2025 IEEE Workshop on Wide Bandgap Power Devices and Applications in Asia (WiPDA Asia 2025)

AA002425

Beijing, China
15-17 August 2025

Pages 520-1036

IEEE Catalog Number: CFP25O09-POD
ISBN: 979-8-3315-1110-4

**Copyright © 2025 by the Institute of Electrical and Electronics Engineers, Inc.
All Rights Reserved**

Copyright and Reprint Permissions: Abstracting is permitted with credit to the source. Libraries are permitted to photocopy beyond the limit of U.S. copyright law for private use of patrons those articles in this volume that carry a code at the bottom of the first page, provided the per-copy fee indicated in the code is paid through Copyright Clearance Center, 222 Rosewood Drive, Danvers, MA 01923.

For other copying, reprint or republication permission, write to IEEE Copyrights Manager, IEEE Service Center, 445 Hoes Lane, Piscataway, NJ 08854. All rights reserved.

*** This is a print representation of what appears in the IEEE Digital Library. Some format issues inherent in the e-media version may also appear in this print version.

IEEE Catalog Number:	CFP25O09-POD
ISBN (Print-On-Demand):	979-8-3315-1110-4
ISBN (Online):	979-8-3315-1109-8
ISSN:	2831-3704

Additional Copies of This Publication Are Available From:

Curran Associates, Inc
57 Morehouse Lane
Red Hook, NY 12571 USA
Phone: (845) 758-0400
Fax: (845) 758-2633
E-mail: curran@proceedings.com
Web: www.proceedings.com

TABLE OF CONTENTS

Design of Short-Circuit and Overload Protection Chips for SiC MOSFETs..1
Jiahui Lv, Yuan Yang, Minmin Zhang

Repetitive Recovery Analysis and Island Width Optimization of 4H-SiC Floating Island Devices.....................6
Runze Xia, Hengyu Wang, Ce Wang, Kuang Sheng

High-Temperature Device Model for GaN-Based MIS-HEMTs...11
Hanlin Cao, Pingyu Cao, Sang Lam, Miao Cui

Dynamic Model of Non-Segmented SiC MOSFETs Considering Switching Behavior and
Temperature Effects...16
Wenzheng Dou, Xin Lan, Ning Zhao

Investigation of the Threshold Voltage Drift in Trench-Type 1200V SiC MOSFETs Under Bipolar
Dynamic Gate Stress..21
Jiaying Cao, Hao Guan, Jiuyang Tang, Liudan Kong, Yifei Chang, Yuhan Duan, Qingchun Zhang, Pan Liu

Analysis of Planar and SBD-Embedded SiC MOSFET Under Avalanche Stress....................................26
Hao Huang, Yuting Jin, Dong Hai, Dingyuan Bao, Taohui Zhang, Haoze Luo, Francesco Iannuzzo, Wuhua Li

Mechanism Analysis of Different Undamped Oscillations for GaN Devices....................................32
Qiang Hu, Jian Chen, Ziyang Wang, Hao Yue, Wensheng Song

Multi-Objective Optimization for Three-Level Power Module Based on Electro-Thermal Co-
Design...38
Runze Wang, Jianing Wang, Shaolin Yu, Xing Li, Xiahao Wang, Baolong Yan, Zhenchun Xia

A Novel Zero-Pressure Packaging Structure of High-Voltage SiC IGBT Device..............................43
Tang Xinling, Wangliang, Du Yujie, Wei Xiaoguang

Design and Characterization of a Novel Embedded SiC Power Module with Double-Sided RDL.................47
Yifei Du, Min Chen, Xinnan Sun, Jie Li, Yucheng Wu, Han Zhou, Feng Jiang

Low-Profile Negative Coupled Inductor for AI Chip Vertical Power Delivery Voltage Regulator.............52
Ruibo Cao, Xiangan You, Fengze Hou, Na Yan, Jian Song, Qidong Wang, Liqiang Cao

Shape Optimization of High-Stress-Buffering Bridge-Typed Spacers for Double-Sided Power
Modules..57
Gaojia Zhu, Ke Xu, Bingru Li, Youzheng Wang, Longnv Li, Yun-Hui Mei

A Study of Multi-Chip Carrier Structure for SiC PCB-Embedded Power Module..............................62
Jia-Jun Lai, Xuelun Zhang, Hao Sun, Ye Zhu, Yunbin Pan, Mingfu Li, Ziliang Shi, Xiong Yang, Xunjin Xu, Min Chen

CNN-Based Rapid Co-Optimization of BV and RON,sp for 4H-SiC SJ MOSFET.................................67
Tiefu Wang, Haoyuan Cheng, Chi Zhang, Hengyu Wang, Kuang Sheng

Thermal Dissipation in GaN-On-Sapphire Power HEMTs: Synergy of Substrate Thinning and
High-Conductivity Thermal Interface Materials..71
Chang Liu, Junsong Jiang, Kun Tan, Jie Lu, Suxia Guo, Cungang Hu, Zhaofu Zhang, Xi Tang, Wenping Cao

Analysis of Influencing Factors on the Thermal Performance of SiC Multi-Chip Parallel Packaging Structure 77
Lei Wang, Rui Jin, Xuebao Li, Peng Zhang, Chao Li

A Synchronously Implanted Termination for 1200V SiC Trench-Gate MOSFET 83
Zhengyun Zhu, Jian Luo, Kai Huang, Qian Wang, Mingyang Gao, Aoxue Xu, Liang Zhou

Research on a Crosstalk Parasitic Oscillation Suppression Method Based on Series Resonance Principle 87
Tiancong Shao, Pei Hu, Yuhan Sun, Kai Wang, Bin Qi, Fangwei Zhao, Zhe Zhang

Effect of Argon-Oxygen Ratios on Crystallization and Surface Morphology of Gallium Oxide Thin Films by Low-Pressure Chemical Vapor Deposition 91
Zhihao Yang, Jichao Hu, Jinhui Yu, Xiaodong Yang, Xiaomin He, Bo Peng

Design of an Active Gate Driver IC for E-Mode GaN HEMTs Based on Adaptive Phased Regulation 95
Wei Liu, Yuan Yang, Wenbin Xing

A Vertical Ultra-Thin Voltage Regulator Module (VRM) with Dual-Phase Integrated Inductor for High-Performance Computing Applications 100
Nianzheng Wang, Hang Kong, Laili Wang

Design of a GaN 1-MHz/48-V-To-1-V/320-A Series-Capacitor Buck Converter 104
Wendi Fan, Yenan Chen

GaN-Based Low-Voltage, High-Current DC-DC Converter for Aerospace Applications 109
Xiao Chen, Zhihao Zhang, Zhongjie Wang, Mingming Ji, Lingxiao Xue

A New Method for Estimating the DC-DC Converter Performance Under Low Voltage and High Current Application 114
Maohang Qiu, Hongbo Ma

A Totem Pole PFC Control Based on Dynamic Root Locus with Wide Output Voltage Range and Low THD 118
Hao Chu, Xiangan Xiao, Xuchen Sun, Jiajia Guan, Cai Chen, Yong Kang

A SiC/Si Hybrid Pulsed Power Supply for High-Current Laser Diode Driving 122
Yihui Tang, Xingye Chen, Yan Su, Helong Li, Zhiqing Yang, Xianbin Qi

Interleaved LLC Resonant Converter with Hybrid PSM/PFM Control for Wide Voltage Gain Applications 128
Jintong Dong, Yijie Wang, Xiaohui Xu, Kai Ji

A Reconfigurable Half-Bridge Buck Based on Dickson SC Converter with DCX-LLC for Wide Input Voltage Range Applications 133
Lichang Man, Yijie Wang, Shanshan Gao, Hongqi Ben, Dianguo Xu

Degradation Evaluation of Single-Event Effects Induced by Heavy-Ion Irradiation on E-Mode p-GaN Gate HEMT Devices 138
Yujie Cheng, Zhangzhe Yan, Jianjun Zhou, Denggui Wang, Zhuangzhuang Hu, Hongfei Wu, Haifeng Cheng

Development of a High-Power-Density SiC Module 142
Jiajun Yang, Xiaoshuang Hui, Puqi Ning

Minimizing Sensor Usage in TSEPs-Based Junction Temperature Estimation of SiC Power Transistor Through Time-Series Analysis 147
Valentyna Afanasenko, Oleksandr Solomakha, Kevin Muñoz Barón, Ingmar Kallfass

Molecular Dynamics Simulations of AgCu Core-Shell Nanoparticle Sintered with Nanoflake 151
Yifeng Chen, Xin Lan, Ziyang Zhang, Xin Li

The Impact of Power Module Layout on Stray Inductance 156
Yuhui Kang, Puqi Ning, Xiaoshuang Hui, Jiajun Yang

Cu/Diamond Composite Electrodes for Low Thermal Resistance Packaging in High-Power IGCTs 160
Tang Xinling, Wei Xiaoguang, Lin Zhongkang, Yang Guang, Yu Kefan, Wang Jingfei

1200V SiC MOSFETs Under Dynamic Reverse Bias Testing in High-Temperature and High-Frequency Environments 164
Liudan Kong, Jiuyang Tang, Jiaying Cao, Yifei Chang, Hao Guan, Yuhan Duan, Qingchun Zhang, Pan Liu

Investigation of Electrical Property and Reliability in AlGaN/GaN MIS-HEMTs Fabricated by Using N2/O2 Composite Plasma Treatment 169
Qingyuan Zuo, Huolin Huang, Yun Lei, Jiayu Zhang, Jianyu Zhao, Jianxun Dai

Investigation on Surge Current Failure in 1200V SiC MOSFETs with Varied Gate Biases in the Third Quadrant 173
Chengyuan Zhou, Hengyu Wang, Yang Zou, Kuang Sheng

Study on SiC Power Module Cooling Technologies Suitable for High-Junction-Temperature and High-Power-Density Applications 177
Baihan Liu, Jianwei Lv, Yipeng Liu, Yifan Zhang, Siqi Dai, Zexiang Zheng, Suhang Wei, Cai Chen, Yong Kang

An Enhanced Over-Current Fault Detection System for a 7.2kV Power Module Using Series-Connected SiC MOSFETs 181
Chunyao Hou, An Lou, Yue Wu, Aozu Luan, Shuai Shao

Modeling of Static Current Sharing of Direct-Paralleled 1.7kV 800A SiC Power Modules 186
Zhongjie Wang, Hui Liao, Lingxiao Xue, Xiangqian Zhang, Liwen Zou, Xiaonan Dong

A 500W 13.56 MHz Amplifier Realized by Two Combining Class-E Circuits 190
Yanfei Ji, Mei Liang, Jiwen Chen, Pengyu Jia, Yihang Zhang

A Method to Enlarge the Soft-Switching Range for the Dual Active Bridge Series Resonant Converter 195
Pengyu Jia, Yimei Xing, Kai Qiu, Mingjun Liu

Bidirectional Energy-Storage Converter Based on Partial Power Processing 201
Zhihao Zhang, Xiao Chen, Fan Zhai, Zhongjie Wang, Lingxiao Xue

3kV/42kW Pump-Back Test for Medium Voltage DC Transformer with Series-Connected SiC MOSFETs 206
Yujian Zong, Shuai Shao, Chaojun Wang, Wentao Cui, Junming Zhang

Thermal Network of Double-Sided Cooling Power Module Based on Cauer Model 211
Jie Li, Min Chen, Xinnan Sun, Jun Huang, Yucheng Wu, Yifei Du, Han Zhou, Feng Jiang

Multi-Objective Automatic Design of Power Module Packaging Based on Artificial Neural Network and Deep Reinforcement Learning ... 215

Weina Mao, Jianing Wang, Shaolin Yu, Zhenchun Xia, Honghong Li, Xiang Pan, Xiahao Wang

A High-Performance Double-Sided Cooling SiC Power Module Packaging Design for EV Inverters ... 221

Haobin Chen, Haidong Yan, Maosheng Zhang, Ji Cheng, Yakun Zhang, Chaohui Liu

Optimal Design of High-Temperature SiC Power Module Based on Gene Algorithm ... 226

Zhenchun Xia, Jianing Wang, Shaolin Yu, Weina Mao, Xiahao Wang, Honghong Li, Runze Wang

Monolithically Integrated Over-Temperature Protection Circuit Based on GaN HEMTs ... 232

Pingyu Cao, Kepeng Zhao, Yihao Xu, Harm Van Zalinge, Ping Zhang, Miao Cui, Fei Xue

Lifetime Modeling of IGBT Devices Based on Micro-Defect Topography Inversion ... 236

Miaomiao Shangguan, Wei Lai, Hao Wang, Yunjie Wu, Yu Liu, Hang Zhao

A 6.78-MHz 2.3-KW Full-Bridge GaN Inverter with Bottom-Side Cooled Transistor GS-065-030-2-L for Wireless Power Transfer ... 241

Jianping Ning, Zhen Sun, Yao Wang, Shuang Zhao

Two-Step Turn-Off Delay Time Control for Efficiency Enhancement in Si/SiC Hybrid Switch-Based Inverters ... 246

Zijie Zheng, Jun Wang, Xuanting Song, Yongzhou Zou, Yuxing Dai, Kamal Al-Haddad

6.78 MHz 2.8 kW Resonant DC-DC Power Conversion Utilizing 650 V GaN HEMT-Based H-Bridge Inverter ... 250

Zhen Sun, Jianping Ning, Yao Wang, Yun Yang

A Passive Cancellation Circuit Incorporating the Motor Stator Windings for Reducing the Common-Mode Noise in Motor Drive Systems ... 255

Lihong Xie, Yuzhen Wu, Xinbo Ruan

A Voltage-Sharing Cascode Switch Structure Using SiC JFET and GaN HEMT ... 260

Yin Fang, Shan Jayamaha, Carl Ngai Man Ho

CLLLC Resonant Converter with Full-Bridge Half-Bridge Switching for Improved Voltage Gain Range and Light-Load Efficiency ... 264

Jinfeng Yu, Sinuo Liu, Yu Gu, Hui Liu

Analysis and Prediction Method of Gate Voltage Peak in Switching Transient of GaN E-HEMT ... 269

Yushan Liu, Yirui Hu, Xiao Li

An Active Gate Charge Controlled Driver for SiC MOSFET with Zero Turn-Off Loss ... 275

Yuze Zheng, Xu Cheng, Fan Zhang, Xiaolu Zhang, Yukun Niu, Xuhui Song

A Modulation Optimization Strategy for Phase Shift Full Bridge with Low Freewheeling Loss ... 280

Enyou Wu, Xuchen Sun, Jiajia Guan, Siyuan Feng, Tianxi Li, Cai Chen, Yong Kang

Optimized Trajectory Control for Soft Start-Up of Multiple-Mode Resonant Switched Capacitor Converter ... 284

Jingjing Qi, Kai Zhang, Haining Zhang

Datasheet-Driven Non-Segmented SPICE Model of SiC MOSFET with Improved Accuracy ... 290

Ziqi Jia, Yu Jiang, Yifan Hu, Hailong He, Chunping Niu, Yi Wu

Investigation and Improvement on Surge Robustness of Double-Trench SiC MOSFETs in Synchronous Rectification Mode 295

Hexin Zhu, Xiaochuan Deng, Haohao Dai, Qian Huang, Xuan Li, Xu Li

A High-Output-Current Radiation Hardened Half Bridge DC-DC Converter for Space Applications 299

Dawei Li, Xiang Zhou, Yeerzhati Nuerdebieke

The Thermal Simulation Study of DC-DC Converters Based on Phase Change Cooling 304

Boyang Liu, Haoyu Zhang, Yu Han, Jialiang Chi, Yao Zhao, Zhiqiang Wang

A 2:1 Capacitive Isolated Resonant Switched-Capacitor DC-DC Converter 308

Xinyu Zhang, Yu Fu, Yucheng Zhao, Shouxiang Li

An Auxiliary Power Supply Utilizing Planar Winding Transformer with High Insulation and Low Coupling Capacitance Considerations 314

Yong Chen, Xuhui Song, Fan Zhang, Yuze Zheng, Xiaolu Zhang, Yukun Niu, Zheyuan Yu, Min Wu, Kaixiang Gong

Gate-Source Voltage Oscillation of Multi-Chip SiC Power Modules Considering Differential Gate Resistance 319

Longnv Li, Chuyuan Liu, Lu Wang, Youzheng Wang, Gaojia Zhu, Yun-Hui Mei

Development and Electrical Characteristics Study of a 10kV/125A SiC MOSFET Module 324

Yaodong Zhang, Yong Chen, Xu Cheng, Xiaotian Zhang

Study of Thermal Stress on Devices in Si/SiC Hybrid Half-Bridge Inverter 328

Yulin Wang, Ruixiao Dong, Jun Wang, Yuxing Dai, Liuchen Chang

A 8:1 Multi-Resonant Switched-Capacitor Converter 334

Yuxin Yan, Yu Fu, Yucheng Zhao, Shouxiang Li

Sustainability of Power Devices: A Perspective on Design for Recycling 339

Jinpeng Cheng, Shuyu Liu, Hao Feng, Li Ran

Cryogenic Output Capacitance Loss of GaN HD-GIT 343

Yudong Wang, Zilong Chen, Yukun Zhang, Chong Dou, Qian Cui, Yuqi Wei

Improved Control Strategy Based on BM Theory for Grid Forming Type VSG System 349

Shuanglong Li, Yajing Zhang, Bin Liu, Baoying Huang, Siyu Pan, Hao Ma

Baseplate Temperature Gradient-Based Health Status Monitoring for Power Module Bonding Wires 355

Yongxin Chen, Lei Xu, Kun Tan, Xi Tang, Cungang Hu, Wenping Cao

Study on Influencing Factors and Mechanisms of Single-Event Gate Rupture in SiC STP-MOSFETs 361

Ying Yang, Zixuan Liu, Chen Wang, Xulong Wang

SiC Power Module with Staggered Terminals Layout Design to Reduce Parasitic Inductance 366

Jiahang Wang Guangzhou, Xi Jiang Guangzhou, Runze Ouyang, Song Yuan, Yuanzhi Zhao, Qingrong Hu, Ying Wang, Xiaowu Gong

Novel Tri-Gate Multichannel Device for Improved Vth Controllability 371

Quanbo He, Hengyu Wang, Florin Udrea

Linear-ESO Based Control for Wide Input Range Partial Power Regulated DCDC Converter 376

Xikun Sang, Shanshan Gao, Yijie Wang, Dianguo Xu

A Multi-Objective Optimization Method Based on NSGA-II Algorithm for Electro-Thermal-Stress Collaborative Design of Intelligent Power Modules 382
Tao Xu, Lei Ming, Ningbo Li, Zihang Gu, Zhiwei Jiao, Zhen Xin

GaN-Based Partial Power DC-DC Converter with Four-Quadrant Operation Capability 388
Chao Liu, Zhe Zhang, Shunqing Wu, Zeqi Yang, Chuang Liu

UIS Ruggedness of Si/SiC Hybrid Switches 393
Chuanqi Zhang, Xuanting Song, Shiwei Liang, Yuxing Dai, Jun Wang, Kamal Al-Haddad

Research on Thermal Conductivity and Mechanical Properties Control of Epoxy Resin in Extreme Environments 398
Zhen Li, Liang Zou, Qingsong Liu, Zhiyun Han, Jinyang Bai

A 3L-ANPC SiC MOSFET Power Module with Low Thermal Coupling and Parasitic Inductance 404
Zedong Xue, Zhiyuan Qi, Zihao Chen, Hao Yuan, Qingwen Song, Yuming Zhang

Research on Transient Electric Field Calculation in Welded Devices Considering Dielectric Relaxation of Insulating Materials 410
Zihan Sang, Zhaocheng Liu, Hao Li, Xuebao Li, Ying Cao, Peng Shu

A Two-Stage Turn-On Gate Driver for SiC MOSFET with Short-Circuit Current Suppression 416
Yong Chen, Xiaolu Zhang, Xu Cheng, Yuze Zheng, Xuhui Song, Yukun Niu, Kaixiang Gong, Fan Zhang

A Zero-Sequence Injection Method to Reduce Electromagnetic Interference 420
Hui Liu, Dong Jiang, Junzhao Zhang

Development of a Novel Analytical Trapped Charge Model for Total Ionizing Dose Effects of SiC MOSFETs 424
Qingmao Hu, Xin Yang, Qingzhong Gui

Method for Estimating Power Loss of IGBT Module in Wind Power Converter Based on Measured Temperature Information 430
Ye Tian, Dawei Chen, Zhijie Zeng, Lixuan Zhu, Guojun Bao, Zhixiang Zou

A Method for Enhancing the Heat Dissipation Performance of Power Devices Through Graphene Coating 436
Xin Li, Jianing Wang, Shaolin Yu, Runze Wang, Xiahao Wang, Baolong Yan, Zhenchun Xia

Research on DC Characteristics of β-(Al0.22Ga0.78)2O3/β-Ga2O3 Modfet 440
Haitao Zhang, Xiaomin He

A Structure-Reconfigurable Electronic Transformer for Renewable Energy DC Distribution Syste 445
Yu Feng, Xianbin Qi, Zhiqing Yang, Jinxiao Wei, Peng Qin, Helong Li, Liu Fang, Lijian Ding

Pyrolysis Process Analysis of Polyimide at High Temperature Based on Molecular Dynamics Simulation 450
Yuteng Jiang, Minglei Xie, Wenzhi He, Zhi Wang, Bingxin Chen, Zhiyun Han, Sixiao Xin, Liang Zou

Analysis of SiC Output Capacitance Effects on Soft-Switching Characteristics in Three-Level Resonant Converters 456
Zhe Shao, Zhiyuan Wang, Binbin Li

On the First Demonstration and Analysis of the HTGB Induced Electrical Degradation of High Voltage SiC IGBT Devices 462
Tuanzhuang Wu, Jiaxing Wei, Junhou Cao, Hao Fu, Zhaoxiang Wei, Desheng Ding, Siyang Liu, Weifeng Sun, Xiaolei Yang, Song Bai

A New Si/SiC Hybrid Interleaved Three-Level ANPC Inverter with Cost and Performance Tradeoff 468
Ruixiao Dong, Jun Wang, Yulin Wang, Yuxing Dai, Liuchen Chang, Chao Zhang

Turn-Off Analysis and Modeling of Releasing Loss in Snubber Capacitor Self-Balancing Circuits for Series-Connected SiC MOSFETs Applied to High-Voltage Pulsed Power Systems 474
Jiaxuan Niu, Xu Cheng, Xu Yang, Yong Chen, Fan Zhang, Kexin Zhao

Optimization and Compensation of Leakage-Induced Deviation of CTTC Magnetic Integrated Structure in CLLC Resonant Converter 480
Liwen Jia, Bodong Li, Yahong Yang, Jianyu Lan, Jiarui Zhang, Kelin Chen, Feng Jiang, Min Chen

Sustained Oscillation Characterization of GaN HEMT at Cryogenic Temperature 485
Zilong Chen, Yuqi Wei, Yanjie He, Yukun Zhang, Chong Dou, Qian Cui

Design of an All-SiC On-Board Auxiliary Inverter for Urban Rail Vehicles 491
Xuefei Li, Yongang Chen, Zixiao Li, Shuiyuan He, Yuwen Qi, Lijun Diao

A Symmetrical Double-Sided Cooled SiC Power Module for Multi-Parallel Applications 495
Guolian Guan, Zhiqiang Zhao, Mingzhi Zhao, Tongyu Zhang, Laili Wang, Dewen Wang

A Transistor Clamp Circuit of On-State Voltage Drop for SiC MOSFET Temperature Monitoring 500
Yixiang Zhao, Hong Li, Xiaofei Hu, Kuang Zhang

A Dynamic Current Balancing Method for Multichip SiC MOSFET Modules with Separate Gate Drive Structures 506
Zicong Li, Zenan Shi, Yifei Luo, Xin Li

A Novel Analytical Physical Model of Gate-Drain Capacitance and Output Characteristics for SiC-MOSFET 512
Ze Tao, Zenan Shi, Yifei Luo, Xin Li

Reliability Analysis for 1200V SiC MOSFETs Under Repetitive Surge Current Operation with Negative Gate-Source Bias 516
Xinbin Zhan, Yanjing He, Jiankun Lai, Xi Jiang, Song Yuan, Hao Yuan, Qingwen Song, Xiaoyan Tang, Xiaowu Gong, Yuming Zhang

A Novel Modulation Method with Simultaneous Reduction of Common-Mode Voltage and Switching Losses for Three-Level SNPC Inverter 520
Shuangxi Zhu, Jiajia Guan, Cai Chen, Yong Kang

An Active Gate Driver of Voltage Overshoot Suppressing for SiC MOSFETs 524
Tingwen Hu, Wensheng Song, Jian Chen, Tao Tang, Hao Yue, Guoyou Liu

A High-Frequency LLC Resonant Converter Incorporating Matrix Transformer 528
Haochen Zhang, Yueshi Guan, Yijie Wang, Dianguo Xu

An Active Gate Driver Addressing GaN HEMT Gate-Source Voltage Overshoots for Both Turn-On and Turn-Off Periods 534
Lurenhang Wang, Yishun Yan, Shuaiqing Zhi, Mingcheng Ma, Xizhi Sun, Dianguo Xu

Application of Wide Bandgap High Frequency Inverter in 10 kW Magnetic Field-Coupled Undersea Wireless Power Transfer Systems 539
  Lei Yang, Jiahua Sun, Yuanfeng Wang

Minimizing Eddy Current Loss for Implanted Wireless Charger Through Phase Difference Optimization 544
  Pengyu Chen, Siyi Yao, Xiyuan Lin, Hongjun Zheng, Congcong Zhang, Minfan Fu

An Accurate Leakage Inductance Model for Magnetic Integrated Planar Transformers in LLC Resonant Converters 550
  Zhili Mo, Xuetong Zhou, Yufei Tian, Li Zheng, Xinhong Cheng

Subharmonic Oscillations in High-Frequency Switched-Mode Power Amplifiers 556
  Wei Liu, Ming Liu

Novel Structure for High-Voltage Vertical β-Ga2O3 Schottky Barrier Diodes: A TCAD Study 562
  Yan Liu, Meng-Qi Fan, Jia-Xiang Chen, Teng Jiao, Mao-Jin Yang, Xian-Hu Zha, Xiao-Ping Wang, Xiang-Jin Ding, Yu-Xi Wan, Dao-Hua Zhang

Efficiency-Oriented Adaptive Dead Time Control for the Dual Active Bridge Converter 566
  Shaoyan Jiang, Xichen Fu, Xingque Xu, Jiabin Ruan, Chuanwei Xiao

Simulation Study of Breakdown Characteristic of Vertical Diamond Schottky Barrier Diodes with Different Drift Layer Parameters and Termination Structures 572
  Tianhe Mi, Peng Wang, Yan Liu, Teng Jiao, Xinchun Cui, Haolin Hu, Yuxi Wan, Daohua Zhang

Multi-Parameter Degradation Modeling Method for Power MOSFETs Integrating Semiconductor Physics Degradation Data 577
  Chenyi Wang, Cen Chen, Weixuan Kong, Haodong Wang, Zhenning Zhou

1.5KW HSC Converter Power Density-Efficiency Advancement: Enabled by Planar Transformer 582
  Pengfei Wang, Vickie Qu, Qianru Shi, Qingchang Liu, Ian Yj Chan, Minfan Fan

Control Strategy of Dual Phase Shifting Dual Active Bridge Converter Based on BM 588
  Yiting Huo, Yajing Zhang, Bin Liu, Baoying Huang, Siyu Pan, Hao Ma

β-Type High-Gain Boost Converter with Diode-Capacitor Cell 594
  Peng Sun, Hong Li, Yidi Liang, Xu Shangguan, Mingbo Wei, Huizhu Zhuang

Soft-Switching Fixed-Frequency Control Strategy for Three-Level Buck-Boost Converter 600
  Fang Li, Yao Xue, Fangwei Zhao, Yajing Zhang, Jun Xu

High-Robust Power Integrated Synchronization Scheme for the Grid-Forming Inverter 606
  Wen Zou, Yuying He, Li Zhang, Dongsheng Yang

Charge Control for Critical Conduction Mode Soft-Switching Grid-Tied Inverters 609
  Zhengzi Lei, Zhongshu Zheng, Wenbo An, Li Zhang, Yuying He

Dynamic Analysis of Hybrid Photovoltaic-Battery System with Dual-Mode Grid-Forming Control 614
  Wenjie Ning, Yiyang Liao, Yaoyu Hu, Shaoze Zhou, Zheng Wei, Yitong Li

Optimization Design of High-Speed and High-Voltage Switching Transient Characteristics for GaN HEMTs Using Multi-Objective Particle Swarm Optimization 619
  Xiao Li, Zhuofan Xiong, Yushan Liu, Qiang Zhou

A Compact EMI Filter Design Method Based on Chaotic SVPWM in Motor Drive System 624
Yanjun Li, Hong Li, Zuoxing Wang, Aojie Liao, Mingxin Shi

A SiC MOSFET Accelerated Degradation Test Platform that Accounts for Turn-On and Turn-Off
Times 630
Bohang Lu, Cen Chen, Zicheng Wang, Xuanyu Lin

Bias Temperature Instability of SiC Trench MOSFET Under DC and AC Gate Stress.................................... 634
Kanghua Yu, Qian Wang, Yuwei Wang, Jun Wang

A Novel Approach to Estimate the Thermal Destruction Point of SiC MOSFETs Under Short
Circuit Conditions 638
Rony Thomas, Zhe Yu, Sebastian Fahlbusch

Investigation on Electrical Properties and Safe-Operating Area for Novel Split-Gate IGBTs 644
Xuanting Song, Jun Wang, Gaoqiang Deng, Zijie Zheng, Yongzhou Zou, Shiwei Liang, Yuxing
Dai, Kamal Al-Haddad

Influencing Factors and Mechanisms of Single-Event Burnout in STP SiC MOSFETs 649
Ying Yang, Xulong Wang, Chen Wang, Zixuan Liu

A Multi-Operation Characterization System for Quantifying Dynamic RON Degradation in p-GaN
Gate GaN HEMTs Under Real-World Switching Conditions.. 655
Junbo Wang, Xiangdong Li, Xi Jiang, Haonan Jiang, Shuzhen You, Yue Hao, Jincheng Zhang

The Establishment of Quasi-3D TCAD Simulation for Bipolar Degradation Occurs in SiC
MOSFET Body Diode 659
Junhou Cao, Xinyu Zhou, Chenlu Wang, Lei Huang, Hao Fu, Zhaoxiang Wei, Tuanzhuang
Wu, Jiaxing Wei, Siyang Liu, Weifeng Sun

Study on AlxGa1-XN Graded Composite Barrier GaN HEMT 663
Ruihao Zhang, Fayu Wan

Chip Screening Strategy for SiC MOSFET Based on Simulated Annealing Algorithm.................................. 667
Zhanshan Zhu, Chuangye Li, Helong Li, Zhiqiang Liu, Lijian Ding

Research on Power Loss Calculation and Optimization of FB-VSCC Active EMI Filter................................ 673
Daozhen He, Hong Li, Yuanheng He, Mingxin Shi

Analysis of Common-Mode EMI in Multi-Converter Systems with EMI Filters 678
Runquan Jiang, Peng Zhou, Guifeng Geng, Xuejun Pei

Research on Key Technology of Low Noise Transformer and Its Application on 110kV Transformer 684
Shouhui Han, Qingsong Liu, Zheng Liu, Liang Zou, Zhiyun Han

Study on Cross-Scale Collaborative Control of Composite Properties for Power Device Packaging.............. 688
Xianfeng Li, Jian Wang, Hanwen Ren, Wei Wang, Chen Chen, Qingmin Li

Partial Discharge Denoising Technology for GIS Equipment Based on CEEMDAN 692
Zhongyue Liu, Pei Cao, Zeyu Li, Hanwen Ren

Improvement of Thermal Conductivity of Epoxy Composites Using the Synergistic Effect of Boron
Nitride Nanosheets and Boron Nitride Whiskers 696
Tiandong Zhang, Chenghai Wang, Xinle Zhang, Tangman Xue, Changhai Zhang, Qingguo
Chi

Study on Surface Discharge Characteristics of Silicone Gel in Salt Fog Environment 700
*Feng Wang, Zhihui Li, Yateng Yany, Shanzhen Fan, Hanwen Ren, Jian Wang, Wei Wang,
Qingmin Li*

Effect of Atomic-Level Roughness on Vertical β-Ga2O3 Schottky Barrier Diode and High-
Temperature Performance 704
*Jiaxiang Chen, Xingye Zhang, Maojin Yang, Xinpeng Lin, Teng Jiao, Yan Liu, Xianhu Zha,
Haolin Hu, Xiangjin Ding, Yuxi Wan, Jun Ma, Mengyuan Hua, Dao-Hua Zhang*

U-Shaped Sloped Field Plate Edge Termination Structure in β-Ga2O3 Vertical SBD 709
*Yanzuo Li, Song Yuan, Xi Jiang, Yanjing He, An Xu, Guorui Mo, Yunxuan Zhao, Zhaoheng
Yan, Qifan Liu, Xinbin Zhan, Ying Wang, Linhai Zhong, Xiaowu Gong*

Enhancement-Mode Vertical (001) β-Ga2O3 Power Transistor Enabled by Dual Ion Implantation 713
*Anjing Luo, Gaofu Guo, Li Zhang, Tiwei Chen, Dengrui Zhao, Zhili Zou, Chunhong Zeng,
Huanyu Zhang, Baoshun Zhang, Zhucheng Li, Xiaodong Zhang, Zhongming Zeng*

Partial Discharge Classification and Intelligent Maintenance Decision-Making Based on
Spikingformer and LLMs 717
*Changdong Wang, Jingli Yang, Xunran Yin, Shuangyan Yin, Tianyu Gao, Yongqi Chang,
Huamin Jie, Zhou Shu, Zhenyu Zhao*

Junction Temperature Fluctuation Suppression Strategy for SiC MOSFETs Based on Equivalent
Gate Resistance Control 721
Ruoyin Wang, Xiaoyong Zhu, Hong Zheng

Study of a High-Sensitivity Recessed-Anode GaN MIS Diode for Temperature Sensing 726
*Yunxuan Zhao, Xi Jiang, Song Yuan, Zhaoheng Yan, Yanzuo Li, Chaofan Deng, Xiangdong Li,
Xiaowu Gong*

Warping Model of SiC Power Module After Reflow Soldering 731
Chang Liu, Yingxin Cui, Yanao Guo, Shoulai Gong, Jisheng Han

A Novel Mechanical Stress Related Failure Mechanism of the State-Of-The-Art SiC Double Trench
MOSFET Under Surge Current Stress 737
*Shikang Xu, Xuan Li, Hanqing Zhao, Yi Wen, Wensong Peng, Zekun Zhou, Xu Li, Xiaochuan
Deng, Bo Zhang*

Impact of Elevated Humidity Conditions on the Thermal and Mechanical Reliability of Silicon
Carbide MOSFETs 741
Jiayu Zhang, Dong Xie, Zhiliang Xu, Zepeng Jiang, Xinglai Ge

Reliability Analysis and Implementation of Silver Sintering Connection in SiC High-Temperature
Package 747
Zizhen Cheng, Wenjie Xu, Fengtao Yang, Zhiqiang Zhao, Dewen Wang, Laili Wang

Impact of Kelvin Source Connection (KSC) Vs. Common Source Connection (CSC) on Power
Cycling Aging Characteristics of Silicon Carbide (SiC) MOSFETs 753
Huaihao Cheng, Yong Chen, Xu Cheng, Xingyu Pei, Jianbiao Li, Hongyuan Wu

SiC VDMOSFET Performance Prediction Method Based on Neural Network 757
Xiamin Hao, Feng He, Rui Jin

Oxidation-Free Sintered Copper Die Attachment for Power Electronics Packaging 762
*Junyang Chen, Haiqiang Zhao, Zewei Zhang, Zhuo Pang, Tsung-Huan Sheng, Yi Chiu, Zhi-
Ying Huang, Yen-Liang Lin, Meiyu Wang*

Advanced Packaging for High-Voltage SiC Module with Excellent Thermal, Mechanical, and Electrical Properties.. 767

Meiyu Wang, Yiting Han, Peng Gao, Zhuo Pang, Yingkun Yang, Haidong Yan

A One-Inductor Two-Switch Three-Port DC-DC Converter for PV-Battery Systems..................................... 772

Yidi Liang, Hong Li, Huizhu Zhuang, Peng Sun, Xu Shangguan

A Parameter Design Method for Bidirectional LLC-C Resonant Converter for Bidirectional On-Board EV Charger Application.. 777

Yiheng Zhang, Pengyu Jia, Mingjun Liu, Yimei Xing

A Phase Shift Control Strategy for the DCX to Realize a Wide Voltage Range by a Variable Mode Transformer 783

Pengyu Jia, Mingjun Liu, Yimei Xing

A Multi-Port Converter-Based Equalization Architecture with Wide Voltage Gain for Long Series Battery Packs... 788

Haipeng Hu, Xianbin Qi, Helong Li, Mingzhu Fang, Peng Qin, Zhiqing Yang, Lijian Ding

Design of Planar Integrated Transformer for LLC Resonant Converters Based on Hybrid Magnetic Materials... 792

Jianguang Yao, Xiaoyi Xu, Yanfang Mao, Zhujian Ou, Runyang Ji

Common-Mode Noise Analysis of Hall-Effect Sensor for SiC Power Converter 798

Guifeng Geng, Peng Zhou, Runquan Jiang, Xuejun Pei

Research on Online Reliability Assessment Technology for IGBT Based on Driver-Side Measurement Fusion 802

Yi Liu, Zhicheng Liu, Yunhui Mei

An Active Current Sharing Strategy Based on Master-Slave Cooperative Control for Paralleled SiC Power Modules... 808

Baolong Yan, Jianing Wang, Shaolin Yu, Honghong Li, Runze Wang, Xin Li

Investigation on SiO2 Gate Formed Through Ultrahigh-Temperature NO Oxidation 813

Yingfeng He, Given Shucheng Chang, Decai Liu, Tao Zhu, Zheyang Li, Rui Jin

The Method for Chip Sorting Based on SiC MOSFET Parameters for Parallel Current Sharing Under a Symmetrical Layout.. 817

Yi Li, Bowen Tian, Rin Zhao, Peng Sun, Zhibin Zhao, Xu Cheng, Yong Chen, Xingyu Pei, Jianbiao Li, Hongyuan Wu

Charaterization of a Nonlinear Conductivity Encapsulant for Electric Field Reduction in High-Voltage Power Module Packaging.. 822

Meiyu Wang, Peng Gao, Yiting Han, Haidong Yan

Configuration Selection for Degradation Trajectory Prediction of Power Modules Based LSTM Model 827

Yichi Zhang, Yi Zhang, Jie Kong, Jiahong Liu, Bo Yao, Huai Wang

A Review and Analysis of Grid-Forming Technologies in Renewable Energy Power Systems..................... 833

Zhicheng Liu, Dezheng Zhang, Yehan Fu, Yuying He, Li Zhang

Switching Oscillation Suppression Based on Embedded SiC Power Module with Low Parasitic Inductance for CLLC Resonant Converter 839

Jiarui Zhang, Bodong Li, Xinnan Sun, Jiahui Wang, Liwen Jia, Kelin Chen, Feng Jiang, Min Chen

Modeling and Design of the Planar Magnetic Integration for Dual-Stage EMI Filters 844
Haiyan Liang, Yitao Liu, Zijian Lu

A Method to Decrease the Submodule Capacitor Voltage Fluctuations in Voltage-Source Modular Multilevel Converter with Wide-Bandgap Power Devices 849
Qian Kang, Tiancong Shao, Trillion Zheng, Yuqing Geng, Yaqi Li, Zhitong Bai, Xiaofeng Yang

Junction Temperature Control and Thermal Stability Enhancement Method for Power Devices Based on Vapor Phase-Change Principle 853
Ruya Song, Shuang Zhao, Jinxiao Wei, Waleed Alhosaini, Lijian Ding

A Self-Clamped L-Shaped Trench Gate SiC MOSFET with Improved Breakdown and Short-Circuit Reliability 857
Xiaobo Cao, Jing Liu, Shaowei Zhang, Qian Zhang, Zhonggang Yin

Degradation of Planar-Gate SiC MOSFETs Under Repetitive Short-Circuit Stress in Different Gate Bias 863
Yifan Wu, Chi Li, Jianwei Liu, Zedong Zheng

Low Roughness and Shape Reforming 4H-SiC Trench Process Optimized by CCP Etching with High Temperature Annealing and Sacrificial Oxidation 867
Qiongyang Zhuang, Xixi Luo, Yu Chen, Caixin Gu, Lei Song, Kaiju Liao, Qin Hu, Jiamin Tian, Yidan Chen, Gang Chen, Jinliang He

Investigation of L-FER ESD Protection Capability on E-Mode p-GaN HEMT 873
Junye Wu, Yitian Gu, Chao Feng, Danfeng Mao, Yanlin Wu, Haolin Hu, Wei Zeng, David Zhou, Yuxi Wan

Study on 4H-SiC Trench Gate Dual-Mode Composite Transistors (T-DCT) Structure to Reduce Ron and Enhance SCWT 877
Wenyu Xi, Cailin Wang, Lei Guan

Super-Junction IGBT with Adaptive Hole Channel Around Stepped Trench Gate for Low On-State Voltage and Low Turn-Off Loss 881
Xuelei Zhou, Hengyu Wang, Yifan Wang, Kuang Sheng

Effect of Voltage Probes on the Characterisation of Switching Processes in Wide-Bandgap Semiconductor Devices 885
Yishun Yan, Lurenhang Wang, Xuchong Cai, Mingcheng Ma, Yanchen Pan, Dianguo Xu

Performance Improvement of SiC n-LTT by Semi-Through Via Structure 890
Yulei Zhang, Xi Wang, Xuhui Pu, Jichao Hu, Hongbin Pu, Yuan Yang

Comparative Investigation of Gate Oxide Degradation in 1.2 kV Planar, Double-Trench, and Asymmetric-Trench SiC MOSFETs 894
Dingkun Zhao, Xin Yang

Understanding the Role of Buffer Traps in GaN HEMTs: A Simulation Study on Dynamic Ron 898
Haiyang Li, Mengqi Fan, Xinyue Dai, Xiaoping Wang, David Zhou, Danfeng Mao, Yan Wang, Yuxi Wan

A 3×1 Silicon Carbide Bidirectional Switch Power Module with Balanced Inductance During Current Commutation 902
Zhiwei Jiao, Lei Ming, Yufeng Cao, Tao Xu, Zihang Gu, Zhen Xin

An Experimental Study on Single Pulse Avalanche Characteristics of Si/SiC Hybrid Switch 907
Hangzhi Liu, Yuming Zhou

Substrate Coupling Considerations for Monolithic Integration of High-Voltage Power Transistors with Low-Voltage Devices and Circuits in GaN-On-Si Technology 911
Rui Ray Yao, Miao Cui, Zhao Wang, Sang Lam, Stephen Taylor

Analysis and Compensation Method of Transient Unbalanced Current in Parallel Connection of SiC MOSFET 916
Yong Chen, Xu Cheng, Xingyu Pei, Jianbiao Li, Hongyuan Wu, Bin Zhao

Design and Fabrication of a SiC Trench MOSFET with Multi-Step P-Type Shielding and Multiple CSL Layers 920
Wei Chen, Fei Guo, Yangyang Wu, Kuan Wang, Zhijie Cheng, Jun Yuan, Rong Zhang, Guoqing Xin, Zhiqiang Wang

A Novel Integrated Power Module Package Method with SiC MOSFETs and Energy Absorber in Solid-State Circuit Breaker Application 924
Dongxin Jin, Jie Gong, Yuchen Wang, Cheng Luo, Xiaojun Dong, Guangyin Lei

Dry Oxidation and SiO$_2$ Deposition Strategies for High-Performance Gate Dielectrics in 3.3 kV SiC Power MOSFETs 929
Zijian Hu, Hongyi Xu, Na Ren, Kuang Sheng

Low Parasitic Repackaging and Integration of Multiple GaN HEMT Devices 933
Yue Chen, Mingrui Zou, Dongjun Jiang, Senhao Liang, Jiakun Gong, Zheng Zeng

Impact of Electroluminescence Spectrum Sampling on SiC MOSFET Junction Temperature and Current Sensing 939
Yuting Jin, Hao Huang, Jingyang Hu, Shengjie Luo, Haoze Luo, Wuhua Li

Research on Threshold Voltage Instability of SiC MOSFETs at High Temperature 944
Xu Cheng, Yong Chen, Xingyu Pei, Jianbiao Li, Hongyuan Wu, Cong Chen

A High-Isolation X-Ray Power Supply with Multi-Transformer Series Configuration 949
Ziyang An, Ye Tian, Jie Ming, Yu Dou, Chushan Li

A High-Speed Dynamic Gate Driver with Low Oscillation for GaN HEMTs 954
Xuetong Zhou, Li Zheng, Xinhong Cheng, Lingyan Shen

Research on an Anti-Offset Wireless Power Transfer System with Auxiliary Resonant Circuit 959
Youzheng Wang, Shengxiu Xu, Shuyu Wang, Hongchen Liu, Longnv Li, Gaojia Zhu, Yunhui Mei

Research on Low Speed Power Boosting Technology for High Speed Maglev Trains 963
Zheyi Zheng, Fuao Chen, Xiaojun Zhang, Haoyun Wang, Yang Chen, Ruikun Mai

Current Overshoot and Oscillation Suppression in SiC MOSFETs Through Variable Gate Capacitance During Turn-On Transient 968
Xuchong Cai, Yishun Yan, Yanchen Pan, Mingcheng Ma, Binbo Xu, Dianguo Xu

An Active Clamped Resonant Ultra-High Frequency Quasi-Square Wave Gate Driver for SiC MOSFET 974
Zhiqing Liang, Zhixing He, Haoyi Sheng, Renfeng Guan, Zhenyuan Ou, Yang Liu, Zongjian Li, Jun Wang

48V-To-0.9V Voltage Regulator with GaN Devices and Integrated Magnetic Design 980
Zikang Li, Jingyang Tan, Yijie Wang, Dianguo Xu

A Bidirectional-Signal Transmission Method for Gate Drive Application Using Single Isolation Transformer 985
Junru Lin, Junming Zhang

A Dual-Voltage 650 V and 100 V GaN Integrated Platform Featuring High-Performance Monolithic Components 990
Yanlin Wu, Junye Wu, Zuoheng Jiang, Danfeng Mao, Keping Wu, Chao Feng, Jiawei Chen, David Zhou, Yuxi Wan

Exploring the Soft-Switching Benefits of TZCM Mode in Three-Level DC-DC Converters Using Wide Bandgap Power Devices 994
Zhigang Yao, Jingrui Liu, Sankun Yao, Bac-Bien Ngo, Ziheng Xiao, Yi Tang

Design and Current-Sharing Study of the TL-Boost Power Unit Based on Discrete Devices in Parallel 1000
Jianing Wang, Honghong Li, Shaolin Yu, Donglei Zhang, Baolong Yan, Zhenchun Xia, Weina Mao

Study on New Structures of EST to Inhibit Snapback Effect and Enhance MCC 1006
Wuhua Yang, Jia Liping, Guo Jiarui, Zhang Chao, Shen Sihao, Wang Cailin

A Transient Interaction Mechanism Analysis Method for the Grid-Forming Voltage Support Device Integrated into LCC-HVDC System 1011
Yanlin Song, Zhichang Yang, Hong Li

A Numerical Model of SiC MOSFET for Electro-Thermal Characteristics Based on TCAD 1017
Yujie Zhang, Yongle Huang, Yifei Luo

The Method to Evaluate the SOA of Drain-Source Voltage for SiC MOSFET 1022
Fengming Yang, Xin Li, Yifei Luo, Lin Liang, Yongle Huang

GaN-Based 1.5 MHz Synchronous Buck Converter with Partial Soft-Switching Control Scheme 1028
Zeqi Yang, Yuan Liu, Zhe Zhang

Design and Demonstration of a Novel SiC Trench MOSFET with Periodically Grounded P Shield Island Based on Secondary Epitaxy Process 1033
Yangyang Wu, Fei Guo, Kuan Wang, Wei Chen, Zhijie Cheng, Yuan Jun, Rong Zhang, Guoqing Xin, Zhiqiang Wang

Author Index

2025 IEEE Workshop on Wide Bandgap Power Devices and Applications in Asia (WiPDA Asia)

A Novel Modulation Method with Simultaneous Reduction of Common-Mode Voltage and Switching Losses for Three-Level SNPC Inverter

Shuangxi Zhu
School of Electrical and Electronic Engineering
Huazhong University of Science and Technology
Wuhan, China
xi_zi@hust.edu.cn

Jiajia Guan
School of Electrical and Electronic Engineering
Huazhong University of Science and Technology
Wuhan, China
jiajiaguan@hust.edu.cn

Cai Chen
School of Electrical and Electronic Engineering
Huazhong University of Science and Technology
Wuhan, China
caichen@hust.edu.cn

Yong Kang
School of Electrical and Electronic Engineering
Huazhong University of Science and Technology
Wuhan, China
ykang@hust.edu.cn

Abstract—The three-phase three-level (3-L) sparse neutral point clamped (SNPC) inverter is representing a promising alternative for motor drives and three-phase PFC rectifier systems. For the operation of these applications, the common-mode (CM) voltage is needed to be reduced. However, conventional modulation schemes with reduced CM voltage leads to high switching losses of the 2-L inverter stage. To address this issue, a novel modulation method based on zero vector transitions is proposed in this article. The proposed method reduces CM voltage amplitude to Vdc/6, and achieves zero switching loss of the 2-L inverter stage. The switching state sequence is determined, and the vectors dwell-time calculation is reconsidered. Finally, the proposed modulation method is verified by the experiment using a 1.5-kW hardware prototype.

Keywords— *Three-phase three-level (3-L) sparse neutral point clamped (SNPC) inverter, common-mode (CM) voltage, zero-voltage switching (ZVS)*

I. INTRODUCTION

The three-level (3-L) sparse neutral point clamped (SNPC) inverter illustrated in Fig. 1 is a particularly promising candidate in three-phase 3-L topologies as lower number of power transistors are employed, i.e., only 10 power transistors, compared to conventional three-phase 3-L inverters [1], [2], [3]. The 3-L SNPC inverter features a cascaded structure composed of a 3-L buck stage and a three-phase 2-L inverter stage. The devices of the 3-L buck stage only need to block half the DC-link voltage, while the voltage rating of the 2-L inverter stage's devices equals the full DC-link voltage. Furthermore, the 3-L buck stage and the 2-L inverter stage can adopt different power semiconductor devices combinations to achieve a better performance with a slight increase in cost [4], e.g., SiC MOSFETs & Si IGBTs, and GaN HEMTs & Si IGBTs.

The operating principle and space vector pulse width modulation (SVPWM) scheme of the SNPC inverter have been analyzed in [1]. Several space vector modulation sequences with different number of switching transitions are identified and compared. the design of SNPC inverter with GaN-Si hybrid structure for motor drive applications is reported in [4]. The mid-point voltage control method based

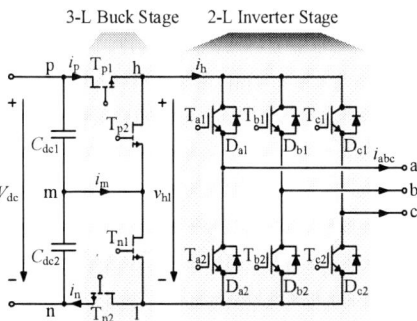

Fig. 1. Schematic of the power circuit of the three-level(3-L) sparse neutral point clamped (SNPC) inverter, which includes the 3-L buck stage and three-phase 2-L inverter stage.

on virtual space vector modulation (VSVM) are analyzed in [5], [6]. A complete analysis of the SNPC inverter is reported and several modulation sequences are investigated in [7]. Among all possible modulation sequences, sequence *U* is the state-of-the-art symmetric modulation sequence. Sequence O generates low switching losses as only six switching transitions occur in a carrier period. However, the switching losses of the 2-L inverter stage remain high when applying the aforementioned sequences, resulting in reduced efficiency. This issue is further exacerbated by the typical use of Si IGBTs in the 2-L inverter stage [4], [8]. Furthermore, SNPC inverter also suffers from high-frequency pulse common-mode (CM) voltage, which incurs leakage currents, and bearing and insulation degradation of the machines for the applications [9], [10].

Thus, to both reduce CM voltage and switching losses for 3-L SNPC inverter, this article proposes a novel modulation method. With the proposed method, the maximum amplitude of CM voltage is reduced to $V_{dc}/6$ and devices of 2-L inverter stage can achieve zero switching loss. The proposed modulation method is verified by the experiment.

979-8-3315-1110-4/25 $31.00 © 2025 IEEE
520

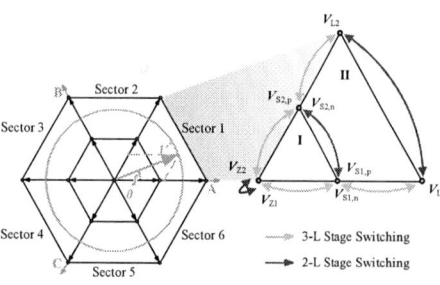

Fig. 3. Output voltage space vector hexagon of 3-L SNPC inverter and available voltage space vectors, including zero, small, and large vectors, in Sector 1.

TABLE II
AVAILABLE VOLTAGE SPACE VECTORS OF SECTOR 1

Vector	$S_{p,n}$	$S_{a,b,c}$	$|V|$	i_m	v_{CM}
V_{Z1}	0,1	1,0,0	0	0	0
V_{Z2}	0,1	1,1,0	0	0	0
$V_{S1,p}$	1,1	1,0,0	$V_{dc}/3$	$-i_a$	$+V_{dc}/6$
$V_{S1,n}$	0,0	1,0,0	$V_{dc}/3$	$+i_a$	$-V_{dc}/3$
$V_{S2,p}$	1,1	1,1,0	$V_{dc}/3$	$+i_c$	$+V_{dc}/3$
$V_{S2,n}$	0,0	1,1,0	$V_{dc}/3$	$-i_c$	$-V_{dc}/6$
V_{L1}	1,0	1,0,0	$2V_{dc}/3$	0	$-V_{dc}/6$
V_{L2}	1,0	1,1,0	$2V_{dc}/3$	0	$+V_{dc}/6$

II. OPERATING PRINCIPLE OF SNPC INVERTER

The 3-L SNPC inverter comprises a 3-L buck stage and a three-phase 2-L inverter stage as shown in Fig. 1. The role of the 3-L buck stage is to create a 3-L switched rail-to-rail voltage v_{hl} including three voltage levels $\{0, V_{dc}/2, V_{dc}\}$. Thus, 3-L switched voltages are available at the three-phase 2-L inverter stage's output terminals.

A. Converter Switching States

The SNPC converter switching states can be identified by the 3-L buck stage and 2-L inverter stage bridge-leg switching functions

$$S_x = \begin{cases} 0, & \text{if } T_{x1} \text{ off, } T_{x2} \text{ on} \\ 1, & \text{if } T_{x1} \text{ on, } T_{x2} \text{ off} \end{cases} \quad x \in p, n, a, b, c. \quad (1)$$

4 switching states exist for the 3-L buck stage, while 2-L inverter stage consists of 8 switching states. In order to operate with 1/3-PWM mode, the 2-L inverter stage zero states are not considered, thus the total count of 2-L inverter stage switching states is 6. The output voltage space vector hexagon is shown in Fig. 2. Three kinds of space vectors are existed and categorized according to their amplitude, i.e., zero vectors V_Z, small vectors V_S, and large vectors V_L. The available space vectors of Sector 1 are listed in TABLE I.

The modulation index is defined as

$$M = \frac{V^*}{V_{dc}/2}, \quad (2)$$

where V^* is the amplitude of the output phase voltage reference. Each sector can be divided in two main areas: area I, and area II. The boundary between two areas is defined as

TABLE I
CONVENTIONAL AND PROPOSED SWITCHING STATE SEQUENCES

Name	Switching State Sequence
U	$V_{S1,p} \leftrightarrows V_{Z1/L1} \leftrightarrows V_{S1,n} \leftrightarrows V_{S2,n} \leftrightarrows V_{Z2/L2} \leftrightarrows V_{S2,p}$
O	$V_{S1,p} \to V_{S2,p} \to V_{Z2/L2} \to V_{S2,n} \to V_{S1,n} \to V_{Z1/L1} \to V_{S1,p}$
N	$V_{L1} \leftrightarrows V_{S1,p} \leftrightarrows V_{Z1} \leftrightarrows V_{Z2} \leftrightarrows V_{S2,n} \leftrightarrows V_{L2}$

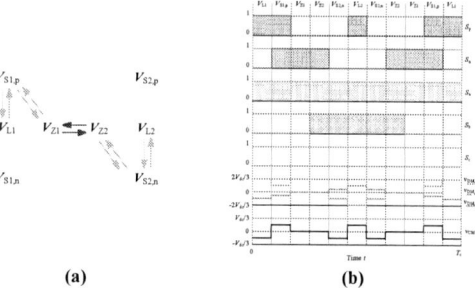

Fig. 2. Switching state sequence N, corresponding (a) graphical representation, and (b) bridge-leg switching functions, DM and CM voltage waveforms.

$$M_{lim} = \frac{1}{\sqrt{3}\cos(\theta - \pi/6)}, \quad (3)$$

i.e., if $M < M_{lim}$, the inverter operates in area I; otherwise it operates in area II.

B. Modulation Strategies

The existing high performing switching state sequences for the SNPC inverter are reported in [7], as listed in TABLE II, where $V_{Z1/L1}$ and $V_{Z2/L2}$ mean that V_{Z1} and V_{Z2} are used in area I and are replaced by V_{L1} and V_{L2} in area II. Sequence U is selected as typical symmetric modulation sequence. Sequence O only has six switching transitions in a carrier period.

III. NOVEL MODULATION METHOD WITH REDUCTION OF CM VOLTAGE AND SWITCHING LOSSES

In order to achieve CM voltage reduction and reduce the losses of the 2-L inverter stage, a novel modulation method based on zero vector transitions is proposed, which reduces the CM voltage to $V_{dc}/6$ and achieves zero switching loss of the 2-L inverter stage.

A. Switching State Sequence Determination

To reduce CM voltage and switching losses, only vectors with lower CM voltage are used, and the bridge-leg of the 2-L inverter stage switches at zero vector transitions. Complying with above limits, the switching state sequence based on zero vector transitions is identified and reported in Fig. 3 and TABLE II. The proposed switching sequence is named sequence N, according to the graphical representation in Fig. 3. Note that the maximum amplitude of CM voltage is reduced to $V_{dc}/6$. Furthermore, the bridge-leg of the 2-L inverter stage switches at zero vector transitions, i.e., switched voltage equals 0, leading to zero switching loss of the 2-L inverter stage.

TABLE III
NOMINAL OPERATING CONDITIONS AND KEY COMPONENTS PARAMETERS OF THE SNPC INVERTER PROTOTYPE

	Description	Value
V_{dc}	DC-link voltage	800 V
M	Modulation index	0.8
I_{acm}	Output phase current amplitude	3.2 A
P	Output power	1.5 kW
f_c	Carrier frequency	20 kHz
T_B	Buck stage transistors	GS66506T
T_I	Inverter stage transistors	IKW08N120CS7
$C_{dc1/2}$	DC-link capacitor	23 uF

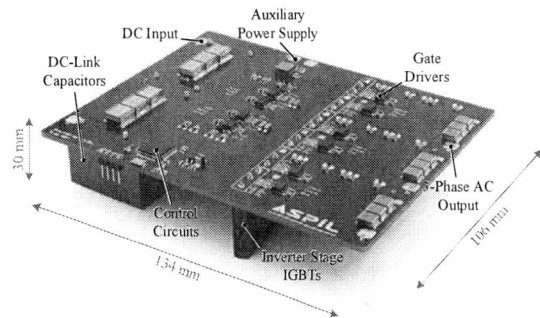

Fig. 4. Hardware prototype of the 1.5-kW 3-phase 3-L SNPC inverter, featuring a boxed volume of $134 \times 106 \times 30$ mm^3.

B. Space Vectors Dwell-time Calculation

Assuming that the reference voltage vector is located in Sector 1, the vectors employed to synthesize the reference voltage vector have to satisfy

$$V^* \angle \theta = (\delta_{S1} + 2\delta_{L1})\frac{V_{dc}}{3} \angle 0° \\ + (\delta_{S2} + 2\delta_{L2})\frac{V_{dc}}{3} \angle 60° + \delta_Z 0 , \quad (4)$$

where δ_{S1} and δ_{S2} represent the duty cycle of the small vectors, δ_{L1} and δ_{L2} represent the duty cycle of the large vectors, δ_Z represents the duty cycle of the zero vectors.

Equation (4) can also be written as

$$M\frac{V_{dc}}{2} \angle \theta = \delta_1 \frac{V_{dc}}{3} \angle 0° \\ + \delta_2 \frac{V_{dc}}{3} \angle 60° + \delta_Z 0 , \quad (5)$$

where

$$\begin{cases} \delta_1 = \delta_{S1} + 2\delta_{L1} \\ \delta_2 = \delta_{S2} + 2\delta_{L2} \end{cases}, \quad (6)$$

δ_1 and δ_2 can be derived as follows

$$\begin{cases} \delta_1 = \sqrt{3}M \sin(\pi/3 - \theta) \\ \delta_2 = \sqrt{3}M \sin(\theta) \end{cases}. \quad (7)$$

Let

$$\begin{cases} \delta_{S1} = 2\lambda\delta_1 \\ \delta_{S2} = 2\lambda\delta_2 \end{cases}. \quad (8)$$

The duty cycles δ_{S1}, δ_{S2}, δ_{L1}, δ_{L2}, and δ_Z can be obtained as follows:

$$\begin{cases} \delta_{S1} = 2\lambda\delta_1 \\ \delta_{S2} = 2\lambda\delta_2 \\ \delta_{L1} = (1/2 - \lambda)\delta_1 \\ \delta_{L2} = (1/2 - \lambda)\delta_2 \\ \delta_Z = 1 - (1/2 + \lambda)(\delta_1 + \delta_2) \end{cases}. \quad (9)$$

λ should satisfy

$$\begin{cases} \lambda \geq \max(0, (\frac{1}{2} - \frac{1}{\delta_1}), (\frac{1}{2} - \frac{1}{\delta_2})) \\ \lambda \leq \min(\frac{1}{2}, \frac{1}{2\delta_1}, \frac{1}{2\delta_2}, (\frac{1}{\delta_1 + \delta_2} - \frac{1}{2})) \end{cases}. \quad (10)$$

The lower and upper limits of parameter λ can be determined based on the range of each term in the functions max() and min() in (10). Thus, the lower and upper limits of parameter λ are obtained as follows:

$$\begin{cases} \lambda \in [0, \frac{1}{2}], & \text{Area I} \\ \lambda \in [0, \frac{1}{\delta_1 + \delta_2} - \frac{1}{2}], & \text{Area II} \end{cases}. \quad (11)$$

The parameter λ increases, δ_S becomes greater while δ_Z becomes smaller. In other words, parameter λ determines the duty cycle ratio of small vectors V_S and zero vectors V_Z. According to the lower and upper limits, parameter λ is defined as

$$\begin{cases} \lambda = \frac{1}{2}k, & \text{Area I} \\ \lambda = k(\frac{1}{\delta_1 + \delta_2} - \frac{1}{2}), & \text{Area II} \end{cases}, \quad (12)$$

where $0 \leq k \leq 1$. Once the parameter k is determined, the duty cycle calculation can be obtained.

IV. EXPERIMENT RESULTS AND ANALYSIS

A hardware prototype of the 3-phase 3-L SNPC inverter shown in Fig. 4 has been built to experimentally quantify the performance of proposed modulation sequence. 650 V GaN HEMTs are employed in the 3-L buck stage and 1200 V Si IGBTs in the 2-L inverter stage.

Fig. 5 verifies the 3-L SNPC inverter prototype operation considering the two conventional modulation sequences and proposed sequence N under the fundamental resistive load operation with a DC-link voltage of 800 V, and modulation index $M = 0.8$. DC-link mid-point voltage v_{mn}, phase switched voltage v_{am}, switched line-to-line voltage v_{ab} are measured. The CM voltage v_{CM} is obtained by averaging three-phase switched voltage. Note that the amplitude of CM voltage decreases obviously to about 133 V ($V_{dc}/6$) for the proposed modulation sequence N, and the switched line-to-line voltage v_{ab} has a zero level near the sinusoidal local average voltage peak, as zero vectors are employed in proposed sequence N.

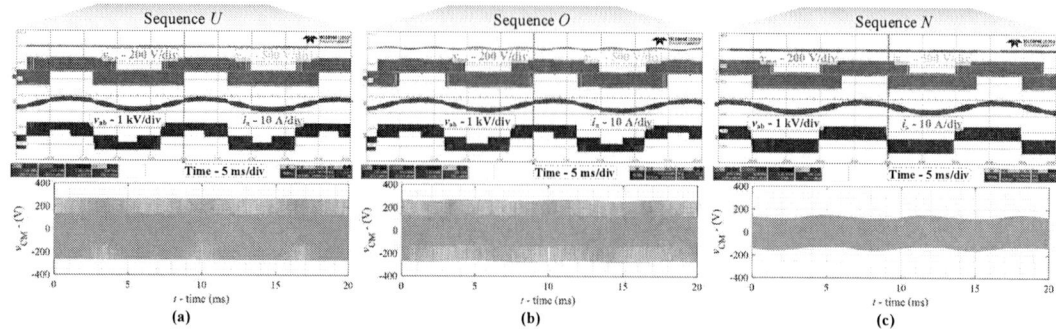

Fig. 6. Experimental waveforms of the 3-L SNPC inverter, under the resistive load operation. Two conventional modulation sequences and proposed sequence are considered, i.e., (a) sequence U, (b) sequence O, and (c) sequence N.

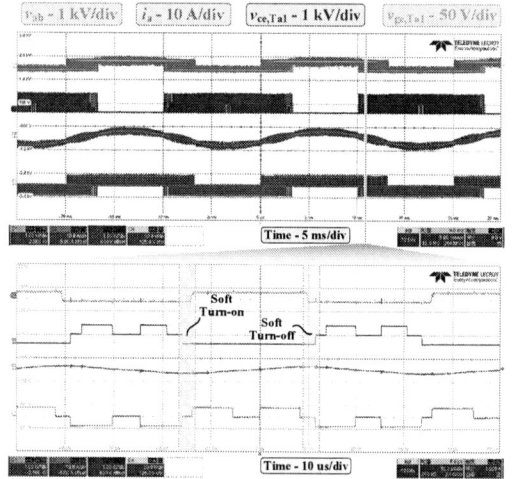

Fig. 5. Zero-voltage turn-on and turn-off waveforms of the inverter stage device T_{a1} under the resistive fundamental load operation with proposed modulation sequence N.

Fig. 6 illustrates switching cycle waveforms under the resistive fundamental load operation with proposed modulation sequence N. Note that the drain-source voltage drops to zero during both turn-on and turn-off processes of 2-L inverter stage devices, thus achieving zero switching loss.

V. Conclusion

This article proposed a novel modulation method for 3-L SNPC inverter to reduce CM voltage and switching losses. The operating principle of SNPC inverter was summarized first, and then detailed analysis of proposed modulation sequence N was provided. The performance of the proposed sequence N was verified with experimental results of a 1.5-kW GaN/Si structure hardware prototype.

References

[1] A. Lange and B. Piepenbreier, "Space vector modulation for three-level simplified neutral point clamped (3L-SNPC) inverter," in *Proc. IEEE Workshop Control Model. Power Electron.*, 2017, pp. 1–8.

[2] M. Schweizer, T. Friedli, and J. W. Kolar, "Comparative Evaluation of Advanced Three-Phase Three-Level Inverter/Converter Topologies Against Two-Level Systems," *IEEE Trans. Ind. Electron.*, vol. 60, no. 12, pp. 5515–5527, Dec. 2013.

[3] D. Cittanti, M. Guacci, S. Miric, R. Bojoi, and J. W. Kolar, "Comparative Evaluation of 800V DC-Link Three-Phase Two/Three-Level SiC Inverter Concepts for Next-Generation Variable Speed Drives," in *Proc. Int. Conf. Elect. Mach. Syst.*, 2020, pp. 1699–1704.

[4] A. Lange, J. Lautner, and B. Piepenbreier, "High Efficiency Three-Level Simplified Neutral Point Clamped (3L-SNPC) Inverter with GaN-Si Hybrid Structure," in *Proc. Int. Power Conv. and Intelligent Motion Conf.*, 2018, pp. 1–7.

[5] C. Qin and X. Li, "Improved Virtual Space Vector Modulation Scheme for the Reduced Switch Count Three-Level Inverter With Balanced and Unbalanced Neutral-Point Voltage Conditions," *IEEE Trans. Power Electron.*, vol. 38, no. 2, pp. 2092–2104, Feb. 2023.

[6] X. Deng *et al.*, "A Novel Virtual Space Vector Modulation With Optimized Neutral-Point Voltage Control Capability for Ten-Switch Three-Phase Three-Level Inverter," *IEEE Trans. Ind. Electron.*, vol. 71, no. 2, pp. 1081–1092, Feb. 2024.

[7] D. Cittanti, M. Guacci, S. Mirić, R. Bojoi, and J. W. Kolar, "Analysis and performance evaluation of a three-phase sparse neutral point clamped converter for industrial variable speed drives," *Electr. Eng.*, vol. 104, no. 2, pp. 623–642, Apr. 2022.

[8] D. Zhang, D. Cittanti, P. Sun, J. Huber, R. I. Bojoi, and J. W. Kolar, "Detailed Modeling and In-Situ Calorimetric Verification of Three-Phase Sparse NPC Converter Power Semiconductor Losses," *IEEE J. Emerg. Sel. Top. Power Electron.*, vol. 11, no. 3, pp. 3409–3423, Jun. 2023.

[9] X. Deng, H. Wang, X. Zhu, H. Wang, W. Zhang, and X. Yue, "Common-Mode Voltage Reduction and Neutral-Point Voltage Control Using Space Vector Modulation for Coupled Ten-Switch Three-Phase Three-Level Inverter," *IEEE Trans. Power Electron.*, vol. 37, no. 6, pp. 6397–6411, Jun. 2022.

[10] F. Wang, Z. Li, and X. Tong, "Modified Predictive Control Method of Three-Level Simplified Neutral Point Clamped Inverter for Common-Mode Voltage Reduction and Neutral-Point Voltage Balance," *IEEE Access*, vol. 7, pp. 119476–119485, Aug. 2019.

An Active Gate Driver of Voltage Overshoot Suppressing for SiC MOSFETs

Tingwen Hu
School of Electrical Engineering
Southwest Jiaotong University
Chengdu, China
hutingwen@my.swjtu.edu.cn

Wensheng Song
School of Integrated Circuits Science
and Engineering
Southwest Jiaotong University
Chengdu, China
songwsh@swjtu.edu.cn

Jian Chen
School of Integrated Circuits Science
and Engineering
Southwest Jiaotong University
Chengdu, China
chenjian@swjtu.edu.cn

Tao Tang
School of Electrical Engineering
Southwest Jiaotong University
Chengdu, China
christophertang@my.swjtu.edu.cn

Hao Yue
School of Electrical Engineering
Southwest Jiaotong University
Chengdu, China
yuehao6866@my.swjtu.edu.cn

Guoyou Liu
School of Integrated Circuits Science
and Engineering
Southwest Jiaotong University
Chengdu, China
liugy@csrzic.com

Abstract—**Compared to silicon-based devices, silicon carbide metal-oxide-semiconductor field-effect transistors (SiC MOSFETs) offer faster switching speeds. However, the fast switching speed of SiC MOSFETs and the inevitable parasitic inductance in the power loop will cause the severe voltage overshoot during the device turn-OFF transient. This paper proposes an active gate driver (AGD) that injects additional gate current into the device during specific stage of the turn-OFF process. This method effectively suppresses v_{ds} overshoot peak under variable load current conditions while exerting less effect on switching speed and switching loss. Finally, an experimental platform is designed to validate the proposed AGD scheme. The experimental results demonstrate that the method achieves a better trade-off between voltage overshoot and turn-OFF loss, while effectively suppressing the v_{ds} overshoot peak under variable load current conditions.**

Keywords—*Active gate driver, overshoot, SiC MOSFETs.*

I. INTRODUCTION

This Wide-bandgap semiconductor power devices, represented by SiC MOSFETs, have been widely utilized in fields such as electric vehicles, renewable energy generation, and aerospace, owing to their fast switching speed, low conduction resistance, and high operating temperature capabilities[1], [2]. However, due to the parasitic inductance in the power loop, the high di/dt of SiC MOSFETs will cause severe overshoot in the drain-source voltage v_{ds}[3].

At present, various approaches have been put forward to suppress the voltage overshoot during device turn-OFF transient, including increasing the gate resistance [4], optimizing the PCB layout [5], adding snubber or damping circuits [6], and employing active gate driver (AGD). AGD technologies have garnered attention for their flexible control over the switching trajectory of SiC MOSFETs. By adjusting driving parameters during the switching process, it offers a highly promising method for suppressing voltage overshoot during the turn-OFF transient of the device [7].

Numerous AGDs have been proposed to optimize the switching performance of SiC MOSFETs. Among them, the open-loop control methods are facile to implement and have a relatively low cost. However, the switching characteristics of SiC MOSFETs exhibit a rather strong dependence on the load current or input voltage. Open-loop control methods

necessitates modeling the device or obtaining the device characteristics under different input voltages and load currents through experiments before designing the driving parameters. Consequently, open-loop control methods are difficult to achieve the optimal performance under varying input conditions.

[1], [3] and [8] establish feedback circuits to detect the stages of the SiC MOSFET switching process, and switch the driving parameters during di/dt or dv/dt stage to optimizing the switching trajectory. However, these methods require setting appropriate trigger thresholds based on the input voltage or load current to ensure that the gate driving parameters switch at the correct switching stage. Under varying load current conditions, the gate driving parameters may switch at incorrect times, resulting in a degradation of the SiC MOSFET's performance.

Therefore, to suppress the v_{ds} overshoot peak of SiC MOSFETs under varying load current conditions, this paper proposes an AGD. The proposed method detects only v_{ds} to identify the stages of the SiC MOSFET's turn-OFF transient. During the di/dt stage, the driving current is adjusted to suppress the v_{ds} overshoot peak. The proposed AGD remains unaffected by changes in load current, thereby enhancing its practicality.

II. PRINCIPLE OF THE PROPOSED AGD

A. The Principle of v_{ds} Overshoot

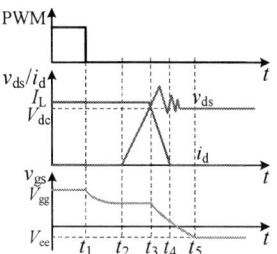

Fig. 1. The turn-OFF waveforms for SiC MOSFETs.

As shown in Fig. 1, the turn-OFF process of a SiC MOSFET can be divided into four stages:

This work was supported in part by the National Natural Science Foundation of China under Grant 52307224.

1). During t_1-t_2, referred to as the turn-OFF delay stage, the gate-source voltage v_{gs} begins to drop, but the drain-source voltage v_{ds} and drain current i_d remain largely unchanged.

2). In the t_2-t_3 stage, the v_{ds} increases significantly as the device transitions into the off-state.

3). In the t_3-t_4 stage, the i_d decreases sharply. The voltage overshoot occurs in this stage due to the interaction of rapid current change (di_d/dt) and parasitic inductance in the power loop. In this stage, di_d/dt can be calculated using (1).

$$\frac{di_d}{dt} = \frac{g_m V_{ee} - g_m \left(I_L + V_{th}\right)}{R_g C_{iss}} \quad (1)$$

Here, g_m represents the transconductance of the device, I_L is the load current, V_{th} is the threshold voltage of the device, R_g is the gate resistance, and C_{iss} is the input capacitance of the device. The peak value of v_{ds} overshoot can be calculated using (2).

$$V_{ds\_peak} = V_{dc} + L_{loop} \left| \frac{di_d}{dt} \right| \quad (2)$$

Here, V_{dc} is the bus voltage, and L_{loop} is the parasitic inductance of the loop.

4). During t_4-t_5, the v_{gs} drops below the threshold voltage, and the i_d decreases to 0 A. At t_5, the v_{gs} stabilizes at V_{ee}, completing the turn-OFF process.

B. The Principle of the Proposed AGD

As shown in Fig. 2, the proposed AGD consists of the following 3 components:

I: RC voltage divider network

II: Stage detection circuit

III: Voltage-controlled current source

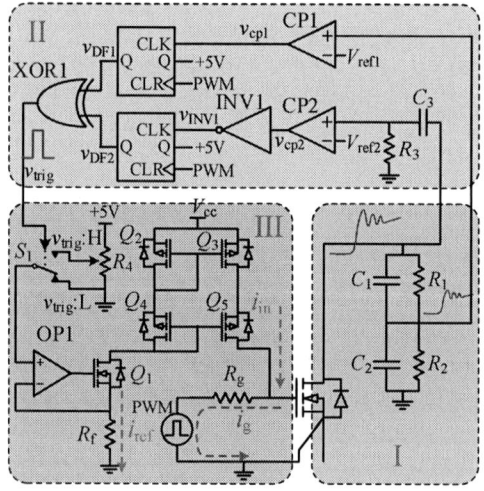

Fig. 2. The proposed AGD topological structure and its hardware implementation.

The proposed AGD uses an RC voltage divider network (Part I in Fig. 2) to step down the hundreds of volts of the SiC MOSFET's v_{ds} to a voltage range acceptable for the subsequent logic circuits. The stage detection circuit (Part II in Fig. 2) processes the output of the RC divider, and through

logical operations, identifies the current decrease stage during the SiC MOSFET turn-OFF transient. During this stage, it generates a trigger pulse v_{trig}. Finally, v_{trig} controls the closing of the switch S_1 at the input of a voltage-controlled current source (Part III in Fig. 2). This allows the current source to output additional current i_{in}, which is injected into the gate of the SiC MOSFET.

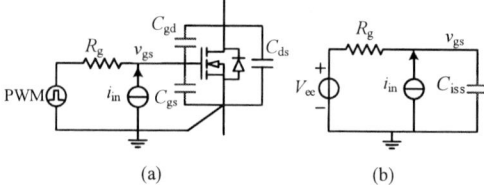

(a)   (b)

Fig. 3. The simplified circuit of the proposed AGD for the id decreasing stage (t_3-t_4). (a) The simplified circuit when the voltage-controlled current source is operating. (b) The further simplified circuit.

Furthermore, the proposed AGD can be simplified to the circuit shown in Fig. 3 when injecting current into the SiC MOSFET gate during the i_d falling stage. When current i_{in} is injected into the gate of the SiC MOSFET, di_d/dt can be calculated using (3).

$$\frac{di_d}{dt} = \frac{V_{ee} - (I_L/g_m + V_{th}) + i_{in} R_g}{(R_g C_{iss})/g_m} \quad (3)$$

Therefore, the expression for the overshoot voltage vos is denoted by:

$$v_{os} = L_{loop} \frac{I_L/g_m + V_{th} - i_{in} R_g - V_{ee}}{(R_g C_{iss})/g_m}. \quad (4)$$

From (4), it can be observed that v_{os} can be regulated by adjusting i_{in}. The larger the i_{in}, the slower the decrease in i_d, resulting in a smaller v_{os}.

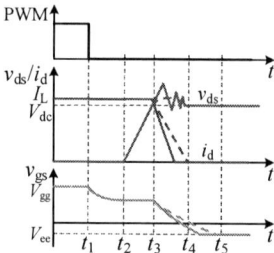

Fig. 4. The turn-OFF waveforms of SiC MOSFETs with and without AGD. (Dashed line represents the turn-OFF waveforms of SiC MOSFETs with AGD; solid line represents the turn-OFF waveforms of SiC MOSFETs without AGD).

As shown in Fig. 4, the proposed AGD controls the di_d/dt during the turn-OFF transient of the SiC MOSFET. The injected current i_{in} adjusts the di/dt during the device turn-OFF process, suppressing the v_{ds} overshoot peak. This method almost does not extend the turn-OFF time of the device and has minimal impact on turn-OFF loss. Fig. 5 illustrates the timing of the key signals in the proposed AGD. It should be noted that in Fig. 2, the reference voltage V_{ref1} of OP1 is theoretically equal to V_{dc}. However, due to the unavoidable delay time t_{delay} in the circuit, it is necessary to appropriately reduce the reference voltage V_{ref1} so that the comparator CP1 is triggered at t'_3 in Fig. 5. This ensures that i_{in}, after t_{delay}, is output during the i_d falling stage. The value of V_{ref1} can be calculated as

$$V_{ref1} = \left(V_{dc} - t_{delay} \frac{dv_{ds}}{dt} \right) \frac{R_2}{R_1 + R_2}. \quad (5)$$

Furthermore, at the end of the i_d falling stage, i_d becomes 0 A. Therefore, even if i_{in} continues to be output after the i_d falling stage, it will not increase the turn-OFF loss of the device. As a result, there is no need for special design considerations for V_{ref2}. In this paper, V_{ref2} is designed as 0.1 V, making CP2 function as a zero-crossing comparator.

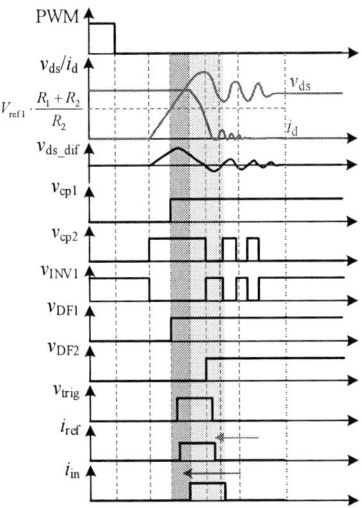

Fig. 5. The signal timing of the proposed AGD.

III. EXPERIMENTAL VALIDATION

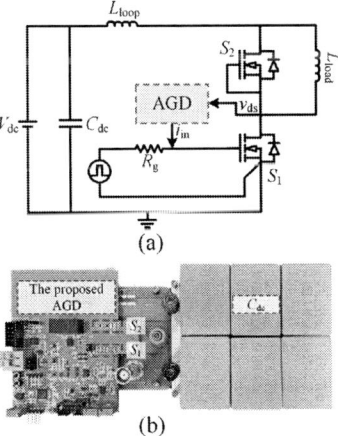

Fig. 6. The experimental platform. (a) Schematic diagram. (b) The proposed AGD.

TABLE I. CIRCUIT PARAMETERS

Symbol	Values
Input voltage V_{dc}	600 V
Load inductance L_{Load}	139 uH
S_1	C3M0040120K
S_2	C3M0040120K
Driver resistor R_g	15 Ω

Based on the above theoretical analysis and the principle of the AGD circuit, an experimental platform is constructed as shown in Fig. 6, with the main parameters provided in TABLE I.

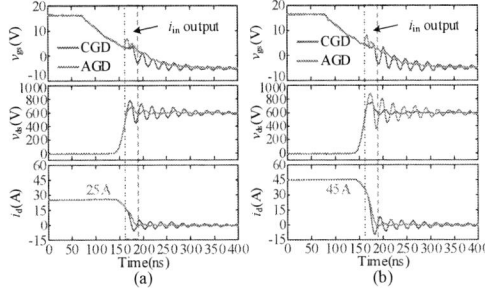

Fig. 7. The turn-OFF waveforms of the SiC MOSFET under different load currents. (a) 600 V/25 A. (b) 600 V/45 A.

Fig. 7 shows the experimental waveforms of the proposed AGD and conventional gate driver (CGD) under test conditions of 25 A and 45 A. During the turn-OFF transient of the SiC MOSFET, the proposed AGD slightly slows down the i_d falling speed, which results in the suppression of the v_{ds} overshoot peak. Moreover, it can be observed that dv_{ds}/dt remains unaffected. Additionally, the oscillations in v_{ds} are mitigated. At the same time, the operation of the proposed AGD does not rely on any signals related to the load current. The proposed method does not require adjusting the circuit parameters based on the load size. Therefore, the circuit parameters of the proposed AGD remain unchanged under both 25 A and 45 A load currents.

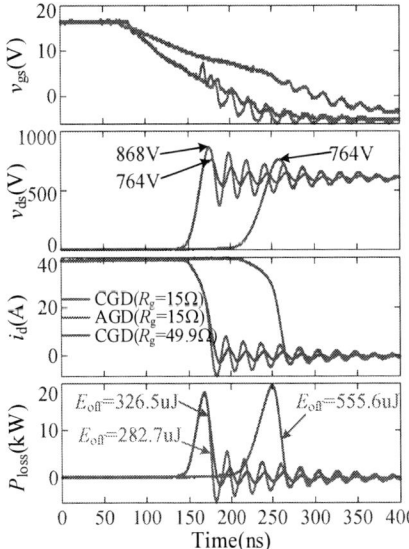

Fig. 8. Waveforms of v_{gs}, v_{ds}, i_d, and energy loss when applying the proposed AGD and the conventional increased gate resistance method, respectively (Test condition: 600V/40A).

Fig. 8 presents the experimental waveforms of the proposed AGD, CGD, and the increased gate resistance method. Under the test condition of 600V/40A, when employing CGD without any voltage overshoot suppression measures, the turn-OFF voltage overshoot peak of the SiC MOSFET reaches 868 V, corresponding to a turn-OFF loss of

282.7 μJ. After increasing the gate resistance to 49.9 Ω with CGD, the voltage overshoot peak decreases to 764 V, representing a 38.8% reduction in overshoot magnitude. However, the turn-OFF loss significantly rises from 282.7 μJ to 555.6 μJ, marking a 96.5% increase. The proposed AGD scheme effectively suppresses voltage overshoot by injecting additional driving current into the gate during the i_d falling stage of device turn-OFF to regulate the current slew rate (di_d/dt). When controlling the voltage overshoot peak to the same 764 V, the AGD scheme achieves a turn-OFF loss of only 326.5 μJ, demonstrating a substantial reduction in turn-OFF loss compared to the traditional increased gate resistance method. Thus, the proposed method achieves a better trade-off between turn-OFF voltage suppression and turn-OFF loss increase. Furthermore, the conventional method of increasing gate resistance significantly prolongs the turn-OFF delay time. In practical applications, this compels the system to extend dead-time periods to prevent shoot-through faults in bridge legs, consequently degrading both the efficiency and output power quality of SiC MOSFET converters. Such compromise ultimately undermines the inherent high-speed switching advantages of SiC devices.

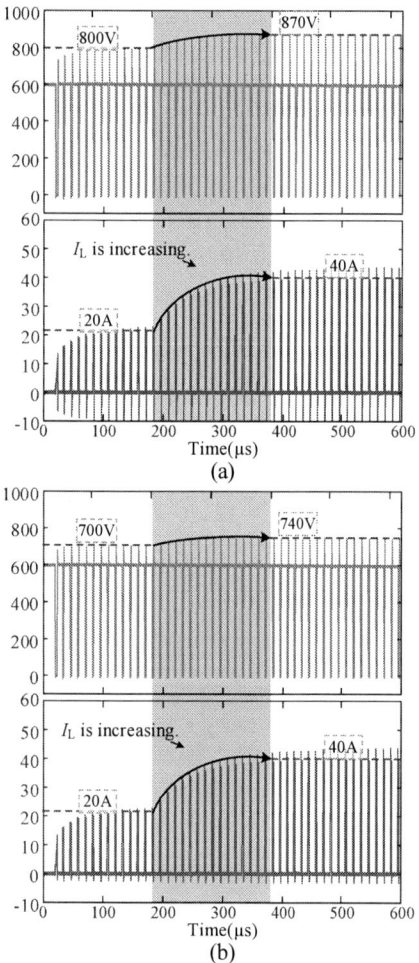

Fig. 9. Comparative experimental waveforms of AGD versus CGD during load current step change from 20A to 40A (DC bus voltage: 600V). (a)

Waveforms under conventional CGD control. (b) Waveforms under the proposed AGD control in this work.

To further validate the effectiveness of the proposed AGD under varying load current conditions, the suppression effect on the v_{ds} overshoot peak is tested under a rapid load current change from 20A to 40A. As shown in Fig. 9(a), when using CGD, the turn-OFF voltage overshoot peak reaches 800 V at a 20A load current. When the load current rapidly increases to 40A, the voltage overshoot peak rises to 870 V. Fig. 9(b) shows that with the proposed AGD, the voltage spikes at 20A and 40A load currents are reduced to 700V and 740, respectively, representing reductions of 50.0% and 48.1%.

IV. CONCLUSION

This paper presents an AGD for suppressing v_{ds} overshoot, featuring a stage recognition circuit and a gate current regulation circuit that do not rely on load current-related signals. The proposed method only adjusts the i_d falling speed during the SiC MOSFET turn-OFF transient to suppress v_{ds} overshoot, with minimal impact on the device turn-OFF speed and turn-OFF loss. An experimental platform is built for verification, and the experimental results demonstrate that the proposed method effectively suppresses v_{ds} overshoot under different load currents.

REFERENCES

[1] A. P. Camacho, V. Sala, H. Ghorbani and J. L. R. Martinez, "A novel active gate driver for improving SiC MOSFET switching trajectory," *IEEE Trans. Ind. Electron*, vol. 64, no. 11, pp. 9032–9042, Nov. 2017.

[2] J. Chen, X. Du, Q. Luo, X. Zhang, P. Sun and L. Zhou, "A review of switching oscillations of wide bandgap semiconductor devices," *IEEE Trans. Power Electron*, vol. 35, no. 12, pp. 13182–13199, Dec. 2020.

[3] P. Nayak and K. Hatua, "Active gate driving technique for a 1200 V SiC MOSFET to minimize detrimental effects of parasitic inductance in the converter layout," *IEEE Trans. Ind. Appl.*, vol. 54, no. 2, pp. 1622–1633, March–April. 2018.

[4] Z. Wu, H. Jiang, Z. Zheng, X. Qi, H. Ma, and L. Liu, L. Ran, "Dynamic dv/dt control strategy of SiC MOSFET for switching loss reduction in the operational power range," *IEEE Trans. Power Electron.*, vol. 37, no. 6, pp. 6237–6241, June. 2022.

[5] N. Zhang, S. Wang, and H. Zhao, "Develop parasitic inductance model for the planar busbar of an IGBT H bridge in a power inverter," *IEEE Trans. Power Electron.*, vol. 30, no. 12, pp. 6924–6933, Dec. 2015.

[6] J. Kim, D. Shin, and S.-K. Sul, "A damping scheme for switching ringing of full SiC MOSFET by air core PCB circuit," *IEEE Trans. Power Electron.*, vol. 33, no. 6, pp. 4605–4615, Jun. 2018.

[7] J. Henn, et al., "Intelligent gate drivers for future power converters," *IEEE Trans. Power Electron.*, vol. 37, no. 3, pp. 3484–3503, Mar. 2022.

[8] J. Cao, Z.-K. Zhou, Y. Shi, and B. Zhang, "An integrated gate driver based on SiC MOSFETs adaptive multi-level control technique," *IEEE Trans. Circuits Syst. I*, vol. 70, no. 4, pp. 1805–1816, Apr. 2023.

2025 IEEE Workshop on Wide Bandgap Power Devices and Applications in Asia (WiPDA Asia)

A High-Frequency LLC Resonant Converter Incorporating Matrix Transformer

Haochen Zhang
Faculty of Electrical Engineering and Automation
Harbin Institute of Technology
Harbin, China
24S106264@stu.hit.edu.cn

Yueshi Guan
Faculty of Electrical Engineering and Automation
Harbin Institute of Technology
Harbin, China
guanyueshi@hit.edu.cn

Yijie Wang
Faculty of Electrical Engineering and Automation
Harbin Institute of Technology
Harbin, China
wangyijie@hit.edu.cn

Dianguo Xu
Faculty of Electrical Engineering and Automation
Harbin Institute of Technology
Harbin, China
xudiang@hit.edu.cn

Abstract—With the development of modern power electronics technology, the miniaturization of switching power supplies and the magnetic integration technology of planar transformers are both topics worth studying. In this paper, a high efficiency and high power density matrix transformer structure for LLC resonant converters is proposed. The matrix transformer helps to reduce the leakage inductance and AC resistance of the winding. It uses the flux offset method to reduce the core size and loss. Next, we select the SPPS type winding form through a comparison process. Finally, in order to verify the correctness of the theory and analysis, the circuit model simulation test is carried out, and a prototype LLC resonant converter with 112.4V/28V and 250W at 1MHz is made.

Keywords—LLC resonant converter, magnetic integrated planar transformer, soft switch, high frequency

I. INTRODUCTION

The LLC resonant converter is a widely used power electronic converter. It plays an important role in many power electronic fields, such as power inverters, power converters, power controllers, etc., with its advantages of high efficiency, low electromagnetic interference, wide load range and high reliability.

At present, the research of LLC resonant converters mainly focuses on the operation, modeling and simulation under resonant point and high power density. Many studies have thoroughly examined the soft switching features and loss analysis of LLC resonant converters, which helps boost the converter's efficiency and lays a strong theoretical groundwork for its use. The LLC resonant converter often operates at high frequency and high buck conditions, which results in several main losses: transformer loss, primary side switch loss, secondary side rectification loss, and termination loss. Transformer loss occupies a large proportion in the high frequency, so a good high frequency and high efficiency transformer is crucial to the LLC resonant converter[1-2].

Furthermore, due to their advantages such as power density, high heat dissipation performance, and small size, planar transformers are currently the subject of extensive research. The planar transformer can use its characteristics to generate leakage and participate in the resonance of the resonator as an inductor, effectively increasing the power density[3-8]. However, this type of planar transformer experiences significant losses when operating at high frequencies and with large currents. In general, the secondary side needs to be connected in parallel to improve the capacity of the load

current, which will further lead to larger space occupation. Therefore, this paper proposes to use an optimized matrix planar transformer, the use of the magnetic integration method, the original need for two ferrite cores into one, reducing 30% of the space and core loss, and finally, effectively improving the efficiency of the LLC resonant converter .

In summary, the structure of this paper is organized as follows: The second part mainly analyzes the principle of the LLC resonant converter; the third part of the matrix planar transformer magnetic integration working principle analysis and winding design; the fourth part is the circuit simulation verification of the prototype and the design of a 1MHz,112.4V/28V, W LLC resonant converter prototype; The fifth part summarizes and looks forward to the future work.

II. THE CURRENT PATH AND MODE ANALYSIS OF LLC RESONANT CONVERTER

This paper uses the most classical half-bridge topology in LLC converter, as shown in Fig. 1. The topology is composed of four modules, which are the primary half bridge inverter, the primary resonator, the planar transformer and the synchronous rectifier. The working process is roughly the input of a DC signal, through the switching tube S_1 and S_2 inverter into AC square wave, then through the primary side resonator into sine wave, and then through the plane transformer to transmit the quantity to the secondary side, and finally through the rectification of the sine into DC output, and then through the large filter capacitor output to the load end.

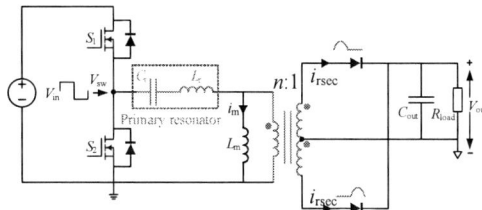

Fig. 1. Topology of Half-bridge LLC resonant converter

For the LLC topology, the planar transformer can be equivalent to two parts of the primary leakage inductance and the excitation inductance in the circuit model, and there will be two resonance points together with the resonant capacitor. When the circuit works under heavy load conditions, the

979-8-3315-1110-4/25 $31.00 © 2025 IEEE   528

current flowing through the primary leakage inductance L_r will be much greater than the current flowing through the primary excitation inductance L_m, then only the primary leakage inductance and the capacitor resonance, called series resonance, the operating point resonance frequency is f_{r1}. When the circuit works under light load conditions, the current flowing through the original side leakage inductance L_r and the current flowing through the original side excitation inductance L_m is approximately in the same order of magnitude, then the excitation inductance L_m, the original side leakage inductance L_r and capacitance C_r resonance, called series parallel resonance, at this time the frequency is recorded as f_{r2}[9].

$$f_{r1} = \frac{1}{2\pi\sqrt{L_r C_r}} \tag{1}$$

$$f_{r2} = \frac{1}{2\pi\sqrt{(L_r + L_m)C_r}} \tag{2}$$

In order to achieve high efficiency and stable operation, the LLC resonant converter must work in the inductive state, and the efficiency of the quasi-resonance is higher than the other two modes, so only the quasi-resonance operating region is analyzed as a representative.

Mode 1, [t_0-t_1]: The working mode is shown in Fig. 2 (a), when the converter works in the dead zone time. Before t_0, the switching tube S_1 is turned off, and the waveform of resonant current i_r is shown in the figure. When entering this mode, the switching tube S_2 is also turned off. At this time, the inductive current can not be mutated, and the current direction in the figure remains unchanged. At this time, the body diode on the switching tube S_2 carries on current charging, and the body diode on the switching tube S_1 carries on current discharge, and the current continues to be transferred by the body diode of S_1 to the DC power supply side after completion. In this mode, both ends of S_1 are always locked by the body diode to 0V, successfully providing conditions for ZVS.

Mode 2, [t_1-t_2]: The working mode of the circuit is shown in Fig. 2 (b). At time t_1, the switch tube S_1 has a driving signal to come. At this time, because the voltage at both ends of the switch tube S_1 is zero in mode 1, S_1 successfully realizes ZVS in opening. At the opening moment of S_1, the input voltage is added to the resonator and both ends of the transformer, at this time i_{Lm} and i_{Lr} drop rapidly, and part of the current in i_{Lm} flows into the transformer from the same end because i_{Lr} drops too fast. At this time, the current on the sub-side switch tube S_3 successfully circulates to complete the power transmission.

Mode 3, [t_2-t_3]: The working mode of the circuit is shown in Fig. 2 (c). In this mode, the switching state of switching tube S_1 and switching tube S_2 does not change. At this time, the switching tube state is consistent with mode 2, and the i_{Lr} current increases in reverse from zero, while the middle i_{Lm} current continues to linearly decrease to zero under the influence of the loading voltage and flows in reverse until it is consistent with i_{Lr}. For the secondary side, the state of the switching tube in mode 3 remains unchanged, completing half a cycle of energy transmission.

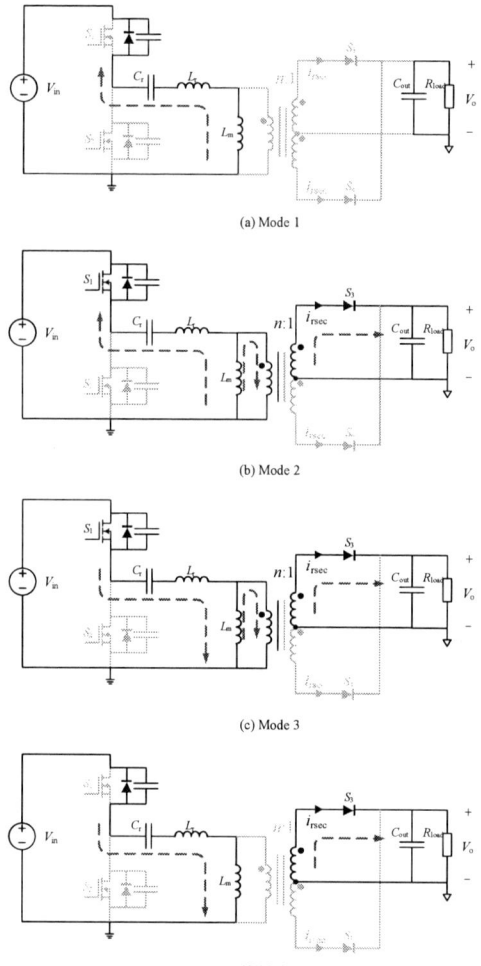

(a) Mode 1

(b) Mode 2

(c) Mode 3

(d) Mode 4

Fig. 2. Working modes of the circuit in the quasi-resonant region

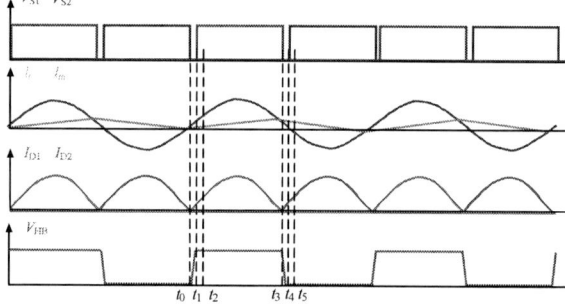

Fig. 3. Key operating waveform of the circuit in the quasi-resonant region

Mode 4, [t_3-t_4]: The working mode of the circuit is shown in Fig. 2 (d). In this mode, both primary switch tubes enter the dead zone, and the on-state of switch tubes S_1 and S_2 in subsequent work is the opposite of that in mode 2. According to the symmetrical structure of Half bridge LLC, the subsequent working mode is exactly the same as that of mode 1-mode 4, except that the change of current flow direction is different from that of the specific working switch tube, which will not be described here.

III. MAGNETIC INTEGRATION DESIGN OF MATRIX TRANSFORMER

A. Structure principle of magnetic integrated matrix transformer

In low-voltage and high-current output scenarios, the LLC resonant converter faces several important challenges:

1) In order to solve the high current application scenario, it is generally necessary to parallel several circuits in the sub-side output so as to improve the capacity with load. However, such a parallel will make it difficult to achieve dynamic current sharing and ultimately reduce the efficiency of the transformer.

2) A large amount of high frequency current flows through the junction of the transformer and diode, which will lead to a large termination loss, and thus the overall efficiency is low.

According to the above design requirements, although the transformer loss of the LLC converter is within a reasonable range, it will also lead to significant winding and termination loss. Therefore, this section proposes an optimized matrix planar transformer structure suitable for high frequency and high current applications[10-11].

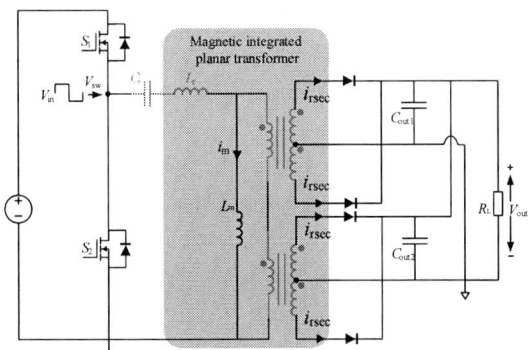

Fig. 4. LLC resonant converter using matrix transformer structure

In this paper, we create the matrix planar transformer by combining two discrete transformers. Each transformer unit has a fixed winding turn ratio, and the final required turn ratio is obtained by connecting the primary side to the secondary side of the two discrete transformer units in series. In the case of high current output, it is best to use one turn for the secondary side winding. Also, because of how the windings are arranged, the matrix transformer works better with heat, and the magnetic force between the windings is low, leading to less AC resistance.

Assuming a turn ratio of $2n$:1, the matrix transformer consists of two discrete transformers. Then the primary side of the two transformers is a series structure, the secondary side is a parallel structure, the primary winding of each discrete transformer unit is n turns, and the secondary winding is 1 turn. At any time in the work, only one secondary side winding leads the current. Such a structure makes the magnetic motive force between the primary and secondary windings smaller; a smaller magnetic motive force will make the transformer produce a smaller AC winding, plus the winding of the primary side is in series, so the secondary side will automatically balance the current, so the matrix

transformer under this structure has smaller loss.

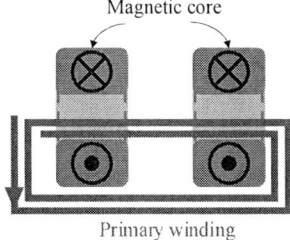

Fig. 5. Primary side winding with two U-I core transformers

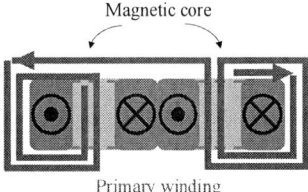

Fig. 6. Primary side windings of two U-I transformers after rotation

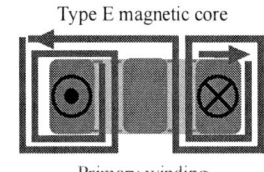

Fig. 7. Type E matrix transformer with flux cancellation

If the matrix transformer uses multiple magnetic cores, this will lead to a large core loss, matrix transformer because each magnetic core has the same magnetic flux, so this paper uses the principle of magnetic flux offset each other for magnetic integration so that the magnetic flux can be completely offset, reducing the core volume and core loss. The specific simplified derivation process is as follows [13-15]:

For a transformer without flux cancellation, in order to realize the circuit structure in Fig. 4, the primary winding pattern of the initial two cores is shown in Fig. 5. In order to achieve flux cancellation and thus reduce the volume of the transformer, the primary side windings must be rearranged to reverse the flux direction of the two cores in Fig. 5, as shown in Fig. 6. Further, the two U-I cores can be rotated 90 degrees and merged to form a single E-I core, as shown in Fig. 7.

The primary side windings are connected in series, and since the magnetic flux density in each core is identical, the magnetic flux in the center leg of the E-I core can effectively cancel itself out. This results in the magnetic flux density of the center column being nearly zero. As a result of this flux cancellation, the two U-I cores typically used in a matrix transformer can be consolidated into a single E-I core. This method can reduce core loss and volume by more than 30%. Consequently, this paper will utilize one E-I magnetic core instead of two U-I magnetic cores for the design of a matrix transformer intended for prototype testing.

B. Winding selection

In the design of planar transformer winding structures, there are two common winding layout schemes: PSSP (Primary-Secondary-Secondary-Primary) and SPPS (Secondary-Primary-Primary-Secondary). The PSSP structure employs a configuration where the primary winding is positioned in the top and bottom layers, while the secondary winding is in the middle. This arrangement facilitates a balanced magnetic field, which enhances magnetic flux coupling. However, a significant issue arises with the terminal connections: the secondary winding must connect to the rectifier and filter capacitor via PCB vias and external traces, leading to an uneven flow of AC current at the connection points. Finite element simulation results reveal that this inadequate connection results in a substantial increase in current density at the terminals. Furthermore, this structure exhibits a critical defect in terminal connections: the secondary winding's link to the rectifier and filter capacitor introduces a serious imbalance in the distribution of AC current at these connection points. The finite element simulation results indicate that this type of inadequate terminal connection leads to a marked increase in current density at the end section. The root cause of this problem is the proximity effect of the reverse current, which causes it to concentrate in the edge region of the conductor. Additionally, the mismatch in leakage flux is further intensified by the incomplete overlap of the primary and secondary windings [16-17].

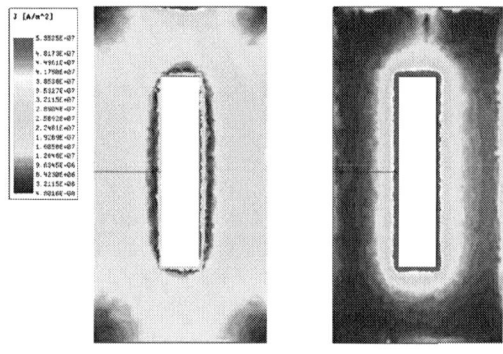

Fig. 8. Side view of matrix transformer of SPPS structure

In contrast, the SPPS structure eliminates terminal connection losses by integrating the rectifier and filter capacitors directly into the secondary winding. The secondary windings are placed in the top and bottom layers, while the primary winding is in the middle layer, and the primary and secondary windings are perfectly aligned to overlap each other. Simulation data show that the current density of the SPPS structure is significantly lower than that of the PSSP structure, and its core advantages are reflected in two aspects: first, the integrated design avoids the current shunt of the uncoupled path so that the secondary current forms a closed-loop flow inside the winding, which significantly reduces the effects of the skinning effect and proximity effect; second, the complete overlap of the primary and secondary windings makes the flux-coupling efficiency increase, effectively suppressing the leakage flux and leading to the edge current aggregation phenomenon. This effectively suppresses the phenomenon of edge current aggregation caused by leakage flux. Even though

the SPPS structure is not as good as the PSSP structure when it comes to how the magnetomotive force is spread out, it has clearer overall benefits in electrical performance.

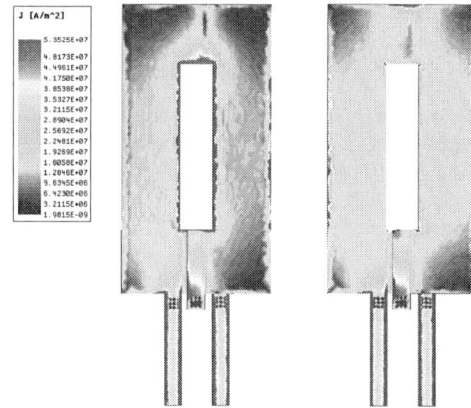

Fig. 9. Side view of matrix transformer of PSSP structure

After the above simulation model construction, the current density distribution diagrams can be obtained as shown in Fig. 8 and Fig. 9[12]. From the figure, it can be learned that although this winding of PSSP has a relatively small magnetomotive force, the fact that it sets the vice-side winding in the inner layer of the PCB board leads to the fact that when connecting the vice-side rectifier part of the transformer, it is necessary to connect the capacitance of the winding and the rectifier part with the SR using an over-hole, which greatly reduces the equivalent resistance of the winding and thus makes the winding terminal loss higher than that of the winding structure of the SPPS.

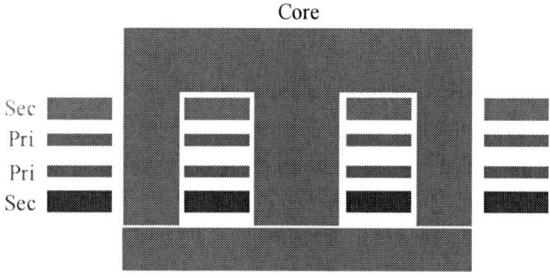

Fig. 10. Side view of matrix transformer of SPPS structure

The quantitative comparison of the two winding structures indicates that the effect of terminal connection loss on system efficiency is significantly greater than the benefits gained from optimizing the winding stacking layout. While the PSSP structure demonstrates superiority in the symmetry of the magnetomotive force, the additional resistance introduced by the terminal connection results in a final winding loss of 0.5502 W. In contrast, the SPPS structure completely eliminates this loss through its integrated design, resulting in a final winding loss of only 0.35618 W in the simulation. Consequently, this paper selects the SPPS-type winding structure as the prototype design for the converter, highlighting its notable advantages in reducing current density and winding losses, which can enhance magnetic energy conversion efficiency in high-power-density converters[19].

IV. EXPERIMENTAL VERIFICATION

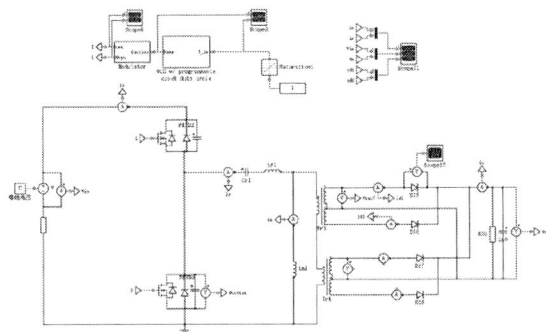

Fig. 11. LLC Circuit simulation diagram of resonant converter

To verify the theoretical feasibility mentioned earlier, the simulation model shown in Fig. 11 is constructed in PLECS. Running the provided simulation circuit diagram yields the simulated waveforms, as illustrated in Fig. 12:

The simulation results indicate that the series structure of the primary side is theoretically feasible; the resonant current exhibits a sinusoidal waveform, and the output voltage aligns with the expected 4:1 step-down ratio. Additionally, the current waveform of the secondary side corresponds to that of the primary side, meeting expectations.

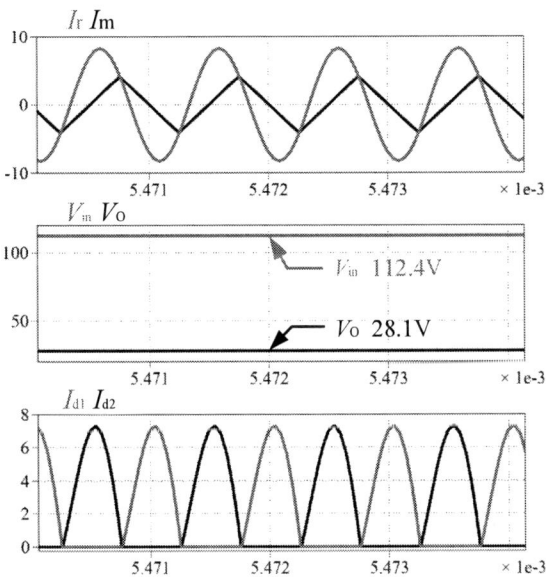

Fig. 12. LLC resonant converter resonant current waveform, input and output waveform, secondary side current waveform

Therefore, build a prototype for test evaluation, as shown in Fig. 13. The test indexes mainly include the resonant current of the prototype, the midpoint voltage waveform of the half-bridge, the output power of the prototype under different loads, and the efficiency of the prototype.

According to the previous analysis, the working frequency selected in this paper is 1 MHz, the magnetic core is 3F46 material, the magnetic core size is E38/25/8, and the switch tube on the original side is GAN device INN650D080BS. The equivalent resonant inductance measured by the prototype is 245nH, and the resonant capacitance is actually adjusted to 73.7nF, which meets the working requirements of the circuit. The input rated voltage is 112.4V and the rated power is 250W.

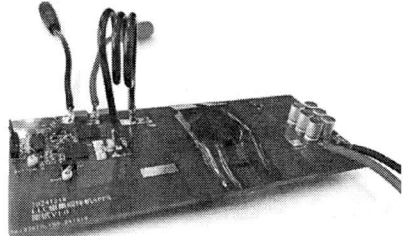

Fig. 13. LLC Schematic of resonant converter prototype

The waveforms under 50% load and 100% load are shown in Fig. 14 and Fig. 15. The two waveforms are similar, except for the resonant current and midpoint voltage amplitudes. Taking full load as an example, under the test conditions of 112.4 V input voltage and 3.1 ohms, the midpoint voltage waveform is a relatively ideal square wave, and the resonant current is an approximate sine wave. At this time, the working output power reaches 242.2 W, and the working efficiency obtained through the test is 92.02%, and the operation condition is good.

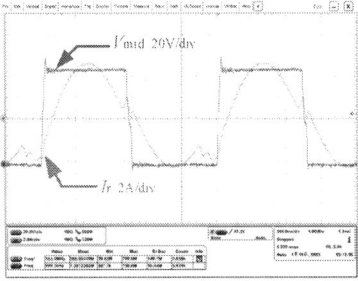

Fig. 14. Resonance current (green) under 50% load and half-bridge midpoint voltage waveform (violet)

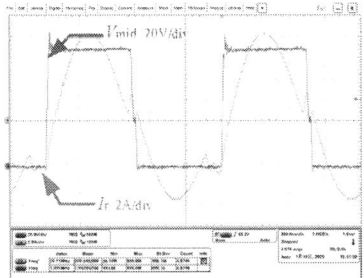

Fig. 15. Resonance current (green) under 100% load and half-bridge midpoint voltage waveform (violet)

Start with a 50V input voltage and gradually increase the input voltage and input power level. For every 10V test, use an oscilloscope to record the midpoint voltage and resonant current time. The following is the recorded experimental data in Table 1:

TABLE I. EXPERIMENTAL DATA RECORDING TABLE

V_{in}(V)	R_{load}(Ω)	P_{out} (W)	P_{in} (W)	efficiency
50	3	47.54	52.8	90.04%
60	4	53.5	58.6	91.30%
70	4	73.39	80	91.74%
80	4	94.1	102.2	92.07%
90	4	122.81	132.7	92.55%
100	4	152.2	164.3	92.64%
112.4	3.1	242.2	263.2	92.02%

As shown in Table 1, as the output power gradually increases, the efficiency of the entire prototype has significantly improved. When the output power exceeds 100 W, the overall efficiency of the prototype rises above 92%, reaching a maximum of 92.64%, indicating that the working conditions are favorable.

V. CONCLUSION

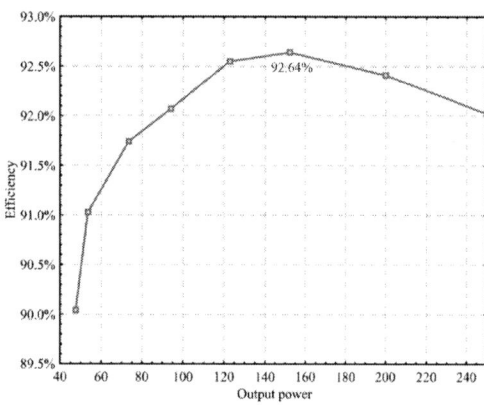

Fig. 16. Variation of output power and overall efficiency of prototype with input voltage

The analysis leads to the following conclusions: Compared to conventional planar transformers, the matrix transformer exhibits reduced core loss. It features low core loss, a simple winding structure, and a highly integrated magnetic design, making it an exceptional planar transformer.

To validate the principles mentioned above, further processing of the tested data yields the efficiency and output power line diagram shown in Fig. 16. The converter's core efficiency exceeds 92%, with a peak efficiency of 92.64%. The operating state is normal, and the working efficiency is high. This data indicates that the matrix transformer is functioning well, and the magnetic integrated transformer holds significant value for practical applications.

REFERENCES

[1] Z. Ouyang, O. C. Thomsen and M. A. E. Andersen, "Optimal Design and Tradeoff Analysis of Planar Transformer in High-Power DC–DC Converters," in IEEE Transactions on Industrial Electronics, vol. 59, no. 7, pp. 2800-2810, July 2012.

[2] Z. Ouyang and M. A. E. Andersen, "Overview of Planar Magnetic Technology—Fundamental Properties," in IEEE Transactions on Power Electronics, vol. 29, no. 9, pp. 4888-4900, Sept. 2014.

[3] B. Yang, Y. Ren, and F. C. Lee, "Integrated magnetic for LLC resonant converter," in Proc. IEEE APEC, 2002, pp. 346–351.

[4] Bo Yang, "Topology Investigation for Front End DC/DC Power Conversion for Distributed Power System," PhD Dissertation, Dept. ECE., Virginia Tech, 2003.

[5] Y. Guan, Z. Wen, Y. Wang and D. Xu, "A Single-Stage Modular DCX with High Voltage Conversion Ratio Based on High Frequency LLC Resonant Converter," 2021 IEEE Workshop WiPDA Asia , Wuhan, China, 2021, pp. 132-137,

[6] Y. Gu, Z. Lu, L. Hang, Z. Qian, and G. Huang, "Three-level LLC series resonant DC/DC converter," IEEE Trans. Power Electron., vol. 20, no.4, pp. 781–789, Jul. 2005.

[7] J. Hu, "Design of a low-voltage low-power dc-dc hf converter", M.S. thesis, Dept. Elect. Eng. Comput. Sci., Massachusetts Institute of Technology(MIT), Cambridge, 2008.

[8] D. Fu, B. Lu, and F. C. Lee, "1 MHz high efficiency LLC resonant converters with synchronous rectifier," in Proc. IEEE Power Spec. Conf., 2007, pp. 2404–2410.

[9] D. Reusch and J. Strydom, ''Understanding the effect of PCB layout on circuit performance in a high-frequency gallium-nitride-based point of load converter,'' IEEE Trans. Power Electron., vol. 29, no.4, pp. 2008–-2015, 2014.

[10] M. He et al., "Analysis of Connection Optimization in Planar Transformer Winding Loss," 2023 IEEE 2nd International Power Electronics and Application Symposium (PEAS), Guangzhou, China, 2023, pp. 604-607.

[11] X. Yang, L. Xiao, Q. Wu, W. Zhao and W. Chen, "An LLC-DAB Hybrid Converter With High Input output Voltage Ratio Based on an Integrated Matrix Transformer," 2024 IEEE International Conference on Electrical Systems for Aircraft, Railway, Ship Propulsion and Road Vehicles & International Transportation Electrification Conference (ESARS-ITEC), Naples, Italy, 2024, pp. 1-5,

[12] A. Jain and I. C. Massimiani, "LCC Resonant Converter Design and Transfer Function Computation Using FHA Analysis," 2021 4th Biennial International Conference on Nascent Technologies in Engineering (ICNTE), Navi Mumbai, India, 2021, pp. 1-5

[13] D. Reusch and F. C Lee, "High frequency bus converter with low loss integrated matrix transformer," in Proc. 27th Annu. IEEE Appl. Power Electron. Conf. Expo., 2012, pp. 1392–1397.

[14] D. Huang, S. Ji, and F. C. Lee, "LLC resonant converter with matrix transformer," IEEE Trans. Power Electron., vol. 29, no. 8, pp. 4339–4347,Aug. 2014.

[15] M. Mu and F. C. Lee, "Design and optimization of a 380V-12V high-frequency, high-current LLC converter with GaN devices and planar matrix transformers," IEEE J. Emerg. Sel. Topics Power Electron.,vol.4, no. 3, pp. 854–862, Sep. 2016.

[16] C. Yan, F. Li, J. Zeng, T. Liu, and J. Ying, ''A novel transformer structure for high power, high frequency converter,'' in Proc. IEEE Power Electron. Specialists Conf., Orlando, FL, USA, Jun. 2007, pp. 214–-218.

[17] A. Nabih and Q. Li, "Low-Profile and High-Efficiency 3 kW 400 V-48 V LLC Converter with a Matrix of Four Transformers and Inductors for 48V Power Architecture for Data Centers," 2021 IEEE Energy Conversion Congress and Exposition (ECCE), Vancouver, BC, Canada, 2021, pp. 1813-1819.

[18] S. -S. Park, M. -S. Jeon, S. -S. Min and R. -Y. Kim, "High-Frequency Planar Transformer Based on Interleaved Serpentine Winding Method With Low Parasitic Capacitance for High-Current Input LLC Resonant Converter," in IEEE Access, vol. 11, pp. 84900-84911, 2023.

[19] M. Dai, X. Zhang, H. Li, D. Zhou, Y. Wang and D. Xu, "LLC Converter With an Integrated Planar Matrix Transformer Based on Variable Width Winding," 2019 22nd International Conference on Electrical Machines and Systems (ICEMS), Harbin, China, 2019, pp. 1-4.

An Active Gate Driver Addressing GaN HEMT Gate-Source Voltage Overshoots for Both Turn-on and Turn-off Periods

1st Lurenhang Wang
School of Electrical Engineering & Automation
Harbin Institute of Technology
Harbin, China
23S006077@stu.hit.edu.cn

2nd Yishun Yan
School of Electrical Engineering & Automation
Harbin Institute of Technology
Harbin, China
24S006068@stu.hit.edu.cn

3rd Shuaiqing Zhi
School of Electrical Engineering & Automation
Harbin Institute of Technology
Harbin, China
21B906032@stu.hit.edu.cn

4th Mingcheng Ma
School of Electrical Engineering & Automation
Harbin Institute of Technology
Harbin, China
24B906005@stu.hit.edu.cn

5th Xizhi Sun
School of Electrical Engineering & Automation
Harbin Institute of Technology
Harbin, China
22S006043@stu.hit.edu.cn

6th Dianguo Xu
School of Electrical Engineering & Automation
Harbin Institute of Technology
Harbin, China
xudiang@hit.edu.cn

Abstract—**Compared with traditional Si MOSFETs, GaN HEMTs have faster turn-on and turn-off speeds and are often used in high-frequency applications. However, the high dv/dt generated by high-frequency switching causes displacement current to flow into the gate drive circuit through the Cgd. High di/dt generates EMI (electromagnetic interference) that also affects the drive circuit, and it is also present in the gate drive circuit. Both of these can cause serious interference in the drive circuit, resulting in positive or negative spikes in the gate-source voltage. These spikes can easily break down the gate, which challenges the safety of GaN HEMTs. To address this problem, we have designed an active gate driver addressing GaN HEMT gate-source voltage overshoots for both turn-on and turn-off periods. The reliability of GaN HEMT applications is improved by connecting a gate capacitor at the right moment to absorb the forward or reverse spikes. In the double pulse test (DPT), at 400 V/16 A, the positive gate-source voltage spike was reduced by 51.5% relative to CGD when the turn-on time was only 3.5 ns. The negative gate-source voltage spike was reduced by 47.6% relative to CGD when the turn-off time was only 5.2 ns. This effectively proves the effectiveness of the designed AGD.**

Keywords—GaN HEMT, gate-source voltage, voltage overshoots suppression, active gate driver

I. INTRODUCTION

GaN HEMT, as a new generation of wide bandgap power semiconductor devices, has attracted widespread attention in fields such as high-frequency communications, power electronics, and automotive electronics due to its significant performance advantages. Compared to Si MOSFET, GaN HEMT has faster switching speeds, lower on-resistance, and smaller parasitic parameters. Therefore, GaN HEMT is often used in medium-power high-frequency applications[1]-[8].

However, the high-frequency characteristics of GaN HEMTs bring the problems of high di/dt and dv/dt. High di/dt will couple with the inevitable parasitic parameters in the power loop, generating EMI in the drive loop. High dv/dt will cause displacement current in C_{gd} to flow into the gate drive circuit, and the gate drive circuit itself also has parasitic

inductance. All three of these will cause the gate-source voltage to oscillate, generating overshoots in the forward or reverse direction[9]-[12].

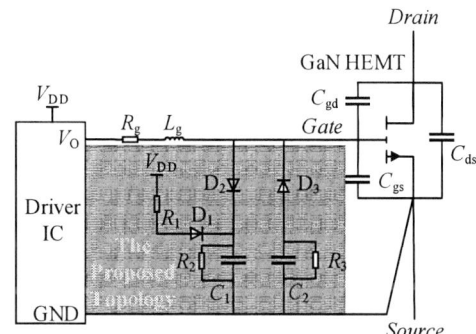

Fig. 1. The proposed topology.

Taking the GS66508B of GaN System Corporation as an example, its maximum transient gate-source voltage is +10 V, and the driving voltage is generally +6 V. Under the influence of displacement current, EMI (Electromagnetic Interference) and stray inductance, the gate-source voltage can easily reach this safety limit value, which is likely to cause breakdown of the gate and pose a challenge to the safety of GaN HEMT (Gallium Nitride High Electron Mobility Transistor). There are generally two methods to suppress the overshoot of the gate-source voltage. One is to increase the gate resistance. A method for predicting the gate voltage overshoot of GaN HEMT is proposed in [13]. By accurately calculating the gate voltage, the value of the gate driving resistance for suppressing the overshoot can be obtained. However, increasing the gate driving resistance will undoubtedly sacrifice the high-speed switching performance of GaN HEMT. A method for suppressing the gate-source voltage oscillation is proposed in [14]. By connecting a pre-charged capacitor in parallel to clamp between the gate and the source, the gate-source voltage oscillation can be effectively suppressed. However, it does not provide a low-impedance

979-8-3315-1110-4/25 $31.00 © 2025 IEEE

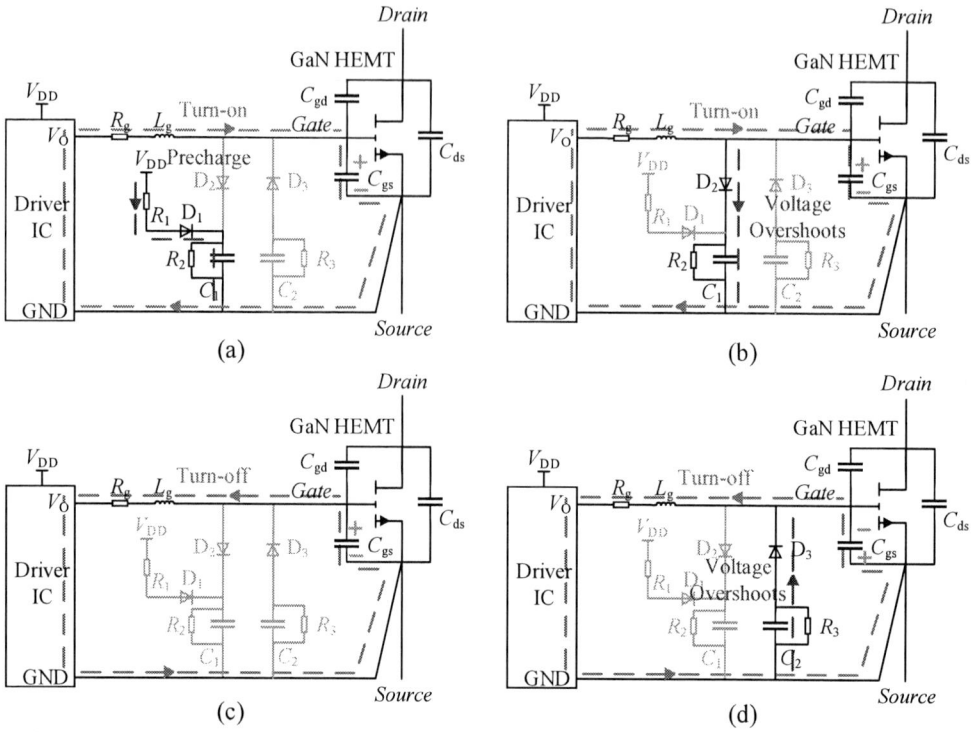

Fig. 2. Operation modes of the proposed topology. (a) Mode I: the process of pre-charging.(b) Mode II: the process of absorbing positive overshoots. (c) Mode III: the process of the initial turn-off period. (d) Mode IV: the process of absorbing negative overshoots.

path for crosstalk, and there may be a misfiring problem in the GaN HEMT bridge application. In [15], the double-pulse circuit of GaN HEMT is modeled. An RC snubber is used to suppress the continuous oscillation, the root locus method is used for analysis, and dipoles are used to completely eliminate the oscillation. However, since this method uses an RC circuit, it will also slow down the turn-on and turn-off speed. For the method of increasing the resistance, it will slow down the turn-on and turn-off speed and increase the switching loss.

The rest of this paper is organized as follows. Section II first introduces the structure and operating mode of the proposed topology. Section III analyzes the transient process of gate-source voltage overshoot suppression, and then introduces the parameter design method. Section IV gives experimental data, and the validity of the design is verified by analyzing the experimental data. Finally, Section V summarizes the entire paper.

II. THE PROPOSED TOPOLOGY AND OPERATION MODE

A. The Proposed Topology

This paper proposes an active gate driver topology that can absorb forward and reverse spikes in the gate-source voltage, as shown in Fig. 1. The topology consists of a precharge module (V_{DD}-R_1-D_1-C_1), a forward voltage spike absorption module (D_2-C_1), and a reverse voltage spike absorption module (C_2-D_3). C_1 is precharged via V_{DD}, and the presence of D_1 ensures that the current does not flow to the power supply side. The purpose of the pre-charge is to ensure that C_1 does not interfere with the normal turn-on process. Only when there is a spike in the gate-source voltage will D_2 conduct, and C_1 absorb the voltage spike. R_2

is present to dissipate the excess charge in C_1 so that the voltage of C_1 returns to its normal value before GaN HEMT is turned on each time. When there is a negative spike in the gate-source voltage during turn-off, D_3 conducts. C_2 absorbs the excess energy and reduces the amplitude of the negative gate-source voltage spike. The presence of R_3 ensures that the charge stored in C_2 can be quickly dissipated, so that there is no charge stored in C_2 before each GaN HEMT turn-off, i.e., the voltage of C_2 is 0, ensuring that D_3 is turned on at the right time. The proposed topology also provides some ideas for further exploring the high-frequency applications of GaN HEMTs.

B. The Operation Modes of the Proposed Topology

The proposed topological operating mode is shown in Fig. 2, and the key voltage and current waveforms are shown in Fig. 3. Figs. 2 and 3 are used as examples to explain in detail how the proposed topology operates during the turn-on and turn-off stages.

Mode I (t_0-t_1): At t_0, the PWM arrives at the rising edge, and the GaN HEMT begins to enter the conduction phase, as shown in Fig. 2(a). At this time, the gate-source voltage begins to rise gradually, but D_2 will not turn on until it reaches the drive voltage, which effectively ensures that the newly added topology will not affect the normal turn-on process. At this time, the GaN HEMT gradually turns on normally, which is the same as CGD.

Mode II(t_1-t_2): At t_1, the gate-source voltage reaches the drive voltage, and the topology operates in the mode shown in Fig. 2(b). At this time, D_2 is turned on, and the gate-source voltage spike is absorbed via D_2-C_1. Excess energy is stored in C_1, which effectively reduces the voltage spike. And because of the presence of D_1, current does not flow into the power supply, ensuring safety.

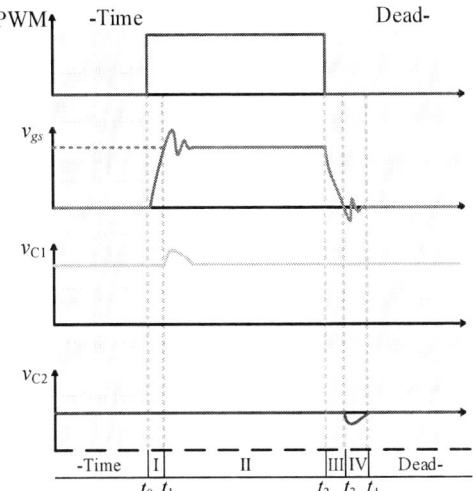

Fig. 3. The key voltage and current timing waveforms of the proposed topology.

Mode III (t_2-t_3): At t_2, the PWM goes to the falling edge, and the GaN HEMT begins to enter the off-state. The topological operation mode is shown in Fig. 2(c). At this time, neither of the two auxiliary branches is turned on, which does not affect the normal off-state process. At this time, the GaN HEMT gradually turns off normally, which is the same as CGD.

Mode IV(t_3-t_4): At t_3, the gate-source voltage begins to oscillate and enter the negative voltage phase, and the topological operation mode is shown in Fig. 2(d). At this time, D_3 is turned on, and C_2 absorbs the negative gate-source voltage spikes. The excess energy is stored in C_2, which effectively reduces the voltage spikes.

III. Transient Analysis and Parameters Design

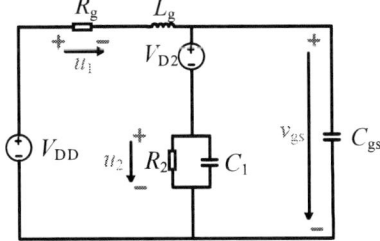

Fig. 4. Equivalent circuit diagram of the turn-on period.

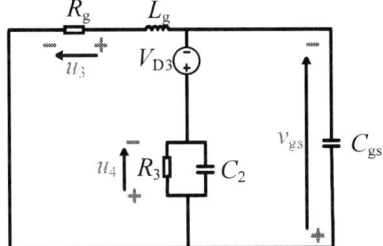

Fig. 5. Equivalent circuit diagram of the turn-off period.

The topology proposed in the initial stage of opening is the same as the CGD operating mode, and will not be analyzed here. The equivalent circuit after the peak suppression module is turned on is shown in Fig. 4 and Fig.5. The analysis of the turn-on and turn-off transient processes is shown in (1) and (2).

In general, C_1 needs to be greater than 10 times C_{gs}, that is, C_1 needs to exceed 2 nF. Considering the suppression of gate-source voltage oscillations and drive losses, it is more appropriate to select C_1, C_2 as 10 nF, R_2 as 50 Ω, and R_3 as 10 Ω.

$$\begin{cases} v_{gs} = V_{D2} + u_2 \\ V_{DD} = u_1 - L_g \dfrac{di_g}{dt} + v_{gs} \\ i_g = \dfrac{u_1}{R_g} \\ \dfrac{u_2}{R_2} + C_1 \dfrac{du_2}{dt} + i_g = C_{gs} \dfrac{dv_{gs}}{dt} \end{cases} \quad (1)$$

$$\begin{cases} v_{gs} = V_{D3} + u_4 \\ u_3 + L_g \dfrac{di_g}{dt} + v_{gs} = 0 \\ i_g = \dfrac{u_3}{R_g} \\ \dfrac{u_4}{R_3} + C_2 \dfrac{du_4}{dt} + i_g = C_{gs} \dfrac{dv_{gs}}{dt} \end{cases} \quad (2)$$

The proposed topology uses three diodes in total. The role of D_1 is to limit the direction of the pre-charge current of C_1, while the role of D_2 and D_3 is to control the gate-source voltage spike absorption module to connect to the gate at the right moment. All three functions require low conduction loss and low reverse recovery, so Schottky diodes NSR0620P2T5G are used for all diodes.

IV. Experimental Verification

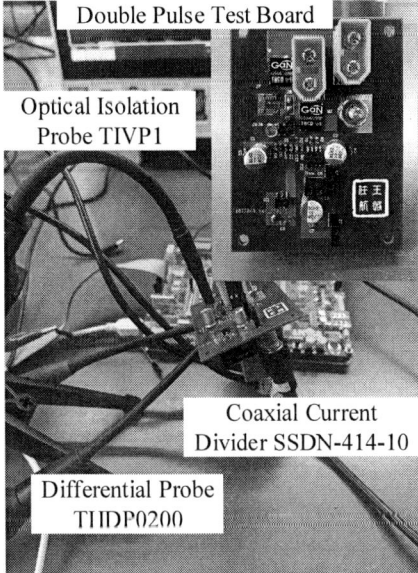

Fig. 6. Experimental platform for double-pulse experiment.

In order to verify the proposed gate-source voltage overshoot suppression method, a double pulse test platform is set up as shown in Fig. 6. The platform uses a TIVP1 optical isolation probe with a bandwidth of 1 GHz to measure the gate-source voltage and a THDP0200 differential probe with a bandwidth of 200 MHz to measure the drain-source voltage. A coaxial current divider SSDN-414-10 with a bandwidth of 2 GHz is used to measure the drain current. The GaN HEMT selected is the GS66508B, which has a Kelvin source and a very small common-source inductance. The driver IC for the GaN HEMT is the 1EDB7275F. The load inductor is 100 μH connected in parallel to the high-side transistor. The IT6018D DC power supply supplies the DC bus voltage. The FPGA provides the GaN HEMT switching signals. The proposed topology was tested with a load current of 16 A and a bus voltage of 400 V.

The turn-on test results are compared with CGD in Fig. 7. The AGD experimental curve in the Fig. 7 is solid, and the CGD is dashed. As can be seen from the AGD and CGD drain-source voltage and drain current curves in Fig. 7, the proposed topology has no effect on the normal turn-on process, and only effectively suppresses the overshoot of the gate-source voltage. A 51.5% reduction of positive gate-source voltage overshoot from 9.3 V to 7.6 V is achieved compared to the CGD. The turn-on time is only 3.5 ns. The turn-off test results are shown in Fig. 8. The proposed topology has no effect on the normal turn-off process as shown above. A 47.6% reduction of negative gate-source voltage overshoot from -4.2 V to -2.2 V is achieved compared to the CGD. The turn-off time is only 5.2 ns.

These two sets of experiments fully verified the advantages of the proposed topology over CGD, that is, without affecting the normal turn-on and turn-off process, the peak value of the v_{gs} overshoot voltage can be greatly reduced, and reliability can be improved.

V. CONCLUSION

This article proposes an active gate driver addressing GaN HEMT gate-source voltage overshoots for both turn-on and turn-off periods. Without affecting the normal turn-on and turn-off process, the gate-source voltage overshoot during turn-on is absorbed by a pre-charge capacitor, and the gate-source voltage overshoot during turn-off is absorbed by an auxiliary capacitance-diode branch. The proposed topology was tested under 400 V/16 A conditions. From the experiment, it can be seen that a 51.5% reduction of positive gate-source voltage overshoot from 9.3 V to 7.6 V is achieved compared to the CGD. The turn-on time is only 3.5 ns. A 47.6% reduction of negative gate-source voltage overshoot from -4.2 V to -2.2 V is achieved compared to the CGD. The turn-off time is only 5.2 ns. The above experimental demonstrates the effectiveness of the proposed topology, which improves the reliability of GaN HEMTs in applications.

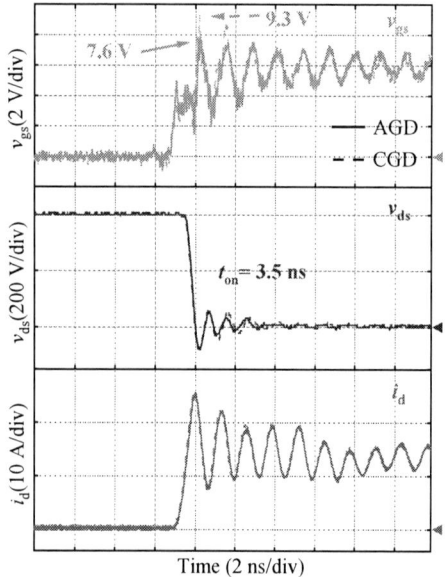

Fig. 7. Double pulse test waveform during turn-on period.

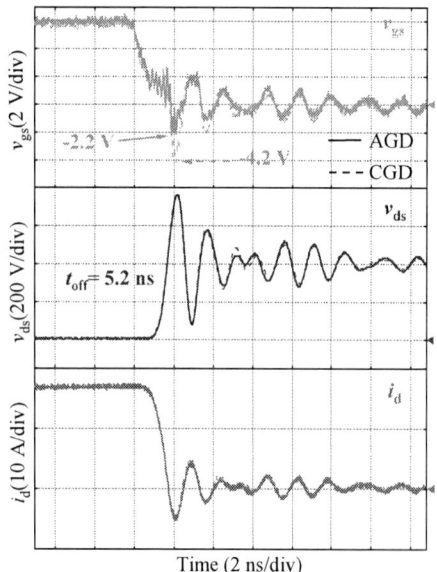

Fig. 8. Double pulse test waveform during turn-off period.

REFERENCES

[1] W. Saito et al., "A 120-W Boost Converter Operation Using a High-Voltage GaN-HEMT," IEEE Electron Device Letters, vol. 29, no. 1, pp. 8-10, 2008.

[2] Z. Liu, X. Huang, F. C. Lee, and Q. Li, "Package Parasitic Inductance Extraction and Simulation Model Development for the High-Voltage Cascode GaN HEMT," IEEE Transactions on Power Electronics, vol. 29, no. 4, pp. 1977-1985, 2014.

[3] Y. Chen, R. Wang, X. Liu, and Y. Kang, "Gate-Drive Power Supply With Decayed Negative Voltage to Solve Crosstalk Problem of GaN Synchronous Buck Converter," IEEE Transactions on Power Electronics, vol. 36, no. 1, pp. 6-11, 2021.

[4] B. Sun, R. Burgos, and D. Boroyevich, "Ultralow Input–Output Capacitance PCB-Embedded Dual-Output Gate-Drive Power Supply for 650 V GaN-Based Half-Bridges," IEEE Transactions on Power Electronics, vol. 34, no. 2, pp. 1382-1393, 2019.

[5] J. Shu, J. Sun, X. Wu and K. J. Chen, "A Dynamic Two-Stage Gate Driver for Unlocking the Fast-Switching Potential of GaN HEMT," in IEEE Transactions on Power Electronics, vol. 40, no. 4, pp. 4752-4756, April 2025.

[6] K. J. Chen et al., "GaN-on-Si Power Technology: Devices and Applications," IEEE Transactions on Electron Devices, vol. 64, no. 3, pp. 779-795, 2017.

[7] Y. Wang, Y. Wang, A. Yu, M. Hu, Q. Wang, C. Pang, H. Xiong, Y. Cheng, J. Qi, Non-Interleaved Shared-Aperture Full-Stokes Metalens via Prior-Knowledge-Driven Inverse Design. Adv. Mater. 2024, 2408978. https://doi.org/10.1002/adma.202408978.

[8] R. Xie et al., "Switching transient analysis for normally-off GaN transistors with p-GaN gate in a phase-leg circuit," in 2017 IEEE

Energy Conversion Congress and Exposition (ECCE), 1-5 Oct. 2017 2017, pp. 399-404.

[9] Y. Wang, A. Yu, Y. Cheng, J. Qi, Matrix Diffractive Deep Neural Networks Merging Polarization into Meta-Devices. Laser Photonics Rev 2024, 18, 2300903. https://doi.org/10.1002/lpor.202300903

[10] S. Yu et al., "A 400-V Half Bridge Gate Driver for Normally-Off GaN HEMTs With Effective Dv/Dt Control and High Dv/Dt Immunity," IEEE Transactions on Industrial Electronics, vol. 70, no. 1, pp. 741-751, 2023.

[11] J. E. Makaran, "Gate Charge Control for MOSFET Turn-Off in PWM Motor Drives Through Empirical Means," IEEE Transactions on Power Electronics, vol. 25, no. 5, pp. 1339-1350, 2010.

[12] J. Wang and H. S.-H. Chung, "Impact of Parasitic Elements on the Spurious Triggering Pulse in Synchronous Buck Converter," IEEE Transactions on Power Electronics, vol. 29, no. 12, pp. 6672-6685, 2014.

[13] J. P. Kozak et al., "An Analytical Model for Predicting Turn-ON Overshoot in Normally-OFF GaN HEMTs," IEEE Journal of Emerging and Selected Topics in Power Electronics, vol. 8, no. 1, pp. 99-110, 2020.

[14] S. Jahdi, O. Alatise, J. A. O. Gonzalez, R. Bonyadi, L. Ran, and P. Mawby, "Temperature and Switching Rate Dependence of Crosstalk in Si-IGBT and SiC Power Modules," IEEE Transactions on Industrial Electronics, vol. 63, no. 2, pp. 849-863, 2016.

[15] Z. Zhang, F. Wang, L. M. Tolbert, and B. J. Blalock, "Active Gate Driver for Crosstalk Suppression of SiC Devices in a Phase-Leg Configuration," IEEE Transactions on Power Electronics, vol. 29, no. 4, pp. 1986-1997, 2014.

Application of Wide Bandgap High Frequency Inverter in 10 kW Magnetic Field-Coupled Undersea Wireless Power Transfer Systems

1st Lei Yang
School of Electrical and Engineering
Xi'an University of Technology
Shaanxi, China
yanglei0930@xaut.edu.cn

2nd Jiahua Sun
School of Electrical and Engineering
Xi'an University of Technology
Shaanxi, China
3201712130@stu.xaut.edu.cn

3rd Yuanfeng Wang
School of Electrical and Engineering
Xi'an University of Technology
Shaanxi, China
924366419@qq.com

Abstract—This article proposes a high-frequency inverter design based on wide bandgap silicon carbide (SiC) material for the high-power demand of 10 kW undersea, combined with magnetic coupling mechanism optimization, to construct a wireless energy transmission system. Breakthrough the performance bottleneck of traditional systems through the high-frequency characteristics of SiC MOSFET devices. By utilizing the high electron mobility, low on resistance, and high temperature resistance of SiC, the system significantly reduces switching losses and temperature rise, while improving power transmission efficiency. Adopting Leeds wire winding and segmented magnetic core design to reduce high-frequency eddy current losses and improve coupling coefficient. Adopting LCC-S type resonant compensation network, combined with finite element simulation optimization of coil structure, to achieve the improvement of coupling coefficient. Achieve a maximum output power of 10 kW at a transmission distance of 10 cm undersea. This design has advantages in transmission efficiency, power capacity, and anti offset capability, providing theoretical basis and technical reference for the engineering application of high-power medium distance UWPT systems.

Keywords—wireless power transfer system, wide bandgap, magnetic field coupling, silicon carbide

I. INTRODUCTION

The rapid development of ocean development and exploration technology has put forward higher requirements for the energy supply of undersea equipment. The traditional cable power supply method is limited by mechanical interface corrosion and poor deployment flexibility, especially in deep-sea high-voltage and high salt environments where the failure rate significantly increases. Magnetic field coupled wireless power transfer technology, through non-contact energy transfer, can effectively avoid the risks caused by physical connections and has become a research hotspot for undersea equipment power supply. However, as a high conductivity medium, seawater can cause severe eddy current losses and electromagnetic shielding effects, leading to a significant decrease in the transmission efficiency of traditional WPT systems. At the same time, the deep-sea high-pressure environment imposes strict requirements on the insulation and heat dissipation design of the system, and existing solutions generally face bottlenecks such as low power levels and short transmission distances.

In recent years, the maturity of wide bandgap silicon carbide power devices has provided a new technological path for high-frequency, high-power WPT systems. Compared to traditional silicon-based devices, SiC MOSFETs have higher switching frequencies and lower conduction losses, which can significantly reduce the size of inverters and improve power density. This is particularly important for space limited undersea applications. When SiC devices operate at high frequencies, the overall system loss will be significantly reduced. However, existing research on the propagation characteristics of high-frequency electromagnetic fields in seawater environments has mostly focused on low-frequency solutions to avoid eddy current problems, and has not fully explored the high-frequency advantages of SiC devices.

In the design of magnetic coupling mechanism, eddy current loss in seawater and coil offset sensitivity are two technical difficulties. The existing schemes mainly reduce the influence of eddy current by adding shielding layer or adopting split magnetic core structure, but it often leads to uneven magnetic field distribution and decreased coupling coefficient. The dynamic displacement of undersea equipment requires the system to have strong anti-deviation ability. By optimizing the topology of the coupling coil, a large offset tolerance can be achieved, but its performance decreases obviously in undersea environment. Therefore, it is urgent to develop a magnetic coupling optimization method that takes into account the high frequency characteristics and seawater adaptability.

Regarding the system control strategy, the characteristic of seawater conductivity changing with temperature and salinity can lead to resonance frequency drift, and traditional fixed frequency tracking algorithms are difficult to adapt to complex marine environments. The adaptive frequency modulation method based on impedance phase detection, but its response speed cannot meet the real-time requirements of high-frequency systems. In addition, thermal management issues under high-power transmission are equally critical. Although SiC devices have high thermal conductivity, the undersea sealing environment limits the heat dissipation path, and temperature rise control needs to be achieved through packaging design and cooling medium optimization. Previous studies have used a combination of aluminum nitride ceramic substrates and oil cooling to control the temperature rise of inverter modules within 40 ° C, but its structural complexity limits the reliability of undersea deployment.

In response to the above issues, this article proposes a WPT system design scheme for undersea 10 kW applications. By designing and protecting SiC high-frequency inverters, improving the loss model of the magnetic coupling mechanism, enhancing the efficiency stability of the system under different operating conditions, and breaking through the limitations of seawater medium on high-frequency magnetic field transmission. Establish a seawater magnetic field coupling model based on COMSOL, design the distribution structure of Litz wire winding, optimize coil parameters using finite element simulation, and ensure efficiency stability under load fluctuations. Through these measures, it is expected to further promote the practical application and technological development of SiC based UWPT systems.

II. THEORY ANALYSIS

High-frequency inverters typically adopt a full-bridge topology structure, which can adjust the output voltage by changing the switch state while keeping the input voltage constant. For the SiC-based inverter in this study, a full-bridge configuration with a working frequency of 100 kHz was used. The full-bridge structure not only improves conversion efficiency but also simplifies control logic and reduces complexity. High-frequency inverters play a crucial role in modern power electronic systems, especially in wireless power transfer systems.

In order to further reduce switching losses, soft-switching technology was introduced in the design. Soft-switching technology can significantly reduce switching losses and improve system efficiency by achieving zero-voltage or zero-current switching during the switching process. This is particularly beneficial in high-frequency applications where traditional hard-switching techniques result in significant power loss. The specific implementation methods include resonant soft switches and quasi-resonant soft switches, each offering unique advantages depending on the application requirements.

As shown in Fig. 1, the experimental platform of this system adopts the LCC-S resonant compensation topology, which has the characteristics of constant coil current on the primary side and constant voltage output on the secondary side. By designing the circuit parameters of the primary-side resonant compensation unit, the system power output becomes independent of the secondary-side conditions, ensuring that the system can stably output the predetermined power regardless of changes in the secondary-side load. At the same time, by optimizing the design on the secondary side, the maximum efficiency of the system no longer depends on the circuit parameters of the primary side, thus achieving a decoupling design between the maximum efficiency and output power of the system. This design method not only enhances the flexibility and stability of the system but also provides a solid foundation for achieving efficient and reliable wireless power transfer. Combining the excellent properties of silicon carbide materials further enhances the overall efficiency and power density of the system.

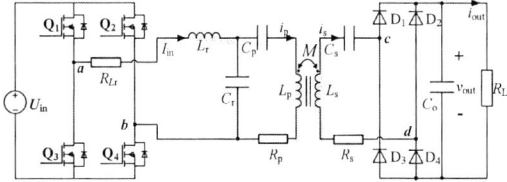

Fig.1. UWPT system LCC-S compensation topology

According to the fundamental equivalent method, the DC source Uin and inverter before point ab can be regarded as equivalent input Vab, while the rectifier and load RL after cd can be regarded as equivalent load Req point. As shown in Fig. 2, this equivalent model simplifies the analysis of the system by consolidating components into more manageable blocks. This approach allows for easier evaluation of system performance under various conditions, making it particularly useful for optimizing power electronics designs.

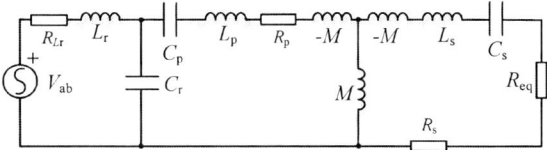

Fig.2. Topological equivalent circuit of resonance compensation

For the above equivalent circuit, when the duty cycle of the inverter bridge control signal is 0.5, the expressions for Vab and Req are as follows.

$$V_{ab} = \frac{2\sqrt{2}}{\pi} V_{in} \tag{1}$$

$$R_{eq} = \frac{8 R_L}{\pi^2} \tag{2}$$

This design adopts split side compensation, with the primary and secondary circuits resonating separately. For the secondary side of the circuit, there are:

$$\omega^2 L_S C_S = 1 \tag{3}$$

Secondly, for the primary side circuit, there are:

$$\omega L_r = \frac{1}{\omega C_r} = \omega L_P - \frac{1}{\omega C_P} \tag{4}$$

For the above design, except for the given angular frequency ω, the independent variables are Lr, Lp, and Ls, and the other variables are used as dependent variables. Therefore, it is necessary to design Lr, Lp, and Ls Convert the impedance on the secondary side to the reflected impedance Zr on the primary side, and when it satisfies the resonance condition, its value is as follows.

$$Z_r = \frac{(\omega M)^2}{R_S + R_{eq} + j\omega L_S + \frac{1}{j\omega C_S}} = \frac{(\omega M)^2}{R_S + R_{eq}} \tag{5}$$

According to circuit theory, when the resonant condition is satisfied, the input impedance Zin exhibits pure impedance, and the reactive power generated by the system is minimized, resulting in the highest system efficiency. From the input end, the input impedance Zin of the entire circuit can be viewed, and the above equation can be obtained.

$$Z_{in} = R_{L_r} + j\omega L_r + \frac{-j\omega L_r \left(Z_r + R_p\right) + \left(\omega L_p - \dfrac{1}{\omega C_p}\right)\omega L_r}{j\left(\omega L_p - \dfrac{1}{\omega C_p} - \omega L_r\right) + Z_r + R_p} = R_{L_r} + \frac{\left(\omega L_r\right)^2}{Z_r + R_p} \quad (6)$$

When the above resonance conditions are met, the secondary output voltage vout is derived as follows. When M and Lr are determined, the characteristic of constant voltage output on the secondary side is satisfied.

$$U_{out} \approx \frac{M}{L_r} U_{in} \quad (7)$$

The input power can be expressed as:

$$P_{in} = \frac{U_{in}^{\ 2}}{Z_{in}} = \frac{U_{in}^{\ 2}}{R_{Lr} + \dfrac{\left(\omega L_r\right)^2}{Z_r + R_p}} \quad (8)$$

The output power can be expressed as:

$$P_{out} = \frac{U_{out}^{\ 2}}{R_{eq}} = \frac{M^2 U_{in}^{\ 2}}{L_r^{\ 2} R_{eq}} \quad (9)$$

When the operating angular frequency of the system is equal to the natural resonant angular power ω_0, the system efficiency expression is obtained as follows. It can be seen that the system efficiency η is a complex function closely related to multiple internal parameter variables of the system.

$$\eta = \frac{P_{out}}{P_{in}} = \frac{M^2 R_{Lr}}{L_r^{\ 2} R_{eq}} + \frac{M^2 \omega^2 \left(R_S + R_{eq}\right)}{R_{eq}\left(\omega^2 M^2 + R_P\left(R_S + R_{eq}\right)\right)} \quad (10)$$

As a specific system, the UWPT system test platform considers the system input voltage, operating frequency, and load conditions as fixed values. The voltage gain is set to a certain proportional relationship based on the system output requirements. While ignoring changes in coil internal resistance, it can be assumed that the system efficiency is only related to the mutual inductance M. The system efficiency increases with an increase in mutual inductance. This assumption simplifies the analysis by focusing on the primary factor affecting efficiency. In practical applications, optimizing mutual inductance through coil design and alignment can significantly enhance the performance of wireless power transfer systems.

With the increase in switching frequency, new problems such as ringing phenomena and high-frequency crosstalk may arise. These issues can lead to increased switch losses or direct bridge arm connections, thereby reducing overall circuit efficiency.

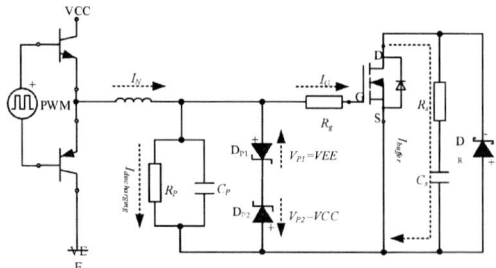

Fig.3. Optimizing RC level shift gate drive structure

In this design, the gate drive structure adopts an optimized RC level shifting scheme. As shown in Figure 3, the traditional level shift structure uses an energy absorption circuit composed of resistors and capacitors to help eliminate gate oscillations, but it will result in energy loss during steady-state operation, which in turn affects circuit efficiency. Therefore, we propose an improvement measure by using magnetic beads instead of parallel circuits composed of Rg and Cg. This design not only effectively suppresses oscillations at specific frequencies, but also reduces switching losses. In addition, the voltage in the on and off states is clamped by diodes DP1 and DP2 respectively to reduce the impact of voltage spikes. Add Rs and Cs between the source and drain to absorb surge currents, ensuring that the switching device always operates within a safe operating range.

The efficiency optimization of underwater wireless power transmission systems requires comprehensive consideration of multiple factors such as electromagnetic coupling characteristics, load matching relationships, and power device stress. By establishing a dynamic correlation model between parameters, it can effectively guide the design of compensation network parameters and the selection of power devices, ultimately achieving the optimization and improvement of overall system performance.

In addition, the conductivity of seawater is much higher than that of air. According to Faraday's law of electromagnetic induction, a rapidly alternating magnetic field will cause significant eddy current losses in seawater. The resonant frequency of underwater power transmission is related to the attenuation constant of seawater. The higher the resonant frequency, the greater the attenuation constant, indicating that the system has higher losses. If the resonant frequency of the system is too low, the system volume will be relatively large. Therefore, if the system volume allows, the resonant frequency should not be too high.

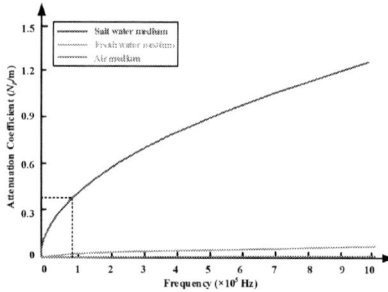

Fig.4. The relationship between resonant frequency and attenuation coefficient

According to previous research, the causes of eddy current losses are actually quite complex. The eddy current loss is related to the magnetic induction intensity B, the magnetic field alternating frequency fs, the water salinity S, and the seawater volume V between the couplers. The eddy current loss caused by seawater can be approximated as:

$$P_{eddy} = f\left(K, B^\alpha, f_s^{\ \beta}, V\right) \quad (11)$$

Among them, K is the attenuation coefficient, representing the influence of seawater salinity and molecular polarization on eddy current losses. α and β respectively represent the positive correlation between eddy current loss and magnetic induction intensity B and switching frequency fs After determining the transmission environment, K remains unchanged. Biot Savart's law states that in a coil with one turn:

$$d\vec{B} = \frac{\mu_0}{4\pi} \times \frac{Idl \times \vec{r}}{r^3} = \frac{\mu_0}{4\pi} \times \frac{Idl \times \sin\theta}{r^2} \quad (12)$$

Among them, μ 0 is the magnetic permeability, dl is the current unit path, r is the radius of a single turn coil, I is the current passing through the coil, and sin θ represents the offset of the current unit path. When dl is small enough, θ=90 °.

When analyzing multi turn coils. In the current circuit L, the magnetic induction intensity B can be expressed as:

$$\oint_L Bdl = \mu_0 NI \quad (13)$$

In a magnetic field constructed by a multi turn coil, the magnetic fields generated by adjacent wires will cancel each other out. In addition, the radius of the coil is much larger than the diameter of the wire. Therefore, the corresponding magnetic field is ignored. The effective length of a wire can be approximately expressed as:

$$L_{eq} = r \quad (14)$$

Therefore, the magnetic induction intensity B can be expressed as follows:

$$B \approx \frac{\mu_0 NI}{r} \quad (15)$$

Based on extensive analysis and experimentation, an empirical equation for eddy current losses has been obtained. The following formula can represent the characteristics of eddy current losses:

$$P_{eddy} = \int K \times B^2 f_s^{2.165} dV, \quad K \approx 3.5 \times 10^{-6} \quad (16)$$

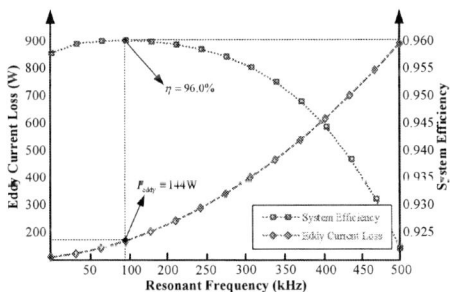

Fig.5. The relationship between system frequency and eddy current loss and system efficiency.

III. EXPERIMENTAL VERIFICATION

A 10KW wireless energy transmission experimental platform was built for basic testing using a high-frequency inverter made of wide bandgap silicon carbide material, as shown in Figure 6. The experimental system consists of a DC power supply with a maximum input voltage of 500 V, a high-frequency inverter with a working frequency of 85 kHz, a rectifier, an electronic load, a magnetic coupling mechanism, and a primary and secondary compensation network. Energy is transmitted to the receiving end through high-frequency magnetic field coupling.

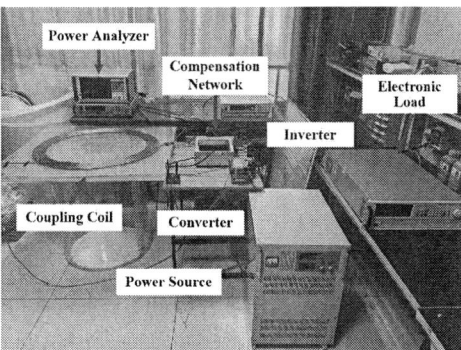

Fig.6. Experimental platform

The resonance state of the primary and secondary circuits is the key to efficient energy transfer in the system. Therefore, the resonance waveforms of the primary and secondary circuits of the UWPT system were measured, as shown in Figure 7. The experimental results show that the output voltage and current phase of the primary inverter are basically the same, and the input voltage and current phase of the secondary rectifier are basically the same. The UWPT system operates in a resonant state. The output voltage amplitude of the primary inverter is equal to the input DC voltage amplitude. The input voltage amplitude of the secondary rectifier is equal to the DC voltage amplitude at both ends of the load. A voltage gain theoretical model that conforms to the LCC-S compensation topology.

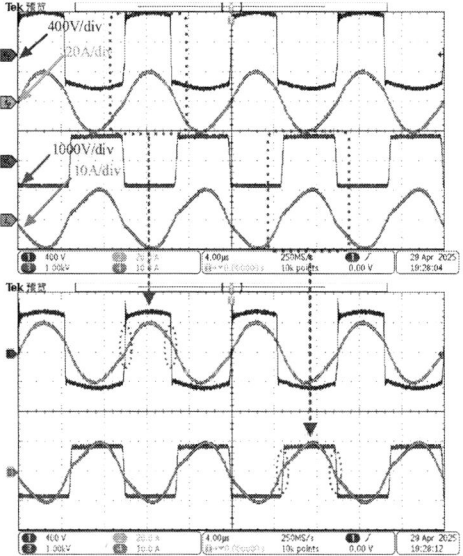

Fig.7. Voltage and current waveforms of primary and secondary sides in resonance state

The fluctuation of load during the charging process of AUV batteries cannot be ignored. To evaluate the constant voltage output capability of the system, as shown in Figure 8, the input-output waveforms of the UWPT system are displayed when the load undergoes a sudden change between 70 Ω and 100 Ω. The experimental data shows that the output voltage remains stable at 700 V with a fluctuation amplitude of less than 2.3%, and the output current is precisely adjusted within the range of 7-10 A as the load changes. The results show that when load fluctuations occur, using open-loop control method can suppress load fluctuations with fast response speed, and the system has the performance of constant voltage output, meeting the constant voltage requirements of AUV battery charging. This feature validates the rationality of the system parameter design.

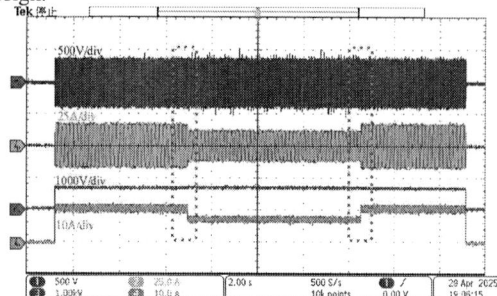

Fig.8. Voltage and current waveforms of primary and secondary sides under load fluctuations

As shown in Figure 9, the power analyzer test results indicate that the system efficiency reaches 96.3% when the output power is increased to 10200 W. This efficiency characteristic confirms the synergistic optimization effect of magnetic coupling mechanism and compensation network. It is worth noting that the efficiency of the system remains above 96% at a rated power of 10 kW, significantly better than similar underwater wireless power transmission systems reported in existing literature.

U_{rms1}	493.095 V
I_{rms1}	21.6137 A
U_{rms2}	0.99840 kV
I_{rms2}	10.2794 A
P_1	10.6576 kW
P_2	10.2630 kW
η_1	96.297 %
$Loss_1$	394.588 W

Fig.9. Maximum output power state

Experimental verification shows that the 10kW UWPT system based on SiC high-frequency inverter has excellent performance, excellent load adaptability, and constant voltage output capability. Verified the effectiveness of reducing losses and improving efficiency of SiC devices, optimizing magnetic coupling mechanisms to enhance coupling coefficients, and overall system design.

IV. CONCUSIONS AND DISCUSSIONS

This article explores an undersea wireless power transfer system with a power level of 10kW. The high-frequency characteristics of SiC MOSFET devices significantly reduce switching losses and temperature rise, while improving power transmission efficiency. Adopting Leeds wire winding and segmented magnetic core design to reduce high-frequency eddy current losses and improve coupling coefficient, and adopting LCC-S type resonant compensation network, combined with finite element simulation optimization of coil structure, to achieve the improvement of coupling coefficient. Achieve a maximum output power of 10 kW at a transmission distance of 10 cm undersea. This design has advantages in transmission efficiency, power capacity, and anti offset capability, providing theoretical basis and technical reference for the

engineering application of high-power medium distance UWPT systems.

REFERENCES

[1] O. C. Onar, M. Chinthavali, S. Campbell, P. Ning, C. P. White and J. M. Miller, "A SiC MOSFET based inverter for wireless power transfer applications," 2014 IEEE Applied Power Electronics Conference and Exposition - APEC 2014, Fort Worth, TX, USA, 2014, pp. 1690-1696.

[2] Y. Wei, L. Du, X. Du, V. S. Machireddy and A. Mantooth, "A Wireless Power Transfer based Gate Driver Design for Medium Voltage SiC MOSFETs," 2021 IEEE International Future Energy Electronics Conference (IFEEC), Taipei, Taiwan, 2021, pp. 1-7.

[3] O. C. Onar, S. Campbell, P. Ning, J. M. Miller and Z. Liang, "Fabrication and evaluation of a high performance SiC inverter for wireless power transfer applications," The 1st IEEE Workshop on Wide Bandgap Power Devices and Applications, Columbus, OH, USA, 2013, pp. 125-130.

[4] K. Matsubara, K. Wada and Y. Suzuki, "Design of magnetic field generator operating at 85 kHz using SiC-MOSFETs for evaluating electromagnetic interference," 2017 IEEE 3rd International Future Energy Electronics Conference and ECCE Asia (IFEEC 2017 - ECCE Asia), Kaohsiung, Taiwan, 2017, pp. 412-416.

[5] C. Da, F. Li, M. Nie, S. Li, C. Tao and L. Wang, "Undersea Capacitive Coupled Simultaneous Wireless Power and Data Transfer for Multiload Applications," IEEE Transactions on Power Electronics, vol. 40, no. 1, pp. 2630-2642, Jan.2025.

[6] R. Hasaba et al., "A Highly Efficient and High Degree of Freedom of Position kW-class Wireless Power Transfer System in Seawater for Small AUVs," 2021 IEEE Wireless Power Transfer Conference (WPTC), San Diego, CA, USA, 2021, pp. 1-4.

[7] E. Rong, P. Sun, K. Qiao, X. Zhang, G. Yang and X. Wu, "Six-Plate and Hybrid-Dielectric Capacitive Coupler for Underwater Wireless Power Transfer," IEEE Transactions on Power Electronics, vol. 39, no. 2, pp. 2867-2881, Feb.2024.

[8] L. Yang et al., "High Power and High Freedom Platform Type Undersea Wireless Power Transfer Station Without Ferrite Core for AUVs," IEEE Journal of Emerging and Selected Topics in Power Electronics, doi:10.1109/JESTPE.2024.3456550.

[9] E. Abramov and M. M. Peretz, "Adaptive Self-Tuned Controller IC for Resonant-Based Wireless Power Transfer Transmitters," IEEE Trans. Power Electron., vol. 36, no. 11, pp. 12413–12431, Nov. 2021, doi: 10. 1109/TPEL.2021.3081018.

[10] C. Yang, M. Lin, and D. Li, "Improving Steady and Starting Characteristics of Wireless Charging for an AUV Docking System," IEEE J. Oceanic Eng., vol. 45, no. 2, pp. 430–441.

[11] S. Pang, J. Xu, H. Li, Q. Ma, and X. Li, "Dual-Frequency Modulation to Achieve Power Independent Regulation for Dual-Load Underwater Wireless Power Connector," IEEE J. Emerg. Sel. Top. Power Electron., vol. 11, no. 2, pp. 2377–2389.

[12] Y. Gu, J. Wang, Z. Liang, and Z. Zhang, "A Wireless In-Flight Charging Range Extended PT-WPT System Using S/Single-Inductor-Double-Capacitor Compensation Network for Drones," IEEE Trans. Power Electron., vol. 38, no. 10, pp. 11847–11858.

[13] S.-J. Jeon and D.-W. Seo, "Capacitance Tuning Method for Maximum Output Power in Multiple-Transmitter Wireless Power Transfer System," IEEE Access, vol. 8, pp. 181674–181682.

[14] S. Ali Khan and D. Ahn, "Automatic Resonance Tuning With ON/OFF Soft Switching for Push–Pull Parallel-Resonant Inverter in Wireless Power Transfer," IEEE Trans. Power Electron., vol. 37, no. 9, pp. 10133–10138.

Minimizing Eddy Current Loss for Implanted Wireless Charger Through Phase Difference Optimization

1st Pengyu Chen
School of Info. Tech.
ShanghaiTech University
Shanghai, China
chenpy2023@shanghaitech.edu.cn

2nd Siyi Yao
School of Info. Tech.
ShanghaiTech University
Shanghai, China
yaosy2022@shanghaitech.edu.cn

3rd Xiyuan Lin
School of Info. Tech.
ShanghaiTech University
Shanghai, China
linxy2023@shanghaitech.edu.cn

4th Hongjun Zheng
Lingang Laboratory
Shanghai, China
zhenghongjun@lglab.ac.cn

5th Congcong Zhang
Lingang Laboratory
Shanghai, China
zhangcc@lglab.ac.cn

6th Minfan Fu*
School of Info. Tech.
ShanghaiTech University
Shanghai, China
fumf@shanghaitech.edu.cn

Abstract—**Rechargeable implantable medical devices (IMDs) inherently incorporate metallic casings, which induce substantial eddy current losses (ECL) and subsequent thermal elevation during wireless charging cycles. This paper presents a phase modulation strategy to mitigate ECL by optimizing the current phase difference ϕ between TX and RX coils. Through analytical modeling of loss mechanisms, we demonstrate that increasing ϕ beyond the conventional 90° operating point significantly suppresses ECL generation. The proposed methodology is rigorously validated via simulations and experimental prototypes. Results confirm the existence of an ECL-minimizing optimal phase ϕ_{opt}, where ϕ_{opt} exhibits strong correlation with TX and RX coils diameter and current distribution.**

Index Terms—**Eddy current losses, implantable device, wireless charger, electromagnetic field modelling.**

I. INTRODUCTION

Over the past five decades, rapid advances in medical technology have spurred the widespread adoption of implantable devices with enhanced functionalities and increased energy demands. For instance, cochlear implants typically operate at around 20 mW [1], while brain-computer interfaces may exceed 500 mW [2], necessitating improved charging capabilities for reliable operation. Wireless power transfer (WPT) [3] is now the preferred charging method for implantable devices. However, for patient safety, implantable devices are encased in metal, which induces eddy current losses (ECL) that lower efficiency and generate harmful heat [4]. Moreover, space constraints force coils to operate above 800 kHz for miniaturization, further exacerbating ECL. Among all types of losses, ECL accounts for the largest share, as shown in Fig.1. Mitigating these losses remains a critical research focus.

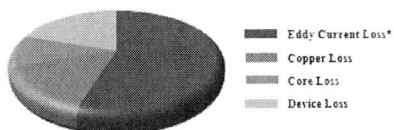

Fig. 1. Loss distribution.

Various strategies have been proposed to mitigate ECL in implantable IPT systems. For example, [5] recommends placing the coil outside the metal casing to distance it from strong magnetic fields, thereby significantly reducing ECL, though this is not clinically viable. In contrast, [6] proposes using a thin, fully enclosed casing that permits magnetic field penetration; despite lowering ECL, this approach compromises efficiency for high-power applications. Similarly, [7] introduces a casing with a circular opening to reduce ECL, yet it still faces inherent efficiency limitations.

In other areas such as consumer electronics or underwater IPT systems, alternative methods have been explored. For instance, [8] employs a control algorithm for maximum efficiency point tracking to reduce eddy currents, but it requires extra receiver circuitry, making it challenging for implantable devices. Likewise, [9] presents a split-casing design that leverages the eddy current effect to enhance magnetic coupling, though its complexity renders it unsuitable for implants. Clearly, most solutions focus on structural rather than circuit-level optimizations. In this paper, we address a method to reduce ECL by optimizing the current phase difference of the coupler coil for implantable WPT systems.

II. Loss Model

The implant structure is shown in Fig.2(a). The RX coil is housed within a metal casing that features a circular opening at its top, sealed with a ceramic piece to allow magnetic field penetration. A ferrite layer is closely attached behind the receiving coil to guides the magnetic field, enhances coupling, and shields the PCB from interference. The metal alloy casing is designed with a smooth, flat surface for biocompatibility, which makes it susceptible to eddy currents.

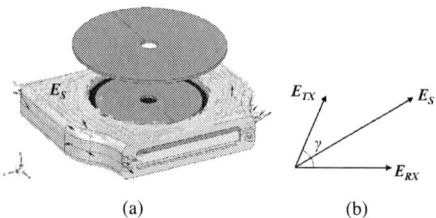

(a) (b)

Fig. 2. Induced electric field. (a) On top casing. (b) Decomposition.

When the WPT system is operating normally, part of the alternating magnetic field generated by the TX and RX coils induces an electric field on the metal casing. Under this induced electric field, eddy currents will be generated in the metal loop. The induced electric field $\mathbf{E_s}$ at each position on the titanium casing can be considered as the vector sum of the electric fields $\mathbf{E_{tx}}$ and $\mathbf{E_{rx}}$ generated individually by the TX and RX coils. Therefore, E can be decomposed as

$$\mathbf{E_s} = \mathbf{E_{tx}} + \mathbf{E_{rx}} \tag{1}$$

According to the derivation in [10], the amplitude of the induced electric field intensity is proportional to the coil current, i.e., $|\mathbf{E}| \propto I$, which holds true for any near-field magnetic coupling WPT system. The currents in the TX and RX coils usually have a certain phase difference ϕ. which can be expressed as:

$$\begin{cases} i_{tx} = I_{txm} \cos(\theta) \\ i_{rx} = I_{rxm} \cos(\theta + \phi). \end{cases} \tag{2}$$

where I_{txm} and I_{rxm} are the peak currents of i_{tx} and i_{rx} respectively, and θ is the phase of i_{tx}, there is $\theta = \omega t$, where ω is the angular frequency. The proportionality coefficient relating electric field intensity to current depends on the metal casing's shape and conductivity and is defined as a_{tx}. According to (2), the induced electric field intensity can be expressed as

$$\mathbf{E_{tx}} = a_{tx} I_{txm} \cos(\theta) \hat{E}_{tx} \tag{3}$$

where $\hat{E}_{tx}$ is the direction vector of $\mathbf{E_{tx}}$. Considering $A_{tx} = a_{tx} I_{txm}$, where A_{tx} is a coefficient related to the TX coil current, the material, and the shape of the metal casing. Similarly, the induced electric field intensity

from RX current depends on $A_{rx} = a_{rx} I_{rxm}$. Hence, the total induced electric field $\mathbf{E_s}$ can be expressed as

$$\mathbf{E_s} = A_{tx} \cos(\theta) \hat{E}_{tx} + A_{rx} \cos(\theta + \phi) \hat{E}_{rx} \tag{4}$$

The magnitude of the induced electric field intensity is

$$\begin{aligned} |\mathbf{E_s}|^2 = &A_{tx}^2 \cos^2(\theta) + A_{rx}^2 \cos^2(\theta + \phi) \\ &+ 2\cos(\theta)\cos(\theta + \phi) A_{tx} A_{rx} \cos(\gamma) \end{aligned} \tag{5}$$

where γ is the angle between $\hat{E}_{tx}$ and $\hat{E}_{rx}$, as shown in Fig.2 (b). By integrating $|\mathbf{E_s}|^2$ with respect to θ from 0 to 2π, the RMS value of $\mathbf{E_s}$ is obtained as

$$\begin{aligned} \int_0^{2\pi} |\mathbf{E_s}|^2 \, d\theta = &\int_0^{2\pi} A_{tx}^2 \cos^2(\theta) \, d\theta \\ &+ \int_0^{2\pi} A_{rx}^2 \cos^2(\theta + \phi) \, d\theta \\ &+ \int_0^{2\pi} 2\cos(\theta)\cos(\theta + \phi) A_{tx} A_{rx} \cos(\gamma) d\theta \end{aligned} \tag{6}$$

By Substituting $|\mathbf{E_{tx}}|^2$ and $|\mathbf{E_{rx}}|^2$ respectively, we get

$$\begin{aligned} \int_0^{2\pi} |\mathbf{E_s}|^2 d\theta = &\int_0^{2\pi} |\mathbf{E_{tx}}|^2 d\theta + \int_0^{2\pi} |\mathbf{E_{rx}}|^2 d\theta \\ &+ k_{cov} \int_0^{2\pi} 2\cos(\theta)\cos(\theta + \phi) d\theta \end{aligned} \tag{7}$$

where $k_{cov} = A_{tx} A_{rx} \cos(\gamma)$. The eddy current loss (ECL) P_e of a conductor is expressed as

$$P_e = \iiint_V \sigma |\mathbf{E}|^2 dV \tag{8}$$

where σ is the electric conductivity of the casing. Substituting (7) into (8), the total ECL is obtained as

$$\begin{aligned} P_{es} &= \iiint_V \sigma \int_0^{2\pi} |\mathbf{E_s}|^2 d\theta dV \\ &= P_{etx} + P_{erx} + P_{cov} \end{aligned} \tag{9}$$

where P_{es} is the total ECL of the casing, and P_{etx} and P_{erx} respectively represent the ECL generated only by i_{tx} and i_{rx}, where $P_{etx} = \iiint_V \sigma \int_0^{2\pi} |\mathbf{E_{tx}}|^2 d\theta dV$, $P_{erx} = \iiint_V \sigma \int_0^{2\pi} |\mathbf{E_{rx}}|^2 d\theta dV$, $k_e = \sigma k_{cov}$, $P_{cov} = \iiint_V k_e \int_0^{2\pi} 2\cos(\theta)\cos(\theta + \phi) d\theta dV$, which is the additional ECL generated through the mutual interaction of E_{tx} and E_{rx}. It can be observed that P_{cov} depends on k_e, ϕ, as well as the volume of the metal casing. When I_{tx}, I_{rx}, the size and position of coils, the structure of titanium casing are all fixed, k_e can be treated as a constant. At this point, the ELC only depends on ϕ.

When the secondary side is not completely compensated, the total ECL equals the sum of the ECL produced individually by TX and RX plus an extra term dependent on ϕ. If this term is negative, the ECL is lower than at $\phi = \pi/2$, suggesting that adjusting ϕ can suppress ECL.

III. CIRCUIT ANALYSIS

This section would analyze the influence of the phase difference between the TX and RX currents on the ECL. Accordingly, an electromagnetic model is built to study the effect of the TX-RX phase difference on eddy current losses. Since the simulation requires both TX and RX coil currents, a circuit is first established [11], with the system architecture shown in Fig.3. This section focuses on circuit analysis, which would offer the design constrain for the simulation.

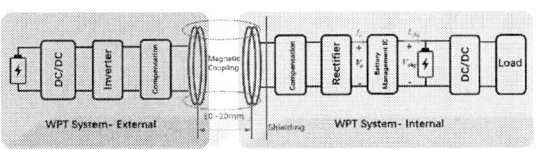

Fig. 3. Architecture of WPT system.

The WPT system uses LCC-S as the compensation network [12] [13], which is shown in Fig.4. Where L_{tx}, L_{rx}, and M represent the self-inductance of the TX coil, the self-inductance of the RX coil, and mutual inductance, respectively. L_t is the compensation inductor, C_{tx}, C_{rx}, and C_t represent the compensation capacitance, and R_{ac} is the equivalent AC load of the rectifier and its subsequent load stage. Specifically, in Fig.3, the maximum output voltage V_{chg} and maximum output current I_{chg} of the charging management IC are constant, and the output voltage of rectifier V_o is also constant [14]. Ignoring chip losses, by power conservation, the rectifier's output current I_o is clamped and can be expressed as

$$I_o = \frac{V_{chg} I_{chg}}{V_o}. \tag{10}$$

From the full-bridge rectifier, the peak current of RX coil I_{rxm} can be expressed as

$$I_{rxm} = \frac{\pi}{2} I_o. \tag{11}$$

Then I_{rxm} needs to be determined. In a practical WPT system, the equivalent load R_{ac} should receive sufficient power to ensure the normal operation. The secondary compensation design typically involves compensating the self-inductance of the RX coil L_{rx} with appropriate capacitor C_{rx} for complete compensation, minimizing reactive power on the receiving side and allowing the load to receive more power [15]. Using the above compensation solution, the phase difference between the TX and RX currents is 90°.

When the phase difference ϕ between the TX and RX currents changes, meaning C_{rx} and L_{rx} are no longer

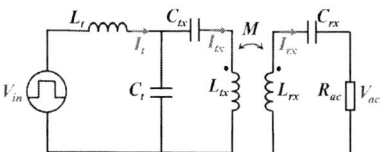

Fig. 4. WPT charging system with LCC-S network.

completely compensated, and the secondary impedance changes. The secondary impedance can be expressed as

$$\begin{aligned} \boldsymbol{Z_{rx}} &= j\omega L_{rx} + \frac{1}{j\omega C_{rx}} + R_{ac} \\ &= jX + R_{ac} = \frac{R_{ac}}{\cos \beta} \angle \beta \end{aligned}. \tag{12}$$

where β is the impedance angle of the secondary side, ranging from $-\pi/2$ to $+\pi/2$, and X is the imaginary part. At this point, the RX coil current $\boldsymbol{I_{rx}}$ (i.e., a phasor form) can be expressed as

$$\boldsymbol{I_{rx}} = \frac{-j\omega M \boldsymbol{I_{tx}}}{(R_{ac}/\cos \beta)\angle \beta} \tag{13}$$

From this equation, the magnitude and phase difference between $\boldsymbol{I_{rx}}$ and $\boldsymbol{I_{tx}}$ can be expressed as

$$\begin{cases} \phi = |-\frac{\pi}{2} - \beta| = \frac{\pi}{2} + \beta \\ \frac{I_{rxm}}{I_{txm}} = \frac{\omega M \cos(\phi - \frac{\pi}{2})}{R_{ac}} \end{cases} \tag{14}$$

From the magnitude relationship between I_{rxm} and I_{txm} in (14), it can be seen that changing ϕ will reduce I_{rxm}, which leads to a decrease in the power received by the load. To maintain a constant power to the load, I_{rxm} must remain unchanged, which requires continuously increasing the TX current I_{txm}. The variation of I_{txm} with ϕ can be expressed as

$$I_{txm} = \frac{R_{ac}}{\omega M \cos(\phi - \frac{\pi}{2})} I_{rxm} = \frac{I_{txm,min}}{\cos(\phi - \frac{\pi}{2})} \tag{15}$$

where $I_{txm,min}$ is the minimum peak current of i_{tx} when $\phi = \pi/2$. $I_{txm,min}$ can be obtained by complete compensating L_{rx}. The following equation can be derived according to [5].

$$\omega M I_{txm,min} = V_{ac,1} = \frac{4}{\pi} V_o \tag{16}$$

where V_o is the output voltage of rectifier mentioned earlier, and $V_{ac,1}$ is the fundamental component of the AC voltage before rectification. Thus, $I_{txm,min}$ can be obtained as

$$I_{txm,min} = \frac{4}{\pi \omega M} V_o = \frac{2}{\pi^2 f M} V_o \tag{17}$$

At this point, all the coil current-related parameters for the electromagnetic field simulation have been obtained.

IV. SIMULATION-BASED LOSS OPTIMIZATION

While the loss model and circuit analysis reveal the phase difference's impact, deriving a specific model for a customized case is challenging. Thus, electromagnetic simulation should be employed to demonstrate the loss reduction effect achieved through phase optimization.

Taking the wireless charger for Brain-Computer Interface (BCI) devices as an example, the operating distance is 2cm, which satisfies the implantation depth requirement. The system operates at 1 MHz to minimize the component size while ensuring compatibility with magnetic materials. The RX coil is placed in a 40 mm-diameter groove, with its diameter limited to 36 mm to accommodate assembly clearance, while the TX coil's outer diameter is generally no more than 60 mm. For BCI implant applications, the battery capacity is typically designed at approximately 2 Wh with an operating voltage range of 3.2–4.2 V. As Li-ion batteries generally restrict charging currents to below 0.5C for cycle life extension, this work selects 0.35C as the charging rate. This configuration yields $V_{chg}I_{chg} = 0.7$ Wh, and V_o is set to 5.5 V (within the 4–7 V range to optimize the charging management IC efficiency). Substituting these into (10) and (11) yields $I_{rxm} = 200mA$, and from which the primary current I_{txm} is determined via (15) and (17).

In practical implantable devices, R_{ac} should receive sufficient power [16]. Thus, when ϕ deviates from $\pi/2$, I_{rxm} must remain constant, which, according to (15), requires I_{txm} to increase. This increase intensifies the varying magnetic field on the metal casing, thereby raising the ECL. The combined effect produces a minimum in the ECL variation with respect to ϕ, as I_{txm} and ϕ are coupled through (15). Figure 5 illustrates the ECL variations for different coil sizes, while Fig. 6 shows the corresponding TX and RX current waveforms at the minimum ECL for each case.

As shown in Fig.5, the ECL curve versus ϕ exhibits a minimum within $[\pi/2, \pi]$, consistent with previous analysis. This implies that increasing ϕ slightly beyond $\pi/2$ effectively suppresses eddy current losses on the metal casing. Furthermore, the ECL reduction with increasing ϕ is strongly influenced by the coil sizes. As seen in Fig. 5(b) and (c), a small D_{tx} with a large D_{rx} results in a limited TX magnetic field and an extensive RX field effect, leading to more pronounced ECL reduction. Additionally, Fig. 5(f) shows that even with a large D_{tx}, a very small I_{rxm} yields significant ECL reduction.

These results indicate that the effectiveness of ECL reduction is linked to the contribution ratio of I_{rx} and I_{tx}: the greater the contribution of I_{rx} relative to I_{tx}, the more significant the reduction achieved by increasing ϕ.

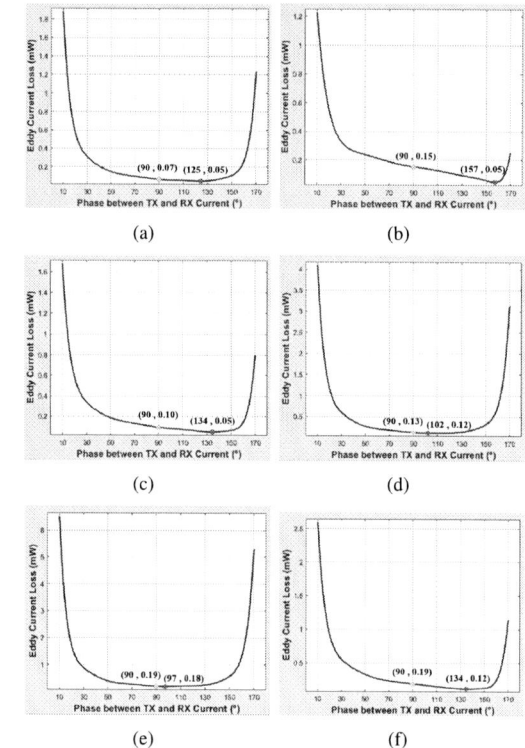

Fig. 5. ECL at different coil sizes. (a) $D_{tx} = 20mm, D_{rx} = 28mm$. (b) $D_{tx} = 20mm, D_{rx} = 36mm$. (c) $D_{tx} = 32mm, D_{rx} = 32mm$. (d) $D_{tx} = 44mm, D_{rx} = 28mm$. (e) $D_{tx} = 56mm, D_{rx} = 28mm$. (f) $D_{tx} = 52mm, D_{rx} = 36mm$

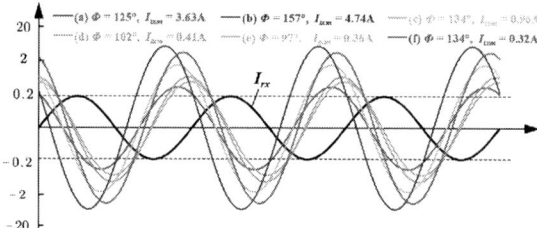

Fig. 6. Waveform of TX and RX current for each case.

V. EXPERIMENT VERIFICATION

Two representative configurations from Fig.5 (d) and (f), designated as experimental Case 1 and Case 2 respectively, were selected for hardware validation. The experimental platform was set up as shown in Fig. 7.

The prototype of this experimental system is shown in Fig.8. The white-colored fixture establishes a standardized 2 cm transmission distance. The upper section places the WPT transmitter containing:

- A PCB integrating full-bridge inverter and LCC compensation network
- TX coil (52 mm diameter)

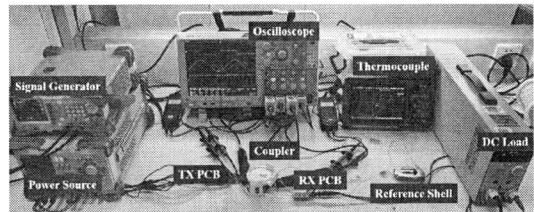

Fig. 7. Experimental setup.

The lower section places the implantable receiver containing:

- A PCB with S-compensation and rectifier
- RX coil (36 mm diameter) with a magnetic-core
- Titanium casing

For the LCC-S compensation network, initial implementation adopted completely compensation of C_{rx} and L_{rx}, achieving approximately 90° phase difference between I_{tx} and I_{rx}. Subsequent adjustment of C_{rx} values through (12) and (14) enabled partial compensation configurations, generating various phase relationships between I_{tx} and I_{rx}.

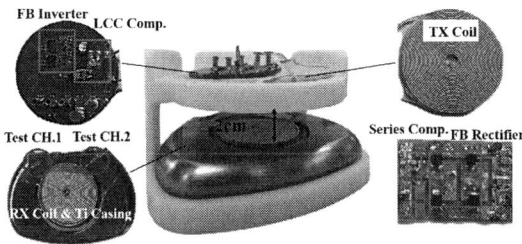

Fig. 8. Prototype of the Case 2 52-36mm WPT system.

Experimental waveforms from Case 1 and Case 2 systems are presented in Fig.9. As shown in Fig.9 (a) and (c), completely compensation yields 90° phase difference between I_{tx} and I_{rx}. After C_{rx} adjustment, Case 1 exhibits 100° phase difference with 0.40 A TX current peak, while Case 2 demonstrates 135° phase difference with 0.36 A TX current peak. These measurements align with the ECL minimum point characteristics in simulation results (see Fig.5 and Fig.6), including both phase differences and I_{txm} magnitudes. This correlation indirectly verifies the accuracy of the theoretical model (15) and simulation, establishing foundational evidence for subsequent thermal validation of ECL suppression effectiveness.

To verify the effectiveness of ECL reduction through currents phase adjustment, thermocouples were deployed to measure temperature rise on the titanium casing, serving as an indirect indicator of ECL generation. Three measurement channels were configured: CH1 and CH2 monitored symmetrical positions on the casing

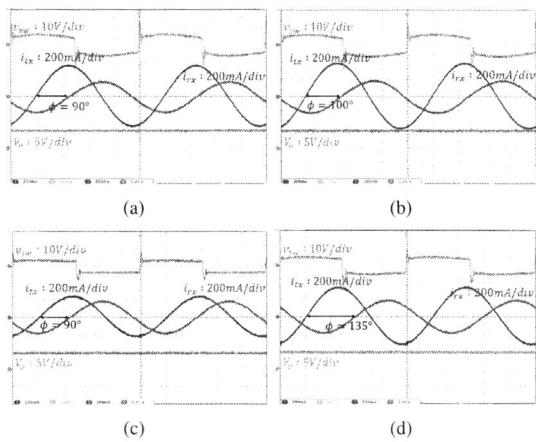

Fig. 9. Waveforms of I_{tx} and I_{rx} with different ϕ. (a) Case 1, $\phi = 90°$. (b) Case 1, $\phi = 100°$. (c) Case 2, $\phi = 90°$. (d) Case 2, $\phi = 135°$.

(shown in Fig.9), while CH3 functioned as a blank reference. Each temperature recording session lasted 6 minutes, with results documented in Fig.10.

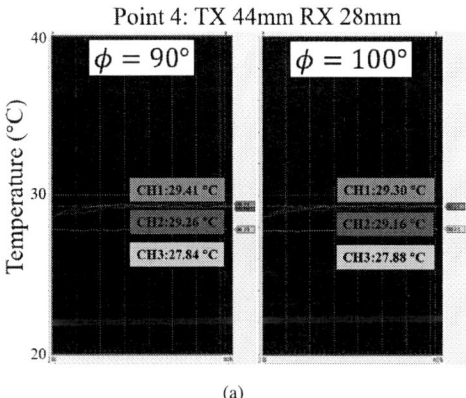

(a)

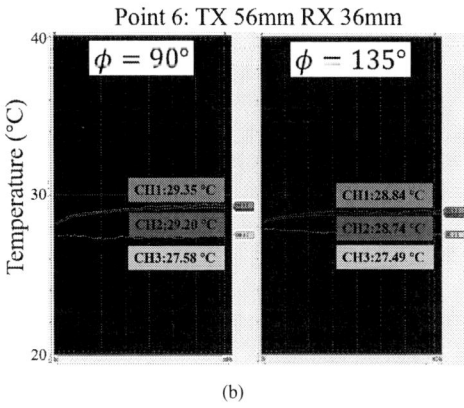

(b)

Fig. 10. Temperature rise characteristics comparison. (a) Temperature rise of Case 1. (b) Temperature rise of Case 2.

As shown in Fig.10 (b), Case 2 exhibits 1.70°C temperature rise at 90° phase difference, decreasing to

1.30°C at 135° phase difference. This 0.4°C reduction corresponds to 70 mW ECL suppression (from 190 mW to 120 mW in Fig. 5(f)), demonstrating significant effectiveness when operating phase between I_{tx} and I_{rx} to the ECL minimum point. Particularly crucial for implantable devices where in-vivo temperature rise must remain below 2°C, this improvement holds substantial clinical relevance.

Fig.10 (a) reveals a more moderate result: Case 1 exhibits 1.49°C temperature rise at 90° phase difference reduces to 1.35°C at 100° phase difference. This marginal 0.14°C improvement aligns with simulated 10 mW ECL reduction (from 130 mW to 120 mW in Fig. 5(d)), further validating the simulation accuracy.

VI. CONCLUSION

This paper optimizes the phase difference between I_{tx} and I_{rx} to reduce eddy current loss of implantable devices, and experimental validation reveals distinct suppression effects across coil configurations. The paper confirms two critical findings from simulations and experiment:

(1) ECL reaches its minimum at an optimal phase shift ϕ_{opt} within $(\pi/2, \pi)$, where ϕ_{opt} is determined by coil diameter ratio D_{tx}/D_{rx}. Adjusting ϕ from $\pi/2$ to ϕ_{opt} achieves ECL reduction.

(2) ECL suppression efficacy correlates strongly with magnitude and distribution of current (I_{tx} and I_{rx}), as evidenced by Case 2's greater suppression than Case 1 when similar adjusting ϕ to ϕ_{opt}.

These findings offer insights into designing efficient wireless power transfer systems, enhancing performance and thermal management in implantable devices.

VII. ACKNOWLEDGMENTS

This work was supported by National Natural Science Foundation of China under Grant 52477013 and Lingang Laboratory under Grant NO. LG-GG202402-06-10.

REFERENCES

[1] M. Saadatzi, M. N. Saadatzi, and S. Banerjee, "Modeling and fabrication of a piezoelectric artificial cochlea electrode array with longitudinal coupling," *IEEE Sensors Journal*, vol. 20, no. 19, pp. 11 163–11 172, 2020.

[2] S. Gao and X. Gao, "The design and implementation of visual braincomputer interfaces," in *2014 International Winter Workshop on Brain-Computer Interface (BCI)*. IEEE, 2014, pp. 1–3.

[3] G. Zheng, T. Li, X. Wang, and M. Fu, "Stability and controller design of a two-stage inductive power transfer system," *IEEE Transactions on Industrial Electronics*, vol. 72, no. 1, pp. 380–389, 2025.

[4] X. Yu, T. Skauli, B. Skauli, S. Sandhu, P. B. Catrysse, and S. Fan, "Wireless power transfer in the presence of metallic plates: Experimental results," *AIP Advances*, vol. 3, no. 6, 2013.

[5] C. Xiao, K. Wei, D. Cheng, and Y. Liu, "Wireless charging system considering eddy current in cardiac pacemaker shell: Theoretical modeling, experiments, and safety simulations," *IEEE Transactions on Industrial Electronics*, vol. 64, no. 5, pp. 3978–3988, 2016.

[6] J. H. Kim and C.-H. Ahn, "Method to reduce metal plate effect between transmitter and receiver in wireless power transfer system," *IEEE Antennas and Wireless Propagation Letters*, vol. 17, no. 4, pp. 587–590, 2018.

[7] H. Mao, B. Yang, Z. Li, S. Song, and X. Zhao, "Flexible and efficient 6.78mhz wireless charging for metal-cased mobile devices using controlled resonance power architecture," in *2017 IEEE Wireless Power Transfer Conference (WPTC)*, 2017, pp. 1–4.

[8] T. Orekan, P. Zhang, and C. Shih, "Analysis, design, and maximum power-efficiency tracking for undersea wireless power transfer," *IEEE Journal of Emerging and Selected Topics in Power Electronics*, vol. 6, no. 2, pp. 843–854, 2018.

[9] N. S. Jeong, S. Kim, H.-J. Lee, and J. H. Kim, "Wireless charging of a metal-encased device," *IEEE Transactions on Antennas and Propagation*, vol. 70, no. 1, pp. 654–663, 2022.

[10] J. Kim, K. Kim, H. Kim, D. Kim, J. Park, and S. Ahn, "An efficient modeling for underwater wireless power transfer using z-parameters," *IEEE Transactions on Electromagnetic Compatibility*, vol. 61, no. 6, pp. 2006–2014, 2019.

[11] X. Wang, R. He, Z. Lin, and M. Fu, "Efficiency control for multi-receiver inductive power transfer systems without knowing realtime coupling," *Wireless Power Transfer*, vol. 12, no. wpt-0024-0019, 2025.

[12] Z. Li, S. Li, H. Deng, Y. Zhang, and W. Hu, "A wireless power transfer based on p-lcc-s compensated topology for artificial catheterization," *Wireless Power Transfer*, vol. 11, no. wpt-0024-0007, 2024.

[13] G. Zheng, C. Qi, Y. Liu, J. Liang, H. Wang, and M. Fu, "Uniform and simplified small-signal model for inductive power transfer systems," *IEEE Transactions on Power Electronics*, vol. 38, no. 2, pp. 2709–2719, 2023.

[14] S. Yao, X. Wang, J. Liang, H. Wang, and M. Fu, "A pv-battery three-port wireless charger for unmanned aerial vehicles," *IEEE Transactions on Industrial Electronics*, pp. 1–5, 2024.

[15] P. Zhao, J. Liang, H. Wang, and M. Fu, "Detuned lcc/s-s compensation for stable-output inductive power transfer system under ultrawide coupling variation," *IEEE Transactions on Power Electronics*, vol. 38, no. 10, pp. 12 342–12 347, 2023.

[16] X. Ji, P. Zhao, H. Wang, H. Yang, and M. Fu, "Multiple-receiver inductive power transfer system based on multiple-coil power relay module," *IEEE Transactions on Circuits and Systems I: Regular Papers*, vol. 70, no. 6, pp. 2625–2634, 2023.

2025 IEEE Workshop on Wide Bandgap Power Devices and Applications in Asia (WiPDA Asia)

An Accurate Leakage Inductance Model for Magnetic Integrated Planar Transformers in LLC Resonant Converters

Zhili Mo
School of Advanced Interdisciplinary Sciences
University of Chinese Academy of Sciences
Beijing, China
mozhili23@mails.ucas.ac.cn

Xuetong Zhou
State Key Laboratory of Materials for Integrated Circuits
Shanghai Institute of Microsystem and Information Technology
Shanghai, China
zhouxuetong@mail.sim.ac.cn

Yufei Tian
State Key Laboratory of Materials for Integrated Circuits
Shanghai Institute of Microsystem and Information Technology
Shanghai, China
tianyufei@mail.sim.ac.cn

Li Zheng
State Key Laboratory of Materials for Integrated Circuits
Shanghai Institute of Microsystem and Information Technology
Shanghai, China
zhengli@mail.sim.ac.cn

Xinhong Cheng
State Key Laboratory of Materials for Integrated Circuits
Shanghai Institute of Microsystem and Information Technology
Shanghai, China
xh_cheng@mail.sim.ac.cn

Abstract—**Magnetic integration effectively enhances the power density of LLC resonant converter through the integration of the resonant inductor and main transformer. Following the technical approach of utilizing inherent leakage inductance of the transformer, precise leakage inductance control becomes critical. In this paper, an accurate leakage inductance model for magnetic integrated planar transformers is proposed. A revised leakage inductance calculation formula is derived, enabling precise computation of air-gapped transformers. Based on the proposed model, a design flow of transformers with customized leakage inductance is proposed and applied to an LLC resonant converter. The resonant frequency accuracy reaches over 99%, demonstrating the precision of the proposed model as well as the feasibility of the design flow.**

Keywords—LLC resonant converter, leakage inductance, planar transformer, magnetic integration

I. INTRODUCTION

LLC resonant converters have gained widespread adoption in high-power applications such as electric vehicles and data centers[1,2]. Due to the zero-voltage-switch (ZVS) and zero-current-switch (ZCS) characteristics, LLC resonant converters exhibit excellent performance in terms of minimizing the switching loss which is magnified proportionally as the operating frequency rises[3]. In the topology of the LLC resonant converter, both the resonant inductor and the main transformer necessitate the utilization of a magnetic core, as shown in Fig. 1(a). These magnetic core components occupy a significant portion of the overall circuit area. To further improve the power density of LLC converter,

recent research is focusing on the integration of its resonant inductor and the main transformer.

Current integrated magnetics follow two technical approaches. The first approach shares the magnetic path to combine the resonant inductor and the main transformer, exemplified by the matrix transformers[4-6], as shown in Fig. 1(b). And the second approach utilizes the inherent leakage inductance of transformer as the resonant inductance L_r[7-9]. Since the equivalent position of L_{lk} coincides precisely with that of L_r within the resonant network, precise leakage inductance control becomes critical for this approach. The magnetic shunts inserting method, as shown in Fig. 1(c), has been proven effective. Furthermore, for transformers with relatively small leakage inductances, magnetic shunts are no longer necessary, and the parasitic parameters can be determined by winding arrangement directly. However, current models for leakage inductance calculation remain limited to transformers without any air gap. Few researchers calculate leakage inductance of air-gapped transformers through analytical formulations and most of them resort to Finite Element Analysis (FEA) instead[9,10].

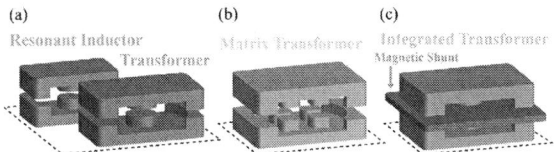

Fig. 1. (a) Traditional magnetic components in the resonant network.
(b) Matrix transformer. (c) Integrated transformer with magnetic shunt.

The inevitable presence of air gaps in LLC transformer designs, essential for magnetic inductance and core saturation prevention, causes challenges in magnetic integration. This work aims to propose a precise calculation of the air-gapped transformers in LLC converter by a correction factor of k_{gap}. In this paper, an accurate leakage inductance model for integrated transformers is proposed. A revised leakage calculation formula is derived, enabling precise computation of air-gapped transformers. A design flow of magnetic-shunt free integrated transformers with customized leakage inductance is proposed, which can effectively shorten the

---

This work was supported by the National Key Research and Development Program of China (Grant No. 2022YFB3604300, 2022YFB3604301, 2022YFB3604303), National Natural Science Foundation of China (Grant No. 11705263), the Science and Technology Commission of Shanghai Municipality (Grant No. 23511102602), Youth Innovation Promotion Association CAS, Autonomous deployment project of State Key Laboratory of Materials for Integrated Circuits (No. SKLJC-Z2024-C02) and Shanghai Post-doctoral Excellence Program (Grant No. 2024697). *(Co-first authors: Zhili Mo and Xuetong Zhou. Co-corresponding authors: Li Zheng and Xinhong Cheng.)*

design cycle of magnetic integrated LLC converters and further increase the power density of them.

This paper is organized as follows. Section II provides a detailed introduction to the proposed model of leakage inductance. Section III presents the design flow of magnetic integrated transformer based on proposed leakage inductance calculation method. In section IV, experimental results are presented, which validate the feasibility of the proposed design flow. Finally, Section V concludes the paper.

II. MODEL OF LEAKAGE INDUCTANCE

A. Prior Theory of Leakage Calculation

Part of the flux generated by primary windings of the transformer will leak from the core, which makes the primary and secondary windings imperfectly coupled and results in leakage inductance. Since most of the leakage flux is stored in the core window, leakage inductance L_{lk} can be calculated according to the stored energy E, where I_p is the current flowing through primary windings, and l_w, b_w, and h are the length, the width and the thickness of each winding layer[11].

$$E_{lk} = \frac{1}{2} L_{lk} I_p^2 \tag{1a}$$

$$E_{lk} = \frac{\mu_0}{2} \sum \int_0^h H^2 \cdot l_w \cdot b_w \cdot dx \tag{1b}$$

According to Ampère's circuital law, magnetic field intensity is determined by the current of windings, thus it varies with the winding layers and results in different leakage inductance for different winding arrangements. (1a) and (1b) have been utilized for calculating leakage inductance of planar transformers with interleaved structure at low frequencies. Following this methodology with the incorporation of more refined physical models, calculation of high-frequency leakage inductance has also been discussed[12].

Theories above remain limited to transformers without any air gap. Conventional leakage inductance calculation by (1a) and (1b) relies on "1-D" assumption, which presumes linear electromagnetic field variation perpendicular to the windings and uniform distribution along winding surfaces[13]. However, this fundamental assumption collapses for practical air-gapped transformers. The air gaps disrupt the uniformity of the magnetic field, as illustrated in Fig. 2, thus the "1-D" assumption turns inapplicable, and the aforementioned formulas become inaccurate. Given the limitation inherent in the conventional model, it becomes essential to undertake model modification efforts aimed at achieving precise customization of the leakage inductance.

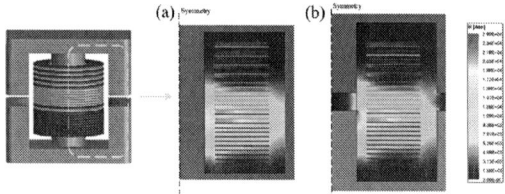

Fig. 2. Distribution of magnetic field intensity in the core window. (a) Transformers without air gap. (b) Air-gapped transformers.

B. Reluctance Model of Leakage Inductance

The precise calculation of this work is based on the magnetic reluctance model of the planar transformer, as shown in Fig. 3, where R_{cs} and R_{cc} are the reluctances of the side legs and the central column, and R_w is reluctance of the air column between primary and secondary windings. Define the unit leakage inductance L_{lk0} as the leakage inductance of the transformer with only one turn of primary winding and secondary winding each, with the unit distance h_0 between the two windings. After calculating L_{lk0}, the model will be extended to transformers with more complex structures.

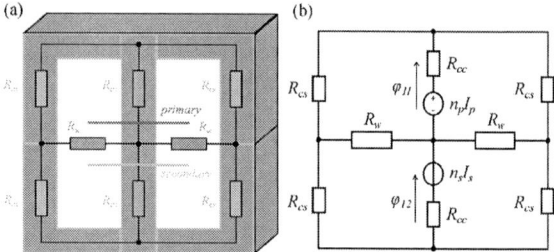

Fig. 3. Reluctance model of planar transformers without air gap.

φ_{11} is the flux generated by the single primary winding and φ_{12} is the flux reaches the secondary winding. Also, φ_{11} represents the total flux in the core while φ_{12} corresponds to the magnetic flux. The difference between them is the leakage flux, which governs the leakage inductance. It is assumed that all leakage flux is constrained to the path where R_w is located, since the flux leaks from the air column between windings into the core window. φ_{11} and φ_{12} are calculated as follows, from which L_{lk0} can be deprived.

$$\varphi_{11} = \frac{2n_p I_p \cdot (R_m + R_w)}{R_m \cdot (R_m + 2R_w)} \tag{2a}$$

$$\varphi_{12} = \frac{2n_p I_p \cdot R_w}{R_m \cdot (R_m + 2R_w)} \tag{2b}$$

$$R_m = 2R_{cc} + R_{cs} \tag{2c}$$

In the formulas above, n_p is the number of primary windings, which turns to be 1 in the model with only one turn of primary winding. I_p is the current that flows through the primary winding. L_{lk0} is deprived as follows. Since the core's permeability μ is much higher than the vacuum permeability μ_0, R_m is much smaller than R_w. Therefore, the magnetic reluctance corresponding to the leakage flux can be approximately expressed as R_w, and L_{lk0} can be likewise approximated.

$$L_{lk0} = \frac{n_p \cdot (\varphi_{11} - \varphi_{12})}{I_p} = \frac{2n_p^2}{R_m + 2R_w} \tag{3a}$$

$$L_{lk0} \approx \frac{1}{R_w} \tag{3b}$$

For air-gapped transformers, shown in Fig. 4, the air gaps introduce additional reluctances: R_{gs} and R_{gc}, which are reluctances of the air gaps on the side legs and on the central column of the core. With the presence of air gaps, the total flux φ_{11}' and the magnetic flux φ_{12}' are calculated as follows.

$$\varphi_{11}' = \frac{2n_p I_p \cdot (R_m + R_w + R_g)}{(R_m + R_g) \cdot (R_m + 2R_w + R_g)} \tag{4a}$$

$$\varphi_{12}' = \frac{2n_p I_p \cdot R_w}{(R_m + R_g) \cdot (R_m + 2R_w + R_g)} \tag{4b}$$

$$R_g = 2R_{gc} + R_{gs}; \ R_w' = R_w + \frac{1}{2}R_g \tag{4c}$$

979-8-3315-1110-4/25 $31.00 © 2025 IEEE

Similar to L_{lk0}, the unit leakage inductance for air-gapped transformers, L_{lk0}', can be expressed and approximated as follows.

$$L_{lk0}' = \frac{n_p \cdot (\varphi_{11}' - \varphi_{12}')}{I_p} = \frac{2n_p^2}{R_m + 2R_w'} \qquad (5a)$$

$$L_{lk0}' \approx \frac{1}{R_w'} \qquad (5b)$$

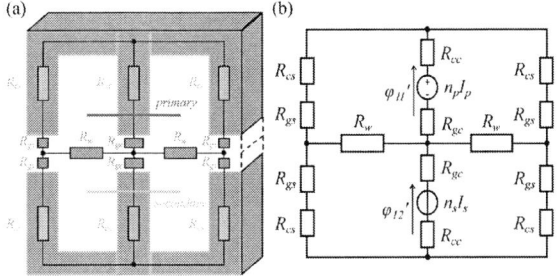

Fig. 4. Reluctance model of air-gapped planar transformers.

The presence of air gaps increases the magnetic reluctance along the leakage flux path. Since the cross-sectional area of the air column between the primary and secondary windings remains unchanged before and after opening the air gap, the effect of the air gap on the transformers can be regarded as that it increases length of the path of leakage flux. The ratio of air-gapped leakage inductance to leakage inductance without any air gap is defined as k_{gap}.

$$k_{gap} = \frac{L_{lk0}'}{L_{lk0}} = \frac{R_w}{R_w'} \qquad (6)$$

For transformers with more than one pair of primary and secondary windings, flux leaks from every air column between each pair of windings. According to Faraday's law of induction and Lenz's law, opposing eddy currents of equal magnitude are induced between adjacent winding layers as shown in Fig. 5, causing the magnetic motive force MMF to vary along the height. MMF quantifies the current's ability to generate a magnetic field and is numerically equal to the product of current and number of turns. Leakage inductances between each pair of adjacent windings can be modeled as unit leakage inductances carrying multiplied current.

For each air column, when calculating its leakage inductance L_{lk}, the leaked flux φ_{11} and φ_{12} at its location are proportional to the local MMF. Leakage inductance for each pair of windings can be determined using (3a), and the total leakage inductance L_{lk} is obtained as the superposition of these individual leakage inductances. If the distance between windings and air gaps is significantly greater than the thickness of air columns, it can be assumed that R_w remains consistent across all air columns, due to their identical cross-sectional area and thickness.

$$L_{lk} = \sum_{\text{air columns}} \frac{n_p^2}{R_w} = \left(\sum_{\text{air columns}} \frac{MMF^2}{I_p^2} \right) \cdot L_{lk0} \qquad (7)$$

Utilizing (5a), the total leakage inductance of air-gapped transformers L_{lk}' can be calculated as follows.

$$L_{lk}' = \left(\sum_{\text{air columns}} \frac{MMF^2}{I_p^2} \right) \cdot L_{lk0}' \qquad (8)$$

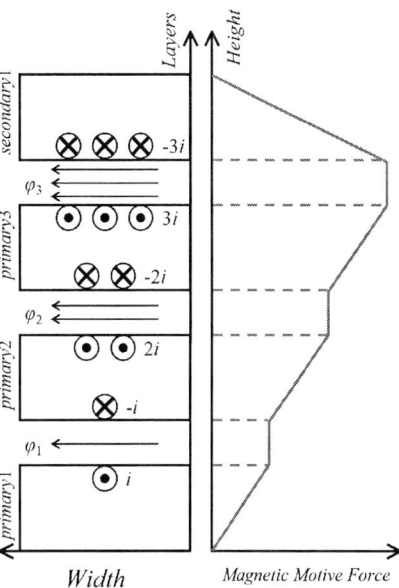

Fig. 5. Distribution of the induced current and magnetic motive force in the winding layers.

The total leakage inductance before and after opening air gaps, L_{lk} and L_{lk}', are both linear superposition of the unit leakage inductance L_{lk0} and L_{lk0}'. Notice that the ratio of them is equal to k_{gap}. k_{gap} is influenced by multiple transformer parameters, including the core shape, core size, and winding layer dimensions, but remains independent of the winding arrangement. When changing arrangement of the windings, the distribution of MMF varies which leads to the alteration of the linear superposition coefficients. Furthermore, leakage inductance can be adjusted by changing the winding arrangement, leaving k_{gap} not affected.

The aforementioned analysis overlooks the thickness of the winding layers. In practical applications, the air columns situated between winding layers are frequently substituted with insulating layers. To build a more accurate model of leakage inductance, thickness of winding layers is no longer ignored, and thickness of air column (insulators) is not strictly equal to the unit distance h_0. MMF of each layer is calculated as follows, where $MMF(x)$ is MMF inside each layer which varies on the direction of thickness, $MMF(0)$ is MMF of the previous adjacent surface, and h_l is the thickness of each layer.

$$MMF(x) = \begin{cases} MMF(0) + I_p \cdot \dfrac{x}{h_l} & \text{In primary windings} \\ MMF(0) & \text{In insulators} \\ MMF(0) - I_s \cdot \dfrac{x}{h_l} & \text{In secondary windings} \end{cases} \qquad (9)$$

The total leakage inductance before and after opening air gap, L_{lk} and L_{lk}', are calculated as follows.

$$L_{lk} = \left(\sum_{\text{layers}} \frac{\int_0^{h_l} MMF(x)^2 dx}{I_p^2 \cdot h_0} \right) \cdot L_{lk0} \qquad (10a)$$

$$L_{lk}' = \left(\sum_{\text{layers}} \frac{\int_0^{h_l} MMF(x)^2 dx}{I_p^2 \cdot h_0} \right) \cdot L_{lk0}' \qquad (10b)$$

The ratio k_{gap} is intrinsic to the transformer's design and remains unchanged regardless of variations in winding arrangements. Therefore, variations in winding arrangements solely modify the coefficients within the linear combination while maintaining k_{gap} invariant. Once k_{gap} is determined, leakage inductance of air-gapped transformers can be obtained by multiplying that of transformers without any air gap by k_{gap}. Using the pre-existing energy model the leakage inductance of transformers without any air gap can be efficiently obtained, hence the prior formula of leakage inductance in [11] can be revised with k_{gap} to make it suitable for air-gapped transformers. The leakage inductance of air-gapped transformers is calculated as follows, where b_w is the average width of windings, l_w is the average perimeter, and h_l is the thickness of each winding layer.

$$L_{lk}' = k_{gap}L_{lk} = k_{gap}\frac{\mu_0 \cdot l_w}{b_w}\sum_{layers}\int_0^{h_l}\frac{MMF^2(x)}{I_p^2}dx \quad (11)$$

In conclusion, for a transformer with specific dimensional parameters, the ratio k_{gap} of its leakage inductance after and before opening the air gap keeps constant regardless of changes in winding arrangements. However, it is difficult to calculate k_{gap} from the existing parameters. Although k_{gap} is nominally derivable from R_w and R_g, its accurate calculation remains challenging due to the fringing effect[14]. Fringing effect expand the effective cross-sectional area of the air gap, thus resulting in reduced magnetic reluctance. A straightforward method to obtain k_{gap} is utilizing FEA to determine the leakage inductance of any given winding arrangement and subsequently calculating k_{gap}. Once the k_{gap} is determined, the leakage inductance can be calculated by (11). This method eliminates the need for performing FEA simulations for every new arrangement, thereby streamlining the calculation process significantly, while the accuracy in determining the leakage inductance across various winding configurations is preserved.

C. Verification by FEA Simulation

The derivation demonstrates that k_{gap} depends on the transformer's structural dimensions and remains independent of the winding arrangement. According to (11), for a specific transformer, its leakage inductance is proportional to the area under the $MMF^2(x)$ curve. Fig. 6 shows the $MMF^2(x)$ curve for a 6:4 transformer with three different winding arrangements: A, non-interleaved structure; B, partly interleaved structure; and C, fully interleaved structure. In the structure notation, 'p' denotes one turn of a primary winding, and 's' stands for a secondary one. Variations in the curve area will cause significant changes in leakage inductance. According to the proposed theory, k_{gap} will remain constant. To validate this deduction, transformers with different parameters were designed, their winding sequences were varied, and k_{gap} was calculated to verify this conclusion. Four planar transformers with different structures were selected and their leakage inductances in different winding arrangements before and after opening the air gap were obtained by FEA simulation. Subsequently, the value of k_{gap} is computed and plotted in Fig. 7.

For transformers with specific structural parameters, the variation of k_{gap} is less than 3% when the winding arrangement varies. When the core type, the winding ratio, or the layer thickness alters, k_{gap} changes correspondingly. The proposed model is effective and precise for air-gapped transformers.

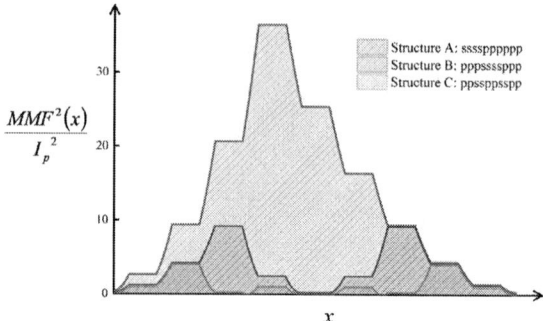

Fig. 6. $MMF^2(x)$ curve for different structures.

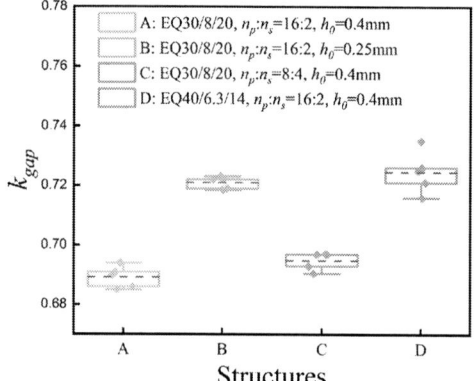

Fig. 7. k_{gap} obtained by FEA Simulation.

III. DESIGN FLOW OF INTEGRATED TRANSFORMERS

Since the value of leakage inductance is closely related to the distribution of magnetic motive force MMF which mainly depends on the winding arrangement, it is feasible to customize the leakage inductance by selecting a proper winding arrangement. The more same-side windings are arranged consecutively, the stronger local magnetic field is obtained, leading to increased energy storage in the core window and consequently higher leakage inductance. Similar methods have been adopted to obtain transformers with low leakage inductance[15]. When it comes to precise leakage control, accurate calculation is needed.

To integrate the resonant inductor L_r with the main transformer in LLC resonant converter, leakage inductance of the transformer L_{lk} needs to match to the value of L_r. The maximum leakage inductance L_{lk_max}, which represents the leakage inductance of the non-interleaved transformer, should be more than L_r. Otherwise, adjustment should be adopted to increase L_{lk_max}, like increasing the layer thickness or reducing the winding width. The value of L_{lk_max} can be obtained by using FEA, while k_{gap} can be determined simultaneously. The target arrangement with the customized leakage inductance can be found simply by traversal calculation of the leakage inductance of each winding arrangement with (11). Shown in Fig. 8, a leakage inductance customization method without using any magnetic shunt is proposed.

Exhaustively evaluating all the possible arrangements provides a straightforward solution. With the derived formula, the computation can be performed efficiently. Meanwhile, alternative methods exist that achieve higher computational efficiency. For planar transformers, reducing the number of intersections between the primary and secondary windings can effectively decrease the winding capacitance[16]. To minimize the computational burden, also to reduce the parasitic capacitance, it is advisable to prioritize the enumeration of winding arrangements that feature a minimal number of intersections between the primary and secondary windings, which is, with either all primary windings or all secondary windings arranged together.

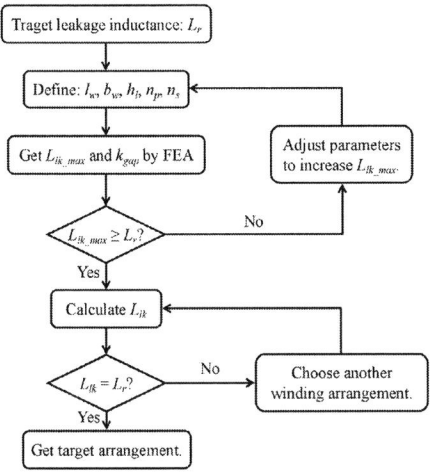

Fig. 8. Design flow of transformers with customized leakage inductance.

IV. EXPERIMENTAL RESULTS

A 400V/48V LLC resonant converter with a 16:2 turn ratio transformer was designed. The operating frequency was set to 500 kHz, and the required resonant inductance was 4 µH, to which value the leakage inductance of the main transformer was customized. The rest of the main transformer parameters are shown in TABLE I. To meet the specified magnetic inductance requirement, the air gap length was set to 1.7 mm. Following the proposed design flow of integrated planar transformers, FEA method was performed on the non-interleaved structure, yielding a calculated k_{gap} as 0.68. After exhaustive calculation of all winding arrangements, the final structure was designed to be '2p+2s+14p'.

TABLE I.   PARAMETERS OF MAIN TRANSFORMER

Parameters	Values
Turn Ratio	16:2
Primary Winding Thickness	70 µm
Secondary Winding Thickness	70 µm
Insulator Thickness	330 µm
Core Type	EQ30/8/20
Average Winding Perimeter	57 mm
Average Winding Width	5 mm

The integrated transformer has been constructed and subjected to testing. Frequency characteristics of its leakage inductance, together with another experimental group, '2s+16p', is shown in Fig. 9. The measured value of its leakage inductance is 3.92 µH at 500 kHz, which deviates by only 2% from the required resonant inductance value. When

the frequency exceeds 500kHz and further increases, the skin effect, the eddy current effect[12] and the shielding effect of the windings[14] cause a reduction in leakage inductance, making the experimental results deviates from the calculation. However, at the rated frequency of 500 kHz, the calculated leakage inductance exhibits high accuracy.

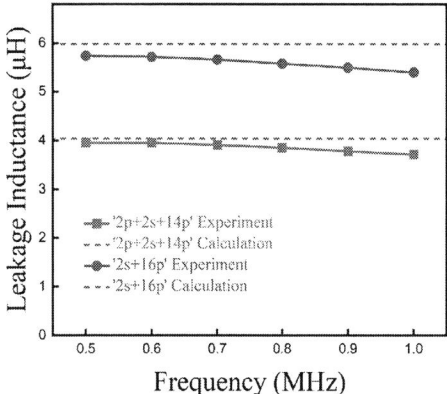

Fig. 9. Frequency characteristics of the leakage inductance.

A test board shown in Fig. 10 was constructed, and its operational functions were examined. With the waveform shown in Fig. 11, the LLC resonant converter worked in the inductive state under the rated operating frequency. ZVS on the primary side is achieved, which meets the need of reducing switching loss.

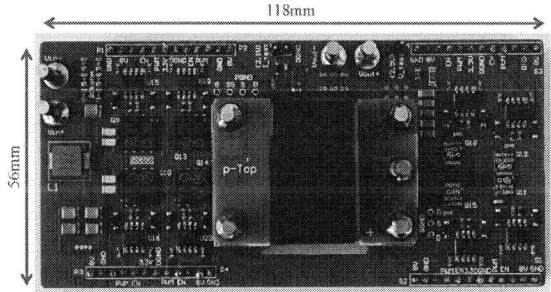

Fig. 10. Test board of LLC resonant converter.

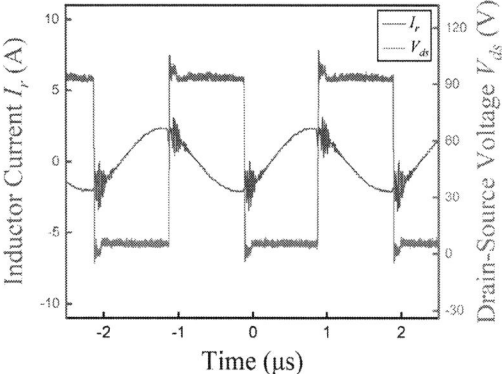

Fig. 11. Waveform of the LLC resonant converter.

During the energy conversion process, the current flowing through the resonant inductor follows a sinusoidal waveform, with a frequency that matches the resonant frequency of the resonant network. Consequently, the resonant frequency can be determined by fitting its current waveform to a sinusoidal curve. By varying the switching frequency f_s, the test board was driven in three distinct conduction modes: discontinuous conduction mode (DCM), boundary conduction mode (BCM) and continuous conduction mode (CCM). The resonant frequency f_r was calculated by performing a sine curve fitting on the measured current waveform. As listed in TABLE II, the average error of f_r in three conduction modes is 3% compared to the expected value 500 kHz. In the commonly adopted mode, BCM, the obtained resonant frequency is 499 kHz, which is close to the rated value. The proposed integrated transformer design method is applicable.

TABLE II. TEST RESULTS OF RESONANT FREQUENCY

Conduction Mode	DCM ($f_s < f_r$)	BCM ($f_s = f_r$)	CCM ($f_s > f_r$)
f_s (kHz)	350	500	700
f_r (kHz)	466	499	490
Error	6.8%	0.2%	2.0%

V. CONCLUSION

In order to integrate the resonant inductor with the main transformer, precise leakage inductance control becomes critical. In this paper, a leakage inductance model for magnetic integrated planar transformers is proposed. A revised leakage calculation formula is derived, enabling precise computation of air-gapped transformers. When the air gap is incorporated to the transformer, its leakage inductance alters in a certain proportion k_{gap} which remains consistent across different winding arrangements. The effect of air gaps on leakage inductance can be attributed to an increase in the path length of the leakage flux. Although k_{gap} appears computable according to the model, its accurate calculation remains challenging due to the fringing effect. One feasible method to obtain k_{gap} is using FEA to determine the leakage inductance of any given winding arrangement and subsequently calculating k_{gap}.

Based on the deprived formula, a design flow of transformers with customized leakage inductance is proposed. An initial FEA simulation should be performed to determine the maximum achievable leakage inductance for the given transformer parameters, while simultaneously calculating k_{gap} from the simulation results. Subsequently, all possible winding arrangements are exhaustively evaluated till the target arrangement is found. This enables the realization of transformers with customized leakage inductance.

The design flow of the integrated transformers was applied to an LLC resonant converter, and a test board was built. The experimental results verified the accuracy of the proposed formula and the feasibility of the design flow. The error of the customized leakage inductance is only 2%, and the measured resonant frequency of LLC converter is 499 kHz, which is close to the rated value. The proposed integrated transformer design method is applicable.

REFERENCE

[1] C. Li, H. Wang and M. Shang, "A Five-Switch Bridge Based Reconfigurable LLC Converter for Deeply Depleted PEV Charging Applications," in *IEEE Transactions on Power Electronics*, vol. 34, no. 5, pp. 4031-4035, May 2019.

[2] B. Xue, H. Wang, J. Liang, Q. Cao and Z. Li, "Phase-Shift Modulated Interleaved LLC Converter With Ultrawide Output Voltage Range," in *IEEE Transactions on Power Electronics*, vol. 36, no. 1, pp. 493-503, Jan. 2021.

[3] J. Xu, L. Gu, Z. Ye, S. Kargarrazi and J. M. Rivas-Davila, "Cascode GaN/SiC: A Wide-Bandgap Heterogenous Power Device for High-Frequency Applications," in *IEEE Transactions on Power Electronics*, vol. 35, no. 6, pp. 6340-6349, June 2020.

[4] S. Gao and Z. Zhao, "Magnetic Integrated LLC Resonant Converter Based on Independent Inductance Winding," in *IEEE Access*, vol. 9, pp. 660-672, 2021.

[5] X. Chen, G. Xu, Q. Shen, Y. Sun and M. Su, "Magnetizing and Leakage Inductance Integration for Split Transformers With Standard UI Cores," in *IEEE Transactions on Power Electronics*, vol. 37, no. 11, pp. 12980-12985, Nov. 2022.

[6] B. Li, X. Huang and J. Zhang, "A High Density 400 W DC/DC Power Module with Integrated Planar Transformer and Half Bridge GaN IC," *2024 IEEE Applied Power Electronics Conference and Exposition (APEC)*, Long Beach, CA, USA, 2024, pp. 94-100.

[7] M. Li, Z. Ouyang and M. A. E. Andersen, "High-Frequency LLC Resonant Converter With Magnetic Shunt Integrated Planar Transformer," in *IEEE Transactions on Power Electronics*, vol. 34, no. 3, pp. 2405-2415, March 2019.

[8] S. A. Ansari, J. N. Davidson and M. P. Foster, "Fully-Integrated Transformer With Asymmetric Primary and Secondary Leakage Inductances for a Bidirectional Resonant Converter," in *IEEE Transactions on Industry Applications*, vol. 59, no. 3, pp. 3674-3685, May-June 2023.

[9] He, P., Mallik, A., Cooke, G. and Khaligh, A., "High-power-density high efficiency LLC converter with an adjustable-leakage-inductance planar transformer for data centers," in *IET Power Electronics*, 12: 303-310.

[10] E. S. Lee, J. H. Park, M. Y. Kim and S. H. Han, "An Integrated Transformer Design With a Center-Core Air-Gap for DAB Converters," in *IEEE Access*, vol. 9, pp. 121263-121278, 2021.

[11] Z. Ouyang, O. C. Thomsen and M. A. E. Andersen, "Optimal Design and Tradeoff Analysis of Planar Transformer in High-Power DC–DC Converters," in *IEEE Transactions on Industrial Electronics*, vol. 59, no. 7, pp. 2800-2810, July 2012.

[12] Z. Ouyang, J. Zhang and W. G. Hurley, "Calculation of Leakage Inductance for High-Frequency Transformers," in *IEEE Transactions on Power Electronics*, vol. 30, no. 10, pp. 5769-5775, Oct. 2015.

[13] M. Chen, M. Araghchini, K. K. Afridi, J. H. Lang, C. R. Sullivan and D. J. Perreault, "A Systematic Approach to Modeling Impedances and Current Distribution in Planar Magnetics," in *IEEE Transactions on Power Electronics*, vol. 31, no. 1, pp. 560-580, Jan. 2016.

[14] Y. Liu, H. Wu and G. Ji, "Inductance Calculation Method Considering the Window Effect of Planarized Magnetic Core," in *IEEE Transactions on Power Electronics*, vol. 38, no. 10, pp. 12999-13007, Oct. 2023.

[15] X. Zhou *et al.*, "A Design Method of Partially Interleaved Winding Structure With Low Leakage Inductance for Planar Transformer Application," in *IEEE Transactions on Power Electronics*, vol. 38, no. 5, pp. 6366-6379, May 2023.

[16] Z. Ouyang and M. A. E. Andersen, "Overview of Planar Magnetic Technology—Fundamental Properties," in *IEEE Transactions on Power Electronics*, vol. 29, no. 9, pp. 4888-4900, Sept. 2014.

Subharmonic Oscillations in High-frequency Switched-Mode Power Amplifiers

Wei Liu [a], Ming Liu [b]

[a] Dept. of Electrical Engineering, Shanghai Jiao Tong University, Shanghai, P.R China,
liuwei78@sjtu.edu.cn

[b] Dept. of Electrical Engineering, Shanghai Jiao Tong University, Shanghai, P.R China,
liuwei78@sjtu.edu.cn

Abstract—Switched Mode Power Amplifiers (SMPA) are preferable in high power applications as its advantage of high-efficiency. While the inevitable effects of parasitic elements hinder its applications in higher frequencies. In this paper, the specific manifestation and mechanism of subharmonic oscillations, which will cause high peak voltage pressure and significant subharmonic distortions, were disclosed and analyzed at the first time in SMPA. Different with linear PA, where the subharmonic oscillation was caused by the input capacitance of transistor, the origination of the phenomenon in SMPA was found to be the non-linear output capacitance of switches in this paper. The time-domain analytical method based on the solutions of Mathieu Equations was carried out to estimate the system stability in Class-E circuit. And it concluded that in Class-E circuit, the parameters of input tank and the bias voltage dominates the phenomenon, which was verified by the experimental results.

Index Terms—*Subharmonic oscillations, parametric oscillations, high-frequency switched mode power amplifier, stability analysis.*

I. INTRODUCTION

Advanced industrial applications are promoting the development of high frequency power electronics (tens of Megahertz or higher), e.g., advanced semiconductor manufacture, plasma generation, and wireless power transfer, all of which are pursuing efficient Power Amplifiers (PA) with higher power capability. Comparing to inefficient linear PA, Switched Mode Power Amplifiers (SMPA) have theoretical 100% efficiency, more desirable in high power applications, but limited in switching frequencies. One of the difficulties of putting SMPA towards higher frequencies is the widespread parasitic elements, e.g., parasitic capacitance and inductance as shown in Fig. 1. In high-frequency applications, the oscillation frequencies induced by parasitic elements are only multiple times of operation frequencies, significantly distorting the normal operation of power converter and causing instability.

Three oscillations types caused by parasitic elements can be generally classified as: natural ringing down oscillations [1], [2], self-sustained oscillations [3], [4], and subharmonic oscillations or parametric oscillations [5], [6]. The former two type belongs to

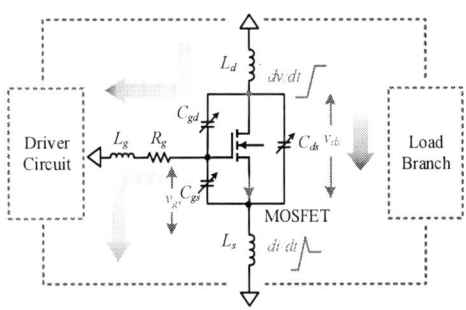

Fig. 1. The effects of parasitic elements in high-frequency power converter.

high-frequency oscillations as their induced oscillation frequencies are higher than the switching frequencies, and are investigated completely recently. The latter one usually causing lower frequencies oscillations, typically at half of switching frequencies, are only reported in linear-PA, but rarely studied in SMPA.

The most commonly studied type is the natural ringing down oscillations, or under-damped natural response, which is triggered by the high *dv/dt* and *di/dt* from the high-frequency switching transient, inducing energy exchange in parasitic elements. Different from this transient natural ring-down oscillations, the self-sustained oscillations will maintain in steady state, and are induced by the negative resistance oscillator. Both two high-frequency oscillations types are likely to induce voltage overshoot, shoot-through, or EMI problems to distort or damage the switches.

Another oscillations type, named subharmonic oscillations or parametric oscillations, acts like the frequency divider and generate half the fundamental frequency for a certain of operation conditions. It has destructive effects due to the increased peak voltage of transistors. And the results of output power drop at some compression points also limits the applications in advancing semiconductor manufacturing, who demands for wide range RF power generation. Different with the former two types, which can be analyzed based on small-signal model, the mechanism of subharmonic oscillation is induced by parametric excitations, and is a large signal issue. In linear PA, as shown in Fig. 2(a), the non-linear input capacitance

C_{iss} is believed to be the major contribution as exhibiting larger variation amplitude than other capacitances. When pumping a non-linear reactance, e.g., C_{iss}, persistently at multiple of oscillation frequencies, negative resistance will be generated when the varied reactance reaching the threshold, as shown in Fig. 2(b). Once there is not sufficient damping, this parametric phenomenon will cause subharmonic frequency to be injection locked to the fundamental.

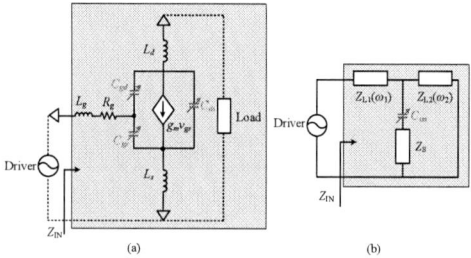

Fig. 2. (a) The origin of subharmonic oscillations in linear PA and (b) its notional model.

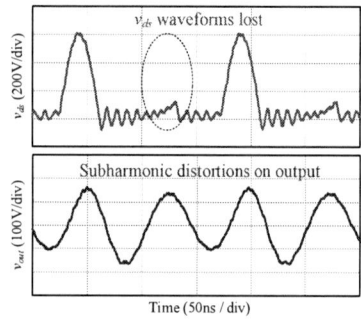

Fig. 3. Experimental waveforms of v_{ds} and v_{out} of MOSFETs in Class-E PA at subharmonic oscillations.

However, when the transistor is fully driven in on-off mode, i.e., acts like switches, the conduction current will not be controlled by the gate signal, and the variations of C_{iss} are no longer dominant. Generally, C_{oss} dominates the non-linear capacitance of transistor, and is considered in this paper as the origin of the subharmonic oscillations in SMPA. When oscillations occur, the waveforms of v_{ds} will swing in half the switching frequencies f_s, as shown in Fig. 3. More seriously, half the waveforms of v_{ds} will lost, i.e., no current charges C_{oss} when the next off driving signal comes, and it seems like that the switch was driven in half the switching frequency with large turning-on duty cycle. This phenomenon will significantly increase the peak voltage pressure, especially for Class-E circuit, which has suffered from high voltage pressure in normal operation. While as to our knowledge, the specific manifestation of subharmonic

oscillations in SMPA have not been disclosed and analyzed at the time of this writing. Therefore, in this paper, the subharmonic oscillation in SMPA was firstly disclosed, and the non-linear output capacitance C_{oss} of switches was discovered to be the origination of the phenomenon. Then, based on the solutions of Mathieu Equations, the explicit condition to estimate the occurrence of subharmonic oscillations was derived. It found that the specific parameters of input tank in Class-E PA were the essential trigger conditions, which is verified by the experiments by changing the value of C_f and bias voltage V_{DD}.

II. MECHANISM OF SUBHARMONIC OSCILLATIONS

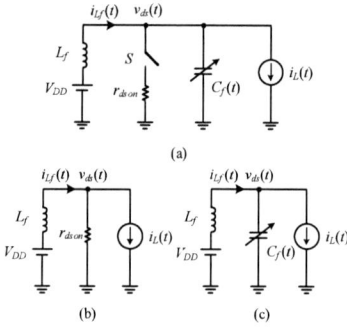

Fig. 4. (a) Schematic of the Class-E circuit with non-linear C_{oss} absorbed in shunt capacitance C_f. (b) S turning on during $[0, DT]$. (c) S turning off during $[DT, T]$.

Fig. 4(a) shows the Class-E topology with non-linear C_{oss}, which have been absorbed in varied shunt capacitance $C_f(t)$, and the load branch is equivalent to the current source $i_L(t)$. As shown in Fig. 5(a), the normal waveforms of v_{ds} should be periodic identical at switching frequency, while the waveforms of v_{ds} in subharmonic oscillations at $f_s/2$ are varied alternately in each period, and in periodic at half of switching frequency, as shown in Fig. 5(b). Hence the idea to diagnose the subharmonic oscillations is through evaluating whether the area of v_{ds} in one period is identical. Here the integral of v_{ds} in nth period can be derived as

$$\int_{(n-1)T}^{nT} v_{ds}dt = \int_{(n-1)T}^{nT} V_{DD} - v_{Lf}(t)dt \\ = V_{DD}T - L_f\Delta i_{Lfn}(T) \quad (1)$$

Where $\Delta i_{Lfn}(T)$ is defined as the increment of the value of i_{Lf} at nT and $(n-1)T$, or the increment of $i_{Lfn}(t)$ at $t = T$ in nth period and $(n-1)$th period as follows.

$$\Delta i_{Lfn}(T) = i_{Lf}(nT) - i_{Lf}((n-1)T) \\ = i_{Lfn}(T) - i_{Lfn\_1}(T) \quad (2)$$

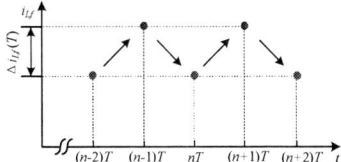

Fig. 5. $\Delta i_{Lfn}(T)$ fluctuates between negative or positive when subharmonic oscillations occurred.

For a stable system at switching frequency, as shown in Fig. 5(a), the value of $\Delta i_{Lfn}(T)$ should be positive or negative first in transient state, then gradually converge to zero in steady state. However, for a system at subharmonic oscillations as shown in Fig. 5(b), the value of $\Delta i_{Lfn}(T)$ is alternately positive and negative in adjacent periods even when system is in steady state as shown in Fig. 5. Considering the integral of v_{ds} during on-time of S is small to be neglected, fluctuating integral of v_{ds} in each period will lead to inconsistent waveforms of v_{ds} in adjacent periods, as shown in Fig. 5(b). Thus, the occurrence of subharmonic oscillation can be estimated by the value of $\Delta i_{Lfn}(T)$. According to Fig. 4(b), when switch S is turning on during $[0, DT]$ in nth period, inductor L_f is charging by V_{DD} through the on-state resistance r_{dson} of switch S, and the current of L_f at time $t = DT$ in nth period can be derived by

$$i_{Lfn}(DT) = (i_{Lfn-1}(T) - \frac{V_{DD}}{r_{dson}})e^{-\frac{r_{dson}}{L_f}DT} + \frac{V_{DD}}{r_{dson}} \quad (3)$$

When S is turning off during $[DT, T]$, the circuit in Fig. 4 can be described by the second-order differential equations with varied parameters as

$$\frac{d^2 q_{Cfn}(t)}{dt^2} + \frac{1}{L_f C_{fn}(t)} q_{Cfn}(t) = \frac{V_{DD}}{L_f} - \frac{di_{Ln}(t)}{dt} \quad (4)$$

Where $i_{Ln}(t)$ is the load current in nth period, $q_{cfn}(t)$ represents the charge of C_f in nth period, $C_{fn}(t)$ is the varied non-linear capacitance C_f in nth period, as defined in the following.

$$C_f(t) = \frac{C_0}{1 - \varepsilon \cos 2\omega_s t} \quad (5)$$

Where the value of ε is defined as the varied amplitude of non-linear capacitance C_f, and ω_s is the switching frequency.

The current of L_f during $[DT, T]$ in nth period satisfy the following expression according to Kirchhoff's law.

$$i_{Lfn}(t) = \frac{dq_{cfn}(t)}{dt} + i_{Ln}(t) \quad (6)$$

The value of $\Delta i_{Lfn}(T)$ is given as

$$\Delta i_{Lfn}(T) = \frac{d\Delta q_{Cfn}(T)}{dt} + \Delta i_{Ln}(T) \quad (7)$$

Where $\Delta i_{Ln}(T) = i_{Ln}(T) - i_{Ln-1}(T)$, $\Delta q_{Cfn}(T) = q_{Cfn}(T) - q_{Cfn-1}(T)$. When the high-$Q$ filter is adopted in the load branch, i.e., the load current $i_{Ln}(t)$ is pure sinusoidal at switching frequency, the value of $\Delta i_{Ln}(T)$ is equal to zero. Hence the value of $\Delta i_{Lfn}(T)$ is only depended on the value of $\Delta q_{Cfn}(T)$. To solve the expression of $\Delta q_{Cfn}(t)$, the second-order differential equation with varied parameters in (4) can be simplified in the following form. ω_{IN} is the resonant frequency of input tank

$$\frac{d^2 \Delta q_{Cfn}(t)}{dt^2} + \frac{1}{L_f C_0}(1 - \varepsilon \cos 2\omega_s(t - DT))\Delta q_{Cfn}(t) = 0 \quad (8)$$

The equation in (7) is known as Mathieu Equations, and combining the expression in (3) and (6), the the value of $\Delta i_{Lfn}(T)$ can be simplified as in the following.

$$\Delta i_{Lfn}(T) = \alpha^{n-1} \Delta i_{Lf1}(T) \quad (9)$$

Where α is given as

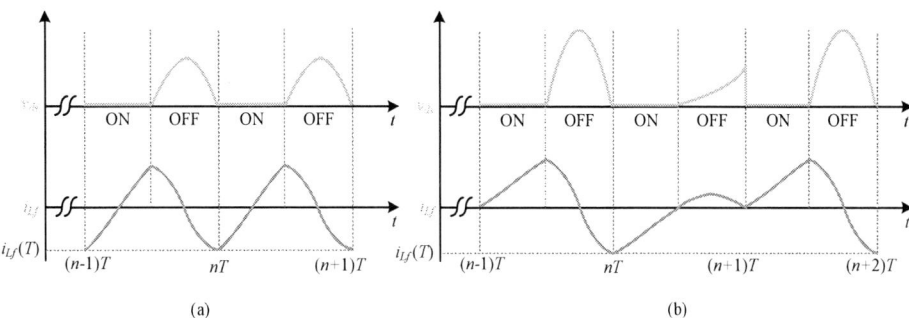

(a)

(b)

Fig. 6. Waveforms of v_{ds} and i_{Lf} under stable operation at the switching frequency f_s and subharmonic oscillations at $f_s/2$. (a) Stable waveforms in the frequency of f_s. (b) Subharmonic oscillations in the frequency of $f_s/2$.

$$\alpha = -\frac{1}{2}e^{-\pi\sigma_{on}}(e^{\frac{1}{2}\lambda_1 T} + e^{\frac{1}{2}\lambda_2 T}) \qquad (10)$$

Where $\lambda_{1,2}$ is given in (11), and σ_{on} is the damping factor of on-resistance as given in (13).

$$\lambda_{1,2} = \pm\frac{1}{2\omega_s}\sqrt{(\frac{1}{2}\omega_{IN}^2\varepsilon)^2 - (\omega_{IN}^2 - \omega_s^2)^2} \qquad (11)$$

$$k_{in} = \frac{1}{\omega_s\sqrt{L_f C_0}} \qquad (12)$$

$$\sigma_{on} = \frac{r_{dson}}{\omega_s L_f} \qquad (13)$$

The system stability is depended on the value of α, and can be summarized as follows:

1) $\alpha \in (-1,1)$. The limit of $\Delta i_{Lfn}(T)$ equals to zero as n approaches infinity, which means the integral of v_{ds} in each switching period is identical [refer to (4)] in steady state, and no subharmonic oscillation occurred in this situation.

2) $\alpha \geq 1$. The limit of $\Delta i_{Lfn}(T)$ equals to $\Delta i_{Lf1}(T)$ ($\alpha = 1$) or infinity ($\alpha > 1$) as n approaches infinity, in either case the current i_{Lf} will keep increasing and the system is unstable.

3) $\alpha \leq -1$. The limit of $\Delta i_{Lfn}(T)$ is either negative or positive depending on the value of n is odd or even. In this situation, v_{ds} will be fluctuated with the period of $2T$ and subharmonic oscillations occurred.

In summary, the system will maintain stable only when α is between -1 and 1. When α is equal or greater than 1, the current of L_f will keep increasing and the system is unstable. When α is equal or less than -1, v_{ds} will fluctuate in the frequency of $f_s/2$, leading to subharmonic oscillations.

Fig. 7 and Fig. 8 illustrate the stability index of a Class-E PA with different nonlinearity index ε, different frequency ratio k_{in}, and different damping factor ζ_{on}. It can be seen from Fig. 7 that when k_{in} is fixed, the stability index α is reversely proportional to the nonlinearity index ε, which means that subharmonic oscillations are more likely to occur with serious non-linearity of switch output capacitance, e.g., the PA with multiple parallel switches or smaller external shunt capacitance. Note the variation range of drain-source voltage will also affect the nonlinearity index ε. The contours ($\alpha = -1$) in Fig. 8 shows that the subharmonic oscillations are more likely to occur when the input tank resonant frequency is close to the switching frequencies, e.g., $k_{in} = 1$. It can be also seen that the switch on-state resistance plays a significant role in damping the subharmonic oscillations, where higher ζ_{on} shows higher tolerance to the non-linearity and achieves larger stable region.

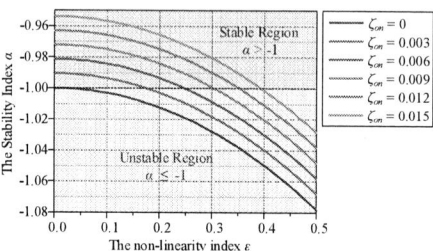

Fig. 7. The stability index α of a Class-E PA with different ε and ζ_{on}.

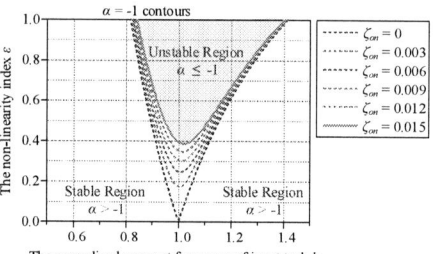

Fig. 8. The contours of stability index α with different ε, k_{in}, and ζ_{on}.

III.   EXPERIMENTAL STUDY

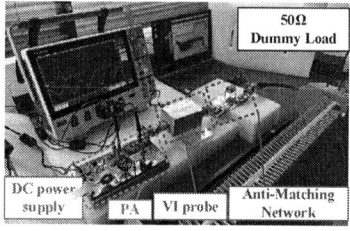

(a)

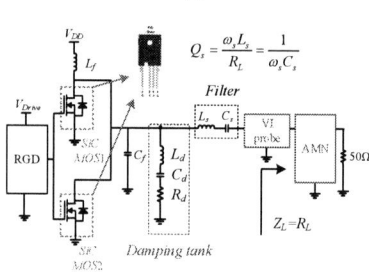

(b)

Fig. 9. The experimental setup. (a) The experimental system; (b) Circuits of the experimental system.

To validate the above analysis about subharmonic oscillations, the experimental prototype based on the 13.56MHz class-E circuit was built as shown in Fig.

9. The experimental system consists of a switched-mode Class E PA, a resonant gate driver, a damping tank a RF VI probe, and an anti-matching network. Here two SiC MOSFETs (C3M0120065K) are connected in paralleling as the switches of the Class E PA. The anti-matching network connecting with a 50Ω dummy load can emulate different load impedances to test the damping effect of load branch. The load filter L_s - C_s is resonant at 13.56MHz, which can be modified to change the Q factor of the load branch. The system efficiency and output power are measured by the VI probe.

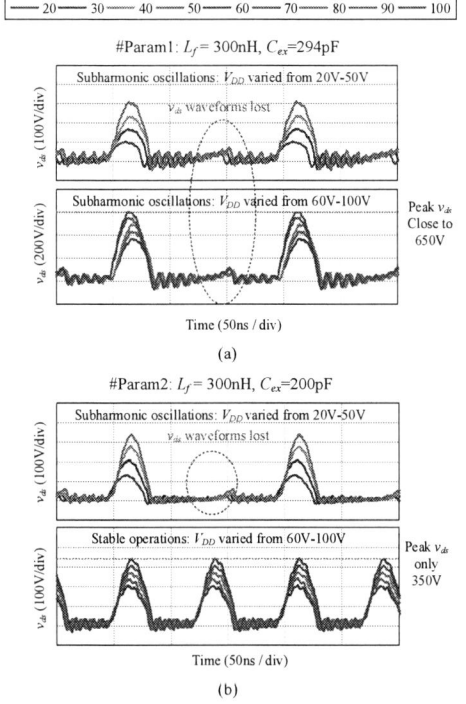

Fig. 10. The experimental waveforms of v_{ds} and v_{out} when V_{DD} is modulated from 20V to 100V. (a) with paremeters#1: L_f = 300nH, C_f = 294pF, R_L = 50Ω, Q_s = 5; (b) with paremeters#2: L_f = 300nH, C_f = 200pF, R_L = 50Ω, Q_s = 5.

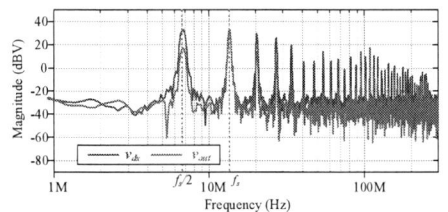

Fig. 11. The measured harmonic contents of v_{ds} and v_{out} when the subharmonic oscillations occurred with paremeters#2: L_f = 300nH, C_f = 200pF, R_L = 50Ω, Q_s = 5.

Fig. 10 shows the experimental waveforms of v_{ds} and v_{out} when V_{DD} is modulated from 20V to 100V. It can be seen from Fig. 10(a) that subharmonic oscillations occurred over the whole V_{DD} modulation range with the parameters #1 (L_f = 300nH, C_f = 294pF, R_L = 50Ω, Q_s = 5). According to the derivation in Section II, k_{in} and ε are calculated as varying from 0.89 to 1.02, and 0.51 to 0.39, respectively, when V_{DD} is sweeping from 20V to 100V. Furthermore, the stability index α is calculated to be smaller than -1 (subharmonic oscillations occurred) within the modulation range of V_{DD}, which matches well with the experimental results. Note that the peak voltages of v_{ds} is almost 6 times of the input dc voltage under subharmonic oscillations and it should be around 3.6 times in the normal operation. Fig. 10 (b) shows that subharmonic oscillations occurred when V_{DD} is modulated from 20V to 50V with the parameter #2 (L_f = 300nH, C_f = 200pF, R_L = 50Ω, Q_s = 5), and no subharmonic oscillations occurred when V_{DD} is larger than 60V. Fig. 11 shows the measured harmonic contents of v_{ds} and v_{out} when the subharmonic oscillations occurred (L_f = 300nH, C_f = 294pF, R_L = 50Ω, Q_s = 5, v_{dd}=100V). It shows that there are significant subharmonic distortions in output voltages of the Class E PA, e.g., at $f_s/2$ and the high-order harmonics of $f_s/2$. The experimental results of Fig. 10 and Fig. 11 validated the influence of circuit parameters, e.g., k_{in} and ε, to subharmonic oscillations.

IV. CONCLUSION

The manifestation and mechanism of subharmonic oscillations have been disclosed and analyzed in this article. It found out that the phenomenon was related to the non-linear output capacitance C_{oss} of MOSFETs and resonant frequency of input tank in Class-E circuit, which is verified in the experiments by changing the value of shunt capacitance C_f and bias voltage V_{DD}. The subharmonic oscillations can be eliminated by properly designing the value of C_f and V_{DD}.

REFERENCES

[1] J. Wang, H. S. Chung, and R. T. Li, "Characterization and Experimental Assessment of the Effects of Parasitic Elements on the MOSFET Switching Performance," *IEEE Trans. Power Electron.*, vol. 28, no. 1, pp. 573–590, Jan. 2013.

[2] Z. Chen, D. Boroyevich, and R. Burgos, "Experimental parametric study of the parasitic inductance influence on MOSFET switching characteristics," in *The 2010 International Power Electronics Conference - ECCE ASIA -*, Sapporo, Japan: IEEE, Jun. 2010, pp. 164–169. Accessed: Feb. 14, 2025. [Online]. Available: http://ieeexplore.ieee.org/document/5543851/

[3] A. Lemmon, M. Mazzola, J. Gafford, and C. Parker, "Instability in Half-Bridge Circuits Switched With Wide Band-Gap Transistors," *IEEE Trans. Power Electron.*, vol. 29, no. 5, pp. 2380–2392, May 2014.

[4] A. Lemmon, M. Mazzola, J. Gafford, and C. Parker, "Stability Considerations for Silicon Carbide Field-Effect

Transistors," *IEEE Trans. Power Electron.*, vol. 28, no. 10, pp. 4453–4459, Oct. 2013.

[5] Sanggeun Jeon, A. Suarez, and D. B. Rutledge, "Global stability analysis and stabilization of a class-E/F amplifier with a distributed active transformer," *IEEE Trans. Microw. Theory Tech.*, vol. 53, no. 12, pp. 3712–3722, Dec. 2005.

[6] J. F. Imbornone, M. T. Murphy, R. S. Donahue, and E. Heaney, "New Insight into Subharmonic Oscillation Mode of GaAs Power Amplifiers Under Severe Output Mismatch Condition".

Novel Structure for High-Voltage Vertical β-Ga₂O₃ Schottky Barrier Diodes: A TCAD Study

Yan Liu, Meng-Qi Fan, Jia-Xiang Chen, Teng Jiao, Mao-Jin Yang, Xian-Hu Zha, Xiao-Ping Wang*, Xiang-Jin Ding, Yu-Xi Wan*, Dao-Hua Zhang*
Shenzhen Pinghu Laboratory, Shenzhen, 518111, China
(*Email: wangxiaoping@phlab.com.cn, wanyuxi@phlab.com.cn, zhangdaohua@phlab.com.cn)

Abstract—The impact of different device structures and parameters on the reverse-bias characteristics of vertical β-Ga₂O₃ Schottky barrier diodes (SBDs) is investigated using calibrated TCAD simulations. The excellent agreement between numerical and experimental results validates the accuracy of our model. The reverse-bias characteristics of β-Ga₂O₃ SBDs with four different termination structures, namely, no termination (SBD 1), mesa termination (SBD 2), stepped-structure mesa termination (SBD 3), and stepped-structure mesa termination with oxide dielectric (SBD 4), are studied. The results show that the stepped-structure mesa termination with proper step structure design could effectively increase the reverse breakdown voltage of β-Ga₂O₃ SBD. Moreover, β-Ga₂O₃ SBD 4 with an optimized structure achieves a breakdown voltage of approximately 1380 V. Compared to the reverse breakdown voltages of SBD 1, SBD 2, and SBD 3, the value of SBD 4 shows an increase of 180%, 48.4%, and 10.9%, respectively. This work will provide theoretical guidance for the preparation of high-voltage β-Ga₂O₃ SBDs.

Keywords—device structure, gallium oxide, reverse-bias characteristics, Schottky barrier diode

I. INTRODUCTION

β-Ga₂O₃ has attracted intensive attention in high-power electronic devices because of its ultra-wide bandgap (~4.8 eV), large breakdown field (~8 MV/cm), and high Baliga's figure of merit (~3444) [1]. A diode is a fundamental device in power electronics systems. For β-Ga₂O₃ diode, the lack of viable p-type doping renders the formation of a traditional p-n diode unfeasible, and a Schottky diode is the attainable device structure at present [2]. Schottky barrier diodes (SBDs) have fast switching speed and reverse recovery characteristics. Moreover, due to the inherent material properties of β-Ga₂O₃, β-Ga₂O₃ SBDs can achieve higher breakdown voltage and lower on-resistance, enabling them to withstand higher power. Actually, the device structure strongly

affects the reverse-bias characteristics of β-Ga₂O₃ SBDs, as a result of a large electric field near the surface of the Schottky contact [3]. This issue can be effectively alleviated by employing the mesa termination, which could improve the device performance. However, the high-quality sample with deep etching depth is still a challenge in the β-Ga₂O₃ SBD fabrication [1]. Deep etching may introduce roughness or micro-cracks, disrupting the smoothness of the metal-semiconductor interface. This would cause localized electric field crowding, increasing leakage current, or reducing barrier homogeneity. Furthermore, the reverse breakdown voltage cannot continually increase with the increase of etching depth. Dhara et al. investigated the deep mesa etch design for efficient edge field termination in β-Ga₂O₃ SBD, and found that the peak electric field near the anode edge reduces significantly with increasing etch depth. For etch depths above 3 μm, the reduction in the peak electric field starts to saturate [4]. With the available etching depth, optimizing device structure and its parameters is an effective approach to improve the reverse-bias characteristics of β-Ga₂O₃ SBDs, but the research is still in its infancy.

In this study, vertical β-Ga₂O₃ SBDs with different termination structures have been designed and studied using technology computer-aided design (TCAD) simulations. The physical models for β-Ga₂O₃ SBDs were established and calibrated, demonstrating excellent agreement between numerical and experimental results, validating the accuracy of our simulations. The influence of structure parameters on the reverse breakdown voltage of proposed SBDs has been analyzed. Furthermore, a feasible approach to enhance the device performance through structure and its parameter optimization has been proposed.

979-8-3315-1110-4/25 $31.00 © 2025 IEEE

II. DEVICE MODELS AND FABRICATION

A. Device structure

Figure 1 shows the schematic cross-section of the vertical β-Ga$_2$O$_3$ SBDs proposed in this work. For devices with different termination structures, the following parameters remain identical. The devices' width and anode diameter are 320 μm and 300 μm, respectively. The substrate is Sn-doped β-Ga$_2$O$_3$, and the doping concentration is 8×10^{18} cm^{-3}. The epitaxial layer is Si-doped β-Ga$_2$O$_3$ with the doping concentration of 9×10^{15} cm^{-3}. The devices have Ni/Au top anode and Ti/Au bottom cathode.

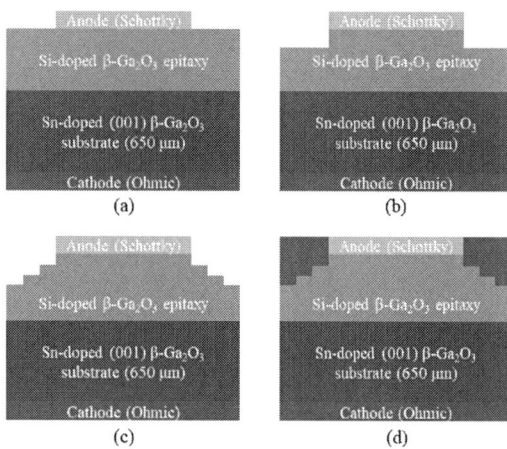

Fig. 1. Schematic cross-sectional structure of (a) β-Ga$_2$O$_3$ SBDs without mesa termination (SBD 1), (b) β-Ga$_2$O$_3$ SBDs with mesa termination (SBD 2), (c) β-Ga$_2$O$_3$ SBDs with stepped-structure mesa termination (SBD 3), (d) β-Ga$_2$O$_3$ SBDs with stepped-structure mesa termination and oxide dielectric (SBD 4) proposed in this work.

B. β-Ga$_2$O$_3$ Calibration

Silvaco TCAD is used in this work. Unlike SiC and GaN, which have established material parameters and simulation models in the TCAD database, Ga$_2$O$_3$, as a relatively new material, has no such data. Therefore, it is essential to develop and calibrate the physical models for the simulation of Ga$_2$O$_3$ devices.

Material parameter: The main material parameters of β-Ga$_2$O$_3$ are listed in Table I.

Table I. Main material parameters used in the simulation [5],[6]

Parameter	β-Ga$_2$O$_3$
Bandgap	4.8 eV
Electron effective mass	0.28 m$_0$
Static dielectric constant	10.2 ε$_0$
Effective conduction band density of states	3.72×10^{18} cm^{-3}
Effective valence band density of states	1.16×10^{19} cm^{-3}
Richardson constant	33.7 A·cm^{-2}·K^{-2}
Critical electric field	8 MV/cm

Mobility: For β-Ga$_2$O$_3$, the room temperature mobility of electrons was found to be 115 cm^2/V·s at the low electric field [7]. Holes are approximately ten times heavier than electrons, which leads to an extremely low mobility of holes. In this work, the hole mobility, which can be considered negligible, is set to 1 cm^2/V·s. At high electric fields, the transport fitting parameters proposed by Barnes are adopted [3].

Incomplete ionization: In Ga$_2$O$_3$ SBDs, incomplete ionization leads to a trade-off between the on-state resistance and breakdown voltage [8], which has been demonstrated in TCAD simulations [9].

Impact ionization: The impact ionization coefficient is determined by the van Overstraeten-de Man model.

Tunneling mechanisms: At high reverse voltage, current traverses the barrier via tunneling, which can be described by the universal Schottky tunneling mechanism. Additionally, band-to-band tunneling is reported to contribute to the current flow and can affect the breakdown [10].

Traps: Zhang et al. [11] performed the deep-level optical spectroscopy and deep-level transient spectroscopy measurements on Ni/β-Ga$_2$O$_3$ Schottky diodes. The traps also affect the tunneling current. Consequently, trap-assisted tunneling has been considered in this work.

Table II. Main model parameters used in the simulation[2],[3],[7]

Model	parameter	value
KLASSEN	μ_{max}	150 cm^2/V·s
	μ_{min}	80 cm^2/V·s
	θ	1.8
	N_{ref}	2×10^{17} cm^{-3}
	α	0.9
Impact ionization	a	7.06×10^5 cm^{-1}
	b	2.1×10^7 V/cm
Traps	e.level	0.6 eV
		0.75 eV
		1.05 eV
	degen	1
	sign/sigp	100

The main models and parameters used for the calibration are listed in Table II. All calibrations were performed under room-temperature conditions. Temperature-dependent testing will be carried out in our future work to further optimize the models.

C. Device fabrication

Fig. 2 shows the schematic cross-section of the Ga$_2$O$_3$ SBD1. The device consists of a 6 μm Si-doped epitaxial drift layer grown by halide vapor phase epitaxy on (001) oriented 650 μm Sn-doped high-conductivity Ga$_2$O$_3$ substrate (from Novel Crystal

Technology, Japan). Samples were first cleaned with solvent and acid to remove contaminants. A back-side ohmic contact of Ti/Au (20/80 nm) was deposited via electron beam evaporation and followed by 1 min 470 °C rapid thermal annealing in N_2 ambient. The anode Ni/Au (20/80 nm) was patterned using photolithography, followed by electron-beam evaporation deposition and a standard lift-off process to form the Schottky metal.

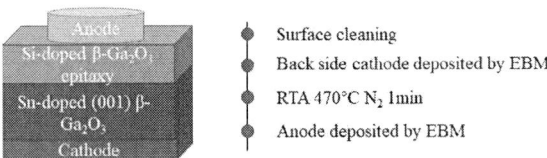

Fig. 2. Schematic cross-sectional structure and main process flow of the SBD1 proposed in this work.

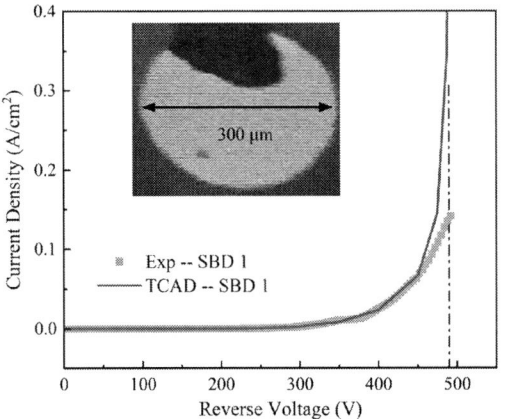

Fig. 3. The reverse voltage of β-Ga$_2$O$_3$ SBD 1 at 300 K. The solid lines and dot scatters are simulated and experimental data, respectively. Inset: the photograph of the fabricated device.

As presented in Fig. 3, the simulated and experimental reverse-bias characteristics of β-Ga$_2$O$_3$ SBD 1 agree well, and the breakdown voltage is approximately 493 V. The good consistency indicates the effectiveness of our simulations.

III. RESULTS AND DISCUSSION

Figure 4 shows the effect of epitaxial layer thickness and etching depth on the reverse-bias breakdown voltage of β-Ga$_2$O$_3$ SBDs. In Fig. 4(a), the breakdown voltage increases with increasing epitaxial layer thickness and starts to saturate when the epitaxial layer thickness reaches 6 µm. Since the carrier concentration of the epitaxial layer remains constant, the depletion region cannot expand further, even with additional increases in epitaxial layer thickness. The maximum depletion width is 7.81 µm,

as presented in Fig. 4(b). In Fig. 4(c), the breakdown voltage increases with the etching depth (d) because the electric field is redistributed, causing the position of the peak electric field to gradually shift toward the trench bottom, as is shown in Fig. 4(d). Related phenomenon has also been observed experimentally, as presented in the insert of Fig. 3: the breakdown point of β-Ga$_2$O$_3$ SBD 1 is located at the edge of the anode, which indicates that the peak electric field is concentrated at the anode edge. Therefore, the mesa termination could directly impact the peak electric field and increase the breakdown voltage.

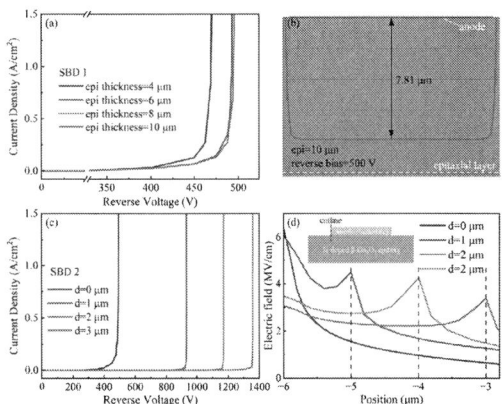

Fig. 4. (a) The reverse voltage of β-Ga$_2$O$_3$ SBD 1 with different thicknesses of the epitaxial layer. (b) The depletion width of β-Ga$_2$O$_3$ SBD 1 with 10 µm epitaxial layer. (c) The reverse voltage and (d) electric field profiles along the cutline of β-Ga$_2$O$_3$ SBD 2 with different etching depths.

Figure 5 shows the reverse voltage and electric field distribution of β-Ga$_2$O$_3$ SBD 3 and SBD 4 with different structure parameters. When the etching depth of SBD 2, as well as that of each step of SBD 3 and SBD 4, is 1 µm, and the sum of w$_1$, w$_2$, and w$_3$ is kept constant, smaller w$_1$ and w$_2$ are beneficial to increasing the breakdown voltage of the device. In Fig. 5(a), the maximum breakdown voltage of β-Ga$_2$O$_3$ SBD 3 is approximately 1245 V, while that of SBD 2 is 930 V. This improvement occurs because the peak electric field could transfer to the second and even the third trench bottom, as can be seen in Fig. 5(c)-(f). In Fig. 5(b), the β-Ga$_2$O$_3$ SBD 4 with an oxide dielectric of a higher dielectric constant could further improve the breakdown voltage, and the highest voltage is about 1380 V. This is due to the fact that the electric field crowding phenomenon at the edge of the device can be alleviated, and the oxide material with a higher dielectric constant will make the electric field distribute more uniformly within it and at the interface, thus reducing the regions where the electric field is concentrated, as illustrated in Fig. 5(g)-(h).

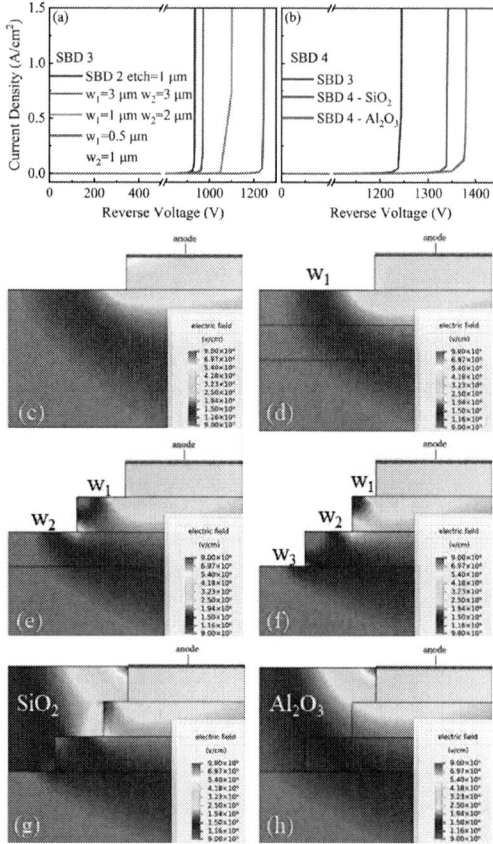

Fig. 5. The room-temperature reverse voltage of (a) β-Ga$_2$O$_3$ SBD 3 with different step widths, (b) β-Ga$_2$O$_3$ SBD 4 with different oxide dielectrics. The electric field distribution of (c) β-Ga$_2$O$_3$ SBD 2 with etching depth of 1 μm, β-Ga$_2$O$_3$ SBD 3 with (d) step width w_1= 3 μm, w_2= 3 μm (e) w_1= 1 μm, w_2= 2 μm (f) w_1= 0.5 μm, w_2= 1 μm, and β-Ga$_2$O$_3$ SBD 4 with (g) SiO$_2$ (dielectric constant of 3.9 ε_0), (h) Al$_2$O$_3$ (dielectric constant of 9.3 ε_0).

IV. CONCLUSION

In summary, β-Ga$_2$O$_3$ SBDs with four different termination structures have been studied by calibrated TCAD simulation. The simulation results agree well with the experimental data, proving the correctness of our simulations. Increasing the epitaxial layer thickness and etching depth can increase the breakdown voltage of β-Ga$_2$O$_3$ SBDs. The stepped-etching mesa termination with optimized structure parameters will effectively enhance the breakdown voltage to 1245 V, and the oxide dielectric can further improve it to 1380 V.

ACKNOWLEDGMENT

This work is supported by the Shenzhen Pinghu Laboratory Project (Grant No. 224110).

REFERENCES

[1] S. Sun, C. Wang, S. Alghamdi, H. Zhou, Y. Hao, and J. Zhang, "Recent advanced ultra-wide bandgap β-Ga$_2$O$_3$ material and device technologies," *Advanced Electronic Materials*, 2025, 11(1), 2300844.

[2] H. Y. Wong, "TCAD simulation models, parameters, and methodologies for β-Ga$_2$O$_3$ power devices," *ECS Journal of Solid State Science and Technology*, 2023, 12(5), 055002.

[3] W. H. Choi, K. Kim, S. G. Jeong, J. H. Han, J. Jang, J. Noh, and J. S. Park, "The significance on structural modulation of buffer and gate insulator for ALD based InGaZnO TFT applications," *IEEE Transactions on Electron Devices*, 2021, 68(12), 6147-6153.

[4] S. Dhara, N. K. Kalarickal, A. Dheenan, C. Joishi, and S. Rajan, "β-Ga$_2$O$_3$ Schottky barrier diodes with 4.1 MV/cm field strength by deep plasma etching field-termination," *Applied Physics Letters*, 2022, 121(20), 203501.

[5] M. Higashiwaki, K. Sasaki, A. Kuramata, T. Masui, and S. Yamakoshi, "Gallium oxide (Ga$_2$O$_3$) metal-semiconductor field-effect transistors on single-crystal β-Ga$_2$O$_3$ (010) substrates," *Applied Physics Letters*, 2012, 100(1), 013504.

[6] J. Yang, S. Ahn, F. Ren, S. J. Pearton, S. Jang, J. Kim, and A. Kuramata, "High reverse breakdown voltage Schottky rectifiers without edge termination on Ga$_2$O$_3$," *Applied Physics Letters*, 2017, 110(19), 192101.

[7] K. Ghosh, "Ab Initio electron transport in monoclinic β-Ga$_2$O$_3$," Ph.D. dissertation, Dept. Elect. Eng., Univ. Buffalo, Buffalo, NY, USA, 2017.

[8] A. T. Neal, S. Mou, R. Lopez, J. V. Li, D. B. Thomson, K. D. Chabak, and G. H. Jessen, "Incomplete ionization of a 110 meV unintentional donor in β-Ga$_2$O$_3$ and its effect on power devices," *Scientific Reports*, 2017, 7(1), 13218.

[9] H. Y. Wong and A. C. F. Tenkeu, "Advanced TCAD simulation and calibration of gallium oxide vertical transistor," *ECS Journal of Solid State Science and Technology*, 2020, 9(3), 035003.

[10] N. A. Moser, "Gallium oxide metal oxide semiconductor field effect transistor analytical modelling & power transistor design trades," Ph.D. dissertation, Dept. Elect. Comput. Eng., George Mason Univ., Fairfax, VA, USA, 2017.

[11] Z. Zhang, E. Farzana, A. R. Arehart, and S. A. Ringel, "Deep level defects throughout the bandgap of (010) β-Ga$_2$O$_3$ detected by optically and thermally stimulated defect spectroscopy," *Applied Physics Letters*, 2016, 108(5), 052105.

Efficiency-Oriented Adaptive Dead Time Control for the Dual Active Bridge Converter

Shaoyan Jiang
Zhongshan Power Supply Bureau of
Guangdong Power Grid Co., Ltd.
Zhongshan, China
773782436@qq.com

Xichen Fu
Zhongshan Power Supply Bureau of
Guangdong Power Grid Co., Ltd.
Zhongshan, China
202321014831@mail.scut.edu.cn

Xingque Xu
Zhongshan Power Supply Bureau of
Guangdong Power Grid Co., Ltd.
Zhongshan, China
710524701@qq.com

Jiabin Ruan
Zhongshan Power Supply Bureau of
Guangdong Power Grid Co., Ltd.
Zhongshan, China
1343392964@qq.com

Chuanwei Xiao
Zhongshan Power Supply Bureau of
Guangdong Power Grid Co., Ltd.
Zhongshan, China
123015452@qq.com

Abstract— **In the full bridge topology, dead time is a crucial parameter that directly affects the energy conversion efficiency and performance of the full bridge topology. Dead time is a period of time without output that is actively added to the control signal of the switching device. Its main purpose is to prevent the upper and lower switching tubes of the bridge arm from conducting at the same time, and to avoid the generation of huge current after short circuit, which may cause device damage. However, the addition of dead time inevitably introduces problems such as energy loss and waveform distortion, which cannot be ignored for DAB converters that pursue high efficiency. This paper will establish a mathematical model of dead time based on GaN HEMT driving circuit, and propose an adaptive dead time control strategy based on this model. The optimal dead time will be obtained through the current operating conditions for control, and finally simulation analysis will be conducted to verify the correctness and effectiveness of the proposed strategy.**

Keywords—Dual active bridge converter, single phase shift control,deadtime.(key words)

I. INTRODUCTION

With the development of distributed generation, energy storage, and electric vehicles, dual active bridge converters have gradually become important power conversion equipment in these fields due to their advantages of high-frequency isolation, symmetrical structure, soft switching characteristics, and high power density. However, the efficiency of DAB converters has always been a research focus[1][2], especially in a wide range of input and output voltages. How to improve their efficiency has become an urgent problem to be solved[3].

.The efficiency of DAB converters is affected by various factors, mainly including switching losses, conduction losses, magnetic core losses, and dead time. Among them, the switching loss is closely related to the characteristics of the switching device, the switching frequency, and the implementation of soft switching; The conduction loss is mainly related to current stress, and the greater the current stress, the higher the conduction loss; The magnetic core loss is related to the design and operating frequency of high-frequency transformers.

Most of the current research on DAB converters focuses on optimizing the modulation method. Reference [4] proposes a dual phase shift modulation method, which, based on SPS modulation, adds uniformly sized internal shifts to both sides of the primary and secondary half bridges simultaneously. Compared with this method, it can achieve countless combinations of reflux power and current stress under the same transmission power, which can cope with the phenomenon of increased reflux power and current stress in the case of input and output voltage mismatch. References [5] and [6] respectively consider the effects of reflux power, current stress, and soft switching on efficiency and optimize them to achieve the goal of improving efficiency. Reference [7] proposes a triple phase shift modulation method, further increasing degrees of freedom to make the optimization effect more comprehensive and effective. More diverse control methods, such as Reference [8] Quantitatively analyzes multiple optimization objectives and propose corresponding optimization strategies, while reducing turn-on and conduction losses to improve efficiency. With the increase of degrees of freedom, although the optimization range expands and the optimization results become more comprehensive, the control difficulty also increases significantly. Therefore, TPS may not be the optimal modulation method in engineering applications [9][10].

There have been many studies on the optimization of modulation strategies for DAB converters, but there has been little research on dead zone time as one of the important factors affecting efficiency. Dead zone time is a key parameter used in DAB converters to prevent switching devices from passing through. However, the existence of dead time can have various adverse effects on the performance of the converter. Reference [11] established a DAB equivalent circuit considering parasitic parameters, obtained a mathematical model of the dead time, and analyzed the high-frequency oscillation phenomenon in the dead time. However, it did not involve how to adjust the control of the dead time. Reference [12] proposed an adaptive dead time method that controls the size of the dead time by detecting whether there is a peak in the leakage source voltage of the switching device. However, errors caused by harmonics can make the judgment inaccurate. Reference [13] proposed a segmented dead time adaptive control strategy based on the mathematical model of the dead time. However, due to the high-frequency characteristics of the DAB converter, a new control strategy still needs to be developed. The study in reference [14] shows that the implementation conditions of zero voltage switching (ZVS) are determined by the dead time and ZVS resonant inductance under resistive load conditions, and a method for setting the dead time range is proposed based on the inductance capacitance parameters in the circuit. However, in

---

This research was funded by China Southern Power Grid Corporation Science and Technology Project under Grant 032000KK52222025.

dynamic load changing operating scenarios, the optimal dead time will significantly vary with load conditions. Therefore, the existing methods still need to be further optimized to establish an adaptive dead time adjustment mechanism suitable for multi condition load changes.

In order to achieve precise and appropriate adaptive control of the dead time of the DAB converter, this paper first analyzes the switching process of the two switching tubes in the bridge arm, obtains the mathematical model of the dead time, and combines it with the high-frequency operating conditions of the DAB converter to provide an adaptive dead time control strategy. Furthermore, the effectiveness and correctness of the proposed method will be verified through simulation experiments.

II. TOPOLOGY AND MODEL OF DEAD TIME

There are four bridge arms in a DAB converter, and each bridge arm has a similar external structure and function as a whole. Taking the common GaN HEMT driving circuit as an example, the bridge arm H and its external circuit are unfolded. The structure is shown in Figure 1, where V_{in} is the input voltage of the entire DAB converter, V_{drive} is the driving voltage, and the PWM control signal output by V_{drive} passes through the gate resistor R_g and the filtering capacitor C_f to reach the GaN HEMT. There are parasitic capacitors C_{gd}, C_{ds}, and C_{gs} between the gate, drain, and source of each GaN HEMT, and the input capacitor $C_{iss}=C_{gs}+C_{gd}$, the output capacitor $C_{oss}=C_{gd}+C_{ds}$, and the reverse transmission capacitor $C_{rss}=C_{gd}$, also known as Miller capacitor; Among them, S and S represent the upper and lower switching tubes of the bridge arm, respectively, to distinguish $Q_1 \sim Q_8$ in the DAB converter; L is the resonant inductance of the DAB converter..

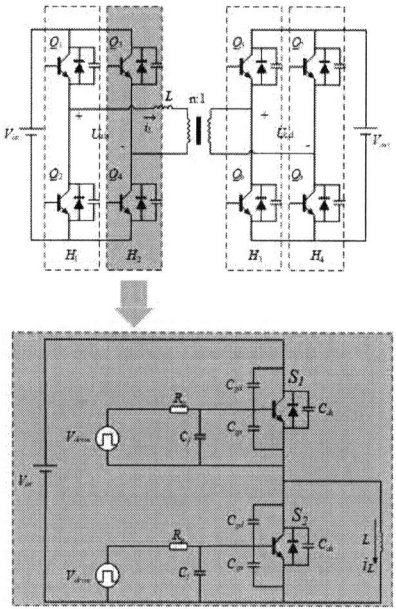

Fig. 1. DAB converter and single bridge arm topology diagram

In order to consider the impact of parasitic capacitance on the switching process, the following will analyze the single bridge arm switching process under non ideal conditions. The current flow direction in Figure 1 is defined as positive. When i $i_L>0$, S_1 is the active switching transistor, and vice versa, S_2

is the active switching transistor. The dead time before the active switching transistor is turned on is defined as the first dead time, and the dead time after turning off is defined as the second dead time. As shown in Figure 2, the opening process of S_1 as the active switching transistor is as follows:

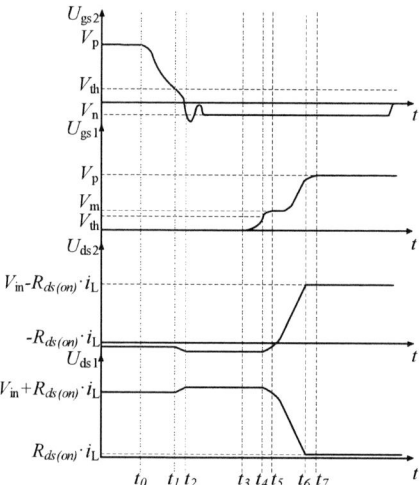

Fig. 2. S1 activation process

stage [t_0-t_1]: Before time t_0, S_2 remains conductive, and the load current i_L passes through S_2. At this time, the drain source voltage U_{ds2} of S_2 is -$R_{sd(on)}$·i_L. At time t_0, the gate source voltage U_{gs2} of S_2 begins to decrease because it has not yet reached the threshold voltage V_{th}, so S_2 remains on. At time t_1, U_{gs2} drops to V_{th}, at which point S_2 is turned off. Throughout the process, S_1 remains turned off, and the drain source voltages of S_1 and S_2 remain unchanged.

stage [t_1-t_2]: The gate source voltage of S_2 decreases from V_{th} to the driving negative voltage V_n In this stage, both the upper and lower transistors are turned off, and the direction of the load current remains unchanged. The load current flows through the body diode of S_2, and the drain source voltage of S_2 is less than -$R_{sd(on)}$·i_L.

stage [t_2-t_3] : During this period, the drain source voltage and gate source voltage remain almost unchanged

Stage [t_3-t_4]: The gate source voltage of S_1 begins to increase and reaches V_{th} at time t_4. Since the upper and lower transistors are still in the off state, the drain source voltage also does not change.

Stage [t_4-t_5]: At time t_4, the load current begins to flow through the S_1 body. At time t_5, the gate source voltage of S_1 rises to the Miller voltage V_m, and the load current completely flows through the S_1 body.

Stage [t_5-t_6]: The gate source voltage of S_1 is maintained at the Miller voltage V_m, which is called the Miller platform. At this time, all gate currents provided by the driver are transferred, thereby charging C_{gd} to achieve rapid voltage changes at the drain source terminal. At time t_6, the drain source voltage of S_2 rises to V_{in}-$R_{sd(on)}$·i_L.

Stage [t_6-t_7]: This process is the process of driving the power supply to charge C_{iss}. At time t_7, the gate source voltage of S_1 is charged to V_p.

In order to prevent S_1 from conducting simultaneously with S_2 during the turn-on process, the gate source voltage of S_2 should decrease from V_{th} to V_{th} before the gate source voltage of S_1 rises to V_{th}. That is, the first dead time t_{d1} should satisfy the following equation:

$$t_{d1} = (t_1 - t_0) - (t_4 - t_3) \geq 0 \qquad (1)$$

In the equation, (t_1-t_0) represents the time when the gate source voltage of S_2 decreases to V_{th}, and (t_4-t_3) represents the time when the gate source voltage of S_1 increases to V_{th}.

If t_{d1} is too large, it will cause the reverse conduction time (t_2 - t_3) of S_2 to be prolonged, increasing the loss. Therefore, the first dead time should be as small as possible. Therefore, there are:

$$t_{d1} = (t_1 - t_0) - (t_4 - t_3) \qquad (2)$$

The equivalent circuit diagram of charging R_g with the input capacitor C_{iss} during the time period t_0 to t_1 is as follows:

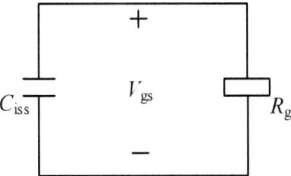

Fig. 3. Equivalent circuit diagram of t0-t1 time period

Can be obtained:

$$t_1 - t_0 = R_g \cdot (C_{gs} + C_{gd}) \cdot \ln(\frac{V_p - V_n}{V_{th} - V_n}) \qquad (3)$$

The input capacitance of GaN devices is charged from 0V to V_{th} during the time period of t_4-t_3. During this period, most of the gate current is used to charge C_{gs} capacitors, and its equivalent circuit diagram is:

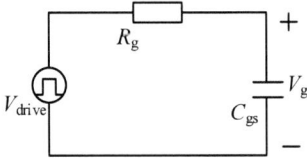

Fig. 4. Equivalent circuit diagram of t3-t4 time period

Can be obtained:

$$t_4 - t_3 = R_g \cdot C_{gs} \cdot \ln(\frac{V_p}{V_p - V_{th}}) \qquad (4)$$

From equations (3) and (4), it can be seen that the first dead time t_d is related to the gate resistance R_g, input capacitance C_{iss}, driving voltage V_{drive}, and threshold voltage V_{th}. These parameters will not change after the driving circuit system is determined. Therefore, the first dead time t_d is a fixed value. In order to preserve a certain margin to prevent the possibility of accidental conduction, a dead time margin coefficient of 1.5

is given. Therefore, the final optimized value for the first dead time segment should be:

$$t_{d1} = 1.5 \cdot [R_g \cdot (C_{gs} + C_{gd}) \cdot \ln(\frac{V_p - V_n}{V_{th} - V_n}) - R_g \cdot C_{gs} \cdot \ln(\frac{V_p}{V_p - V_{th}})] \qquad (5)$$

Normally, the second dead time t_{d2} should also satisfy equations (1), but the impact of output capacitance needs to be considered during the S_1 shutdown process. If t_{d2} is too small, when the drain source voltage of S2 has not yet dropped to 0, the S_2 body is already conducting. Not only will the output capacitor discharge the S_2 body, but the input voltage will also discharge the S_2 body. This phenomenon increases losses and also easily leads to forward fluctuations in U_{gs1}, causing the bridge arm to misguide conduction.

In order to prevent the occurrence of the above phenomenon, the second dead time t_{d2} should be set longer. It should meet the requirement that after the drain source voltage U_{ds2} of S_2 completely drops to 0, the gate source voltage U_{gs1} of S_2 only reaches the threshold voltage. Therefore, the second dead time t_{d2} adds a discharge process of the output capacitor C_{oss2} on the basis of the first dead time t_{d1}, that is:

$$t_{d2} = 1.5 \cdot [R_g \cdot (C_{gs} + C_{gd}) \cdot \ln(\frac{V_p - V_n}{V_{th} - V_n})$$
$$- R_g \cdot C_{gs} \cdot \ln(\frac{V_p}{V_p - V_{th}}) + \frac{2 \cdot Q_{oss}}{i_L}] \qquad (6)$$

In the equation, Q_{oss} is the output charge of GaN devices, which can be found in the data manual. It is not difficult to find that the second dead time t_{d2} is not only related to the parameters of the driving circuit, but also to the magnitude of the load current. When the load current is small, the output capacitor charges slowly, and the dead time should be set larger; On the contrary, when the load current is high, the output capacitor charges faster, allowing the dead time to be set smaller and infinitely close to the first dead time t_{d1}.

III. ADAPTIVE DEAD TIME CONTROL STRATEGY

In order to prevent the bridge arm from passing through and minimize the distortion of the output waveform, appropriate modulation of the dead time should be carried out - under the premise of the bridge arm not passing through, the dead time should be minimized as much as possible. Through the mathematical modeling of the dead time mentioned above, it can be concluded that after the circuit system is determined, the dead time is only related to the load current. Therefore, it is only necessary to obtain the load current to calculate the appropriate dead time.

In a DAB converter, the load current of the first two sets of bridge arms is the inductor current, and the load current of the last two sets of bridge arms is the current induced by the inductor current through the transformer. By calculating the size of the inductor current, the appropriate dead time can be adaptively controlled. As shown in the figure, it is a typical curve of the inductor current and control signal of the DAB converter. It can be seen that the frequency of the inductor current transformation is of the same order of magnitude as the operating frequency of the DAB converter. Therefore, determining the first and second dead time sequence by the direction of the inductor current and controlling the size of the two dead times separately will greatly increase the burden on the control system. Therefore, a more reasonable control strategy should be selected.

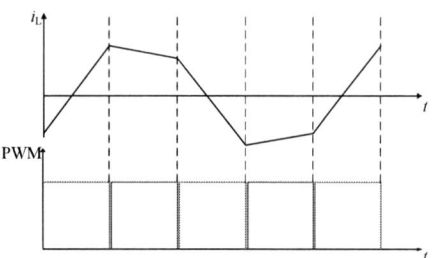

Fig. 5. Typical waveform of inductance current in DAB converter

The traditional fixed dead time value process is to make two dead time periods equal to the second dead time period when the load is small. However, as the load increases, the size of the second dead time period gradually decreases, and the traditional fixed dead time period gradually deviates from the optimal value, resulting in losses. Therefore, the dual dead time control strategy that can be adjusted according to the inductance current under different loads can improve efficiency compared to traditional fixed dead time periods, and do not consider the control system design problems caused by complex measurement of inductance current direction.

Simultaneously control two dead time periods, with the setting of the second dead time period as the main focus. In the expression of the second dead time period, the load current iL is in the denominator, which means that the smaller the load current, the larger the dead time period, resulting in increasingly severe waveform distortion. Therefore, harmonic limiting measures need to be added; In addition, in order to ensure the safety of the switch tube, the minimum value of the dead time should also be set with a limit; The dead time within the intermediate range should be set according to the expression, and the final selection of the dead time should be:

$$t_{d1} = t_{d2} = t_d = \begin{cases} T_{\min} & i_{L\max} < i_L \\ 1.5 \cdot (A + \dfrac{2 \cdot Q_{oss}}{i_L}) & i_{L\min} < i_L < i_{L\max} \\ T_{\max} & i_L < i_{L\min} \end{cases}$$

(7)

In the equation, A is a constant related to the driving circuit, $T_{\min}$ and $T_{\max}$ are the minimum and maximum values of the limited dead time, corresponding to the maximum and minimum values of the load current $i_{L\max}$ and $i_{L\min}$, respectively. Both can be determined by actual operating conditions, where $i_{L\max}$ is set based on the maximum current that the system can achieve, and $i_{L\min}$ is determined based on whether there is a peak in the switch current under the minimum operating conditions.

Taking SPS modulation as an example, the inductor current is constantly changing, as shown in Figure 5. However, the inductor current corresponding to the switching process is the inflection point of the inductor current, and the size of this inflection point can be calculated through the parameters of the main circuit. Therefore, the changes in the main circuit can be reflected in the dead time through the inductor current. By controlling the dead time, efficiency improvement can be

achieved under different operating conditions. The inductance current inflection point under SPS modulation is:

$$i_{Lt0} = \frac{nV_2(1 - 2D - k)}{4fL}$$

(8)

$$i_{Lt2} = \frac{nV_2[1 - k(1 - 2D)]}{4fL}$$

(9)

In the formula, n is the transformer ratio, V_2 is the output voltage, D is the external shift ratio, k is the voltage ratio, and its value is V_1/nV_2. f is the operating frequency of the DAB converter.

Similarly, in order to ensure the safety of the switching transistor, a smaller inductor current is selected for control, as shown in the adaptive control block diagram of the entire dead time:

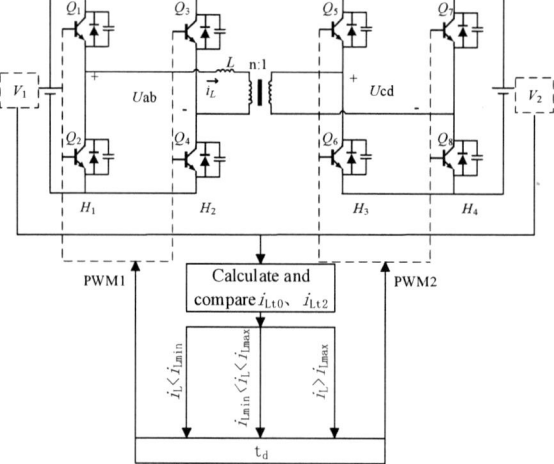

Fig. 6. Dead time control diagram

IV. SIMULATION ANALYSIS

In order to verify the correctness and effectiveness of the dead time mathematical model and control strategy mentioned above, this paper will build a DAB converter simulation example in Matlab/Simulink and conduct simulation experiments for research. The detailed calculation parameters are shown in the table below. The GaN switch tube model in this article is based on TI's LMG3422R030 as an example. According to its data manual, the output charge Q_{oss} is 175nC (V_{DS} = 0V to 400V), the switch tube output capacitance is 170pF, and the switch tube output capacitance C_{oss} is 170pF.

TABLE I. SIMULATION PARAMETERS

Main circuit parameters	value
switching frequency f/Hz	10^5
Transformer ratio n	1
Input voltage V_1/V	500
output voltage V_2/V	500
Inductance L/H	5×10^{-6}
Drive gate level resistor R_g/Ω	0.03
Gate level filtering capacitor C_f/F	10^{-9}
Range of driving voltage V_{drive}/V	15
Output charge Q_{oss}/C	175×10^{-9}

Switching tube output capacitor C_{oss}/F	170×10^{-12}

To verify the correctness of the dead time formula, control variable methods were applied to Cgd, Rg, and iLC to obtain the optimal dead time and corresponding loss power under different simulation conditions. As shown in Figure 7, as the size of the gate drain capacitance increases, the discharge time of the input capacitance becomes longer, and therefore the optimal dead time gradually increases. The relationship between the gate resistance and the optimal dead time is shown in Figure 8. Similar to the gate drain capacitance, the time constant is the product of capacitance and resistance. As the gate resistance increases, the optimal dead time also gradually increases. Since the variable is almost only related to the switching loss, the loss power presented in Figures 7 and 8 also gradually increases with the resistance and capacitance. Figure 9 shows the relationship between the optimal dead time, loss power, and inductor current. The optimal dead time decreases as the inductor current increases, confirming the process described earlier where the output capacitor needs to be discharged through the inductor current. During this process, the larger the inductor current, the shorter the discharge time, which corresponds to the optimal dead time formula. In addition, due to the involvement of inductor current, changes in power loss are not only caused by switch losses. There is also the influence of conduction loss, and as the inductor current increases, the power loss will gradually increase.

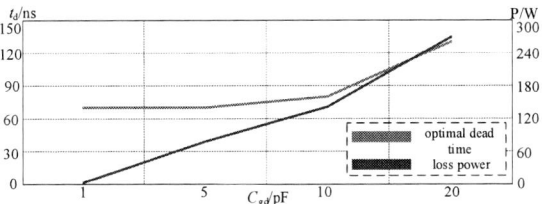

Fig. 7. The relationship between optimal dead time, loss power, and gate drain capacitance

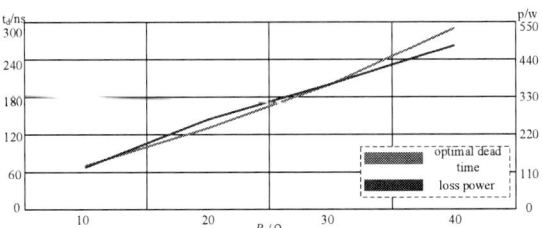

Fig. 8. The relationship between optimal dead time, power loss, and gate resistance

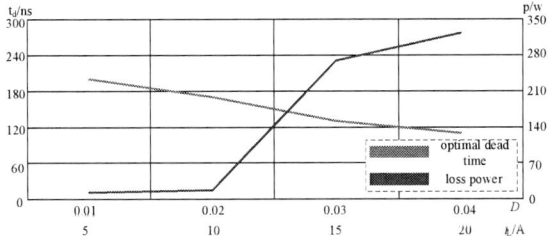

Fig. 9. The relationship between optimal dead time, loss power, and inductor current

The above validates the correctness of the optimal dead time formula, and the following will simulate and verify the adaptive control strategy for dead time.

a) Within the adjustable dead zone range: Based on the above example, an additional operating condition variation is added - the output voltage decreases from 500V to 250V at 400us (running time 800us), and the effectiveness of the adaptive dead zone time control strategy is verified by observing the changes in inductor current and power. As shown in Figure 10(a), it is the waveform diagram of the input and output voltage. In order to observe the changes in inductor current and power before and after the dead time change more clearly, the updated dead time output delay updates the dead time after the power stabilizes at 700us. Figure 10(b) shows the changes in the dead time. According to the ideal state, changes in this operating condition will cause an increase in the instantaneous value of the inductor current during the dead zone phase, thereby reducing the optimal dead zone time. During the time period of 400us-700us, as it still continues the optimal dead zone time of the previous operating condition and deviates from the optimal dead zone time of the current operating condition, the loss power will be greater than that after 700us. Figure 11(a-d) shows the changes in inductor current, input power, output power, and loss power throughout the entire control process. According to Figure 11(d), it can be seen that in the case of operating condition changes, the control strategy will update the dead zone time and reduce losses, which verifies the correctness and effectiveness of this strategy.

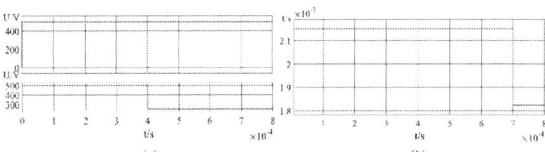

Fig. 10. Changes in Input/Output Voltage and Dead Time under Adaptive Dead Zone Control Strategy

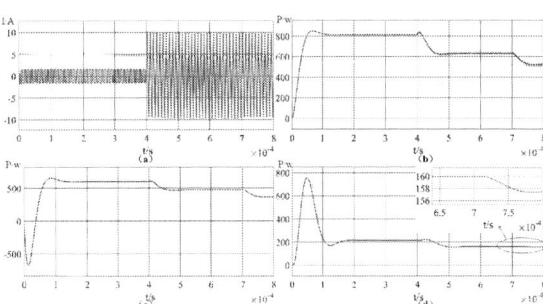

Fig.11. Changes in inductor current and power under adaptive dead zone control strategy when output voltage decreases

b) At the dead zone boundary: Based on the above example, two situations of whether to limit the inductor current value are analyzed. Figure 12 shows the variation of the dead zone time and the waveform of the drain current of the switch under the control strategy without current limitation. Figure 13 shows the variation of the dead zone time and the waveform of the drain current of the switch under the current limitation control strategy. It can be seen that compared to the control strategy without limiting the inductor current in Figure 12(a), the minimum dead time in figure 13(a) is limited to 150ns, and in 12(b), the drain current of the switch tube exhibits a peak phenomenon due to the small dead time, indicating that there may be a misleading conduction phenomenon. Therefore, it is necessary to add control that limits the inductor current to ensure the safe operation of the entire DAB converter, and the change in dead time also verifies the effectiveness of the adaptive dead time control strategy.

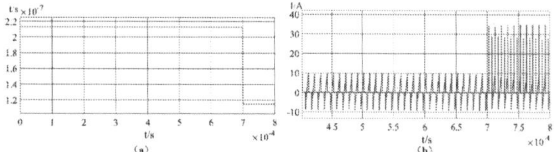

Fig.12. Changes in Dead Time without Current Limitation Control Strategy and Waveform of Drain Current of Switching Tube

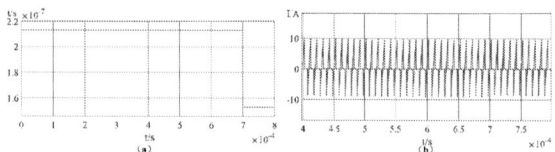

Fig.13. shows the variation of dead time under the current limiting control strategy and the waveform of the drain current of the switching transistor

V. CONCLUSION

This paper introduces the principle and impact of dead time in DAB converters. By analyzing the switching process of the two switching tubes in the bridge arm, a mathematical model of the dead time is obtained. In order to prevent the distortion of output waveform caused by too small dead time and too large dead time, the dead time is limited to a special range. Combined with the high-frequency operating conditions of DAB converters, an adaptive dead time control strategy is proposed. Furthermore, in the simulation experiment, the correctness of the parameters in the dead time formula was verified through the control variable method, and simulation analysis was conducted on the internal and external boundaries of the dead time selection range. It was found that within the dead time selection range, the proposed control strategy can effectively reduce losses, and the setting of the dead time selection boundary can not only reduce losses but also avoid the phenomenon of bridge arm through.

REFERENCE

[1] WANG Panpan,XU Zehan,WANG Li, et al A Hybrid Optimization Control Strategy of Efficiency and Dynamic Performance of Dual-Active-Bridge DC-DC Converter Based on Triple-Phase-Shift [J].Transactions of China Electrotechnical Society,2022,37(18):4720-4731.DOI:10.19595/j.cnki.1000-6753.tces.211525

[2] ZHANG Xiaowei,SU Xingyu,ZHOU Jinghua.Optimization Strategy of DAB Converter Based on Asymmetric Duty Modulation [J].Power Electronics,2023,57(12):117-120.

[3] BAN Guobang,ZENG Baobao,YUAN Xufeng,et al. Optimal Control of Dual Active Bridge Converters under Dual Phase Shift Control [J]. Electrical Automation,2023,45(05):96-98+102.

[4] Hua Bai,Mi, C.Eliminate Reactive Power and Increase System Efficiency of Isolated Bidirectional Dual-Active-Bridge DC–DC Converters Using Novel Dual-Phase-Shift Control[J].IEEE Transactions on Power Electronics, 2008, 23(6):2905-2914.DOI:10.1109/TPEL.2008.2005103.

[5] WANG Qi,ZHANG Zeke,LIU Ke,et al. Global Optimization Control of Minimum Backflow Power for Dual-active-bridge Converters Under Dual-phase-shift Control [J].Power System technology ,2024,48(09):3921-3930.DOI:10.13335/j.1000-3673.pst.2023.1593

[6] SUN Biaoguang,LI Jingzheng,DENG Xuzhe. Optimal control of minimum current stress in a dual-active-bridge DC-DC converter under dual phase shift control [J]. Power System Protection and Control,2023,51(20):107-118.DOI:10.19783/j.cnki.pspc.230247

[7] Wu K , De Silva C W , Dunford W G .Stability Analysis of Isolated Bidirectional Dual Active Full-Bridge DC–DC Converter With Triple Phase-Shift Control[J].IEEE Transactions on Power Electronics, 2012, 27(4):2007-2017.DOI:10.1109/TPEL.2011.2167243.

[8] GONG Linxiao,LI Wenhui,XU Junzhong,et al. Hybrid Phase Shift Control Strategy for High-frequency DAB Converter Based on Multi-objective Optimization [J]. Proceedings of the CSEE,2024,44(04):1517-1534.DOI:10.13334/j.0258-8013.pcsee.223036

[9] HUANG Yunfei,ZHONG Qihao,OU Yangyoupeng,et al.Overview of Topologies and Control Strategies for Dual-active-bridge Converters[J].Journal of Power Supply,2024,22(04):53-65.DOI:10.13234/j.issn.2095-2805.2024.4.53

[10] Juan,David,Páez,et al.Overview of DC–DC Converters Dedicated to HVdc Grids[J].IEEE Transactions on Power Delivery, 2018.DOI:10.1109/TPWRD.2018.2846408.

[11] HU Yujie,LI Zixin,ZHAO Cong,et al. Mechanism Analysis and Suppression of Oscillation in Dead Time of Series Resonant Dual Active Bridge Based on MOSFET [J].Transactions of China Electrotechnical Society,2022,37(10):2549-2558.DOI:10.19595/j.cnki.1000-6753.tces.201485

[12] LI Jin,LIU Jinjun,Dushan Boroyevich.Adaptive Dead-time Control Scheme for High-switching-frequency Dual-active-bridge Converter[J].Power Eletronics,2017,51(12):41-45.

[13] QIN Haihong,WANG Wenlu,XIE Sixuan,et al. A Novel Dead-time Adaptive Control Method for GaN-based Motor Drive [J]. Proceedings of the CSEE,2023,43(11):4422-4434.DOI:10.13334/j.0258-8013.pcsee.213277

[14] ZHI Qianjin,ZHANG Yanxin,LI Hongchang,et al. Dead Time Optimization of ZVS Full-bridge Converter With Resonant Loads[J]. Power Electronics,2023,57(05):119-122+137.

Simulation study of breakdown characteristic of vertical diamond Schottky barrier diodes with different drift layer parameters and termination structures

Tianhe Mi
Shenzhen Pinghu
Laboratory
Shenzhen, China
mitianhe@phlab.com.cn

Peng Wang
Shenzhen Pinghu
Laboratory
Shenzhen, China
wangpeng@phlab.com.cn

Yan Liu
Shenzhen Pinghu
Laboratory
Shenzhen, China
liuyan@phlab.com.cn

Teng Jiao
Shenzhen Pinghu
Laboratory
Shenzhen, China
jiaoteng@phlab.com.cn

Xinchun Cui
Shenzhen Pinghu
Laboratory
Shenzhen, China
cuixinchun@phlab.com.cn

Haolin Hu
Shenzhen Pinghu
Laboratory
Shenzhen, China
huhaolin@phlab.com.cn

Yuxi Wan
Shenzhen Pinghu
Laboratory
Shenzhen, China
wanyuxi@phlab.com.cn

Daohua Zhang
Shenzhen Pinghu
Laboratory
Shenzhen, China
zhangdaohua@phlab.com.cn

Abstract—This paper reports the effects of drift layer thickness, doping concentration, and terminal structures on the breakdown voltage and on-resistance of the vertical diamond Schottky barrier diodes (SBDs) via simulation. The results reveal that the breakdown voltage increases with the increase of the drift layer thickness and the decrease of the doping concentration until it approaches saturation. Meanwhile, the optimal drift layer length is obtained to balance the breakdown voltage and on-resistance. Through meticulous design and optimization of various terminal structures, the breakdown voltage of the SBD is significantly increased from the initial 490 V to 1980 V, while the maximum electric field in the device is effectively reduced to 3.1 MV/cm. This work provides crucial theoretical basis and technical support for the design and manufacturing of high performance power devices.

Keywords—Diamond, Schottky barrier diode, edge termination, breakdown, on-resistance, TCAD

I. INTRODUCTION

In modern power electronics, power devices with higher performance based on various semiconductor materials have been developed to achieve higher energy utilization efficiency[1]. Among diverse materials, diamond stands out for its remarkable properties. With a wide bandgap of around 5.5 eV, much larger than Silicon (1.12 eV), Silicon carbide (3.26 eV) and Gallium nitride (3.4 eV)[2],[3], diamond offers excellent electrical insulation, enabling high voltage operation with low leakage. In addition, diamond offers extremely high thermal conductivity, up to 2200 W/(m·K)[4], ensuring efficient heat dissipation and improving device reliability under high power conditions[5],[6].

Schottky barrier diodes (SBDs), with low on-resistance, fast switching speed and reverse recovery, are widely applied in power systems. The optimization of SBD parameters and structure to achieve the enhancement of breakdown voltage and internal breakdown electric field is essential for improving the utilization of the epitaxial layer while ensuring the high-voltage application. Wherein, the epitaxial layer thickness and doping concentration need to be carefully adjusted to realize the trade-off between the device breakdown voltage and on-resistance. Further, to alleviate the electric field concentration effect at the anode edge of the SBDs,

termination structures need to be employed. The use of TCAD in the device design can greatly reduce the experiment cost and promote the shift from experience-driven to data-driven device design.

Thus, in this study, the effects of drift layer thickness, doping concentration, and different terminal structures on the breakdown voltage and on-resistance of diamond SBDs were comprehensively investigated via TCAD to deepen the understanding of the breakdown mechanisms in diamond SBDs and contribute to the development of more efficient and reliable high voltage diamond based power devices.

II. MAIN MODELS AND PARAMETERS

In our simulations, various physically based models and their parameters, which are applicable to diamond materials, have been implemented first. The physical models adopted in the simulation mainly include the effective density of states, energy band model, mobility model, SRH recombination model, incomplete ionization model, and most importantly, the impact ionization model. The electron-hole pair generation rate G_{impact} from impact ionization at high electric fields is given by :

$$G_{impact} = \alpha_n \times n \times v_n + \alpha_p \times p \times v_p \qquad (1)$$

where v_n (v_p) is the electron (hole) velocity and n (p) is

TABLE I.   IMPACT IONIZATION COEFFICIENTS FOR DIAMOND

Parameters	Values
$a_n(\text{cm}^{-1})$	1.89E5
$a_p(\text{cm}^{-1})$	5.48E6
$b_n(\text{V/cm})$	1.7E7
$b_p(\text{V/cm})$	1.42E7
c_n	1
c_p	1

the concentration of electron (hole), and α_n (α_p) represents the ionization rate for electron (hole). Besides, based on the

979-8-3315-1110-4/25 $31.00 © 2025 IEEE

following relationship between the impact ionization coefficient of electrons $\alpha_{n,p}$ and electric field F_{ava}:

$$\alpha_{n,p}(F_{ava}) = a_{n,p}exp[-\left(\frac{b_{n,p}}{F_{ava}}\right)^{c_{n,p}}] \qquad (2)$$

the change of $\alpha_{n,p}$ with F_{ava} can be determined. There are various options for the correlation coefficient, such as Rashid coefficients[7], Isberg coefficients[8], Watanabe coefficients[9], Hiraiwa coefficients[10], Kamakura coefficients[11] etc. In this work, the Rashid's coefficient is adopted, for it's fitting parameters were obtained via extrapolation of coefficient parameters from silicon and silicon carbide based on the bandgap of diamond, and the breakdown field strength obtained by simulation calculations using this coefficient, which is about 2MV/cm, is very close to that reported in many literatures[12][13][14]. The relevant parameters are shown in Table 1.

III. RESULTS AND DISCUSSION

Simulations were conducted on diamond SBDs with different drift layer parameters and different terminal structures.

A. The influence of drift layer parameters on device characteristics

The drift layer thickness and doping concentration need to be carefully designed before the devices fabrication since they have a significant impact on the reverse blocking and forward conduction characteristics. Based on the simulation models in section II, the effects of drift layer parameters including drift layer length and doping concentration on the breakdown voltage of ideal parallel-plane diamond SBD are studied. The simulation structure and parameter Settings are shown in Fig. 1. The simulation results of breakdown voltage as a function of drift layer length under different doping concentrations and the electric field strength distribution varying with drift layer length at a fixed doping concentration are shown in Fig. 2.

As shown in Fig. 2, the breakdown voltage increase with the thickness of the drift layer within a certain range. In particular, when the doping concentration is set to 5E15 cm^{-3}, the breakdown voltage rises from 720 V to 1380 V and gradually reaches saturation as the thickness of the drift layer increases from 4 μm to 14 μm, which becomes more pronounced as the doping concentration of the drift layer decreases. This is due to the transition of the device from a punch through to a non-punch through mode, as shown in Fig. 2(b). In the non-punch through mode, as the drift layer thickness is further increased, the electric field no longer extends to the heavily doped substrate, which results in the device's breakdown voltage no

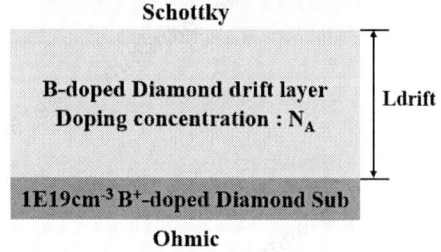

Fig. 1. The schematic structure and parameters of the SBD for simulation.

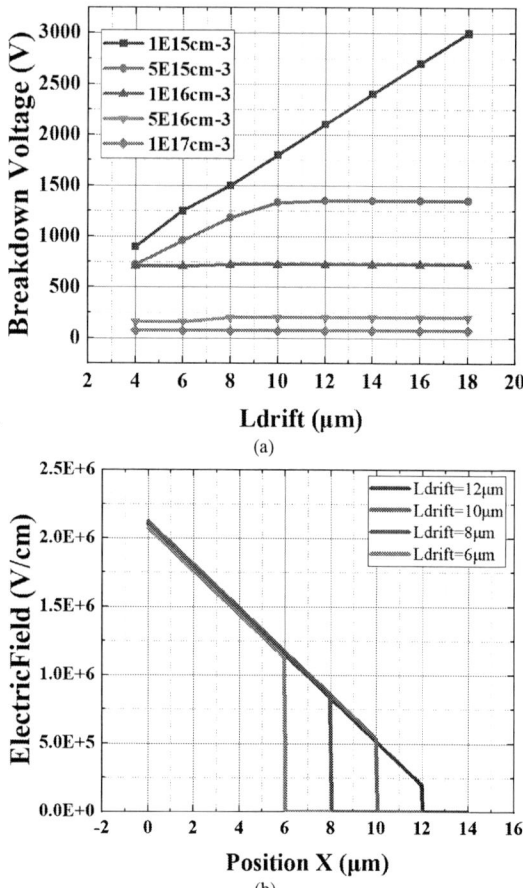

Fig. 2. The variation of breakdown voltage with thickness of drift layer under different doping concentrations(a) and the variation of electric field distribution with the thickness of the drift layer at doping concentration=5E15 cm^{-3}(b).

longer increasing with the increase of drift layer thickness. When designing a device, the breakdown voltage should be considered first. As shown in Fig. 2, the influence of doping concentration on the breakdown voltage is significantly higher than that of thickness, therefore, for high-voltage devices, the doping concentration of the drift layer, which directly determines the upper limit of the breakdown voltage, should be determined first. The thickness should then be designed to meet the voltage requirement while considering its effect on the on-resistance. From a comprehensive consideration, there must be a "perfect" drift layer thickness to achieve the minimum on-resistance while meeting the voltage requirements, and further, to achieve the maximum Baliga's Figure of Merit (BFOM) under a specific doping concentration of the drift layer. The influence of the drift layer thickness on the on-resistance was simulated. Based on this, BFOM of the device was calculated by combining the breakdown voltage and on-resistance, the results are shown in Fig. 3.

As shown in Fig.3, the on-resistance of the device increases approximately linearly as the thickness of the drift layer increases. However, the breakdown voltage does not exhibit a linear relationship with the drift layer thickness as shown in Fig. 2. Consequently, there exists a unique drift layer thickness at which the BFOM of the device reaches its

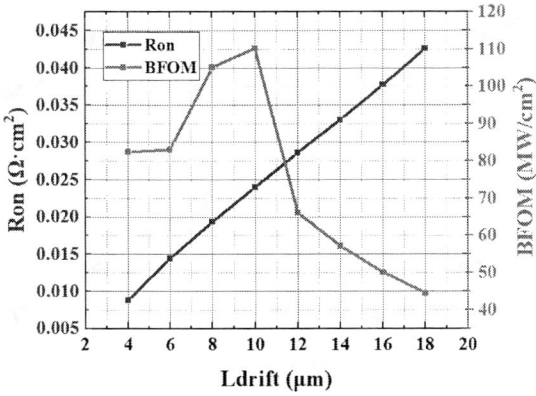

Fig. 3. The variation of on-resistance and BFOM of the device with thickness of the drift layer at doping concentration=5E15 cm-3.

maximum value. According to the simulation results, this optimal thickness is determined to be 10 μm when the doping concentration of the drift layer is 5E15 cm^{-3}, and the BFOM reaches 110 MW/cm^2. This conclusion is consistent with the results of some previous studies[10][15], which demonstrates the correctness of our simulation. Therefore, the doping concentration and thickness of the drift layer need to be elaborately designed to achieve optimal performance.

B. The influence of terminal structure on breakdown voltage

The terminal structure can effectively alleviate the electric field crowding effect at the anode edge and thus significantly increase the device breakdown voltage. Therefore, in this study, the effect of the mesa and field plate structure on the device breakdown voltage is investigated, as shown in Fig. 4, where the dielectric used in the field plate is SiO2, with the field plate length set to 20 μm and the thickness of 1 μm. Throughout the simulation, the drift layer thickness and doping concentration are set to 10 μm and 1E15 cm-3, respectively. The breakdown voltages corresponding to different device structures obtained by simulation are shown in Fig. 5. It should be noted that a series resistor with a resistance of 1E13 Ω was connected in series at one end of the SBD device to ensure the convergence of the simulation, and the voltage was applied across both the series resistor and the SBD device. Consequently, after breakdown, the internal current of the device stabilized at a relatively low order of 1E-

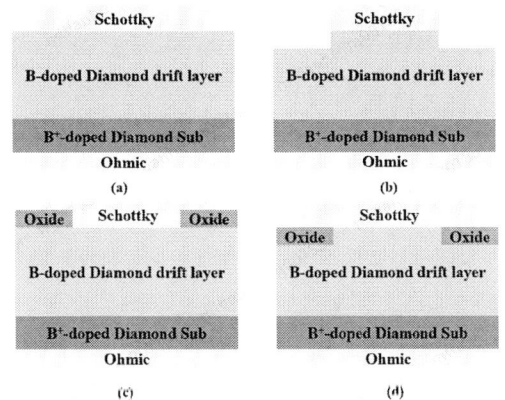

Fig. 4. The schematic structures of the traditional planar SBD (a), the SBD with mesa (b), the SBD with SiO2 field plate (c), and the SBD with planar mesa (d), respectively.

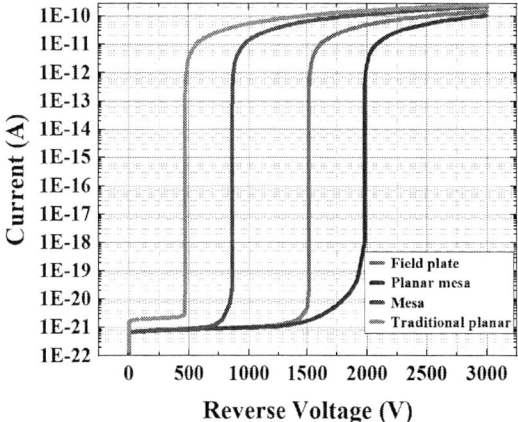

Fig. 5. The breakdown curves of the four structures and the distribution of internal electric fields.

10 A, while its magnitude increased by approximately 11 orders of magnitude compared to that before breakdown.

As depicted in Fig. 5, the breakdown voltages have been elevated from 490 V to 800 V, 1500 V, and 1980 V, respectively, upon the application of termination structures, which indicates the effectiveness of the termination structure in increasing the breakdown voltage of the device, attributed to the reduction or shifting of the peak electric field at the anode edge, as illustrated in Fig. 6.

In the structure depicted in Fig. 6, the device with the

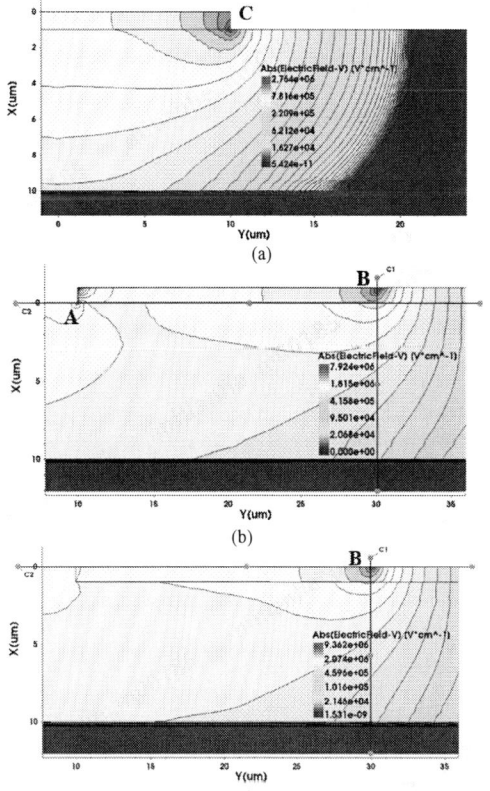

Fig. 6. The Electric field distribution in mesa structure(a), field plate structure(b), and planar mesa structure(c) at the bias of 1200 V.

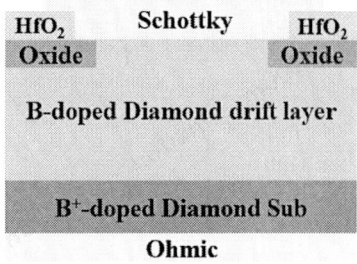

Fig. 7. The schematic structure of the SBD with a composite structure consisting of a field plate and a planar mesa.

planar mesa configuration achieves the maximum breakdown voltage of 1980 V. This is attributed to the fact that in the mesa and field plate structures, electric field peaks exist at Points C and A on the diamond surface, respectively. By contrast, in the planar mesa structure, there is no longer an electric field peak on the diamond surface, which has shifted to to the point B on the surface of SiO_2 with a low dielectric constant. However, it should be noted that the peak electric field at point B in the dielectric is as high as 9.36 MV/cm in this device as shown in Fig. 6(c), which may lead to premature breakdown of the dielectric since the theoretical breakdown field of diamond is higher than that of most dielectrics, which is up to 10 MV/cm. Because the normal component of the displacement vector is required to be continuous at the interface of different media, according to:

$$\varepsilon_1 E_1 = \varepsilon_2 E_2 \qquad (3)$$

where ε and E are the dielectric constant and electric field of the two materials, respectively. Therefore, to reduce the risk of premature breakdown of SiO_2, HfO_2 with higher relative permittivity (approximately 24) was chosen to be deposited on the surface of SiO_2, and Schottky contact is deposited to cover part of SiO_2 and HfO_2 to form a composite structure consisting of a field plate and a planar mesa as shown in Fig. 7.The SBD with a composite structure consisting of a field plate and a planar mesa shown in Figure 7 was simulated, and its internal electric field distribution are shown in Fig. 8.

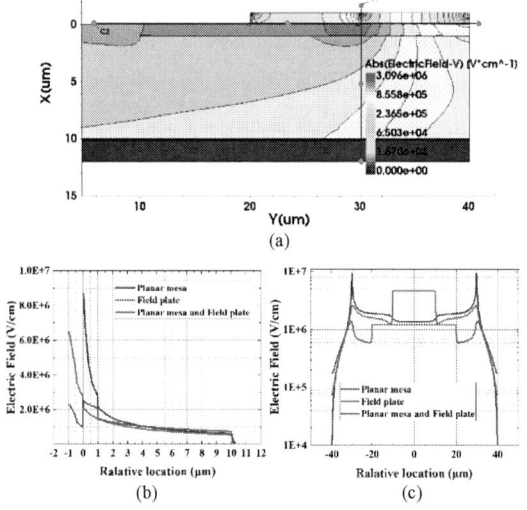

Fig. 8. The Electric field distribution in device with a composite structure (a) and the curves of electric field distribution along C1(Y=30 μm) (b) and C2(X=0 μm) (c) in different structure at the bias of 1200 V.

From Fig. 8, by adopting the composite structure, the position of the maximum electric field of the device is shifted from the edge of the Schottky contact to the inside of the dielectric with lower dielectric constant at 1200 V reverse bias according to (3). Meanwhile, the peak electric field decreasing from 9.36 MV/cm to 3.1 MV/cm. as well as the distribution of the electric field becomes more uniform.

Unfortunately, two-dimensional (2D) simulation cannot model three-dimensional corner effects (such as electric field concentration at device edges), which may overestimate the breakdown voltage to a certain extent. For example, premature breakdown may occur at the corner of the field oxide layer in planar devices due to three-dimensional electric field distortion, but this phenomenon cannot be captured by 2D simulation. We will conduct 3D simulations and supplement them with relevant experiments to calibrate and validate the simulation results in future work. The related findings will be presented in the following publications.

IV. CONCLUSION

In summary, the breakdown characteristics and the BFOM of diamond SBDs with different drift layer parameters and termination structures are studied by TCAD simulation. Breakdown voltages and BFOM are obtained for devices with different drift layer parameters. Besides, by employing different termination structures, a significant improve in breakdown voltage is realized.

ACKNOWLEDGMENT

This work is supported by the Shenzhen Pinghu Laboratory Project (Grant No. 225030).

REFERENCES

[1] J. Millán, P. Godignon, X. Perpiñà, A. Pérez-Tomás, J. Rebollo, "A Survey of Wide Bandgap Power Semiconductor Devices," in IEEE Transactions on Power Electronics, vol. 29, pp. 2155-2163, May 2014.

[2] B. Ozpineci, "Comparison of Wide-Bandgap Semiconductors for Power Electronics Applications," , Jan. 2004.

[3] A. Denisenko, E. Kohn, "Diamond power devices. Concepts and limits," Diamond and related materials, vol. 14, pp. 491-498, 2005.

[4] N. Donato, N. Rouger, J. Pernot, G. Longobardi, F. Udrea, "Diamond power devices: state of the art, modelling, figures of merit and future perspective," Journal of Physics D: Applied Physics, vol. 53, 093001.

[5] H. Umezawa, M. Nagase, Y. Kato, and S. I. Shikata, "High temperature application of diamond power device." Diamond and related materials, vol. 24, pp. 201-205, 2012.

[6] H. Umezawa, "Diamond Semiconductor Devices for harsh environmental applications," 2022 6th IEEE Electron Devices Technology & Manufacturing Conference (EDTM), Oita, Japan, pp. 297-299, March, 2022

[7] S. J. Rashid, A. Tajani, D. J. Twitchen, L. Coulbeck, F. Udrea, T. Butler, "Numerical Parameterization of Chemical-Vapor-Deposited (CVD) Single-Crystal Diamond for Device Simulation and Analysis," in IEEE Transactions on Electron Devices, vol. 55, pp. 2744-2756, Oct. 2008

[8] J Isberg, M Gabrysch, A Tajani, DJ Twitchen, "Transient current electric field profiling of single crystal CVD diamond," Semiconductor science and technology, vol. 21, pp. 1193, July, 2006

[9] T Watanabe, M Irie, T Teraji, T Ito, Y Kamakura, K Taniguchi, "Impact excitation of carriers in diamond under extremely high electric fields." Japanese Journal of Applied Physics, vol. 40, pp. L715, 2001.

[10] A Hiraiwa, H Kawarada, "Blocking characteristics of diamond junctions with a punch-through design." J. Appl. Phys. Vol. 117, pp. 124503, 2015.

[11] Y Kamakura, T Kotani, K Konaga, N Minamitani, G Wakimura, N Mori, "Ab initio study of avalanche breakdown in diamond for power device applications." 2015 IEEE International Electron Devices Meeting (IEDM). IEEE, pp. 5.2.1-5.2.4., 2015.

[12] P Reinke, F Benkhelifa, L Kirste, H Czap, L Pinti, V Zürbig, V Cimalla, C Nebel, O Ambacher, "Influence of Different Surface Morphologies on the Performance of High-Voltage, Low-Resistance Diamond Schottky Diodes," in IEEE Transactions on Electron Devices, vol. 67, pp. 2471-2477, June 2020.

[13] X.X. Yu, R.Z. Wang, B. Qiao, Z.H. Li, R. Sheen, J.J. Zhou, L.K. Zhou, "High Current Diamond Power Schottky Barrier Diode." VACCUM ELECTRONICS, vol. 5, pp. 59-63, May 2024.

[14] D. Zhao, Z.C. Liu, J. Wang, W.Y. Yi, R.Z. Wang, K.Y. Wang, "Performance Improved Vertical Diamond Schottky Barrier Diode With Fluorination-Termination Structure," in IEEE Electron Device Letters, vol. 40, pp. 1229-1232, Aug. 2019.

[15] H. Huang, J. Huang, H. Hu, J. Cheng and B. Yi, "Analytical Models of Breakdown Voltage and Specific On-Resistance for Vertical GaN Unipolar Devices," in IEEE Access, vol. 7, pp. 140383-140390, 2019.

Multi-parameter Degradation Modeling Method for Power MOSFETs Integrating Semiconductor Physics Degradation Data

Chenyi Wang
Department of Electrical Engineering
& Automation
Harbin Institute of Technology
Harbin, China
23b906054@stu.hit.edu.cn

Cen Chen
Department of Electrical Engineering
& Automation
Harbin Institute of Technology
Harbin, China
macchan_ee@hit.edu.cn

Weixuan Kong
Department of Electrical Engineering
& Automation
Harbin Institute of Technology
Harbin, China
2021113187@stu.hit.edu.cn

Haodong Wang
Department of Electrical Engineering
& Automation
Harbin Institute of Technology
Harbin, China
23S006027@stu.hit.edu.cn

Zhenning Zhou
Department of Electrical Engineering
& Automation
Harbin Institute of Technology
Harbin, China
2022111538@stu.hit.edu.cn

Abstract—**Power MOSFETs is the most important component of electronic system, and their reliability directly determines the service effectiveness of electronic system. In order to evaluate reliability of power MOSFETs, some power MOSFETs degradation modelling methods have been proposed. However, most of these methods focus only on the degradation law of a single external parameter of the power MOSFETs, but ignore the relationship between the degradation of multi-parameter of the power MOSFETs and the physical nature behind the parameter degradation. The reliability assessment results based on the single parameter degradation model have a large deviation from the actual situation. This paper proposes a multi-parameter degradation modeling method for power MOSFETs that integrates semiconductor physics and multi-dimensional degradation data. The physical properties of power MOSFETs are analyzed by combining semiconductor physics principles with test data. This approach allows for the extraction of degradation information related to multiple parameters of power MOSFETs using the PSO optimization algorithm. Subsequently, a multi-parameter degradation model of power MOSFETs is established. The method proposed in this article considers the physical nature of the multi-parameter degradation process of power MOSFET, and obtains the multi-parameter degradation information of power MOSFET through the multi-dimensional degradation data of power MOSFET and the PSO optimization algorithm, so as to perform degradation modeling. The established degradation model describes the overall degradation trend of the power MOSFET through the degradation laws of multiple parameters. This research contributes to reliability assessment and maintenance planning of power MOSFETs from the perspective of actual physical information.**

Keywords—*Power MOSFET, Semiconductor Physics, Degradation Model, Multi-parameter, Particle Swarm Optimization*

I. INTRODUCTION (HEADING 1)

Power MOSFETs are crucial in high-power, high-frequency power electronic converters [1]. As key components, their parameter drift or failure can lead to severe consequences, especially since these converters often operate in demanding high-frequency, high-temperature, and high-

voltage environments [2]. Consequently, Power MOSFETs face significant reliability challenges from operational strains.

The predominant failure modes in MOSFETs are categorized into package and chip issues [3][4]. While package failures involve aspects like bonding wire fractures, chip failures include mechanisms such as Time Dependent Dielectric Breakdown (TDDB) [5], Bias Temperature Instability (BTI) [6], and Hot Carrier Injection (HCI) [7]. For power MOSFETs, with their vertical conductive channels and thick gate oxides, BTI is a particularly prevalent failure mechanism due to prolonged exposure to gate bias and high-temperature stress during practical applications. This article focuses on the degradation of N-type power MOSFETs under the BTI mechanism.

To characterize MOSFET behavior, various digital models, such as PSpice-based [8] or electrothermal models [9], have been developed to describe the relationship between external performance (e.g., voltage, current) and internal physical information. However, these digital models, while effective for performance simulation, have rarely been utilized to analyze or describe degradation processes. For describing product aging, degradation modeling methods are typically either data-driven or based on failure physics.

Data-driven degradation modeling generally relies on accelerated test data or historical data, often employing stochastic processes like Wiener or inverse Gaussian processes [10][11], to establish mathematical models. While these methods can track parameter drift, they often lack a deep consideration of the physical meaning behind the product's degradation process. This limitation is significant for power MOSFETs, where underlying physical information greatly impacts output performance, and multiple parameters degrade concurrently. Correlation analysis alone is insufficient to capture the intrinsic nature of this multi-parameter degradation.

Conversely, physics-of-failure modeling analyzes the product's failure mechanism to explore the relationship between underlying physical properties and external performance parameters [12][13]. This approach can

979-8-3315-1110-4/25 $31.00 © 2025 IEEE

characterize the degradation process from its fundamental nature and is suitable for multi-parameter co-degradation products like MOSFETs. However, developing such models solely through analytical physical principles can be exceedingly complex and lengthy.

In summary, current research often focuses on the drift of a single parameter for MOSFET degradation modeling, or establishes mathematical models lacking a description of actual internal physical parameter degradation. This can lead to reliability assessments that deviate significantly from the actual device conditions. Therefore, this paper proposes a multi-parameter degradation modeling method for power MOSFETs that integrates semiconductor physics and multi-dimensional degradation data. By incorporating multi-dimensional degradation data into a physics-informed digital model of the MOSFET and utilizing Particle Swarm Optimization (PSO) to extract key internal physical information, we develop a multi-parameter degradation model. This model considers both the multi-dimensional degradation data and the actual physical properties of the power MOSFET, aiming to better describe the overall aging state. The subsequent sections detail the physics-based digital model and parameter identification (Section 2), the experimental setup and multi-parameter degradation modeling results (Section 3), followed by conclusions and future prospects (Section 4).

II. Digital model of power MOSFET Considering Semiconductor Physics

This section first introduces the physical structure of the power MOSFET, establishes a numerical model linking its internal parameters to external behavioral characteristics, and proposes a parameter estimation method based on measured data.

A. Digital Modeling of Power MOSFETs

Using the vertical double-diffusion MOSFET (VDMOSFET) as an example, Fig. 1(left) illustrates its microstructure, while the right side depicts the key electrical topology, including equivalent parasitic components (resistance, capacitance, inductance) and the body diode. These elements collectively influence device performance through distinct mechanisms.

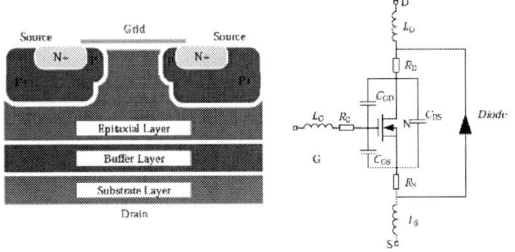

Fig. 1. The microphysical structure and electrical topology of power MOSFET.

The relationship between gate-source voltage (V_{gs}) and threshold voltage (V_{th}) governs the power MOSFET's operational states. The Shichman-Hodges model typically characterizes its voltage-current behavior: when $V_{gs} < V_{th}$, the device remains off ($I_d=0$); when $V_{gs} > V_{th}$, conduction initiates, with performance becoming $V_{ds(sat)}$-dependent.

Equation 1 defines the supersaturation voltage $V_{ds(sat)}$:

$$V_{ds}(sat) = V_{gs} - V_{th} \tag{1}$$

When the drain-source voltage V_{ds} is less than $V_{ds(sat)}$, the power MOSFET operates in the "linear region". When the drain-source voltage V_{ds} is greater than $V_{ds(sat)}$, the power MOSFET operates in the "saturation region". Therefore, the drain-source voltage can be expressed as Equation 2:

$$\begin{cases} I_d = 0, V_{gs} < V_{th} \\ I_d = K_p(V_{gs} - V_{th} - 0.5V_{ds})V_{ds}, V_{gs} \geq V_{th}, V_{ds} \leq V_{ds}(sat) \\ I_d = 0.5K_p(V_{gs} - V_{th})^2, V_{gs} \geq V_{th}, V_{ds} > V_{ds}(sat) \end{cases} \tag{2}$$

K_P is a parameter related to the power MOSFET process manufacturing, such as Equation 3：

$$K_P = \frac{W\mu_n C_{ox}}{2L} \tag{3}$$

Where W is the channel width, L is the channel length, C_{ox} is the oxide capacitance, and μ_n is the mobility. Equation 4 shows: Since the channel modulation effect increases with the increase of V_{ds}, the effective length of the channel decreases.

$$L' = \frac{L - \Delta L}{L} \tag{4}$$

Therefore, λ is introduced to adjust the reduction of the effective length of the channel in Equation 5:

$$\begin{cases} I_d = 0, V_{gs} < V_{th} \\ I_d = K_p(1 + \lambda V_{ds})(V_{gs} - V_{th} - 0.5V_{ds})V_{ds}, V_{gs} \geq V_{th}, V_{ds} \leq V_{ds}(sat) \\ I_d = 0.5K_p(1 + \lambda V_{ds})(V_{gs} - V_{th})^2, V_{gs} \geq V_{th}, V_{ds} > V_{ds}(sat) \end{cases} \tag{5}$$

The output characteristics of the power MOSFET can be well described by Equation 5, as shown in Fig. 2.

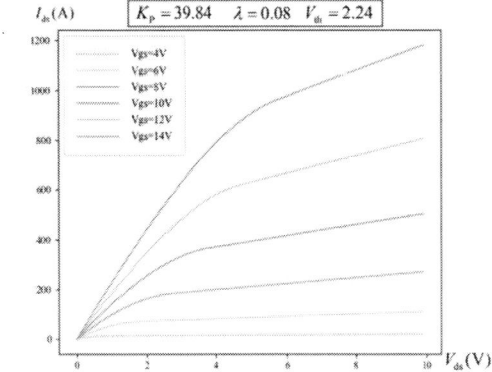

Fig. 2. Output characteristics of the power MOSFET obtained from digital model.

B. Parameter Identification Method Based on PSO

This paper proposes a Particle Swarm Optimization (PSO)-based parameter estimation method: Experimental I_{ds}-V_{ds} curves are acquired using a B1505A tester under three selected V_{gs} voltages (subject to instrument current limits).

The measured (V_{ds}, V_{gs}) values are input into the model to generate simulated I_{ds} (Fig. 3). The PSO algorithm iteratively optimizes model parameters within a reasonable range by minimizing the root mean square error (RMSE) between simulated and measured I_{ds} (Equation 6). PSO initializes a particle swarm with randomized positions (parameter solutions) and velocities (search capability), calculates the fitness value (f_{obj}, corresponding to RMSE in Equation 6) for each particle position during iterations, updates the global optimal solution P_{gd}, and maintains search dimensions consistent with the number of model parameters.

$$f_{obj} = RMSE(I_{ds}) = \sqrt{\frac{\sum_{j=1}^{N}[i_{ds(digitalmodel)} - i_{ds(realtest)}]^2}{N}} \quad (6)$$

According to the position of the optimal solution, the particle updates its speed and position according to the Equation 7. Among them, ω is called the inertia factor, C_1 and C_2 are called the acceleration constants, generally taken as [0, 4]. $random(0,1)$ represents a random number on the interval [0,1]. P_{id} represents the d_{th} dimension of the individual extreme value of the i_{th} variable. P_{gd} represents the d_{th} dimension of the global optimal solution.

$$V_{id} = \omega V_{id} + C_1 random(0,1)(P_{id} - X_{id}) + C_2 random(0,1)(P_{gd} - X_{id})$$
$$X_{id} = X_{id} + V_{id} \quad (7)$$

A numerical model is constructed for a power MOSFET to correlate the physical properties of the power MOSFET with its output characteristics. A numerical model parameter estimation method based on the PSO algorithm is also proposed. This method combines the power MOSFET digital model with the measured output characteristic data. It can estimate the intrinsic physical parameter information of the power MOSFET and provide support for power MOSFET degradation modeling.

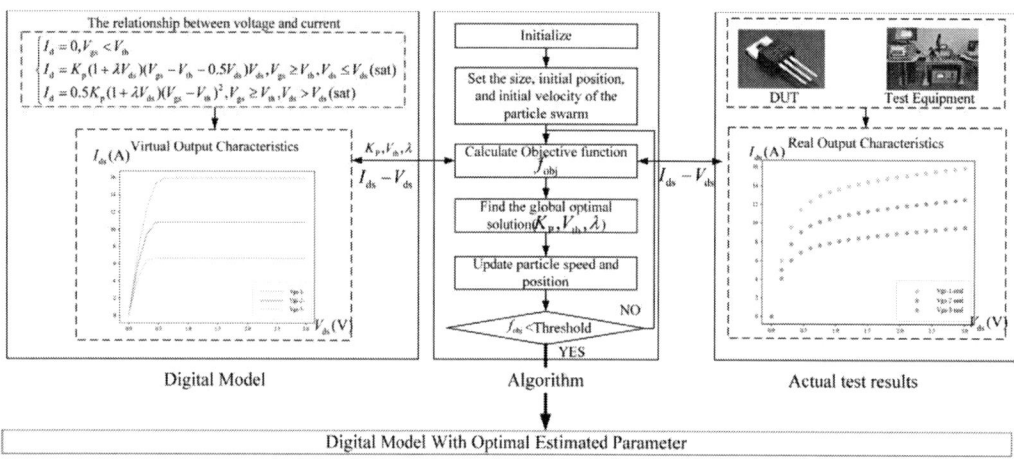

Fig. 3. Method route to integrate actual test data and digital models of power MOSFETs through PSO algorithm.

III. DIGITAL MODEL OF POWER MOSFET CONSIDERING SEMICONDUCTOR PHYSICS

In this section, an accelerated degradation test is conducted on the power MOSFET to obtain multi-dimensional degradation data. Through the parameter estimation method mentioned in the previous section, the degradation of the actual physical information of the power MOSFETs is obtained, and multi-parameter degradation modeling is performed. The model is finally represented by a numerical model of the power MOSFET, which can depict the overall degradation trend of the power MOSFET.

A. BTI Aging Experiment Setup

Electro-thermal stress (V_{gs} = 29V, 120°C) was applied to 19 identical power MOSFETs using a DC power supply and thermal chamber (Fig. 4). Output characteristics (I_{ds}-V_{ds}) were measured every 24 hours after a 1-hour recovery period to isolate permanent degradation. Parameter estimation integrated multi-dimensional aging data (voltage, current, temperature) with the MOSFET numerical model, enabling multi-parameter degradation trend analysis. Automated testing systems ensured rapid device characterization. This approach validated the interdependencies between physical parameter drift (K_P, λ, and V_{th}) and electro-thermal aging, supporting robust degradation modeling.

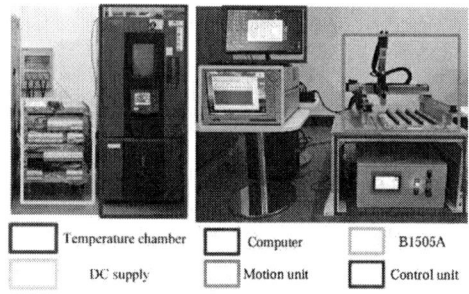

Fig. 4. Experimental and test devices.

B. Accelerated Test and Degradation Modeling

Fig. 5 gives the testing and modeling ideas during the aging process of power MOSFET. First, obtain multi dimensional power MOSFET degradation information (I_{ds}-V_{ds}) through the aging and testing equipment. Then, the power MOSFET degradation information is combined with the digital model, and the parameters (K_P, λ, and V_{th}) in the digital model are obtained through the PSO optimization algorithm.

These parameters are also internal physical information that characterizes the aging of the device.

For a single parameter, a power form degradation model is used to describe the degradation behavior. Combining the numerical model of the power MOSFET with the distribution characteristics of the parameter P, the degradation model of the power MOSFET output characteristics can be obtained, as shown in Equation 8.

After 43 days of continuous testing, the degradation trend of parameters KP, λ, and Vth are obtained as shown in Fig. 6.

After fitting, the model parameters are obtained as shown in Table 1. The prediction effect of the model under 95% confidence interval is shown in Fig. 7.

$$
\begin{cases}
\Delta V_{th} = A_1 t^{P_1} (P_1 \sim N(\mu_1, \theta_1^2)) \\
\Delta K_P = A_2 t^{P_2} (P_2 \sim N(\mu_2, \theta_2^2)) \\
\Delta \lambda = A_3 t^{P_3} (P_3 \sim N(\mu_3, \theta_3^2)) \\
OutputCharacteristic \begin{cases}
I_d = 0, V_{gs} < V_{th} \\
I_d = K_P(1 + \lambda V_{ds})(V_{gs} - V_{th} - 0.5V_{ds})V_{ds}, V_{gs} \geq V_{th}, V_{ds} \leq V_{ds}(\text{sat}) \\
I_d = 0.5K_P(1 + \lambda V_{ds})(V_{gs} - V_{th})^2, V_{gs} \geq V_{th}, V_{ds} > V_{ds}(\text{sat})
\end{cases}
\end{cases}
\tag{8}
$$

TABLE I.   TABLE TYPE STYLES

Parameter	A1	A2	A3	P1	P2	P3
Value	0.0566	-2.3784	0.0125	N(0.2153,0.257^2)	N(0.2074,0.0353^2)	N(0.2300,0.0436^2)

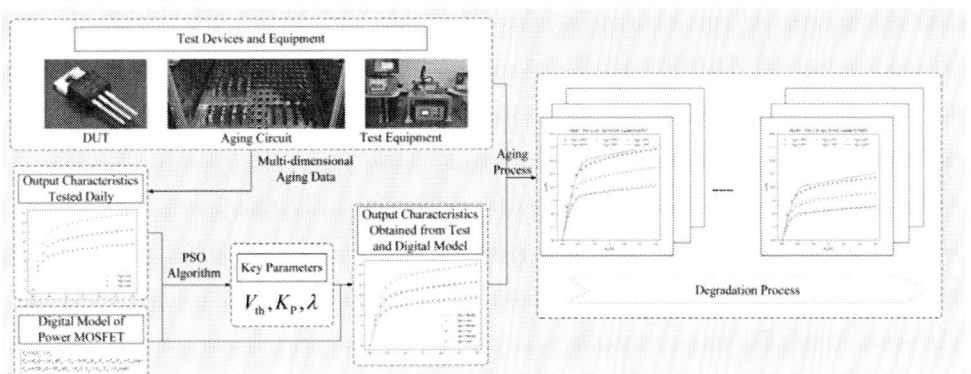

Fig. 5.   Degradation Modeling Process.

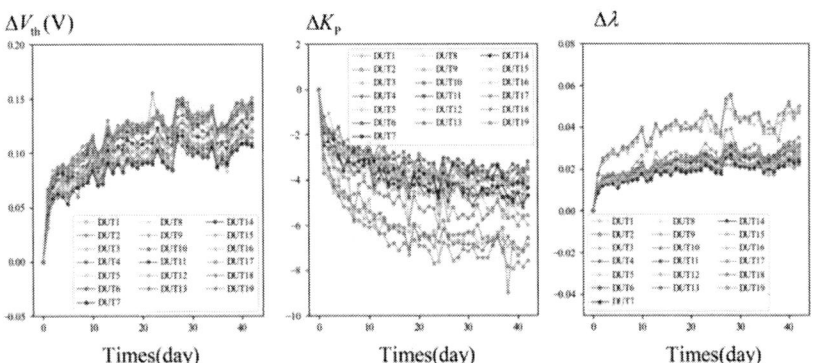

Fig. 6.   Degradation Modeling Process.

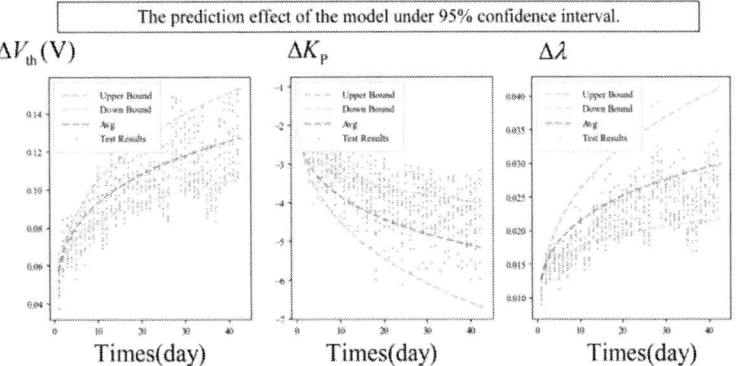

Fig. 7.   Degradation Modeling Process.

IV. Conclusions

This study presents a physics-informed approach to describe power MOSFET aging by integrating semiconductor physics with multi-dimensional aging data. By employing the Particle Swarm Optimization (PSO) algorithm, key physical parameters (K_P, λ, V_{th}) are estimated through the MOSFET digital model and experimental data (I_{ds}-V_{ds}-V_{gs}). The model accounts for time-dependent parameter distributions, enabling accurate characterization of multi-parameter degradation under electro-thermal stress. Experimental results validate the method's effectiveness in capturing interdependencies among degradation mechanisms and output characteristics. This framework establishes a synergy between failure physics and data-driven modeling, providing multidimensional reliability metrics that outperform traditional single-parameter evaluation methods.

References

[1] F. Yang, X. Tang, X. Wei, L. Sang, R. Liu, S. Bai, T. Peng, G. Zhao, P. Yang, T. Yang, and G. Tang, "Demonstrations of high voltage sic materials, devices and applications in the solid state transformer," *J. Cryst. Growth*, vol. 604, p. 127059, 2023.

[2] J. Wang and X. Jiang, "Review and analysis of sic mosfets' ruggedness and reliability," *IET Power Electron.*, vol. 13, no. 3, pp. 445–455, 2020.

[3] Y. Wang, Y. Ding, and Y. Yin, "Reliability of wide band gap power electronic semiconductor and packaging: A review," *Energies*, vol. 15, no. 18, 2022.

[4] B. J. Nel and S. Perinpanayagam, "A brief overview of sic mosfet failure modes and design reliability," *Procedia CIRP*, vol. 59, pp. 280–285, 2017.

[5] M. Kimura, "Field and temperature acceleration model for time-dependent dielectric breakdown," *IEEE Trans. Electron Devices*, vol. 46, no. 1, pp. 220–229, Jan. 1999.

[6] J. F. Zhang and W. Eccleston, "Positive bias temperature instability in mosfets," *IEEE Trans. Electron Devices*, vol. 45, no. 1, pp. 116–124, Jan. 1998.

[7] K.-L. Chen, S. A. Saller, I. A. Groves, and D. B. Scott, "Reliability effects on mos transistors due to hot-carrier injection," *IEEE Trans. Electron Devices*, vol. 32, no. 2, pp. 386–393, Feb. 1985.

[8] C. Leonardi, A. Raciti, F. Frisina, R. Letor, and S. Musumeci, "A new power mosfet model including the variation of parameters with the temperature," in *Proc. Unspecified Conf.*, Mar. 1998, pp. 261–266. (Conference name not provided in source)

[9] C. Sintamarean, F. Blaabjerg, and H. Wang, "A novel electro-thermal model for wide bandgap semiconductor based devices," in *Proc. Unspecified Conf.*, Sep. 2013, pp. 1–10. (Conference name not provided in source)

[10] K. A. Doksum and A. Hoyland, "Models for variable-stress accelerated life testing experiments based on wiener processes and the inverse gaussian distribution," *Theory Probab. Appl.*, vol. 37, no. 1, pp. 137–139, 1993.

[11] B. Yan, H. Wang, and X. Ma, "Correlation-driven multivariate degradation modeling and rul prediction based on wiener process model," *Qual. Reliab. Eng. Int.*, vol. 39, no. 8, pp. 3203–3229, 2023.

[12] D. E. Helling and B. Wong, "Applying a physics-of-failure model to predicting surface mount solder joint reliability," *Qual. Reliab. Eng. Int.*, vol. 7, no. 5, pp. 403–410, 1991.

[13] K. Upadhyayula and A. Dasgupta, "Physics-of-failure guidelines for accelerated qualification of electronic systems," *Qual. Reliab. Eng. Int.*, vol. 14, no. 6, pp. 433–447, 1998.

1.5KW HSC Converter Power Density-Efficiency Advancement: Enabled by Planar Transformer

Pengfei Wang
Delta Electronics
AEPCL-ShanghaiTech University
Shanghai, China
Pengfei.Wang@deltaww.com;
wangpf2024@shanghaitech.edu.cn

Vickie Qu
Delta Electronics
Shanghai, China
Vickie.Qu@deltaww.com

Qianru Shi
Delta Electronics
Shanghai, China
Qianru.Shi@deltaww.com

Qingchang Liu
Delta Electronics
Shanghai, China
Qingchang.Liu@deltaww.com

IAN YJ CHAN
Cyntec Tech-Delta Electronics
Shanghai, China
IAN.YJ.CHAN@Cyntec.com

Minfan Fan
AEPCL-ShanghaiTech University
Shanghai, China
fumf@shanghaitech.edu.cn

Abstract—Amid growing demands for high - efficiency power in data centers and enterprise servers, this paper presents a novel 1.5kW non - isolated, unregulated step - down Hybrid Switched Capacitor (HSC) converter. The HSC converter, with a unique topology using only two - winding for 4:1 voltage conversion, reduces design complexity. The proposed planar transformer, featuring an offset core and optimized windings, cuts losses effectively. Experiments show a peak efficiency of ~98.2% and full - load efficiency of 97.4%. A 60ns dead time enables MOSFET zero - voltage switching (ZVS), and experimental waveforms match simulations, validating the approach's practicality for high power density applications.

Keywords—1.5KW HSC converter, Planar transformer, Efficiency, ZVS.

I. INTRODUCTION

In the current digital era, data centers and enterprise servers are growing exponentially, which has led to an urgent demand for ultra - efficient power supplies. Given the surging power requirements, along with strict energy - efficiency and spatial constraints [1-3], understanding the technological driving forces and trends in computer/telecom power supplies is crucial for formulating advanced power - conversion strategies. In distributed power systems, especially those centered on data centers, a variety of topological architectures and design considerations have emerged to optimize power delivery, where different converter configurations play a vital role.

Numerous previous research efforts have propelled the development of power - conversion technology. The 1MHz LLC converter in [4], integrating GaN devices and a refined transformer, achieves high efficiency and power density. GaN devices, with their excellent electron mobility and fast switching speed, enable higher frequency operation, reducing the size of passive components and enhancing power density. However, their high cost limits applications in cost - sensitive scenarios. The LLC resonant converter with a matrix transformer [5] enables efficient magnetic - flux distribution, but its complex design and control circuitry pose challenges in manufacturing and reliability. The LLC converter in [6] with an integrated planar matrix transformer for high - output - current applications benefits from good thermal performance, yet its production requires advanced skills, potentially increasing costs. The 3KW 400V - 48V LLC converter in [7], designed for data - center 48V power architecture, has circuit complexity issues. The high - density hybrid switched - capacitor converter in [8], novel non - isolated LLC resonant

converters in [9], and multi - phase buck converters with extended duty - cycle in [10] share common features that serve as a basis for innovation. Switched - tank converters [11] offer a compact solution but face capacitor - voltage - balancing problems. The 48V voltage - regulator module in [12] with a PCB - winding matrix transformer emphasizes innovative winding methods, but its performance depends on PCB - material quality. The adoption of gallium nitride (GaN) devices significantly enhances high-frequency performance in two-phase DC/DC converters, particularly at switching frequencies exceeding 2 MHz, thereby enabling the widespread application of planar transformers in high-frequency power modules [13-14]. For two-stage 48V voltage regulation architectures, the integration of resonant capacitors with series-resonant converters forms a hybrid DC transformer (DCX), while the intermediate bus converter (IBC) achieves wide input voltage adaptation (40V-60V range) through phase-shift regulation. Notably, both topologies leverage zero-voltage switching (ZVS) mechanisms independent of transformer magnetizing current, allowing simplification from conventional 4n:1 to n:1 transformer configurations, which collectively contribute to improved peak efficiency [15-16].

The 1.5KW HSC Converter represents a milestone in this field. This paper will conduct an in - depth design analysis, comprehensive evaluation, and experimental verification. The goal is to reveal how it improves power density and efficiency, contributing to the progress of power - converter - technology. The following sections will explore this topic in detail, from basic principles to practical implementation and performance assessment.

II. HSC TOPOLOGY ANALYSIS

In the hybrid switched capacitor converter (HSC) proposed in [8], when the voltage conversion ratio is 4:1, as shown in the circuit diagram of *Figure 1*, compared with the traditional multi - tapped autotransformer (MTA) adopted in [8], the structure of the transformer has changed from four sets of windings to only two sets of wind which are composed of an interleaved resonant capacitor Cr connected to a coupled - type transformer (CT). This change significantly reduces the design and PCB wiring complexity of the transformer, as well as the overall circuit complexity.

In this interleaved parallel circuit, two sets of resonant capacitors Cr with equal parameters and resonant inductors Lr work alternately within one cycle. It's worth noting that Lr is composed of the transformer leakage inductance and PCB parasitic inductance. This makes the resonant current flowing through the two sets completely symmetric, so the circuit can well ensure symmetry.

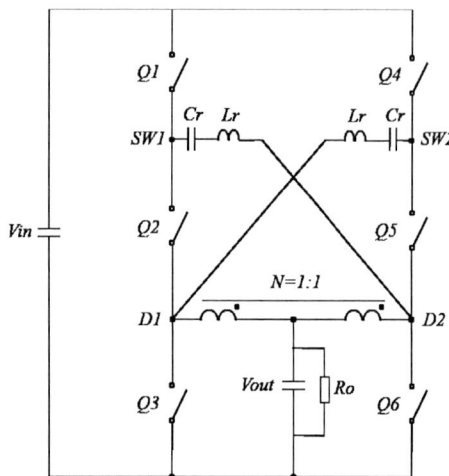

Figure. 1. HSC topology converter with coupled transformer

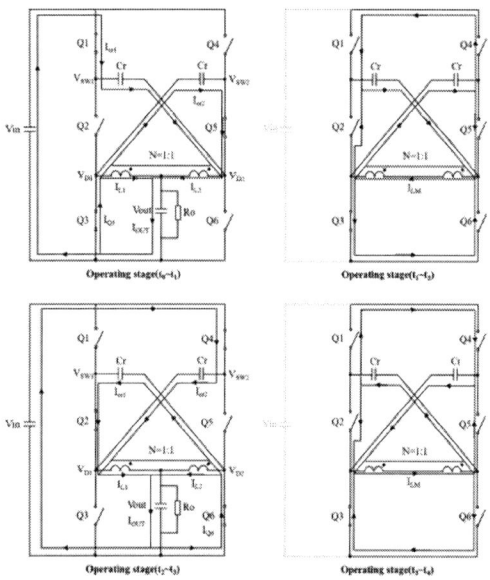

Figure. 2. Topological stages with coupled transformer

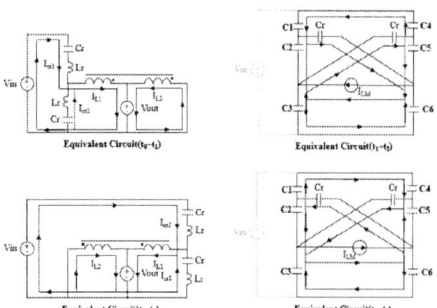

Figure. 4. Equivalent circuits for four subintervals

Taking into account the converter illustrated in *Figure 2* and the converter waveforms demonstrated in *Figure 3*, the operation mode of the hybrid switched capacitor converter (HSC) with a coupled transformer within one switching cycle from t_0 to t_4 will be analyzed through four sub-intervals. Meanwhile, modal analysis will be performed with reference to the equivalent circuit depicted in *Figure 4*, and state equations will be provided to calculate the voltage conversion ratio. The detailed description of the four sub-intervals is as follows:

a) $t_0 - t_1$: At time t_0, switches *Q1*, *Q3*, and *Q5* are turned on. As a result, there are three branches working simultaneously. Branch *1* is: $Vin \rightarrow Q1 \rightarrow Cr_1 \rightarrow Lr_1 \rightarrow L_2 \rightarrow Vout \rightarrow GND$. Branch *2* is: $Q3 \rightarrow Lr_2 \rightarrow Cr_2 \rightarrow Q5 \rightarrow L_2 \rightarrow Vout \rightarrow GND$. Branch *3* is: $Q3 \rightarrow L_1 \rightarrow Vout \rightarrow GND$. When the parameters of the resonant capacitors and resonant inductors in branches *1* and *2* are equal, the symmetry of the currents in these two branches can be ensured, that is, $I_{cr1} = I_{cr2}$. Consequently, the coupled transformer can maintain magnetic balance. Therefore, in practical applications, it is advisable to ensure that the resonant parameters of the two branches are as equal as possible.

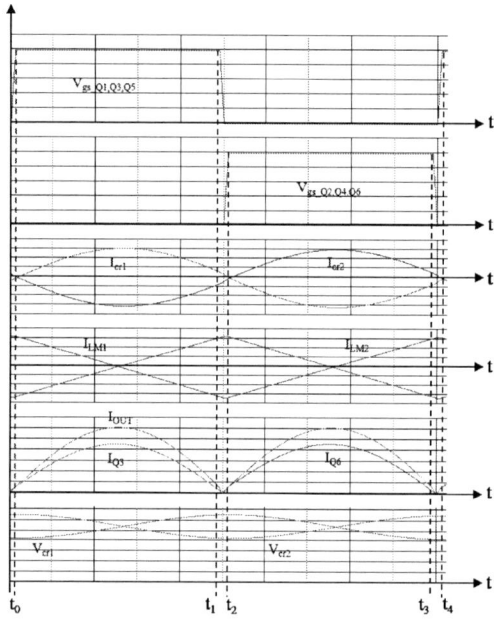

Figure. 3. Operation key waveform of HSC topology with coupled transformer

Since the turns ratio of the two windings of the coupling transformer is *1:1*, according to Kirchhoff's Voltage Law *(KVL)*, the winding voltage of the equivalent branch *3* is equal to *Vout*. Therefore, the winding voltages of branch *1* and branch *2* are also *Vout*. The derivation of the state equation for branch *1* is as follows:

$$Vin - 2Vout = vcr1 + Lr1\frac{dicr1}{dt} \qquad (1)$$

$$Cr1\frac{dvcr1}{dt} = icr1 \qquad (2)$$

Set the initial states in advance: $i_{cr1}|_{t=t0} = i_{cr}$, $v_{cr1}|_{t=t0} = v_{Cr}$.

$$w0 = \frac{1}{\sqrt{Lr1 \cdot Cr1}} \qquad (3)$$

$$Z0 = \sqrt{\frac{Lr1}{Cr1}} \qquad (4)$$

w_0 and Z_0 act as the characteristic parameters of this resonant circuit.

Establish a second-order linear ordinary differential equation based on the given state equations *(1)* and *(2)*. By substituting the initial parameters and equations *(3)* and *(4)*, the state equations of i_{cr1} and v_{cr1} can be obtained respectively as follows:

$$icr1 = \sqrt{icr^2 + \left(\frac{vcr - (Vin - 2 \cdot Vout)}{Z0}\right)^2} \cdot cos[w0 \cdot (t - t0)] \qquad (5)$$

$$vcr1 = \sqrt{(icr \cdot Z0)^2 + \left(vcr - (Vin - 2 \cdot Vout)\right)^2} \cdot sin[w0 \cdot (t - t0)] + Vin - 2 \cdot Vout \qquad (6)$$

Similarly, Establish the state equations for equivalent branch 2, and thus the following two state equations can be obtained:

$$2 \cdot Vout = vcr2 + Lr2\frac{dicr2}{dt} \qquad (7)$$

$$cr2\frac{dvcr2}{dt} = icr2 \qquad (8)$$

Simultaneously solve state equations *(7)* and *(8)* to establish a second-order linear ordinary differential equation. By substituting the initial parameters and equations *(3)* and *(4)*, the state equations of i_{cr2} and v_{cr2} can also be obtained respectively as follows:

$$icr2 = \sqrt{icr^2 + (\frac{2 \cdot Vout - vcr}{Z0})^2} \cdot cos[w0 \cdot (t - t0)] \qquad (9)$$

$$vcr2 = \sqrt{(icr \cdot Z0)^2 + (2 \cdot Vout - vcr)^2} \cdot sin[w0 \cdot (t - t0)] + 2 \cdot Vout \qquad (10)$$

Assume that $i_{cr1} = i_{cr2}$ and $v_{cr1} = v_{cr2}$. Then, by simultaneously solving equations *(5)* and *(9)*, as well as equations *(6)* and *(10)* respectively, the relational expressions between the resonant capacitor voltage, the output voltage and the input voltage can be derived respectively as follows:

$$Vin = 2 \cdot Vcr \qquad (11)$$

$$Vin = 4 \cdot Vout \qquad (12)$$

In the equivalent circuit within the time interval from t_0 to t_1, according to Kirchhoff's Current Law *(KCL)*, we can obtain that $i_{cr1} + i_{cr2} = i_{L1}$. Moreover, since $i_{L1} = i_{L2}$ and $i_{cr1} = i_{cr2}$, it follows that $i_{out} = 4 \cdot i_{cr}$, which means that the output current within one period is four times the input current.

b) $t_1 - t_2$:

During the time interval from t_1 to t_2, the dead-time period t_{dead} begins. Since the magnetizing inductance is much larger than the resonant inductance, the magnetizing current I_{LM} charges the parasitic capacitance C_1, C_3, and C_5, and discharges the parasitic capacitance C_2, C_4, and C_6. At t_2, the parasitic capacitance C_1 will be charged to $\frac{vin}{2}$, the parasitic capacitance C_3 will be charged to $2 \cdot Vout$, and the parasitic capacitance C_5 will be charged to Vin. Meanwhile, the parasitic capacitances C_2, C_4, and C_6 will be discharged to $0V$, which prepares for achieving Zero-Voltage Switching *(ZVS)* in the next moment. At this time, the peak value of the magnetizing current I_{Lm} that creates the conditions for realizing the *ZVS* of the MOSFETs is $\frac{Vout}{2 \cdot fs \cdot Lm}$, and the

magnetizing inductance Lm needs to satisfy the condition that $Lm \leq \frac{Ts \cdot tdead}{32 \cdot Vin \cdot Ceq}$.

c) $t_2 - t_3$: During the period from t_2 to t_3, when t equals t_2, switches Q_2, Q_4, and Q_6 are turned on under the condition of Zero-Voltage Switching *(ZVS)*. Similar to the working state in the first phase *(t_0 - t_1)*, there are also three branches, which are described as follows: Branch *1*: $Vin \rightarrow Q_4 \rightarrow C_{r2} \rightarrow L_{r2} \rightarrow L_1 \rightarrow Vout \rightarrow GND$. Branch *2*: $Q_6 \rightarrow L_{r1} \rightarrow C_{r1} \rightarrow Q_2 \rightarrow L_1 \rightarrow Vout \rightarrow GND$. Branch *3*: $Q_6 \rightarrow L_2 \rightarrow Vout \rightarrow GND$. It can be clearly observed from the simulation waveforms in Figure *6* that $i_{Cr2} = -i_{Crl}$, indicating that they have the same magnitude but opposite directions. The same situation holds for the magnetizing currents of the two sets of windings as well as the resonant capacitors on both sides. Consequently, the structure is symmetrical, and the equations derived in the first phase remain applicable. In other words, equations *(11)* and *(12)* can also be derived accordingly.

d) $t_3 - t_4$:

During this stage, the equivalent circuit in *Figure 4* is in the dead-time period, just like the second stage. However, the current directions for charging and discharging the parasitic capacitances are reversed. The parasitic capacitances C_2, C_4, C_6 start to be charged, while the parasitic capacitances C_1, C_3, C_5 start to be discharged until they reach 0V, which provides the conditions for achieving Zero-Voltage Switching *(ZVS)* of Q_1, Q_3, Q_5 in the next stage. At the end of the dead-time period, that is, at t4, the parasitic capacitance C_2 is charged to Vin, C_4 is charged to $\frac{Vin}{2}$, and C_6 is charged to $2 \cdot Vout$. Meanwhile, by observing the current waveform of Q_6 from *Figure 3*, it can be seen that the natural turn-off, namely Zero-Current Switching *(ZCS)*, is fully achieved.

Considering the working states of the four stages in total, Q_1, Q_2, Q_4, Q_5 can achieve *ZVS* but not *ZCS*. Q_3 and Q_6 can achieve both *ZVS* and *ZCS*. Moreover, for a voltage conversion ratio of *4:1*, the Hybrid Switched Capacitor *(HSC)* topology has only two sets of windings, which significantly reduces the complexity of transformer design. Throughout the entire cycle, the currents of the two sets of windings are always equal, effectively reducing the winding losses. Compared with the traditional *LLC* with the same conversion ratio, it has obvious advantages.

III. PROPOSED PLANAR TRANSFORMER DESIGN

Magnetic cores commonly use high - frequency power ferrite materials, which have low hysteresis and eddy current losses. When magnetic materials are repeatedly flipped in a magnetic field, energy is consumed during magnetic flipping, causing magnetic hysteresis loss. The hysteresis loop of ferrite materials has a special shape. Its narrow shape means that magnetic domain flipping during magnetization is relatively easy, with lower remaining magnetic induction strength and coercive force. This effectively reduces hysteresis loss. When the magnetic field passes through the magnetic core, an induced current (eddy current) is generated. Due to the high resistance of ferrite materials, the eddy current strength generated is relatively small, effectively reducing the eddy current loss.

Figure 5a presents the layout diagram of the column within the coil and core. When the circumference of the coil remains constant, the magnetic field generated by the core can penetrate the coil uniformly. As a result, the magnetic - flux density is consistent throughout the entire coil. This characteristic contributes to reducing energy losses and hysteresis, thereby enhancing the efficiency of the transformer.

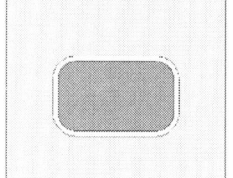 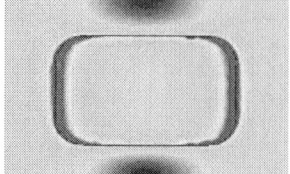

Figure. 5a. Planning diagram of the coil

Figure. 5b. B field cloud image of the core column

Figure 5b shows the B - field cloud image of the column in the magnetic core. Owing to the concentration of magnetic flux, the column in the magnetic core serves as the primary magnetic - flux path. According to the magnetic - circuit Ohm's law *(13)*, the magnetic resistance of the magnetic core

$$\phi = \frac{NI}{Rm} \qquad (13)$$

is significantly smaller than that of the surrounding air. When the magnetic flux passes through a region of concentration with a fixed cross - sectional area, the value of the magnetic induction intensity B in the surrounding area is relatively large.

Existing low - voltage products demonstrate high - current output characteristics, with the critical reference factor being the thickness of the coil. *Figure 6* shows the parallel layers under different core and coil thicknesses. As the number of layers' increases, the current distribution per layer decreases, thereby reducing the corresponding winding losses. The thickness of the core above and below is adjusted to maximize the number of parallel layers. *Table 1* shows the loss simulations for three scenarios indicate that subject to the height constraints of the product, when the thickness of the core above and below is set at *2.2mm* and the number of parallel layers reaches six, the loss can be minimized.

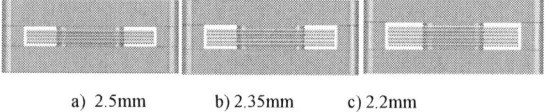

a) 2.5mm   b) 2.35mm   c) 2.2mm

Figure. 6. Model diagram of different pendulum thicknesses

Table. 1. Loss distribution of different pendulum thicknesses

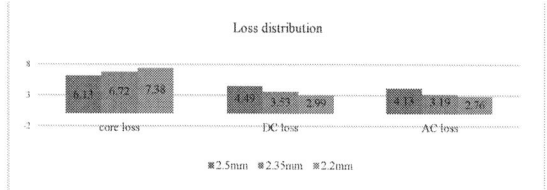

Figure 7 elucidates the internal windings of the transformer. The primary winding and the secondary winding are arranged in an alternating pattern, forming a sandwich - like structure with a turn ratio of *1:1*. This configuration effectively mitigates the winding losses and diminishes the leakage flux. The transformer employs a custom - tailored "UI" - type magnetic core design. The secondary winding is connected to the SR MOSFETs and the output capacitor. Thus, the output winding is typically designed as a single - turn, which streamlines the winding design and fabrication of the PCB, attaining the HSC design with a *12V* output.

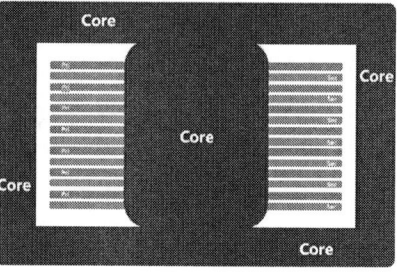

Figure. 7. Idealized model diagram of transformer

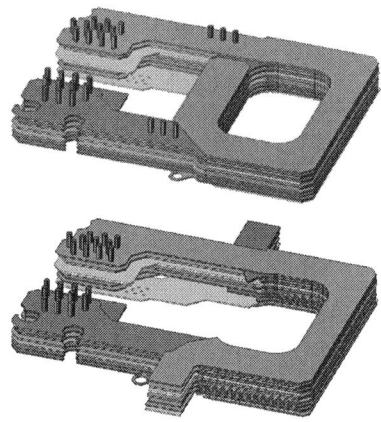

Figure. 8a. The winding structure of the transformer
Figure. 8b: Model diagram of short terminals

In the simplified model shown in *Figure 8a*, the current directions of the primary and secondary windings are opposite, as are the magnetic flux directions. This causes the magnetic flux to cancel out, achieving coupling. Maxwell simulation data in *Table 2* shows a significant increase in DC loss compared to the ideal model. To meet design requirements, the coil is connected to MOSFETs, whose position affects the coil length and thus the DC loss. Power module efficiency is mainly affected by transformer losses. Transformer AC losses are mainly caused by the skin and proximity effects, while DC losses come from winding resistance. *Figure 8b* shows an optimization of the original design. By shortening the column winding in the core, the DC loss is effectively reduced. *Table 2* shows both AC and DC losses have decreased. The long coil has a higher loss than the short one. *Table 2* shows the DC loss difference between the two cases is *1.17w*, indicating that during the transition from short to long, the DC loss increases more significantly than AC loss. This discovery is significant for power system design and efficiency improvement.

Table. 2 The loss distribution of the ideal model and the actual model

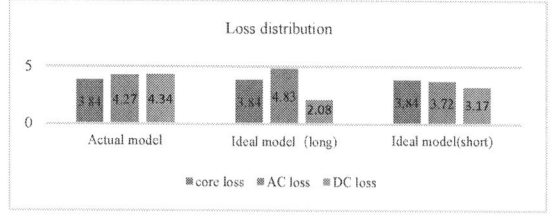

979-8-3315-1110-4/25 $31.00 © 2025 IEEE

IV. EXPERIMENTAL RESULTS

To validate the proposed approach, a *1500W*、 *4:1* module has been constructed and must be soldered onto a *230×120* mm evaluation board (EVB) for testing, as depicted in *Figure 9*. For *Q1, Q2, Q4,* and *Q5,* two *QE046N08LM5CGSC* double - sided heat − dissipating MOSFETs from Infineon are connected in parallel, while for *Q4* and *Q6,* three *IQE013N04LM6CGSC* double - sided heat - dissipating MOSFETs from Infineon are paralleled. To guarantee zero - voltage switching (ZVS) for all MOSFETs, a *100μm* air gap is introduced into the central magnetic core of the transformer, yielding an excitation inductance of *280nH. 6pcs GRM21BZ71H475KE15L X7R 0805* multilayer

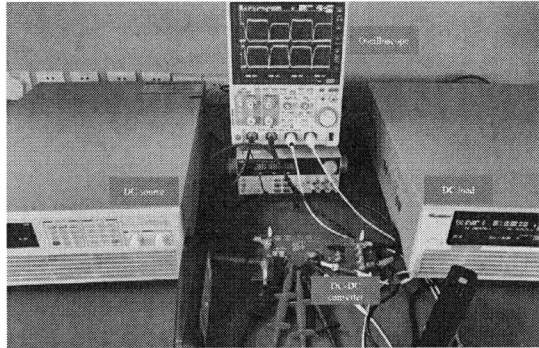

Figure. 9. Experiment Platform

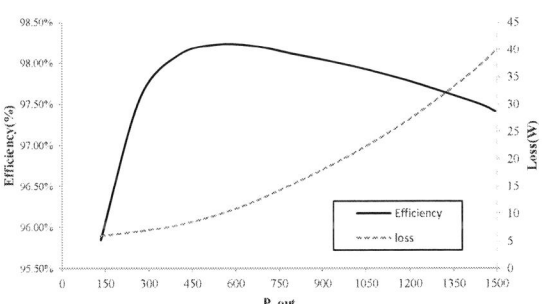

Figure. 10. Efficiency and power losses trends for the proposed 4:1 converter

ceramic capacitors (MLCCs) are respectively connected in parallel as resonant capacitors C_{r1} and C_{r2}. The two sets of windings, with a turns ratio of *1:1* and each having *7* turns, are wound in an interleaved parallel configuration. The printed circuit board (PCB) of this module is fabricated using a *16 - layer* board with a *3 - ounce* copper thickness. Experimentally, the peak efficiency reaches approximately *98.2%,* and the full - load efficiency is *97.4%.* The losses, encompassing MOSFET losses, driving losses, transformer losses, auxiliary power losses, and other losses (as shown in *Figure 10*), remain stable across the entire load spectrum.

Figure 11 showcases the primary experimental waveforms of the proposed module. These include the driving signals of *Q1, Q2, Q3* and the resonant waveform of the resonant capacitor Cr1. The experimental waveforms bear a remarkable resemblance to the simulated counterparts, validating the practical viability of the proposed theory. The dead - time is configured to be *60* ns. As depicted in the driving signals of *Q1* and *Q3,* along with the drain − source

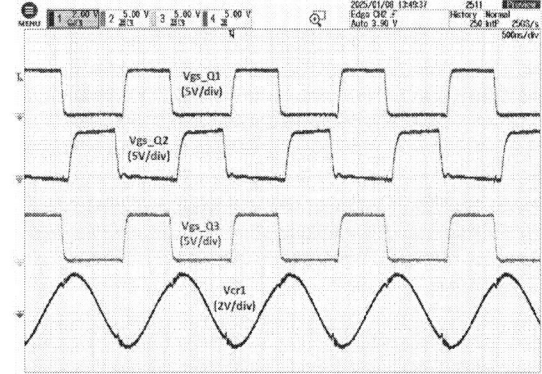

Figure. 11. *Q1、 Q2、 Q3* driver signal and resonant capacitor C_{r1} waveform at *Vin = 54V* with *Iout = 20A*

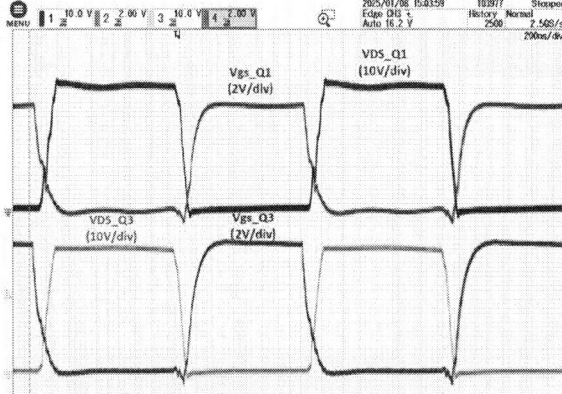

Figure. 12. *Vds、 Vgs* of *Q1* and *Vds、 Vgs* of *Q3* at *Vin = 54V* with *Iout = 116A*

(DS) voltages in *Figure 12*, it is evident that there is no intersection between the MOS driving signals and the DS voltages. Thus, a dead - time of *60* ns is entirely adequate for the MOS to accomplish zero - voltage switching (ZVS).

V. CONCLUSIONS

This paper explores the design, analysis, and experimental validation of a *1.5kW* HSC converter. Topological analysis shows it can achieve efficient power conversion with fewer windings and ensure ZVS for most switches. The optimized planar transformer boosts efficiency. Experimental results verify high efficiency performance across loads. The match between experimental and simulated waveforms validates the theory.

REFERENCES

[1] Efficient power supplies for data center and enterprise servers [Online]. Available: www.80plus.org.

[2] M. M. Jovanovic, "Technology drivers and trends in power supplies for computer/telecom," in Proc. APEC, 2006, Plenary session presentation.

[3] F. C. Lee, P. Barbosa, P. Xu, J. Zhang, B.Yang, and F.Canales, "Topologiesand design considerations for distributed power system applications," in Proc. IEEE, Jun. 2001, vol. 89, no. 6, pp. 939–950.

[4] Runruo Chen, Sheng-yang Yu. "A High-efficiency High-power-density 1MHz LLC Converter with GaN Devices and Integrated Transformer." in Proc. IEEE, 2018, pp. 791 - 796

[5] Daocheng Huang, Shu Ji, and Fred C. Lee. "LLC Resonant Converter With Matrix Transformer." IEEE TRANSACTIONS ON POWER ELECTRONICS, VOL. 29, NO. 8, AUGUST 2014, pp. 4339 - 4347.

[6] Chao Fei, Fred C. Lee, Qiang Li. "High-Efficiency High-Power-

Density LLC Converter With an Integrated Planar Matrix Transformer for High-Output Current Applications." IEEE TRANSACTIONS ON INDUSTRIAL ELECTRONICS, VOL. 64, NO. 11, NOVEMBER 2017, pp. 9072 - 9082.

[7] Ahmed Nabih, Qiang Li. "Low-Profile and High-Efficiency 3kW 400 V-48V LLC Converter with a Matrix of Four Transformers and Inductors for 48V Power Architecture for Data Centers." in Proc. IEEE 2021, pp. 1813 - 1819.

[8] Roberto Rizzolattiy, Christian Rainery, Stefano Saggini, Mario Ursino. "High Density Hybrid Switched Capacitor Converter for Data-Center Application." in Proc. IEEE 2021, pp. 1288 - 1293.

[9] Daocheng Huang, Xinke Wu, Fred. C. Lee. "Novel Non-isolated LLC Resonant Converters." in Proc. IEEE 2021, pp. 1373 - 1380.

[10] Yungtaek Jang, Milan M. Jovanović, and Yuri Panov. "Multi-Phase Buck Converters with Extended Duty Cycle." in Proc. IEEE 2006, pp. 38 - 44.

[11] Shuai Jiang, Chenhao Nan, Xin Li, Chee Chung, Mobashar Yazdani. "Switched Tank Converters." in Proc. IEEE 2018, pp. 81 - 90.

[12] Mohamed H. Ahmed, Chao Fei, Fred C. Lee, Qiang Li. "48-V Voltage Regulator Module With PCB Winding Matrix Transformer for Future Data Centers." IEEE TRANSACTIONS ON INDUSTRIAL ELECTRONICS, VOL. 64, NO. 12, DECEMBER 2017, pp. 9302 – 9310.

[13] Minfan Fu, Chao Fei, Yuchen Yang, Qiang Li, Fred C Lee."A GaN-Based DC/DC Module For Railway Applications:Design Consideration and High-Frequency Digital Control. " IEEE Transactions on Industrial Electronics, VOL. 67 ,Issue:2, February 2020, pp. 1638-1647.

[14] Minfan Fu, Chao Fei, Yuchen Yang, Qiang Li, Fred C Lee. "Optional Design of Planer Magnetic Components For a Two-Stage GaN-Based DC-DC Converter." IEEE Transactions on Power Electronics, VOL,34, Issue:4, April 2019, pp.3329-3338.

[15] Jiawei Liang, Yao Qin, Yu Liu, Minfan Fu, Haoyu Wang. "Phase Shift Regulated Resonant Switched-Capacitor-Based Intermediate Bus Converter for 48V Date Center Power System." IEEE Transactions on Industrial Electronics, VOL. 72 , No.2, February 2025, pp. 1475-1485.

[16] Jiawei Liang, Liang Wang, Junrui Liang, Minfan Fu, Haoyu Wang. "A Switched-Capacitor and Series-Resonant Hybrid MHz DCX in Data Center Applictions." IEEE Transactions on Power Electronics, VOL,39, Issue:10, Octorber 2024, pp.99:1-12.

Control Strategy of Dual Phase Shifting Dual Active Bridge Converter Based on BM

Yiting Huo
School of Automation,Beijing InformationScience & Technology University
Beijing,China
Email: 2024020393@bistu.edu.cn

Yajing Zhang*
School of Automation, Beijing Information Science & Technology University
Beijing,China
Email: zhangyajing@bistu.edu.cn

Bin Liu
State Grid Economic and Technological Research Institute Co
Beijing,China
Email:
liubin@chinasperi.sgcc.com.cn

Baoying Huang
State Grid Economic and Technological Research Institute Co
Beijing, China
Email:
huangbaoying@chinasperi.sgcc.com.cn

Siyu Pan
State Grid Economic and Technological Research Institute Co
Beijing, China
Email:
pansiyu@chinasperi.sgcc.com.cn

Hao Ma
School of Automation, Beijing Information Science & Technology University
Beijing, China
Email: 2022020463@bistu.edu.cn

Abstract—Dual Active Bridge (DAB) converters are pivotal in renewable energy, electric vehicles, and energy storage systems due to bidirectional power transfer, galvanic isolation, and zero-voltage switching capabilities. While Silicon Carbide (SiC) devices enhance efficiency through reduced conduction losses and high-frequency operation, conventional Single-Phase-Shift (SPS) control under non-unity voltage ratios induces excessive circulating power and current stress, degrading system efficiency and reliability. The Brayton-Moser (BM) form provides a powerful nonlinear control method through energy-based port Hamiltonian system modeling. To address the aforementioned issues, this study proposes an enhanced Brayton-Moser theory-based nonlinear control strategy integrated with Dual-Phase-Shift (DPS) modulation and minimum circulating power optimization. To enhance system performance under passive conditions and optimize efficiency, a Brayton-Moser theory-guided DAB-based DC power conversion system has been architected and experimentally substantiated via simulation platforms.

Keywords—Dual Active Bridge; Wide Bandgap; Improved BM control; PI control; Minimum circulating power control

I. INTRODUCTION

With the escalating severity of environmental pollution and global warming, contemporary society demands increasingly reliable energy supplies [1]. Concurrently, the growing infrastructure charging requirements, renewable energy integration, and electric vehicle adoption necessitate robust solutions. To enable seamless transition capabilities between diverse loads and energy sources while reducing system costs [2], the Dual Active Bridge (DAB) converter has emerged as a critical power conversion solution in renewable energy generation, electric vehicle charging, and DC microgrid applications. Recent advancements in Silicon Carbide (SiC) power MOSFET technology have focused on minimizing the on-resistance of the device to reduce conduction losses, thereby significantly improving energy efficiency. Consequently, this paper employs silicon carbide MOSFETs, Wide-Bandgap (WBG) semiconductor devices, to leverage their superior electrical properties [3]. The DAB converter's prominence stems from its bidirectional power transfer capability, high-frequency electrical isolation, and soft-switching techniques. Nevertheless, its performance is highly dependent on the selected modulation strategy. For bidirectional isolated DC-DC converters, phase shift modulation strategies are predominantly adopted, where marked differences exist between conventional Single Phase Shift (SPS) control and enhanced Dual Phase Shift (DPS) control in terms of efficiency, dynamic response, and device stress. Unlike SPS modulation, which is effective only when the input-to-output voltage ratio is unity, DPS modulation offers superior flexibility in power regulation [4]. A hybrid modulation strategy proposed in [5] reduces the RMS current in the inductors. Some other advancements in control strategies include model predictive methods with optimized regenerative power [6], novel controllers addressing input-side voltage imbalance [7], solutions for power imbalance mitigation [8], and near full-range Zero Voltage Switching (ZVS) implementations [9].

To further enhance system robustness, R.K. Brayton and J.K. Moser proposed the Brayton-Moser (BM) control strategy, which eliminates system overshoot and delivers advantages such as high stability and rapid transient response. As demonstrated in [10], the application of this control strategy to DAB converters enables voltage regulation across wide operating ranges through an energy-based methodology. Further studies demonstrate that integrating BM controllers with PI controllers improves dynamic performance [11], with additional benefits such as: Steady-state error reduction via optimized BM control [12]; Enhanced system immunity to disturbances [13]; Simplified control model structures [14]; Improved dynamic performance in DAB applications through BM model implementation [15].

The rest of this paper is organized as follows: Section II briefly introduces the topology of the DAB DC-DC converter under DPS modulation and the theoretical framework for minimum circulating power control under DPS operation. Section III presents the improved Brayton-Moser control theory based on mixed potential function analysis. Section IV conducts simulation-based experiments for the DAB converter, validating the effectiveness of the proposed control strategy. Finally, section V concludes this article.

Project supported by the National Natural Science Foundation of China (52237008) and the Headquarters Technology Projects of State Grid Corporation of China (5200-202456095A-1-1-ZN).

II. DUAL ACTIVE BRIDGE CONVERTER WITH DUAL PHASE SHIFT MODULATION

A. Topology of the Dual Active Bridge DC-DC Converter

The topology of the Dual Active Bridge (DAB) DC-DC converter is illustrated in Fig. 1. It comprises two full-bridge converters, designated as H_1 (primary side) and H_2 (secondary side), which are constructed using Silicon Carbide (SiC) Wide-Bandgap (WBG) semiconductor devices. Leveraging the bidirectional power transfer capability, high-frequency electrical isolation, and soft-switching technology, the DAB converter is widely adopted as a pivotal power conversion solution in renewable energy generation, electric vehicle charging, and DC microgrid applications. The converter achieves a stable voltage step-up ratio of 7; however, to accommodate long-term aging, temperature variations, and unexpected overvoltage conditions, industrial standards recommend constraining the step-up range within 5.83.

In Fig. 1, Q_1-Q_{12} are SiC MOSFETs; C_1 is the input capacitance and C_2 is the output capacitance; C_{fcA} and C_{fcB} are flying capacitances; V_{in} and V_{out} are voltage voltages; V_1 and V_2 are the voltages of primary side and secondary side; I_{out} is the output current; L_1 is the inductance; i_{L1} is the current of L_1; i_2 is the current of secondary side; R_{in} and R_{out} are input resistance and output resistance.

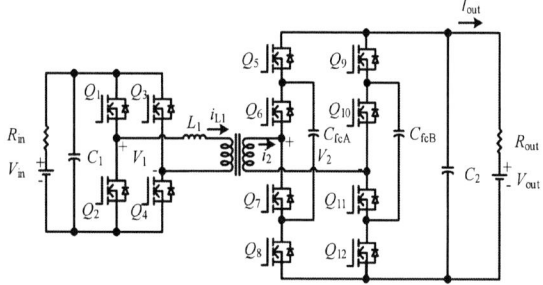

Fig. 1. Topology structure of dual active bridge

This paper employs a Dual Phase Shift (DPS) modulation strategy. Compared to conventional Single Phase Shift (SPS) modulation, the DPS approach introduces an additional internal phase shift angle D_1 alongside the external phase shift angle D_2. Through coordinated adjustment of D_1 and D_2, this strategy enables multi-objective optimization of power transfer, circulating power, and soft-switching range is achieved. Under varying voltage ratios $K=V_1/(NV_2)$, the proposed strategy minimizes circulating power, reduces current stress, and enhances efficiency. The primary/secondary voltage waveforms and inductor current dynamics illustrated in Fig. 2, while the operating modes of the switching devices are depicted in Fig. 3. Where V_{gs} is the working status of the switch tubes.

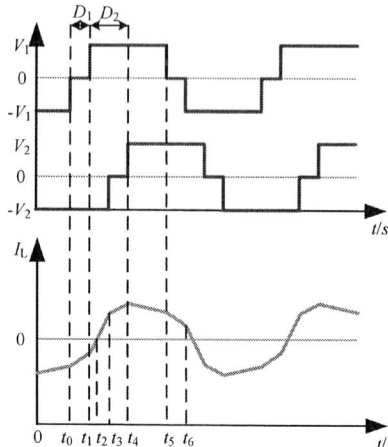

Fig. 2. Voltage and current waveforms of inductors

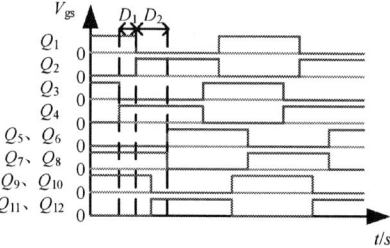

Fig. 3. Working mode of switch tube

B. Minimum Circulating Power Control Theory

Under Dual Phase Shift modulation, the waveforms of V_1 and V_2 are jointly determined by D_1 and D_2, exhibiting multi-level voltage characteristics. The expressions for inductor voltage and inductor current remain identical to those under Single Phase Shift modulation. The power transfer equation can be expressed as:

$$P = \frac{NV_1V_2}{8L_1f_s}[4D_2(1-D_2)-2D_1^2] \tag{1}$$

where the power transfer magnitude is predominantly governed by D_2, while D_1 optimizes the inductor current waveform. In the DPS operation, the impact of circulating power must be addressed. Circulating power arises when the voltage across the inductor and the polarity of the inductor current are mismatched. This phenomenon complicates the regulation of system current and voltage, compromises system stability, and necessitates additional current paths for reverse energy flow. These paths introduce higher resistive and switching losses. In terms of output performance, under high load currents or elevated load voltages, circulating power exacerbates output voltage and current fluctuations, adversely affecting the normal operation of connected loads. To minimize circulating power across the converter's full operating range, this paper proposes a segmented optimization strategy derived based on theoretical analysis of the circulating power expression. This strategy identifies the optimal phase shift angles D_1 and D_2. The per-unitized power transfer equation for the DAB DC-DC converter under DPS modulation is given by:

$$P_D^* = \frac{P_D}{P_B} = 4D_2 - 4D_2^2 - 2D_1^2 \qquad (2)$$

Where P_D^* represents the normalized power transfer relative to the converter's maximum capacity.

When $0 \leqslant P_D^* \leqslant 1/2$, D_1 and D_2 can be expressed by:

$$D_1 = D_2 = \frac{2 - \sqrt{4 - 6P_D^*}}{6} \qquad (3)$$

The minimum circulating power can be expressed by:

$$P_{cir}^* = \frac{(1-K)^2(4 + \sqrt{4 - 6P_D^*})}{72(K+1)} \qquad (4)$$

When $1/2 < P_D^* < 2/3$, D_1 and D_2 can be expressed by:

$$D_1 = D_2 = \frac{2 + \sqrt{4 - 6P_D^*}}{6} \qquad (5)$$

The minimum circulating power can be expressed by:

$$P_{cir}^* = \frac{(1-K)^2(4 - \sqrt{4 - 6P_D^*})}{72(K+1)} \qquad (6)$$

When $2/3 \leqslant P_D^* \leqslant 1$, D_1 and D_2 can be expressed by:

$$D_1 = (K+1)\sqrt{\frac{1 - P_D^*}{2K^2 + 4K + 6}}$$
$$D_2 = \frac{1}{2} - \frac{\sqrt{2}}{2}\sqrt{\frac{1 - P_D^*}{K^2 + 2K + 3}} \qquad (7)$$

The minimum circulating power can be expressed by:

$$P_{cir}^* = \frac{[K - (K^2 + 2K + 3)\sqrt{\dfrac{1 - P_D^*}{2K^2 + 4K + 6}}]^2}{2(K+1)} \qquad (8)$$

III. BRAYTON-MOSER MODEL AND CONTROLLER FOR THE DUAL ACTIVE BRIDGE CONVERTER

The Brayton-Moser (BM) formalism provides a powerful nonlinear control methodology through energy-based port-Hamiltonian system modeling. For electrical networks like the Dual Active Bridge converter, this approach enables systematic energy shaping via the Mixed Potential Function (MPF), which provides a unified representation that combines circuit stored energy and dissipation characteristics. The developed BM-based controller explicitly addresses the inherent nonlinearities in DAB dynamics while achieving superior transient performance through direct energy trajectory regulation. In this work, an improved BM-based controller is developed to enhance transient response and minimize overshoot.

For the Dual Active Bridge DC-DC converter topology shown in Fig.1, the power transfer under lossless conditions

is governed by the phase shift modulation between the primary and secondary bridges. To establish the Brayton-Moser model, the following steps are adopted:

According to the Brayton-Moser formalism, the differential equations describing the nonlinear circuit dynamics are derived as follows:

$$\begin{cases} -L_1 \dfrac{di_{L_1}}{dt} = \dfrac{\partial P}{\partial i_{L_1}}(i_{L_1}, V_{C_2}) \\ C_2 \dfrac{dV_{C_2}}{dt} = \dfrac{\partial P}{\partial V_{C_2}}(i_{L_1}, V_{C_2}) \end{cases} \qquad (9)$$

The Kirchhoff's Voltage Law (KVL) relationships for the primary full-bridge circuit (H_1) within the DAB converter configuration are derived as follows:

$$L_1 \frac{di_{L_1}(t)}{dt} = V_1 - NV_2 \qquad (10)$$

According to the transmission power expression of Equation (1), the Kirchhoff's Current Law (KCL) relationships for the secondary full-bridge circuit (H_2) within the DAB converter configuration are derived as follows:

$$C_2 \frac{dV_2(t)}{dt} = \frac{NV_1}{8L_1 f_s}[4D_2(1-D_2) - 2D_1^2] - \frac{V_2}{R} \qquad (11)$$

Combining the KVL equation with the KCL equation:

$$\begin{cases} L_1 \dfrac{di_{L_1}(t)}{dt} = V_1 - NV_2 \\ C_2 \dfrac{dV_2(t)}{dt} = \dfrac{NV_1}{8L_1 f_s}[4D_2(1-D_2) - 2D_1^2] - \dfrac{V_2}{R} \end{cases} \qquad (12)$$

The state-space equation derived from the Brayton-Moser (BM) formulation is established as follows:

$$\begin{bmatrix} L_1 \dfrac{di_{L_1}(t)}{dt} \\ C_2 \dfrac{dV_2(t)}{dt} \end{bmatrix} = \begin{bmatrix} 0 & 1 \\ 0 & \dfrac{N[4D_2(1-D_2) - 2D_1^2]}{8L_1 f_s} \end{bmatrix} \begin{bmatrix} i_{L_1} \\ V_1 \end{bmatrix} - \begin{bmatrix} N \\ \dfrac{1}{R} \end{bmatrix} V_2 \quad (13)$$

State space equation based on Brayton Moser model:

$$\begin{cases} -L_1 \dfrac{di_{L_1}}{dt} = \dfrac{\partial P}{\partial i_{L_1}}(i_{L_1}, V_2) \\ C_2 \dfrac{dV_2}{dt} = \dfrac{\partial P}{\partial V_2}(i_{L_1}, V_2) \end{cases} \qquad (14)$$

Let i_{Ld} and V_{2d} denote the desired trajectories of the averaged inductor current and averaged output voltage over one switching period, respectively. The error trajectories of the state variables are defined as:

$$\begin{cases} -L_1 \dfrac{di_{Ld}}{dt} = \dfrac{\partial P}{\partial i_{Ld}}(i_{Ld}, V_{2d}) \\[2mm] C_2 \dfrac{dV_{2d}}{dt} = \dfrac{\partial P}{\partial V_{2d}}(i_{Ld}, V_{2d}) \end{cases} \quad (15)$$

And the error dynamics equations are derived as:

$$\begin{cases} -L_1 \dfrac{d\tilde{i}_{L_1}}{dt} = \dfrac{\partial P}{\partial i_{L_1}}(i_{L_1}, V_2) - \dfrac{\partial P}{\partial i_{Ld}}(i_{Ld}, V_{2d}) \\[2mm] C_2 \dfrac{d\tilde{V}_2}{dt} = \dfrac{\partial P}{\partial V_2}(i_{L_1}, V_2) - \dfrac{\partial P}{\partial V_{2d}}(i_{Ld}, V_{2d}) \end{cases} \quad (16)$$

Incorporate Dissipative Factors of Series and Shunt Damping into the Error Dynamics Equations:

$$\begin{cases} -L_1 \dfrac{d\tilde{i}_{L_1}}{dt} = \dfrac{\partial P}{\partial \tilde{i}_{L_1}}(\tilde{i}_{L_1}, \tilde{V}_2) + \dfrac{\partial P_{Ri}}{\partial \tilde{i}_{L_1}}(\tilde{i}_{L_1}, \tilde{V}_2) \\[2mm] C_2 \dfrac{d\tilde{V}_2}{dt} = \dfrac{\partial P}{\partial \tilde{V}_2}(\tilde{i}_{L_1}, \tilde{V}_2) + \dfrac{\partial P_{Gi}}{\partial \tilde{V}_2}(\tilde{i}_{L_1}, \tilde{V}_2) \end{cases} \quad (17)$$

Where R_i and G_i are the series damping injected in series with the inductor and the shunt damping injected in parallel with the capacitor, respectively. These terms $P_{Ri}=1/2R_i i_L{}^2$ 、 $P_{Gi}=1/2G_i V_2{}^2$ represent the injected dissipative components.

Construct the Lyapunov function as follows:

$$V(x) = \frac{1}{2}L_1 \tilde{i}_L{}^2 + \frac{1}{2}C_2 \tilde{V}_2{}^2 \quad (18)$$

If and only if $i_L=0$ and when $V_2=0$, $V(x)=0$. Otherwise, $V(x)>0$. the derivative is as follows:

$$\dot{V}(x) = L_1 \tilde{i}_L \dot{\tilde{i}}_L + C_2 \tilde{V}_2 \dot{\tilde{V}}_2 \quad (19)$$

The error dynamic equation can be expressed as:

$$\begin{cases} L_1 \dot{\tilde{i}}_{L_1} = -N\tilde{V}_2 - R_i \tilde{i}_{L_1} \\[2mm] C_2 \dot{\tilde{V}}_2 = -\dfrac{1}{R}\tilde{V}_2 \end{cases} \quad (20)$$

Therefore:

$$\dot{V}(x) = -N\tilde{V}_2 \tilde{i}_{L_1} - R_i \tilde{i}_{L_1}{}^2 - \frac{1}{R}\tilde{V}_2{}^2 < 0 \quad (21)$$

Thus, the system is stable in the Lyapunov sense, and energy reshaping can be accomplished by injecting virtual damping into the system. The virtual damping factor can be inserted either in series with the input inductor or in parallel with the output capacitor. Given that series placement of the virtual damping resistance R_i with the input inductor creates a duty cycle independent of the damping parameters, the preferred strategy involves parallel integration of the virtual

damping conductance G_i with the output capacitor. The implementation procedure is outlined below:

Let $P_{R_i} = 0$, the closed-loop dynamic equation can be expressed as:

$$\begin{cases} -L_1 \dfrac{d\tilde{i}_{L_1}}{dt} = \dfrac{\partial P}{\partial \tilde{i}_{L_1}}(\tilde{i}_{L_1}, \tilde{V}_2) \\[2mm] C_2 \dfrac{d\tilde{V}_2}{dt} = \dfrac{\partial P}{\partial \tilde{V}_2}(\tilde{i}_{L_1}, \tilde{V}_2) + \dfrac{\partial P_{G_i}}{\partial \tilde{V}_2}(\tilde{i}_{L_1}, \tilde{V}_2) \end{cases} \quad (22)$$

Subtracting the above equation from the state equation of the Brayton-Moser model yields:

$$\begin{cases} -L_1 \dfrac{di_{Ld}}{dt} = \dfrac{\partial P}{\partial i_{Ld}}(i_{Ld}, V_{2d}) \\[2mm] C_2 \dfrac{dV_{2d}}{dt} = \dfrac{\partial P}{\partial V_{2d}}(i_{Ld}, V_{2d}) - \dfrac{\partial P_{G_i}}{\partial \tilde{V}_2}(\tilde{i}_{L_1}, \tilde{V}_2) \end{cases} \quad (23)$$

Substituting the original equation of state yields:

$$\begin{cases} -L_1 \dfrac{di_{Ld}}{dt} = -V_1 + NV_{2d} \\[2mm] C_2 \dfrac{dV_{2d}}{dt} = \dfrac{N[4D_2(1-D_2)-2D_1^2]}{8L_1 f_s}V_1 - GV_{2d} - G_i \tilde{V}_{2d} \end{cases} \quad (24)$$

Subsequently, the converter control law under the condition of shunt-connected virtual damping with the output capacitor is derived:

$$\begin{cases} \dot{V}_{2d} = \dfrac{1}{C_2}\left[\dfrac{N[4D_2(1-D_2)-2D_1^2]}{8L_1 f_s}V_1 - GV_{2d} - G_i \tilde{V}_{2d}\right] \\[2mm] 4D_2(1-D_2)-2D_1^2 = \dfrac{8L_1 f_s}{NV_1}\left(C_2 \dfrac{dV_{2d}}{dt} + GV_{2d} + G_i \tilde{V}_{2d}\right) \end{cases} \quad (25)$$

Considering the aforementioned minimum circulating power control, the internal phase shift angle D_1 varies across different segmented circulating power ranges, and the corresponding external phase shift angle D_2 is defined as follows:

$$D_2 = \frac{1}{2} - \frac{1}{2}\sqrt{1 - \left[2D_1^2 + \dfrac{8L_1 f_s}{NV_1}\left(C_2 \dfrac{dV_{2d}}{dt} + GV_{2d} + G_i \tilde{V}_{2d}\right)\right]} \quad (26)$$

The improved BM control block diagram is shown in the Fig. 4. Where V_{2d} is expected trajectory of secondary side voltage, f_s is switching frequency, N is transformer ratio, G_i is parallel damping factor.

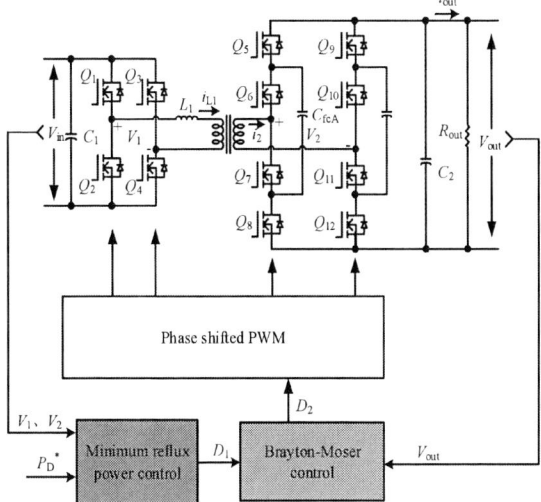

Fig. 4. Improved BM control block diagram

IV. SIMULATION VERIFICATION

To validate the feasibility of the proposed control strategy, a simulation model corresponding to Fig. 1 was constructed in MATLAB/Simulink, with simulation parameters detailed in Table 1.

TABLE I.   SIMULATION PARAMETERS OF DAB CONVERTER

Parameters	Value
Input Voltage V_1/V	400
Output Voltage V_2/V	750
Inductance f/μH	30
Switching Frequency f/kHz	100
Transformer ration N	1:2
Primary full-bridge capacitor C_1/μF	50
Secondary full-bridge capacitor C_2/μF	50
Secondary half-bridge capacitor C_{fcA}/mF	1.5

A. Impact of Minimum Circulating Power Control on System Performance

Under PI control, the input voltage is set to 300 V and the output voltage to 750 V. A comparative analysis of inductor current, output voltage, and output current is conducted between the system without minimum circulating power control and the system with minimum circulating power control. The output power is illustrated in the Fig. 5.

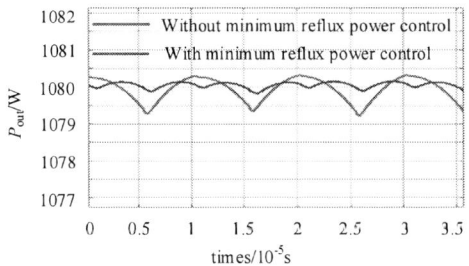

Fig. 5. The impact of minimum reflux power control on ripple

It can be observed that although the system employing minimum circulating power control exhibits a slight increase in rise time, it achieves reduced overshoot, shorter settling time, and significantly diminished output voltage and current ripple under steady-state conditions compared to the uncontrolled system.

B. Comparative analysis of improved BM and PI control during transient state

In this study, the output voltage is set to 750 V. For the improved Brayton-Moser based controller, the injected damping parameters are configured as G=0.02082 Ω and G_i=0.0098 Ω. The PI controller parameters are set to K_p=0.0008 and K_i=0.05. Comparative analyses between the BM controller and PI controller are conducted under three segmented operating conditions $0 \le P_D^* \le 1/2$, $1/2 < P_D^* < 2/3$, $2/3 \le P_D^* \le 1$, with the best results illustrated in the Fig. 6.

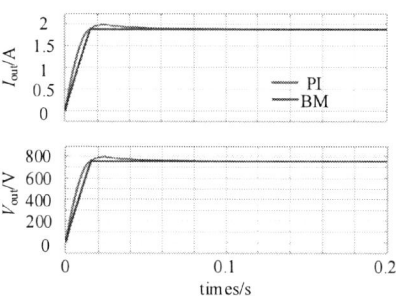

Fig. 6. Output voltage and output current under PI control and BM control

When $0 \le P_D^* \le 1/2$, under improved BM-based control, the output voltage stabilizes at 0.03s with no overshoot. In contrast, under PI control, the output voltage achieves stabilization at 0.065s with an overshoot of approximately 3.33%. When $1/2 < P_D^* < 2/3$, under improved BM-based control, the output voltage stabilizes at 0.0135s with no overshoot. In contrast, under PI control, the output voltage achieves stabilization at 0.085s with an overshoot of approximately 5.20%. And when $2/3 \le P_D^* \le 1$, under improved BM-based control, the output voltage stabilizes at 0.046s with no overshoot. In contrast, under PI control, the output voltage achieves stabilization at 0.12s with an overshoot of approximately 5.53%. The results demonstrate that the optimal minimum circulating power is achieved when the normalized power P_D^*=0.64. Waveforms of output voltage and current under each condition are shown in Fig. 6. The PI control model attains stability at 0.105s, while the improved BM control model achieves steady-state operation at 0.015s. This comparison clearly indicates that the enhanced BM control strategy exhibits significantly superior performance over the conventional PI control approach.

At 0.15 seconds, a resistive load with a value of 1600Ω was connected in parallel to the converter, resulting in a step change in the output current. A comparative analysis of the transient responses of improved BM and PI control strategy was performed, with the specific simulation results illustrated in Fig. 7. Furthermore, at the same moment, the desired voltage was altered to 800V, which caused changes in both the output voltage and the output current. The detailed simulation results are presented in Fig. 8.

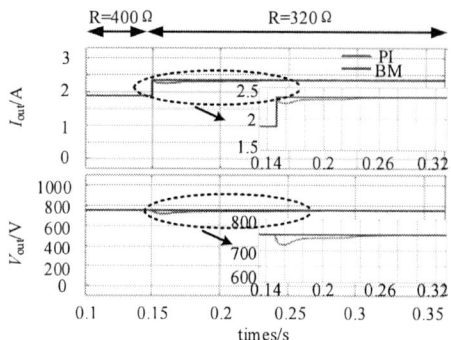

Fig. 7. Comparison of output waveforms under conditions of abrupt current value changes

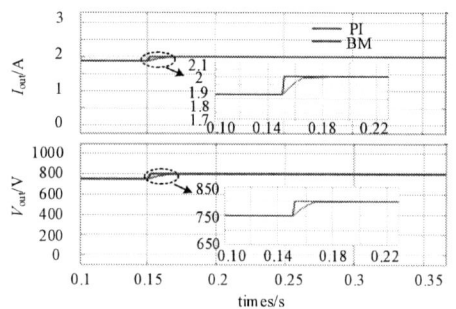

Fig. 8. Comparison of output waveforms under conditions of abrupt voltage value changes

Based on the comparative analysis of improved BM control and PI control under the aforementioned scenarios, it is evident that BM control effectively improves the system's settling time while eliminating the adverse effects of overshoot. When applied to Dual Active Bridge DC-DC converters, the improved BM control strategy achieves enhanced operational efficiency, reduced device stress, and improved dynamic stability, thereby offering significant advantages for high-performance power conversion systems.

V. CONCLUSION

This paper investigates the Dual Active Bridge DC-DC converter with Dual Phase Shift modulation using Brayton-Moser theory. Through theoretical derivation and simulation experiments, the following conclusions are drawn:

First, this paper introduce the minimum circulating power control into the DAB converter under PI control significantly reduces output current and voltage ripple.

In addition, The minimum circulating power controller enhances system stability and improves dynamic performance by mitigating reactive power losses and current stress.

The proposed improved BM control strategy eliminates overshoot while maintaining rapid transient response, as validated by simulation results.

Finally, compared to PI control, the improved BM strategy demonstrates superior stability and fast dynamics across segmented power transfer conditions, achieving: Zero overshoot during load/voltage transients and twice faster settling time.

REFERENCES

[1] M. Zhang, Y. Peng, Y. Zeng and Y. Chen, "Neural Network-Based Analysis of Local Extreme Cold Events in the Context of Global Warming: Causes and Predictions," Wuhu, China, 2024, pp. 142-145.

[2] A. Panchbhai, G. Chilkalpudi and A. Kumar, "Analysis of Circulating Power in Triple Active Bridge Converter: Impact of SPS and DPS Control," Trivandrum, India, 2023, pp.1-6.

[3] J. J. Kim, J. -H. Park, S. Sabri, B. Fetzer, B. Hull and S. -H. Ryu, "Investigation into Relationship of the Switching Performance and Short-Circuit Withstand Time on 1.2 kV 4H-SiC Power MOSFETs," Bremen, Germany, 2024, pp. 148-151.

[4] S. Chaurasiya and B. Singh, "A Load Adaptive DPS Control for DAB with Reduced Current Stress for Wide Load and Voltage Range," Jaipur, India, 2020, pp.1-6.

[5] S. Fang, P. Dai and S. Liu, "A Hybrid Modulation for Multilevel DAB Converter Based on Asymmetrical Duty Modulation," Guangzhou, China, 2023, pp. 68-73.

[6] S. Huang, J. Xing, N. Wang and F. Song, "TPS-MPC Method with Backflow Power Optimization for Series Resonant DAB Converter," Chengdu, China, 2024, pp. 2356-2359.

[7] S. -H. Park, I. -D. Kim, S. -M. Song and J. Kim, "Design of Battery Charger and Discharger using Series-input and Parallel-output connected DAB Converter," Chiang Mai, Thailand, 2022, pp. 1-5.

[8] Z. Chen, Z. Zhang, X. Sun, Z. Li and X. Liu, "An Optimized Return Power Control for DAB Converter Cluster with ISOP Configuration," Guangzhou, China, 2023, pp. 321-326.

[9] M. Zhang, H. Zou, S. Farzamkia, C. Chen and A. Q. Huang, "Three Phase High-Frequency-Link-Y-Configuration AC-DC DAB Converter with Monolithic Bidirectional GaN Switch," Phoenix, AZ, USA, 2024, pp. 1130-1136.

[10] H. Zhou, A. M. Khambadkone and X. Kong, "Passivity-Based Control for an Interleaved Current-Fed Full-Bridge Converter With a Wide Operating Range Using the Brayton–Moser Form," in *IEEE Transactions on Power Electronics*, vol. 24, no. 9, pp. 2047-2056, Sept. 2009.

[11] Shair J, Xie X, Li H, et al. A grid-side multi-modal adaptive damping control of super- /sub-synchronous oscillations in type-4 wind farms connected to weak AC grid[J]. Electric Power Systems Research, 2023,215:108963.

[12] K. Shipra, R. Maurya and S. N. Sharma, "Brayton-moser passivity based controller for electric vehicle battery charger," in *CPSS Transactions on Power Electronics and Applications*, vol. 6, no. 1, pp. 40-51.

[13] F. Yumin, L. Jianguo, Z. Yajing and W. Jiuhe, "Power Shaping Control of Single-Phase Grid-Connected Converter Based on Brayton-Moser Model," Beijing, China, 2023, pp. 12-18.

[14] Z. Wu *et al.*, "A Novel Method for Estimating the Region of Attraction for DC Microgrids via Brayton-Moser's Mixed Potential Theory," in *IEEE Transactions on Smart Grid*, vol. 14, no. 4, pp. 3313-3316, July 2023.

[15] H. Ma, Y. -J. Zhang, J. -G. Li and J. -H. Wang, "Control Strategy of DAB Converter Based on Brayton-Moser Model," Singapore, Singapore, 2023, pp. 1-5.

2025 IEEE Workshop on Wide Bandgap Power Devices and Applications in Asia (WiPDA Asia)

β-Type High-Gain Boost Converter with Diode-Capacitor Cell

Peng Sun
School of Electrical Engineering
Beijing Jiaotong University
Beijing, China
23126350@bjtu.edu.cn

Hong Li*
College of Electrical Engineering
Zhejiang University
Hangzhou, China
hong_li@zju.edu.cn

Yidi Liang
School of Electrical Engineering
Beijing Jiaotong University
Beijing, China
23111448@bjtu.edu.cn

Xu Shangguan
School of Electrical Engineering
Beijing Jiaotong University
Beijing, China
22110467@bjtu.edu.cn

Mingbo Wei
School of Electrical Engineering
Beijing Jiaotong University
Beijing, China
23111436@bjtu.edu.cn

Huizhu Zhuang
School of Electrical Engineering
Beijing Jiaotong University
Beijing, China
24126364@bjtu.edu.cn

Abstract—**In modern microgrids and energy storage systems, high step-up DC converters are often required to facilitate power transfer between low-voltage sources and high-voltage DC buses. In this paper, a β-type high-gain boost converter with a diode-capacitor cell (β-DCBC) is proposed. The β-DCBC achieves a high voltage gain of 3/(1-2D), where D is the same duty cycle applied to both switches. The paper first introduces the operational principles of the β-DCBC, followed by an analysis and comparison of the voltage and current stresses of its components with those of existing high step-up DC converters in literature for similar applications. Furthermore, the practical voltage gain and efficiency—including the effects of parasitic parameters and power losses—are thoroughly evaluated. Finally, circuit simulations demonstrate the low stress and cost-effective advantages of the β-DCBC. This topology provides a promising solution for photovoltaic (PV) systems in DC microgrids that require high voltage gains.**

Keywords—DC-DC boost converter, lower voltage stress, high voltage gain

I. Introduction

Renewable energy systems are rapidly advancing to meet the increasing global energy demand while promoting environmental sustainability [6]. DC microgrids facilitate the integration of renewable energy sources and loads. Typically, renewable sources such as photovoltaics (PV) and fuel cells generate low and variable DC voltages. Consequently, DC-DC boost converters with high voltage gain are often necessary to satisfy load or grid voltage requirements [7]. As a result, DC-DC converters play a crucial role in enabling efficient energy transfer within DC microgrids and have been extensively studied and widely implemented [8-9].

In recent years, various methods have been developed to construct high-gain converter topologies, including quadratic type [10-11], switched-inductor [12-13], switched-capacitor [14-15], voltage multiplier [16-17], and coupled inductor [18] approaches. Due to their two-stage structures, the efficiency of quadratic boost converters tends to be lower compared to single-stage designs. The voltage boosting method based on switched-inductor cells is simple in topology and easy to control; however, the core size of the inductor limits the power density of the converter. Both voltage multiplier and switched-capacitor topologies employ diode-capacitor boost cells to

enhance voltage gain, offering excellent scalability. Nonetheless, higher-order configurations can cause excessive current spikes in the switching devices. Additionally, voltage gain can be further increased by combining voltage multiplier cells with interleaved parallel structures [19-20], though their duty cycle is inherently limited. The converter topology proposed in [21] effectively addresses this issue. Using a coupled inductor, the voltage at both ends can be increased by adjusting the turns ratio, thereby enhancing the high-gain capability of the DC-DC converter. However, leakage inductance in coupled inductors can generate voltage spikes at both ends of the main switch, leading to additional electromagnetic interference.

In this paper, a β-type high-gain boost converter with a diode-capacitor cell (β-DCBC) designed for PV systems in DC microgrids is proposed. The main advantages of the β-DCBC are its simple structure, featuring only a single inductor, and its ability to maintain a continuous input current. Additionally, it achieves high voltage gain with a low duty cycle and exhibits extremely low voltage stress on the components. Furthermore, the practical voltage gain and efficiency of the β-DCBC are analyzed theoretically. Finally, the converter's performance is validated through theoretical analysis and simulation.

II. Operation Principle

The β-DCBC topology consists of β-type structure and diode-capacitor cell is shown in Fig. 1. It includes an input source V_{in}, two switches S_1 and S_2, five diodes $D_1 \sim D_5$, a single inductor L, four capacitors $C_1 \sim C_4$.

The theoretical waveforms of the β-DCBC include CCM and DCM, as shown in Fig. 2. In CCM, there are two operation modes during one switching period in the steady-state. The corresponding current path for each operation modes is Fig. 3. Besides, v_g is the gate-source voltage of S_1 and S_2, $v_{S1,2}$ is the drain-source voltage of S_1 and S_2, $v_{D1\sim5}$ is the voltage of $D_1 \sim D_5$.

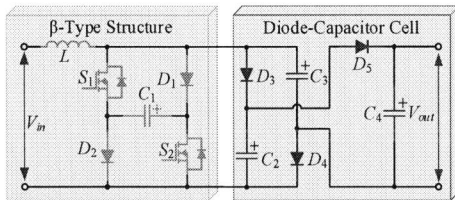

Fig. 1. β-DCBC Topology.

---

* Corresponding Author, E-mail: Hong Li, hong_li@zju.edu.cn
* Supported in part by the National Science Fund for Distinguished Young Scholars 52325704, in part by the Key Program of National Natural Science Foundation of China 52237008.

979-8-3315-1110-4/25 $31.00 © 2025 IEEE

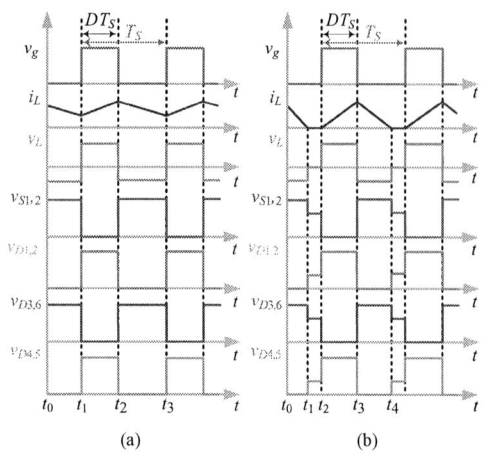

Fig. 2. Theoretical waveforms of the proposed β-DCBC, (a)CCM, (b)DCM.

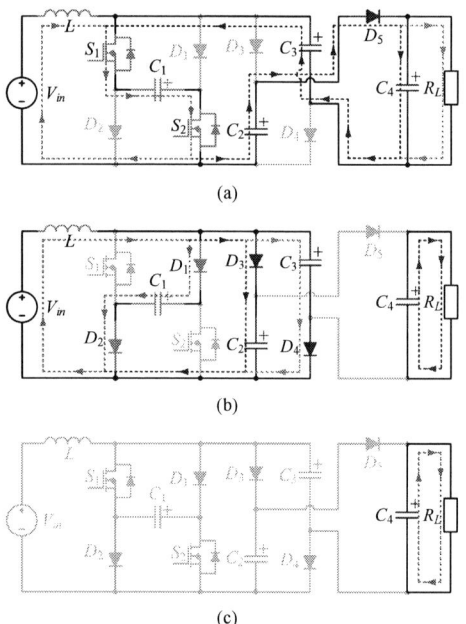

Fig. 3. Current path of β-DCBC, (a) Model I, (b) Model II, (c) Model III.

Mode I (t_1-t_2): As shown in Fig. 3(a), at $t=t_1$, S_1 and S_2 are turned on. L is charged linearly by V_{in} and C_1 in series. D_1, D_2, D_3 and D_4 are reverse bias condition while D_5 is forward biased. C_4 is charged by C_1, C_2 and C_3 in series. Furthermore, C_1, C_2 and C_3 are discharged to R_L in series.

$$\begin{cases} V_{in} + V_{C1} = L\dfrac{di_L}{dt} \\ V_{C1} + V_{C3} + V_{C3} = V_{C4} = V_{out} \end{cases} \tag{1}$$

Mode II (t_2-t_3): As shown in Fig. 3(b), at $t=t_2$, S_1 and S_2 are turned off. C_1 is charged by V_{in} and L in series. D_5 is reverse bias condition while D_1, D_2, D_3 and D_4 are forward biased. C_2 and C_3 are charged in series by V_{in} and L respectively. C_4 discharges to R_L.

$$\begin{cases} V_{in} - V_{C1} = L\dfrac{di_L}{dt} \\ V_{C1} = V_{C2} = V_{C3} \\ V_{C4} = V_{out} \end{cases} \tag{2}$$

This mode ends S_1 and S_2 when is turned on under the condition of CCM, and the duration of this mode is $(1-D)T_S$. Under the condition of DCM, this mode ends when i_L drops to zero. The duration of this mode for ΔT_{M2_DCM}, which can be calculated by (4), where i_{L_max} is the maximum current of the inductor L.

$$\begin{cases} V_{in} + V_{C1} = \dfrac{L\left(i_{L\_max} - 0\right)}{DT_S} \\ V_{in} - V_{C1} = \dfrac{L\left(0 - i_{L\_max}\right)}{\Delta T_{M2\_DCM}} \end{cases} \tag{3}$$

$$\Delta T_{M2\_DCM} = \dfrac{\left(V_{in} + V_{C1}\right)T_S}{V_{C1} - V_{in}} \tag{4}$$

Mode III (t_3-t_4): As shown in Fig. 3(c), at $t=t_3$, the current of L is equal to zero, and all diodes are reverse biased and turned off. C_4 discharges to R_L. This mode starts when i_L drops to zero, and ends when S_1 and S_2 are turned off. The duration of this mode ΔT_{M3} can be obtained by (6).

$$V_{C4} = V_{out} \tag{5}$$

$$\Delta T_{M3} = \left(1-D\right)T_S - \Delta T_{M2\_DCM} \tag{6}$$

The detailed operation principles of every stage are presented in Fig. 2. Under the different modes, the operation modes of proposed β–DCBC are shown in Table I.

TABLE I – DISTRIBUTION OF OPERATION MODES UNDER CCM AND DCM

Converters modes	Operating modes
CCM	M1, M2
DCM	M1, M2, M3

III. STYLING STEADY ANALYSIS

A. Voltage Gain

According to volt-second balance principle on L, (7) can be obtained:

$$\left(V_{in} + V_{C1}\right)DT_S + \left(V_{in} - V_{C1}\right)\left(1-D\right)T_S = 0 \tag{7}$$

V_{C1}, V_{C2}, V_{C3} and V_{C4} can be calculated by simultaneous solution of (8) and the working mode of Fig. 3:

$$\begin{cases} V_{C1} = \dfrac{1}{1-2D}V_{in} & V_{C2} = \dfrac{1}{1-2D}V_{in} \\ V_{C3} = \dfrac{1}{1-2D}V_{in} & V_{C4} = \dfrac{3}{1-2D}V_{in} \end{cases} \tag{8}$$

Therefore, the voltage gain of the proposed converter is given by (9).

$$M_{CCM} = \frac{V_{C4}}{V_{in}} = \frac{V_{out}}{V_{in}} = \frac{3}{1-2D} \qquad (9)$$

B. Voltage Stress and Current Stress

Based on the working mode of Fig. 3 and (8), the voltage stresses of S_1, S_2, D_1, D_2, D_3, D_4 and D_5 are calculated by:

$$\begin{cases} V_{S1} = V_{S2} = V_{D1} = V_{D2} = \dfrac{1}{1-2D}V_{in} = \dfrac{1}{3}V_{out} \\[2mm] V_{D3} = V_{D4} = V_{D5} = \dfrac{2}{1-2D}V_{in} = \dfrac{2}{3}V_{out} \end{cases} \qquad (10)$$

According to the ampere second equilibrium principle, the average currents flow through each capacitor during each mode are shown in the Table II.

TABLE II - AVERAGE VALUE OF CURRENT IN MODE I AND MODE II

Capacitors	Model I	Model II
C_1	$-\dfrac{D+1}{(1-2D)D}I_{out}$	$\dfrac{D+1}{(1-2D)(1-D)}I_{out}$
C_2	$-\dfrac{1}{D}I_{out}$	$\dfrac{1}{1-D}I_{out}$
C_3	$-\dfrac{1}{D}I_{out}$	$\dfrac{1}{1-D}I_{out}$
C_4	$\dfrac{1-D}{D}I_{out}$	$-I_{out}$

Based on Table I and the current paths for each mode depicted in Fig. 3, the current stresses of the semiconductor devices are:

$$\begin{cases} I_{S1} = I_{S2} = \dfrac{D+1}{(1-2D)D}I_{out} \\[2mm] I_{D1} = I_{D2} = \dfrac{D+1}{(1-2D)(1-D)}I_{out} \\[2mm] I_{D3} = I_{D4} = \dfrac{1}{1-D}I_{out} \\[2mm] I_{D5} = \dfrac{1}{D}I_{out} \end{cases} \qquad (11)$$

Based on Table II, the current stress on the capacitors can be derived from $0 < D < 0.5$:

$$\begin{cases} I_{C1} = \dfrac{D+1}{(1-2D)D}I_{out} \\[2mm] I_{C2} = \dfrac{1}{D}I_{out} \\[2mm] I_{C3} = \dfrac{1}{D}I_{out} \\[2mm] I_{C4} = \dfrac{1-D}{D}I_{out} \end{cases} \qquad (12)$$

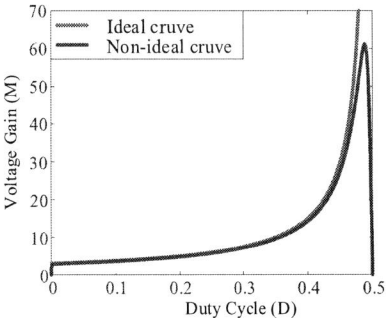

Fig. 4. The comparison of the idea and non-idea voltage gain.

Form (8) ~ (10), the voltage stresses of S_1, S_2 in β-DCBC are one-third of the output voltage, so do as the voltage stresses of D_1, D_2 and C_1~C_3. The voltage stresses of D_3, D_4 and D_5 are two-thirds of the output voltage.

C. Effect on voltage gain and efficiency

The non-ideal factors, i.e., parasitic resistor of inductor and switches, will degrade the voltage gain and efficiency of the β-DCBC. To analyze the effect of parasitic parameters, the voltage gain and efficiency are calculated by considering the parasitic resistors of L, S_1, S_2, C_1~C_4 and forward voltages of D_1~D_5. The sizes of the parasitic parameters are r_L=10mΩ, r_S=40mΩ, r_C=10mΩ and V_{FD}=0.7V.

Based on (11) and the current paths for each mode depicted in Fig. 3. Similar to the previous voltage gain analysis, based on the volt-second balance of inductor L, the non-ideal voltage gain of the β-DCBC, considering parasitic parameters, can be expressed as (13).

$$M_{real} = \frac{M_{CCM} - \left(\dfrac{5-4D}{1-2D}\right)\dfrac{V_{FD}}{V_{in}}}{1 + \dfrac{1}{(1-2D)R_L}\left(\dfrac{9r_L}{1-2D} + ar_C + br_S\right)} \qquad (13)$$

where

$$\begin{cases} a = \dfrac{2D^2 - 6D + 4}{D} + \dfrac{6D+6}{1-2D} + \dfrac{4-6D}{1-D} \\[2mm] b = \dfrac{6D+6}{1-2D} + \dfrac{2D+2}{D} \end{cases}.$$

According to (9) and (13), the comparison of the voltage gains for the idea and non-idea β-DCBC can be provided as Fig. 4, where R_L=640Ω and V_{in}=30V. Based on the comparison results shown in Fig. 4, the voltage gain is degraded rapidly when duty cycle higher than 0.4.

In addition to conduction losses, the efficiency is also influenced by the switching losses of S_1 and S_2. Based on (13), the efficiency of the β-DCBC can be determined using (14).

$$\eta = \frac{M_{real}}{M_{CCM}} - \frac{D+1}{27D}M_{real}\left(t_r + t_f\right)f_S \qquad (14)$$

According to (14), the calculated efficiency versus the duty cycle of the β-DCBC considering conduction losses and switching losses is shown in Fig. 5, where t_r=11ns, t_f=9 ns and f_S=100kHz. According to Fig. 5, the efficiency is degraded rapidly when duty cycle higher than 0.4.

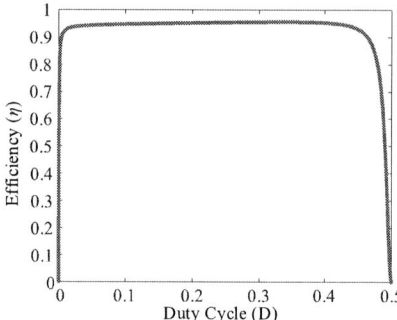

Fig. 5. The calculated efficiency of the proposed β-DCBC considering conduction losses and switching losses.

D. Performance Comparison

To verify the advantages of the β-DCBC, the converters proposed in [1-5] are selected to make a performance comparison is shown in Table III. These include component counts, voltage gain, voltage stress and current stress of switches and diodes, and input current comparisons. Besides, Fig. 6 to Fig. 10 illustrate the comparison of the voltage gain, voltage stress and current stress of switches and diodes graphically.

Fig. 6 illustrates the comparison of voltage gain, with the β-DCBC exhibiting the highest gain among the proposed converters in [1–5]. Its voltage gain is equivalent to that of the converter in [1], but the diode current stress in the β-DCBC is lower. This underscores the β-DCBC's advantages in achieving high voltage gain with reduced stress on the components.

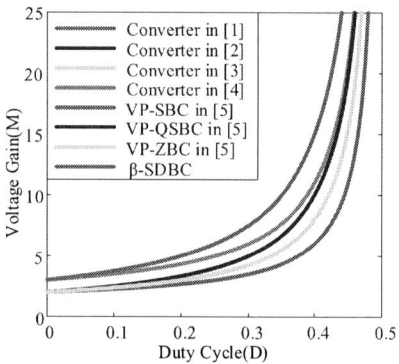

Fig. 6. The comparison of voltage gain.

Fig. 7 compares the maximum voltage stresses on each converter switch. The β-DCBC exhibits the lowest switching voltage stress among the converters proposed in [1–5]. While its maximum switch voltage stress is comparable to that of the converter in [1], the β-DCBC benefits from a lower capacitance voltage stress, highlighting its advantage in reducing switching voltage stresses.

Fig. 8 compares the maximum diode voltage stress across different converters. The results show that, although the diode voltage stress in the proposed converters in [3] and [5] is lower than that of the β-DCBC, their voltage gains are also lower than the β-DCBC. Additionally, the converters in [3] and VP-ZBC in [5] require a higher number of inductors, which could lead to a bulkier design. Moreover, the input current ripples of VP-SBC and VP-ZBC in [5] are pulsating, potentially affecting their overall performance.

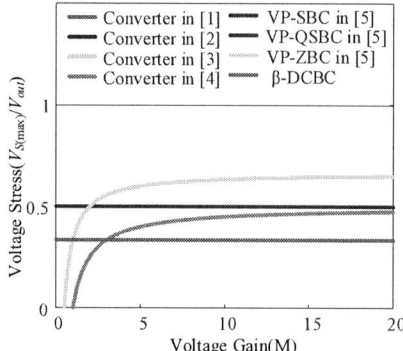

Fig. 7. The comparison of voltage stress of switches.

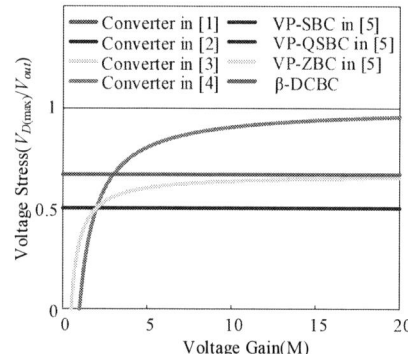

Fig. 8. The comparison of voltage stress of diodes.

TABLE III - PERFORMANCE COMPARISON OF THE PROPOSED CONVERER WITH OTHER HIGH GAIN CONVERERS

Converter	[1]	[2]	[3]	[4]	VP-SBC in [5]	VP-QSBC in [5]	VP-ZBC in [5]	β-DCBC
No.of *L/C/S/D*	1/4/2/5	2/5/1/4	2/4/1/3	1/3/2/4	1/3/2/4	1/3/2/4	2/4/1/3	1/4/2/5
Voltage Gain(M)	$\dfrac{3}{1-2D}$	$\dfrac{2}{1-2D}$	$\dfrac{2-D}{1-2D}$	$\dfrac{3-2D}{1-2D}$	$\dfrac{2-2D}{1-2D}$	$\dfrac{2}{1-2D}$	$\dfrac{2-D}{1-2D}$	$\dfrac{3}{1-2D}$
Voltage stress of switches($V_{S(max)}/V_{out}$)	$\dfrac{1}{3}$	$\dfrac{1}{2}$	$\dfrac{2M-1}{3M}$	$\dfrac{M-1}{2M}$	$\dfrac{1}{2}$	$\dfrac{1}{2}$	$\dfrac{2M-1}{3M}$	$\dfrac{1}{3}$
Voltage stress of diodes($V_{D(max)}/V_{out}$)	$\dfrac{2}{3}$	$\dfrac{1}{2}$	$\dfrac{2M-1}{3M}$	$\dfrac{M-1}{M}$	$\dfrac{1}{2}$	$\dfrac{1}{2}$	$\dfrac{2M-1}{3M}$	$\dfrac{2}{3}$
Current stress of switches ($I_{S(max)}/I_{out}$)	$\dfrac{M^2-M}{M-3}$	$\dfrac{2M^2-2M}{M-2}$	$\dfrac{2M^2-3M+1}{M-2}$	$\dfrac{(M-1)^2}{M-3}$	$\dfrac{M^2-M}{M-2}$	$\dfrac{M^2+2M}{2M-4}$	$\dfrac{2M^2-3M+1}{M-2}$	$\dfrac{M^2-M}{M-3}$
Current stress of switches ($I_{D(max)}/I_{out}$)	$\dfrac{M^2+M}{M+3}$	$\dfrac{2M^2}{2+M}$	$\dfrac{2M^2-M}{M+1}$	$\dfrac{(M-1)^2}{M+1}$	$M-1$	$\dfrac{M^2-2M}{2M+4}$	$\dfrac{2M(M-1)}{M+1}$	$\dfrac{M^2-M}{M+3}$
Continuous input current	Yes	Yes	Yes	No	No	Yes	No	Yes

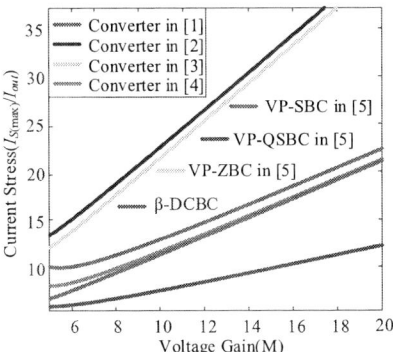

Fig. 9. The comparison of voltage gain.

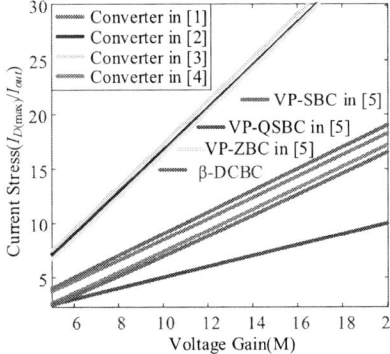

Fig. 10. The comparison of voltage gain.

Fig. 9 compares the maximum current stress on the switches for each converter. The results indicate that, although the maximum switch current stress of the converter in [4] and the VP-SBC and VP-QSBC in [5] is lower than that of the β-SCBC, the peak currents in [4] and VP-SBC are very close to the β-SCBC. Additionally, the voltage gain of the converter in [4] is lower than that of the β-SCBC, and its input current exhibits pulsating behavior. The disadvantages of the converter in [5] relative to the β-SCBC will not be discussed further.

Fig 10 compares the maximum diode current stress across different converters. The results show that only the VP-QSBC proposed in [5] exhibits a lower diode current stress than the β-DCBC. However, its voltage gain is lower than that of the β-DCBC, and its switch voltage stress is higher. This highlights the advantage of the β-DCBC's lower current stress on its diodes.

In summary, the excellent overall performance of β-DCBC has been proved.

IV. SIMULATION ANALYSIS

A 400W PSIM simulation platform was built to verify the correctness of the theoretical analysis. 30V/400V voltage conversion in PV system. Parameters are shown in Table IV. The simulated results are shown in Fig. 11 to Fig. 15.

The waveforms v_{out} and v_{in} are shown in Fig. 11. The voltage of output voltage is 400V, while the input voltage is 30V, which means the voltage gain is 13.3. The characteristic of high voltage gain is verified by the simulated results.

Fig.12 depicts the waveforms of i_{in} and v_L. It is proved that the input current of β-DCBC is characterized as continuous.

TABLE IV - SIMULATION PARAMETERS

Parameters or devices	Symbol	Value
Input Voltage	V_{in}	30V
Output Voltage	V_{out}	400V
Output Power	P_{out}	400W
Switching Frequency	f_S	100kHz
Inductor	L	320uH
Capacitors	C_1	40uF
	$C_2\sim C_4$	20uF

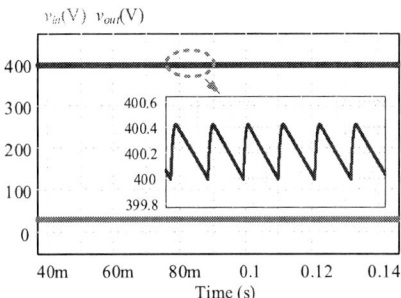

Fig. 11. Waveform of v_{in} and v_{out}.

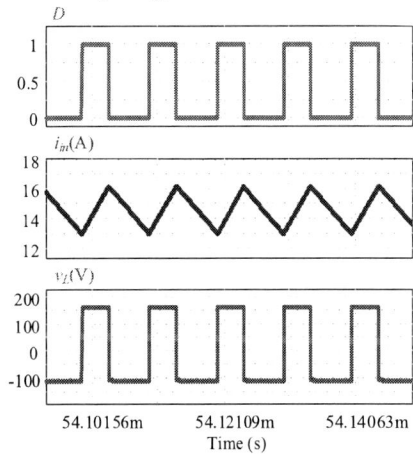

Fig. 12. Waveform of i_{in} and v_l.

The waveforms of the v_{C1}, v_{C2} and v_{C3} are shown in Fig. 13. The voltage stresses of C_1, C_2 and C_3 are only 100V, and that of C_4 is 400V, while the output voltage is 400V. The maximum capacitive voltage stress of C_1, C_2 and C_3 is only one third of the output voltage, verifying the correctness of the theoretical analysis.

The waveforms of v_{S1}, v_{S1}, v_{D1}, v_{D2}, v_{D3}, v_{D4} and v_{D5} are shown in Fig. 14. The voltage stresses of S_1, S_2, D_1, D_2 are only

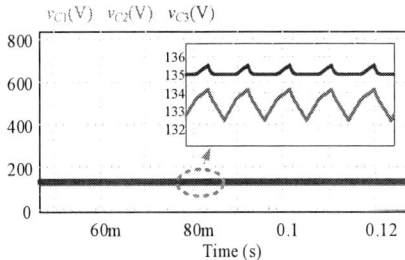

Fig. 13. Waveforms of v_{C1}, v_{C2} and v_{C3}.

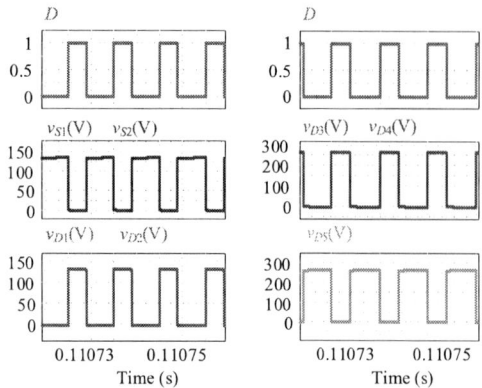

Fig. 14. Waveforms of v_{S1}, v_{S2} and $v_{D1}\sim v_{D5}$.

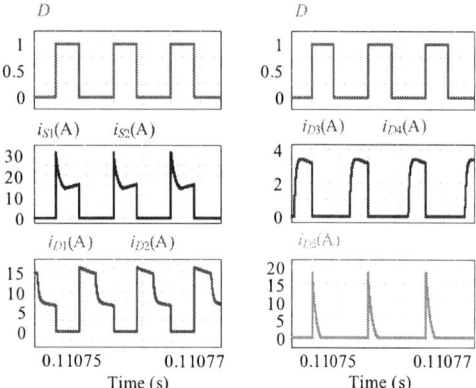

Fig. 15. Waveforms of i_{S1}, i_{S2} and $i_{D1}\sim i_{D5}$.

100V, and that of D_3, D_4 are 200V, while the output voltage is 400V. Hence the low voltage switches with low conduction resistance can be used in the converter to improve the efficiency.

The waveforms of I_{S1}, I_{S2}, I_{D1}, I_{D2}, I_{D3}, I_{D4} and I_{D5} are shown Fig.15. The peak currents of S_1 and S_2 are around 30A, while the peak currents of D_1 and D_2 are around 15A, while the peak currents of D_3, D_4 and D_5 are 4A, 4A and 20A respectively.

CONCLUSIONS

In this paper, the β-DCBC achieves a voltage gain of 3/(1-2D), with simulation analysis verifying its advantages in PV systems. A voltage gain of 13.3 can be obtained with a duty cycle of only D=0.39. The switch voltage stress is just one-third of the output voltage, which similarly applies to the voltage stresses on diodes D_1, D_2, and capacitors $C_1\sim C_3$. Compared to existing high-gain topologies in [1–5], the β-DCBC offers the benefits of high voltage gain and low device voltage stress. Furthermore, detailed power loss analysis indicates that the converter can achieve an efficiency of up to 96%. Therefore, the proposed β-DCBC provides a promising high step-up converter topology for PV systems in DC microgrids.

REFERENCES

[1] S. Miao, W. Liu and J. Gao, "Single-Inductor Boost Converter With Ultrahigh Step-Up Gain, Lower Switches Voltage Stress, Continuous Input Current, and Common Grounded Structure," *IEEE Trans. Power Electron.*, vol. 36, no. 7, pp. 7841-7852, July 2021.

[2] Y. Zhang, C. Fu, M. Sumner and P. Wang, "A Wide Input-Voltage Range Quasi-Z-Source Boost DC–DC Converter With High-Voltage

Gain for Fuel Cell Vehicles," *IEEE Trans. Ind. Electron.*, 65, no. 6, pp. 5201-5212, June 2018.

[3] M. Veerachary and P. Kumar, "Analysis and Design of Quasi-Z-Source Equivalent DC-DC Boost Converters," in IEEE Transactions on Industry Applications, vol. 56, no. 6, pp. 6642-6656, Nov.-Dec. 2020.

[4] M. -K. Nguyen, T. -D. Duong and Y. -C. Lim, "Switched-Capacitor-Based Dual-Switch High-Boost DC–DC Converter," *IEEE Trans. Power Electron.*, vol. 33, no. 5, pp. 4181-4189, May 2018.

[5] G. Zhang et al., "A Generalized Additional Voltage Pumping Solution for High-Step-Up Converters," *IEEE Trans. Power Electron.*, vol. 34, no. 7, pp. 6456-6467, July 2019.

[6] D. Yu, J. Yang, R. Xu, Z. Xia, H. H. -C. Iu and T. Fernando, "A Family of Module-Integrated High Step-Up Converters With Dual Coupled Inductors," *IEEE Access,*, vol. 6, pp. 16256-16266, 2018.

[7] M. Das and V. Agarwal, "Design and Analysis of a High-Efficiency DC–DC Converter With Soft Switching Capability for Renewable Energy Applications Requiring High Voltage Gain," *IEEE Trans. Ind. Electron.*, vol. 63, no. 5, pp. 2936-2944, May 2016.

[8] H. Tarzamni, H. S. Gohari, M. Sabahi and J. Kyyrä, "Nonisolated High Step-Up DC–DC Converters: Comparative Review and Metrics Applicability," *IEEE Trans. Power Electron.*, vol. 39, no. 1, pp. 582-625, Jan. 2024.

[9] Q. Zhang et al., "Output Impedance Modeling and High-Frequency Impedance Shaping Method for Distributed Bidirectional DC–DC Converters in DC Microgrids," *IEEE Trans. Power Electron.*, vol. 35, no. 7, pp. 7001-7014, July 2020.

[10] M. Hajilou and H. Farzanehfard, "Single Switch Ultra-High Step-Up Quadratic Converter With Low Input Current Ripple," *IEEE Trans. Ind. Electron.*, vol. 72, no. 1, pp. 411-418, Jan. 2025.

[11] C. -Y. Chan, S. Chincholkar and W. Jiang, "A Modified Fixed Current-Mode Controller for Improved Performance in Quadratic Boost Converters," in *IEEE Transactions on Circuits and Systems II: Express Briefs*, vol. 67, no. 10, pp. 2014-2018, Oct. 2020.

[12] C. Li, H. Li, N. Wang, X. Sun and L. Cheng, "A Full Soft-Switching High Step-Up DC/DC Converter With Active Switched Inductor and Three-Winding Coupled Inductor," *IEEE Trans. Power Electron.*, vol. 38, no. 10, pp. 13133-13146, Oct. 2023.

[13] R. Fani, Z. Akhlaghi and E. Adib, "High Step-Up DC-DC Converter by Integration of Active Switched Inductors, Built in Transformer, and Multipliers," *IEEE Trans. Power Electron.*, vol. 39, no. 2, pp. 2468-2477, Feb. 2024.

[14] I. -B. Kong, W. -S. Kim and S. -W. Lee, "A Novel High-Voltage-Gain Quasi-Resonant DC–DC Converter With Active-Clamp and Switched-Capacitor Techniques," *IEEE Trans. Power Electron.*, vol. 38, no. 6, pp. 7810-7820, June 2023.

[15] G. Wu, X. Ruan and Z. Ye, "Nonisolated High Step-Up DC–DC Converters Adopting Switched-Capacitor Cell," *IEEE Trans. Ind. Electron.*, vol. 62, no. 1, pp. 383-393, Jan. 2015.

[16] M. R. S. de Carvalho, E. A. O. Barbosa, F. Bradaschia, L. R. Limongi and M. C. Cavalcanti, "Soft-Switching High Step-Up DC–DC Converter Based on Switched-Capacitor and Autotransformer Voltage Multiplier Cell for PV Systems," *IEEE Trans. Ind. Electron.*, vol. 69, no. 12, pp. 12886-12897, Dec. 2022.

[17] P. Mohseni, S. Mohammadsalehian, M. R. Islam, K. M. Muttaqi, D. Sutanto and P. Alavi, "Ultrahigh Voltage Gain DC–DC Boost Converter With ZVS Switching Realization and Coupled Inductor Extendable Voltage Multiplier Cell Techniques," *IEEE Trans. Ind. Electron.*, vol. 69, no. 1, pp. 323-335, Jan. 2022.

[18] T. Liu, M. Lin and J. Ai, "High Step-Up Interleaved dc–dc Converter With Asymmetric Voltage Multiplier Cell and Coupled Inductor," in IEEE Journal of Emerging and Selected Topics in Power Electronics, vol. 8, no. 4, pp. 4209-4222, Dec. 2020.

[19] M. Meraj, M. S. Bhaskar, A. Iqbal, N. Al-Emadi and S. Rahman, "Interleaved Multilevel Boost Converter With Minimal Voltage Multiplier Components for High-Voltage Step-Up Applications," *IEEE Trans. Power Electron.*, vol. 35, no. 12, pp. 12816-12833, Dec. 2020.

[20] A. Alzahrani, M. Ferdowsi and P. Shamsi, "High-Voltage-Gain DC–DC Step-Up Converter With Bifold Dickson Voltage Multiplier Cells," *IEEE Trans. Power Electron.*, vol. 34, no. 10, pp. 9732-9742, Oct. 2019.

[21] H. Li, P. Sun, M. Wei, X. Shangguan, J. Hu and Y. Zeng, "A β-type Ultrahigh-Gain Boost Converter," *2024 IEEE Energy Conversion Congress and Exposition (ECCE)*, Phoenix, AZ, USA, 2024.

2025 IEEE Workshop on Wide Bandgap Power Devices and Applications in Asia (WiPDA Asia)

Soft-Switching Fixed-Frequency Control Strategy for Three-Level Buck-Boost Converter

Fang Li
School of Automation
Beijing Information Science and
Technology University
Beijing, China
lifang@bistu.edu.cn

Yao Xue
China Huaneng Clean Energy
Research Institute
Beijing, China
y_xue@qny.chng.com.cn

Fangwei Zhao
China Huaneng Clean Energy
Research Institute
Beijing, China
fw_zhao@qny.chng.com.cn

Yajing Zhang
School of Automation
Beijing Information Science and
Technology University
Beijing, China
zhangyajing@bistu.edu.cn

Jun Xu
School of Automation
Beijing Information Science and
Technology University
Beijing, China
13552848601@163.com

Abstract— A soft-switching fixed-frequency control strategy based on the non-inverting three-level Buck-Boost converter is proposed in this paper. Soft-switching for all switches in the whole domain can be achieved by the specific inductor current waveform trajectories. Limiting conditions are imposed on the three duty cycle control variables, and the mathematical calculation results of the maximum output power point are provided. The linear modulation method that each duty cycle increases proportionally when the load current increases is proposed. The simple fixed frequency control strategy can be used easily, and can achieve ultra-wide range voltage regulation with smooth transitions between different operating conditions. Simulation results demonstrate the feasibility of this control strategy, and experimental validation will be conducted in the future.

Keywords—non-inverting three-level Buck-Boost converter, zero-voltage-switching (ZVS), fixed-frequency, control strategy, multiple control variables

I. INTRODUCTION

The four-switch Buck-Boost (FSBB) converter has attracted much attention and has become more and more popular due to its voltage regulation ability, low inductor current and soft switching. In order to improve the efficiency, some control methods for the FSBB converter have been presented in the publications, the simplest control scheme that switches ON and OFF simultaneously has been improved to the two-mode modulation scheme and the quadrangle control method[1-3]. The inductor current is shaped into quadrangle to achieve the zero-voltage-switching (ZVS) of all switches[1]. It is suitable operating at high switching frequency to achieve high power density for the switching noise can be significantly reduced. The output power and efficiency can be further increased by introducing multi-frequency control strategy, while the control complexity increases at the same time[4]. When there are two or three control variables, many papers used the look-up-table to realize the control strategy, which is difficult to reproduce[5].

In the efficiency test of FSBB converter, the peak efficiency often appears when the output voltage is closed to the input voltage. That's because the voltage across the inductor is very small, the time interval that power directly conduct form input to output is large. However, when the difference between the input and output voltage is increased,

the larger inductor peak current lead to larger RMS value, and the efficiency decreased clearly. Therefore, FSBB converter is not perfect in the application with ultra-wide voltage regulation range [6-7].

The topology of three-level Buck-Boost converter is shown in Figure 1, which replace the primary bridge of FSBB to a three-level bridge. The voltage of inductor left terminal is v_{AB}, can be Vin, Vin/2 and 0. The input voltage regulation range can be widened by the increasing of the voltage level.

A simple fixed-frequency control strategy for three-level Buck-Boost converter has been proposed in [8]. In Buck mode, only the left bridge is adjusted. While in Boost mode, only the right bridge is adjusted. Moreover, an improvement mode transition state is proposed for the output voltage is closed to the input voltage. The fixed-frequency control is relatively simple, and effectively reducing the number of switch actions. However, hard switching of part switches under continuous conduction mode (CCM) limits the switching frequency and efficiency.

Three-level Buck-Boost converter as a derived topology of FSBB, the operational principle and the improved control methods can be implemented similarly. The quadrangle control method with little negative current to achieve the ZVS of all switches should be introduced and improved in the converter [9-11]. As the number of switches increases in the three-level Buck-Boost converter, the control variables also increase, and the complexity of the control strategy also increases, further research is needed.

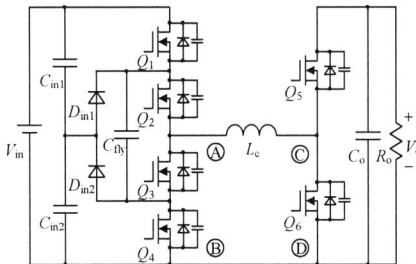

Fig. 1. Topology of three-level Buck-Boost converter.

In order to optimize the control of the three-level Buck-Boost converter and achieve a wider voltage regulation range for renewable energy systems, this paper proposes a

---

This work has been supported by the National Natural Science Foundation of China (52237008) and the Headquarters Technology Projects of State Grid Corporation of China (5200-202456095A-1-1-ZN).

979-8-3315-1110-4/25 $31.00 © 2025 IEEE

simplified fixed-frequency control strategy that ensuring ZVS of all switches in the whole operating range. The control strategy is suitable for silicon MOSFETs, and the better efficiency performance can be achieved by the wide bandgap devices. The operational principles are given in Section II, the key waveforms and theoretical analysis calculations are given in Section III, and the closed-loop control strategy and simulation results based on PSIM software are given in Section IV, at last is the conclusion.

II. OPERATIONAL PRINCIPLES

Operational modes of three-level Buck-Boost converter are shown in Fig. 2. The voltage across the inductor are labeled, too. V_{in} is the input voltage, and V_{out} is the output voltage. The converter works as the FSBB when Q_1 and Q_2, Q_3 and Q_4 turn ON and OFF synchronously, as Fig. 2(a)-(d). Two switches in series are used as one, which will increase the conduction loss. When Q_1 and Q_3, Q_2 and Q_4 are ON, the voltage of port AB is $V_{in}/2$ in ideal case. The voltages across the inductor are the same in Fig.2 (e) and (f), Fig.2 (g) and (h), and the equivalent operation modes are the same. As the number of input side switches increases, the voltages across the switches drop to $V_{in}/2$, new voltage level appears, the control variables increase, and the operating voltage range of the converter also increases.

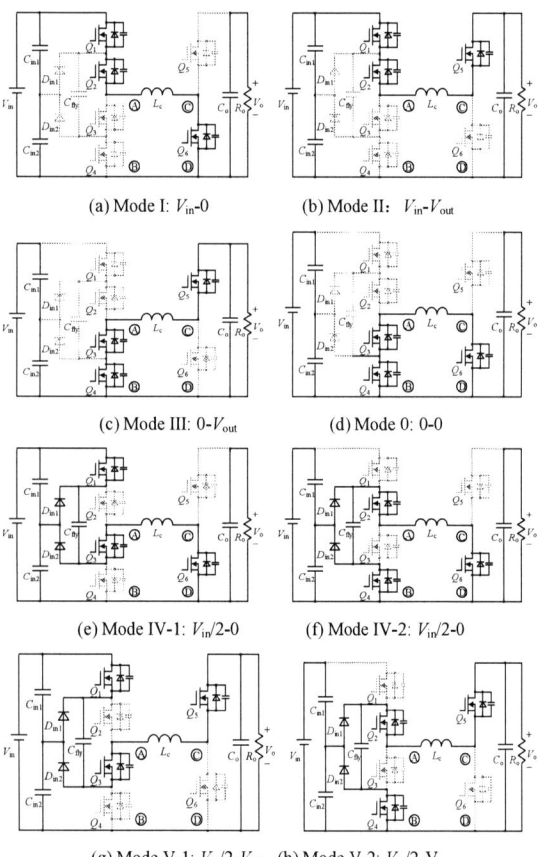

(a) Mode I: V_{in}-0   (b) Mode II: V_{in}-V_{out}

(c) Mode III: 0-V_{out} (d) Mode 0: 0-0

(e) Mode IV-1: $V_{in}/2$-0   (f) Mode IV-2: $V_{in}/2$-0

(g) Mode V-1: $V_{in}/2$-V_{out} (h) Mode V-2: $V_{in}/2$-V_{out}

Fig. 2. Operational modes and the voltage across the inductor.

The voltage ratio of output and input port is M, i.e.

$$M = \frac{V_{out}}{V_{in}}. \tag{1}$$

The changing slopes of inductor current are different with different M, as shown in Table I. It can be assumed that the inductor current remains constant in Mode 0 ignoring the loss ideally. ↑ indicates current is rising, ↑↑ indicates current is rising rapidly, while ↓ and ↓↓ indicate current is decreasing gently or rapidly. The inductor current first increases and then decreases, and repeat periodically.

TABLE I.   THE CHANGING SLOPES OF INDUCTOR CURRENT

Operational Modes	Voltage Ratio		
	M>1	*0.5≤M≤1*	*M<0.5*
I		↑↑	
II	↓	↑	↑
III		↓↓	
IV		↑	
V	↓	↓	↑
0	Remain constant		

Assuming the circuit operates under ideal conditions, neglecting parasitic parameters in various components and wires, the input and output voltage are constant during one switching period. The output capacitances of switching devices are the same as C_{oss}.

The dead times in switching legs are the same as t_{ead}. And the ZVS condition are written as follows:

$$\left| I_V \right| \geq \frac{2C_{oss} \, \mathbf{max}\left(V_{in}/2, V_{out}\right)}{t_{dead}}. \tag{2}$$

In the analysis below, we assume dead times is small enough that it can be ignored during the equivalent period T.

The quadrangle control method with little negative current can realize ZVS of all power switches of FSBB. That can be inherited for the three-level Buck-Boost converter. Q_1 and Q_4, Q_2 and Q_3, Q_5 and Q_6 are turned on complementary correspondingly. It can be obtained that the conditions for the implementation of ZVS soft switching are: the inductor current is larger than I_v when Q_3, Q_4 or Q_5 is turned ON, and the inductor current is lower than - I_v when Q_1, Q_2 or Q_6 is turned ON.

III. KEY WAVEFORMS AND MATHEMATICAL ANALYSIS

In order to realize ZVS of all switches, a control strategy is proposed considering the variation of inductor current. The key waveforms of three-level Buck-Boost converter at different M interval are given in Fig.3. I_v and -I_v are represented by two dashed lines. Each operating waveform contains mode 0, where the inductor current remains constant as -I_v. Thus the converter operates in the discontinuous conduction mode (DCM).

The moment when Q_1 turning ON is the beginning of the entire switching cycle. The definitions of the control variables shown in Fig. 3 are: D_1 represents the duty cycle when $v_{AB}=V_{in}$, D_2 represents the duty cycle when $v_{AB}=V_{in}/2$, D_3 represents the duty cycle that from the beginning to Q_5 turning ON. d represents the duty cycle corresponding to operational mode III, and satisfying

$$M = \frac{V_{out}}{V_{in}} = \frac{D_1 + D_2 / 2}{D_1 + D_2 + d - D_3} .$$ (3)

In actual control, $-I_v$ current detection is used to turn OFF Q_5. When $D_1+D_2+d=1$, operational mode 0 no longer appears, that is the boundary conduction mode (BCM).

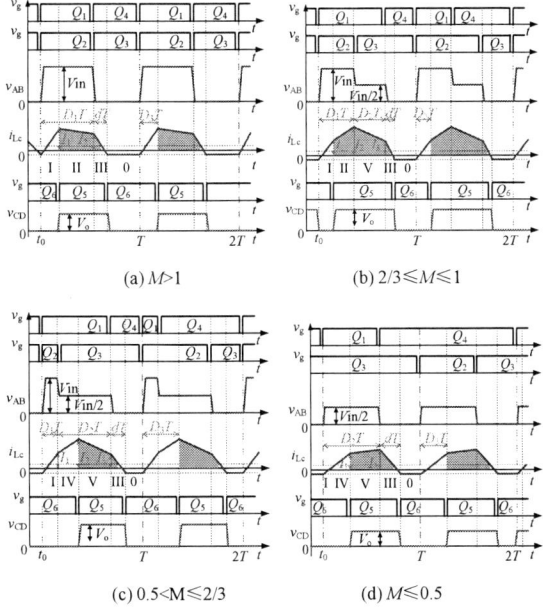

(a) $M>1$ (b) $2/3 \leqslant M \leqslant 1$

(c) $0.5<M\leqslant 2/3$ (d) $M\leqslant 0.5$

Fig. 3. Key waveforms of the converter at different voltage intervals.

There are three duty cycle control variables at fixed switching frequency, D_1, D_2 and D_3. The feasible combinations of control variables corresponding to any operating condition are innumerable. Therefore, various feasible control strategies exist.

This paper proposes a relatively straightforward control strategy by set limiting conditions for the duty cycles. As shown in Table II, when $M>1$, $V_{in}/2$ is not suitable to boost the output voltage, so $D_2=0$; when $M\leqslant 0.5$, V_{in} is not required for $V_{in}/2$ is even more appropriate, so $D_1=0$. When $0.5<M\leqslant 1$, it need smooth transition among the above intervals, and splits into two different situations based on the relationship between D_1 and D_3. The waveforms are different as Fig. 3(b) and Fig.3(c) shown.

TABLE II. THE LIMITING CONDITIONS FOR THE DUTY CYCLES

Voltage Ratio	The limiting conditions	
$M>1$	$D_2=0$	
$2/3<M\leqslant 1$	$\dfrac{D_2}{D_1}=\dfrac{2-2M}{2M-1}$	$D_3<D_1$
$0.5<M<2/3$	$\dfrac{D_1}{D_2}=\dfrac{2M-1}{2-2M}$	$D_1\leqslant D_3$
$M\leqslant 0.5$	$D_1=0$	

It is necessary to ensure ZVS soft switching with inductor current regardless of the operating conditions. To simplify the calculation, I_v is represented by the duty cycle D' as

$$D' = \frac{2I_v L}{V_{in}T} .$$ (4)

Find the maximum output current value based on the limiting conditions in Table II. When $M>1$, the operational principles are the same to the step-up modes in FSBB[12], the control variables and the output current of the maximum point are

$$D_1 = \frac{M^2 + M - MD'/2}{M^2 + M + 1}, D_2 = 0, D_3 = \frac{M^2 + D'/2}{M^2 + M + 1}.$$ (5)

$$I_{out\_MAX} = \frac{V_{in}T}{2L} \frac{M-(M+1)D'+D'^2/4}{M^2+M+1} .$$ (6)

When $M\leqslant 0.5$, the results are similar as

$$\begin{aligned} D_1 &= 0 \\ D_2 &= \frac{4M^2 + 2M - 2MD'}{4M^2 + 2M + 1} \\ D_3 &= \frac{4M^2 + D'}{4M^2 + 2M + 1} \end{aligned}$$ (7)

$$I_{out\_MAX} = \frac{V_{in}T}{2L} \frac{M-(2M+1)D'+D'^2/2}{4M^2+2M+1} .$$ (8)

When $0.5<M\leqslant 1$, there are four operational modes besides Mode 0, and the maximum output current always occurs in BCM, so the formula is satisfied:

$$\begin{cases} V_{in}\left(D_1 + \dfrac{D_2}{2}\right) = V_{out}\left(1-D_3\right) \\ \dfrac{D_2}{D_1} = \dfrac{2-2M}{2M-1} \; or \; \dfrac{D_1}{D_2} = \dfrac{2M-1}{2-2M} \end{cases} .$$ (9)

When $D_3<D_1$, the results of maximum output point are

$$\begin{aligned} D_1 &= \frac{2M^2 + M - 1 - MD' + D'/2}{2M^2 - M + 2} \\ D_2 &= \frac{2 - 2M^2 + MD' - D'}{2M^2 - M + 2} \\ D_3 &= \frac{2M^2 - 2M + 1 + D'/2}{2M^2 - M + 2} \end{aligned}$$ (10)

$$I_{out\_MAX} = \frac{V_{in}T}{2L} \frac{-M^2 + 3M - 1 - MD' - D' + D'^2/4}{2M^2 - M + 2} .$$ (11)

When $D_1\leqslant D_3$, the results of maximum output point are

$$D_1 = \frac{2M - MD' - 1 + D'/2}{2 - M}$$

$$D_2 = \frac{2 - 2M - D' + MD'}{2 - M} \qquad . \qquad (12)$$

$$D_3 = \frac{1 - M + D'/2}{2 - M}$$

$$I_{out\_MAX} = \frac{V_{in}T}{2L} \frac{M/2 - D' + D'^2/4}{2 - M} . \qquad (13)$$

The boundary condition between Fig. 3(b) and Fig.3(c) is $D_3 = D_1$, and the results of (10) and (12) are consistent, both being $M = 2/(3-D')$. And the corresponding output current values are also equal by (11) and (13). Due to the small value of D', the simplified approximation column for the critical point in the previous discussion is written as $M = 2/3$.

The variation of the duty cycles corresponding to the maximum output point when $D' = 0.05$ is as shown as Fig.4.

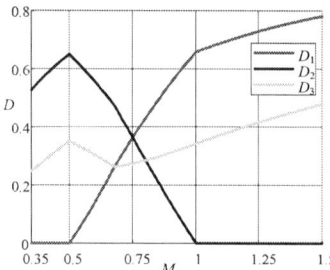

Fig. 4. Duty cycles corresponding to the maximum output point.

Since the control object of the converter is keeping the output voltage constant, the actual output current value is divided to the per-unit-value for simple analysis:

$$I_{out}^* = I_{out} / \frac{V_{out}T_s}{2L} \qquad (14)$$

Draw the variation curve of maximum output current per-unit-value under different M when $D' = 0.05$, as the black line shown in Fig. 5.

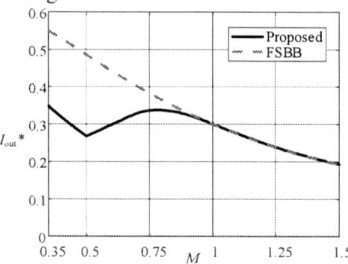

Fig. 5. The maximum output current per-unit-value under different M.

The maximum output current per-unit-value under different M of FSBB is the purple dashed line in Fig.5[12]. It is also the maximum output current of three-level Buck-Boost converter only contains operational modes I, II and III. However, with the introduction of operational modes IV and V, as well as the limiting conditions for the duty cycles, the output power is significantly decreased. Since the rated power of the load is parallel to the x-axis line in Fig.5, the proposed

control strategy in this paper is more suitable for wide input voltage range.

IV. CLOSED LOOP CONTROL STRATEGY AND SIMULATION RESULTS

Most of the state-of-the-art modulation schemes require the 2-D or 3-D look-up-table (LUT) to achieve high efficiency and other optimal performances. Those control strategies are complex and cannot be directly used. This paper proposes a linear modulation method, as shown in Fig.6. When the load current increases, each duty cycle increases proportionally until reaching the BCM mode. The heavier the load, the shorter the time interval of Mode 0. Although the efficiency is not the most perfect, it is acceptable with ZVS soft switching of all power devices.

In the BCM state, variable frequency control can further increase the output power. However, the control complexity increases at the same time. This paper adopts the simple fixed frequency control that can be used easily, and the closed-loop control system is shown in Fig. 7.

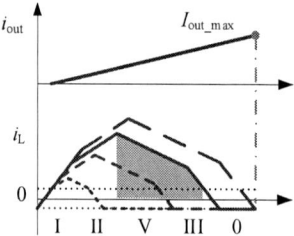

Fig. 6. Principle of the proposed modulation scheme using $2/3 \leqslant M \leqslant 1$ as an example.

The variables sampled in the main circuit are V_{in}, V_{out}, and i_L. The voltage ratio M can be calculated. D' should be obtained when the parameters are determined. In the proposed trajectory module, the values of maximum output point are calculated, and combined with the output of the PI controller v_{con} to determine the real-time duty ratio D_1, D_2 and D_3 linearly. Once i_L dropping to $-I_v$ is detected, signal Q_{5end} will turn OFF Q_5 and turn ON Q_6 after the dead time. Next, the gate signal module transforms the duty cycles and Q_{5end} signal into actual gate signals to the converter. There is no mode transition over voltage range, and the linear calculation is capable using the real-time computation.

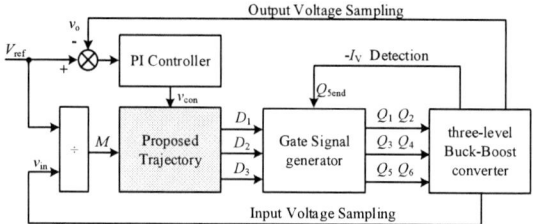

Fig. 7. Closed-loop control system of the three-level Buck-Boost converter.

In order to validate the feasibility of the proposed control strategy in this paper, a three-level Buck-Boost converter simulation model was established using PSIM software. The calculation of duty ratios can be realized in the Simplified C Block of PSIM. The model parameters are as follows: inductance is 4μH, output voltage is 48V, and input voltages

979-8-3315-1110-4/25 $31.00 © 2025 IEEE

are 35V, 60V, 85V and 160V. It can realize ZVS of all power switches with expected current waveform trajectories. In Fig. 8, the ZVS of Q_1 and Q_6 with $-I_v$ are shown.

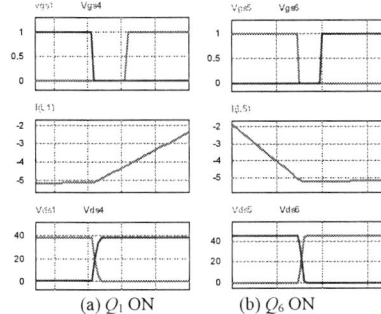

(a) Q_1 ON   (b) Q_6 ON

Fig. 8.   ZVS of Q_1 and Q_6 with $-I_v$.

The load step waveforms are given in Fig. 9 at V_{in}=35V (M>1) and V_{in}=140V (M<0.5) respectively. From the zoom in waveform, the lowest negative current is maintained at $-I_v$. And when the load changes heavy, the time interval of Mode 0 is decreasing correspondingly.

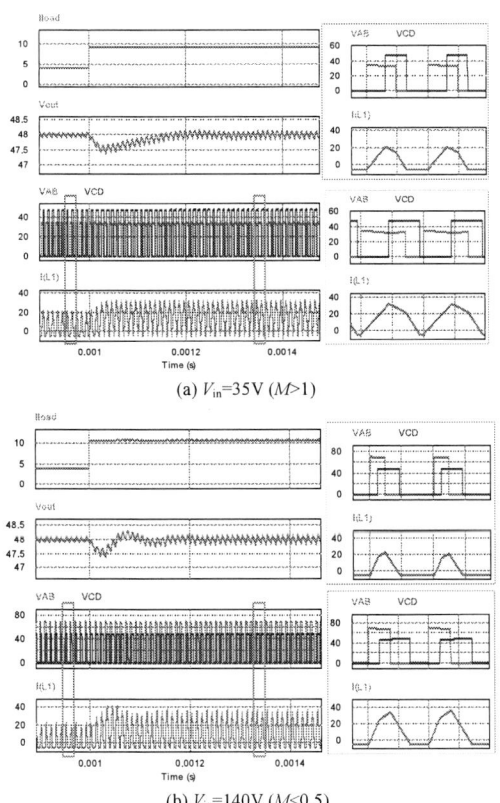

(a) V_{in}=35V (M>1)

(b) V_{in}=140V (M<0.5)

Fig. 9.   Waveforms of load step.

Fig. 10 shows that the input voltage jumps from 85 V to 60V while the load is constant. From the zoom in waveform, v_{AB} and v_{CD} changes as the proposed operational principles shown in Fig.3 (b) and (c). The output voltage can quickly be stabilized in many periods. The converter can work stably under different working conditions with smooth transition.

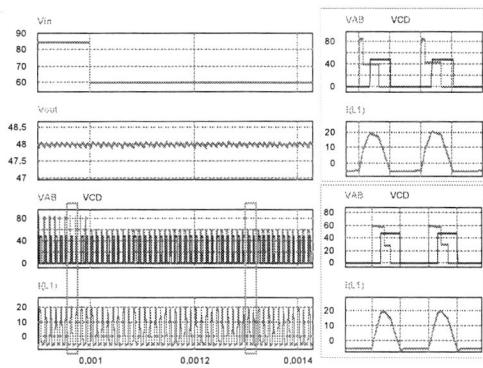

Fig. 10. Waveforms of input voltage step.

V.   CONCLUSIONS

This paper proposes a soft-switching fixed-frequency control strategy for wide-voltage-range, non-inverting three-level buck-boost converter. The mathematical analysis and calculation are given based on the operational principle with the limiting conditions. The linear control strategy is simple that can be calculate in real time，while ZVS of all switches ensure the efficiency. It is easy and convenient for engineering applications. From the simulation waveforms, the converter operates stably under different working conditions. In the future, more detailed analysis and control strategy optimization will be provided, and the prototype should be built and the actual ZVS effect and efficiency will be tested.

ACKNOWLEDGMENT

This work has been supported by the National Natural Science Foundation of China (52237008) and the Headquarters Technology Projects of State Grid Corporation of China (5200-202456095A-1-1-ZN).

REFERENCES

[1] P.Vinciarelli, "Buck-boost dc-dc switching conversion," U.S. Patent US006788033B2,Sep.7,2004.

[2] Xiao. H , and S. Xie . "Interleaving double-switch buck–boost converter," IET Power Electronics 5.6(2012):899-908.

[3] X. Ren, X. Ruan, H. Qian, M. Li and Q. Chen, "Three-Mode Dual-Frequency Two-Edge Modulation Scheme for Four-Switch Buck–Boost Converter," IEEE Transactions on Power Electronics. 24.2(2009):499-509.

[4] F. Liu, J. Xu, Z. Chen, P. Yang, K. Deng and X. Chen, "A Multi-Frequency PCCM ZVS Modulation Scheme for Optimizing Overall Efficiency of Four-Switch Buck–Boost Converter With Wide Input and Output Voltage Ranges, " IEEE Transactions on Industrial Electronics, 70.12:12431-12441, Dec. 2023.

[5] F. Liu, J. Xu, Z. Chen, R. Huang and X. Chen, "A Constant Frequency ZVS Modulation Scheme for Four-Switch Buck–Boost Converter With Wide Input and Output Voltage Ranges and Reduced Inductor Current, " IEEE Transactions on Industrial Electronics, 70.5:4931-4941, May 2023.

[6] Y. Shi. , and Yang. X . "Zero-Voltage Switching PWM Three-Level Full-Bridge DC–DC Converter With Wide ZVS Load Range," IEEE Transactions on Power Electronics. 10(2013):28.

[7] Y. Cao, Y. Bai, AU- V. Mitrovic, AU - B. Fan, AU - D. Dong, AU - R. Burgos, et al., "A Three-Level Buck–Boost Converter With Planar Coupled Inductor and Common-Mode Noise Suppression," IEEE Transactions on Power Electronics, 38. 9(2023):10483-10500.

[8] F. Li, R. Hao, H. Lei, X. You, C. Ke and J. Wang, "Non-Inverting Three-Level Buck-Boost Converter for Wide Voltage Range Application," 2018 IEEE Energy Conversion Congress and Exposition (ECCE) IEEE, 2018. *(references)*

[9] Q. Liu, Q. Qian, M. Zheng, S. Xu, W. Sun, and T. Wang, "An Improved Quadrangle Control Method for Four-Switch Buck-Boost Converter With Reduced Loss and Decoupling Strategy," IEEE Trans. Power Electron., vol. 36(9):10827-10841, 2021.

[10] Z. Zhou, H. Li, and X. Wu, "A Constant Frequency ZVS Control System for the Four-Switch Buck–Boost DC–DC Converter With Reduced Inductor Current," IEEE Trans. Power Electron., 34.7:5996-6003, 2019.

[11] L. Tian, X. Wu, C. Jiang, and J. Yang, "A Simplified Real-Time Digital Control Scheme for ZVS Four-Switch Buck–Boost With Low Inductor Current," IEEE Trans. Ind. Electron, 69.8:7920-7929, 2022.

[12] Z. Fu, F. Li, F. Zhang, Y. Zhang, J. Xu and M. Qiu, "A Simplified Quadrangle Control Strategy for Four-Switch Buck-Boost Converter, " 2024 IEEE 10th International Power Electronics and Motion Control Conference (IPEMC2024-ECCE Asia), Chengdu, China: 4522-4527, 2024.

979-8-3315-1110-4/25 $31.00 © 2025 IEEE

2025 IEEE Workshop on Wide Bandgap Power Devices and Applications in Asia (WiPDA Asia)

High-Robust Power Integrated Synchronization Scheme for the Grid-forming Inverter

Wen Zou
School of Electrical and Power Engineering
Hohai University
Nanjing, China
wen.zw@hhu.edu.cn

Yuying He
School of Electrical and Power Engineering
Hohai University
Nanjing, China
heyuying@hhu.edu.cn

Li Zhang
School of Electrical and Power Engineering
Hohai University
Nanjing, China
zhanglinuaa@hhu.edu.cn

Dongsheng Yang
Department *of Electrical Engineering*
Eindhoven University of Technology
Eindhoven, Netherlands
d.yangl@tue.nl

Abstract—**Synchronization stability is a critical foundation for both grid-following (GFL) and grid-forming (GFM) inverter systems. However, the widely adopted phase-locked loop (PLL) in GFL inverters and power synchronization control (PSC) in GFM inverters demonstrate complementary characteristics in terms of grid strength adaptability. From a control structure perspective, their inherent similarity and duality can be clearly identified, offering new insights into their potential integration. This paper proposes a power integrated synchronization for the robust operation of grid-forming inverters. This method synergistically integrates PLL and PSC technologies within the GFM inverter framework, thus enabling the system to achieve plug-and-play functionality irrespective of grid strength variations. Extensive experimental results validate the theoretical analysis and demonstrate the effectiveness of the proposed methodology.**

Keywords—*PLL, PSC, synchronization, stability.*

I. Introduction

Grid-tied inverters, acting as a crucial interface between distributed renewable energy generation systems and the utility grid, hold significant importance in modern power systems. To accommodate the diverse penetration levels of power electronics in modern power systems, the integration of both voltage-source inverters and current-source inverters is essential [1]. Voltage-source inverters, which can establish grid frequency and voltage amplitude, are known as grid-forming inverters, while current-source inverters, characterized by their ability to synchronize with grid voltage and frequency, are referred to as grid-following inverters [2].

Grid-following inverters exhibit reduced robustness under stiff grid conditions but enhanced robustness under weak grid conditions, whereas grid-forming inverters demonstrate weaker robustness under stiff grid conditions and stronger robustness under weak grid conditions [3]. Given the complementary stability characteristics of the two types of inverters under varying grid strengths, many scholars have explored the integration of the outer synchronization control strategies of the two types of inverters, a methodology termed hybrid synchronization control. Ref. [4] and [5] elucidated that the hybrid synchronization control strategy functions analogously to a damping winding in grid-forming inverters, thereby enhancing the transient stability of the system. On the

other hand, Ref. [6] validated through impedance modeling that the hybrid synchronization strategy can improve the small-signal stability of grid-following inverters under weak grid conditions. Although the aforementioned studies have offered comprehensive analyses of hybrid synchronization control strategies, they primarily focused on the direct superposition of the two synchronization methods, overlooking the intrinsic structural relationships between GFL and GFM inverters.

With the deepening understanding of synchronization principle, stability characteristics, dynamic behaviors, and physical properties of both GFL and GFM inverters, significant advancements have been achieved in exploring the intrinsic connections between these two types of inverters. Early work in Ref. [7] highlighted the structural similarities between the PLL synchronization in GFL inverters and the PSC in GFM inverters. Subsequently, Ref. [8] comprehensively summarized and delineated the similarities and differences in the control structures of GFL and GFM inverters. Furthermore, Ref. [9] provided a detailed analysis of the dual characteristics of GFL and GFM inverters, focusing on their control structures, swing characteristics, and adaptability ranges to grid strength.

Building upon the comprehensive analysis of GFL and GFM inverters presented in the references, this study introduces a novel integrated power synchronization control strategy within the GFM inverter paradigm. This approach significantly bolsters the operational stability of GFM inverters in robust grid environments, thereby facilitating a seamless plug-and-play capability for these inverters.

II. Power Synchronization for the Robust Operation of Grid-forming Inverters

A. Duality of Inner-loop

For inner-loop control of grid-following and grid-forming inverters, GFL inverters employ vector current control to regulate the inverter as a current source, whereas GFM inverters utilize dual-loop voltage-current control to operate the inverter as a voltage source. The equivalent models of both inverter types are illustrated in Fig. 1., where i_{GFL} represents the equivalent current source of the GFL inverter, Y_{GFL} denotes the equivalent admittance of the GFL inverter, u_{GFM} signifies the equivalent voltage source of the GFM inverter, and Z_{GFM} corresponds to the equivalent impedance of the GFM inverter. Additionally, i_g represents the grid-connected current, u is the voltage at the point of common coupling (PCC), Z_g is the grid impedance, and v_g is the grid voltage. This clearly

This work is supported in part by National Natural Foundation of China under Grant 524036711, in part by the Natural Science Foundation of Jiangsu Province under Grant SBK2023045379, and in part by "Chunhui Plan" Collaborative Research Project of the Ministry of Education under Grant HZKY20220141.

979-8-3315-1110-4/25 $31.00 © 2025 IEEE

demonstrates the dual characteristics of their inner-loop control structures.

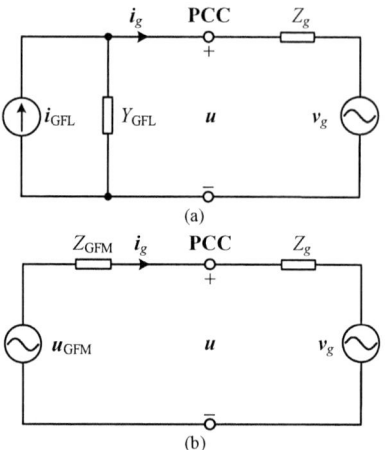

(a)

(b)

Fig. 1. Duality between the equivalent model of the GFL and GFM inverters. (a) Equivalent model of GFL inverter. (b) Equivalent model of GFM inverter.

B. Power integrated Synchronizations

The GFL inverter is synchronized with the grid through the PLL, where θ denotes the synchronization phase angle. In contrast, the GFM inverter is synchronized by power loops via the widely-used P-ω droop method, where the output active power P are calculated by the instantaneous power theory and the low-pass filter (LPF) is usually embedded to emulate the inertia. The synchronization parts of the GFL and GFM inverter seem to be irrelevant. In fact, Ref. [9] has already demonstrated the duality of the synchronization parts in GFL and GFM inverters, as illustrated in Fig.2.

The original block diagram of the PLL is shown in Fig. 2(a), where the q-axis voltage u_q is regulated to zero via the PLL regulator $G_{PLL}(s) = K_{PLLp} + K_{PLLi}/s$, where K_{PLLp} is the proportional gain and K_{PLLi} is the integral gain. The original block diagram of the PSC is shown in Fig. 2(b), where m_p is gain of P-ω droop. Observing Fig. 2, the u_q PLL for the GFL inverter is equivalent to the Q-ω droop(QSC) as shown in Fig. 2(c). This gives the duality relationship between the synchronization of GFL and GFM inverters.

Inspired by the duality of the GFL and GFM inverters from the perspectives of their structure and synchronization, we attempt to integrate the two kinds of synchronization so as to make the best of them. Thus, the power integrated synchronization for the robust operation of grid-forming inverters is proposed based on the duality theorem hereinafter, as shown in Fig. 2(d), ensuring the grid-synchronization stability in stiff grids. This actually aligns with intuition, as the instability of GFM inverters under stiff grid conditions typically manifests as sub-synchronous or super-synchronous oscillations, which primarily fall within the frequency range dominated by the synchronization control loop.

III. STABILITY ANALYSIS FOR GFM INVERTERS WITH INTEGRATED SYNCHRONIZATION

This section will analyze the power integrated synchronization stability. In Ref. [10], a SISO model is developed to describe the grid-synchronization characteristics of the PLL-based inverter precisely, and intuitively show how the PLL interact with the other parts. On the basis of this work, this section will establish the SISO models for GFM inverter of power integrated synchronization, and thereby unfold its grid-synchronization dynamics.

Due to space limitations, the detailed small-signal modeling of the system is not fully elaborated in this paper. Based on the analysis in the previous section, the closed-loop transfer function block diagrams of the SISO models for power integrated synchronization are directly presented, as illustrated in Fig. 3. Where $\tilde{\delta}$ are the perturbed value of the power angle δ. The power angle δ is also the phase angle between the inverter's dq frame and the grid voltage, i.e., $\delta = \theta - \theta_g$.

Based on Fig. 3, the bode diagrams of the SISO open-loop gains of the power integrated synchronization under stiff grid can be depicted, as shown in Fig. 4. By comparing with the original loop gains, owing to the introduced $f_{\delta\_GFMq}(s)QSC(s)$ in the loop gain, the phase plot of open-loop gains can also be raised to be located above the line of $-180°$, resulting in sufficient phase margin under stiff grid. Therefore, the system stability of the GFM inverter could be enhanced by the proposed power integrated synchronization.

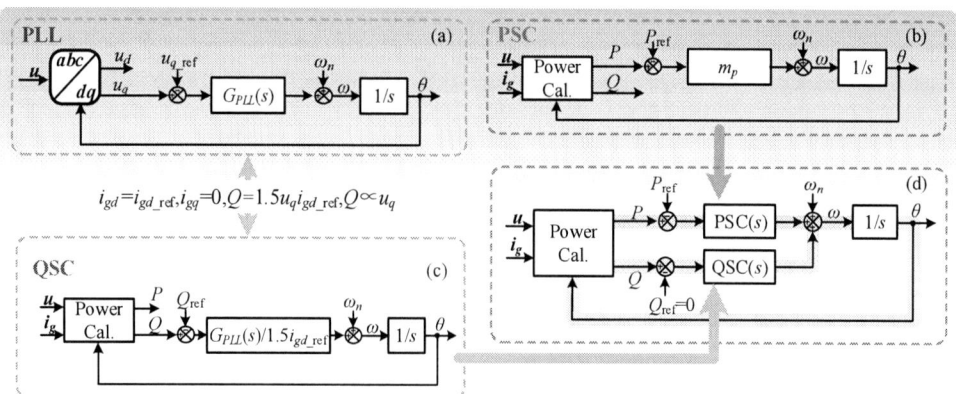

Fig. 2. Power integrated Synchronizations

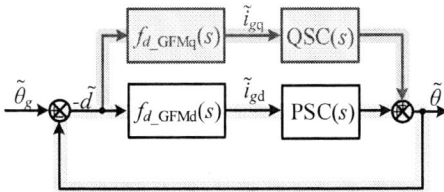

Fig. 3.   Closed-loop SISO model of the GFM inverter with the power integrated synchronization.

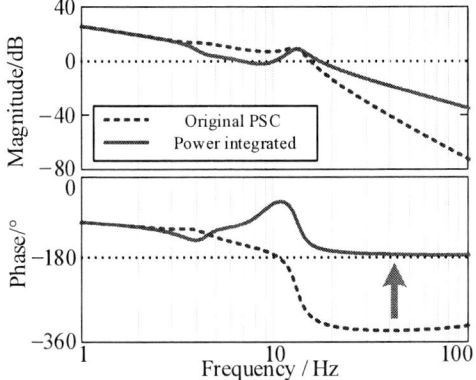

Fig. 4.   Bode diagrams of the GFM inverter with the conventional PSC and the proposed power integrated synchronization.

IV.   VERIFICATION

To verify the accuracy of the proposed power integrated synchronization and the correctness of the theoretical analysis, experiments are performed. The main circuit of the GFM inverter and the stiff grid are implemented on the OPAL-RT Target Real-time Platform. The control strategy of the inverter is implemented through digital control. The output voltage and current of the inverter are measured by the oscilloscope Tektronix MDO34. The parameters for these experiments are listed in Table I.

TABLE I.   TABLE TYPE STYLES

System Parameters			
Parameter	Value	Parameter	Value
Input voltage V_{dc}	700 V	Fundamental frequency f_o	50 Hz
Grid voltage V_g (RMS)	220 V	Switching frequency f_{sw}	20 kHz
Output power P_o	10 kW	Sampling frequency f_s	20 kHz
Inverter-side inductor L_f	3.2 mH	Filter capacitor C_f	10 µF

Fig. 5 shows the experimental waveforms of the GFM inverters under the stiff grid condition of SCR = 29. As shown, obvious oscillation exists in the GFM inverters with the conventional PSC, while the system with the proposed power integrated synchronization remains stable.

The above results confirm that the power integrated synchronization method is effective to enhance the system stability of GFM inverter under stiff grid conditions.

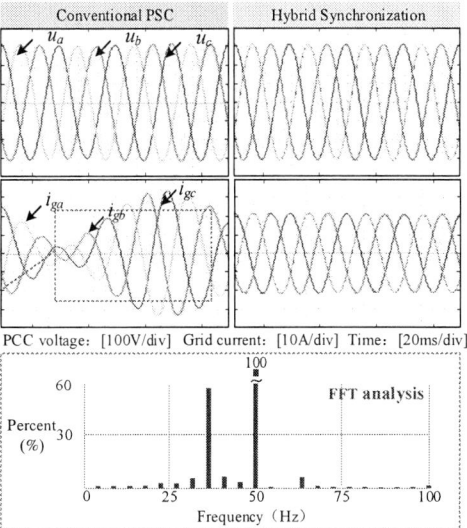

Fig. 5.   Experimental results of the GFM inverter with the conventional PSC and the proposed power integrated synchronization under SCR = 29.

V.   CONCLUSION

This paper proposes a novel power integrated synchronization control method for GFM inverters, which combines the QSC transformed from the PLL with the conventional PSC of GFM inverters. The proposed method significantly enhances the stability of GFM inverters under stiff grid conditions.

REFERENCES

[1]   J. Matevosyan et al., "Grid-Forming Inverters: Are They the Key for High Renewable Penetration?," *IEEE Power and Energy Magazine*, vol. 21, no. 2, pp. 77-86, March-April 2023.

[2]   Y. Lin et al., "Pathways to the Next-Generation Power System With Inverter-Based Resources: Challenges and recommendations," *IEEE Electrification Magazine*, vol. 10, no. 1, pp. 10-21, March 2022.

[3]   Bahrani, "Power-Synchronized Grid-Following Inverter Without a Phase-Locked Loop,"*IEEE Access* vol. 9, pp. 112163-112176, 2021.

[4]   T. Liu and X. Wang, "Physical Insight Into Hybrid-Synchronization-Controlled Grid-Forming Inverters Under Large Disturbances," *IEEE Transactions on Power Electronics*, vol. 37, no. 10, pp. 11475-11480, Oct. 2022.

[5]   L. Harnefors, J. Kukkola, M. Routimo, M. Hinkkanen and X. Wang, "A Universal Controller for Grid-Connected Voltage-Source Converters," *IEEE Journal of Emerging and Selected Topics in Power Electronics*, vol. 9, no. 5, pp. 5761-5770, Oct. 2021.

[6]   F. Chen, L. Zhao, L. Harnefors and X. Wang, "Impedance Modeling for Quadrature-Axis Active Damping of PLL Dynamics," *2022 IEEE 23rd Workshop on Control and Modeling for Power Electronics (COMPEL)*, Tel Aviv, Israel, 2022, pp. 1-7.

[7]   Q. -C. Zhong and D. Boroyevich, "Structural resemblance between droop controllers and phase-locked loops," *IEEE Access*, vol. 4, pp. 5733–5741, 2016.

[8]   X. Wang, H. Wu, F. Zhao, "Stability and control of grid-forming converter," IEEE ECCE Europe 2023 Tutorial presentation.

[9]   Y. Li, Y. Gu and T. C. Green, "Revisiting Grid-Forming and Grid-Following Inverters: A Duality Theory," *IEEE Transactions on Power Systems*, vol. 37, no. 6, pp. 4541-4554.

[10] L. Huang et al., "Grid-Synchronization Stability Analysis and Loop Shaping for PLL-Based Power Converters With Different Reactive Power Control," *IEEE Transactions on Smart Grid*, vol. 11, no. 1, pp. 501-516, Jan. 2020.

2025 IEEE Workshop on Wide Bandgap Power Devices and Applications in Asia (WiPDA Asia)

Charge Control for Critical Conduction Mode Soft-switching Grid-Tied Inverters

Zhengzi Lei
School of Electrical and Power Engineering
Hohai University
Nanjing, China
hhulzz@hhu.edu.cn

Zhongshu Zheng
School of Electrical and Power Engineering
Hohai University
Nanjing, China
hhuzzs@hhu.edu.cn

Wenbo An
School of Electrical and Power Engineering
Hohai University
Nanjing, China
anwenbo@hhu.edu.cn

Li Zhang*
School of Electrical and Power Engineering
Hohai University
Nanjing, China
zhanglinuaa@hhu.edu.cn

Yuying He
School of Electrical and Power Engineering
Hohai University
Nanjing, China
heyuying@hhu.edu.cn

Abstract—Critical conduction mode (CRM) based soft switching inverters are beneficial for achieving efficient grid connection of photovoltaics. In order to improve system performance, a charge control scheme is proposed in this paper. By taking the output charge quantity of the bridge arm as the control target, the system model based on the output charge quantity of the bridge arm is established. The specific implementation processes of the proposed charge control scheme are provided, and the analysis is conducted in the discrete-domain. Finally, a 1-kW single-phase LCL-type grid-tied inverter is built for experiment verification. Experimental results demonstrate that compared to the average current control scheme, the proposed scheme significantly improves dynamic response capabilities during current fluctuations.

Keywords— one cycle control, soft-switching grid-tied inverter, digital control, dynamic response

I. Introduction

Over the past decade, researchers have been dedicated to enhancing and achieving the efficiency and compactness of LCL-type grid-tied inverters through the utilization of high-frequency silicon carbide (SiC) MOSFETs [1]. Furthermore, zero-voltage switching (ZVS) techniques, when applied to grid-tied inverters operating in CRM [2], can effectively minimize switching losses and overcome the limitations of switching frequency. In the realm of LCL grid-tied inverters employing CRM soft-switching technology, the average current control scheme is commonly adopted for grid-tied current control. In [3], a state-plane plot is combined with average current controls to calculate the ZVS turn-ON delay. However, this scheme necessitates a zero-crossing detection (ZCD) circuit and adjusts control dynamically based on switching cycle changes, resulting in increased implementation complexity. Recent studies have paid much attention to the analysis and design of the one cycle control (OCC) in the fields such as active Power Factor Correction (PFC), DC-DC converter [4]. With its advantages of simple implementation, small computational requirements and robust stability, an effective solution is offered by OCC to address the above deficiencies. However, as a nonlinear algorithm, traditional OCC also has its limitations. Due to its reliance on a fixed switching frequency, it cannot be directly applied to CRM grid-tied inverters, where the switching frequency varies dynamically.

This work was supported by the National Natural Science Foundation of China under Grant 52322705, in part by the Natural Science Foundation of Jiangsu Province under Grant BK20230037.

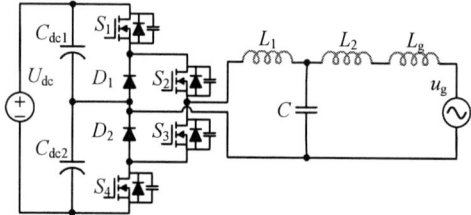

Fig. 1. The topology of CRM grid-tied inverter.

This paper draws inspiration from the concept of OCC and optimizes it to propose an improved charge control scheme tailored to the charge characteristics of CRM grid-tied inverters. A control model based on the output charge quantity Q_{inv} of the bridge arm is derived in this paper, the proposed charge control scheme is both simple and effective. Unlike the control scheme typically used in CRM grid-tied inverters, which focus on the control with the output voltage u_{inv} of the bridge arm. The control target of the charge control scheme is changed to the output charge Q_{inv} of the bridge arm, thereby enabling the indirect control of the inductance current on the inverter side. Compared to average current control, the dynamic response is faster. Besides, the burden on the digital controller of the proposed scheme is reduced, which is beneficial for reducing the control costs.

II. Review of the Conventional Average Current Control Model

A. Review of CRM Grid-tied Inverter with the Conventional Average Current Control

In principle, the conventional control model can be applied to various topologies. In this paper, the topology employed for the CRM grid-tied inverter is the three-level neutral-point-clamped (3L-NPC) topology, as depicted in Fig. 1, where C_{dc1} and C_{dc2} are dc-link capacitors. The LCL filter is composed of inverter-side inductor L_1, capacitor C, and grid-side inductor L_2.

The soft-switching CRM operation with conventional average current control is as follows. When the inverter-side current reaches zero, the inverter-side inductor and switch junction capacitors resonate, reducing the switch drain-to-source voltage to zero and enabling ZVS turn-ON. The soft-switching CRM operation has two typical implementation approaches. One is based on the Zero-Crossing Distortion (ZCD) circuit. The circuit is employed to generate the ZCD

979-8-3315-1110-4/25 $31.00 © 2025 IEEE   609

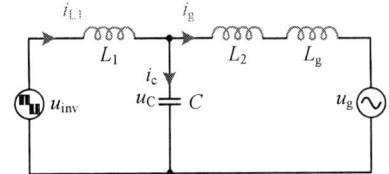

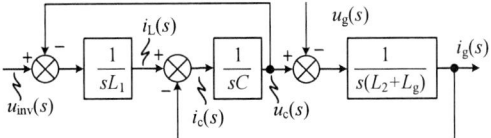

Fig. 2. The circuit of CRM grid-tied inverter based on the voltage among the bridge arms u_{inv}.

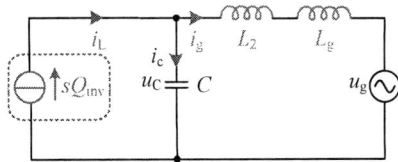

Fig. 4. The circuit of CRM grid-tied inverter based on the output charge quantity Q_{inv} of the bridge arm.

Fig. 3. The conventional control model of CRM grid-tied inverter.

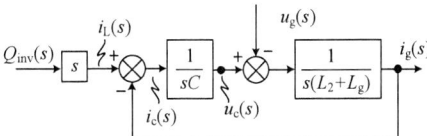

Fig. 5. The charge control model of CRM grid-tied inverter in the continuous-domain.

signal for synchronous switches, and the turn-OFF extension after the ZCD signal is used to generate the reverse inductor current. The other one is based on the direct digital calculation, plotting state-plane trajectories to determine off-time t_{off} and extension time t_{ext}. [5] proposed the state-plane plot based on a constant digital control frequency to calculate the ZVS turn-ON delay. This method avoids the ZCD circuit and reduces the performance demands on digital controllers.

B. Control Model of CRM Grid-tied Inverter Based on the Voltage among the Bridge Arms u_{inv}

Since the conventional CRM grid-tied inverter modeling primarily pays more attention to the relationship between u_{inv} and i_g, the circuit can be further simplified to the configuration shown in Fig. 2. Generally, the grid impedance at the PCC can be equivalent to an ideal voltage source u_g in series with a grid impedance Z_g. A pure inductor L_g is considered here to emulate the worst case which can be regarded as one part of the LCL filter. Consequently, the grid impedance Z_g is simplified as the line reactance L_g.

Based on the principles of KCL and KVL, the circuit equation can be formulated to indicate the relationship between the grid current i_g, the voltage among the bridge arms u_{inv} and the grid voltage u_g, as given in (1) and (2):

$$i_g(s) = G_1(s)u_{\text{inv}}(s) + G_2(s)u_g(s) \qquad (1)$$

Where

$$\begin{cases} G_1(s) = \dfrac{1}{L_1(L_2+L_g)Cs^3 + (L_1+L_2+L_g)s} \\ G_2(s) = \dfrac{-(L_1Cs^2+1)}{L_1(L_2+L_g)Cs^3 + (L_1+L_2+L_g)s} \end{cases} \qquad (2)$$

According to (1) and (2), the conventional control model of the CRM grid-tied inverter can be obtained in the continuous-domain, as illustrated in Fig. 3. It is a typical three-order model, which can be widely used in conventional control such as average current control.

III. PROPOSED CHARGE CONTROL FOR CRM GRID-TIED INVERTERS

Since the inductance current of inverter side i_{L1} returns to

zero at each switching cycle, there must be a certain mathematical relationship between the charge quantity flowing through the inductor and the turn-ON time of switching within one cycle. In this section, the control model based on the output charge quantity Q_{inv} of the bridge arm is constructed, and the specific implementation processes of the proposed charge control scheme are provided.

A. The Control Model Based on the Output Charge Quantity Q_{inv} of the Bridge Arm

Due to the integral relationship between the current i_{L1} of the inverter side and charge quantity Q_{L1} flowing through the inductor L_1 of the inverter side.

$$Q_{\text{inv}} = \frac{i_{\text{L1}}}{s} \qquad (3)$$

The output charge quantity Q_{inv} flows through the inductance of inverter side L_1, and subsequently is diverted to the capacitor C, the network side inductor L_2 and the grid impedance L_g. The output charge quantity Q_{inv} of the bridge arm is equal to the charge quantity Q_{L1} flowing through the inductor L_1 of the inverter side. Thus, the current i_{L1} can be expressed by the differential sQ_{inv} of the output charge quantity Q_{inv} from the bridge arm.

$$i_{\text{L}} = sQ_{\text{L1}} = sQ_{\text{inv}} \qquad (4)$$

Furthermore, the LCL circuit model, reformulated based on the output charge of the bridge arm, is shown in Fig 4. In this representation, the voltage u_{inv} of the bridge arm and the inductance L_1 of the inverter side can be equivalent to the current source sQ_{inv}, derived from the output charge of the bridge arm. From Fig. 4, the charge control model for the CRM grid-tied inverter can be obtained in the continuous-domain, as shown in Fig. 5. Then, the circuit equations can also be formulated to indicate the relationship between the grid current i_g, the voltage among the output charge quantity Q_{inv} of the bridge arm and the grid voltage u_g as follows:

$$i_g(s) = G_Q(s)Q_{\text{inv}}(s) + G_u(s)u_g(s) \qquad (5)$$

Where

979-8-3315-1110-4/25 $31.00 © 2025 IEEE

$$\begin{cases} G_Q(s) = \dfrac{s}{(L_2 + L_g)Cs^2 + 1} \\[2mm] G_u(s) = \dfrac{-sC}{(L_2 + L_g)Cs^2 + 1} \end{cases} \quad (6)$$

Here, the equivalence of the current source sQ_{inv}, leads to the model and circuit equations no longer containing the inductance L_1. Compared with the third-order control model based on the voltage among the bridge arms u_{inv}, the model based on the output charge quantity Q_{inv} of the bridge arm simplifies to a second-order model, achieving the reduction of the control system, which reduces the overall complexity of the system. Essentially, it is an optimized idea of OCC, by controlling the output charge quantity Q_{inv} of the bridge arm, the inductance current of inverter side i_{L1} is indirectly controlled.

B. Charge Control

As digital control is employed by the actual control system, the charge control of the CRM grid-tied inverter is further analyzed in the discrete-domain. According to the law of conservation of electric charge, within the kth control cycle, the following requirements must be satisfied by the total output charge quantity $Q_{inv}(k)$, the total charge quantity $Q_C(k)$ flowing through the filter capacitor C, and the total charge quantity $Q_2(k)$ flowing through the grid impedance ($L_2 + L_g$) of the grid side.

$$Q_{inv}(k) = Q_C(k) + Q_2(k) \quad (7)$$

Where

$$\begin{cases} Q_C(k) = C[u_C(k) - u_C(k-1)] \\[2mm] Q_2(k) = i_g(k)T_c \end{cases} \quad (8)$$

Then $Q_{inv}(k)$ can be obtained as

$$Q_{inv}(k) = C[u_C(k) - u_C(k-1)] + i_g(k)T_c \quad (9)$$

In the (k-1)th and kth switching cycles, the relationship between the capacitor voltage u_C, grid voltage u_g, and grid current i_g can be expressed as

$$\begin{cases} u_C(k) = (L_2 + L_g)\dfrac{i_g(k) - i_g(k-1)}{T_c} + u_g(k) \\[3mm] u_C(k-1) = (L_2 + L_g)\dfrac{i_g(k-1) - i_g(k-2)}{T_c} + u_g(k-1) \end{cases} \quad (10)$$

Based on (9) and (10), the following relationship can be deduced as

$$Q_{inv}(k) = C[u_g(k) - u_g(k-1)] + \left(\dfrac{L_2 C}{T_c} + \dfrac{L_g C}{T_c} + T_c\right)i_g(k)$$
$$- \left(\dfrac{2L_2 C}{T_c} + \dfrac{2L_g C}{T_c}\right)i_g(k-1) + \left(\dfrac{L_2 C}{T_c} + \dfrac{L_g C}{T_c}\right)i_g(k-2) \quad (11)$$

The flow diagram illustrating the specific implementation of charge control is depicted in Fig. 6. First, within the k-th

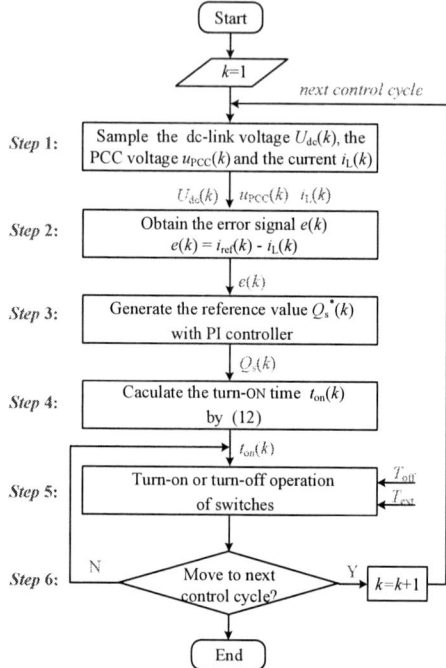

Fig. 6. The flow diagrams of charge control in the time-domain.

control cycle, the sampled inverter current $i_{L1}(k)$ is compared with the referenced current $i_{ref}(k)$, and the error signal $e(k)$ is obtained. After passing the PI controller, the command value $Q_s*(k)$ of the output charge of the bridge arm within a single switching cycle is generated. Next, through switching mode determination and switching time calculation module, the turn-ON time $t_{on}(k)$ of the kth control cycle are generated according to (12).

$$t_{on}(k) = \sqrt{\dfrac{4L_1 u_C(k)}{U_{dc}\left[\dfrac{U_{dc}}{2} - u_C(k)\right]}\left[Q_s(k) + \pi\sqrt{L_1 C_{oss}}\right]} \quad (12)$$

Upon application of the turn-ON time $t_{on}(k)$ to the drive circuit, the total charge output of the grid-tied CRM inverter in the kth control cycle is achieved, which can be represented as actually $Q_{inv}(k)$. Finally, bring $Q_{inv}(k)$ into the second-order control model under charge control, and then the charge control of CRM grid-tied inverter is successfully realized.

IV. EXPERIMENTAL RRSULTS

A 1-kW CRM grid-tied inverter prototype was built to verify the effectiveness of proposed scheme, as represented in Fig. 7 and the specifications of the prototype are presented in TABLE I. The leakage inductor of the transformer T is taken as the grid-side inductor L_2. For comparison, the average current control is operated for comparison, as referenced in [5].

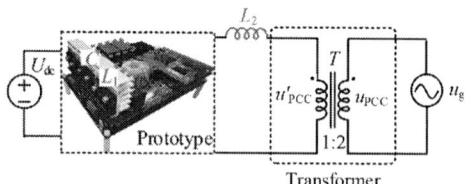

Fig. 7. The flow diagrams of charge control in the time-domain.

TABLE I. TABLE TYPE STYLES

Parameter	Value
DC-link (U_{dc})	500 V
Transformer ratio	1:2
Secondary side voltage (u_{PCC})	220 V / 50 Hz
L_1 / C_f / L_2	40 μH / 6.6 μF / 208 μH
Rated Power	1 kW
Control frequency (f_c)	60 kHz

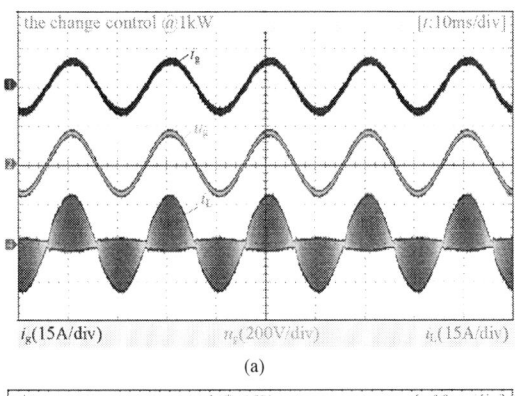

(a)

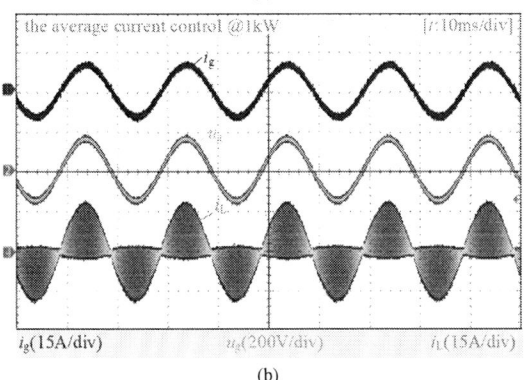

(b)

Fig. 8. The steady-state waveforms of CRM grid-tied inverter. (a) with the charge control scheme. (b) with the average current control scheme.

The steady-state waveforms with the charge control scheme and the average current control scheme are shown in Fig. 8 (a) and (b). It can be seen that the quality of the grid-tied current i_g exhibits good sinuosity when using the proposed charge control scheme. Similar to the average current control scheme, there is no obvious distortion of the grid-tied current i_g. The total harmonic distortion (THD) of the grid-tied current i_g with the charge control scheme is 2.35%, only slightly higher than the average current control scheme (2.15%). This proves that the proposed charge control scheme can also operate the inverter in CRM.

Additionally, the dynamic performances of the proposed charge control and the average current control schemes were evaluated. In order to emulate the worst-case scenario, step changes occurred at the peak of u_{PCC}. The output power of the grid-tied inverter was abruptly changed from 0.5kW to 1kW and then returned to 0.5kW within 0.1 seconds. As depicted in Fig. 9 (a) and (b), when the grid-tied current reference I_{ref} is step-changed, oscillations of grid-tied current i_g appears with the average current control scheme, while the grid-tied current i_g overshoot with the charge control scheme is smaller.

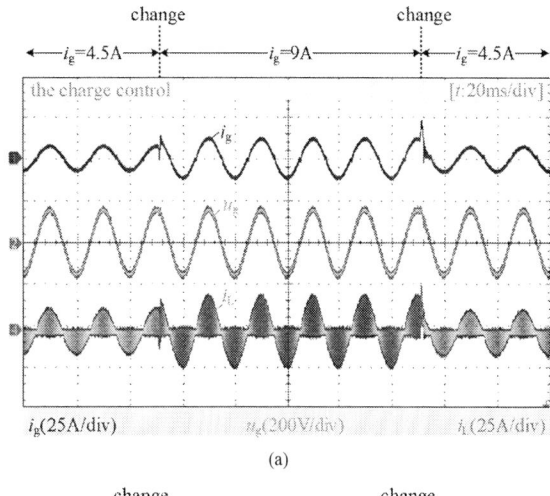

(a)

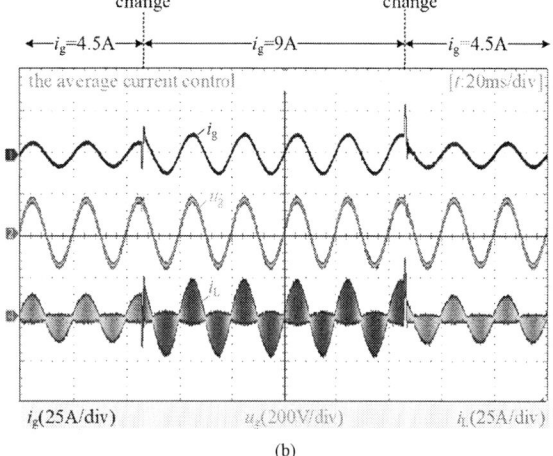

(b)

Fig. 9. The steady-state waveforms of CRM grid-tied inverter. (a) with the charge control scheme. (b) with the average current control scheme.

Therefore, the charge control scheme enhances the dynamic performance of the grid-tied inverter.

V. CONCLUSION

This paper proposed a charge control scheme for CRM-based soft-switching grid-tied inverters. Theoretical analysis and experimental results demonstrate that the charge control is practical, feasible, and straightforward to implement, effectively reducing the order of the model while maintaining satisfactory control performance with smaller calculation requirements. In future research, the application of proposed charge control scheme will be further extended to three-phase system.

REFERENCES

[1] C. Liu, K. Zhuang, Z. Pei, D. Zhu, X. Li, Q. Yu and H. Xin, "Hybrid SiC-Si DC–AC Topology: SHEPWM Si-IGBT Master Unit Handling High Power Integrated With Partial-Power SiC-MOSFET Slave Unit Improving Performance," IEEE Trans. Power Electron., vol. 37, no. 3, pp. 3085-3098, March 2022.

[2] Y. Yan, H. Gui and H. Bai, "Complete ZVS analysis in dual active bridge," IEEE Trans. Power Electron., vol. 36, no. 2, pp. 1247 1252, Feb. 2021.

[3] G. Son, Z. Huang, Q. Li, and F. Lee, "Analysis and control of critical conduction mode high-frequency single-phase transformerless PV

inverter," IEEE Trans. Power Electron., vol. 36, no. 11, pp. 13188–13199, Nov. 2021.

[4] C. Wang, H. Hu, H. Cheng, Z. Zhao and J. Liu, "Voltage Balancing Control of Cascaded Single-Phase VIENNA Converter Based on One Cycle Control With Unbalanced Loads," IEEE Access, vol. 8, pp. 95126-95136, 2020.

[5] Z. Zheng, L. Zhang, C. Wu, Y. Wang, Z. Lei and K. Sun, "Variable OFF-Time and Deadtime Scheme With Optimized Control Frequency for Soft-Switching Single-Phase Inverters," IEEE Trans. Power Electron., vol. 38, no. 4, pp. 4972-4987, April 2023.

Dynamic Analysis of Hybrid Photovoltaic-Battery System with Dual-Mode Grid-Forming Control

Wenjie Ning[1], Yiyang Liao[1*], Yaoyu Hu[1], Shaoze Zhou[2], Zheng Wei[2], Yitong Li[1]

*liaoyiyang2003@stu.xjtu.edu.cn

[1]School of Electrical Engineering, Xi'an Jiaotong University, China

[2]NARI Technology Co., Ltd., China

Abstract—To address the stability challenges of renewable energy integration under low-carbon energy transitions, this paper proposes a hybrid photovoltaic-battery system with dual-mode grid-forming control. The system dynamically switches between constant-power mode and frequency-supporting mode to adapt to different grid-supporting requirements. For the constant-power mode, the battery storage compensates the fluctuations of photovoltaic (PV) power generation. Meanwhile, the voltage-forming-frequency-following control is employed to the dc-ac inverter to support the ac grid voltage and regulate the dc bus voltage. For the frequency-supporting mode, the battery energy storage system (BESS) actively stabilizes the dc-bus voltage, while the dc-ac inverter actively supports the grid frequency and voltage magnitude through droop control. Small-signal stability analysis and time-domain validations demonstrate robust dynamic performance in both modes. Mode transitions induce only a 3% transient frequency deviation on the ac grid, confirming system stability and mode compatibility. The findings provide theoretical and technical insights for stable operation of high-penetration renewable energy systems.

Index Terms—hybrid PV-battery system, grid-forming inverter, small-signal stability, mode transition.

I. INTRODUCTION

The global imperative for low-carbon energy transitions has significantly elevated the role of renewable energy sources in modern power systems [1]–[3]. This shift has driven the proliferation of power electronic converters while precipitating stability challenges due to the phased retirement of conventional synchronous generators [4]. To maximize solar energy in photovoltaic (PV) generation systems, a novel hybrid PV-battery system has been developed. Unlike conventional renewable energy systems employing grid-following converters, the proposed architecture operates in two distinct grid-forming modes: constant-power mode and frequency-supporting mode [5]–[7]. This dual-mode capability enables adaptive power management under varying conditions.

This work was supported by the Science and Technology Project of NARI Technology Co., Ltd. under grant SGN-RGF00XAJS2412906.

The innovation of the hybrid system lies in its mode-switching capability based on battery energy storage system (BESS) dynamics and different grid-supporting requirements. When operating in constant-power mode the inverter employs a voltage-forming-frequency-following control [5] to maximize PV power utilization while performing dual voltage regulation (ac grid voltage stabilization and dc bus voltage support). When operating in frequency-supporting mode, the inverter adopts conventional active power droop control [8] to stabilize grid voltage amplitude and frequency and the BESS actively regulates dc bus voltage through boost converter.

II. OPERATION PRINCIPLES

The entire hybrid PV-battery system is depicted in Fig. 1. The system's energy supply is composed of two components: a PV generation system and a BESS. The two modules are individually interfaced via their respective boost converters, with their outputs coupled to form unified dc buses [9]. The dc bus is regulated to exhibit constant-power source characteristics through closed-loop control of the boost converters. And it is interfaced with an inverter to deliver energy to the ac grid. This inverter operates in two distinct control modes: voltage-forming-frequency-following control [5] and conventional droop control [6]. The two control modes can be dynamically transitioned between varying operating conditions to ensure system stability.

A. Constant-Power Mode

In this mode, the PV array operates as the dominant energy source, while the BESS provides active power compensation to maintain the dc bus's constant-power source characteristics under varying load conditions. According to [5], the synchronization mechanism of constant-power mode is depicted in the lower-right section of Fig. 2a. Building upon conventional matching control strategies [10], the voltage-forming-frequency-supporting incorporates a virtual resistor

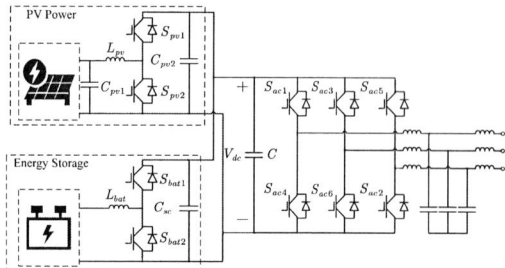

Fig. 1. Circuit diagram of the hybrid PV-battery system.

and a virtual pole-pair number to achieve ac side voltage formation capability and grid frequency tracking functionality while actively regulating dc bus voltage stability. The operating frequency of the inverter is governed by (1).

$$\omega = \omega^* + N\left(R\frac{p_s - p}{v_{\mathrm{dc}}} + v_{\mathrm{dc}} - v_{\mathrm{dc}}^*\right) \qquad (1)$$

where, ω is the angular frequency, ω^* is the angular frequency reference, N is the matching ratio, R is the virtual damping resistance, p_s is the ac-side measured active power, p is the dc-side measured active power, v_{dc} is the dc bus voltage, v_{dc}^* is the dc bus voltage reference.

The inverter that employs this control strategy exhibits stable and rapid dynamic responses to external grid frequency variations while enhancing the stiffness of the grid voltage. Concurrently, its dc-side voltage regulation capability enables adaptation to constant-power input.

Due to the dc bus voltage regulation capability inherent in the voltage-forming-frequency-following control mode, the PV boost converter could operate in MPPT mode [11]. The boost converter of the BESS employs a power compensation control scheme to maintain the constant-power source characteristics of the dc bus [12]. This is why the mode is called constant-power mode. The detailed architecture of the system is illustrated in Fig. 2a, which delineates the coordinated control strategies of the boost converters and the grid-tied inverter controlled by the voltage-forming-frequency-following control, highlighting their hierarchical interaction to enforce the regulation of dc bus voltage and the synchronization of the ac grid.

B. Frequency-Supporting Mode

In this mode, the BESS operates as the main energy source, keeping the dc bus's voltage constant. In terms of frequency-supporting capability, the power-flow control theory indicates that active power is primarily connected to the power angle and system frequency

[6]. So the output frequency of the ac voltage of the inverter can be expressed as

$$\omega = \omega^* - m\left(p^* - p\frac{1}{1 + s/\omega_f}\right) \qquad (2)$$

where ω is the angular frequency, ω^* is the angular frequency reference, m is the frequency droop gain, p^* is the active power reference, p is the measured active power, and ω_f is the bandwidth of the low-pass filter.

The block diagram illustrating the implementation of the frequency-droop equation is shown in the lower-right section of Fig. 2b [13]. In practical applications, a low-pass filter (LPF) is typically added to the measurement and feedback channel of the grid-side power signal. This filter has been proven to be a necessary condition for stabilizing the control loop [6]. Its introduction helps suppress disturbances and oscillations present in each power measurement while also providing a certain level of inertia to the system [13].

In contrast to the control modes discussed in Section II-A, droop control, while providing ac-side voltage magnitude and frequency regulation capabilities, inherently lacks dc bus voltage regulation functionality [4]. The BESS boost converter assumes responsibility for dc bus voltage stabilization [8].The detailed system architecture is illustrated in Fig. 2b.

C. Mode Selection

The hybrid PV-battery system could automatically select the most suitable operating mode based on the state of charge (SOC) of the battery energy storage system (BESS) and real-time grid requirements.

* Frequency-Supporting Mode: When the BESS possesses sufficient energy reserves (i.e., SOC> 60%), which can effectively stabilize the dc bus voltage, the droop control method (discussed in Section II-B) dominated by the BESS will be selected.
* Constant-Power mode: When the BESS has insufficient energy reserves (i.e., SOC< 60%) to adequately regulate the dc bus voltage, while photovoltaic (PV) generation remains robust, the PV-dominated voltage-forming-frequency-supporting (discussed in Section II-A) control strategy can be employed.

The dynamic characteristics and small-signal stability analysis of these two modes will be discussed in the next section.

III. CASE STUDIES

In this section, the small-signal stability and time-domain response characteristics of the two operational modes will be discussed. A comparative stability analysis is conducted to evaluate system performance in different operational modes and the switching process.

979-8-3315-1110-4/25 $31.00 © 2025 IEEE

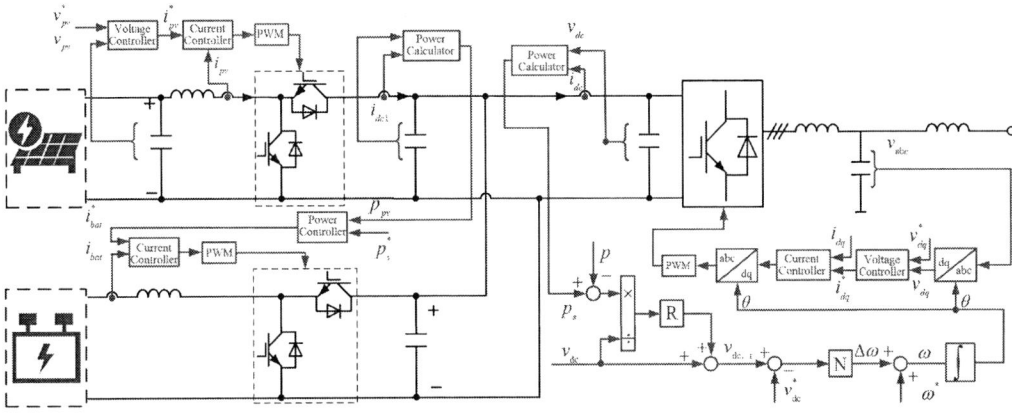

(a) Control structure of constant-power mode

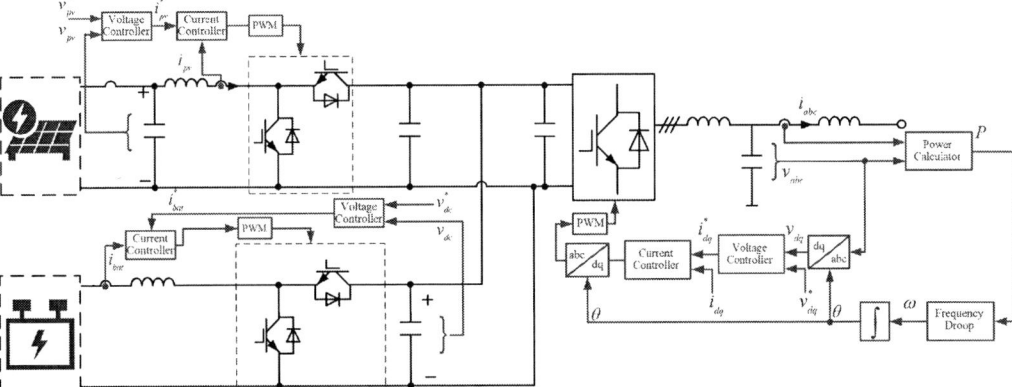

(b) Control structure of frequency-supporting mode

Fig. 2. Control structure comparison of two operating modes.

The results obtained from both analytical approaches are systematically cross-verified to ensure methodological consistency and conclusion validity [14].

A. Small-Signal Stability

Small-signal stability analysis of the system is conducted via a state-space modeling approach, yielding a pole plot of the system at its steady-state operating point [15]. The detailed pole map of the two operating mode is shown in Fig. 3. The two operational modes exhibit a closely aligned pole distribution, distinguishing by two critical aspects:

* DC Mode Emergence: The frequency support mode introduces an additional dc-dominant eigenvalue (λ_{dc}), reflecting the enhanced low-frequency voltage regulation dynamics inherent to this configuration.
* Oscillation Amplitude Discrepancy: The damping ratios (ξ) of oscillatory modes differ marginally between modes, with the constant-power mode demonstrating increased amplitude overshoot compared to the frequency-supporting case.

The detailed transient processes induced by small-signal disturbances will be analyzed in the subsequent section through time-domain response evaluations.

B. Time-Domain Responses

The small-signal stability of the system is validated through time-domain response analysis, yielding distinct transient response waveforms. These waveforms enable identification of dynamic modes, which are cross-validated against the system poles derived in the Section III-A to confirm analytical coherence. At $t = 1.5s$, a step disturbance is introduced to the grid-side voltage frequency, followed by a voltage magnitude perturbation at $t = 2.0s$, to evaluate the system's transient stability and dynamic mode interactions. Fig. 4 presents transient responses under constant-power mode and Fig. 5 presents responses under frequency-supporting mode.

The transient responses of the two operational modes demonstrate approximate equivalence under grid-side voltage disturbance conditions. The distinction lies in the dc bus voltage behavior during grid frequency perturbations. In constant power mode, the

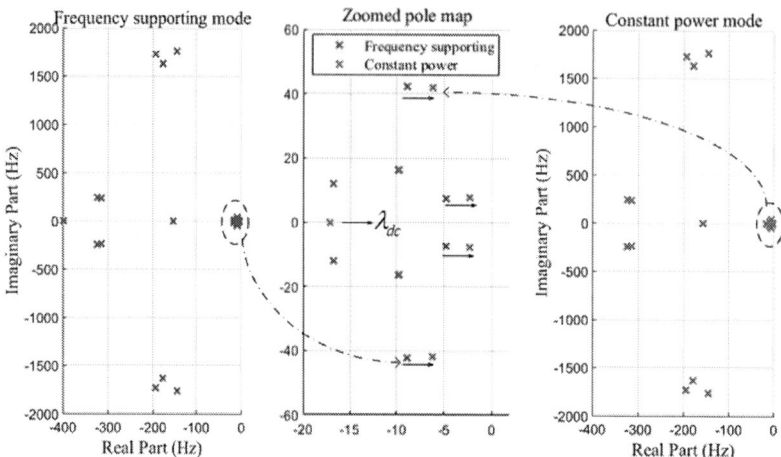

Fig. 3. Pole map of two operating modes

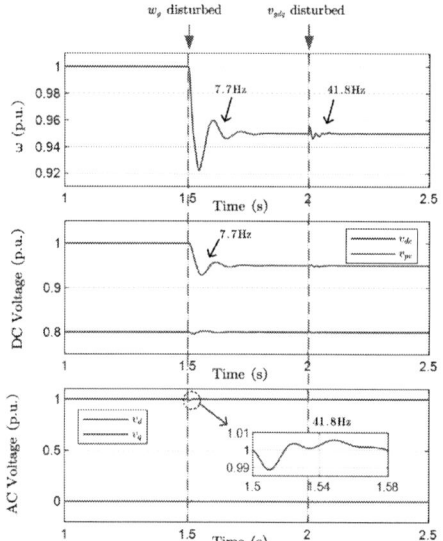

Fig. 4. Small-signal waveform of constant-power mode.

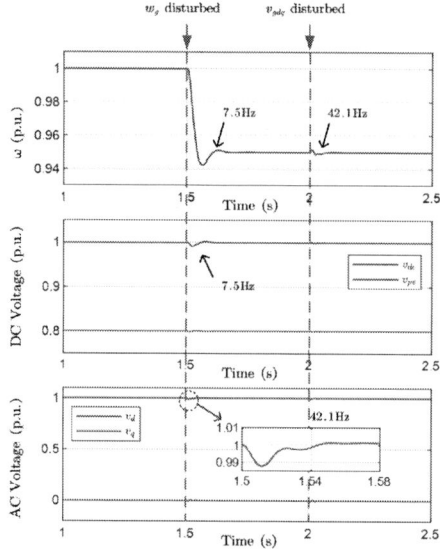

Fig. 5. Small-signal waveform of frequency-supporting mode.

dc bus voltage operating point varies with ac frequency fluctuations due to the inherent coupling relationship between dc voltage and ac frequency in this control strategy(as described in equation (1)). Conversely, in frequency support mode, the BESS actively regulates the dc bus voltage, thereby maintaining its operating point invariant regardless of ac frequency variations.

Fig. 6 demonstrates the system's operational states under both control modes. When PV output power fluctuates, the BESS actively compensates by adjusting its power injection to maintain stable energy delivery to the grid. At $t = 3.5s$, a mode transition is executed, during which the transient process induces only a 3% frequency deviation on the ac grid side. This result conclusively validates the operational compatibility between the two control modes.

IV. CONCLUSIONS

This paper proposes a hybrid PV-battery system with dual-mode grid-forming control. The state-space model is developed to evaluate small-signal stability of the whole system, revealing closely aligned pole distributions between two modes with minor damping ratio difference. Time-domain simulations demonstrate the robust dynamic performance of the whole system under disturbances and mode transitions. The key contributions include:

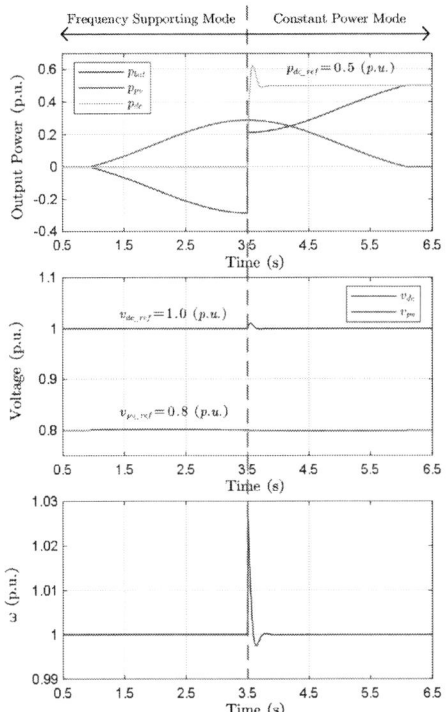

Fig. 6. Time response of different mode.

* A dual-mode control framework enabling adaptive operation under varying battery energy storage system (BESS) dynamics and ac grid conditions;
* The controller design can be guided by the established full-order state-space model;
* Validation of mode-switching feasibility with minimal transient impacts.

This work provides new insights into deploying hybrid PV-battery systems in future power girds with high penetration level of renewable energy.

ACKNOWLEDGEMENT

This work was supported by the Science and Technology Project of NARI Technology Co., Ltd. under grant SGNRGF00XAJS2412906.

REFERENCES

[1] R. Rosso, X. Wang, M. Liserre, X. Lu, and S. Engelken, "Grid-forming converters: Control approaches, grid-synchronization, and future trends—a review," *IEEE Open Journal of Industry Applications*, 2021.

[2] J. Cao and A. Emadi, "A new battery/ultracapacitor hybrid energy storage system for electric, hybrid, and plug-in hybrid electric vehicles," *IEEE Transactions on Power Electronics*, 2012.

[3] J. Fang, Y. Tang, H. Li, and X. Li, "A battery/ultracapacitor hybrid energy storage system for implementing the power management of virtual synchronous generators," *IEEE Transactions on Power Electronics*, 2018.

[4] C. Sun, S. Q. Ali, G. Joos, and F. Bouffard, "Design of hybrid-storage-based virtual synchronous machine with energy recovery control considering energy consumed in inertial and damping support," *IEEE Transactions on Power Electronics*, 2022.

[5] C. Ai, Y. Li, Z. Zhao, Y. Gu, and J. Liu, "An extension of grid-forming: A frequency-following voltage-forming inverter," *IEEE Transactions on Power Electronics*, 2024.

[6] J. Guerrero, L. G. de Vicuna, J. Matas, M. Castilla, and J. Miret, "Output impedance design of parallel-connected ups inverters with wireless load-sharing control," *IEEE Transactions on Industrial Electronics*, 2005.

[7] Y. Li, Y. Gu, Y. Zhu, A. Junyent-Ferré, X. Xiang, and T. C. Green, "Impedance circuit model of grid-forming inverter: Visualizing control algorithms as circuit elements," *IEEE Transactions on Power Electronics*, 2021.

[8] Y. Li, Y. Gu, and T. Green, "Revisiting grid-forming and grid-following inverters: A duality theory," *IEEE Trans. on Power Syst.*, 2022.

[9] J. M. Guerrero, J. C. Vasquez, J. Matas, M. Castilla, and L. Garcia de Vicuna, "Control strategy for flexible microgrid based on parallel line-interactive ups systems," *IEEE Transactions on Industrial Electronics*, 2009.

[10] C. Arghir and F. Dörfler, "The electronic realization of synchronous machines: Model matching, angle tracking, and energy shaping techniques," *IEEE Transactions on Power Electronics*, 2020.

[11] J. Rocabert, A. Luna, F. Blaabjerg, and P. Rodríguez, "Control of power converters in ac microgrids," *IEEE Transactions on Power Electronics*, 2012.

[12] H. S. Rizi, Z. Chen, A. Q. Huang, and P. Rodriguez, "A comprehensive strategy for grid forming control in dc coupled photovoltaic and battery energy storage inverters," *2024 IEEE 15th International Symposium on Power Electronics for Distributed Generation Systems (PEDG)*, 2024.

[13] S. D'Arco and J. A. Suul, "Equivalence of virtual synchronous machines and frequency-droops for converter-based microgrids," *IEEE Transactions on Smart Grid*, 2014.

[14] X. Li, Y. Hu, Y. Shao, and G. Chen, *Mechanism analysis and suppression strategies of power oscillation for virtual synchronous generator*. IEEE, 2017.

[15] P. Kundur, *Power system stability and control*. McGraw-hill New York, 1994, vol. 7.

2025 IEEE Workshop on Wide Bandgap Power Devices and Applications in Asia (WiPDA Asia)

Optimization Design of High-Speed and High-Voltage Switching Transient Characteristics for GaN HEMTs Using Multi-Objective Particle Swarm Optimization

Xiao Li, Zhuofan Xiong, Yushan Liu, Qiang Zhou
School of Automation Science and Electrical Engineering
Beihang University
Beijing, China
li_xiao@buaa.edu.cn

Abstract—**Wide-bandgap (WBG) devices, such as Gallium Nitride High Electron Mobility Transistors (GaN HEMT), have become core components in efficient power conversion systems due to their high switching speed and low conduction resistance. While WBG devices can significantly reduce switching losses and improve system efficiency, their high-speed switching introduces transient issues like overcurrent and overvoltage during switching process, which can affect the device's safety. Therefore, a balance between high efficiency and reliability must be found. To predict the high-speed switching characteristics of GaN HEMT, an accurate analysis model based on the device's datasheet is proposed in this paper. Building on this model, a Multi-Objective Particle Swarm Optimization (MOPSO) algorithm is used to optimize design parameters including gate resistance and parasitic inductance, ensuring the best balance between switching loss and device safe operation. The MOPSO algorithm improves system efficiency and suppresses overcurrent and overvoltage issues through global search and the generation of Pareto-optimal solutions. The findings provide a theoretical foundation and design guidance for the reliable application of GaN HEMT in efficient power conversion systems.**

Keywords—component, formatting, style, styling, insert (key words)

I. INTRODUCTION

With the rapid development of power electronics technology, wide-bandgap (WBG) devices, such as gallium nitride high-electron-mobility transistors (GaN HEMTs) and silicon carbide metal-oxide-semiconductor field-effect transistors (SiC MOSFETs), have become core components of high-efficiency power conversion systems due to their advantages of high switching speed, low on-resistance, and high power density [1-3]. Compared to traditional silicon-based devices, WBG devices can significantly reduce switching losses, thereby improving system efficiency and power density. However, the fast-switching speed also introduces issues such as turn-on overcurrent and turn-off overvoltage, which may threaten the safe operation of the devices. Therefore, while ensuring that the devices operate within the safe operating area (SOA) and EMI limits, it is necessary to balance efficiency and reliability and minimize switching losses [4-5].

To address these challenges, it is essential to find suitable methods for predicting switching characteristics. Currently, there are three main approaches for predicting switching losses, as well as turn-on overvoltage and overcurrent: 1) empirical formula-based calculations, 2) simulation-based calculations, and 3) analytical model-based calculations. Empirical formula-based methods are simple but less accurate

over a wide operating range, especially for GaN HEMTs. Due to their low gate charge and high transconductance, traditional assumptions (such as separating current commutation from voltage changes and assuming a constant Miller plateau) are not applicable. The high-speed switching of GaN HEMTs makes the effects of parasitic inductance and capacitance non-negligible, particularly the significance of common-source inductance in low-voltage, high-current applications [6]. Although analytical models have advantages in modeling nonlinearities and covering wide operating ranges [7-8], existing models still have limitations, especially in modeling the channel current (i_{ch}) and nonlinear capacitance of GaN HEMTs [9-11]. The lack of consideration for short-channel effects leads to inaccurate i_{ch} estimation, particularly during the current rise phase. Additionally, modeling nonlinear capacitance based on C-V curves has limitations. The reverse transfer capacitance C_{rss} (i.e., C_{gd}) cannot be directly applied to switching loss models, and the dynamic gate-drain charge Q_{gd} needs to be characterized more accurately. Therefore, existing analytical models require improvement, and more suitable datasheet-based methods are needed to accurately predict the switching transients of GaN HEMTs. Recent studies [12] have proposed extraction methods that combine C-V curves, gate charge curves, and transfer characteristic curves to establish more accurate switching transient analysis models. However, these studies mostly focus on turn-off losses, which account for only a small portion of total switching losses. Moreover, at high switching speeds, slower switching and larger parasitic inductance exacerbate overcurrent and overvoltage issues, compromising device safety. To address this, in Paper [13], an overvoltage suppression circuit (OVSC) with switching loss optimization and clamping energy feedback for SiC MOSFETs has been proposed. However, the additional circuit design increases complexity in practical applications.

To balance switching losses and safe operation, gate resistance (R_g) and parasitic inductance (L_{loop}) in PCB layout become critical design parameters. Gate resistance directly affects switching speed, while parasitic inductance is closely related to transient overshoot and oscillations. Therefore, selecting appropriate gate resistance and optimizing PCB layout to control parasitic inductance are key to achieving efficient and reliable switching operations.

Considering the above issues, an accurate analysis model based on the device's datasheet is proposed in this paper for GaN HEMT switching transients in high-speed and high-voltage applications, combined with a multi-objective particle swarm optimization (MOPSO) algorithm for design optimization. As shown in Table I, MOPSO can

979-8-3315-1110-4/25 $31.00 © 2025 IEEE

simultaneously optimize conflicting objectives such as switching losses and safe operation, effectively improving efficiency and ensuring device reliability through global search and Pareto optimal solutions [14-15]. The algorithm optimizes parameters such as gate resistance and parasitic inductance, achieving an optimal balance between switching losses and safe operation. This provides a theoretical foundation and design guidance for the reliable application of GaN HEMTs in high-efficiency power conversion systems.

TABLE I   COMPARISON OF MOPSO WITH NSGA-II, MOEA/D

Characterise	MOPSO	NSGA-II	MOEA/D
Implementation complexity	simple	medium	medium
Convergence rate:	fast	medium	medium
Parallelism	strong	medium	medium
Adaptability	strong	medium	dependency decomposition strategy

II. ANALYTICAL MODEL OF SWITCHING LOSS

As shown in Fig .1, this paper presents an equivalent circuit model for the switching transient characteristics of GaN HEMTs, incorporating key parasitic parameters. The model is based on device-level characteristics, such as channel current and parasitic capacitances (C_{gs}, C_{gd}, C_{ds}), and includes the equivalent circuits of the gate driver, coaxial current splitter, load inductance, and decoupling capacitors. Parasitic parameters of the PCB are extracted using Ansys, and the basic equations are derived using Kirchhoff's Current Law (KCL) and Kirchhoff's Voltage Law (KVL). The switching transient of SW_1 is analyzed in detail, while SW_2 is treated as a passive transistor in parallel with the load inductance, providing a theoretical foundation for circuit optimization.

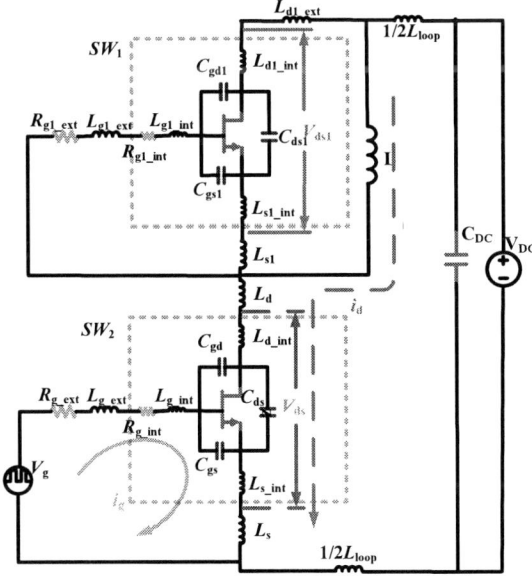

Fig.1. GaN HEMTs half-bridge equivalent circuit model

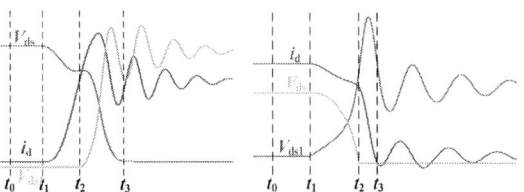

Fig.2. Typical waveforms of switching transition.

$$
\begin{cases}
i_d = i_{ch} + C_{ds}(V_{ds}) \cdot \dfrac{dV_{ds}}{dt} - C_{gd}(V_{gd}) \cdot \dfrac{dV_{gd}}{dt} \\
V_g = V_{gs} + R_g\, i_g + L_g \dfrac{di_g}{dt} + L_s \left(\dfrac{di_d}{dt} + \dfrac{di_g}{dt} \right) \\
i_g = C_{gs}(V_{gs}, V_{ds}) \cdot \dfrac{dV_{gs}}{dt} - C_{gd}(V_{gd}) \cdot \dfrac{dV_{gd}}{dt} \\
V_{DC} = V_{ds} + V_{ds1} + L_{loop} \dfrac{di_d}{dt} + R_{loop}\, i_d \\
i_d = I_L + C_{gd1}(V_{gd1}) \cdot \dfrac{dV_{ds1}}{dt} + C_{ds1}(V_{ds1}) \cdot \dfrac{dV_{ds1}}{dt} + i_{ch1}
\end{cases}
\tag{1}
$$

Among them:

$$
\begin{cases}
R_g = R_{g\_ext} + R_{g\_int} \\
R_{loop} = R_C + R_B \\
L_{loop} = L_{d\_int} + L_{s\_int} + L_{d\_int1} + L_{s\_int1} + L_{shunt} + L_B + L_C \\
L_g = L_{g\_ext} + L_{g\_int} \\
L_s = L_{s\_ext} + L_{s\_int} \\
C_{oss1} = C_{ds1} + C_{gd1}
\end{cases}
\tag{2}
$$

The channel current i_{ch} of a GaN HEMT is determined by V_{gs} and V_{ds}, as described by Equation (3), and is fitted and constructed as shown in Fig.3 based on datasheet's transfer characteristic curve data and output characteristic curve data. For nonlinear capacitance, they are traditionally described as functions of V_{ds}, but in reality, V_{gs} also significantly influences their behavior. Therefore, Equations (4) and (5) are used to model them, with additional construction based on Q-V curve data from the datasheet, as shown in Fig.4(a) and 4(b). Based on the above analysis, an accurate datasheet-based full-characteristics analytical model of GaN HEMTs for switching transient prediction is developed.

$$
i_{ch} = A_1 \cdot \ln\left(1 + e^{\frac{V_{gs}-k2 \cdot V_{TH}}{k3}}\right) \cdot \frac{V_{ds} \cdot (m+nV_{gs})}{1 + V_{ds} \cdot (x0\_0 + x0\_1 \cdot V_{gs})}
\tag{3}
$$

$$
C_{gs} = r_1 \cdot \frac{e^{\frac{Vgs-r2}{r3}}}{1 + e^{\frac{Vgs-r2}{r3}}} \mid r_4 \cdot \frac{e^{\frac{Vds-r5}{r6}}}{1 + e^{\frac{Vds-r5}{r6}}}
\tag{4}
$$

$$
C_{gd} = m_1 \cdot \frac{e^{\frac{Vgs-m2}{r3}}}{1 + e^{\frac{Vgs-m2}{m3}}} + m_4 \cdot \frac{e^{\frac{Vds-m5}{m6}}}{1 + e^{\frac{Vds-m5}{m6}}}
\tag{5}
$$

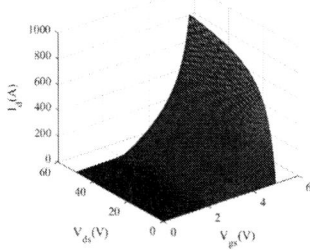

Fig.3. Fitted three-dimensional channel current model.

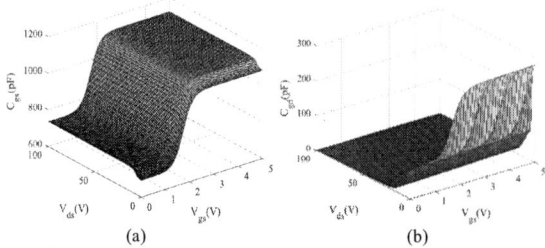

Fig.4. (a) Fitted $C_{gs}(V_{gs}, V_{ds})$ model; (b) Fitted $C_{gd}(V_{gs}, V_{ds})$ model.

Based on the above analysis, an accurate analytical model of switching loss for GaN HEMTs' switching transient prediction is developed. And the waveform of i_d and V_{ds} can be obtained. The turn-on and turn-off loss (P_{on}/P_{off}) can be represented by the following equations:

$$P_{on} = f_s \int_{t_1}^{t_3} i_{ch} v_{ds} dt = f_s \int_{t_1}^{t_3} (i_d + i_{Coss}) v_{ds} dt$$

$$= f_s \int_{t_1}^{t_3} i_d v_{ds} dt + P_{Coss} \tag{6}$$

$$P_{off} = f_s \int_{t_1}^{t_3} i_{ch} v_{ds} dt = f_s \int_{t_1}^{t_3} (i_d - i_{Coss}) v_{ds} dt$$

$$= f_s \int_{t_1}^{t_3} i_d v_{ds} dt - P_{Coss} \tag{7}$$

where i_{Coss} is the current discharging the output capacitance and P_{Coss} is the loss of the output capacitance. P_{Coss} can be obtained by the following equation:

$$P_{Coss} = f_s \int_0^{V_{DC}} v_{ds} C_{oss}(v_{ds}) dv_{ds} \tag{8}$$

To verify the proposed model's effectiveness, a double-pulse test platform was set up as shown in Fig.5. EPC2045 was used as the switching device, with LMG1205 as the driver IC. A DC power supply adjusted the DC link voltage, and a function generator provided PWM signals to the gate driver. V_{gs} and V_{ds} were measured using a Tektronix TPP1000 probe, while i_d was measured using a 2 GHz coaxial current shunt. The switching transient waveform were measured under different V_{ds} and I_L. The proposed model and the experimental results were compared as shown in Figs. 6-7. Therefore, it can be seen that the model proposed in this paper can more accurately describe the switching transient behavior of GaN HEMTs under high-speed and high-voltage conditions.

Based on this model, a design method is proposed by optimizing R_g and parasitic inductance (L_{loop}) in PCB layout, providing theoretical support for the reliable operation of GaN HEMT in high-speed and high-voltage applications. Detailed descriptions are given in the next sections.

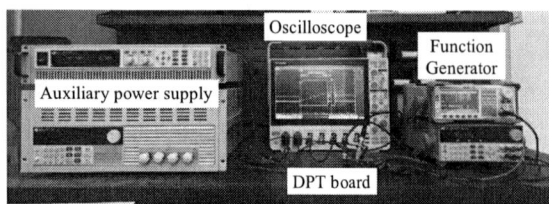

Fig.5. Experiment platform

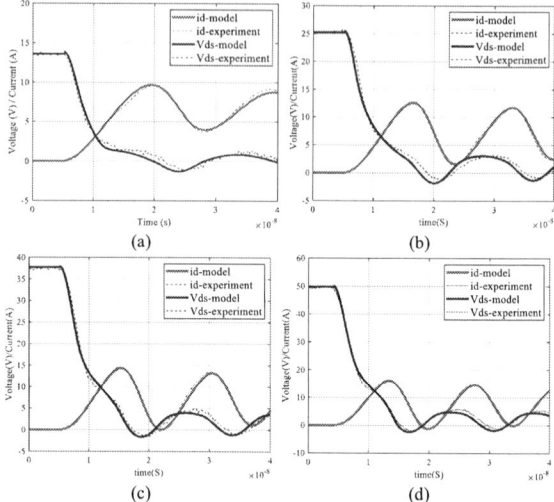

Fig. 6. Comparison results of turn-on process at R_g=10Ω, I_L=7A: (a) for V_{in}=12V; (b) for V_{in}=24V;(c) for V_{in}=36V; (b) for V_{in}=48V

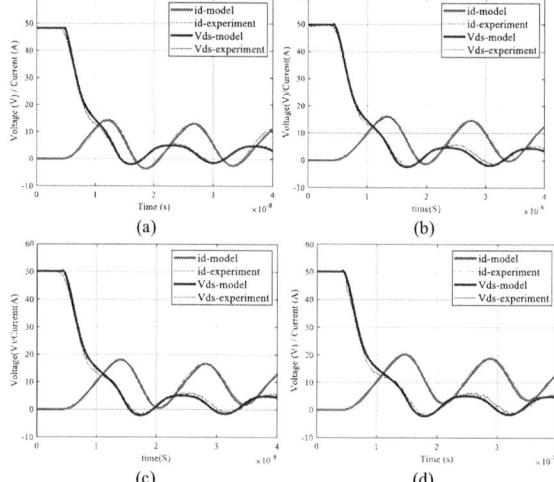

Fig. 7. Comparison results of turn-on process at R_g=10Ω, V_{in}=48A: (a) for I_L=5A; (b) for I_L=7A;(c) for I_L=9A; (b) for I_L=11A

III. OPTIMIZATION OF GaN HEMT SWITCHING TRANSIENT CHARACTERISTICS USING MOPSO

In this section, a method utilizing Multi-Objective Particle Swarm Optimization (MOPSO) is introduced to optimize the switching transient characteristics of GaN HEMTs. The method involves adjusting the gate resistance (R_g) and parasitic inductance (L_{loop}) in the PCB layout to minimize switching losses (P_{on}/P_{off}), peak conduction current (i_{dmax}), maximum transient gate-source voltage during turn-on (V_{gsmax}), and voltage overshoot during turn-off (V_{dsmax}).

In engineering practice, , i_{dmax}, V_{gsmax}, and V_{dsmax} should be kept below 80% of their respective maximum ratings ($i_{d\_max\ rating}$, $V_{gs\_max\ rating}$ and $V_{ds\_max\ rating}$) to ensure device safety and reliability. Therefore, the optimization problem is formulated as a multi-objective optimization problem (MOOP), with the objective function defined as:

$$\min F(x) = [P_{on}(x), i_{dmax}(x), V_{gsmax}(x),$$
$$P_{off}(x), V_{dsmax}(x)] \tag{9}$$

$$\begin{cases} i_{dmax} < 0.8 \times i_{d\_max\ rating} \\ V_{gsmax} < 0.8 \times V_{gs\_max\ rating} \\ V_{dsmax} < 0.8 \times V_{ds\_max\ rating} \end{cases} \quad (10)$$

where $x = [R_g, L_{loop}]$ represents the decision variables. Constraints include the feasible ranges for R_g and L_{loop}.

MOPSO is employed to solve this problem due to its ability to handle multiple objectives and generate Pareto-optimal solutions. The flowchart of the algorithm is shown in Fig.8. The algorithm updates particle positions and velocities using the following equations:

$$v_i(t+1) = w \cdot v_i(t) + c_1 \cdot r_1 \cdot (pbest_i - x_i(t)) + c_2 \cdot r_2 \cdot (gbest - x_i(t)) \quad (10)$$

$$x_i(t+1) = x_i(t) + v_i(t+1) \quad (11)$$

An external archive is used to store non-dominated solutions, and crowding distance calculation ensures diversity among solutions:

$$crowding\_distance(i) = \sum_{m=1}^{k} \frac{f_m(i+1) - f_m(i-1)}{f_m^{max} - f_m^{min}} \quad (12)$$

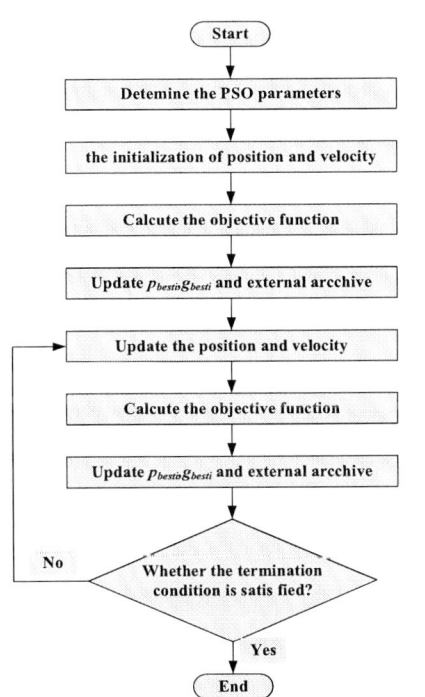

Fig.8. Fitted three-dimensional channel current model.

As shown in Fig.9, MOPSO optimization was performed for the operating condition of V_{ds}= 48V and I_L= 9 A. The set R_g ranges from 1Ω to 100Ω and the L_{loop} ranges from 5nH to 50nH. The optimization provided multiple Pareto-optimal solutions for R_g and L_{loop}, illustrating the trade-offs between switching losses, i_{dmax}, V_{gsmax} and V_{dsmax}. For the switching transient, lower R_g and higher L_{loop} resulted in lower switching losses but led to higher values of i_{dmax}, V_{gsmax} and V_{dsmax}. Therefore, considering these trade-offs, the design values of R_g=18.42 Ω and L_{loop}= 33.42nH were ultimately selected.

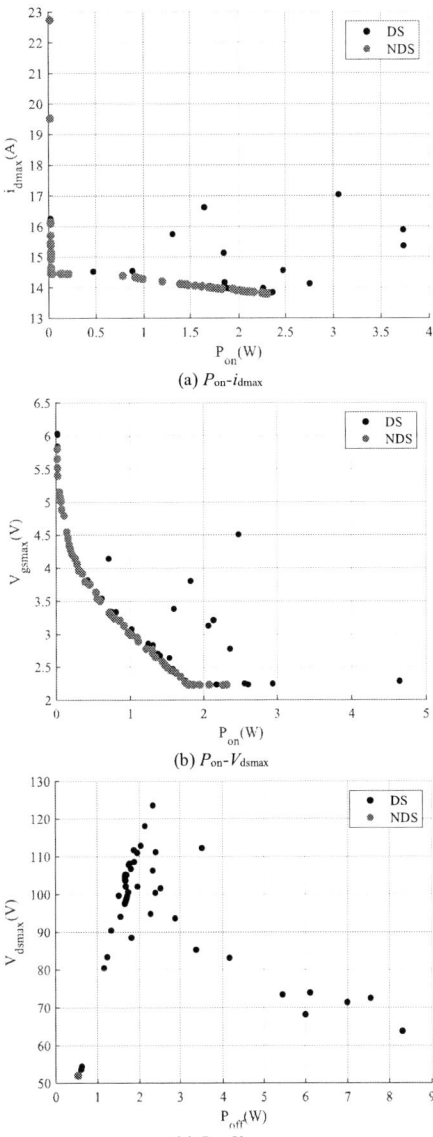

(a) P_{on}-i_{dmax}

(b) P_{on}-V_{dsmax}

(c) P_{off}-V_{dsmax}

Fig.9. Optimization results after using the MOPSO algorithm.

IV. CONCLUSION

A design method based on Multi-Objective Particle Swarm Optimization (MOPSO) is proposed to optimize the switching transient characteristics of GaN HEMTs in high-speed and high-voltage applications. By integrating a datasheet-driven analytical model with the MOPSO algorithm, the multi-objective conflict between switching loss minimization and device safety is effectively balanced. First, the proposed analytical model is validated through double-pulse tests, demonstrating significantly improved prediction accuracy for switching transient waveforms under different voltage and load current conditions compared to conventional models. Based on this model, R_g and L_{loop} are optimized, achieving reduced switching losses ((P_{on}/P_{off})) while suppressing transient overcurrent ((i_{dmax}), gate-source voltage overshoot (V_{gsmax}), and drain-source voltage overshoot (V_{dsmax}). These optimizations ensure device operation within 80% of maximum ratings. The MOPSO algorithm generates Pareto-optimal solutions through global search, explicitly revealing trade-offs between parameters and providing flexible design guidelines for engineering applications. Future work will focus on temperature-dependent modeling and algorithm refinement to further enhance the comprehensive performance and practical applicability of GaN HEMTs.

ACKNOWLEDGMENT

This work was supported in part by the National Natural Science Foundation of China under Grant 52407193, and in part by Chunhui Project Foundation of the Education Department of China under Grant 202200504.

REFERENCES

[1] D. Han and B. Sarlioglu, "Deadtime Effect on GaN-Based Synchronous Boost Converter and Analytical Model for Optimal Deadtime Selection," in *IEEE Transactions on Power Electronics*, vol. 31, no. 1, pp. 601-612, Jan. 2016.

[2] E. A. Jones, F. F. Wang and D. Costinett, "Review of Commercial GaN Power Devices and GaN-Based Converter Design Challenges," in *IEEE Journal of Emerging and Selected Topics in Power Electronics*, vol. 4, no. 3, pp. 707-719, Sept. 2016.

[3] J. Chen, Q. Luo, J. Huang, Q. He and X. Du, "A Complete Switching Analytical Model of Low-Voltage eGaN HEMTs and Its Application in Loss Analysis," in IEEE Transactions on Industrial Electronics, vol. 67, no. 2, pp. 1615-1625, Feb. 2020.

[4] P. Yi, Y. Cui, A. Vang and L. Wei, "Investigation and evaluation of high power SiC MOSFETs switching performance and overshoot voltage," 2018 IEEE Applied Power Electronics Conference and Exposition (APEC), San Antonio, TX, USA, 2018, pp. 2589-2592.

[5] Y. Liu, X. Liu, X. Li and H. Yuan, "Analytical Model and Safe-Operation-Area Analysis of Bridge-Leg Crosstalk of GaN E-HEMT Considering Correlation Effect of Multi-Parameters," in IEEE Transactions on Power Electronics, vol. 39, no. 7, pp. 8146-8161, July 2024.

[6] Y. Liu, J. Cao and X. Li, "Switching Loss Model for Fast-Switching GaN HEMT in Half-Bridge Circuit Considering Parasitic Inductance and Temperature Effect," in IEEE Transactions on Circuits and Systems I: Regular Papers, vol. 71, no. 12, pp. 6128-6137, Dec. 2024.

[7] K. Wang, X. Yang, H. Li, H. Ma, X. Zeng, and W. Chen, "An analytical switching process model of low-voltage eGaN HEMTs for loss calculation, " *IEEE Trans. Power Electron.*, vol. 31, no. 1, pp. 635–647, Jan. 2016.

[8] J. Chen, Q. Luo, J. Huang, Q. He, and X. Du, "A complete switching analytical model of low-voltage eGaN HEMTs and its application in loss analysis, " *IEEE Trans. Ind. Electron.*, vol. 67, no. 2, pp. 1615-1625, Feb. 2020.

[9] Z. Qi et al., "An Accurate Datasheet-Based Full-Characteristics Analytical Model of GaN HEMTs for Deadtime Optimization," in IEEE Transactions on Power Electronics, vol. 36, no. 7, pp. 7942-7955, July 2021.

[10] Z. Dong, X. Wu, H. Xu, N. Ren and K. Sheng, "Accurate Analytical Switching-On Loss Model of SiC MOSFET Considering Dynamic Transfer Characteristic and Qgd," in *IEEE Transactions on Power Electronics*, vol. 35, no. 11, pp. 12264-12273, Nov. 2020.

[11] X. Li, Z. Xiong, J. Cao and Y. Liu, "Analytical Model of Fast-Switching GaN HEMTs in Bridge-Leg Considering Nonlinear Characteristics," *2023 IEEE Energy Conversion Congress and Exposition (ECCE)*, Nashville, TN, USA, 2023, pp. 5939-5945.

[12] S. Song, H. Peng, X. Chen, Q. Xin and Y. Kang, "Determination and Implementation of SiC MOSFETs Zero Turn-off Loss Transition Considering No Miller Plateau," in *IEEE Transactions on Power Electronics*, vol. 38, no. 12, pp. 15509-15521, Dec. 2023.

[13] C. Yang, Y. Pei, L. Wang, L. Yu, F. Zhang and B. Ferreira, "Overvoltage and Oscillation Suppression Circuit With Switching Losses Optimization and Clamping Energy Feedback for SiC MOSFET," in IEEE Transactions on Power Electronics, vol. 36, no. 12, pp. 14207-14219, Dec. 2021.

[14] Q. Zhu et al., "An External Archive-Guided Multiobjective Particle Swarm Optimization Algorithm," in IEEE Transactions on Cybernetics, vol. 47, no. 9, pp. 2794-2808, Sept. 2017.

[15] Q. Lin et al., "Particle Swarm Optimization With a Balanceable Fitness Estimation for Many-Objective Optimization Problems," in IEEE Transactions on Evolutionary Computation, vol. 22, no. 1, pp. 32-46, Feb. 2018.

A Compact EMI Filter Design Method Based on Chaotic SVPWM in Motor Drive System

Yanjun Li
The School of Electrical Engineering
Beijing Jiaotong University
Beijing, China
yanjunli@bjtu.edu.cn

Hong Li*
The College of Electrical Engineering
Zhejiang University
Hangzhou, China
hong_li@zju.edu.cn

Zuoxing Wang
The School of Electrical Engineering
Beijing Jiaotong University
Beijing, China
19117029@bjtu.edu.cn

Aojie Liao
The School of Electrical Engineering
Beijing Jiaotong University
Beijing, China
24121300@bjtu.edu.cn

Mingxin Shi
The School of Electrical Engineering
Beijing Jiaotong University
Beijing, China
24121327@bjtu.edu.cn

Abstract—As motor drive systems evolve towards higher voltage, higher frequency, and higher power density, the issue of electromagnetic interference (EMI) becomes increasingly critical. Consequently, the size of the passive EMI filters (PEFs) used to suppress EMI tends to increase, which restricts the integration of motor drive systems. To address this issue, this paper proposes a method to reduce the volume of PEF in motor drive systems based on chaotic space vector pulse width modulation (SVPWM). First, the mechanism of EMI source in the system is analyzed using the double Fourier transform method. From a mathematical perspective, the effectiveness of chaotic modulation in suppressing EMI sources is demonstrated. On this basis, the suppression effects of various chaotic SVPWM—generated by various chaotic mappings—on EMI sources are compared. Finally, the performance of these different chaotic SVPWM methods in suppressing common-mode (CM) EMI is evaluated to determine their effectiveness in reducing the size of CM EMI filters. This paper offers a novel approach for reducing the size of EMI filter and enhancing overall power density in motor drive systems.

Keywords—space vector pulse width modulation (SVPWM), chaotic modulation, electromagnetic interference (EMI), passive EMI filter.

I. INTRODUCTION

Motor drive system, typically composed of an inverter and a motor, is the power core of electric vehicles [1]. However, as the motor drive system evolves towards higher voltage, higher frequency, and higher power density, electromagnetic interference (EMI) issues arise [2]. These EMI issues not only affect the operation of the motor drive system itself but may also interfere with the functioning of nearby in-vehicle devices, posing a significant challenge to the reliability of electric vehicles. Currently, relevant standards such as CISPR25 and CISPR32 set amplitude limits for EMI during various frequency bands, providing a reference for EMC design [3], [4]. Compliance with these standards is necessary for motor drive systems to be marketable. Therefore, EMI suppression design for motor drive systems is essential.

The main method for suppressing EMI in motor drive systems is the design of passive EMI filter (PEF). However, PEFs occupy more than 30% of the system's total volume, typically in common-mode (CM) inductance, hindering the development of high-power-density motor drive systems [5]. Therefore, most research aimed at reducing the volume of PEF focuses on lowering the CM inductance volume. In [6], hybrid EMI filter is used to reduce the size of EMI filter. In [7], magnetic integration technology is used to design planar integrated EMI filter. In [8], amorphous iron and nanocrystalline materials are used as the core material for PEF. However, the above solutions are relatively expensive, which is not conducive to low-cost design for motor drive systems.

In recent years, low-EMI modulations have attracted widespread attention from researchers due to their low-cost advantage [9]. One such modulation is chaotic modulation. However, because the EMI suppression effect of chaotic modulation is typically less than 10dBµV, using chaotic modulation alone is often insufficient to bring the EMI into compliance with standards [10]. Therefore, chaotic modulation needs to be combined with PEF. In [11], a method to reduce the size of PEF for Boost converter and grid-connected inverter is proposed by using chaotic modulation, providing a reference for this study. However, the suppression effect of chaotic SVPWM on EMI sources in motor drive systems has not yet been quantitatively evaluated using mathematical methods, and there is also a lack of quantitative studies on its impact in reducing the size of system filters.

This paper proposes a method to reduce the size of PEF in motor drive systems by introducing chaotic modulation in space vector pulse width modulation (SVPWM), while ensuring compliance with CISPR32 standard. It also compares the extent of size reduction of PEF under different chaotic mappings. The main innovations of the article are as follows:

- Introduction of chaotic SVPWM into the design of PEF in motor drive systems.

- Analysis of EMI sources in motor drive systems using the double Fourier transform method and validation of the effectiveness of chaotic SVPWM in suppressing system EMI sources.

---

*This work is supported in part by the National Science Fund for Distinguished Young Scholars of China under Grant 52325704, in part by the Shunyi Innovation Joint Fund of Beijing Natural Science Foundation under Grant L247004, and in part by the National Natural Science Foundation of China under Grant 52237008.

*Corresponding Author, Email: Hong Li, hong_li@bjtu.edu.cn

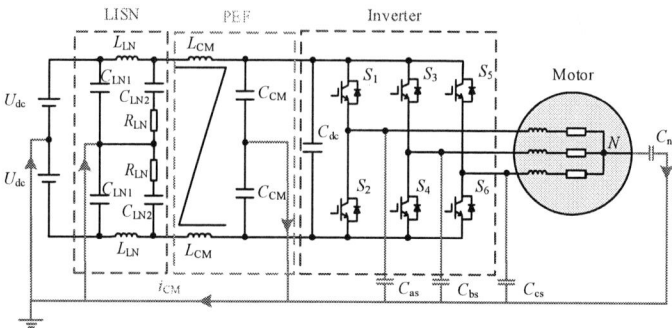

Fig. 1. CM equivalent circuit of the motor drive system.

- Quantification of the size reduction of PEF under different chaotic SVPWM, providing a reference for selecting the most suitable chaotic SVPWM.

This paper is structured as follows: Section II introduces the mathematical analysis method for EMI sources in motor drive systems based on the double Fourier transform. Section III describes chaotic SVPWM and its different chaotic mappings based on the mathematical analytical expression of the system's EMI source. Section IV calculates the parameter values of PEF based on the EMI simulation results of chaotic SVPWM. verifies the accuracy of the calculations through simulations. Section V concludes the paper.

II. ANALYSIS OF EMI SOURCES IN MOTOR DRIVE SYSTEM

Since the volume proportion of the CM PEF is relatively large in the overall PEF, this paper takes the CM PEF as an example to demonstrate the effectiveness of chaotic SVPWM in reducing the volume of the EMI filter. CM equivalent circuit of the motor drive system with the inclusion of a CM PEF and Line Impedance Stabilization Network (LISN) is shown in Fig.1. Z_{LN} is the impedance of the LISN connected to the motor drive system, The network consists of inductors L_{LN}, capacitors C_{LN1} and C_{LN2}, and resistors R_{LN}. The source impedance mainly consists of the parasitic capacitance of the switching devices to ground C_{as}, C_{bs} and C_{cs}, typically ranging from 10pF to 100pF. When the load impedance is low and the source impedance is high, an LC EMI filter is selected. L_{CM} represents the inductor used in the PEF, while C_{CM} represents the capacitor used in the PEF.

In motor drive systems, the source of EMI is the switching devices within the inverter. When these switching devices are in operation, their switching actions generate dv/dt and di/dt, which through circuit components, the motor, and parasitic capacitance formed by parasitic effects, lead to electromagnetic interference. Therefore, the modulation method of the inverter in a motor drive system directly determines the trend and magnitude of the EMI in the system. This thesis begins with modulation and uses the double Fourier series method to express the EMI sources in the system through numerical analytical methods.

The commonly used modulation method in current inverters is SVPWM with a constant switching frequency. For this type of modulation, it is possible to derive normalized expressions for the equivalent modulation wave and carrier wave [12]. The expression for the modulation wave is shown in (1), while the expression for the carrier wave is given in (2).

$$
u_r(t) = \begin{cases} \sqrt{3}M\cos\left(\omega_0 t - \dfrac{\pi}{6}\right) & 0,\, \omega_0 t < \dfrac{\pi}{3},\, \pi,,\, \omega_0 t < \dfrac{4\pi}{3} \\[2mm] 3M\cos\omega_0 t & \dfrac{\pi}{3},,\, \omega_0 t < \dfrac{2\pi}{3},\, \dfrac{4\pi}{3},,\, \omega_0 t < \dfrac{5\pi}{3} \\[2mm] \sqrt{3}\cos\left(\omega_0 t + \dfrac{\pi}{6}\right) & \dfrac{2\pi}{3},,\, \omega_0 t < \pi,\, \dfrac{5\pi}{3},,\, \omega_0 t < 2\pi \end{cases} \quad (1)
$$

$$
u_s(t) = \begin{cases} -\left[\omega_s t - 2k\pi\right]\dfrac{2}{\pi} - 1 & 2\left(k-\dfrac{1}{2}\right)\pi,,\, \omega_s t < 2k\pi \\[2mm] \left(\omega_s t - 2k\pi\right)\dfrac{2}{\pi} - 1 & 2k\pi,,\, \omega_s t,, 2\left(k+\dfrac{1}{2}\right)\pi \end{cases} \quad (2)
$$

Here, $u_r(t)$ represents the modulation wave, $u_s(t)$ represents the carrier wave, ω_0 represents the angular frequency of the modulation wave, ω_s represents the angular frequency of the carrier wave, and M is the modulation index.

When $u_r(t)$ equals $u_s(t)$, the switching instants can be determined, from which the switching function can be obtained. By defining $x = \omega_s t$, $y = \omega_o t$, the double Fourier transform result of the switching function $s_a(\omega_s t, \omega_o t)$ can then be expressed as follows:

$$
s_a(x, y) = \frac{A_{00}}{2} + \sum_{n=1}^{\infty}\left(A_{0n}\cos(nx) + B_{0n}\sin(ny)\right)
$$
$$
+ \sum_{m=1}^{\infty}\left(A_{m0}\cos(mx) + B_{m0}\sin(my)\right) + \quad (3)
$$
$$
\sum_{m=1}^{\infty}\sum_{n=\pm 1}^{\pm\infty}\left[A_{mn}\cos(mx + ny) + B_{mn}\sin(mx + ny)\right]
$$

$$
\begin{cases} A_{mn} = \dfrac{1}{2\pi^2}\displaystyle\int_{-\pi}^{\pi}\int_{-\pi}^{\pi} s_a(x, y)\cos(mx + ny)\,\mathrm{d}x\mathrm{d}y \\[3mm] B_{mn} = \dfrac{1}{2\pi^2}\displaystyle\int_{-\pi}^{\pi}\int_{-\pi}^{\pi} s_a(x, y)\sin(mx + ny)\,\mathrm{d}x\mathrm{d}y \end{cases} \quad (4)
$$

Where the index m represents the harmonic order corresponding to the switching frequency and its multiples, while n represents the harmonic order around the switching frequency.

979-8-3315-1110-4/25 $31.00 © 2025 IEEE

$$s_a(\omega_s t, \omega_o t) = U_{dc} + U_{dc} M \cos(\omega_o t) + \sum_{n=3,9,15,\cdots}^{\infty} \frac{3\sqrt{3}MU_{dc}}{\pi(n^2-1)} \sin\left(\frac{n\pi}{6}\right)\sin\left(\frac{n\pi}{2}\right)\cos(n\omega_o t) +$$

$$\frac{8U_{dc}}{m\pi^2}\sum_{m=1}^{\infty}\sum_{n=-\infty}^{+\infty}
\begin{bmatrix}
\dfrac{\pi}{6}\sin\left[(m+n)\dfrac{\pi}{2}\right]\cdot\left[J_n\left(m\dfrac{3\pi}{4}M\right)+2\cos\left(\dfrac{n\pi}{6}\right)\cdot J_n\left(m\dfrac{\sqrt{3}\pi}{4}M\right)\right]+ \\[2ex]
\dfrac{1}{n}\sin\left(\dfrac{m\pi}{2}\right)\cos\left(\dfrac{n\pi}{2}\right)\sin\left(\dfrac{n\pi}{6}\right)\cdot\left[J_n\left(m\dfrac{3\pi}{4}M\right)-J_n\left(m\dfrac{\sqrt{3}\pi}{4}M\right)\right]\Bigg|_{n\neq0} + \\[2ex]
\sum_{k=1(k\neq-n)}^{\infty}
\begin{bmatrix}
\dfrac{1}{n+k}\sin\left[(m+k)\dfrac{\pi}{2}\right]\cos\left[(n+k)\dfrac{\pi}{2}\right]\sin\left[(n+k)\dfrac{\pi}{6}\right] \\[1ex]
\left\{J_k\left(m\dfrac{3\pi}{4}M\right)+2\cos\left[(2n+3k)\dfrac{\pi}{6}\right]\cdot J_n\left(m\dfrac{\sqrt{3}\pi}{4}M\right)\right\}
\end{bmatrix} + \\[3ex]
\sum_{k=1(k\neq n)}^{\infty}
\begin{bmatrix}
\dfrac{1}{n-k}\sin\left[(m+k)\dfrac{\pi}{2}\right]\cos\left[(n-k)\dfrac{\pi}{2}\right]\sin\left[(n-k)\dfrac{\pi}{6}\right]\times \\[1ex]
\left\{J_k\left(m\dfrac{3\pi}{4}M\right)+2\cos\left[(2n-3k)\dfrac{\pi}{6}\right]\cdot J_n\left(m\dfrac{\sqrt{3}\pi}{4}M\right)\right\}
\end{bmatrix}
\end{bmatrix}\cos(m\omega_s t + n\omega_o t) \quad (5)$$

The final expression is given in (5). This formula can be used to calculate the spectrum of the inverter's switching voltage u_{ds} [13].

To validate (5), the traditional SVPWM (T-SVPWM) is employed. The spectrum of u_{dsS1} in the range of 20 kHz to 100 kHz is analyzed through both calculation and simulation, as shown in Fig. 2, and a comparison between the simulation and the analytical results in 20kHz is shown in Fig. 3, with the simulation parameters listed in Table 1. It can be observed that the energy of the EMI source is concentrated at the switching frequency and its harmonics. At the switching frequency of 20 kHz and its harmonics, the analytical results for the switching device u_{dsS1}, obtained via double Fourier transform, u_{dsS1} are in good agreement with the FFT results directly applied to in the SVPWM simulation. This confirms that the proposed (5) effectively reflects the EMI distribution characteristics of SVPWM. Therefore, by varying the switching frequency, the distribution of the EMI source can be spread out, thereby achieving the goal of suppressing EMI.

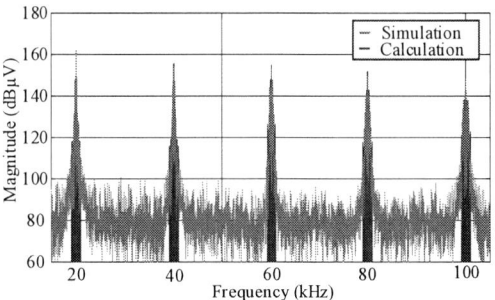

Fig. 2. Verification of SVPWM double Fourier transform result in 20kHz-100kHz.

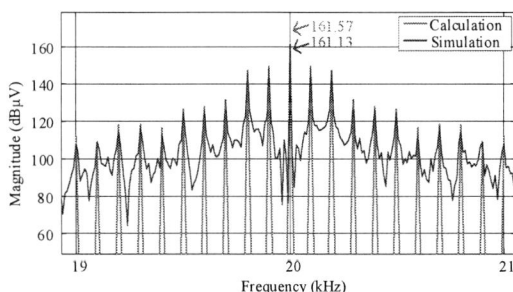

Fig. 3. Verification of SVPWM double Fourier transform result in 20kHz.

TABLE I
PARAMETERS OF MOTOR DRIVE SYSTEM

Parameter	Symbol	Value
Input dc voltage	U_{dc}	175V
Rated power	P	400W
Output line voltage RMS	U_{orms}	220V
Reference switching frequency	f_r	20kHz
Maximum frequency offset	k_r	0.2
Switching frequency range	f_c	16kHz-24kHz

III. MECHANISM OF CHAOTIC SVPWM AND CHAOTIC MAPPINGG

The generation method of chaotic SVPWM involves using chaotic mapping to disturb the switching frequency of T-SVPWM, causing chaotic variations in the inverter's switching frequency within a certain range. Since the EMI energy is concentrated around the switching period, by varying the frequency, the EMI energy of traditional SVPWM can be spread over a wider frequency range, thus achieving the goal of reducing EMI spikes. Its EMI suppression effect is shown in Fig.4. The expression for the chaotic switching frequency is given by (6):

$$f_c = f_r + k_r x_i f_r \quad i = 1, 2, \ldots \quad (5)$$

Where f_c is the chaotic carrier frequency, f_r is the reference carrier frequency, k_f is the maximum frequency offset, and x_i is the iterative values of chaotic mapping.

From (6), it can be seen that the chaotic mapping has a significant impact on the distribution of the chaotic frequency, which in turn affects the EMI suppression effect. This study compares the EMI suppression effects of traditional Chebyshev mapping and the novel proposed Logistic-Chebyshev mapping for the motor drive system, and calculate the CM inductance value of the PEF under both chaotic mappings [14].

The expression for the Chebyshev mapping is shown in (7).

$$x_{i+1} = \cos[k \cdot \arccos(x_i)] \quad x_i \in (-1,1) \; i = 0,1,2\dots \quad (6)$$

Where k is the control parameter, and x_i represents the iterative values of Chebyshev mapping.

And the expression for the Logistic-Chebyshev mapping is shown in (8).

$$\begin{cases} x_{i+1} = \mu x_i(1-x_i) \quad x_i \in (0,1) \quad i = 0,1,2,\dots \\ y_{i+1} = \cos[(k+2x_{i+1}) \cdot \arccos(z_i)] \quad y_i \in (-1,1) \\ z_{i+1} = [2(x_{i+1}-0.5)+y_{i+1}]/2 \end{cases} \quad (7)$$

where μ is the control parameter for the Logistic mapping, x_i is the iterative value of the Logistic mapping, k is the control parameter for the Chebyshev mapping, y_i is the iterative value of the Chebyshev mapping, and z_i is the final output value after the perturbation.

Using (5), the double Fourier transform result of u_{dsS1} for the switch S1 of the inverter is carried out under two modulations: traditional chaotic SVPWM (T-SVPWM) and Logistic-Chebyshev chaotic SVPWM (LCC-CSVPWM). This allows for a quantitative comparison of the EMI suppression effectiveness of the two methods in motor drive systems, which is shown in Fig.6.

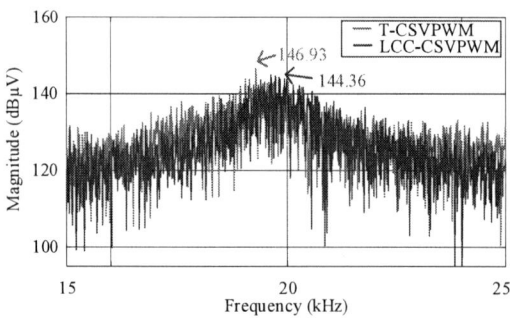

Fig. 6. Comparison of EMI source between two types of chaotic SVPWM in 20kHz.

From the computed EMI distribution, it can be observed that the EMI energy generated by LCC-CSVPWM is concentrated around the switching frequency, with a peak of 144.36dBµV. This represents a reduction of 2.5dBµV compared to the peak EMI of T-CSVPWM, and approximately 18dBµV lower than the sharp EMI peak observed at the switching frequency in T-SVPWM. Therefore, LCC-CSVPWM demonstrates a superior ability to

disperse EMI energy concentrated at the switching frequency.

The parameters of the motor drive system are set as shown in Table I, and simulation result of the initial EMI spectrum under three different modulations are obtained. The comparison of their CM EMI spectrum is shown in the Fig.7 and Fig.8.

TABLE II
COMPARISON OF PASSIVE EMI FILTER IN MOTOR DRIVE SYSTEMS UNDER DIFFERENT MODULATIONS

Modulation	C_{CM}	L_{CM}	Percentage
T-SVPWM	2200pF	6.12mH	100%
T-CSVPWM	2200pF	4.94mH	80.7%
LCC-CSVPWM	2200pF	4.22mH	68.9%

IV. DESIGN OF EMI FILTERS WITH CHAOTIC SVPWM

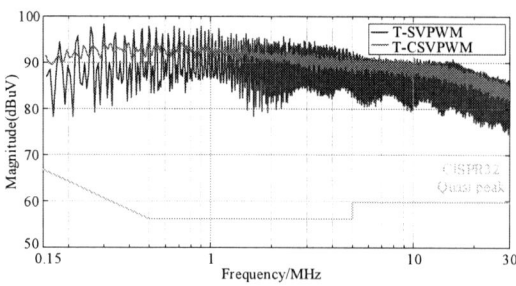

Fig. 7. Effect of T-CSVPWM for CM EMI suppression.

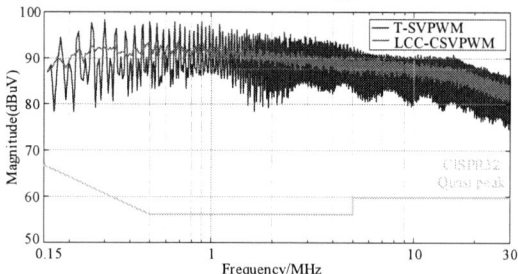

Since the volume proportion of the CM PEF is relatively large in the overall PEF, this paper takes the CM PEF as an example to demonstrate the effectiveness of chaotic SVPWM in reducing the volume of the EMI filter. The detailed design process of the PEF can be found in [11].

The relationship between the cut-off frequency $f_{\text{cut-off}}$ and the value of CM inductance and capacitance is shown in (9):

$$L_{CM} \cdot C_{CM} = \frac{1}{8\pi^2 f_{\text{cut-off}}^2} \quad (8)$$

In the filter design process, the CM capacitance value is typically determined based on the leakage current requirements, and then the CM inductance value is calculated to meet the EMI suppression requirements.

After obtaining the CM EMI spectrum, the parameters of the CM PEF for the motor drive system are calculated according to the CM PEF calculation method in Section II. The CM capacitor in the filter, which needs to meet the system's leakage current requirements, is set to 2200pF. Based on this, the CM inductance value is calculated to ensure compliance, and the required inductance value to meet the CISPR32 standard for the motor drive system is shown in the Table II.

To verify the inductance values calculated in Table II, simulations are conducted using MATLAB/Simulink for the proposed motor drive system. Fig.10, Fig.11 and Fig.12 show the CM EMI spectra after adding the inductance values derived in Section III for T-SVPWM, T-CSVPWM, and

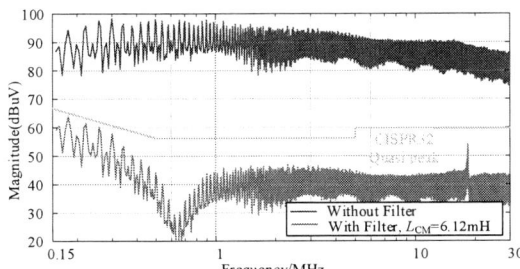

Fig. 10. Comparison of EMI without filter and with filter in T-SVPWM.

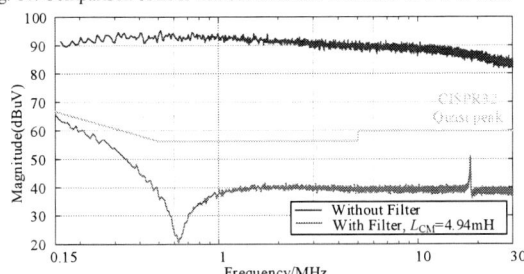

Fig. 11. Comparison of EMI without filter and with filter in T-CSVPWM.

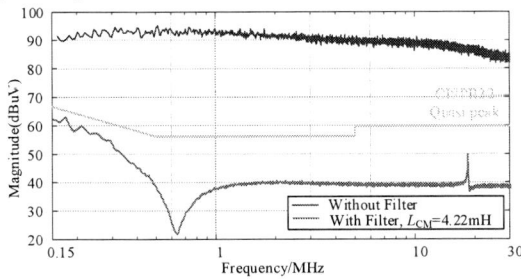

Fig. 12. Comparison of EMI without filter and with filter in LCC-CSVPWM.

LCC-CSVPWM.

From the simulation results, it can be observed that the inductance values calculated earlier are sufficient to meet the EMI suppression requirements. The inductance value is approximately proportional to the inductance volume [11]. Therefore, the degree of reduction in the inductance value can be used as an estimate for the corresponding reduction in the inductance volume.

Based on the estimation, it can be seen that under the traditional chaotic algorithm, the inductance value is reduced by 19.3% compared to traditional SVPWM, while under the

LCC chaotic algorithm, the inductance value is reduced by 31.1% compared to traditional SVPWM, as shown in Table II.

After determining the required inductance value, the inductor volume is calculated based on the method presented in [11], using the inductance formula. The inductor structure is illustrated in fig. 13.

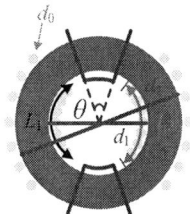

Fig. 13. CM inductors structures in motor drive systems.

Given the parameters of the magnetic core, the formula for calculating the inductance is shown in (10).

$$L = \frac{\mu_0 \mu_r N^2 A_e}{l_e} \tag{10}$$

As illustrated in the figure, an angular separation of 80° is maintained between the positive and negative windings to ensure system safety. Under this condition, the expression for the number of turns can be obtained in (11)

$$N \cdot d_0 = \frac{140°}{180°} \pi \frac{d_1}{2} \tag{11}$$

By combining Equations (10) and (11), the expression for D_{out} can be obtained as (12):

$$L = \frac{49 \mu_0 \mu_r h \pi d_1^2}{162 d_0^2} \cdot \frac{d_2 - d_1}{d_2 + d_1} \tag{12}$$

where L is the inductance, μ_0 is the permeability of free space, μ_r is the relative permeability, h is the height of the magnetic core, D_0 is the diameter of the winding wire, D_{out} is the outer diameter of the core, and D_{in} is the inner diameter of the core.

Equation (12) is solved numerically in MATLAB, and the parameters along with the results are summarized in Table III.

TABLE III
COMPARISON OF CM INDUCTOR CORE VOLUME

Modulation	T-SVPWM	T-CSVPWM	LCC-CSVPWM
L_{CM}	6.12mH	4.94mH	4.22mH
μ_r	7000	7000	7000
h	15mm	15mm	15mm
d_0	1.5mm	1.5mm	1.5mm
d_1	20mm	20mm	20mm
d_2	38.6mm	20.2mm	20mm
V_{LCM}	14.547cm³	11.608cm³	10.143cm³
V_{LCM} Percentage	100%	79.79%	69.72%

According to the calculation results, under LCC-CSVPWM, the volume of the common-mode choke is reduced by 30%, which is approximately 10% more than the reduction achieved under T-CSVPWM.

V. CONCLUSION

This paper proposes a method to reduce the volume of the passive EMI filter in motor drive systems based on chaotic SVPWM. This approach reduces the EMI magnitude that needs to be suppressed and decreases the volume of the CM inductance in the PEF. The paper also compares the suppression effectiveness of two different chaotic mappings for the chaotic SVPWM on the PEF volume. Based on simulation results, the LCC-CSVPWM achieves the best suppression effect on the inductance volume in PEF, leading to a 31.1% reduction in the required inductance value, which is approximately 11% more effective than T-CSVPWM. Furthermore, according to the calculation results, under LCC-CSVPWM, the volume of the common-mode choke is reduced by 30%, which is about 10% more than the reduction achieved under T-CSVPWM. This method offers an effective solution for reducing the EMI filter volume in motor drive systems and improving their power density.

REFERENCES

[1] L. Zhou and M. Preindl, "Variable Switching Frequency Techniques for Power Converters: Review and Future Trends," *IEEE Trans. Power Electron.*, vol. 38, no. 12, pp. 15603-15619, Dec. 2023.

[2] F. A. Kharanaq, A. Emadi and B. Bilgin, "Modeling of conducted emissions for EMI analysis of power converters: State-of-the-art review," *IEEE Access*, vol. 8, pp. 189313-189325, 2020.

[3] Vehicles, boats and internal combustion engines - Radio disturbance characteristics - Limits and methods of measurement for the protection of on-board receivers, CISPR25-Edition 5.0, Dec. 2021.

[4] Electromagnetic compatibility of multimedia equipment - Emission requirements, CISPR32-Edition 2.0, Mar. 2015.

[5] How Active EMI Filter ICs Mitigate Common-Mode Emissions and Increase Power Density in Single- and Three-Phase Power Systems,

Texas Instruments, Mar. 2023. [Online]. Available: https://www.ti.com/lit/pdf/slvafj9.

[6] H. Li, S. Wang, D. He, Z. Zhao and W. Su, "A Hybrid EMI Filter Incorporating Active Y-Capacitor for Common-Mode Noise Mitigation," *IEEE Trans. Power Electron.*, vol. 40, no. 4, pp. 5252-5264, April 2025.

[7] Y. Liu, X. Zhang, Z. Lu and J. Yin, "Design of Planar Magnetic Integrated LCL-EMI Hybrid Filter for the Grid-Connected Inverter," *IEEE Trans. Ind. Appl.*, vol. 60, no. 3, pp. 4280-4291, May-June 2024.

[8] M. K ˌacki, M. S. Rylko, J. G. Hayes, and C. R. Sullivan, "Magnetic materialselection for EMI filters," in Proc. IEEE Energy Convers. Congr. Expo.,Cincinnati, OH, USA, 2017, pp. 2350–2356.

[9] B. Wunsch, I. Stevanović and S. Skibin, "Length-Scalable Multiconductor Cable Modeling for EMI Simulations in Power Electronics," *IEEE Trans. Power Electron.*, vol. 32, no. 3, pp. 1908-1916, March 2017.

[10] Z. Zhang, Y. Hu, X. Chen, G. W. Jewell and H. Li, "A Review on Conductive Common-Mode EMI Suppression Methods in Inverter Fed Motor Drives," *IEEE Access*, vol. 9, pp. 18345-18360, 2021.

[11] H. Li, Y. Ding, C. Zhang, Z. Yang, Z. Yang and B. Zhang, "A Compact EMI Filter Design by Reducing the Common-Mode Inductance With Chaotic PWM Technique," *IEEE Trans. Power Electron.*, vol. 37, no. 1, pp. 473-484, Jan. 2022.

[12] J. Meng, W. Ma, L. Zhang, Q. Pang and Z. Zhao, " DM and CM EMI Sources Modeling for Inverters Considering the PWM Strategies," *Tranactions of. China Electrotechnical.Society* , vol. 22, no. 12, pp. 92-97, Dec. 2007.

[13] N. Li, W. Zhou, and Q. Liu, " Research on Output Common Mode Voltage of aThree-Phase Inverter Based on the AnalysisMethod of Double Fourier Series," *Chinese Journal of Electron Devices* , vol. 46, no. 3, pp. 697-704, Jun. 2023.

[14] H. Li, Y. Li, Z. Wang, M. Zhou, B. Zhang and Z. Yang, "Logistic-Chebyshev Cascaded Chaotic Space Vector Pulse Width Modulation for Common-Mode EMI Suppression in Motor Drive System," *2024 2nd China Power Supply Society Electromagnetic Compatibility Conference (CPEMC)*, Hangzhou, China, 2024, pp. 299-304,

A SiC MOSFET accelerated degradation test platform that accounts for turn-on and turn-off times

Bohang Lu [a], Cen Chen [a], Zicheng Wang [a] and Xuanyu Lin [a]

[a] Harbin Institute of Technology, Harbin 150001, China, E-mail 2021113137@stu.hit.edu.cn
[a] Harbin Institute of Technology, Harbin 150001, China, E-mail mac-chan ee@hit.edu.cn
[a] Harbin Institute of Technology, Harbin 150001, China, E-mail 22B906030@stu.hit.edu.cn
[a] Harbin Institute of Technology, Harbin 150001, China, E-mail 2022111796@stu.hit.edu.cn

Abstract—As the demand for applications and technological advancements continues to rise, the performance of silicon carbide (SiC) metal-oxide-semiconductor field-effect transistors (MOSFETs) has been subject to ongoing optimization. Nevertheless, in practical implementations, SiC MOSFETs encounter various reliability challenges, one of the most pressing being the issue of alternating current bias temperature instability (AC BTI). The turn-on and turn-off times of SiC MOSFETs exhibit sensitivity to AC BTI. During degradation assessments, significant alterations in these timing parameters are observed throughout the degradation process, which subsequently influences the degradation rate and presents difficulties regarding stress stability in accelerated testing scenarios. To mitigate this issue, the present study proposes the development of a self-regulating gate drive resistance platform aimed at maintaining the stability of the turn-on and turn-off times of SiC MOSFETs during accelerated testing conditions. An examination of strategies for preserving the consistency of on-off timing in accelerated degradation testing.

Keywords—AC BTI, SiC MOSFETs, reliability, Degradation, Accelerated degradation testing

I. INTRODUCTION

SiC MOSFETs are widely employed in various power electronic converters. As application requirements evolve and processing technologies advance rapidly, SiC MOSFETs have undergone substantial optimization in key performance metrics, such as blocking voltage and switching loss, thereby approaching the theoretical limits established for wide bandgap semiconductor devices [1]. However, in practical applications, SiC MOSFETs remain vulnerable to electrical, thermal, and various complex environmental stresses, which raise numerous reliability concerns [2]. A particularly urgent reliability issue that requires immediate attention is the AC bias temperature instability (AC BTI), which is closely associated with the degradation of the gate oxide layer in SiC MOSFETs. This phenomenon is characterized by the drift of several performance parameters, particularly the threshold voltage of SiC MOSFETs, when subjected to prolonged AC bipolar gate bias voltage and elevated temperatures. Numerous studies have indicated that the turn-on and turn-off times of SiC MOSFETs are critical stress factors that influence their AC BTI[3].

However, during degradation tests, the turn-on and turn-off times demonstrate considerable variability throughout the degradation process, which subsequently affects the degradation rate. This variability poses a challenge to maintaining stability in stress conditions during accelerated testing[4].

This article discusses the development of a self-regulating platform for gate drive resistors, which is intended to ensure the stability of the turn-on and turn-off times of SiC MOSFETs during accelerated testing. Chapter 2 offers a succinct analysis of the correlation between degradation and the turn-on and turn-off times. Chapter 3 elucidates the fundamental principles governing the self-regulating gate drive resistor platform, while Chapter 4 describes the experimental setup established to validate the platform's functionality.

II. THE DEPENDENCE OF DEGRADATION ON TRANSITION TIMES

This section seeks to explore the necessity of developing a self-regulating platform for gate drive resistance by analyzing the relationship between degradation phenomena and the turn-on and turn-off times. This investigation will be supported by experimental validation.

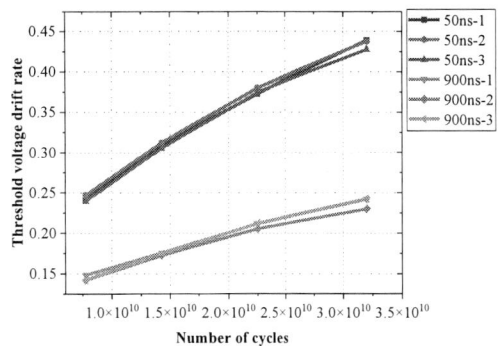

Fig. 1. Threshold voltage drifts under different turn-on and turn-off times

To investigate the influence of turn-on and turn-off times on AC BTI, two distinct groups of devices

979-8-3315-1110-4/25 $31.00 © 2025 IEEE

underwent accelerated testing. The first group was characterized by turn-on and turn-off times of 50 nanoseconds, whereas the second group exhibited both times of 900 nanoseconds. The gate drive voltage was established at 20V/-10V, with a duty cycle consistently maintained at 50% and a switching frequency of 100 kHz. The outcomes of the testing are presented in Figure 1.

AC BTI is attributed to the rapid alternation of the gate bias voltage polarity in SiC MOSFETs[5]. The transition of the gate bias voltage polarity typically occurs within a timeframe of 300 nanoseconds. However, the process of charge trapping and subsequent emission back into the channel transpires at a considerably slower rate, lagging behind the polarity transition of the gate bias[6]. Consequently, the charges that become trapped in defects induced by the bias voltage during the initial half of the cycle do not return to the channel in a timely manner. As a result, when the new gate bias voltage is applied, it induces additional charges into these defects, facilitating the recombination of electrons and holes[7]. The energy released during this recombination process excites the local vibrational modes of the defects, thereby activating further receptor-like interface defects. These newly formed defects are challenging to recover and tend to adopt a negative charge state by capturing electrons during the threshold voltage extraction process[8].This phenomenon ultimately leads to a permanent positive shift in the threshold voltage, as illustrated in Figure 2.

The turn-on and turn-off times of SiC MOSFETs represent a significant factor influencing AC BTI. Prolonged turn-on and turn-off times allow a greater number of previously trapped charge carriers to escape prior to recombination with carriers of opposite polarity, consequently diminishing the drift in threshold voltage.

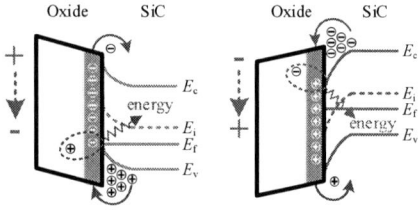

Fig. 2. The influence of turn-on and turn-off times on AC BTI.

To investigate the effects of AC BTI on the turn-on and turn-off times of semiconductor devices, a series of devices characterized by a nominal turn-on and turn-off time of 50 ns were subjected to accelerated degradation testing through the manipulation of gate resistance. The gate drive voltage was established at 20 volts for the turn-on phase and -10 volts for the turn-off phase, while maintaining a duty cycle of 50% and a switching frequency of 100 kHz. To facilitate a substantial observation of variations in turn-on and t

urn-off times, these parameters were monitored in a circuit configuration with a gate drive resistance of 200 ohms. The findings of this experimental investigation are presented in Figure 3.

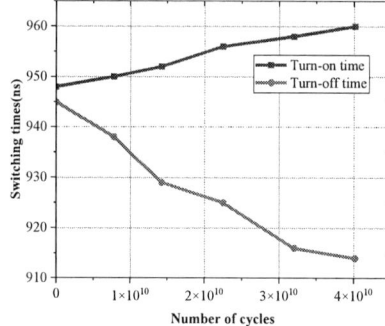

Fig. 3. the effects of AC BTI on the turn-on and turn-off times of semiconductor devices

As the accelerated testing advances, the trapped charges demonstrate increasingly significant negative charge properties. This phenomenon results in a more substantial positive shift in the threshold voltage and causes the repulsion of electrons upon the application of a positive gate voltage, thereby impeding the turn-on process. Conversely, the application of a negative gate voltage attracts holes, thereby promoting the turn-off process. Consequently, with the progression of accelerated testing, the turn-on time of the SiC MOSFET progressively lengthens, while the turn-off time correspondingly shortens[9].

III. SET UP

The self-regulating platform delineated in this article operates on the principle of modulating the gate drive resistance. This modulation is contingent upon the measured turn-on and turn-off times, allowing for an adaptive adjustment of the gate drive resistance.

Fig. 4. The programmable resistor board

The self-regulating platform employs a programmable resistor board to modify the gate resistance, as illustrated in Figure 4. This board is developed using a modular design to facilitate the functionality of digital resistance adjustment. The primary control unit comprises several programmable resistor modules, each incorporating solid-state relays and resistor components arranged in parallel. By managing the activatio n and deactivation of the solid-state relays, it is possible to adjust the resistance value.

In this study, a microcontroller is employed to regulate a solid-state relay, with each input/output (I/O) port of the microcontroller dedicated to controlling a single solid-state relay. When the I/O port is set to a high logic level, the solid-state relay activates, resulting in a low resistance state within the resistance module. Conversely, when the I/O port is at a low logic level, the solid-state relay deactivates, allowing the resistance module to output the combined value of the parallel resistors. The overall output resistance of the programmable resistance board is determined by the cumulative output resistances of all individual modules.

The oscilloscope is utilized to identify the devices involved in the accelerated testing process and facilitates communication with the host computer. Subsequently, the host computer regulates the output of the microcontroller's input/output ports based on the data acquired. When the host computer analyzes the on/off time relayed by the oscilloscope, it determines whether the received data falls within the predetermined threshold range. If the data is within this range, the host computer promptly exits the loop to finalize the feedback process. Conversely, if the data exceeds the established threshold, the host computer directs the microcontroller to decrease the resistance value of the programmable resistor board. In instances where the data is below the threshold, the host computer instructs the microcontroller to increase the resistance value of the programmable resistor board.

IV. TEST

To assess the functionality of the self-regulating platform, the subsequent experiment was devised.

Upon integrating the potentiometer in series with the programmable resistor board, it functions as the gate drive resistor within the device's drive circuit. By varying the resistance of the potentiometer, the changes in the turn-on and turn-off speeds during the degradation process are effectively simulated. Subsequently, a feedback platform is employed to fine-tune the turn-on and turn-off durations to predetermined specifications. The turn-on time is calibrated to fall within the range of 269 ns to 279 ns, while the turn-off time is adjusted to lie between 254 ns and 264 ns. The experimental configuration is illustrated in Figure 5, and the corresponding results are detailed in Table 1.

Upon modiying the resistance value of the variable resistor, the programmable resistor board is capable of adjusting the turn-on and turn-off times to the predetermined range, thereby providing comprehensive validation of the operational efficacy of the self-feedback test platform.

Fig. 5. The experimental configuration

TABLE I THE RESULT OF TEST

Potentiometer turn-on resistance	Potentiometer turn-off resistance	Programmable resistor board turn-on resistance	Programmable resistor board turn-off resistance	Turn-on time	Turn-off time
20Ω	20Ω	20Ω	20Ω	274ns	259ns
23Ω	17Ω	20Ω	20Ω	290ns	233ns
23Ω	17Ω	17Ω	23Ω	269ns	255ns
26Ω	14Ω	17Ω	23Ω	294ns	234ns
26Ω	14Ω	14Ω	27Ω	275ns	260ns
29Ω	11Ω	14Ω	27Ω	292ns	238ns
29Ω	11Ω	11Ω	29Ω	271ns	256ns

V. CONCLUSION

This article examines and substantiates the impact of turn-on and turn-off times on the AC BTI of SiC MOSFETs, highlighting that these temporal parameters evolve as accelerated testing progresses. As the duration of accelerated testing increases, the turn-on and turn-off times of the devices tend to lengthen, which significantly affects the stress stability throughout the testing process. To address this issue, the article proposes the development of a self-adjusting gate drive resistance platform aimed at maintaining the stability of the turn-on and turn-off times of SiC MOSFETs during accelerated testing. Additionally, experimental investigations were conducted to validate the effectiveness of this platform.

REFERENCES

[1] J. Chen, X. Du, Q. Luo, X. Zhang, P. Sun, and L. Zhou, "A review of switching oscillations of wide bandgap semiconductor devices," IEEE Trans. Power Electron., vol. 35, no. 12, pp. 13 182–13 199, 2020.

[2] Wang J, Jiang X. Review and analysis of SiC MOSFETs' ruggedness and reliability[J]. IET Power Electronics, 2020, 13(3): 445-455.

[3] M. W. Feil et al., "Towards understanding the physics of gate switching

instability in silicon carbide MOSFETs," in Proc. IEEE Int. Rel. Phys.

ymp. (IRPS), Mar. 2023, pp. 1–10.

[4] Jiang H, Zhong X, Qiu G, et al. Dynamic gate stress induced threshold voltage drift of silicon carbide MOSFET[J]. IEEE Electron Device Letters, 2020, 41(9): 1284-1287.

[5] Cai Y, Sun P, Chen C, et al. Investigation on Gate Oxide Degradation of SiC MOSFET in Switching Operation[J]. IEEE Transactions on Power Electronics, 2024.

[6] Cai Y, Chen C, Zhao Z, et al. Characterization of gate-oxide degradation location for SiC MOSFETs based on the split C–V method under bias temperature instability conditions[J]. IEEE Transactions on Power Electronics, 2023, 38(5): 6081-6093.

[7] Feil M W, Waschneck K, Reisinger H, et al. Towards understanding the physics of gate switching instability in silicon carbide MOSFETs[C]//2023 IEEE International Reliability Physics Symposium (IRPS). IEEE, 2023: 1-10.

[8] Lelis A J, Habersat D B. AC-stress degradation in SiC MOSFETs[C]//Materials Science Forum. Trans Tech Publications Ltd, 2023, 1092: 151-155.

[9] Habersat D B, Lelis A J. AC-stress degradation and its anneal in SiC MOSFETs[J]. IEEE Transactions on Electron Devices, 2022, 69(9): 5068-5073.

Bias Temperature Instability of SiC Trench MOSFET under DC and AC gate stress

1st Kanghua Yu
College of Electrical and
Information Engineering
Hunan University
Changsha, China
kanghuayu@hnu.edu.cn

2nd Qian Wang
College of Electrical and
Information Engineering
Hunan University
Changsha, China
Wang_qian@hnu.edu.cn

3rd Yuwei Wang
College of Electrical and
Information Engineering
Hunan University
Changsha, China
yuweiwang@hnu.edu.cn

4th Jun Wang
College of Electrical and
Information Engineering
Hunan University
Changsha, China
junwang@hnu.edu.cn

Abstract—This study investigates the threshold voltage (V_{th}) instability and capacitance-voltage (C_g-V_g) degradation mechanisms in commercial SiC trench MOSFETs under DC and AC bias temperature instability (BTI) stresses. By analyzing ΔV_{th} and C_g-V_g characteristics under varying voltage biases, temperatures, and stress modes (DC, unipolar AC, and bipolar AC), distinct degradation is identified. High-bias DC BTI (e.g., 35 V) induces ΔV_{th} via Fowler-Nordheim tunneling at near-interface traps, aligning with conventional models. In contrast, bipolar AC stress (20 V/−10 V) triggers severe ΔV_{th} degradation (up to 5 V) and C_g-V_g drifts in both depletion and inversion regions at room temperature, which may attribute to bulk defect generation driven by recombination-enhanced defect reactions during polarity switching. Elevated temperatures (175°C) suppress AC-induced degradation, while unipolar AC stress shows opposite effects in ΔV_{th} degradation, highlighting the critical role of dynamic switching in defect kinetics. These findings reveal that AC BTI degradation stems from carrier recombination during rapid switching, fundamentally differing from the field-driven trapping mechanisms in DC BTI.

Keywords—Trench MOSFET, DC BTI, AC BTI

I. Introduction

Due to the characteristic of high switching frequencies, high blocking voltages, and low loss, SiC MOSFETs offer a widespread application in electric vehicle drives, photovoltaic converters, and aerospace power supplies[1, 2]. However, the issue of instability of threshold voltage (V_{th}) inevitably results in the degradation of device lifetime, and thus hinder the anticipated applications, which remains to be of great concern in the academia and industry[3-7].

The study of threshold voltage instability is characterized by DC bias temperature instability (DC BTI) and gate switching instability (GSI, also known as AC BTI). The mechanisms of DC BTI in SiC MOSFETs have been extensively studied[7-10], the degradation physics under AC BTI remain insufficient understood. Recent studies revealed that AC BTI is related to interface defect creation via recombination-enhanced defect reactions (REDRs)[11], while existing models based on electric-field-enhanced trapping fail to fully align with observations[12]. Therefore, the understanding of defect dynamics and the mechanism of threshold drift driven by the AC BTI is not clear. In addition, experimental further reveals that AC BTI degradation in MOSFET devices exhibits stronger correlation with switching cycles rather than stress time[12-15]. Particularly, the reliability is more serious in the trench SiC MOSFETs owing to the inhomogeneity of SiC oxidation process and the gate oxide electric field crowding at the corner of the trench bottom[16, 17].

Furthermore, a comprehensive characterization of the degradation mechanisms in C_g-V_g characteristics of trench MOSFETs remains insufficiently explored. Consequently, it is of utmost importance to understand the behavior of the threshold voltage and C-V characteristic under different operating condition.

This work investigates the DC and AC BTI in commercial SiC trench MOSFETs. In this study, the DC field effect is decoupled from the dynamic switching contribution by ΔV_{th} and C_g-V_g characteristics for different voltage bias and temperature under multiple stress modes (DC, unipolar AC, and bipolar AC stresses). Under high-bias DC BTI, charge trapping is dominated by Fowler-Nordheim tunneling at near-interface defects, aligning with conventional bias temperature instability models. In contrast, bipolar AC stress induces a fundamentally different mechanism, where rapid carrier recombination during polarity switching generates bulk defects in the SiC substrate and oxide, accounting for the exacerbated drifts of threshold voltage and C_g-V_g curve. Additionally, elevated temperatures suppress AC-induced degradation, indicating interactions between thermal activation and dynamic defect generation kinetics. The findings provide critical insights for developing AC BTI models and optimizing trench MOSFET designs for high-frequency applications.

II. Experiments

In this study, the commercially available trench SiC MOSFETs was selected as the device under test (DUT). The experiment relies on the BTI automated test system and the DUTs testing board, which can well automate the AC BTI experiment stress stage and test stage. The drain and source electrodes of the DUT are shorted to the ground during the stress stage to eliminate the drain stress. As shown in Fig.1, the measurement-test-measurement (MSM) sequence diagram is employed to evaluate the threshold voltage (V_{th}) shift under negative gate bias, positive gate bias and gate switching stress, with threshold voltage sweeps performed at certain stress cycle intervals to monitor degradation patterns. Three consecutive sweeps are performed as preconditioning after the end of each stress period to exclude the influence of interface traps with very short capture-emission time, and ensure V_{th} of MOSFETs are extracted at a defined interface energy state. The V_{th} in SiC MOSFETs is extracted from third sweep of I_g-V_g curve at the V_d bias of 10 V and I_d current density of 5 mA. The threshold voltage drift (ΔV_{th}) is defined as the difference between the measured value and the initial value of the V_{th}, i.e. $\Delta V_{th} (t_S) = V_{th} (t_S) - V_{th} (0)$, where t_S refers to the stress time. The AC BTI is conducted at 100 kHz with a gate stress waveform alternating between three switching patterns: 20

V/−10 V, 0 V/20 V, and 0 V/−10 V. The transfer and C_g-V_g characteristics are measured at room temperature using the power device analyzer (Agilent B1505A). The C_g-V_g measurement is conducted at the frequency of 10 kHz with the drain and source electrodes shorted.

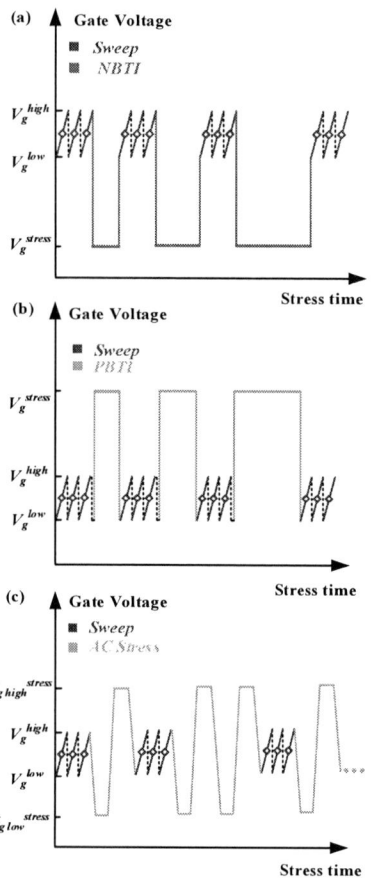

Fig. 1. The MSM sequence used to evaluate V_{th} shift of devices under (a) NBTI, (b) PBTI and (c) AC BTI.

III. RESULTS AND DISCUSSION

A. Comparison of Threshold voltage shift

Fig.2 shows the results of positive bias temperature instability (PBTI), negative bias temperature instability (NBTI), and AC BTI. After 10^5 s stress of PBTI and NBTI, small ΔV_{th} is observed when V_g=20 V (ΔV_{th}=0.5 V) and -10 V (ΔV_{th}=-0.1 V), which is ascribe to the charging of traps at or near the SiC/SiO$_2$ interface. When V_g is up to 35 V, a comparatively larger ΔV_{th} (ΔV_{th}=1.70 V) is observed, which directly correlates with the FN tunneling becoming the dominant charge injection mechanism, indicating that FN tunneling significantly assisted the charging of the traps. Furthermore, the notable V_{th} drift, up to 5V, is observed only when the gate is biased at bipolar AC BTI ($V^{AC}_{g\,low}$ = -10 V and $V^{AC}_{g\,high}$ =20 V), which is twice the initial V_{th}. The difference in degradation characteristics between AC and DC stress reveals fundamentally distinct degradation mechanisms.

In order to further investigate the degradation mechanism, the temperature dependence of different stress

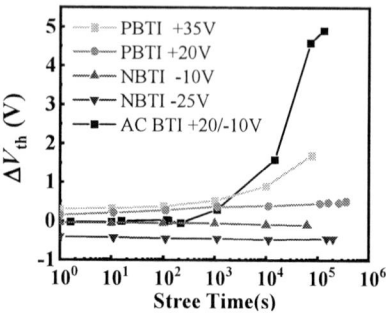

Fig. 2. ΔV_{th} as a function of stress time under various gate stress of DC BTI and AC BTI at room temperature.

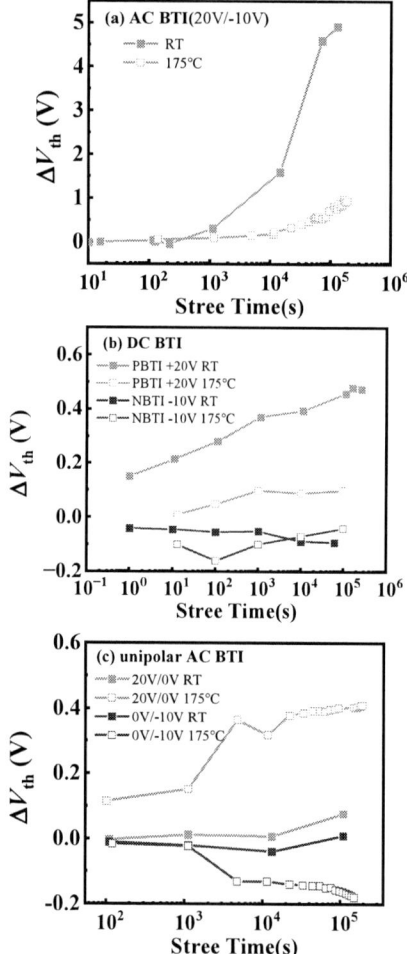

Fig. 3. Comparison of ΔV_{th} as a function of cumulative stress time at room temperature and 175°C for (a) AC BTI (+20 V/-10 V), (b) DC BTI and (c) unipolar (20V/0V and 0 V/-10 V) AC BTI.

condition is also investigated in Fig.3. During the high-temperature degradation, the V_{th} is measured at high temperature after certain stress cycles. The device was then removed from the heating stage for room-temperature characterization using the B1505A analyzer, before being returned to the heating stage for continued aging. Ultimately, the calculated ΔV_{th} between room-temperature and high-temperature measurements demonstrated consistent results. As depicted in Fig.3(a) and (b), the drift of the V_{th} value decreases significantly at higher temperatures for both PBTI and bipolar AC stress (+20 V/-10 V). The AC stress is composed with switching part and DC stress part. To investigate the DC stress part of the AC stress, unipolar AC stress is also conducted with $V_{g\,low}^{AC}$ =0 V (-10 V) and $V_{g\,high}^{AC}$ =20 V (0 V). Different from bipolar AC stress, V_{th} exhibit higher shift with the higher temperature under 20V/0V and 0 V/-10 V stress, suggesting that the decrease in ΔV_{th} with increasing temperature under 20 V/-10 V stress is caused by switching part. For the dependence of temperature, recent literature does not yet allow for a clear conclusion, some devices show increasing V_{th} drift with increasing temperature[18], whereas others show decreasing V_{th} with increasing temperature[13, 18].

B. Comparison of C_g-V_g characteristics

In order to further study the behavior of traps, C_g-V_g measurements are conducted between the gate and source, where drain and source of the DUT are shorted. As shown in Fig.4(a), the C_g-V_g curve including depletion and inversion region. During the region I, holes in the channel region are gradually depleted. While in the region II, electrons from the n well gradually slide into the channel to form an inversion layer. Under the DC BTI, only high stresses introduce shift in region II of CV curve, where NBTI at V_g =-25 V causes a negative shift and PBTI at V_g =35 V a positive shift, which is align to the behavior of the ΔV_{th}. Under low stress bias, the small band bending predominantly facilitates direct tunneling. It can be concluded that the limited energy provided by low applied voltages proves insufficient to enable substantial direct tunneling, consequently leading to minimal shifts in both threshold voltage and C_g-V_g characteristics. In contrast, FN tunneling significantly enhances carrier injection probabilities into donor-like and acceptor-like traps at near interface (NITs), ultimately introducing the negative and positive shift, respectively.

Different from the DC BTI, Fig.5(a) illustrates the C_g-V_g characteristic under the bipolar AC stress, where the notable shifts both in region I and II are observed at room temperature. The shift in region II, which corresponds to the inversion region, shows similar behavior to DC BTI. This similarity suggests that the primary mechanism of the shift in region II under AC BTI involves the charging of acceptor-like near-interface traps (NITs) and interface states. However, the positive shift observed in region I (depletion region) presents a unique characteristic of AC stress that distinguishes it from DC BTI behavior, which may originate from the fundamental differences in carrier dynamics under bipolar AC stress. During high-frequency AC stress, the SiC channel experiences rapid alternations between electron and hole dominance. This dynamic environment creates conditions for enhanced carrier recombination events. The energy released during the non-radiative recombination could provide sufficient activation energy to generate new hole-related defects in the SiC bulk material. The bulk defects would manifest as additional positive charge centers in the depletion region, which effectively increase the

required gate voltage to fully deplete holes from the channel region, explaining the observed rightward shift in region I. While, as shown in Fig.5(b), there is only small shift in both region I and II under the bipolar AC stress at 175°C, which align with the degradation behavior of ΔV_{th}, indicating that the high temperature suppressed the trap charge. However, the underlying physical mechanisms require further systematic characterization. Furthermore, no significant shift is observed under unipolar stress at both room and high temperature, as illustrated in Fig.5(c), (d), (e) and (f), suggesting that the DC stress part of the AC stress is not the dominant contributor to AC BTI degradation. It further demonstrates that the trap capturing occurred in bipolar AC BTI is mainly caused by the switching part between the accumulation of holes and inversion of electrons during the polarity transitions.

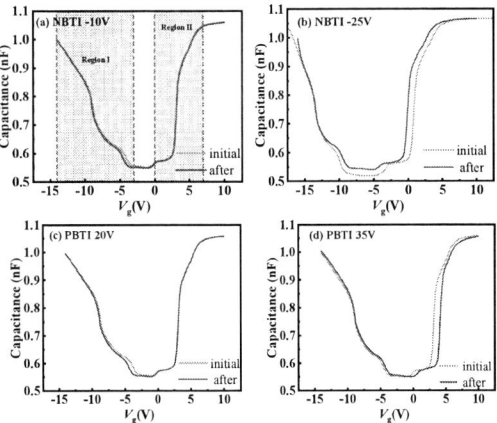

Fig. 4. Comparison of C_g-V_g curves before and after (a)-10 V NBTI, (b) -25 V NBTI, (c) 20 V PBTI and (d)35 V PBTI at room temperature.

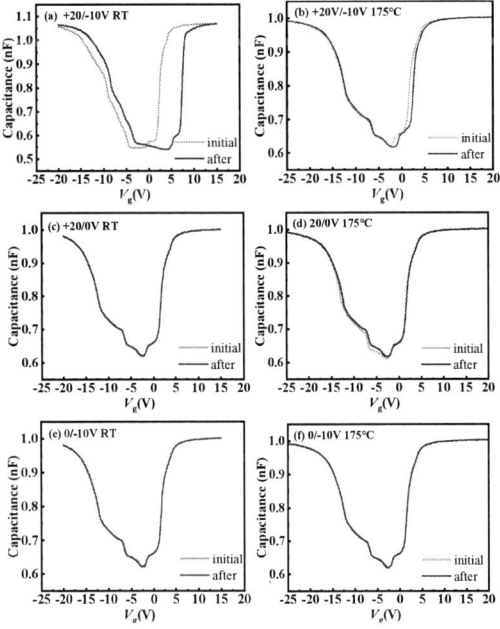

Fig. 5. Comparison of C_g-V_g curves before and after AC BTI with the V_g = (a) 20 V/-10 V, (c) 0 V/-10 V, (e) 20 V/0 V at room temperature, and V_g = (b) 20 V/-10 V, (d) 0 V/-10 V, (f) 20 V/0 V at 175°C.

IV. Conclusion

This study systematically investigates bias temperature instability in SiC trench MOSFETs through DC and AC stress tests, analyzing bias voltage and temperature effects on ΔV_{th} and C_g-V_g characteristics. Results demonstrate that ΔV_{th} induced by DC BTI at high biases (V_g=35 V) is dominated by FN tunneling-assisted trap charging. In contrast, bipolar AC stress causes severe ΔV_{th} degradation (up to 5 V), accompanied by positive drifts in both depletion and inversion region of the C_g-V_g at room temperature. These AC-specific effects diminish at 175°C, while unipolar AC stress shows negligible degradation. The results conclusively demonstrate that DC BTI primarily activates existing traps through sustained electric fields, whereas AC BTI degradation originates from dynamic switching processes during polarity transitions. The switching between the accumulation of holes and inversion of electrons in AC stress facilitate the recombination, which may provide sufficient activation energy to generate defects. However, the exact physical mechanisms governing this defect generation process require further systematic characterization.

Acknowledgment

This work was supported by Yuelushan University Science & Technology Park under Grant H202291400539.

References

[1] T. Kimoto, "Material science and device physics in SiC technology for high-voltage power devices," *Jpn. J. Appl. Phys.,* vol. 54, no. 4, p. 040103, Mar. 2015.

[2] X. She, A. Q. Huang, O. Lucia, and B. Ozpineci, "Review of Silicon Carbide Power Devices and Their Applications," *IEEE Trans. Ind. Electron.,* vol. 64, no. 10, pp. 8193-8205, 2017.

[3] T. Grasser *et al.,* "The Paradigm Shift in Understanding the Bias Temperature Instability: From Reaction–Diffusion to Switching Oxide Traps," *IEEE Trans. Electron Devices,* vol. 58, no. 11, pp. 3652-3666, 2011.

[4] S. Mahapatra *et al.,* "A Generic Trap Generation Framework for MOSFET Reliability—Part I: Gate Only Stress–BTI, SILC, and TDDB," *IEEE Trans. Electron Devices,* vol. 71, no. 1, pp. 114-125, 2024.

[5] T. Kimoto and H. Watanabe, "Defect engineering in SiC technology for high-voltage power devices," *Appl. Phys. Express,* vol. 13, no. 12, p. 120101, Nov. 2020.

[6] C. Yang, S. Wei, and D. Wang, "Bias temperature instability in SiC metal oxide semiconductor devices," *J. Phys. D: Appl. Phys.,* vol. 54, no. 12, 2021.

[7] K. Puschkarsky, T. Grasser, T. Aichinger, W. Gustin, and H. Reisinger, "Review on SiC MOSFETs High-Voltage Device Reliability Focusing on Threshold Voltage Instability," *IEEE Trans. Electron Devices,* vol. 66, no. 11, pp. 4604-4616, 2019.

[8] K. Puschkarsky, H. Reisinger, T. Aichinger, W. Gustin, and T. Grasser, "Understanding BTI in SiC MOSFETs and Its Impact on Circuit Operation," *IEEE Transactions on Device and Materials Reliability,* vol. 18, no. 2, pp. 144-153, 2018.

[9] T. Watanabe, Y. Fukui, S. Hino, S. Tomohisa, N. Miura, and K. Nishikawa, "Categorization of PBTI Mechanisms on 4H-SiC MOSFETs by the Stress Gate Voltage and Channel Plane Orientation," *IEEE Transactions on Device and Materials Reliability,* vol. 23, no. 1, pp. 99-108, 2023.

[10] T. Aichinger, G. Rescher, and G. Pobegen, "Threshold voltage peculiarities and bias temperature instabilities of SiC MOSFETs," *Microelectron. Rel.,* vol. 80, pp. 68-78, Jan. 2018.

[11] T. Grasser *et al.,* "A Recombination-Enhanced-Defect-Reaction-Based Model for the Gate Switching Instability in SiC MOSFETs," presented at the 2024 IEEE International Reliability Physics Symposium (IRPS), 2024.

[12] M. W. Feil *et al.,* "Gate Switching Instability in Silicon Carbide MOSFETs—Part I: Experimental," *IEEE Trans. Electron Devices,* vol. 71, no. 7, pp. 4210-4217, 2024.

[13] D. B. Habersat and A. J. Lelis, "AC-Stress Degradation and Its Anneal in SiC MOSFETs," *IEEE Trans. Electron Devices,* vol. 69, no. 9, pp. 5068-5073, 2022.

[14] M. W. Feil *et al.,* "Towards Understanding the Physics of Gate Switching Instability in Silicon Carbide MOSFETs," presented at the 2023 IEEE International Reliability Physics Symposium (IRPS), 2023.

[15] H. Jiang, X. Zhong, G. Qiu, L. Tang, X. Qi, and L. Ran, "Dynamic Gate Stress Induced Threshold Voltage Drift of Silicon Carbide MOSFET," *IEEE Electron Device Letters,* vol. 41, no. 9, pp. 1284-1287, 2020.

[16] X. Deng *et al.,* "Short-Circuit Capability Prediction and Failure Mode of Asymmetric and Double Trench SiC MOSFETs," *IEEE Trans. Power Electron.,* vol. 36, no. 7, pp. 8300-8307, 2021.

[17] M. Sagawa, H. Miki, Y. Mori, H. Shimizu, and A. Shima, "Evaluation of gate oxide reliability in 3.3 kV 4H-SiC DMOSFET with J-Ramp TDDB methods," in *2018 IEEE 30th International Symposium on Power Semiconductor Devices and ICs (ISPSD)*, 13-17 May 2018 2018, pp. 363-366, doi: 10.1109/ISPSD.2018.8393678.

[18] H. Jiang *et al.,* "A Physical Explanation of Threshold Voltage Drift of SiC MOSFET Induced by Gate Switching," *IEEE Trans. Power Electron.,* vol. 37, no. 8, pp. 8830-8834, 2022.

A Novel Approach to Estimate the Thermal Destruction Point of SiC MOSFETs under Short Circuit Conditions

Rony Thomas
Application Engineering SiC
Nexperia Gmbh
Hamburg, Germany
rony.thomas@nexperia.com

Zhe Yu
Application Engineering SiC
Nexperia Gmbh
Hamburg, Germany
zhe.yu@nexperia.com

Sebastian Fahlbusch
Application Engineering SiC
Nexperia Gmbh
Hamburg, Germany
sebastian.fahlbusch@nexperia.com

Abstract—**Short circuit events in silicon carbide (SiC) MOS-FETs can create critical reliability issues because of rapid temperature rise, which frequently results in thermal failure of the device. This paper discusses a new approach for the estimation of thermal destruction point during short circuit conditions by integrating experimentally measured power dissipation data into thermal models created in simulation software. The proposed methodology enables an accurate prediction of the thermal destruction limits, giving more insight about the device operation in highly stressful environments. The works also focus on the impact and changes in destruction temperature while varying different electrical parameters.**

Index Terms—**Short circuit, thermal destruction point, power dissipation, thermal models, simulation**

I. INTRODUCTION

There are practical difficulties in determining the precise junction temperature at which a MOSFET fails during a short circuit event. The extreme nature of the short circuit condition causing abrupt and localized junction temperature rise, making direct temperature measurement difficult. Conventional thermal sensors, such as infrared cameras or thermocouples, lack the required spatial and temporal resolution to capture the temperature at the precise moment of device failure. Additionally, the destruction process is highly dynamic, involving combined interactions of self-heating, carrier mobility degradation, package inductances, and threshold voltage shifts, all occurring quickly in a few microseconds [2].

The electrical measurements can impacts the short circuit withstand time (SCWT) and the energy dissipated before failure. they do not reveal the exact junction temperature at which failure occurs. Similarly, most modern simulation models lack in their predictive capabilities when it comes to short circuit destruction. Most of the thermal models often do not incorporate detailed physics-based mechanisms governing thermal runaway and failure, leading to inaccuracies in predicting both the withstand time and the temperature during short circuit failure.

A method based on thermal networks is presented to get around these restrictions. When the MOSFET's thermal behavior is modeled and included into the experimental analysis,

the junction temperature during a short circuit event can be estimated. This method provides a more accurate estimation of the temperature at failure. For the initial evaluation, 1200 V SiC MOSFET from Nexperia (NSF040120L4A0) is used for experiment and modelling. In this paper, the short circuit destruction time (SCDT) refers to the time until the device physically fails under short circuit conditions, which is slightly different from the short circuit withstand time that defines the maximum safe duration before failure begins.

II. METHODOLOGY

The methodology used in this paper is illustrated as a flowchart in Fig. 1. It consists of two parallel paths: one based on experimental measurements and the other derived from simulation.

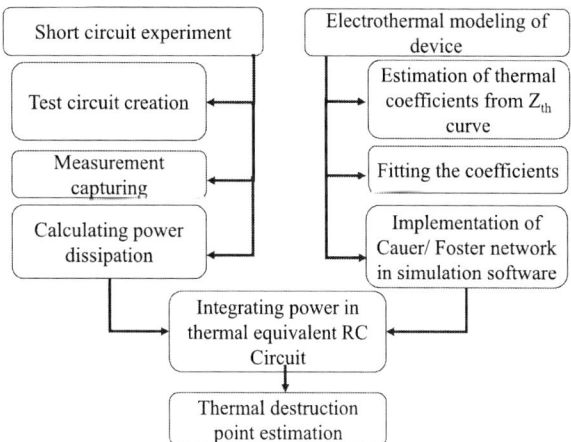

Fig. 1. Proposed Methodology

Unlike typical MOSFET operation, a short circuit occurs under high drain source voltage (V_{ds}) and drain current (I_d), resulting in extremely high-power dissipation. This results in both severe thermal and electrical stress, causing device failure within microseconds.

In most cases, the SCDT of SiC devices is only a few microseconds, making it difficult to accurately determine the exact junction temperature at the critical moment of failure. To build the hardware setup, a Type 1 short circuit configuration is used, where the Device Under Test (DUT) is connected across DC link capacitors [1].

During the short circuit event, the V_{ds} and I_d are measured, as shown in Fig. 2. This data is then used to calculate the instantaneous power dissipation until the device fails. In particular, this is not the typical Type 1 configuration in which both switches have similar R_{dson} levels. In this case, different switches are intentionally selected, including a low R_{dson} high side switch. As a result, the voltage and current stress are predominantly imposed across the DUT. In simulation software, power dissipation can be incorporated into a behavioral current or voltage source, which drives the thermal network. Thermal modeling can be performed using either the Foster network or the Cauer network. In the proposed methodology, the thermal network is derived from the thermal impedance graphs (Z_{th}) in the device datasheets using curve-fitting tools.

III. SHORT CIRCUIT MEASUREMENTS

A. Test Setup

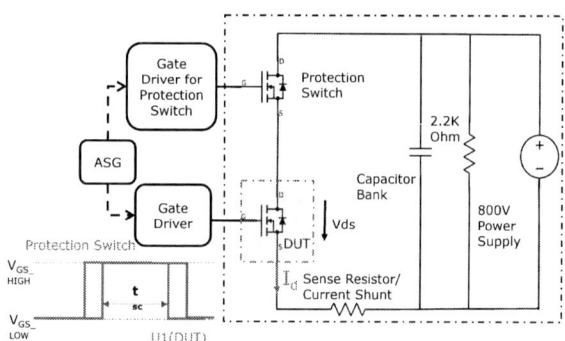

Fig. 2. Test Circuit for Type 1 Short Circuit [[3]]

To prepare for the designated short circuit measurements, a test bench was developed specifically for studying 1200V SiC MOSFETs. While the Type 1 short circuit, which involves direct turn-on into a short circuit, can be initiated using just the device-under-test and a capacitor bank, this employs a half-bridge setup [1]. This setup contains a high current capable SiC MOSFET functioning as both a protective device and a high-side switch. The circuit is designed for short circuit tests with a DC-link voltage of 800 V. The circuit diagram for the short circuit testing and the gate pulse conditions are shown in Fig. 2. An arbitrary function generator (AFG) is used to turn on the protection switch before the SC pulse is applied to the system and to turn it off afterwards. Short circuit dynamics are captured using an oscilloscope. The complete photograph of the test setup is shown in Fig. 3.

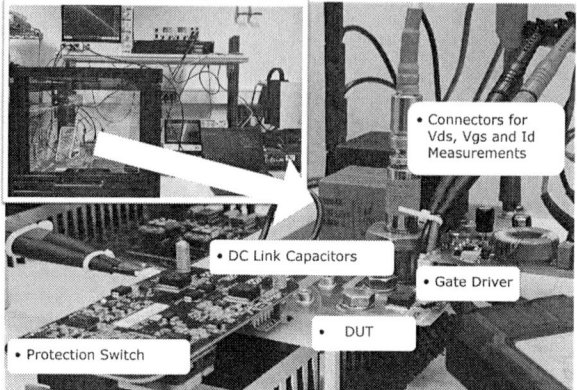

Fig. 3. Photograph of Testbench

The DUT is mounted on a detachable external PCB (Fig3) that can be easily disconnected from the half bridge PCB for testing. The data captured during the experiments through oscilloscope were extracted by using MATLAB script.

B. Measurements

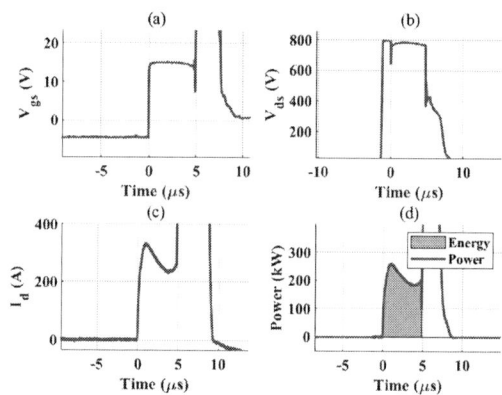

Fig. 4. Calculating power and energy (d) from short circuit waveforms , for TO247-4, 40 mΩ devices, V_{ds} =800 V (b) , Vgs=15 V (a)

The instantaneous power during the short circuit event is calculated by multiplying drain source voltage (V_{ds}) with drain current (I_d).

$$P_{sc} = V_{ds} \times I_d \tag{1}$$

The short circuit waveforms corresponding to the SiC MOSFET (NSF040120L4A0) under defined conditions are shown in Fig. 4. Here, the device undergoes a Type 1 short circuit condition and was destroyed at 5.1 μs under V_{ds} = 800 V and V_{gs} = 15 V. The peak current during the short circuit event exceeded 320 A. The entire system is controlled via MATLAB, ensuring synchronized operation and the ability to automate repetitive testing tasks. The data captured during the experiments were stored in an excel file and the power dissipation across the MOSFET during actual operation is given across the defined RC network. LTspice has an inherent feature of defining the behavioural sources with values from pwl file. The file format '.pwl' can be used to define the signal of voltage/current sources in simulation.

IV. MODELLING OF THERMAL NETWORKS

Thermal networks simplify the heat equation, enabling to predict temperature changes and heat flow during power dissipation events. Power dissipation can be directly related to temperature rise through thermal impedance. Cauer and Foster networks are two equivalent circuit representations used for thermal modeling [5]. The construction of a Cauer network involves discretizing the thermal structure into layers, each representing a distinct material layer [4]. This requires material properties and dimensions, which manufacturers often do not provide. In contrast, thermal resistance (R_{th}) and thermal

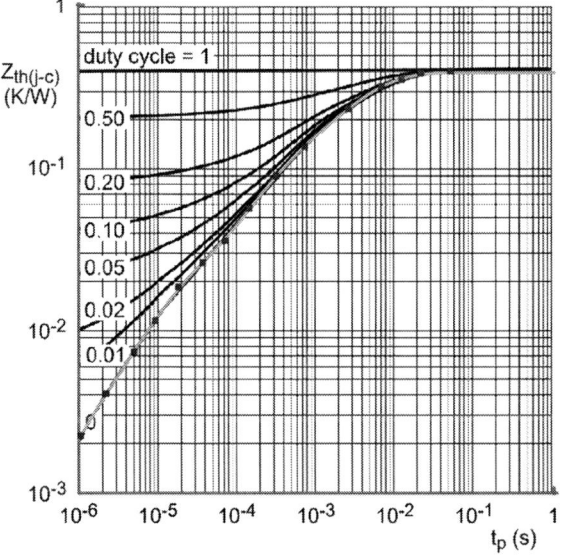

Fig. 5. Curve Fitting using PLECS [7]

capacitance (C_{th}) corresponding to Foster network can be derived numerically from thermal impedance curves (Z_{th}) in data sheets, offers a more practical alternative, especially for time-domain analysis. There are various methods available to extract thermal capacitance (C_{th}) and thermal resistance (R_{th}) by fitting the thermal impedance (Z_{th}) curve. Among these, simulation tools such as PLECS provide an intuitive and streamlined approach for curve fitting and thermal model extraction. Alternatively, MATLAB's curve fitting toolbox can also be used for this purpose, although it may be less convenient for application engineers due to its more manual setup process.

The thermal chain feature of the PLECS software is used to fit the Z_{th} curve. The transient impedance curve fitting of NSF040120L4A0 is shown in Fig. 5. This tool enables accurate curve fitting with precise convergence and provides the equivalent thermal network in either Foster or Cauer model representations. The transient thermal impedance is modeled using a Foster network consisting of multiple RC elements. For five sets, the equation can be expressed as

$$Z_{th}(t) = \sum_{i=1}^{5} R_i \left(1 - e^{-\frac{t}{\tau_i}} \right) \tag{2}$$

$$\tau_i = R_i * C_i \tag{3}$$

Each term in the summation corresponds to a thermal RC pair, where:

R_i is the thermal resistance of the i-th stage (in K/W),

C_i is the thermal capacitance of the i-th stage (in J/K),

τ_i is the thermal time constant of the i-th stage, t is the elapsed time (in seconds),

$Z_{th}(t)$ is the time-dependent thermal impedance (in K/W).

TABLE I
FOSTER EQUIVALENT THERMAL PARAMETERS FROM CURVE FITTING

Param.	1	2	3	4	5
R (K/W)	0.0009	0.0140	0.0783	0.1660	0.1244
τ (s)	8.1×10^{-7}	1.2×10^{-5}	3.0×10^{-4}	2.7×10^{-3}	8.4×10^{-3}

Table I presents the equivalent set of thermal resistance (R) and time constant (τ) values obtained by fitting the Z_{th} curve using PLECS software.

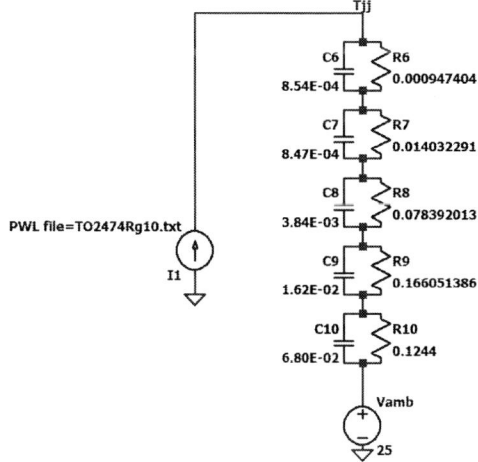

Fig. 6. Simulation circuit with Foster network in LTspice

The Foster model can be converted back into a Cauer network. However, when derived directly from a given thermal impedance curve without additional crystal or structural information, the physical interpretability of the Cauer network becomes limited. Although the Cauer model is often considered more physical representative, in such cases it does not provide meaningful insights into the actual device construction. The equivalent simulation circuit, which is used for initial analysis, is shown in Fig. 6. Here, the equivalent junction-to-case temperature (T_{jc}) is obtained by calculating the voltage drop across the equivalent thermal network, which is represented using resistance and capacitance. The behavioral current source I1 has a magnitude equivalent to the power dissipation of the MOSFET tested.

V. INCORPORATING EXPERIMENTAL DATA INTO ELECTRO-THERMAL MODELS

An approach was taken to predict the temperature variations observed in the experiment by comparing them with simulation results using LTspice. It has an inherent feature of defining the behavioral models with predefined magnitude. As discussed earlier, the voltage across the junction to ambient (across the network) represents the equivalent junction temperature of the system. It is feasible to apply precise electrical stress to specified thermal models, enabling the prediction of the system's thermal stress.

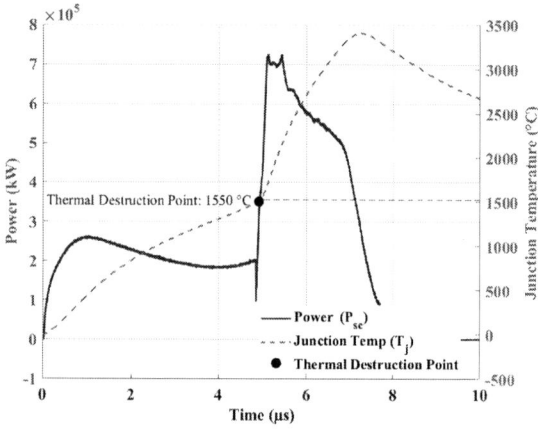

Fig. 7. Illustration of thermal destruction point estimation from defined methodology

The thermal destruction point (1550°C) occurring during the Type 1 short circuit fault is illustrated in Fig. 7 by combining the measurement and simulation methods. The x-axis represents time in μs, while the dual y-axes show the instantaneous power dissipated across the DUT and the simulated junction temperature in °C, respectively. The device failure point is clearly visible in the short circuit waveform, where the current abruptly shoots and the power rises sharply. For the 40 m TO-247 4-pin device, the short circuit destruction occurred at approximately 5.1 μs. By intersecting this time point with the junction temperature curve, the thermal destruction point of the MOSFET can be identified. All figures discussed in this section follow the same methodology, with variations only in the electrical parameters.

VI. IMPACT OF ELECTRICAL PARAMETERS ON JUNCTION TEMPERATURE

This method of determining the junction temperature enables the analysis of how variations in several electrical parameters influence device behavior under stress conditions. The electrical testing involves systematically varying the following parameters:

- Drain-source voltage (V_{ds})
- Gate-source voltage (V_{gs})
- Threshold voltage (V_{th})
- Gate resistance (R_g)
- Drain source on resistance (R_{dson})

The power dissipation across the SiC MOSFET during the short circuit Type 1 fault was applied as an input to the thermal network in LTspice to analyze junction temperature variations. The impact of thermal stress while increasing V_{ds} is illustrated in Fig. 8.

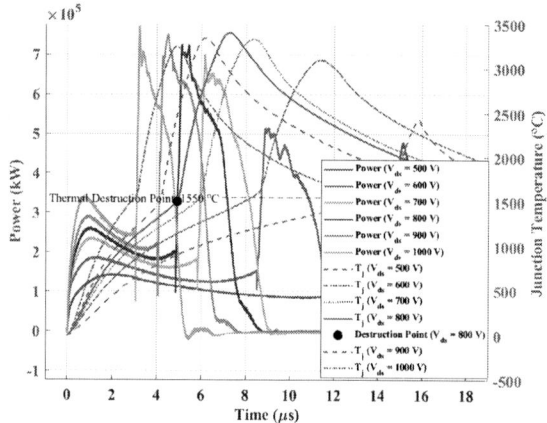

Fig. 8. Effect of V_{ds} on thermal stress for TO247-4, 40 mΩ devices, V_{ds} =500 V - 1000 V, V_{gs}=15 V, R_g=5.6 Ω

The drain source voltage has been varied between 500 V and 1000 V, and the impact is analyzed here. It is clear from the waveform that, although the short circuit withstanding time and the instantaneous power dissipation change with increase in V_{ds} the device fails at similar thermal conditions. This method can be used to interpolate the trajectory along which short circuit events occur with respect to a consistent junction temperature, enabling the prediction of SCDT at any given voltage.

The effect of V_{gs} on the thermal destruction point while changing from 15 V to 18 V is depicted in Fig. 9. Even though the SCWT is reduced from 5.1 μs to 3.2 μs with increasing V_{gs}, the junction temperature at which failure occurs remains the same for the given device.

979-8-3315-1110-4/25 $31.00 © 2025 IEEE

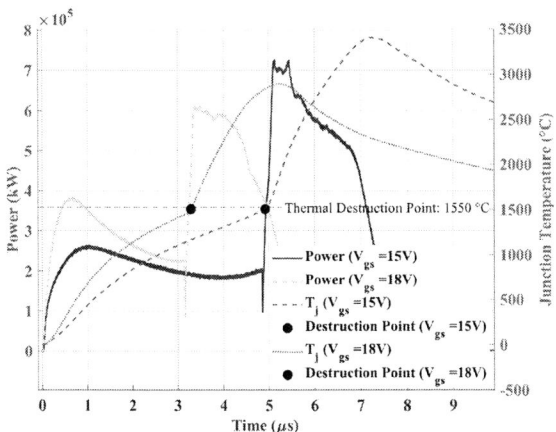

Fig. 9. Effect of V_{gs} on thermal stress for TO247-4, 40 m Ω devices, V_{ds} =, 800 V , V_{gs}=15 V and 18 V, R_g=5.6 Ω

A set of SiC MOSFETs were characterized to evaluate the influence of threshold voltage variation on short circuit behavior. The V_{th} values were measured using a curve tracer by selecting V_{gs} under a 4 mA current. For each package, two devices from 20 samples were selected: one with the highest V_{th} (V_{thMAX}), one with the lowest V_{th} (V_{thMIN}). Devices with minimum and maximum V_{th} values,showing a spread of approximately 5%, were selected for analysis. These devices were subjected to identical short circuit conditions to examine the impact on junction temperature. As shown in Fig. 10, despite the variation in V_{th}, the resulting junction temperatures at the point of thermal failure showed almost no significant deviation.

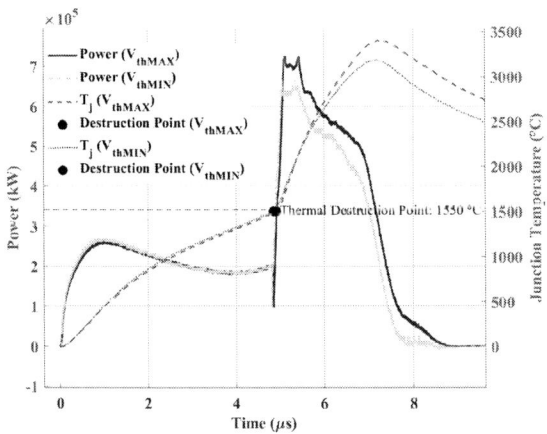

Fig. 10. Effect of V_{th} on thermal stress for TO247-4, 40 m Ω devices, V_{ds} =800 V, V_{gs}=15 V, R_g=5.6 Ω

Additionally, the influence of gate R_g was investigated using values of 5.6 Ω, 10 Ω, and 30 Ω. While an increase in R_g led to minor changes in short circuit duration and

switching speed, the ultimate junction temperature at which the MOSFET underwent thermal destruction remained nearly unchanged. This effect is illustrated in Fig. 11.

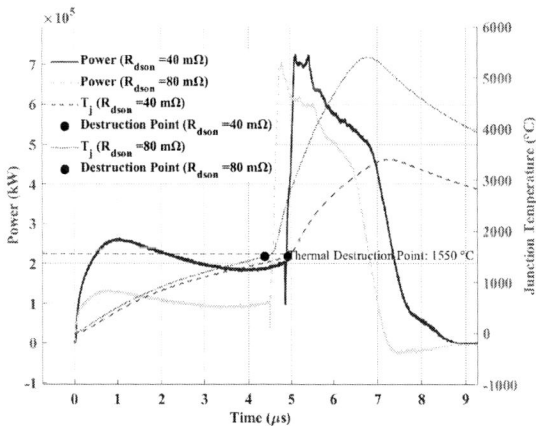

Fig. 11. Effect of R_g on thermal stress for TO247-4, 40 m Ω devices, V_{ds} =800 V, V_{gs}=15 V,R_g=5.6 Ω

The device NSF040120L4A0 was compared with NSF080120L4A0, which features a higher R_{dson} value. Since R_{dson} directly correlates with the crystal size, a notable difference in thermal behavior was expected. Although the instantaneous power dissipation of the 80 m Ω device is roughly half that of the 40 m Ω device, the thermal destruction occurred at similar temperature range with difference in short circuit destruction times. The results are illustrated on Fig. 12.

Fig. 12. Fig. 12. Effect of R_{dson} on thermal stress for TO247-4, 40 m Ω and TO247-4, 80 m Ω devices, V_{ds} =800V, V_{gs}=15 V, R_g=5.6 Ω

VII. Accuracy and Validation of Method

The accuracy and reliability of the proposed method for estimating junction temperature from the thermal network primarily depend on two key aspects: the accuracy of the

Z_{th} curve and the precision of the testing setup. While the test method can be carefully controlled by using consistent equipment, questions arise regarding the accuracy of the Z_{th} curves from which the thermal network is derived.

According to JEDEC standards, Z_{th} values in the sub-100μs region are typically extrapolated due to inherent measurement delays and inconsistencies at very short time intervals [6]. This makes the accuracy of the early transient region questionable. Since this method fully relies on datasheet-provided parameters, the accuracy of the results is directly tied to the fidelity of the Z_{th} data. Although the calculated absolute temperature may deviate slightly, the comparative analysis between different cases remains valid. This implies that, once a baseline measurement is established, it is possible to accurately predict device behavior under other operating conditions, such as varying V_{ds} or V_{gs}. As a next step, validation of this methodology using TCAD simulations is planned to further assess its robustness and improve confidence in its predictive capability.

VIII. CONCLUSION

Here in this paper, a novel approach combining measurement and simulation provides strong insight into the destruction temperature during short circuit events. One of the main findings from the initial results is that although variations in V_{ds} or V_{gs} directly influence the short circuit destruction time of the SiC MOSFET, device destruction consistently occurs at the same junction temperature. This work also analyzes the influence of R_g, V_{th}, and R_{dson} on thermal destruction levels and junction temperature variations during short circuit operation. It was observed that variations in V_{th} do not significantly affect either SCDT or the junction temperature. However, changes in R_g and R_{dson} do affect SDWT, yet the thermal destruction still consistently occurs around 1550°C. Multiple devices were examined to determine their thermal destruction points. Although the melting point of silicon carbide (SiC) is approximately 2730°C, failure occurs much earlier, around 1550°C. This early failure can be attributed to gate oxide degradation, which occurs at lower temperatures. The study finds that changes in electrical parameters do not alter the ultimate thermal destruction temperature, a conclusion verified across different device packages. It should be noted that the accuracy of these findings depends heavily on the precision of datasheet parameters. .

REFERENCES

[1] A. Engel, "Short circuit behaviour and short circuit robustness of silicon carbide MOSFETs – A review," 2021.

[2] A. Perez-Tomas, P. Brosselard, P. Godignon, J. Millan, and N. Mestres, "Field effect mobility temperature modeling of 4H SiC metal oxide semiconductor transistors," *J. Appl. Phys.*, vol. 100, no. 11, p. 114508, 2006.

[3] J. Lutz, H. Schlangenotto, U. Scheuermann, and R. De Doncker, *Semiconductor Power Devices: Physics, Characteristics, Reliability.* Springer, 2011.

[4] M. März and P. Nance, "Thermal modeling of power-electronic systems," Application Note, Infineon Technologies AG, Munich, Germany, Jul. 6, 2000. [Online]. Available: http://www.infineon.com/products/power/pdf/mmpn$_e$ng.pdf

[5] Infineon Technologies AG, "Thermal equivalent circuit models," Application Note AN2015-10, Version 1.00, Aug. 2015. [Online]. Available: https://www.infineon.com/dgdl/Infineon-Thermal$_e$quivalent$_c$ircuit$_m$odels $- ApplicationNotes - v01_02 - EN.pdf?fileId = db3a30431a5c32f2011aa65358394dd2$

[6] JEDEC Solid State Technology Association, "JESD51-14: Transient dual interface test method for the measurement of the thermal resistance junction-to-case of semiconductor devices with heat flow through a single path," Arlington, VA, USA, Oct. 2010.

[7] Nexperia, "NSF040120L4A0 – 1200 V, 40 mΩ, N-channel SiC MOSFET," Datasheet, Aug. 15, 2024. [Online]. Available: https://assets.nexperia.com/documents/data-sheet/NSF040120L4A0.pdf

Investigation on Electrical Properties and Safe-Operating Area for Novel Split-gate IGBTs

Xuanting Song
College of Electrical and Information Engineering
Hunan University
Changsha, China
songxt@hnu.edu.cn

Jun Wang
College of Electrical and Information Engineering
Hunan University
Changsha, China
junwang@hnu.edu.cn

Gaoqiang Deng
School of Integrated Circuit Science and Engineering
University of Electronic Science and Technology of China
Chengdu, China
gqdeng@uestc.edu.cn

Zijie Zheng
College of Electrical and Information Engineering
Hunan University
Changsha, China
zijiezheng@hnu.edu.cn

Yongzhou Zou
College of Electrical and Information Engineering
Hunan University
Changsha, China
yongzhouzou990183@hnu.edu.cn

Shiwei Liang
College of Electrical and Information Engineering
Hunan University
Changsha, China
swliang@hnu.edu.cn

Yuxing Dai
College of Electrical and Electronic Engineering
Wenzhou University
Wenzhou, China
daiyx@hnu.edu.cn

Kamal Al-Haddad
Department of Electrical Engineering
École de Technologie Supérieure
Montreal, Canada
kamal.al-haddad@etsmtl.ca

Abstract—In this work, the electrical properties and safe-operating area (SOA) of two novel split-gate insulated gate bipolar transistors (IGBTs): the interlaced gate IGBT(IGT IGBT) and the segmented gate IGBT(SGT IGBT) are systematically investigated. Through TCAD Sentaurus simulations, the impact of the poly gate-to-emitter area ratio (W_G/W_E) on device performance is analyzed. Both structures feature a built-in p-MOSFET to enhance conductivity modulation and reduce turn-off loss(E_{off}). The simulation results demonstrate distinct performance advantages between the two designs: the IGT IGBT achieves superior forward conduction characteristics and enhanced electromagnetic interference (EMI) suppression compared to the SGT IGBT, while the latter exhibits lower switching losses and reduced input capacitance. Increasing the W_G/W_E ratio amplifies channel density, thereby improving forward conduction capability. However, this enhancement inversely degrades switching efficiency and compromises short-circuit withstand robustness.

Keywords—*IGBT, safe-operating area(SOA), on-voltage drop(V_{on}), turn-off loss(E_{off})*

I. Introduction

Insulated gate bipolar transistor (IGBT) is one of crucial components in medium- to high-voltage power conversion systems, enabling critical applications such as smart grids, industrial motor drives, and automotive electronics[1-4]. A central challenge in IGBT advancement lies in optimizing the trade-off between the on-state voltage drop ($V_{on}/V_{CE(sat)}$) and turn-off loss (E_{off}). Widely adopted techniques, including carrier storage (CS) layers[5], injection enhancement[6-7], and wafer thinning[8], improve conductivity modulation to

reduce V_{on} while maintaining low E_{off}. However, highly doped or thick CS layers degrade breakdown voltage (BV) and safe-operating area (SOA). To mitigate this, the p-ring IGBT structure was introduced[9], employing floating p-type layers at the trench bottom to balance charge and shield the gate trench from electric field stress, thereby preserving BV. However, SOA limitations persist due to dynamic avalanche and latch-up in p-ring IGBT.

To achieve low loss and wide SOA, the novel interlaced gate IGBT(IGT IGBT) and the segmented gate IGBT(SGT IGBT) have been proposed and preliminary verified their superior performance in our previous work[10-11]. However, the area ratio between the poly gate and the poly emitter will critically influences the performance of these IGBT designs. The comparison between these two designs is also worth investigating. This paper will systematically analyze the impact of this area ratio and further investigate the performance and SOA differences between the two IGBTs.

II. Device Structure and Mechanism

The schematic view of IGT IGBT and SGT IGBT are showed in Fig.1. The poly gate and emitter are interleaved in the trench along the z-axis. In the back side of these IGBTs, the built-in p-MOSFET is formed by p-body, n-CS layer,p-ring and trench emitter. In the off-state, the p-ring structure maintains the high blocking voltage. In the initiall on-state, the potential of collector is relatively low. The built-in p-MOSFET is closed, thus enhanced the conductivity modulation of the n-drift region. When the collector potential increases, the p-MOSFET becomes conductive. This action provides a path that facilitates rapid hole extraction.

This work was supported in part by the National Natural Science Foundation of China under Grant U21A20499 and in part by the Yuelu Mountain National University Science and Technology City under Grant H202291400539.

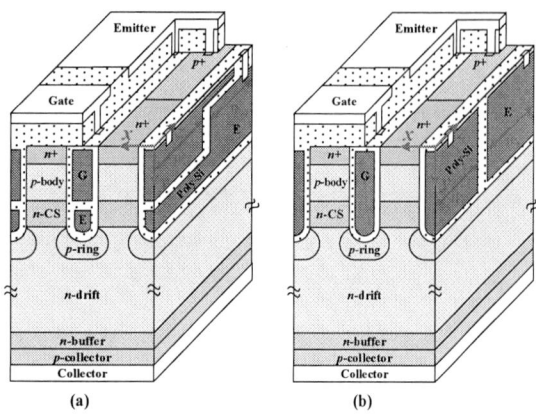

Fig.1 Schematic view of (a) SGT IGBT and (b) IGT IGBT

TABLE I. KEY PARAMETERS IN SIMULATIONS

Key Parameters	Devices	
	IGT IGBT	SGT IGBT
Cell width(μm)	6	6
Mesa width(μm)	4	4
Trench width(μm)	2	2
P-ring interval(μm)	1	1
Gate oxide thickness(μm)	0.1	0.1
N-drift thickness(μm)	112	112
N-CS thickness(μm)	2	2
N-buffer thickness(μm)	1	1
N-drift doping(cm^{-3})	8×10^{13}	8×10^{13}
N-CS doping(cm^{-3})	2×10^{17}	2×10^{17}
N-buffer doping(cm^{-3})	2×10^{16}	8×10^{16}
P-ring doping(cm^{-3})	2×10^{17}	2×10^{17}

Consequently, this leads to a reduction in both the saturation current and the turn-off loss.

For all LIGBTs in this work, the key parameters are listed in Table I. The TCAD Sentaurus [12] is used to investigate the characteristics of IGBTs. The excessive carriers' lifetime is set to 1μs. The physical models are used including high field saturation mobility model, Enormal mobility model, Phumob mobility model, doping and temperature dependence Shockley-Read-Hall recombination, Auger recombination, Unibo avalanche recombination, effective intrinsic density (OldSlotboom) model and self-heating model (only used in short-circuit and UIS simulations)

III. RESULTS AND DISCUSSION

Firstly, the forward blocking characteristic of IGT and SGT IGBT is investigated in Fig.2. Due to the p-ring region under the trench oxide, the electric field is concentrated at the junction between the p-ring and n-drift layer, effectively protecting the corner of the trench gate oxide. It is observed that the area ratio of the poly gate to the emitter(W_G/W_E) does not influence the blocking capability of both IGBTs. All the IGBTs achieve a breakdown voltage (BV) of 1368V.

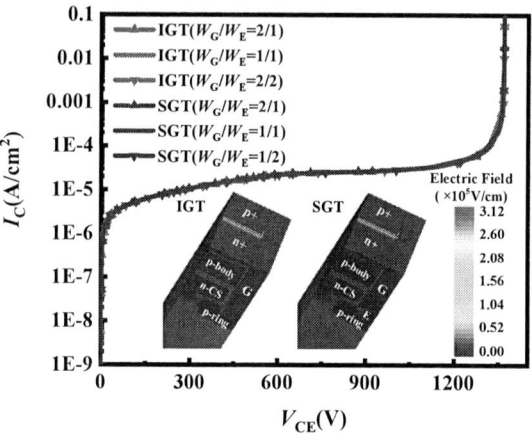

Fig.2 Breakdown voltage characteristics of IGT and SGT IGBT

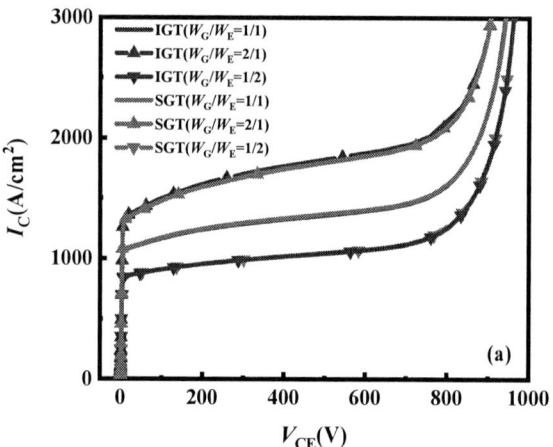

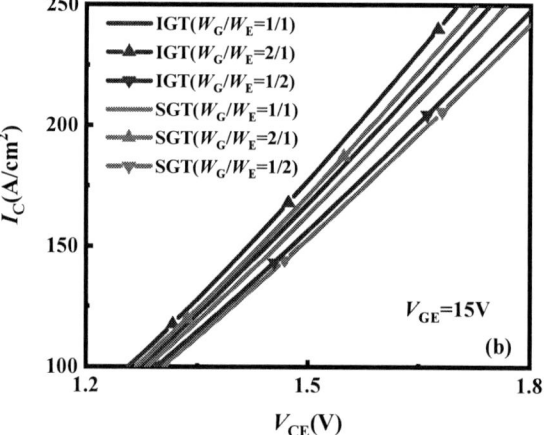

Fig.3 The forward output characteristics of both IGBTs with different W_G/W_E, (a) saturation current, (b) forward voltage drop(V_{on}) when I_c above 100A/cm^2.

Fig.3 illustrates the comparison of IGT and SGT IGBT in forward output characteristics. Also the impact of W_G/W_E is also investigated. As shown in Fig.3(a), the two IGBTs exhibit nearly identical saturation currents, owing to their similar doping profiles and the inclusion of a built-in p-MOSFET structure. Fig.3(b) shows the V_{on} above 100A/cm^2,

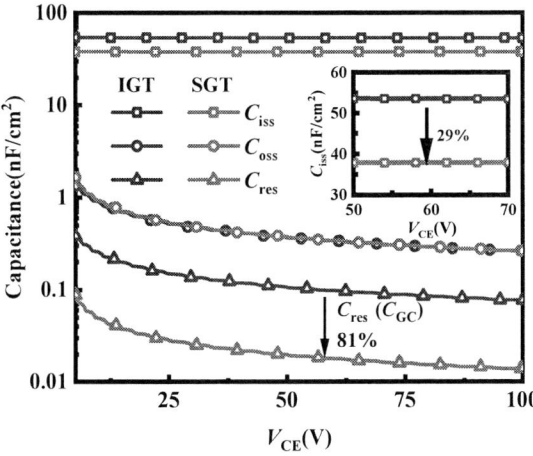

Fig.4 Capacitance as functions of V_{CE} for IGT and SGT IGBT.

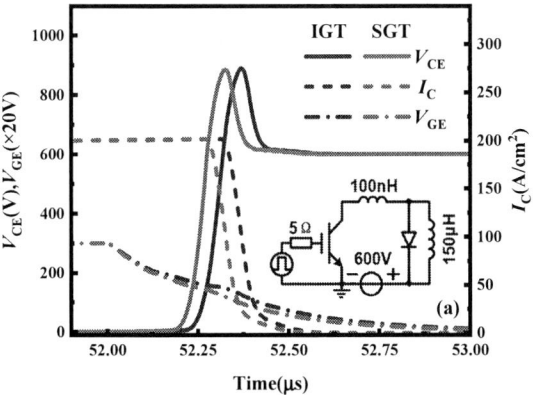

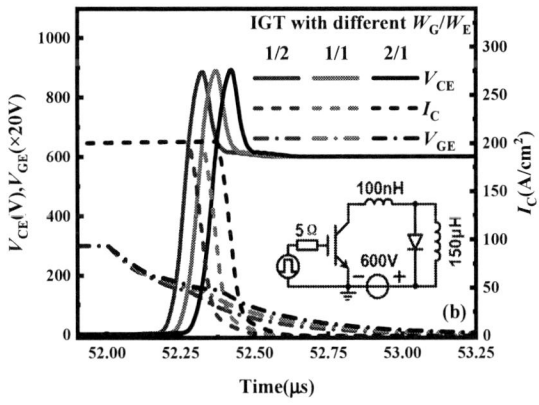

Fig.5 Turn-off waveforms of (a) IGT and SGT IGBT, (b) IGT IGBT with different W_G/W_E.

the IGT IGBT has lower V_{on} under the same W_G/W_E. This improvement is attributed to the deeper gate poly, which enhances the electron channel within the device.When W_G/W_E increases, the channel density also increases, leading to strengthened conductivity modulation and consequently an increase in the saturation current. Additionally, the latch-up phenomenon induced by the parasitic n+/p-body/n-CS transistor is more likely to occur earlier under these conditions. Conversely, decreasing the W_G/W_E enhances the p-MOSFET part while weakening conductivity modulation,

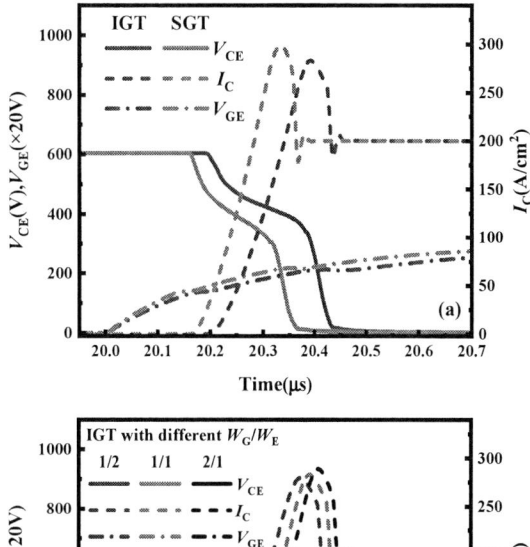

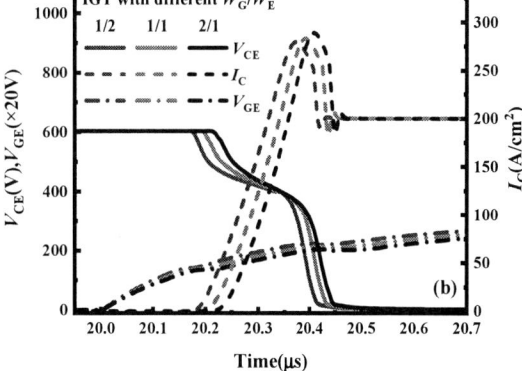

Fig.6 Turn-on waveforms of (a) IGT and SGT IGBT, (b) IGT IGBT with different W_G/W_E.

leading to a reduction in saturation current and a delay in latch-up occurrence.

Fig.4 compares the characteristic capacitances of the SGT and IGT IGBT. Due to the poly emitter under poly gate, the overlap between gate and collector for SGT IGBT is greatly reduced. Therefore, the transfer capacitance(C_{res}) of SGT IGBT is reduced by 81% compared to IGT IGBT. Moreover, the input capacitance(C_{iss}) is also reduced by 29%, which is beneficial to the switching performance of devices. Additionally, as W_G/W_E increases, the overlap between gate and collector increases, which leads to an increase of C_{res}.

Both IGBTs in this work are designed for fast switching. As shown in Fig.5(a), the turn-off waveforms of the IGT and SGT IGBT are compared. The test circuit and parameters are shown in the inset. Due to its lower C_{res}, the SGT IGBT exhibits a significantly shorter Miller plateau, resulting in a reduced turn-off delay time(t_{d_off}) of 0.28μs with a 5Ω gate resistor (R_G). In contrast, the IGT IGBT demonstrates a longer t_{d_off} of 0.33μs under the same conditions. Additionally, the E_{off} for the SGT IGBT is 11.56mJ/cm², compared to 12.07mJ/cm² for the IGT IGBT. The SGT IGBT shows more superior turn-off performance than IGT IGBT. When varying W_G/W_E, the characteristic capacitance of IGBTs is changed, which will affect the turn-off performance. Fig.5(b) presents the turn-off waveforms of the IGT IGBT under varying W_G/W_E ratios. As W_G/W_E increases, the gate-collector overlap region expands, leading to an elevation in C_{res}. This would prolong the t_{d_off} as a longer capacitive discharge process

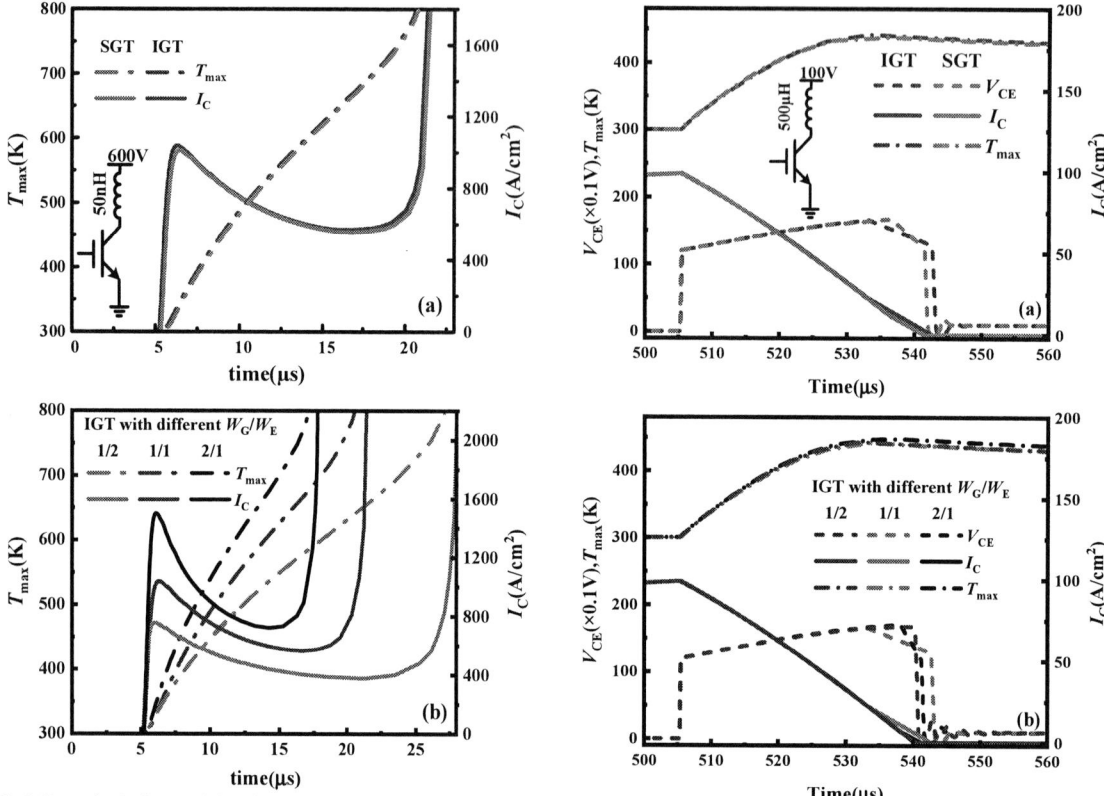

Fig.7 Short-circuit characteristics of (a) IGT and SGT IGBT, (b) IGT IGBT with different W_G/W_E.

Fig.8 UIS characteristics of (a) IGT and SGT IGBT, (b) IGT IGBT with different W_G/W_E.

required(SGT IGBT has the similar principle when changing W_G/W_E).

Fig.6(a) compares the turn-on characteristics of the two IGBTs. Owing to its lower lower C_{iss}, the SGT IGBT achieves a reduced turn-on delay time(t_{d_on}) of 0.16μs, whereas the IGT IGBT exhibits a longer t_{d_on} of 0.2 μs. Furthermore, the SGT IGBT demonstrates a steeper current rise rate(dI_C/dt) of 2500 A/μs, substantially exceeding the 1700 A/μs value of the IGT IGBT. This allows SGT IGBT to have a faster turn-on speed, but it simultaneously intensifies electromagnetic interference (EMI) risks. Additionally, the turn-on surge current(I_{surge}) is also larger in SGT IGBT. For the turn-on loss(E_{on}), with the larger dI_C/dt and smaller C_{iss}, the SGT IGBT reduces the E_{on} to 12.25mJ/cm², compared to 14.93 mJ/cm² for the IGT IGBT. These results indicate a performance trade-off: the SGT IGBT prioritizes switching loss reduction, whereas the IGT IGBT offers superior EMI suppression and lower I_{surge}. Fig.6(b) exhibits the impact of W_G/W_E. Taking IGT IGBT as an example, when enlarging W_G/W_E, the C_{iss} increases and further leads to longer t_{d_on} and larger overshoot current.

The SOA has been a major concern in IGBT design. Fig.7(a) quantitatively compares the short-circuit (SC) characteristics of the two IGBTs under 600V DC-link voltage, revealing the waveform of I_c and maximum temperature inside cell (T_{max}). A key mechanism governing SC withstand capability lies in the parasitic n+/p-body/n-CS transistor triggering-an effect analogous to thyristor latch-up. For both IGBTs in this work, the integrated p-MOSFET structure effectively suppresses carrier injection at the parasitic transistor's emitter-base junction, delaying latch-up initiation. This enables both IGBTs to achieve 20μs SC withstand time, meeting the industrial requirements, Fig.7(b) compares the impact of W_G/W_E on SC robustness(Taking IGT IGBT as an example). Under a larger W_G/W_E, the saturation current is enhanced. Consequently, the temperature rises faster under larger W_G/W_E, which eventually makes a premature failure. The current capability and thermal stability need to be balanced in these IGBTs design.

The UIS capability of both IGBTs is also investigated for the SOA evaluation Since it is widely used to examine the device turn-off capability. Fig.8(a) shows the comparison of the UIS waveform for IGT and SGT IGBTs. The test condition is shown in the inset. When devices turn off at 100A/cm², the waveforms are similar for both IGBTs. With the similar integrated p-MOSFET for enhanced latch-up immunity, basically the UIS robustness is similar for both IGBTs. Fig.8(b) discusses the effect of W_G/W_E on UIS(Taking IGT IGBT as an example). It seems no significant impact are made by varying the W_G/W_E, showing the structural insensitivity of UIS robustness for both IGBTs

IV. CONCLUSION

This study investigates two innovative IGBT configurations through structural modifications in poly gate and poly emitter ratio for performance comparison. The IGT IGBT shows a better V_{on} and EMI suppression characterstics, while SGT IGBT exhibits superior switching losses and lower

C_{iss} and C_{res}. Investigations of the W_G/W_E ratio further indicates that increased values induce trade-off: While improving forward conduction characteristics, they simultaneously degrade switching performance and short-circuit withstand capability.

REFERENCES

[1] N. Iwamuro and T. Laska, "IGBT History, State-of-the-Art, and Future Prospects," in IEEE Transactions on Electron Devices, vol. 64, no. 3, pp. 741-752, March 2017.

[2] S. Rezwan, S. Hossain, S. Tasnim and M. M. Rahman, "H2D4-Type Single Phase Transformer-Less Inverter with Reactive Power Control for Grid-tied PV System," 2018 International Conference on Smart Grid and Clean Energy Technologies (ICSGCE), Kajang, Malaysia, 2018, pp. 112-118.

[3] J. Yang, Z. He, J. Ke and M. Xie, "A New Hybrid Multilevel DC–AC Converter With Reduced Energy Storage Requirement and Power Losses for HVDC Applications," in IEEE Transactions on Power Electronics, vol. 34, no. 3, pp. 2082-2096, March 2019.

[4] I. -H. Ji et al., "A New Soft Self-Clamping Scheme for Improving the Self-Clamped Inductive Switching (SCIS) Capability of Automotive Ignition IGBT," Proceedings of the 19th International Symposium on Power Semiconductor Devices and IC's, Jeju, Korea (South), 2007, pp. 145-148.

[5] H. Takahashi, H. Haruguchi, H. Hagino, and T. Yamada, "Car rier stored trench-gate bipolar transistor (CSTBT)—A novel power device

[6] M. Kitagawa, I. Omura, S. Hasegawa, T. Inoue, and A. Nakagawa, "A 4500 V injection enhanced insulated gate bipolar transistor (IEGT) operating in a mode similar to a thyristor," in IEDM Tech. Dig., Washington, DC, USA, Dec. 1993, pp. 679–682.

[7] M. Sumitomo, J. Asai, H. Sakane, K. Arakawa, Y. Higuchi, and M. Matsui, "Low loss IGBT with partially narrow mesa structure (PNM IGBT)," in Proc. 24th Int. Symp. Power Semiconductor Devices ICs, Bruges, Belgium, Jun. 2012, pp. 17–20.

[8] T. Laska, M. Munzer, F. Pfirsch, C. Schaeffer, and T. Schmidt, "The field stop IGBT (FS IGBT). A new power device concept with a great improvement potential," in Proc. 12th Int. Symp. Power Semiconductor Devices ICs, Toulouse, France, 2000, pp. 355–358.

[9] M. Antoniou et al., "Experimental demonstration of the p-ring FS+ Trench IGBT concept: A new design for minimizing the conduction losses," in Proc. IEEE 27th Int. Symp. Power Semiconductor Devices IC's (ISPSD), May 2015, pp. 21–24.

[10] Y. Wu, G. Deng et al., "Improving dynamic characteristics for IGBTs by using interleaved trench gate," in Chinese Phys. B, Dec. 2023, vol 32, no 12, pp.128503(1)- 128503(5).

[11] G. Deng, J. Wang, Y. Wu, C. Tan and S. Liang, "3-D Segmented Gate Concept: A New IGBT Solution for Reduced Loss and Improved Safe-Operating Area," in IEEE Transactions on Electron Devices, vol. 70, no. 6, pp. 3172-3178, June 2023.

[12] Sentaurus Device User Guide, Synopsys, Mountain View, CA, USA, 2018.

Influencing Factors and Mechanisms of Single-Event Burnout in STP SiC MOSFETs

* Ying Yang
Department of Electronic Engineering
Xi'an University of Technology
Xi'an, Shaanxi, China
* Corresponding author: yangy@xaut.edu.cn

Xulong Wang
Department of Electronic Engineering
Xi'an University of Technology
Xi'an, Shaanxi, China
2230321215@stu.xaut.edu.cn

Chen Wang
Department of Electronic Engineering
Xi'an University of Technology
Xi'an, Shaanxi, China
863416316@qq.com

Zixuan Liu
Department of Electronic Engineering
Xi'an University of Technology
Xi'an, Shaanxi, China
2565932926@qq.com

Abstract—The excellent performance of SiC MOSFETs makes them highly promising for aerospace applications. However, single-event burnout (SEB) induced by particle radiation in space limits their use in space stations. To provide a theoretical basis for single-event protection, this study investigates the factors influencing SEB in 1200V SiC MOSFETs. The results show that when the linear energy transfer (LET) of the incident particle exceeds 0.1 pC/μm, the SEB threshold voltage is 424 V. The most sensitive region of the device to single-event effects is located 0.9 μm from the center of the cell, near the vertical PN junction formed by the P-body and the drift region. This is because the damage to the depletion region of the PN junction after single-particle incidence is more extensive in this area, and the holes generated by single-particle collisions with the lattice in the drift region are more likely to drift and accumulate in the P-body region. As a result, the potential of the P-body increases, and when it exceeds the turn-on voltage of the PN junction, the parasitic bipolar junction transistor (BJT) turns on, leading to a sharp increase in electron concentration in the drift region. This causes avalanche breakdown at the interface between the drift region and the substrate, triggering single-event effects. This study provides important data support and design guidance for optimizing the SEB resistance of shallow trench planar-gate MOSFETs (STP-MOS).

Keywords—shallow trench planar-gate MOSFETs, single-event burnout, single-particle strike

I. INTRODUCTION

SiC power devices are susceptible to single-event burnout (SEB) under heavy-ion irradiation, which can cause significant damage to the devices [1],[2],[3],[4],[5]. In recent years, extensive research has been conducted on the SEB effects in planar-gate MOSFETs. The factors influencing SEB include the linear energy transfer (LET) of the incident particles, the strike location, and the particle types [6][7]. Studies have shown that higher LET values, strike locations closer to the parasitic bipolar junction transistor (BJT) region, and deeper penetration depths result in more severe damage to the devices. By analyzing the electron and hole density

distributions and lattice temperature changes after particle strikes, researchers have elucidated the mechanisms of SEB [8]. The primary cause of SEB in SiC MOSFETs is the turn-on of the parasitic BJT (with the N+ source region as the emitter, the P-body region as the base, and the N- drift region as the collector), which triggers avalanche multiplication and ultimately leads to device burnout.

In planar-gate MOSFETs, the JFET region formed between the P-body and the drift region beneath the gate introduces additional JFET resistance. To reduce the on-resistance, the shallow trench planar-gate structure removes a portion of the SiC beneath the gate and fills it with SiO2, significantly lowering the drift region resistance and thereby improving conduction performance [9]. Currently, research on SEB has primarily focused on planar-gate MOSFETs, while the factors influencing SEB in shallow trench planar-gate MOSFETs remain unclear. This paper conducts an in-depth investigation into the factors affecting SEB in 1200 V SiC shallow trench planar-gate MOSFETs.

II. DEVICE STRUCTURE AND PARAMETER SETTINGS

The shallow trench planar-gate MOSFETs (STP-MOS) features a vertical structure, with the gate and source located at the top of the device and the drain at the bottom, as illustrated in Fig. 1(a). Fig. 1(b) compares the output characteristics of the shallow trench planar-gate MOSFET (STP-MOS) with those of a conventional planar-gate MOSFET (Con-MOS). From Fig.1(b), it is evident that the output characteristics of the STP-MOS are superior to those of the Con-MOS, with a 10.71% reduction in specific on-resistance [9].

The single-event strike locations are set at 0.5 μm, 0.9 μm, 1.5 μm, 2.2 μm, and 4.6 μm from the center of the cell, with vertical penetration as shown in Fig. 2. The strike depth is 13 μm, and the linear energy transfer (LET) values are set to 0.05 pc/μm、0.07 pc/μm、0.09 pc/μm、0.1 pc/μm.

When conducting SEB studies, the gate-source voltage bias is set to 0 V, and the drain-source bias voltage is maintained below the device's breakdown voltage.

This work was supported by the Shaanxi Provincial Key Research and Development Program (Grant No.2025CY-YBXM-146) and the National Natural Science Foundation of China (Grant No. 62174134) .

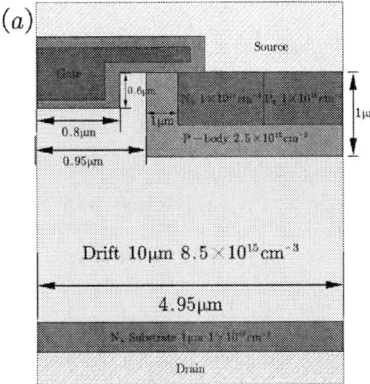

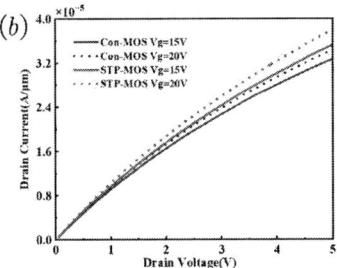

Fig. 1. illustrates the schematic diagram of the shallow trench planar-gate structure and a comparison of the output characteristic curves between STP-MOS and Con-MOS. (a) Schematic diagram of the trench planar-gate device structure; (b) Comparison of the output characteristic curves.

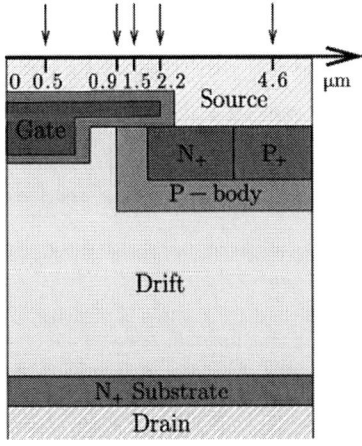

Fig. 2. illustrates the single-event strike locations.

III. ANALYSIS OF SINGLE-EVENT BURNOUT RESULTS AND INFLUENCING FACTORS

A. Analysis of Single-Event Burnout Threshold Voltage and Sensitive Regions

Fig. 3 presents the single-event burnout (SEB) threshold voltages of STP-MOS and Con-MOS devices under single-particle strikes at different incident positions relative to the cell center, with a fixed LET value of 0.1 pC/μm. The SEB threshold voltage is defined as the minimum drain voltage required to elevate the lattice temperature to 2000 K after single-particle irradiation. Both devices exhibit the lowest

SEB threshold voltage at 0.9 μm from the cell center, with values of 350 V for STP-MOS and 320 V for Con-MOS. The most sensitive region is consistently located at 0.9 μm from the cell center. This sensitive position aligns with [10], corresponding to the vertical PN junction interface in the SiC bulk region, indicating that the shallow trench structure of STP-MOS does not alter the SEB-sensitive region of the device.

Fig. 4 shows the single-event burnout (SEB) threshold voltage and the lattice temperature versus time curves of STP-MOS under single-particle strikes at different positions. The Fig. reveals that the lattice temperature rise rate varies with strike location. The fastest temperature rise occurs when particles strike at 2.2 μm from the cell center, while the slowest occurs at 0.5 μm. The temperature rise rates at 0.9 μm, 1.5 μm, and 2.2 μm are similar and higher than those at 0.5 μm and 4.6 μm. This phenomenon occurs because the three strike positions (0.9 μm, 1.5 μm, and 2.2 μm) are closest to the parasitic bipolar transistor. When holes generated by single-particle strikes quickly enter the P-body region, they turn on the parasitic bipolar transistor faster, thereby triggering SEB more rapidly.

The significance of the 0.9 μm position lies in its proximity to the vertical PN junction interface formed between the P-body region and the drift region beneath the gate. When a single particle strikes at this location, the PN junction formed by the P-body region and the N- drift region is reverse-biased due to the drain-source voltage bias being greater than 0 V. After the single-particle strike, the large number of electron-hole pairs generated by collisions with the lattice causes the depletion region of the vertical PN junction to rapidly collapse. Holes in the drift region drift toward the P-body region under the influence of the electric field, leading to the turn-on of the parasitic bipolar junction transistor (BJT) (with the N+ source region as the emitter, the P-body region as the base, and the drift region as the collector). Compared to other strike locations, the unique position at 0.9 μm from the cell center results in a more extensive disruption of the depletion region of the PN junction formed by the P-body region and the drift region after a single-particle strike. This makes it easier for holes to drift from the drift region into the P-body region under

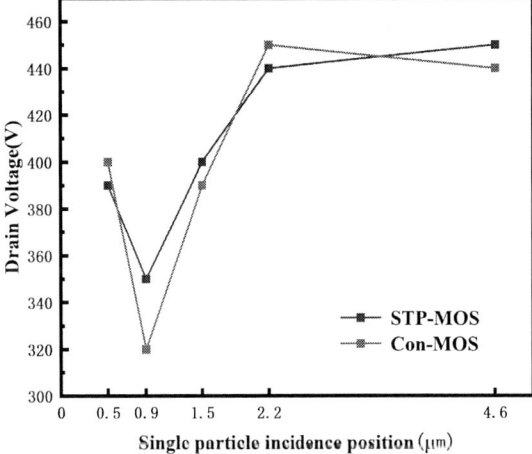

Fig. 3. SEB threshold voltage of STP-MOS and Con-MOS under single-particle strikes at different incident positions.

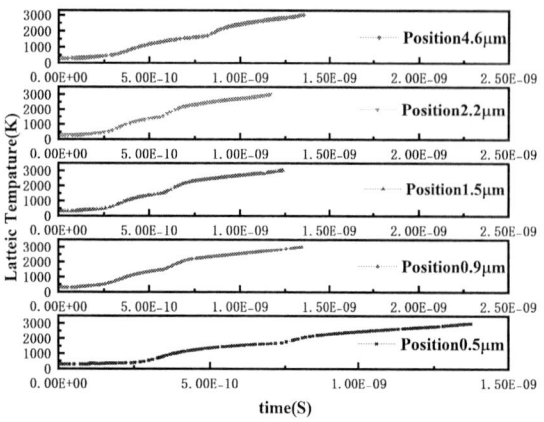

Fig. 4. Lattice temperature versus time curves for single-event burnout (SEB) induced by single-particle strikes at different locations.

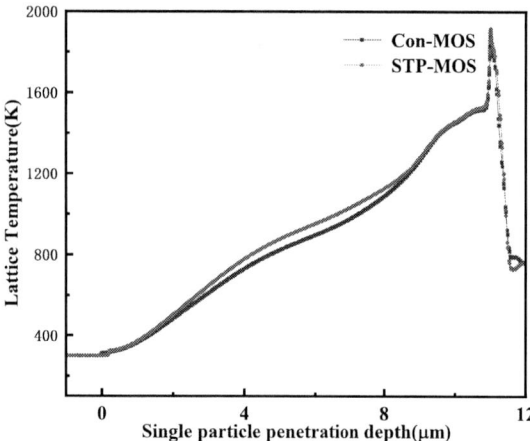

Fig. 5. Lattice temperature distribution along the strike path for STP-MOS and Con-MOS at 0.5 μm (0 denotes the device surface, with increasing coordinates representing the substrate direction).

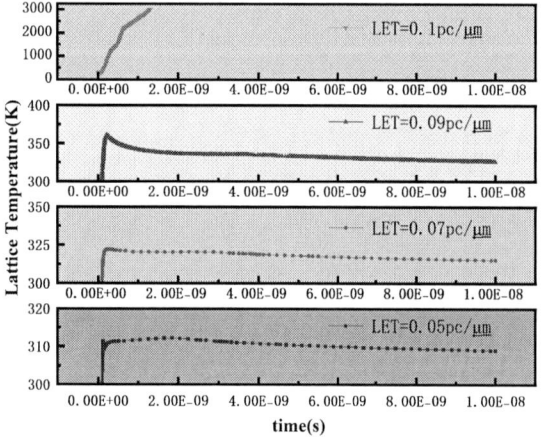

Fig. 6. Lattice temperature versus time curves for devices under different LET values.

the electric field, facilitating the turn-on of the parasitic BJT. Consequently, the 0.9 μm position becomes the most susceptible region to SEB, resulting in a significantly lower SEB threshold voltage at this location compared to others.

The structural difference between STP-MOS and Con-MOS lies in their gate positions, prompting further investigation of single-particle strikes at 0.5 μm from the cell center. Fig. 3 reveals that the SEB threshold voltages for STP-MOS and Con-MOS at 0.5 μm are 390 V and 400 V respectively, showing no significant difference between the two devices. Fig. 5 displays the lattice temperature distribution within the devices when struck at 0.5 μm from the cell center. Both devices exhibit their peak lattice temperatures at 11 μm depth (the interface between drift region and substrate), with STP-MOS reaching 1915 K and Con-MOS reaching 1881 K. The STP-MOS structure features a trenched SiC body region under the gate center, resulting in a slightly thinner central body region thickness. This leads to a marginally more non-uniform lattice temperature distribution in STP-MOS compared to Con-MOS.In conclusion, the gate structure of STP-MOS does not significantly affect the device's single-event effects.

B. Influence of Different LET Values on Single-Event Burnout

Fig. 6 shows the variation of lattice temperature with time after single-particle incidence at different LET values (0.05 pc/μm, 0.07 pc/μm, 0.09 pc/μm, and 0.1 pc/μm), with the incidence position being 0.9 μm away from the center of the cell. It can be observed from the Fig. that when the LET value is 0.1 pc/μm, the lattice temperature gradually increases after single-particle incidence and reaches 2000 K at 0.5 ns, resulting in single-particle burnout of the device. When the LET value is less than 0.1 pc/μm, the lattice temperature of the device also gradually increases after single-particle incidence, but the highest temperature it reaches is only 360 K (LET = 0.09 pc/μm). After reaching the peak temperature, the temperature of the curve slowly decreases, and the device does not undergo single-particle burnout. The reason for this is that when the LET value of the single particle is greater than or equal to 0.1 pc/μm, a sufficient number of holes are generated in the device to turn on the parasitic transistor. When the LET value is less than 0.1 pc/μm, the holes generated by the single-particle collision with the lattice are not enough to turn on the parasitic transistor, and thus the device does not experience single-particle effects.

C. Single-Event Burnout (SEB) Effects in Devices at Different Operating Temperatures

In order to investigate the single-particle burnout effect of the device at different junction temperatures, the junction temperature variation of the device after single-particle incidence at a junction temperature of 425 K (150°C) was studied and compared with the single-particle effect of the device at room temperature of 300 K (25°C). Fig. 7 shows the single-particle burnout threshold voltage at different incidence positions of single-particle incidence when the device is at junction temperatures of 300 K and 425 K. It can be seen from the Fig. that the increase in device junction temperature will significantly lead to a decrease in the single-particle burnout threshold voltage. The reason for this phenomenon is analyzed as follows: With the increase in temperature, the intrinsic carrier concentration of the material increases significantly, and the impact ionization rate also rises. After single-particle incidence, due to the higher junction temperature, which causes an increase in the impact ionization rate and the intrinsic carrier concentration, more electron-hole pairs will be generated within the device, making it easier to trigger single-particle effects. That is to say, single-particle burnout

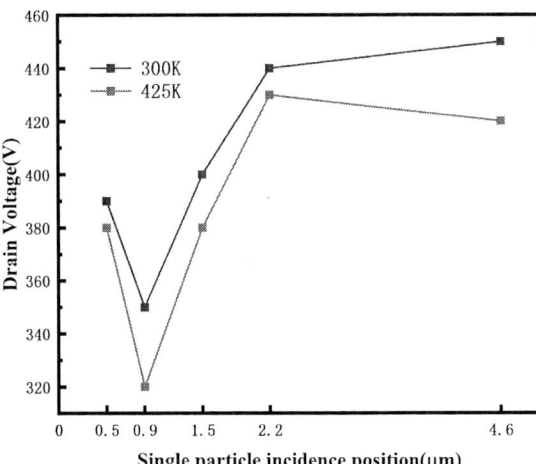

Fig. 7. Single-Particle Burnout Threshold Voltage at Different Incidence Positions of the Device under Various Operating Temperatures.

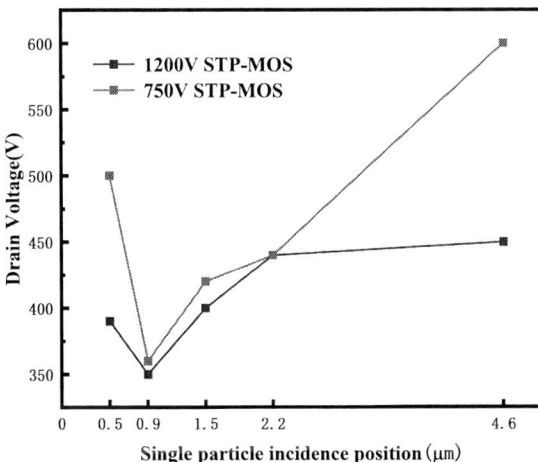

Fig. 8. Single-Particle Burnout Threshold Voltage at Different Incidence Positions for Devices with Rated Voltages of 750 V and 1200 V.

will occur at a lower drain-source voltage. Therefore, as the device junction temperature rises, the single-particle burnout threshold voltage decreases.

D. Single-Particle Burnout Effects of Devices with Different Rated Voltages

To investigate the single-particle effect of devices with different rated voltage levels, devices with a rated voltage of 750 V were used for the experiment. The cell structure of the devices used in the experiment remained unchanged, with only the drift region thickness adjusted to 7 μm and the drift region doping concentration set at 2×10^{16} cm^{-3}. Fig. 8. shows the minimum drain voltage at which single-particle burnout occurs at different single-particle incidence positions for 750 V and 1200 V rated STP-MOS devices. It can be clearly observed from Fig. 8 that the minimum drain voltage at which single-particle burnout occurs, namely the single-particle burnout threshold voltage, is 360 V, corresponding to a single-particle incidence position of 0.9 μm away from the cell center. The sensitive position for single-particle burnout is the same. Meanwhile, the single-particle burnout threshold voltage of the 750 V rated device is generally higher than that of the 1200 V rated device at different positions. This is because the drift

region doping concentration of the 750 V rated device is relatively higher, resulting in a higher recombination rate of electron-hole pairs generated by impact ionization in the drift region. Therefore, the 750 V device requires a higher drain voltage to generate a stronger electric field, producing more electron-hole pairs to turn on the parasitic transistor and trigger the single-particle effect. In summary, the sensitive position for single-particle burnout is the same for both 1200 V and 750 V rated STP-MOS devices, but the single-particle burnout threshold voltage is higher for the 750 V device. Without changing the device structure, the sensitive area for single-particle burnout will not change for devices with different rated voltage levels.

IV. ANALYSIS OF SINGLE-EVENT BURNOUT MECHANISMS

Fig. 9 and Fig. 10 show the distributions of hole current density and electron current density within the device at different time points (0.1 ns, 0.15 ns, and 0.5 ns) after a single-particle strike vertically incident at 0.9 μm from the center of the cell gate. Fig. 9 demonstrates that after the single-particle strike, due to the forward bias of the drain-source voltage, holes along the strike path drift from the drift region toward the P-body region under the influence of the electric field, leading to a significant increase in hole concentration near the P-body region. Fig. 10 shows that electrons drift toward the substrate and accumulate at the interface between the drift region and the substrate.

After the single-particle strike, collisions with the lattice along the strike path generate a large number of electron-hole pairs. The newly generated electrons and holes move in opposite directions under the influence of the forward drain-source voltage: electrons move toward the substrate, while holes drift toward the P-body region, causing a significant

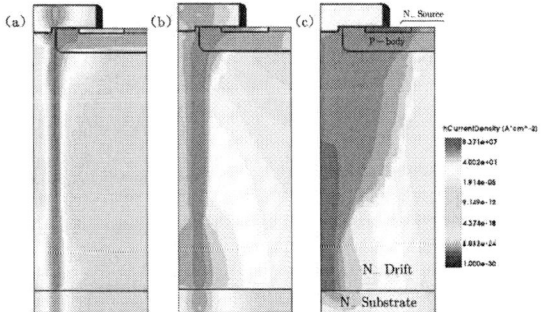

Fig. 9. Distribution of Hole Current Density at Different Times After Single-Particle Incidence. (a) 0.1 ns, (b) 0.15ns and (c) 0.5 ns.

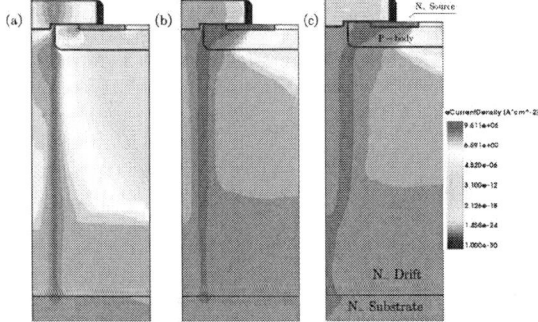

Fig. 10. Distribution of Electron Current Density at Different Times After Single-Particle Incidence. (a) 0.1 ns, (b) 0.15ns and (c) 0.5 ns.

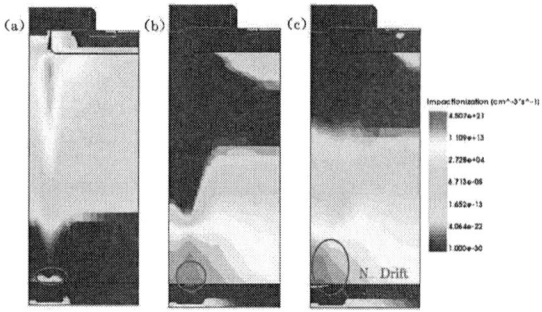

Fig. 11. illustrates the impact ionization distribution within the device at different times following a single-particle incidence: (a) 0.1 ns, (b) 0.15 ns, and (c) 0.5 ns.

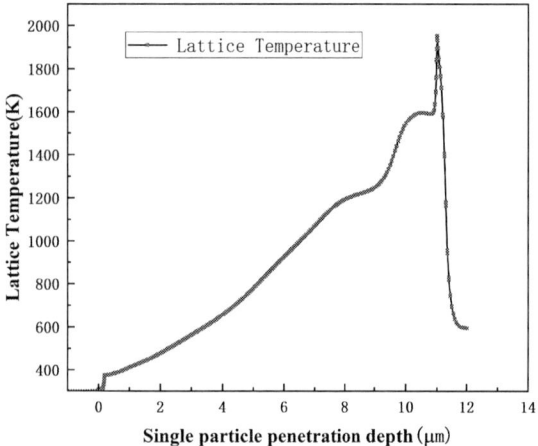

Fig. 12. Lattice Temperature Distribution Along the Incidence Path at 0.9 μm When Single-Particle Burnout Occurs.

increase in the potential of the P-body region. When the potential difference between the P-body region and the N+ source region exceeds the turn-on voltage of the PN junction, the parasitic bipolar junction transistor (BJT) is triggered and turns on.

Observe the electron current density distribution at the moment of 0.15 ns in Fig. 10(b). There is a high electron density in the P-body region, while in the electron current density distribution at the moment of 0.1 ns in Fig. 10(a), no high electron density is found in the P-body region. Therefore, the high electron density in the P-body region at the moment of 0.15 ns in Fig. 10(b) is not caused by the single-particle collision with the lattice, but is the net electron density after subtracting the recombined electrons in the P-body region from the electrons diffused from the N+ source emitter into the P-body region when the parasitic transistor is in the conductive state. Thus, at 0.15 ns, the parasitic transistor has already been in the conductive state.

Under the applied forward bias, electrons drift toward the bottom of the drift region and accumulate at the N-/N+ interface between the drift layer and substrate. This carrier accumulation leads to a substantial increase in electron concentration at the hetero junction interface. When the local electric field exceeds the critical breakdown field strength of 4H-SiC material, avalanche multiplication generates copious electron-hole pairs. The generated holes continue migrating toward the P-body region, further enhancing the forward bias across the emitter-base junction of the parasitic transistor. This positive feedback mechanism substantially increases electron population in the drift region, causing abrupt escalation of drain current and ultimately triggering single-event burnout (SEB).

The proposed SEB mechanism shows consistency with [10], wherein parasitic bipolar transistor turn-on was likewise identified as the root cause of SEB failure.

Fig. 11 shows the distribution of impact ionization at different times (0.1 ns, 0.15 ns, and 0.5 ns) after single-particle incidence on the device. It can be observed from the Fig. that as time increases, the high impact ionization region at the interface between the substrate and the drift region continues to expand, but the expanding area is only within the drift region. The only difference between the substrate and the drift region is the high doping concentration of the substrate ($2 \times 10^{19} \text{cm}^{-3}$). This is because the high substrate doping concentration causes the space charge region to expand in the drift region, resulting in a high electric field mainly distributed in the drift region with the peak electric field located at the interface between the drift region and the substrate. Therefore, the impact ionization rate in the drift region is significantly higher. As seen in Fig. 11(b), at 0.15 ns, a high impact ionization rate has already appeared at the interface between the drift region and the substrate of the device, and the device has already undergone avalanche breakdown with the formation of the positive feedback mechanism of the parasitic transistor. As time goes on, the high impact ionization region at the interface between the substrate and the drift region continues to expand, and so does the avalanche region, eventually leading to single-particle burnout of the device.

Fig. 12 shows the lattice temperature distribution inside the device along the incidence path at 0.9 μm where single-particle burnout occurs. The coordinate 0 represents the device surface, and the direction of increasing coordinates is towards the substrate. It can be observed from the Fig. that during single-particle burnout, the lattice temperature along the incidence path increases towards the substrate direction and reaches a peak at 11 μm, which is precisely located at the interface between the drift region and the substrate, corresponding to the region with high impact ionization rate in Fig. 11. Therefore, combining Fig.s 10 and 11, the high electron density at the bottom of the drift region significantly increases the impact ionization rate at the interface between the drift region and the substrate. The intense lattice collisions lead to an increase in lattice temperature. At the same time, the electrons generated by the conducting parasitic transistor and the electrons produced by lattice collisions drift to the interface between the substrate and the drift region under the forward drain-source voltage, causing the impact ionization at the interface to continue increasing, which in turn leads to a continuous rise in lattice temperature and eventually results in single-particle burnout.

V. CONCLUSION

This paper investigates the single-particle burnout effect of a 1200 V trench planar gate MOSFET, which has a single-particle burnout threshold voltage of 350 V, while the threshold voltage for single-particle burnout of Con-MOS is 320 V. The sensitive area for single-particle burnout in both devices is located at a position 0.9 μm away from the center of the cell. This is because the destruction of the PN junction depletion region after particle incidence is more severe, causing holes to more easily drift towards the P-body region.

When the LET value of the particle exceeds 0.1 pC/μm, the device begins to experience single-particle burnout. The gate structure of the STP-MOS does not affect the single-particle burnout effect of the device. The single-particle burnout effect of the device under different operating temperatures was also studied, and it was found that an increase in operating temperature leads to a decrease in the threshold voltage for single-particle burnout. The single-particle burnout threshold voltage also varies significantly among devices with different voltage ratings; the device with a rated voltage of 750 V has a higher single-particle burnout threshold voltage of 350 V. Analysis reveals that the mechanism of single-particle burnout in devices is that the holes generated by particle incidence trigger the parasitic transistor to turn on, the electron concentration in the drift region continues to rise, and the electric field at the interface between the drift region and the substrate reaches the avalanche breakdown field of the device, ultimately leading to single-particle burnout. The research results of this paper indicate that the burnout mechanism of trench planar gate devices is consistent with that of conventional planar gate devices, and the shallow trench structure has no significant impact on the single-particle burnout effect of the device.

ACKNOWLEDGMENT

This work was supported by the Shaanxi Provincial Key Research and Development Program (Grant No.2025CY-YBXM-146) and the National Natural Science Foundation of China (Grant No. 62174134).

REFERENCES

[1] W. Huang, C. Fu, Y. Ma, M. Huang, X. Dong and Q. Yu, "Study on Single Event Effect of SiC MOSFET by Proton Irradiation," 2024 IEEE 17th International Conference on Solid-State & Integrated Circuit Technology (ICSICT), Zhuhai, China, 2024,pp. 1-3.

[2] J.-C. Zhou, Y. Wang, M.-T Bao, X.-J. Li, J.-Q. Yang, and F. Cao, "Simulation study of single-event-burnout reliability for 1.7-kv 4H-SiC VDMOSFET," IEEE Trans. Device Mater. Rel., vol. 22, no. 3, 2022, pp. 431-437.

[3] K. Liu et al., "Sensitivity and mechanism study of single- event burnout in 4H-SiC devices with FLRs termination," IEEE Trans. Electron Devices, vol. 70, no. 6, 2023, pp. 3196-3201.

[4] K. Liu et al., "Experimental and Simulation Study of Single- Event Leakage Current Degradation and Damage Mechanism in 4H-SiC PiN Diodes," IEEE Trans. Electron Devices, vol. 71, no. 8, 2024, pp. 4891-4896.

[5] J. -K. Shi et al., "Simulation Study on Single-Event Burnout Reliability of 900V 4H-SiC Quasi Vertical Double Diffused MOSFET," IEEE Access, vol. 13, 2024, pp. 5023-5031.

[6] L. Mo et al.,"Single event burnout of SiC MOSFET induced by atmospheric neutrons," Microelectronics Reliability, vol. 146, 2023, pp. 114997.

[7] R. Yang et al., "An Improved Single-Event Effect Performance SiC MOSFET of Hole Extraction Pillar Combined With Multilayer P-Shield Structure," IEEE Transactions on Electron Devices, vol. 71, no. 2, 2024,pp.1018-1023.

[8] F. -K. Liu, Z. -L. Liu and X. -J. Li, "Impact of 6 º Co- γ Irradiation Pre-Treatment o Single-Event Burnout in N- Channel Power VDMOS Transistors," IEEE Electron Device Letters, vol. 45, no. 7, 2024, pp. 1105-1108.

[9] C. Wang, Y. Yang, Z. Liu and Q. Zhao, "Shallow Trench Design for Silicon Carbide Planar-Gate MOSFETs," 2024 3rd International Symposium on Semiconductor and Electronic Technology (ISSET), Xi'an, China, 2024, pp. 43- 48.

[10] A. F. Witulski et al., "Single-Event Burnout Mechanism in SiC Power MOSFETs," IEEE Transactions on Nuclear Science, vol. 65, no. 8, 2018, pp. 1951-1955.

2025 IEEE Workshop on Wide Bandgap Power Devices and Applications in Asia (WiPDA Asia)

A Multi-Operation Characterization System for Quantifying Dynamic R_{ON} Degradation in p-GaN gate GaN HEMTs under Real-World Switching Conditions

Junbo Wang[1], Xiangdong Li[1,2*], Xi Jiang[1,2], Haonan Jiang[1], Shuzhen You[1,2], Yue Hao[1,2], and Jincheng Zhang[1,2*]

[1]Guangzhou Wide Bandgap Semiconductor Innovation Center, Guangzhou Institute of Technology, Xidian University, Guangzhou, China, *E-mail: xdli@xidian.edu.cn

[2]State Key Laboratory of Wide Bandgap Semiconductor Devices and Integrated Technology, School of Microelectronics, Xidian University, Xi'an, China, *E-mail: jchzhang@xidian.edu.cn

Abstract—The dynamic on-resistance (dR_{ON}) degradation of commercial p-GaN gate HEMTs poses a significant challenge to their broader adoption in the power electronics market, highlighting the critical need for accurate and effective evaluation methods. In this work, a precise evaluation system for measuring dR_{ON} of the device is established, enabling a comprehensive assessment of commercial HEMTs under various operating conditions. The device exhibits intensified dR_{ON} degradation with increasing bus voltage, reaching 1.2× dR_{ON} at the V_{BUS} of 500 V in double-pulse test (DPT). During DC stress test, the device gradually degrades to 1.3× dR_{ON} as stress time increases. Finally, in the dynamic high-temperature operating life (DHTOL) test, the device achieves the 1.6× dR_{ON} under continuous high-temperature, high-voltage, and hard-switching coupling stress. These results provide critical insights into the degradation of p-GaN gate HEMTs and validate the robustness of the proposed evaluation system.

Keywords—GaN HEMTs, dynamic R_{ON}, R_{ON} evaluation system, reliability

I. INTRODUCTION

GaN HEMTs, owing to their excellent switching characteristics, have enabled high-efficiency and high-power-density power systems, such as fast charging and data centers, etc., achieving great success in the consumer electronic market [1][2][3][4][5]. However, the dR_{ON} degradation of GaN HEMTs, resulting from the combined effects of current collapse under high-voltage stress and self-heating under high-current injection, poses a significant challenge to their further applications [6][7][8].

To address these challenges, extensive research has been conducted to optimize the dynamic characteristics of GaN HEMTs through epitaxial engineering [9], structural design [10], and surface passivation [11], with subsequent board-level validation under various hard-switching stress conditions [12][13]. However, the lack of standardized evaluation protocols hinders the comparison of the power devices.

Following the release of the JEDEC standard for switching reliability evaluation, many studies have adopted its unified test methodology to assess GaN power devices using in-situ platforms with clamping circuits [14][15][16]. Nevertheless, as shown in Fig. 1, the existing clamping circuits can only monitor degradation trends, precluding the precise value of the real dR_{ON} of devices. This limitation prevents the determination of whether the device degradation level meets acceptable criteria for practical applications.

In this work, a novel circuit system capable of precisely characterizing the dR_{ON} of GaN power devices is proposed, which is employed to comprehensively evaluate the dR_{ON} degradation of the commercial 700-V p-GaN gate HEMTs (INN700D240B) from innoscience under multiple operating conditions, including double-pulse test (DPT), DC stress, and dynamic high-temperature operating life (DHTOL).

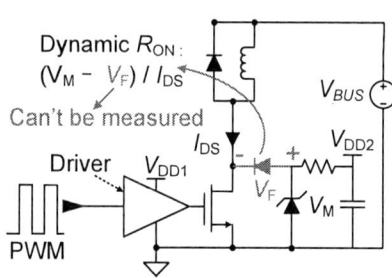

Fig. 1. Problems in the existing clamping circuit.

II. DYNAMIC RON EVALUATION SYSTEM AND OPERATION PRINCIPLE

The schematic diagram of the proposed circuit system and the experimental setup are illustrated in Fig. 2. The core component of the proposed system is the clamping circuit, which plays a critical role in accurately characterizing the dR_{ON} degradation of GaN power devices. Unlike conventional clamping circuits, the proposed design features two series-connected 1200-V high-voltage fast recovery diodes D_1 and D_2 to withstand the bus voltage when the device is turn-OFF. Next, a current source circuit is employed for controlling voltage drops V_{D1} and V_{D2} across the diodes when the device is turn-ON.

The voltage drops V_{D1} and V_{D2} can be given by:

$$V_{D1} = V_A - V_D \tag{1}$$

$$V_{D2} = V_B - V_A \tag{2}$$

It should be noted that V_A, V_B, and V_D denote the voltages at nodes A, B, and D with respect to ground, respectively. At this stage, since V_{D1} equals V_{D2}, the V_D can be expressed as:

$$V_D = 2V_A - V_B \tag{3}$$

At this stage, when R_1 equals R_2, the V_{OUT} of the OPAx388-based operational amplifier circuit is given by:

$$V_{OUT} = 2V_A - V_B \tag{4}$$

Therefore, V_{OUT} equals V_D during turn-ON phase. Finally, the dR_{ON} of the device can be represented as

$$dR_{ON} = (V_{OUT} \div I_{SHUNT}) - R_{SHUNT} \tag{5}$$

Where I_{SHUNT} can be measured using a current shunt (SSDN-414-10) with a R_{SHUNT} of 0.1 Ω.

979-8-3315-1110-4/25 $31.00 © 2025 IEEE   655

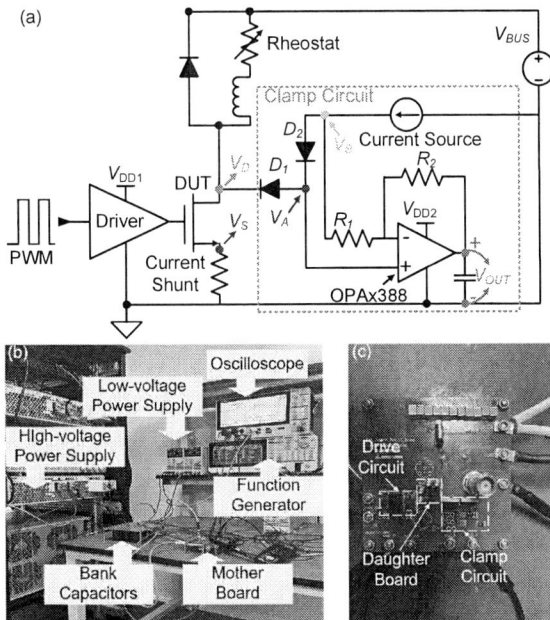

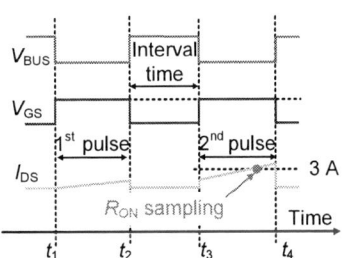

Fig. 2. (a) Schematic diagram of dR_{ON} measurement circuit. Photograph of (b) measurement setup and (c) test board.

III. RESULTS AND DISCUSSION

A. DPT

Fig. 3. Switching waveforms under double-pulse testing. dR_{ON} is extracted when I_{DS} reaches 3 A.

TABLE I. SUMMARY OF IMPORTANT PARAMETERS FOR DPT

V_{BUS}(V)	1st pulse width (µs)	2nd pulse width (µs)	interval time (µs)
50	45	3	2
100	21	3	2
200	9	3	2
300	6	3	2
400	3	3	2
500	3	3	2

Fig. 3 depicts the waveform of the DPT measurement, which is commonly used to assess the impact of hard switch stress on the device under test [17]. During the 1st pulse, the DUT conducts to charge the load inductance, enabling the power circuit to establish the expected operating current. Upon completion of the 2nd conduction, the constant current method, as specified in the device's datasheet, is employed to determine the dR_{ON} of the device. Table I shows the width and interval time of the 1st and 2nd pulses at various bus voltages.

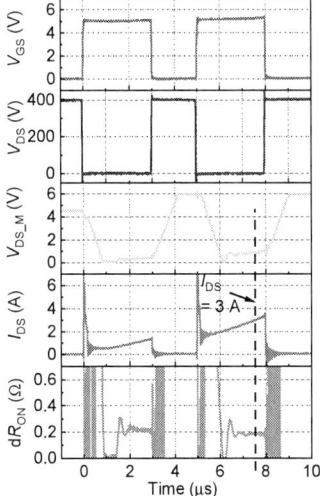

Fig. 4. Measured waveforms of the DPT under the V_{BUS} of 400 V.

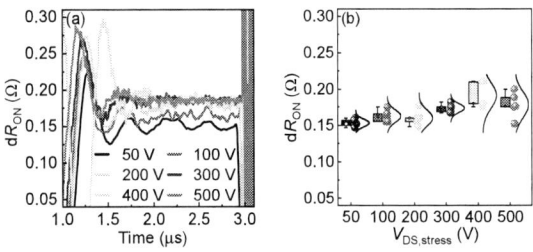

Fig. 5. (a) Measured waveforms of dR_{ON} with various bus voltages during the 2nd pulse under double-pulse testing (b) Extracted dR_{ON} values of multiple devices under various voltages.

As shown in Fig. 4, at the V_{BUS} of 400 V, the extracted dR_{ON} value is approximately 0.2 Ω, which closely matches the R_{ON} value specified in the device datasheet [18]. Furthermore, tests conducted on multiple devices at different V_{BUS} revealed that the dR_{ON} increases with higher bus voltages. However, most devices remain below 1.2× dR_{ON} till 500 V as illustrated in Fig. 5.

B. DC Stress

The DC stress mode was introduced to investigate the dR_{ON} degradation of devices under long-term high-voltage reverse bias stress and to assess the recovery degree of devices after stress removal [19]. The corresponding test setup waveform is illustrated in Fig. 6.

Fig. 7(a) illustrates the test waveform for extracting the dR_{ON} during the DC stress test of the device at a bus voltage of 400 V. Fig. 7(b) shows the test waveform for extracting dR_{ON} from the recovery characteristics of the device after removing long-term high-voltage stress.

Fig. 8(a) illustrates the time-resolved degradation characteristics of the dR_{ON}. It can be observed that under long-term high-voltage stress at a bus voltage of 400 V, the device exhibits rapid degradation within the first few tens of seconds, followed by gradual saturation over several hundred seconds. However, as the stress time extends to several thousand seconds, further degradation occurs, eventually reaching approximately 1.3× dR_{ON}.

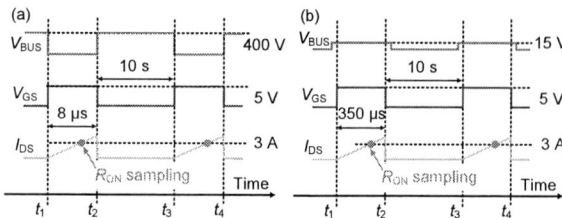

Fig. 6. Measurement sequences of the DC stress test during (a) stress phase and (b) recovery phase.

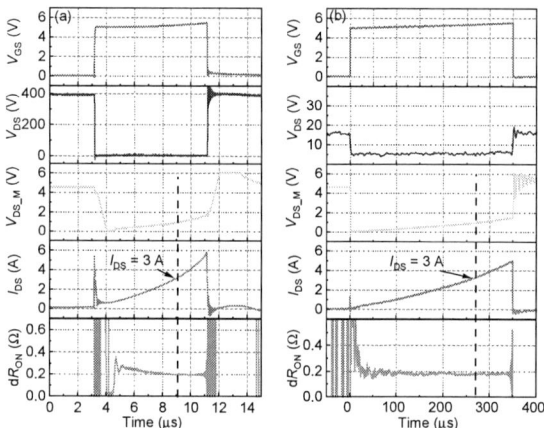

Fig. 7. Measured switching waveforms of the DC stress mode during (a) stress phase and (b) recovery phase.

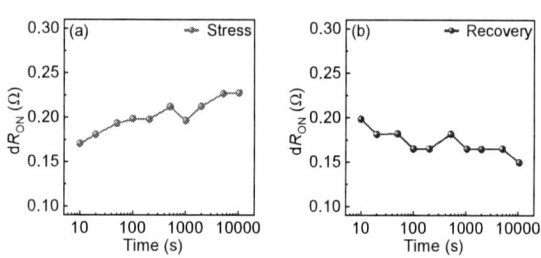

Fig. 8. Time-resolved dR_{ON} under DC stress mode during (a) stress phase and (b) recovery phase.

As depicted in Fig. 8(b), it is evident that the device recovers to its initial state within a few seconds after stress removal. Notably, at the end of the recovery phase, the dR_{ON} of the device remains lower than its value during the initial stress stage, which may be attributed to the degradation induced by the initial 10-s high-voltage stress.

C. DHTOL

DHTOL test, as defined in the JEDEC standard JEP 180.01, is a critical method for evaluating the reliability of GaN power devices under continuous hard-switching stress at high temperatures and high voltages [20]. As shown in Fig. 9, the waveform of the dR_{ON} extracted from the device during DHTOL testing is depicted. Fig. 10 illustrates the test waveform under operating conditions with V_{BUS} of 400 V, V_{GS} of 5 V, average I_{DS} of 3 A, switching frequency of 100 kHz, and duty cycle of 50 %.

Fig. 11 illustrates the time-resolved dR_{ON} characteristics of the device. Under high-temperature and high-voltage stress conditions, the dR_{ON} degrades significantly, reaching

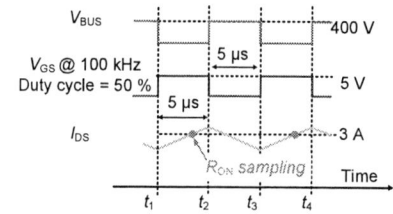

Fig. 9. Key waveforms for measuring dR_{ON} under DHTOL test.

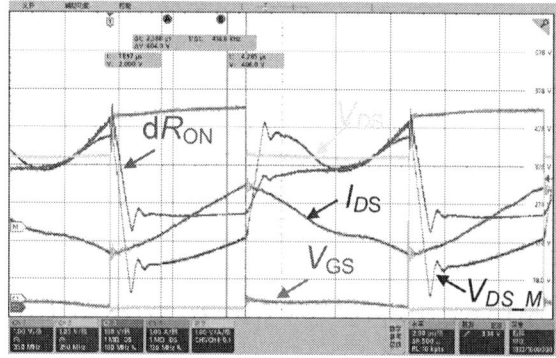

Fig. 10. Switching waveforms of the DHTOL test at V_{BUS} = 400 V, average I_{DS} = 3 A.

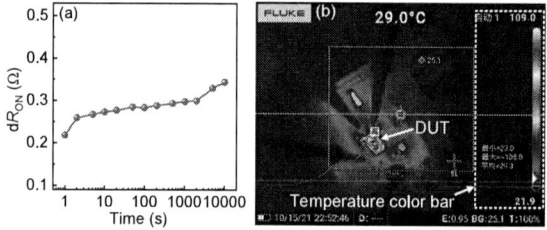

Fig. 11. (a) Time-resolved dR_{ON} and (b) temperature distribution of R_{ON} evaluation system under DHTOL test.

0.34 Ω at the end of the stress stage, which corresponds to a 1.6× increase compared to the initial value. The results align closely with the device's datasheet, which specifies a junction temperature (T_j) of 150 °C for the R_{ON}. This phenomenon not only validates the accuracy of the dR_{ON} testing system but also confirms that the observed degradation trend aligns with theoretical expectations. As shown in Fig. 11(b), during the DHTOL testing, the tested device achieved an average temperature of approximately 108 °C under stable system conditions. This result further validates the reliability and stability of current commercial p-GaN gate HEMTs.

IV. CONCLUSION

In this work, the dR_{ON} degradation of commercial p-GaN gate HEMTs was systematically evaluated under various stress conditions using a precise measurement system. The DPT test revealed that dR_{ON} degradation intensifies with increasing bus voltage, reaching 1.2× dR_{ON} at the highest voltage. Under DC stress testing, the device gradually degraded to 1.3× dR_{ON} with prolonged stress time. Finally, the DHTOL test demonstrated that the device achieved the expected 1.6× dR_{ON} under continuous high-temperature, high-voltage, and hard-switching coupling stress. These inspiring new findings not only provide critical insights into the degradation of p-GaN gate HEMTs, but also validate the robustness of the proposed evaluation system.

ACKNOWLEDGMENT

This work was supported in part by the National Key Research and Development Program of China under Grant 2021YFB3600900, in part by Natural Science Foundation of Guangzhou under Grant 2023A04J1063, in part by Zhuhai Industry University Research Cooperation Project under Grant 2320004002835, and in part by Natural Science Basic Research Program of Shaanxi under Grant 2024JC-YBQN-0621.

REFERENCES

[1] K. J. Chen, O. Häberlen, A. Lidow, C. L. Tsai, T. Ueda, Y. Uemoto, and Y. Wu, "GaN-on-Si power technology: Devices and applications," *IEEE Trans. Electron Devices*, vol. 64, no. 3, pp. 779–795, Mar. 2017.

[2] H. Amano, Y. Baines, E. Beam, M. Borga, T. Bouchet, P. R. Chalker, M. Charles, K. J. Chen, N. Chowdhury, and R. Chu, "The 2018 GaN power electronics roadmap," *Journal of Physics D: Applied Physics,* vol. 51, no. 16, pp. 163001, Mar. 2018.

[3] K. J. Chen, O. Häberlen, A. Lidow, C. L. Tsai, T. Ueda, Y. Uemoto, and Y. Wu, "GaN-on-Si power technology: Devices and applications," *IEEE Trans. Electron Devices*, vol. 64, no. 3, pp. 779–795, Mar. 2017.

[4] B. Bakeroot, A. Stockman, N. Posthuma, S. Stoffels, and S. Decoutere, "Analytical model for the threshold voltage of p-(Al)GaN high-electronmobility transistors," *IEEE Trans. Electron Devices*, vol. 65, no. 1, pp. 79–86, Jan. 2018.

[5] Roy K. Y. Wong, H. C. Chiu, J. H. Zhang, C.Zhou, T. Zhao, Y. B. Wu, H. Liao, S. He, A. B. Zhang, Y. B. Zou, S. Li, M. Zhang, M. Wu, J. Lee, P. W. Chen, A. Xie, and J. Zhang, "High-Performance GaN-on-Si Power Devices with Ultralow Specific On-resistance Using Novel Strain Method Fabricated on 200 mm CMOS-Compatible Process Platform," in *2019 International Symposium on Power Semiconductor Devices and ICs (ISPSD)*, Shanghai, China, 2019, pp. 67-70.

[6] M. Meneghini, P. Vanmeerbeek, R. Silvestri, S. Dalcanale, A. Banerjee, D. Bisi, E. Zanoni, G. Meneghesso, and P. Moens, "Temperature-Dependent Dynamic R_{ON} in GaN-Based MIS-HEMTs: Role of Surface Traps and Buffer Leakage," *IEEE Trans. Electron Devices*, vol. 62, no. 3, pp. 782-787, Mar. 2015.

[7] R. Vetury, N. Q. Zhang, S. Keller, and U. K. Mishra, "The impact of surface states on the DC and RF characteristics of AlGaN/GaN HFETs," *IEEE Trans. Electron. Devices*, vol. 48, no. 3, pp. 560–566, Mar. 2001.

[8] L. Efthymiou, K. Murukesan, G. Longobardi, F. Udrea, A. Shibib, and K. Terrill, "Understanding the threshold voltage instability during OFF-state stress in p-GaN HEMTs," *IEEE Electron Device Lett.*, vol. 40, no. 8, pp. 1253–1256, Aug. 2019.

[9] J. Yang, J. Wei, Y. Wu, M. Nuo, Z. Chen, X. Yang, M. Wang, and B. Shen, "600-V p-GaN Gate HEMT With Buried Hole Spreading Channel Demonstrating Immunity Against Buffer Trapping Effects," *IEEE Electron Device Lett.*, vol. 44, no. 2, pp. 225-228, Feb. 2023.

[10] W. Saito, T. Nitta, Y. Kakiuchi, Y. Saito, K. Tsuda, I. Omura, and M. Yamaguchi, "Suppression of dynamic on-resistance increase and gate charge measurements in high-voltage GaN-HEMTs with optimized field-plate structure," *IEEE Trans. Electron Devices*, vol. 54, no. 8, pp. 1825–1830, Aug. 2007.

[11] X. Li, N. Posthuma, B. Bakeroot, H. Liang, S. You, Z. Wu, M. Zhao, G. Groeseneken, and S. Decoutere, "Investigating the Current Collapse Mechanisms of p-GaN Gate HEMTs by Different Passivation Dielectrics," *IEEE Tran. Power Electronics*, vol. 36, no. 5, pp. 4927-4930, May 2021.

[12] F. Zhou, W. Xu, F. Ren, Y. Xia, L. Wu, T. Zhu, D. Chen, R. Zhang, Y. Zheng, and H. Lu, "1.2 kV/25 A Normally off P-N Junction/AlGaN/GaN HEMTs With Nanosecond Switching Characteristics and Robust Overvoltage Capability," *IEEE Trans. Power Electronics*, vol. 37, no. 1, pp. 26-30, Jan. 2022.

[13] Y. Huang, Q. Jiang, S. Huang, Z. Ji, X. Wang, and X. Liu, "Investigation of Saturated R_{ON} on GaN Power HEMTs by a Re-Configurable Continuous Switching Platform," *IEEE Trans. Electron Devices*, vol. 71, no. 8, pp. 4879-4884, Aug. 2024.

[14] Guideline For Switching Reliability Evaluation Procedures For Gallium Nitride Power Conversion Devices, document JEP 180.01, JEDEC Solid State Technol. Assoc., Arlington County, VA, USA, Jan. 2021.

[15] S. R. Bahl, F. Baltazar, and Y. Xie, "A Generalized Approach to Determine the Switching Lifetime of a GaN FET," in *2020 IEEE International Reliability Physics Symposium (IRPS)*, Dallas, TX, USA, 2020, pp. 1-6.

[16] S. Yin, Y. Lin, R. Hao, S. Jin, C. He, W. Yao, X. Li, Q. He, X. Pu, X. Su, Y. Zou, H. Cai, K. -J. Lee, M. Wang, H. Guo, K. Shen, F. Wang, H. -C. Chiu, L. Chen, D. Marcon, and R. K. . -Y. Wong, "Evaluation of Reliability and Lifetime of 650-V GaN-on-Si Power Devices Fabricated on 200-mm CMOS-Compatible Process Platform for High-Density Power Converter Application," in *2022 IEEE International Symposium on Power Semiconductor Devices and ICs (ISPSD)*, Vancouver, BC, Canada, 2022, pp. 93-96.

[17] Z. Jiang, M. Hua, X. Huang, L. Li, C. Wang, J. Chen, and K. J. Chen, "Negative Gate Bias Induced Dynamic ON-Resistance Degradation in Schottky-Type p-Gan Gate HEMTs," *IEEE Trans. Power Electronics*, vol. 37, no. 5, pp. 6018-6025, May 2022.

[18] Innoscience, INN700D240B Datasheet. (2024). [Online]. Available: https://www.innoscience.com/.

[19] Y. Huang, Q. Jiang, S. Huang, X. Wang, and X. Liu, "Characterization of Electrical Switching Safe Operation Area on Schottky-Type P-GaN Gate HEMTs," *IEEE Trans. Power Electronics*, vol. 38, no. 7, pp. 8977-8989, Jul. 2023.

[20] F. Hosseinabadi, S. Chakraborty, S. K. Bhoi, G. Prochart, D. Hrvanovic, and O. Hegazy, "A Comprehensive Overview of Reliability Assessment Strategies and Testing of Power Electronics Converters," *IEEE Open Journal of Power Electronics*, vol. 5, pp. 473-512, 2024.

2025 IEEE Workshop on Wide Bandgap Power Devices and Applications in Asia (WiPDA Asia)

The Establishment of Quasi-3D TCAD Simulation for Bipolar Degradation Occurs in SiC MOSFET Body Diode

Junhou Cao, Xinyu Zhou, Chenlu Wang, Lei Huang, Hao Fu, Zhaoxiang Wei, Tuanzhuang Wu
Jiaxing Wei*, Siyang Liu*, Weifeng Sun
National ASIC System Engineering Research Center
Southeast University
Nanjing, China
*Email: jiaxingwei@seu.edu.cn; liusy2017@seu.edu.cn;

Abstract—**Bipolar degradation (BD) in SiC MOSFETs is one of the reliability issues that limit the long-term operation of the devices. Under bipolar conduction conditions, as the Stacking Faults (SF) in the epitaxial layer expanding, the device resistance and body diode voltage drop increase significantly. This paper determines the correlation between the degree of electrical parameter degradation and the expansion of the SF area through constant current electrical stress loading and Photoluminescence (PL) defect distribution characterization. In addition, by introducing a 0.75nm-thick 3C-SiC layer at a 4° angle and controlling its expansion length, a quasi-3D simulation to describe the BD process was constructed for the first time. The changes in various parameters in the epitaxial layer obtained from the simulation were consistent with existing research conclusions, and the simulated body diode V_f shift achieves a high degree of consistency with the actual measurements. This work contributes to the study of the mechanism of bipolar degradation in SiC and the reliability assessment of devices under long-term stress in applications.**

Index Terms—**Reliability, Bipolar Degradation, SiC MOSFET, TCAD Simulation**

I. Introduction

Bipolar Degradation (BD) is a typical degradation mechanism occurring in bipolar SiC devices. It significantly increases the voltage drop of the PiN diodes and the ON-resistance of the MOSFETs, making it one of the most critical reliability issues during the long-term operation of SiC MOSFETs and PiN diodes [1]. Existing studies indicate that BD is related to the Basal Plane Dislocations (BPDs) generated during the SiC substrate growth process. During the process of bipolar conduction, the energy generated by the recombination of electrons and holes promotes the propagation of the BPDs, and leads to an increase in the area of Stacking Fault (SF) regions, which subsequently causes a series of degradations, such as increased resistance in the epitaxial layer, reduced carrier lifetime, and enlarged leakage paths [2].

Numerous studies have extensively investigated the changes in defect regions during bipolar degradation. Through the Photoluminescence(PL) characterization of degraded devices, two different shapes of the SF region were observed [3]. This phenomenon is interpreted as a result of BPD propagation along different crystallographic directions [4]. Besides, the lattice dislocation of the SF region is also observed through the Transmission Electron Microscopy(TEM) characterization [5]. based on these results, the 2D TCAD simulation has been proposed and used for mechanism studies [6]. However, the established simulations use a horizontal line to model the SF region, which can only achieve qualitative study, and differs significantly from the actual distribution of defects.

In this study, a quasi-3D TCAD simulation established for qualitative study of BD occurs in SiC MOSFETs is propsed for the first time. To gain a deeper understanding of the BD mechanism, the electrical characteristics degradation of a 1700V SiC MOSFET after 3A current stress were extracted. The corresponding expansion of the SF region was characterized through PL image. Through electrical simulation and physical parameter analysis, more details during the BD process within the epitaxial layer was observed. A comparison between the simulation and actual measurement degradation results shows a high degree of consistency in the degradation of the body diode voltage drop.

II. Test Setup and Results

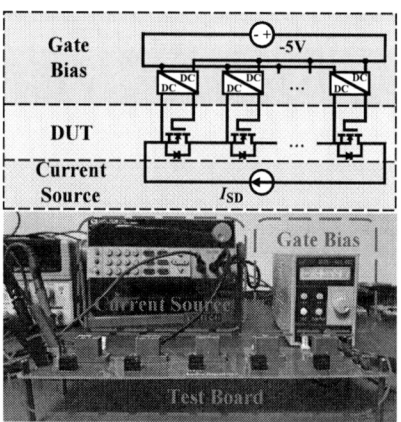

Fig. 1. The Schematic of the constant current stress applying and the testing platform

979-8-3315-1110-4/25 $31.00 © 2025 IEEE

A commercial SiC MOSFET rating at 1700V, 3.7A was selected as the Device Under Test (DUT). A constant current testing platform was set up for degradation evaluation, as shown in Fig.1. To ensure the channel was remained in a completely OFF-state, -5V bias voltage was applied to the gate. During the I_{SD}=3A stress degradation process, two DUTs were connected in series, the forward conduction and the body diode characteristics of two DUTs(#1 and #2) were extracted respectively to record degradation process as shown in Fig.2. Where the I_{max} and V_f represents the forward conduction current and the forward voltage drop on the body diodes measured after devices return to room temperature. Two devices show similar degradation trends, a rapidly degradation was observed in 4 minutes and followed by saturation. After 1 hour stress application, bipolar degradation of two DUTs no longer occur. Comparing the saturation state with the original state, the increase in V_f is about 4%(Fig.3).

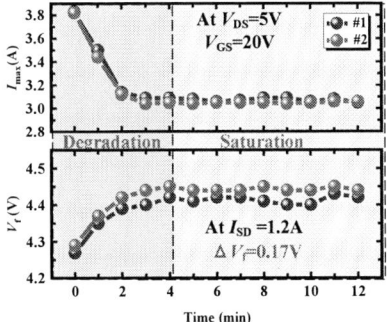

Fig. 2. The degradation and saturation of I_{max} and V_f during the 3A current stress

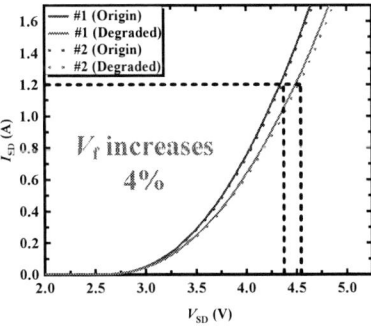

Fig. 3. The shift of body diode characteristics curves before and after stress

To observe the defect regions expanded during the stress process more clearly, PL characterization was carried out on the degraded device #1. The distribution of SF areas was obtained by exciting at a wavelength of 325nm and photographing with a camera equipped with a 420nm bandpass filter [5]. A strip-shaped signal region was observed in the active area of the DUT as shown in Fig.4. Which was confirmed to be Bar-Shaped Stacking Fault (BSSF) based on comparison with previously reported studies [7].

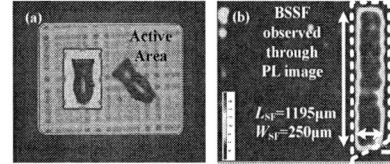

Fig. 4. (a)The de-packaged device #1 and (b) SF distribution observed in the PL image.

III. SIMULATION AND VERIFICATION

To investigate the impact of SF defect regions on the device conduction process, TCAD simulation was employed for further analysis. Fig.5(a) illustrates the expansion of BSSF in the epitaxial layer of the SiC MOSFET. The TEM results indicate that, a noticeable lattice misalignment was observed in the SF region within the device cross-section [5]. Which exhibits the same lattice ordering as that of 3C-SiC and thus forming a thin heterojunction structure(Fig.5(b)). Besides, the angle between the inserted defect layer and the substrate was measured tobe 4° [4]. The PL image reveals that the width of the SF Region(W_{SF}) is approximately 250μm. According to (1), the estimated epitaxial thickness T_{epi} is about 17.5μm, which is also consistent with the standard epitaxial specifications for 1700V-rated SiC MOSFET devices.

$$T_{epi} = W_{SF} \times \tan(4°) \tag{1}$$

Finally, a 1700V-rated SiC MOSFET device was constructed in Sentaurus TCAD software. The epitaxial layer was set to 17.5μm thickness with a doping concentration of 1E16cm^{-3}. The device area in the simulation was configured to match the active area of the DUT at 0.0095 cm^2. To model the degraded device, the simulated structure was devided into two distinct regions: the region with defect and the region without defect (as shown in Fig.6(a)). For accurate modeling of the extended stacking fault (SF) region, a 0.75 nm-thick 3C-SiC layer was inserted into the epitaxial layer of the region with defect at a 4° tilt angle (Fig.6(b)). The length and width of this layer were set to 1195μm and 250μm, respectively, based on PL image measurements. To maintain the total device area, the corresponding undegraded region was adjusted to 0.0062 cm^2.

Comparative simulation of the device's bulk band structure reveals that the inserted 3C-SiC defect region creates a potential well at the heterointerface with adjacent 4H-SiC due to the band offset(Fig.7). This potential well induces localized carrier trapping and forms a depletion region that impedes carrier transport [8], ultimately leading to a significant increase in the epitaxial layer resistance.

To further investigate the impact of the SF region on conduction capability, body diode current conduction at 3A was simulated for both origin and degraded devices(Fig.8). Current density distribution analysis demonstrates that nearly all current is blocked by the potential barrier near the 3C-SiC layer, forcing the parallel-connected region without defect to carry substantially higher current density. Under actual operating conditions, this current crowding

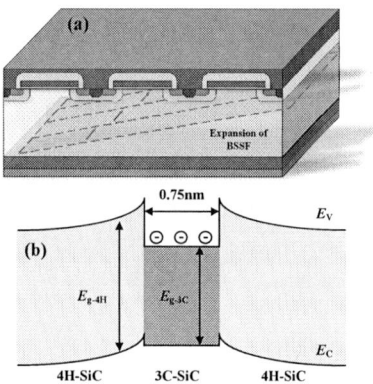

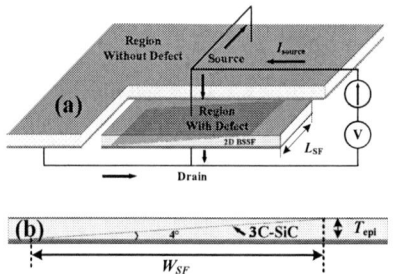

Fig. 5. The schematic diagram of (a) the BSSF expansion in the epitaxial layer and (b) the localized defect heterojunction interface.

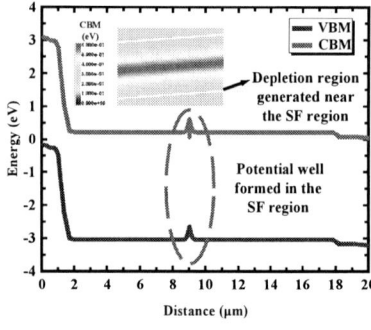

Fig. 6. (a) Implementing quasi-3D simulation of BD by paralleling the structure with and without the introduced SF region. (b) The 2D simulation structure established for degraded device by introducing 0.75nm 3C-SiC layer as the SF region.

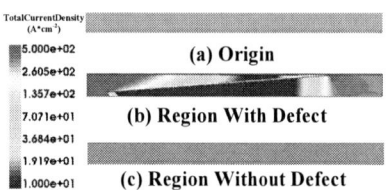

Fig. 7. Vertical energy band diagram of the region with defect.

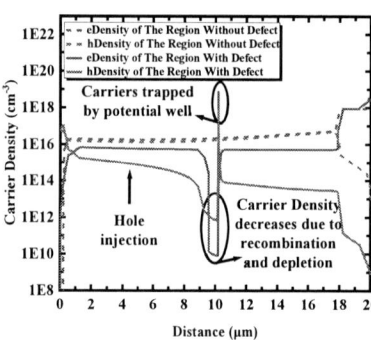

Fig. 8. The current density distribution of (a) Origin device, (b) the region with defect and (c) the region without defect in degraded device under the condition of I_{SD}=3A.

effect may lead intensified carrier recombination in healthy regions [9]. and result in generation of other potential dislocations. These observations suggest a positive feedback mechanism in BD.

By analyzing the carrier distribution within the device under the on-state condition, the trapping and depletion effects of the defect layer on carriers can be more intuitively observed as shown in Fig.9. Under a conduction current of 3A, the concentrations of electrons and holes within the potential well of the defect region reach 1E19cm^{-3} and 1E18cm^{-3} respectively, significantly higher than the carrier concentration under normal conduction conditions($\sim$ 1E16cm^{-3}). This elevated carrier concentration leads to an increased recombination rate within the potential well, thereby providing additional energy for defect propagation, which further accelerates the expansion of the existing SF region. Moreover, the carrier concentration in other regions of the epitaxial layer is notably reduced due to recombination and depletion effects, resulting in a decrease in current density within the defect region.

The body diode characteristics of the region with and without defect are shown in Fig.10. The results show that the conduction in the region without defect is dominant in a degraded device, and the current is nearly completely blocked in he region with defect. Comparing the device before and after degradation, V_f increased by 6.9%, which achieves a good alignment with the actual measured result.

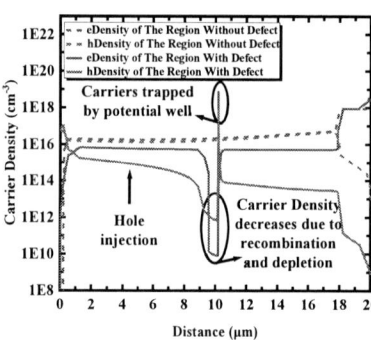

Fig. 9. The carrier density in the whole epitaxial layer of the region without defect (in dashed line) and the region with defect (in solid line).

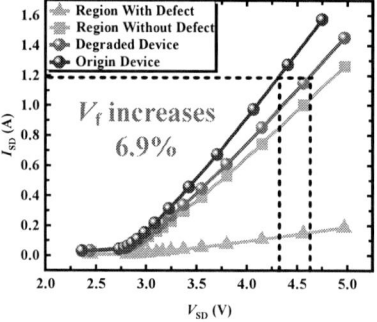

Fig. 10. Shift in the characteristic curves of the DUT's body diode before and after degradation in simulation.

IV. CONCLUSION

A quasi-3D simulation suitable for bipolar degradation of the body diode in SiC MOSFET devices has been constructed in this paper. The simulation was built based on the actual measurement results of a commercial 1700V device, and the expansion of the BSSF area in the epitaxial layer was simulated by inserting a 0.75nm thick layer of 3C-SiC at a 4-degree angle, with the area of the SF region set according to the actual measured PL image. The parameters of the epitaxial layer in the electrical simulation are consistent with existing research, and the simulated drift of the body diode drop voltage after degradation also shows good agreement with the actual measurement results.

Under the on-state condition, the heterojunction barrier near the SF region restricts carrier transport and induces current crowding in the undegraded areas, which may promote the initiation of new defect regions. Moreover, carrier concentration analysis reveals enhanced recombination activity within the SF region, potentially providing the energy necessary for the propagation of existing defects. Taken together, these results suggest that BD may exhibit a positive feedback mechanism. Therefore, fundamentally reducing the substrate defect density remains an effective strategy to mitigate this issue[]. This research helps to understand the mechanism of bipolar degradation and correspondingly improves the reliability of SiC MOSFETs under long-term working conditions.

ACKNOWLEDGMENT

This work is supported in part by the National Natural Science Foundation of China under grant 62434002 and grant 62174029; in part by the Fundamental Research Funds for the Central Universities under grant 2242024RCB0028; in part by the Science and Technology Major Project of Jiangsu Province under grant BG 2024001; in part by the Fund for Transformation of Scientific and Technological Achievements of Jiangsu Province under grant BA2023001; in part by the Natural Science Foundation of Jiangsu Province under grant BK20232027; in part by the Distinguished Young Scientists Foundation of Jiangsu Province under grant BK20230025; and in part by the Fund for Transformation of Scientific and Technological Achievements of Wuxi City under grant C20231021.

REFERENCES

[1] A. Agarwal, H. Fatima, S. Haney, and S.-H. Ryu, "A new degradation mechanism in high-voltage sic power mosfets," *IEEE Electron Device Letters*, vol. 28, no. 7, pp. 587–589, 2007.

[2] T. Kimoto and H. Watanabe, "Defect engineering in sic technology for high-voltage power devices," *Applied Physics Express*, vol. 13, no. 12, p. 120101, 2020.

[3] A. Okada, J. Nishio, R. Iijima, C. Ota, A. Goryu, M. Miyazato, M. Ryo, T. Shinohe, M. Miyajima, T. Kato, *et al.*, "Dependences of contraction/expansion of stacking faults on temperature and current density in 4h-sic p–i–n diodes," *Japanese Journal of Applied Physics*, vol. 57, no. 6, p. 061301, 2018.

[4] T. Kimoto, A. Iijima, H. Tsuchida, T. Miyazawa, T. Tawara, A. Otsuki, T. Kato, and Y. Yonezawa, "Understanding and reduction of degradation phenomena in sic power devices," in *2017 IEEE International Reliability Physics Symposium (IRPS)*, pp. 2A–1, IEEE, 2017.

[5] G. Feng, J. Suda, and T. Kimoto, "Characterization of major in-grown stacking faults in 4h-sic epilayers," *Physica B: Condensed Matter*, vol. 404, no. 23-24, pp. 4745–4748, 2009.

[6] A. Lachichi and P. Mawby, "Modeling of bipolar degradations in 4h-sic power mosfet devices by a 3c-sic inclusive layer consideration in the drift region," *IEEE Transactions on Power Electronics*, vol. 37, no. 3, pp. 2959–2969, 2021.

[7] H. Tsuchida, K. Murata, T. Tawara, M. Miyazato, T. Miyazawa, and K. Maeda, "Suppression of bipolar degradation in 4h-sic power devices by carrier lifetime control," in *2019 IEEE International Electron Devices Meeting (IEDM)*, pp. 20–1, IEEE, 2019.

[8] C. Taniguchi, A. Ichimura, N. Ohtani, M. Katsuno, T. Fujimoto, S. Sato, H. Tsuge, and T. Yano, "Theoretical investigation of the formation of basal plane stacking faults in heavily nitrogen-doped 4h-sic crystals," *Journal of Applied Physics*, vol. 119, no. 14, 2016.

[9] A. Iijima and T. Kimoto, "Electronic energy model for single shockley stacking fault formation in 4h-sic crystals," *Journal of Applied Physics*, vol. 126, no. 10, 2019.

Study on $Al_xGa_{1-x}N$ Graded Composite Barrier GaN HEMT

Ruihao Zhang and Fayu Wan

Nanjing University of Information Science and Technology, Nanjing, China, ruihao.zhang@nuist.edu.cn

Abstract— This work designs an $Al_xGa_{1-x}N$ graded composite barrier GaN HEMT (GB-HEMT) and compares it with a fixed component barrier GaN HEMT (FB-HEMT). Through simulation with Silvaco TCAD, it is found that the GB-HEMT has wider electron gas and flatter transconductance (g_m). Its saturation output current reaches 1.73 A/mm and the turn-off characteristics is also greatly improved. In addition, due to the exceptional direct current (DC) performance, the radio frequency (RF) performance of GB-HEMT is also excellent. In terms of small-signal characteristics, the f_t/f_{max} of GB-HEMT are 138.4/271.6 GHz. For large-signal characteristics, its power added efficiency (*PAE*) and gain are 18.7 dB and 44.4%, respectively, with the power density of 3.8 W/mm.

Keywords— GaN HEMT; Graded composite barrier (GB); Large-signal performance; Three-dimensional electron gas (3DEG).

I. INTRODUCTION

GaN High Electron Mobility Transistors (HEMTs) are widely used in many fields due to their excellent characteristics [1-4] Currently, in reported AlGaN/GaN HEMTs, the maximum output power has reached 10 W/mm [5,6], and the cut-off frequency exceeds 100 GHz [7]. Compared to traditional narrow bandgap semiconductors, third-generation semiconductors represented by GaN exhibit superior electrical properties, including high electron saturation velocity, high thermal stability, and high breakdown voltage, making them widely applicable in communication systems, aerospace, and other fields.

Compared with traditional fixed composition barrier GaN HEMT (FB-HEMT), the gradient composite barrier GaN HEMT (GB-HEMT) offers many advantages. FB-HEMTs with fixed composition barrier layers can lead to higher interface scattering and parasitic parameters, which become more severe as the aluminum content in AlGaN increases. This work proposes a GaN HEMT with a graded composite barrier structure which can effectively improve interface quality, reduce interface scattering, and thus enhancing the HEMT's electrical performance [8].

Currently, GaN HEMTs exhibit very high power density in practical applications. However, existing AlGaN/GaN HEMT technology faces issues such as gain compression [9] and significant nonlinearity [10] at higher frequencies. The decline in transconductance (g_m) of HEMTs is related to saturation velocity and nonlinear source resistance. One method to suppress the nonlinearity caused by the decline in g_m is to use a three-dimensional electron gas (3DEG) channel instead of a two-dimensional electron gas (2DEG) channel. FB-HEMTs only form 2DEG under the AlGaN/GaN interface near the side of GaN, while a graded channel can form electron gas in the $Al_xGa_{1-x}N$ graded barrier layer. This results in better transconductance flatness and improving the linearity of HEMT [11,12]. Owing to the exceptional direct current (DC) performance, the radio frequency (RF) performance of GB-HEMT is also excellent compared with FB-HEMT.

II. DEVICE STRUCTURE

As shown in Fig. 1(a), the device structure of the FB-HEMT from top to bottom is as follows: a T-shaped gate with a gate length of 100 nm, a 60 nm thick Si_3N_4 passivation layer, a 20 nm thick $Al_{0.25}Ga_{0.75}N$ fixed component barrier layer, a 400 nm thick GaN channel layer, an AlGaN back-barrier layer (with the fixed molar composition of 0.1), a 1.2 μm thick unintentionally doped GaN buffer layer, and a 1 μm thick SiC substrate. The gate-to-source distance is 1.1 μm, and the gate-to-drain distance is 1.6 μm.

As shown in Fig. 1(b), for GB-HEMT, the 20 nm thick $Al_xGa_{1-x}N$ gradient composite barrier layer includes a 5 nm thick fixed composition barrier layer with a fixed component $Al_{0.4}Ga_{0.6}N$ barrier layer and a 15 nm thick graded composition barrier layer. Since it is difficult to achieve an ideal linear gradient of the Al component in the back-barrier layer with existing device processes, the AlGaN barrier is divided into multiple thin layers, with the Al component uniformly decreasing from 0.4 to 0 in steps of 1 nm. Since the Al composition between each thin layer is similar, the interface scattering is relatively small and can be ignored. Besides, the barrier with ideal linear gradient Al component or with step-by-step gradient Al component have little impact on the device performance. The Al component of the barrier layers of these two HEMTs is approximately the same for comparison.

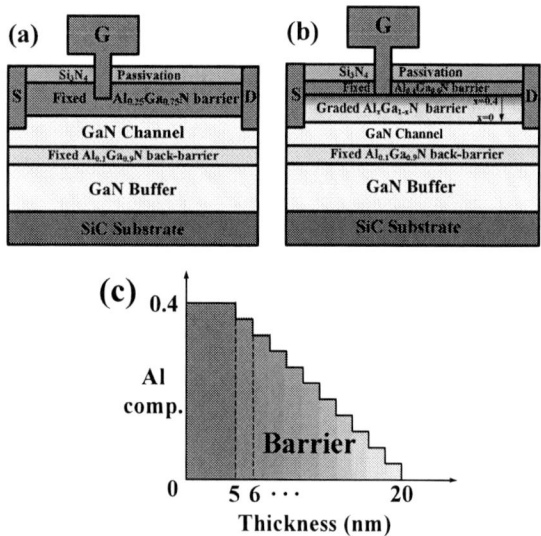

Fig. 1. (a) Device structure of FB-HEMT; (b) Device structure of GB-HEMT; (c) The Al composition distribution of GB-HEMT barrier.

This work uses Silvaco TCAD software for simulation. The simulation specifies the electric field dependent mobility (fldmob) and the concentration dependent mobility (conmob). Shockley-Read-Hall (SRH) along with temperature mobility model (lat.temp) are also employed in the simulation.

In order to reduce the deviation between simulation results and actual results, bulk trap model (trap) and the interface trap model (inttrap) are used in the simulation [13-15]. For the bulk trap, the concentration is 1×10^{16} cm^{-3} in the buffer layer, located 0.36 eV underneath the conduction band. Besides, FB-HEMT has acceptor trap on the interface between the Al$_{0.25}$Ga$_{0.75}$N barrier and the GaN channel, the trap concentration is set to be 6×10^{12} cm^{-2} and is located at level 0.3 eV below the conduction band. Since the mole fraction of barrier in GB-HEMT gradually increases and no mutation occurs at the interface [16]. As a result, the interface trap of the GB-HEMTs is ignored.

III. RESULTS AND DISCUSSION

Fig. 2(a) shows the variation of energy band with depth of FB-HEMT and GB-HEMT, and depth is the distance from the surface of the device. The potential well in the heterostructure is extended wider from the graded AlGaN barrier (y = 5 nm) to the heterointerface (y = 20 nm). Therefore, 3DEG is formed in the graded AlGaN barrier, which can be called polarization doping. Fig. 2(b) shows the variation of electron concentration with depth. It is found that for traditional FB-HEMT, the peak electron concentration in the channel reaches 2.3×10^{20} cm^{-3}, followed by a sharp decline, which forms a triangular distribution. While for GB-HEMT, the electrons gather in a wide potential well. Hence, the carriers have a certain distribution thickness in the graded barrier, with the electron gas concentration exceeding 2×10^{19} cm^{-3} over a width of around 15 nm. The 2DEG in the channel has become 3DEG. The electron volume density can be translated into electron sheet density (n_s) by integrating n along the depth. It can be calculated that the n_s of FB-HEMT is 1.52×10^{13} cm^{-2}, while the n_s of GB-HEMT reaches 3.03×10^{13} cm^{-2}. The higher n_s enables GB-HEMT to have larger drain-source current (I_{DS}) than FB-HEMT.

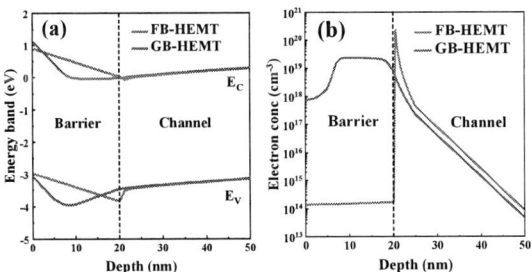

Fig. 2. (a) Variation of energy band with depth of FB-HEMT and GB-HEMT; (b) Variation of electron concentration with depth of FB-HEMT and GB-HEMT.

Alloy scattering is one of the key scattering mechanisms that affects the electron transport properties of AlGaN/GaN HEMTs. In the Al$_x$Ga$_{1-x}$N barrier layer, due to the differences in atomic size and electronegativity of Al and Ga atoms, the random distribution causes local fluctuations in the conduction band edge, leading to scattering effects and reducing the electron mobility (μ_n) along with the electron saturation velocity (v_{sat}). Since the 2DEG for FB-HEMT is primarily concentrated in the GaN channel layer, the alloy scattering has little impact on it. Therefore, as shown in Fig. 3(a) and (b), μ_n and v_{sat} of 2DEG remain unchanged. In contrast, for GB-HEMTs, the 3DEG is mainly concentrated in the graded AlGaN barrier, so μ_n and v_{sat} of 3DEG are lower for alloy scattering.

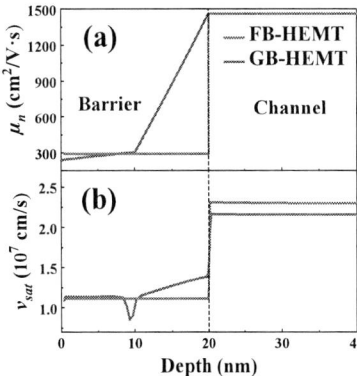

Fig. 3. (a) Variation of electron mobility with depth of FB-HEMT and GB-HEMT; (b) Variation of electron saturation velocity with depth of FB-HEMT and GB-HEMT.

To achieve excellent small-signal characteristics, the gate length of the HEMT is generally short, which easily leads to short-channel effects, deteriorating the electrical properties. Considering these issues, this work employs a recessed gate structure [17,18] and a back-barrier structure [19,20] to enhance the confinement of the electron gas and gate control capability. The transfer characteristic curve can be used to determine the quality of the gate control capability.

Setting the drain-source voltage (V_{DS}) to 10 V, as shown in Fig. 4(a), after structural optimization, the on/off ratio of the FB-HEMT reaches 3×10^7, while the on/off ratio of the GB-HEMT reaches 7×10^{10}. The GB-HEMT has a higher on/off ratio. The subthreshold swing (SS) of the FB-HEMT is 386 mV/dec, while the SS of the GB-HEMT is 375 mV/dec. The GB-HEMT also has a lower subthreshold swing, indicating better off-state characteristics. The improvement in off-state characteristics may be related to the reduction in peak electron concentration in GB-HEMT.

Transconductance flatness is an important performance indicator of HEMT. Replacing the 2DEG channel with a 3DEG channel can significantly improve the linearity of the HEMT, increase the saturation current, and enhance output power. The GaN HEMT with a graded composite barrier structure proposed in this work has a higher and wider carrier channel and a flatter transconductance curve.

$$I_{DS} = \begin{cases} qn_s\mu_nE, & \text{Linear region} \\ qn_sv_{sat}, & \text{Saturation region} \end{cases} \quad (1)$$

Eq. (1) shows the calculation of I_{DS} under ideal conditions, where E is the electric field strength and q is the electron charge. When V_{DS} remains constant, E is only affected by V_{GS}. It can be seen that when the HEMT operates in the linear region, I_{DS} should increase linearly with the increase of E, and the transconductance should remain constant in the linear region. However, for conventional FB-HEMT, there is a certain transition from the linear region to the saturation region, which prevents the transconductance from staying at a high level.

Refer to Fig. 4(b), which shows the transconductance curves of the FB-HEMT and the GB-HEMT. With V_{DS} set to 10 V, it can be seen that the peak transconductance (g_{mmax}) of the FB-HEMT is 570.4 mS. When V_{DS} remains unchanged, the g_m will first rise rapidly to reach g_{mmax} as the gate bias (V_{GS}) increases, then drop rapidly, and g_m cannot maintain a high level for long. In contrast, GB-HEMT has a higher and flatter transconductance, with g_m greater than 500 mS from -2.5 V to 2 V, and g_{mmax} reaching 655.3 mS.

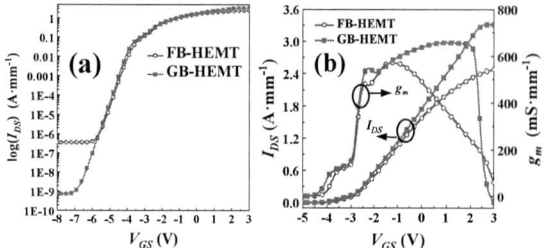

Fig. 4. Transfer characteristics and transconductance curves of FB-HEMT and GB-HEMT: (a) Transfer characteristic curve with a logarithmic scale on the Y-axis; (b) Transfer characteristic and transconductance curves with a linear scale on the Y-axis.

The 3DEG of GB-HEMT brings large saturation output current. It can be seen from Fig. 5 that the saturation output current at V_{GS}=0 V reaches 1.73 A/mm, which is 0.22 A/mm higher than that of FB-HEMT. The n_s of GB-HEMT is around twice than that of FB-HEMT, but I_{DS} is only increased by 12.5%. According to Eq. (1), when the electric field remains unchanged, n_s and μ_n/v_{sat} jointly determine I_{DS}. The 2DEG of FB-HEMT is concentrated in the buffer layer with large μ_n/v_{sat}, while 3DEG of GB-HEMT is mainly concentrated in the graded barrier layer with lower μ_n/v_{sat}, which limits the increase of I_{DS}.

In addition, GB-HEMT can suppress the current collapse. According to Fig. 5, when V_{GS}=0 V, after changing the fixed Al component barrier to the gradient composite barrier, the current collapse rate drops from 10.6% to 2.9%, a decrease of 7.3%.

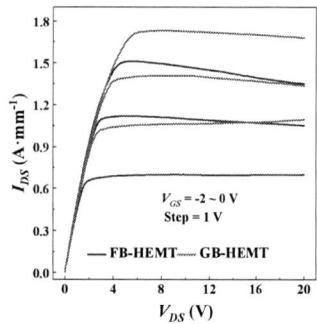

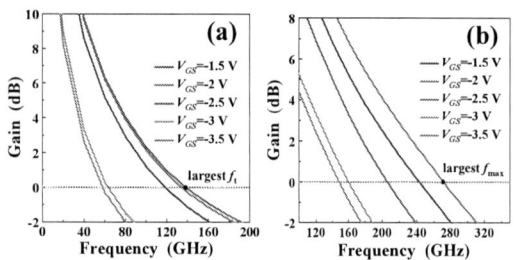

Fig. 5. Output characteristic curves of FB-HEMT and GB-HEMT.

Fig. 6. Small-signal characteristic curves of the GB-HEMT under different gate voltages: (a) f_t; (b) f_{max}.

The GB-HEMT has excellent small-signal characteristics [21,22]. Small-signal simulations are conducted on the GB-HEMT with V_{DS} =15 V. To find the maximum cut-off frequency (f_t) and maximum oscillation frequency (f_{max}), the V_{GS} of the GB-HEMT is linearly reduced from -1.5 V to -3.5 V in steps of -0.5 V. Fig. 6(a) and (b) show the small-signal characteristic curves of the HEMT under different gate voltages. From Fig. 6(a), it can be seen that when V_{GS}=-2.5 V, f_t reaches a maximum of 138.4 GHz; from Fig. 6(b), it can be seen that when V_{GS}=-2 V, f_{max} reaches a maximum of 271.6 GHz.

5G communication system has high requirements for the linearity of the power amplifier. Fig. 7 is a schematic diagram of the nonlinear distortion of the amplifier circuit. It can be seen that great linearity of the HEMT can make the linear working area wider. Hence, the circuit design margin can also be larger. On the contrary, if the linearity is poor, the input signal swing will be limited, and nonlinear distortion is likely to occur. The linearity of the HEMT is related to the transconductance. If the transconductance cannot be stabilized at a high level, the linear amplification area will be narrow, seriously affecting the performance of the amplifier. Since the proposed GB-HEMT has higher and flatter transconductance, the GB-HEMT has better large-signal performance than FB-HEMT.

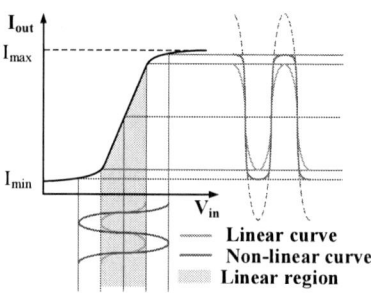

Fig. 7. The mechanism of GaN HEMT nonlinear distortion.

As shown in Fig. 8, after changing the fixed barrier to the graded composite barrier, the large-signal characteristics of GaN HEMT working at 8 GHz are improved, especially the power gain, which is 4.7 dB higher than FB-HEMT. Besides, the output power density of GB-HEMT is 3.8 W/mm, an increase of 0.64 W/mm; the power added efficiency (PAE) is 44.4%, an increase of 2.1%. The GB-HEMT has flatter transconductance [Fig. 4(b)] and larger saturated output current [Fig. 5], which is an important reason for the improvement of the large-signal performance.

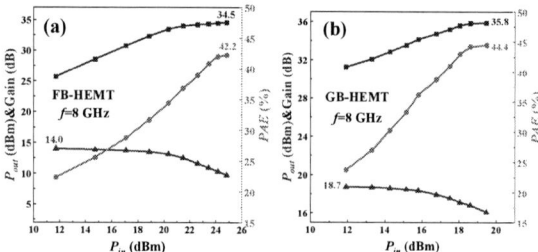

Fig. 8. The large-signal performance of: (a) FB-HEMT; (b) GB-HEMT.

IV. CONCLUSION

This work designs a graded composite barrier GaN HEMT. Compared to the FB- HEMT, the GB-HEMT features a wider potential well as well as thicker electron gas. The 2DEG in the channel transforms into 3DEG which results in higher and flatter transconductance and higher output saturation current. By analyzing the factors affecting I_{DS}, the reason why GB-HEMT has a larger I_{DS} is explored. The GB-HEMT also exhibits stronger off-state characteristics and better small-signal performance, with f_t and f_{max} reaching 138.4 GHz and 271.6 GHz respectively. Besides, for large-signal characteristics, its PAE and gain are 18.7 dB and 44.4 %, respectively, with the power density of 3.8 W/mm, indicating broad application prospects.

REFERENCES

[1] "AlGaN/GaN Based Diodes for Liquid Sensing," *Chinese Physics Letters,* vol. 30, no. 3, p. 037301, 2013/03/01 2013.

[2] N. Moultif *et al.,* "S-band pulsed-RF operating life test on AlGaN/GaN HEMT devices for radar application," *Microelectronics Reliability,* vol. 100-101, p. 113434, 2019/09/01/ 2019.

[3] J.-M. Zeng *et al.,* "Effect of Barrier Temperature on Photoelectric Properties of GaN-Based Yellow LEDs*," *Chinese Physics Letters,* vol. 37, no. 3, p. 038502, 2020/03/01 2020.

[4] H. Guan, W. Li, X. Tong, G. Shen, F. Li, and H. Zhang, "DC and RF performance of HR Si(111)-based AlGaN/GaN MIS-HEMT with a symmetrical multi-finger grid array structure for 5G N28 700MHz low-bias-control applications," *Materials Science in Semiconductor Processing,* vol. 164, p. 107619, 2023/09/01/ 2023.

[5] M. W. Lee *et al.,* "Over 10W/mm High Power Density AlGaN/GaN HEMTs With Small Gate Length by the Stepper Lithography for Ka-Band Applications," *IEEE Journal of the Electron Devices Society,* vol. 11, pp. 311-318, 2023.

[6] T. Ohki *et al.,* "An Over 20-W/mm S-Band InAlGaN/GaN HEMT With SiC/Diamond-Bonded Heat Spreader," *IEEE Electron Device Letters,* vol. 40, no. 2, pp. 287-290, 2019.

[7] L. Li *et al.,* "GaN HEMTs on Si With Regrown Contacts and Cutoff/Maximum Oscillation Frequencies of 250/204 GHz," *IEEE Electron Device Letters,* vol. 41, no. 5, pp. 689-692, 2020.

[8] T. Fang, R. Wang, H. Xing, S. Rajan, and D. Jena, "Effect of Optical Phonon Scattering on the Performance of GaN Transistors," *IEEE Electron Device Letters,* vol. 33, no. 5, pp. 709-711, 2012.

[9] A. Ahmed, S. S. Islam, and A. F. M. Anwar, "Frequency and temperature dependence of gain compression in GaN/AlGaN HEMT amplifiers," *Solid-State Electronics,* vol. 47, no. 2, pp. 339-344, 2003/02/01/ 2003.

[10] S. K. Dhar *et al.,* "Impact of Input Nonlinearity on Efficiency, Power, and Linearity Performance of GaN RF Power Amplifiers," in *2020 IEEE/MTT-S International Microwave Symposium (IMS),* 2020, pp. 281-284.

[11] M. G. Ancona, J. P. Calame, D. J. Meyer, S. Rajan, and B. P. Downey, "Compositionally Graded III-N HEMTs for Improved Linearity: A Simulation Study," *IEEE Transactions on Electron Devices,* vol. 66, no. 5, pp. 2151-2157, 2019.

[12] S. H. Sohel *et al.,* "X-Band Power and Linearity Performance of Compositionally Graded AlGaN Channel Transistors," *IEEE Electron Device Letters,* vol. 39, no. 12, pp. 1884-1887, 2018.

[13] P. Vigneshwara Raja, J.-C. Nallatamby, N. DasGupta, and A. DasGupta, "Trapping effects on AlGaN/GaN HEMT characteristics," *Solid-State Electronics,* vol. 176, p. 107929, 2021/02/01/ 2021.

[14] M. Meneghini *et al.,* "Buffer Traps in Fe-Doped AlGaN/GaN HEMTs: Investigation of the Physical Properties Based on Pulsed and Transient Measurements," *IEEE Transactions on Electron Devices,* vol. 61, no. 12, pp. 4070-4077, 2014.

[15] W. Zhang, Y. Zhang, W. Mao, X. Ma, J. Zhang, and Y. Hao, "Influence of the Interface Acceptor-Like Traps on the Transient Response of AlGaN/GaN HEMTs," *IEEE Electron Device Letters,* vol. 34, no. 1, pp. 45-47, 2013.

[16] Kim J-G, "Optimization of Epitaxial Structures on GaN-on-Si(111) HEMTs with Step-Graded AlGaN Buffer Layer and AlGaN Back Barrier," *Coatings,* vol. 14, no. 6, pp. 700, 2024.

[17] S. Wu *et al.,* "A Millimeter-Wave AlGaN/GaN HEMT Fabricated With Transitional-Recessed-Gate Technology for High-Gain and High- Linearity Applications," *IEEE Electron Device Letters,* vol. 40, no. 6, pp. 846-849, 2019.

[18] Y. Zhang *et al.,* "High-Temperature-Recessed Millimeter-Wave AlGaN/GaN HEMTs With 42.8% Power-Added-Efficiency at 35 GHz," *IEEE Electron Device Letters,* vol. 39, no. 5, pp. 727-730, 2018.

[19] M. Micovic *et al.,* "GaN double heterojunction field effect transistor for microwave and millimeterwave power applications," in *IEDM Technical Digest. IEEE International Electron Devices Meeting, 2004.,* 2004, pp. 807-810.

[20] L. Yang *et al.,* "The DC Performance and RF Characteristics of GaN-Based HEMTs Improvement Using Graded AlGaN Back Barrier and Fe/C Co-Doped Buffer," *IEEE Transactions on Electron Devices,* vol. 69, no. 8, pp. 4170-4174, 2022.

[21] J. S. Moon *et al.,* "Power Scaling of Graded-Channel GaN HEMTs With Mini-Field-Plate T-gate and 156 GHz fT," *IEEE Electron Device Letters,* vol. 42, no. 6, pp. 796-799, 2021.

[22] J. S. Moon *et al.,* "360 GHz fMAX Graded-Channel AlGaN/GaN HEMTs for mmW Low-Noise Applications," *IEEE Electron Device Letters,* vol. 41, no. 8, pp. 1173-1176, 2020.

Chip Screening Strategy for SiC MOSFET Based on Simulated Annealing Algorithm

1st Zhanshan Zhu
State Key Laboratory of Advanced Vehicle Integration and Control China FAW Group Co.,LTD.
Changchun, China
zhuzhanshan@faw.com.cn

2nd Chuangye Li
School of Electrical Engineering Hefei University of Technology.
Hefei, China
chaungye.li@hfut.edu.cn

3rd Helong Li
School of Electrical Engineering Hefei University of Technology.
Hefei, China
helong.li@hfut.edu.cn

4th Zhiqiang Liu
State Key Laboratory of Advanced Vehicle Integration and Control China FAW Group Co.,LTD.
Changchun, China
liuzhiqiang@faw.com.cn

5th Lijian Ding
School of Electrical Engineering Hefei University of Technology.
Hefei, China
lijian.ding@hfut.edu.cn

Abstract—This paper studies the spread of key electrical parameters of Silicon Carbide (SiC) MOSFETs and its effect on the parallel current sharing of the devices. This paper proposes a screening method for parallel SiC MOSFETs based on the simulated annealing algorithm. Compared with traditional methods, this screening method can significantly reduce the current imbalance degree of parallel devices and narrow the difference in switching losses. The imbalance degree of the turn-on loss is 29.7%, while after optimization, the imbalance degree of the turn-on loss is only 7.5%. This paper aims to provide guidelines on device screening for paralleling SiC MOSFETs.

Keywords—Electrical parameters; SiC MOSFET; screen; parallel-connection.

I. INTRODUCTION

Compared with silicon materials, silicon carbide materials (SiC) possess a higher bandgap, a greater breakdown field strength and better thermal stability [1]-[2]. In addition, SiC metal-oxide-field-effect-transistors (MOSFETs) do not have the tail current during the turn off process due to the characteristics of unipolar devices. Therefore, SiC MOSFETs have fast switching speed and lower switching loss compared to Si insualted-gate-bipolar-ransistors (IGBT) with same voltage and current rating,which make it possible to penetrate in various high-voltage, high-frequency and high-power application scenarios [3]-[4].For example, in fields such as inverters in renewable energy power generation systems, power drive systems of electric vehicles.

Due to the relatively low current capacity of single SiC MOSFET chip, the multichip power module with paralleled SiC MOSFETs is widely used in the aforementioned systems. However, the discrepancy in chip parameters of SiC MOSFETs is more severe due to the limitation of manufacturing process. even for SiC MOSFETs of the same batch, parameters such as on-resistance (R_{dson}), threshold voltage (V_{th}) and parasitic capacitance will also have a certain discreteness[5]-[6]. This discreteness will lead to an unbalanced current distribution among various devices when they work in parallel. The R_{dson} affects the on-state current among the devices, while the V_{th} affects the sharing of transient current[7].The phenomenon of unbalanced current distribution will not only reduce the overall efficiency of the parallel system, but also cause overheating due to excessive current flowing through some devices, accelerate the aging

and even damage of the devices, and seriously threaten the reliability and stability of the system.

To alleviate this problem, a variety of device screening strategies have been proposed for application in previous research and engineering practices. Traditional screening methods based on a single parameter range, such as screening operations carried out according to the preset limited ranges of on-resistance or threshold voltage [8], have the advantage that the implementation process is relatively simple and convenient and does not require the support of complex calculation models and algorithms. However, they ignore the interrelationships and synergistic effects among parameters and it is difficult to achieve the ideal current sharing effect under the comprehensive optimization of multiple parameters.In[9], a new classification criterion for multi-device screening is developed based on the distance coefficient of the transfer curve (TC). This classification method can suppress the transient unbalanced current among parallel devices. However, the calculation of the TC distance coefficient is rather complicated and it is difficult to meet the needs of the industry.The clustering screening method views device parameters as multi-dimensional data points and uses clustering algorithms (such as K-means) to classify devices according to parameter similarity. It takes key parameters as multi-dimensional coordinates for grouping [10]. However, the results of clustering algorithms are highly sensitive to the selection of initial cluster centers and the setting of the number of clusters. Different initial settings may lead to completely different clustering results, which in turn will affect the accuracy and stability of screening.

This paper proposes a chip screening strategy based on the simulated annealing algorithm. This screening strategy can effectively reduce the parameter differences among parallel chips and further reduce the imbalance of current and junction temperature. Compared with traditional screening methods, the simulated annealing algorithm can comprehensively consider the complex relationships among multiple device parameters. Through the temperature control and the adjustment of the state transition probability in the simulated annealing process, it gradually finds the screening scheme that can achieve the best current sharing effect for parallel devices.

This paper is organized as follows: In Section II, the influence of V_{th} discrepancy and R_{dson} discrepancy on the current sharing of paralleled SiC MOSFETs are analyzed. The proposed screening method based on the simulated annealing algorithm is proposed in Section III. In Section IV, the feasibility of this screening strategy is verified through experiments. The conclusion are given in Section V.

II. THE INFLUENCE OF DEVICE MISMATCH ON THE CURRENT SHARING OF PARALLELED SIC MOSFETS

During the operation of parallel devices, the current-sharing effect is closely related to device parameters. Among them, the on-resistance and the threshold voltage are the two most critical parameters that affect the parallel current-sharing performance. [11]

A. Device Parameters Test and Hardware Setup

Thirty single-chip SiC MOSFETs from the same batch is selected, and the static characteristics of these 30 devices is tested in detail used a power device analyzer. V_{th} and R_{on} variations of these devices are shown in Fig.1. A parallel double-pulse test platform for single-chip silicon carbide MOSFET devices as shown in Fig.2 is set up.In the simulation and experimental study in this paper,the gate-source voltage bias is 18V and -5 V , the driving resistance is 15Ω, and the bus voltage is 600V.

The MOSFET drain current is measured with a two-stage current measurement method, which includes a ten-turn current transformer at the first stage and a Pearson Current Monitor in the second stage as shown in Fig.2. A Pearson Current Monitor is usually equipped with a wide bandwidth and currents can be accurately measured by it within a wide frequency range.

B. Influence of V_{th} Mismatch on Current Sharing

During the device switching process, the drain current I_d satisfies the following relationship：

$$I_d = g_m(V_{gs} - V_{th}), V_{gs} > V_{th} \qquad (0)$$

where g_m is the transconductance of SiC MOSFET ,and

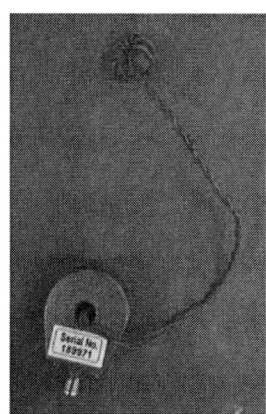

Fig. 2. Double-pulse test platform

$$g_m = \frac{\mu C_{OX} W}{L_{CH}}[V_{gs} - V_{th}] \qquad (2)$$

In the formula, μ represents the effective carrier mobility; C_{OX} is the capacitance per unit area of the gate oxide layer; and W and L_{CH} are the channel width and length, respectively[12].

For the switching state, as can be seen from (1), the drain current is controlled by the gate-source voltage and the threshold voltage. Devices with a smaller threshold voltage will turn on preferentially during the turn-on process and share more current; however, they will turn off with a delay during the turn-off process and share more current. Fig.3 shows the switching transient current with V_{th} mismatch.

C. Influence of R_{dson} Mismatch on Current Sharing

Generally speaking, the on-resistance is regarded as a parameter that plays a key role only in the steady-state current sharing. When the parallel devices are in the conduction stage,

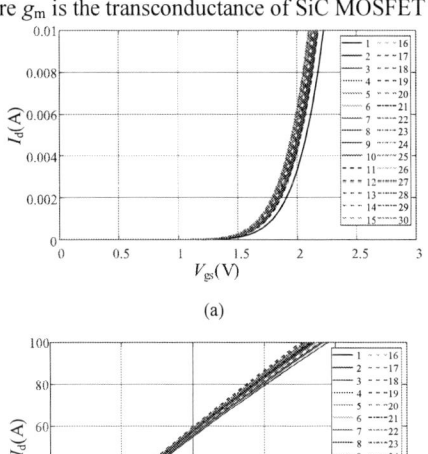

(a)

(b)

Fig. 1. MOSFETs parameters variation. (a) V_{th} variation.(b) R_{dson} variation.

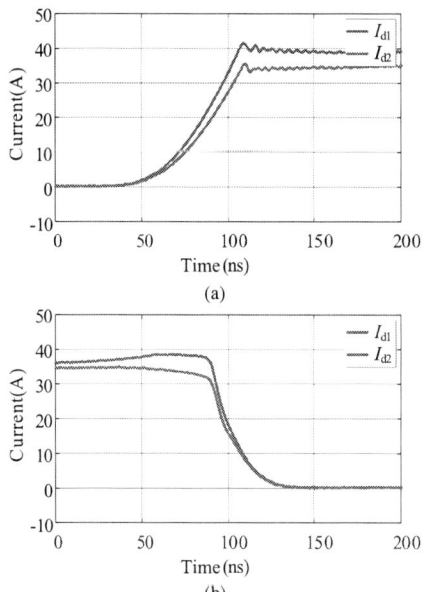

(a)

(b)

Fig. 3. Current sharing performance with V_{th1}=2.4V,V_{th2}=2.6V.(a) Turn-on. (b) Turn-off.

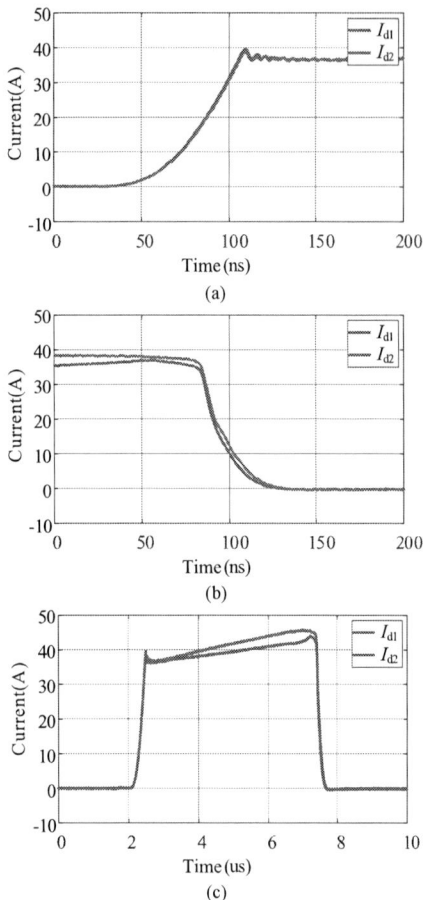

Fig. 4. Current sharing performance with R_{dson1}=18.7mΩ , R_{dson2}=17.3mΩ.(a) Turn-on. (b) Turn-off. (c) On state.

as shown in (3), the on-resistance of each device affects the distribution of the steady-state current[13]. For the conduction state, assuming that the on-resistances of two SiC MOSFETs are R_{dson1} and R_{dson2} respectively. Then, the currents of the two devices can be expressed as:

$$\begin{cases} I_{d1} = \dfrac{R_{dson2}}{R_{dson1} + R_{dson2}} I_L \\ I_{d2} = \dfrac{R_{dson1}}{R_{dson1} + R_{dson2}} I_L \end{cases} \quad (3)$$

In the formula, $I_L=I_{d1}+I_{d2}$ represents the sum of the output currents of the two devices.

It can be obtained from (3) that the current sharing ability of the devices is affected by the discreteness of the on-resistance. The switching transient and conduction state waveforms with differences in on-resistance are shown in Fig.4. From the results in the figure, it can be seen that the differences in the on-resistance among the parallel SiC MOSFETs do not have an impact on the current distribution characteristics during the turn-on and turn-off stages of the devices, but a lower conduction state current is resulted in by a higher on-resistance.

III. SCREENING BASED ON THE SIMULATED ANNEALING ALGORITHM

This paper proposes a SiC MOSFET chip screening strategy based on the Simulated Annealing (SA) algorithm. It ensures that the parameters among parallel chips are as close as possible, thereby reducing the imbalance of current and junction temperature and improving the working stability and performance of the power module.

A. Theoretical foundation

The SA algorithm is rooted in the physical phenomenon of the annealing process of solid substances in statistical mechanics. Under the framework of statistical mechanics, the probability of a system being in a specific state follows the Boltzmann distribution. By analogy in the field of optimization problems, the value of the objective function of an optimization problem can be regarded as the energy in a physical system, and the various feasible solutions in the search space correspond to different states of the system. The SA algorithm explores the optimal solution in the solution space by simulating the characteristic that the system state evolves with the change of temperature during the solid annealing process, and it is used to solve combinatorial optimization problems.

The SA algorithm has asymptotic convergence. Based on the Markov chain theory, if the simulated annealing algorithm is regarded as a Markov chain, under certain conditions (such as the temperature dropping slowly enough and the state transition probability satisfying the detailed balance condition), this Markov chain is ergodic. As the number of iterations tends to infinity, the simulated annealing algorithm will converge to the global optimal solution with a probability of 1. However, in practical applications, due to the limitations of computational resources and time, it is difficult to achieve theoretically infinite iterations and extremely low temperatures. Therefore, appropriate parameters must be chosen to have a sufficiently good approximate optimal solution found within a reasonable time.

B. Parameter Definition

Take the different screening results of chips as the solution space. Each screening result among them represents a feasible solution x. In order to conduct effective searches and solution updates in the solution space, define the neighborhood structure $N(x)$. In this algorithm, the neighborhood structure is defined by exchanging the positions of two random chips.

In order to accurately describe the differences in device parameters, the device parameters involved in the screening process have been standardized to eliminate differences in parameter dimensions and orders of magnitude. The linear normalization formula used in this paper is as follows:

$$x_{nom} = \frac{x - x_{min}}{x_{max} - x_{min}} \quad (4)$$

where x_{min} and x_{max} are the minimum and maximum values of a device parameter among all chips on the wafer, respectively; x is the value of the device parameter for the chip and x_{nom} is the normalized device parameter.

To measure the superiority and inferiority of different screening results, the objective function f is defined as the sum of the variances of the chip parameters within each parallel group. A better screening result is indicated by a small value of the objective function. There are two cases for calculation. When the number of chips n can be evenly divided by the number of parallel chips m, the objective function is defined as (5), where d_{ij} is the device parameter vector of the j_{th} chip in the i_{th} group, and d includes the threshold voltage and on-resistance parameters of the device; when the number of chips cannot be evenly divided by the number of parallel chips m, let $k = mod(n/m)$, then the objective function is defined as (6). Apply the objective function to each solution and calculate its objective function value.

$$f(x) = \sum_{i=1}^{n/m} \sum_{j=1}^{m} \frac{1}{m}(d_{ij} - \overline{d_i})^2 \qquad (5)$$

$$f(x) = \sum_{i=1}^{(n-k)/m} \sum_{j=1}^{m} \frac{1}{m}(d_{ij} - \overline{d_i})^2 \qquad (6)$$

Set the parameters of the simulated annealing algorithm: the initial solution X_0, the initial chip sorting method; the initial temperature T_0, and T_0 should be large enough to ensure that the algorithm can fully explore the solution space in the initial stage and avoid falling into a local optimal solution prematurely; the temperature decay coefficient α ($0<\alpha<1$), which determines the attenuation rate of the temperature with the progress of iterations; the termination temperature T_{min}. When the temperature drops below T_{min}, the algorithm is regarded as converged and terminates its operation; the length L of the Markov chain at each temperature stage, which represents the number of neighborhood search and solution update operations carried out at a fixed temperature.

After the initialization is completed, the iteration begins, divided into an inner loop and an outer loop.

Inner loop: Under each temperature T, start the iterative process of the Markov chain with a length of L. According to the already defined neighborhood structure $N(x)$, randomly select a neighborhood solution xnew from the neighborhood of the current solution x. Calculate the values of the objective function (energy) of the new solution and the current solution $\Delta E = f(xnew)-f(x)$, that is, judge the superiority or inferiority before and after screening, and then judge whether to accept the screening result according to the Metropolis criterion, accept the screening result with probability P. The expression of the Metropolis criterion is shown in (7).Outer loop: When the L iterations of the Markov chain under the current temperature are completed, update the temperature and enter the next temperature stage to continue the iteration.

$$P = \begin{cases} 1, \Delta E < 0 \\ e^{-\frac{\Delta E}{T}}, \Delta E \geq 0 \end{cases} \qquad (7)$$

The algorithm continuously performs outer loop iterations, constantly updates the temperature and executes inner loop operations at each temperature stage until the temperature T drops to be less than or equal to the termination temperature, at which point the algorithm terminates its operation.

At the beginning of the algorithm, a relatively high initial temperature is set, which is similar to the state where a solid is at a high temperature. At this time, the system is in a relatively active state and is allowed to accept inferior solutions, so that it can jump out of the local optimal solution with a relatively high probability. As the algorithm progresses, the temperature gradually decreases, just like the gradual drop in the temperature of a solid during the annealing process.

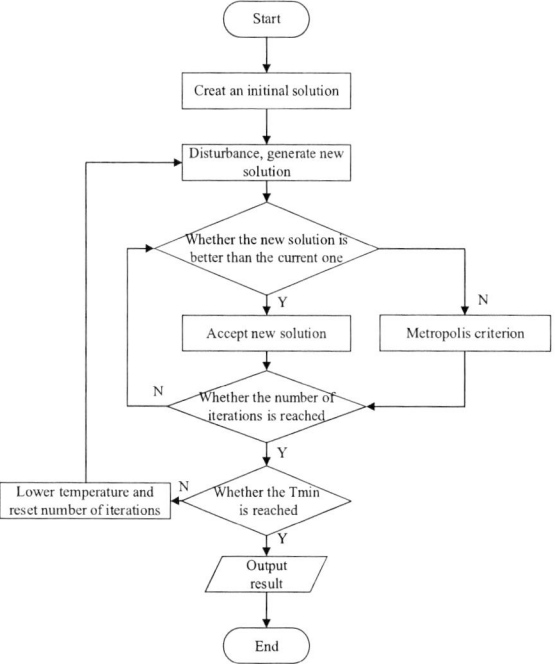

Fig. 5. Flowchart of the Simulated Annealing Algorithm.

Meanwhile, the probability of accepting inferior solutions also gradually decreases, and the system gradually becomes stable. Eventually, an approximate optimal solution is found.

The process flow of the chip screening method based on the SA algorithm is shown in Fig.5.

C. Algorithm verification

In this work, static tests are carried out on 30 discrete SiC MOSFETs of the same batch, and the test results are shown in Fig.6. In this paper, 30 devices are screened, and 28 devices are selected for grouping, with 4 devices in each group. The grouping before and after optimization is shown in Fig.7. The parameters of each group of devices are more dispersed before optimization, while the parameters of each group of devices are similar after optimization.

In terms of measuring the differences among chips, the sum of the variances within the group is selected as the evaluation criterion. This criterion can reflect the degree of dispersion among the individuals within the same group of

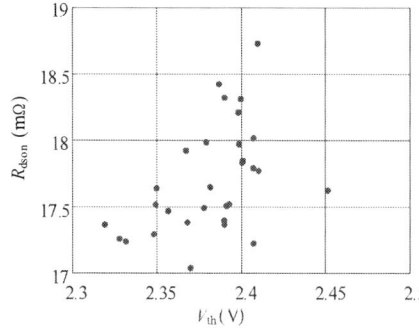

Fig. 6. Distribution of device parameters.

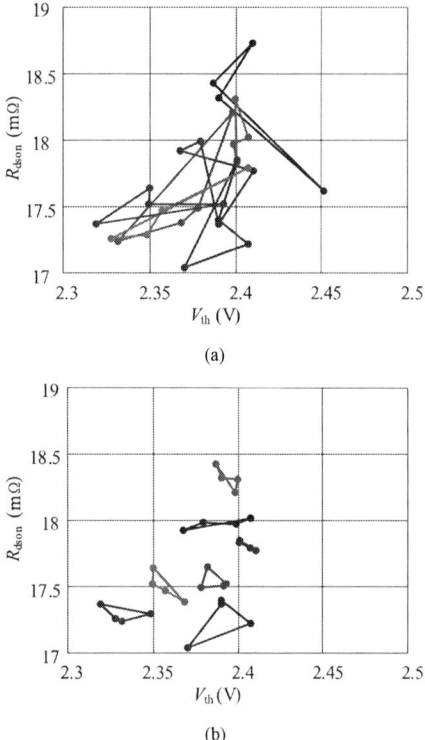

Fig. 7. Distribution of device parameters.(a) Original. (b) Optimized.

TABLE I.   SUM OF VARIANCES BEFORE AND AFTER OPTIMIZATION

Screen method	The sum of variances
Original	0.72
Optimized	0.058

chips as a whole. The larger the sum of the variances is, the greater the differences among the chips will be.Therefore, the sums of the variances before and after optimization are shown in Table I. After optimization by this algorithm, the differences among the parameters of the chip devices in each group are reduced.

IV. EXPERIMENTAL VERIFICATION

In order to verify the effectiveness of this algorithm, on the basis of the above-mentioned test platform, two SiC MOSFETs are packaged in a flip-chip manner to reduce the impact of circuit mismatch and realize the parallel connection of four SiC MOSFETs. Dual-pulse tests are carried out respectively for a group of four devices without screening and a group of four devices after being screened by this algorithm, to conduct a comparative analysis of the current sharing performance of the devices in the dual-pulse tests before and after screening.

The test results of the devices without screening are shown in Fig.8. There are significant differences in both the transient current and the steady-state current. The test results of the devices screened by this algorithm are shown in Fig.9, and both the transient current and the steady-state current are improved.

The losses of the paralleled SiC MOSFETs are calculated respectively, and the results are shown in Table II. In order to evaluate the differences in the losses of the paralleled SiC

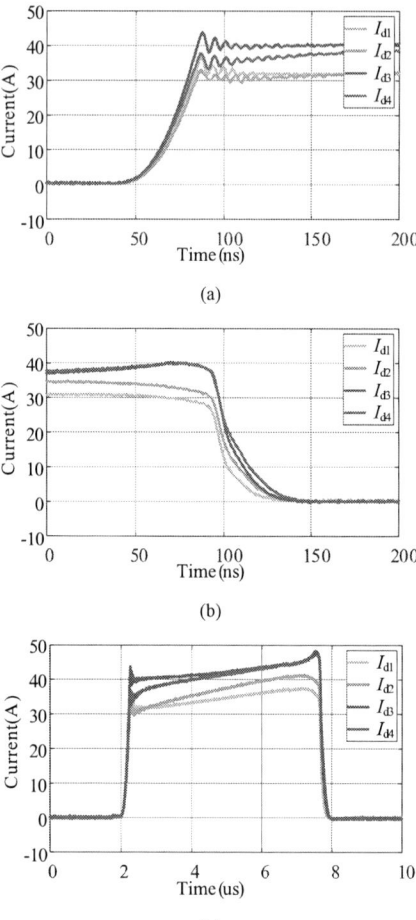

Fig. 8. Waveform of device in original order. (a)Turn-on. (b) Turn-off. (c) On state.

MOSFETs, the device loss imbalance degree E_U defined as shown in (8) is introduced.

$$E_{\mathrm{U}} = \frac{E_{\max} - E_{\min}}{E_{\max}} \qquad (8)$$

Before optimization, the imbalance degree of the turn-on loss is 29.7%, while after optimization, the imbalance degree of the turn-on loss is only 7.5%. Before optimization, the imbalance degree of the sum loss is 24.3%, while after optimization, the imbalance degree of the sum loss is only 7.2%.

This result confirms the feasibility of the screening by this algorithm. It can effectively reduce the current imbalance degree of the parallel-connected devices and significantly decrease the loss differences among the parallel devices. The

TABLE II.   UM OF VARIANCES BEFORE AND AFTER OPTIMIZATION

		1	2	3	4
Loss before optimization /mJ	Turn-on	2.911	2.621	3.726	3.153
	Turn-off	2.401	2.956	3.291	3.445
	Sum	5.312	5.577	7.017	6.598
Loss after optimization /mJ	Turn-on	3.445	3.386	3.183	3.158
	Turn-off	2.943	2.9562	2.746	2.815
	Sum	6.388	6.3422	5.929	5.973

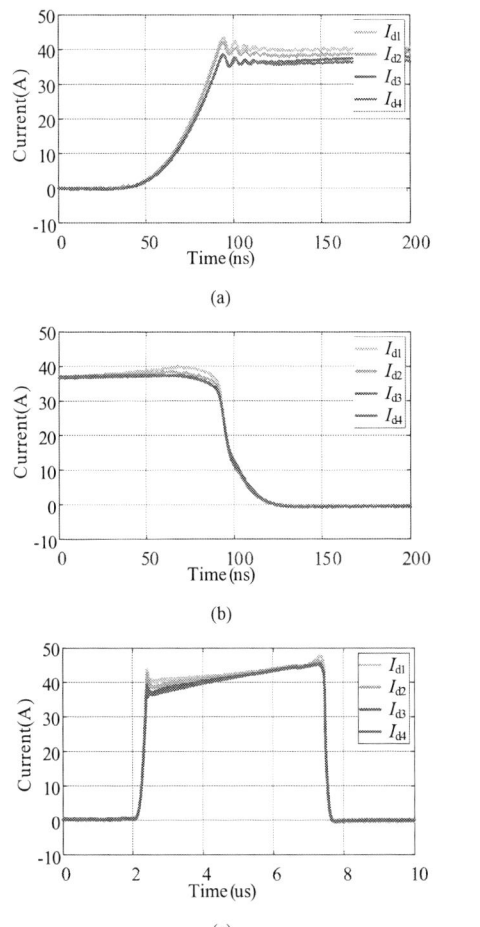

(a)

(b)

(c)

Fig. 9. Waveform of device after optimization. (a)Turn-on. (b) Turn-off. (c) On state.

thermal stress and electrical stress borne by each device are more uniform, thus reducing the risk of individual devices failing prematurely due to overheating.

V. CONCLUSION

This research conducts an in-depth exploration into the problem of uneven current sharing in parallel-connected SiC MOSFETs. Firstly, the impact of chip parameter mismatch is analyzed. The on-resistance dominates the steady-state current distribution, while the threshold voltage affects the switching transient current. Subsequently, a screening strategy based on the simulated annealing algorithm is proposed. This algorithm can significantly reduce the parameter differences among the parallel-connected devices. Finally, through the parallel dual-pulse tests on four devices, it is effectively verified that this algorithm can effectively reduce the uneven current sharing degree of the parallel-connected devices.

REFERENCES

[1] K. Sheng, Q. Guo, J. Zhang et al, "Development and Prospect of SiC Power Devices in Power Grid," Proceedings of the CSEE, 2012, 32(30): 1-7.

[2] J. Millán, P. Godignon, X. Perpiñà, A. Pérez-Tomás and J. Rebollo, "A Survey of Wide Bandgap Power Semiconductor Devices," in IEEE Transactions on Power Electronics, vol. 29, no. 5, pp. 2155-2163, May 2014.

[3] S. Jahdi, O. Alatise, J. A. Ortiz Gonzalez, R. Bonyadi, L. Ran and P. Mawby, "Temperature and Switching Rate Dependence of Crosstalk in Si-IGBT and SiC Power Modules," in IEEE Transactions on Industrial Electronics, vol. 63, no. 2, pp. 849-863, Feb. 2016.

[4] X. She, A. Q. Huang and R. Burgos, "Review of Solid-State Transformer Technologies and Their Application in Power Distribution Systems," in IEEE Journal of Emerging and Selected Topics in Power Electronics, vol. 1, no. 3, pp. 186-198, Sept. 2013.

[5] D. Peftitsis, R. Baburske, J. Rabkowski, J. Lutz, G. Tolstoy and H. -P. Nee, "Challenges Regarding Parallel Connection of SiC JFETs," in IEEE Transactions on Power Electronics, vol. 28, no. 3, pp. 1449-1463, March 2013.

[6] H. Li, S. Zhao, X. Wang, L. Ding and H. A. Mantooth, "Parallel Connection of Silicon Carbide MOSFETs—Challenges, Mechanism, and Solutions," in IEEE Transactions on Power Electronics, vol. 38, no. 8, pp. 9731-9749, Aug. 2023.

[7] H. Li et al., "Influences of Device and Circuit Mismatches on Paralleling Silicon Carbide MOSFETs," in IEEE Transactions on Power Electronics, vol. 31, no. 1, pp. 621-634, Jan. 2016, doi: 10.1109/TPEL.2015.2408054.

[8] M. Riccio et al., "Analysis of Device and Circuit Parameters Variability in SiC MOSFETs-Based Multichip Power Module," 2018 20th European Conference on Power Electronics and Applications (EPE'18 ECCE Europe), Riga, Latvia, 2018, pp. P.1-P.9.

[9] J. Ke, Z. Zhao, P. Sun, H. Huang, J. Abuogo, and X. Cui, "Chips Classification for Suppressing Transient Current Imbalance of Parallel-Connected Silicon Carbide MOSFETs," IEEE Trans. Power Electron.,vol. 35, no. 4, pp. 3963-3972, Apr 2020.

[10] Y. Liu et al., "A New Screening Method for Alleviating Transient Current Imbalance of Paralleled SiC MOSFETs," 2020 IEEE 1st China International Youth Conference on Electrical Engineering (CIYCEE), Wuhan, China, 2020, pp. 1-6.

[11] H. Li, S. Munk-Nielsen, S. Bęczkowski and X. Wang, "A Novel DBC Layout for Current Imbalance Mitigation in SiC MOSFET Multichip Power Modules," in IEEE Transactions on Power Electronics, vol. 31, no. 12, pp. 8042-8045, Dec. 2016, doi: 10.1109/TPEL.2016.2562030.

[12] Z. Ceng, W. Shao, B. Hu et al, "Active Current Sharing of Paralleled SiC MOSFETs by Coupling Inductors," Proceedings of the CSEE, 2017, 37(07): 2068-2081.

[13] P.Sun,Z.Zhao,Y.Cai et al, "Chip Screening for Parallel Silicon Carbide MOSFET Based on Switching Energy Balancing," Proceedings of the CSEE, 2019, 39(19): 2068-2081.

Research on Power Loss Calculation and Optimization of FB-VSCC Active EMI Filter

Daozhen He
School of Electrical Engineering
Beijing Jiaotong University
Beijing, China
22121452@bjtu.edu.cn

Hong Li*
College of Electrical Engineering
Zhejiang University
Hangzhou, China
hong_li@zju.edu.cn

Yuanheng He
School of Electrical Engineering
Beijing Jiaotong University
Beijing, China
21291072@bjtu.edu.cn

Mingxin Shi
School of Electrical Engineering
Beijing Jiaotong University
Beijing, China
24121327@bjtu.edu.cn

Abstract—**The use of wide-bandgap semiconductor devices, with their high switching speed and efficiency, has gained widespread application in power electronics. However, their application also exacerbates electromagnetic interference (EMI) issues, posing challenges to traditional suppression strategies and creating an urgent need for advanced solutions to enhance system electromagnetic compatibility (EMC). Although passive EMI filters are widely used in engineering, their design for miniaturization and lightweighting under high power density conditions still faces certain limitations. This has spurred in-depth research into active EMI filters (AEFs). AEFs feature various topologies and can achieve interference suppression within specific frequency ranges with a compact size. However, existing research often overlooks the power supply and power loss issues that limit AEF effectiveness. To address this, this paper takes the commonly used feedback voltage-sensing current-compensation (FB-VSCC) AEF as an example. The analysis first identifies the sources of power loss within the AEF and proposes a detailed method for calculating AEF power loss. Based on this, the relationship between power loss and EMI attenuation is explored. Optimization strategies for power loss, aimed at balancing insertion loss with system efficiency, are then proposed. Finally, the effectiveness of the optimization method was verified through simulation.**

Keywords—*active EMI filter, power loss, voltage-sensing current- compensation*

I. INTRODUCTION

High-speed switching operations in power semiconductor devices produce severe electromagnetic interference (EMI) noise, adversely affecting surrounding electronic equipment and potentially causing malfunctions. With the wide application of power converters in electric vehicles, industrial automation and consumer electronics, the requirements for their performance, efficiency and reliability continue to increase, and EMI problems become more and more prominent, becoming a key factor that cannot be ignored in the design. Therefore, in order to ensure the normal operation of the equipment and meet the relevant electromagnetic compatibility standards, techniques such as filtering, optimized layout and soft switching are often used to reduce EMI [1]. In general, installing passive EMI filters (PEFs) is the most convenient and effective method in engineering, but with the development trend of high power density, traditional

PEF faces certain challenges in the miniaturization and lightweight design of the converter. In contrast, active EMI filters (AEFs), with its advantages of small size and light weight, have gradually attracted wide attention from researchers and shown good application prospects [2].

AEF is mainly composed of noise sensing, noise amplification and noise compensation, and is divided into six basic topologies according to different noise sensing and compensation methods. Among them, feedforward AEF has more stringent requirements on circuit gain and component components, which makes it difficult to achieve the ideal EMI suppression effect [3]. The feedback (FB) AEF is designed to be more flexible, mainly using two typical topologies of voltage-sensing current-compensation (VSCC) and current-sensing current-compensation (CSCC) [4-8]. However, these two AEF have their own advantages and disadvantages. FB-CSCC AEF relies on current transformer for noise sensing, which may lead to volume increase and efficiency decrease in high power density systems. FB-VSCC AEF uses only capacitors for noise sensing and compensation. Although the structure design is the most compact, its stability is obviously inferior to that of FB-CSCC AEF[9].

At present, remarkable progress has been made in the research of AEF, mainly in the aspects of topology optimization, modeling and stability analysis. Besides, Texas Instruments has launched FB-VSCC AEF products and successfully applied them in AC-DC converters to improve the power density of the system [10]. However, the supply and power losses that limit AEF performance and applications are rarely discussed. Most studies only show the filtering effect of AEF, and only a few analyze and quantify the power loss of AEF [4,8,11-15]. But in fact, optimizing the power loss of AEF is of great significance, which can improve system efficiency, reduce heat generation, extend equipment life and alleviate heat dissipation problems.

This paper aims to analyze the source of AEF power loss and its influence on the overall performance by calculating the AEF power loss, and put forward the corresponding optimization strategy. Firstly, taking the FB-VSCC AEF in [16] as an example, the calculation method of AEF power loss was introduced in detail, and then the relationship between power loss and EMI attenuation effect was discussed. Finally, the optimization method of AEF power loss was verified through simulation.

This work was supported in part by the National Science Fund for Distinguished Young Scholars of China under Grant 52325704, and in part by the Shunyi Innovation Joint Fund of Beijing Natural Science Foundation under Grant L247004.

II. POWER LOSS CALCULATION AND OPTIMIZATION METHOD OF ACTIVE EMI FILTER

A. Loss calculation method

AEF generally uses passive components such as inductors or capacitors to sample noise signals, which are amplified by operational amplifiers or class AB push-pull power amplifiers and re-injected into the main circuit to cancel each other with the original noise signal to reduce noise flowing through the EMI victim end. According to the components and working principle of AEF, the loss of AEF mainly includes the power loss to maintain the static operating point of the amplifier circuit, the power loss of the cancellation current generated by the amplifier circuit, and the power loss caused by the resistance used for noise sensing and compensation [2,15].

The first two types of losses are the power dissipation generated by the AEF amplification part, while the third type stems from energy consumption in the sensing and compensation part of the AEF. Since the inductance and capacitance components in this part do not consume energy, the main energy loss occurs in the resistive elements. Moreover, the voltage and current across these resistive components are relatively small, so this power loss can be considered negligible [4]. This paper takes the FB-VSCC AEF proposed in [16] as shown in Fig. 1 as an example to analyze the power loss of AEF.

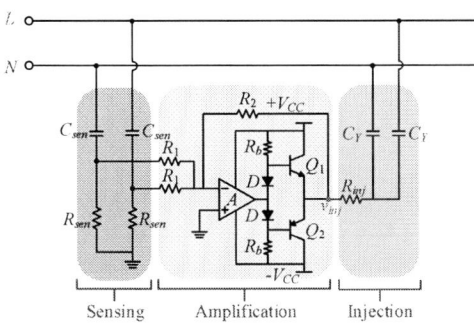

Fig. 1. Topology of FB-VSCC AEF [16].

Firstly, for the static power loss of AEF, the detailed calculation process of the static current of the operational amplifier was derived in [17]. Due to its complexity, this paper selects (1) for approximate estimation. Among them, V_{CC} is the voltage of the AEF power supply, and I_q is the quiescent current of the operational amplifier. Their typical values can be found in the relevant datasheet. The typical static current value of the AD826 operational amplifier selected in this paper is 6.6mA at ±5V and 6.8mA at ±15V.

$$P_q = V_{CC} \cdot \left(I_q + I_{BQ}\right) \tag{1}$$

For the quiescent current I_{BQ} of a transistor, since the two transistors in the class AB push-pull amplification circuit are complementary and symmetrical, analyzing half of them is sufficient. Based on the common collector amplification circuit shown in Fig.1, the quiescent current of the transistor can be calculated according to (2). Here, β represents the current amplification factor of the transistor, and R_b is the bias resistor of the transistor, which is used to set the appropriate static operating point of the transistor. V_{BEQ} is the PN junction voltage between the base and emitter of a transistor, and it is usually 0.7V for a silicon transistor.

$$I_{BQ} = \frac{V_{CC} - V_{BEQ}}{R_b + \left(1 + \beta\right)R_e} \tag{2}$$

Regarding the calculation method of power loss that generates cancellation current, there are mainly two analysis perspectives: the AEF itself and the perspective of the filtered system.

From the perspective of AEF, class AB push-pull power amplifiers are often added because amplifier circuits using only operational amplifiers are limited in current injection capacity. The output current of this power amplifier directly affects the ability of AEF to counteract EMI. In addition, class AB amplifiers have a large output power, which means that their power consumption to the power supply is correspondingly high, accounting for a major part of the AEF loss. Assuming that V_{CC} is the voltage of the AEF power supply and I_{inj} is the average injection current generated by the class AB push-pull circuit, the power loss caused by the cancellation current is shown as follows [12].

$$P_{cancel} = V_{CC} \cdot I_{inj} \tag{3}$$

From the perspective of the filtered system, the attenuation of the EMI spectrum measured before and after adding AEF can reflect the magnitude of the cancellation current of AEF. As shown in Fig. 2, the variation of voltage in the EMI spectrum is converted into the variation of current, then calculate the power required by each frequency point, and finally perform superposition to obtain the total power, as shown below [8,11].

$$P_{cancel} = \sum_{f=kf_s} V_{CC} \triangle i_L(f) \tag{4}$$

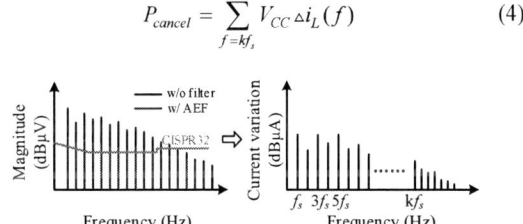

Fig. 2. The cancelling current spectrum of the converter with AEF.

Here, $\triangle i_L$ represents the change in current, fs represents the switching frequency, and f is the multiple of the switching frequency. For sensing and compensation circuit in the AEF the power loss caused by the resistance element, can be calculated by measuring the voltage of the each resistance, however, this part of the loss is relatively small, only accounts for what percentage of total power dissipation AEF so negligible.

In summary, the total power loss expression of AEF is as follows.

$$P_{AEF} = P_q + P_{cancel} \tag{5}$$

B. Loss optimization method

According to the current distribution relationship of the BJT, the average injection current generated by the class AB push-pull amplifier circuit is：

$$I_{inj} = I_e = \left(1 + \beta\right)I_b \tag{6}$$

Where I_b is the base current, which is equal to the static current I_{BQ} of the transistor in (2).

Substituting (6) into (3) can obtain the following expression.

$$P_{cancel} = (1+\beta) \cdot \frac{V_{CC}(V_{CC}-V_{BE})}{R_b} \quad (7)$$

Then the total power loss of the AEF is shown as (8).

$$P_{AEF} = V_{CC}I_q + \frac{V_{CC}(V_{CC}-V_{BE})(2+\beta)}{R_b} \quad (8)$$

From the above expression, it can be seen that once the models of the operational amplifier and the transistor are determined, the power loss of the AEF is decided by the supply voltage V_{CC} and the bias resistor R_b. Specifically, the supply voltage V_{CC} is positively correlated with the power loss of AEF. When the supply voltage V_{CC} rises, both the injection current and the power loss increase accordingly. On the contrary, the bias resistance R_b is negatively correlated with the power loss. As the bias resistance increases, the injection current and power loss decrease instead.

However, the magnitude of the injected current not only affects the power loss of the AEF, but also limits its insertion loss. In fact, insertion loss can be regarded as a form of power loss, which reflects the power loss of the noise signal after passing through the EMI filter. The schematic diagram of the insertion loss calculation for AEF is shown in Fig. 3. Since the noise source impedance Z_n is much larger than the load impedance Z_{LISN}, the current I_{Zn} flowing to the noise source impedance in Fig. 3 can be ignored.

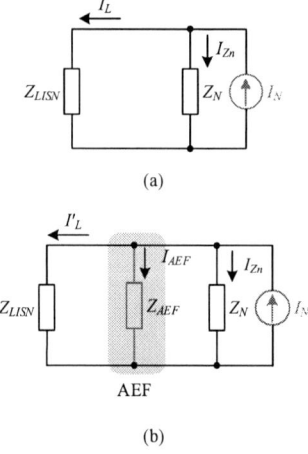

(a)

AEF

(b)

Fig. 3. EMI filter insertion loss diagram. (a) Before inserting the filter. (b) After inserting the filter.

Thus, the insertion loss of AEF can be expressed as:

$$\begin{cases} IL = 20\lg\frac{I_L}{I'_L} = 20\lg\left(\frac{I_N}{I_N - I_{AEF}}\right) \\ I_{AEF} \approx I_{inj} = (1+\beta)\frac{V_{CC}-V_{BE}}{R_b} \end{cases} \quad (9)$$

It can be known from (9) that the supply voltage V_{CC} and the bias resistor R_b are also related to the insertion loss. Therefore, it is necessary to analyze the influence of the above two parameters on the insertion loss of AEF, so as to achieve

a reasonable balance between filtering performance and power loss in the design of circuit parameters.

Fig. 4 shows the variation of the AEF insertion loss under different supply voltages and bias resistors. It can be seen from Fig. 4(a) that with the increase of the supply voltage, the insertion loss in the high-frequency band (5-30MHz) gradually increases. This indicates that increasing the supply voltage enhances the filtering ability of the filter to a certain extent. This might stem from the fact that the higher supply voltage increases the output voltage and current of the amplifier circuit, and at the same time, the unit gain bandwidth of the operational amplifier is relatively high, thereby enhancing the ability to suppress high-frequency interference. By observing Fig. 4(b), it can be found that as the bias resistor Rb decreases from 10kΩ to 1kΩ, the insertion loss also shows a certain increase in the high-frequency region, and its principle of action is the same as increasing the supply voltage.

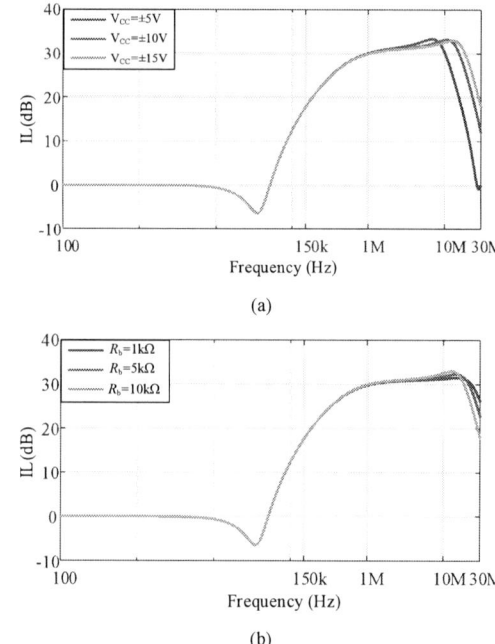

(a)

(b)

Fig. 4. Insertion loss of AEF at different supply voltages and bias resistors. (a) Different power supply voltages. (b) Different bias resistors.

In conclusion, the supply voltage and bias resistance have significant influences on insertion loss and power loss. By reasonably selecting the supply voltage and bias resistor, an optimized balance between power consumption and the overall filtering effect can be achieved during the design process.

Therefore, the optimized design method for AEF power loss is as follows: First, the goal must be clearly defined, that is, to minimize the power loss of AEF to the greatest extent under the premise of meeting EMI standards. During the design stage, amplifiers with low quiescent current and low power consumption should be given priority, and the required circuit gain and bandwidth can be achieved by adjusting the bias resistance value. During this process, combined with simulation analysis and measured results, the bias resistance value that not only meets the filtering requirements but also minimizes power loss is found. Finally, the lowest operating voltage range should be selected based on the filtering target to reduce unnecessary power loss. The entire optimization process needs to go through multiple simulation verifications and experimental tests, constantly adjusting parameters, and finally achieving the optimal design scheme. Next, the above analysis will be verified through simulation.

III. SIMULATION VERIFICATION

The EMI simulation circuit of buck converter is built based on Ltspice software platform, and the AEF shown in Fig. 1 is applied to the converter. The EMI suppression effect of AEF under different power supply voltages is shown in Fig. 5, and the corresponding power loss calculation is shown in Tab. I. It can be seen from the result that when the power supply voltage is smaller, the power loss is smaller, but the EMI suppression effect is also weakened. Therefore, under the premise of meeting the EMI standard, the power loss can be reduced by appropriately reducing the power supply voltage.

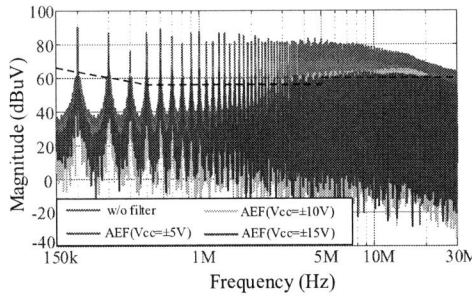

Fig. 5. EMI spectrum of buck converter with AEF at different supply voltages.

TABLE I. OUTPUT CURRENT AND POWER LOSS OF AEF AT DIFFERENT SUPPLY VOLTAGES

supply voltage V_{CC} /V	output current I_{inj} /mA	power loss P_{AEF} /mW
5	7.63	71.15
10	13.00	198.00
15	16.62	351.30

The EMI suppression effect of AEF under different bias resistors is shown in Fig. 6, and the corresponding power loss calculation is shown in Tab. II. As can be seen from the result, when the bias resistance is smaller, the injection current is larger, and the power loss is greater, but the EMI suppression effect is better. Therefore, when the EMI attenuation requirement is met, the power loss can be reduced by appropriately increasing the bias resistance. In summary, the above simulation results are consistent with the theoretical analysis in Section II.

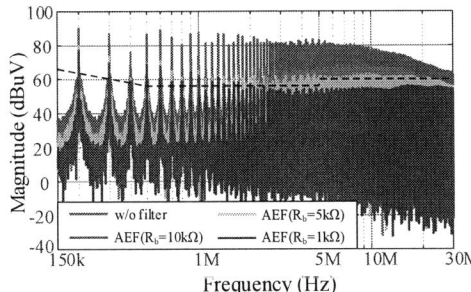

Fig. 6. EMI spectrum of buck converter with AEF with different bias resistors.

TABLE II. OUTPUT CURRENT AND POWER LOSS OF AEF WITH DIFFERENT BIAS RESISTORS

bias resistance R_b /kΩ	output current I_{inj} /mA	power loss P_{AEF} /mW
1	36.77	653.55
5	22.20	435.00
10	16.62	351.30

IV. CONCLUSION

This paper presents a comprehensive analysis of power loss in AEF and explores optimization methods to enhance their performance. The loss calculation method reveals that the primary contributor to AEF power loss is the current required to counteract EMI, significantly influenced by the design of class AB push-pull amplifier. Simulation results demonstrate that reducing the supply voltage can effectively decrease power loss while maintaining acceptable EMI suppression levels. Additionally, adjusting the bias resistance shows a trade-off between power loss and EMI attenuation, where higher resistance can lead to lower losses without compromising performance. Overall, the findings provide valuable insights for optimizing AEF design to achieve a balance between efficiency and EMI mitigation.

REFERENCES

[1] K. Mainali and R. Oruganti, "Conducted EMI Mitigation Techniques for Switch-Mode Power Converters: A Survey," in IEEE Transactions on Power Electronics, vol. 25, no. 9, pp. 2344-2356, Sept. 2010.

[2] B. Narayanasamy and F. Luo, "A Survey of Active EMI Filters for Conducted EMI Noise Reduction in Power Electronic Converters," in IEEE Transactions on Electromagnetic Compatibility, vol. 61, no. 6, pp. 2040-2049, Dec. 2019.

[3] W. Chen, W. Zhang, X. Yang, Z. Sheng and Z. Wang, "An Experimental Study of Common- and Differential-Mode Active EMI Filter Compensation Characteristics," in IEEE Transactions on Electromagnetic Compatibility, vol. 51, no. 3, pp. 683-691, Aug. 2009.

[4] D. Shin, S. Jeong and J. Kim, "Quantified Design Guidelines of a Compact Transformerless Active EMI Filter for Performance, Stability, and High Voltage Immunity," IEEE Trans. Power Electron., vol. 33, no. 8, pp. 6723-6737, Aug. 2018.

[5] K. Zhang, K. -W. Wang and H. S. -H. Chung, "High-Attenuation Wideband Active Common-Mode EMI Filter Section," in IEEE Transactions on Power Electronics, vol. 37, no. 5, pp. 5479-5490, May 2022.

[6] H. Li, C. Zhang, Y. Ding, Z. Yang and B. Zhang. "Optimization of High Frequency Noise Suppression Effect of Active EMI Filter Based on Chaotic Spread Spectrum PWM Method," Proceedings of the CSEE,2022,42(13):4642-4651.

[7] R. Goswami and S. Wang, "Modeling and Stability Analysis of Active Differential-Mode EMI Filters for AC/DC Power Converters," in IEEE Transactions on Power Electronics, vol. 33, no. 12, pp. 10277-10291, Dec. 2018.

[8] B. Narayanasamy, H. Peng, Z. Yuan, A. I. Emon and F. Luo, "Modeling and Analysis of a Differential Mode Active EMI Filter With an Analog Twin Circuit," in IEEE Transactions on Electromagnetic Compatibility, vol. 62, no. 4, pp. 1591-1600, Aug. 2020.

[9] D. Shin, S. Jeong, Y. Baek, C. Park, G. Park and J. Kim, "A Balanced Feedforward Current-Sense Current-Compensation Active EMI Filter for Common-Mode Noise Reduction," in IEEE Transactions on Electromagnetic Compatibility, vol. 62, no. 2, pp. 386-397, April 2020.

[10] How Active EMI Filter ICs Mitigate Common-Mode Emissions and Increase Power Density in Single- and Three-Phase Power Systems, Texas Instruments, Mar. 2023. [Online]. Available: https://www.ti.com/lit/pdf/slvafj9.

[11] R. Goswami, S. Wang, E. Solodovnik and K. J. Karimi, "Differential Mode Active EMI Filter Design for a Boost Power Factor Correction AC/DC Converter," in IEEE Journal of Emerging and Selected Topics in Power Electronics, vol. 7, no. 1, pp. 576-590, March 2019.

[12] S. Wang, Y. Y. Maillet, F. Wang, D. Boroyevich and R. Burgos, "Investigation of Hybrid EMI Filters for Common-Mode EMI

Suppression in a Motor Drive System," IEEE Trans. Power Electron., vol. 25, no. 4, pp. 1034-1045, April 2010.

[13] M. C. Di Piazza, A. Ragusa and G. Vitale, "Design of Grid-Side Electromagnetic Interference Filters in AC Motor Drives with Motor-Side Common Mode Active Compensation," in IEEE Transactions on Electromagnetic Compatibility, vol. 51, no. 3, pp. 673-682, Aug. 2009.

[14] M. C. Di Piazza, A. Ragusa and G. Vitale, "Power-Loss Evaluation in CM Active EMI Filters for Bearing Current Suppression," in IEEE Transactions on Industrial Electronics, vol. 58, no. 11, pp. 5142-5153, Nov. 2011.

[15] Yu Zhang, "Research on the mechanism of noise filtering and power estimation in AEF power supply," M.S. thesis, Dept. Elect. Eng., Xi'an University of technology, China, 2023.

[16] H. Li, D. He, S. Wang and Z. Zhao, "Active X-Y Capacitors Based Hybrid EMI Filter Design," 2024 2nd China Power Supply Society Electromagnetic Compatibility Conference (CPEMC), Hangzhou, China, 2024, pp. 338-342.

[17] J. Ji, X. He, J. Lu. "A study of power consumption prediction models for active EMI filters power supplies. 2023 1st China Power Supply Society Electromagnetic Compatibility Conference (CPEMC)," 2023.

Analysis of Common-mode EMI in Multi-converter Systems with EMI filters

Runquan Jiang
State Key Laboratory of High Density Electrical Energy Conversion
Huazhong University of Science and Technology
Wuhan, China
M202372276@hust.edu.cn

Peng Zhou
State Key Laboratory of High Density Electrical Energy Conversion
Huazhong University of Science and Technology
Wuhan, China
zhou_p@hust.edu.cn

Guifeng Geng
State Key Laboratory of High Density Electrical Energy Conversion
Huazhong University of Science and Technology
Wuhan, China
M202372291@hust.edu.cn

Xuejun Pei
State Key Laboratory of High Density Electrical Energy Conversion
Huazhong University of Science and Technology
Wuhan, China
ppei215@hust.edu.cn

Abstract—The DC power system with multiple converters is widely used in multi-electric aircraft, which complicates the analysis of conducted electromagnetic interference (EMI). Although EMI filters are currently the most effective method for suppressing conducted EMI, the interaction influence among multiple EMI filters has not been studied yet, seriously threatening the electromagnetic safety of multi-converter systems. Firstly, based on the behavioral model, this paper proposes a general analysis method for common-mode (CM) EMI in the multi-converter system. Then, based on this method, the influence of the interaction between EMI filters on the suppression performance of multi-converter system containing distributed EMI filters is analyzed. It is found that the EMI filter with high cut-off frequency will weaken the performance of other filters with low cut-off frequency under some frequencies. Finally, the behavior frequency-domain model is compared with the time-domain model in simulation, and shows good consistency.

Keywords—Common-mode system noise prediction, electromagnetic interference (EMI), behavioral model (BM), muti-converter system, EMI filter.

I. INTRODUCTION

To achieve sustainable development in aviation, the electrification of future aircraft becomes imminent. which makes a variety of electronic converters with different topologies and power levels appear inside the aircraft[1]. Fig.1 shows the power system of a more electric aircraft. The alternator is connected to the DC bus through the power electronic rectifier, and supplies power to different loads through different power converters[2]. The DC power supply system of the aircraft brings more complex conducted electromagnetic interference issues. Even when individual power electronic converters meet electromagnetic compatibility (EMC) standards, the assembled system may still experience excessive conducted EMI[3]. This may cause severe interference to sensitive equipment on the aircraft, seriously threatening the safe and reliable operation of the

aircraft. Therefore, it is urgent to study the conduction mechanism of EMI and its suppression in DC power system.

Currently, there is no comprehensive theoretical analysis method for conducted EMI in DC power systems. Existing research mainly focuses on modeling and predicting the EMI of individual power electronic converters[4], using primarily time-domain and frequency-domain methods. The time-domain method is easy to implement but not convenient for analysis[5]. The frequency-domain method includes the detailed model and the behavioral model (BM). The white box model focuses on the generation and propagation of noise inside the converter[6-7], which requires the detailed structure of the converter and the high frequency parameters of the circuit. Therefore, it lacks universality. In [3], the high-frequency interactive resonance in a DC power system composed of two identical converters is studied based on their detailed model. However, in more-electric aircraft, the topologies and numbers of converters are not fixed[8]. Using detailed model to study EMI paths is difficult and lacks generality. In contrast, the behavioral model uses an impedance and source to model a converter[9]. It has demonstrated good accuracy in studies that do not concentrate on the internal characteristics of the converter and is expected to be used in EMI modeling of aircraft DC power supply system.

Passive EMI filters are widely used in different electronic converters as the most effective method to suppress conducted EMI[10]. In a DC power system, different EMI filters are connected on the same DC bus, as shown in Fig.2 (a), whose mutual influence mechanism has not been explained clearly. In [11], the insertion loss of filters in a system consisting of three identical converters and their EMI filters is studied. It finds that filters with the same parameters will cause the resonance point of the low frequency to shift, resulting in the total EMI exceeding the standard when connected together. However, the interaction between EMI filters with different parameters has not been studied further.

Recently, centralized EMI filter has been proposed to replace the traditional distributed EMI filter in DC power systems, as shown in Fig.2 (b). In [12], the volume of centralized EMI filter is analyzed and has more advantages

---

This work was supported in part by the National Natural Science Foundation of China under Grant 52377187, in part supported by the China Postdoctoral Science Foundation under Grant Number 2024M761010, and in part of the stably supported project of the State Key Laboratory of High Density Electrical Energy Conversion under Grant Number DN2024-02. The authors are with the State Key Laboratory of High Density Electrical Energy Conversion.

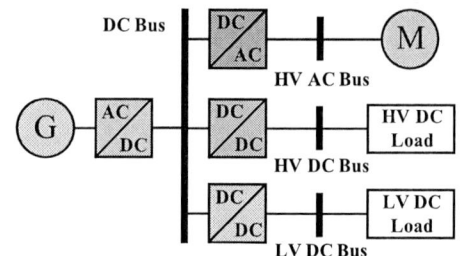

Fig. 1 Multi-electric aircraft's electrical system

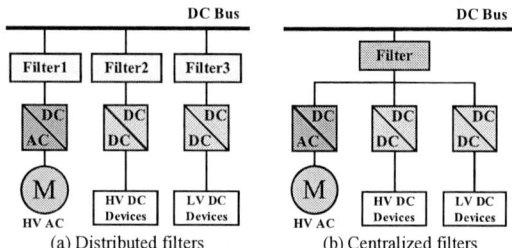

(a) Distributed filters   (b) Centralized filters

Fig. 2. Distributed and centralized EMI filters

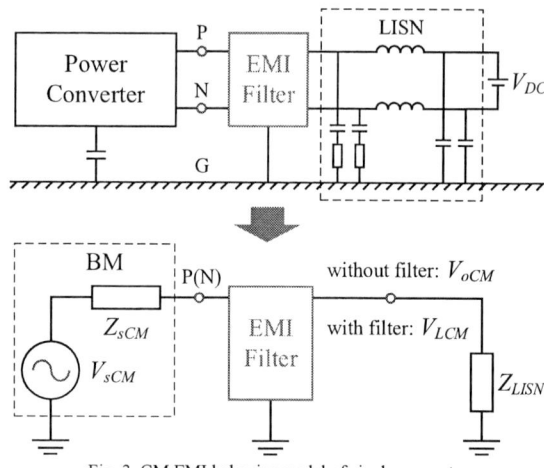

Fig. 3. CM EMI behavior model of single converter

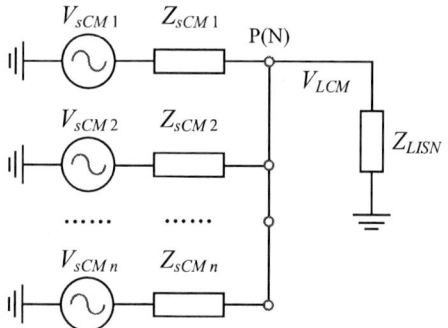

Fig. 4. CM EMI behavior model of multi-converter system

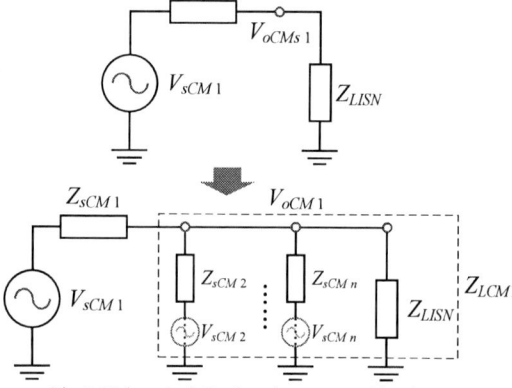

Fig. 5. Noise calculation based on superposition theorem

than that of traditional distributed filters. However, there is a lack of a systematic noise path modeling basis as the theoretical support for filter design.

Therefore, this paper studies the CM EMI of the DC power supply systems with three types of converters. Firstly, a CM frequency-domain model of the multi-converter system is established based on the behavioral model. The concept of parallel loss is introduced to analyze the mutual influence of noise among the multi-converter systems. Then, a simulation model of a DC multi-converter system containing three converters is built, and the accuracy of this behavioral model is verified. Finally, based on the behavioral model, the CM EMI of the system after adding the distributed CM EMI filter is analyzed. The interactions among the EMI filters in the system are analyzed and compared with the time-domain simulation.

II. EMI BEHAVIOR MODEL OF MULTI-CONVERTER SYSTEM WITH DIFFERENT CONVERTERS

A. EMI behavior model of single converter

The EMI behavior frequency-domain model of power electronic converter focuses on its external characteristics. According to Thevenin's theorem, power electronic equipment is equivalent to an active port network. The equivalent CM source impedance Z_{sCM} is used to represent the conducted CM impedance characteristics of the converter, while the equivalent CM interference source V_{sCM} is used to represent its CM interference characteristics. Therefore, a group (V_{sCM}, Z_{sCM}) can model the CM interference emission characteristics of a converter. Basing on BM, Fig. 3 shows the conducted CM noise modeling process of a power converter that includes an external impedance (e.g. EMI filter) and a linear impedance stabilization network (LISN), where the external impedance is not necessary.

In the BM, Z_{sCM} can be obtained with an impedance measurement tool, while V_{sCM} cannot be measured directly. It can be calculated through formula (1) by measuring the CM interference (V_{oCM}) on the LISN without EMI filter.

$$V_{sCM} = V_{oCM} \frac{Z_{LISN} + Z_{sCM}}{Z_{LISN}} \quad (1)$$

V_{LCM} is the CM EMI on the LISN after adding an EMI filter. Then, considering the source impedance and load impedance of the converter, the CM insertion loss IL_{CM} of the filter can be calculated by formula (2).

$$IL_{CM} = 20\log(\frac{V_{LCM}}{V_{oCM}}) \quad (2)$$

979-8-3315-1110-4/25 $31.00 © 2025 IEEE   679

B. EMI behavior model of multi-converter system

The frequency-domain BM are extended from a single converter to the DC power supply system that composed of multiple converters. Fig. 4 shows a model of a power supply system with n converters connected to the same DC bus based on BM, where n groups (V_{sCMi}, Z_{sCMi}) are connected in parallel to the LISN.

Then the superposition theorem is applied to predict the CM EMI V_{LCM} on the LISN of the converter system. Each interference source is analyzed separately. Taking V_{sCM1} as an example, when considering the effect of V_{sCM1}, other voltage interference sources (V_{sCM2} to V_{sCMn}) are short. Therefore, the load changes from LISN to the impedance Z_{LCM1} which is parallel of LISN, Z_{sCM2}, ..., Z_{sCMn}, as shown in Fig. 5. The calculation formula of Z_{LCM1} is (3).

$$Z_{LCM1} = Z_{LISN} \parallel Z_{sCM2} \parallel ... \parallel Z_{sCMn} \quad (3)$$

V_{oCMs1} is the CM EMI on the LISN generated by V_{sCM1} in the single-converter system, and V_{oCM1} is the CM EMI on the LISN generated by V_{sCM1} in the multi-converter system. V_{oCMs1} and V_{oCM1} can be calculated by formula (4) and (5).

$$V_{oCMs1} = V_{sCM1} \frac{Z_{LISN}}{Z_{LSIN} + Z_{sCM1}} \quad (4)$$

$$V_{oCM1} = V_{sCM1} \frac{Z_{LCM1}}{Z_{LCM1} + Z_{sCM1}} \quad (5)$$

It can be seen from formula (4) and (5) that when considering the effect of V_{sCM1} acting alone in multi-converter system, Z_{sCM2}, ..., Z_{sCMn} provide additional paths for the CM EMI, changing the load impedance form LISN to Z_{LCM1}. Furthermore, it changes the voltage division between Z_{sCM1} and the load impedance. According to the superposition theorem, the total CM EMI of the multi-converter system V_{oCM} is the sum of the interference V_{oCMi} generated by each source V_{sCMi}, and it can be calculated by formula (6).

$$V_{oCM} = \sum_{i=1}^{n} V_{oCMi} = \sum_{i=1}^{n} V_{sCMi} \frac{Z_{LCMi}}{Z_{LCMi} + Z_{sCMi}} \quad (6)$$

C. Influence of CM EMI between multiple converters

The CM EMI flows between multiple converters through their source impedance and mutually influences each other. In order to measure the impact of the addition of other converters on EMI, following the definition of insertion loss IL of EMI filters, the ratio of the CM EMI generated by a single source in a multi-converter system V_{oCMi} and an independent system V_{oCMsi} is defined as parallel loss PL. PL can be calculated by formula (7), which reflects the influence of the source impedance of different converters on the impedance of LISN.

$$PL_i = 20\log(\frac{V_{oCMi}}{V_{oCMsi}}) = 20\log(\frac{Z_{LCMi}}{Z_{LISN}} \frac{Z_{LISN} + Z_{sCMi}}{Z_{LCMi} + Z_{sCMi}}) \quad (7)$$

III. A SIMULATION OF A DC POWER SUPPLY SYSTEM CONTAINING THREE CONVERTERS

In this section, a simulation of a DC power supply system with three different converters was constructed to demonstrate the mutual influence of noise among them.

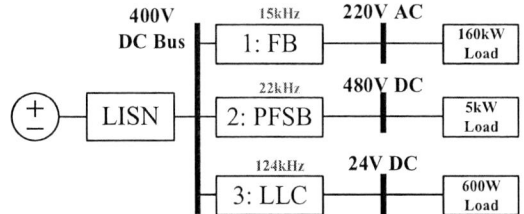

Fig. 6. Simulation structure of the multi-converter system

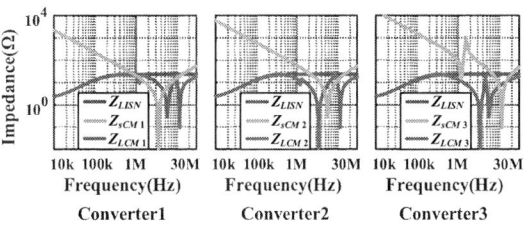

Fig. 7. Comparison of impedance

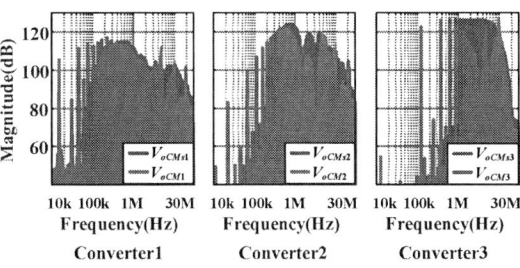

(a) CM EMI of each converter in single- converter and multi-converter system

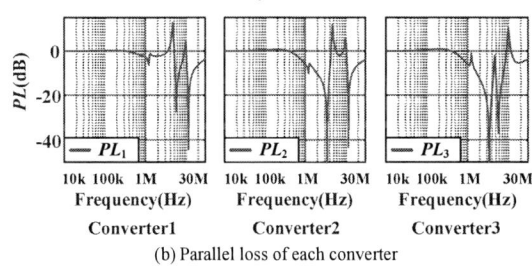

(b) Parallel loss of each converter

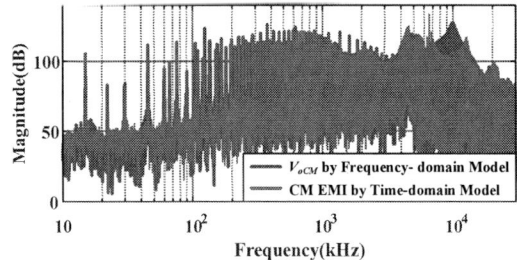

(c) Results of frequency-domain and time-domain model

Fig. 8. Comparison of interference

A. The simulation structure of the system

Fig. 6 is a simplified schematic diagram of a 400V DC power supply system containing three power converters. It is composed of a full-bridge converter, a phase-shifted full-bridge converter, and an LLC converter. The output voltages and load powers of each converter are shown in the figure. The switching frequencies of the three converters are 15kHz, 22kHz and 124kHz respectively. The simulation was designed and implemented in MATLAB 2024, which allows for the estimation and processing of the input EMI of three converters and the combined EMI of the multi-converter system. In the simulation, the CM source impedance and the CM interference source of the three converters are extracted, and the frequency-domain BM of the system is built.

B. Comparison of impedance

Fig. 7 shows the comparison of the CM source impedance Z_{sCMi} and load impedance of each converter in the independent converter (Z_{LISN}) and multi-converter systems (Z_{LCMi}). It can be seen that the CM source impedance of each converter is capacitive at low frequencies and turns to inductive at high frequencies. The CM load impedance of the independent converter Z_{LISN} is half of the standard LISN impedance, approximately 25Ω. In the multi-converter system, the CM load impedance of each converter Z_{LCMi} is the parallel connection of the LISN and the source impedance Z_{sCMi} of the other two converters. At low frequencies, the value of capacitive impedance is much greater than that of Z_{LISN}, so the parallel load impedance Z_{LCMi} after the three are connected in parallel is dominated by LISN. However, at high frequencies, the impedance of capacitive sources drops sharply with frequency, causing significant changes in the impedance of parallel loads Z_{LCMi}. It can be seen that Z_{LCMi} is less than Z_{LISN} over 1MHz.

C. Comparison of CM EMI

Fig. 8 (a) shows the comparison of CM EMI generated by each converter on the LISN in the independent system V_{oCMsi} and the multi-converter system V_{oCMi}. It can be seen that due to the change in load impedance Z_{LCMi}, V_{oCMi} increases at some frequencies compared to V_{oCMi} but decreases in most frequency bands. Fig. 8 (b) shows the PL of each converter. The part greater than 0dB is the noise amplification caused by the interactive resonance between the converters. Meanwhile, the part less than 0dB is that the source impedance improves the Z_{LISN}, causing the noise that originally flowed into the LISN to flow into other converters. Fig. 8 (c) shows the comparison of CM EMI calculated by the formula (6) and that in the time-domain simulation. It can be seen that BM of the multi-converter system can predict the CM EMI well. The introduction of PL visually presents the mutual influence of the noise flow paths between different converters.

IV. EMI BEHAVIOR MODEL OF MULTI-CONVERTER SYSTEM WITH DISTRIBUTED EMI FILTERS

According to the design process of EMI filter, EMI filters are designed respectively for the three converters in Fig. 6, whose topologies are all CL filters. The parameters of the three EMI filters are shown in Table 1. After adding EMI filters to the three independent converters, the noise generated by the converters will be attenuated according to the insertion loss IL of the EMI filters after considering the source and load impedance.

However, in a multi-converter system, as analyzed in Section II, when considering the influence of the EMI filter of one converter, the other EMI filters will also affect the load impedance in the form of PL. For example, Fig. 9 is the CM BM of the multi-converter system after adding the CM EMI filters in Fig. 6. When considering the effect of V_{sCM1}, filters 2 and 3 provide extra flow paths for the CM EMI to the ground.

Table 1. Table Type Styles

Filter	Parameters of C and L		Cut-off frequency f_r
1	C_1=4.4uF	L_1=200uH	5.3kHz
2	C_2=2uF	L_2=4uH	56.3kHz
3	C_3=0.4uF	L_3=1500uH	6.5kHz

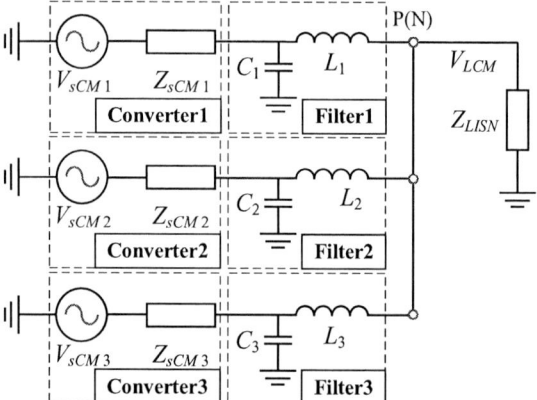

Fig. 9. The BM of multi-converter system with distributed EMI filters

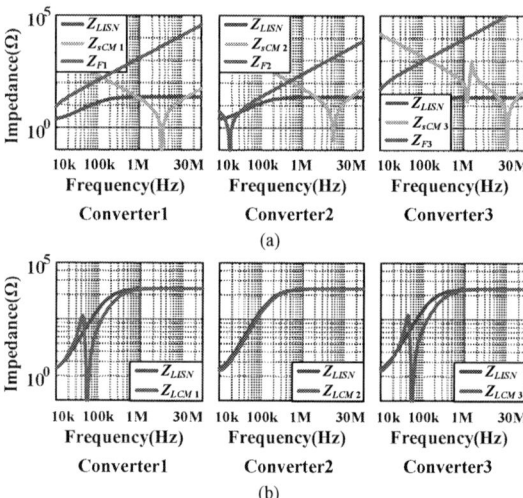

Fig. 10. Comparison of impedance in multi-converter system

This means that the insertion loss IL of EMI filter 1 cannot be used as the evaluation criterion for its attenuation ability of the CM EMI of converter 1.

A. Analysis of PL caused by other filters

In Fig.9, from P towards the interference source, the impedance Z_i of each converter with EMI filters is calculated by formula (8). Since the CM source impedance of the converter is generally the ground capacitance of the switching device, which is much smaller than the CM capacitance in filters, Z_i is approximately equal to the impedance of the filter Z_{Fi} and can be further simplified into formula (9).

$$Z_i = Z_{Li} + Z_{Ci} \parallel Z_{sCMi} \tag{8}$$

$$Z_i \approx Z_{Li} + Z_{Ci} = Z_{Fi} \tag{9}$$

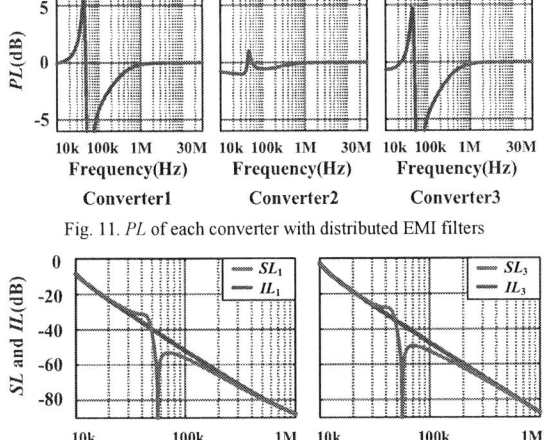

Fig. 11. *PL* of each converter with distributed EMI filters

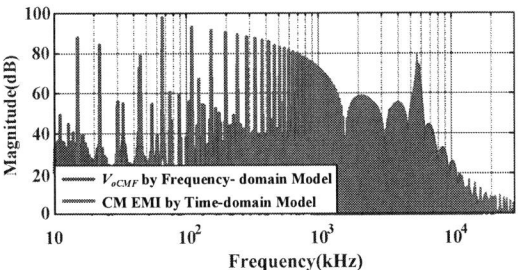

Fig. 12. *SL* and *PL* of filter 1 and 3

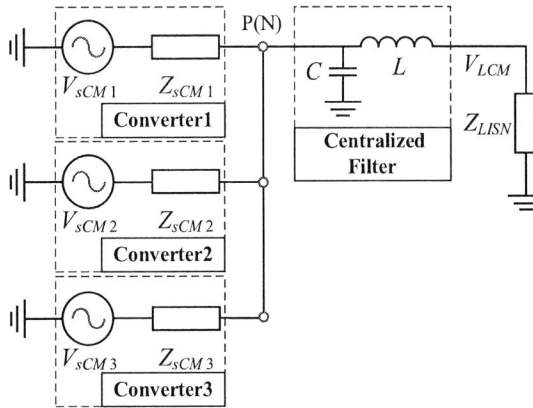

Fig. 14 The BM of multi-converter system with centralized EMI filter

Fig.13. Comparison of CM EMI in the time-frequency domain after adding EMI filters

Therefore, after adding the distributed filters, the parallel load impedance Z_{LCM1} changes from formula (3) to (10), and the *PL* of each converter changes from formula (7) to (11).

$$Z_{LCM1} = Z_{LISN} \| Z_{f2} \| Z_{f3} \qquad (10)$$

$$PLF_i = 20\log(\frac{Z_{LCMi}}{Z_{LISN}} \frac{Z_{LISN} + Z_{fi}}{Z_{LCMi} \| Z_{fi}}) \qquad (11)$$

The comparison of Z_{Fi} and Z_{sCMi} is shown in Fig. 10(a). Since the cut-off frequency of the CL filter f_r is generally set low to obtain a higher insertion loss at high frequencies. Therefore, f_r is generally lower than the switching frequency of the converter. Z_{Fi} is capacitive in the frequency band before f_r and inductive after f_r, which is much greater than the CM impedance of LISN. However, around f_r, the value of ZFi becomes very small due to the resonance between C and L, which is of the same order of magnitude as Z_{LISN}. However, according to Section II, when Z_{Fi} and Z_{LISN} are similar, the parallel load impedance Z_{LCMi} will change significantly, as shown in Fig. 10 (b).

Therefore, different from the situation of each converter without filters in Section II, affected by other EMI filters, the *PL* of each converter are mainly concentrated at low frequencies and almost zero at high frequencies, as shown in Fig. 11. The third branch is taken as an example to explain this phenomenon with the flow path of CM EMI: At low frequencies, the impedance value of the inductor is low. C_1 and C_2 can be regarded as directly connected to P (N). Together with filter 3, they form a Π-type filter which has better suppression than the CL filter, so the performance of filter 3 is optimized. Therefore, the parallel loss of converter 3 PF_3 is less than 0dB, as shown in Fig. 11. At high frequencies, the impedance of the inductor is very large, so C_1 and C_2 can hardly provide any additional interference flow paths. Therefore, the PF_3 at high frequencies is 0dB. Near the cut-off frequency of filter 2 f_{r2}, as the impedance of filter 2 switches from capacitive to inductive, PF_1 reaches its maximum and minimum values respectively on the left and right sides of f_{r2}. As shown in Fig. 11, PF_1 and PF_3 reach around 5 dB at f_{r2}, which means that in the multi-converter system, filters 1 and 3 will lose 5dB suppression effect on the interference noise.

B. SL analysis of multi-converter systems

In order to measure the noise suppression ability of filter i in the system for the converter i, formula (12) defines the sum noise loss SL_i on branch i. SL_i is the sum of the insertion loss IL_i of the EMI filter i and the PL_i caused by the parallel connection of EMI filters of other converters and LISN. Fig.12 shows the comparison of IL and SL of filter 1 and 3. It can be seen that in converters 1 and 3, due to the influence of filter 2, the total loss at 40-50 KHZ is about 5dB higher than the insertion loss. This may result in EMI filters 1 and 3 designed in accordance with the standard process being unable to reduce the CM EMI of converters 1 and 3 below the standard limit in the multi-converter system, even if they pass the EMC standard in the single converter system. This will bring extremely unstable factors to the entire power supply system.

$$SL_i = PL_i + IL_i \qquad (12)$$

Meanwhile, the *SL* of each filter can be applied to calculate the CM EMI generated by the multi-converter system after adding the distributed EMI filter. Formula (6) can be further simplified into Formula (13). Fig. 13 is the comparison of the CM EMI calculated based on formula (13) and the time-domain simulation of the system that contains EMI filters. It can be seen that the results in the time and frequency model are highly consistent, which means that this method can predict the total EMI of the system after adding the distributed EMI filter well.

$$V_{oCM} = \sum_{i=1}^{n} V_{sCMi} \cdot 10^{\frac{SL_i}{20}} \qquad (13)$$

C. Limitations of distributed EMI filters

In the multi-converter system, distributed EMI filters provide each other with additional flow paths to the ground, causing a significant change in SL caused by the filter with a high cut-off frequency f_{rh}. Generally, the frequency band that conducts the most severe EMI in a power converter is the first few harmonic points f_{kn} of its switching frequency[13]. Therefore, when f_{rh} and f_{kn} are close, it may cause the noise at this frequency to exceed the EMC standard, which brings more uncontrollable factors to the design of distributed EMI filters in the system.

However, centralized EMI filters combine multiple distributed EMI filters into one filter, avoiding the interactive influence among them. Fig. 14 is a diagram of the centralized EMI filter of this multi-converter system. According to the superposition theorem, taking V_{sCM1} as an example. When V_{sCM1} acts alone, the equivalent capacitance C_e formed by the parallel connection of the source impedances Z_{sCM2} and Z_{sCM2} of other converters and the capacitance C of the centralized filter can be calculated by formula (14). Since the equivalent capacitance of the CM source impedance is generally much smaller than the CM capacitance of the filter, it can be ignored. It can be seen that the large capacitor C in the centralized EMI filter blocks the interaction between the converters. Therefore, when designing centralized EMI filters, the interaction influence between the converter and its EMI filters can be avoided. Moreover, the redundancy of passive components can also be avoided and improve the power density of the system. The specific design method for centralized filters is not introduced in detail due to the space limit.

$$C_e = C \parallel Z_{sCM2} \parallel Z_{sCM3} \approx C \qquad (14)$$

V. CONCLUSION

Firstly, a frequency-domain modeling method for CM EMI of multi-converters system based on the behavioral model is proposed, and the concept of parallel loss is introduced to measure the interaction influence of noise between different converters. Then, a simulation model of a DC multi-converter system containing three different converters was built to verify the accuracy of the behavioral model. On this basis, the influence of the interaction between distributed EMI filters on the CM EMI suppression of the system is analyzed. It is found that EMI filters with high cut-off frequencies will weaken the performance of certain frequency bands of filters with low cut-off frequencies. Then,

the behavioral frequency-domain model with distributed EMI filters was compared with the time-domain simulation to verify its accuracy. Finally, the advantages of centralized EMI filters in multi-converter systems are briefly expounded.

REFERENCES

[1] B. Sarlioglu and C. T. Morris, "More Electric Aircraft: Review, Challenges, and Opportunities for Commercial Transport Aircraft," in IEEE Transactions on Transportation Electrification, vol. 1, no. 1, pp. 54-64, June 2015.

[2] D. Wang et al., "Multilevel Inverters for Electric Aircraft Applications: Current Status and Future Trends," in IEEE Transactions on Transportation Electrification, vol. 10, no. 2, pp. 3258-3282, June 2024.

[3] P. Zhou, X. Pei, Q. Chen, Y. Zhang and H. Fan, "EMI Behavioral Model Based CM Noise Prediction Method for DC Power System Considering Multi-Noise Coupling," in IEEE Transactions on Power Electronics, vol. 38, no. 4, pp. 4658-4667, April 2023.

[4] P. Zhou, X. Pei, K. Zhang and Y. Shan, "Improved EMI Behavioral Modeling Method of Three-Phase Inverter Based on the Noise-Source Phase Alignment," in IEEE Transactions on Power Electronics, vol. 37, no. 8, pp. 9333-9344, Aug. 2022.

[5] L. Ran, S. Gokani, J. Clare, K. Bradley, and C. Christopoulos, "Conducted electromagnetic emissions in induction motor drive systems part I: Time domain analysis and identification of dominant modes," IEEE Trans. Power Electron., vol. 13, no. 4, pp. 757-767, Jul. 1998.

[6] L. Ran, S. Gokani, J. Clare, K. J. Bradley, and C. Christopoulos, "Conducted electromagnetic emissions in induction motor drive systems. II. Frequency domain models," IEEE Trans. Power Electron., vol. 13, no. 4, pp. 768-776, Jul. 1998.

[7] M. Jin and M. Weiming, "Power converter EMI analysis including IGBT nonlinear switching transient model," IEEE Trans. Ind. Electron., vol. 53, no. 5, pp. 1577-1583, Oct. 2006.

[8] B. J. Brelje and J. R. R. A. Martins, "Electric, hybrid, and turboelectric fixed-wing aircraft: A review of concepts, models, and design approaches," Prog. Aerosp. Sci., vol. 104, pp. 1-19, Jan. 2019.

[9] F.-Y. Shih, D. Y. Chen, Y.-P. Wu, and Y.-T. Chen, "A procedure for designing EMI filters for AC line applications," IEEE Trans. Power Electron., vol. 11, no. 1, pp. 170-181, Jan. 1996.

[10] S. Gulur, V. M. Iyer and S. Bhattacharya, "Passive CM Filter Configuration for a Multistage Grid-Tied Solid State Transformer," in IEEE Journal of Emerging and Selected Topics in Industrial Electronics, vol. 4, no. 3, pp. 710-717, July 2023, doi: 10.1109/JESTIE.2023.3268623.

[11] Michio Tamate, Akio Toba, K. Wada and T. Shimizu, "Analysis of EMI filter attenuation characteristics for a multi-converter system," 2009 IEEE 6th International Power Electronics and Motion Control Conference, Wuhan, China, 2009, pp. 957-962, doi: 10.1109/IPEMC.2009.5157522.

[12] L. Malburg, N. Moonen and F. Leferink, "Analysis of Multi-Filter EMI Mitigation for Weight and Volume Optimization," 2023 International Symposium on Electromagnetic Compatibility - EMC Europe, Krakow, Poland, 2023, pp. 1-6, doi: 10.1109/EMCEurope57790.2023.10274223.

[13] X. Zhao et al., "Planar Common-Mode EMI Filter Design and Optimization for High-Altitude 100-kW SiC Inverter/Rectifier System," in IEEE Journal of Emerging and Selected Topics in Power Electronics, vol. 10, no. 5, pp. 5290-5303, Oct. 2022.

Research on key technology of low noise transformer and its application on 110kV transformer

Shouhui Han
School of Electrical Engineering
Shandong University
Jinan, China
hanshoushui@mail.sdu.edu.cn

Qingsong Liu
CSG EHV Power Transmission
Company
Electric Power Research Institute
Guangzhou, China
595083554@qq.com

Zheng Liu
School of Electrical Engineering
Shandong University
Jinan, China
1253145632@qq.com

Liang Zou
School of Electrical Engineering
Shandong University
Jinan, China
zouliang@sdu.edu.cn

Zhiyun Han
School of Electrical Engineering
Shandong University
Jinan, China
hanzhiyun@sdu.edu.cn

Abstract—**Aiming at the problem of 110kV transformer operation noise exceeding the standard, this study systematically analyzes the noise mechanism under the action of electromagnetic-mechanical-acoustic multi-physical field coupling, and puts forward a multi-objective optimization design method based on finite element simulation. By constructing a coupled electromagnetic-mechanical-acoustic field model, the nonlinear influence of rated capacity, magnetic flux density and other parameters on the noise is revealed, which breaks through the limitations of the traditional single physical field simulation. The composite noise reduction structure of double-layer magnetic shielding sandwiched by corrugated paperboard is innovatively proposed, combining the dual mechanisms of sound absorption and vibration isolation, and the simulation verifies that its noise reduction effect is significantly better than that of the traditional single-layer magnetic shielding scheme. The study further quantifies the threshold values of passive noise reduction measures such as clamping force optimization (0.11-0.13 MPa is the best range) and core material replacement, and amends the applicable boundaries of engineering empirical formulas, which provides theoretical support and engineering guidance for the standardized design of low-noise transformers.**

Keywords—*Low-noise transformer, Multi-physical field coupling model, Magnetostrictive effects, Passive noise reduction technique, Vibration isolation, Correction of engineering equations*

I. INTRODUCTION

As the core equipment of the power grid system, power transformers play a key role in the transmission and distribution of power. The noise generated during its operation not only affects the reliability of the equipment itself (such as accelerating the aging of insulation materials, resulting in partial discharge), but also causes significant interference with the living environment of the surrounding residents.[1] According to the International Electrotechnical Commission (IEC) standards, 110kV transformer operation noise needs to be controlled at 55 dB (A) or less, while the actual engineering often due to electromagnetic-mechanical coupling effect of complexity, design parameter redundancy and other issues lead to noise exceeding the standard.[2] Traditional research focuses on a single means of noise reduction (e.g. acoustic enclosure, sound-absorbing materials) or a single physical field (electromagnetic field, mechanical vibration) analysis, and lacks a systematic exploration of the synergistic effect of multiple parameters.[3] In addition, the existing engineering empirical formulas are mostly based on simplified assumptions, which are difficult to adapt to the dynamic characteristics of the new transformer structure.[4] Therefore, it is of great theoretical and practical significance to carry out in-depth analysis of the noise mechanism and multi-objective optimization design of 110kV transformers.

II. RESEARCH STATUS

Existing research on the 110kV transformer noise and core density, magnetostriction and other core parameters of the dynamic correlation law lack of systematic quantitative analysis, especially for the multi-physical field coupling under the vibration noise propagation mechanism has not yet established reproducible mathematical models.[5] The simulation results are compared and verified with the results calculated by the engineering formula to verify whether the engineering formula of the core intrinsic frequency is in accordance with the engineering reality.[6]

Current passive noise reduction measures (such as clamping force optimization, positioning structure improvement) mostly rely on empirical design, the lack of quantitative assessment of the noise reduction effect of the system.[7] The industry urgently needs to explore both theoretical support and engineering feasibility of new noise reduction programs to meet the increasingly stringent environmental standards on the transformer acoustic performance challenges.

The design of low-noise transformers has not yet formed a complete technical closed loop from simulation modeling, parameter optimization to effect verification, which restricts the standardization and improvement of product performance.[8] By constructing a multi-physical field coupling model of electromagnetic-mechanical-acoustic field, this study systematically reveals the nonlinear influence of rated capacity and other operating parameters on noise, which will provide the industry with a generalizable simulation analysis paradigm.

In recent years, with the improvement of environmental protection standards and the densification of urban power grids, the transformer noise pollution problem has received increasing attention. The current research mainly focuses on the optimization of passive noise reduction technology, multi-physical field coupling modeling and new materials and structural design of three major directions.[9]

In the traditional noise reduction methods, the improvement of the vibration characteristics of the iron core is the core direction.[10] Research shows that the core magnetostrictive effect is one of the main sources of noise, through the optimization of the core material (such as the use of amorphous alloy or high permeability silicon steel) can reduce the hysteresis loss and vibration amplitude; at the same time, the clamping force regulation and optimization of the positioning structure can inhibit the mechanical resonance of the iron core and the windings, the effect of noise reduction up to 3-5dB. In addition, the vibration isolation design of the tank structure (such as elastic shims, damping coatings) has also been widely proven to reduce the noise propagation. can reduce noise propagation.[11]

With the development of numerical simulation technology, the finite element-based electromagnetic-mechanical-acoustic multi-physical field coupling model has gradually become a mainstream research tool.[12] By establishing a coupled model of transformer vibration and noise, scholars have analyzed the nonlinear effects of rated capacity, magnetic flux density and other parameters on noise, and revealed the positive correlation between magnetostriction and noise intensity. However, the existing models mostly simplify the boundary conditions, and the adaptability to complex working conditions still needs to be improved.

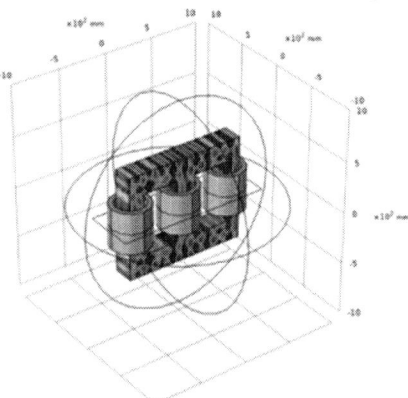

Fig 1. Electromagnetic-mechanical-acoustic multi-physics field coupling modeling

In recent years, hybrid noise reduction schemes have become a new trend, e.g., combining active noise cancellation (ANC) with passive structural optimization, or introducing new composite materials (e.g., magnetic-acoustic bifunctional shielding). Some studies have attempted to design the core shape or the internal acoustic structure of the fuel tank through topology optimization, but engineering applications still face the challenges of cost and reliability.

III. MODELING AND DATA ANALYSIS

The core and winding parts, which need to be analyzed, are dissected with refined free tetrahedral meshes t

o ensure their accuracy in the solving process, while the external air domain is dissected with conventional meshes to improve the solving efficiency and ensure the solving accuracy at the same time. A total of 48759 cells are dissected.

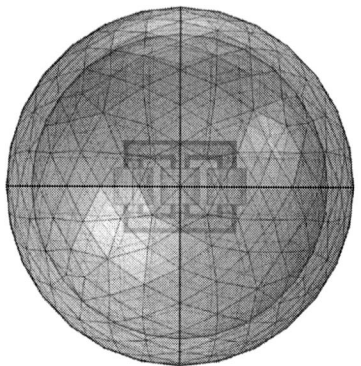

Fig 2. Solution Domain Mesh Segmentation Results

Take the scaling multiplier as the horizontal coordinate and the characteristic frequency as the vertical coordinate to draw the line graph of the frequency obtained from the formula calculation and the frequency obtained from the simulation as shown in Figs. 3-4. It can be seen that, under the premise of the core width to maintain a certain, the core intrinsic frequency and its height of the relationship between the trend in line with the trend of the engineering formula, are inversely proportional to the relationship.

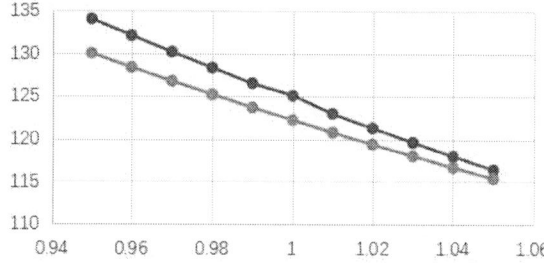

Fig 3. Variation curves of simulated and calculated values of core eigenfrequency at different lengths

According to the transformer vibration noise generation mechanism and simulation results analysis, power transformer passive noise reduction can be roughly divided into two ideas: the first idea is to reduce the vibration of the core, windings and cooling system in the process of manufacturing and assembling the transformer; the second idea is to isolate the vibration by vibration occurring within the transformer, the transformer tank and the fixtures of the connecting jie, and in the specific space area of the placement of transformer Sound insulation or anechoic treatment to reduce noise pollution.

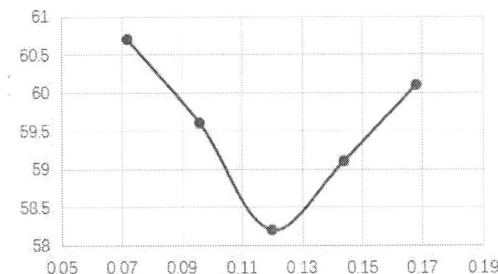

Fig 4. Variation curve of sound pressure level under different core clamping force

IV. RESEARCH CONTENT

A. coupled multi-physics field modeling methods

Innovative model construction:

Based on the finite element simulation platform, a multi-physical field coupling model of electromagnetic-mechanical-acoustic field is constructed, which realizes the whole process of dynamic analysis from electromagnetic excitation to mechanical vibration, and then to acoustic field propagation. The model breaks through the limitations of traditional single-field simulation and significantly improves the accuracy of transformer noise prediction.[13]

Coupling process optimization:

A step-by-step coupling strategy (electromagnetic→mechanical→acoustic field) is proposed to solve the computational efficiency problem of complex physical field interactions through boundary acceleration transfer and pressure load feedback mechanisms, which provides a methodological reference for the multidisciplinary joint simulation of similar devices.[14]

B. noise reduction technology programs

New structural design:

A composite noise reduction measure of double-layer magnetic shielding sandwiched by wavy cardboard is proposed to be installed on the inner wall of the fuel tank. Through the multi-level structural design (magnetic shielding + wavy cardboard), combined with the dual mechanism of sound absorption and vibration isolation, effectively reduce the noise transmission, simulation verification of its noise reduction effect is significantly better than the traditional single-layer magnetic shielding program.

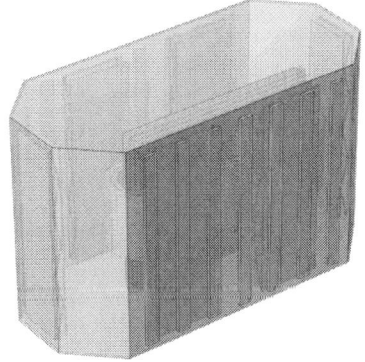

Fig 5. Double-layer magnetic shielding clip wavy cardboard simulation model

Quantitative analysis of passive noise reduction:

For the first time, a variety of passive noise reduction measures (clamping force optimization, core material replacement, upper and lower positioning reinforcement, etc.) are systematically simulated and quantified, and the noise reduction thresholds of different measures (e. g., the range of optimal values of the clamping force is 0.11-0.13 MPa) are clarified to provide precise guidance for engineering practice.

C. Theoretical validation and correction of engineering formulas

Engineering validation of the intrinsic frequency formula:

Through the simulation experiments of scaling the core size (overall enlargement/reduction, single dimension adjustment), the linear inverse relationship between the core intrinsic frequency and the size is verified, and the applicable boundaries of the formula in specific working conditions are revealed. This result provides a reliable basis for rapid prediction of core vibration characteristics in engineering design.[15]

Energy-equivalent analysis of magnetostrictive effects:

The innovative equivalent of magnetostriction rate to strain energy volume density bypasses the complexity of the traditional permeability-density model, simplifies the process of magnetostriction force calculation, and provides new ideas for subsequent research.

V. CONCLUSION

In this study, the dynamic propagation mechanism of 110kV transformer noise is systematically revealed based on the electromagnetic-mechanical-acoustic multi-physical field coupling modeling method. By constructing a finite element simulation model, the whole process of dynamic analysis from electromagnetic excitation to mechanical vibration and then to acoustic field radiation is realized for the first time. The model breaks through the limitations of the traditional single field simulation, comprehensively considers the core magnetostrictive effect, the fluctuation of electromagnetic force of the winding and the vibration transmission of the tank structure and other key factors, and optimizes computational efficiency through the boundary acceleration transmission and pressure load feedback mechanism. Simulation results show that the nonlinear relationship between rated capacity and magnetic flux density significantly affects the noise intensity, for example, when the magnetic flux density is increased from 1.6T to 1.8T, the increase in sound pressure level can reach 4.2dB, which reveals the prediction bias of the traditional empirical formulae under the high magnetic density working condition. On this basis, the research team innovatively proposed a double-layer magnetic shielding sandwich corrugated cardboard composite noise reduction structure: the inner layer adopts highly permeable silicon steel sheet to suppress the leakage flux, and the outer layer forms an acoustic wave scattering cavity through the corrugated cardboard, which combines the magnetic shielding material's permeability and the cardboard's porous sound-absorbing properties, realizing the dual suppression of electromagnetic noise and mechanical vibration. Experimental data show that the structure in the 1250kVA transformer on the application of the average sound pressure level at 1m from 58dB (A) to 52dB (A), compared with the traditional single-layer magnetic shielding scheme to reduce the 3.2dB, and frequency characteristics analysis shows that the attenuation effect of the low-frequency noise of the following 500Hz is particularly significant.

The study further quantifies the synergistic effect of passive noise reduction measures, and determines the best parameter combinations of clamping force optimization, core material replacement, and positioning structure reinforcement by orthogonal test method. When the core clamping force is controlled in the range of 0.11-0.13MPa, the micro-displacement between the core stacks can be effectively suppressed, and with the application of amorphous alloy materials, the magnetostrictive coefficient is reduced to 60% of that of the conventional silicon steel wafers, and the overall noise reduction threshold is measured to be 4.8dB. At the same time, the inverse linear relationship between the core intrinsic frequency and geometric dimensions has been verified through the size-scaling experiments, for example, by reducing the core height from 1200mm to 1200mm, and the core height is reduced to 4.5mm. For example, when the core height is scaled down from 1200mm to 1000mm, the fundamental frequency rises from 215Hz to 258Hz. The result not only corrects the scope of application of the size coefficient in the engineering formula (the error rate is reduced from 12% to 3.5% by the traditional formula), but also provides a theoretical basis for the rapid prediction of the vibration characteristics of transformers of different capacities. It is worth noting that the research team for the first time the magnetostriction rate is equivalent to the strain energy body density, through the principle of conservation of energy to simplify the magnetic - force coupling calculation process, this method for the subsequent quantitative analysis of magnetoelastic effect opens up a new path.

VI. FUTURE AND PROSPECTS

Looking toward the future, the study suggests deepening the low-noise transformer technology from three aspects: firstly, exploring the deep integration of active noise reduction (ANC) system and passive optimization structure, such as deploying piezoelectric sensor arrays inside the tank and generating inverted acoustic waves to achieve dynamic noise offset through adaptive algorithms; secondly, in view of the complex electromagnetic environment faced by the transformers in the urban power grid (e.g., harmonic contamination, transient over-voltages, etc.), it is necessary to Secondly, for the complex electromagnetic environment faced by transformers in urban power grids (such as harmonic pollution, transient overvoltage, etc.), it is necessary to establish an adaptive validation system for the full working conditions of the multi-physical field coupling model; and finally, promote the formulation of standardized design guidelines, combine the simulation and optimization process with intelligent manufacturing technology, and develop a transformer noise prediction platform based on digital twin. These technological breakthroughs will help realize the paradigm shift from "empirical trial and error" to "model-driven" transformer noise control, and provide key technical support for the construction of green smart grid.

REFERENCES

[1] Y. Niu, G. Wang and H. Wang, "Research on Transformer Noise and Noise Reduction Measures Based on Multi-physical Field Coupling," 2024 International Conference on Electrical Power Systems and Intelligent Control (EPSIC), Changsha City, China, 2024, pp. 40-45, doi: 10.1109/EPSIC63429.2024.00014.

[2] F. Qiu, "Analysis on Vibration and Noise Characteristics of Epoxy Resin Dry-Type Transformer and Noise Reduction Measures," 2023 IEEE 4th International Conference on Electrical Materials and Power Equipment (ICEMPE), Shanghai, China, 2023, pp. 1-4, doi: 10.1109/ICEMPE57831.2023.10139447.

[3] J. Liang, T. Zhao, L. Zou, L. Zhang and Z. Li, "Adaptive Active Noise Control System of Power Transformer," 2015 Fifth International Conference on Instrumentation and Measurement, Computer, Communication and Control (IMCCC), Qinhuangdao, China, 2015, pp. 1394-1397, doi: 10.1109/IMCCC.2015.298.

[4] R. S. Girgis, M. Bernesjo and J. Anger, "Comprehensive analysis of load noise of power transformers," 2009 IEEE Power & Energy Society General Meeting, Calgary, AB, Canada, 2009, pp. 1-7, doi: 10.1109/PES.2009.5275883.

[5] P. Peng et al., "Design and optimization of noise reduction patch on transformer based on acoustic array technology," 2018 Chinese Automation Congress (CAC), Xi'an, China, 2018, pp. 1365-1368, doi: 10.1109/CAC.2018.8623603.

[6] W. W. L. Keerthipala, Zhou RuJing, Tan Eu Leong and Chionh Chang Jinn, "Electronic circuits for active control of acoustic noise generated by high voltage transformers," 1998 International Conference on Power Electronic Drives and Energy Systems for Industrial Growth, 1998. Proceedings., Perth, WA, Australia, 1998, pp. 243-248 Vol.1, doi: 10.1109/PEDES.1998.1330021.

[7] B. He, Y. Chen, W. Liu, B. Sheng, Y. -F. Liu and P. C. Sen, "Investigation and Reduction of Common Mode Current in Center-Tapped Transformer of LLC Resonant Converters," 2022 IEEE Applied Power Electronics Conference and Exposition (APEC), Houston, TX, USA, 2022, pp. 1686-1691, doi: 10.1109/APEC43599.2022.9773439.

[8] S. Wang, S. Lu and K. Ren, "ME-TransNet: A Modal-Enhanced Transformer Denoising Network for Attenuating Low-Frequency Swell Noise," in IEEE Geoscience and Remote Sensing Letters, vol. 21, pp. 1-5, 2024, Art no. 7501805, doi: 10.1109/LGRS.2024.3358746.

[9] R. S. Girgis, M. S. Bernesjö, S. Thomas, J. Anger, D. Chu and H. R. Moore, "Development of Ultra-Low-Noise Transformer Technology," in IEEE Transactions on Power Delivery, vol. 26, no. 1, pp. 228-234, Jan. 2011, doi: 10.1109/TPWRD.2010.2070812.

[10] R. Girgis, M. Bernesjö, S. Thomas, J. Anger, D. Chu and H. Moore, "Development of ultra — Low noise transformer technology," 2011 IEEE Power and Energy Society General Meeting, Detroit, MI, USA, 2011, pp. 1-8, doi: 10.1109/PES.2011.6038896.

[11] R. S. Girgis, M. Bernesjo and J. Anger, "Comprehensive analysis of load noise of power transformers," 2009 IEEE Power & Energy Society General Meeting, Calgary, AB, Canada, 2009, pp. 1-7, doi: 10.1109/PES.2009.5275883.

[12] H. Kishimoto, K. Komoku, J. Furuta and N. Itoh, "A Study on 23-GHz Low-Phase-Noise VCO with Transformer Output," 2024 IEEE Asia-Pacific Microwave Conference (APMC), Bali, Indonesia, 2024, pp. 348-350, doi: 10.1109/APMC60911.2024.10867767.

[13] R. S. Girgis and M. Bernesjö, "Appropriate test conditions proposed for Industry Standards of measuring transformer noise," IEEE PES General Meeting, Minneapolis, MN, USA, 2010, pp. 1-7, doi: 10.1109/PES.2010.5590027.

[14] Q. Zhang et al., "A W-Band Low Noise Amplifier with High Gain and Low Noise Figure in 65-nm CMOS," 2022 IEEE MTT-S International Microwave Workshop Series on Advanced Materials and Processes for RF and THz Applications (IMWS-AMP), Guangzhou, China, 2022, pp. 1-3, doi: 10.1109/IMWS-AMP54652.2022.10107105.

[15] S. -L. Jang, Y. -P. Hsieh, M. -H. Juang, J. -Y. Sung and W. -C. Lai, "Low-Voltage Quadrature Voltage-Controlled-Oscillator Using Twisted Transformer," in IEEE Microwave and Wireless Technology Letters, vol. 34, no. 6, pp. 667-670, June 2024, doi: 10.1109/LMWT.2024.3386332

Study on Cross-scale Collaborative Control of Composite Properties for Power Device Packaging

Xianfeng Li
School of Electrical and Electronic
Engineering
North China Electric Power University
Beijing, China
Leepioneer@126.com

Jian Wang
School of Electrical and Electronic
Engineering
North China Electric Power University
Beijing, China
wangjian31791@ncepu.edu.cn

Hanwen Ren
School of Electrical and Electronic
Engineering
North China Electric Power University
Beijing, China
rhwncepu@ncepu.edu.cn

Wei Wang
School of Electrical and Electronic
Engineering
North China Electric Power University
Beijing, China
wwncepu@163.com

Chen Chen
Wuxi power supply company
State Grid Jiangsu Electric Power Co.,
Ltd.
Wuxi, China
chenchen921197@163.com

Qingmin Li
School of Electrical and Electronic
Engineering
North China Electric Power University
Beijing, China
lqmeee@ncepu.edu.cn

Abstract—Aiming at the deterioration of electrothermal properties of silicone gel, an h-BN doped silicone gel composite material was proposed in this paper. The h-BN/ silicon gel material preparation platform and the material electrical breakdown experimental platform were set up, and the thermal conductivity and insulation properties were co-optimized by adjusting the dispersion of the filler. The experimental results show that the introduction of h-BN significantly improves the thermal conductivity of the material, which reaches 0.329 W/(m·K) when doped with 30 wt%, which is 2.12 times that of the pure matrix. The thermal stability of the composite was further verified by thermogravimetric analysis. In terms of electrical properties, the breakdown field strength of 10 wt% doping reaches 28.62 kV/mm, which is 25% higher than that of pure silicon gel. At 30 wt% doping, the breakdown field strength decreased by 20% due to packing agglomeration. This study provides a new idea for improving the electro-thermal performance of packaging materials for new energy high frequency power devices.

Keywords—*Power devices, Package insulation, thermal conductivity, Breakdown field strength, h-BN, silicone gel*

I. INTRODUCTION

With the advancement of the "dual carbon" strategy, power electronic equipment is rapidly developing towards high-frequency and high power density[1]-[4]. As core components, the packaged insulation materials of power devices face the challenge of extreme working conditions of electro-thermal multi-field coupling[5]. On the one hand, the local electric field distortion and dielectric loss of the device are intensified, and the material aging threatens the reliability of the insulation[6]. On the other hand, the junction temperature of the device is high, while the low thermal conductivity of traditional silica gel cannot meet the demand for efficient heat dissipation[7]. How to simultaneously enhance the thermal conductivity and electrical breakdown resistance of packaging materials has become a key challenge in the reliability design of new energy equipment.

In this field, many researchers have conducted extensive explorations. Muhammad Khan found that organic and inorganic nanofillers have synergistic effects on the dielectric and mechanical properties of epoxy resin composites[8]. Hexagonal boron nitride (h-BN) is considered as an ideal functional filler because of its wide band gap, high thermal conductivity and low dielectric loss. Ting Wang prepared a high thermal conductivity polyimide (PI) composite film containing BNNSs fillers, in which the in-plane thermal conductivity of PI composite film with BNNSs of 7 wt% was as high as 2.95 W/mK[9]. Bolin Tang synthesized KH550-modified h-BN nano-filler by one-step ball milling method. The results showed that the thermal conductivity of the composite was increased by 195.7% compared with that of pure epoxy resin at 10 wt%, but this study only focused on the thermal conductivity of the material[10]. Tianze Wang studied boron nitride nanosheets/polyetherimide composites with volume resistivity up to 8.16×10^{15} Ω·cm, but did not conduct specific research on electrical properties[11]. In summary, most of the existing studies focus on the single regulation of electrical or thermal properties, while the influence of h-BN doping amount and dispersion state on the electro-thermal properties of composites is non-linear, so it is urgent to reveal the performance optimization mechanism from the perspective of multi-scale interface regulation.

This paper focuses on the packaging requirements of new energy power devices, takes h-BN/ silicone gel composite material as the research object, conducts characterization and characterization by SEM, builds an electrical breakdown experiment platform, and systematically explores the influence of filler doping amount on microstructure, thermal properties and electrical insulation characteristics. This paper provides a theoretical basis for the development of high thermal conductivity and breakdown resistant packaging materials. Through h-BN gradient doping and interface modification strategies, it is expected to solve the collaborative optimization problem of heat dissipation and insulation of new energy power devices.

II. MATERIAL PREPARATION AND ELECTRICAL AGING EXPERIMENT

A. Preparation of h-BN/ silicone gel composites

In order to carry out various physical and chemical properties tests, h-BN/ silicone gel composite samples were prepared. The h-BN pre-treatment and composite preparation process were shown in Fig. 1.

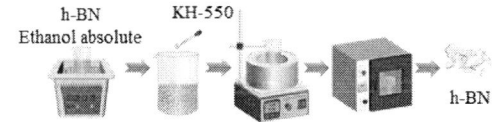

This work was supported by the National Key R&D Program of China-Key special project of intergovernmental international Science and Technology Innovation cooperation (2025YFE0106300).

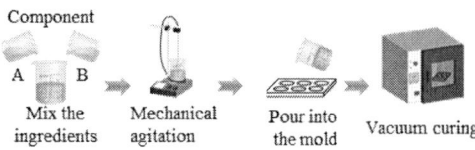

Fig.1. Platform for the preparation of h-BN/ silicone gel composites

The surface pretreatment of hexagonal boron nitride was carried out with silane coupling agent KH-550. (1)h-BN was added to anhydrous ethanol, dispersed by ultrasound for 60 minutes, and then KH-550 was added to the anhydrous ethanol. (2)Stir the mixed solution in a water bath at 80 ° C for 4 hours. (3)The solution is dried in a vacuum drying oven at 130℃. (4)The modified KH-550/h-BN powder is ground, sifted and dried. The macro and micro morphologies of the modified h-BN powder are shown in Fig. 2.

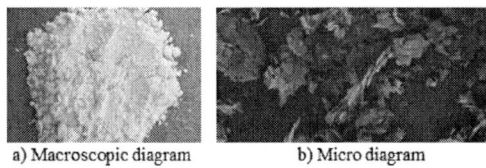

a) Macroscopic diagram   b) Micro diagram

Fig. 2 Morphology of h-BN powder modified by KH-550

h-BN/silicon gel composites were prepared by mechanical blending modification. (1) The A component, h-BN, B components are added to the container in turn, where the mass ratio of A and B components is 1:1. (2) Using mechanical stirring, the substrate and inorganic filler are fully mixed for 20min. (3) The material is degassed and then poured into the mold. (4) The mold was placed in a vacuum oven and cured at 60℃ for 12 h to obtain H-BN/silicon gel composite samples with H-BN mass fractions of 0%, 10%, 20% and 30%, respectively.

B. Electrical breakdown experiment

In order to study the electrical breakdown characteristics of silicone gel under high frequency voltage, this paper built an experimental platform as shown in Fig. 3, which was composed of high frequency voltage source, high voltage probe, digital oscilloscope, microscopic imaging system and UHF partial discharge monitoring system.The output waveform of high voltage and high frequency AC power supply is sine wave, the output voltage is continuously adjustable in the range of 0-30 kV, and the output frequency range is 1 kHz-50 kHz. Digital oscilloscope sampling rate is 10 GS/s.

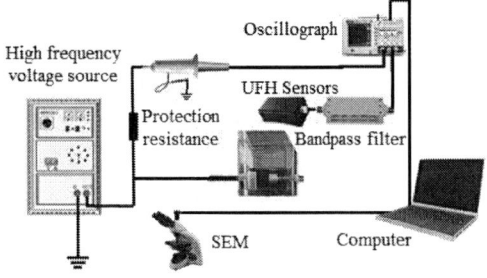

Fig. 3 Schematic diagram of the electrical breakdown experimental platform

III. EXPERIMENTAL RESULTS AND ANALYSIS

A. The microstructure of composite materials

After gold spraying treatment, SEM observation was carried out on the cross section of the composite material, and the microscopic morphology of the cross section of the composite material under different h-BN doping amounts was obtained. The dispersion of h-BN was observed, and the results were shown in Fig.4.

The cross-section of pure silicone gel is smooth and flat under the electron microscope. When the doping amount is 10 wt%, h-BN disperses more uniformly in silicon gel. When the doping amount reaches 20 wt%, h-BN may be connected. With the further increase of doping amount, h-BN agglomeration occurs in some regions, which will affect the properties of the composites.

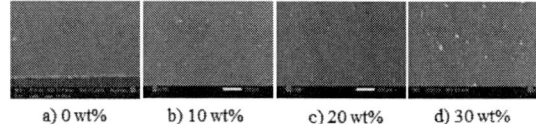

a) 0 wt%   b) 10 wt%   c) 20 wt%   d) 30 wt%

Fig. 4 Microstructure of composites with different doping amounts

B. Thermal properties of modified silicon gel

1. Thermal conductivity analysis

The thermal conductivity of h-BN/ silicone gel composites was tested, and the effect of introducing h-BN on the thermal properties of silicone gel was studied. In this paper, the thermal conductivity of silicone gel composites was tested at 60℃, 80℃ and 100℃, and the thermal conductivity of the composites under different doping amounts was obtained. In order to further analyze the efficiency of improving the thermal conductivity of the materials by h-BN doping amount, the average value of the thermal conductivity at each temperature was taken, as shown in Fig. 5.

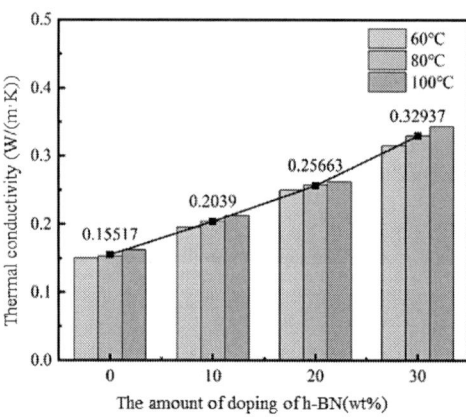

Fig. 5 Thermal conductivity of composites with different doping amounts

The thermal conductivity of pure silicone gel is low, only 0.15517 W/(m·K), and the thermal conductivity of the composite increases with the increase of h-BN doping amount. At the same time, with the increase of temperature, the thermal conductivity of the composite material will also increase to a certain extent. The thermal conductivity of pure silicone gel and h-BN composites with 10 wt%, 20 wt% and 30 wt% doping is 0.2039 W/(m·K), 0.25663 W/(m·K) and 0.32937 W/(m·K), respectively, and the thermal conductivity is nonlinear. At 10wt%, the thermal conductivity is 1.31 times that of pure silicone gel, and at 30wt%, the thermal conductivity is 2.12 times that of pure silicone gel.

The introduction of h-BN can effectively improve the thermal conductivity of the composite material. h-BN has good thermal conductivity, and its addition will improve the thermal conductivity efficiency of phonons, thus improving the thermal conductivity of the composite material.

2. Thermogravimetric analysis

Thermogravimetric analysis is one of the most direct methods to quickly evaluate the thermal stability of insulating materials. The change of mass fraction percentage and thermogravimetric rate of composite samples during thermal aging can reveal the change of internal structure of composite materials during thermal aging.

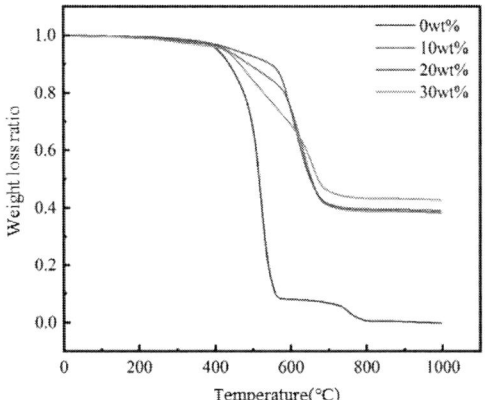

Fig. 6 Thermogravimetric curve

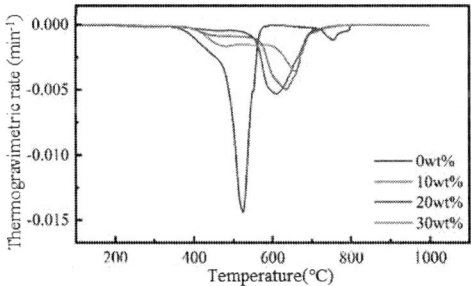

Fig. 7 Thermogravimetric rate

TGA and DTG curves of different materials under nitrogen atmosphere are shown in Fig. 4 and Fig. 5 respectively. According to the TGA curve, the thermal decomposition of pure silicone gel can be divided into two stages, the first thermal decomposition stage starts at about 410°C, and the degradation rate reaches the maximum at 460°C. The second thermal decomposition stage begins at 750°C, and the degradation rate reaches its maximum at 790°C. When the temperature reaches 800°C, the pure silicone gel has basically no residue.

For h-BN/ silicon gel composites, the decomposition is divided into two stages. With the increase of h-BN doping amount, the decomposition degree decreases gradually, and the improvement effect is very obvious. When the doping amount is 30 wt%, the decomposition rate of the first stage of thermal decomposition is higher, but the initial temperature of

the second stage is higher. In general, h-BN can effectively improve the thermal stability of silicon gel. With the increase of doping amount, the maximum degradation rate, degradation degree and degradation initial temperature are increased. However, when the doping amount is higher, the decomposition rate shows a different trend, which is partly because the addition of h-BN affects the addition reaction of methyl silicone oil and vinyl silicone oil in silica gel, resulting in a decrease in polymerization degree. On the other hand, when the content of h-BN is high, agglomeration will occur and defects such as holes and cracks will occur in the matrix, which will have a negative impact on the pyrolysis rate.

3. Electrical performance analysis

(1) Breakdown voltage analysis

The breakdown voltage is measured at 10 kHz AC voltage, and the breakdown field strength and its distribution characteristics can be described by obtaining a two-parameter Weibull distribution. The Weibull distribution function is shown in (1):

$$P(x) = 1 - \exp[-(\frac{E}{\alpha})^{\beta}]$$

Where, $P(x)$ represents the cumulative failure probability, E represents the breakdown field strength under the corresponding probability, β represents the shape parameter, and α represents the scale parameter. The two-parameter Weibull distribution of the breakdown field strength of the composite is shown in Fig. 8. The cumulative failure probability of 63.2% is selected as the breakdown field strength of the composite. h-BN doping has a significant effect on the breakdown field strength of the composites. With the increase of h-BN doping, the breakdown voltage increases first and then decreases. The breakdown voltage of pure silicon gel is 22.89 kV/mm. When the doped h-BN is 10 wt%, the breakdown voltage is about 28.62 kV/mm, reaching the highest. When the doping amount is 20 wt%, the breakdown voltage of the composite decreases to 26.47 kV/mm, which begins to decline, but it is still higher than that of pure silicone gel. When the doping amount continues to increase to 30 wt%, the breakdown voltage of the composite material drops to about 20.36 kV/mm, which is lower than that of pure silicon gel.

h-BN enhances the breakdown field strength of the composite material, which is related to its inhibition effect on electrical branches. 1) h-BN makes it easy for carriers to collide and decelerate when moving in the electric field, and the average free travel of carriers becomes shorter, which makes it impossible for carriers to accumulate more energy in the electric field, and it is difficult to form an electric branch channel; 2) h-BN has a high breakdown field strength, and when the electrical branch channel reaches its vicinity, it cannot pass through, so it needs to bypass h-BN, resulting in the electrical branch channel becoming tortuous, and its development speed will be slowed down accordingly. The further increase of the doping amount of h-BN will lead to the decrease of the breakdown field strength. This is because when the doped h-BN content increases to a certain level, agglomeration or the introduction of some defects may occur in the composite. Moreover, with the increase in h-BN density, h-BN will be connected, and the carriers will migrate along the h-BN surface to the opposite side, due to the existence of interface barrier effect, the carriers will be forced to move only along the direction of the electric field line, shortening the time for the electrical branch to reach the electrode, so the breakdown is more likely to occur.

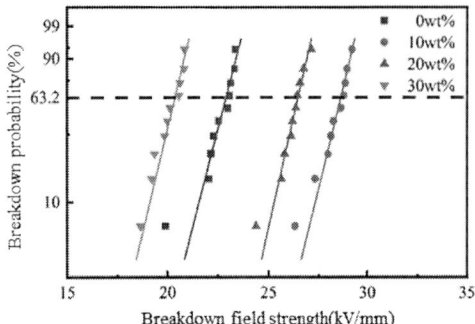

Fig. 8 Weibull distribution of breakdown field strength

(2) Dielectric spectrum in frequency domain

The dielectric loss were measured to investigate the effect of h-BN doping on the dielectric properties of silicon-gel composites. The dielectric properties are tested at 25°C and the test frequency ranges from 10 to 10^6 Hz. The results are shown in Fig.9.

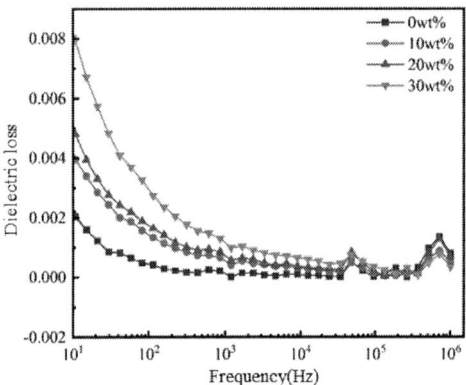

Fig. 9 The trend of dielectric loss with frequency of electric field

The trend of dielectric loss of composites with different doping amounts decreases with the increase of frequency. At low frequency, the dielectric loss increases significantly, and at high frequency, the loss decreases significantly, which is close to the pure silicone gel sample. The interface between h-BN and silicone gel and the polarity of h-BN will strengthen the relaxation polarization of the composite material, resulting in greater polarization loss, so the dielectric loss will increase. At low frequency, the dipole steering can keep up with the change of frequency. With the increase of frequency, the polarization process of the dipole is suppressed, resulting in a smaller relaxation loss, and therefore the dielectric loss will decrease with the increase of frequency.

IV. CONCLUSIONS

In this paper, an electrical breakdown experiment platform was built and h-BN/ silicone gel composite was prepared. SEM observation, thermal conductivity test, thermogravimetric test, breakdown voltage test and dielectric property test were carried out, and the corresponding test results were analyzed. The main conclusions of this paper are as follows:

1) Scanning electron microscopy (SEM) showed that when the doping amount of h-BN reached 30 wt%, the filler formed a continuous thermal conductivity network, and the thermal conductivity of the composite was increased to 0.3294 W/(m·K), an increase of 112% compared with the matrix.

2) Thermogravimetric analysis (TGA) showed that with the increase of h-BN doping amount, the maximum degradation rate, degradation degree and degradation starting temperature of the composite increased with the increase of h-BN doping amount.

3) The electrical test shows that the breakdown field strength changes non-monotonically with h-BN doping: when the doping is 10 wt%, the breakdown field strength reaches 28.6 kV/mm, which increases by 25%; When doped with 30 wt%, the interfacial defect is caused by packing agglomeration, and the breakdown field strength decreases by 11%.

4) Frequency domain dielectric spectrum shows that the trend of dielectric loss is to decrease, and the dielectric loss increases greatly at low frequency.

REFERENCES

[1] J. Wang, W. Wu, Y. Song, et al., "Trap Distribution and Along-surface Discharge Characterization of Aromatic Compound-modified Silicone Gel," IEEE Transactions on Dielectrics and Electrical Insulation, 2024.

[2] H.Y. Chen, V.V. Ginzburg, J. Yang, et al., "Thermal Conductivity of Polymer-Based Composites: Fundamentals and Applications," Progress in Polymer Science, vol. 59, pp. 41-85, 2016.

[3] L. Li, B. Zhou, G.J. Han, et al., "Understanding the effect of interfacial engineering on interfacial thermal resistance in nacre-like cellulose nanofiber/graphene film," Composites Science and Technology, vol. 197, 2020.

[4] Hannan MA, Al-Shetwi AQ, Mollik MS, et al., "Wind Energy Conversions, Controls, and Applications: A Review for Sustainable Technologies and Directions,"Sustainability,vol.15, no. 5: pp. 3986, 2023.

[5] S. S. Ang and H. Zhang, "High temperature power electronic module packaging," 2015 China Semiconductor Technology International Conference, Shanghai, China, pp. 1-3, 2015.

[6] W. Wang et al., "Interfacial Discharge Characteristics and Insulation Life Analysis of Package Insulation Under Square Voltage Coupled With High Frequency and Steep dv/dt," IEEE Transactions on Electron Devices, vol. 72, no. 1, pp. 350-356, Jan. 2025,

[7] J. Wang, C. Chen, H. Yan, et al., "Electrical Trees of Silicone Gel Encapsulation Materials in Power Electronic Modules Self-Healing Properties and Influencing Factors," IEEE Transactions on Industry Applications, vol. 60, no. 1, pp. 1288-1297,2024.

[8] Muhammad Khan, Aqeel A. Khurram, Tiehu Li, et al., "Synergistic effect of organic and inorganic nano fillers on the dielectric and mechanical properties of epoxy composites," Journal of Materials Science & Technology, vol. 34, no. 12, pp. 2424-2430, 2018.

[9] T. Wang, M. Wang, L. Fu, et al., "Enhanced Thermal Conductivity of Polyimide Composites with Boron Nitride Nanosheets," Sci Rep 8, 1557, 2018.

[10] B Tang, M Cao, Y Yang, et al., "Synthesis of KH550-Modified Hexagonal Boron Nitride Nanofillers for Improving Thermal Conductivity of Epoxy Nanocomposites," Polymers, vol. 15, no.6, pp. 1415, 2023.

[11] Tianze Wang, Hui Chi, Danying Zhao, et al., "A highly thermally conductive yet electrically insulating boron nitride nanosheets/polyetherimide composite with oriented structure," Composites Part A:Applied Science and Manufacturing, vol. 188, 2025.

Partial Discharge Denoising Technology for GIS Equipment Based on CEEMDAN

Zhongyue Liu
State Grid Shanghai Electric
Research Institute
Shanghai, China
972563484@qq.com

Pei Cao
State Grid Shanghai Electric
Research Institute
Shanghai, China
caopei2009@163.com

Zeyu Li
North China Electric Power
University
Beijing, China
1350364621@qq.com

Hanwen Ren
North China Electric Power
University
Beijing, China
rhwncepu@ncepu.edu.cn

Abstract—Addressing the problems with noise interference in partial discharge (PD) signals of gas - insulated switchgear (GIS) and the shortcomings of traditional denoising methods, such as mode mixing and residual noise, the article introduces a dual - threshold collaborative denoising approach that utilizes Complete Ensemble Empirical Mode Decomposition with Adaptive Noise (CEEMDAN). To enhance the signal decomposition process, Gaussian white noise is incorporated, which significantly alleviates the modal aliasing issue. Subsequently, a two - stage filtering framework is established. In the initial stage, components dominated by noise are eliminated using a correlation coefficient threshold. Then, in the subsequent stage, multiple intrinsic mode functions (IMFs) are chosen based on a variance contribution threshold. In the single and double exponential attenuation and oscillatory PD model experiments, the signal quality is significantly improved after noise reduction. The technique offers high - quality signals for GIS partial discharge detection, aiding in the enhancement of diagnostic accuracy and dependability for faults.

Keywords—*partial discharge, CEEMDAN, modal aliasing, noise reduction*

I. INTRODUCTION

Gas-insulated switchgear (GIS), a critical high-voltage equipment in power systems, relies on SF6 gas insulation and encapsulated components within a metal casing. However, partial discharges may occur due to manufacturing/assembly defects or long-term aging, which not only waste energy but progressively damage insulation systems, potentially triggering severe equipment failures or accidents, posing significant risks to grid stability and personnel safety.

Recent advancements in signal processing have witnessed significant methodological developments in localized discharge denoising research. Extensive research efforts by both domestic and international researchers have yielded multiple denoising frameworks through systematic investigations. Conventional methodological frameworks have predominantly employed three principal techniques: Fourier Transform (FT) for frequency-domain analysis, Wavelet Transform (WT) for time-frequency localization, and Empirical Mode Decomposition (EMD) for adaptive signal processing. Although FT can effectively remove periodic noise, it is less effective in processing non-smooth signals[1]. The wavelet transform offers certain advantages when dealing with non - smooth signals. However, the lack of adaptability in threshold selection and in determining the number of decomposition levels leads to unstable noise reduction outcomes[2]. J. Xie proposed a denoising approach for localized signal denoising grounded in sparse representation theory. By constructing an overcomplete dictionary of atoms that are well - matched to localized pulses yet orthogonal to noise components, and integrating a fast searching algorithm to expedite the selection of optimal matching atoms, this method achieves enhanced denoising performance compared to conventional wavelet - based techniques[3]. However, this method relies on the establishment of a high-quality training sample library and the design of an efficient optimization solver, so there may be some obstacles to its generalization in practical applications. The EMD technique has the ability to adaptively split signals into multiple Intrinsic Mode Functions (IMFs), however, it is susceptible to mode mixing and endpoint effects, which can impair its noise reduction effectiveness.

The Ensemble Empirical Mode Decomposition (EEMD) approach is employed to address the mode - related issues of the EMD technique. It manages to circumvent the mode - mixing issue through the inclusion of white noise in the signal. Regardless, this method is computationally intensive, and the introduced white noise might linger, thereby impacting the signal's reconstruction precision[4]. The Complementary Ensemble Empirical Mode Decomposition (CEEMD) method addresses limitations in conventional EEMD by strategically incorporating complementary white noise pairs into the original signal. This innovative approach enables mutual cancellation of residual noise components during signal reconstruction while maintaining decomposition accuracy. Compared with traditional EEMD implementations, CEEMD achieves enhanced computational efficiency through optimized iteration requirements, effectively balancing signal fidelity and processing economy[5]. In addition, some scholars have also proposed hybrid noise reduction strategies combining multiple methods, such as a hybrid noise reduction method based on CEEMD and wavelet thresholding, which decomposes the signal by CEEMD and then utilizes the wavelet thresholding method to reduce the noise of the decomposed IMFs, and achieves a better noise reduction effect. However, the method still has some limitations when dealing with complex signals, such as the lack of adaptivity in threshold selection[6].

To address complex noise interference in partial discharge (PD) signals, this study proposes a CEEMDAN-based denoising method. The approach decomposes contaminated signals into intrinsic mode functions (IMFs), selects PD-related components through correlation coefficients and variance contribution analysis, and reconstructs them to suppress noise. Simulation results

979-8-3315-1110-4/25 $31.00 © 2025 IEEE

demonstrate that this method effectively reduces interference and waveform distortion while mitigating mode aliasing, thereby enhancing PD detection accuracy and reliability for subsequent feature extraction and pattern recognition.[7][8].

II. FUNDAMENTAL PRINCIPLE

A. Introduction to the CEEMDAN Principle

EMD, being an adaptive means for nonlinear and nonsmooth signals, has the capacity to dissect the signal into a collection of IMFs, but it is plagued by the issues of mode aliasing and endpoint effects. To address these issues, EEMD has been proposed to reduce mode aliasing by adding white noise and applying EMD multiple times. However, this approach introduces the new problem of residual auxiliary noise.

For further optimization, CEEMDAN has been developed, which not only inherits the advantages of EEMD, but also reduces unnecessary noise residuals by adding adaptive noise to the residuals at each stage more accurately and achieves the goal of almost zero reconstruction error. Compared to EEMD, CEEMDAN completes signal decomposition with fewer averaging times, providing more efficient and accurate results. The detailed procedures will be outlined below.

(1) A Gaussian white noise with amplitude ε is added to $x(t)$ for $i=1, \cdots, I$ times consecutively to obtain I noise sequences $x_i(t) = x(t) + \varepsilon n_i(t)$. $x_i(t)$ is decomposed into a series of IMFs by EMD to obtain a total of I sets of decompositions. The first mode obtained from the CEEMDAN decomposition can be regarded as the average of the first order IMF in this group I decomposition. IMF1 can be expressed as follows.

$$IMF1 = \frac{1}{I} \sum_{i=1}^{I} IMF1_i \qquad (1)$$

(2) Update the residual value $r(t)$ using the formula provided below.

$$r(t) = x(t) - IMF1 \qquad (2)$$

(3) $x(t)$ is converted into the residual value $r(t)$, and then similar to step (1), Gaussian noise is introduced into the residual value $r(t)$,generating I new noise sequences $r(t) + \varepsilon n_i(t)$, and each of the noise sequences is subjected to EMD decomposition separately to obtain I sets of decompositions, and the average of the first-order IMFs of each set of decompositions is denoted as the second-order IMF of the CEEMDAN decomposition with the following equation Shown.

$$IMF2 = \frac{1}{I} \sum_{i=1}^{I} E_1\bigl(r(t)+\varepsilon n_i(t)\bigr) \qquad (3)$$

(4) Recalculate the residual value in accordance with the following formula.

$$r_k(t) = r_{k-1}(t) - IMF_k \qquad (4)$$

(5) For the next order IMF solution of CEEMDAN decomposition, return to step (3) and loop through steps (3) and (4) until the residuals can no longer be obtained by EMD decomposition. Assuming that the signal $x(t)$ is decomposed by CEEMDAN into k IMFs, it can be represented as the sum of the eigenmode functions and the residuals.

$$x(t) = \sum_{k=1}^{K} IMF_k + r_k(t) \qquad (5)$$

In addition, if eventually $r_k(t)$ is counted as an IMF, then it can be decomposed to obtain $k+1$ IMFs. This study decomposes it according to this method. It is shown in the following equation.

$$x(t) = \sum_{k=1}^{K+1} IMF_k \qquad (6)$$

B. Correlation Coefficient

The correlation coefficient is a dimensionless numerical indicator used to measure the similarity between two signals, ranging from -1 to 1. A value close to 1 indicates a strong morphological similarity, while a value near -1 suggests a significant inverse relationship. If the coefficient approaches 0, it signifies no notable linear connection between the signals. The correlation coefficient plays a crucial role in identifying which IMFs have strong correlation with the original signals, and then filtering out the components that contain PD feature information. The detailed calculation formula is presented below.

$$r = \frac{\sum\limits_{i=1}^{n} (x_i-\bar{x})(y_i-\bar{y})}{\sqrt{\sum\limits_{i=1}^{n} (x_i-\bar{x})^2}\sqrt{\sum\limits_{i=1}^{n} (y_i-\bar{y})^2}} \qquad (7)$$

In Equation (7), x_i and y_i denote the instantaneous amplitude of the two signals at the ith sampling point, respectively. $\bar{x}$ and $\bar{y}$ represent the arithmetic means of the two signals, respectively. Additionally, n signifies the total number of sampling points for the signals.

The selection of correlation coefficient thresholds for IMF filtering requires balancing signal preservation and noise removal: excessively high thresholds risk losing valid signal components, while overly low thresholds retain excessive noise. This study adopts a threshold of 0.3, retaining IMFs with coefficients above this value as valid components and discarding those below as noise.

C. Variance Contribution Rate

The variance contribution rate quantifies the proportion of variance that a particular IMF contributes to the overall variance of the signal.

$$e(j) = \frac{D(j)}{\sum\limits_{i=1}^{N} D(j)} \times 100\% \qquad (8)$$

In the formula, $D(j)$ mathematically quantifies the dispersion measure for the jth Intrinsic Mode Function (IMF), computed through the averaging process of squared deviations between each IMF's discrete sampling data and its central tendency. Specifically, this statistical parameter is derived from the following relationship.

$$D(j) = \frac{1}{n} \sum_{i=1}^{n} C_j(i)^2 - \left(\frac{1}{n}\sum_{i=1}^{n} C_j(i)\right)^2 \qquad (9)$$

In the given context, $C_j(i)$ signifies the amplitude of the jth IMF at the ith sampling point. The variable n denotes the aggregate count of signal sampling points, and N represents the overall quantity of IMFs.

In the present research, the calculation procedure for the variance contribution rate is as follows. For each IMF, the variance $D(j)$ is determined by the equation. Based on $D(j)$ obtained in step (1), calculate the variance contribution rate $e(j)$ for each IMF using the equation. A threshold is set based on the distribution of variance contributions. IMFs are considered to have a greater impact on the original data compared to other components when, and only when, the variance contribution rate of these components exceeds the threshold value. Combine the IMFs screened by B and take the common parts to further enhance the denoising capability.

This study implements dual-criteria filtration (correlation-variance) to identify IMFs with significant energy-spectral alignment to source signals, effectively isolating physically meaningful components while discarding low-correspondence elements representing stochastic interference.

III. SIMULATION AND ANALYSIS

The UHF signals generated in the process of partial discharge have a very fast rate of change and a very short duration, usually only 2 to 100 nanoseconds, coupled with the fact that there are significant differences in the presentation of different devices and different types of PD signals. To confirm the noise - reduction effectiveness of the approach put forward in this study, it's essential to faithfully depict PD signals with exact mathematical representations. Empirical observations from industrial applications and scholarly investigations reveal that PD phenomena manifest through four canonical discharge modalities, grouped into two principal attenuation modalities: mono-exponential (with non-oscillatory and oscillatory variants) and bi-exponential (similarly demonstrating fundamental and oscillation-enhanced configurations). Based on this, the following four mathematical models are selected to simulate the partial discharge signal.

Equation for the mono-exponential decay model.

$$D_1 = Ae^{\frac{-(t-t_0)}{\tau}} \tag{10}$$

Equation for the bi-exponential decay model.

$$D_2 = B\left(e^{\frac{-1.3(t-t_0)}{\tau}} - e^{\frac{-2.2(t-t_0)}{\tau}}\right) \tag{11}$$

The equation for oscillatory mono-exponential decay phenomena is expressed.

$$D_3 = Ce^{\frac{-(t-t_0)}{\tau}} \sin\left(2\pi f(t-t_0)\right) \tag{12}$$

The equation for oscillatory bi-exponential decay phenomena is expressed.

$$D_4 = D\left(e^{\frac{-1.3(t-t_0)}{\tau}} - e^{\frac{-2.2(t-t_0)}{\tau}}\right) \sin\left(2\pi f(t-t_0)\right) \tag{13}$$

The mathematical form uses the A-D coefficient as the amplitude parameter corresponding to each signal. The parameter τ signifies the rate of signal decay, while f indicates the frequency of the oscillatory decay. The four discharge pulses mentioned above are simulated using MATLAB. Field-deployed UHF systems for partial discharge detection capture signals containing broadband random interference with uniform power spectral density across the electromagnetic spectrum. To realistically simulate field-acquired PD signals, MATLAB's awgn function is employed to add noise to clean PD signals, generating colored-noise PD signals with 0dB SNR. The time-domain waveform of the PD signal after superimposed white noise is presented in Fig. 1.

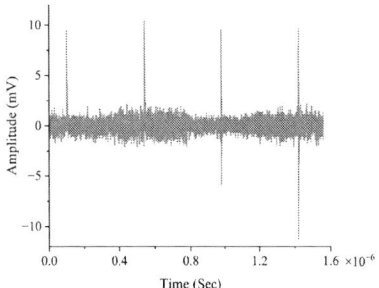

Fig. 1. Dye-noise PD signal

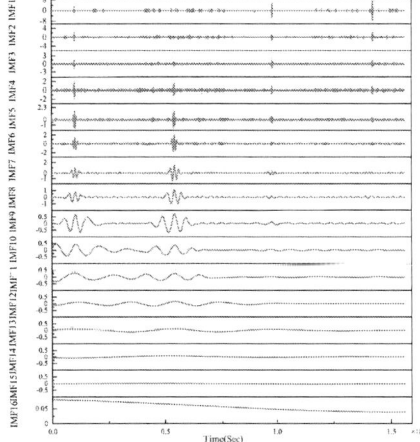

Fig. 2. Results of CEEMDAN decomposition

As depicted in Fig. 2, the dyed - noise PD signal has been decomposed into 16 IMFs via CEEMDAN, with the signals containing different frequency components being effectively separated into the corresponding IMFs. To enhance identification accuracy of PD-related IMF components, statistical metrics including correlation coefficients and variance contributions were systematically quantified, with the analytical results cataloged in Tables 1 and 2.

TABLE I.   EACH IMF CORRELATION COEFFICIENT

IMF	Correlation coefficients	IMF	Correlation coefficients

IMF		IMF	
IMF1	0.2316	IMF9	0.1746
IMF2	0.3440	IMF10	0.1036
IMF3	0.6861	IMF11	0.0148
IMF4	0.5374	IMF12	0.0914
IMF5	0.3963	IMF13	0.0667
IMF6	0.3064	IMF14	0.0088
IMF7	0.2977	IMF15	0.0345
IMF8	0.2055	IMF16	0.0345

TABLE II.   CONTRIBUTION OF EACH IMF VARIANCE

IMF	Variance contribution ratio	IMF	Variance contribution ratio
IMF1	3.39%	IMF9	2.47%
IMF2	6.68%	IMF10	2.44%
IMF3	9.43%	IMF11	2.33%
IMF4	5.79%	IMF12	2.51%
IMF5	6.83%	IMF13	2.24%
IMF6	3.58%	IMF14	2.36%
IMF7	3.25%	IMF15	2.05%
IMF8	3.45%	IMF16	1.05%

This study establishes correlation coefficient and variance contribution rate thresholds at 0.3 and 3.74% respectively. As illustrated in Table 1, IMF2 through IMF6 exhibit correlation coefficients exceeding the specified 0.3. This quantitative evidence demonstrates that these particular IMFs contain the principal PD signal components, with IMF1 falling below both the correlation coefficient criteria and the variance contribution threshold of 3.74%. However, from Table 2, it is concluded that IMF2, IMF3, IMF4, and IMF5 have a greater impact on the original data compared to other IMFs, and IMF2, IMF3, IMF4, and IMF5 are finally selected for signal reconstruction. As presented in Figure 4, the signal reconstruction results exhibit distinct waveform characteristics following empirical mode decomposition processing. The graphical representation demonstrates effective preservation of critical PD signal components while eliminating baseline interference, with complete morphological details visible in the processed waveform.

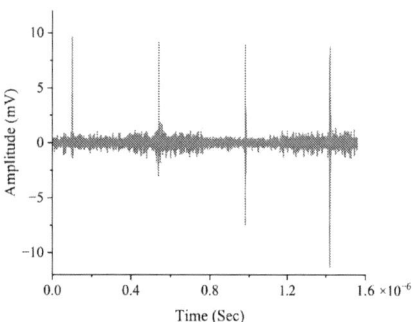

Fig. 3.   Reconstructed signal

This study evaluates denoising performance using three metrics: SNR, NCC, and RMSE. Higher SNR/NCC values and lower RMSE indicate superior noise reduction and fidelity. Quantitative results in Table 3 validate the method's effectiveness through balanced signal preservation and interference removal.

TABLE III.   EVALUATION OF NOISE REDUCTION

Evaluation indicators	SNR	NCC	RMSE
Methodology of this paper	9.0814	0.8251	0.2392

IV. CONCLUSION

In this study, a CEEMDAN - based noise reduction method is proposed to address the susceptibility of GIS partial discharge (PD) signals to noise interference.

A dual - threshold screening mechanism using the correlation coefficient and variance contribution rate is incorporated. This overcomes the limitations of traditional modal decomposition methods, such as modal aliasing, noise residuals, and subjective component selection. Simulation results show that the method significantly improves key performance metrics. It enhances the signal - to - noise ratio (SNR), indicating better signal distinguishability from noise, increases the normalized correlation coefficient, suggesting a closer resemblance to the original signal, and reduces the root - mean - square error (RMSE), meaning more accurate signal reconstruction.

The method can accurately extract PD feature components while suppressing high - and low - frequency noise. It thus provides a high - quality signal for subsequent fault diagnosis in GIS.

V. ACKNOWLEDGMENT

The research presented in this paper received financial support from the Science and Technology Project of State Grid Shanghai, identified by the project number B3094024000C. This project focuses on the extraction of time - frequency features and the optimal localization of Ultra - High Frequency attenuated partial discharge signals in Gas - Insulated Switchgear within a real interference environment.

REFERENCES

[1] S. S. Agaian, Mei-Ching Chen and C. L. P. Chen, "Noise reduction algorithms using Fibonacci Fourier transforms," 2008 IEEE International Conference on Systems, Man and Cybernetics, Singapore, 2008, pp. 1048-1052.

[2] I. Sornsen, C. Suppitaksakul and R. Kitpaiboontawee, "Partial Discharge Signal Detection in Generators Using Wavelet Transforms," 2021 International Conference on Power, Energy and Innovations (ICPEI), Nakhon Ratchasima, Thailand, 2021, pp. 195-198.

[3] F. Law, J. Xie, and Y. Wang, "Partial discharge signal sparse representation denoising method," Proceedings of the CSEE, vol. 35, no. 10, pp. 2625-2633, 2015.

[4] M. V. Rojas-Moreno, G. Robles, J. M. Martínez-Tarifa and J. M. Fresno, "Ensemble Empirical Mode Decomposition for the denoising of partial discharges measured in UHF," 2016 IEEE International Conference on Dielectrics (ICD), Montpellier, France, 2016, pp. 963-966.

[5] P. Chen, D. -L. Han, Q. -F. Cai and H. Liu, "Application of wavelet transform and EEMD in electromagnetic acoustic signal de-noising," 2013 International Conference on Quality, Reliability, Risk, Maintenance, and Safety Engineering (QR2MSE), Chengdu, China, 2013, pp. 1734-1737.

[6] J. R. Yeh, J. S. Shieh, and N. E. Huang, "Complementary ensemble empirical mode decomposition: A novel noise enhanced data analysis method," Adv. in Adaptive Data Analysis, vol. 2, no. 2, pp. 135-156, 2010.

[7] J. C. Chan, H. Ma, T. K. Saha and C. Ekanayake, "Self-adaptive partial discharge signal de-noising based on ensemble empirical mode decomposition and automatic morphological thresholding," in IEEE Transactions on Dielectrics and Electrical Insulation, vol. 21, no. 1, pp. 294-303, February 2014.

[8] L. Haifei, H. Guofang, Z. Qingwei, C. Houcheng and Z. Wenqiang, "Partial Discharge Denoising Method for Switchgear Based on CEEMD-Wavelet Threshold," 2019 2nd World Symposium on Communication Engineering (WSCE), Nagoya, Japan, 2019, pp. 31-37.

Improvement of thermal conductivity of epoxy composites using the synergistic effect of boron nitride nanosheets and boron nitride whiskers

Tiandong Zhang
School of Electrical and Electronic Engineering, Harbin University of Science and Technology
Harbin, China
tdzhang@hrbust.edu.cn

Chenghai Wang
School of Electrical and Electronic Engineering, Harbin University of Science and Technology
Harbin, China
1920300043@stu.hrbust.edu.cn

Xinle Zhang
School of Electrical and Electronic Engineering, Harbin University of Science and Technology
Harbin, China
2310303055@stu.hrbust.edu.cn

Tangman Xue
School of Electrical and Electronic Engineering, Harbin University of Science and Technology
Harbin, China
xbwang@hrbust.edu.cn

Changhai Zhang
School of Electrical and Electronic Engineering, Harbin University of Science and Technology
Harbin, China
chzhang@hrbust.edu.cn

Qingguo Chi
School of Electrical and Electronic Engineering, Harbin University of Science and Technology
Harbin, China
qgchi@hrbust.edu.cn

Abstract—Using epoxy resin (EP) composites with high thermal conductivity as electronic packaging materials is an effective means to solve the heat dissipation problem of electronic components. In this study, BNWKs-BNNSs/EP composites were prepared by mixing two-dimensional BNNSs and one-dimensional boron nitride whiskers (BNWKs) with different mass ratios at a fixed filling content of 30wt%, in order to achieve the effect of "1+1>2". The experimental results show that adding a small amount of BNWKs can effectively improve the thermal conductivity of the composite, but when the content exceeds a certain level, the thermal conductivity of the composite will decrease significantly. The thermal conductivity of 3.75wt% BNWKs-26.25wt% BNNSs/EP composite is the highest, which can reach 1.497 W/(m·K), which is 1.20 times that of 30wt% BNNSs/EP. At the same time, at room temperature, its volume resistivity and breakdown field strength are still as high as 1.15×10^{14} $\Omega \cdot$m and 70.98 kV/mm, which has relatively excellent electrical insulation performance and meets the application requirements of packaging materials. This study provides an effective method for preparing epoxy resin matrix composites with high thermal conductivity.

Keywords— Epoxy resin, boron nitride nanosheets, boron nitride whiskers, thermal conductivity

I. INTRODUCTION

With the continuous innovation and breakthroughs in science and technology, the integration and power density of electronic components inside electrical equipment have been constantly increasing. Although this has enhanced the processing speed of electrical equipment and diversified its functions, making people's lives more convenient, the heat accumulated in a unit space has also risen sharply [1-3]. According to relevant experimental statistics, for every 2°C increase in the operating environment temperature inside electrical equipment, its safety and reliability decrease by 10%, and its working life is also greatly reduced [4]. This seriously affects the safe and normal operation and service life of the equipment, and also restricts the further application of electronic components. Therefore, developing packaging technologies and materials with excellent performance has become a key scientific research issue that needs to be urgently addressed [5].

At present, scholars at home and abroad mainly improve the thermal conductivity of epoxy resin matrix composites through two aspects and means of intrinsic type and filled type. The intrinsic mode control of polymer preparation process is complex and cumbersome, and the preparation cycle is long. At the same time, the improvement of thermal conductivity of polymer is limited. Compared with optimizing the polymer chain at the molecular level to improve the thermal conductivity, directly adding high thermal conductivity fillers can often improve the thermal conductivity of polymers more effectively. Compared with the method of using a single filler, the method of using two or more different forms or sizes of thermal conductive filler mixed filler often has better improvement effect. Zeng prepared a series of epoxy resin-based materials by incorporating 20 μm and 70 μm alumina spherical particles into the epoxy resin matrix at different mass ratios [6]. When the content of Al_2O_3 filler reaches 75 wt% (20 μm:70 μm=1:2 wt%), the thermal conductivity of the composite reaches 1.91 W/(m·K), which is significantly higher than that of the composite doped with 20 μm alumina or 70 μm alumina alone. Chen successfully prepared a series of ZnOs@T-ZnOw/EP composites with a mixture of spherical ZnOs and tetrapod-like ZnO whiskers (T-ZnOw), and the most suitable ratio was found by varying the mass ratio of ZnOs and T-ZnOw. The thermal conductivity of 40 wt% ZnO and 60 wt% T-ZnOw composites reaches 0.52 W/(m·K), which is 2.76 times higher than that of the pure epoxy resin, when the filler content is 11.08 vol%.

In this paper, two-dimensional BNNSs and one-dimensional boron nitride whiskers (BNWKs) were selected as high thermal conductivity fillers, and the synergistic effect of the two was used to further improve the thermal conductivity of the composites. When the doping amount of BNWKs is 3.75 wt% and BNNSs is 26.25 wt%, the thermal conductivity of BNWKs-BNNSs/EP composite is the highest, which is 1.497 W/(m·K), 19.57% higher than that of 30 wt% BNNSs/EP. The synergy effect calculation results show that the synergy effect rate is the highest, which is 1.044, with strong synergy effect. At the same time, the 3.75 wt% BNWKs-26.25 wt% BNNSs/EP composite has high thermal conductivity, and its volume resistivity and breakdown field strength are still as high as 1.15×10^{14} Ω·m and 70.98 kV/mm, which has good electrical insulation performance and meets the application requirements of packaging materials.

II. EXPERIMENTAL SECTION

A. Materials

Bisphenol A epoxy resin (EP, E51) was purchased from Nantong Xingchen synthetic materials Co., Ltd; Methyl hexahydrophthalic anhydride (MHHPA) was purchased from Changzhou Runxiang Chemical Co., Ltd; 2,4,6-tris (Dimethylaminomethyl) phenol (DMP-30) was purchased from Shanghai McLean Biochemical Technology Co., Ltd; Hexagonal boron nitride (h-BN) was purchased from Dandong Rijin Technology Co., Ltd; Boron nitride whiskers (BNWKs) were purchased from Nangong Fenghui nanotechnology Co., Ltd.

B. Preparation of BNWKs-BNNSs/EP composites

The preparation flow chart of BNWKs-BNNSs/EP composites is shown in Fig. 1. BNNSs and BNWKs were magnetically stirred in a beaker for 2 h according to different mass ratios (1:0, 1:1, 1:3, 1:5, 1:7 and 0:1), and then mixed evenly. The mixed solution was dried in a vacuum dryer to obtain BNNSs-BNWKs uniform powder. 10 g EP, 8 g MHHPA, 0.1 g DMP-30 and 7.7 g BNNSs-BNWKs uniform powders with different mass ratios were put into the beaker, and the mass fraction of filler was fixed at 30 wt%. Put the beaker in the vacuum mixing box and mix for 2 h. Then drop the mixed rubber liquid into the mold and put it into the flat vulcanizer for pressure curing. BNWKs-BNNSs/EP composites were successfully prepared after curing at 120 °C for 2 h, then curing at 150 °C for 4 h, and then natural cooling.

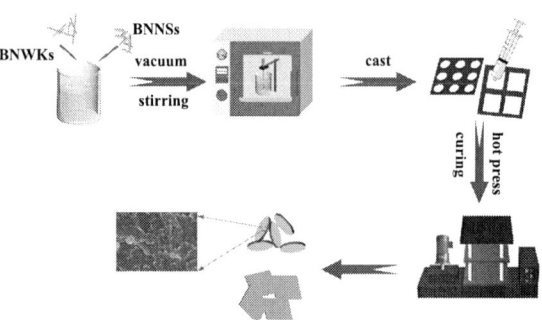

Fig. 1. Bschematic diagram of preparation of BNNSs-BNWKs/EP Composites.

III. RESULTS AND DISCUSSION

Fig. 2 shows the SEM of BNWKs and BNWKs-BNNSs/EP composites. From Fig. 2(a), it can be seen that BNWKs is a typical rod structure, with a length of about 50 μm and a diameter of about 3 μm. Fig. 2(b) shows the cross-sectional SEM of 3.75 wt% BNWKs-26.25 wt% BNNSs/EP. It can be found that the filler is evenly distributed without obvious agglomeration. At the same time, it can be observed that BNWKs are overlapped between BNNSs, forming an obvious bridging effect, which can provide a faster path for heat transfer and form a more perfect heat conduction network, which is helpful to improve the thermal conductivity of composites.

Fig. 2(c) shows the cross-section SEM of 7.5 wt% BNWKs-22.5 wt% BNNSs/EP. It can be found that BNWKs are disorderly distributed in the EP matrix, and its dispersion

uniformity is reduced. It can be seen from Figure 2(d) that this uneven phenomenon is more obvious. From the red mark in the figure, it can be seen that BNWKs have obvious agglomeration phenomenon, which will seriously weaken the "bridge" effect of BNWKs, which is not conducive to the establishment of thermal conduction path in the composite, on the contrary, it will also reduce the thermal conductivity of the composite.

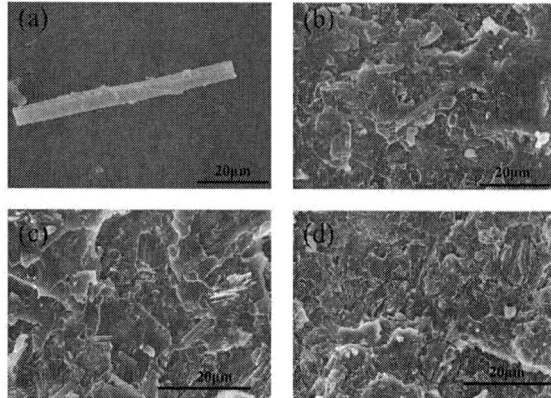

Fig. 2. SEM image of (a) BNWKs, (b) 3.75 wt% BNWKs-26.25 wt% BNNSs/EP, (c)7.5 wt% BNWKs-22.5 wt% BNNSs/EP (d)15 wt% BNWKs-15 wt% BNNSs/EP.

Fig. 3 shows the XRD pattern of BNWKs-BNNSs/EP composite. It can be seen from the figure that there is an obvious "drum like" diffraction peak at 10°~20°, which is the characteristic diffraction peak of amorphous EP. At the same time, the characteristic diffraction peaks of BN appeared near 26.7°, 41.6° and 55.1°, corresponding to its (002), (100) and (004) crystal planes, respectively. In addition, there is no additional impurity peak in the composite, which means that the preparation process of BNWKs-BNNSs/EP composite is only physical doping, and there are no other substances and chemical reactions.

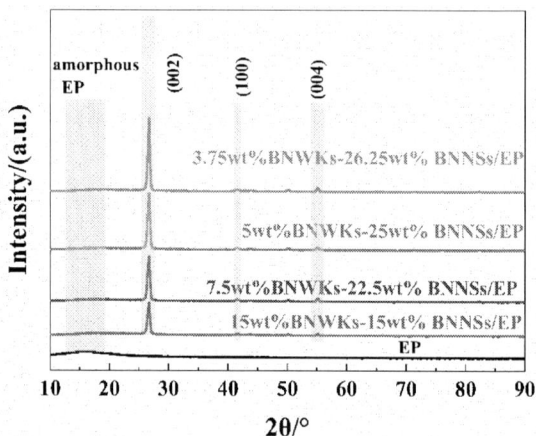

Fig. 3. XRD pattern

Fig. 4 shows the thermal conductivity of the composite. It can be found from the figure that the thermal conductivity of 3.75 wt% BNWKs-26.25 wt% BNNSs/EP composite is the highest, which is 1.497 W/(m·K), 8.32 times that of EP, and 19.57% higher than that of 30wt% BNNSs/EP; The thermal conductivity of 30wt% BNWKs/EP was the lowest, which was 0.37 W/(m·K), only 2.06 times that of EP. Compared with 30 wt% BNNSs/EP, the thermal conductivity decreased by 70.45%. On the whole, it can be found that with the increase of BNWKs content, the thermal conductivity of the composites increases first and then decreases. When the content of BNWKs in BNWKs-BNNSs/EP composites is low (≤5 wt%), the thermal conductivity of the composites is still superior to that of BNNSs/EP, but when the content of BNWKs is too high (≥7.5 wt%), the thermal conductivity of the composites is seriously deteriorated. The reason for this phenomenon is that the addition of BNWKs has two effects on the composites: (1) BNWKs is rod-shaped and has a high aspect ratio, which is easy to form a thermal bridge between BNNSs, so as to speed up the phonon transmission, reduce the phonon scattering and improve the thermal conductivity path, which is very beneficial to improve the thermal conductivity of the composites; (2) The contact between one-dimensional BNWKs is point contact, and the interface thermal resistance is large. At the same time, the high aspect ratio is also prone to agglomeration, which is unfavorable to the improvement of the thermal conductivity of the composite. When the content of BNWKs is low, BNWKs can play a good role in bridging, providing a fast channel for heat transfer. At this time, the influence of (1) is greater than (2), which is conducive to the improvement of the thermal conductivity of the composite; However, when the content of BNWKs is too high, the bridging effect of BNWKs is gradually weakened. At the same time, due to the characteristics of high aspect ratio and easy agglomeration, BNWKs form a point contact surface, which increases the thermal resistance of the composite, reduces the phonon transmission rate, and further weakens the synergistic effect between BNWKs and BNNSs, resulting in the further reduction of the thermal conductivity of the composite. At this time, the influence of (2) is greater than (1), which hinders the improvement of the thermal conductivity of the composite. Fig. 4 (b) shows the change curve of the thermal conductivity of each composite material at different temperatures. It is not difficult to see that with the rise of temperature, the thermal conductivity of each composite material also increases slightly, which may be caused by the mismatch between the thermal expansion coefficient of the filler and the matrix. The thermal expansion coefficient of the matrix is higher than that of the filler. In the process of temperature rise, the continuous extrusion of the filler leads to closer contact between the filler and the filler and between the filler and the matrix, thus reducing the interface thermal resistance of the composite material and speeding up the phonon transmission.

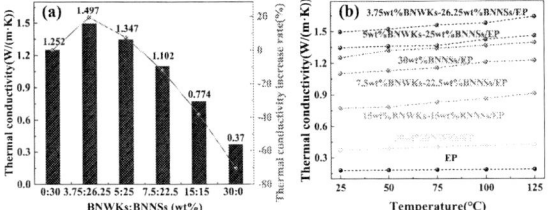

Fig. 4. Thermal conductivity of BNWKs-BNNSs/EP composites.

In order to further verify the synergistic effect between BNWKs and BNNSs, the synergistic effect rate f_X is used to analyze the thermal conductivity of the composite. The specific calculation formula is as follows:

$$f_X = \frac{K - K_m}{\left(K_1 - K_m\right) + \left(K_2 - K_m\right)} \quad (1)$$

Where: f_X is the synergistic effect rate, %; K、K_m、K_1 and K_2 are the thermal conductivity of BNWKs-BNNSs/EP composites, EP, BNNSs/EP composites and BNWKs/EP composites, respectively, W/(m·K)。 The higher the value of F_X, the more significant the synergistic effect between BNNSs and BNWKs. The specific calculation results are shown in Fig. 5. It can be seen from the figure that the synergistic effect rate of the two fillers gradually decreases with the increase of BNWKs content. The synergistic effect rate of 3.75wt% BNWKs-26.25wt% BNNSs/EP composite is the highest, which is 1.044, showing a strong synergistic effect. The synergistic effect rate of 15 wt% BNWKs-15wt% BNNSs/EP composite is the lowest, which is only 0.471, indicating that the synergistic effect is weak, which is consistent with the thermal conductivity test.

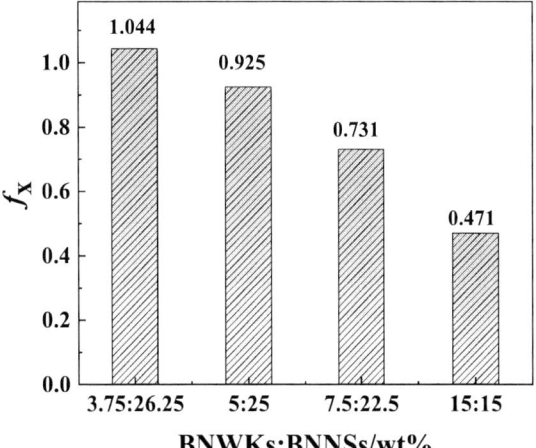

Fig. 5. Analyses of synergistic effect of BNWKs BNNSs/EP composites.

It can be seen from Fig. 6 (a) that the dielectric constant of the composite material gradually decreases with the increase of frequency, because the establishment of slow polarization takes a long time, and with the rapid change of electric field frequency, the slow polarization gradually fails to keep up with the change of electric field. At the same time, through comparison, it can be found that with the increase of BNWKs content, the dielectric constant of the composite also gradually increases. At the frequency of 10^6 Hz, The dielectric constant of EP is only 4.08, while the dielectric constant of 30 wt% BNWKs/EP is 5.96. There may be two reasons for this: 1) the increase of BNWKs content leads to more interface polarization under the same filling mass fraction; 2) The rod-shaped structure of BNWKs is easy to form a needle plate electrode structure near both ends of the positive and negative plates, which leads to the concentration of electric field at both ends of the electrode and the increase of the dielectric constant of the composite. At the same time, it can also be found that the dielectric loss of the composites increases with the increase of BNWKs content. At 10^6 Hz, the dielectric loss of EP is 0.0155, while that of 30 wt% BNWKs/EP is 0.03927, which is much higher than that of EP. It can be seen that the introduction of BNWKs leads to the deterioration of the dielectric properties of the composites. Fortunately, at the frequency of 10^6 Hz, the dielectric constant and dielectric loss of 3.75 wt% BNWKs-26.25 wt% BNNSs/EP with the highest thermal conductivity are 4.5 and 0.01977, respectively, which meet the application requirements.

In order to further study the dielectric properties of 3.75 wt% BNWKs-26.25 wt% BNNSs/EP composites at high temperature, the relevant measurements were carried out, and the results are shown in Fig. 6 (b). It can be seen from the figure that the dielectric constant and dielectric loss of the composite increase with the increase of temperature. In high frequency environment (10^6 Hz), the dielectric constant of the composite is 4.49 at 25 °C, the dielectric loss is 0.003261, the dielectric constant is 4.97 at 150 °C, and the dielectric loss is 0.0188, which still meets the application requirements of packaging materials, reflecting its application prospect in high frequency and high temperature environment.

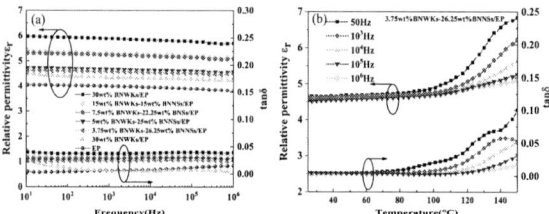

Fig. 6. Dielectrics of BNWKs-BNNSs/EP composites.

One of the basic requirements of electronic components for packaging materials is to require materials with high volume resistivity. The higher the volume resistance of packaging materials, the better the protection of internal components and reduce the impact of high electric field on components. Fig. 7 (a) shows the volume resistivity of the composite. It can be found from the figure that compared with EP, the volume resistivity of the composite is seriously deteriorated. At the same time, with the increase of BNWKs content, the deterioration is also gradually aggravated. In particular, the volume resistivity of 30wt% BNWKs/EP composite decreases to 9.04×10^{12} $\Omega \cdot$m, which may be due to the following two reasons: 1) there are inevitable minor defects or cracks in the preparation process of the composite, resulting in the decline of physical resistivity; 2) The introduction of rod-shaped BNWKs results in the concentration of electric field at both ends of the whisker, which further reduces the volume resistivity of the composite. Fortunately, although the volume resistivity of the composite decreased significantly compared with that of EP matrix, it was still higher than 10^9 $\Omega \cdot$m, meeting the application requirements of packaging materials.

The breakdown field strength is another important characteristic parameter that determines the insulation characteristics of composite materials. Fig. 7(b) shows the breakdown field strength of the composite. It can be seen that the breakdown strength of 3.75 wt% BNWKs-26.25 wt% BNNSs/EP composite is 70.98 kV/mm, which is only 15.94% lower than that of EP, and still maintains a high breakdown strength. This shows that when the doping content of BNWKs is 3.75 wt% and BNNSs is 26.25 wt%, the epoxy resin composite still has good breakdown strength under the premise of ensuring relatively excellent thermal conductivity, which can meet the application requirements.

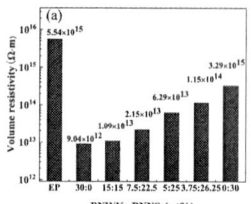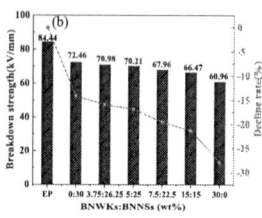

Fig. 7. Volume resistivity and breakdown field strength of composites.

IV. CONCLUSIONS

In this paper, the total filler mass fraction is kept constant by 30 wt%, and BNWKs-BNNSs/EP composite materials are prepared in different mass ratios of BNWKs and BNNSs. When the doping content of BNWKs is 3.75 wt% and the doping content of BNNSs is 26.25 wt%, the thermal conductivity of BNWKs-BNNSs/EP composites is the highest, which is 1.497 W/(m·K), which is 731.7% higher than that of EP and 1.2 times the thermal conductivity of 30 wt% BNNSs/EP. At the same time, synergistic effect calculation proves that it shows excellent synergistic effect. Although the introduction of one-dimensional BNWKs reduces the electrical insulation performance of composite materials to a certain extent, 3.75 wt% BNWKs-26.25 wt% BNNSs/EP composite materials can still ensure that their volume resistivity (1.15×10^{14} $\Omega \cdot$m) and breakdown field strength (70.98 kV/mm) can still meet the application requirements of packaging materials with excellent thermal conductivity.

ACKNOWLEDGMENT

This work was supported by the National Natural Science Foundation of China (No. 52277024). Heilongjiang Provincial Natural Science Foundation of China (No. ZD2024E008). Heilongjiang Chunyan Innovation Team Program (CYQN24003).

REFERENCES

[1] Y. F. Wen, C. Chen, Y. S. Ye, Z. G. Xue, H. Y. Liu, X. P. Zhou, Y. Zhang, D. Q. Li, X. L. Xie, and Y. W. Mai, "Advances on thermally conductive epoxy-based composites as electronic packaging underfill materials—a review, " Adv. Mater., vol. 34, Dec 2022.

[2] X. W. Cao, L. T. Ge, X. C. Yin, and G. J. He, "Preparation of noncovalent functionalized Boron Nitride and its 3D collaborative thermal conduction network with epoxy, " Surf. Interfaces., vol. 34, Nov 2022.

[3] Y. Hu, Ch. Chen, Y. F. Wen, Zh. G. Xue, X.P. Zhou, D. A. Shi, G. H. Hu, and X. L. Xie, "Novel micro-nano epoxy composites for electronic packaging application: Balance of thermal conductivity and processability, " Compos. Sci. Technol., vol. 209, June 2022.

[4] Y. M. Yao, J. J. Sun, X. L. Zeng, R. Sun, J. B. Xu, and Ch. P. Wong, "Construction of 3D skeleton for polymer composites achieving a high thermal conductivity, " SMALL. vol. 14, March 2018.

[5] Zh. M. Chen, J. Xie, Y. H. Fu, X. L. Wang, X. L. Zhang, Sh. Wang, Ch. X. Xiong, and Sh. P. Zhu, "Enhanced thermal conductivity of epoxy resin by incorporating pitch-based carbon fiber modified by Diels-Alder reaction, " Diamond Relat. Mater. vol. 127, Aug. 2022.

[6] X. Y. Zeng, Zh. G. Zhang, Y. Q. Pan, Y. L. Zhang, L. and X. Hou, "Epoxy-based composites with enhanced thermal properties through collective effect of different particle size fillers, " POLYM POLYM COMPOS. vol. 30, June 2022.

[7] G. Chen, L. F. Li, P. F. Guo, and X. Ma, " Preparation and performance of epoxy resin-based thermal conductive composites with different morphologies of ZnO, "J. Sol-Gel Sci. Technol. vol. 107 May 2023.

Study on Surface Discharge Characteristics of Silicone Gel in Salt Fog Environment

Feng Wang
North China Electric Power University
Beijing, China
wangfeng990109@163.com

Zhihui Li
Guangdong Power Grid Company of Limited Liability
Guangzhou, China
lzh_ncepu@163.com

Yateng Yany
North China Electric Power University
Beijing, China
yangyateng2002@163.com

Shanzhen Fan
North China Electric Power University
Beijing, China
szfan_bj@163.com

Hanwen Ren
North China Electric Power University
Beijing, China
rhwncepu@ncepu.edu.cn

Jian Wang
North China Electric Power University
Beijing, China
wangjian31791@ncepu.edu.cn

Wei Wang
North China Electric Power University
Beijing, China
wwncepu@163.com

Qingmin Li
North China Electric Power University
Beijing, China
lqmeee@ncepu.edu.cn

Abstract—IGBT (Insulated Gate Bipolar Transistor) power modules are the core components of offshore wind power converters, and the rapid development of the offshore power generation industry has driven a significant increase in their demand. Under long-term operating conditions, IGBT modules are susceptible to corrosion from high salt spray penetration, leading to module failure. As the main insulation material for IGBT devices, silicone gel's insulation performance is also affected by salt fog particles, making it necessary to study its insulation performance in a salt fog environment. In this study, silicone gel was subjected to surface discharge testing at different frequencies under varying salt fog concentrations using a controlled variable method, and surface damage characteristics were investigated. The research findings indicate that the starting voltage and flashover voltage of surface discharge are minimally affected by salt fog concentration. the measurement finds that there is a special frequency-induced inflection point phenomenon of surface breakdown voltage at 15 kHz under the salt fog environment. The surface damage morphology of the silicone gel varies at different frequencies, with the most severe damage observed at 15 kHz frequency. These research results can provide valuable insights for the insulation design of IGBT packaging.

Keywords—Offshore wind power, insulated gate bipolar transistor (IGBT), organic silicone gel, salt spray, gas-solid surface discharge

I. INTRODUCTION

With the increasing prominence of global energy shortages, environmental pollution, and extreme weather conditions, clean energy sources such as wind power have garnered significant attention. In recent years, China's offshore wind power industry has experienced rapid growth, driven by policy support, market demand, and technological advancements. The country has consecutively led global rankings in annual installed capacity for three years, consequently spurring a sharp increase in demand for equipment such as offshore wind power converters[1]. Insulated Gate Bipolar Transistor (IGBT), the core components responsible for energy conversion and transmission in wind power converters, typically employ organic silicone gel as an encapsulation material, enveloping critical components such as power device chips, bonding wires, and Direct Copper Bond (DCB) substrates[2]. Over extended service in marine environments, gaps between the silicone gel and substrates widen, allowing high-concentration salt fog to permeate the module through interface voids. Additionally, temperature fluctuations can lead to the formation of condensation droplets on the surface of the organic silicone gel, causing it to penetrate the module[3]. These factors pose higher demands on the electrical performance of encapsulation materials in salt fog environments.

Current research on organic silicone gel both domestically and internationally primarily focuses on the impact of electrical-thermal stress on insulation performance, with limited studies investigating the performance of silicone gel in salt fog environments. Analysis by Ka. Fischer et al. on global wind turbine failure statistics reveals that submodule failures (IGBT modules) constitute the largest portion of inverter component failure modes (22%)[4]. In a study on corrosion of IGBT modules in offshore wind power converters, Zhao Shilin et al. discovered significant amounts of sulfides and chlorides at the chip surfaces, DCB board copper layers, and output terminals of faulty and long-serving IGBT modules, significantly reducing their service life[5]. Research by J. Kiilunen et al. on the reliability of inverters in salt fog environments found that

This work was supported in part by National Key R&D Program of China (2025YFE0106300), National Natural Science Foundation of China (No. U24B2092, 52207153), Beijing Natural Science Foundation (L241043).

half of the faulty samples exhibited IGBT module failures, all attributed to short-circuit failures caused by salt fog corrosion[6]. Therefore, investigating the electrical insulation performance of organic silicone gel in salt fog environments holds significant implications for the development of IGBT technology.

To address the aforementioned issues, this study proposes a needle-plate electrode model at the organic silicone gel-salt fog interface for surface discharge testing. Samples are subjected to partial discharge inception voltage (PDIV) and repetitive flashover voltage tests at different frequencies in varying concentrations of salt fog environments, considering the influence of factors such as temperature and humidity on surface discharge. Utilizing scanning electron microscopy and other electrical equipment, the surface morphology and generated substances of organic silicone gel after flashovers at different frequencies in N3 salt fog are examined, providing a foundational and theoretical basis for further research on partial discharge along power module surfaces and insulation optimization design.

II. EXPERIMENTAL PLATFORM AND SAMPLE PREPARATION

The experimental platform used in this study, as shown in Fig. 1, mainly consists of a high-frequency positive square wave power supply, a needle-plate electrode mold, a fog control box, a temperature and humidity controller, a high-voltage probe, and a high-speed digital oscilloscope. The high-frequency positive square wave power supply can output continuously adjustable voltages ranging from 0 to 50 kV with a frequency range of 0 to 50 kHz. The high-voltage probe has a voltage division ratio of 1000:1 for measuring the high-frequency positive square wave voltage. The high-speed digital oscilloscope has a maximum sampling rate of 5 GS/s and a sampling bandwidth of 1.5 GHz. Salt fog is generated using an ultrasonic atomizer, transported to the fog chamber through a gas hose, and the humidity inside the fog chamber is controlled to be 96% and the temperature at room temperature using an intelligent humidity controller and dehumidifier. The needle electrode of the electrode model is made of 0.8 mm stainless steel, and both positive and negative electrode plates are made of stainless steel. 70mm×40mm×5mm acrylic mold is used in the experiment, with organic silicone gel encapsulated inside the mold. The distance between the stainless steel needle tip and the negative electrode plate is 3mm, with an angle of 15° between the needle and the sample[7]. The experiment is conducted by applying pressure on the positive electrode plate.

To avoid unnecessary influences of contamination, moisture, and other factors on the insulation performance of samples and test results, it is necessary to clean and dry the mold before testing, and the organic silicone gel sample needs to undergo degassing and curing treatment during preparation. The mold is cleaned with anhydrous

ethanol and dried in a 60°C oven for 2 hours. Prior to filling, the organic silicone gel is prepared. 40g both of components A and B are weighed separately using a digital balance, mixed with a mechanical stirrer for 25 minutes until evenly mixed, and then poured into the acrylic mold so that the needle tip slightly touches the silicone gel surface. To eliminate air bubbles formed during filling, a 30-minute vacuum deaeration is conducted, followed by curing the sample in a 60°C constant temperature drying oven for 12 hours to form a solid structure[8].

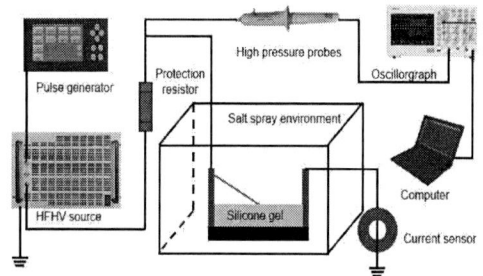

Fig.1. Experimental platform for surface discharge testing

Salt fog solutions with concentrations of 1% NaCl, 3% NaCl, and 5% NaCl are denoted as N1, N3, and N5, respectively. The organic silicone gel is subjected to surface insulation strength testing at frequencies of 5 kHz, 10 kHz, and 5 kHz under different concentrations of salt fog[9]: the voltage is uniformly increased at a rate of 100V/s until intermittent discharge and surface flashover occur, as shown in Fig. 2. The voltage at which partial discharge initiates (Partial Discharge Inception Voltage, PDIV) and the flashover voltage (FV) of the sample are recorded, followed by repetitive flashover voltage testing after surface flashover. Ten tests are conducted at different frequencies, and the average data is taken as the final result.

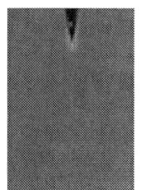

Fig. 2. Development of surface discharge (a) Partial discharge inception (b) Intermittent discharge (c) Surface flashover

III. ANALYSIS OF SURFACE DISCHARGE TEST RESULTS

In order to study the effect of different concentrations of salt fog on the insulating strength of the surface of silicone gel, select the starting discharge voltage and three salt fog concentrations: N1, N3 and N5. The starting flashover voltage and the repeated flashover voltage of the surface of the silicone gel are tested, the test results are shown in Fig. 3.

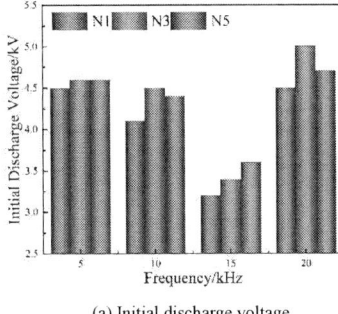

(a) Initial discharge voltage

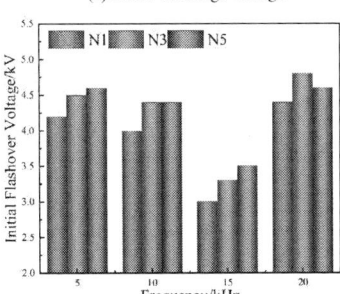

(b) Initial flashover voltage

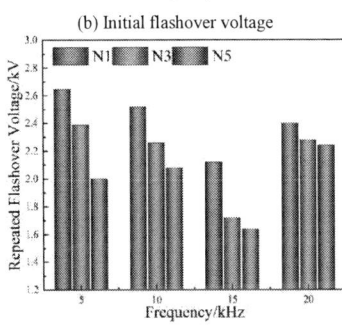

(c) Repeated flashover voltage

Fig.3. Electrical insulation strength at different frequencies

At a constant salt spray concentration, the starting voltage, flashover voltage, and repeated flashover voltage of silicone gel initially decline as the frequency decreases but then rise as the frequency increases, exhibiting an inflection point at 15 kHz.

The variation in breakdown voltage along the surface of the silicone gel can be attributed to the dissipation of tip charge at different frequencies and the high-frequency thermal effects associated with the "frequency-induced inflection point" phenomenon. When the frequency is below 15 kHz, the thermal effects dominate the discharge process. As the frequency rises, the motion of the tip charge and free particles in the salt spray intensifies, making the tip more prone to corona discharge. Once the frequency exceeds 15 kHz, the high-frequency thermal effect reaches saturation. The continuous injection of charge into the needle tip

dissipates and repels with frequency variations, hindering charge accumulation.

IV. SURFACE DAMAGE CHARACTERISTICS OF SILICA GEL

To investigate the surface morphology changes during surface flashover processes of silicone gel under high-frequency electrical stress in a salt fog environment, optical microscopy was employed to test and analyze the surface appearance of samples after surface flashover at different frequencies.

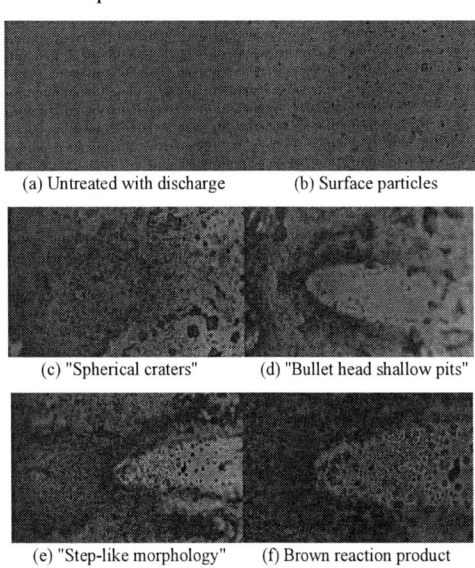

(a) Untreated with discharge	(b) Surface particles
(c) "Spherical craters"	(d) "Bullet head shallow pits"
(e) "Step-like morphology"	(f) Brown reaction product

Fig. 4. Surface discharge damage appearance1

The microstructure of the surface of the silicone gel before flashover treatment and under three different frequency damage conditions is shown in the following figures. As seen in Fig. 4(a), the surface of the sample without flashover treatment is smooth, but there are some impurities sticking to the surface of the silica gel, which is more pronounced in the salt fog environment. After evaporation of moisture, NaCl particles adhere to the surface in a relatively dispersed form, as shown in Fig. 4(b). At a frequency of 5 kHz, the accumulation of charge at the needle tip increases the field strength between the needle tip and the negative substrate, forming a discharge path and causing surface flashover, leaving behind "spherical craters," brown deposits, and NaCl particles on the surface, as shown in Fig. 4(c). This may be attributed to the main accumulation of charge at the needle tip and the insufficient clarity of the high-frequency heating effect. At 10 kHz frequency, the surface of the material after flashover exhibits a different microstructure. The material does not form the penetration craters as shown in Fig. 4(c) but instead leaves behind "bullet head shallow pits" enclosing the needle tip, with a depth slightly smaller than the "spherical craters" in Fig. 4(c). Additionally, brown substances are generated at the tip of the shallow pits, as shown in Fig. 4(d). This could be due to the increased high-frequency heating effect with higher frequency, leading to an increased accumulation of charge at the needle tip, intensified movement of salt fog particles, making surface flashover easier. Observing the overall surface damage of the material at 15 kHz frequency, it is found that the material surface forms more pronounced discharge craters, including both "spherical craters" and "bullet head shallow pits," with a layered structure resembling "steps" between the craters, as shown in Fig. 4(e). Furthermore, more pronounced brown substances appear around the craters, possibly due to the high-frequency joule heating effect causing the decomposition of the silica gel, leading to free NaCl micro-particles directly contacting the decomposed substances and undergoing cross-linking reactions, as shown in Fig. 4(f). Materials subjected to surface flashover at different frequencies exhibit distinct morphological features, reflecting the insulation performance of silica gel under high-frequency electrical stress.

V. Conclusions

This paper established a salt fog environment gas-solid surface discharge test system to study the high-frequency surface discharge characteristics of the silicone gel-salt fog interface under different influencing factors. Through theoretical analysis and experimental research, the phenomenon of rapid voltage drop during the flashover process was investigated, and the surface damage of the material after repeated flashovers was characterized. The following conclusions were drawn: (1) The concentration of salt fog has little effect on the starting voltage and flashover voltage of surface discharge. (2) The repetitive flashover voltage is inversely proportional to the concentration of salt spray, and the above parameters of surface discharge show a special phenomenon of "frequency inflection point" at 15 kHz frequency. (3) Surface damage is lighter at 5 kHz and 10 kHz frequencies, accompanied by "spherical craters" and "bullet head shallow pits." (4) Surface damage is most severe at 15 kHz, exhibiting a combination of the two previous morphologies, forming a "step-like" appearance, and generating more brown substances in the surrounding area.

The research in this paper was supported by the National Key Research and Development Program of China (2021YFB2601404), the National Natural Science Foundation of China (52207153), and the Natural Science Foundation of Beijing Municipality (3232053). The project focuses on the surface discharge characteristics of insulating silica gel for IGBT encapsulation under salt spray environment.

REFEREFCES

[1] H. Wang, F. Blaabjerg, K. Ma and R. Wu, "Design for reliability in power electronics in renewable energy systems – status and future," 4th International Conference on Power Engineering, Energy and Electrical Drives, Istanbul, Turkey, 2013, pp. 1846-1851, doi: 10.1109/PowerEng.2013.6889108.

[2] H. Jain, B. Mather, A. K. Jain and S. F. Baldwin, "Grid-Supportive Loads—A New Approach to Increasing Renewable Energy in Power Systems," in IEEE Transactions on Smart Grid, vol. 13, no. 4, pp. 2959-2972, July 2022, doi: 10.1109/TSG.2022.3153230.

[3] R. Kumar, F. Ke, D. England, A. Summers and L. Young, "A New Halogen-Free Vapor Phase Coating for High Reliability & Protection of Electronics in Corrosive and Other Harsh Environments," 2022 International Conference on Electronics Packaging (ICEP), Sapporo, Japan, 2022, pp. 87-88, doi: 10.23919/ICEP55381.2022.9795465.

[4] Fischer et al., "Reliability of Power Converters in Wind Turbines: Exploratory Analysis of Failure and Operating Data From a Worldwide Turbine Fleet," in IEEE Transactions on Power Electronics, vol. 34, no. 7, pp. 6332-6344, July 2019, doi: 10.1109/TPEL.2018.2875005.

[5] S. Zhao, J. Yang, Y. Tong and E. Yao,"Progress on Corrosion and Protection of IGBT Modules Used in Offshore Wind Power Converter ,"in Material Reports, vol 37, pp. 407-413, July 2023.

[6] J. Kiilunen and L. Frisk, "Reliability testing of frequency converters with salt spray and temperature humidity tests," 2009 European Microelectronics and Packaging Conference, Rimini, Italy, 2009, pp. 1-5.

[7] S. Li, W. Si and Q. Li, "Partition and recognition of partial discharge development stages in oil-pressboard insulation with needle-plate electrodes under combined AC-DC voltage stress," in IEEE Transactions on Dielectrics and Electrical Insulation, vol. 24, no. 3, pp. 1781-1793, June 2017, doi: 10.1109/TDEI.2017.006361.

[8] F. Zeng, D. Su, R. Chen, Q. Yao, L. Li and J. Tang, "Effect of Thermal Oxidative Aging on Cross-Linking Network and Electrical Property of Silicone Gel for IGBT Packaging," in IEEE Transactions on Dielectrics and Electrical Insulation, vol. 31, no. 2, pp. 1012-1019, April 2024, doi: 10.1109/TDEI.2023.3345259.

[9] T. Yamashita, K. Iwanaga, T. Furusato, H. Koreeda, T. Fujishima and J. Sato, "Improvement of insulation performance of solid/gas composite insulation with embedded electrode," in IEEE Transactions on Dielectrics and Electrical Insulation, vol. 23, no. 2, pp. 787-794, April 2016, doi: 10.1109/TDEI.2015

Effect of Atomic-Level Roughness on Vertical β-Ga$_2$O$_3$ Schottky Barrier Diode and High-Temperature Performance

Jiaxiang Chen[a], Xingye Zhang[a], Maojin Yang[a], Xinpeng Lin[a], Teng Jiao[a], Yan Liu[a], Xianhu Zha[a], Haolin Hu[a], Xiangjin Ding[a], Yuxi Wan[a,*], Jun Ma[b], Mengyuan Hua[b,*], Dao-Hua Zhang[a,*]

[a] Shenzhen Pinghu Laboratory, Shenzhen, China
[b] Department of Electronic and Electrical Engineering, Southern University of Science and Technology, Shenzhen, China

(*Email: wanyuxi@phlab.com.cn; huamy@sustech.edu.cn; zhangdaohua@phlab.com.cn)

Abstract—We demonstrate the atomic scale surface roughness of β-Ga$_2$O$_3$ epitaxial layer by developing a two-step inductively coupled plasma and wet-repair treatment. The superior surface roughness of 0.12 nm and close-to-unity ideal factor (*n*) of 1.05 were achieved using optimized surface treatment on vertical β-Ga$_2$O$_3$ Schottky barrier diodes (SBDs). Compared to untreated samples, the blocking capability was enhanced with increasing breakdown voltage (BV) above 1.6 times. Furthermore, a comprehensive analysis of the electrical properties at high temperature was analyzed to verify that the improved Ni/β-Ga$_2$O$_3$ interface quality can lead to an enhancement of the device performance. When subjected to an off-state stress of -100 V at 175℃, the surface-treated SBDs perform stable forward characteristics with slight current-density degradation (ΔJ_{max}<0.01%). The results pave the way for further improving the surface quality of β-Ga$_2$O$_3$ devices and validate the potential of SBDs in high-temperature power application scenarios.

Keywords—Gallium oxide, Schottky barrier diode, smooth surface roughness, high-temperature, reliability

I. INTRODUCTION

β-phase Ga$_2$O$_3$ has emerged as a highly promising ultra-wide bandgap semiconductor (~4.8 eV) for next-generation power electronics, exhibiting superior material characteristics, enabling revolutionary performance in high-voltage rectifiers, power conversion systems, and extreme-environment electronic devices.[1, 2] β-Ga$_2$O$_3$ Schottky barrier diode (SBD) as rectifier ensures robust operation at high-temperature and in harsh environment.[3] Moreover, the compatibility of Ga$_2$O$_3$ with cost-effective melt-growth technique further enhances its scalability for power electronics.

The air-exposed surface of β-Ga$_2$O$_3$ epi-wafers may contain impurities and contaminations, such as adsorbed carbon, surface voids, and native oxides.[4]

Interface states and defects may account for conduction and blocking properties in Ga$_2$O$_3$ power devices. The high-quality interface is critical to device performance.[5]Therefore, surface pre-treatments are needed to optimize the surface roughness and quality. The roughness of the Ga$_2$O$_3$ surface is also the key aspect for Schottky barrier height (SBH) homogeneity.[6] In recent years, numerous surface treatment strategies for β-Ga$_2$O$_3$ Schottky diodes, such as annealing, wet chemical, F-plasma and dry etching, have been used to enhance the interface quality and reduce roughness.[7, 8] Among them, inductively coupled plasma (ICP) etching can effectively improve the surface quality. However, atomic-scale surface roughness of β-Ga$_2$O$_3$ wafer is still a challenge.

In this work, the vertical β-Ga$_2$O$_3$ SBD with self-aligned mesa termination was fabricated, which can eliminate alignment deviation and simplify the fabrication process. The unreliable surface on the top of (001) Ga$_2$O$_3$ epi-layer was removed by two-step inductively coupled plasma (ICP) dry etching and wet repair treatment. The surface roughness can be suppressed to the atomic-scale of 0.12 nm, which is the minimum value so far. Moreover, the electrical performance was investigated from room temperature (RT) up to 175 ℃.

II. DEVICE FABRICATION AND MEASUREMENT

A. Device structure and fabrication

Fig.1 shows the schematic cross-section of the self-aligned Ga$_2$O$_3$ SBDs in this study. The device consists of a 6 μm Si-doped epitaxial drift layer grown by halide vapor phase epitaxy (HVPE) on (001) oriented 650 μm Sn-doped high-conductivity Ga$_2$O$_3$ substrate (from Novel Crystal Technology, Japan).

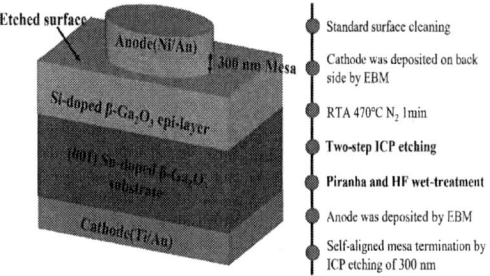

Fig. 1. Schematic cross-sectional structure and main process flow of the self-aligned SBDs used in this work.

The samples were first processed by solvent and acid cleaning. Then, the backside ohmic contact Ti/Au (20/80nm) was deposited by e-beam evaporation followed by 1min 470°C rapid thermal annealing in N_2 ambient. Then, the 3 min fast ICP etching with source power of 300 W, chamber pressure of 10 mTorr, and a mixture of $BCl_3/Cl_2/Ar$ was used to etch off the front face of wafer. To reduce surface roughness, 1 min slow ICP etching with source power of 200 W and same $BCl_3/Cl_2/Ar$ mixture was used on samples. Following two-step dry etching, the surfaces of samples were repaired by Piranha solution ($H_2O_2/H_2SO_4=1:3$) and hydrofluoric acid (HF), sequentially. After that, Ni/Au/Cr (20/80/170 nm) as Schottky metal and hard mask was subsequently deposited using electron-beam evaporation with standard lift-off process. Finally, the shallow-groove mesa termination was conducted by $BCl_3/Cl_2/Ar$ gas combinations with an etched depth of around 300 nm.

B. Characterization methodology

The I-V characteristics of the devices were measured by Keysight B1505A Semiconductor Device Analyzer. The breakdown measurements of diodes were carried out submerged in Fluorinert to rule out air arcing.

III. RESULTS AND DISCUSSION

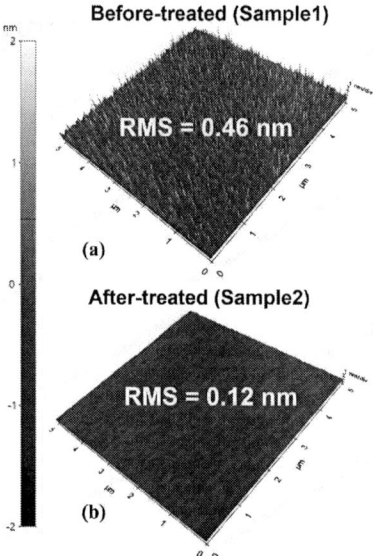

Fig. 2. Measured three-dimensional surface-roughness profile using AFM of (a) before dry-etched sample 1, and (b) after dry-etched sample 2.

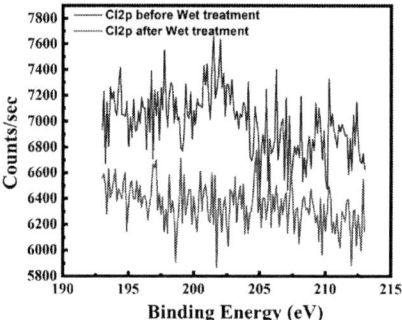

Fig. 3. XPS spectra of Cl for surface by-product induced by dry etching: without wet-solution process (black line), and with wet-solution process (blue line).

To verify the effect of etching treatment on the surface uniformity, AFM was performed over a 5×5 μm^2 scanning area, as shown in Fig. 2. For comparison, the RMS roughness of the sample Sample 1 (before ICP etching and wet-treatment) reached 0.46 nm, and that of Sample 2 (after surface treatment) was only 0.12 nm, reaching atomically flat surface. In addition, the repaired film showed a more uniform step-flow morphology. During the ICP dry-etching, the excessive by-products can be induced by BCl_3 plasma, such as $GaCl_3$. By X-ray photoelectron spectroscopy (XPS) analysis, the types of surface compounds (Cl 2p) after ICP dry-etching without wet-treatement were determined at level of 0.3% in Fig 3. However, the presence of a minimal amount of chlorine was identified on the after wet-etching surface, suggesting that the appropriate dry etching and wet-etching processes could effectively reduce the by-products adsorbed on the surface. This demonstrates the effectiveness of the combining dry-etching and wet-solution treatment for restoring damaged surfaces and reduceing the surface roughness.

As shown in Fig.4 (a-b), *J–V* characteristics of devices with/without (w/wo) surface treatment at room temperature were plotted. Both SBDs exhibit the high on/off current ratio ($J_{on/off}$) of 10^9, indicating devices have the outstanding charge modulation. The on-resistance ($R_{on,sp}$) of two SBDs are 5.56 and 5.23 $m\Omega \cdot cm^2$, respectively. Further, the barrier height of the SBD can be extracted from the equation shown as follows:

$$J_0 = A^* T^2 \exp\left(\frac{-q\Phi_B}{kT}\right)$$

where J_0 is the saturation reverse current density, A^* is Richardson's constant, T is measured temperature, and Φ_B presents the barrier height. Φ_B is extracted from the linear fit of the Richardson plot. After the surface treatment, Schottky barrier height (SBH) was increased from 1.02 eV to 1.12 eV, which favors the reduction of the leakage current under reverse bias. Besides, the *n* can be extracted from *I-V* curves. The surface-treated SBD (sample 2) exhibited an ideal factor (*n*) of 1.05, which is much lower than the corresponding value of 1.65. For sample 2, the value of *n* is around unity (1.05), suggesting the surface-treated SBDs have homogeneous SBH distribution.[9]

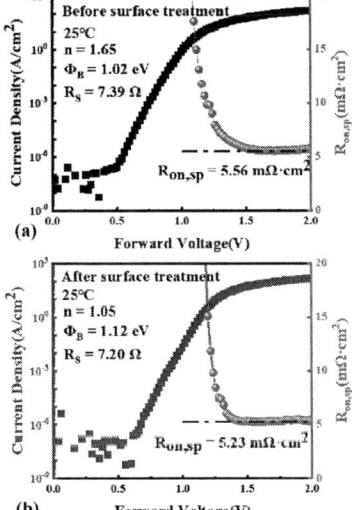

(a)

(b)

Fig. 4. Forward characteristics of (a) before and (b) after surface treatment SBDs.

Nevertheless, the high value of ideal factor (1.65) of untreated sample 1 may originate from the interfacial trap-state and Gaussian distribution of barrier height inhomogeneity.[10, 11] In Fig. 4(a-b), the series resistance (R_s) exhibits a slight decrease from 7.39 to 7.20 Ω, indicating the enhanced conduction properties after surface-treatment.

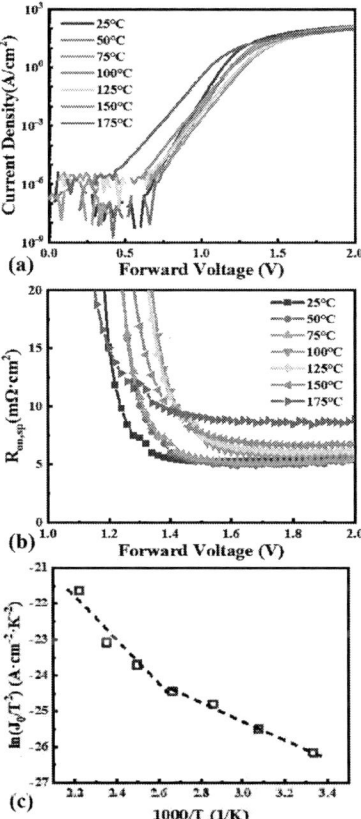

(a)

(b)

(c)

Fig. 5. Temperature-dependent (a) forward J-V and (b) $R_{on,sp}$ characteristics in logarithmic scale from 25 to 175 °C . (c) Temperature-dependent Richardson plot.

Figure 5(a-b) summarize the temperature-dependent forward *J–V* curves of sample 2 over the range from 25 to 175 °C. As the temperature increases, the current density of the device increases slightly in the subthreshold region and decreases after conduction. With increasing temperature, although the on-resistance ($R_{on,sp}$) of the device increases from 5.2 $m\Omega \cdot cm^2$ to 8.3 $m\Omega \cdot cm^2$ and the $J_{on/off}$ decreases from 10^9 to 10^8, the devices still maintain excellent operating conditions over a wide temperature range. [12] Moreover, the Φ_B was determined using I-V curves, which increased from

1.12 to 1.34 eV with elevating temperature. In Fig. 5(c), the temperature-dependent Richardson plot was showed. Also, the deviation in the Richardson plots may be due to the spatially inhomogeneous SBH and potential fluctuations at the interface.[13]

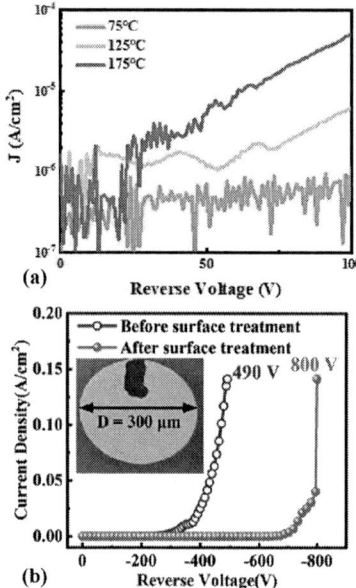

Fig. 6. (a) Temperature-dependent reverse current characteristics of SBD. (b) Breakdown voltage of SBDs with/without surface treatment at room temperature.

Fig. 6(a) shows the leakage current density for erverse bias up to 100 V in the range of 75 ℃ to 175 ℃. The breakdown characteristics of the two devices at room temperature are illustrated in Fig. 6(b). The reverse breakdown voltage (BV) of the surface-treated sample 2 can reach 800 V, much higher than that of sample 1 (490 V), owing to better surface quality. It should be noted that the reverse BV of sample 1 is above 1.6 times of magnitude higher than that of sample 2, which is due to the much lower Φ_B as well as the possible presence of surface-state assisted tunneling effect. Moreover, from the breakdown photograph shown in the inset of Fig. 6(b), the breakdown point of the device is located at the edge of the anode, which is attributed to the electric field effect at the edge of the anode. This indicates that the peak electric field of the device is not fully transferred to the bottom of the trench at this etching depth. It is promising to continue to deepen the mesa-termination to further increase the BV.

Fig.7 shows the forward characteristics before and after applying a reverse bias of -100V with stress time of 10 seconds. In Fig.7 (a-c), the threshold voltage shift is smaller when increasing temperature, owing to thermally enhanced carrier trapping process.[14, 15] With increasing temperature from 75 to 175 ℃, the degradation of forward current-density at 2 V is suppressed from 7% to 0.01%, indicating that surface-treated SBDs show the enhanced reliability at high temperature.

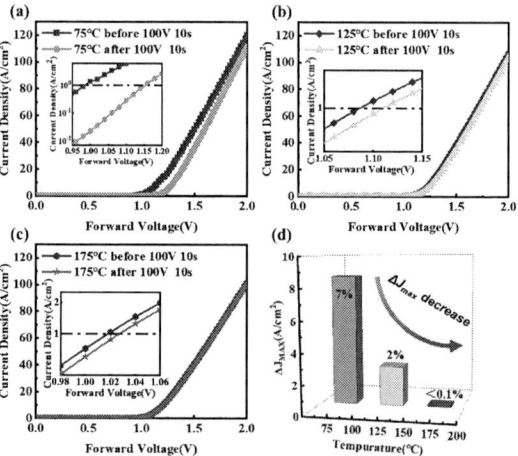

Fig. 7. Stress-reliability of SBD in high temperature range from 75 to 175 ℃ (a-c). Inset figure: J shift around 1A/cm². (d) Degradation of saturation J at 2V with different temperature.

IV. CONCLUSION

In summary, we have achieved a smooth surface with a roughness of 0.12 nm by developing the two-step ICP etching and wet-repair surface treatment. Enhanced surface quality can significantly improve the BV above 1.6 times of SBDs. In addition, the ideal factor (n=1.05) was obtained, which was attributed to the low surface density. Off-state stress measurements reveal reliable dynamic performance of surface-treated SBDs under harsh high-temperature conditions.

ACKNOWLEDGMENT

This work is supported by the Shenzhen Pinghu Laboratory Project (Grant No. 224110).

REFERENCES

[1]   H. Qu, W. Huang, Y. Zhang, J. Sui, J. Chen, B. Chen, D. W. Zhang, Y. Wang, Y. Lv, Z. Feng, and X. Zou, "Reliable electrical performance of β-Ga₂O₃ Schottky barrier diode at cryogenic temperatures," *Journal of Vacuum Science & Technology A,* 2024, 42 (2), 023418.

[2]   A. Sircar, S. Saha, U. Singisetti, and X. Yao, "Performance Analysis of a two-stage Ga2O3 Voltage Multiplier," in *2024 IEEE 11th Workshop on Wide Bandgap Power Devices & Applications (WiPDA),* 2024, pp. 1-4.

[3] B. Wang, M. Xiao, X. Yan, H. Y. Wong, J. Ma, K. Sasaki, H. Wang, and Y. Zhang, "High-voltage vertical Ga_2O_3 power rectifiers operational at high temperatures up to 600 K," *Applied Physics Letters,* 2019, 115 (26) 263503.

[4] J. Wen, W. Hao, Z. Han, F. Wu, Q. Li, J. Liu, Q. Liu, X. Zhou, G. Xu, S. Yang, and S. Long, "Vertical β-Ga_2O_3 Power Diodes: From Interface Engineering to Edge Termination," *IEEE Transactions on Electron Devices,* 2024, 71 (3), 1606-1617.

[5] A. Elwailly, M. Xiao, Y. Zhang, and H. Y. Wong, "Design Space of Vertical Ga_2O_3 Junctionless FinFET and its Enhancement with Gradual Channel Doping," in *2020 IEEE Workshop on Wide Bandgap Power Devices and Applications in Asia (WiPDA Asia),* 2020, pp. 1-5.

[6] L. A. Lyle, "Critical review of Ohmic and Schottky contacts to β-Ga_2O_3," *Journal of Vacuum Science & Technology A,* 2022, 40 (6), 060802.

[7] X. Xu, Y. Deng, S. Ye, D. Chen, T. Li, M. Zhu, and H. Zhang, "Demonstration of Surface Treatment for beta-Ga_2O_3 Schottky Barrier Diode with High Breakdown Voltage and Low Specific On-Resistance," in *2024 36th International Symposium on Power Semiconductor Devices and ICs (ISPSD),* 2024, pp. 240-243.

[8] Y.-H. Hong, X.-F. Zheng, Y.-L. He, F. Zhang, X.-Y. Zhang, X.-C. Wang, J.-N. Li, D.-P. Wang, X.-L. Lu, H.-B. Han, X.-H. Ma, and Y. Hao, "The optimized interface characteristics of β-Ga_2O_3 Schottky barrier diode with low temperature annealing," *Applied Physics Letters,* 2021, 119 (13), 132103.

[9] V. Mikhelashvili, G. Eisenstein, and R. Uzdin, "Extraction of Schottky diode parameters with a bias dependent barrier height," *Solid-State Electronics,* 2001, 45 (1), 143-148.

[10] H. Qu, J. Chen, Y. Zhang, J. Sui, Y. Gu, Y. Deng, D. Su, R. Zhang, X. Lu, and X. Zou, "Emission and capture characteristics of electron trap (E_{emi} = 0.8 eV) in Si-doped β-Ga_2O_3 epilayer," *Semiconductor Science and Technology,* 2023, 38 (1), 015001.

[11] H. Xue, Q. He, G. Jian, S. Long, T. Pang, and M. Liu, "An Overview of the Ultrawide Bandgap Ga_2O_3 Semiconductor-Based Schottky Barrier Diode for Power Electronics Application," *Nanoscale Research Letters,* 2018, 13 (1), 290.

[12] M. K. Yadav, A. Mondal, S. K. Sharma, and A. Bag, "Unveiling Thermal Effects on Sn-Doped β-Ga_2O_3 Schottky Barrier Diodes on Sapphire for High-Temperature Power Electronics," *IEEE Transactions on Electron Devices,* 2024, 71 (3), 1529-1534.

[13] Şükrü Karataş, Muzaffer Çakar, Abdülmecit Türüt, "On the electrical characteristics of the Al/rhodamine-101/p-Si MS structure at low temperatures," *Materials Science in Semiconductor Processing,* 2014, 28, 135-143.

[14] J. Chen, H. Luo, H. Qu, M. Zhu, H. Guo, B. Chen, Y. Lv, and X. Zou, "Single-trap emission kinetics of vertical β-Ga_2O_3 Schottky diodes by deep-level transient spectroscopy," *Semiconductor Science and Technology,* 2021, 36 (5), 055015.

[15] S. Bu, Y. Wang, X. Zheng, S. Yue, D. Lin, L. Yi, V. Melikyan, X. Ma, and Y. Hao; "Forward bias stress-induced degradation mechanism in β-Ga_2O_3 SBDs: A trap-centric perspective." *Applied Physics Letters.* 2025; 126 (12), 122104.

U-Shaped Sloped Field Plate Edge Termination Structure in β-Ga₂O₃ Vertical SBD

Yanzuo Li
State Key Laboratory of Wide-Bandgap
Semiconductor Devices and Integrated
Technology, School of
Microelectronics, Xidian University,
Xi'an 710071, China
liyanzuo@stu.xidian.edu.cn

Song Yuan
Guangzhou Institute of Technology,
Xidian University,
Guangzhou, 510555, China
syuan@xidian.edu.cn

Xi Jiang
Guangzhou Institute of Technology,
Xidian University,
Guangzhou, 510555, China
xjiang@xidian.edu.cn

Yanjing He
Guangzhou Institute of Technology,
Xidian University,
Guangzhou, 510555, China
hyj@xidian.edu.cn

An Xu
Guangzhou Institute of Technology,
Xidian University,
Guangzhou, 510555, China
xuan@stu.xidian.edu.cn

Guorui Mo
Guangzhou Institute of Technology,
Xidian University,
Guangzhou, 510555, China
23111213631@stu.xidian.edu.cn

Yunxuan Zhao
Guangzhou Institute of Technology,
Xidian University,
Guangzhou, 510555, China
23111213638@stu.xidian.edu.cn

Zhaoheng Yan
State Key Laboratory of Wide-Bandgap
Semiconductor Devices and Integrated
Technology, School of
Microelectronics, Xidian University,
Xi'an 710071, China
zhyan_2@stu.xidian.edu.cn

Qifan Liu
State Key Laboratory of Wide-Bandgap
Semiconductor Devices and Integrated
Technology, School of
Microelectronics, Xidian University,
Xi'an 710071, China
qifanliu@stu.xidian.edu.cn

Xinbin Zhan
State Key Laboratory of Wide-Bandgap
Semiconductor Devices and Integrated
Technology, School of
Microelectronics, Xidian University,
Xi'an 710071, China
xbzhan@stu.xidian.edu.cn

Ying Wang
Xi'an University of Posts and
Telecommunication, School of
Electronic Engineering,
Xi'an, 710121, China
ywang1@xupt.edu.cn

Linhai Zhong
Guangzhou Institute of Technology,
Xidian University,
Guangzhou, 510555, China
23111213614@stu.xidian.edu.cn

Xiaowu Gong
State Key Laboratory of Wide-Bandgap
Semiconductor Devices and Integrated
Technology, School of
Microelectronics, Xidian University,
Xi'an 710071, China
xwgong@xidian.edu.cn

Abstract—In this work, we developed a β-Ga₂O₃ Vertical Schottky barrier diode (SBD) device with a U-shaped Sloped field plate (U FP) and established its corresponding fabrication process flow. The deep mesa termination with a depth of approximately 5 μm serves as the basis for the entire termination structure. Next, a 100nm-thick Al₂O₃ layer and BCB dielectric were deposited in the trench region. Using the associated "puddle development" technique, arc-shaped pits were fabricated. Subsequently, metal field plates connected to the anode were deposited in these pits to weaken and redistribute the electric field peak at the edge of the mesa termination structure. TCAD simulations demonstrated that the introduction of arc-shaped field plates effectively weakened and redistributed the original electric field peaks. Furthermore, annealing and wet chemical treatment were incorporated into the device fabrication process to address etching-induced defects and mitigate the sharp decline in forward current density post-etching. Currently, we have fabricated devices with an on-resistance of 9.5 mΩ·cm² and a breakdown voltage of approximately 1.8 kV. This work will provide valuable insights for the development of high-voltage β-Ga₂O₃ vertical diodes in kilovolt-class applications.

Keywords— β-Ga₂O₃, power devices, vertical Schottky diode, termination.

I. INTRODUCTION

β-Ga₂O₃ has garnered significant attention from both academia and industry in power semiconductor applications due to its outstanding properties: an ultra-wide bandgap of 4.5-4.9 eV, a theoretical critical field strength of ~8 MV/cm[1], controllable n- type doping capabilities, and the potential for low-cost fabrication[2], [3]. Currently, researchers have conducted extensive research on β-Ga₂O₃ high-voltage vertical diodes, but the device performance still falls far short of the material's theoretical potential. This is primarily due to the excessive electric field concentration at the edges of the device. Due to the current inability to achieve effective doping in β-Ga₂O₃, many edge termination structures widely used in traditional semiconductor materials like Si, SiC, and GaN cannot be effectively utilized in β-Ga₂O₃. In β-Ga₂O₃ diodes, the edge termination structures reported to date primarily include field plates (FP) [4], recessed surface etching [5], mesa structures [6], and junction termination extension (JTE)

979-8-3315-1110-4/25 $31.00 © 2025 IEEE

[7]. All these edge termination structures have demonstrated effectiveness in improving the reverse breakdown voltage characteristics of the devices to varying degrees.

In this work, a U-Shaped Sloped Field Plate (U FP) termination has been proposed in β-Ga₂O₃ Vertical SBD. The similar Sloped field plate has been reported in both Si and SiC [8]-[11]. However, due to material differences, β-Ga₂O₃ lacks etching approaches analogous to the Bosch process [12]. In this work, a novel process design distinct from those used in Si and SiC has been proposed. TCAD simulations demonstrate that the introduction of a U-shaped field plate on the mesa termination can mitigate the electric field spike at the corner of the mesa edge.

II. DEVICE STRUCTURE AND FABRICATION

Figure 1(a)-(c) shows the β-Ga₂O₃ SBD with unterminated, mesa and U-shaped sloped field plate edge termination. This terminal structure consists of deep mesa etching and a U-shaped field plate. Deep mesa structures are processed by ICP etching, and BCB dielectric is subsequently filled into the

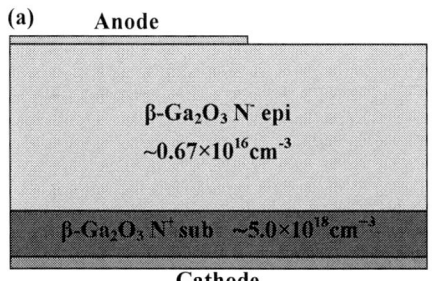

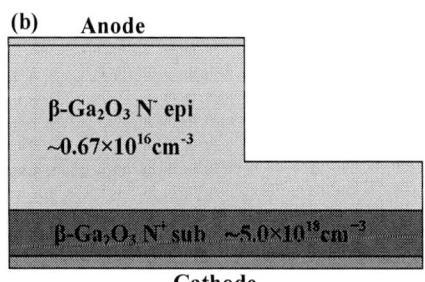

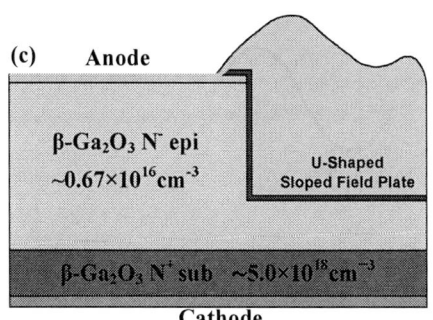

Fig. 1(a). Schematic view of unterminated Schottky Barrier Diode (UT-SBD). (b) Schematic view of mesa Schottky Barrier Diode (mesa-SBD). (c) Schematic view of U-Shaped Sloped Field Plate Schottky Barrier Diode (U FP-SBD)

trenches. Furthermore, a layer of oxide is placed between the β-Ga₂O₃ and BCB to reduce the density of interface states. The BCB dielectric has a high breakdown electric field strength (5.3 MV/cm), enabling the terminal structure to withstand high voltage. In addition, The U-shaped field plate is located in the deep trench. The electric field peaks in the trench terminal structure appear at the sidewalls and corners. U-shaped field plate can reduce and redistribute the electric field peaks in the trench terminal structure. In summary, the introduction of a U-shaped field plate in a semiconductor device effectively redistributes electric field peaks. Instead of these peaks occurring at the edges of the Schottky junction, they now appear inside the junction and at the end of the field plate. This redistribution enhances the device's ability to withstand reverse voltage without breakdown, improving its reverse breakdown voltage. The U-shaped field plate likely contributes to this improvement by influencing the electric field distribution, potentially creating a more gradual field gradient or shielding vulnerable areas from high stress. This advancement could lead to more reliable and higher-voltage semiconductor devices, which is particularly beneficial for applications in power electronics.

For the process of the β-Ga₂O₃ SBD, the epitaxial wafers we used are provided by Novel Crystal Technology. The thickness of the β-Ga₂O₃ (001) epitaxial wafer is approximately 630 μm (Sn doping: $\sim 5.0 \times 10^{18} \text{cm}^{-3}$), along with a 10μm epitaxial layer(Si doping: $\sim 0.67 \times 10^{16} \text{cm}^{-3}$). The net doping of $0.67 \times 10^{16} \text{cm}^{-3}$ is extracted from the C-V characteristics at 1 MHz, as shown in Fig. 2(a) and (b).

TABLE I.   MAJOR PARAMETERS OF THE FABRICATED DEVICES.

Device parameters	Value
Mesa depth (μm)	~5
U-shaped field plate depth (μm)	25
U-shaped field plate length (μm)	~1.8
N⁻-epi layer concentration (cm⁻³)	0.67e16
N⁻-epi layer thickness (μm)	~10
N⁺ substrate thickness (μm)	~630

The fabrication process flow started with an organic and acid cleaning step to remove the particles from the wafer's surface. Then, the Ni/Pt (100nm/100nm) metal etching mask was deposited on the front side by electron beam evaporation. A deep mesa structure of approximately 5 um is achieved using BCl$_3$/Ar-based ICP etching process. The wafer was subsequently treated with piranha for 30 minutes to remove metal etching mask and improve the interface quality. Subsequently, rapid thermal annealing (RTA) was performed at 500°C for 5 minutes in N$_2$. Next, the Ti/Au (60/200nm) was deposited on the backside as the cathode, and a rapid thermal annealing is performed for one minute at 475°C under a N$_2$ atmosphere. The Ni/Au (50nm/100nm) was deposited on the top region of mesa to serve as Schottky contact for SBD. After that, Al$_2$O$_3$ is deposited to a thickness of 100 nm using ALD at 300°C on the wafer. The BCB is filled into the grooves and processed into suitable curved pits using puddle development technology. The dielectrics were opened by ICP dry etching process. Subsequently, the Ni/Au (50/100nm) metal stack was formed by electron beam evaporation and lift-off as field plate. The remaining gaps are filled with BCB dielectric. Finally, RIE etching is used to create openings in the anode area for testing purposes. The radius of the devices are designed to be 70μm. The field plate length and depth are approximately 25 μm and 1.8 μm, respectively. Some major parameters of the fabricated devices are summarized in Table I.

III. EXPERIMENTAL RESULT AND DISCUSSIONS

Three types of devices were fabricated: UT-SBD, mesa-SBD, and U FP-SBD. After completing device fabrication, we used Keysight B1500 and B1505 systems to test device performance. Fig. 3(a) and (b) shows the forward I-V characteristics of the SBD.

As shown in Fig. 3(a) and (b), we can observe that under materials with similar doping concentrations, the forward characteristics of the three samples are very close. Firstly, regarding the turn-on voltage, both U FP-SBD and mesa-SBD exhibit slightly higher values compared to UT-SBD. Using a forward current density of 1 mA/cm^2 as the reference criterion, the turn-on voltages of UT-SBD, mesa-SBD, and U FP-SBD were extracted as 0.5 V, 0.58 V, and 0.64 V, respectively. The observed difference is likely attributed to the combined effects of extended thermal soaking during etching and the Al$_2$O$_3$ deposition at 300°C. This thermal history, analogous to low-temperature annealing, could induce structural modifications at the Schottky interface. Additionally, the specific on-resistances extracted from them are approximately 7.5 mΩ·cm^2, 11 mΩ·cm^2, and 9.6 mΩ·cm^2, respectively. The ideality factors of the three samples are all very close, measuring 1.05. This observation, to some extent, indicates that all three SBDs exhibit high-quality Schottky interfaces.

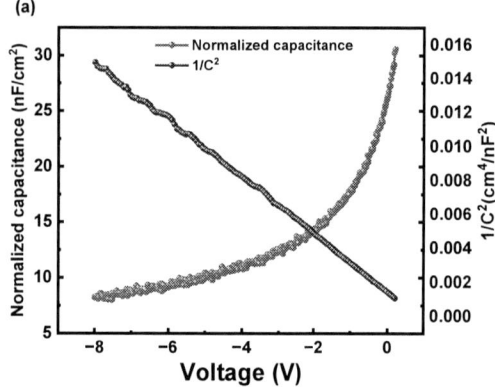

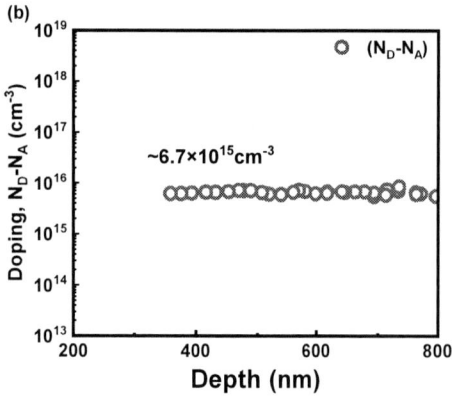

Fig.2. (a) The 1 MHz CV characteristic of SBD and carrier concentration extraction from 1/C^2 fitting plot of SBD. (b) Measured net doping concentration of the (001) β-Ga$_2$O$_3$ drift layer extracted from the C–V measurement.

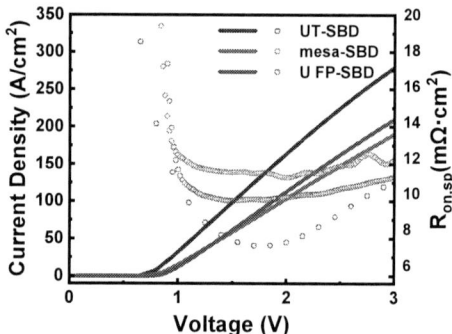

Fig. 3(a) Forward I-V characteristics of the fabricated unterminated SBD (UT-SBD), mesa-SBD and U FP-SBD in the linear scale.

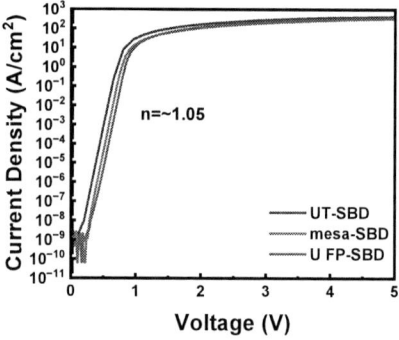

Fig. 3(b) Forward I-V characteristics of the fabricated unterminated SBD (UT-SBD), mesa-SBD and U FP-SBD in the semi-log scale.

TCAD simulations preliminarily demonstrate the location of the peak electric field in conventional Schottky barrier

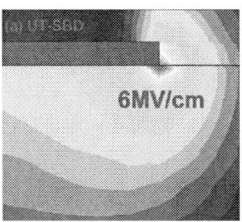
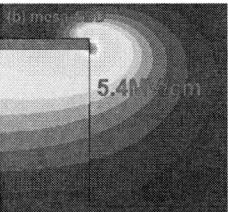
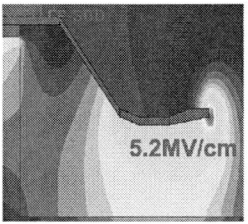

Fig. 4(a) Extracted E-field profile of UT-SBD. (b)Extracted E-field profile of mesa-SBD. (c) Extracted E-field profile of U FP-SBD. The above simulations are at 1000V reverse bias.

diodes (SBDs). Furthermore, they reveal the mitigation effect of mesa termination on the peak electric field and identify the inherent peak electric field location within the mesa-terminated structure itself. Furthermore, the simulations illustrate the redistribution effect of U-shaped field plates on the peak electric field in mesa-terminated structures and provide preliminary validation of the effectiveness of the U-shaped field plate termination.

From Fig. 5 of breakdown voltage, the breakdown voltages of UT-SBD, mesa-SBD, and U FP-SBD are 925 V, 1389 V, and 1819 V, respectively. By comparison, their device breakdown voltages have been improved, which to some extent demonstrates the effectiveness of the terminal structure. The U FP-SBD exhibits a Baliga's Figure of Merit (BFOM) of 0.34GW/cm^2. This value is not competitive compared to existing literature reports, indicating that further optimization and refinement of our device fabrication process are required to enhance performance.

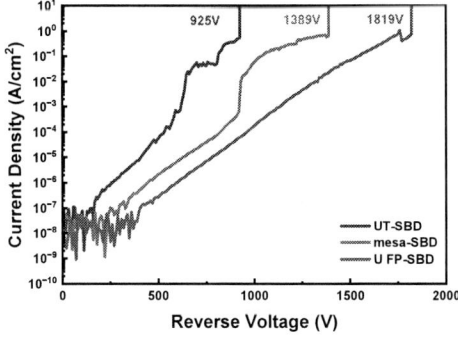

Fig. 5. Reverse I–V characteristics of UT-SBD, Mesa-SBD, FPDM-SBD and FPDM-HJD.

IV. CONCLUSIONS

In summary, this work combines field plate termination with mesa termination to form a downward-curved arc field plate termination structure. We have designed and developed compatible fabrication processes for β-Ga$_2$O$_3$ materials, fabricated corresponding devices, and conducted preliminary electrical testing. However, the performance metrics of the devices fabricated in our current experimental work still fall short of expectations. In subsequent efforts, we will focus on optimizing the fabrication processes to further enhance the device performance metrics and refine the relevant testing procedures for the devices. The results partially validate the feasibility of the proposed design approach.

ACKNOWLEDGMENT

The authors would like to thank the National Science Foundation for Young Scientists of China (NSFC for Young Scientists) for their support and funding (Grant No. 62204194 and No. 62404116).

REFERENCES

[1] M. Higashiwaki, K. Sasaki, H. Murakami, Y. Kumagai, A. Koukitu, A. Kuramata, T. Masui, and S. Yamakoshi, "Recent progress in Ga2O3 power devices," Semicond. Sci. Technol., vol. 31, no. 3, pp. 34001–34011, Jan. 2016.

[2] A. Kuramata, K. Koshi, S. Watanabe, Y. Yamaoka, T. Masui, and S. Yamakoshi, "High-quality β-Ga2O3 single crystals grown by edge defined film-fed growth," Jpn. J. Appl. Phys., vol. 55, no. 12, Nov. 2016, Art. no. 1202A2.

[3] M. Higashiwaki and G. H. Jessen, "Guest editorial: The dawn of gallium oxide microelectronics," Appl. Phys. Lett., vol. 112, no. 6, pp. 60401–60404, Feb. 2018.

[4] J. Yang et al., "Reverse breakdown in large area, field-plated, vertical β-Ga2O3 Rectifiers," ECS J. Solid State Sci. Technol., vol. 8, no. 7, pp. 3159–3164, Mar. 2019.

[5] W. Guo, Z. Han, X. Zhao, G. Xu, and S. Long, "Large-area β Ga2O3 Schottky barrier diode and its application in DC–DC converters," J. Semiconductors, vol. 44, no. 7, Jul. 2023, Art. no. 072805.

[6] F. Otsuka, H. Miyamoto, A. Takatsuka, S. Kunori, K. Sasaki, and A. Kuramata, "Large-size (1.7 × 1.7 mm2) β-Ga2O3 field-plated trench MOS-type Schottky barrier diodes with 1.2 kV breakdown voltage and 109 high on/off current ratio," Appl. Phys. Exp., vol. 15, no. 1, Dec. 2021, Art. no. 016501.

[7] W. Hao et al., "High-Performance Vertical β-Ga2 O3 Schottky Barrier Diodes Featuring P-NiO JTE with Adjustable Conductivity," 2022 International Electron Devices Meeting (IEDM), San Francisco, CA, USA, 2022, pp. 9.5.1-9.5.4.

[8] Yang W, Feng H, Fang X, et al. A novel sloped field plate-enhanced ultra-short edge termination structure[J]. IEEE Electron Device Letters, 2016, 37(4): 471-473.

[9] Yang W, Feng H, Fang X, et al. Design and characterization of sloped-field-plate enhanced trench edge termination[J]. IEEE Transactions on Electron Devices, 2016, 64(3): 728-734.

[10] Liu Y, Yang W, Feng H, et al. Trench field plate engineering for high efficient edge termination of 1200 V-class SiC devices[C]//2019 31st International Symposium on Power Semiconductor Devices and ICs (ISPSD). IEEE, 2019: 143-146.

[11] Liu Y, Yang W, Feng H, et al. Design and characterization of the deep-trench, U-shaped field-plate edge termination for 1200-V-Class SiC devices[J]. IEEE Transactions on Electron Devices, 2019, 66(10): 4251-4257.

[12] Han C, Zhang Y, Song Q, et al. An improved ICP etching for mesa-terminated 4H-SiC pin diodes[J]. IEEE transactions on electron devices, 2015, 62(4): 1223-1229.

2025 IEEE Workshop on Wide Bandgap Power Devices and Applications in Asia (WiPDA Asia)

Enhancement-Mode Vertical (001) β-Ga₂O₃ Power Transistor Enabled by Dual Ion Implantation

Anjing Luo
School of Nano-Tech and Nano-Bionics
University of Science and Technology of China
Hefei , China

Gaofu Guo
Nanofabrication facility, Suzhou Institute of Nano-Tech and Nano-Bionics
ChineseAcademy of Sciences
Suzhou, Jiangsu , China

Li Zhang
Nanofabrication facility, Suzhou Institute of Nano-Tech and Nano-Bionics
ChineseAcademy of Sciences
Suzhou, Jiangsu , China

Tiwei Chen
School of Nano-Tech and Nano-Bionics
University of Science and Technology of China
Hefei , China

Dengrui Zhao
Nanofabrication facility, Suzhou Institute of Nano-Tech and Nano-Bionics
ChineseAcademy of Sciences
Suzhou, Jiangsu , China

Zhili Zou
School of Nano-Tech and Nano-Bionics
University of Science and Technology of China
Hefei , China

Chunhong Zeng*
Nanofabrication facility, Suzhou Institute of Nano-Tech and Nano-Bionics
ChineseAcademy of Sciences
Suzhou, Jiangsu , China
chzeng2007@sinano.ac.cn

Huanyu Zhang
School of Nano-Tech and Nano-Bionics
University of Science and Technology of China
Hefei , China

Baoshun Zhang
Nanofabrication facility, Suzhou Institute of Nano-Tech and Nano-Bionics
ChineseAcademy of Sciences
Suzhou, Jiangsu , China

Zhucheng Li
Nanofabrication facility, Suzhou Institute of Nano-Tech and Nano-Bionics
ChineseAcademy of Sciences
Suzhou, Jiangsu , China

Xiaodong Zhang*
Nanofabrication facility, Suzhou Institute of Nano-Tech and Nano-Bionics
ChineseAcademy of Sciences
Suzhou, Jiangsu , China
xdzhang2007@sinano.ac.cn

Zhongming Zeng
Nanofabrication facility, Suzhou Institute of Nano-Tech and Nano-Bionics
ChineseAcademy of Sciences
Suzhou, Jiangsu , China

Abstract—**This study presents a high-performance enhancement-mode (E-mode) β-Ga₂O₃ vertical field-effect transistor (DI-FET) featuring a planar gate structure, fabricated through sequential nitrogen and silicon ion implantation. The device demonstrates a maximum output current density of 123.5 A/cm² at a drain-source voltage (V_{DS}) of 10 V, a peak transconductance of 49 S/cm², a specific on-resistance ($R_{on,sp}$) of 54.6 mΩ·cm², and a high threshold voltage (V_{th}) of 5.7 V. Furthermore, a breakdown voltage (BV) of 812 V is achieved at a gate-source voltage (V_{GS}) of 0 V. These results underscore the considerable potential of nitrogen/silicon dual-ion-implanted β-Ga₂O₃ DI-FETs for next-generation high-performance power electronics.**

Keywords—enhancement-mode, vertical β-Ga₂O₃ power transistor, dual-ion implantation

I. Introduction

Compared with gallium nitride (GaN), silicon carbide (SiC), and other wide-bandgap semiconductors, β-Ga₂O₃ exhibits superior electronic properties. Its ultra-wide bandgap of approximately 4.9 eV suppresses leakage currents under high-voltage operation, while its theoretical critical breakdown field of 8 MV/cm—substantially exceeding those of GaN and SiC—enables device miniaturization without compromising voltage-handling capability. Furthermore, β-Ga₂O₃ devices achieve a Baliga's figure of merit (BFOM) of 3444, indicating significantly reduced conduction losses during power conversion. The monoclinic crystal structure endows the β phase with exceptional thermodynamic and chemical stability, and melt-based growth techniques facilitate the fabrication of large-area, high-quality single-crystal substrates. These substrates can be produced cost-effectively and at scale, maintaining excellent performance under high electric fields and elevated temperatures.Large-area β-Ga₂O₃ substrates maintain excellent performance under high electric fields and elevated temperatures, providing a robust material platform for next-generation power and RF electronic devices[1],[2],[3],[4].

Power devices are classified as vertical or lateral according to their current-flow direction[5],[6],[7],[8],. To fully exploit the advantages of β-Ga₂O₃, vertical architectures have become predominant. In contrast to lateral devices, the breakdown voltage of a vertical device scales with its drift-layer thickness, enabling both high current and high breakdown voltage without enlarging the lateral footprint, thereby achieving superior power density[9],[10],[11],[12].

The scarcity of p-type doping techniques has constrained the development of bipolar β-Ga₂O₃ power devices, resulting in limited research on enhancement-mode vertical β-Ga₂O₃ transistors[13]. To date, depletion-mode vertical Ga₂O₃ FETs are classified into geometry-limited FinFETs and current blocking layer-type (CBL- type) transistors. Although Ga₂O₃ FinFETs have demonstrated kilovolt-level breakdown voltages, their reliance on electron-beam lithography impedes device throughput and yield

979-8-3315-1110-4/25 $31.00 © 2025 IEEE   713

[14],[15],[16],[22]. Therefore, devices employing current-blocking layers have attracted significant interest owing to their considerable potential. Notable examples include current-aperture vertical electron transistors (CAVETs), vertical diffusion barrier FETs (VDBFETs), dual-ion-implanted FETs (DI-FETs), and U-shaped trench gate metal–oxide–semiconductor field-effect transistors (UMOSFETs)[17],[18],[19],[20],[21],[22].

The planar-gate β-Ga$_2$O$_3$ vertical dual-ion-implanted field-effect transistor (DI-FET) is intrinsically a normally-off (enhancement-mode) device. Ion implantation enables precise, controllable doping, which is highly advantageous for device engineering. Moreover, the planar fabrication process provides a cost-effective, scalable production route, positioning DI-FETs as a promising platform for next-generation power devices[23]. In this work, we report an enhancement-mode planar-gate β-Ga$_2$O$_3$ DI-FET fabricated via sequential nitrogen and silicon ion implantation, achieving a breakdown voltage (BV) of 812 V, a specific on-resistance ($R_{on,sp}$) of 54.6 mΩ·cm^2, and a threshold voltage (V_{th}) of 5.7 V.

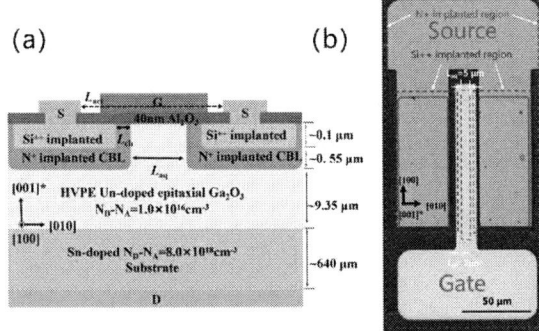

Fig.1. (a) Schematic structure of the Ga$_2$O$_3$ vertical double-implantation field-effect transistor (DI-FET). (b) Optical microscope image of the Ga$_2$O$_3$ vertical double-implantation field-effect transistor (DI-FET). The devices reported in this work had L_{ap} = 5 μm and L_{ch} = 2 μm.

II. DEVICE FABRICATION

A β-Ga$_2$O$_3$ DI-FET was fabricated on a (001)-oriented Sn-doped β-Ga$_2$O$_3$ substrate,with a 10 μm n-Ga$_2$O$_3$ drift layer grown by halide vapor phase epitaxy (HVPE). Figure. 1(a)-(b) illustrates the device structure and optical microscopy images of the Ga$_2$O$_3$ vertical double-implantation field-effect transistor. The MOSFETs featured a source width of 2 × 40 μm, an aperture length (L_{ap}) of 5 μm, and an aperture width of 100 μm. The effective gate length of 2 μm corresponded to the channel length (L_{ch}).

The complete fabrication sequence for the β-Ga$_2$O$_3$ DI-FET is depicted in Fig. 2. A 100 nm SiO$_2$ hard mask was first deposited by PECVD to define the nitrogen-implantation regions. Nitrogen ions were implanted using an NV-GSD-HE implanter to yield a box-profile doping concentration of 5 × 10^{18} cm^{-3} at ~0.65 μm depth. Post-implantation activation was conducted in an oxygen ambient at 900 °C for 30 min to activate dopants and anneal implantation damage. Subsequently, silicon ions were implanted to form an n$^+$ surface layer and activated by a 15 min anneal at 900 °C in nitrogen, promoting low-resistance ohmic contacts. Figure. 3 presents the simulated depth profiles of the nitrogen and silicon implants. Ti/Au drain electrodes were then deposited by electron-beam

evaporation. A 40 nm Al$_2$O$_3$ gate dielectric was grown via PEALD and patterned by inductively coupled plasma etching, followed by Ti/Au source metallization. Rapid thermal annealing at 475 °C for 60 s completed the formation of low-resistance source/drain contacts. Finally, Ni/Au was deposited by electron-beam evaporation to define the gate electrode.

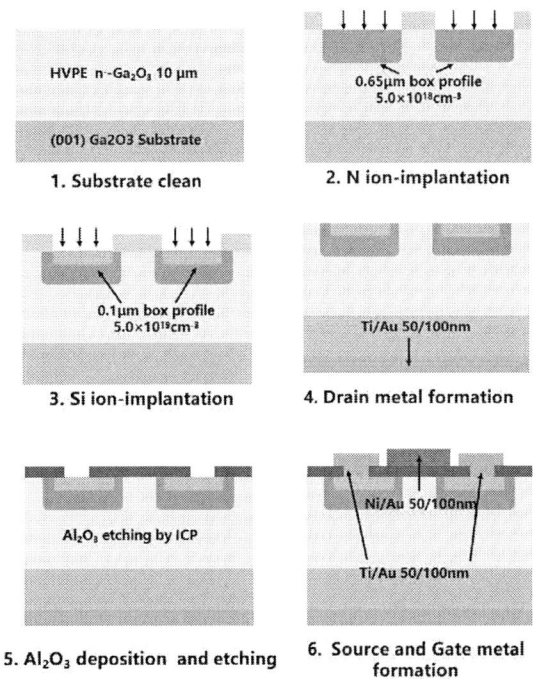

Fig. 2. Baseline process flow of Ga$_2$O$_3$ vertical DI-FET.

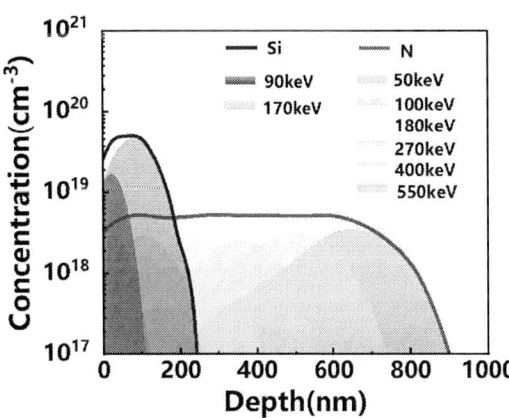

Fig. 3. Simulated Si and N depth profile after implantation.

III. RESULTS AND DISCUSSION

The electrical performance of the DI-FET was evaluated at room temperature using a Keysight B1505A Power Device Analyzer. Representative DC I–V characteristics are presented in Figure 4. The output characteristics (Figure 4(a)) were measured by sweeping the gate voltage (V_G) from 0 to 10 V in 1 V steps. The device achieved a maximum drain current density of 123.5 A/cm^2 and a specific on-resistance ($R_{on,sp}$) of 54.6 mΩ·cm². The half-cell current density was normalized using the area defined as $Wc \times (L_{act}/2 + L_t)$, where Wc = 40 μm is the Ti/Au contact width, L_{act} = 19 μm is the active region length (Figure 1(a)), and L_t = 5.8 μm is the transfer length extracted from TLM measurements. A diode-like conduction behavior was observed, consistent with previous reports, and is attributed to point-defect diffusion during ion implantation annealing, which forms an electron barrier at the current aperture, thereby increasing the turn-on resistance[24].

Due to the deep-level nature of nitrogen traps, electrons in the gate region (L_g) are fully neutralized under zero bias, resulting in normally-off operation. The transfer characteristics (Figure 4(b)) at V_{DS} = 10 V yield a threshold voltage (V_{th}) of 5.7 V (defined at I_D = 1 A/cm²) and a peak transconductance ($g_{m,max}$) of 49 S/cm², indicating excellent gate control.

Figure 4(c) shows the breakdown characteristics, where a breakdown voltage (BV) of 812 V is achieved at V_{GS} = 0 V. This demonstrates that annealing at 900 °C for 30 minutes in an oxygen atmosphere successfully activated a portion of the implanted dopants, enabling the formation of a current-blocking layer with effective current suppression and high-voltage tolerance. As illustrated in Figure 5, the β-Ga_2O_3 DI-FET surpasses the theoretical material limits of silicon and exhibits excellent performance among current-blocking-layer-based devices.

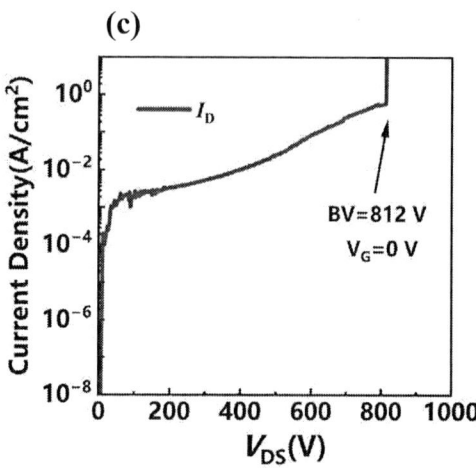

Fig.4. Electrical Performance of the DI-FET. (a)Transfer I_D-V_{GS} characteristics (V_{DS}=10 V). (b) Output I_D-V_{GS} characteristics. (c) Reverse breakdown characteristics (V_{GS} = 0 V).

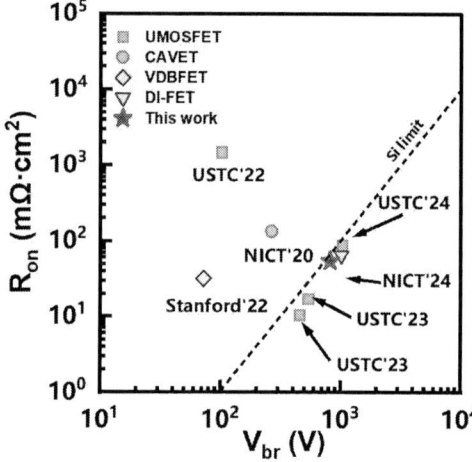

Fig.5. Benchmark plot of V_{br} and $R_{(on,sp)}$ for state-of-the-art vertical β-Ga_2O_3 MOSFETs.

IV. CONCLUSION

In this work, we present a high-performance, enhancement-mode β-Ga_2O_3 dual-ion-implanted vertical FET (DI-FET) with a planar-gate architecture, fabricated via sequential nitrogen and silicon ion implantation. The device achieved a breakdown voltage (BV) of 812 V and, at V_{GS} = 10 V, exhibited a specific on-resistance ($R_{on,sp}$) of 54.6 mΩ·cm² and a threshold voltage (V_{th}) of 5.7 V. Because the nitrogen implant dose and annealing conditions were not fully optimized, the current-blocking layer's effectiveness was limited, resulting in relatively high off-state current and a breakdown voltage below the intrinsic Ga_2O_3 limit. Nonetheless, these findings demonstrate the considerable promise of N/Si dual-ion implantation for the development of high-performance Ga_2O_3 power devices.

ACKNOWLEDGMENT

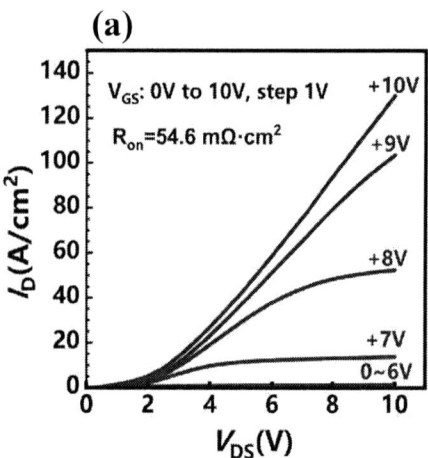

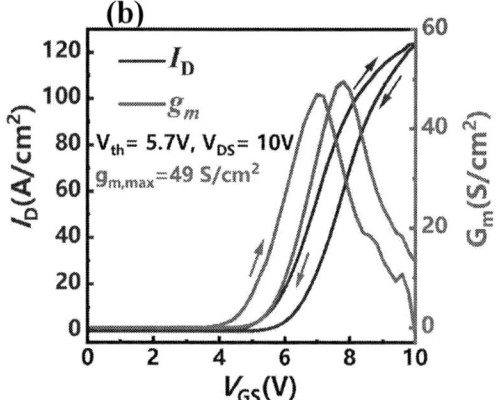

The authors would like to thank Nano Fabrication Facility and Vacuum Interconnected Nanotech Workstation (NANO-X) of Suzhou Institute of Nano-Tech and Nano-Bionics, Chinese Academy of Sciences for their technical support.

REFERENCES

[1] M. Higashiwaki, K. Sasaki, H. Murakami, Y. Kumagai, A. Koukitu, A. Kuramata, T. Masui, and S. Yamakoshi, "Recent progress in Ga$_2$O$_3$ power devices," Semicond. Sci. Technol., vol. 31, no. 3, Jan. 2016,Art. no. 034001.

[2] X. Wei, X. Zhang, W. Tang, W. Liu, X. Zhou, W. Tang, Y. Ma, T. Chen, X. Huang, H. Qian, G. Yu, X. Zhang, W. Shen, Y. Fan, Z. Zeng, H. Wang, Y. Cai, and B. Zhang, "2.0 kV/2.1 m ·cm^2 lateral p-GaN/AlGaN/GaN hybrid anode diodes with hydrogen plasma treatment," IEEE Electron Device Lett., vol. 43, no. 5, pp. 693–696, May 2022.

[3] Y. Ma, T. Chen, X. Zhang, W. Tang, B. Feng, Y. Hu, L. Zhang, X. Zhou, X. Wei, K. Xu, D. Mudiyanselage, H. Fu, and B. Zhang, "High- photoresponsivity self-powered a-, ε-, and β-Ga$_2$O$_3$/p-GaN heterojunc-

[4] tion UV photodetectors with an in situ GaON layer by MOCVD," ACS Appl. Mater. Interface, vol. 14, no. 30, pp. 35194–35204, Aug. 2022.

[5] W. Tang, Y. Ma, X. Zhang, X. Zhou, L. Zhang, X. Zhang, T. Chen, X. Wei, W. Lin, D. H. Mudiyanselage, H. Fu, and B. Zhang, "High-quality (001) β-Ga$_2$O$_3$ homoepitaxial growth by metalorganic chemical vapor deposition enabled by in situ indium surfactant," Appl. Phys. Lett.,vol. 120, no. 21, May 2022, Art. no. 212103.

[6] S. Sharma, K. Zeng, S. Saha, and U. Singisetti, "Field-plated lateral Ga$_2$O$_3$ MOSFETs with polymer passivation and 8.03 kV breakdown voltage," IEEE Electron Device Lett., vol. 41, no. 6, pp. 836–839, Jun. 2020.

[7] H. Zhou, K. Maize, G. Qiu, A. Shakouri, and P. D. Ye, "β-Ga$_2$O$_3$ on insulator field-effect transistors with drain currents exceeding 1.5 A/mm and their self-heating effect," Appl. Phys. Lett., vol. 111, no. 9, Aug. 2017, Art. no. 092102.

[8] Y. Lv, H. Liu, X. Zhou, Y. Wang, X. Song, Y. Cai, Q. Yan, C. Wang, S. Liang, J. Zhang, Z. Feng, H. Zhou, S. Cai, and Y. Hao, "Lateral β-Ga$_2$O$_3$ MOSFETs with high power figure of merit of 277 MW/cm^2," IEEE Electron Device Lett., vol. 41, no. 4, pp. 537–540, Apr. 2020.

[9] Y. Lv, X. Zhou, S. Long, X. Song, Y. Wang, S. Liang, Z. He, T. Han, X. Tan, Z. Feng, H. Dong, X. Zhou, Y. Yu, S. Cai, and M. Liu, "Source- field-plated β-Ga$_2$O$_3$ MOSFET with record power figure of merit of 50.4 MW/cm^2," IEEE Electron Device Lett., vol. 40, no. 1, pp. 83–86, Jan. 2019.

[10] K. Zeng, A. Vaidya, and U. Singisetti, "1.85 kV breakdown voltage in lateral field-plated Ga$_2$O$_3$ MOSFETs," IEEE Electron Device Lett., vol. 39, no. 9, pp. 1385–1388, Sep. 2018.

[11] C. Wang, H. Zhou, J. Zhang, W. Mu, J. Wei, Z. Jia, X. Zheng,X. Luo, X. Tao, and Y. Hao, "Hysteresis-free and μs-switching of D/E-modes Ga$_2$O$_3$ hetero-junction FETs with the BV^2/R$_{on,sp}$ of 0.74/0.28 GW/cm^2," Appl. Phys. Lett., vol. 120, no. 11, Mar. 2022, Art. no. 112101.

[12] M. H. Wong and M. Higashiwaki, "Vertical β-Ga$_2$O$_3$ power transistors: A review," IEEE Trans. Electron Devices, vol. 67, no. 10,pp. 3925–3937, Oct. 2020.

[13] R.H. Horng, X.Y. Tsai, F.G. Tarntair, J.M. Shieh, S.H. Hsu, J.P. Singh, G.C. Su, P. L. Liu, P-type conductive Ga$_2$O$_3$ epilayers grown on sapphire substrate by phosphorus-ion implantation technology, Mater. Today Adv. 20 (2023) 100436,

[14] W. Li, H. G. Xing, K. Nomoto, K. Lee, S. M. Islam, Z. Hu, M. Zhu, X. Gao, M. Pilla, and D. Jena, "Development of GaN vertical trench-MOSFET with MBE regrown channel," IEEE Trans. Electron Devices, vol. 65, no. 6, pp. 2558–2564, Jun. 2018.

[15] Z. Hu, K. Nomoto, W. Li, Z. Zhang, N. Tanen, Q. T. Thieu, K. Sasaki,A. Kuramata, T. Nakamura, D. Jena, and H. G. Xing, "Breakdown mech- anism in 1 kA/cm^2 and 960 V E-mode β-Ga$_2$O$_3$ vertical transistors,"Appl. Phys. Lett., vol. 113, no. 12, Sep. 2018, Art. no. 122103.

[16] Z. Hu, K. Nomoto, W. Li, R. Jinno, T. Nakamura, D. Jena, and H. Xing, "1.6 kV vertical Ga$_2$O$_3$ FinFETs with source-connected field plates and normally-off operation," in Proc. 31st Int. Symp. Power Semicond. Devices ICs (ISPSD), May 2019, pp. 483–486.

[17] W. Li, K. Nomoto, Z. Hu, T. Nakamura, D. Jena, and H. G. Xing, "Single and multi-fin normally-off Ga$_2$O$_3$ vertical transistors with a breakdown voltage over 2.6 kV," in IEDM Tech. Dig., Dec. 2019, pp. 270–273.

[18] K. Zeng, R. Soman, Z. Bian, S. Jeong, and S. Chowdhury, "Vertical Ga$_2$O$_3$ MOSFET with magnesium diffused current blocking layer,"IEEE Electron Device Lett., vol. 43, no. 9, pp. 1527–1530, Sep. 2022.

[19] M. H. Wong, K. Goto, Y. Morikawa, A. Kuramata, S. Yamakoshi, H. Murakami, Y. Kumagai, and M. Higashiwaki, "All-ion-implanted planar-gate current aperture vertical Ga$_2$O$_3$ MOSFETs with Mg-doped blocking layer," Appl. Phys. Exp., vol. 11, no. 6, May 2018, Art. no. 064102.

[20] Y. Ma, X. Z. Zhou , W. B. Tang , X. D. Zhang , G. W. Xu, Z. M. Zeng , and S. B. Long , "702.3 A·cm^{-2}/10.4 mΩ·cm^2 β-Ga$_2$O$_3$ U-Shape Trench Gate MOSFET With N-Ion Implantation,"IEEE Electron Device Lett., vol. 44, no. 3, pp. 384–387,Mar. 2023, doi: 10.1109/LED.2023.3235777.

[21] M. H. Wong, K. Goto, H. Murakami, Y. Kumagai, and M. Higashiwaki, "Current aperture vertical β-Ga$_2$O$_3$ MOSFETs fabricated by N- and Si- ion implantation doping," IEEE Electron Device Lett., vol. 40, no. 3, pp. 431–434, Dec. 2018.

[22] M. H. Wong, H. Murakami, Y. Kumagai, and M. Higashiwaki, "Enhancement-mode β-Ga$_2$O$_3$ current aperture vertical MOSFETs with N-ion-implanted blocker," IEEE Electron Device Lett., vol. 41, no. 2, pp. 296–299, Dec. 2019.

[23] M. H. Wong, H. Murakami, Y. Kumagai, and M. Higashiwaki "Aperture-limited conduction and its possible mechanism in ion-implanted current aperture vertical β-Ga$_2$O$_3$ MOSFETs," Appl. Phys. Lett., vol. 118, no. 1, Jan. 2021, Art. no. 012102.

[24] M. H. Wong, H. S. Murakami, Y.S. Kumagai; M. K Higashiwaki, "Enhancement-Mode β-Ga$_2$O$_3$ Current Aperture Vertical MOSFETs With N-Ion-Implanted Blocker," in IEEE Electron Device Letters, vol. 41, no. 2, pp. 296-299, Feb. 2020.

Partial Discharge Classification and Intelligent Maintenance Decision-Making Based on Spikingformer and LLMs

Changdong Wang
School of Electronic and Information
Engineering
Harbin Institue of Technology
Harbin, China
hitwcd@stu.hit.edu.cn

Jingli Yang*
School of Electronic and Information
Engineering
Harbin Institue of Technology
Harbin, China
Technological Innovation Center of
Littoral Test, China
jinglidg@hit.edu.cn
*Corresponding author

Xunran Yin
School of Electronic and Information
Engineering
Harbin Institue of Technology
Harbin, China
yinxunran@stu.hit.edu.cn

Shuangyan Yin
School of Electronic and Information
Engineering
Harbin Institue of Technology
Harbin, China
21B905048@stu.hit.edu.cn

Tianyu Gao
School of Electronic and Information
Engineering
Harbin Institue of Technology
Harbin, China
gaotianyu0714@hit.edu.cn

Yongqi Chang
School of Electronic and Information
Engineering
Harbin Institue of Technology
Harbin, China
changyq@hit.edu.cn

Huamin Jie
School of Electrical and Electronic
Engineering,
Nanyang Technological University,
Singapore
Jieh0002@e.ntu.edu.sg

Zhou Shu
Department of Electrical and Computer
Engineering, National University of
Singapore, Singapore
shuzhou@nus.edu.sg

Zhenyu Zhao*
Department of Electrical and Computer
Engineering at National University of
Singapore, Singapore
zhaozy@nus.edu.sg
*Corresponding author

Abstract—Partial discharge (PD) classification is an important part of power equipment health management, and accurately identifying PD types and their development trends is crucial for intelligent operation and maintenance of equipment. This article proposes an intelligent PD diagnosis framework that integrates Spikingformer and large language model (LLM). Firstly, event driven PD feature extraction is carried out using Spikingformer combined with spiking neural network (SNN), and the dynamic adjustment mechanism of membrane potential is adopted to improve classification accuracy. Subsequently, based on the SNN classification results, LLM combines device operation data with historical maintenance records to make intelligent operation and maintenance decisions. In addition, we introduce an optimization mechanism based on physics knowledge to enable LLM to consider PD development laws and device physical models when providing decision recommendations, thereby improving the interpretability and adaptability of the system. The experimental results show that this method achieves efficient and accurate PD recognition under various working conditions, and can provide reasonable intelligent operation and maintenance strategies, providing a new solution for equipment health management in fields such as smart grid, transformer monitoring, and wind power operation and maintenance

Keywords—*Partial discharge classification, Intelligent maintenance, Spiking neural network, LLMs*

I. INTRODUCTION

Partial discharge (PD) classification plays a crucial role in the health management of high-voltage power systems, transformers, and major mechanical equipment [1-3]. PD is an early indicator of insulation degradation in electrical equipment, and failure to detect and analyze it in a timely manner may lead to severe equipment malfunctions or even catastrophic failures [4-6]. Traditional PD recognition methods rely on feature engineering and machine learning models, yet they often suffer from low recognition accuracy and high computational energy consumption, making deployment challenging in real-world applications [7].

To address these challenges, spiking neural networks (SNNs) have gained significant attention in the field of fault diagnosis due to their excellent energy efficiency [8-10]. While these approaches mitigate the computational burden to some extent, intelligent decision-making and maintenance management still rely heavily on expert intervention. In recent years, the introduction of large language models (LLMs) in the fault diagnosis domain has opened new directions for enhancing intelligent maintenance decision-making for critical equipment [11-13].

Inspired by these advancements, this study proposes an event-driven PD classification and intelligent decision-making framework that integrates Spikingformer, a spiking self-attention network, with LLMs. Spikingformer combines the temporal encoding capabilities of SNNs with the global context modeling power of transformers, enabling efficient PD feature extraction. By leveraging spiking self-attention mechanisms, Spikingformer processes PD signals in a biologically inspired, low-power computing manner

979-8-3315-1110-4/25 $31.00 © 2025 IEEE

while capturing both local transient patterns and long-term dependencies, thereby improving classification performance. After PD classification, LLMs serve as intelligent decision-making agents, generating precise maintenance strategies based on PD recognition results, historical maintenance records, equipment operating parameters, and physics-informed models. By integrating Spikingformer's event-driven PD classification capability with LLM-powered predictive analytics, this study develops an efficient and intelligent PD monitoring and decision-making system, enabling real-time PD detection and data-driven intelligent maintenance decisions. This framework can be widely applied in smart grids, high-voltage insulation monitoring, and industrial equipment health assessment, contributing significantly to enhancing equipment lifespan and operational efficiency.

The structure of this paper is as follows: Section II presents the methodologies developed in this study. Section III details the experimental setup and results analysis. Finally, Section IV concludes with key insights and discusses possible directions for future research.

II. PROPOSED METHOD

As shown in **Fig. 1**, This framework integrates the event-driven Spikingformer model (ESNN) with Large Language Models (LLMs) to achieve efficient and low-power Partial Discharge (PD) classification and intelligent maintenance decision-making. The first phase (Phase 1) focuses on PD data acquisition and preprocessing. Sensors are installed on high-voltage electrical equipment (such as transmission towers and transformers) to collect PD signals in real time, which are then transmitted to a data acquisition system for storage and processing. The acquired signals are subsequently divided into a training set and a testing set, where the training set is used for model construction, and the testing set is employed for PD classification and intelligent maintenance analysis. After appropriate preprocessing, the data is forwarded to the next phase for modeling. The second phase utilizes an event-driven

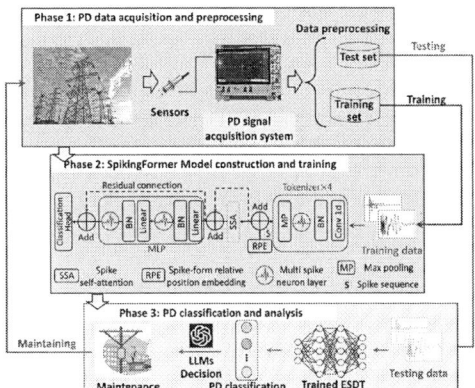

Fig. 1. The PD classification framework based on the proposed SpikingFormer.

spiking neural network to model PD signals and train an efficient PD classifier. The Spikingformer serves as the core of the model framework, featuring Spike Self-Attention (SSA) to enhance the recognition of key patterns in PD signals, Spike-form Relative Position Embedding (RPE) to improve temporal information capture, and a multi-spike neuron layer that simulates the dynamic behavior of biological neurons. Additionally, MLP (Multi-Layer Perceptron), Max Pooling (MP), and 1D Convolution further optimize feature representations and enhance PD classification performance. With residual connections, the model effectively processes complex PD signals and, through training, produces an efficient Event-Driven Spiking Transformer model for PD classification. The third phase is responsible for PD recognition and intelligent maintenance analysis. First, the trained model classifies PD signals from the testing data, identifying the PD type and severity. Subsequently, LLMs act as intelligent decision-making agents, integrating PD classification results, historical maintenance records, equipment operating parameters, and physics-informed models to generate precise maintenance strategies. These strategies include fault diagnosis, maintenance recommendations (such as component replacement or parameter adjustments), and operational optimization under physical constraints. Finally, based on LLM-

TABLE III

AVERAGE PD CLASSIFICATION RESULTS OF THE DIFFERENT METHODS ON DIFFERENT SAMPLE SET.

Methods	Metrics	300 samples	Test time(s)↓	150 samples	Test time(s)↓
Resnet [14]	Acc↑	94.95±0.82	0.45	93.63±0.51	0.36
	Recall↑	94.56±0.77		92.48±0.74	
	F1↑	94.92±0.49		92.72±0.39	
Vision Transformer [15]	Acc↑	94.51±0.70	0.36	92.48±0.51	0.34
	Recall↑	93.03±0.49		91.15±0.49	
	F1↑	93.21±0.48		91.28±0.44	
Proposed	Acc↑	**95.13±0.22**	**0.25**	**94.10±0.17**	**0.19**
	Recall↑	**94.36±0.34**		**93.13±0.51**	
	F1↑	**94.75±0.47**		**93.65±0.34**	

driven decisions, the system can execute different levels of maintenance actions, including continuous monitoring, parameter adjustments, or immediate shutdown for inspection and repair. This framework can be widely applied in smart grids, high-voltage insulation monitoring, wind power maintenance, and industrial equipment health assessment, significantly improving the accuracy, real-time performance, and intelligence level of PD detection while providing robust support for efficient equipment maintenance and lifecycle management.

III. EXPERIMENTAL STUDY

To thoroughly evaluate the effectiveness of the proposed approach, a series of comparative experiments are conducted under the following conditions: Benchmark Models – The evaluation includes traditional deep learning architectures such as ResNet [14] and Vision Transformer [15]. Experimental Setup – Our method utilizes a batch size of 64 and is optimized using the Adam optimizer within the PyTorch framework. Computational Environment – The experiments are executed on a system equipped with an 11th Gen Intel(R) Core (TM) i7-11800H @ 2.30GHz CPU and an NVIDIA GeForce RTX 3050 GPU, operating in a Python 3.7 environment. Table 1 shows the average classification performance of different methods in partial discharge (PD) fault diagnosis tasks, including ResNet, Vision Transformer, and proposed methods, evaluated on 300 and 150 sample datasets, respectively. The indicators in the table include accuracy (Acc), recall (Recall), F1 score (F1), and inference time (Test time, s). The results indicate that the proposed method outperforms the baseline model in all performance metrics, particularly in terms of accuracy (Acc: 95.13% vs 94.95%) and F1 score (94.75% vs 94.92%). In addition, the inference time of the proposed method was significantly reduced (0.25s vs 0.45s), indicating its higher computational efficiency. As the sample size decreases, the classification performance of all methods decreases, but the proposed method has the smallest decrease, indicating its strong robustness. Example of large model analysis results: Mild partial discharge (such as mild corona discharge): Maintenance suggestion: Continue monitoring, regularly conduct partial discharge detection, and observe discharge trends. Optimization measures: Adjust the operating voltage of the equipment appropriately to avoid excessive

IV. CONCLUSION

In this study, we proposed an event-driven PD classification and intelligent maintenance framework by integrating Spikingformer with Large Language Models (LLMs) to enhance fault diagnosis accuracy and computational efficiency. Experimental results demonstrate that our method outperforms ResNet and Vision Transformer, achieving higher accuracy (95.13%) and F1-score (94.75%) while significantly reducing inference time (0.25s vs. 0.45s). Furthermore, it maintains robust performance even with a reduced dataset, demonstrating strong generalization ability. These findings suggest that the proposed framework is well-suited for real-time PD monitoring and large-scale industrial deployment, offering an effective solution for smart grids, high-voltage insulation monitoring, and predictive maintenance applications. Future work will focus on developing interpretability for this model to enhance its industrial application potential.

V. ACKNOWLEDGEMENT:

This work was supported by the National Natural Science Foundation of China (No. 62171157), Chinese Scholarships Council of China (grant number: 202306120133), and China·Association·for Science·and Technology Youth Talent·Support·Engineering Doctoral Program.

REFERENCES

[1] C. Zhang et al., "Combining multi-level feature extraction algorithm with residual graph convolutional neural network for partial discharge detection," Measurement, vol. 242, p. 116151, 2025.

[2] H. Wang et al., "High-Frequency Current Transformer Design and Analysis for Partial Discharge Detection of Power Electronic Modules," IEEE Trans. Power Electron., 2025.

[3] C. Wang et al., "Learning to Imbalanced Open Set Generalize: A Meta-Learning Framework for Enhanced Mechanical Diagnosis," IEEE Trans. Cybern., Early Access, 2025, doi: 10.1109/TCYB.2025.3531494.

[4] Q. Khan et al., "Partial discharge detection and diagnosis in gas insulated switchgear: State of the art," IEEE Electr. Insul. Mag. , vol. 35, no. 4, pp. 16-33, 2019.

[5] L. F. Freitas-Gutierres et al., "Advancing substation inspection: The Hilbert–Huang transform approach for partial discharge recognition and assessment," Measurement , p. 116846, 2025.

[6] C. Wang, et. al., "An Uncertainty Perception Metric Network for Machinery Fault Diagnosis Under Limited Noisy Source Domain and Scarce Noisy Unknown Domain," Advanced Engineering Informatics, vol. 62, Oct. 2024, Art. no. 102682.

[7] Z. Shu et al., "Partial discharge detection and classification using low-noise UHF sensing frontend and wavelet scattering feature extraction network," IEEE Trans. Microw. Theory Tech. , 2024.

[8] C. Wang et al., "An energy-efficient mechanical fault diagnosis method based on neural dynamics-inspired metric SpikingFormer for insufficient samples in industrial Internet of Things," IEEE Internet Things J. , vol. 12, no. 1, pp. 1081-1097, Jan. 2025, doi: 10.1109/JIOT.2024.3476034.

[9] C. Wang et al., "Neural-transformer: A brain-inspired lightweight mechanical fault diagnosis method under noise," Reliab. Eng. Syst. Saf. , vol. 251, p. 110409, Aug. 2024.

[10] R. Gast, S. A. Solla, and A. Kennedy, "Neural heterogeneity controls computations in spiking neural networks," Proc. Natl. Acad. Sci. , vol. 121, no. 3, p. e2311885121, 2024.

[11] H. A. A. M. Qaid et al., "FD-LLM: Large Language Model for Fault Diagnosis of Machines," arXiv preprint arXiv:2412.01218 , 2024.

[12] R. Gitzel et al., "Towards Cognitive Assistance and Prognosis Systems in Power Distribution Grids–Open Issues, Suitable Technologies, and Implementation Concepts," IEEE Access , 2024.

[13] X. Zhang et al., "A novel method for intelligent operation and maintenance of transformers using deep visual large model DETR+ X and digital twin," Sci. Rep. , vol. 15, no. 1, p. 98, 2025.

[14] Z. Shu et al., "Partial Discharge Detection and Classification Using Low-Noise UHF Sensing Frontend and Wavelet Scattering Feature Extraction Network," IEEE Trans. Microw. Theory Tech. , 2024, doi: 10.1109/TMTT.2024.3393993.

[15] K. He, X. Zhang, S. Ren, and J. Sun, "Deep residual learning for image recognition," in Proc. IEEE Conf. Comput. Vis. Pattern Recognit. (CVPR) , pp. 770-778, Jun. 2016

Junction Temperature Fluctuation Suppression Strategy for SiC MOSFETs Based on Equivalent Gate Resistance Control

Ruoyin Wang[a], Xiaoyong Zhu[a], and Hong Zheng[a]

[a] School of Electrical and Information Engineering, Jiangsu University, Zhenjiang, China,
1000006773@ujs.edu.cn

I. INTRODUCTION

Power electronic devices are critical in the conversion, transmission, and control of electric energy, and they have an impact on power network expansion and performance in the future. Wind power [1] and wireless power transfer (WPT) technology [2] are two examples of new energy technologies that are intrinsically tied to the support of power electronic devices. Researching how to increase the reliability of power devices, and consequently improving the safety of power electronic systems, is of practical value and significance. Temperature is the key factor limiting the dependability of SiC MOSFETs, accounting for roughly 55% [3].

The reliability of SiC devices is seen as a significant area of concern. As a result of direct contact between materials with different thermal expansion coefficients, the packed components are subjected to thermal stress [4]. Thermal stress will deteriorate the weaker components in the encased device, resulting in device failure [5]. In general, converters work in non-stationary situations, creating random power fluctuations in the SiC MOSFETs. Increased thermal conductivity and Young's modulus of SiC [6] may create higher stress in the molded solder layer during power variation. In addition, power fluctuations generate temperature fluctuations in SiC MOSFETs [7], resulting in welding fatigue [8]. The service life of SiC solder layers is just one-third that of ordinary silicon devices [6]. As a result, temperature has a significant impact on SiC MOSFET reliability. Active thermal management (ATM) technique for SiC MOSFETs is critical for lowering thermal stress shock and delaying the aging rate of power devices [9]. ATM technology is mainly used to control the amount of heat produced by limiting the amplitude of junction temperature changes, as "Peak-clipping and valley filling". Currently, the common method for junction temperature management is to change the switching loss by varying the switching frequency [10]. This strategy cannot change the junction temperature of a single semiconductor device. For particular systems, such as magnetically coupled resonant WPT systems, frequency is a critical parameter.

This letter proposes an equivalent gate resistance control(EGRC)-based ATM for SiC MOSFETs. Driving resistance can dynamically adjust switching loss. Because online resistance cannot be changed,

parallel resistors must be used. This approach requires a large number of resistors to adjust the resistance, resulting in poor temperature regulation. This letter proposes a drive resistance adjustment technique. Changing the delay time of the delay switch achieves continuous drive resistance adjustment. A full-bridge inverter prototype shows the practicality and superiority of the suggested technique, which reduces temperature fluctuations.

In this letter, we detail the system architecture in Section II, briefly discuss the control structure in Section III, and the experimental prototype and experimental results are presented in Section IV before concluding the letter in Section V.

II. MATHEMATICAL MODEL

The control circuit based on a full-bridge inverter is built to validate the effectiveness, as shown in Fig.1. In this letter, the SiC MOSFET's gate circuit is simplified, as shown in Fig.2.

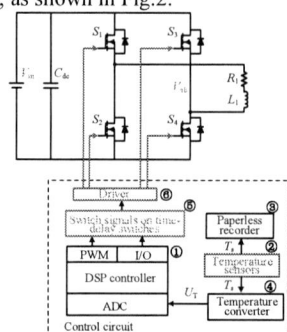

Fig.1 System architecture

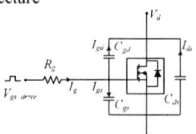

Fig.2 Simplified model of SiC MOSFET.

There are three stages in the turn-on and turn-off processes respectively [2].

$$\begin{cases} T_1 = \left(C_{gs} + C_{gd}\right) \times R_g \times \ln\left(V_{gs\_drive} / \left(V_{gs\_drive} - V_{th}\right)\right) \\ T_2 = \left(C_{gs} + C_{gd}\right) \times R_g \times \ln\left(V_{gs\_drive} / \left(V_{gs\_drive} - V_{gs\_miller}\right)\right) \\ T_3 = Q_{gd} \times R_g / \left(V_{gs\_drive} - V_{gs\_miller}\right) \\ T_4 = \left(C_{gd} + C_{gs}\right) \times R_g \times \ln\left(V_{gs\_drive} / V_{gs\_miller}\right) \\ T_5 = Q_{gd} \times R_g / V_{gs\_miller} \\ T_6 = \left(C_{gd} + C_{gs}\right) \times R_g \times \ln\left(V_{gs\_miller} / V_{th}\right) \end{cases} \quad (1)$$

where V_{gs_drive} is driving voltage, V_{gs_miller} is Miller voltage, V_{th} is threshold voltage, R_g is driving resistance, I_g is gate driving current, C_{gd} and C_{gs} are parasitic capacitance, the voltage is V_{gd} and V_{gs}, respectively, Q_{gd} is gate drain charge. Switching loss is expressed as follows.

$$E_{on} = \int_{t_1}^{t_2} V_d \frac{I_d}{t_2-t_1}(t-t_1)dt + \int_{t_2}^{t_3} I_d \frac{V_d}{t_3-t_2}(t-t_2)dt = \frac{I_d V_d (T_2+T_3)}{2} \quad (2)$$

As T_2 and T_3 increase, so does turn-on loss. Turn-off loss is:

$$E_{off} = \int_{t_5}^{t_6} I_d \frac{V_d}{t_6-t_5}(t-t_5)dt + \int_{t_6}^{t_7} V_d \frac{I_d}{t_7-t_6}(t-t_6)dt = \frac{I_d V_d (T_5+T_6)}{2} \quad (3)$$

Then E_{sw} is approximately proportional to switching time T_i

$$E_{sw} = E_{on} + E_{off} \approx \frac{I_d V_d}{2}\sum T_i = \frac{I_d V_d}{2}\sum\left(\frac{Q_1}{I_{g\_T_2}} + \frac{Q_2}{I_{g\_T_3}} + \frac{Q_3}{I_{g\_T_5}} + \frac{Q_4}{I_{g\_T_6}}\right) \quad (4)$$

where Q is the gate charge. According to this, the gate current can be used to control the switching loss.

$$I_g = \left(V_{gs\_drive} - V_{gs}\right)/R_g \quad (5)$$

Therefore, this letter adjusts R_g, thus realizes the adjustment of E_{sw}, as shown in Fig. 3. Further, the switching loss is

$$P_{sw} = f_{sw}k_I k_U E_{sw} = f_{sw} I_d V_d E_{sw}/I_{ref}V_{ref} \quad (6)$$

where f_{sw} is switching frequency, k_I and k_U are conversion coefficients of voltage and current respectively, I_{ref} and V_{ref} are reference values.

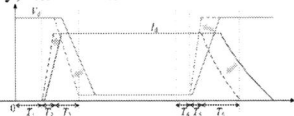

Fig.3 Adjustment diagram schematic.

Then T_i can be regarded as a function of R_g:

$$P_{sw} = \frac{f_{sw}I_d^2 V_d^2}{2I_{ref}V_{ref}}f(R_g) = P_{sw}(R_g) \quad (7)$$

$$\begin{cases} T_{J\_sw}(R_g) = P_{sw}(R_g)\cdot R_{J-c} \\ \Delta T_{J\_sw}(R_g) = \left[P_{sw}(R_{g1}) - P_{sw}(R_{g2})\right]\cdot R_{J-c} \end{cases} \quad (8)$$

where $T_{J\_sw}(R_g)$ is the junction temperature corresponding to the switching loss part under R_g, $\Delta T_{J\_sw}(R_g)$ is the adjustment range when the driving resistance changes from R_{g1} to R_{g2}, and R_{J-c} is the thermal resistance between junction and case. In summary, (7) and (8) can be used to calculate the range of R_g under varying load I_d.

III. ATM METHOD BASED ON EGRC

To adjust the overall resistance, change the number of driving resistors as shown in Fig.4(a). This method necessitates the use of more equipment. As shown in Fig.4(b), this letter introduces the delay switch concept. The voltage and current waveform with a large driving resistance R_{gmax} is shown in Fig.5. The turn-on signal is employed as a reference signal. A minor driving resistance R_{gmin} is connected when S_{w1} is turned off after T_{d1}. In Fig.5, the waveform becomes a dashed

line at this point. The procedure for turning off the device is the same. R_{gmin} is connected to modify the current and voltage waveform after the delay T_{d2}.

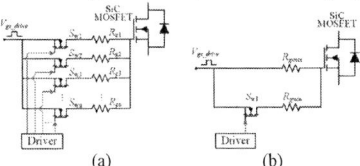

Fig.4(a)Conventional methods, (b)EGRC.

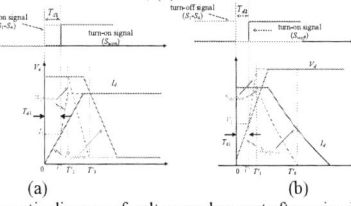

Fig.5 Schematic diagram of voltage and current after using EGRC.

As shown in Fig. 5(a), re-establish the relationship:

$$E_{on}(T_{d1}) = \frac{V_d I_d}{2T_2^2}\left(T_2 - T_2'\right)T_{d1}^2 + \frac{V_d I_d}{2}\left(T_2' + T_3'\right) \quad (9)$$

Then derivation

$$E_{on}'(T_{d1}) = V_d I_d \left(T_2 - T_2'\right)T_{d1}/T_2^2 \quad (10)$$

And because of

$$\begin{cases} T_2 = \left(C_{gs} + C_{gd}\right)\times R_{gmax}\times\ln\left(\frac{V_{gs\_drive}}{V_{gs\_drive} - V_{gs\_miller}}\right) \\ T_2' = \left(C_{gs} + C_{gd}\right)\times R_{gmin}\times\ln\left(\frac{V_{gs\_drive}}{V_{gs\_drive} - V_{gs\_miller}}\right) \end{cases} \quad (11)$$

$R_{gmax} > R_{gmin}$, therefore $T_2 - T_2' > 0$, it means $E_{on}'(T_{d1}) > 0$, the turn-on loss E_{on} is a monotone increment function with respect to T_{d1}. The analysis of the turn-off stage is consistent with the turn-on process, as shown in Fig.5(b).

$$E_{off}(T_{d2}) = \frac{V_d I_d}{2T_5^2}\left(T_5 - T_5'\right)T_{d1}^2 + \frac{V_d I_d}{2}\left(T_5' + T_6'\right) \quad (12)$$

Then derivation

$$E_{off}'(T_{d2}) = V_d I_d \left(T_5 - T_5'\right)T_{d2}/T_5^2 \quad (13)$$

And because of

$$\begin{cases} T_5 = Q_{gd}R_{gmax}/V_{gs\_miller} \\ T_5' = Q_{gd}R_{gmin}/V_{gs\_miller} \end{cases} \quad (14)$$

$R_{gmax} > R_{gmin}$, it means $E_{off}'(T_{d2}) > 0$, the E_{off} is a monotone increment function with respect to T_{d2}. In conclusion, the junction temperature of SiC MOSFET can be controlled by the delay time T_d. Fig. 6 depicts the design of the ATC control system based on EGRC proposed in this letter.

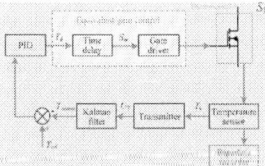

Fig.6 Control block diagram.

IV. EXPERIMENTAL VERIFICATION

An experimental platform is developed as shown in Fig. 7 to verify the feasibility of the proposed strategy. It primarily consists of a SiC MOSFET-based full-bridge inverter, a DC source, a rectifier bridge, and an electronic load.

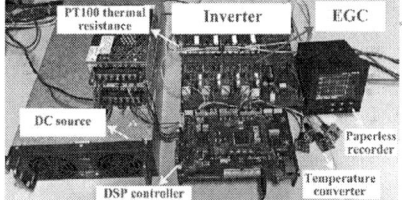

Fig.7 The experiment platform.

In the experiment, the junction temperature measurement method based on the RC thermal network model is used. This method requires the measurement of the case temperature T_c, the thermal resistance θ_{jc} between the junction and the case, combined with the power consumption P_{sw}. The junction temperature T_j and case temperature T_c can be expressed:

$$T_j = P_{sw} \cdot \theta_{jc} + T_c \qquad (15)$$

where P_{sw} is the switching loss and θ_{jc} is the steady-state thermal resistance. The Foster thermal network model belongs to a series structure which is easier to be extracted by curve fitting technology. θ_{jc} reflects the thermal properties of the package structure and materials. The final value represents the magnitude of the stable crust thermal resistance $Z_{th}(t = \infty)$.

$$Z_{th}(t=\infty) = \sum_{i=1}^{3} \theta_{thi} = \theta_{jc} \qquad (16)$$

Joint Electron Device Engineering Council presented a transient thermal resistance test standard:

$$Z_{th}(t) = \left(T_j(t) - T_j(t=0) \right) / P_H \qquad (17)$$

The thermal resistance curves under different measurement conditions will be separated from the contact point of the case surface. Two different working conditions can be set as coated with thermal grease and without grease. The impedance curve test platform built in this letter is shown in Fig. 8. The cooling curve is used to measure θ_{jc}, the SiC MOSFET to be tested is turned on, and the constant voltage source V_{in} is connected to the load resistance R_L to heat it until the junction temperature reached stability, the power consumption is

$$P_H = V_{in}I_{in} - I_{in}^2 R_L \qquad (18)$$

where I_{in} indicates the heating current. In each working condition, the voltage source is set as 200V and 300V respectively, and R_L is 20Ω. The test results are shown in Fig.9. The cooling curve and transient impedance curve without thermal conductivity grease are shown in Fig.9(a), and the cooling curve and transient impedance curve with thermal conductivity grease are shown in Fig.9(b). Fig.10 shows the thermal resistance curve with and without silicone grease. The thermal resistance θ_{jc} is the separation point of the two curves, which is about 0.592°C/W.

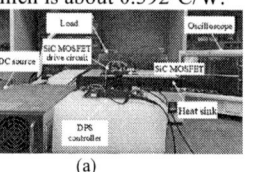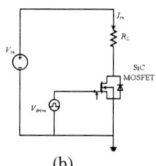

Fig.8(a)The impedance curve test platform, (b) the test principle.

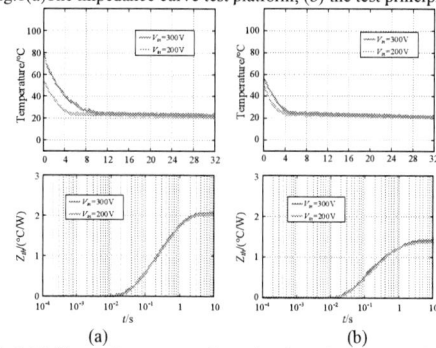

Fig.9 (a) The cooling curve and transient impedance curve of SiC MOSFET without thermal conductivity grease, (b) with thermal conductivity grease.

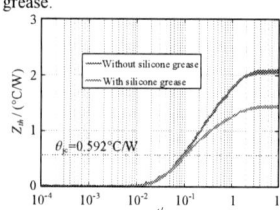

Fig.10 Test results of thermal resistance.

Fig. 11(a) and (b) show the voltage and current waveforms at the turn-on stage, respectively. Fig. 12(a) and (b) show waveforms at the turn-off stage. The larger R_g at the turn-on stage, the longer the rise time of I_d and the longer the drop time of V_d, implying a higher switching loss. The turn-off stage is the same as the turn-on stage. If the auxiliary switch's conducting time is too short, it's possible that the auxiliary switch will have been turned off before the main switch's turn-on (or turn-off) stage is complete. The only remedy that can be offered is that the auxiliary switch's conducting duration should equal the main switch's longest turn-on (or turn-off) period. After testing, SiC MOSFET with 30Ω gate resistance has the longest turn-on and turn-off time, which are 132ns and 113ns, respectively. In conclusion, the conducting time of the auxiliary switch is fixed at 140 ns, which completely covers the turn-on or turn-off stage and prevents the auxiliary switch from being turned off too soon. The voltage and current waveforms are shown in Fig.13. Then the auxiliary switches are tested under various T_d. The ability of the auxiliary switch to vary the turn-on and turn-off trajectories with different delay times is demonstrated. The turn-off signal is used as the reference signal when the SiC MOSFET is in the turn-off stage (Fig. 13a). Following the delay T_{d2}, the auxiliary switch's access causes V_d to rapidly climb, changing the trajectory of turn-off voltage and current. The turn-on signal is used as the reference signal when in the turn-on stage (Fig. 13b). The auxiliary switch's access causes the I_d to rapidly rise after the delay T_{d1}, changing the trajectory of the turn-on waveform. The voltage and current curves differ depending on the T_d.

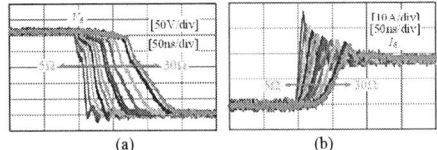

(a) (b)

Fig.11(a) The voltage waveform, (b) current waveform at the turn-on stage

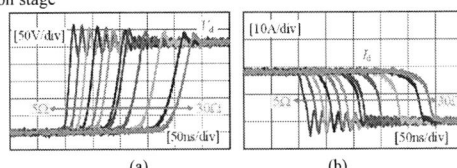

(a) (b)

Fig.12(a) The voltage waveform, (b) current waveform at the turn-off stage

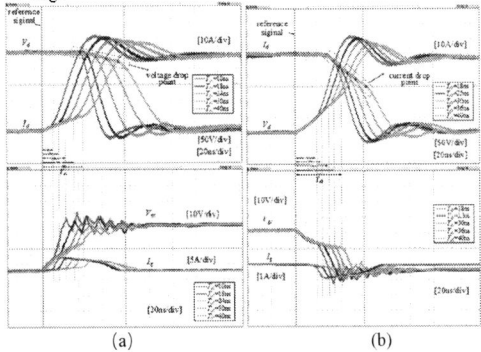

(a) (b)

Fig.13 Voltage and current waveforms at (a) turn-off stage,(b)turn-on stage.

TABLE I lists the system parameters. The power fluctuation is defined as 45A-43A-25A-45A-30A-25A-43A. Furthermore, with the rated current I_R=43A, the temperature is around 70°C, which is taken as the target temperature T_{ref}. The results are shown in Fig. 14. The largest temperature fluctuation is reduced from 18.93°C to 9.94°C. Then it is required to compare the estimated and experimental results further. IXFN50N120SIC is the SiC MOSFET used in this letter, the auxiliary switch is BSC052N03LS. As

shown in Fig. 15(a), the calculated value of temperature regulation range is obtained through (7) and (8), resulting in the regulation range that can be accomplished by altering R_g under various I_d. This letter compares the experimental and computed results shown in Fig. 15(b). The highest error is only 0.45°C.

V. CONCLUSION

In this letter, an EGRC-based ATM strategy for SiC MOSFETs is proposed. Modifying the driving resistance can reduce the switching loss, which is subsequently followed by adjusting junction temperature. To complete the design, a comparable switching mechanism is provided. At the same time, by adjusting the switching loss, this letter provides the control range. Using the full-bridge inverter as an example, the recommended ATM system can significantly reduce junction temperature fluctuations when power varies.

TABLE I
PARAMETERS OF INVERTER

Parameters	Value
V_{in}	250V
f_w	85kHz
T_a	28°C
I_R	43A
R_{gmin}	6Ω
R_{gmax}	30Ω

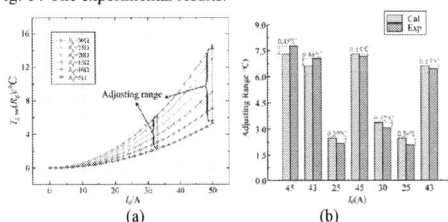

Fig. 14 The experimental results.

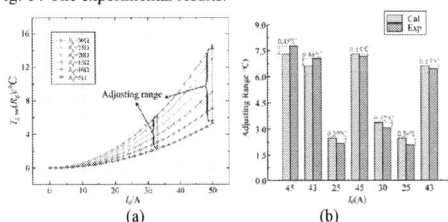

(a) (b)

Fig.15 (a)Adjustable junction temperature range; (b)comparison of calculated results with experimental results.

REFERENCE

[1] Q. Zhang, "a new method for monitoring the junction temperature of SiC MOSFET on line based on on-state resistance," 2019 22nd International Conference on electrical machines and systems (ICEMS), Harbin, China, 2019

[2] R. Wang, L. Tan, C. Li, T. Huang, H. Li and X. Huang, "Analysis, Design, and Implementation of Junction Temperature Fluctuation Tracking Suppression Strategy for SiC MOSFETs in Wireless High-Power Transfer,"

in IEEE Transactions on Power Electronics, vol. 36, no. 1, pp. 1193-1204, Jan. 2021

[3] Ma K , Blaabjerg F . Thermal optimised modulation methods of three-level neutral-point-clamped inverter for 10 MW wind turbines under low-voltage ride through[J]. Iet Power Electronics, 2012, 5(6):920-927.

[4] B. Hu., "Failure and Reliability Analysis of a SiC Power Module Based on Stress Comparison to a Si Device," IEEE Trans. Device Mater. Relib., Dec. 2017.

[5] M. Ciappa, "Selected failure mechanisms of modern power modules," Microelectronics Reliability, vol.42, no.4–5, pp. 653–667, Apr. 2002.

[6] C. Herold, M. Schaefer, F. Sauerland, T. Poller, and J. Lutz, "Power cycling capability of Modules with SiC-Diodes," CIPS 2014; 8th International Conference on Integrated Power Electronics Systems, Nuremberg, Germany, pp. 1-6, 2014.

[7] S. Liebig and J. Lutz, "Efficiency and lifetime of an active power filter with SiC-MOSFETs for aerospace application," PCIM Europe 2014; International Exhibition and Conference for Power Electronics, Intelligent Motion, Renewable Energy and Energy Management, Nuremberg, Germany, pp. 1-9, 2014.

[8] F. P. McCluskey, et al., "Reliability of high temperature solder alternatives," Microelectronics Reliability, vol. 46, no. 9–11, pp. 1910–1914, Sep. 2006.

[9] M. Riccio, A. Irace, G. Breglio, P. Spirito, V. Kosel, M. Glavanovics, and A. Satka, "Thermal simulation and ultrafast IR temperature mapping of a smart power switch for automotive applications," in Proc.21stInt.Symp. PowerSemicond.DevicesIC's, Barcelona,Spain,Jun.14 –18,2009, pp. 200–203

[10] T. A. Polom, and B. Wang, "Control of Junction Temperature and Its Rate of Change at Thermal Boundaries via Precise Loss Manipulation," IEEE Trans. on Ind. Applicat., vol. 53, no. 5, pp. 4796–4806, Sep. 2017.

Study of a High-Sensitivity Recessed-Anode GaN MIS Diode for Temperature Sensing

Yunxuan Zhao
Guangzhou Institute of Technology,
Xidian University,
Guangzhou, 510555, China
23111213638@stu.xidian.edu.cn

Xi Jiang
Guangzhou Institute of Technology,
Xidian University,
Guangzhou, 510555, China
xjiang@xidian.edu.cn

Song Yuan
Guangzhou Institute of Technology,
Xidian University,
Guangzhou, 510555, China
syuan@xidian.edu.cn

Zhaoheng Yan
State Key Laboratory of Wide-Bandgap
Semiconductor Devices and Integrated
Technology, School of
Microelectronics, Xidian University,
Xi'an 710071, China
zhyan_2@stu.xidian.edu.cn

Yanzuo Li
State Key Laboratory of Wide-Bandgap
Semiconductor Devices and Integrated
Technology, School of
Microelectronics, Xidian University,
Xi'an 710071, China
liyanzuo@stu.xidian.edu.cn

Chaofan Deng
Guangzhou Institute of Technology,
Xidian University,
Guangzhou, 510555, China
22111212540@stu.xidian.edu.cn

Xiangdong Li
Guangzhou Institute of Technology,
Xidian University,
Guangzhou, 510555, China
xdli@xidian.edu.cn

Xiaowu Gong
State Key Laboratory of Wide-Bandgap
Semiconductor Devices and Integrated
Technology, School of
Microelectronics, Xidian University,
Xi'an 710071, China
xwgong@xidian.edu.cn

Abstract—In this paper, a temperature sensor with a partial P-GaN cap layer and a semicircular recessed anode AlGaN/GaN MIS diode is presented. The sensor utilizes the depletion effect of the P-GaN layer and the passivation effect of the insulating dielectric layer to reduce the diode's forward current. Additionally, the recessed anode structure helps mitigate longitudinal dislocation defects within the epitaxial layers. As a result, the sensor exhibits excellent temperature sensitivity across a wide temperature range from 25 °C to 200 °C. The proposed sensor achieves a maximum sensitivity of 4.77 mV/K at a current density of 0.00013 A/cm², which is approximately 3.2 times higher than that of conventional GaN SBD temperature sensors. The device also maintains a low reverse leakage current due to the insulating dielectric layer.

Keywords—GaN power devices, temperature sensors, temperature sensitivity.

I. INTRODUCTION

Gallium nitride (GaN) power devices offer low conduction losses and high switching speeds[1], making them well-suited for applications such as electric vehicles, server power supplies, and aerospace power systems[2]. However, under high-frequency and high-power operating conditions, increased power dissipation leads to a rise in junction temperature. The high temperature leads to several undesirable effects, including increased leakage current, reduced electron mobility and accelerated device degradation. These factors significantly impair the performance, reliability, and lifespan of the device[3].

As power densities continue to increase, there is a growing need for real-time junction temperature monitoring to ensure thermal stability and system safety. The integration of temperature sensors provides a compact and reliable approach for enabling thermal feedback control in GaN-based power systems.

The recessed anode AlGaN/GaN Schottky barrier diode (SBD) has been demonstrated with a sensitivity of 1.05 mV/K [4], and can be laterally integrated with HEMT structures. In contrast, the vertical NiN-anode GaN SBD exhibits a higher sensitivity of 2.54 mV/K[5], but is not compatible with lateral integration.

In this paper, a GaN MIS diode temperature sensor with a P-GaN cap layer and a semicircular T-shaped recessed anode is proposed. The sensor utilizes the depletion effect of the P-GaN layer and the surface passivation of the dielectric layer effectively suppress the forward current. Meanwhile, the T-shaped recessed structure reduces the influence of dislocation-induced leakage paths, thus improving the temperature sensitivity. At a current of 0.00013 A cm⁻², the sensitivity reaches a maximum value of 4.77 mV/K while maintaining good linearity and low reverse leakage current. In addition, the sensor can be laterally integrated with P-GaN/AlGaN/GaN HEMT to improve the device junction temperature measurement accuracy.

II. DEVICE STRUCTURE

The three-dimensional structure of the GaN recessed-anode MIS diode is shown in Fig. 1(a). A corresponding two-dimensional plan view is shown in Fig. 1(b), while Fig. 1(c) provides a microscopic image of the fabricated device. The device was fabricated by first etching the P-GaN using an inductively coupled plasma etching (ICP) process to etch away the unwanted P-GaN layer. This was followed by mesa etching to achieve device isolation. Then, the anode recess was etched to a depth of approximately 100 nm below the trench layer. A 4.4 nm Al₂O₃ dielectric layer was then deposited using atomic layer deposition (ALD) to form the MIS structure.

For cathode formation, the underlying region was etched to 100 nm below the AlGaN buffer layer. A Ti/Al/Ni/Au (20/140/50/40 nm) metal stack was deposited to create the

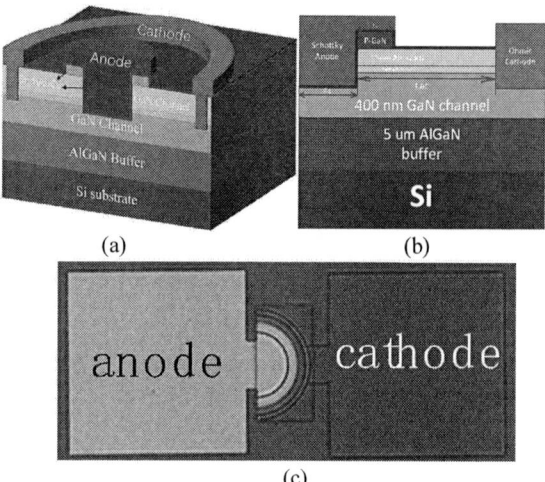

(a)   (b)

(c)

Fig. 1 Structure of recessed anode MIS diode (a) 3D structural view (b) 2D cross-sectional view (c) Microscopic view of GaN recessed-anode MIS diode.

cathode electrode, which adopts a semicircular ring geometry with a width of 10 μm. A semicircular T-shape anode electrode with a radius of R_a = 30, 35, 40, 45, 50, and 55 μm, the distance between the anode electrode and the cathode electrode L_{ac}=10, 15, 20, 25, 30 μm and the width of the P-GaN L_{pgan}=10 μm.

Based on the practical trade-off between integration and leakage performance, the structure with R_a=50 μm and L_{ac}=20 μm was selected for detailed analysis in this work.

III. OPERATING PRINCIPLE OF TEMPERATURE DIODE

The operation of the device is based on two dominant carrier transport mechanisms under forward bias: direct electron tunneling (DT) and thermionic emission (TE). Owing to the thin oxide layer (4.4 nm), direct tunneling becomes feasible, with the effective barrier height determined by the conduction band offset between the dielectric layer and GaN. Simultaneously, electrons can also overcome the barrier through the thermionic emission process. As a result, the forward current is primarily governed by the combined contributions of DT and TE mechanisms.

The direct tunneling current density equation is:

$$
\begin{aligned}
J_{DT} &= A^* \exp(-\alpha_T d \sqrt{q \varphi_T}) \\
&\exp(-\frac{q \varphi_{B0}}{kT})[\exp(\frac{qV}{\eta_T kT}) - 1]
\end{aligned}
\tag{1}
$$

where A^*, α_T, d, φ_T, q, φ_{B0}, k, η_T represent effective Richardson constant (A^* is related to the effective mass of the electrons, which in this paper takes the value of 26.9 A cm^{-2} K^{-2}). a constant of 1.01 eV$^{-1/2}$ A^{-1}, oxide thickness, effective barrier height in direct tunneling (1.1 V), electron charge, Schottky barrier height extracted from the same batch of GaN SBDs without the 2nm Al$_2$O$_3$ insertion layer (approximately linearly positively correlated with the temperature), the Boltzmann constant, ideal factor in direct tunneling. The equation is processed by ignoring the effect of Schottky barrier height on direct tunneling, and considering the forward voltage V >> kT, the equation simplifies to a form similar to that of the hot electron emission model:

$$
I_D = AeA^* T^2 \exp(\frac{-q\phi_B}{kT}) \exp[\frac{q(V_F - I_D R_S)}{nkT}]
\tag{2}
$$

The on-state voltage is:

$$
V_F = n\phi_B + \frac{nkT}{q}[\ln(\frac{I_D}{AeA^*}) - 2\ln(T)] + I_D R_S
\tag{3}
$$

where n, Φ_B, T, A_e and R_S represent ideal factor, the Schottky barrier height, Kelvin temperature, area of the Schottky anode, and series resistance.

The derivation of the on-state voltage V_F with respect to the temperature T leads to the formula for the sensitivity of the device:

$$
\frac{dV_F}{dT} = \frac{nk}{q}\left(\ln(\frac{I_D}{AeA^* T^2}) - 2\right)
\tag{4}
$$

It is obtained from equation (4):

$$
\frac{dV_F}{dT} \propto \frac{nk}{q}\ln(\frac{I_D}{AeA^*})
\tag{5}
$$

From Eq. (5), it can be seen that the temperature sensitivity of the device can be increased by decreasing the forward current or increasing the anode area. However, increasing the anode area significantly raises fabrication complexity and cost. Therefore, the sensitivity is primarily improved by minimizing the forward current.

The P-GaN layer depletes the two-dimensional electron gas beneath it, which increases the resistance in that region and reduces the forward current. Simultaneously, the presence of the dielectric layer provides surface passivation and further suppresses current conduction. In addition, the leakage current caused by dislocation defects in the epitaxial layer can be mitigated by the anode recessed structure.

IV. TEMPERATURE SENSOR DIODE SIMULATION

In order to investigate the operating mechanism of the recessed anode GaN MIS diode temperature sensor, the forward conduction characteristics of the device are evaluated using Sentaurus TCAD software. Aluminum nitride (AlN) was employed as the passivation layer material. The key parameters of the simulated device structure are in agreement with the experiment results. The anode radius R_a=50 μm and the cathode-anode spacing L_{ac} is 20 μm.

Fig. 2(a)~(c)show the simulation results of I-V characteristics of the recessed anode GaN MIS diode temperature sensor under varying dielectric layer thicknesses. As the thickness of the dielectric layer increases, the temperature-dependent variation in the on-state voltage becomes more significant at the fixed current level. These results indicate that the thickness of the dielectric layer plays a critical role in determining the temperature sensitivity of the device.

When the device operates in the subthreshold region, the current is positively correlated with the temperature, and the presence of notches as well as the dielectric layer reduces the

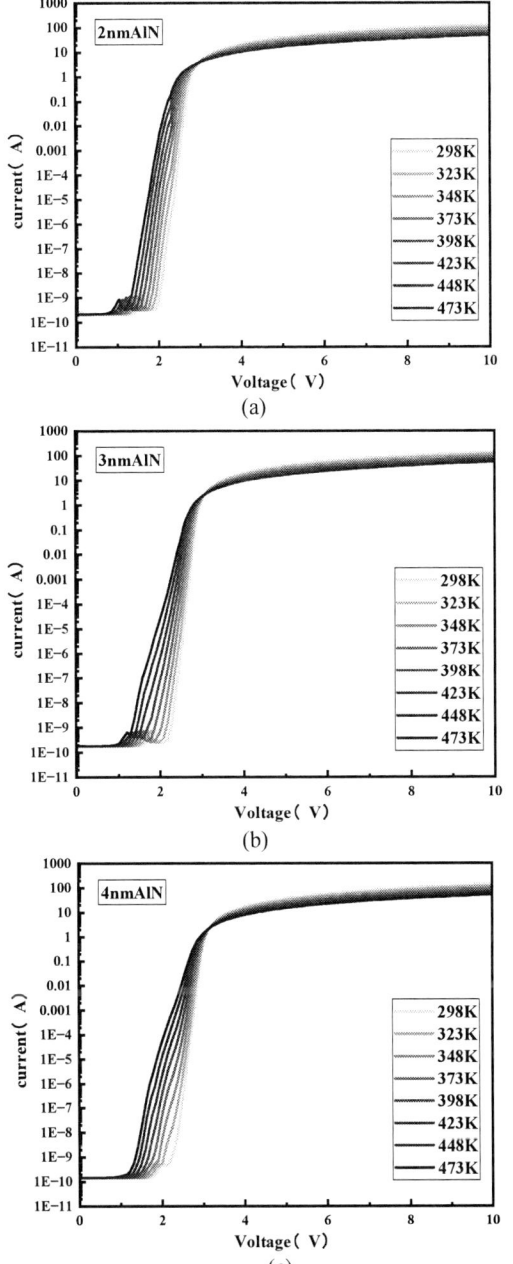

Fig. 2 Positive conduction IV characteristic curves for different dielectric layer thicknesses (a) 2nmAlN (b) 3nmAlN (c) 4nmAlN

longitudinal dislocation defects and interface states, and thus the positive conduction current is reduced, improving the temperature sensitivity.

Fig. 3 shows the relationship between the voltage drop and the temperature for GaN recessed-anode MIS diodes with different thicknesses of the dielectric layer under a bias current of 1e-7 A. Specifically, the extracted temperature sensitivity is calculated to be 3.57 mV/K (R^2=0.99995) for 2 nm dielectric layer thickness, 4.83 mV/K (R^2=0.98937) for 3nm dielectric layer thickness, and 5.17 mV/K (R^2=0.9948) for 4 nm dielectric layer thickness. These results indicate that

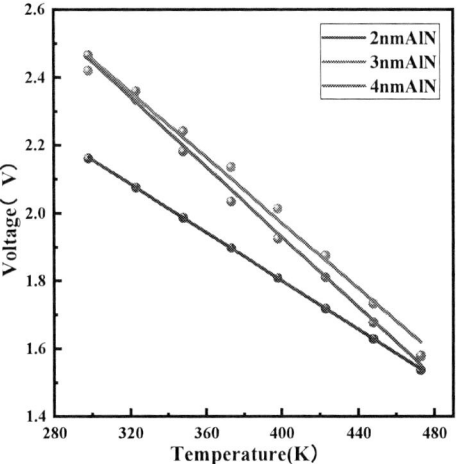

Fig. 3 Temperature sensitivity at different dielectric layer thicknesses (forward current of 1e-7 A)

increasing the dielectric thickness up to 4 nm significantly improves temperature sensitivity while maintaining near-ideal linearity.

The observed improvement is primarily attributed to the suppression of subthreshold leakage current, which is influenced by interface states introduced during the etching process. The presence of the dielectric layer reduces the leakage caused by the interface state after etching process, which lowers the current value in the subthreshold region. Therefore, the device sensitivity increases with the thickness of the dielectric layer.

V. EXPERIMENT TEST RESULT AND ANALYSIS

From Fig. 4(a), it can be observed that, when the cathode-anode distance is 20 μm, the reverse leakage current depends on both the anode radius and temperature. As the anode radius and temperature increase, the reverse leakage current also increases. Fig. 4(b) shows that, when the anode radius is 50 μm, the reverse leakage varies with the cathode-anode distance. It can be clearly seen that, under the room temperature, the change of the cathode-anode distance will not affect the reverse leakage too much. Therefore, suitable anode radius and cathode-anode distance should be selected according to the actual situation in order to obtain good reverse characteristics.

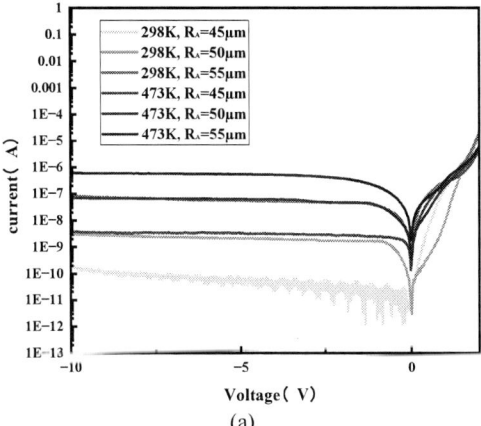

(a)

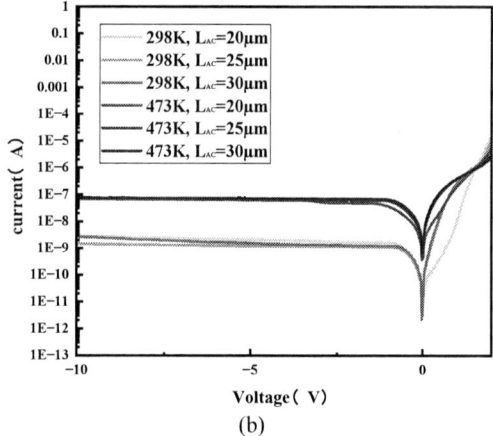

(b)

Fig. 4 (a) Variation of device reverse leakage with cathode-anode distance for anode radius equal to 50 μm (b) Variation of device reverse leakage with anode radius for device cathode-anode distance equal to 20 μm.

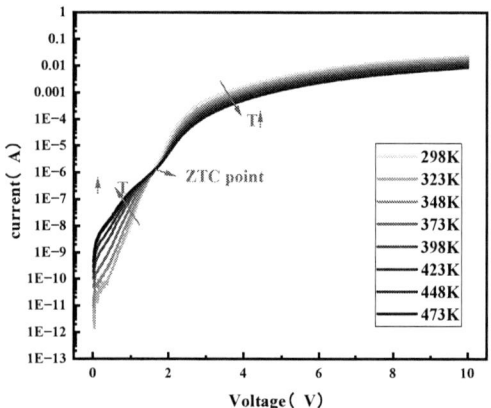

Fig. 5 I-V characteristics of recessed anode GaN MIS diode with temperature (semi-logarithmic coordinates).

The temperature-dependent I–V curves shown in Fig. 5 demonstrate two distinct regions. In the subthreshold regime, the forward voltage exhibits a positive temperature coefficient due to reduced barrier height and increased thermally activated carriers. Beyond the zero-temperature coefficient point, in the full conduction region, the I–V curves exhibit a negative shift with increasing temperature. This behavior is consistent with the thermionic emission model typically observed in SBD.

As shown in Fig. 6(a), the voltage drop of the recessed anode MIS diode has a good temperature dependence in the subthreshold region, for the current range from 4e-8 to 5e-9 A. In addition, Fig. 6(b) illustrates the sensing capability of a notch anode GaN MIS diode temperature sensor at current densities of 0.00013 A cm^{-2} to 0.00102 A cm^{-2}, which shows a gradual increase in the sensitivity with decreasing current, from 3.35 mV/K ($R^2 = 0.98284$) to 4.77 mV/K ($R^2 = 0.98322$). The sensitivity increases substantially and the linearity is close to the ideal case. The presence of the insulating dielectric layer decreases the forward leakage associated with the interface states, while the presence of the recessed anode reduces the forward leakage of the longitudinal dislocations of the epitaxial structure, and the temperature sensitivity increases with the decrease of current. In addition, Fig. 6(b) depicts the linear relationship

between sensitivity and current density, in agreement with theoretical calculations.

As shown in Fig. 7, the recessed-anode GaN MIS diode exhibits suppressed interface states and an increased barrier height, primarily attributed to the presence of the alumina dielectric layer. This results in reduced reverse leakage current.

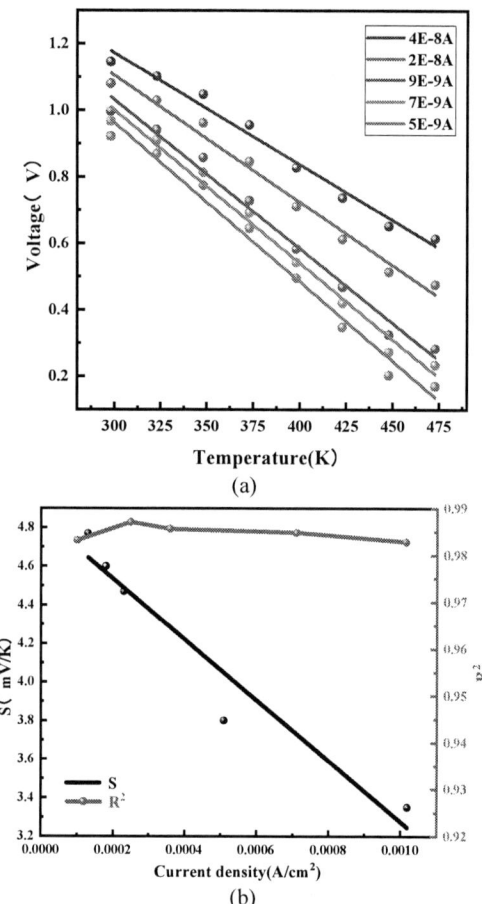

(a)

(b)

Fig. 6 (a) Forward voltage versus temperature at different currents. (b) Temperature sensitivity (S) and linearity of fit (R^2) versus current density.

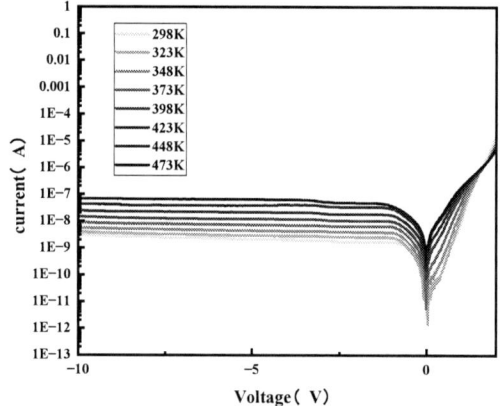

Fig. (7) Temperature-dependent reverse leakage current of the recessed-Anode GaN MIS Diode.

TABLE I. COMPARISON OF THE SENSITIVITY OF THE DIFFERENT GAN DIODE TEMPERATURE SENSORS

Device type	Current(A cm^{-2})	S （mV/K）	Ref
GaN SBD	0.56	1.13	[6]
Vertical GaN SBD	0.06	1.35	[7]
NiN GaN SBD	0.0000032	2.54	[5]
GaN-on-GaN SBD(ET)	0.0001	2.022	[8]
Lateral GaN HAD	/	1.64	[9]
RA-GaN SBD	/	1.05	[4]
PCT-SBD	2.04	2.54	[10]
TiN GaN SBD	/	1.20	[11]
GaN recessed-anode MIS diode	0.00013	4.77	This work

Table I shows the performance of the recessed anode MIS diode compared to previously reported GaN temperature sensors. The sensitivity of the recessed anode MIS diode is significantly better than previously reported GaN Schottky diode temperature sensors. Compared with the vertical devices in the table, such as vertical GaN SBD, NiN SBD, GaN-ON-GaN SBD(ET), the recessed anode MIS diodes not only have higher sensitivity, but also are easy to be integrated horizontally, enabling monolithic integration with GaN power devices. The recessed anode MIS diode temperature sensor achieves 322% higher sensitivity than standard GaN SBD sensors and 231% higher than vertical GaN SBD temperature sensor. Compared to devices with the same lateral integration capability like the RA-GaN SBD, the proposed device exhibits a 326% increase in sensitivity, primarily due to the dielectric layer and the partial P-GaN cap.

VI. CONCLUSION

In this paper, a recessed-anode GaN MIS diode temperature sensor was fabricated. The sensor utilizes the depletion effect of the P-GaN layer and the passivation effect of the insulating dielectric layer to reduce the diode's forward current. The T-shaped recessed anode structure further reduces the influence of dislocation defects, thus improving the temperature sensitivity. The result showed good temperature dependence in the temperature range of 298 K-473 K, and drastically improved the temperature sensitivity of the device.

The recessed-anode GaN MIS diode's temperature sensitivity reaches up to 4.77 mV/K (R^2=0.98322). The recessed-anode GaN MIS diode can also be monolithically integrated with matched P-GaN/AlGaN/GaN HEMT to improve the temperature measurement accuracy of AlGaN/GaN heterojunction.

ACKNOWLEDGMENT

This work was supported in part by the Young Scientists Fund of the National Natural Science Foundation of China under Grants 62404166 and 62204194, and in part by the Guangdong Basic and Applied Basic Research Fund under Grant 2025A1515011926.

REFERENCES

[1] Y. Zhang, A. Dadgar, T .Palacios. "Gallium nitride vertical power devices on foreign substrates: a review and outlook , " Journal of Physics D: Applied Physics, 2018, 51.doi:10.1088/1361-6463/aac8aa.

[2] X. Zou, X. Zhang, X. Lu, C. W. Tang and K. M. Lau, "Breakdown Ruggedness of Quasi-Vertical GaN-Based p-i-n Diodes on Si Substrates," in IEEE Electron Device Letters, vol. 37, no. 9, pp. 1158-1161, Sept. 2016, doi: 10.1109/LED.2016.2594821.

[3] I. Saidi, Y. Cordier, M. Chmielowska, H. Mejri, and H. Maaref, "Thermal effects in AlGaN/GaN/Si high electron mobility transistors," Solid-State Electronic., vol. 61, no.1, pp. 1-6, Jul. 2011, doi:10.1016/j.sse.2011.02.008.

[4] T. Pu et al., "Recessed Anode AlGaN/GaN Schottky Barrier Diode for Temperature Sensor Application," in IEEE Transactions on Electron Devices, vol. 68, no. 10, pp. 5162-5166, Oct. 2021, doi: 10.1109/TED.2021.3105498.

[5] L. Li, T. Pu, X. Li and J. -P. Ao, "Effect of Anode Material on the Sensitivity of GaN Schottky Barrier Diode Temperature Sensor," in IEEE Sensors Journal, vol. 22, no. 3, pp. 1933-1938, 1 Feb.1, 2022, doi: 10.1109/JSEN.2021.3133895.

[6] X. Li, T. Pu, X. Li, L. Li and J. -P. Ao, "Correlation Between Anode Area and Sensitivity for the TiN/GaN Schottky Barrier Diode Temperature Sensor," in IEEE Transactions on Electron Devices, vol. 67, no. 3, pp. 1171-1175, March 2020, doi: 10.1109/TED.2020.2968358.

[7] L. Li, X. Li, T. Pu, S. Cheng, H. Li and J. -P. Ao, "Vertical GaN-Based Temperature Sensor by Using TiN Anode Schottky Barrier Diode," in IEEE Sensors Journal, vol. 21, no. 2, pp. 1273-1278, 15 Jan.15, 2021, doi: 10.1109/JSEN.2020.3018330.

[8] X. Li et al., "Effect of Helium-Implanted Edge Termination on GaN-on-GaN Schottky Barrier Diode Temperature Sensors," in IEEE Sensors Journal, vol. 23, no. 24, pp. 30112-30118, 15 Dec.15, 2023, doi: 10.1109/JSEN.2023.3325663.

[9] X. Wei et al., "Dual Current and Voltage Sensitivity Temperature Sensor Based on Lateral p-GaN/AlGaN/GaN Hybrid Anode Diode," in IEEE Sensors Journal, vol. 21, no. 20, pp. 22459-22463, 15 Oct.15, 2021, doi: 10.1109/JSEN.2021.3109915.

[10] Z. Yan et al., "A Novel AlGaN/GaN-Based Schottky Barrier Diode With Partial P-GaN Cap Layer and Semicircular T-Anode for Temperature Sensors," in IEEE Transactions on Electron Devices, vol. 70, no. 10, pp. 5087-5091, Oct. 2023, doi: 10.1109/TED.2023.3306736.

[11] L. Li, J. Chen, X. Gu, X. Li, T. Pu, and J.-P. Ao, "et al.Temperature sensor using thermally stable TiN anode GaN Schottky barrier diode for high power device application," Superlattices and Microstructures, vol. 123, pp. 274-279, Nov. 2018, doi:10.1016/j.spmi.2018.09.007.

Warping model of SiC Power Module after Reflow Soldering

Chang Liu
Institute of Novel Semiconductors
Shandong University
Jinan,China
202234091@mail.sdu

Yingxin Cui*
Institute of Novel Semiconductors
Shandong University
Jinan,China
cuiyingxin@sdu.edu.cn

Yanao Guo
Institute of Novel Semiconductors
Shandong University
Jinan,China

Shoulai Gong
Institute of Novel Semiconductors
Shandong University
Jinan,China

Jisheng Han
Institute of Novel Semiconductors
Shandong University
Jinan,China

Abstract—The mismatch in coefficients of thermal expansion (CTE) among materials in silicon carbide (SiC) power modules during reflow soldering leads to significant warpage of the Direct Bonded Copper (DBC) substrate and then increases residual stresse between the DBC and the SiC device, adversely affecting device and module stability and reliability. However, quantitative analyses of warpage behavior of the SiC module and effective mitigation strategies remain limited. In this paper, a thermo-mechanical coupled finite element model of a 62 mm, 1200 V/300 A SiC half-bridge power module is established to simulate the warpage phenomenon under realistic reflow soldering conditions. Simulation results indicate a typical warpage profile with a raised center and lowered edges, matching experimental measurements within a maximum error of 4.7%. The maximum stress value of the solder layer reached 564 MPa. To mitigate warpage and solder-layer stresses, a mechanical pre-warping method is proposed and evaluated. Simulation and experimental analyses show that applying pre-warping between 250 µm and 500 µm can significantly reduce the warpage of DBC substrate up to 27.3% and the stress of solder-layer up to 44%. However, further increases in pre-warp magnitude yield diminishing improvements and introduce substantial initial stresses. This paper quantitatively clarifies the impact of pre-warping magnitude on the suppression of DBC substrate warpage and the stress reduction of solder-layer, providing practical guidance for optimal SiC power module packaging design and improved reliability.

Keywords—SiC Power Module, Warping, Reflow Soldering

I. Introduction

Silicon carbide (SiC) power modules have emerged as highly attractive devices in power electronics applications due to their outstanding properties, including high power density, high-temperature operational capability, and reduced switching losses. However, during the packaging process, especially in the reflow soldering stage, warpage of the Direct Bonded Copper (DBC) substrate frequently occurs, driven by significant mismatches in coefficients of thermal expansion (CTE) among the constituent materials[1-2] This warpage not only compromises the uniformity and integrity of the soldered interfaces but also leads to increase the thermal contact resistance, thereby diminishing heat dissipation efficiency and negatively impacting electrical performance and reliability[3-4]. Consequently, understanding the mechanisms underlying

warpage formation during the reflow soldering process, precisely predicting the warpage of DBC substrate, and developing effective suppression methods are critical for enhancing the quality and performance of SiC power module packaging.

This paper presents an in-depth investigation into the warpage behaviors of SiC power module subjected to typical reflow soldering conditions. A fully coupled thermo-mechanical finite element simulation model is constructed, closely replicating actual assembly conditions, to predict and analyze the warpage profiles and residual stress distributions. The coordinate measuring equipment is utilized to test the experimental warpage of the DBC substrate, and the test results are used for calibration the validate the simulation results, ensuring the accuracy and credibility of the proposed numerical approach.

Furthermore, based on the insights obtained from simulation and experiment, this study proposes a novel warpage suppression strategy, which is called mechanical pre-warping. By varying pre-warp magnitudes systematically, the research quantifies the corresponding improvements in warpage reduction and solder-layer stress mitigation. Ultimately, this comprehensive analysis not only deepens the fundamental understanding of the warpage phenomena associated with SiC power module packaging, but also provides valuable theoretical and experimental guidance toward optimized design and reliability of SiC module.

II. Thermal-Mechanical Coupling Simulation Model of Reflow Soldering

A. *Warpage phenomenon of the copper baseplate after reflow soldering*

Fig.1 illustrates the packaging structure of a SiC half-bridge power module with a 62 mm footprint, rated at 1200 V and 300 A, which serves as the basis for this study.

979-8-3315-1110-4/25 $31.00 © 2025 IEEE

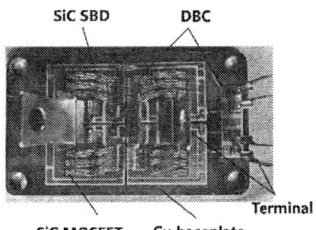

Fig. 1 Top view of the internal structure of a 62 mm packaged 1200 V/300 A SiC MOSFET module after removal of the plastic encapsulation

In the module packaging process, the DBC substrate were bonded to the copper baseplate through a vacuum reflow soldering method. Initially, a 100 μm thick layer of PbSn₅Ag₂.₅ solder was printed on the top surface of a 3 mm thick copper baseplate. Subsequently, four power loop DBC substrates and one gate loop DBC substrate were mounted on the solder-coated surface. Each DBC substrate consisted of a 0.63 mm thick alumina (Al_2O_3) ceramic layer with 96% purity, sandwiched between two 0.3 mm thick copper layers. After assembly, vacuum reflow soldering was performed to ensure the formation of high-quality solder joints layer. At the preheating and heating stage, the structure of the DBC substrate and copper baseplate expanded freely with the temperature rising，as shown in Figure 2 (a),[5]. However, during the cooling stage, as the temperature drops, the mismatch in coefficients of thermal expansion (CTEs) among the different materials and the solidification-induced constraints of the solder led to warpage and the development of residual stresses, as shown in Fig 2(b). The initially flat baseplate exhibited noticeable warpage after reflow soldering. It shows a visible gap between the center of the baseplate and the reference horizontal line, indicating a significant warpage phenomenon after reflow soldering.

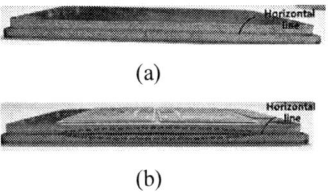

(a)

(b)

Fig. 2. Baseplate profiles before and after reflow soldering: (a) Flat copper baseplate; (b) Warped copper baseplate

B. 3D model of the SiC power module

Based on the actual module, the SiC power module geometry model established in this study consists of three parts: the DBC substrate, the solder layer, and the copper baseplate, as shown in Fig. 3.

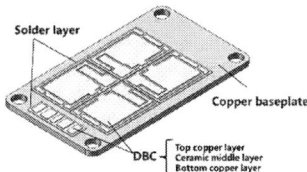

Fig. 3 Reflow Soldering Simulation Model

The critical geometrical dimensions of the model are presented in TABLE I.

TABLE I. THE CRITICAL GEOMETRICAL DIMENSIONS OF THE MODEL

Component	Dimension L×W×H (mm)
Ceramic layer of power loop DBC	49.5 × 24.5 × 0.38
Copper layer of power loop DBC	31.6 × 22.5 × 0.3
Solder layer of power loop DBC	31.6 × 22.5 × 0.1
Ceramic layer of gate loop DBC	29 × 9 × 0.38
Copper layer of gate loop DBC	27 × 7 × 0.3
Solder layer of gate loop DBC	31.6 × 22.5 × 0.1
Cu Baseplate	104.5 × 59.5 × 3

C. Material properties

During the reflow soldering process, the solder material undergoes a temperature cycle from room temperature 25 ℃ to above its liquidus temperature 287-294 ℃, followed by cooling back to room temperature. The solder used in this study is PbSn₅Ag₂.₅, with a melting (liquidus) temperature range of 287–294 °C. During this thermal process, the solder experiences creep elastic deformation, and plastic deformation, all of which are closely related to the temperature profile and heating/cooling rates. To effectively characterize the thermos-mechanical behavior of solder during the reflow process, the Anand viscoelastic constitutive model is widely employed. The mathematical formulation of the Anand model is given as follows [6]:

$$\dot{\varepsilon}_p = A \exp\left(-\frac{Q}{RT}\right)\left[\sinh\left(\xi\frac{\sigma}{s}\right)\right]^{1/m} \quad (1)$$

$$\dot{s} = \left\{h_0\left|1 - \frac{s}{s^*}\right|^{\alpha} \text{sign}\left(1 - \frac{s}{s^*}\right)\right\}\dot{\varepsilon}_p \quad (2)$$

$$s^* = \hat{s}\left[\frac{\varepsilon_p}{A}\exp\left(\frac{Q}{RT}\right)\right]^n \quad (3)$$

where $\dot{\varepsilon}_p$ is the inelastic strain rate, A is the pre-exponential factor, Q is the activation energy, R is the gas constant, T is the absolute temperature, ξ is the stress multiplier, σ is the equivalent stress, s is the deformation resistance, ṡ is the rate of deformation resistance, m is the strain rate sensitivity factor, h₀ is the hardening constant, s* is the saturation value of deformation resistance, α is the hardening sensitivity parameter, and n is the strain rate sensitivity factor of the saturation resistance.

The model parameters can be obtained by fitting stress–strain curves from material tensile or shear tests. The Anand constitutive model parameters for PbSn₅Ag₂.₅ solder are listed in TABLE II [7]

TABLE II. ANAND CONSTITUTIVE MODEL PARAMETERS FOR PbSn₅AG₂.₅ SOLDER

Symbol	Value	Symbol	Value
m	0.307	α	3.96
A	26400	$\hat{S}$(MPa)	35.75
Q	90697	s(MPa)	30
h_0(MPa)	36700	n	0.03
ξ	13		

The other material parameters are listed in Table III.

The table presents key thermophysical and mechanical properties of materials, including the specific heat capacity (C_p, J/kg·K), thermal conductivity (K, W/m·K), coefficient of thermal expansion (CTE, 1/K), density (ρ, kg/m³), Young's modulus (E, GPa), and Poisson's ratio (ν).

TABLE III.   MATERIAL PARAMETERS

Material	Cp (J/kg·K)	K (W/m·K)	CTE (1/K)	ρ (kg/m³)	E (GPa)	ν
Al_2O_3	900	490	4.3 e-6	3216	748	0.45
Cu	385	400	17 e-6	8960	120	0.35
solder	130	40	2.9 e-5	11110	24.7	0.35

D. FE model

The reflow soldering process was modeled through a coupled thermo-mechanical analysis, with the thermal load applied to the entire model based on the actual process profile, as shown in Fig. 4.

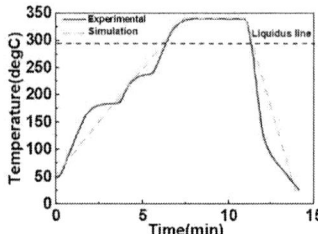

Fig. 4. Reflow soldering temperature curve

The soldering behavior can be divided into three main stages: heating, reflow, and cooling. In the heating stage, when the temperature exceeds the melting point of the solder alloy, the solder enters intoa molten state, and the mechanical interaction between the DBC plates and the baseplate can be neglected. Upon cooling to the solidus temperature, the solder solidifies and forms a mechanical bond between the DBC plates and the baseplate. This physical phenomenon was modeled using the "activation" function available in COMSOL Multiphysics. [8]

During the reflow stage, before solder solidification, the solder joints are unable to support mechanical loads, and the DBC plates are in a free-expansion state, unaffected by the solder joints. In this stage, internal stresses within the DBC plates remain close to zero. In the simulation, the activation function was implemented by applying a scaling factor to drastically reduce the elastic modulus and density of the solder material, effectively modeling it as a "nonexistent" state. When the temperature decreases to the solidus point and the activation conditions are satisfied, the solder material is reactivated from a stress-free state, establishing mechanical constraints between the DBC plates and the baseplate, thereby inducing stress.

To ensure numerical convergence during simulation, appropriate mechanical boundary conditions were applied. A rigid body motion suppression constraint was imposed on the baseplate to prevent global translation or rotation of the model, thereby ensuring the stability and accuracy of the simulation results.

E. Analysis of Experimental and Simulation Results

The reflow soldering simulation result is shown in Fig. 5, where the baseplate exhibits a warpage pattern characterized by a "high center and low edges", consistent with the experimental morphology of the sample shown in Fig. 2. By

extracting the warpage data from the bottom surface of the baseplate, the maximum warpage was found to be 533 μm.

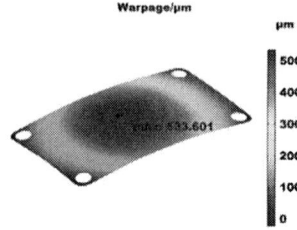

Fig. 5. Simulated warpage of the flat baseplate after reflow soldering.

For the experimental sample shown in Fig. 2, the bottom-side warpage of the baseplate after the reflow soldering process was measured using a coordinate measuring machine (CMM). The measurement setup is illustrated in Fig. 6(a), and the contact-type probe used for the measurement is shown in Fig. 6(b). The probe started scanning from the midpoint of the short edge on the backside of the baseplate and moved along the long edge, with a point interval of 2 mm, to acquire the warpage data.

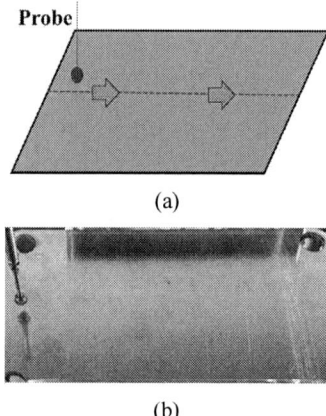

Fig.6. Warpage measurement: (a) Schematic of CMM measurement setup; (b) CMM measurement result

As the experimental sample is shown in Fig. 2, the bottom-side warpage of the baseplate after reflow soldering was measured using a coordinate measuring machine (CMM). The measurement setup is illustrated in Fig. 5(a), and the contact-type probe used for the measurement is shown in Fig. 5(b). The probe started scanning from the midpoint of the short edge on the backside of the baseplate and moved along the long edge, with a point interval of 2 mm, to acquire the warpage data. At the same time, the warpage values from the simulation results after reflow soldering were also extracted for comparison.

Fig. 7 presents the comparison between the simulation and experimental results. The maximum measured warpage is 509 μm. Using the experimental data as a reference, the simulation exhibits a 4.7% error at the location of maximum warpage.

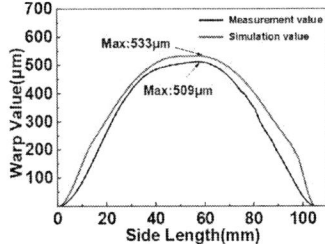

Fig. 7. Comparison between experimental and simulation results.

The other material parameters are listed in Table IV.

TABLE IV.   COMPARISON OF WARPAGE AT DIFFERENT POSITIONS.

Parameters	20 mm	40 mm	60 mm	80 mm
Experimental value (μm)	263	485	510	346
Simulation value (μm)	305	507	529	402
Error (%)	15.9	4	3.7	16.1

The errors between the simulation results and the measured values within these intervals are all within 20%, with the minimum error of 3.7% occurring at the 60 mm position. The observed discrepancies may be attributed to several non-ideal factors: first, uneven heating within the reflow furnace chamber, where the top and bottom heating methods may cause non-uniform thermal distribution across the sample, second, slight displacement of the DBC substrate during the reflow process, which affects the final warpage results, finally, material property variations, where local inhomogeneities in material properties can influence the magnitude of warpage.[9]

Despite these potential sources of error, the simulated warpage data shows good agreement with the experimental measurements, thereby validating the accuracy of the established simulation model and providing a foundation for further research and optimization of warpage behavior.

III.   WARPAGE SUPPRESSION AND IMPROVEMENT REPARE

In industrial applications, excessive warpage of SiC power modules beyond a specific threshold can render them unsuitable for practical use. Since the flat baseplate exhibits a "high center and low edges" warpage pattern after reflow soldering, a pretreatment strategy was proposed to mitigate this effect. Prior to reflow, a counter warpage was mechanically induced in the baseplate to compensate for the anticipated deformation. To evaluate the effectiveness of this approach, finite element simulations were conducted.

A.   Effectiveness Verification of Pre-Warping in Warpage Suppression

As illustrated in Fig. 8(a), a finite element model of a baseplate with an initial counter warpage of 500 μm was established. The geometric dimensions, thermos-physical properties, and boundary conditions were maintained consistent with the previous simulation model, and the simulation procedures followed the same methodology as previously described.

Subsequently, based on the simulation design, an actual baseplate with the corresponding pre-applied warpage was fabricated. A mechanical pre-bending process was employed to induce controlled deformation in the copper baseplate, resulting in a free-state configuration with the desired warpage. The fabricated baseplate is shown in Fig. 8(b). To verify the warpage accuracy, a coordinate measuring machine (CMM) was utilized to scan the baseplate surface. The measured maximum warpage was 480 μm, presenting a small deviation from the target value of 500 μm, thereby confirming that the fabricated baseplate satisfied the experimental requirements.

Following the reflow soldering process, the assembled sample with the pre-warped baseplate was obtained, as shown in Fig. 8(c). Morphological analysis revealed that the overall warpage of the sample was significantly reduced compared to the sample without pre-treatment. The gap between the baseplate and the reference horizontal line was considerably smaller, demonstrating that the pre-applied counter warpage effectively mitigated the warpage induced during the reflow process.

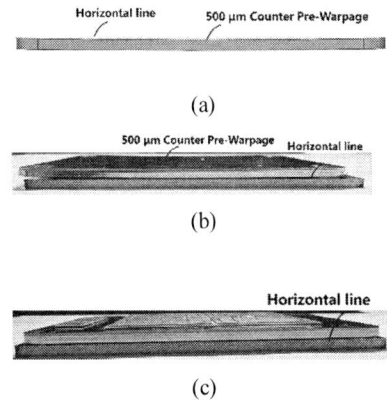

Fig. 7 Baseplate profiles: (a) Simulation with 500 μm pre-warp; (b) Fabricated baseplate with 480 μm pre-warp; (c) Baseplate after reflow soldering.

To quantitatively evaluate the results, the bottom-side warpage of the baseplate after reflow soldering was measured using a coordinate measuring machine (CMM) and compared with the simulation results, as shown in Fig. 8. The experimental measurement indicated a maximum warpage of 312.5 μm, while the simulated value was 338.5 μm, resulting in an 8% error at the location of maximum warpage.

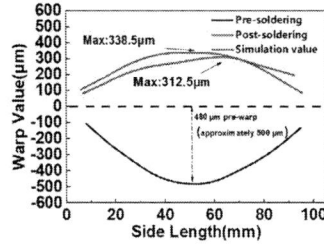

Fig. 8. Comparison between experimental and simulation results.

By comparing the simulation and experimental results, it was observed that the simulation error increased. This discrepancy may be attributed to the non-uniform distribution of solder during the printing process, caused by the pre-applied warpage, where a thicker solder layer tends to accumulate in the bent regions. Additionally, the simulation model could not fully capture the dynamic melting and solidification behavior of the solder, leading to additional errors. Nevertheless, the error in the maximum warpage remains relatively small, and the results still provide significant reference value and can serve as an effective predictive tool.

B. Effect of Different Pre-Warping Magnitudes on Warpage Behavior

Further, three-dimensional models of baseplate subcomponents with initial pre-warp magnitudes of 250 μm, 750 μm, 1000 μm, 1250 μm, and 1500 μm were constructed. The overall warpage morphologies after reflow soldering simulations for different pre-warped baseplates are shown in Fig. 9. From the morphological analysis, as illustrated in Fig. 9(a), the baseplate with an initial pre-warp of 250 μm exhibited significant warpage phenomenon after reflow soldering, with the maximum deformation near the center reaching approximately 450 μm. As the magnitude of the applied counter pre-warp increased, the overall deformation of the structure, as shown in Fig. 9(b)–(e), progressively decreased, approaching approximately 300 μm. As shown in Fig. 9(f), the maximum deformation of the baseplate was reduced to less than 300 μm when the initial pre-warp reached 1500 μm.

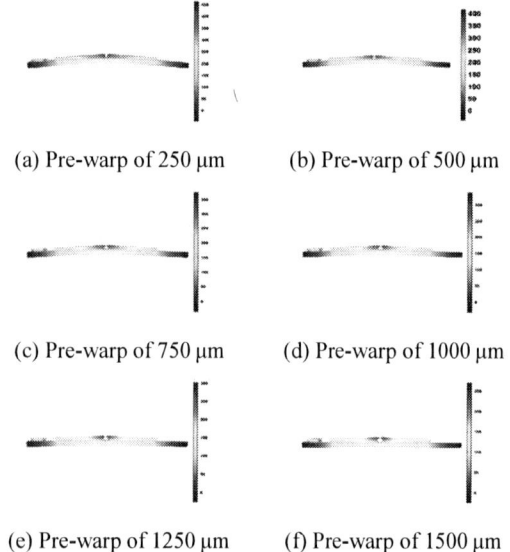

(a) Pre-warp of 250 μm   (b) Pre-warp of 500 μm

(c) Pre-warp of 750 μm   (d) Pre-warp of 1000 μm

(e) Pre-warp of 1250 μm   (f) Pre-warp of 1500 μm

Fig. 9. Baseplate warpage profiles after reflow soldering with different pre-warp magnitudes: (a) 250 μm pre-warp; (b) 500 μm pre-warp; (c) 750 μm pre-warp; (d) 1000 μm pre-warp; (e) 1250 μm pre-warp; (f) 1500 μm pre-warp.

C. Analysis of Simulation Results

The stress within the solder layer after reflow soldering has a significant impact on the reliability of power modules. Stress concentration positions are prone to initiating cracks, which can gradually propagate and eventually lead to delamination or failure under repeated stress cycles [10-11]. The pre-warping treatment significantly suppressed the deformation after reflow soldering, resulting in a baseplate profile closer to a flat state. To provide a consistent characterization, the maximum warpage in the z-direction at the bottom of the baseplate and the maximum stress in the solder layer of the power circuit were extracted, as shown in Fig. 10.

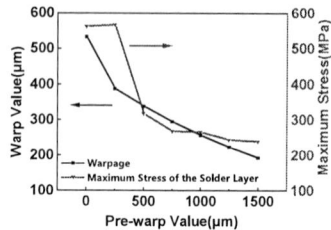

Fig. 10. Simulation results of the maximum z-direction warpage at the bottom of the baseplate and the maximum stress in the solder layer of the power circuit under different pre-warping conditions.

When comparing the flat baseplate with the baseplate having a pre-warp of 250 μm, the warpage decreased from 533 μm to 388 μm, representing a reduction of 145 μm and a decrease of 27.3%. This stage exhibited a significant reduction in warpage, indicating that applying a low pre-warp magnitude is highly effective in suppressing warpage. As the pre-warp value continued to increase from 250 μm to 1500 μm, the trend of warpage reduction gradually slowed. The warpage decreased from 388 μm to 193.2 μm, although the reduction persisted, the rate of decrease diminished, falling from 38.2% to approximately 50%. This suggests that beyond a certain pre-warp magnitude, the improvement in warpage suppression tends to stabilize, and further increasing the pre-warp provides diminishing returns.

Regarding the change in the maximum stress of the solder layer, after reflow soldering, the maximum stress increased slightly from 563 MPa for the flat baseplate to 567 MPa for the baseplate with a 250 μm pre-warp, indicating minimal variation. However, between 250 μm and 500 μm pre-warp values, the maximum stress dropped significantly from 567 MPa to 318 MPa, a reduction of 249 MPa, corresponding to a 44% decrease, demonstrating that pre-warping notably reduced the solder layer's maximum stress. As the pre-warp further increased from 500 μm to 1500 μm, the reduction trend became more gradual, with the maximum stress decreasing from 318 MPa to 239 MPa, a further reduction of only 24%. This indicates that beyond a certain pre-warp magnitude, the improvement in reducing the solder layer stress also tends to diminish.

Through reflow soldering simulations of baseplates with different pre-warp magnitudes, the study demonstrates that applying an appropriate counter pre-warp can effectively suppress warpage after reflow soldering, particularly at lower pre-warp levels where the reduction in warpage is most significant. However, as the pre-warp magnitude continues to increase, the reduction in warpage gradually becomes less pronounced, indicating a diminishing improvement effect.

Regarding the solder layer stress, pre-warping can significantly reduce the maximum stress, with the most notable reduction occurring between 250 μm and 500 μm of pre-warp. As the pre-warp magnitude increases further, the decrease in maximum stress becomes progressively smaller. Although pre-warping is effective in minimizing both post-reflow warpage and solder layer stress, an excessively high pre-warp value can introduce substantial initial stress into the baseplate, adversely affecting the reflow soldering quality.

Therefore, it is necessary to comprehensively balance the pre-warp magnitude in practical design, the initial stress within the baseplate, the post-reflow warpage, and the maximum stress in the solder layer to achieve optimal assembly quality and long-term reliability.

IV. CONCLUSION

In this study, a thermo-mechanical coupled simulation accurately modeled the warpage behavior of a 1200 V, 300 A SiC half-bridge power module with a 62 mm footprint during reflow soldering, matching experimental results within 4% deviation. A mechanical pre-warping technique effectively reduced post-reflow deformation and solder stress. Pre-warp magnitudes between 250 μm and 500 μm provided optimal results, with a 250 μm pre-warp reducing warpage by 27.3% and a 500 μm pre-warp further decreasing solder stress by approximately 44%. However, benefits plateaued beyond 500 μm, as higher pre-warp introduced significant internal stresses, potentially harming solder joint quality. Thus, balancing pre-warp magnitude, internal stress, and warpage is essential for achieving optimal reliability. These findings offer practical guidelines for enhancing the reliability of SiC power module assemblies.

ACKNOWLEDGEMENTS

This work was supported by the Key R&D Program of Shandong Province (Grant 2022ZLGX02 and Grant 2022CXGC010103), Taishan Scholars Program of Shandong Province (Grant tsqn202306069), Natural Science Foundation of Shandong Province (Grant ZR2022QF089).

REFERENCES

[1] Chuang W C, Chen W L. Study on the Strip Warpage Issues Encountered in the Flip-Chip Process [J]. Materials, 2022, 15(1).

[2] Guo Y, Liu M, Yin M, et al. Reliability Sensibility Analysis of the PCB Assembly concerning Warpage during the Reflow Soldering Process [J]. Mathematics, 2022, 10(17)

[3] Le Henaff F, Azzopardi S, Woirgard E, et al. Lifetime Evaluation of Nanoscale Silver Sintered Power Modules for Automotive Application Based on Experiments and Finite-Element Modeling [J]. Ieee Transactions on Device and Materials Reliability, 2015, 15(3): 326-334.

[4] Zhou Y, Xu L, Liu S. Optimization for warpage and residual stress due to reflow process in IGBT modules based on pre-warped substrate [J]. Microelectronic Engineering, 2015, 136: 63-70.

[5] Baek J-H, Park D-W, Oh G-H, et al. Effect of cure shrinkage of epoxy molding compound on warpage behavior of semiconductor package [J]. Materials Science in Semiconductor Processing, 2022, 148: 106758.

[6] Long X, Chen Z, Wang W, et al. Parameterized Anand constitutive model under a wide range of temperature and strain rate: experimental and theoretical studies [J]. Journal of Materials Science, 2020, 55(24): 10811-10823.

[7] Wang G Z, Cheng Z N, Becker K, et al. Applying Anand Model to Represent the Viscoplastic Deformation Behavior of Solder Alloys [J]. Journal of Electronic Packaging, 1998, 123(3): 247-253.

[8] Khatibi G, Kotas A B, Lederer M. Effect of aging on mechanical properties of high temperature Pb-rich solder joints [J]. Microelectronics Reliability, 2018, 85: 1-11.

[9] Matvienko O, Daneyko O, Kovalevskaya T, et al. Investigation of Stresses Induced Due to the Mismatch of the Coefficients of Thermal Expansion of the Matrix and the Strengthening Particle in Aluminum-Based Composites [J]. Metals, 2021, 11(2).

[10] Xie L, Deng E, Gu D, et al. Remaining Useful Lifetime Prediction Method of Power Modules Based on the Aging Characteristic Parameters [J]. IEEE Transactions on Power Electronics, 2025, 40(1): 2086-2098.

[11] Xiang D, Ran L, Tavner P, et al. Condition Monitoring Power Module Solder Fatigue Using Inverter Harmonic Identification [J]. IEEE Transactions on Power Electronics, 2012, 27(1): 235-247.

A novel mechanical stress related failure mechanism of the state-of-the-art SiC double trench MOSFET under surge current stress

Shikang Xu
State Key Laboratory of Electronic Thin Films and Integrated Devices University of Electronic Science and Technology of China
Chengdu, China
shikangxu@std.uestc.edu.cn

Xuan Li*
State Key Laboratory of Electronic Thin Films and Integrated Devices University of Electronic Science and Technology of China
Chengdu, China
xuanli@uestc.edu.cn

Hanqing Zhao
State Key Laboratory of Electronic Thin Films and Integrated Devices University of Electronic Science and Technology of China
Chengdu, China
hanqingzhao@std.uestc.edu.cn

Yi Wen
State Key Laboratory of Electronic Thin Films and Integrated Devicesg University of Electronic Science and Technology of China
Chengdu, China
wenyi169@163.com

Wensong Peng
Shenzhen Institute for Advanced Study University of Electronic Science and Technology of China
Shenzhen, China
pengwensong_uestc@163.com

Zekun Zhou
State Key Laboratory of Electronic Thin Films and Integrated Devices University of Electronic Science and Technology of China
Chengdu, China
zkzhou@uestc.edu.cn

Xu Li
School of Integrated Circuit Science and Engineering Southwest Jiaotong university
Chengdu, China
xuli@swjtu.edu.cn

Xiaochuan Deng
State Key Laboratory of Electronic Thin Films and Integrated Devices University of Electronic Science and Technology of China
Chengdu, China
xcdeng@uestc.edu.cn

Bo Zhang
State Key Laboratory of Electronic Thin Films and Integrated Devices University of Electronic Science and Technology of China
Chengdu, China
zhangbo@uestc.edu.cn

Abstract—In this paper, the surge reliability of the state-of-the-art reinforced double trench silicon carbide (SiC) metal-oxide-semiconductor field effect transistors (RDT-MOSFETs) featuring the deep source trench is investigated. A previously unreported failure mechanism is identified, which leads to the short circuit between the drain and source terminals. After microcosmic failure analysis, a 10 μm-long crack connecting the poly Si in source trench and the SiC in drift region is found. The device-level electro-thermal-mechanical analysis is employed to explain the formation of the crack. The results indicate that a novel mechanical stress concentration is introduced by the deep source trench and primarily focuses at its corner. Once the crack forms, expansion occurs easily due to the inherently low fracture toughness of SiC. This failure mechanism highlights the necessity for device designers to carefully consider reliability risk related to mechanical stress.

Keywords—SiC MOSFET, reinforced double trench, failure mechanism, mechanical stress, surge current

I. INTRODUCTION

Due to low on-resistance, fast switching speed and high junction temperature, the high-voltage silicon carbide metal-oxide-semiconductor field effect transistors (SiC MOSFETs) emerge rapidly in power applications, including electric vehicles, solar inverters and traction inverters[1], [2], [3], [4]. The SiC MOSFET exhibits attracting trade-off between on-resistance and switching loss, which is competitive to its counterpart Si IGBT. However, the reliability of SiC MOSFET still requires further verification, one critical aspect of which is surge reliability. Regularly, a freewheeling diode is needed to conduct the reverse current during the dead time in applications like synchronous rectifiers. Moreover, the device itself on the AC side also suffers large surge current when a short-circuit fault occurs in AC/DC converter. A fast

recovery diode is connected in anti-parallel with the IGBT to conduct the reverse surge current[5], while the body diode of SiC MOSFET is expected to replace the aforementioned freewheeling diode in terms of lower cost and higher integration density[6], [7], [8]. Therefore, the high surge current withstand capability is essential for SiC MOSFET when operating in the third quadrant to avoid irreversible failure.[9].

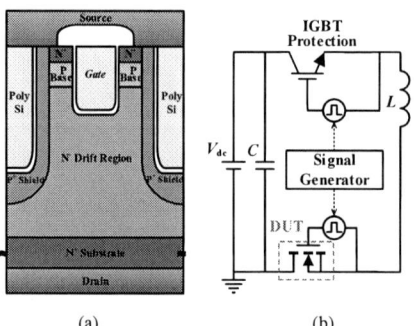

(a)   (b)

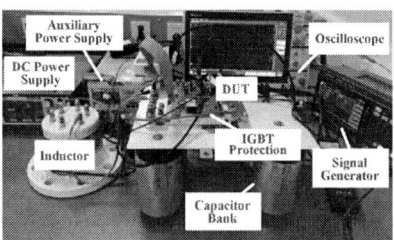

(c)

Fig. 1. Surge test setups. (a) Cell structure of the DUT, (b) topology of the test circuit, and (c) demonstration of the surge test.

The double trench MOSFET (DT-MOSFET) with gate trench and source trench is one of the widely-used trench gate SiC MOSFETs. Several researchers study its reliability under surge current stress, and the failure reasons are thermal burnout[10], [11], [12], [13] and fragile gate rupture[14]. The thermal burnout is typically due to the high junction temperature induced breakdown of the body diode or melting of aluminum, which eventually results in short circuit between three terminals or between gate and source terminals. The gate rupture, on the other hand, is typically due to the hot electron injection into the gate oxide, which eventually results in the short circuit between gate and source terminals. Based on previous generation DT-MOSFET, the state-of-the-art reinforced double trench SiC MOSFET (RDT-MOSFET) features the deeper source trench and smaller cell pitch exhibiting higher power density[15]. However, its surge reliability, particularly with respect to its failure mechanisms, is not investigated up till now.

In this paper, the surge current reliability of RDT-MOSFET operating in the third quadrant is investigated experimentally. Furthermore, a novel failure mechanism beyond those aforementioned is identified which causes short circuit between drain and source terminals. The root cause is the mechanical stress concentration at the corner of the deep source trench through microcosmic failure analysis and device-level electro-thermal-mechanical analysis.

II. EXPERIMENTAL SETUP AND RESULT

To test the surge current reliability of RDT-MOSFET, the 1200 V-rated SiC MOSFET SCT40346KL manufactured by Rohm is selected as the device under test (DUT) and the experiment setup is designed as shown in Fig 1. The typical specific on-state resistance $R_{on,sp}$ ($R_{ds(on)} \times$active area, @V_{GS}=18 V) and rated current I_{rate} (@T_c=100 °C) of DUT are 2.2 mΩ·cm² and 30 A, respectively[16]. The cell pitch and depth of source trench are measured as 2.1 μm and 1.9 μm, respectively. The capacitor bank C composed of four paralleled 500 μF/1200 V capacitors and the inductor of 5 mH generate 10 ms half sinusoidal wave to simulate the surge current in actual applications[13], [17]. A 1700 V/600 A IGBT module is connected in series with DUT to control the surge current generation and protect the test platform. The gate-source voltage (V_{GS}) is set as 0 V(the recommended turn-off gate-source drive voltage) to turn off the channel.

The peak surge current ($I_{surge-peak}$) is increased from 125 A with a step of 10 A through the increasing of DC bus voltage (V_{dc}) until the DUT fails as shown in Fig. 2. The source-drain voltage (V_{SD}) and surge current stress increase at the same time. The DUT fails at the $I_{surge-peak}$ of 145 A, which is five times I_{rate} ($5 \times I_{rate}$). The V_{SD} increases from surge current begins as I_{surge}. However, there is an abnormal increase at 6 ms which indicates failure occurs. The three terminal resistance of failed devices indicates the loss of blocking capability and the short circuit between drain and source terminals as presented in Table I. This signifies that an additional conduction path is generated during the surge process, apart from the body diode.

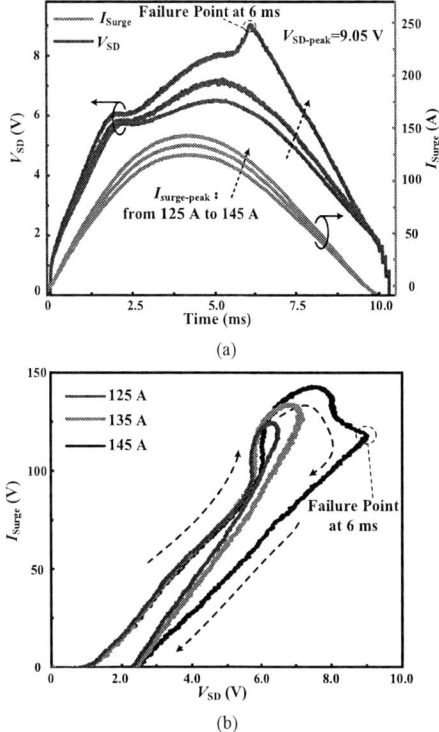

(a)

(b)

Fig. 2. Surge waveforms of DUT from 125 A to 145 A. The DUT fails at 6 ms indicated by a sharp increase of V_{SD} when $I_{surge-peak}$ of 145.0 A.

TABLE I.
TYPICAL THREE-TERMINAL RESISTANCE OF FRESH AND FAILED RDT-MOSFET UNDER SURGE CURRENT STRESS

Device	Three-Terminal Resistance		
	R_{gs}	R_{ds}	R_{gd}
Fresh	∞	∞	∞
Failed	∞	0.3 Ω	∞

III. NOVEL FAILURE MECHINAME

To uncover the underlying failure mechanism, microcosmic failure analysis is applied to the failed DUT. Obvious melted surface source aluminum from the top view is shown in Fig. 3(a) which indicates that the junction temperature of DUT reaches 933K (the melting point of aluminum). Furthermore, there is a significant crack around the source trench measuring up to 10 μm and crossing four cells as exhibited in the scanning electron microscope (SEM) image along a-a' cross-section in Fig. 3(b). The crack connecting the SiC in drift region and poly Si in source trench provides an electrical path between the drain and source terminals, which explains the failure.

 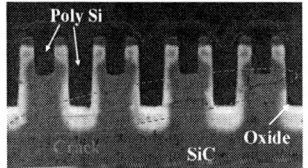

(a) (b)

Fig. 3. Image of the failed DUT. (a) Top view, and (b) SEB images along a-a' cross-section view.

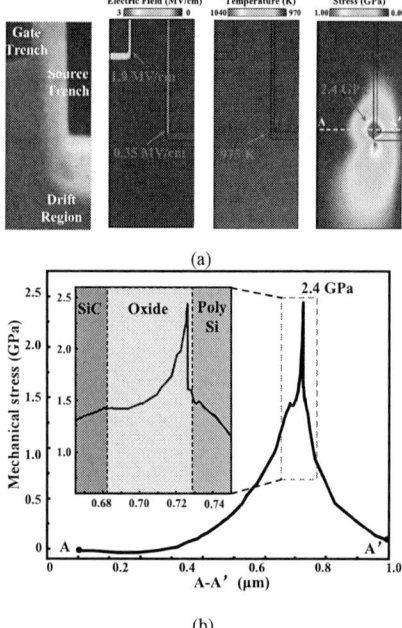

(a)

(b)

Fig. 4. (a) Electric field, junction temperature and mechanical stress around the source trench, and (b) mechanical stress along A-A' cutline of RDT-MOSFET.

Device-level electro-thermal-mechanical finite-element simulation is performed using Sentaurus TCAD. The simulation results including the distribution of electric field, junction temperature and mechanical stress around the corner of the source trench are demonstrated in Fig. 4. Electric field is eliminated from the possible causes of crack for that it is too low to reach the breakdown electric field of SiO_2 (~10MV/cm) due to the protection of p-shield region[18]. Moreover, the electric field around the source trench is lower than that around the gate trench. The junction temperature is relatively high reaching up to 975 K, but still remains below the melting point of SiO_2. However, it brings significant stress concentrations at the corner and the surrounding of source trench as shown in Fig 4(a).

The mechanical stress is concentrated at the interface between SiC, SiO_2 and poly Si as shown in the extracted stress along A-A' cutline in Fig. 4(b). The peak stress reaches 2.4 Gpa and locates within the oxide layer, which far exceeds its strength of 1.4 GPa. Consequently, the crack is generated by significant mechanical stress. Connecting the source and drain terminals, the crack brings localized increase in current density. Furthermore, the temperature as well as mechanical stress are further increased. Once initiated in the SiC drift region, the expansion of crack is facilitated by the intrinsically low toughness of SiC.

To investigate the formation process of the crack, the distributions of mechanical stress and temperature at different time point after the surge begins are presented in Fig. 5(a). There is obvious mechanical stress in the source oxide layer at 4 ms (point b) where the peak value reaches the strength of SiO_2. However, this small stress concentration point is insufficient to form crack causing device failure. Until 6 ms after surge begins (point c), the high mechanical stress region completely covers the oxide and crack forms.

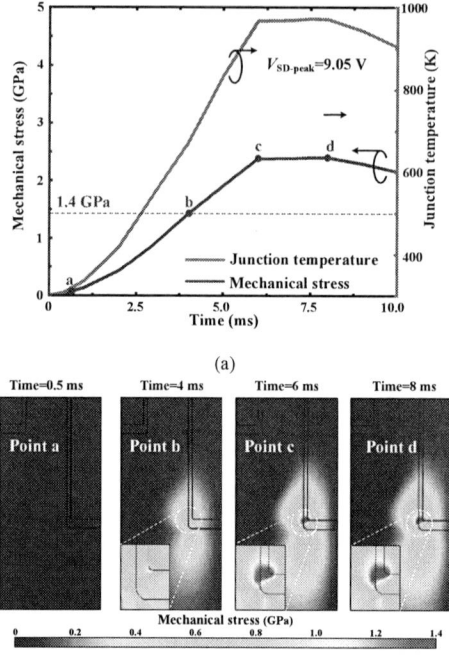

(a)

(b)

Fig. 5. (a) Temperature and peak mechanical stress of RDT-MOSFET during the surge current process, and (b) distribution of mechanical stress around the source trench at 0.5 ms, 4 ms, 6ms and 8 ms after suffering surge current.

TABLE II.
KEY MECHANICAL PARAMETERS OF SiC, POLY SI AND SiO_2

Material	Key mechanical parameters			
	Strength (GPa)	CTE (10⁻⁶/K)	Young's Modulus (GPa)	Poisson's Ratio
SiC	21[a]	4.3	500	0.157
Poly Si	2.0	2.6	169	0.22
SiO_2	1.4	0.5	75	0.17

[a] 0.3 GPa at 1000 °C

The schematic illustration of mechanical stress generation within the oxide layer in the source trench is illustrated in Fig. 6. It is considered that no mechanical stress between different materials at the reference temperature(T_{ref}). As the temperature increases, mechanical stress is produced by thermal expansion. Besides the high junction temperature, significant mechanical parameters difference between multiple material layers (SiC, oxide and poly Si) also contributes to the mechanical stress. The key mechanical parameters of SiC, SiO_2 and poly Si are listed in Table II[19]. The mismatch of coefficient of thermal expansion (CTE) between these materials mainly facilitates generation of crack. On the one hand, due to the mismatch between SiO_2 and SiC($\alpha_{SiC}/\alpha_{Oxide}$=8.6), the more expanded SiC generates the compressive stress($\sigma_{SiC/Oxide}$) on the adjacent oxide. On the other hand, the mismatch between SiO_2 and Poly Si($\alpha_{Oxide}/\alpha_{Poly\ Si}$=5.2) also leads to the compressive stress($\sigma_{Poly\ Si/Oxide}$). Finally, crack forms when the stress at SiO_2 exceeds its strength.

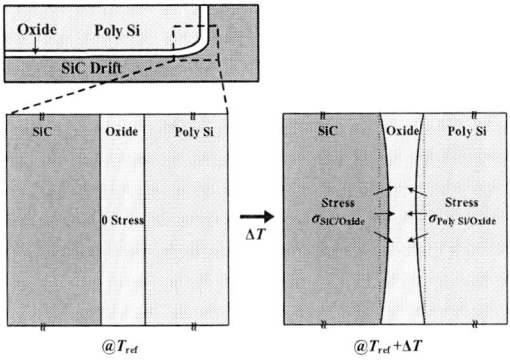

Fig. 6. Schematic illustration of mechanical stress generation within the oxide layer, induced by the thermal expansion mismatch between SiC, poly Si, and oxide.

IV. CONCLUSION

The surge reliability of RDT-MOSFET is studied in depth. An unreported failure mechanism is revealed for the first time. The RDT-MOSFET withstands 135 A peak surge current and fails at $I_{surge-peak}$ of 145 A when the V_{GS} is 0V. The drain and source terminals of the failed device is shorted. This failure mechanism attributes to the extremely high mechanical stress around the corner of the source trench through microcosmic failure analysis and device-level electro-thermal-mechanical finite-element simulation. The mismatch between materials in source trench that leads to the mechanical stress. It is essential for designers to take good balance between performance and reliability in trench gate type SiC MOSFETs.

V. ACKONWLEDGMENT

This work was supported by the National Natural Science Foundation of China under Grant U2469204 and Grant 62434002. (Corresponding author: Xuan Li.)

REFERENCES

[1] X. She, A. Q. Huang, Ó. Lucía, and B. Ozpineci, "Review of silicon carbide power devices and their applications," IEEE Trans. Ind. Electron., vol. 64, no. 10, pp. 8193 – 8205, Oct. 2017, doi: 10.1109/TIE.2017.2652401.

[2] M. Buffolo et al., "Review and outlook on GaN and SiC power devices: Industrial state-of-the-art, applications, and perspectives," IEEE Trans. Electron Devices, vol. 71, no. 3, pp. 1344–1355, Mar. 2024, doi: 10.1109/TED.2023.3346369.

[3] X. Li et al., "Achieving zero switching loss in silicon carbide MOSFET," IEEE Trans. Power Electron., vol. 34, no. 12, pp. 12193–12199, Dec. 2019, doi: 10.1109/TPEL.2019.2906352.

[4] X. Li et al., "A SiC power MOSFET loss model suitable for high-frequency applications," IEEE Trans. Ind. Electron., vol. 64, no. 10, pp. 8268–8276, Oct. 2017, doi: 10.1109/TIE.2017.2703910.

[5] F. Carastro, J. Mari, T. Zoels, B. Rowden, P. Losee, and L. Stevanovic, "Investigation on diode surge forward current ruggedness of si and SiC power modules," in 2016 18th European Conference on Power Electronics and Applications (EPE'16 ECCE Europe), Sep. 2016, pp. 1–10. doi: 10.1109/EPE.2016.7695685.

[6] S. Palanisamy, T. Basler, J. Lutz, C. Künzel, L. Wehrhahn-Kilian, and R. Elpelt, "Investigation of the bipolar degradation of SiC MOSFET body diodes and the influence of current density," in 2021 IEEE International Reliability Physics Symposium (IRPS), Mar. 2021, pp. 1–6. doi: 10.1109/IRPS46558.2021.9405183.

[7] B. Liang et al., "Surge characteristics of planar-gate silicon carbide MOSFET in the third quadrant," IEEE Transactions on Power Electronics, pp. 1–14, 2025, doi: 10.1109/TPEL.2025.3561810.

[8] D. Ma et al., "Degradation evaluation and defects analysis for 1.2-kV planar-gate SiC MOSFETs under repetitive surge current stress," IEEE Trans. Electron Devices, vol. 70, no. 12, pp. 6473–6479, Dec. 2023, doi: 10.1109/TED.2023.3323912.

[9] Y. Ye, A. Hensler, T. Basler, and J. Lutz, "Surge current test of SiC MOSFET with planar assembling and joining technology," in ISPS' 23 Proceedings, Czech Technical University in Prague, 2023, pp. 126–130. doi: 10.14311/ISPS.2023.020.

[10] Y. Zhao et al., "Electrical-thermal coupling modeling of SiC MOSFETs based on field-circuit coupling and its application in junction temperature calculation during surges," IEEE Trans. Power Electron., vol. 40, no. 2, pp. 2651 – 2667, Feb. 2025, doi: 10.1109/TPEL.2024.3493382.

[11] Z. Wang et al., "Reliability investigation on SiC trench MOSFET under repetitive surge current stress of body diode," in 2020 IEEE Workshop on Wide Bandgap Power Devices and Applications in Asia (WiPDA Asia), Suita, Japan: IEEE, Sep. 2020, pp. 1 – 4. doi: 10.1109/WiPDAAsia49671.2020.9360249.

[12] W. Huang et al., "Investigation of surge current reliability of 1200V planar and trench SiC MOSFET," in 2020 IEEE 15th International Conference on Solid-State & Integrated Circuit Technology (ICSICT), Nov. 2020, pp. 1–3. doi: 10.1109/ICSICT49897.2020.9278173.

[13] X. Zhan et al., "Investigation on degradation of 1200-V planar and trench SiC MOSFET under surge current stress of body diode," IEEE Transactions on Electron Devices, vol. 71, no. 1, pp. 709–714, Jan. 2024, doi: 10.1109/TED.2023.3335892.

[14] Z. Zhu, H. Xu, L. Liu, N. Ren, and K. Sheng, "Investigation on surge current capability of 4H-SiC trench-gate MOSFETs in third quadrant under various V_{GS} biases," IEEE J. Emerg. Sel. Top. Power Electron., vol. 9, no. 5, pp. 6361 – 6369, Oct. 2021, doi: 10.1109/JESTPE.2020.3028094.

[15] X. Li et al., "An in-depth investigation into short-circuit failure mechanisms of state-of-the-art 1200 V double trench SiC MOSFETs," IEEE Trans. Power Electron., pp. 1 – 8, 2024, doi: 10.1109/TPEL.2024.3431296.

[16] Rohm, "SCT4036KL-silicon carbide power MOSFET datasheet," 2022. [Online]. Available: http://www.rohm.com/

[17] M. Alaluss, C. Böhm, C. Herrmann, T. Basler, R. Elpelt, and G. Zeng, "3rd quadrant surge current SOA of SiC MOSFETs with different voltage class," Solid State Phenom., vol. 360, pp. 1–8, Aug. 2024, doi: 10.4028/p-fOWA4y.

[18] K. Yao, H. Yano, H. Tadano, and N. Iwamuro, "Investigations of SiC MOSFET short-circuit failure mechanisms using electrical, thermal, and mechanical stress analyses," IEEE Trans. Electron Devices, vol. 67, no. 10, Art. no. 10, Oct. 2020, doi: 10.1109/TED.2020.3013192.

[19] W. N. Sharpe, Mechanical Properties of MEMS Materials, The MEMS Handbook. Boca Raton, FL, USA: CRC Press, 2002.

Impact of Elevated Humidity Conditions on the Thermal and Mechanical Reliability of Silicon Carbide MOSFETs

Jiayu Zhang
School of Electrical Engineering

Southwest Jiaotong University
Chengdu, China
zjy26@my.swjtu.edu.cn

Dong Xie
School of Integrated Circuits Science
and Engineering
Southwest Jiaotong University
Chengdu, China
xiedong@my.swjtu.edu.cn

Zhiliang Xu
School of Electrical Engineering

Southwest Jiaotong University
Chengdu, China
xzl@my.swjtu.edu.cn

Zepeng Jiang
School of Electrical Engineering
Southwest Jiaotong University
Chengdu, China
Jiangzp@my.swjtu.edu.cn

Xinglai Ge*
School of Electrical Engineering
Southwest Jiaotong University
Chengdu, China
xlge@swjtu.edu.cn

Abstract—Silicon carbide metal oxide semiconductor field effect transistors(SiC MOSFETs)are widely used in new energy vehicles, photovoltaic systems, and other outdoor working conditions, and their long-term operational reliability is affected by factors such as ambient temperature and humidity. High humidity environments will cause degradation of SiC MOSFET packages, but the associated aging mechanism is unclear. Thus, to better understand these effects, this work simulates environmental stresses using two testing setups: a high voltage-high humidity high temperature reverse bias (HV-H3TRB) test platform and a power cycling test (PCT) platform. Test results show that moisture causes corrosion at the bonding wire and solder interface, which leads to an increase in overall device thermal resistance and a reduction in bonding wire connection. The combination of these two conditions resulted in a 32% reduction of the power cycling lifetime, guiding the reliability evaluations of power devices in outdoor applications.

Keywords—SiC MOSFETs, high voltage-high humidity high temperature reverse bias test (HV-H3TRB), Corrosion, Reliability.Introduction (Heading 1)

I. INTRODUCTION

Metal-oxide-semiconductor field effect transistor (SiC MOSFET) devices have become an indispensable part of renewable energy generation[1], aerospace, and electric vehicles due to their low switching losses and high switching frequency[2]. With the gradual commercialization of SiC MOSFETs, the issue of their long-term operational reliability has begun to receive attention from both academia and industry. However, for power devices applied in new energy vehicles, wind turbines, photovoltaic power generation, and other outdoor operating conditions, they are simultaneously affected by temperature[3], humidity[4], salt spray, and other factors, and these environments pose serious challenges to the long-term reliability of SiC MOSFETs[5]. Therefore, it is crucial to study the reliability of devices in high-humidity environments, which is of great value to improve the reliability of power devices[6].

SiC MOSFET reliability assessments typically involve two key tests: the high-humidity high-temperature reverse

bias (H3TRB) and power cycling tests (PCT)[7],[8]. While these evaluations provide valuable insights, current research primarily examines these stressors in isolation. Emerging evidence suggests a concerning trend—PCT preconditioning can intensify degradation during subsequent H3TRB testing, revealing unexpected interactions between these protocols[5],[10]. In practical applications, SiC MOSFETs encounter complex, simultaneous stressors that challenge their durability. For instance, sustained exposure to high humidity levels can threaten device performance even after they are stored in outdoor environments. The above studies have not deeply explored the intrinsic mechanism of humidity affecting device reliability. The question of whether relative humidity affects device package reliability is still unclear and needs more adequate experimental arguments, and the uncertainty of this question brings troubles to the assessment of device reliability in practical engineering applications.

Therefore, in this work, we developed a specialized test platform to evaluate the performance of SiC MOSFETs under high-temperature and high-humidity environment. Through comprehensive failure analysis involving decapsulation techniques, digital optical microscopy, and SEM examination, we identified moisture penetration as a critical factor affecting device encapsulation, particularly for the bonding wires and solder layers. Finally, it was found that moisture intrusion has a significant effect on encapsulation such as bonding wires and solder layer. Test results show that moisture causes corrosion at the bonding wire and solder interface, which leads to an increase in overall device thermal resistance and a reduction in bonding wire connection strength. The combination of these two conditions resulted in a 32% reduction of the power cycling lifetime, guiding the reliability evaluations of power devices in outdoor applications.

II. EXPERIMENTAL SETUP

A. Device Used in Tests and Test Method

SiC MOSFETS are available in standard TO-247 packages from Infineon asymmetric trench .For DC characterization, we used Agilent B1505A power device analyzers, while PCT were conducted using an HD Semi-PCT-1000 A-6 system.

This work was supported in part by the Joint Fund for Railway Basic Research of the National Natural Science Foundation of China under Grant U2468228, in part by the Fundamental Research Funds for the Central Universities under Grant 2682025CX055, and in part by the Sichuan Science and Technology Program under Grant MZGC20240128.

979-8-3315-1110-4/25 $31.00 © 2025 IEEE

The experimental setup involved dividing the test devices into two distinct groups. Group 1 underwent the HV-H3TRB aging tests under harsh conditions—85°C temperature, 85% relative humidity, and 80% of the rated drain-source voltage (V_{DSS}) for 1,000 hours, before proceeding to the PCT. Group 2 serving as the control, skipped the aging phase and went straight into power cycling. This approach allowed us to evaluate how environmental humidity impacts the device reliability. In order to deeply compare the differences in thermal stresses and failure modes caused by moisture infiltration, this paper sets up comparative tests under the same power cycling experimental conditions (t_{on} = 2s, $\Delta T_j \approx$ 80 K and $T_{jmax} \approx 125$ °C). Test circuits and schematic for the PCT are shown in Fig. 1.

B. Test System

In Group 1, the devices under tests (DUTs) undergo the HV-H3TRB evaluation before proceeding to the PCT. To provide context, the H3TRB test involves exposing all DUTs to an 85°C, 85% relative humidity chamber while wiring them in parallel, as illustrated in Fig. 2. Under HV-H3TRB conditions, 80% of the rated blocking voltage (V_{DSS}) is applied across the drain and source terminals, with the gate and source short-circuited. A precision resistor monitors the drain-source leakage current (I_{DSS}), and if this current exceeds the predefined failure threshold, an automated high-voltage switch immediately isolates the defective unit from the test setup.

In each PCT, at least six devices of each device type are used. The saturation voltage (V_{ds}), junction temperature (T_j), and thermal resistance (R_{th}) are the online monitoring parameters. The DC parameters are measured only once before and after the HV-H3TRB, including the transfer characteristic curve, the output curve, the resistance, and the gate leakage current.

The failure criterion is defined as $\Delta V_{ds} > 5\%$ or $\Delta R_{th} > 20\%$. Total PCT time is longer than the lifetime defined by the failure criterion.

III. Experimental Results

A. HV-H3TRB Performance

Group 1 conducted a 1000h HV-H3TRB aging test at 85°C/85%RH and 80% of the rated voltage V_{DSS} (960V), and then removed the devices for power cycling after all devices had reached the test time. This test simulates the performance of power devices that have been dormant for years, often in a high humidity setting like offshore wind farms. Group 2 was used as a control group and was subjected to direct the PCTs (in comparison with Group 1 of devices, the role of ambient humidity can be analyzed). Static parameters were compared at room temperature before and after the HV-H3TRB testing to assess its chip degradation.

The results showed that the leakage current I_{DSS} and blocking voltage V_{DSS} were basically unchanged after the HV-H3TRB test. In terms of gate performance, the gate leakage current I_{GSS} was essentially unchanged, and the threshold voltage V_{gsth} decreased by 2.84% on average. It can be seen that there is essentially no degradation in the static parameters of the device. The device chip has good resistance to moisture and no chip failure has occurred. This shows that the blocking ability of silicon carbide chips does not decrease in high humidity environments.

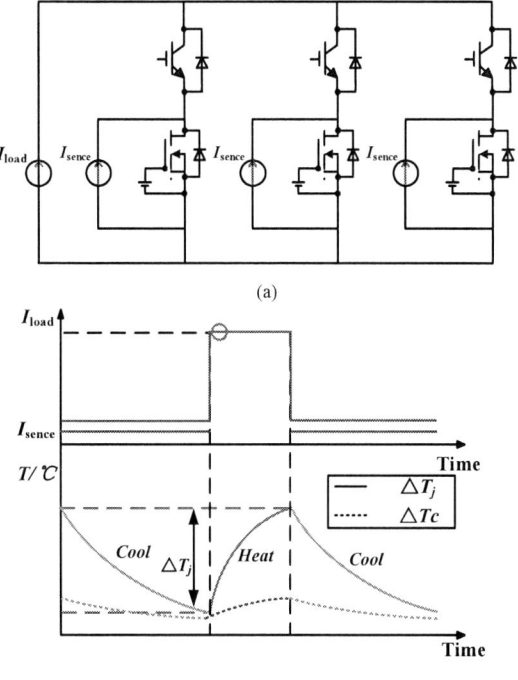

(a)

(b)

Fig. 1. Test circuits and test schematic. (a) Test circuit with three phases and one devices in each phase. (b) Test schematic for the PCT

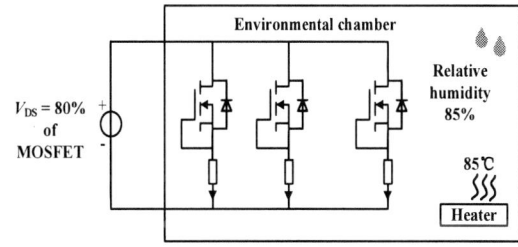

Fig. 2. Test circuit of the HV-H3TRB.

TABLE I.   STATIC PARAMETER

static parameter	After H3TRB
I_{DSS}	unchanged
V_{DSS}	unchanged
I_{GSS}	unchanged
V_{gsth}	decreased by 2.84%
$R_{DS(on)}$	increases by 2.7%

At the same time, the pass-state impedance parameter ($R_{DS(on)}$) of the device under humidity stress showed a deterioration trend, and this parameter drift phenomenon mainly originates from the electrochemical erosion process at the metal interface triggered by the penetration of ambient moisture. The impedance of the bonding wire connection part increased by 2.7%. In order to judge the instrumentation stability under close application conditions, a double-pulse test was also performed. The test results from Fig. 3 have shown that, after 1000h aging, no significant changes are observed in the voltage waveform and the value of dv/dt does not affect the normal operation of the device.

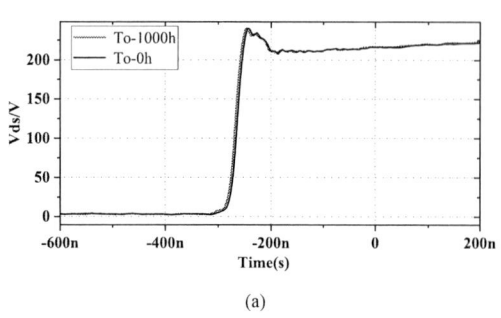

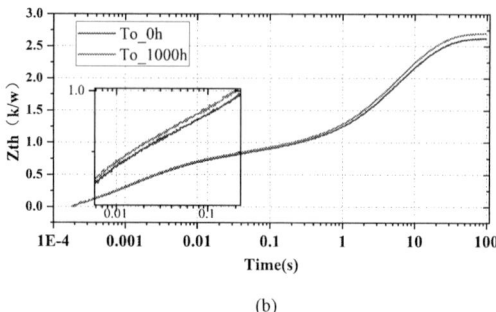

(a)   (b)

Fig. 3. H3TRB Performance (a) turn-off behavior of V_{ds}. (b) Z_{thjs} characteristics

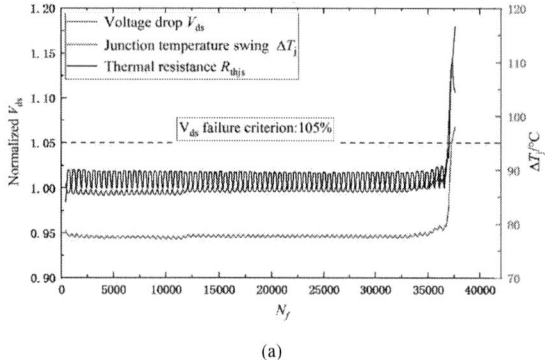

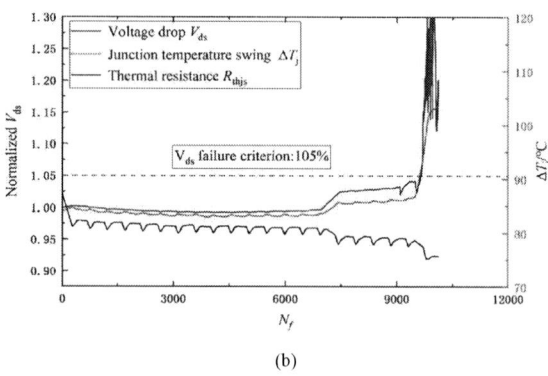

(a)   (b)

Fig. 4. Measured results in saturation mode. (a) Measured V_{ds}, R_{th}, ΔT_j change trend for one DUT in Group 2. (b) Measured V_{ds}, R_{th}, ΔT_j change trend for one DUT in Group 1.

However, humidity diffuses from the environment into the device, which in turn causes a change in the transient thermal resistance (R_{th}) of the device junction heatsink from Fig. 3(b). This results in a larger junction temperature due to the mismatch in the Coefficient of Thermal Expansion (CTE) between the materials. Meanwhile, the CTE mismatch between materials could result in higher stresses and accelerate the aging process.

B. PCT Performance

Following the HV-H3TRB test, the PCT was performed on both sets of DUT devices, in which V_{DS}, R_{th}, and $T_{vj\,min}$ and $T_{vj\,max}$ were monitored for each device. The test results of the DUTs are shown in Fig. 4.

The V_{DS} of Group 1 increases slowly and linearly in the early stage, and then starts to increase exponentially in the late stage of the test, and all the bonding wires are broken in a short time, and the R_{th} is relatively stable during the PCT as a whole. The development of V_{DS} in Group 2 is completely different from that of Group 1 DUT, with a stepwise increase in the early stage of aging, and after reaching the failure criterion, it shows a long time continuous oscillation, remains stable for a long time, and then finally the V_{DS} increases abruptly.

This result indicates that the package is significantly more fragile after the HV-H3TRB test where the contact between the bonding wires and the aluminum layer on the chip surface is not strong, which can affect the package reliability of the devices. Considering the differences in thermal resistance and device characteristics during device mounting, the device lifetimes of the two groups were normalized to a junction temperature fluctuation of ΔT_j=80K and a minimum junction temperature of $T_{j\,min}$=45°C according to the CIPS08 model for a more accurate lifetime comparison. From Fig. 5, when the maximum junction temperature is 125°C and the junction temperature fluctuation is 80 K, the average lifetime of the new device is 36806 times, and the average lifetime of the moisture-aged device is 23372 times, which is a reduction of about 30%, and the moisture intrusion has a significant effect on the device lifetime.

The Weibull probability plot was used to linearize the failure data from the two groups, and the angle of the resulting straight line directly relates to the β morphology parameter value. As depicted in Fig. 6, the blue dots and their solid lines illustrate the reliability profile for the first batch of samples, whereas the red dots and their broken lines signify the lifetime distribution for the second batch. The shape and scale parameters of the Weibull distribution for the second set of device lifetimes are 35.7 and 37897, respectively. According to a numerical breakdown, the Weibull parameters (shape factor = 3.27, scale factor = 26546) for the first group of devices dropped to 91% and 32% of their baseline counterparts in the second group.

It can be seen that water vapor intrusion has a significant impact on device package reliability. Moisture intrusion affects the weakening of the device bondline connection strength and the increase of thermal resistance, both of which can reduce the power cycle lifetime of the device. In the case of outdoor operation of the device, adequate protection should be taken to prevent water vapor from intruding and affecting the power devices.

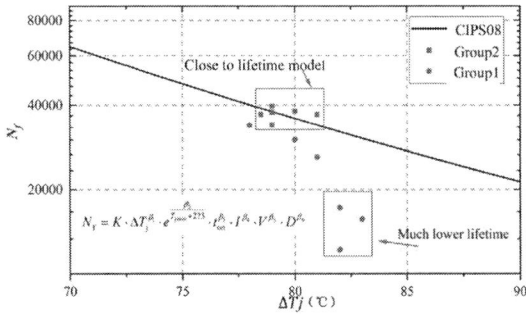

Fig. 5. Lifetime comparison of Group 1 and Group 2

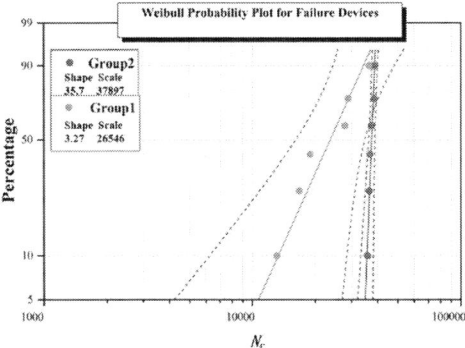

Fig. 6. Weibull distribution of Group 1 and Group 2

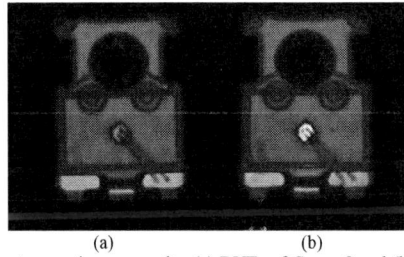

Fig. 7. Front acoustic scan results. (a) DUTs of Group 2 and (b) DUTs of Group 1.

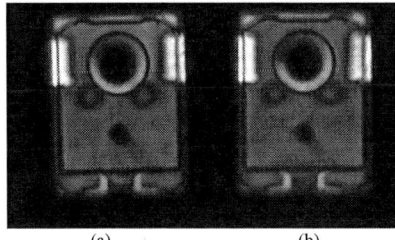

Fig. 8. Acoustic scan result. (a) DUTs of Group 2 and (b) DUTs of Group 1.

Fig. 9. Optical inspection result.(a) DUTs of Group 2(b) DUTs of Group1.

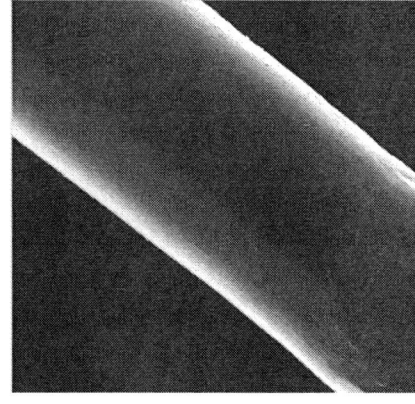

(a)

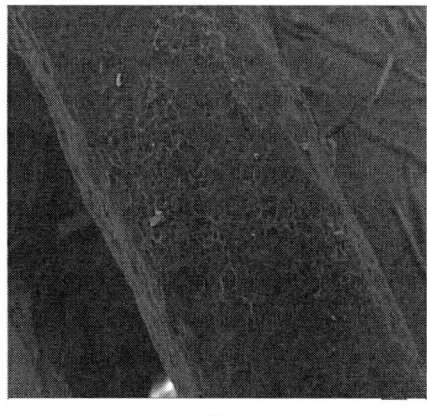

(b)

Fig. 10. SEM of bond wire (a) DUTs of Group 2 (b) DUTs of Group 1.

IV. FAILURE MECHANISM ANALYSIS

In order to systematically reveal the mechanism of water vapor permeation on the degradation of power devices, deeply analyzing the causes of the significant reduction of device lifetime, the device after the PCT was opened, the surface of the chip was first optically inspected and analyzed using an optical microscope, followed by observation of the device bonding lines for corrosion.

Figs. 7 and 8 showed significant damage to the source bonding lines of the devices in the front side acoustic sweep of the devices. Figs. 7(b) revealed delamination of the chip and the plastic sealing material, which was not observed for healthy devices. Microscopic observation results from Figs. 9 show that the chip surface exposed to humidity has typical humidity erosion characteristics. This indicates that the bonding wires and the aluminum layer on the chip surface have been corroded by water vapor.

More importantly, the lead bonding surfaces in Figs. 10 showed a significant roughening trend, which originates from the electrochemical corrosion triggered by moisture penetration. This interfacial degradation process directly weakens the mechanical stability of the package system, leading to the impairment of the structural integrity of the package, and ultimately resulting in the lifetime degradation under power cycling conditions.

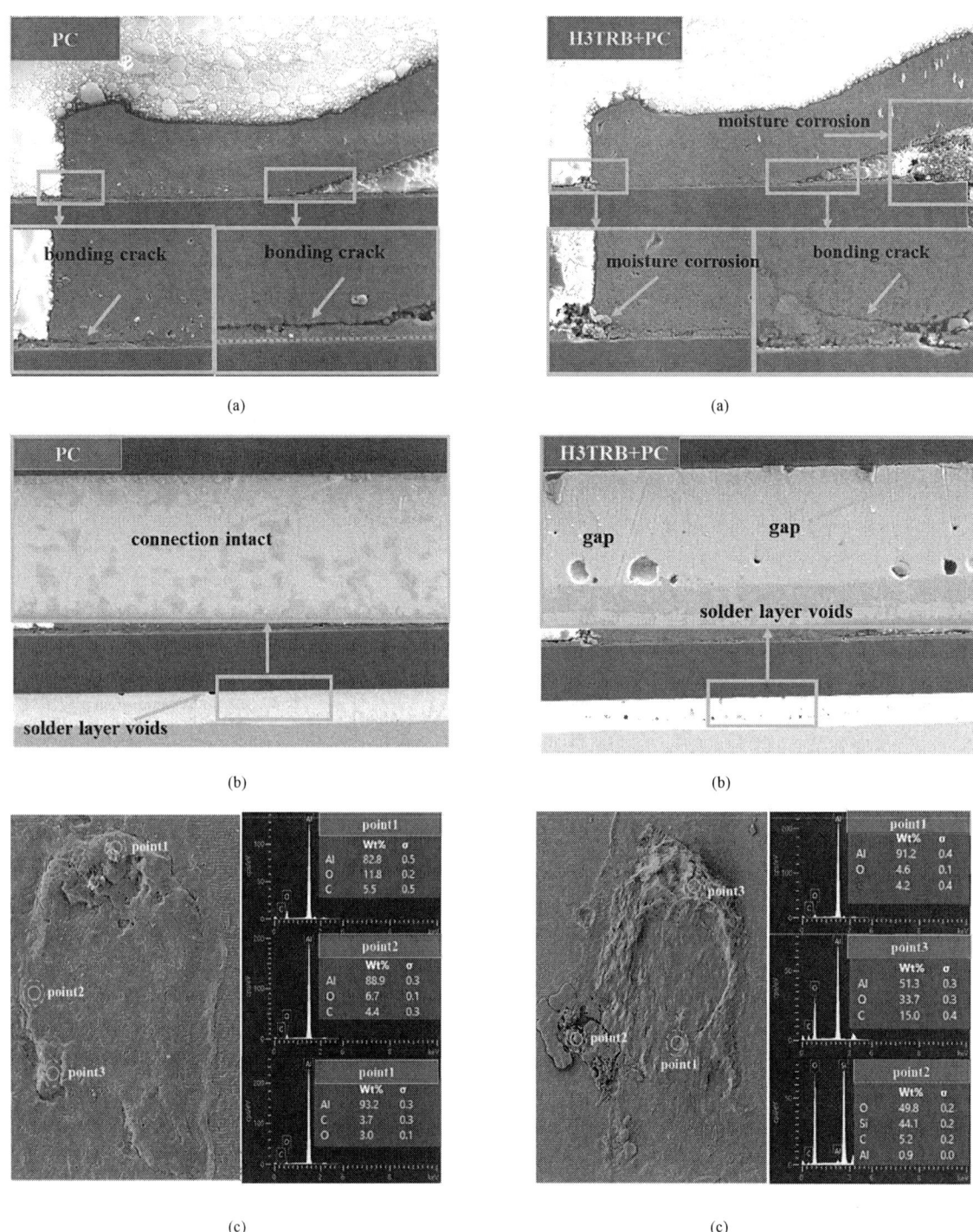

Fig. 11. SEM results (EDS mode) of the normal signs. (a) Bond wire feet. (b) Solder layer. (c) Result of EDS

Fig. 12. SEM results (EDS mode) of the corrosion signs. (a) Bond wire feet. (b) Solder layer. (c) Result of EDS

To elucidate the mechanistic correlation between humidity ingress levels and accelerated power cycling endurance degradation, targeted decapsulation was performed on representative specimens from the test population, enabling systematic failure mode characterization through cross-sectional analysis. Figs. 11 and 12 show normal and moisture corroded devices, respectively. The failure can be observed in more detail by metallographic microscopy and SEM analysis. Fig. 12(a) demonstrates moisture induced corrosion at both bonding wire head and root interfaces, directly compromising contact integrity and package structural stability. Fig. 12(b) reveals delamination at the molding compound interface, creating interfacial gaps where moisture condensation would accelerate the Al layer corrosion.

From Fig. 12, the moisture corrosion analysis reveals pronounced void formation and interfacial gaps within the solder layer of Group 2, constituting a primary contributing factor to the observed R_{th} elevation. EDX characterization in Fig. 12(c) demonstrates significant corrosion progression at both the aluminum surface layer and bonding wire interfaces in Group 1 specimens (a absent phenomenon absent in Group 2). Therefore, the delamination, cracks and voids in the device after moisture invasion will affect the chip heat dissipation and accelerate the failure of the device.

V. CONCLUSIONS

In this work, the influence of humidity on the reliability of SiC MOSFETs in the PCT is analyzed through experimental results, statistical regularity analysis, and SEM with EDX, and the preliminary conclusions are as follows.

1) SiC MOSFET terminals have good humidity resistance, and after 1000h of H3TRB test, the blocking voltage basically does not drop and dynamic performance is intact and undamaged.

2) Moisture intrusion will make the thermal resistance of SiC MOSFET increase, and corrode bonding wires and surface aluminum layer. This causes the V_{ds} to rise step by step in the early stage of the power cycle, and then to oscillate for a long time after reaching the failure criterion, resulting in a significant reduction in the device lifetime..

3) The moisture absorption expansion of the package causes solder deformation, and the layering between solder and chip is the main reason leading to the change of R_{th}.References

REFERENCES

[1] H. Wang, M. Liserre, and F. Blaabjerg, "Toward Reliable Power Electronics: Challenges, Design Tools, and Opportunities," IEEE Industrial Electronics Magazine, vol. 7, no. 2, pp. 17–26, Jun. 2013.

[2] U. -M. Choi, F. Blaabjerg, and K. -B. Lee, "Study and Handling Methods of Power IGBT Module Failures in Power Electronic Converter Systems," IEEE Transactions on Power Electronics, vol. 30, no. 5, pp. 2517–2533, May 2015.

[3] Z. Li et al., "The Influence of Special Environments on SiC MOSFETs," Materials, vol. 16, p. 6193, Sep. 2023.

[4] L. Ooi, D. Goh, and V. C. Ngwan, "High Temperature Reverse Bias (HTRB) & Temperature Humidity Bias (THB) Reliability Failure Mechanisms and Improvements in Trench Power MOSFET and IGBT," IEEE Journal of the Electron Devices Society, vol. 9, pp. 1181–1187, 2021.

[5] F. Hoffmann, N. Kaminski, and S. Schmitt, "Investigation on the Impact of Environmental Stress on the Thermo-Mechanical Reliability of IGBTs by Means of Consecutive H3TRB and PCT Testing," in 2021 33rd International Symposium on Power Semiconductor Devices and ICs (ISPSD), Jun. 2021, pp. 371–374.

[6] Y. Wang, E. Deng, L. Wu, Y. Yan, Y. Zhao, and Y. Huang, "Influence of Humidity on the Power Cycling Lifetime of SiC MOSFETs," IEEE Transactions on Components, Packaging and Manufacturing Technology, vol. 12, no. 11, pp. 1781–1790, Nov. 2022.

[7] H. Matsushima, R. Yamada, and A. Shima, "Two Mechanisms of Charge Accumulation in Edge Termination of 4H-SiC Diodes Caused by High-Temperature Bias Stress and High-Temperature and High-Humidity Bias Stress," IEEE Transactions on Electron Devices, vol. 65, no. 8, pp. 3318–3325, Aug. 2018.

[8] M. Wang, Y. Chen, Z. He, Z. Wu, and B. Li, "Comparative Investigation on Aging Precursor and Failure Mechanism of Commercial SiC MOSFETs Under Different Power Cycling Conduction Modes," IEEE Transactions on Power Electronics, vol. 38, no. 6, pp. 7142–7155, Jun. 2023.

[9] T. Gao, L. Ding, J. Wang, W. Chen, and X. Fan, "Analysis of Blocking Capability Failure Mechanism in IGBT Module Under High Salt Spray Environment," IEEE Transactions on Dielectrics and Electrical Insulation, vol. 30, no. 6, pp. 2914–2922, Dec. 2023.

[10] A. Brunko, W. Holzke, H. Groke, B. Orlik, and N. Kaminski, "Model-Based Condition Monitoring of Power Semiconductor Devices in Wind Turbines," in 2019 21st European Conference on Power Electronics and Applications (EPE '19 ECCE Europe), Sep. 2019, p. P.1-P.9.

Reliability Analysis and Implementation of Silver Sintering Connection in SiC High-temperature Package

Zizhen Cheng
State Key Laboratory of Electrical Insulation and Power Equipment
Xi'an Jiaotong University
Xi'an, China
zzcheng@stu.xjtu.edu.cn

Wenjie Xu
State Key Laboratory of Electrical Insulation and Power Equipment
Xi'an Jiaotong University
Xi'an, China
317311026@stu.xjtu.edu.cn

Fengtao Yang
State Key Laboratory of Electrical Insulation and Power Equipment
Xi'an Jiaotong University
Xi'an, China
yangfengtao@stu.xjtu.edu.cn

Zhiqiang Zhao
State Key Laboratory of Electrical Insulation and Power Equipment
Xi'an Jiaotong University
Xi'an, China
zhaozq@stu.xjtu.edu.cn

Dewen Wang
China North Vehicle Research Institute
China North Vehicle Research Institute
Beijing, China
dwwangbfcl@163.com

Laili Wang
State Key Laboratory of Electrical Insulation and Power Equipment
Xi'an Jiaotong University
Xi'an, China
llwang@mail.xjtu.edu.cn

Abstract—**This paper focuses on the nano-silver sintering technology in Silicon Carbide (SiC) high-temperature packaging. It comprehensively investigates and analyses the failure modes and high-temperature ageing characteristics of sintered nano-silver from both macro and microscopic perspectives. In addition, this work realizes the package of SiC high-temperature power devices based on the sintered nano-silver interconnection technology, and validates the high-temperature operation capability of the SiC device under the junction temperature up to 488°C.**

Keywords—Silicon Carbide, High-temperature package, Silver sintering

I. INTRODUCTION

The requirements for high-temperature operation capability of power semiconductor devices in high-tech fields such as electric vehicles and aerospace urgently promote the application of SiC power semiconductor devices [1]. However, the insufficiency of high-temperature packaging technology, especially the interconnection technology, tends to brings thermal failure, microcracking and fracture to SiC devices [2], which severely restricts the performance of SiC devices.

As shown in Fig. 1, in the package structure, the interconnection layer locates in the core of the power device heat dissipation path, and directly faces the concentrated stress. For this reason, it is necessary to conduct research on the good electrical and thermal conductivity, high mechanical strength, high melting point and matched thermal expansion coefficient interconnection materials and technology, to enhance the electrical and thermal characteristics of the

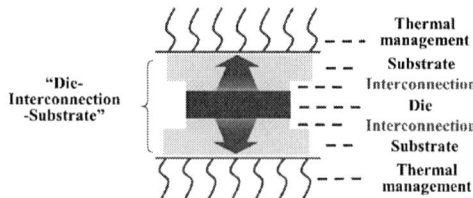

Fig. 1 Partial schematic of packaged power devices.

packaged power devices and the reliability of high-temperature operation.

Nano-silver sintering technology is a high-temperature connection technology with great application potential. The technology utilizes the size effect of nano-silver particles to significantly reduce the melting point of the material, while after the sintering process the silver particles are densified to form a silver sintering layer. Through this process, the sintering interconnection structure obtains similar performance to silver such as high melting point, excellent electrical, thermal and mechanical properties.

The interfacial interconnection strength of nano-silver sintering is often the most critical factor affecting the chip interconnection strength [3]. For this reason, researchers have mainly carried out studies from two perspectives: interfacial interconnection mechanism and solder joint failure mechanism. Researches have been conducted to investigate the interconnection strength between sintering nano-silver and silver-plated, gold-plated and copper-plated substrate. The findings indicate the connection with silver-plated substrate is strong due

979-8-3315-1110-4/25 $31.00 © 2025 IEEE

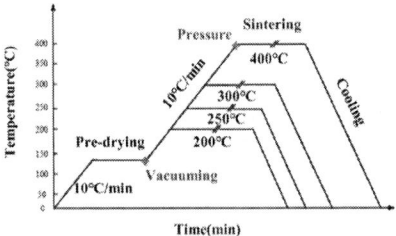

Fig. 2 Sintering process.

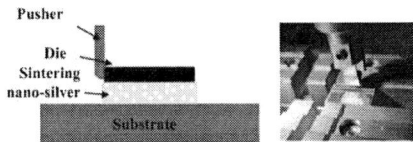

Fig. 3 Schematic diagram and test procedure of push-pull force test.

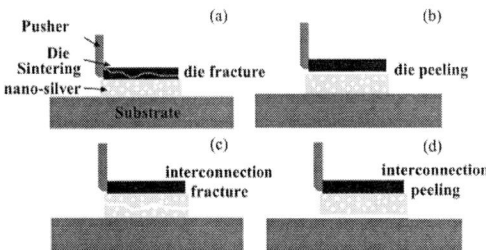

Fig. 4 Faliure form of "sandwich" connection structures: (a) Form I, die fracture; (b) Form II, die peeling;. (c) Form III, interconnection fracture; (d) Form IV, interconnection peeling.

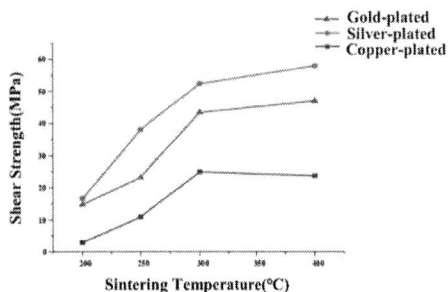

Fig. 5 Shear strength as a function of sintering temperature.

to great diffusion between the same atoms in [4]; Xiaomin Wang [5] studied the densification and interfacial diffusion behavior of sintering nano-silver on gold-plated substrate; the connection with copper-plated substrate exist in two forms based on the oxygen content and therefore have instable interfacial strength [6].

Although there are several studies on the sintering mechanism of nano-silver, the characteristics and applications of sintering silver for SiC power device packaging are not yet well fully studied. This paper conducts research in the field of SiC high-temperature power device packaging. First of all, from the micro-scale and macro-scale aspects, the failure modes and high temperature ageing characteristics of the sintering silver were investigated. Then, the package of high-temperature power devices was achieved based on nano-silver sintering, and the high-temperature working capability was tested, verifying that the highest operation temperature could reach 488°C.

II. STUDY ON THE FAILURE OF SINTERING NANO-SILVER CONNECTIONS

This section focuses on exploring the failure modes of sintering nano-silver interconnecting layers under different sintering temperatures and with substrates of three types plated metals (i.e. silver, gold and copper). Based on the combination of macroscopic and microscopic result, this section analyzed the failure forms of nano-silver sintering.

A. Experiment Settlement

According to the test provisions of GB/T4937.19-2018 on chip shear strength, this work carries out the test of sintered nano-silver connection failure by preparing a "sandwich" connection

structure which consists of die, sintering interconnection and substrate. The parameters of the sintering process in this work are as follows: nano-silver paste printed with the size of 5 mm × 5 mm and a thickness of 100 μm, a constant heating rate of 10 °C/min, an auxiliary pressure set at 1 MPa, and a sintering lasting time of 2h. It also should be noted that the sintering temperature is one of the parameters for the study of failure in this section, so the temperature is set as 200°C, 250°C, 300°C and 400°C, respectively. The sintering process in this study is shown in Fig. 2.

After sintering, the shear strength of the sample is tested through a push-pull experiment as shown in Fig. 3, and the microscopic sintering structure is observed through scanning electron microscope (SEM).

B. Analysis on the Failure

In general, there are four types of shear test failure forms in "sandwich" connection structures (shown in Fig. 4). In case of Form I, the interconnection strength of the interconnection structure is higher than the fracture strength of the die, so the test can not accurately reflect the perfomance of sintering techonology. In case of Form II, direct peeling of the die occurs during the shear test, which indicates that the metallurgical connection is weak at the die/solder interface. In this situation, the strength of the interconnection structure is determined by the peeling strength at the die/nano-silver interface. In failure Form III, the shear failure occours within the sintered

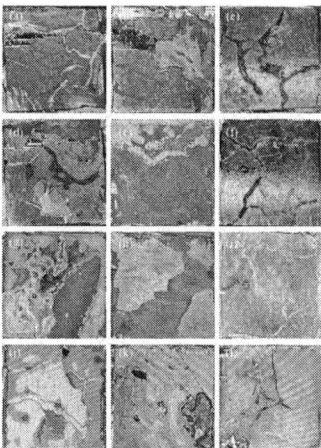

Sintering temperature: (a)~(c): 200°C, (d)~(f): 250°C, (g)~(h): 300°C, (j)~(l): 400°C.

Substrate metallization: (a)(d)(g)(j): gold-plated, (b)(e)(h)(k): silver-plated, (e)(f)(i)(l): copper-plated.

Fig. 6 Sintering nano-silver connecting layer residual connection interface under different sintering temperature and metallization.

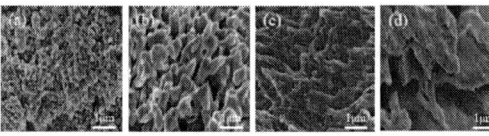

Fig. 7 Microstructure and morphology of sintered nano-silver residuals.

nano-silver layer. This indicates that a strong metallurgical interconnection between the sintered layer, the die and ceramic substrate have formed, and the fracture strength of the sintered nano-sliver is the weakness of the interconnection structure. When failure Form IV happens, the sintered layer and ceramic substrate interface is completely peeled off, and it demonstrates the metallurgical connection is the weak spot.

The shear strength test curve of the sintered nano-silver interconnection structure is shown in Fig. 5. The shear strength of the three types of samples increases with the increase of sintering temperature in the temperature range from 200°C to 300°C. The interconnection strength of the structure on silver-plated substrate exceeds 35 MPa at the sintering temperature of 250°C, and further rises over 50 MPa when the sintering temperature reaches 300°C. The sintered nano-silver connection structure has the best strength performance among the three test samples. This is mainly due to the fact that the self-diffusion between homogeneous atoms is significantly faster than the mutual diffusion between heterogeneous atoms. Thus, the metallurgical connection between the sintered silver layer and the silver-plated substrate is likely to be strong. The shear strength of interconnection structure on gold-plated substrate is less than 20 MPa when sintering temperatures is below 250°C. But when the sintering temperature exceeds 300°C, the interconnection strength rises to more than 40MPa. However, the strength of interconnection structure on copper-plated substrate remains lower than 20MPa under different sintering temperatures. This is mainly because that the surface of the copper-plated substrate is more prone to oxidation at high temperatures, leading to the aggravation of connecting strength and CTE mismatch between the connecting layers.

After the push-pull force test, the residual connection interfaces of the sintered silver connection layer with different sintering temperatures and metal interfaces are obtained and shown in Fig. 6. The microscopic morphology of the sintering silver residual sturctures on the silver-plated substrate are shown in Fig. 7. It can be seen from that the failure locations of the interconnection structure of gold-plated substrate and the silver-plated substrate change continuously with different sintering temperatures, whereas that of the copper-plated substrate is always located at the connection layer/substrate interface, and there is almost no residual sintered nano-silver on the substrate.

When the sintering temperature is lower than 250°C, the failure of the connection structures of both gold-plated and silver-plated substrates is mainly located in the sintering silver layer, which indicates that the strength of the sintering strcuture is lower than that of the connection interface. With the increasing sintering temperature, the sintering between Ag particles is more adequate and the sintering strcuture is more robust, so the connection strength is also rising. When the sintering temperature is further increased to 300°C, the failure positions on the gold-plated substrate and silver-plated substrate move to interface. Thus, the connection strength is co-determined by the interface interconnection strength and the fracture strength of the nano-silver sintering structure. The fracture form of the residual sintering nano-silver structure changes from brittle fracture under sintering temperature of 250°C to violent ductile fracture in 300°C, which shows that there is a significant increase in the fracture strength of the connection structure. When the sintering temperature rises to 400°C, the failure location of interconnection structure on gold-plated and silver-plated substrate are almost completely transferred to the interface, at this time the interconnection strength of the solder joints mainly depends on the interfacial bonding strength. For the connection structure on copper-plated substrate, the failure form is always located at the interface. With the sintering temperature rising, the bond strength between the solder layer and the substrate has been enhanced, but remains at low value.

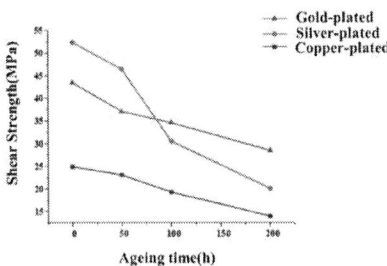

Fig. 8 Shear strength curve with ageing time.

In conclusion, the interconnection strength and failure modes of sintered nano-sliver interconnection structure are related to the sintering temperature and substrate metallization. For the specific type of nano-silver paste studied in this work —— fully sintered paste, the interconnection is formed by the fuction of sintering of Ag particles and diffusion between metal atoms with few effect of organic solvent. Therefore, the type of plated metal on the substrate instead of sintering temperature is critical for the strength of interconnection structure. Among them, the interconnection structure on the silver-plated substrate is significantly stronger than the other two kinds of structure due to greater metal diffusion. One notable result is that the strength of sintered interconnection structure hardly increases when the sintering temperature exceeds 300°C. As a result, we set the sintering temperature as 300°C in the following research for balancing the interconnection performance and cost-efficiency.

III. CHARACTERISTICS OF SINTERED SILVER HIGH-TEMPERATURE AGEING

In this section, the samples of sintereing nano-silver connections are prepared based on the sintering process shown in Fig. 2 (sintering temperature of 300°C) for high-temperature ageing experiments. The experimental temperature of high-temperature ageing is selected as 200°C, and the ageing time are 50h, 100h and 200h, respectively.

A. Analysis on Macro-mechanical Properties of Ageing Sintering Nano-silver Connection

The curves of the shear strength of sintering nano-silver connection with gold-plated, silver-plated and copper-plated substrates are shown in Fig. 8. In general, the interconnect strength of all three connections decreases with increasing ageing time, but the magnitude of the decrease and the reliability are quite different.

Among them, the initial shear strength of the connection structure on silver-plated substrate before ageing is the highest, reaching 52.4MPa. However, the decrease of the shear strength is obviously aggravated with ageing prolonging. When the ageing time reaches 200h, the shear strength is only 20.1MPa,

which is much lower than the initial shear strength by 61.64%. The connection on gold-plated substrate connection has the best high-temperature ageing characteristics. After 50h, the decreasing trend of shear strength with the increasing ageing time has slowed down. After 200h's ageing, the shear strength still maintained at 65.6% of the initial shear strength. The connection structure on copper-plated substrate has the worst high-temperature ageing characteristics. Although the decrease is not large, the reliability of connection on the copper-plated substrate after 200h of ageing is extremely low due to the low initial shear strength.

Due to the limitation of time and lack of equipments, the longer ageing tests and temperature cycle tests have not been conducted. However, we will carry out research on that in the feature to further impove our work.

B. Microscopic Analysis on the High-temperature Ageing Failure Mechanism

Due to the limitation of space, this section analyzes the high-temperature ageing failure mechanism of sintering nano-silver by taking that on silver-plated substrates as the example.

The residual connection structures on silver-plated substrate after ageing and push-pull force test are shown in Fig. 9, and the elemental composition of the selected position after ageing 100h is shown in Fig. 10. It can be clearly seen that when the ageing time is 50h, the residual interface does not appear obvious traces of oxidation or copper substrate, and the residual sintered silver on the substrate increased significantly. This is mainly due to the the coarsening of the sintered structure by ageing, and the Kirkendall holes combines and forms the initiation of microcracks. When the ageing time is increased to 100 h, the residual interface starts to show obvious oxidation traces and the original silver plating layer is barely recognizable. Elemental analysis indicates that high temperature oxidation has occurred at region 1 and 3; while the elements at region 2 are mainly Ag, indicateing that the Cu-Ag interface has not been fully oxidized after ageing for 100h. When the ageing time further increases to 200h, the connection strength drops to a lower level, and the failure location almost entirely locates at the interface of the copper substrate/silver-plated layer. This is because as the ageing time continues to increase, the oxides will gradually extend from the edge of the connection to the center, and ultimately penetrate through the entire connection interface, leading to complete failure.

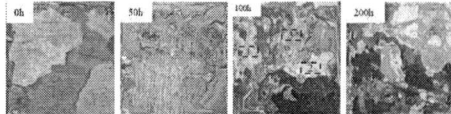

Fig. 9 Residual interfaces on aged silver-plated substrates after shear strength tests.

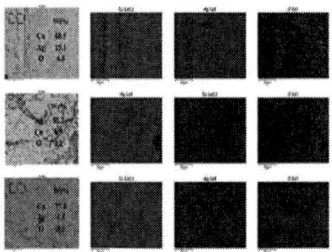

Fig. 10 Analysis of elemental composition at the selection locations of residual interfaces after 100h of ageing.

Fig. 11 Produced high-temperature SiC devices.

IV. SiC High-temperature Packaging Devices Based on Sintering Nano-silver

In this section, the packaging of SiC high-temperature power devices is realized on the basis of the previous study, and the high-temperature operation capability of the power devices is tested and evaluated.

A standard TO-254 pakage is selected as the basic package form in this work. The gold-plated SiC MOSFET die is sintered following the process curve as shown in Fig. 2 (sintering temperature of 300°C), and aluminum wire bonding is used for electrical interconnections. The completed power device is shown in Fig. 11, which can stably work under 500V/70A conditions (shown in Fig. 12).

High-temperature experiments based on the self-conducting characteristics of the SiC MOSFET body diode are conducted to examine the working capability of the sintering nano-silver connection under extreme high temperature. As the current flow through the body diode continues to increase, the temperature of the device also continues to rise. Ultimately, when the junction temperature of the device reaches as high as 488°C (as shown in Fig. 13), the power device can still operate, which has far exceeded the maximum tolerance temperature of commercial power devices.

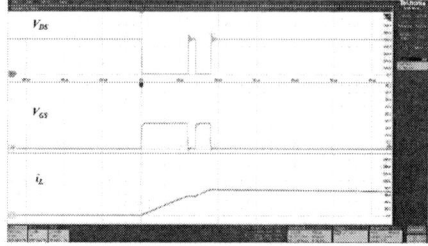

Fig. 12 Switching waveform of the device.

Fig. 13 Infrared images of high-temperature experiments at junction temperature of 488°C.

V. Conclusion

This paper focuses on the high-temperature interconnection technology challenges in SiC packaging applications. Aimed at advanced sintering nano-silver technology, researches have been conducted on the high-temperature reliability of sintering nano-silver connection from the micro-scale and macro-scale aspects, and in-depth analysis have been developed on the high-temperature ageing failure mechanism. The analysis indicates the effect of sintering temperature and plated metal on interconnection strength and failure mode. In addition, the experiment results proves that silver-plated substrate has the best connection performance, and the optimal sintering temperature is 300°C in this work. At last, based on the sintering nano-silver interconnection, the package of SiC high temperature power device is realized, and the high temperature serviceability is examined at the junction temperature up to 488°C.

References

[1] H. Lee, V. Smet, and R. Tummala, "A review of SiC power module packaging technologies: Challenges, advances, and emerging issues," IEEE J. Emerg. Sel. Top. Power Electron., vol. 8, no. 1, pp. 239–255, Mar. 2019.

[2] X. She, A. Q. Huang, Ó. Lucía, and B. Ozpineci, "Review of silicon carbide power devices and their applications," IEEE Trans. Ind. Electron., vol. 64, no. 10, pp. 8193–8205, Oct. 2017.

[3] H. G. Zheng, D. Berry, K. D. T. Ngo, et al., "Chip-bonding on copper by pressureless sintering of nano-silver paste under controlled atmosphere," IEEE Trans. Compon. Packag. Manuf. Technol., vol. 4, no. 3, pp. 377–384, Mar. 2014.

[4] J. G. F. Bai and G. Q. Lu, "Thermomechanical reliability of low-temperature sintered silver die attached SiC power

device assembly," IEEE Trans. Device Mater. Reliab., vol. 6, no. 3, pp. 436–441, Sep. 2006.

[5] X. Wang, Y. Mei, X. Li, M. Wang, Z. Cui, and G.-Q. Lu, "Pressureless sintering of nano-silver paste as die attachment on substrates with ENIG finish for semiconductor applications," J. Alloys Compd., vol. 777, pp. 578–585, Mar. 2019.

[6] F. Yang, W. B. Zhu, W. Z. Wu, et al., "Microstructural evolution and degradation mechanism of SiC-Cu chip attachment using sintered nano-Ag paste during high-temperature ageing," J. Alloys Compd., vol. 844, pp. 156442, Oct. 2020.

Impact of Kelvin Source Connection (KSC) vs. Common Source Connection (CSC) on Power Cycling Aging Characteristics of Silicon Carbide (SiC) MOSFETs

1st Huaihao Cheng
Zhuhai Power Supply Bureau of Guangdong Power Grid Co., Ltd.
Zhuhai, China
865693953@qq.com

2nd Yong Chen
Zhuhai Power Supply Bureau of Guangdong Power Grid Co., Ltd.
DC Power Distribution and Consumption Technology Research Center of Guangdong Power Grid Co., Ltd.
Zhuhai, China
35665035@qq.com

3rd Xu Cheng
Zhuhai Power Supply Bureau of Guangdong Power Grid Co., Ltd.
DC Power Distribution and Consumption Technology Research Center of Guangdong Power Grid Co., Ltd.
Zhuhai, China
664157102@qq.com

4th Xingyu Pei
Zhuhai Power Supply Bureau of Guangdong Power Grid Co., Ltd.
DC Power Distribution and Consumption Technology Research Center of Guangdong Power Grid Co., Ltd.
Zhuhai, China
31977550@qq.com

5th Jianbiao Li
Zhuhai Power Supply Bureau of Guangdong Power Grid Co., Ltd.
DC Power Distribution and Consumption Technology Research Center of Guangdong Power Grid Co., Ltd.
Zhuhai, China
zhlijianbiao@126.com

6th Hongyuan Wu
Zhuhai Power Supply Bureau of Guangdong Power Grid Co., Ltd.
DC Power Distribution and Consumption Technology Research Center of Guangdong Power Grid Co., Ltd.
Zhuhai, China
whyqsxddc@163.com

Abstract—This study investigates the reliability of Silicon Carbide (SiC) MOSFETs under high-temperature power cycling, focusing on the degradation mechanisms induced by Kelvin Source Connection (KSC) and Common Source Connection (CSC) configurations. Experiments were conducted on commercial 1200V/30A devices through 28,000 power cycles under junction temperatures of 175-220°C and load currents of 18-20A. Results demonstrate that under high-temperature (220°C) and high-current (20A) stress, KSC devices exhibit significantly greater threshold voltage drift (15% for DUT2) compared to CSC devices (~10% for DUT3-4). Notably, KSC devices show a 20% abrupt increase in on-resistance after 27,000 cycles, while CSC devices display earlier but milder packaging degradation. The work reveals the accelerated gate oxide degradation in KSC structures caused by insufficient drive-power loop decoupling, and highlights CSC's reliability advantages through optimized bond wire stress distribution. This study provides theoretical guidance for high-reliability SiC device packaging design and holds substantial engineering significance for SiC MOSFET durability in high-temperature power applications such as electric vehicles and photovoltaic inverters.

Keywords—Silicon Carbide (SiC) MOSFET, Power cycling reliability, Kelvin Source Connection (KSC), Common Source Connection (CSC)

I. INTRODUCTION

With the accelerating global energy transition, third-generation wide-bandgap semiconductor material silicon carbide (SiC) has emerged as a core material to replace silicon-based power devices due to its superior physical properties: higher bandgap (3.3eVvs.1.1 eV), breakdown field strength (2.8 kV/cm vs. 0.3 kV/cm), and thermal conductivity (4.7 W/m·Kvs.1.23 W/m·K) [1]. In high-power-density applications such as photovoltaic inverters and electric

vehicles, SiC MOSFETs have achieved significant commercial progress, yet their long-term operational reliability remains inadequately addressed. Studies reveal that under identical operating conditions, the power cycling lifetime of SiC devices is only one-third of silicon-based counterparts [2], with primary failure mechanisms including packaging degradation and gate oxide degradation.

Existing accelerated aging methods, such as High-Temperature Gate Bias (HTGB) [3-5], High-Temperature Gate Switching (HTGS) [6], and Chopper Mode Bias (CMB) [7], exhibit notable limitations: HTGB and HTGS focus on gate oxide degradation but fail to account for packaging stress; Thermal Cycling (TC) [8] addresses packaging failures while neglecting gate oxide damage; Direct Current Power Cycling (DCPC) [9] applies dual stresses simultaneously but suffers from insufficient on-state voltage drop measurement accuracy [10]. To address these challenges, F. Yang et al. [11] proposed optimized on-state voltage monitoring using bypass devices and delayed turn-off strategies, while S.Dusmez et al.[12] employed multi-device parallel configurations for stress distribution balancing. However, the former introduces voltage overshoot risks, and the latter incurs high sample costs. Notably, the gate oxide in SiC MOSFETs exhibits heightened sensitivity to positive bias temperature instability, while packaging degradation correlates closely with bond wire stress, making multi-parameter correlation modeling critical for decoupling failure mechanisms.

Researchers worldwide have extensively explored power cycling test methodology improvements. For source terminal optimization, H. Li et al. [13] demonstrated that Kelvin Source Connection (KSC) eliminates driver-power loop coupling in single-chip Common Source Connection (CSC) configurations but introduces auxiliary source coupling paths in multi-chip parallel setups. J. Lehmann et al. [14] proposed a Kelvin emitter detection method for silicon IGBTs, which proves incompatible with SiC devices. A. Fayyaz et al. [15] developed a three-phase inverter-based ACPC platform utilizing body diode freewheeling for stress application, yet its practicality is limited by on-state voltage oscillation. In test platform development, F. Yang etal.[16] compared silicon and SiC device testing methodologies, proposing three enhanced modes: body diode reverse conduction heating (requiring mitigation of stress decay at high junction temperatures), reverse channel conduction heating (suffering from partial current shunting through body diodes), and forward conduction heating (enabling on-state voltage measurement via auxiliary switches but requiring voltage overshoot suppression).

Addressing these challenges, this study constructs a multi-device parallel high-temperature power cycling test platform for SiC MOSFETs. High-temperature power cycling aging tests are performed on TO-packaged SiC MOSFET devices with both Common Source Connection (CSC) and Kelvin Source Connection (KSC) configurations. By monitoring threshold voltage drift and body diode forward voltage degradation, the analysis concludes that CSC-configured SiC devices demonstrate superior reliability in high-stress environments..

II. DEVELOPMENT OF MULTI-DEVICE PARALLEL HIGH-TEMPERATURE POWER CYCLING TEST PLATFORM

To achieve aging experimental results reflecting the combined effects of multiple degradation mechanisms in silicon carbide(SiC)MOSFETs while enabling accelerated aging without overstressing specific parameters (e.g.,excessive maximum junction temperature $T_{j,max}$ that exacerbates packaging degradation), this study proposes a fixed junction temperature differential mode based on a thermal network model. The junction temperature (Tj) is calculated from case temperature (Tc), and junction-to-case 1thermal resistance (Rth) is determined solely before power cycling tests to avoid recalibration during aging. This thermal network-based junction temperature cycling control compensates for the impact of conduction loss increases (caused by gate oxide degradation) on turn-on time (ton).

Power cycling test time efficiency and platform scalability are critical for multi-device parallel testing. A multi-device parallel alternating conduction strategy is implemented, as shown in the test circuit schematic (Fig.1). Parallel devices share a single power source, where the turn-on time (ton) and load current of each device influence the turn-off time (toff) of others. When a device cools to the minimum junction temperature (Tjmin,typically ambient), it remains idle until its next cycling phase. During the cooling phase of all devices, the power source current is stabilized through a bypass device to maintain output continuity.

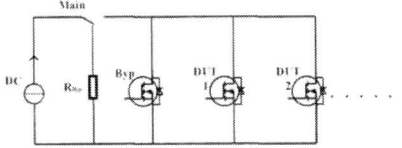

Fig. 1. Power Cycling Test Circuit Schematic Diagram

Taking three tested devices as an example, the control timing logic for each switch and device under test (DUT) in the test circuit schematic is illustrated in Fig.2.

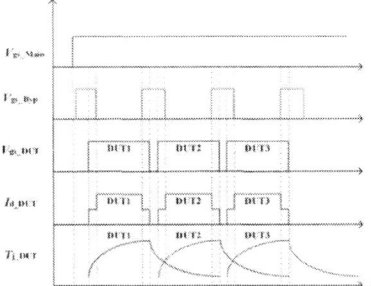

Fig. 2. Power Cycling Control Timing Logic

As shown in Figure 4, after the power cycling test begins, the main switch relay is switched to the DUT loop. Upon confirming reliable connection, the cycling process initiates. At the start of a cycle, the bypass device is first activated, and the power supply outputs the preset current value for DUT1. After the power supply stabilizes, DUT1 is driven to turn on, and the bypass device is deactivated. DUT1 then begins self-heating through conduction losses. The bypass device during startup prevents oscillations during the power supply transition from voltage source to current source. When DUT1 reaches the preset junction temperature (Tj), the bypass device is reactivated, and DUT1 is turned off. During this bypass freewheeling phase, the power supply current is reprogrammed to the preset value for DUT2. After output stabilization, DUT2 is driven to turn on, and the bypass device is deactivated, initiating DUT2 heating. During DUT switching, the bypass device serves as a freewheeling path, allowing modification of the power supply current to apply different stress levels across parallel DUTs.

III. EXPERIMENTAL ANALYSIS OF SOURCE TERMINAL CONFIGURATION EFFECTS ON SiC MOSFET AGING CHARACTERISTICS

A. Experimental Samples and Condition Setup

This study utilizes commercial 1200V/30 A SiC MOSFETs as tested devices, available in two source terminal configurations: **TO-247-4 packages with Kelvin Source Connection (KSC) and conventional TO-247-3 packages with Common Source Connection (CSC)(Fig. 3). Both packages employ identical dies to eliminate chip-related variability. Four samples were tested: DUT1-2 (TO-247-4) and DUT3-4 (TO-247-3). After 19,000 cycles, DUT1 failed and was replaced with DUT5 (TO-247-4).

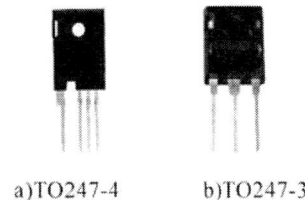

a)TO247-4   b)TO247-3

Fig. 3. SiC MOSFET Discrete Devices with Different Source Terminal Configurations

Thermal resistance parameters were set to Tj,min =27∘C and Tj,max=175∘C. The load current (Iload) was 18 A. Under these conditions, each power cycle lasted approximately 2 minutes, with a turn-on time (*t*on) of ~30 s and turn-off time (*t*off) of ~90 s. Static and dynamic characteristic tests were performed every 1,000 cycles.

To compare degradation mechanisms under varying stress levels,after 18,000 cycles, *I*load was increased to 20 A and *Tj*,max to 220∘C (while *Tj*,min remained 27∘C). This adjustment reduced the cycle duration to 70 s (*t*on≈20 s, *t*off ≈50 s). Post-stress characterization continued every 1,000 cycles.The experiment spanned 8 weeks, accumulating 28,000 cycles.

B. Analysis of Aging Effects on Static Characteristics of Silicon Carbide (SiC) MOSFETs

(1) Threshold Voltage Degradation Analysis
from left to right and then moving down to the next line. This is the author sequence that will be used in future citations and by indexing services. Names should not be listed in columns nor group by affiliation.

Gate oxide traps are identified as the primary cause of threshold voltage (*V*th) instability in SiC MOSFETs. As shown in Figure 4, the threshold voltage drift percentage (Δ*V*th) is defined as the relative change during aging. Under the initial aging condition of *Tj*,max=175∘C, DUT1 exhibited a gradual increase in *V*th, rising by 37% after 16,000 cycles, while other devices remained stable with fluctuations within 10%.

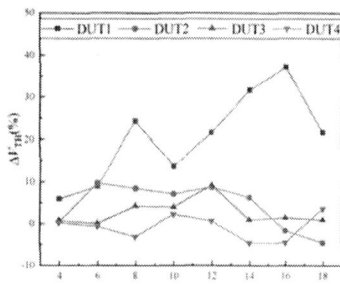

Fig. 4. L_{Load}=18A,T_{Jmax}=175°C Percentage Change of Threshold Voltage (Δ*V*th) During Aging

After increasing the stress to *Tj*,max=220∘C (Fig.5), all tested devices showed positive *V*th drift with eventual saturation. Devices with Kelvin Source Connection (KSC) demonstrated greater *V*th shifts compared to those with Common Source Connection (CSC). DUT1 was removed from testing after 19,000 cycles due to gate-source short-circuit failure. DUT2 exhibited a 15% *V*th increase after 28,000 cycles, DUT3-4 showed ~10% increases, and DUT5 (newly added) displayed a 35% rise after 9,000 cycles.

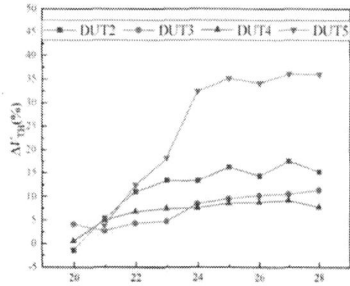

Fig. 5. L_{Load}=20A,T_{Jmax}=220°C Percentage Change of Threshold Voltage (Δ*V*th) During Aging

(2) Body Diode Forward Voltage Drop Degradation Analysis

Under negative gate bias conditions, when the SiC MOSFET is reliably turned off, current flows solely through the intrinsic body diode within the chip. Since aging-induced device degradation does not affect this structure, monitoring the body diode forward voltage drop (*V*SD) under negative gate bias enables complete decoupling of gate oxide degradation effects, allowing independent monitoring of packaging degradation. Conversely, *V*SD changes under zero gate bias also reflect gate oxide degradation.

As shown in Figure 12, under the *Tj*,max=175∘C stress condition, DUT1-4 exhibited no significant *V*SD variations or minor increases during the initial 18,000 cycles. Based on threshold voltage (*V*th) and on-resistance (*R*on) degradation trends, DUT1 primarily experienced gate oxide degradation, while DUT2-3 showed minor packaging degradation.

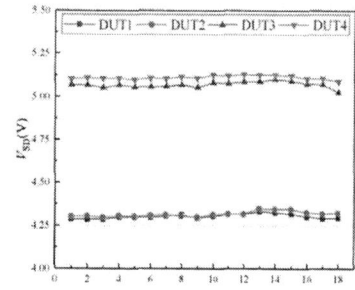

Fig. 6. L_{Load}=18A,T_{Jmax}=175°C ,V_{GS}=-4V,I_D=10A Body Diode Forward Voltage Drop (*V*SD) Variation During Aging

With continued aging, DUT2 displayed an abrupt *V*SD shift after 27,000 cycles, DUT3 after 22,000 cycles, DUT4 exhibited gradual *V*SD increases, and DUT5 remained stable. These observations align with *R*on degradation analysis: DUT2 and DUT3 experienced packaging degradation at 27,000 and 22,000 cycles, respectively. DUT4's *R*on increase arose from combined gate oxide and packaging degradation.

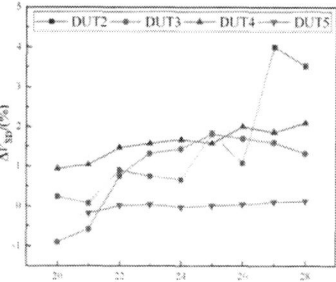

Fig. 7. L_{Load}=20A,T_{Jmax}=220°C ,V_{GS}=-4V,I_D=10A Percentage Change of Diode Forward Voltage Drop (Δ*V*SD) During Aging

Figure 13 illustrates *V*SD evolution under *Tj*,max=220∘C. Under zero gate bias (Figure 14), *V*SD increases reflect contributions from both gate oxide and packaging degradation. For example, DUT5's steady *V*SD rise indicates gate oxide degradation.

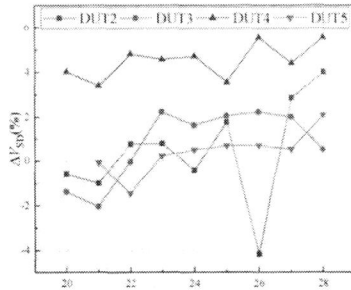

Fig. 8. L_{Load}=20A,T_{Jmax}=220°C ,V_{GS}=0V,I_D=10A Percentage Change of Diode Forward Voltage Drop (ΔVSD) During Aging

In summary, VSD monitoring under negative gate bias effectively isolates packaging degradation from gate oxide effects, demonstrating high sensitivity to abrupt packaging failures but limited response to gradual degradation. VSD trends under different biases corroborate degradation mechanisms inferred from Vth and Ron analyses.

IV. CONCLUSIONS

This study systematically investigates the aging characteristics of silicon carbide (SiC) MOSFETs with Kelvin Source Connection (KSC) and Common Source Connection (CSC) configurations through high-temperature power cycling experiments. Experimental results demonstrate that under high-stress conditions (220°C/20A), KSC devices exhibit significantly higher threshold voltage drift (15%) caused by gate oxide degradation compared to CSC devices (10%). Additionally, KSC devices show a 20% abrupt increase in on-resistance after 27,000 cycles, while CSC devices experience earlier but milder packaging degradation. The study reveals that the accelerated gate oxide degradation in KSC structures stems from insufficient decoupling of the driver-power loop, whereas CSC configurations demonstrate superior reliability due to optimized bond wire stress distribution. By monitoring the body diode forward voltage drop, effective decoupling of packaging degradation from gate oxide degradation is achieved, validating the dynamic evolution of failure mechanisms under multi-stress coupling effects. These findings provide critical theoretical support for high-reliability SiC device packaging design and lifetime prediction, offering practical guidance for device selection in high-temperature, high-power applications such as electric vehicles and photovoltaic inverters.

V. ACKNOWLEDGMENTS

This work was supported by science and technology project of China Southern Power Grid under Grant 030400KC23090017(GDKJXM20231033).

REFERENCES

[1] Zhao, B., Qin, H. H., Ma, C. Y., Yuan, Y., & Zhong, Z. Y. (2014). Research on Switching Characteristics of SiC Power Devices. Advanced Technology of Electrical Engineering and Energy, 33(3), 18-22.

[2] J. Chen, E. Deng, Z. Zhao, Y. Wu and Y. Huang, "Power Cycling Capability Comparison of Si and SiC MOSFETs under Different Conduction Modes," 2020 32nd International Symposium on Power Semiconductor Devices and ICs (ISPSD), Vienna, Austria, 2020, pp. 541-544.

[3] U. Karki and F. Z. Peng, "Effect of gate-oxide degradation on electrical parameters of power MOSFETs," IEEE Trans. Power Electron., vol. 33, no. 12, pp. 10764-10773, Dec.

[4] JT. Aichinger, G. Rescher, and G. Pobegen, "Threshold voltage peculiarities and bias temperature instabilities of SiC MOSFETs," Microelectron. Reliab., vol. 80. pp. 68-78, 2018.

[5] A. J. Lelis, R. Green, D. B. Habersat, and M. El, "Basic mechanisms of threshold-voltage instability and implications for reliability testing of SiC MOSFETs," IEEE Trans. Electron Devices, vol. 62, no. 2, pp. 316-323, Feb. 2015..

[6] K. Puschkarsky, H. Reisinger, T. Aichinger, W. Gustin, and T. Grasser. "Understanding BTI in SiC MOSFETs and its impact on circuit operation, " IEEE Trans. Device Mater. Reliab., vol. 18, no. 2, pp. 144-153, Jun.2018.

[7] A. Bolotnikov et al., "Utilization of SiC MOSFET body diode in hard switching applications," Mater. Sci. Forum, vol. 778-780, pp. 947-950,2014.

[8] P. Ning et al., "SiC wirebond multichip phase-leg module packaging design and testing for harsh environment," IEEE Trans. Power Electron., vol. 25, no. 1, pp.16-23, Jan. 2010.

[9] T. Hung, L. Liao, C. C. Wang, W. H. Chi, and K. Chiang, "Life prediction of high-cycle fatigue in aluminum bonding wires under power cycling test," IEEE Trans. Device Mater. Reliab., vol. 14, no. 1, pp. 484-492, Mar. 2014.

[10] Li, H., Munk-Nielsen, S., Wang, X., Beczkowski, S., Jones, S. R., and Dai, X. "Effects of Auxiliary-Source Connections in Multichip Power Module", IEEE Transactions on Power Electronics, vol. 32, no, 10, pp. 7816-7823, 2016.

[11] F. Yang, E. Ugur, B. Akin and G. Wang, "Design Methodology of DC Power Cycling Test Setup for SiC MOSFETs," in IEEE Journal of Emerging and Selected Topics in Power Electronics.doi: 10.1109/JESTPE.2019.2914419

[12] S. Dusmez and B. Akin, "An accelerated thermal aging platform to monitor fault precursor on-state resistance," 2015 IEEE International Electric Machines& Drives Conference (IEMDC), Coeur d'Alene, ID, USA, 2015, pp. 1352-1358.

[13] Li, H., Munk-Nielsen, S., Wang, X., Beczkowski, S., Jones, S. R., and Dai, X,"Effects of Auxiliary-Source Connections in Multichip Power Module", IEEE Transactions on Power Electronics, vol. 32, no, 10, pp. 7816-7823, 2016

[14] J. Lehmann, M. Netzel, R. Herzer, and S. Pawel, "Method for Electrical Detection of Bond Wire Lift-Off for Power Semiconductors," IEEE 15th International Symposium on Power Semiconductor Devices and ICs, 2003,pp.333-336.

[15] A. Fayyaz, G. Romano, and A. Castellazzi, "Body diode reliability investigation of SiC power MOSFETs," Microelectron. Reliab., vol. 64, pp.530-534, 2016.

[16] F. Yang, E. Ugur, B. Akin and G. Wang, "Design Methodology of DC Power Cycling Test Setup for SiC MOSFETs," in IEEE Journal of Emerging and Selected Topics in Power Electronics, doi: 10.1109/JESTPE.2019.2914419

2025 IEEE Workshop on Wide Bandgap Power Devices and Applications in Asia (WiPDA Asia)

SiC VDMOSFET Performance Prediction Method Based on Neural Network

Xiamin Hao
Beijing Institute of Smart Energy,
Beijing Huairou Laboratory
Beijing, China
haoxiamin0515@126.com

Feng He
Beijing Institute of Smart Energy,
Beijing Huairou Laboratory
Beijing, China
hefeng@bise.hrl.ac.cn

Rui Jin
Beijing Institute of Smart Energy,
Beijing Huairou Laboratory
Beijing, China
jinrui@bise.hrl.ac.cn

Abstract— Silicon carbide (SiC) vertical double-diffusion MOSFETs (VDMOSFETs) exhibit excellent performance in power electronics. This study employs TCAD simulations to model SiC VDMOSFETs, generating a dataset of 2000 device structures and their simulation results. A neural network model is trained on this dataset to predict key device parameters. Prediction accuracy is evaluated using mean squared error (MSE), mean absolute error (MAE), and R-squared (R^2). Furthermore, the trained model facilitates parameter sensitivity analysis. Results demonstrate that appropriate model selection (e.g., CNN) enables accurate prediction of electrical characteristics and effective sensitivity analysis.

Keywords—SiC VDMOSFET, neural network, power device, machine learning, TCAD

I. INTRODUCTION

Silicon carbide (SiC) MOSFETs have become pivotal components in next-generation high-power electronic devices owing to their superior material properties, such as high breakdown field, excellent thermal conductivity, and high electron saturation velocity [1–4]. Among various structures, the vertical double-diffused MOSFET (VDMOSFET) offers significant advantages. Its design effectively mitigates electric field concentration at the trench gate corner through a relatively simplified fabrication process while maintaining high breakdown voltage capability. Consequently, SiC VDMOSFETs have demonstrated excellent performance and have emerged as the dominant architecture for high-voltage SiC power MOSFETs in recent years[5].

However, the electrical characteristics of SiC VDMOSFET, including on resistance ($R_{on, sp}$), breakdown voltage (BV), and threshold voltage (V_{th}), are highly sentive to numerous structural and process parameters. Traditional design heavily relies on Technology Computer-Aided Design (TCAD) simulation. This approach often entails lengthy design cycles, substantial computational costs, and faces challenges related to numerical convergence. The modeling and simulation process consumes a lot of resources and time, so how to improve simulation efficiency has become the focus of research.

The rapid advancement of neural network (NN) technology presents a promising avenue for predicting device structure-performance relationships. Machine learning (ML) algorithms excel at capturing complex, nonlinear parameter dependencies and have gained significant traction in the microelectronics device domain [6–10]. For instance, ML-based methods have been successfully applied to predict key

electrical characteristics in integrated Schottky barrier diode (SBD)-MOSFETs and high-electron-mobility transistors (HEMTs) [11–13].

This paper proposes a NN-based framework for predicting the electrical characteristics of VDMOSFET and conducting multi-parameter sensitivity analysis. Considering the numerous parameters and complex mechanisms that affect the performance of VDMOSFET devices, it is imperative to select and optimize the NN to improve its accuracy.

II. METHOD

While TCAD simulations provide valuable insights, they are inherently limited to extrapolating results based on the implemented physical models. To enhance the fidelity of our training data, we rigorously incorporate comprehensive physical models within the TCAD software environment. Furthermore, although NN predictions do not explicitly model the underlying physical processes, employing advanced network architectures can maximize prediction accuracy within this inherent limitation [6].

In this study, we first utilize TCAD to model and simulate VDMOSFET devices, generating a dataset of 2000 distinct device structures and their corresponding electrical characteristics to serve as the NN training and validation set. Input parameters comprise epitaxial layer doping concentration, JFET region width, channel length, P$^+$ region width, and N$^+$ region width. Output parameters include the forward IV curve, specific on resistance $R_{on,sp}$, transfer characteristic curve, as well as the extracted V_{th}, reverse bias IV curve, and the extracted BV. These outputs comprehensively characterize device performance and exhibit complex dependencies on the input parameters, enabling both performance prediction and sensitivity analysis via the NN. Following dataset generation, the NN is trained. The accuracy of the predictions is rigorously evaluated using established metrics: mean square error (MSE), mean absolute error (MAE), and R-squared (R^2). Furthermore, SHapley Additive exPlanations (SHAP) will be used to analyze parameter sensitivity mechanisms.

III. RESULTS AND DISCUSSION

A. Device Modeling and Dataset Generation

The simulated VDMOSFET device structure is illustrated in Fig. 1. Key structural dimensions include a 0.05μm gate oxide, 0.3μm deep N$^+$ and P$^+$ regions, a 0.9μm thick P$_{well}$

979-8-3315-1110-4/25 $31.00 © 2025 IEEE

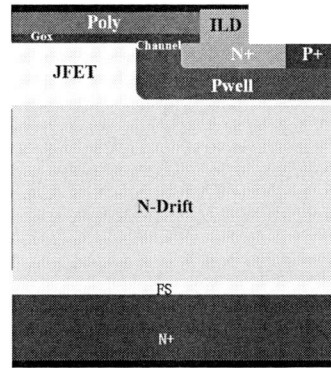

Fig. 1. Schematics of VDMOSFET

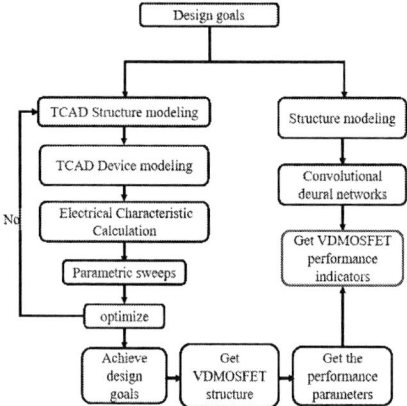

Fig. 2. Flowchart comparing traditional TCAD-based and proposed NN-based design methodologies for VDMOSFETs

region, a 12μm thick N⁻ epitaxial layer, and a 0.2μm thick N⁺ substrate. To ensure simulation fidelity, critical physical models were incorporated within TCAD, accounting for phenomena such as avalanche ionization, high-field and surface vertical field mobility degradation, incomplete ionization, and Auger recombination.

TABLE II.   VALUES OF INPUT PARAMETERS

Input Parameters	Values
N_{epi} (cm⁻³)	1E13~1E18
W_{jfet} (μm)	1~1.4
$W_{channel}$ (μm)	0.4~1
W_{P+} (μm)	0.4~2
W_{N+} (μm)	1.5~2.1

Focusing on the primary electrical characteristics of MOSFETs—threshold voltage (V_{th}), specific on-resistance ($R_{on,sp}$), and breakdown voltage (BV)—as the key output metrics, we identified five major input parameters governing these properties: epitaxial layer doping concentration (N_{epi}), JFET region width (W_{jfet}), channel width ($W_{channel}$), P⁺ region

width (W_{P+}), and N⁺ region width (W_{N+}) [14]. Table I details the parameter ranges employed in the bias sweep simulations to ensure device setting rationality.

Systematic TCAD simulations were performed across this parameter space, generating the forward Id-Vg (transfer), Id-Vd (output), and reverse bias I-V (breakdown) curves for each device configuration. From these curves, the target metrics (V_{th}, $R_{on,sp}$, BV) were extracted. This process yielded a comprehensive dataset of 2000 distinct samples, enhancing data generalization capability. The dataset was partitioned into training (1600 samples), validation (200 samples), and testing (200 samples) sets. Fig. 2 schematically compares the traditional TCAD-based design flow with the proposed NN-based approach, highlighting the latter's direct prediction capability.

B. Structural Parameter Sensitivity Analysis

To quantitatively evaluate the impact of structural parameters on device performance, conventional statistical analysis and SHAP were applied to the TCAD-generated dataset.

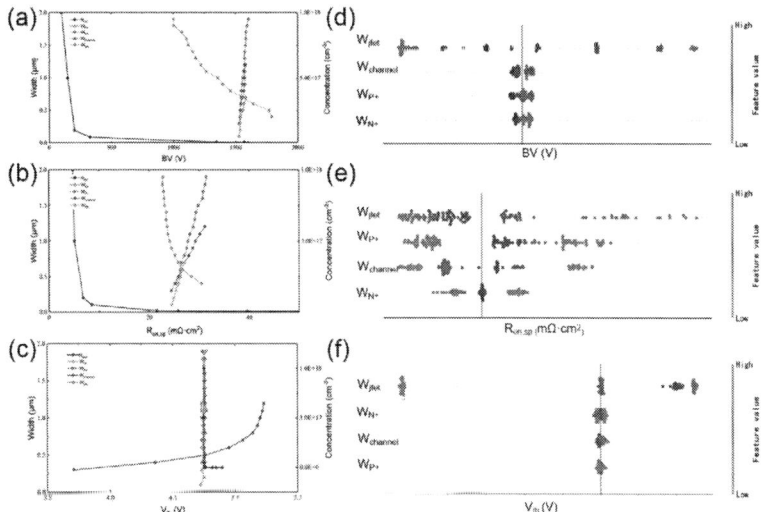

Fig. 3. Analysis of individual structural parameter impacts on key device metrics: (a) Conventional statistical analysis; (b) SHAP-based sensitivity analysis.

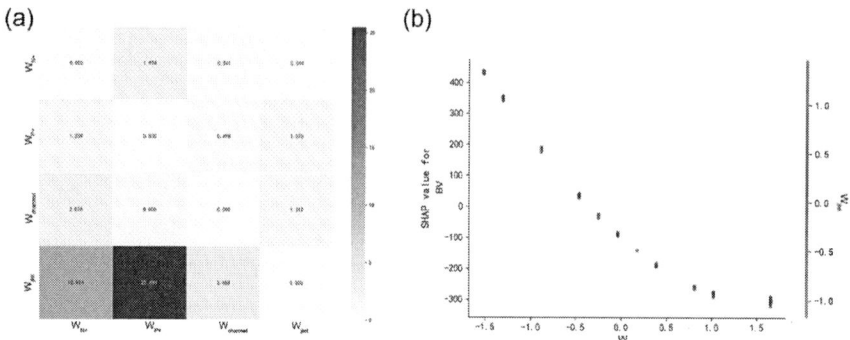

Fig. 4. SHAP analysis of multi-parameter influence mechanisms on BV: (a) Mutual influence mechanism of W_{N+}, W_{P+}, $W_{channel}$, and W_{jfet}; (b) Impact mechanism of the strong coupling between W_{jfet} and W_{P+}.

Fig. 3 presents the combined results. Both BV and $R_{on,sp}$ exhibit a pronounced negative correlation with the ratio N_{epi}/W_{jfet} (SHAP coefficient φ = -0.18 ± 0.02, p < 0.001). Notably, the influence of N_{epi} alone could not be robustly assessed via SHAP due to its limited effective sample size within the dataset, potentially affecting the interaction analysis. Conversely, BV and $R_{on,sp}$ increase with the lateral extension of both W_{P+} and W_{N+} (SHAP interaction index δ = 0.22). The threshold voltage Vth demonstrates a strong, positive dependence on Lchannel (φ = 0.41, 95% CI: [0.38, 0.45]), while showing negligible dependence on N_{epi}, W_{jfet}, W_{P+} and W_{N+} ($|\varphi|$ < 0.05).

Beyond individual effects, SHAP dependency plots (Fig. 4) effectively decouple synergistic interactions between parameters, corroborating and extending insights from TCAD simulations. Due to the aforementioned limitations with Nepi, the decoupling analysis focused on W_{N+}, W_{jfet}, W_{P+} and $W_{channel}$.

Taking BV as an example, SHAP revealed the strongest mutual influence mechanism between W_{P+} and W_{jfet}, while the coupling effect of W_{N+} with W_{jfet} was relatively weaker. BV generally decreases as W_{P+} increases. However, when W_{P+}

exceeds ~8.5μm, a further increase in BV necessitates increasing W_{jfet}. Conversely, for W_{P+} below ~5.5μm, increasing BV requires decreasing W_{jfet}.

The strong agreement between TCAD-derived trends and SHAP interpretability establishes a robust framework for predicting VDMOSFET performance. Crucially, the parameter decoupling achieved via SHAP provides fundamental insights for the precise adjustment of key electrical characteristics like BV, V_{th}, and $R_{on,sp}$.

C. NN Model Performance

To effectively extract features from the complex dataset, two NN architectures were implemented and compared: a Fully Connected Neural Network (FCNN) and a Convolutional Neural Network (CNN), as depicted in Fig. 5.

Results demonstrate the superior predictive capability of the CNN architecture for modeling the intricate relationships in VDMOSFET devices. Despite extensive hyperparameter tuning of the FCNN (e.g., 5 fully connected layers: 512 → 512 → 256 → 256 → 128 neurons; Scaled Exponential Linear Unit (SELU) activation and layer normalization; progressive dropout rates: 0.5 → 0.3), its fitting performance remained

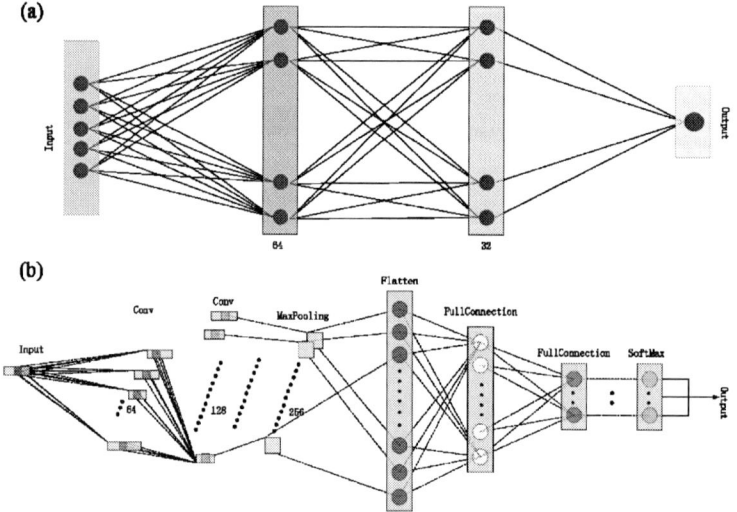

Fig. 5. Architectures of the implemented neural networks: (a) Fully Connected Neural Network (FCNN); (b) Convolutional Neural Network (CNN).

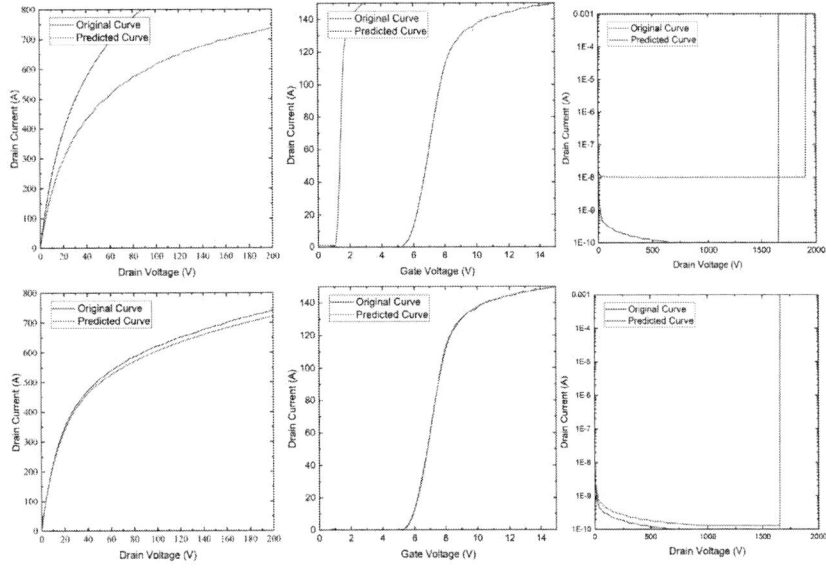

Fig.6. Curve fitting results: (a-c) Id-Vd, Id-Vg, and BV curves based on the FCNN model; (d-f) Corresponding curves based on the CNN model.

TABLE II.   VALUES OF EVALUATION INDICATORS

	IdVd Curves (Ron,sp)		IdVg Curves (Vth)		BV Curves (BV)	
	FCNN	CNN	FCNN	CNN	FCNN	CNN
MSE	0.016	0.003	0.036	0.002	0.011	0.001
MAE	0.244	0.042	0.526	0.036	0.153	0.028
R2	0.561	0.987	0.263	0.995	0.737	0.996

unsatisfactory, as evidenced by the poor agreement between predicted and actual Id-Vd, Id-Vg, and BV curves shown in Fig. 6(a)-(c). The CNN architecture, designed to leverage spatial relationships within the input data, comprises an input layer, a feature expansion module, a feature extraction module (including three convolutional blocks: Conv1d channels: 64→128→256, kernel_size=3, padding=1; BatchNorm1d; ReLU; MaxPool1d kernel_size=2), and an output layer utilizing global average pooling to predict the characteristic curves.

In contrast to the FCNN, the CNN achieved excellent curve fitting, as shown in Fig. 6(d)-(f). Quantitative evaluation metrics (Table II) further confirm the CNN's superiority. For Ron,sp, Vth, and BV predictions, the CNN achieved MAE <

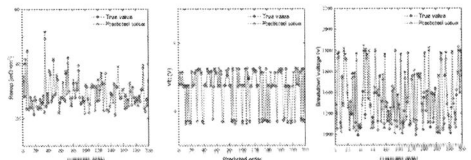

Fig.7. Comparison of CNN model predictions (red dots) versus true TCAD values (black dots) for device characteristics.

0.1, MSE < 0.01, and R^2 > 95% across all curve types. Conversely, the FCNN results were markedly inferior (e.g., MSE often > 0.01, MAE > 0.1, R^2< 80%).

The high accuracy of the CNN predictions is visually confirmed in Fig. 7, which compares predicted values (red dots) against actual TCAD-simulated values (black dots) for a representative sample. The close proximity of the data points and their consistent trendlines demonstrate the model's high predictive fidelity.

IV. CONCLUSION

In a word, the method of training and predicting neural network algorithm based on TCAD software simulation data is a method to quickly design VDMOSFETs with specific performance goals. The results show that there is a strong correlation between the predicted value and the actual measured value. This method greatly reduces the manpower, material resources, and time usually required when using TCAD simulation to analyze equipment performance. A neural network effectively establishes a fast relationship between device parameters and performance, which promotes researchers to carry out device optimization research and accelerates the progress of VDMOSFET performance prediction.

ACKNOWLEDGMENT

This work was supported by the Management Technology Project of State Grid Corporation of China (SGCC) Headquarters, No. 5500-202399661A-3-2-ZN.

REFERENCES

[1] D. Johannesson, K. Jacobs, S. Norrga, A. Hallén, M. Nawaz, and H. P. Nee, "Wide-range prediction of ultra-high voltage sic igbt static performance using calibrated tcad model," Mater. Sci. Forum, vol. 1004 MSF, pp. 911–916, 2020, doi: 10.4028/www.scientific.net/MSF.1004.911.

[2] T. Kimoto, "Bulk and epitaxial growth of silicon carbide," Prog. Cryst. Growth Charact. Mater., vol. 62, no. 2, pp. 329–351, 2016, doi: 10.1016/j.pcrysgrow.2016.04.018.

[3] T. Kimoto, "Material science and device physics in SiC technology for high-voltage power devices," Jpn. J. Appl. Phys., vol. 54, no. 4, p. 40103, 2015, doi: 10.7567/jjap.54.040103.

[4] A. Vasilev et al., "TCAD Modeling of Temperature Activation of the Hysteresis Characteristics of Lateral 4H-SiC MOSFETs," IEEE Trans. Electron Devices, vol. 69, no. 6, pp. 3290–3295, Jun. 2022, doi: 10.1109/TED.2022.3166123.

[5] B. Yi, J. Cheng, M. Kong, B. Zhang, and X. B. Chen, "A high-voltage p-LDMOS with enhanced current capability comparable to double RESURF n-LDMOS," Proc. Int. Symp. Power Semicond. Devices ICs, vol. 2018-May, no. 51237001, pp. 148–151, 2018, doi: 10.1109/ISPSD.2018.8393624.

[6] A. M. Massimo Orazio Spata, Sebastiano Battiato, Alessandro Ortis, Francesco Rundo, Michele Calabretta, Carmelo Pino, "Deep Learning Algorithm for Advanced Level-3 Inverse- Modeling of Silicon-Carbide Power MOSFET Devices." doi: https://arxiv.org/pdf/2310.17657.

[7] Z. Ni, X. Lyu, O. P. Yadav, B. N. Singh, S. Zheng, and D. Cao, "Overview of Real-Time Lifetime Prediction and Extension for SiC Power Converters," IEEE Trans. Power Electron., vol. 35, no. 8, pp. 7765–7794, Aug. 2020, doi: 10.1109/TPEL.2019.2962503.

[8] Z. Zhang, A. Mehrabi, W. D. Van Driel, and R. H. Poelma, "The Potential of Machine Learning for Thermal Modelling of SiC Power Modules - A Review," in 2024 IEEE 10th Electronics System-Integration Technology Conference (ESTC), IEEE, Sep. 2024, pp. 1–8. doi: 10.1109/ESTC60143.2024.10712111.

[9] X. Niu et al., "SiC MOSFET with Integrated SBD Device Performance Prediction Method Based on Neural Network," Micromachines, vol. 16, no. 1, p. 55, Dec. 2024, doi: 10.3390/mi16010055.

[10] L. Lin, X. Wang, and J. Hu, "Prediction of JTE breakdown performance in SiC PiN diode radiation detectors using TCAD augmented machine learning," Nucl. Instruments Methods Phys. Res. Sect. A Accel. Spectrometers, Detect. Assoc. Equip., vol. 1061, p. 169102, Apr. 2024, doi: 10.1016/j.nima.2024.169102.

[11] S. Bin Kutub, H.-J. Jiang, N.-Y. Chen, W.-J. Lee, C.-Y. Jui, and T.-L. Wu, "Artificial Neural Network-Based (ANN) Approach for Characteristics Modeling and Prediction in GaN-on-Si Power Devices," in 2020 32nd International Symposium on Power Semiconductor Devices and ICs (ISPSD), IEEE, Sep. 2020, pp. 529–532. doi: 10.1109/ISPSD46842.2020.9170110.

[12] A. Kilic, R. Yildirim, and D. Eroglu, "Machine Learning Analysis of Ni/SiC Electrodeposition Using Association Rule Mining and Artificial Neural Network," J. Electrochem. Soc., vol. 168, no. 6, p. 062514, Jun. 2021, doi: 10.1149/1945-7111/ac0aaa.

[13] J. Wei, H. Wang, T. Zhao, Y.-L. Jiang, and J. Wan, "A New Compact MOSFET Model Based on Artificial Neural Network With Unique Data Preprocessing and Sampling Techniques," IEEE Trans. Comput. Des. Integr. Circuits Syst., vol. 42, no. 4, pp. 1250–1254, Apr. 2023, doi: 10.1109/TCAD.2022.3193330.

[14] M. Torky and T. P. Chow, "Comparative Performance Evaluation of Conventional and Superjunction Vertical 4H-SiC High-Voltage Power MOSFETs," Mater. Sci. Forum, vol. 1062 MSF, pp. 565–569, 2022, doi: 10.4028/p-37r645.

979-8-3315-1110-4/25 $31.00 © 2025 IEEE

Oxidation-free Sintered Copper Die Attachment for Power Electronics Packaging

Junyang Chen[a], Haiqiang Zhao[a], Zewei Zhang[a], Zhuo Pang[a], Tsung-Huan Sheng [b], Yi Chiu [b], Zhi-Ying Huang [b], Yen-Liang Lin [b] and Meiyu Wang [a,c,*]

[a] College of Electronic Information and Optical Engineering, Nankai University, Tianjin 300350, China;

b Geckos Technology Corp., No. 31, Xingkeyuan Rd., Houli Dist., Taichung City 421007, Taiwan

[c] Shenzhen Research Institute of Nankai University, Shenzhen 518083, China;

* Corresponding author: meiyuwang@nankai.edu.cn

Abstract—The rapid development of wide-bandgap semiconductors imposed stringent requirements for die attachment for power electronic packaging. Among which, sintered copper is becoming an advanced die attachment technology, since it offers much higher melting point, thermal conductivity, and reliability than traditional solders; lower cost and better resistance to electrochemical migration than sintered silver. However, the inherent property of copper oxidation during sintering in the air is an obstacle to applications. In this study, an antioxidative copper paste was developed by encapsulating 600 nm submicron copper particles with reducing solvents. The SEM and EDS results verified oxidation-free and delamination-free Cu-Cu joints. The effects of sintering temperature, pressure, and time were investigated in detail. Die-shear strength of 48.1 MPa, electrical resistivity of 5.19×10^{-7} $\Omega\cdot m$, and thermal conductivity of 209.36 W/m·K were obtained, indicating excellent mechanical, electrical, and thermal properties. This study provides a more economical and advanced packaging technology for power electronics packaging.

Keywords—Electronics packaging, sintered copper, die attachment, shear strength, electrical resistivity, thermal conductivity

I. INTRODUCTION

With the rapid development of the wide bandgap (WBG) semiconductors, higher requirements have been placed on the die attachment materials for power electronics packaging. The high operating temperature, large breakdown electric field, and excellent thermal conductivity of power semiconductor devices demand that packaging materials possess good mechanical strength, thermal conductivity, electrical conductivity, and reliability [1-3].

Sintered nanosilver materials, due to their 3~5 times better thermal and electrical conductivities and

reliability to solders, as well as the high melting point (961°C), have become an advanced die attachment technology in the power electronics packaging [4, 5]. However, the high cost of silver nanoparticles and its susceptibility to electrochemical migration, limit its widespread application [6]. In contrast, copper (Cu) offers 1000 times lower costs, higher melting point (1083°C), and stronger resistance to electrochemical migration [7]. Therefore, sintered-Cu hold great promise as advanced die attachment technology for applications in WBG semiconductors packaging [8].

Currently, the main challenge of sintered copper processes is the tendency of copper to oxidation during sintering in the air [9]. To address this issue, this study employs submicron Cu particles encapsulated with reducing solvents to enhance its oxidation resistance while promoting densification sintering. When sintered at 260°C under 5 MPa for 30 min, the die-shear strength of 48.1 MPa, electrical resistivity of 5.19×10^{-7} $\Omega\cdot m$, and thermal conductivity of 209.36 W/m·K indicates excellent mechanical, electrical, and thermal properties. The microstructure characterization and mechanism explanation will be provided in the full paper later.

II. MATERIALS AND EXPERIMENTAL PROCEDURES

The Cu paste was made by mixing submicron Cu particles (diameter of 600 nm, Geckos Technology Corp., Ncp06) with reducing solvents including thinners, binders, and dispersants. For the die-shear strength tests, the Cu-Cu joint samples were fabricated by attaching 3 mm × 3 mm × 0.3 mm Cu dummy devices on 20 mm × 20 mm × 1 mm Cu substrate by stencil-printing 80 μm-thick Cu paste. For the electrical resistivity tests, the resistor samples were made by 80 μm-thick copper paste stencil-printed in a serpentine pattern on an Al_2O_3-ceramic substrate. For the thermal conductivity tests, the bulk samples were made to 10 mm × 10 mm × 3 mm. After samples fabrication, the sintering was conducted at temperature of 220~260°C under pressures of 1~5 MPa holding for 10~30 min. The effects of sintering parameters such as temperature, pressure, and time on the microstructure and mechanical, electrical, and thermal performance were evaluated in detail.

III. RESULTS AND DISCUSSION

A. Microstructure and chemistry

Fig. 1 shows the thermogravimetric analysis (TGA) and differential scanning calorimetry (DSC) test results of the Cu paste. In the temperature range of 100-175°C, the TGA curve drops sharply, while the DSC curve shows an endothermic peak. This is due to the decomposition and volatilization of thinners. In the 175-275°C range, dispersants and binders continue to volatilize, resulting in a further weight loss in the TGA and exothermic peak in the DSC. When heated to 400°C, the TGA curve reaches steady and the solid loading is calculated as 79.4%.

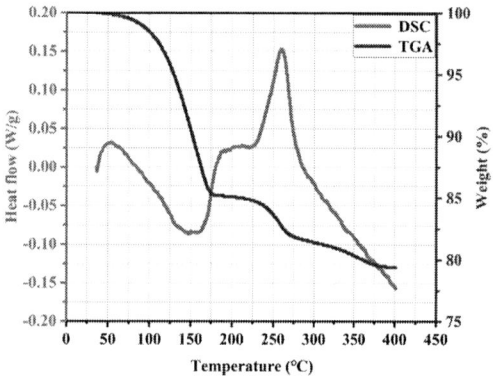

Fig. 1. The TGA and DSC test results of the Cu paste.

Fig. 2 shows the microstructural morphology of the as-printed copper paste powder. The observation was carried out using a scanning electron microscope (SEM) at a magnification of 50,000. The powder particles predominantly exhibit spherical and polyhedral shapes, which are beneficial for achieving high packing density and uniform shrinkage during the sintering process. The submicron particle size is 600 nm in average, with size distribution ranging from approximately 200 nm to 800 nm, allowing smaller particles to fill the interspaces between larger ones and thus enhancing the overall densification. The particle surfaces appear relatively smooth, with no evident oxidation layers or porous structures, indicating that oxidation did not occur during the powder preparation process.

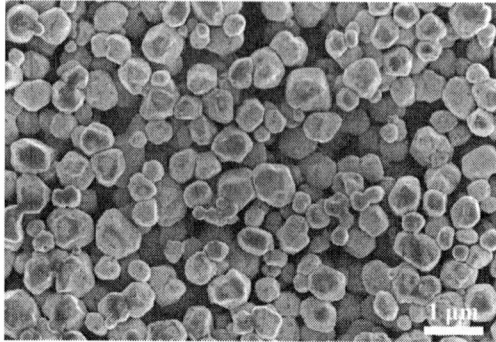

Fig. 2. SEM images of as-print Cu paste.

Fig. 3 shows the cross-sectional SEM image of the Cu–Cu interconnection joint sintered at 260 °C under a pressure of 5 MPa for 30 minutes. The sintered copper layer exhibits a dense and uniform microstructure, and the Cu–Cu joint shows good interfacial bonding. No obvious voids or cracks are observed, and there is no evidence of delamination or debonding. The copper particles have bonded effectively, forming a continuous and well-integrated metallic phase.

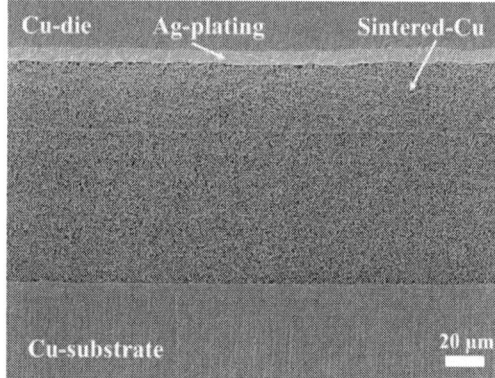

Fig. 3. Cross-sectional SEM image of sample.

Fig. 4 and Fig. 5 show the SEM image of magnified bonding interface between sintered copper and Cu substrate, as well as the corresponding EDS line scan elemental analysis across the interface. The SEM image shows well-bonded interface without delamination or crack. The EDS results indicate that Cu is uniformly distributed at the bonding interface, with no evident elemental segregation or compositional inhomogeneity. Moreover, no significant oxygen signal or peak was detected, indicating that no noticeable oxidation occurred during the sintering process. The oxidation-free Cu-Cu joints were achieved successfully.

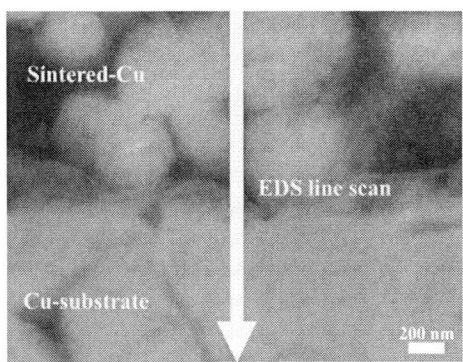

Fig. 4. Magnified SEM image of bonding interface between sintered-Cu and Cu-substrate.

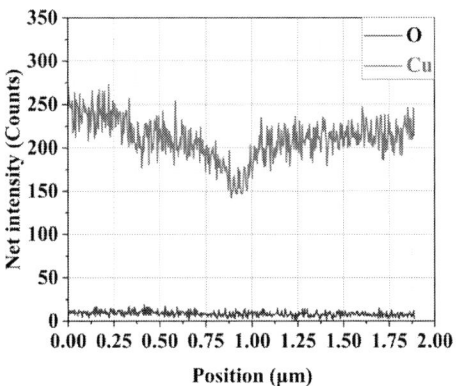

Fig. 5. EDS line scan element analysis at bonding interface between sintered-Cu and Cu-substrate.

B. Sintering thermodynamic and kinetic mechanisms

The sintering parameters, i.e., temperature (T), time (t), and pressure (P_a) have significant influences on the performance of the sintered-Cu joints. This is because that the pressure sintering rate ($d\rho/dt$) is a function of the driving force (kinetics) and the mobility (thermodynamics) expressed as follows:

$$\frac{d\rho}{dt} = A\left(\frac{P_a}{3} + \frac{\gamma}{3r}\right) \times \left(\frac{D}{k_B T}\right) \qquad (1)$$

$$D = D_0 e^{\left(\frac{-Q}{RT}\right)} \qquad (2)$$

where ρ is the relative density of a sintering powder compact, γ is the surface free energy of the powder particles, r is the average particle size (r=200 nm in this study), γ/r is the sintering potential, k_B is the Boltzmann constant, A is constant. D is the atomic diffusivity controlling the sintering. Q is the activation energy, D_0 is a term determined by atomic vibration frequency and jump distance, and R is the gas constant.

The higher sintering temperature (T) will exponentially accelerate the sintering mobility and promotes densification sintering rate. The longer sintering time (t) will result in greater densification in the bondline and longer inter-diffusion length at the bonding interface. The applied external pressure (P_a) has three functions in promoting sintering: the first adds to the driving-force term, as expressed in Eq. (1), the second causes deformation of the particles to promote neck growth, and the third increases the powder compact's contact area at the two bonding interfaces to promote adhesion.

C. Die-shear strength

Fig. 6 shows the die-shear strength test results of Cu-Cu joints under different sintering parameters. Overall, the die-shear strengths almost linearly increase with higher sintering temperature, pressure, and time. This is because, as expressed in Eq. (1) and (2), the higher sintering parameters increase both the cohesive strength in the sintered-Cu bondline and the adhesive strength at the Cu-Cu bonding interface. Strong bonding strength of higher than the criterion of strong bonding (30 MPa) was obtained. Specifically, the die-shear strength reached 48.1 MPa when sintering at 260°C, 33.8 MPa when sintering for 20 min, and 35.9 MPa when sintering under pressure of 3 MPa.

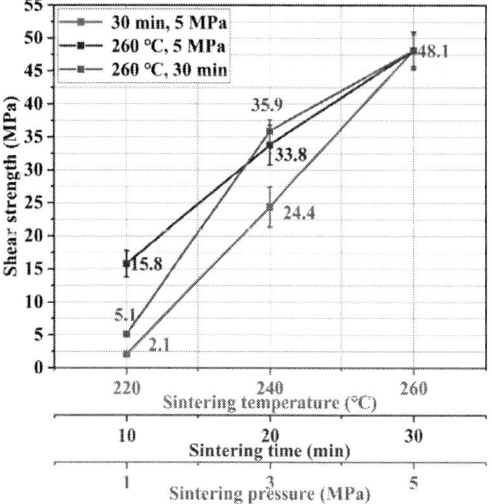

Fig. 6. Effect of sintering parameters on die-shear strength.

D. Electrical resistivity

The electrical resistivity was measured using four-terminal method. Fig. 7 shows the measured results of the resistor samples sintered under different sintering conditions. As shown by the red line in Fig. 7, the sintering temperature has the most significant influence on the electrical resistivity. As the sintering temperature increased from 220°C to 260°C, the electrical resistivity significantly decreased by 12 times, from 62.7×10^{-7} $\Omega \cdot m$ to 5.19×10^{-7} $\Omega \cdot m$, indicating that the electrical conductivity improves greatly with increasing sintering temperature. When sintering at temperature of 220°C, the large electrical resistivity agreed well with the low die-shear strength of 2.1 MPa, indicating 220°C is too low to achieve densification sintering. The blue line and the green line in Fig. 7 show the effects of sintering time and pressure. The results indicate that electrical resistivity decreases with an increase in sintering time, while not much affected by the pressure variation. Overall, when sintering at 260°C for 20~30 min under 1~5 MPa, the resistivity is about 30 times larger than that of the bulk copper (1.75×10^{-8} $\Omega \cdot m$), and similar with the reported values, indicating a good electrical property.

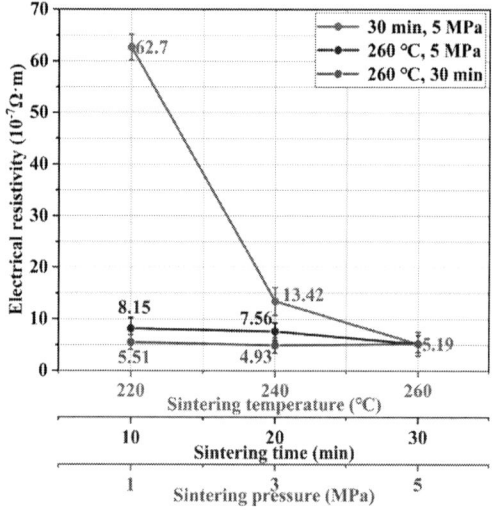

Fig. 7. Effect of sintering parameters on electrical resistivity.

E. Thermal conductivity

The thermal conductivity of the bulk samples was measured using a laser flash method. Fig. 8 shows the variation in thermal conductivity under different sintering conditions. As shown by the red line in Fig. 8, similar with the test results of die-shear strength and electrical resistivity, the increase in sintering temperature leads to a significant improvement in thermal conductivity, reaching 209.36 W/(m·K) when sintering at 260°C, demonstrating excellent thermal conduction

properties. The blue line and the green line in Fig. 8 show that as the sintering time and pressure increase, the thermal conductivity also gradually improves. When sintering at temperature ≥240°C holding for ≥20 min under ≥3 MPa, the thermal conductivities were higher than 170 W/(m·K), demonstrating excellent thermal properties.

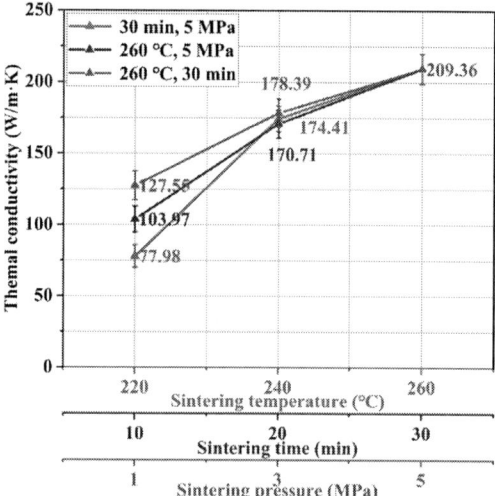

Fig. 8. Effect of sintering parameters on thermal conductivity.

IV. CONCLUSIONS

This study investigated sintered copper as die attachment technology for power electronics packaging. An antioxidative copper paste was developed by encapsulating 600 nm submicron copper particles with reducing solvents not only to enhance its oxidation resistance but also to promote the densification sintering. The SEM and EDS results verified oxidation-free and delamination-free Cu-Cu joints. The effects of sintering temperature, pressure, and time were investigated in detail. Among which, the sintering temperature has the most significant effects on the sintered quality. When sintering at 260°C under 5 MPa for 30 min, die-shear strength of 48.1 MPa, electrical resistivity of 5.19×10^{-7} $\Omega \cdot m$, and thermal conductivity of 209.36 W/m·K were obtained, indicating excellent mechanical, electrical, and thermal properties. This study provides a more economical and advanced packaging technology for power electronics packaging.

REFERENCES

[1] X. Li, L. Lin, C.-J. Du *et al.*, "Effect of Oxygen Content on Bonding Performance of Sintered Silver Joint on Bare Copper Substrate," *IEEE Transactions on Components, Packaging and Manufacturing Technology,* vol. 13, no. 3, pp. 391-398, 2023.

[2] H. Yan, Y.-H. Mei, X. Li *et al.*, "A Multichip Phase-Leg IGBT Module Using Nanosilver Paste by Pressureless Sintering in Formic Acid Atmosphere," *IEEE Transactions on Electron Devices,* vol. 65, no. 10, pp. 4499-4505, 2018.

[3] T. Suzuki, Y. Yasuda, T. Terasaki *et al.*, "Macro- and Micro-Deformation Behavior of Sintered-Copper Die-Attach Material," *IEEE Transactions on Device and Materials Reliability,* vol. 18, no. 1, pp. 54-63, 2018.

[4] M. Wang, Y. Mei, X. Li *et al.*, "Pressureless Silver Sintering on Nickel for Power Module Packaging," *IEEE Transactions on Power Electronics,* vol. 34, no. 8, pp. 7121-7125, 2019.

[5] F. Yu, J. Cui, Z. Zhou *et al.*, "Reliability of Ag Sintering for Power Semiconductor Die Attach in High-Temperature Applications," *IEEE Transactions on Power Electronics,* vol. 32, no. 9, pp. 7083-7095, 2017.

[6] W. Lv, J. Liu, Y. Mou *et al.*, "Fabrication and Sintering Behavior of Nano Cu–Ag Composite Paste for High-Power Device," *IEEE Transactions on Electron Devices,* vol. 70, no. 6, pp. 3202-3207, 2023.

[7] L. Wu, J. Qian, F. Zhang *et al.*, "Low-Temperature Sintering of Cu/Functionalized Multiwalled Carbon Nanotubes Composite Paste for Power Electronic Packaging," *IEEE Transactions on Power Electronics,* vol. 37, no. 2, pp. 1234-1243, 2022.

[8] J. Alptekin, K. Antony Jesu Durai, D. K. Kumaravel *et al.*, "A Cu–Cu Wire-Bonding Enabled by a Cu-Selective Passivation Coating to Enhance Packaging Reliability," *IEEE Transactions on Components, Packaging and Manufacturing Technology,* vol. 13, no. 12, pp. 1923-1928, 2023.

[9] J. Kahler, N. Heuck, A. Wagner *et al.*, "Sintering of Copper Particles for Die Attach," *IEEE Transactions on Components, Packaging and Manufacturing Technology,* vol. 2, no. 10, pp. 1587-1591, 2012.

Advanced Packaging for High-voltage SiC Module with Excellent Thermal, Mechanical, and Electrical Properties

Meiyu Wang [a, b], Yiting Han [a], Peng Gao [a], Zhuo Pang [a], Yingkun Yang [c, d*], Haidong Yan [e*]

[a] College of Electronic Information and Optical Engineering, Nankai University, Tianjin 300350, China
[b] Shenzhen Research Institute of Nankai University, Shenzhen 518083, China
[c] Microsystem and Terahertz Research Center, China Academy of Engineering Physics, Chengdu 610200, China
[d] Institute of Electronic Engineering, China Academy of Engineering Physics, Mianyang 621999, China
[e] College of Electrical Engineering, Zhejiang University, Hangzhou 311200, China
[*] Corresponding author: yangyingkun_mtrc@caep.cn, haidong_yan@zju.edu.cn

Abstract—To address the critical challenges of thermal management, thermo-mechanical stress concentration, and electric field distortion in high-voltage SiC power modules, this study proposes a double-sided packaging architecture incorporating sintered-silver interconnects and through-ceramic-vias (TCV). Leveraging multi-physics simulations using Ansys Workbench and COMSOL Multiphysics software, the proposed design demonstrates superior performance across thermal, mechanical, and electrical domains. Compared to conventional single-sided modules, the TCV-enhanced architecture achieves a 34°C reduction in junction temperature from 198.12°C to 164.09°C at a power dissipation of 100 W, 50%, 66.6%, and 71.2% lower von-Mises stress in terms of overall package, SiC die, and sintered-silver interconnects, respectively. Furthermore, electric field simulations reveal a 20.6% reduction from 70.3 kV/mm to 55.8 kV/mm in peak field intensity at triple points when applying a DC voltage of 10 kV , attributed to optimized geometric symmetry, TCV-based current redistribution, and thermal-electrical coupling effects. These advancements validate the architecture's capability to simultaneously enhance power density, reliability, and insulation performance for 10 kV-class SiC modules, paving the way for next-generation high-voltage power electronics.

Keywords—Power Module Packaging, High Voltage, Multiphysics Simulation, Thermal Property, Thermo-Mechanical Stress, Electric Field.

I. INTRODUCTION

Compared to conventional silicon (Si), wide-bandgap semiconductors such as silicon carbide (SiC) exhibit superior performance advantages, primarily manifested in their larger bandgap width, higher critical breakdown field strength, elevated saturation drift velocity, and enhanced thermal conductivity [1, 2]. These intrinsic properties enable SiC power devices to demonstrate exceptional capabilities, including high-temperature and high-voltage tolerance, reduced switching losses, and low on-resistance [3, 4]. Consequently, SiC-based devices have been progressively implemented in high-voltage and high-power applications, showcasing unique technological merits and promising development prospects. However, significant challenges remain in the integration and large-scale deployment of high-voltage SiC power modules (≥10 kV). The primary limitations include: (1) existing high-voltage SiC modules predominantly adopt packaging materials, structures, and processes originally developed for low-voltage Si-based devices. This approach induces electric field concentration at triple points, such as ceramic-copper-encapsulant interfaces, leading to partial discharge breakdown and localized dielectric degradation, thereby severely limiting voltage endurance; (2) conventional

single-sided cooling modules with wire-bonded interconnects exhibit inadequate thermal dissipation efficiency, failing to mitigate heat accumulation under high power density; (3) while double-sided cooling architectures improve thermal management, coefficient of thermal expansion (CTE) mismatch-induced thermo-mechanical stress concentration at material interfaces compromises module reliability and operational lifespan.

To address the inherent limitations of current SiC packaging technologies in terms of high-temperature reliability, manufacturing complexity, and cost-effectiveness, the research community has achieved significant progress through coordinated multi-dimensional innovations encompassing structural optimization such as 3D stacking [5] and dual-side cooling [6], and material advancements such as silver sintering [6] and copper interconnects [7]. A representative example is Tesla's Model Y, which implements TPAK-packaged SiC modules with dual-side cooling architecture and sintered silver interconnects. This configuration demonstrates a 30% improvement in power density and extended driving range when supporting 1 kV high-voltage platforms, thereby validating the effectiveness of structure-material co-design strategies in high-voltage applications [8]. However, as device voltage ratings advance beyond 10 kV, conventional packaging solutions increasingly exhibit critical challenges including localized electric field distortion, cumulative thermomechanical stress, and insulation performance limitations.

Consequently, this study proposes a double-sided interconnection architecture with through-ceramic-vias (TCV) in the upper direct-plated-copper (DPC) to enhance thermal dissipation efficiency and mitigate thermos-mechanical stress. Multiphysics simulations encompassing thermal dissipation, thermo-mechanical stress, and electric field analyses were conducted using ANSYS Workbench and COMSOL Multiphysics software, providing advanced packaging solutions for 10kV high-voltage SiC diode module with excellent thermal, thermo-mechanical, and insulation performance.

II. POWER MODULE PACKAGING ARCHITECTURE DESIGN AND METERIAL DESIGN

This study proposed two SiC diode power module architectures with distinct structural configurations, as shown in the cross-sectional schematics of Fig. 1. Module A features a double-sided interconnection design, integrating symmetrically aligned copper (Cu) spacers adjacent to the SiC die and a precisely arranged array of Cu-plated TCV in the upper DPC substrate. In contrast, module B adopts a single-sided interconnection approach, utilizing two-tiered Cu spacers of varying heights to maintain equivalent module dimensions. To ensure high-voltage insulation reliability, 0.25 mm radius fillets were implemented at the corner regions of both upper and lower DPC copper layers. To enhance the thermal management, highly thermal conductive sintered-silver was used as interconnection material and AlN ceramic was used as the insulation material in the DPC substrates, respectively [9]. Epoxy resin of Henkel FP4531 was used as the encapsulant material. This insulation material combination of AlN and epoxy leverages superior electrical insulation properties—including high dielectric strength (>20 kV/mm) and volume resistivity (>1×10^{15} Ω·cm)—to effectively suppress arcing and electromagnetic interference, thereby ensuring stable operation of SiC power modules under high-voltage conditions.

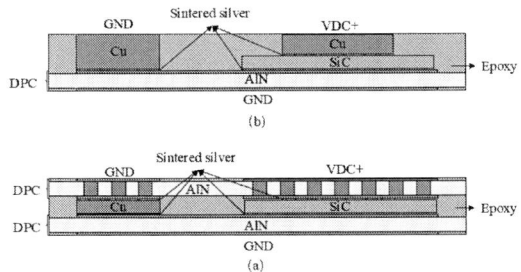

Fig. 1 Schematic cross-sectional schematics of (a) module A and (b) module B.

III. THERMAL, MECHANICAL, AND ELECTRICAL SIMULATION ANALYSIS

Prior to conducting the Multiphysics simulations including the thermal, mechanical, and electrical analyses, the thermophysical, mechanical, and dielectric properties of constituent materials were rigorously characterized. Table I lists the material parameters used in the Multiphysics simulations. It is worth noting that the Young's modulus of sintered-silver depends greatly on the temperature. The measured Young's modulus profile of sintered silver versus temperature is shown in Fig. 2. The Anand model was used

for describing the viscoelasticity of sintered-silver. Table II lists the parameters of the Anand Model for sintered-silver.

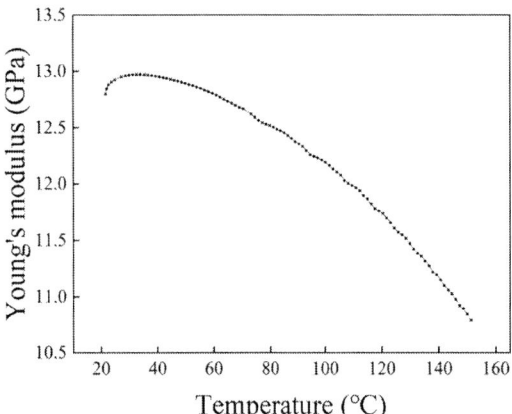

Fig. 2 Mesured Young's modulus profile of sintered silver versus temperature.

A. Thermal dissipation simulation

In power electronics applications, excessive junction temperatures may induce device degradation through thermal runaway mechanisms, thus necessitating optimized thermal management to maintain operational temperatures below critical thresholds. Thermal dissipation simulations of the two modules were conducted using Ansys Workbench. This study employs an advanced material configuration comprising an AlN ceramic substrate, sintered silver interconnection layers, epoxy encapsulant, and Cu spacer layers, with precisely defined thermophysical parameters including thermal conductivity and CTE for each constituent material. A hybrid meshing strategy was implemented to optimize computational accuracy-efficiency balance. Specifically, conventional free tetrahedral elements were applied to regular geometric regions such as AlN substrates and Cu layers, while unstructured refined tetrahedral meshing was specifically adopted for complex geometries such as TCV arrays and die peripheries. Emphasis was placed on generating boundary-layer meshes with elevated nodal density at critical interfaces between die and sintered-silver to precisely resolve interfacial thermal resistance characteristics.

TABLE I. Material parameters used in the Multiphysics simulations.

Material	Thermal Conductivity (W/m·K)	Young's Modulus (GPa)	Poisson's ratio	Coefficient Thermal Expansion ($\times 10^{-6}$ /K)	Electrical Conductivity (S/m)	Relative dielectric constant
AlN Ceramic	120	320	0.22	4.6	1.0×10^{-10}	9
Cu (Spacer)	400	128	0.34	16.5	5.81×10^{7}	1
Sintered-silver	175	Fig. 2	0.38	19	2.05×10^{7}	6.9
Epoxy (Henkel FP4531)	0.75	37.8	0.38	56	5.88×10^{-15}	3.29
SiC die	370	410	0.45	4.5	1.0×10^{-15}	9.7

Table II. Parameters of the Anand model for sintered-silver [10]

Saturation stress S_0 (S_0MPa)	Activation energy Q/R (Q/RK)	Pre-exponential factor A (s^{-1})	Hardening/softening coefficient ξ	Strain rate sensitivity exponent m	Initial hardening modulus h_0 (MPa)	Modified Saturation Stress $\hat{s}$ ($\hat{s}$MPa)	Stress exponent n	Strain rate sensitivity of hardening a
2.768	5709	9.81	11	0.6572	15800	67.389	0.00326	1

For thermal analysis, as shown in Fig. 3, a power dissipation of 100 W was applied to the SiC die, and a heat convection coefficient of 5000 W/m²·K was imposed on the bottom substrate surfaces. Fig. 4 shows the temperature distribution of the two modules. Module A achieves a junction temperature of 164.09°C, demonstrating about 34°C reduction compared to that of module B (198.12°C). This significant thermal performance enhancement (17.2%) validates the superior heat dissipation capability of the proposed double-sided architecture interconnected by TCV.

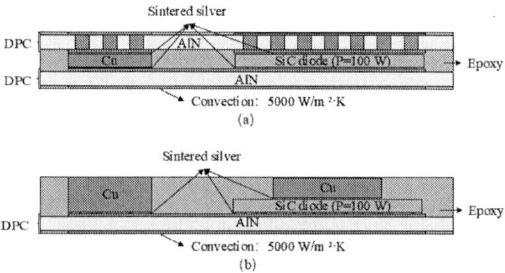

Fig. 3. Load setting in the thermal simulations of (a) module A and (b) module B.

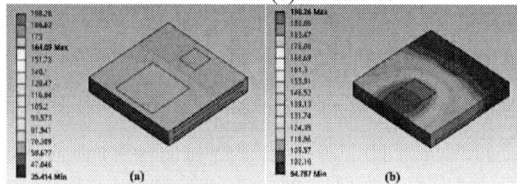

Fig. 4 Simulation results of temperature distributions in (a) module A and (b) module B.

B. Thermo-mechanical stress simulation

Thermal induces volumetric expansion or contraction in packaging materials, generating thermo-mechanical stresses due to CTE mismatch across material layers. Excessive thermo-mechanical stresses may exceed the ultimate strength thresholds of materials and interconnect interfaces, potentially initiating crack propagation or catastrophic failure. The von Mises equivalent stress serves as a critical metric for evaluating material yielding susceptibility. In power module design, minimizing stress under high-voltage operational loads is essential to prevent irreversible plastic deformation. Reduced stress correlates with enhanced resistance to fatigue damage and localized yielding, thereby improving structural reliability and operational lifespan.

In thermo-mechanical stress analysis, unlike conventional thermal simulations that primarily focus on the independent aspects of structural design, material property definition, and meshing strategies, coupled thermo-stress analysis necessitates bidirectional interactions between thermal and structural fields through Multiphysics coordination. Specifically, in the thermal stress simulation of the two modules, the temperature load is the same as that in the temperature simulation of section III.A. That is, a power dissipation of 100 W was applied to the SiC die, and a heat convection coefficient of 5000 W/m² · K was

imposed on the bottom substrate surfaces. Furthermore, the bottom substrate surfaces were set as fixed surfaces, i.e., zero displacement.

The comparative thermo-mechanical stress analysis of the two modules at working conditions were simulated. As illustrated in Figs. 5~7 and quantified in Table III, the two modules demonstrate significant stress performance differences. Specifically, module A exhibits a 50% (806.43 MPa) lower overall stress compared to Module B. For the SiC die, module A achieves a stress of 216.32 MPa, representing a 66.6% decrease from that of module B (647.48 MPa). For the sintered-silver interconnects, module A shows a remarkable stress reduction to 38.89 MPa, which is 71.2% lower than that of Module B (134.96 MPa). Overall, a significant decrease in stress was obtained in module A by utilizing the double-sided DBC interconnection architecture.

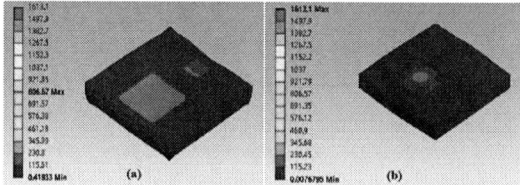

Fig.5 Overall von-Mises stress simulation results of (a) module A and (b) module B.

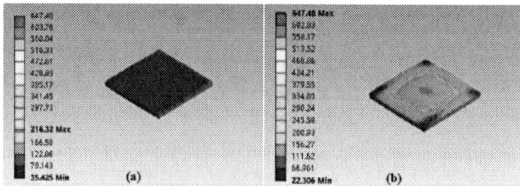

Fig.6 Von-Mises stress simulation results of SiC diode in (a) module A and (b) module B.

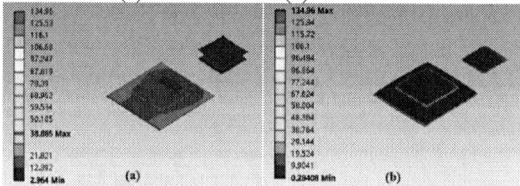

Fig. 7 Von-Mises stress simulation results of sintered-silver layer in (a) module A and (b) module B.

TABLE III. Stress comparison of critical components in module A and module B

Von Mises stress (MPa)	Module A	Module B	Reduction
Overall	806.67	1613.10	50.0%
SiC die	216.32	647.48	66.6%
Sintered-silver	38.89	134.96	71.2%

C. Electric field simulation

In high-voltage module packaging, the spatial distribution of electric field (E-field) intensity is critically influenced by packaging geometries and material dielectric properties. Excessive E-field concentrations exceeding material breakdown thresholds may induce localized dielectric failure, compromising device reliability and operational safety. This study conducted electric field simulations using COMSOL Multiphysics software, following a systematic workflow comprising four principal phases: (1) three-dimensional (3D) geometric modeling based on actual dimensions, (2) parameterized material database configuration, (3) hybrid structured and unstructured meshing discretization, and (4) electrostatic boundary condition application with numerical solving. There are three critical implementation details worth emphasizing. First, the meshing strategy for distinct components strictly adhered to the previously established thermal-mechanical simulation protocol, thereby maintaining computational consistency. Second, pure metallic domains were systematically excluded from the electric field computation domain to eliminate spurious field distortions. Third, as illustrated in Fig. 8, the potential boundary conditions were defined by applying a DC voltage of 10 kV (VDC+) to the high-voltage terminal while maintaining all other electrical contacts at ground potential. The simulation results revealed that Module A exhibited a maximum electric field intensity of 55.8 kV/mm at the triple junction of die-epoxy-silver interface, representing a 20.6% (14.5 kV/mm) reduction compared to the 70.3 kV/mm observed at the equivalent location in Module B.

The reason leading to the superior E-field performance of Module A relative to Module B stems from two critical design innovations. Firstly, the structural optimization including the TCVs integration and symmetrical architecture suppress field distortion. Module A employs TCVs and a symmetrical dual-side Cu spacer configuration as depicted in Fig. 1(a), which establish 3D current path dispersion to redistribute E-field lines. The TCV structure vertically channels current from the die's upper surface to the underlying substrate, thereby eliminating asymmetric current path-induced field convergence inherent in Module B's single-side design as depicted in Fig. 1(b). Furthermore, the optimized TCVs array arrangement combined with 0.25 mm edge fillets on Cu layers effectively mitigate E-field intensification caused by geometric discontinuities.

Secondly, the gradient modulation reduces the partial discharge and dielectric breakdown susceptibility. Multiphysics simulations conducted in COMSOL reveal distinct E-field characteristics between the modules. As shown in Fig. 8(a), module A exhibits radially symmetric field distribution around TCV regions despite localized high-field zones (>50 kV/mm), accompanied by gradient E-field transitions. This behavior arises from the TCV's 3D current redistribution mechanism — the vertically interconnected Cu-plated vias enable efficient current transfer from the chip's upper surface to the substrate, circumventing edge field distortion caused by unidirectional current convergence in conventional single-side packaging as demonstrated in Fig. 8(b). In contrast, Module B's single-layer interconnection concentrates extreme field intensities of 70.3 kV/mm at die-edge transition zones, where abrupt geometric features such as right-angle edges and material interface discontinuities collectively induce pronounced field line distortion.

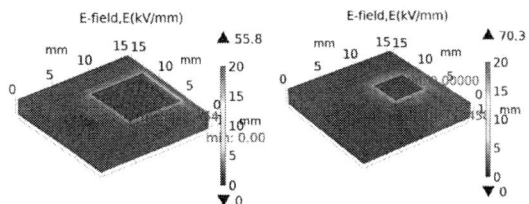

Fig. 8 Electric field simulation results under high voltage of 10 kV in (a) module A and (b) module B.

IV. CONCLUSION

High-voltage SiC power modules face intrinsic challenges in thermal management, thermo-mechanical stress, and electric field uniformity. This study improves these limitations through a double-sided DPC architecture integrated with sintered-silver interconnects and TCV. Compared with the single-sided module, the double-sided module lowers the junction temperature by 34°C from 198.12°C to 164.09°C at a power dissipation of 100 W, enhancing material stability and operational lifespan under high power density. Simultaneously, by leveraging symmetric TCV layouts and geometric optimization, the proposed architecture reduces von-Mises stress by 50.0–71.2% across critical components, including the SiC die and sintered-silver interconnects, effectively mitigating interfacial delamination and fatigue failure risks. When applying a DC voltage of 10 kV, the electric field simulations further validate a 20.6% reduction in peak field intensity at triple points (55.8 kV/mm vs. 70.3 kV/mm), attributed to TVC-driven current redistribution and thermal-electrical coupling effects that suppress dielectric degradation. By integrating structural symmetry and multi-physics co-design, this work provides a scalable solution for 10 kV-class SiC modules, advancing their application in electric vehicles, renewable energy systems, and industrial drives.

V. REFERENCES

[1] Z. Chen and A. Q. Huang, "High Performance SiC Power Module Based on Repackaging of Discrete SiC Devices," *IEEE Transactions on Power Electronics,* vol. 38, no. 8, pp. 9306-9310, 2023.

[2] L. Zhang, X. Yuan, X. Wu, C. Shi, J. Zhang, and Y. Zhang, "Performance Evaluation of High-Power SiC MOSFET Modules in Comparison to Si IGBT Modules," *IEEE Transactions on Power Electronics,* vol. 34, no. 2, pp. 1181-1196, 2019.

[3] X. Sun *et al.*, "Design and Evaluation of a Face-Down Embedded SiC Power Module With Low Parasitic Inductance and Low Thermal Resistance," *IEEE Transactions on Power Electronics,* vol. 38, no. 3, pp. 2799-2804, 2023.

[4] R. Paul, R. Alizadeh, X. Li, H. Chen, Y. Wang, and H. A. Mantooth, "A Double-Sided Cooled SiC MOSFET Power Module for EV Inverters," *IEEE Transactions on Power Electronics,* vol. 39, no. 9, pp. 11047-11059, 2024.

[5] J. Shu, J. Sun, M. Tao, Y. Du, S. W. R. Lee, and K. J. Chen, "Unlocking the Full Potential of GaN/SiC Cascode Device With 3D Co-

Packaging and Enhanced dv/dt Control Capability," *IEEE Transactions on Power Electronics,* vol. 40, no. 5, pp. 6874-6882, 2025.

[6] Y. Yan *et al.*, "A Novel Double-Sided Cooling Silicon Carbide Power Module With Ultralow Parasitic Inductance Based on an Interleaved Power Loop," *IEEE Transactions on Power Electronics,* vol. 39, no. 10, pp. 12570-12588, 2024.

[7] L. Wu *et al.*, "Low-Temperature Sintering of Cu/Functionalized Multiwalled Carbon Nanotubes Composite Paste for Power Electronic Packaging," *IEEE Transactions on Power Electronics,* vol. 37, no. 2, pp. 1234-1243, 2022.

[8] A. K. Morya *et al.*, "Wide Bandgap Devices in AC Electric Drives: Opportunities and Challenges," *IEEE Transactions on Transportation Electrification,* vol. 5, no. 1, pp. 3-20, 2019.

[9] C. Kuring *et al.*, "GaN-Based Multichip Half-Bridge Power Module Integrated on High-Voltage AlN Ceramic Substrate," *IEEE Transactions on Power Electronics,* vol. 37, no. 10, pp. 11896-11910, 2022.

[10] D. J. Yu, X. Chen, G. Chen, G. Q. Lu, Z. Q. J. M. Wang, and Design, "Applying Anand model to low-temperature sintered nanoscale silver paste chip attachment," vol. 30, no. 10, pp. 4574-4579, 2009.

2025 IEEE Workshop on Wide Bandgap Power Devices and Applications in Asia (WiPDA Asia)

A One-Inductor Two-Switch Three-Port DC-DC Converter for PV-Battery Systems

Yidi Liang
School of Electrical Engineering
Beijing Jiaotong University
Beijing, China
23111448@bjtu.edu.cn

Hong Li*
College of Electrical Engineering
Zhejiang University
Hangzhou, China
hong_li@zju.edu.cn

Huizhu Zhuang
School of Electrical Engineering
Beijing Jiaotong University
Beijing, China
24126364@bjtu.edu.cn

Peng Sun
School of Electrical Engineering
Beijing Jiaotong University
Beijing, China
23126350@bjtu.edu.cn

Xu Shangguan
School of Electrical Engineering
Beijing Jiaotong University
Beijing, China
22110467@bjtu.edu.cn

Abstract—**The increasing demand for photovoltaic battery systems with low cost, high reliability, and high efficiency drives the requirement for efficient and compact three-port converters (TPCs), necessitating simpler topologies and a reduced number of components. In this paper, a new TPC topology based on one inductor and two switches was proposed. Only one inductor, two switches, and two diodes are required by this topology, resulting in a reduction in the number of components from 7 to 5, corresponding to a reduction of 29%, thereby decreasing the system size and cost. The operating modes and states of the proposed topology were analyzed theoretically, and its feasibility was verified through simulations and experiments. Experimental results demonstrated that the efficiency of the topology reached 89%, 95.3%, and 91% under SIDO, DISO, and SISO operating states, respectively. Compared with the existing TPCs, the proposed topology was found to possess advantages of a simpler structure and a reduced number of components, making it suitable for applications where common ground is not required and strict cost limitations are imposed.**

Keywords—*photovoltaic battery systems, three-port converter, one inductor, two switches, topology*

I. INTRODUCTION

With the rapid development of renewable energy technologies [1], the demands for high power density [2], low cost [3], and high reliability [4] in photovoltaic (PV)-battery systems have steadily increased. Significant attention has been drawn to Three-Port Converters (TPCs) due to their capability to integrally manage a PV array, an energy storage battery, and the DC bus. Through the reduction in the number of components and interconnection lines, both energy conversion efficiency enhancement and significant reduction in system complexity and overall cost can be achieved in TPCs [5-6], as illustrated in Fig. 1. By this integrated architecture, efficient coordinated management of PV generation and battery energy storage is enabled, thereby fully meeting the three core requirements of high efficiency, high reliability, and low cost for PV-battery systems.

At present, research on TPC topology has been focused on three aspects: combinations of basic DC-DC converters, topological structure optimization, and new topology designs. In Ref. [7], a boost-sepic TPC topology comprising 9 components was proposed. Although the basic power

conversion function was achieved by the topology, a large number of components resulted in a bulky system and elevated cost. In Ref. [8], a boost TPC comprising 7 components (6 of which are switching components) was proposed by improving the boost converter. Although the topological structure was optimized, the number of switching components remained relatively high. To reduce the number of switching components, a PWM TPC was proposed in Ref. [9]. Although the total number of components was still 7, the number of switching components was successfully reduced to 4, thereby effectively reducing the complexity and cost of the system. However, these topologies still suffer from excessive component counts, high system complexity, and the associated cost increases, which limit their large-scale deployment in distributed PV-battery or residential scenarios.

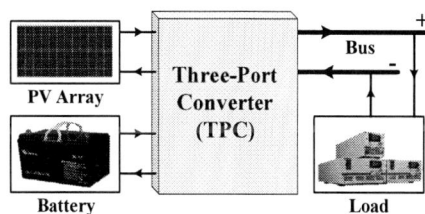

Fig. 1. Integrated management of PV array, battery, and bus based on TPC.

Therefore, a new TPC based on one inductor and two switches is proposed. System size and cost are reduced while buck-boost functions are maintained by simplifying the number of components to only 1 inductor, 2 switches, and 2 diodes. Furthermore, adaptability to various operating states is provided through a flexible power allocation mechanism, thereby enabling the energy management requirements of PV-battery systems under different operating conditions to be met.

II. OPERATION PRINCIPLE

For the PV-battery system application (60V PV input, 24V battery, and 100V DC bus) [10-11], a new TPC based on one inductor and two switches was proposed, named 123C, as illustrated in Fig. 2. The proposed topology is composed of 3 functional ports: the PV input port (V_{PV}), the battery port (V_B), and the DC bus output port, which consists of a filter capacitor (C_O) and a load resistor (R_O). The topology is constituted by 2 switches (S_1, S_2), 2 diodes (D_1, D_2), and 1 inductor (L). Energy distribution among the PV input, battery, and DC bus is effectively enabled by this compact structure.

---

* Corresponding Author, E-mail: Hong Li, hong_li@zju.edu.cn
* Supported in part by the National Science Fund for Distinguished Young Scholars 52325704, in part by the Key Program of National Natural Science Foundation of China 52237008.

979-8-3315-1110-4/25 $31.00 © 2025 IEEE

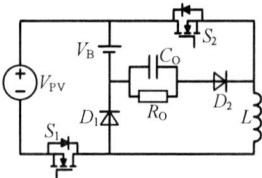

Fig. 2. One-inductor two-switch three-port converter (123C).

Based on different combinations of switching states, four operating modes are defined for the proposed 123C topology. As shown in Table I, the switching states and corresponding energy transfer paths for each mode are described as follows.

TABLE I.   FOUR OPERATING MODES OF 123C TOPOLOGY

Mode	S_1	S_2	Power flow
Mode 1	ON	ON	$V_{PV} \rightarrow L$
Mode 2	OFF	ON	$V_B \rightarrow L$
Mode 3	ON	OFF	$V_{PV}+L \rightarrow R_O+V_B$
Mode 4	OFF	OFF	$L \rightarrow R_O$

Mode 1: as shown in Fig. 3(a), S_1 and S_2 are turned on, D_1 and D_2 are reverse biased, V_{PV} supplies power to L.

Mode 2: as shown in Fig. 3(b), S_2 is turned on, S_1 are turned off, D_1 is forward biased, D_2 is reverse biased, V_B supplies power to L.

Mode 3: as shown in Fig. 3(c), S_1 is turned on, S_2 is turned off, D_1 is reverse biased, D_2 is forward biased, V_{PV} and L supply power to V_B and R_O.

Mode 4: as shown in Fig. 3(d), S_1 and S_2 are turned off, D_1 and D_2 are forward biased, L supplies power to R_O.

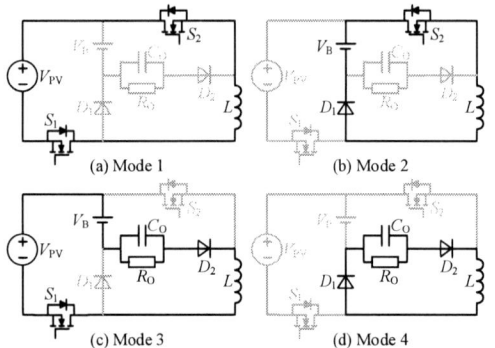

(a) Mode 1   (b) Mode 2

(c) Mode 3   (d) Mode 4

Fig. 3. Four operating modes of 123C topology.

Based on the analysis of the four operating modes of the 123C topology, three typical operating states have been identified, each corresponding to a distinct energy transfer path and power distribution strategy.

(1) Single-Input Dual-Output (SIDO) state: In this state, V_{PV} simultaneously supplies power to V_B and R_O. This state is ideal for conditions with ample sunlight and high load demand, effectively enhancing PV-battery system energy utilization.

(2) Dual-Input Single-Output (DISO) state: In this state, V_{PV} and V_B work together to supply power to R_O. This state typically occurs under low sunlight conditions or sudden load increases, where the energy storage unit provides supplementary power to ensure stable PV-battery system operation.

Single-Input Single-Output (SISO) state: In this state, V_{PV} directly supplies power to R_O. This state is suitable for scenarios with stable sunlight conditions and moderate load demand, enabling efficient and direct energy transfer.

A. SIDO state

Within a complete switching cycle, this operating state comprises three consecutive operating modes: Mode 1, Mode 3, and Mode 4. The corresponding theoretical waveforms is illustrated in Fig. 4, where v_{gs1}, v_{gs2} represent the gate drive signals for switches S_1, S_2. According to volt-second balance principle on L, (1) can be obtained, where, d_1, d_2 denote the duty cycles of S_1, S_2, T denote the switching period, V_{in}, V_b, and V_o represent the voltages across V_{PV}, V_B, and R_O, respectively. Then, (2) can be calculated from (1).

$$V_{in}d_2T + \left(V_{in} - V_b - V_o\right)\left(d_1 - d_2\right)T + \left(-V_o\right)\left(1 - d_1\right)T = 0 \quad (1)$$

$$V_o = \frac{d_1 V_{in} - (d_1 - d_2)V_b}{1 - d_2} \quad (2)$$

Fig. 4. The theoretical waveform of 123C in SIDO state.

B. DISO state

Within a complete switching cycle, this operating state comprises three consecutive operating modes: Mode 1, Mode 2, and Mode 4. The corresponding theoretical waveforms is illustrated in Fig. 5. According to volt-second balance principle on L, (3) can be obtained. Then, (4) can be calculated from (3).

$$V_{in}d_1T + V_b\left(d_2 - d_1\right)T + \left(-V_o\right)\left(1 - d_2\right)T = 0 \quad (3)$$

$$V_o = \frac{d_1 V_{in} + (d_2 - d_1)V_b}{1 - d_2} \quad (4)$$

Fig. 5. The theoretical waveform of 123C in DISO state.

C. SISO state

Within a complete switching cycle, this operating state comprises two consecutive operating modes: Mode 1, and Mode 4. The corresponding theoretical waveform is illustrated in Fig. 6. According to volt-second balance principle on L, (5) can be obtained. Then, (6) can be calculated from (5).

$$V_{in}d_1T + \left(-V_o\right)\left(1 - d_1\right)T = 0 \quad (5)$$

$$V_o = \frac{d_1 V_{in}}{1 - d_1} \quad (6)$$

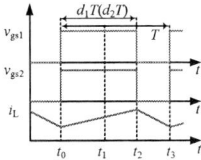

Fig. 6. The theoretical waveform of 123C in SISO state.

Based on the above three operating states, the voltage and current stresses endured by each switching components can be quickly obtained, as shown in Table II.

TABLE II. COMPONENTS STRESS

Component	Voltage stress	Current stress
S_1	$V_{in}-V_b$	I_o
S_2	V_o+V_b	I_o
D_1	$V_{in}-V_b$	I_o
D_2	V_o+V_b	I_o

III. SIMULATION AND EXPERIMENTAL VERIFICATION

To verify the feasibility of the 123C topology, based on the parameters shown in Table III, a simulation and experimental platform with a power level of 100 W (the average power level is less than 100 W, and the instantaneous peak power can reach 100 W) was built.

TABLE III. SIMULATION AND EXPERIMENTAL PARAMETERS

Parameter	Value
Input port voltage V_{in} (V)	60
Energy storage port voltage V_b (V)	24
Output port voltage V_o (V)	100
Inductor L (mH)	1
Switching frequency f_S (kHz)	100
Output load R_O (Ω)	300
Output capacitor C_O (μF)	10
Switch S_1, S_2	IMZA65R027M1H
Diode D_1, D_2	IDP08E65D1

The simulated waveform of the 123C topology at different operating states is shown in Fig. 7 to Fig. 9. i_L is the current of the L, $V_{S1}, V_{S2}, V_{D1}, V_{D2}$ are the voltage stresses of S_1, S_2, D_1, D_2, respectively. Fig. 7 is the waveform of the 123C work in the SIDO state. In this state, V_{PV} simultaneously supplies power to V_B and R_O. Fig. 8 is the waveform of the 123C work in the DISO state. In this state, V_{PV} and V_B work together to supply power to R_O. Fig. 9 is the waveform of the 123C work in the DISO state. In this state, V_{PV} directly supplies power to R_O.

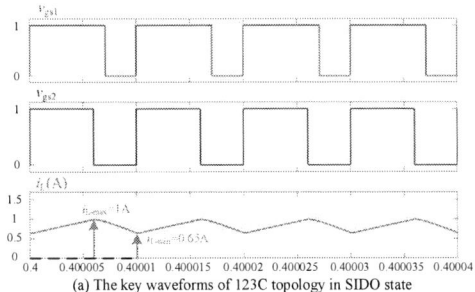

(a) The key waveforms of 123C topology in SIDO state

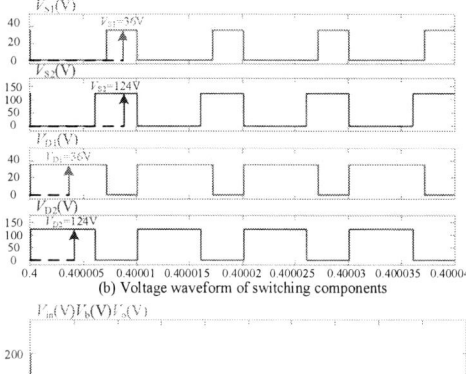

(b) Voltage waveform of switching components

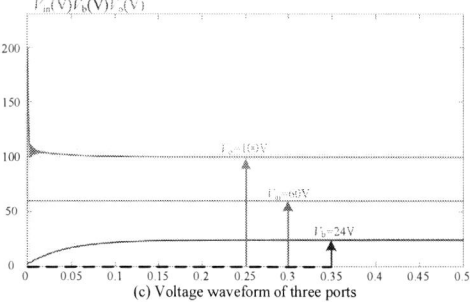

(c) Voltage waveform of three ports

Fig. 7. The simulated waveforms of 123C in SIDO state.

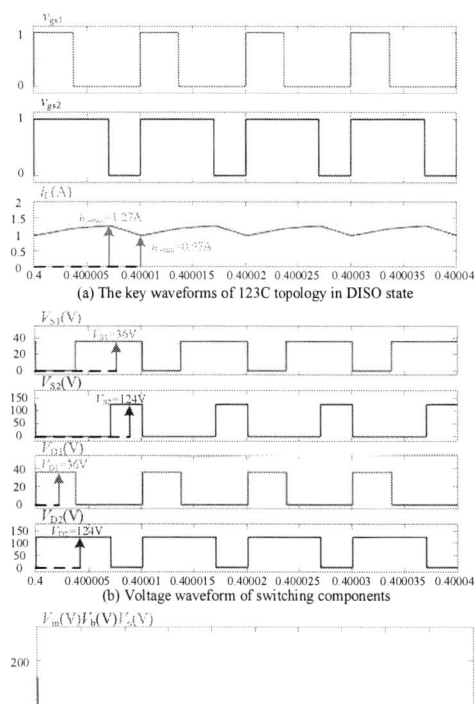

(a) The key waveforms of 123C topology in DISO state

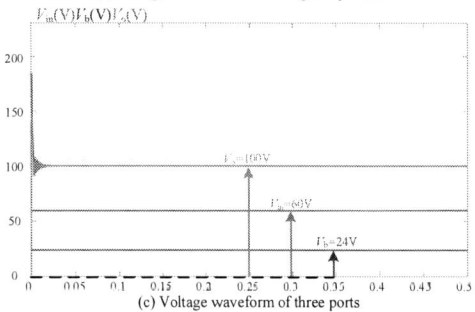

(b) Voltage waveform of switching components

(c) Voltage waveform of three ports

Fig. 8. The simulated waveforms of 123C in DISO state.

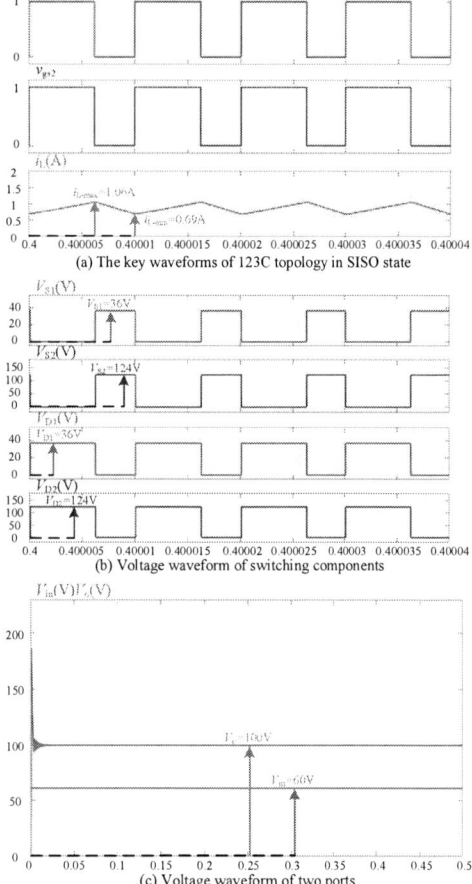

(a) The key waveforms of 123C topology in SISO state

(b) Voltage waveform of switching components

(c) Voltage waveform of two ports

Fig. 9. The simulated waveforms of 123C in SISO state.

The experimental platform is shown in Fig. 10, and the corresponding experimental results are shown in Fig. 11 to Fig. 13. The voltage stresses of switches S_1 and S_2 and diodes D_1 and D_2 are 36 V, 124 V, 36 V, and 124 V respectively, which is consistent with the theoretical analysis in Table II. The efficiencies of the converter under the three operating states of SIDO, DISO and SISO are 89%, 95.3% and 91% respectively. The theoretical analysis, simulation and experimental results are consistent, verifying the feasibility of the 123C topology and indicating that it can operate stably and meet the requirements of the PV-battery system.

(a)123C

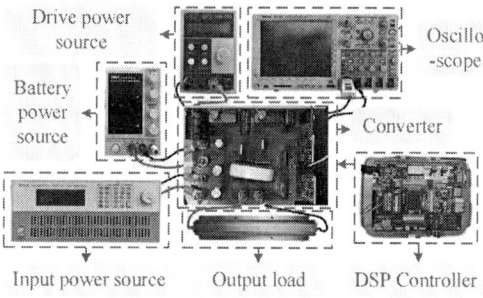

(b)Schematic of the experimental platform

Fig. 10. Schematic of the 123C and experimental platform.

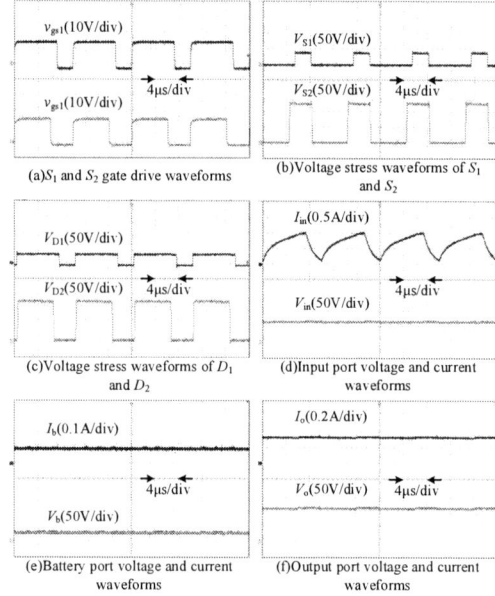

(a)S_1 and S_2 gate drive waveforms

(b)Voltage stress waveforms of S_1 and S_2

(c)Voltage stress waveforms of D_1 and D_2

(d)Input port voltage and current waveforms

(e)Battery port voltage and current waveforms

(f)Output port voltage and current waveforms

Fig. 11. The experimental waveforms of 123C in SIDO state.

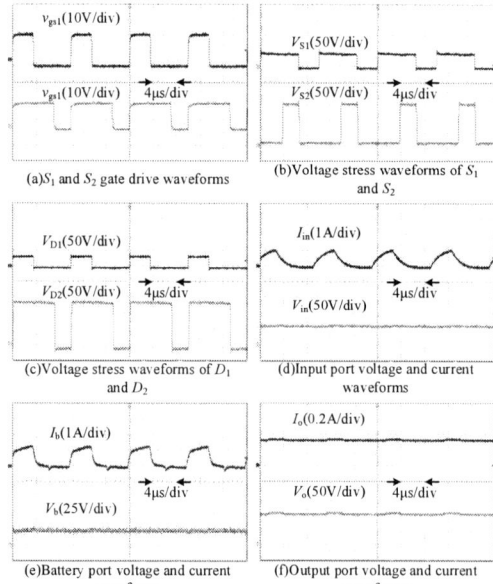

(a)S_1 and S_2 gate drive waveforms

(b)Voltage stress waveforms of S_1 and S_2

(c)Voltage stress waveforms of D_1 and D_2

(d)Input port voltage and current waveforms

(e)Battery port voltage and current waveforms

(f)Output port voltage and current waveforms

Fig. 12. The experimental waveforms of 123C in DISO state.

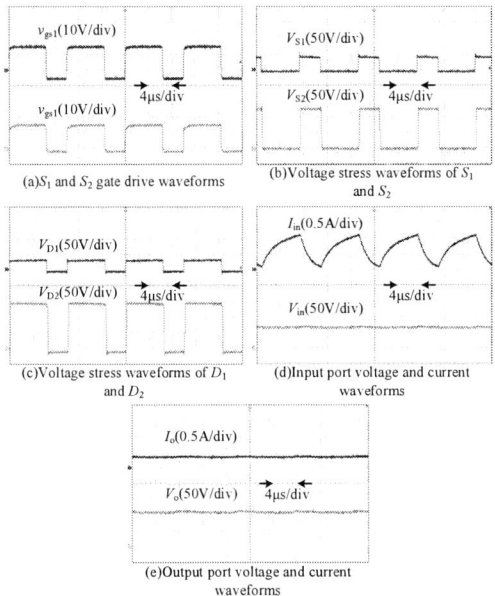

(a) S_1 and S_2 gate drive waveforms

(b) Voltage stress waveforms of S_1 and S_2

(c) Voltage stress waveforms of D_1 and D_2

(d) Input port voltage and current waveforms

(e) Output port voltage and current waveforms

Fig. 13. The experimental waveforms of 123C in SISO state.

As shown in Table IV, compared to existing TPC topologies [5, 12-16], the topology proposed in this paper reduces the number of components from 7 to 5, a 29% reduction, while offering high reliability and high power density. The voltage stress on the switching components varies little across the topologies, but 123C uses a non-common-ground configuration, rendering it suitable only for cost sensitive applications such as PV-battery system where common grounding is not strictly required.

TABLE IV. PERFORMANCE COMPARISON OF THE PROPOSED CONVERTER WITH OTHER CONVERTERS

Parameter	Converter						
	[5]	[12]	[13]	[14]	[15]	[16]	123C
No. of S/D	3/3	3/1	4/2	3/3	2/2	3/3	2/2
No. of L/C	1/0	2/1	3/1	1/0	2/1	1/0	1/0
Voltage stress of switching components	V_o	V_o	V_o	V_{in}	$V_o \mid V_b$	V_o	$V_o + V_b$
Common ground	YES	NO	YES	NO	NO	YES	NO

IV. CONCLUSIONS

In this paper, a new TPC based on one inductor and two switches, named 123C, was proposed. Efficient energy management among the PV input port, battery port, and output port was achieved by this topology through a structure comprising 1 inductor, 2 switches, and 2 diodes. Compared to existing TPCs, the number of components in 123C was reduced from 7 to 5, corresponding to a reduction of 29%, thus improving power density and system reliability while effectively reducing overall cost. However, as a non-common-ground configuration was utilized by 123C, this topology is only suitable for applications where common grounding is not required and strict cost constraints exist.

REFERENCES

[1] A. Ashoornezhad, Q. Asadi, R. Saberi and H. Falaghi, "Considering the Impact of PV-Battery Systems on Long-Term Distribution Network Planning," 2023 27th International Electrical Power Distribution Networks Conference (EPDC), Mashhad, Iran, Islamic Republic of, 2023, pp. 161-166.

[2] N. Sivasankar and K. Devabalaji, "Smart Multiport Bidirectional Non-Isolated DC-DC Converter for Solar PV-Battery Systems," 2022 First International Conference on Electrical, Electronics, Information and Communication Technologies (ICEEICT), Trichy, India, 2022, pp. 1-8.

[3] F. Härtel and T. Bocklisch, "Minimizing Energy Cost in PV Battery Storage Systems Using Reinforcement Learning," in IEEE Access, vol. 11, pp. 39855-39865, 2023.

[4] M. Deckers, L. Van Cappellen, G. Emmers, F. Poormohammadi and J. Driesen, "Cost Comparison for Different PV-Battery System Architectures Including Power Converter Reliability," 2022 24th European Conference on Power Electronics and Applications (EPE'22 ECCE Europe), Hanover, Germany, 2022, pp. 1-11.

[5] T. Cheng, D. D. C. Lu, A. Gong and D. Verstraete, "Analysis of a three-port DC-DC converter for PV-battery system using DISO boost and SISO buck converters," 2015 Australasian Universities Power Engineering Conference (AUPEC), Wollongong, NSW, Australia, 2015, pp. 1-6.

[6] C. Yin, H. Li, Y. Li, W. Su and T. Q. Zheng, "A New Family of Non-Isolated Single-Inductor Three-Port Converter Based on A Storage Port Switch-Commutated Unit," 2022 International Power Electronics Conference (IPEC-Himeji 2022- ECCE Asia), Himeji, Japan, 2022, pp. 2307-2311.

[7] A. Ruhela and K. A. Chinmaya, "A Novel Boost-SEPIC based Three-Port DC-DC Converter for Solar PV Integrated E-Boat Applications," 2022 International Conference on Smart Energy Systems and Technologies (SEST), Eindhoven, Netherlands, 2022, pp. 1-6.

[8] Z. Wei and C. Tang, "A Non-Isolated Three-Port DC-DC Converter," 2024 IEEE 7th Advanced Information Technology, Electronic and Automation Control Conference (IAEAC), Chongqing, China, 2024, pp. 1369-1374.

[9] S. S. Nair and M. Rajeev, "A Novel High Gain Non-Isolated Three-port Converter for Stand-Alone PV Applications," 2023 International Conference on Computer, Electronics & Electrical Engineering & their Applications (IC2E3), Srinagar Garhwal, India, 2023, pp. 1-6.

[10] H. Khoramikia, M. Heydari and S. M. Dehghan, "A New Three-Port Non-Isolated DC-DC Converter for Renewable Energy Sources Application," Electrical Engineering (ICEE), Iranian Conference on, Mashhad, Iran, 2018, pp. 1101-1106.

[11] H. Zhu, D. Zhang, B. Zhang and Z. Zhou, "A Nonisolated Three-Port DC-DC Converter and Three-Domain Control Method for PV-Battery Power Systems," in IEEE Transactions on Industrial Electronics, vol. 62, no. 8, pp. 4937-4947, Aug. 2015.

[12] H. Nagata and M. Uno, "Nonisolated PWM Three-Port Converter Realizing Reduced Circuit Volume for Satellite Electrical Power Systems," in IEEE Transactions on Aerospace and Electronic Systems, vol. 56, no. 5, pp. 3394-3408, Oct. 2020.

[13] H. Liu, D. Zhang, L. Qu and Q. Tong, "A Non-isolated Three-Port Converter Interfacing Composite Bus in Power Conditioning Unit for Deep Space Detector," 2022 IEEE 5th International Conference on Electronics Technology (ICET), Chengdu, China, 2022, pp. 220-226.

[14] P. Zhang, Y. Chen and Y. Kang, "Nonisolated Wide Operation Range Three-Port Converters With Variable Structures," in IEEE Journal of Emerging and Selected Topics in Power Electronics, vol. 5, no. 2, pp. 854-869, June 2017.

[15] Q. Tian, G. Zhou, R. Liu, X. Zhang and M. Leng, "Topology Synthesis of a Family of Integrated Three-port Converters for Renewable Energy System Applications," in IEEE Transactions on Industrial Electronics, vol. 68, no. 7, pp. 5833-5846, July 2021.

[16] H. Wu, Y. Xing, Y. Xia and K. Sun, "A family of non-isolated three-port converters for stand-alone renewable power system," IECON 2011 - 37th Annual Conference of the IEEE Industrial Electronics Society, Melbourne, VIC, Australia, 2011, pp. 1030-1035.

A Parameter Design Method for Bidirectional LLC-C Resonant Converter for Bidirectional On-Board EV Charger Application

Yiheng Zhang
College of Electrical and Control Engineering
North China University of Technology
Beijing, China
2022312080114@mail.ncut.edu.cn

Pengyu Jia
College of Electrical and Control Engineering
North China University of Technology
Beijing, China
jiapengyu@ncut.edu.cn

Mingjun Liu
College of Electrical and Control Engineering
North China University of Technology
Beijing, China
2023312080102@mail.ncut.edu.cn

Yimei Xing
College of Electrical and Control Engineering
North China University of Technology
Beijing, China
2023312080119@mail.ncut.edu.cn

Abstract—A detail design process is presented for a bidirectional LLC-C resonant converter which is applied in on-board charger (OBC). The switching frequency range of charge and discharge mode is very close by the proposed design method. Firstly, the fundamental harmonic approximation (FHA) and resonant network equivalence were used to rebuilt the gain model in both of directions. Based on the model, a design variable *m* represented the normalized frequency ratio in both of direction is introduced to simplify gain expression and benefit analysis. After a detail analysis for the effect of design variables, a parameter design method is presented. A nominal charge power 6.6 kW prototype is bult to verify the effectiveness and feasibility of theoretical analysis.

Keywords—Bidirectional LLC-C resonant converter, on-board charger, variable frequency control

I. INTRODUCTION

Recently, the electric vehicles (EVs) and plug-in hybrid electric vehicles (PHEVs) have become increasingly popular in global automative market. More requirements have raised for on-board charger (OBC). In vehicle-to-grid (V2G), the OBC is not only satisfy the demand of charge to battery but also is equipped with the ability of transferring power back to the grid [1]-[2]. Therefore, bidirectional power transmission is an important metric for OBC.

The typical scheme of OBC is a two-stage structure. The topology, as shown in Fig. 1, is made up of a bidirectional power factor correction (PFC) and a bidirectional DC-DC converter. In the DC-DC converter, bidirectional CLLC resonant converter or bidirectional LLC-C resonant converter is a good choice due to its soft-switching and bidirectional power transmission capability [3]. In order to reduce the variation range of the switching frequency with a wide voltage range, [4] proposed an asymmetric parameter design method for bidirectional CLLC resonant converter. However, it is necessary to iterate repeatedly from two directions. To reduce the complexity of the design process, the symmetrical design method is usually utilized in bidirectional CLLC resonant converter[5]-[7]. Therefore, the design process is carried out only in one power flow direction.

This work was supported in part by the Beijing Natural Science Foundation under Grant 3232044, and in part by the R&D Program of Beijing Municipal Education Commission under Grant KM202110009011.

It is clear that the number of components in a CLLC converter is more compared with the LLC-C resonant converter. For the LLC-C resonant converter, the gain model of both directions is bult and analyzed by fundamental harmonic approximation (FHA) in [8]-[9]. However, there is a lack of detailed parameter design guidelines.

In this article, FHA and a resonant network equivalence were used to rebuilt the gain model in both of direction. Then, A detailed parameter design method is presented to reduce the whole switching frequency range of the converter. The charge and discharge mode gain curves designed by the proposed method are shown in red and bule line of Fig. 2, respectively. It is clearly seen that the whole switching frequency range of converter is reduced.

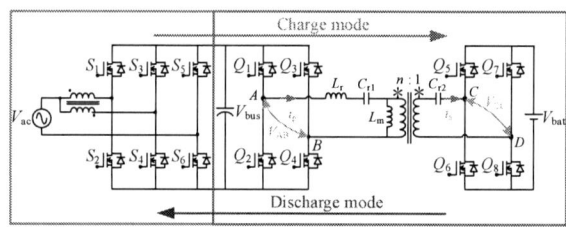

Fig. 1. Conventional two-stage on-board charger topology.

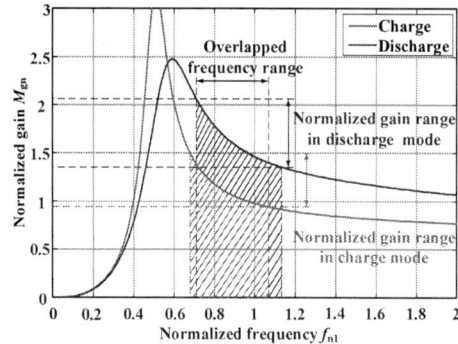

Fig. 2. The normalized gain curves of charge and discharge mode under proposed method.

II. THE FHA MODEL OF LLC-L RESONANT CONVERTER

A. Normalized gain in the charge and discharge mode

Based on FHA, the equivalent circuit in the charge mode is obtained as shown in Fig. 3. The expression of the normalized gain M_{gn1} in charge mode is given in (1).

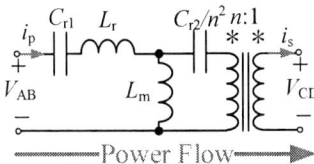

Fig. 3.   The FHA equivalent circuit in the charge mode.

$$M_{gn1} = \cfrac{1}{\sqrt{[1 + \cfrac{1}{L_{n1}}(1 - \cfrac{1}{f_{n1}^2})]^2 + Q_1^2[(f_{n1} - \cfrac{1}{f_{n1}})* \atop \sqrt{(1 - \cfrac{C_{n1}}{L_{n1}f_{n1}^2}) - \cfrac{C_{n1}}{f_{n1}}]^2}}} \quad (1)$$

where, f_{r1}, R_{01}, Q_1, L_{n1}, C_{n1} and f_{n1} respectively denote the resonant frequency, the characteristic impedance, quality factor, inductance ratio and the normalized frequency in the charge mode, defined as (2)-(8).

$$M_{gn1} = \frac{nV_{bat}}{V_{bus}}, f_{r1} = \frac{1}{2\pi\sqrt{L_r C_{r1}}}$$

$$R_{01} = \sqrt{\frac{L_r}{C_{r1}}}, \quad Q_1 = \frac{R_{01}}{8n^2 R_{L1}/\pi^2}$$

$$L_{n1} = \frac{L_m}{L_r}, \quad C_{n1} = \frac{C_{r1}}{C_{r2}/n^2} \quad (2)\text{-}(8)$$

$$f_{n1} = \frac{f_s}{f_{r1}}$$

The equivalent circuit in the discharge mode is shown in Fig. 4(a). Because the two mutually equivalent two-port networks have the same port characteristics. Rf.[10] is proved that the circuit shown in Fig. 4(a) can be completely equivalent to the circuit shown in Fig. 4(b) where the parameters relationship is given in (9)-(11). As a result, the expression of the normalized gain M_{gn2} in discharge mode is given in (12).

$$L_q = kL_q, \ L_d = kL_r$$

$$k = \frac{L_m}{L_m + L_r} = \frac{L_{n1}}{1 + L_{n1}} \quad (9)\text{-}(11)$$

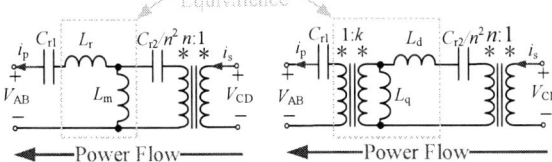

Fig. 4.   The FHA equivalent circuit in the discharge mode (a)LLC-C (b)C-T-LLC.

$$M_{gn2} = \cfrac{1}{k\sqrt{[1 + \cfrac{1}{L_{n2}}(1 - \cfrac{1}{f_{n2}^2})]^2 + Q_2^2[(f_{n2} - \cfrac{1}{f_{n2}})' \atop \sqrt{(1 - \cfrac{C_{n2}}{L_{n2}f_{n2}^2}) - \cfrac{C_{n2}}{f_{n2}}]^2}}} \quad (12)$$

where, f_{r2}, R_{02}, Q_2, L_{n2}, C_{n2} and f_{n2} respectively denote the resonant frequency, the characteristic impedance, quality factor, inductance ratio and the normalized frequency in the discharge mode, defined as (13)-(19).

$$M_{gn2} = \frac{V_{bus}}{nV_{bat}}, \ f_{r2} = \frac{1}{2\pi\sqrt{L_d C_{r2}/n^2}}$$

$$R_{02} = \sqrt{\frac{L_d}{C_{r2}/n^2}}, \quad Q_2 = \frac{R_{02}}{8k^2 R_{L2}/\pi^2} \quad (13)\text{-}(19)$$

$$L_{n2} = \frac{L_q}{L_d}, \quad C_{n2} = \frac{C_{r2}/n^2}{C_{r1}/k^2}, f_{n2} = \frac{f_s}{f_{r2}}$$

Substitute (9)-(11) into (17) and (18). The relationship between the parameters of charge and discharge mode is given in (20)-(21).

$$L_{n2} = L_{n1}$$

$$C_{n2} = \frac{L_{n1}^2}{(1 + L_{n1})^2 C_{n1}} \quad (20)\text{-}(21)$$

$$m = \frac{f_{n2}}{f_{n1}} \quad (22)$$

In order to make the switching frequency range in charging and discharging modes overlaps as well as possible, as shown in Fig. 2. By a normalized frequency ratio m given in (22), the parameter relationship between charge and discharge mode can be derived as shown in (23)-(25). Therefore, the gain curve of both modes can place in a plane to facilitate design. The new expression of normalized gain of both modes is respectively given in (26) and (27).

$$\frac{f_{r2}}{f_{r1}} = \frac{1}{m}$$

$$C_{n1} = \frac{L_{n1}}{m^2(1 + L_{n1})} \quad (23)\text{-}(25)$$

$$C_{n2}'' = \frac{m^2 L_{n1}}{1 + L_{n1}}$$

$$M_{gn1} = \cfrac{1}{\sqrt{[1 + \cfrac{1}{L_{n1}}(1 - \cfrac{1}{f_{n1}^2})]^2 + Q_1^2[(f_{n1} - \cfrac{1}{f_{n1}})* \atop \sqrt{(1 - \cfrac{1}{(1+L_{n1})m^2 f_{n1}^2}) - \cfrac{L_{n1}}{(1+L_{n1})m^2 f_{n1}}]^2}}} \quad (26)$$

$$M_{gn2} = \cfrac{1}{\cfrac{L_{n1}}{1+L_{n1}}\sqrt{[1 + \cfrac{1}{L_{n1}}(1 - \cfrac{1}{m^2 f_{n1}^2})]^2 + Q_2^2[(mf_{n1} - \cfrac{1}{mf_{n1}})* \atop \sqrt{(1 - \cfrac{1}{(1+L_{n1})f_{n1}^2}) - \cfrac{mL_{n1}}{(1+L_{n1})f_{n1}}]^2}}}$$

$$(27)$$

B Analysis about the gain with respect to design variables

1)   *the effect of maximum quality factor Q_{max} for gain:*

Fig. 5 shows the cluster of normalized gain curve in charge mode under different quality factors Q_1. The cluster of gain curve in discharge mode is similar to the charge mode. It is clearly seen from Fig. 5 that the gain monotonicity is changed when the quality factor Q_1 is taken a large value. It will be unfavorable to the closed loop control system design. As a result, it is not appropriate to take too large when designing the maximum value of the quality factor Q_{max}.

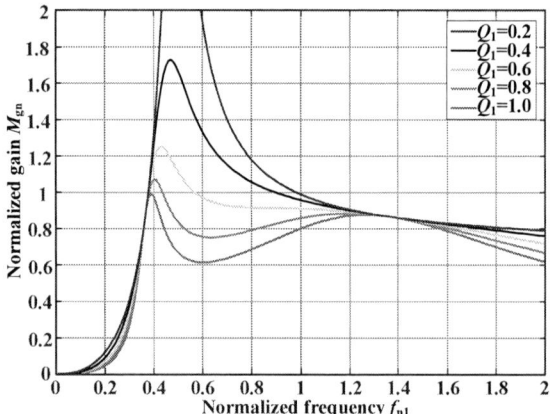

Fig. 5.   The normalized gain curves in different Q_1 with L_{n1}=3, m=1.

2)   the effect of inductance ratio L_{n1} for gain:

Fig. 6 shows a 3D plot of the charge normalized gain with respect to the inductance ratio L_{n1} and normalized frequency f_{n1}. The 3D plot of discharge gain is similar to the charge mode. The normalized frequency f_{n1} corresponding to the peak gain is far away from unity when the inductance ratio L_{n1} is taken a larger value. Then, the switching frequency is far away from the resonant frequency result in more harmonics and a low efficiency. On the other hand, the frequency variation range is wide, which is not benefit to the design of transformer.

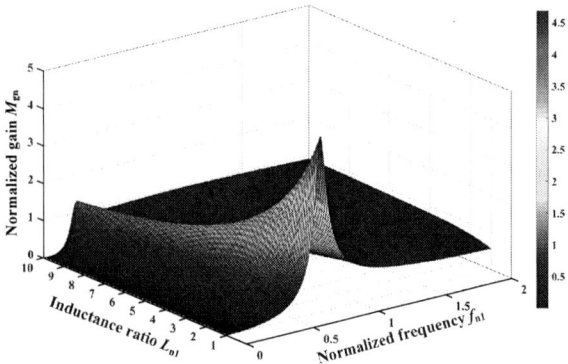

Fig. 6.   The 3D plot of M_{gn}-f_{n1}-L_{n1} with Q_1=0.3, m=1.

3)   the effect of normalized frequency ratio m for gain:

Fig. 7 shows the cluster of normalized gain curve in both modes under different normalized frequency ratios m. When $m>1$, the larger m is, the steeper the gain curve of discharge mode is. But it is more difficult to realize a step-down condition and it not be possible to realize zero voltage switching (ZVS). When $m<1$, the smaller m is, the steeper the gain curve of charge mode is. And the gain curve of discharge mode moves towards the high frequency region result in a

wide switching frequency. With a carefully comparison between $m>1$ and $m<1$, it can be found that the gain curve of charge mode is steep and the overlapped frequency is wide under m<1. Therefore, in order to satisfy the gain demand in both directions and realization of ZVS with a narrow range of frequency variation, the value of m should be satisfy $m \leqslant 1$ and as close to the 1 as possible.

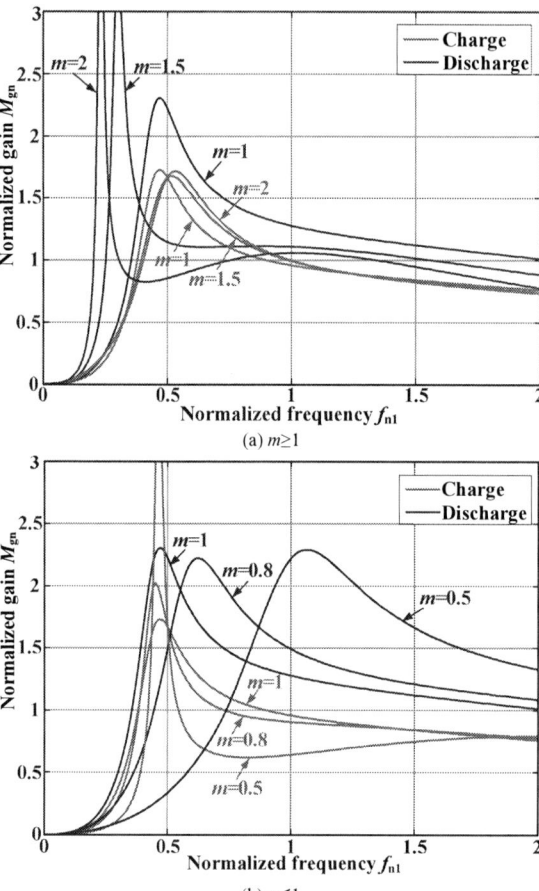

(a) $m \geq 1$

(b) $m \leq 1$

Fig. 7.   The normalized gain curves of both modes in different m with L_{n1}=3, Q_1=Q_2=0.4.

III.   The Circuit Parameter Design

1)   Determination for the transformer turns ratio n:

The LLC-C resonant converter has a good step-up ability in discharge mode [8]. However, f_{n1} has to increase to more than 2 to realize a step-down gain requirement. In order to avoid the above problems, converter in the discharge mode is more suitable for operating in the step-up region. Therefore, the transformer ratio n should be designed in step-down mode as given in (28).

$$n_s = \frac{V_{bus}}{V_{bat\text{-}max}}(1.2 \sim 1.5) \qquad (28)$$

2) Calculation for the normalized voltage gain range:

When calculating the normalized gain range, it is necessary to consider the practical requirement in V2G. The nominal output power in discharge mode is specified equal to half of that in charge mode in our prototype. For example, the nominal output power in charge and discharge mode is respectively 6.6kW and 3.3kW. The converter structure could reconfigure by the power level. In discharge mode, the inverter stage of converter could change from original full bridge to half bridge to avoid the difficult of step-down in discharge mode. The corresponding relationship of voltage between charge and discharge mode is shown in Fig. 8, where the bus voltage is always 400V and the battery voltage ranges from 280V to 450V. The maximum and minimum values of the normalized gain in both directions can be calculated according to (29)-(32).

$$M_{gn1-max} = \frac{nV_{bat-max}}{V_{bus}}, M_{gn1-min} = \frac{nV_{bat-min}}{V_{bus}}$$
$$M_{gn2-max} = \frac{2V_{bus}}{nV_{bat-min}}, M_{gn2-min} = \frac{2V_{bus}}{nV_{bat-max}} \quad (29)\text{-}(32)$$

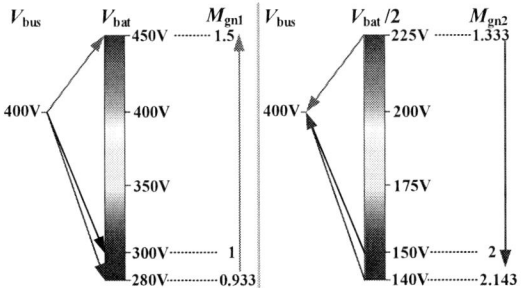

Fig. 8. The corresponding relationship between DC-bus voltage and battery voltage in both modes.

3) Selection for the maximum quality factor Q_{max}:

From Section II, Q_{max} should not be designed too large. The suggested value is between 0.5 and 0.6.

4) Selection for an inductance ratio L_{n1}:

From Section II, L_{n1} should not be designed too large. This value in the range of 1~3 is recommended as initial conditions for design.

5) Selection for the normalized frequency ratio m :

From Section II, m should be less than 1 and as close to 1 as possible. This value in the range of 0.8~1 is recommended as initial conditions for design.

6) Calculation for the characteristic impedance R_{01}:

The battery voltage is generally a variable range, so Q_{max} corresponds to the minimum battery voltage at the nominal output power. R_{01} is calculated by (33) after the L_{n1}, Q_{max} and m are determined.

$$R_{01} = Q_{max} \frac{8n^2}{\pi^2} \frac{V_{bat-min}^2}{P_{o1}} \quad (33)$$

7) Selection for the resonant frequency f_r:

The resonant frequency f_r is selected according to the desired switching frequency range so to reduce the harmonics.

8) Calculation for the resonant tank parameters:

With the mentioned design results, the resonant tank parameters can be calculated according to (34)-(37).

$$L_r = \frac{R_{01}}{2\pi f_{r1}}, L_m = L_{n1}L_{r1}$$
$$C_{r1} = \frac{1}{R_0 2\pi f_{r1}}, C_{r2} = \frac{(1+L_{n1})n^2}{m^2 L_{n1}R_{01} 2\pi f_{r1}} \quad (34)\text{-}(37)$$

The design steps are summarized to a flowchart as shown in Fig. 9.

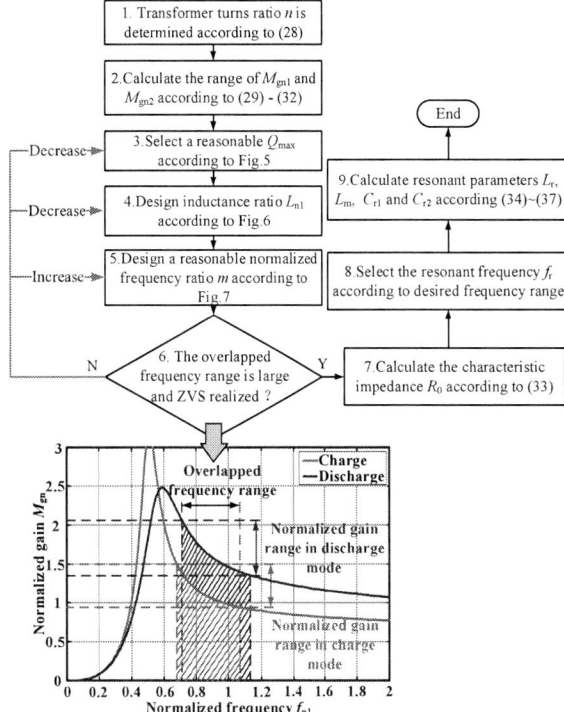

Fig. 9. The flowchart of parameter design process.

IV. EXPERIMENTAL RESULT

In order to verify the feasibility and effectiveness of the proposed parameter design method, a 6.6kW (3.3kW in discharge mode) prototype with GaN switches is built and tested. The main circuit parameters of the LLC-C resonant converter can be calculated according to the proposed design method, which are given in TABLE I.

TABLE I
MAIN CIRCUIT PARAMETERS IN THE PROTOTYPE OF THE LLC-C RESONANT CONVERTER

Parameters	Value
DC bus voltage V_{bus}	400V
Battery voltage V_{bat}	280V-450V
Charge nominal output power P_{o1}	6.6kW
Discharge nominal output power P_{o2}	3.3kW
Transformer turns ratio n_4 (N_p:N_s)	16:12
Charge mode resonant frequency f_{r1}	180kHz
Primary resonant capacitor C_{r1}	87nF
Primary resonant inductor L_r	9μH
Magnetizing inductor L_m	24μH
Secondary resonant capacitor C_{r2}	161nF
Switches Q_1~Q_8	GAN039-650NTBA (two in parallel)

The experimental waveforms under nominal output power in charge mode are given in Fig. 10. The experimental waveforms under nominal output power in discharge mode are given in Fig. 11. The key waveforms of voltage and current are measured at the DC-DC converter of the OBC, where the DC bus voltage V_{bus} is fixed to 400V. Moreover, the efficiency curves of converter in the charge and discharge mode are given in Fig. 12 and Fig. 13. The peak efficiency of charge and discharge mode is 97.55% and 97.47%, respectively.

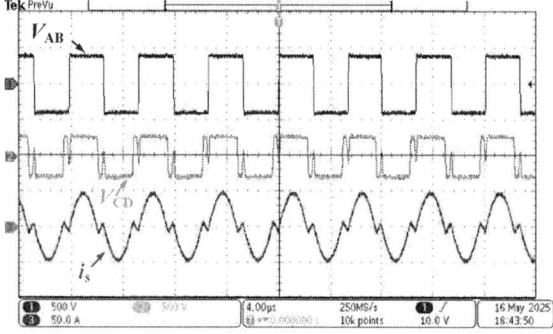

(a) V_{bus}=400V, V_{bat}=280V

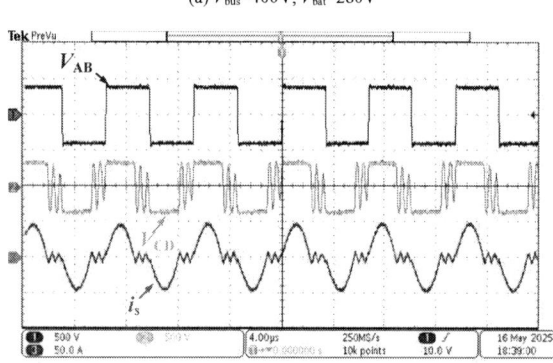

(b) V_{bus}=400V, V_{bat}=350V

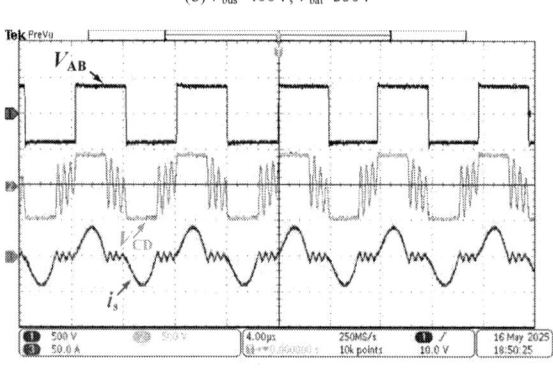

(c) V_{bus}=400V, V_{bat}=450V

Fig. 10. The experimental waveforms under nominal output power in the charge mode.

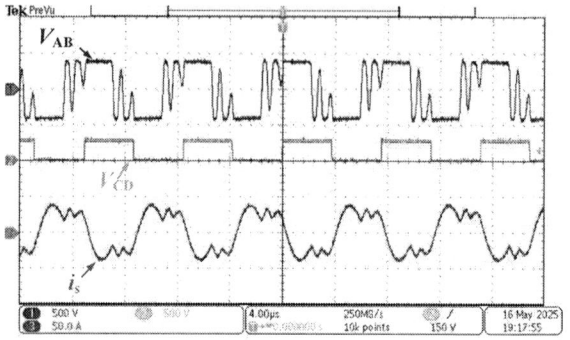

(a) V_{bat}=280V, V_{bus}=400V

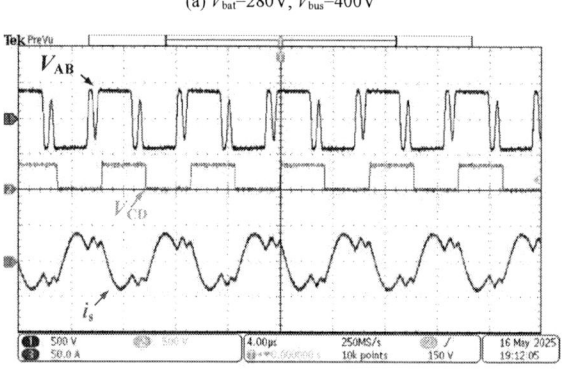

(b) V_{bat}=350V, V_{bus}=400V

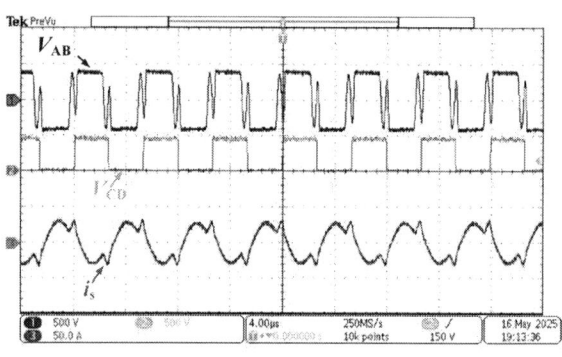

(c) V_{bat}=450V, V_{bus}=400V

Fig. 11. The experimental waveforms under nominal output power in the discharge mode.

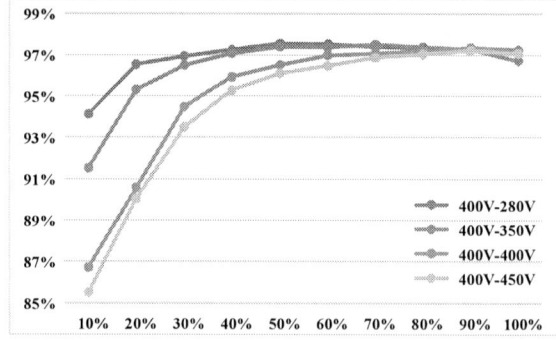

Fig. 12. The efficiency performances in the charge mode with different power levels and battery voltages.

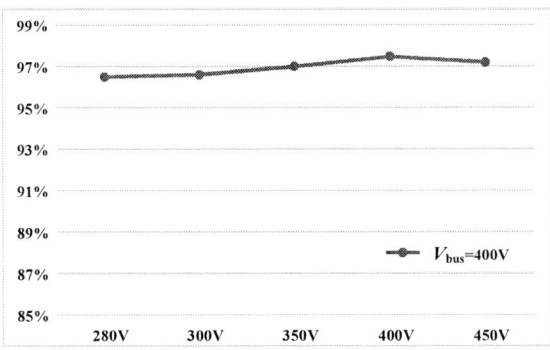

Fig. 13. The efficiency performances in discharge mode with different battery port voltages under 3.3kW output power.

V. CONCLUSIONS

For the LLC-C resonant converter, the FHA and resonant network equivalence were used to rebuilt the gain model in both directions. Then, a detail parameter design method is presented to reduce the whole switching frequency range of the converter. The experimental results show the feasibility of proposed method. The peak efficiency of charge and discharge mode is 97.55% and 97.47%, respectively.

REFERENCES

[1] Z. U. Zahid, Z. M. Dalala, R. Chen, B. Chen and J. -S. Lai, "Design of Bidirectional DC–DC Resonant Converter for Vehicle-to-Grid (V2G) Applications," in IEEE Transactions on Transportation Electrification, vol. 1, no. 3, pp. 232-244, Oct. 2015.

[2] R. P. Upputuri and B. Subudhi, "A Comprehensive Review and Performance Evaluation of Bidirectional Charger Topologies for V2G/G2V Operations in EV Applications," in IEEE Transactions on Transportation Electrification, vol. 10, no. 1, pp. 583-595, March 2024.

[3] C. Zhang, P. Li, Z. Kan, X. Chai and X. Guo, "Integrated Half-Bridge CLLC Bidirectional Converter for Energy Storage Systems," in IEEE Transactions on Industrial Electronics, vol. 65, no. 5, pp. 3879-3889, May 2018.

[4] J. Min and M. Ordonez, "Bidirectional Resonant CLLC Charger for Wide Battery Voltage Range: Asymmetric Parameters Methodology," in IEEE Transactions on Power Electronics, vol. 36, no. 6, pp. 6662-6673, June 2021.

[5] J. -H. Jung, H. -S. Kim, M. -H. Ryu and J. -W. Baek, "Design Methodology of Bidirectional CLLC Resonant Converter for High-Frequency Isolation of DC Distribution Systems," in IEEE Transactions on Power Electronics, vol. 28, no. 4, pp. 1741-1755, April 2013.

[6] S. Zou, J. Lu, A. Mallik and A. Khaligh, "Bi-Directional CLLC Converter With Synchronous Rectification for Plug-In Electric Vehicles," in IEEE Transactions on Industry Applications, vol. 54, no. 2, pp. 998-1005, March-April 2018.

[7] P. He, A. Mallik, A. Sankar and A. Khaligh, "Design of a 1-MHz High-Efficiency High-Power-Density Bidirectional GaN-Based CLLC Converter for Electric Vehicles," in IEEE Transactions on Vehicular Technology, vol. 68, no. 1, pp. 213-223, Jan. 2019.

[8] W. Chen, P. Rong and Z. Lu, "Snubberless Bidirectional DC–DC Converter With New CLLC Resonant Tank Featuring Minimized Switching Loss," in IEEE Transactions on Industrial Electronics, vol. 57, no. 9, pp. 3075-3086, Sept. 2010.

[9] C. -W. Lin, H. -J. Chiu, M. -S. Tzeng, J. -W. Yeh and C. -H. Huang, "Novel Bidirectional On-Board Charger for G2V and V2X Applications on Wide-Range Batteries," in IEEE Journal of Emerging and Selected Topics in Power Electronics, vol. 12, no. 2, pp. 2292-2305, April 2024.

[10] X. Tan and X. Ruan, "Equivalence Relations of Resonant Tanks: A New Perspective for Selection and Design of Resonant Converters," in IEEE Transactions on Industrial Electronics, vol. 63, no. 4, pp. 2111-2123, April 2016.

A Phase Shift Control Strategy for the DCX to Realize a Wide Voltage Range by a Variable Mode Transformer

Pengyu Jia
College of Electrical and Control
Engineering
North China University of Technology
Beijing, China
jiapengyu@ncut.edu.cn

Mingjun Liu
College of Electrical and Control
Engineering
North China University of Technology
Beijing, China
2023312080102@mail.ncut.edu.cn

Yimei Xing
College of Electrical and Control
Engineering
North China University of Technology
Beijing, China
2023312080119@mail.ncut.edu.cn

Abstract—A phase shift control strategy for the variable mode DC/DC transformer (DCX) to realize a wide voltage range is proposed. The DCX is realized by the series resonant converter (SRC), which employs a variable mode transformer. Different from a regular full bridge SRC-DCX, the circuit topology is realized by some certain numbers of the half bridge (HB) SRC modules. By continuously changing the phase shift angles between the HB SRC modules, the effective turns ratio of the variable mode transformer can vary. As a result, the voltage conversion ratio of the DCX can be adjusted by the phase shift angle even if the frequency is fixed. The gain model of the proposed DCX is derived and the voltage conversion model is verified by the experimental results.

Keywords—DC/DC transformer, series resonant converter, soft switching, phase shift control

I. Introduction

Resonant converters are easy to realize the soft-switching compared with the traditional chopper converters. However, different resonant converters are with different features. A traditional series resonant converter (SRC) is hard to decrease its output voltage in a light load while a parallel resonant converter (PRC) is hard to boost its voltage in a heavy load condition. Different from them, LLC resonant converter becomes more and more popular because it can achieve the soft-switching in the whole load range. By adjusting the inductance ratio or the quality factor, the boost capability can be improved. Besides, by setting the converter operating in a proper frequency range, the output voltage can be easily decreased by slightly changing the switching frequency. Therefore, it is easy to realize a high output efficiency performance.

For the resonant converters, the output voltage is usually controlled by adjusting the switching frequency, known as the pulse frequency mode (PFM). In the SRC or the LLC converter, if the switching frequency f_s is equal to the resonant frequency f_r, the normalized voltage gain of the converter is always equal to unity. In this condition, the harmonic of the resonant current can be limited in a very tiny range. In this case, the operation status of the LLC converter is equivalent to the SRC. Besides, the primary-side switches always work with a zero-voltage-switching (ZVS) on and the rectifier stage is in a critical discontinuous current mode. Correspondingly, the reverse recovery current of the diode is suppressed and the

converter can achieve a very high efficiency. In this case, the voltage conversion ratio is independent with the load. By utilizing this feature, the converter can be applied as a high frequency DC-DC transformer (DCX) to achieve the isolation function[1-4].

In order to make the voltage conversion ratio can be adjusted, the intuitive approach is operating the DCX as a regular LLC resonant converter [5-10] or a dual bridge (DB) - series resonant converter (SRC) [11-13], where only the control strategy needs to be changed but no main circuit components need to be changed. However, a big challenge to achieve a wide voltage range by a conventional LLC converter is the realization of synchronous rectification. Under the condition of step-up mode, it is not easy to obtain an analytical solution for the rectifier switch conducting phase and duration, so additional circuits such as the sampling circuit by a current transformer or the synchronous rectifier chip is generally used to drive the secondary-side switches [8-10], which increase the cost and control complexity. Moreover, a wide voltage range generally requires a very small inductance ratio, which is defined as the value of the magnetizing inductance divided by resonant inductance. As a result, this normally requires a small value for the magnetizing inductance, as well as a long air gap in the magnetic core. Correspondingly, it will make the design of the transformer difficult, moreover, the circulating current will increase and the efficiency will decrease.

Synchronous rectification can be naturally satisfied in a DB-SRC because each bridge leg is normally driven with a pair of complementary pulses, so that the secondary-side winding current always flows through the channels of the switches. Moreover, the converter can achieve wide voltage range by increasing the control freedom degrees besides switching frequency, such as the phase shift angle between the primary side and secondary side [11-13]. However, the control strategy becomes complicated when the light-load condition appears.

In [14-17], a pre-regulation converter is combined with a DCX so as to achieve a wide voltage range. However, the switch count of the pre-regulation leads to an increased cost of the whole circuit. Moreover, the efficiency performance is also limited by the power loss of the two stages. Different from a two-stage configuration, a regulation converter which handles partial of the total power is inserted into the circuit [18-20], in order to compensate the voltage variation range so to make the DCX be able to operate at the optimal efficiency point. Compared with the two-stage configuration, the power

This work was supported in part by the Beijing Natural Science Foundation under Grant 3232044, and in part by the R&D Program of Beijing Municipal Education Commission under Grant KM202110009011.

stress distributed into the regulation converter becomes much smaller. Therefore, the efficiency is improved. However, the cost is still increased compared with a regular DCX.

II. THE PROPOSED SRC-DCX

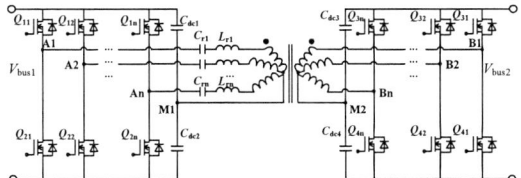

Fig. 1. Circuit schematic of the variable-mode SRC

Fig.1 shows a schematic of the proposed DCX. Different from a regular full bridge SRC-DCX, it can be considered as multiple phase HB-SRC submodules connected in parallel, where the module number n can be selected equal to 2 or larger. The switching frequency (f_s) of the HB SRC modules are set equal to the natural resonant frequency (f_r) and the output gain of the converter is regulated by adjusting the phase shift angles of each HB SRC modules. If the module number is n, the range of the phase angle of between each two modules ranges from 0 to $2\pi/n$. The control signals of the switches are shown in Fig.2.

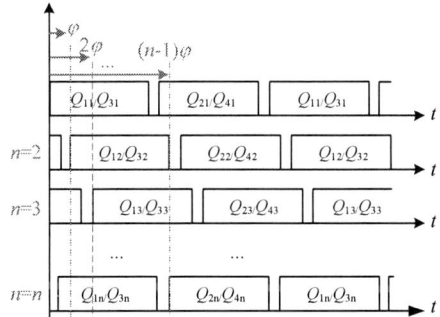

Fig. 2. typical control signal waveforms corresponding to n submodules.

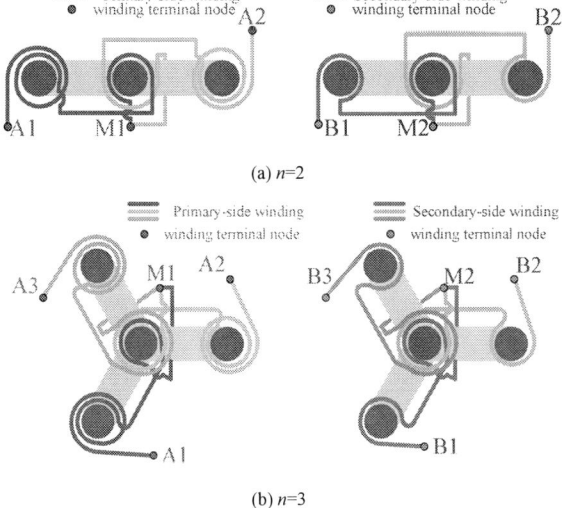

(b) n=3

Fig. 3. Schematic of the variable mode transformer realization method in a top view

Fig.3 shows the schematic of the transformer realization method in a top view, where the module number n is respectively selected equal to 2 and 3 for examples. All the windings are aligned origin-symmetric. For the two HB-SRC module case, the coils around the center column of the core is with no flux when the phase difference of the two modules is equal to π. Similarly, if the module number n is equal to 3, the coil around the center column is with no flux when the phase difference of the three modules is equal to $2\pi/3$. However, when the phase difference of all the HB-SRC modules is equal to 0, which means they are with the same operating phase, the flux of the center column exists and gets maximum. By utilizing this feature, the windings around the center column can be enabled or disabled, and as a result, turns ratio can be controlled by the phase shift angle of the multiple HB-SRC modules.

Moreover, when the number of the submodules equals 2, the windings can be arranged symmetrically about the Y-axis, which is given in Fig.4. By this means, the terminal nodes of all the windings can be set in the same side of the core so as to facilitate the manufacture of the transformer and simplify the layout of the PCB. Here only the primary-side windings are given as an example and the secondary-side windings are arranged in the same way. When comparing Fig.4(a) and Fig.4(b), it is clear that the phase shift angle between the two bridge legs of the converter should be equal to $(\pi-\varphi)$ in Fig.4(b) so that the transformer is equivalent to the case of Fig.4(a) with a phase shift angle equal to φ.

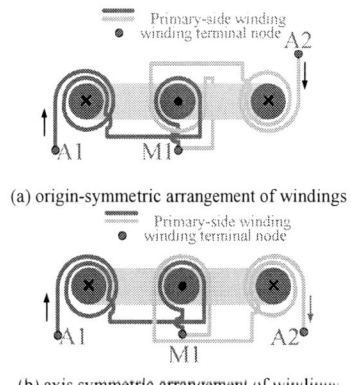

(a) origin-symmetric arrangement of windings

(b) axis symmetric arrangement of windings

Fig. 4. Two equivalent winding methods of a two-phase variable mode transformer in a top view

The magnetic circuit of transformer in the two-phase HB-SRC module case are analyzed in the following so to derive the voltage gain of the DCX, which use the phase shift angle φ as the only control variable. In this case, the phase shift angle between the two modules changes from 0 to π. For a transformer, the designers always introduce an air gap to obtain a stable inductance value. Therefore, air gaps are considered and they are assumed existing in all the columns of the core. In the core center column, there may be an air gap of which the length is equal to l_{gm}, and meanwhile, the length of the air gap in the edge column is assumed equal to l_{gs}. The winding turns in the center column and the winding turns in the edge columns are respectively defined as k_1, n_1 for the primary-side winding, and defined as k_2, n_2 for the secondary-side winding. For the case in Fig.3(a), it is obvious that k_1=2, n_1=2, k_2=2 and n_2=1.

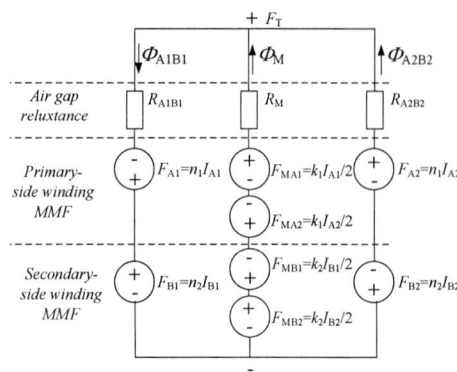

Fig. 5. Magnetic circuit model in a two-phase variable mode transformer

The magnetic circuit of the two-phase variable model transformer is given in Fig.5, where the reference positive directions for the magnetomotive force (F), flux (Φ) are marked. It should be noted that the model of Fig.5 is corresponding to the case of axis symmetric arrangement of windings, such as Fig.4(b). Since the core material permeability is much larger than the permeability of free space (μ_0), the reluctances of the core material is omitted and only the ones caused by air gaps are considered for simplification. The air gaps of two core edge columns are with the same length, therefore, the air gap reluctances R_{A1B1} and R_{A2B2} are equal to each other as shown in (1). The reluctance of the air gap in the center column is R_M. This can be either different with or equal to R_{A1B1} (R_{A2B2}).

$$R_{A1B1} = R_{A2B2} \quad (1)$$

According to Fig.5, the following equations can be derived according to Fig.5, where F_T denotes the magnetomotive force from the bottom to the top of the core center column.

$$\begin{cases} F_{A1} - F_{B1} + F_T = \phi_{A1B1}R_{A1B1} \\ F_{MA1} + F_{MB2} - F_{MB1} - F_{MA2} - F_T = \phi_M R_M \\ F_{A2} - F_{B2} - F_T = \phi_{A2B2}R_{A2B2} \\ \phi_{A1B1} - \phi_{A2B2} = \phi_M \end{cases} \quad (2)\sim(5)$$

By derivation and simplification, the relationship between winding voltages and currents can be obtained. The equations are complicated and shows a coupled relationship between all the four winding ports, which are A1M1, A2M1, B1M2 and B2M2. It is clear that the effective transformer turns ratios for each HB modules, that are $N_{ratio\_A}=V_{A1M1}/V_{B1M2}$ and $N_{ratio\_B}=V_{A2M1}/V_{B2M2}$, directly affect the voltage gain. Since the outputs of the two HB modules are connected in parallel, these two effective turns ratios must be proved equal to each other. Therefore, the function of N_{ratio_A} as well as N_{ratio_B} with respect to the phase shift angle φ is derived and obtained as follows.

The winding voltages of the primary side (or the secondary side) is controlled with the phase shift angle φ between the two submodules. Here, the primary-side winding voltages, V_{A1M1} and V_{A2M1}, are described by (6) and (7), where V_m denotes the amplitude of a given sinusoidal voltage.

$$\begin{cases} V_{A1M1} = V_m \sin(2\pi f_s t) \\ V_{A2M1} = V_m \sin(2\pi f_s t - \varphi) \end{cases} \quad (6)(7)$$

Then the secondary-side winding voltages can be obtained as follows.

$$\begin{cases} V_{B2M2} = \dfrac{(k_1 n_2 - k_2 n_1)V_m \sin(2\pi f_s t)}{2n_1(k_1 + n_1)} \\ \qquad + \dfrac{(k_2 n_1 + k_1 n_2 + 2n_1 n_2)V_m \sin(2\pi f_s t - \varphi)}{2n_1(k_1 + n_1)} \\ V_{B1M2} = \dfrac{(k_2 n_1 + k_1 n_2 + 2n_1 n_2)V_m \sin(2\pi f_s t)}{2n_1(k_1 + n_1)} \\ \qquad + \dfrac{(k_1 n_2 - k_2 n_1)V_m \sin(2\pi f_s t - \varphi)}{2n_1(k_1 + n_1)} \end{cases} \quad (8)(9)$$

Therefore, the winding voltages of the secondary side can be calculated, which are given in (10) and (11).

$$\begin{cases} V_{AMP\_B2M2} = V_m \sqrt{\left[\dfrac{(k_1 n_2 - k_2 n_1) + (k_2 n_1 + k_1 n_2 + 2n_1 n_2)\cos(\varphi)}{2n_1(k_1 + n_1)}\right]^2 + \left[\dfrac{(k_2 n_1 + k_1 n_2 + 2n_1 n_2)\sin(\varphi)}{2n_1(k_1 + n_1)}\right]^2} \\ V_{AMP\_B1M2} = V_m \sqrt{\left[\dfrac{(k_2 n_1 + k_1 n_2 + 2n_1 n_2) + (k_1 n_2 - k_2 n_1)\cos(\varphi)}{2n_1(k_1 + n_1)}\right]^2 + \left[\dfrac{(k_1 n_2 - k_2 n_1)\sin(\varphi)}{2n_1(k_1 + n_1)}\right]^2} \end{cases} \quad (10)(11)$$

According to (8)~(11), the effective transformer turns ratios N_{ratio_A} and N_{ratio_B} can be solved based on Fig.5, which are always equal to each other and the analytical solution with respect to φ is given in (12).

$$\begin{aligned} N_{ratio\_A} &= N_{ratio\_B} \\ &= \sqrt{\dfrac{2n_1^2(k_1 + n_1)^2}{\begin{array}{l} k_2^2 n_1^2 + 2k_2 n_1^2 n_2 + (k_1^2 + 2k_1 n_1 + 2n_1^2)n_2^2 \\ + (k_1 n_2 - k_2 n_1)(k_2 n_1 + k_1 n_2 + 2n_1 n_2)\cos\varphi \end{array}}} \end{aligned} \quad (12)$$

The SRC shows a unity voltage conversion ratio when the switching frequency is equal to the natural resonant frequency. Therefore, voltage gain of the DCX can be adjusted by the phase shift angle φ according to (12).

III. EXPERIMENT VERIFICATION

As known, the output is independent of the load when the DCX operates at the resonant frequency. For each HB submodule of the converter, if the primary side and secondary side are operated in phase under the condition of $f_n=1$, the normalized voltage gain is always equal to 1. The output voltage can be changed according to the variation of the effective turns ratio N_{ratio_A} (or N_{ratio_B}). It is clear that the range of the effective turns ratio for the DCX can be determined by (12), hence, this can be applied to design an isolated DC/DC converter corresponding to a given voltage range, which can be also considered as a DCX with adjustable voltage conversion ratio.

TABLE I.   MAIN CIRCUIT PARAMETERS

Port voltage 1	V_{bus1}	200V
Switching frequency	f_s	98kHz
Effective transformer ratio 1	$N_{ratio}(\varphi=\pi)$	4:1
Effective transformer ratio 2	$N_{ratio}(\varphi=0)$	2:1
Transformer core	Core T_1	E58/11/38-3C95
Resonant inductor in the HB module	L_r	43μH
Resonant capacitor in the HB module	C_r	60nF
Resonant inductor core	Core L_r	PQ32/32-3C95
Power switch type	$Q_1\sim Q_8$	GS66508T

Fig. 6. The variable mode transformer and the resonant tanks

A prototype of two-phase HB-SRC DCX is built to make the verification, where the main circuit parameters and the specifications are given as shown in TABLE I. In order to reduce the parasitic capacitance from the active switches, the enhanced GaN switches are selected as the power switches. The transformer photo is given in Fig.6, besides, the PCB winding layout is given in Fig.7 which is realized by a four-layer structure. It is clear that the axis-symmetrical arrangement for each submodule is adopted. Therefore, when the phase shift angle φ is equal to 0, the winding in the core center column is disabled.

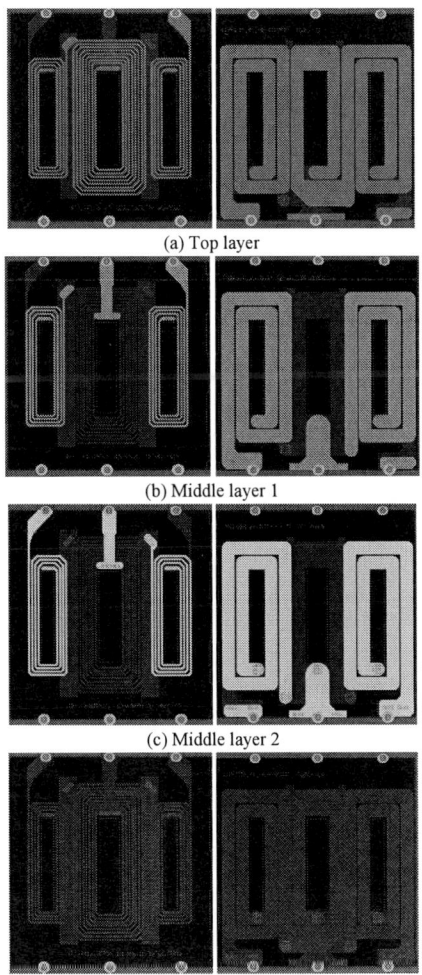

(a) Top layer

(b) Middle layer 1

(c) Middle layer 2

(d) Bottom layer

Fig. 7. PCB winding layout

The voltage conversion ratio (V_{bus1}/V_{bus2}) is predicted by the effective turns ratio of the transformer (N_{ratio}) and the curve is given in Fig.8. Meanwhile, the tested results are marked in the same picture. The voltage ratio is adjusted only by controlling the phase shift angle φ of the two HB-SRC modules. In Fig.8, it is clear that the predicted error approaches its maximum value when the phase shift angle gets $\pi/2$, but the prediction error gradually decreases when the phase shift angle deviate from $\pi/2$. Moreover, when the phase shift angle is around $\pi/2$, the load change slightly affects the voltage gain. When the phase shift angle gets close to 0 or π, the predicted results show a good accordance with tested ones. Correspondingly, the load nearly has no effect on the output gain.

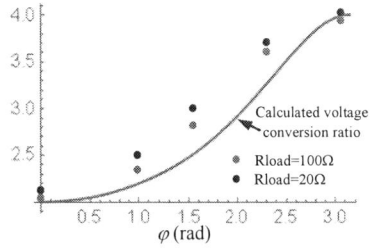

Fig. 8. Experimental tested voltage conversion ratios (V_{bus1}/V_{bus2}) and the theoretical results for the DCX

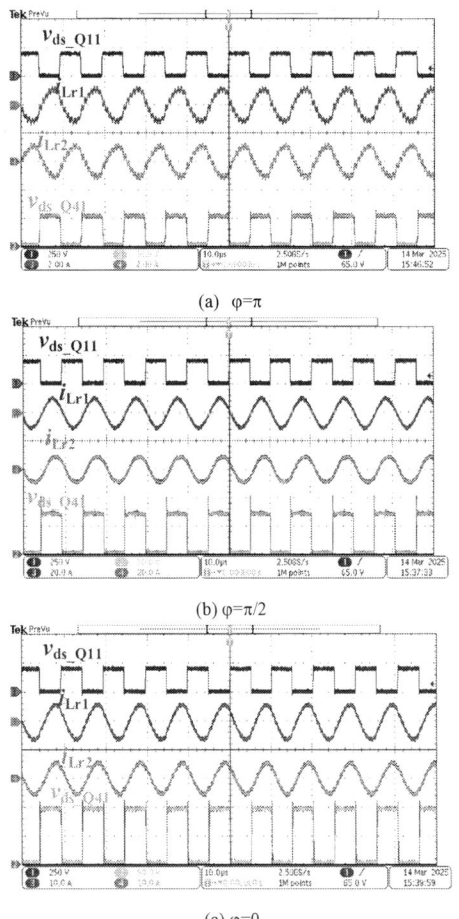

(a) $\varphi=\pi$

(b) $\varphi=\pi/2$

(c) $\varphi=0$

Fig. 9. Waveforms of the converter with phase shifted control

Fig.9 shows the steady-state waveforms of the converter operating at $\varphi=0$, $\pi/2$ and π, respectively. It is clear that the resonant currents are balanced at $\varphi=0$ and $\varphi=\pi$. However, due to the existence of the power exchange with the phase shift angel between different windings, the resonant currents in two modules are unbalanced at $\varphi=\pi/2$. In fact, only the two cases of $\varphi=0$ and $\varphi=\pi$ can realize a symmetrical power distribution on the two submodules. If the phase shift angle deviates from these two values, the unbalanced phenomenon is always present because the two module is coupled by the variable mode transformer.

It should be noted that there exists some high frequency oscillation, such as Fig.9(a). This can be explained that the PCB winding capacitance forms high frequency resonant path with the inductor and this phenomenon can be alleviated by increasing the distance between each piece of the PCB winding.

IV. CONCLUSIONS

An SRC-DCX with variable mode transformer is proposed in this paper, which can be considered as a combination of multiple HB-SRC submodules in parallel. The output gain of DCX is regulated only by adjusting the phase shift angel of each module and the frequency is always fixed near the resonant frequency. Besides, the number of submodules can be selected according to the power level. The voltage gain model of the DCX is derived and it is verified by the experimental results.

REFERENCE

[1] X. Wu, H. Chen, and Z. Qian, "1-MHz LLC resonant dc transformer (DCX) with regulating capability," IEEE Trans. Ind. Electron., vol. 63, no. 5, pp. 2904–2912, May 2016.

[2] W. Feng, P. Mattavelli, and F. C. Lee, "Pulsewidth locked loop (PWLL) for automatic resonant frequency tracking in LLC DC–DC transformer (LLC-DCX)," IEEE Trans. Power Electron., vol. 28, no. 4, pp. 1862 1869, Apr. 2013.

[3] Ü. , S.Alemdar and O.Keysan,"Design and implementation of an unregulated DC-DC transformer (DCX) module using LLC resonant converter," in Proc. 8th IET Int. Conf. Power Electron., Mach. Drives, Glasgow, Scotland, 2016, pp. 1–6.

[4] D.Gu,Z.Zhang,Y.Wu,D.Wang,H.Gui,andL.Wang,"High efficiency LLC DCX battery chargers with sinusoidal power decoupling control,"in Proc. IEEE Energy Convers. Congr. Expo., Milwaukee, WI, USA, 2016, pp. 1–7.

[5] J. Deng, S. Li, S. Hu, C. C. Mi and R. Ma, "Design methodology of LLC resonant converters for electric vehicle battery chargers", IEEE Trans. Veh. Technol., vol. 63, no. 4, pp. 1581-1592, May 2014.

[6] Y. Shen, W. Zhao, Z. Chen and C. Cai, "Full-bridge LLC resonant converter with series-parallel connected transformers for electric vehicle on-board charger", IEEE Access, vol. 6, pp. 13490-13500, 2018.

[7] Bo Yang, F. C. Lee, A. J. Zhang and G. Huang, "LLC resonant converter for front end DC/DC conversion", Proc. 17th Annu. IEEE Appl. Power Electron. Conf. Expo., pp. 1108-1112, 2002.

[8] D. Fu, Y. Liu, F. C. Lee and M. Xu, "A Novel Driving Scheme for Synchronous Rectifiers in LLC Resonant Converters," in IEEE Transactions on Power Electronics, vol. 24, no. 5, pp. 1321-1329, May 2009.

[9] W. Feng, F. C. Lee, P. Mattavelli and D. Huang, "A Universal Adaptive Driving Scheme for Synchronous Rectification in LLC Resonant Converters," in IEEE Transactions on Power Electronics, vol. 27, no. 8, pp. 3775-3781, Aug. 2012.

[10] C. Fei, Q. Li and F. C. Lee, "Digital Implementation of Adaptive Synchronous Rectifier (SR) Driving Scheme for High-Frequency LLC Converters With Microcontroller," in IEEE Transactions on Power Electronics, vol. 33, no. 6, pp. 5351-5361, June 2018.

[11] P. Jia, K. Qiu, M. Liang, T. Shao and Y. Zhang, "A Simple Variable Frequency and Single-Phase-Shifted Control for the Dual Active Bridge Series Resonant Converter," in IEEE Journal of Emerging and Selected Topics in Industrial Electronics, vol. 6, no. 1, pp. 308-326, Jan. 2025.

[12] C. Sun et al., "Generalized Multiphase-Shift Transient Modulation for Dual-Active-Bridge Series-Resonant Converter," in IEEE Transactions on Power Electronics, vol. 38, no. 7, pp. 8291-8309, July 2023.

[13] F. M. Ibanez, J. M. Echeverria, J. Vadillo and L. Fontan, "A Step-Up Bidirectional Series Resonant DC/DC Converter Using a Continuous Current Mode," in IEEE Transactions on Power Electronics, vol. 30, no. 3, pp. 1393-1402, March 2015.

[14] J. Lee, Y. Jeong, and B. Han, "An isolated DC/DC converter using high frequency unregulated LLC resonant converter for fuel cell applications," IEEE Trans. Ind. Electron., vol. 58, no. 7, pp. 2926–2934, Jul. 2011.

[15] M. H. Ahmed, C. Fei, F. C. Lee, andQ. Li, "48-V voltage regulator module with PCB winding matrix transformer for future data centers," IEEE Trans. Ind. Electron., vol. 64, no. 12, pp. 9302–9310, Dec. 2017.

[16] F. Liu, G. Zhou, X. Ruan, S. Ji, Q. Zhao, and X. Zhang, "An input-series output-parallel converter system exhibiting natural input-voltage-sharing and output-current-sharing," IEEE Trans. Ind. Electron., vol. 68, no. 2, pp. 1166–1177, Feb. 2021.

[17] R. Gu, J. Duan, D. Zhang and H. Liu, "Regulated Series Hybrid Converter With DC Transformer (DCX) for Step-Up Power Conversion," in IEEE Transactions on Industrial Electronics, vol. 69, no. 9, pp. 8961-8971, Sept. 2022.

[18] V. Li, M. H. Ahmed, Q. Li and F. C. Lee, "Modeling and control of sigma converter for 48V voltage regulator application", Proc. IEEE. Energy. Convers. Congr. Expo., pp. 1199-1204, 2018.

[19] T. Liu, X. Wu and S. Yang, "1 MHz 48–12 V regulated DCX with single transformer", IEEE J. Emerg. Sel. Topics Power Electron., vol. 9, no. 1, pp. 38-47, Feb. 2021.

[20] D. Neumayr, M. Vöhringer, N. Chrysogelos, G. Deboy and J. W. Kolar, " p 3 DCT—Partial-power pre-regulated dc transformer ", IEEE Trans. Power Electron., vol. 34, no. 7, pp. 6036-6047, Jul. 2019.

2025 IEEE Workshop on Wide Bandgap Power Devices and Applications in Asia (WiPDA Asia)

A Multi-Port Converter-Based Equalization Architecture with Wide Voltage Gain for Long Series Battery Packs

Haipeng Hu
School of Electrical Engineering
Hefei University of Technology
Hefei, China
haipeng.hu@mail.hfut.edu.cn

Xianbin Qi*
School of Electrical Engineering
Hefei University of Technology
Hefei, China
xianbin_qi@hfut.edu.cn

Helong Li
School of Electrical Engineering
Hefei University of Technology
Hefei, China
helong.li@hfut.edu.cn

Mingzhu Fang
National Key Laboratory of Deep
Space Exploration
Hefei, China
mingzhu_fang@163.com

Peng Qin
School of Electrical Engineering
Hefei University of Technology
Hefei, China
pengqin@hfut.edu.cn

Zhiqing Yang
School of Electrical Engineering
Hefei University of Technology
Hefei, China
zhiqing.yang@hfut.edu.cn

Lijian Ding
School of Electrical Engineering
Hefei University of Technology
Hefei, China
ljding@hfut.edu.cn

Abstract—To address the challenges of insufficient voltage gain, structural complexity, and efficiency limitations in energy management of long-series battery strings, this study proposes a wide-voltage-gain equilibrium architecture based on multi-port converters. The architecture enables flexible substitution of arbitrary multi-port converters with port multiplexing capability, maintaining compact configuration while supporting intelligent power path reconfiguration. Validation through PLECS simulations integrated with GaN power devices demonstrates that the proposed control strategy achieves efficient conversion from 4-16V input to over 64V output, confirming its technical feasibility in both wide voltage gain and rapid cell balancing. These findings provide an innovative technical pathway for high-efficiency energy management in high-density energy storage systems.

Keywords—Multiport converters; Wide voltage gain; GaN; Battery string balancing

I. INTRODUCTION

Lithium-ion batteries have become critical power sources in electric vehicles, energy storage systems, and unmanned aerial vehicles due to their high energy density and long cycle life. However, battery pack operation faces challenges such as cell capacity variations, internal resistance drift, and inconsistent self-discharge rates [1], leading to reduced usable capacity and increased thermal runaway risks. Battery balancing technology, which dynamically manages cell uniformity through energy redistribution, serves as a core approach to enhance safety and longevity. Existing passive balancing methods suffer from low efficiency and excessive heat generation. While active balancing methods (classified into four categories by energy transfer types: C2C [2], [3] C2S [4] -[7],S2C [8], [9]and C2S2C[10],[11]) achieve higher efficiency, they remain constrained by topological complexity, limited voltage gain range, and high-frequency losses in silicon-based devices, hindering adaptation to wide-voltage scenarios. Current methods still fail to balance equilibrium speed, efficiency and practicality, necessitating improvements in circuit architectures and energy transfer pathways.

To address the challenges of topological complexity and limited efficiency in traditional active balancing technologies, this paper introduces the following innovations to overcome existing technical bottlenecks: (1) a multi-port converter-based balancing architecture with an integrated switching matrix to enable diverse energy transfer paths; (2) a wide-voltage gain control strategy achieving a voltage conversion range from 4 V–16 V to over 64 V; and (3) the incorporation

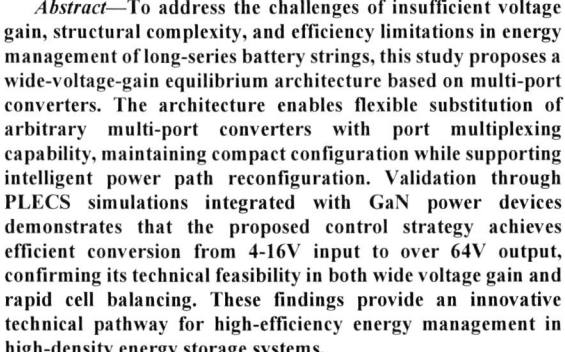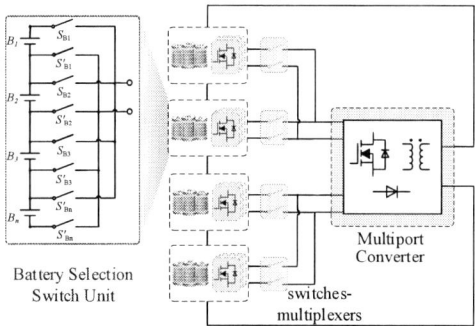

Fig. 1. The proposed architecture

of GaN devices into the multi-port converter to enhance high-frequency performance. Section II details the operational states and working principles of the proposed balancer. Section III presents the design methodology and parameter calculations. Section IV provides experimental validation results, and Section V concludes the study.

II. OPERATION AND ANALYSIS OF THE PROPOSED CONVERTER

A. Introduction to Topological Structure

To address the cell capacity discrepancies in long-series battery pack operation, this paper proposes an efficient battery balancing scheme based on a multi-port converter and its implementation method. The topology is shown in Fig. 1.

Specifically, the long-series battery pack is divided into m independent battery sub-modules, each consisting of n serially connected battery cells. Each cell is equipped with two independent selection switches for connecting to either the primary-side port of the multi-port converter or a bypass branch. By controlling the conduction states of the switch array, the following functions can be achieved: Cell energy extraction: When a cell requires energy replenishment, battery selection switch closes to connect it to the converter's primary side; Multi-cell cooperative balancing: Multiple cells can be paralleled to the converter through battery selection switch for cross-cell energy transfer; Flexible topology reconfiguration: Dynamic switching combinations enable cascaded expansion or fault isolation of battery sub-modules. The secondary side of the multi-port converter is directly connected to the entire battery pack's bus. The specific experimental topology is shown in Fig. 2

979-8-3315-1110-4/25 $31.00 © 2025 IEEE

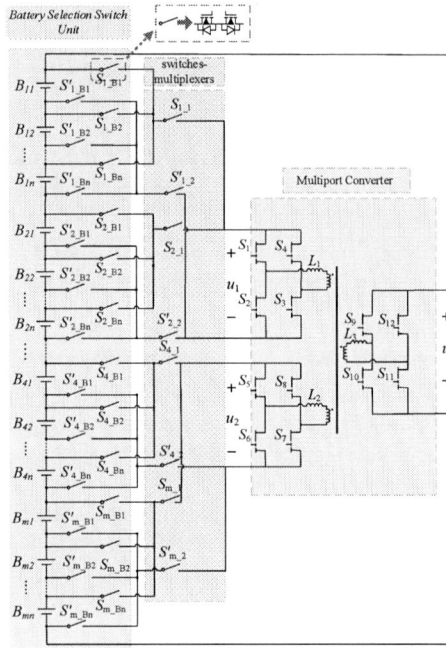

Fig. 2. The basic topology of the proposed converter.

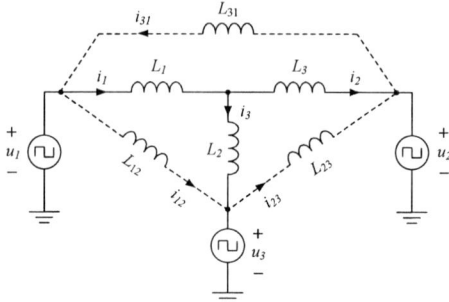

Fig. 3. Equivalent circuit of three-winding transformer.

Table I PARAMETERS FOR THE CIRCUIT

Parameter	Value
Low-voltage side VLV	4 -16 V
High-voltage side VHV	64 V
Rated Power P_{rated}	12W-48W
Switching frequency f_S	500k Hz
Transformer T_1, T_2, T_3	1:1:4
Primary Side Switch Model	Epc2024
Secondary Side Switch Model	Epc2065
Inductors L_1, L_2, L_3	1.13μH

B. Operation Principles of the Proposed Converte

The most fundamental modulation strategy for Proposed converters is phase-shift modulation. Fig. 3 illustrates three equivalent circuit diagrams of the three-winding transformer referred to Port 1, while Fig. 4 demonstrates the operational current schematic during one switching cycle of the three-port converter. Based on the delta equivalent circuit of the high-frequency transformer and instantaneous voltage values, the instantaneous current in the delta-equivalent circuit can be calculated using the following equation:

$$
\begin{cases}
i_{12}(t) = i_{12}(t_0) + \dfrac{1}{L}\int_{t_0}^{t} u_1(\tau) - u_2(\tau) d\tau \\[2mm]
i_{23}(t) = i_{23}(t_0) + \dfrac{1}{L}\int_{t_0}^{t} u_2(\tau) - u_3(\tau) d\tau \\[2mm]
i_{31}(t) = i_{31}(t_0) + \dfrac{1}{L}\int_{t_0}^{t} u_3(\tau) - u_1(\tau) d\tau
\end{cases}
\tag{1}
$$

The inter-port power transfers P_{12}, P_{23}, and P_{31} are defined as the power flowing through branch inductors L_{12}, L_{23}, and L_{31} in the delta-equivalent circuit, respectively. These power transfers can be calculated using Equation (2):

$$
\begin{cases}
P_{12} = \dfrac{1}{T_s}\int_{t_0}^{t_0+T_s} \left[u_1(t) - u_2(t)\right] i_{12}(t) dt = \dfrac{V_1 V_2 D_{12}}{2 f_s L_{12}}(1 - |D_{12}|) \\[2mm]
P_{23} = \dfrac{1}{T_s}\int_{t_0}^{t_0+T_s} \left[u_2(t) - u_3(t)\right] i_{23}(t) dt = \dfrac{V_2 V_3 D_{23}}{2 f_s L_{23}}(1 - |D_{23}|) \\[2mm]
P_{31} = \dfrac{1}{T_s}\int_{t_0}^{t_0+T_s} \left[u_3(t) - u_1(t)\right] i_{31}(t) dt = -\dfrac{V_3 V_1 D_{13}}{2 f_s L_{31}}(1 - |D_{13}|)
\end{cases}
\tag{2}
$$

Where $D_{23}=D_{13}-D_{12}=0$. Defining P_1, P_2, and P_3 as the average power at each transformer winding port over one switching period, the port power of the proposed converter can be calculated using Equation (3):

$$
\begin{cases}
P_1 = P_{12} - P_{31} = \dfrac{V_1 V_2 D_{12}}{2 f_s L_{12}}(1 - |D_{12}|) + \dfrac{V_1 V_3 D_{13}}{2 f_s L_{31}}(1 - |D_{13}|) \\[2mm]
P_2 = P_{12} - P_{23} = \dfrac{V_1 V_2 D_{12}}{2 f_s L_{12}}(1 - |D_{12}|) - \dfrac{V_2 V_3 D_{23}}{2 f_s L_{23}}(1 - |D_{23}|) \\[2mm]
P_3 = P_{23} - P_{31} = \dfrac{V_2 V_3 D_{23}}{2 f_s L_{23}}(1 - |D_{23}|) + \dfrac{V_1 V_3 D_{13}}{2 f_s L_{31}}(1 - |D_{13}|)
\end{cases}
\tag{3}
$$

Through power analysis of the proposed converter, the switching-period-averaged currents at each port can be idealized as I_{dc1}, I_{dc2} and I_{dc3}

$$
\begin{cases}
I_{dc1} = \dfrac{P_1}{V_1} = \dfrac{V_3 D_{13}}{2 f_s L_{31}}(1 - |D_{13}|) \\[2mm]
I_{dc2} = \dfrac{P_2}{V_2} = \dfrac{V_3 D_{23}}{2 f_s L_{23}}(1 - |D_{23}|) \\[2mm]
I_{dc3} = \dfrac{P_3}{V_3} = \dfrac{V_2 D_{23}}{2 f_s L_{23}}(1 - |D_{23}|) + \dfrac{V_1 D_{13}}{2 f_s L_{31}}(1 - |D_{13}|)
\end{cases}
\tag{4}
$$

where $D_{13}=D_{23}$. This derivation neglects all switching device losses and the high-frequency transformer's copper and core losses.

III. PARAMETER DESIGN

A. Parameter calculation

The design of the *Proposed* converter should consider the following principles:

1) Design the voltage ratio of the three-winding high-frequency transformer to be equal to the rated voltage ratio of the three DC voltage ports, $1:n_2:n_3=V_1:V_2:V_3$

2) Design the transformer leakage inductance $L_1=L_2=L_3$ based on the principle of equal inductance in each branch of the delta equivalent circuit, i.e., $L=L_{12}=L_{23}=L_{31}$. This aims to enable the three ports of the converter to have identical power transfer capabilities. The formula for inductance L can be obtained by transforming Formula 3.

979-8-3315-1110-4/25 $31.00 © 2025 IEEE

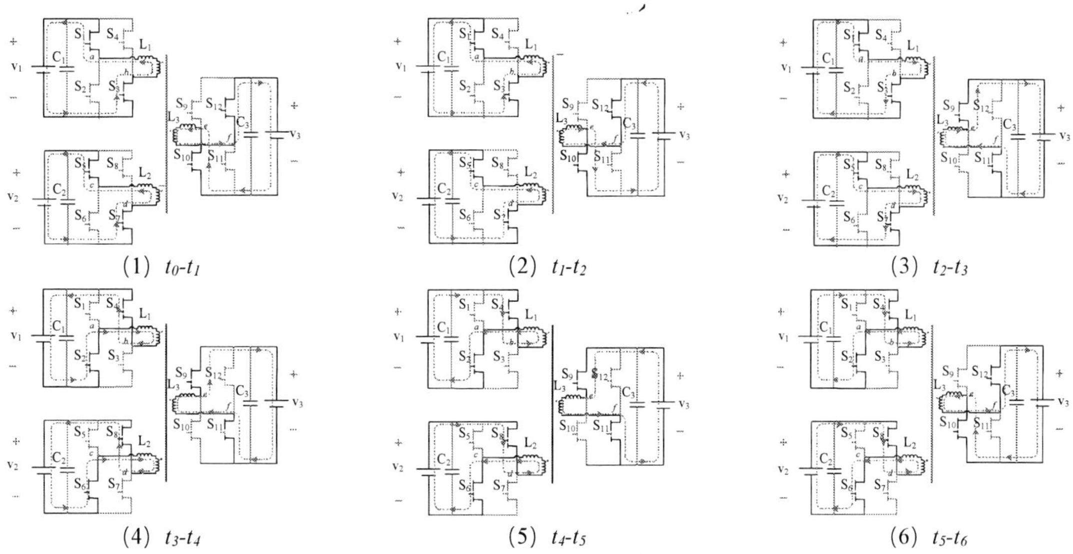

(1) t_0-t_1 (2) t_1-t_2 (3) t_2-t_3

(4) t_3-t_4 (5) t_4-t_5 (6) t_5-t_6

Fig. 4. The working waveform under SPS control

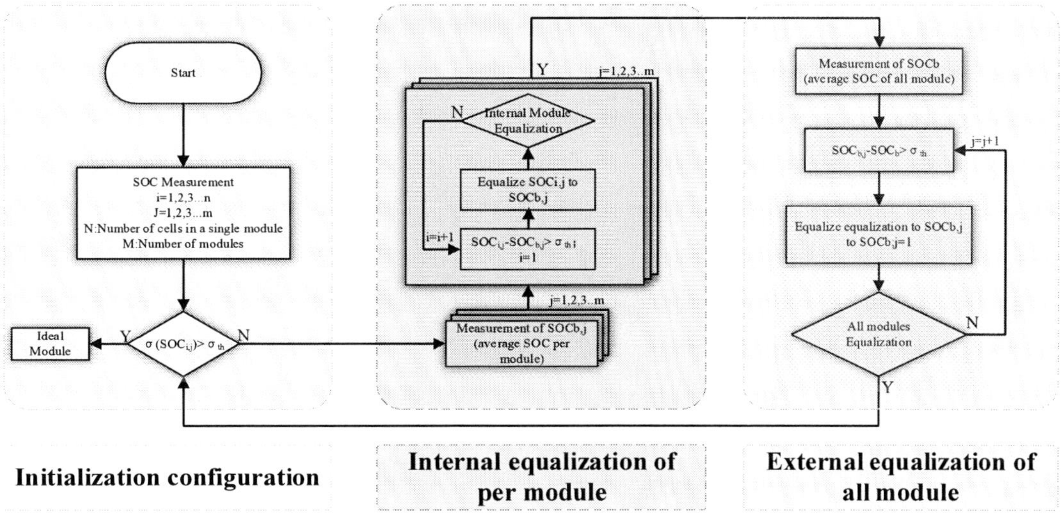

Fig. 5. Algorithm flow chart

B. Control algorithm

The long-series retired lithium battery pack is divided into m modules, with each module comprising n cells. The cells in the j-th module are labeled with indices i (cell index) and j (module index), where its state of charge is denoted as $SOC_{i,j}$. The battery balancing process comprises three operational stages: initialization configuration, internal equalization of per module and external equalization of all module, with the specific control algorithm depicted in Fig. 5.

The process initiates with initialization detection of cell imbalance in each battery unit. According to the converter port quantity, corresponding battery groups are allocated for simultaneous equalization. During this phase, the cell with maximum SOC discharges to its adjacent cell, then both cells are connected for discharge. This operation repeats until all battery groups achieve internal equalization. Subsequently, modules exceeding the average SOC undergo discharge equalization to attain global balance in the long-series battery .

IV. VERIFICATION RESULTS

The PLECS simulation results shown in Fig. 5 demonstrate the SOC variation trends of 16 battery cells during the balancing process. Equilibrium is achieved at time

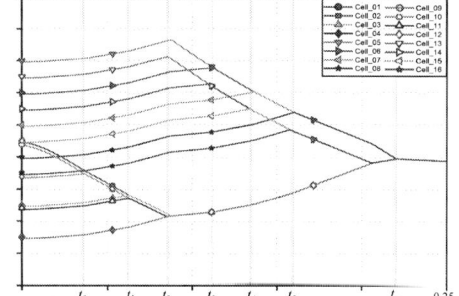

Fig.6. Results of equalization for lithium-ion battery cells.

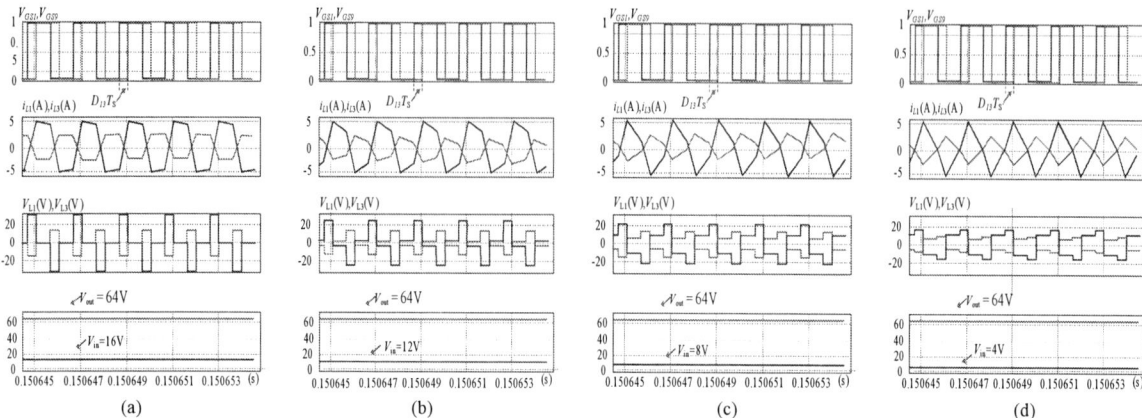

Fig. 7. Voltage and current waveforms in the equalization stage. (a) 0 to t_0 mode. (b) t_0 to t_1 mode. (c) t_1 to t_2 mode. (d) t_5 to t_6 mode.

t_0, which verifies the control logic in Fig. 5. The process is divided into intra-pack balancing and inter-module energy redistribution phases. Corresponding voltage and current waveforms of the converter under different operating conditions are provided in Fig. 6.

During the time interval from 0 to t_0, a selected battery is connected to the primary side of the converter to perform battery equalization. At the instant of t_0, the equalization process continues until the capacity of the battery matches that of the most closely - capacitated battery.

From t_0 to t_1, two selected batteries are connected to the primary side of the converter for battery equalization. At t_1, equalization proceeds until the batteries' capacities are brought into close proximity.

Between t_1 and t_2, three selected batteries are connected to the primary side of the converter for equalization. At themoment of t_2, the equalization within the module is finalized. Subsequently, the reversing switch is toggled, and the connection is shifted to the non-equalized module. The process from 0 to t_2 is then repeated until, at t_5, the internal equalization of all modules is completed.In the time span from t_5 to t_6, battery modules with capacities exceeding the average value are selected for discharging, until the overall equalization of the entire battery pack is achieved.

V. CONCLUSION

To address insufficient voltage gain and structural complexity, in energy management of long-series battery strings, this paper proposes a wide-voltage-gain equilibrium architecture based on multi-port converters. The architecture innovatively integrates port-multiplexed multi-port converter topology, supporting modular replacement and implementing hierarchical balancing strategy to achieve efficient step-up conversion from 4-16V input to over 64V output, while maintaining dynamic power path reconfiguration capability. Notably, the battery balancing mechanism is decoupled into intra-module and inter-module collaborative modes. PLECS simulations validate the technical advantages of GaN devices in enabling rapid balancing within compact configuration. Through co-optimization of topology and wide-bandgap devices, this solution establishes a new paradigm for high-efficiency energy management in high-density energy storage systems.

ACKNOWLEDGMENT

This work was supported in part by the Anhui Provincial Natural Science Foundation under Grant 2408085QE151, in part by the Open Fund of the State Key Laboratory of High-Efficiency and High-Quality Conversion for Electric Power under Grant 2024KF010.

REFERENCES

[1] Q. Wang, B. Mao, S. Stoliarov, and J. Sun, "A review of lithium-ionbattery failure mechanisms and fire prevention strategies," Prog. EnergyCombust. Sci., vol. 73, pp. 95–131, Jul. 2019.

[2] VARDHAN R K,SELVATHAI T,REGINALD R,et al.Modeling of single inductor based battery balancing circuit for hybrid electric vehicles[C]//Proceedings of IECON 2017-43rd Annual Conference of the IEEE Industrial Electronics Society.Beijing:IEEE,2017.

[3] P. A. Cassani and S. S. Williamson, "Feasibility analysis of a novel cellequalizer topology for plug-in hybrid electric vehicle energy-storagesystems," IEEE Trans. Veh. Technol., vol. 58, no. 8, pp. 3938–3946,Oct. 2009.

[4] X. Qi, Y. Wang and M. Fang, "An Integrated Cascade Structure-Based Isolated Bidirectional DC–DC Converter for Battery Charge Equalization," in IEEE Transactions on Power Electronics, vol. 35, no. 11, pp. 12003-12021, Nov. 2020,

[5] S. Li, C. C.Mi, andM. Zhang, "A high-efficiency active battery-balancingcircuit using multiwinding transformer," IEEE Trans. Ind. Appl., vol. 49,no. 1, pp. 198–207, Jan./Feb. 2013.

[6] X. Qi, Y. Wang, Y. Wang and Z. Chen, "Optimization of Centralized Equalization Systems Based on an Integrated Cascade Bidirectional DC–DC Converter," in IEEE Transactions on Industrial Electronics, vol. 69, no. 1, pp. 249-259, Jan. 2022.

[7] X. Qi, Y. Wang, M. Fang, Y. Wang and Z. Chen, "Principle and Topology Derivation of Integrated Cascade Bidirectional Converters for Centralized Charge Equalization Systems," in IEEE Transactions on Power Electronics, vol. 37, no. 2, pp. 1852-1869, Feb. 2022.

[8] M. Einhorn, W. Roessler, and J. Fleig, "Improved performance of serially connected Li-ion batteries with active cell balancing in electricvehicles," IEEE Trans. Veh. Technol., vol. 60, no. 6, pp. 2448–2457,Jul. 2011.

[9] X. Qi, Y. Wang, M. Fang, Y. Wang and Z. Chen, "Multiport DC–DC Converter With Integrated Cascaded Structure for Optimizing Centralized Battery Equalization System," in IEEE Transactions on Power Electronics, vol. 37, no. 12, pp. 15111-15126, Dec. 2022.

[10] Y. Shang, B. Xia, C. Zhang, N. Cui, J. Yang, and C. Mi, "Amodularizationmethod for battery equalizers using multiwinding transformers," IEEETrans. Veh. Technol., vol. 66, no. 10, pp. 8710–8722, Oct. 2017.

[11] A. Tavakoli, S. A. Khajehoddin and J. Salmon, "A Modular Battery Voltage-Balancing System Using a Series-Connected Topology," in *IEEE Transactions on Power Electronics*, vol. 35, no. 6, pp. 5952-5964, June 2020,

Design of Planar Integrated Transformer for LLC Resonant Converters Based on Hybrid Magnetic Materials

Jianguang Yao, Xiaoyi Xu, Yanfang Mao, Zhujian Ou, Runyang Ji
State Grid Nantong Power Supply Company
Nantong, Jiangsu Province, China
615589521@qq.com
5235799@qq.com
ntmaoyanfang2014@163.com
zj_ou89@163.com
jjrunyang@sina.com

Abstract—**LLC resonant converters are widely utilized in applications such as server power supplies, data center power systems, LED drivers, and electric vehicle onboard chargers. Traditional LLC converters employ wire-wound transformers and additional magnetic components as resonant inductors, significantly increasing the overall size of the converter. Planar magnetic integration technology enables the integration of resonant inductors and transformers, thereby enhancing power density. In various magnetic integration schemes using ferrite cores, air gaps are typically unavoidable, and the resulting fringing flux exacerbates conductor eddy current losses. This paper proposes a magnetic integration scheme for LLC resonant converters based on an E-core structure, where magnetic powder cores replace traditional air gaps to form a gapless integrated magnetic component with hybrid magnetic materials. The proposed design significantly reduces winding losses caused by fringing flux from air gaps, improves overall converter efficiency, and demonstrates more pronounced benefits at higher power levels. In an LLC resonant converter operating at 3.5 kW output power, the integrated transformer with the hybrid magnetic material solution achieves a 49.6% reduction in copper loss and 35.76% decrease in total magnetic component loss compared to the conventional air-gap design. The proposed hybrid magnetic material-based LLC converter attains a peak efficiency of 97.13%, demonstrating a 0.86% efficiency enhancement over the air-gap counterpart.**

Keywords—*LLC resonant converter, planar magnetic integration, fringing flux, hybrid magnetic materials*

I. INTRODUCTION

The growing demands of the energy conversion market drive the development of power converters toward higher efficiency and power density. High-performance power semiconductor devices have become indispensable for meeting these stringent requirements for enhanced efficiency and miniaturization. With the advent of wide bandgap (WBG) power semiconductor devices, it is feasible to elevate switching frequencies to hundreds of kilohertz (kHz) or even megahertz (MHz) ranges. At such high switching frequencies, the size of passive components can be drastically minimized. Additionally, the adoption of soft-switching topologies effectively mitigates excessive switching losses[1],[2]. The LLC converter exhibits soft-switching characteristics, enabling high efficiency in high power density conditions. Specifically, it can realize zero voltage switching (ZVS) for primary side switches, as well as the secondary rectifier (SR) switches. In addition, thanks to the sinusoidal-like secondary current, the SR turns off at nearly zero current (nearly ZCS) and turn off loss for the SR is mitigated. Therefore, for the LLC resonant converter, the device switching related loss only incorporates primary device turn off loss and gate loss for primary and SR devices[3].

In recent years, planar transformers have gained increasing popularity in high-frequency power converters due to their distinct advantages, including low-height design, excellent thermal performance, modularity, manufacturing simplicity, enhanced reliability, and higher power density. Planar magnetic components and PCB windings, characterized by their high integration density, are gradually replacing conventional magnetic components and Litz windings, serving as an effective solution for high-frequency, high-power-density LLC converters. Additionally, magnetic integration offers another viable approach to significantly reduce the volume and losses of magnetic components. Extensive research has been conducted on magnetic integration technologies. The integration of two transformers into a single core, achieved by leveraging the magnetic flux cancellation principle, effectively reduces the volume and losses of magnetic components [5]. A topology-based integration of resonant inductors and transformers has been demonstrated to minimize winding and core losses [6]. For CLLC resonant tanks, a configuration where windings are arranged on one side column of the core, with the central column forming a leakage flux path to enhance leakage inductance, enables the integration of three magnetic elements [7]. Inductance tuning via horizontal air gap adjustments has been proposed by positioning primary windings on core side columns and secondary windings on the central column [8]. The introduction of magnetic shunts between primary and secondary windings creates low-reluctance paths for leakage flux, increasing leakage inductance to facilitate inductor-transformer integration [9].

In high-frequency magnetics, ac winding losses are affected by skin and proximity effects, including uneven current distribution due to fringing magnetic fields around air gaps. Many efforts have addressed the problem of fringing field losses at high frequency for planar magnetics. It is well known that fringing effects can be mitigated using distributed air gaps. The open-circuit copper screen acts as a flux barrier, reducing the fringing effect through eddy currents that generate counteracting flux opposing the fringing flux [10]. Additionally, shaping copper windings to form "keep-away regions" has been shown to effectively minimize fringing loss [11]. Positioning conductors at a distance from air gaps has been shown to significantly mitigate the fringing flux impact [12]. An orthogonal air gap approach—a distributed technique involving gaps in core segments parallel and perpendicular to windings has been proposed to optimize flux distribution [13].

Parameter	Value
$N_\mathrm{p}{:}N_\mathrm{s}$	13:10
f_{r1}	300kHz
S_1-S_4	CREE C3M0015065D
D_1-D_4	CREE C6D16065H

This paper proposes an E-core-based hybrid magnetic material integration scheme tailored for LLC converters, where magnetic powder cores replace air gaps in conventional integrated planar transformers. By maintaining equivalent magnetic reluctance between the magnetic powder core and the air gap, the proposed hybrid design significantly suppresses fringing flux from air gaps, thereby reducing winding eddy current losses. Finite element simulations demonstrate a 23.5% reduction in winding losses compared to traditional air-gap configurations, with the efficiency improvement becoming more pronounced at higher power levels.

The paper is structured as follows: Section II presents the design considerations and circuit parameters of the LLC resonant converter. Section III elaborates on the integrated transformer design, including practical model parameters, loss analysis, magnetic circuit evaluation, and core/winding configurations. Section IV provides a comparative analysis of simulation results. Section V concludes the study.

II. DESIGN OF LLC RESONANT CONVERTER

A. Parameter Design of the LLC Resonant Converter

Fig. 1 shows the circuit topology of LLC converter, where L_r represents the resonant inductance, L_m denotes the magnetizing inductance, and C_r is the resonant capacitor. When energy is transferred from the primary to the secondary side of the transformer, the magnetizing inductance becomes clamped, leaving only L_r and C_r to resonate at frequency f_{r1}.

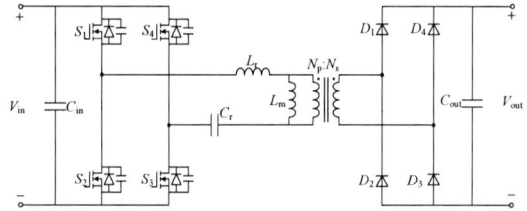

Fig. 1. Circuit diagram of LLC resonant converter.

Conversely, when the transformer ceases energy transformer and the output capacitor supplies the load, L_m participates in the resonance, resulting in a modified resonant frequency f_{r2}. To achieve high efficiency in the LLC converter, meticulous parameter design is essential [14].

$$f_{r1} = \frac{1}{2\pi\sqrt{L_r C_r}} \tag{1}$$

$$f_{r2} = \frac{1}{2\pi\sqrt{(L_m + L_r) C_r}} \tag{2}$$

The specifications and detailed parameters of the proposed converter are summarized in Table 1.

TABLE I. LLC CONVERTER SPECIFICATIONS AND PARAMETERS

Parameter	Values
V_in	400V
V_out	200-480V
L_r	8uH
L_m	50uH
C_r	35nF

B. Loss Analysis of the LLC Resonant Converter

The power losses in the LLC resonant converter include conduction and switching losses of semiconductor devices, core losses, and winding losses of transformers and inductors. This paper employs a transformer with integrated resonant inductance, and the loss analysis of the integrated magnetic components will be discussed in the following section. When operating near the resonant point, the LLC converter achieves ZVS for the secondary-side power MOSFETs and ZCS for the secondary-side diodes. The conduction and switching losses of the power MOSFETs, as well as the conduction losses of the diodes, can be quantified through thermal simulation in PLECS.

III. DESIGN OF PLANAR INTEGRATED TRANSFORMER

A. Equivalent Circuit Modeling and Loss Analysis of Planar Transformers

Fig. 2 illustrates the practical equivalent circuit model of the transformer, where L_m, L_kp and L_ks denote the magnetizing inductance, primary-side leakage inductance, and secondary-side leakage inductance, respectively. In this study, the proposed magnetic integration scheme employs magnetic circuit design to utilize the primary-side leakage inductance as the resonant inductor of the LLC converter while eliminating the secondary-side leakage inductance. The parasitic resistances of the primary and secondary windings are denoted as R_p and R_s, respectively, whereas R_s models the core loss resistance. The parasitic capacitances C_p, C_s, and C_ps correspond to the intra-winding capacitances of the primary and secondary sides, as well as the inter-winding capacitance between them. Compared to conventional wire-wound magnetics, planar magnetic components have higher parasitic capacitance, which can disrupt the normal operation of LLC resonant converters. It should be noted that the inter-winding capacitance C_ps provides a low-impedance path for common-mode noise, leading to electromagnetic interference (EMI) issues. Additionally, the intra-winding capacitances (C_p, C_s) can resonate with the magnetizing inductance and resonant inductance, causing oscillations in the transformer output voltage and resonant current waveforms. These parasitic capacitances are highly influenced by factors such as the winding overlap area, inter-winding spacing, and winding geometry. Therefore, minimizing parasitic capacitance is a critical design objective.

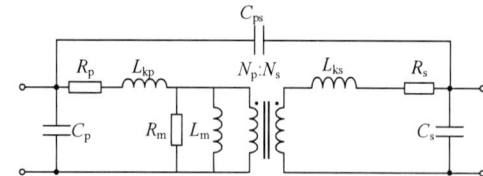

Fig. 2. Planar transformer equivalent circuit model.

The losses in planar transformers consist of core loss and winding loss. Core loss primarily arise from hysteresis loss, eddy current loss, and residual loss. The conventional core loss model employs the Steinmetz equation

$$P = Cf^{\alpha}B_m^{\beta}V \qquad (3)$$

C, α, β expressed as empirical coefficients typically provided by the core manufacturer's datasheet, B_m is the peak magnetic density, V is the core volume. However, this formulation is accurate only under sinusoidal excitation and exhibits inherent limitations for non-sinusoidal waveforms.

Winding losses in transformers dramatically increase with high frequency due to eddy current effects. Eddy current losses, including skin effect and proximity effect losses, seriously impair the performance of transformers in high-frequency power conversion applications. Both the skin effect and the proximity effect cause the current density to be nonuniform in the cross section of the conductor and thus cause a higher winding resistance at higher frequency. Planar winding structures feature flat rectangle copper conductors with higher width-to-thickness ratios. The most commonly used expression for the ac resistance of the mth layer is derived as follows

$$\frac{R_{ac,m}}{R_{dc,m}} = \frac{\varepsilon}{2}\left[\frac{\sinh\varepsilon + \sin\varepsilon}{\cosh\varepsilon - \cos\varepsilon} + (2m-1)^2 \frac{\sinh\varepsilon - \sin\varepsilon}{\cosh\varepsilon + \cos\varepsilon}\right] \qquad (4)$$

where ε is an effective thickness of the skin depth, which is defined as h/δ, h is the thickness of conductor, and δ is the skin depth at a given frequency. The variable m represents a winding portion and defined as a ratio:

$$m = \frac{F(h)}{F(h) - F(0)} \qquad (5)$$

where $F(0)$ and $F(h)$ are the MMFs at the limits of a layer surface. The first term in (3) describes the skin effect factor, and the second term represents the proximity effect factor. The proximity effect loss in a multilayer winding may strongly dominate the skin effect loss, depending on the value of m, which relates to the winding arrangements. Calculate the AC/DC resistance ratio of each layer, multiply it by the square of the RMS value of the current to obtain the winding loss.

B. Magnetic Circuit Modeling of Planar Integrated Transformers

The proposed magnetic integration scheme adopts a conventional E-core structure, with primary and secondary windings wound on the left and right legs, while the center leg is designed to provide a leakage flux path. As illustrated in Fig. 3(a), the primary winding comprises N_{p1} turns on the left leg with upward magnetomotive force (MMF) polarity and N_{p2} turns on the right leg with downward MMF polarity. The secondary windings, wound on both legs with N_{s1} and N_{s2} turns, generate opposing MMFs relative to the primary windings. When $N_{p1} > N_{p2}$ and $N_{s1} < N_{s2}$, the flux paths are depicted in Fig. 3(c). The red path represents the main excitation flux, the yellow path denotes primary-side leakage flux, and the blue path corresponds to secondary-side leakage flux. This configuration retains both primary and secondary leakage inductances, making it suitable for CLLC resonant converters. When $N_{p1} > N_{p2}$ and $N_{s1} = N_{s2}$, the secondary windings on the left and right legs generate counteracting fluxes in the center leg, resulting in zero net secondary leakage inductance. This configuration is specifically optimized for magnetic integration in LLC resonant converter.

Fig. 3(b) shows the equivalent magnetic circuit of the proposed integrated magnetic structure. By applying the superposition theorem to the magnetic circuit, the flux values in the three branches can be determined as:

$$\phi_1 = \frac{\left(N_{p1}R_1 + N_{p1}R_2 + N_{p2}R_1\right)i_p - \left(N_{s1}R_1 + N_{s1}R_2 + N_{s2}R_1\right)i_s}{R_2^2 + 2R_1R_2} \qquad (6)$$

$$\phi_2 = \frac{\left(N_{p1} - N_{p2}\right)i_p + \left(N_{s2} - N_{s1}\right)i_s}{R_2 + 2R_1} \qquad (7)$$

$$\phi_3 = \frac{\left(N_{p1}R_1 + N_{p2}R_1 + N_{p2}R_2\right)i_p - \left(N_{s1}R_1 + N_{s2}R_1 + N_{s2}R_2\right)i_s}{R_2^2 + 2R_1R_2} \qquad (8)$$

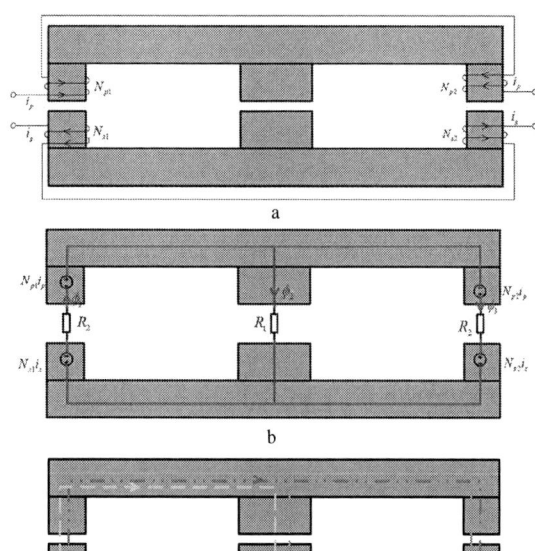

Fig. 3. Integrated magnetic structure. (a) Winding configuration. (b) Equivalent magnetic circuit model. (c) Magnetic flux path.

The reluctance of the ferrite core is negligible compared to that of the air gaps; thus, only the air gap reluctances are considered in the analysis. Here, R_1 and R_2 represent the air gap reluctances of the center leg and side legs, respectively.

$$R_1 = \frac{l_1}{\mu_0 A_1} \qquad (9)$$

$$R_2 = \frac{l_2}{\mu_0 A_2} \qquad (10)$$

A_1 and A_2 denote the cross-sectional areas of the center leg and side legs, respectively, and l_1, l_2 are the air gap lengths corresponding to the center leg and side legs. Based on the definitions of self-inductance and mutual inductance, the primary self-inductance L_1, secondary self-inductance L_2, and mutual inductance M can be derived as given in Eq. (11-13).

$$L_1 = \frac{\left(N_{p1}^2 + N_{p2}^2 + 2N_{p1}N_{p2}\right)R_1 + \left(N_{p1}^2 + N_{p2}^2\right)R_2}{R_2^2 + 2R_1R_2} \tag{11}$$

$$L_2 = \frac{\left(N_{s1}^2 + N_{s2}^2 + 2N_{s1}N_{s2}\right)R_1 + \left(N_{s1}^2 + N_{s2}^2\right)R_2}{R_2^2 + 2R_1R_2} \tag{12}$$

$$M = \frac{\left(N_{p1}N_{s1} + N_{p1}N_{s2} + N_{p2}N_{s1} + N_{p2}N_{s2}\right)R_1 + \left(N_{p1}N_{s1} + N_{p2}N_{s2}\right)R_2}{R_2^2 + 2R_1R_2} \tag{13}$$

By equating the input-output port characteristics of the mutual inductance model and the physical magnetic structure, the magnetizing inductance L_m and leakage inductances L_{kp}, L_{ks} can be derived. Theoretically, arbitrary inductance values can be achieved through magnetic integration by rationally allocating winding turns and precisely controlling reluctance values. When the secondary windings are evenly distributed across both the primary and secondary sides ($N_{s1}=N_{s2}=N_{s/2}$) and identical air gaps are applied to the center leg and side legs ($R_2=2R_1$), secondary leakage inductance is zero, the expressions for the magnetizing inductance, primary leakage inductance, and the inductance ratio are given by

$$L_{pm} = \frac{N_p^2}{4R_1} \tag{14}$$

$$L_{kp} = \frac{\left(N_{p1} - N_{p2}\right)^2}{8R_1} \tag{15}$$

$$L_n = \frac{2N_p^2}{\left(N_{p1} - N_{p2}\right)^2} \tag{16}$$

Magnetic Integration Procedure for LLC Resonant Converters:

Divide the secondary winding into even turns and wind them equally on both side legs. This ensures cancellation of secondary leakage flux in the center leg. Tune the primary winding turns on the left and right legs to achieve the desired inductance ratio. Adjust the air gap lengths in the center leg and side legs to precisely control reluctances and thereby finalize absolute inductance values.

C. Specific Structure of Planar Integrated Transformer

In this design, the transformer employs a turn ratio of 1.3:1. To prevent excessive core flux density and avoid saturation, the primary winding is configured with 13 turns. while the secondary winding comprises 10 turns equally split into on the side legs. This symmetrical secondary winding arrangement eliminates secondary leakage inductance, enabling leakage inductance integration for the LLC resonant converter. To align the magnetizing-to-leakage inductance ratio with the LLC resonant tank requirements, the primary winding is asymmetrically distributed with $N_{p1}=10$ and $N_{p2}=3$. This configuration ensures that closely approximates the inductance ratio of the LLC resonant cavity, optimizing converter performance.

An E-core (model E64_10_50) with LIANFENG_NH9A material was selected for the magnetic structure. Based on the designed circuit parameters and winding distribution, the initial air gap length was calculated as 0.52 mm using Eq. 5. However, due to fringing flux effects at the air gap edges,

the effective cross-sectional area exceeds the core's physical cross-sectional area, introducing a discrepancy in the theoretical calculation.

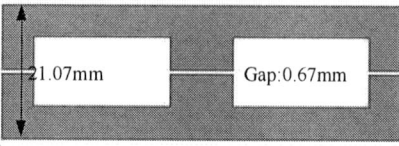

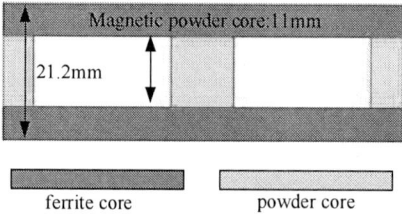

Fig. 4. Core structures of two Solutions. (a) Air gap. (b) Hybrid magnetic materials.

To obtain more accurate air gap value, finite element analysis (FEA) was performed via Ansys Maxwell, refining the air gap to 0.67 mm for optimal flux density distribution and inductance accuracy. To mitigate the eddy current effects induced by fringing flux at the air gap edges, a hybrid magnetic core structure was developed by replacing the air gaps in the three legs with powder cores. The final design adopts NPX19 powder cores (manufactured by POCO) for their stable permeability and low core loss characteristics. With a relative permeability of 19, the powder core length was optimized to 11 mm through FEA to maintain equivalent reluctance to the original air gap configuration. The core structures of the traditional air-gap design and the hybrid magnetic material design are compared in Fig. 4.

The parasitic capacitance of planar transformers is heavily influenced by their PCB winding structures. Conventional PCB substrates, such as FR-4, exhibit a relative permittivity of 4.2-4.6, resulting in significant parasitic capacitance due to enhanced electric field coupling between conductive layers. To mitigate this issue, a zero voltage gradient winding layout is proposed[15]. The key idea here is that the top and bottom layers of each PCB should be identical and connected in parallel to achieve zero voltage gradient and minimize the parasitic capacitance. The winding structure and simulation model are illustrated in Fig. 5. The primary winding, comprising 13 turns, is implemented using a dual-PCB configuration: a 2-layers PCB(PCB1) positioned at the top of the core window and a 4-layers PCB(PCB2) at the bottom. The 10 turns allocated to the left-side leg are equally distributed across both PCBs, with 5 turns wound symmetrically on the top and bottom layers of each PCB. These layers are connected in parallel to enforce a zero voltage gradient. The 3 turns allocated to the right-side leg are distributed across both PCBs as follows:1 turn is wound on the top PCB, with 1 turn per layer (top and bottom) connected in parallel.2 turns are implemented on the bottom PCB, featuring 2 turns per layer (top and bottom) in a parallel configuration. The two PCBs are electrically and mechanically interconnected via a central connector to ensure synchronized current sharing and structural rigidity. To further suppress intra-winding parasitic capacitance, the inner layers of the four-layer PCB employ a Z-type winding pattern, which staggers conductive traces to disrupt capacitive coupling paths between adjacent turns. The secondary winding, consisting of 10 turns, is implemented using a single four-layer PCB(PCB2) centrally positioned within the core window. The winding is equally divided between the left and right legs, with 5 turns wound on each leg. Each leg's 5 turns are distributed across the top and bottom layers of the PCB (5 turns per layer) and connected in parallel. The inner layers of the four-layer PCB are routed to ensure dot convention alignment. All three PCBs employ a standard thickness of 1.6 mm, and a 1.6 mm spacing is maintained between adjacent PCBs to minimize inter-winding parasitic capacitance.

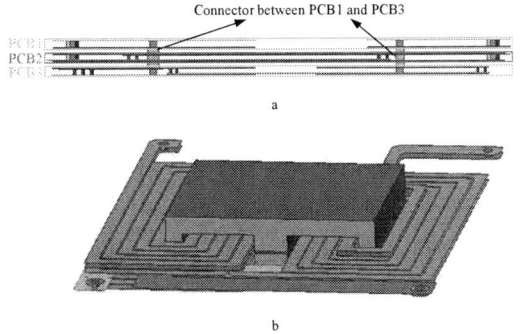

Fig. 5. Winding structure and simulation model. (a)Winding structure. (b)Simulation model

IV. SIMULATION RESULTS ANALYSIS

A. Comparison of simulation results

The power device losses are obtained through PLECS thermal simulations. By exporting the primary and secondary current waveforms from PLECS and applying them as excitation sources in Ansys transient field simulations, the core losses and winding copper losses of the integrated magnetic components are derived.

When the LLC resonant converter operates at the resonant frequency with an output power of 3.5 kW, the losses of the integrated transformers for both design schemes, as simulated in Maxwell transient field analysis, are shown in Fig. 6, and the overall loss distribution is presented in Fig. 7.

As illustrated in Figs. 6, 7, the hybrid magnetic material integrated transformer (composed of a magnetic powder core replacing the air gap) demonstrates comparable semiconductor losses to the conventional air-gap design.

However, it achieves a significant reduction in copper losses (49.6% average reduction) with only a moderate increase in core losses, resulting in a 35.73% total loss reduction for the integrated transformer. Figure 8 compares the flux density distributions of the two schemes at 3.5 kW output power, both remaining below the saturation flux density of the core materials.

B. Comparison of Efficiency Curves

Fig. 9 illustrates the efficiency curves of the two magnetic core configurations. The LLC converter employing the hybrid magnetic material integrated transformer (with a magnetic powder core replacing the air gap) demonstrates

superior efficiency across the full power range compared to the conventional air-gap design. The proposed hybrid solution achieves a peak efficiency of 97.15%, representing a 0.86% improvement over the air-gap counterpart.

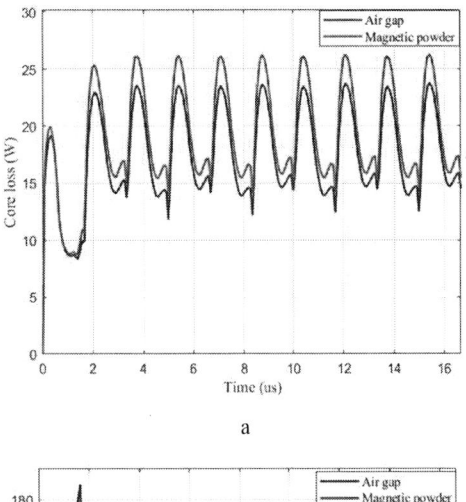

a

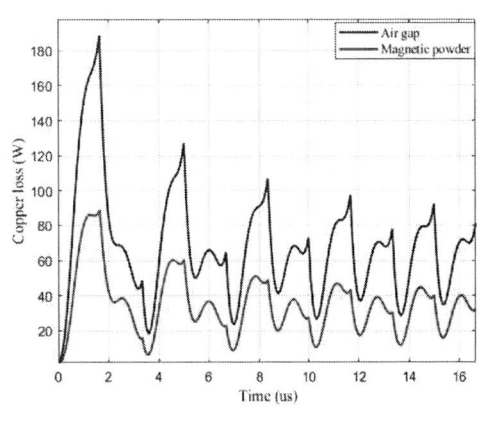

b

Fig. 6. Core loss and copper loss of two integrated transformers at 3.5 kW. (a) Core loss. (b) Copper loss.

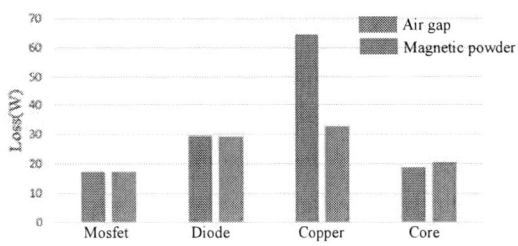

Fig. 7. Overall loss distribution of the LLC resonant converter at 3.5 kW.

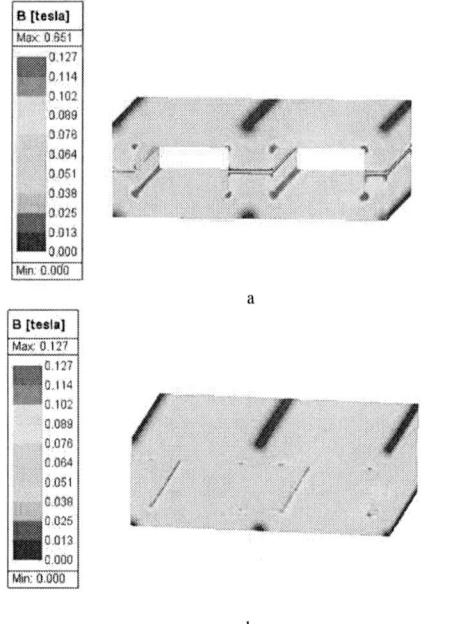

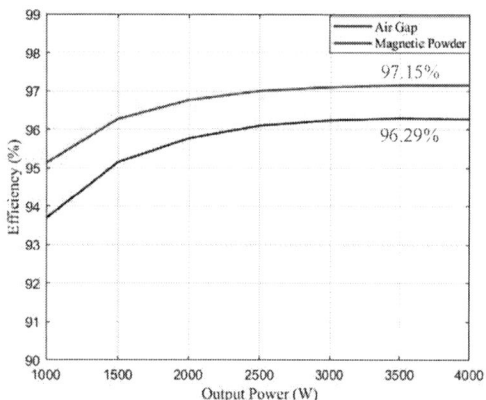

Fig. 8. Flux density distribution of the two magnetic core designs: (a) Air-gap design, (b) Magnetic powder core design.

Fig. 9. Efficiency curves of the two designs.

V. CONCLUSION

This paper proposes a planar integrated transformer employing a hybrid magnetic material composed of a magnetic powder core to replace the airgap, forming an airgap-free structure. The PCB winding configuration utilizes a zero-voltage-gradient method to minimize parasitic capacitance. Compared to the conventional airgap-based integrated transformer, the proposed hybrid magnetic design achieves significant copper loss reduction while maintaining equivalent magnetizing inductance and leakage inductance. In the LLC resonant converter operating at 3.5 kW output power, the integrated transformer with the hybrid magnetic material solution achieves a 49.6% reduction in copper loss and 35.76% decrease in total magnetic component loss compared

to the conventional air-gap design. The proposed hybrid magnetic material-based LLC converter attains a peak efficiency of 97.13%, demonstrating a 0.86% efficiency enhancement over the air-gap counterpart.

ACKNOWLEDGMENT

This project received support from the Key Technology Project of State Grid Nantong Power Supply Company. Research on Energy Management and Power Generation Optimization Strategies for Multi-Port Converters Applied to Residential Photovoltaic Energy Storage Systems (Grant No.NT2024001).

REFERENCES

[1] X. Wang, S. Zhao, Z. Yang, Y. Liu, H. A. Mantooth, and L. Ding, "Comprehensive analysis of synchronous rectifying signal delay of high-frequency LLC resonant DC/DC converter," *IEEE Trans. Power Electron.*, vol. 39, no. 10, pp. 13348–13364, Oct. 2024.

[2] S. Zhao, A. Kempitiya, W. T. Chou, V. Palija, and C. Bonfiglio, "Variable DC-link voltage LLC resonant DC/DC converter with wide bandgap power devices," *IEEE Trans. Ind. Appl.*, vol. 58, no. 3, pp. 2965–2977, May-Jun. 2022.

[3] F. C. Lee, Q. Li, and A. Nabih, "High frequency resonant converters: An overview on the magnetic design and control methods," *IEEE J. Emerg. Sel. Topics Power Electron.*, vol. 9, no. 1, pp. 11–23, Feb. 2021.

[4] Z. Ouyang, O. C. Thomsen, and M. A. E. Andersen, "Optimal design and tradeoff analysis of planar transformer in high-power DC–DC converters," *IEEE Trans. Ind. Electron.*, vol. 59, no. 7, pp. 2800–2810, Jul. 2012.

[5] C. Fei, F. C. Lee, and Q. Li, "High-efficiency high-power-density LLC converter with an integrated planar matrix transformer for high-output current applications," *IEEE Trans. Ind. Electron.*, vol. 64, no. 11, pp. 9072–9082, Nov. 2017.

[6] Y. Liu, H. Wu, J. Zou, J. Hu, and C. Zhao, "CLL resonant converter with secondary side resonant inductor and integrated magnetics," *IEEE Trans. Power Electron.*, vol. 36, no. 10, pp. 11316–11325, Oct. 2021.

[7] B. Li, Q. Li, and F. C. Lee, "High-frequency PCB winding transformer with integrated inductors for a bi-directional resonant converter," *IEEE Trans. Power Electron.*, vol. 34, no. 7, pp. 6123–6135, Jul. 2019.

[8] M. D'Antonio, S. Chakraborty, and A. Khaligh, "Planar transformer with asymmetric integrated leakage inductance using horizontal air gap," *IEEE Trans. Power Electron.*, vol. 36, no. 12, pp. 14014–14028, Dec. 2021.

[9] M. Li, Z. Ouyang, and M. A. E. Andersen, "High-frequency LLC resonant converter with magnetic shunt integrated planar transformer," *IEEE Trans. Power Electron.*, vol. 34, no. 3, pp. 2405–2415, Mar. 2019.

[10] J. Fletcher, B. Williams, and M. Mahmoud, "Air gap fringing flux reduction in inductors using open-circuit copper screens," in *Proc. Electr. Power Appl.*, vol. 152, no. 4, pp. 990–996, Jul. 2005.

[11] L. Ye, G. R. Skutt, R. Wolf, and F. C. Lee, "Improved winding design for planar inductors," in *Proc. IEEE PESC*, 1997, pp. 1561–1567.

[12] Z. Ouyang, G. Sen, O. C. Thomsen, and M. A. E. Andersen, "Analysis and design of fully integrated planar magnetics for primary-parallel isolated boost converter," *IEEE Trans. Ind. Electron.*, vol. 60, no. 2, pp. 494–508, Feb. 2013.

[13] S. Mukherjee, Y. Gao, and D. Maksimović, "Reduction of AC winding losses due to fringing-field effects in high-frequency inductors with orthogonal air gaps," *IEEE Trans. Power Electron.*, vol. 36, no. 1, pp. 815–828, Jan. 2021.

[14] J. Deng, S. Li, S. Hu, C. C. Mi, and R. Ma, "Design methodology of LLC resonant converters for electric vehicle battery chargers," *IEEE Trans. Veh. Technol.*, vol. 63, no. 4, pp. 1581–1592, May 2014.

[15] M. A. Saket, N. Shafiei, and M. Ordonez, "LLC converters with planar transformers: Issues and mitigation," *IEEE Trans. Power Electron.*, vol. 32, no. 6, pp. 4524–4542, Jun. 2017.

Common-mode Noise Analysis of Hall-Effect Sensor for SiC Power Converter

Guifeng Geng
State Key Laboratory of High Density Electrical Energy Conversion
Huazhong University of Science and Technology
Wuhan, China
M202372291@hust.edu.cn

Peng Zhou
State Key Laboratory of High Density Electrical Energy Conversion
Huazhong University of Science and Technology
Wuhan, China
zhou_p@hust.edu.cn

Runquan Jiang
State Key Laboratory of High Density Electrical Energy Conversion
Huazhong University of Science and Technology
Wuhan, China
M202372276@hust.edu.cn

Xuejun Pei
State Key Laboratory of High Density Electrical Energy Conversion
Huazhong University of Science and Technology
Wuhan, China
ppei215@hust.edu.cn

Abstract—**The fast switching capability of SiC-MOSFET devices enables significant improvements in power density for power electronic converters. However, this technological advantage introduces substantial challenges in electromagnetic compatibility due to high dv/dt-induced noise effects, particularly impacting the accuracy of sensor sampling circuits. This paper systematically investigates the cross-domain electromagnetic coupling phenomena between power circuits and measurement systems in closed-loop hall effect sensor sampling configurations. Through comprehensive characterization of parasitic capacitance parameters in typical hall effect sensor primary-secondary structures, we reveal the multi-path conduction mechanisms by which power circuit dv/dt transients propagate as common-mode interference to sensitive sampling circuits. The study establishes a detailed equivalent circuit model elucidating the common-mode interference transmission path, supported by rigorous analysis of the parasitic capacitance coupling effects between high-voltage power components and low-voltage measurement circuits. These findings provide critical insights for developing effective electromagnetic interference suppression strategies in high-switching-speed power electronic systems.**

Keywords—*hall effect sensor, current measurement, voltage measurement, electromagnetic interference (EMI), SiC-MOSFET, common-mode noise*

I. INTRODUCTION

The advent of wide bandgap semiconductor devices, particularly silicon carbide (SiC) metal oxide semiconductor field effect transistors (MOSFETs), has revolutionized power electronics by enabling high switching frequencies exceeding traditional silicon-based solutions. However, the high switching speed characteristics of SiC-MOSFET bring about high dv/dt noise issues[1], posing new challenges to the hardware circuit design of power electronic converters.

As the power density of power electronic converters increases, parasitic parameters inevitably exist between the power circuit and control circuit of the converter[2]. The high dv/dt of the SiC MOSFET switching process in the power circuit will excite displacement current on the parasitic

capacitance between the power circuit and control circuit, which will couple to the control system in the form of common mode current[3], affecting the integrity of the PWM signal and the voltage and current sampling results. For example, in 2022, SUNGROW pointed out at "TI Green Energy Semiconductor Technology Innovation Summit" that there is a serious phenomenon of EMI crosstalk between power circuit and control circuit in SST, which seriously affects communication cables and sampling circuits.

Currently, research on EMI crosstalk between power circuit and control circuit in power electronic converters focuses on gate crosstalk[4]-[7], while research on the common mode interference mechanism of isolated sampling circuits is insufficient, lacking theoretical basis and feasible suppression measures. Therefore, analyzing the common mode interference coupling mechanism of isolated sampling circuits and proposing feasible suppression methods are of great significance for improving the reliability of SiC power converters.

Reference [8] analyzed the problem of high-frequency noise in the load current of a three-phase voltage source inverter with R-L load. The high-frequency noise component in the load current is believed to be caused by dv/dt generated by switching actions and parasitic capacitance of the load inductance. Compared with the fundamental component, the amplitude of the high-frequency noise component is high, close to the unit signal-to-noise ratio, which may affect the overcurrent protection circuit.

The parasitic capacitance parameter between the power circuit and the isolated sampling circuit is crucial. Reference [9] presents a method for extracting the parasitic capacitance of the primary and secondary sides of a current hall effect sensor. High frequency and low voltage sine excitation are used on the primary side, and the voltage on the secondary side detection resistor R is measured. By calculating, the parasitic capacitance of the primary and secondary sides of the LA-55-P current hall effect sensor is approximately 0.31pF. The coupling path between the high dv/dt of the power circuit and the sampling circuit of the current hall effect sensor is analyzed. Reference [10] proposed a grounding shielding strategy for current hall effect sensors in inverters, connecting the shielding layer of the output cable to the potential dead point on the DC side of the inverter. Experimental results

---

This work was supported in part by the National Natural Science Foundation of China under Grant 52377187, in part supported by the China Postdoctoral Science Foundation under Grant Number 2024M761010, and in part of the stably supported project of the State Key Laboratory of High Density Electrical Energy Conversion under Grant Number DN2024-02.

showed that the noise suppression performance of this shielding strategy was superior to traditional grounding shielding. Reference [11] analyzed the equivalent common mode conduction impedance of various forms of voltage and current sampling circuits, and provided corresponding expressions for calculating the common mode conduction impedance. The CMTI performance of the sampling circuit was evaluated based on the magnitude of the common mode conduction impedance of the sampling circuit.

This article takes a typical Buck circuit as the research object and studies the closed-loop voltage and current hall effect sensor sampling circuit. The parasitic capacitance parameters of the hall effect sensor's primary and secondary sides are extracted through experiments. The common mode interference conduction path from high dv/dt of the power circuit to the secondary side of the voltage and current hall effect sensor is analyzed, and an equivalent circuit model of common mode interference is established. The calculation expression for the interference voltage in the sampling results is given.

II. PARASITIC CAPACITANCE EXTRACTION OF CLOSED-LOOP HALL-EFFECT SENSOR

A. Structure of Hall-effect Sensor

The closed-loop hall effect sensors used in power electronic converters include closed-loop voltage hall effect sensors and closed-loop current hall effect sensors. This article selects CHB-20L (closed-loop voltage hall effect sensor) and LA-50P (closed-loop current hall effect sensor) as research objects, and their internal structure diagrams are shown in Fig. 1 and Fig. 2. It can be seen that the main difference between the two lies in the number of turns of the primary winding and the rated current value of the primary winding.

Fig. 1. Internal structure of CHB-20L

Fig. 2. Internal structure of LA-50P

B. Parasitic Capacitance Extraction

The experimental scheme for measuring parasitic capacitance used in this article refers to the scheme proposed in reference [9]. A signal generator is used to apply a 20MHz, ±10V sine wave excitation to the primary winding, measure the sine voltage VR on the secondary resistor, calculate the current value in the circuit, and obtain the impedance of the circuit. Subtracting the impedance of the measured resistor can obtain the impedance of the parasitic capacitance. The results in this paper are shown in TABLE I. .

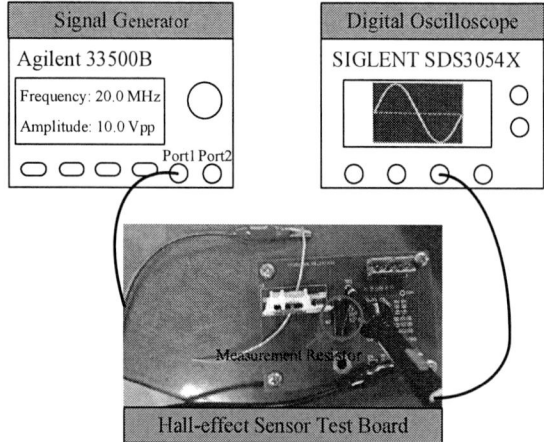

Fig. 3. Schematic diagram of experimental scheme for extracting parasitic parameters of hall effect sensor

TABLE I. PARASITIC CAPACITANCE EXTRACTION RESULTS OF HALL EFFECT SENSOR

	Type	Test Result
Current hall effect sensor	LA-50P	0.32pF
Voltage hall effect sensor	CHB-20L	4.78pF

III. COMMON MODE INTERFERENCE ANALYSIS

The parasitic capacitance between the power circuit and the sampling circuit is an important path for the high dv/dt of the power circuit to conduct to the sampling circuit. The following equation shows the principle of coupling dv/dt and parasitic capacitance to generate common mode noise current. The peak value of the common mode noise current is proportional to the capacitance of the parasitic capacitance Cp and dv/dt.

$$i_{CM\_niose} = C_p \frac{dv}{dt}$$

According to experimental tests, the switching process of SiC MOSFET can generate a dv/dt of up to 40kV/us. Even if the parasitic capacitance is only a few pF, such a high dv/dt can generate noise currents of tens of mA, which has adverse effects on the sampling circuit.

The power circuit and control circuit structure of Buck converter considering parasitic parameters is shown in Fig. 4 and Fig. 5, and parasitic capacitance value is shown in TABLE II.

TABLE II. PARASITIC CAPACITANCE VALUE OF BUCK CONVERTER

Symbol	Description	Value
Cp1	Parasitic capacitance of midpoint node of SiC-MOSFET bridge to PE	50pF
Cp2	Parasitic capacitance of Vdc+ to PE	60.8pF
Cp3	Parasitic capacitance of Vdc- to PE	21.5pF
Cp4	Parasitic capacitance of AGND- to PGND	1nF

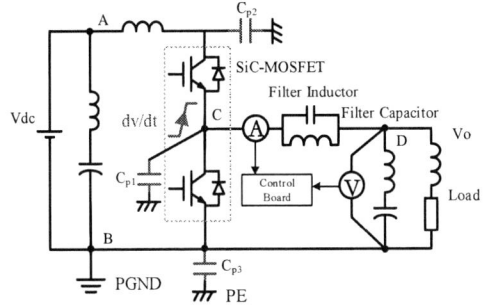

Fig. 4. Power circuit structure of Buck converter considering parasitic parameters

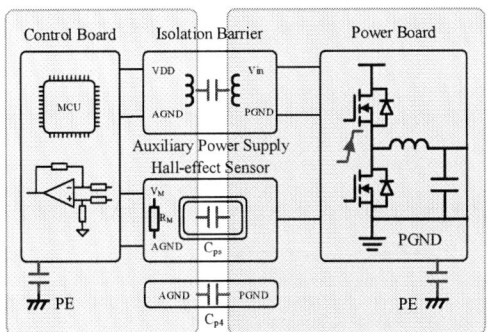

Fig. 5. Control circuit structure of Buck converter considering parasitic parameters

The midpoint node of the half bridge of the Buck converter is the emission source of dv/dt, which enters the control circuit in the form of common mode interference through the parasitic capacitance of the Hall sensor. In the sampling stage of the control circuit, dv/dt returns through the parasitic capacitance between the control circuit ground and the power circuit ground, forming a closed-loop conduction path.

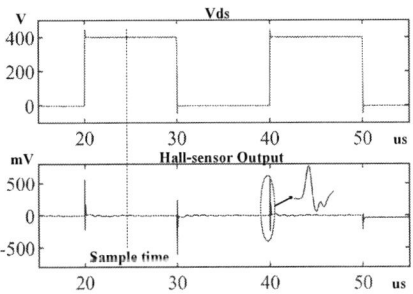

Fig. 6. Interference components in sampled signals

Fig. 6 shows the interference spikes in the sampled signal, indicating that voltage spikes occur during the switching process and have a short duration. For commonly used average current control, the sampling point is usually set at the middle position of the duty cycle, and the sampling results are almost unaffected by interference spikes. For the overcurrent protection circuit of the converter, the protection threshold time is compared with the sampling signal, so interference spikes can cause the overcurrent protection circuit to malfunction, affecting the normal operation of the converter.

IV. MODEL OF COMMON MODE INTERFERENCE

A. Interference Source Model

The third part of the analysis shows that the dv/dt caused by the high-speed switching of SiC MOSFETs is a common mode interference source. Many scholars have conducted research on mathematical models of interference sources[12][13], established segmented linear models of switch transient processes, and calculated the time-domain waveform of interference sources. The mathematical model of the interference source is not the focus of this article, so we will not delve into the mathematical model of the interference source in depth. Therefore, this article uses experimental methods to obtain the waveform of the interference source, and the extracted time-domain waveform of the interference source is shown in Fig. 7.

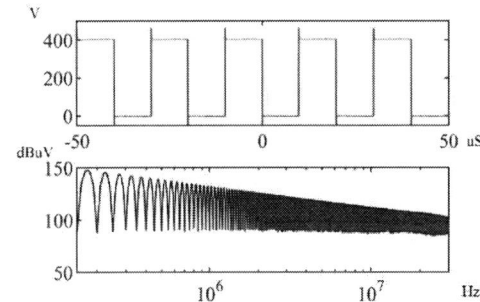

Fig. 7. Waveform and spectrum of interference source

B. Common-mode Equivalent Circuit Model

The current Hall sensor sampling circuit and the voltage Hall sensor sampling circuit have different measurement points in the power circuit, and their corresponding interference paths are also different. This section will analyze the common mode interference equivalent circuit models of the two separately. In order to facilitate the establishment of an equivalent model for common mode interference, this article makes the following assumptions: different sampling circuits are independent of each other and have no field circuit coupling.

For the current Hall sensor sampling circuit as shown in Fig. 8, the common mode interference current is coupled to the sampling circuit through the parasitic capacitance of the Hall sensor at point C, flows into the signal conditioning circuit to form a differential mode voltage, and then returns to the power side through the capacitor set between the analog ground (AGND) of the sampling circuit and the power ground (PGND) of the power circuit.

$$V_{c\_o} = I_{CM\_c} \times Z_M = \frac{V_{ds}}{Z_{ps\_c} + Z_M + Z_G} \times Z_M$$

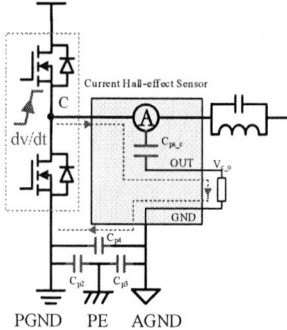

Fig. 8. Schematic diagram of interference path in current hall effect sensor sampling circuit

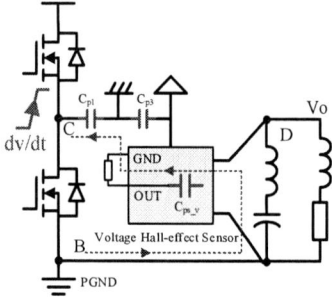

Fig. 9. Schematic diagram of interference path in voltage hall effect sensor sampling circuit

For the voltage Hall sensor sampling circuit as shown in Fig. 9. The common-mode interference current in the Hall-effect voltage sensor sampling circuit establishes a closed-loop conduction path through parasitic coupling mechanisms, as illustrated in Fig. 9. This current originates from the power ground (PGND) of the power conversion unit and couples into the sensing circuit via the parasitic capacitance of the Hall sensor. Subsequently, the interference current propagates through the printed circuit board's stray capacitance to protective earth (PE) formed by the signal conditioning circuitry. Finally, the current completes its circulation path by returning to the power supply domain through the ground-referenced parasitic capacitance at the midpoint node of the half-bridge.

$$V_{v\_o} = I_{CM\_v} \times Z_R = \frac{V_{ds}}{Z_{ps\_v} + Z_M + Z_G + Z_o} \times Z_R$$

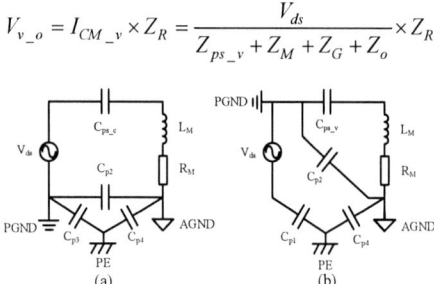

Fig. 10. Common-mode equivalent circuit model of hall effect sensor, (a) Current hall effect sensor, (b) Voltage hall effect sensor

Fig. 10 shows the common mode interference equivalent circuit models of the current Hall sensor sampling circuit and the voltage Hall sensor sampling circuit. The common mode

current generates a voltage drop across the sampling resistor, causing disturbance to the sampling signal.

V. CONCLUSION

This paper analyzes the common-mode interference mechanism of hall effect sensor sampling circuit in SiC power converter. Parasitic capacitance between primary side and secondary side of the hall effect sensor are extracted through experiments and the voltage hall effect sensor has larger parasitic capacitance. Then the interference source of the power circuit and the conduction path of common mode interference coupled to the sampling circuit are located. The interference mechanism and possible influence of the sampling circuit are analyzed, and the common-mode interference equivalent model is established.

REFERENCES

[1] S. Ji, S. Zheng, F. Wang and L. M. Tolbert, "Temperature-dependent characterization, modeling, and switching speed-limitation analysis of third-generation 10-kV SiC MOSFET," in IEEE Transactions on Power Electronics, vol. 33, no. 5, pp. 4317-4327, May 2018.

[2] B. F. Kjærsgaard et al., "Parasitic capacitive couplings in medium voltage power electronic systems: an overview," in IEEE Transactions on Power Electronics, vol. 38, no. 8, pp. 9793-9817, Aug. 2023.

[3] .K. Mainali, S. Madhusoodhanan, A. Tripathi, K. Vechalapu, A. De and S. Bhattacharya, "Design and evaluation of isolated gate driver power supply for medium voltage converter applications," 2016 IEEE Applied Power Electronics Conference and Exposition (APEC), Long Beach, CA, USA, 2016, pp. 1632-1639.

[4] P. Wang, L. Zhang, X. Lu, H. Sun, W. Wang and D. Xu, "An improved active crosstalk suppression method for high-speed SiC MOSFETs," in IEEE Transactions on Industry Applications, vol. 55, no. 6, pp. 7736-7744, Nov.-Dec. 2019.

[5] S. Jahdi, O. Alatise, P. Alexakis, L. Ran and P. Mawby, "The impact of temperature and switching rate on the dynamic characteristics of silicon carbide schottky barrier diodes and MOSFETs," in IEEE Transactions on Industrial Electronics, vol. 62, no. 1, pp. 163-171, Jan. 2015.

[6] S. Xu, W. Sun and D. Sun, "Analysis and design optimization of brushless DC motor's driving circuit considering the Cdv/dt induced effect," 2010 IEEE Energy Conversion Congress and Exposition, Atlanta, GA, USA, 2010, pp. 2091-2095.

[7] J. Wang and H. S. -h. Chung, "Impact of parasitic elements on the spurious triggering pulse in synchronous buck converter," 2013 IEEE Energy Conversion Congress and Exposition, Denver, CO, USA, 2013, pp. 480-487.

[8] A. Kumar, S. Parashar and S. Bhtattacharya, "Continuous heat run test of latest generation power modules for 10 kV 4H-SiC MOSFETs in medium voltage power converters," 2018 IEEE Energy Conversion Congress and Exposition (ECCE), Portland, OR, USA, 2018, pp. 1949-1955.

[9] M. R. Nielsen, M. Kirkeby, H. Zhao, D. N. Dalal, M. Møller Bech and S. Munk-Nielsen, "Noise Analysis of current sensor for medium voltage power converter enabled by Silicon-Carbide MOSFETs," 2022 IEEE 9th Workshop on Wide Bandgap Power Devices & Applications (WiPDA), Redondo Beach, CA, USA, 2022, pp. 180-185.

[10] J. Yoo, Y. -R. Lee, H. Kim and S. -K. Sul, "Shielding technique for noise reduction in hall-effect current sensor of voltage source inverter," 2022 IEEE Energy Conversion Congress and Exposition (ECCE), Detroit, MI, USA, 2022, pp. 1-5.

[11] W. Meng, F. Zhang, Z. Fu and G. Dong, "High dv/dt noise modeling and reduction on control circuits of GaN-based full bridge inverters," in IEEE Transactions on Power Electronics, vol. 34, no. 12, pp. 12246-12261, Dec. 2019.

[12] M. R. Ahmed, R. Todd and A. J. Forsyth, "Predicting SiC MOSFET behavior under hard-switching, soft-switching, and false turn-on conditions," in IEEE Transactions on Industrial Electronics, vol. 64, no. 11, pp. 9001-9011, Nov. 2017.

[13] Y. Xie, C. Chen, Y. Yan, Z. Huang and Y. Kang, "Investigation on ultralow turn-off losses phenomenon for SiC MOSFETs with improved switching model," in IEEE Transactions on Power Electronics, vol. 36, no. 8, pp. 9382-9397, Aug. 2021.

979-8-3315-1110-4/25 $31.00 © 2025 IEEE

Research on Online Reliability Assessment Technology for IGBT Based on Driver-Side Measurement Fusion

Yi Liu
School of Electrical Engineering
Tiangong University
Tianjin, China
yiliu@tiangong.edu.cn

Zhicheng Liu
School of Electrical Engineering
Tiangong University
Tianjin, China
2431070897@tiangong.edu.cn

Yunhui Mei
School of Electrical Engineering
Tiangong University
Tianjin, China
meiyunhui@163.com

Abstract—**Insulated Gate Bipolar Transistor (IGBT) have been widely adopted in energy conversion systems, and their reliability has become a critical focus in industrial applications. Temperature is the primary cause of failure in power electronic converters, and accurate junction temperature estimation of IGBT modules serves as a fundamental basis for reliability assessment. This paper proposes an online junction temperature evaluation method for IGBT based on multi-drive monitoring quantity fusion. By integrating gate current (I_g), Miller voltage (V_{GP}), and plateau duration during turn-off ($t_{MP,off}$), a three-domain Kalman filter spanning voltage, time, and current is constructed. Furthermore, adaptive Kalman filtering is employed to achieve dynamic noise suppression. Experimental results demonstrate that this method achieves a fusion weight adaptation accuracy of ±1.5%, enabling precise measurement of junction temperature.**

Keywords—*IGBT junction temperature monitoring, driver-side sampling, multi-parameter fusion, Kalman filtering, adaptive noise suppression*

I. INTRODUCTION

High-voltage and high-power power electronic devices have been comprehensively applied in fields such as new energy power generation, flexible DC transmission, electric locomotive traction, and electric vehicles. In these applications, power electronic devices typically operate in harsh operating environments and face severe reliability challenges. Their health management and reliability assessment have increasingly attracted the attention of both the academic and industrial communities.The Insulated Gate Bipolar Transistor (IGBT), as a core component in modern power electronic systems, has its junction temperature directly impacting module lifespan and system reliability. Statistics indicate that approximately 55% of IGBT failures are caused by excessive junction temperatures[1]. High-power IGBT modules are internally composed of multiple material layers, including aluminum bond wires, power chips, solder layers, copper layers, ceramic substrates, and baseplates. The two most common long-term failure mechanisms induced by temperature are the aging of aluminum bond wires and solder layer degradation. Due to the mismatch in coefficients of thermal expansion (CTE) among the internal layers of IGBT modules, long-term thermo-mechanical fatigue accumulation leads to void formation and delamination between layers, ultimately resulting in permanent failure of the power devices[2]–[3],as shown in Fig.2. For short-term thermal breakdown failures, IGBT modules are prone to transient overtemperature failures under operating conditions such as overcurrent, short-circuit, and current imbalance. Therefore, junction temperature monitoring technologies for IGBTs, particularly real-time monitoring under practical operating conditions in converters, are crucial for enhancing the performance, reliability, and longevity of power conversion systems[4].

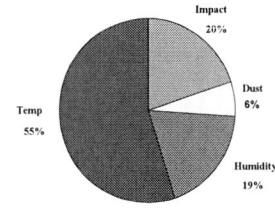

Fig. 1.   Proportion of IGBT failure

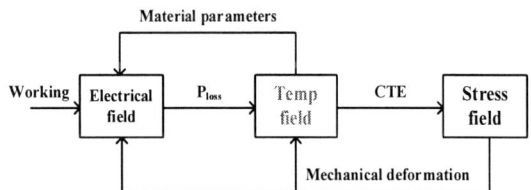

Fig. 2.   IGBT failure mechanism

Junction temperature measurement methods for power semiconductor devices can be broadly categorized into four classes: physical contact methods, optical techniques, thermal network models, and thermo-sensitive electrical parameters (TSEPs)[5]–[7]. While physical contact and optical methods are straightforward to implement, they require destructive modifications to the device packaging to create measurement channels, making them highly intrusive. Thermal network models rely on computationally intensive calculations and are constrained by significant limitations. In contrast, the thermo-sensitive electrical parameters method enables non-invasive temperature estimation by leveraging external electrical parameter measurements without altering the device packaging. It achieves microsecond-level response times, making it suitable for online applications.

In recent years, TSEPs-based monitoring methods have emerged as a research hotspot.Temperature influences three fundamental properties of semiconductor materials: intrinsic carrier concentration, carrier mobility, and carrier lifetime,

This work was supported by the National Natural Science Foundation of China under Grant 52177189, and also by the Tianjin Municipal Science and Technology Bureau under Grants 24JCZXJC00130 and 21JCJQJC00150.

which collectively govern the electrical behavior of emiconductor devices. This relationship has led to the identification of numerous TSEPs, including collector-emitter voltage Vce[8], rate of collector-emitter voltage change[9], threshold voltage V_{th}[10], Miller plateau voltage V_{MP}[11]–[13], and gate current I_g[14]–[16]. These parameters are typically converted into junction temperature (T_j) using pre-calibrated look-up tables. However, measurement inaccuracies and the low temperature sensitivity of TSEPs often result in significant noise in T_j.Furthermore, the aging-induced degradation of power modules introduces additional complexities, as certain TSEPs exhibit drift characteristics correlated with material fatigue, thereby exacerbating estimation inaccuracies.

This paper presents a novel online evaluation framework based on multi-sensor fusion from the gate-drive interface. The methodology synergistically integrates multiple TSEPs measurement models to overcome the limitations of single-parameter approaches. Direct extraction of T_j from individual thermo-sensitive electrical parameters remains challenging due to parameter interdependencies. For instance, studies on the Miller plateau voltage (V_{MP})[11], demonstrate its pronounced dependence on collector current:elevated I_C levels induce nonlinear V_{MP}-T_j relationships in high-temperature regimes, primarily due to accelerated self-heating effects. Studies on the gate peak current (($I_{g,peak}$)[16],investigations into $I_{g,peak}$ reveal its strong correlation with localized thermal conditions near the gate pad and robustness against bond wire degradation. Nevertheless,Ig,peak-based methods exhibit degraded accuracy in estimating average chip temperature under uniform thermal gradients.

To address these challenges, a Kalman filter (KF)-based algorithm is implemented for dynamic T_j estimation. The KF framework, renowned for its real-time operability and noise suppression capabilities[17], aligns with the stringent requirements of IGBT thermal monitoring. By incorporating a state-space thermal model and a power loss estimator driven by load current measurements, the KF adaptively compensates for aging-induced parameter drift and cooling condition variations [18]. A weighted fusion strategy is applied to gate current ($I_{g,peak}$), Miller voltage (V_{MP}), and turn-off delay time ($t_{d,off}$), augmented by wavelet-based denoising and dual-variable voltage-time modeling. This hybrid approach mitigates the intrinsic limitations of single-TSEPs methods—including noise susceptibility, operational dependency, and calibration errors—achieving a T_j estimation accuracy of ±1.3°C across the entire operational temperature range.

The paper is structured as follows: Section II elaborates on the physical principles governing the selected TSEPs. Section III details the gate-drive-side sampling circuitry design. Section IV formulates the Kalman filter algorithm integrated with the thermal model. Section V validates the methodology through experimental characterization of multichip IGBT modules. Section VI concludes the study.

II. MODEL ANALYSIS OF THE RELATIONSHIP BETWEEN TSEP AND JUNCTION TEMPERRATURE

The Insulated Gate Bipolar Transistor (IGBT) is a high-power semiconductor device characterized by a composite structure combining a bipolar junction transistor (BJT) with a MOS-gated topology. A cross-sectional view of its typical internal architecture is illustrated in Fig.3. When the gate-to-emitter voltage exceeds the threshold voltage , an inversion layer forms at the surface of the p-type emitter region, establishing an n-channel. This enables electron injection from the N+ emitter region into the N- drift region. Concurrently, the conduction of the pnp bipolar transistor between the collector and emitter drives the IGBT into its on-state. Based on this operational principle, the equivalent circuit of a practical IGBT is depicted in Fig.3.

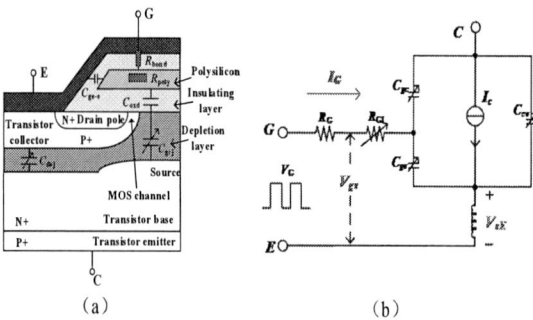

Fig. 3. (a)The internal structure of IGBT (b)The simplified model on the drive side

Fig.4 illustrates the switching waveforms of an IGBT,Where V_{CE} is the gate drive voltage, V_{CE} is the collector-emitter voltage, V_{TH} is the threshold voltage, V_{MP} is the Miller voltage level, and I_C is the on-state current. This process is generally divided into 8 stages, namely t_0 to t_8.

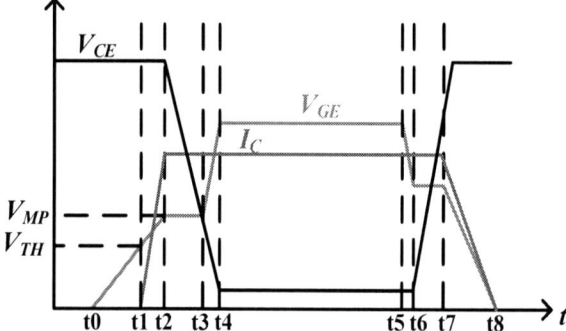

Fig. 4. IGBT Conduction Waveform

A. Gate peak Current $I_{g,peak}$

Elevated temperatures modify both the gate oxide capacitance and carrier mobility, thereby influencing the charging and discharging dynamics of the gate. Gate current exhibits an approximately linear rise before reaching its peak (linear charging phase), expressed as:

$$I_G(t) = \frac{V_G}{R_G + R_{G,int}(T)} + \frac{V_G}{L_S}t \tag{1}$$

Here, V_G is the gate voltage, R_G is the external gate resistor, and $R_{G,int}$ is the temperature-dependent internal gate resistor. Junction temperature (T_j) is derived from $R_{G,int}$ as:

$$R_{G,int}(T_j) = \alpha_1 T_j(k) + \beta_1 \tag{2}$$

979-8-3315-1110-4/25 $31.00 © 2025 IEEE

B. Miller Voltage V_{MP}

The formation of Miller voltage is related to the parasitic capacitance (C_{GC}) of the IGBT, and C_{GC} is affected by temperature. An increase in temperature will reduce the dielectric constant of the semiconductor material, leading to changes in C_{GC}, which in turn affects V_{MP}:

$$V_{\mathrm{MP}}(T) = \sqrt{\frac{2I_d L}{\mu(T)C_{\mathrm{ox}}W}} + V_T(T) \qquad (3)$$

In the formula, I_g represents the gate current, and R_G,int is the internal gate resistance of the IGBT, which increases linearly with temperature.

By decoupling the relationship between I_C and V_{MP} in reference [4], the relationship between the Miller voltage and the junction temperature can be expressed as:

$$V_{\mathrm{MP}}(T) = I_g R_{\mathrm{G,int}}(T) + \sqrt{\frac{2\left(\cosh\dfrac{W}{L}-1\right)I_c L}{\mu(T)C_{\mathrm{ox}}W}} + V_T(T) \qquad (4)$$

Experimental results show that V_{MP} decreases approximately linearly with temperature and can be modeled as:

$$V_{\mathrm{MP}}(T) = I_g R_{\mathrm{G,int}}(T) + V_T(T) \qquad (5)$$

C. Turn-off Miller Voltage Delay Time $t_{MP,off}$

The turn-off delay time is defined as the interval between the initiation of the gate-emitter voltage (V_{GE}) decay and the commencement of the collector-emitter voltage (V_{CE}) rise. The Miller plateau duration during turn-off ($t_{MP,off}$) reflects the carrier transit velocity within the IGBT. Elevated temperatures reduce carrier mobility, thereby prolonging the charge accumulation time during switching transients.

The onset of the $t_{MP,off}$ rise (the end of the turn-off delay) coincides with the V_{GE} approaching the Miller voltage (V_{Miller}), which is a function of the threshold voltage (V_{th}), transconductance (g_{fs}), and load current (I_L), as expressed by:

$$V_{\mathrm{Miller}} = V_{\mathrm{th}} + \frac{I_L}{g_{fs}} \qquad (7)$$

where g_{fs} denotes the IGBT transconductance, typically derived from the transfer characteristics in datasheets at a specific junction temperature (T_j). For a given device, both V_{th} and g_{fs} exhibit temperature dependency. Consequently, as shown in the equation above, the Miller plateau duration during turn-off is a temperature-sensitive parameter. However, it is critical to note that load current (I_L) influences V_{Miller}, thereby modulating $t_{MP,off}$.

Studies in [19] demonstrate that $t_{MP,off}$ increases with rising T_j. For a fixed T_j, $t_{MP,off}$ elongates with higher DC bus voltages (V_{dc}), while under identical T_j and V_{dc}, it decreases with increasing load current (I_L).

Experimental results indicate that $t_{MP,off}$ exhibits an approximately linear dependence on temperature, which can be modeled as:

$$t_{MP,off}(T_j) = \alpha_3 T_j(k) + \beta_3 \qquad (8)$$

III. THE DESIGN OF THE DRIVER SAMPLING CIRCUIT

A. Circuit Architecture Design

The drive-side sampling circuit needs to accurately capture I_g, V_{MP}, and $t_{MP,off}$ without interfering with the switching characteristics of the IGBT. The overall architecture is shown in Fig.5, which includes the following modules:

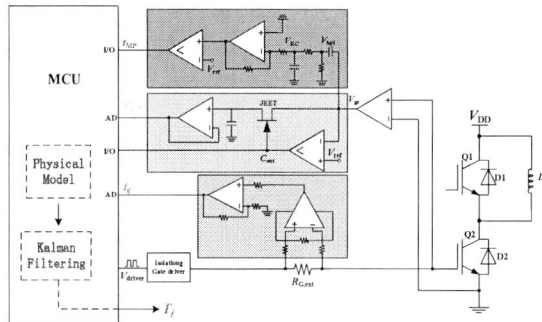

Fig. 5. The architecture of the drive-side sampling circuit

(1) Gate current detection: The peak voltage of the external gate resistor $V_{R,gext}$ can be measured to monitor the peak current of the gate. An instrumentation amplifier is used to collect the voltage across the gate drive resistor and convert it into current.

(2) Miller voltage measurement module: This module consists of a differential input section, a comparison section, and a sample-and-hold section.

(3) Miller platform duration measurement module: Since the derivative of the Miller platform with respect to time is zero, a high-pass filter is used to differentiate the gate drive waveform of the IGBT. By using a comparator, a low-level signal can be generated, triggering the acquisition of the Miller platform signal.

IV. IMPLEMENTATION OF MULTI-SENSOR DATA FUSION

A. Construction of the Observation Model

In practical systems, where data is sampled in discrete form (e.g.at a switching frequency of f_{sw}=10 kHz), the continuous-time model must be discretized. The discretization is performed using a time step k, which corresponds to one or multiple switching periods , depending on the system's sampling frequency and application requirements.

The template is designed for, but not limited to, six authors. Based on the relationship between temperature-sensitive parameters and junction temperature established in Chapter 3, the gate current (I_g), Miller voltage V_{MP}, and turn-off Miller plateau duration $t_{MP,off}$ are integrated into the observation vector z_k, which has a nonlinear relationship with the junction temperature T_j as follows:

$$\mathbf{z}_k = \begin{bmatrix} I_g(k) \\ V_{MP}(k) \\ t_{MP,off}(k) \end{bmatrix} = \begin{bmatrix} \alpha_1 T_j(k) + \beta_1 \\ \alpha_2 T_j(k) + \beta_2 \\ \alpha_3 T_j(k) + \beta_3 \end{bmatrix} + \mathbf{v}_k \qquad (9)$$

Where:

- α_i, β_i : Coefficients of the linear model

- V_k: Observation noise vector with an initial covariance matrix, and the noise variance is calibrated through steady-state experiments.

B. Design of State Space Model

To describe the dynamic characteristics of the junction temperature, the state variables are selected as the junction temperature and its rate of change:

$$\mathbf{x}_k = \begin{bmatrix} T_j(k) \\ \dot{T}_j(k) \end{bmatrix} \qquad (10)$$

Where $\dot{T}_j(k)$ is the rate of change of the junction temperature, reflecting the acceleration of temperature change.

C. Construction of the State Equation Construction of the State Equation

Based on the first-order thermodynamic model, the dynamic evolution equation of the junction temperature is:

$$\mathbf{x}_k = \mathbf{A}\mathbf{x}_{k-1} + \mathbf{B}P_{\text{loss}}(k) + \mathbf{w}_k \qquad (11)$$

D. Sage-Husa Adaptive Kalman Filter Implementation

a) Positioning

Set the initial state estimate error covariance matrix $P_0 =$ diag(1,0.1).

Define forgetting factors $\alpha = 0.01$ and $\beta = 0.01$ to balance the influence of historical data and new observations:

b) Prediction Step:

$$\hat{\mathbf{x}}_k^- = \mathbf{A}\hat{\mathbf{x}}_{k-1} + \mathbf{B}P_{\text{loss}}(k)$$
$$P_k^- = \mathbf{A}P_{k-1}\mathbf{A}^T + Q \qquad (12)$$

c) Update Step:

Calculate the Kalman gain:

$$\mathbf{K}_k = P_k^- \mathbf{H}_k^T (\mathbf{H}_k P_k^- \mathbf{H}_k^T + R)^{-1} \qquad (13)$$

Update the state estimate and covariance.

d) Adaptive adjustment of noise covariance:

Online update of process noise covariance Q and observation noise covariance R:

$$\hat{\mathbf{x}}_k = \hat{\mathbf{x}}_k^- + \mathbf{K}_k(\mathbf{z}_k - \mathbf{h}(\hat{\mathbf{x}}_k^-))$$
$$P_k = (I - \mathbf{K}_k \mathbf{H}_k) P_k^- \qquad (15)$$

V. EXPERIMENTAL VERIFICATION

The experimental verification scheme of this paper is based on the single-phase full-bridge inverter platform, with a focus on the measurement method of the junction temperature of the IGBT in the lower arm of the half-bridge structure. The FF450R12ME4 IGBT half-bridge module was adopted as the test object in the experiment. After opening the cover of the module and removing the surface silicone grease, since each bridge arm of this model of IGBT module is composed of three chips in parallel, in order to obtain the average junction temperature of each bridge arm, a fiber optic temperature sensor (with a temperature variation range of 0-200°C, a sampling rate of 10Hz, and an accuracy of ±2°C) is attached to the surface of each chip. Establish a quantitative mapping model between junction temperature and driving electrical quantity. The inverter test device is shown in Fig. 6.

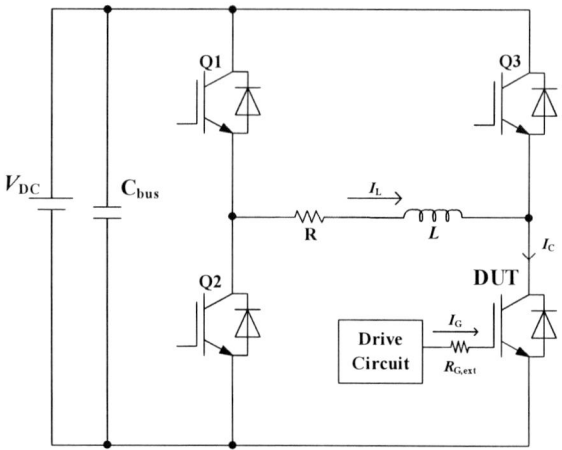

Fig. 6. Single-phase inverter circuit topology

A. The relationship between the driving monitoring quantity and Tj was established

By applying a constant conduction current to the IGBT under test to generate steady-state power loss, a controlled thermal boundary condition is constructed using a water-cooled heat sink. When the module reaches a thermal equilibrium state, the temperature field distribution on the chip surface is monitored in real time using the fiber optic sensor. After processing the data, the effective average temperature of the module is calculated. After the junction temperature stabilizes at the preset reference value, the drive monitoring quantities are collected synchronously to establish a mapping model between junction temperature and drive monitoring quantities.

This experiment conducted pulse tests within the range of 20 to 125°C, and a total of 12 temperature points were used. Take 10 samples at each temperature point and calculate the average value. The relationship between the driving monitoring quantity and the temperature was calculated by using the simple linear regression method, and the linear model of the driving monitoring quantity and the junction temperature was established. Fig.7 to Fig.9 show the experimental results of the fitting.

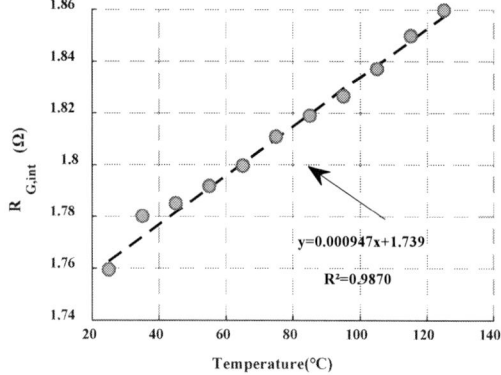

Fig. 7. Relationship between $R_{G,\text{int}}$ and T_j.

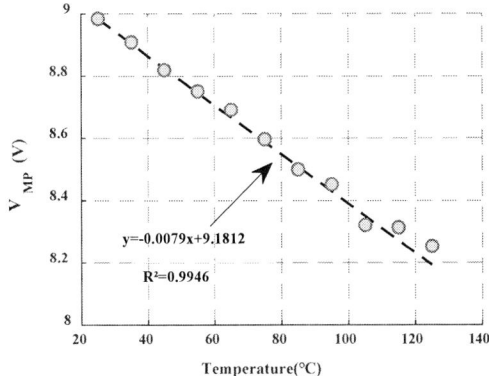

Fig. 8. Relationship between V_{MP} and T_j.

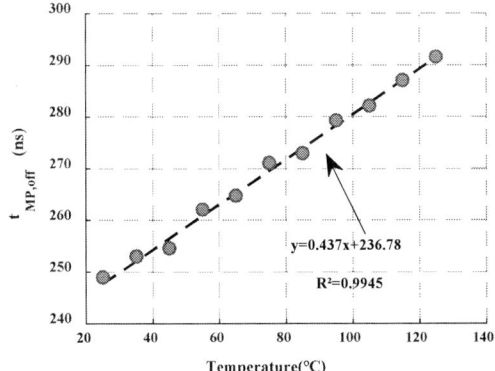

Fig. 9. Relationship between $t_{MP,off}$ and T_j.

The above experiments show that the temperature sensitivity of a single drive monitoring quantity presents a relatively obvious segmented feature. That is, at lower temperatures, the linearity of Miller voltage sensitivity to temperature is better, but as the temperature increases, the linearity deteriorates. In the medium-temperature region, the linearity of the peak current at the gate should be better. Although the Miller platform has good sensitivity in the high-temperature range for turn-off time, it changes relatively little at low temperatures and has poor linearity. The problem of insufficient linearity of a single temperature-sensitive parameter throughout the entire temperature range can be solved by introducing the variable weight fusion mechanism based on Kalman filtering. The following will verify this method through specific experiments.

B. *Verification of Kalman Filter Estimate of Junction Temperature*

Subsequent dynamic validation experiments are conducted as follows: The calibrated module is re-encapsulated and integrated into the inverter system, and the half-bridge circuit under test is operated under rated conditions. Based on the gate drive characteristic monitoring circuit designed in Section 3, the characteristic quantities of the drive waveform are collected in real time through a 16-bit high precision ADC (resolution 0.1mV), and combined with the previously established temperature-electrical model, non-intrusive junction temperature online estimation is achieved. This experimental scheme can effectively verify the dynamic

accuracy and engineering practicability of the proposed junction temperature monitoring method by comparing the direct measurement values from the fiber optic sensor with the model inversion results.

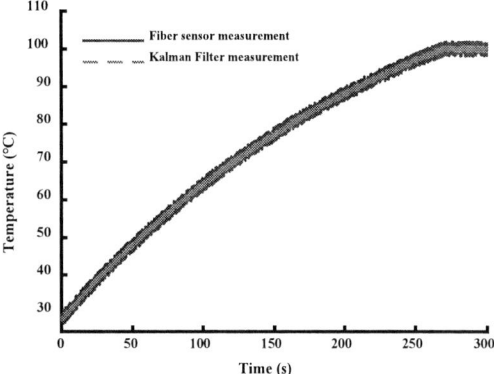

Fig. 10. Comparison of Kalman filter junction temperature measurement and fiber optic sensor measurement

Fig.10 shows the comparison between the estimated T_j value given by the Kalman filter and the measured T_j value obtained by the fiber optic sensor. When the T_j estimation module given by Kalman filtering continuously rises, it can accurately track the measurement results of the optical fiber sensor, and the measured noise is less than that of the optical fiber sensor. The results show that the studied method for measuring the junction temperature of IGBT based on the fusion of driving monitoring quantities is feasible and effective.

VI. CONCLUSION

Realizing the online junction temperature monitoring of IGBT modules is an important basis for the reliability assessment of them. This paper proposes an online reliability assessment method for IGBT based on the fusion of multiple temperature-sensitive parameters. Compared with the single-parameter assessment method, it can evaluate the junction temperature of IGBT more accurately, thereby improving the accuracy of IGBT reliability assessment. Multi-parameter fusion utilizes the complementary information of multiple temperature-sensitive parameters, which can more comprehensively reflect the working status of IGBTs, effectively reduce the evaluation errors caused by external interference or self-characteristic limitations of a single parameter, significantly improve the full-temperature range accuracy and real-time performance of IGBT junction temperature monitoring, and provide theoretical support and technical guarantee for the reliability management of power electronic systems. Subsequently, the multi-parameter fusion algorithm can be further optimized and the experimental conditions expanded to enhance the practicability and universality of this method.

REFERENCES

[1] H. Wang, M. Liserre and F. Blaabjerg, "Toward Reliable Power Electronics: Challenges, Design Tools, and Opportunities," in IEEE Industrial Electronics Magazine, vol. 7, no. 2, pp. 17-26, June 2013.

[2] U. -M. Choi, F. Blaabjerg and S. Jørgensen, "Power Cycling Test Methods for Reliability Assessment of Power Device Modules in Respect to Temperature Stress," in IEEE Transactions on Power Electronics, vol. 33, no. 3, pp. 2531-2551, March 2018.

[3] J.Lutz, H. Schlangenotto, U. Scheuermann, and R. D. Doncker, Semiconductor Power Device—Physics, Characteristic, Reliability. New York, NY, USA: Springer-Verlag, 2011, ch. 11.

[4] S. Yang, A. Bryant, P. Mawby, D. Xiang, L. Ran and P. Tavner, "An industry-based survey of reliability in power electronic converters," 2009 IEEE Energy Conversion Congress and Exposition, San Jose, CA, USA, 2009.

[5] Y. Avenas, L. Dupont and Z. Khatir, "Temperature Measurement of Power Semiconductor Devices by Thermo-Sensitive Electrical Parameters—A Review," in IEEE Transactions on Power Electronics, vol. 27, no. 6, pp. 3081-3092, June 2012.

[6] N. Baker, M. Liserre, L. Dupont and Y. Avenas, "Improved Reliability of Power Modules: A Review of Online Junction Temperature Measurement Methods," in IEEE Industrial Electronics Magazine, vol. 8, no. 3, pp. 17-27, Sept. 2014.

[7] E. R. Motto and J. F. Donlon, "IGBT module with user accessible on-chip current and temperature sensors," 2012 Twenty-Seventh Annual IEEE Applied Power Electronics Conference and Exposition (APEC), Orlando, FL, USA, 2012, pp. 176-181.

[8] Y. Peng, Q. Wang, H. Wang and H. Wang, "An On-Line Calibration Method for TSEP-Based Junction Temperature Estimation," in IEEE Transactions on Industrial Electronics, vol. 69, no. 12, pp. 13616-13624, Dec. 2022.

[9] A. Bryant et al., "Investigation into IGBT dv/dt during turn-off and its temperature dependence," IEEE Trans. Power Electron., vol. 26, no. 10, pp. 3019–3031, Mar. 2011.

[10] M. Du, Y. Tang, M. Gao, Z. Ouyang, K. Wei and W. G. Hurley, "Online Estimation of the Junction Temperature Based on the Gate Pre-Threshold Voltage in High-Power IGBT Modules," in IEEE Transactions on Device and Materials Reliability, vol. 19, no. 3, pp. 501-508, Sept. 2019.

[11] S. Cheng, Y. Hu, C. Xiang, J. Liu, X. Wu and J. Yao, "An Online Condition Monitor Method for IGBT Independent of Collector Current," in IEEE Transactions on Transportation Electrification, vol. 8, no. 4, pp. 4607-4621, Dec. 2022.

[12] Y. Quan et al., "Online Junction Temperature Monitoring Method for SiC MOSFETs Based on Turn-Off Miller Plateau Voltage," in IEEE Transactions on Power Electronics, vol. 39, no. 12, pp. 15800-15810, Dec. 2024.

[13] C. H. van der Broeck, A. Gospodinov and R. W. De Doncker, "IGBT Junction Temperature Estimation via Gate Voltage Plateau Sensing," in IEEE Transactions on Industry Applications, vol. 54, no. 5, pp. 4752-4763, Sept.-Oct. 2018.

[14] N. Baker, S. Munk-Nielsen, F. Iannuzzo and M. Liserre, "IGBT Junction Temperature Measurement via Peak Gate Current," in IEEE Transactions on Power Electronics, vol. 31, no. 5, pp. 3784-3793, May 2016.

[15] Q. Zhang and P. Zhang, "An Online Junction Temperature Monitoring Method for SiC MOSFETs Based on a Novel Gate Conduction Model," in IEEE Transactions on Power Electronics, vol. 36, no. 10, pp. 11087-11096, Oct. 2021.

[16] N. Baker, L. Dupont, S. Munk-Nielsen, F. Iannuzzo and M. Liserre, "IR Camera Validation of IGBT Junction Temperature Measurement via Peak Gate Current," in IEEE Transactions on Power Electronics, vol. 32, no. 4, pp. 3099-3111, April 2017.

[17] M. S. Grewal and A. P. Andrews, Kalman Filtering: Theory and Practice Using MATLAB. New York, NY, USA: Wiley, 2011.

[18] Z. Sun, M. Ma, M. Zhan and J. Wang, "Junction temperature estimation in IGBT power modules based on Kalman filter," 2017 IEEE Conference on Energy Internet and Energy System Integration (EI2), Beijing, China, 2017, pp. 1-6.

[19] H. Luo, Y. Chen, P. Sun, W. Li and X. He, "Junction Temperature Extraction Approach With Turn-Off Delay Time for High-Voltage High-Power IGBT Modules," in IEEE Transactions on Power Electronics, vol. 31, no. 7, pp. 5122-5132, July 2016.

979-8-3315-1110-4/25 $31.00 © 2025 IEEE

An Active Current Sharing Strategy Based on Master-Slave Cooperative Control for paralleled SiC Power Modules

Baolong Yan[a], Jianing Wang[b], Member, IEEE, Shaolin Yu[b], Honghong Li[b], Runze Wang[a], Xin Li[a]

[a] School of Elechical and Information Engineering, Anhui University of Science and Technology Huainan, China,
2023201832@aust.edu.cn

[b] Hefei Comprehensive National Science Center, (Anhui Energy Laboratory), the Institute of Energy, Hefei, China,

Abstract—**SiC MOSFETs offer high switching speeds and low power losses, which make them widely used in the power electronics with the requirements of higher power density and efficiency. Due to the limited current capacity of a single module, multiple modules should be connected in parallel to meet the demands of higher power applications. However, achieving effective current sharing remains a critical challenge. This paper presents a Master-Slave Cooperative Control Strategy to balance the current among parallelled SiC MOSFET modules in high-power applications. The strategy decouples the factors affecting dynamic current imbalance, specifically switching delay and current change rate, and automatically optimizes these parameters for the selected master module. This approach eliminates the excessive adjustment cycles associated with traditional dynamic master-slave control schemes that rely on a single master module for multi-parameter adjustments. The proposed Master-Slave Cooperative Control Strategy implements closed-loop current regulation to achieve current sharing between SiC power modules. The feasibility of the strategy is verified through simulations for theoretical verification.**

Keywords—*Master-Slave Cooperative Control, Active gate driver, current sharing, SiC MOSFET, paralleled power modules.*

I. INTRODUCTION

As power electronics advance towards higher density, frequency, and reliability, Silicon Carbide (SiC) power devices are increasingly replacing traditional silicon devices in areas like renewable energy, electric vehicles, and industrial drives, thanks to their high voltage tolerance, high-temperature operation, and fast switching speeds. To meet the required current capacity for higher power applications, two common methods are typically used: paralleling multiple SiC MOSFET modules or paralleling multiple power converters. However, the latter method requires derating, which increases the system's size and cost while reducing its power density. Therefore, paralleling multiple SiC MOSFET modules is a cost-effective solution. However, several factors can cause both dynamic and static current imbalances in paralleled SiC MOSFET modules, potentially leading to system failures.

In existing methods for mitigating unbalanced dynamic currents in paralleled SiC devices, in addition to optimizing the power circuit structure [1, 2], screening the paralleled devices for uniformity [3], and incorporating chokes [4, 5], active gate drivers (AGDs) featuring online closed-loop adjustment and high-precision compensation have garnered increasing attention. AGDs regulate current sharing by real-time detection of dynamic current imbalances in parallel modules, such as turn-on/turn-off delay mismatch ($\Delta T_{on/off\text{-}delay}$) and current change rate mismatch ($\Delta di_{ds}/dt$), and dynamically adjusting gate drive parameters, including gate resistance (R_g)

This work was supported by Anhui Province Key Research and Development Program Project, Project Number:JZ2024AKKG0057.

and gate voltage (V_{gs}). The traditional master-slave control strategy, due to the fixed master module selection, introduces the risk of compromising the current-sharing mechanism if the master fails. The PI control strategy, with its fixed proportional-integral coefficients, lacks flexibility to adapt to transitions between multiple operating modes, reducing its effectiveness in dynamic conditions. Neural network controllers, while powerful, require higher hardware costs because of the increased computational resources needed for both training and operation. Fuzzy logic requires a pre-established rule base and may experience decision-making ambiguities or gaps when faced with unknown operating conditions.

To address the challenges mentioned above, this paper proposes a master-slave cooperative control strategy to resolve the current imbalance issue in the parallel configuration of three SiC power modules. By decoupling the switching delay and current change rate, and optimizing the selection of the master module for both parameters, the dynamic adjustment step size for both parameters is reduced. This ensures that their adjustment ranges are kept within optimal limits, thus reducing the current sharing cycle duration and enhancing adjustment efficiency.

II. THE PROPOSED MASTER-SLAVE COOPERATIVE CONTROL STRATEGY

To achieve the sharing of dynamic current ($T_{on/off\text{-}delay}$ and di_{ds}/dt) among power modules, two conditions must be met: first, compensating for $\Delta T_{on/off\text{-}delay}$ to synchronize the turn-on/off events across all parallel modules, and second, ensuring uniformity of $\Delta di_{ds}/dt$ across modules for consistent current distribution. The proposed AGD scheme adjusts dynamic current by regulating the $T_{on/off\text{-}delay}$ through pulse edge timing and modifying gate resistance (R_g) to control $\Delta di_{ds}/dt$, thereby achieving balanced currents. However, the adjustment of these parameters is interdependent, meaning that modifying one affects the other, which could reduce control efficiency or cause instability. To improve system performance and stability, decoupling the control of switching delay and current change rate is necessary.

The AGD scheme presented in this paper is designed to parallel three SiC MOSFET power modules, as illustrated in Fig.1, employing a master-slave cooperative framework. Dynamic current deviation compensation in a closed-loop manner is managed by a DSP. Each AGD collects the dynamic current deviations ($\Delta T_{on/off\text{-}delay}$ and $\Delta di_{ds}/dt$). By processing the information from each AGD, the master AGD module is identified, and the gate drive parameters are adjusted to regulate the $T_{on/off\text{-}delay}$ and di_{ds}/dt for the next pulse. During the current pulse, the master AGD module only samples the dynamic current information of the active device without actively modifying the drive parameters.

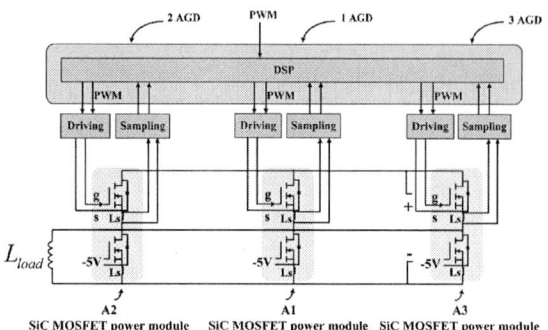

Fig. 1. The structure of the proposed AGD scheme

A. The decoupling process of switching delay and current change rate in the proposed AGD scheme

The hardware structure of the AGD for each parallel device is shown in Fig.2. It mainly consists of three parts.

1) Current Sampling Circuit: The dynamic current information is sampled by the induced voltage generated by the parasitic inductance at the source terminal. During this sampling process, the dynamic current data is converted into voltage trigger signals (V_{cp1}, V_{cp2}, V_{cp3}, V_{cp4}), which are then accessible to the DSP.

2) DSP-based Control Circuit: This circuit is responsible for processing the dynamic current information between the parallel devices. It controls the gate driver circuit and receives trigger signals from the current sampling circuit.

3) Gate Driver Circuit: Composed of three driver circuits (Q_1, Q_2, Q_3), this system controls the turn-on gate resistance (R_{on} and R_{on_aux}) and the turn-off gate resistance (R_{off} and R_{off_aux}) to regulate the dynamic current of the parallel SiC MOSFETs.

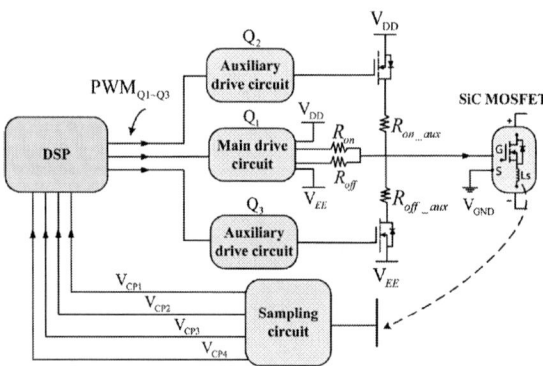

Fig. 2. The hardware structure of AGD of each paralleled SiC MOSFET

The sampling circuit [6] comprises three stages: a passive integrator circuit, an operational amplifier circuit, and a comparator circuit. The passive integrator circuit detects the induced voltage across the source parasitic inductance of the SiC module. Through Equation (1), it converts dynamic current information into voltage representation. As derived from the equation, the initiation time of V_{in} corresponds to the

current's starting moment, while the voltage variation rate directly reflects the current's rate of change.

$$Vin = \frac{1}{R \cdot C} \cdot \int_0^t \frac{di}{dt} \cdot L \cdot dt \qquad (1)$$

The operational amplifier circuit amplifies the output voltage V_{in} from the integrator circuit to facilitate subsequent processing by the comparator circuit. As illustrated in Fig.3, the comparator circuit generates V_{cp1} and V_{cp2} by comparing the amplified signal with predefined reference voltages V_{ref1} and V_{ref2}. The DSP calculates the dynamic current characteristics of the module by measuring two critical intervals:

1) The count value (cnt_{on_delay}) between the PWM rising edge and the V_{cp1} rising edge, representing the turn-on delay.

2) The count value (cnt_{on_slope}) between the V_{cp1} and V_{cp2} rising edges, corresponding to the current slew rate.

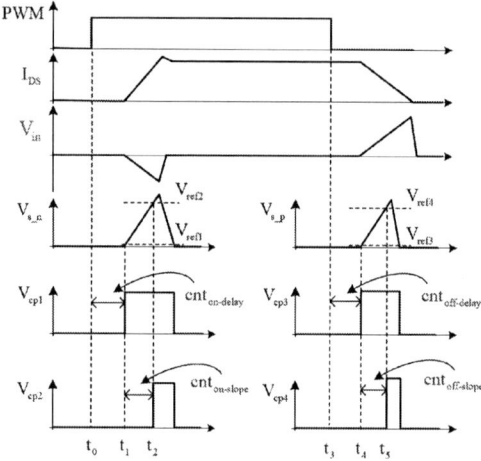

Fig. 3. The waveforms of the current sampling circuit.

The operating principle of the gate driver circuit is illustrated in Fig.2. The turn-on process is governed by the activation of Q_1 and Q_2, while the turn-off process is controlled by Q_1 and Q_3. Taking the turn-on process as an example (waveforms shown in Fig.4), Q_1 serves as the reference for both SiC module turn-on and turn-off operations, with Q_2 maintaining synchronization to the rising edge of Q_1.

When Q_2 is enabled, the equivalent gate resistance R_g becomes the parallel combination of R_{on} and R_{on_aux}. Conversely, when Q_2 is disabled, R_g equals R_{on} alone. This configuration enables dynamic adjustment of the current slew rate through modulation of R_g. The turn-on delay (T_{on_delay}) and turn-off delay (T_{off_delay}) are independently regulated by advancing or delaying the rising edges of Q_1 and Q_2.

The principle of independently adjusting $T_{on/off-delay}$ and di_{ds}/dt in the AGD is presented. The turn-on and turn-off processes controlled by the proposed AGD scheme are shown in Fig.4. The stages of these processes are outlined as follows. The emphasis is placed on the decoupling process.

Turn-on Process:

S_1: Compensation stage of $T_{on-delay}$

979-8-3315-1110-4/25 $31.00 © 2025 IEEE

This stage is a preset compensation phase designed to offset the $\Delta T_{\text{on-delay}}$ between parallel devices. Upon receiving the rising edge of the PWM signal, the DSP introduces a delay of T_{S1} to alter the switching state of the gate driver circuit (Q_1-Q_3). By adjusting T_{S1}, this ensures that all parallel devices turn on simultaneously.

S_3: Control stage of di_{ds}/dt

This stage corresponds to the rising phase of the drain-source current. The proposed AGD controls the change rate of V_{gs} by adjusting the gate drive resistances, thereby regulating the current slew rate (di_{ds}/dt). The S_3 stage is divided into two sub-stages: S_{3-1} and S_{3-2}. In the S_{3-1} sub-stage, the switching state configuration is as follows: Q_1 and Q_2 are ON, while Q_3 is OFF. At this point, the gate resistance (R_g) is the parallel combination of R_{on} and R_{on_aux}. In the S_{3-2} sub-stage, only Q_2 switches from ON to OFF, and the gate resistance equals R_{on}.

The turn-off process follows a similar approach. Through the analysis of the stages in both the turn-on and turn-off processes, it becomes evident that, through the cooperation of the proposed gate driver circuits (Q_1-Q_3), the decoupling of switching delay and current change rate can be achieved.

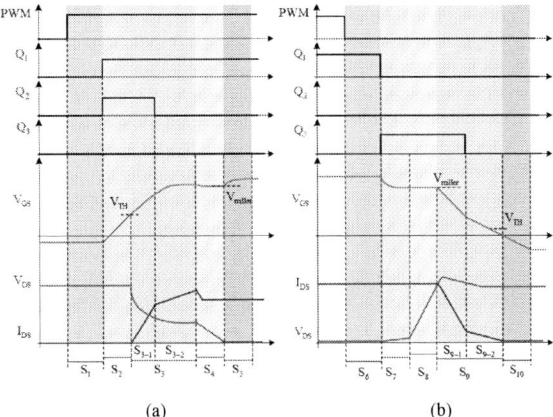

(a)   (b)

Fig. 4. The turn-on and turn-off processes controlled by the proposed AGD. (a) The turn-on processes. (b) The turn-off processes.

B. The control strategy of the proposed AGD scheme

Based on the sampling circuit described in reference [6], the output signals (V_{cp1}, V_{cp2}, V_{cp3}, V_{cp4}) are processed to obtain the dynamic current data for each SiC module, yielding the counting values: $cnt_{\text{on-delay}}$ and $cnt_{\text{on-slope}}$.

Turn-on Process: The control system collects dynamic current information from the three modules to compute the matrices D1 = [$cnt_{\text{on-delay-1}}$; $cnt_{\text{on-delay-2}}$; $cnt_{\text{on-delay-3}}$], which correspond to the turn-on delay, and D2 = [$cnt_{\text{on-slope-1}}$; $cnt_{\text{on-slope-2}}$; $cnt_{\text{on-slope-3}}$], which correspond to the current change rate. Matrices D1 and D2 are compared to determine the median values within each matrix. The SiC module corresponding to the median value is selected as the master module, while the other modules are designated as slave modules. The deviation values are then determined (as shown in the equation below). Based on the deviation values, each AGD determines whether the corresponding module serves as the master module for switching delay or current change rate. Subsequently, when the next pulse arrives, the PWM waveform for the slave

module's driver circuit is adjusted. The turn-off process is carried out in a similar manner.

The compensation associated with the S1 stage, is provided by equations (2) and (3).

$$\Delta cnt_{\text{on/off-delay1}} = cnt_{\text{on/off-delay-master}} - cnt_{\text{on/off-delay-slave1}} \quad (2)$$

$$\Delta cnt_{\text{on/off-delay2}} = cnt_{\text{on/off-delay-master}} - cnt_{\text{on/off-delay-slave2}} \quad (3)$$

The turn-on delay associated with the AGD is given by equation (4).

$$T_{S1\_current} = T_{S1\_last} + \Delta cnt_{\text{on-delay-last}} \cdot T_{\text{working}} \quad (4)$$

The compensation associated with the S1 stage, is provided Tworking, represents the DSP's resolution.

Current change rate compensation corresponding to the S3-1 stage is provided by equations (5) and (6).

$$\Delta cnt_{\text{on/off-slope1}} = cnt_{\text{on/off-slope-master}} - cnt_{\text{on/off-slope-slave1}} \quad (5)$$

$$\Delta cnt_{\text{on/off-slope2}} = cnt_{\text{on/off-slope-master}} - cnt_{\text{on/off-slope-slave2}} \quad (6)$$

The current change rate associated with the AGD is expressed in equation (7).

$$T_{S3-1\_current} = T_{S3-1\_last} \pm T_{\text{working}} \quad (7)$$

Fig.5 presents the logic flow diagram of the proposed AGD scheme, illustrating how the AGDs from each module cooperate to collect and process dynamic current information during the rising and falling edges of the pulse signal. The DSP identifies the optimal master module for switching delay and the optimal master module for current change rate based on the deviation values of dynamic current information from the different modules.

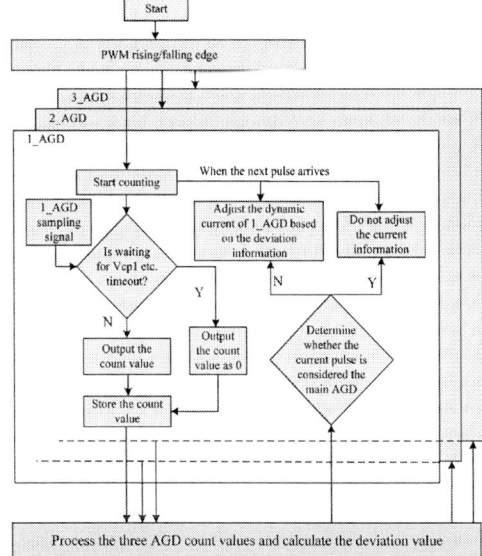

Fig. 5. The logic flow chart of the proposed AGD scheme.

III. SIMULATION VERIFICATION

The feasibility of the proposed active current sharing strategy is validated through simulation studies in this section. According to the structure of main circuit shown in Figure.6. The upper MOSFETs of the parallel power modules remain in the off state and are connected in parallel with the load inductor (L_{load}) to ensure continuous current. The lower MOSFETs in the parallel configuration are controlled by the respective AGD to verify the proposed scheme.

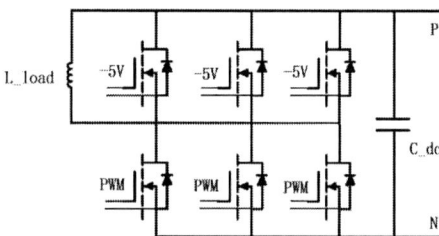

Fig. 6. The structure of main circuit of the proposed AGD scheme.

To validate the feasibility of the proposed control scheme, simulation analyses were conducted on the dynamic current acquisition and processing capabilities of the aforementioned sampling circuit for SiC modules. The current slew rate regulation through variable gate resistance Rg in the driver circuit was experimentally demonstrated, as depicted in Fig. 7, with the following test conditions: gate resistance of 10 Ω, DC bus voltage of 600 V, and switching current of 160 A.

Fig.7 illustrates the dynamic current extraction performance of the sampling circuit. The passive integrator detects the induced voltage across the SiC module's source parasitic inductance, translating dynamic current characteristics (T_{on/off_delay}, di_{ds}/dt) into the output voltage V_{in}. The waveforms in Fig.7 sequentially represent:

- Gate drive voltage V_{gs} (black trace)
- Conducting current I_d (blue trace)
- Passive integrator output V_{in} (red trace)
- Comparator outputs V_{cp1} and V_{cp2} (blue and purple traces, respectively)

As evidenced in Fig.7, the initiation of V_{in} precisely coincides with the turn-on instant of the SiC module's drain-source current I_{ds}. During switching transients, the linear proportionality between Vin and Ids, as derived in Equation (1), is confirmed by the congruent waveform trends of I_{ds} and V_{in}.

Simulation results have conclusively demonstrated the sampling circuit's effectiveness in capturing and processing dynamic current parameters (T_{on/off_delay}, di_{ds}/dt) of SiC modules.

The switching delay adjustment in the gate driver circuit is achieved by modulating the initiation timing of the PWM rising edge. Fig.8 demonstrates the current slew rate regulation methodology. For three paralleled SiC modules with identical specifications, progressive increases in the auxiliary gate driver Q₂'s activation timing (i.e., extending the T_{S3-1} duration) induce clearly observable modifications in the

drain-source current slew rate (di_{ds}/dt) of the corresponding modules. As shown in Fig.8, the waveforms sequentially depict:

- Primary gate drive signal V_{gs_main} of Q_1
- Auxiliary gate drive signal Vgs_aux of Q_2
- Conducting current I_{ds} of the module

The green, yellow, and red traces correspond to the waveforms of the three paralleled SiC modules, respectively.

The simulation validates the driver circuit's capability to dynamically modulate the equivalent gate resistance R_g, enabling controlled adjustment of the I_{ds} slew rate.

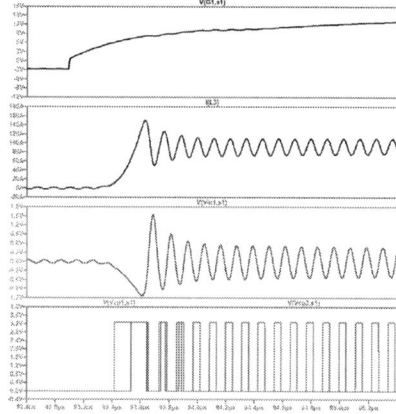

Fig.7. The simulated waveforms of the current sampling circuit

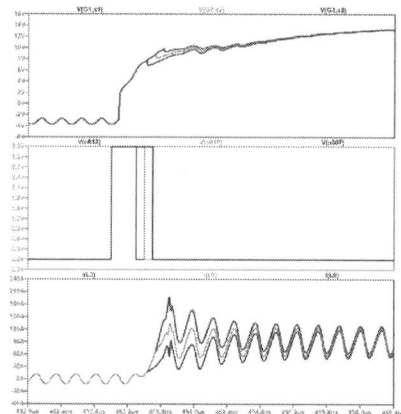

Fig.8. The waveforms of the proposed AGD scheme for adjusting di_{ds}/dt.

For current imbalance regulation among three paralleled modules, the initial uneven current distribution between SiC modules under the simulation conditions specified at the beginning of this section is shown in Fig. 9 (a). Through the proposed sampling circuit, dynamic current parameters (T_{on/off_delay}, di_{ds}/dt) of the SiC modules are acquired and processed. The rising edge timings of Q_1 and Q_2 in the driver circuit are adjusted based on the deviation values ($\Delta cnt_{on/off\_delay}$), while the conduction period of Q_2 is modulated according to $\Delta cnt_{on/off\_slope}$, thereby achieving current sharing optimization.

The rising edge synchronization adjustment is demonstrated in Fig.9 (b), where the gate signal phasing compensates for switching delay mismatches. Correspondingly, Fig.9 (c) illustrates the current slew rate (di_{ds}/dt) adaptation through dynamic gate resistance (R_g) tuning, thereby demonstrating successful current sharing regulation across paralleled modules. In each figure, the red, green, and yellow traces correspond to the conducting currents of the three paralleled SiC modules, respectively.

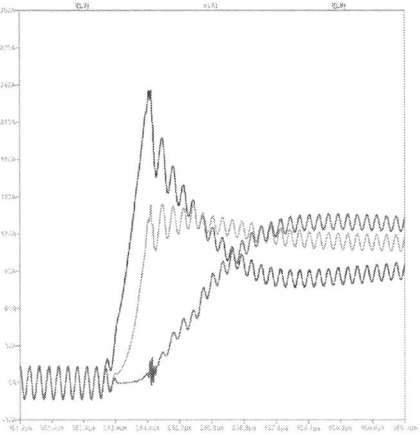

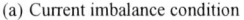

(a) Current imbalance condition

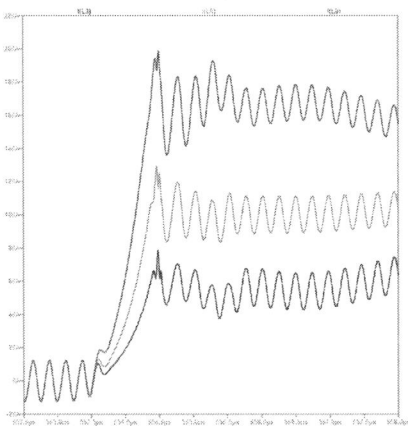

(b) Turn-on delay (T_{on_delay}) regulation

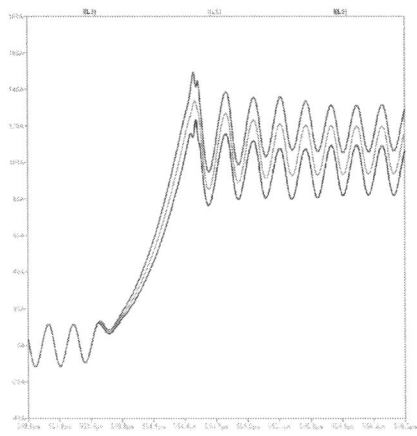

(c) Current slew rate (di_{ds}/dt) adjustment

Fig.8. Simulation waveforms of the proposed AGD scheme

IV. CONCLUSION

This paper proposes a Master-Slave Cooperative Control Strategy for high-power SiC MOSFETs to achieve current balancing among paralleled SiC MOSFET modules. The principal innovations are summarized as follows:

1) Decoupled Dynamic Parameter Control: The proposed Active Gate Driver (AGD), based on a reconfigurable gate resistor architecture, enables independent regulation of switching delay (T_{on/off_delay}) and current slew rate (di_{ds}/dt) during transient operations. This decoupling mechanism eliminates mutual interference between parameters, ensuring precise adjustment of dynamic current imbalances..

2)Adaptive Multi-Module Coordination: Unlike conventional master-slave schemes constrained by fixed master modules, the proposed strategy dynamically identifies optimal reference modules for switching delay and slew rate based on real-time parameter deviations. This adaptive coordination effectively mitigates current imbalances in paralleled SiC modules, even under asymmetrical operating conditions.

3)Scalability for Multi-Module Systems: The modular design of the AGD and distributed control logic ensure robust current sharing in expanded configurations with four or more paralleled modules. By autonomously distributing regulation tasks among local controllers and dynamically updating master-slave relationships, the strategy maintains synchronization and current balance across large-scale parallel systems without compromising stability.

The simulation results validate the strategy's capability to achieve precise current sharing through hardware-driven parameter decoupling and adaptive control.

REFERENCES

[1] J. Qu, Q. Zhang, X. Yuan, and S. Cui, "Design of a Paralleled SiC MOSFET Half-Bridge Unit with Distributed Arrangement of DC Capacitors," IEEE Trans. Power Electron., vol. 35, no. 10, pp. 1087910891, Oct. 2020.

[2] B. Zhang, R. Wan g, P. Barbosa, Q. Cheng, Y. Tsai, W. Wan g, W. Lai, and F. Shih, "Common Source Inductance Compensation Technique for Dynamic Current Balancing in SiC MOSFETs Parallel Operations ," IEEE Trans. Power Electron., vol.38, no.11, pp.13944-13956, 2023.

[3] B. Zhao, Q. Yu, P. Sun, Y. Cai, and Z. Zhao, "Device Screening Strategy for Suppressing Current Imbalance in Parallel-connected SiC MOSFETs," IEEE Trans. Device Mater. Rel., vol. 21, no. 4, pp. 556–568, Dec. 2021.

[4] Y. Mao, Z. Miao, C. -M. Wang and K. D. T. Ngo, "Balancing of Peak Currents Between Paralleled SiC MOSFETs by Drive-Source Resistors and Coupled Power-Source Inductors," IEEE Trans. Ind. Electron., vol. 67, no. 10, pp. 8334-8343, Oct. 2017.

[5] Z. Zeng, X. Zhang, and Z. Zhang, "Imbalance Current Analysis and Its Suppression Methodology for Parallel SiC MOSFETs with Aid of a Differential Mode Choke," IEEE Trans. Ind. Electron., vol. 67, no. 2, pp. 1508-1519, Feb. 2020.

[6] Y. Li et al., "An Online Unified Delay and Slew Rate Regulation for Current Sharing in Paralleled SiC Power Modules With Active Gate Drivers," in IEEE Transactions on Industrial Electronics, vol. 72, no. 4, pp. 3246-3256, April 2025.

Investigation on SiO$_2$ Gate Formed through Ultrahigh-temperature NO Oxidation

line 1: 1st YingFeng He
line 2: *Beijing Institute of Smart Energy*
line 3: *Beijing Huairou Laboratory*
line 4: Beijing, China
line 5: heyingfeng@bise.hrl.ac.cn

line 1: 2nd Given Shucheng Chang
line 2: *Beijing Institute of Smart Energy*
line 3: *Beijing Huairou Laboratory*
line 4: Beijing, China
line 5: changshucheng@bise.hrl.ac.cn

line 1: 3rd Decai Liu
line 2: *Beijing Institute of Smart Energy*
line 3: *Beijing Huairou Laboratory*
line 4: Beijing, China
line 5: liudecai@bise.hrl.ac.cn

line 1: 4th Tao Zhu
line 2: *Beijing Institute of Smart Energy*
line 3: *Beijing Huairou Laboratory*
line 4: Beijing, China
line 5: zhutao@bise.hrl.ac.cn

line 1: 5th ZheYang Li
line 2: *Beijing Institute of Smart Energy*
line 3: *Beijing Huairou Laboratory*
line 4: Beijing, China
line 5: lizheyang@bise.hrl.ac.cn

line 1: 6th Rui Jin
line 2: *Beijing Institute of Smart Energy*
line 3: *Beijing Huairou Laboratory*
line 4: Beijing, China
line 5: jinrui@bise.hrl.ac.cn

Abstract—A novel thermal oxidation process combined nitric-oxide and high temperature is reported to improve the SiO$_2$ gate quality for 4H-silicon carbide metal-oxide-semiconductor field effect transistors. The structure and electrical properties of the SiO$_2$ films formed by the method were studied in detail. The results showed that ultrahigh temperature nitric-oxide oxidation method can achieve high-quality SiO$_2$ gate. The 4H-silicon carbide metal-oxide-semiconductor field effect transistors with the SiO$_2$ gate formed through this method can work properly which proves that the method is compatible with existing SiC manufacturing processes.

Keywords—4H-SiC, nitric-oxide oxidation, MOSFET, Critical electric field, interface

I. Introduction

Silicon carbide (SiC) possesses excellent physical and electrical properties (wide-bandgap, high field breakdown, high thermal conductivity and so on), which is considered a promising candidate for power device applications. High voltage controllable silicon carbide power electronic devices are the best choice for intelligent electric power equipment [1-2]. Unlike other third-generation semiconductor materials, SiC power devices can get the gate oxide through thermal oxidation. The fabrication process is similar to the Si devices which is helpful with the manufacture of the SiC power devices. However, the quality of the SiO$_2$/SiC interface formed by standard dry O$_2$ oxidation is poor which affects device loss characteristics and long-term reliability [3-4]. The post-oxidation annealing with nitric-oxide (NO) or Nitrous oxide (N$_2$O) gas has been proved to be an effective way to improve the interface quality while not significantly degrading the oxide reliability [5-6]. However, the improvement is still far from satisfactory. Many alternative oxidation and interface passivation methods have been proposed to further improve the quality of SiO$_2$/4H-SiC interface and gate oxide.

So far, a post-oxidation annealing (POA) in NO or N$_2$O ambient is a standard method to improve the quality of gate oxide and the interface [5-6]. However, the channel mobility of SiC MOSFET is limited to typically about 25–35 cm^2/Vs even after the post-oxidation annealing [7]. For further improving the quality of SiO$_2$/4H-SiC interface, some impurities were incorporated into the interface. Previous work has reported the incorporation of phosphorus [8], boron [9],

barium [10] and so on. However, these passivation methods reduce the reliability of gate-oxide. Due to this, the study on the improvement of the NO or N$_2$O passivation method has attracted great interest of researchers [11-12].

In this work, a novel thermal oxidation process combined nitric-oxide and high temperature is reported to improve the SiO$_2$ gate quality for 4H-SiC metal-oxide-semiconductor field effect transistors (MOSFETs). The results showed that the method can achieve high-quality SiO$_2$ gate while it slightly reduces the improvement effect on the SiO$_2$/SiC interface. The SiO$_2$ film grown via the method has a larger refractive index value. The sample with the SiO$_2$ has a lower leakage current and a higher oxide critical electric field. In addition, the 4H-SiC MOSFETs with SiO$_2$ gate formed through the method can work properly which proves that the method is compatible with existing SiC manufacturing processes.

II. Experiment Details

The n-MOSFETs are implemented using a vertical MOSFET (VMOSFET) on 4H-SiC (0001) epilayers with 12 μm and 8 ×10^{15} cm^{-3} doping concentration. Besides, the vertical n-type MOS capacitor (MOSCAP) structure is used to form the test pattern for material testing area on the same wafer.

The 4H-SiC (0001) epilayers are divided into two groups (A and B). The thermal oxidation conditions for sample A and B are shown in Table I.

Atomic force microscopy (Dimension ICON, Bruker, USA) was employed to investigate the surface morphology of SiO$_2$. The refractive index of the SiO$_2$ was measured using Spectroscopic ellipsometer (SE). And the thickness of SiO$_2$ films were determined through the SE data. All electrical characteristic measurements were performed on the Keysight B1505A semiconductor parameter analyzer. The interface state density (D_{it}) was evaluated by the high (100 kHz)-low (quasi-static) method from the C–V characteristics.

TABLE I. THE THERMAL OXIDATION CONDITIONGS FOR SAMPLE A AND B

Sample	Oxidation	Annealing
A	1450 °C NO	—
B	1250 °C O$_2$	1250 °C NO 1 h

This work was supported by Scientific and Technology Project of State Grid Corporation of China "Research on the Control Electrode Technology of Silicon Carbide Chips Based on Polycrystalline Silicon Material", the project number is "5500-202399656A-3-2-ZN".

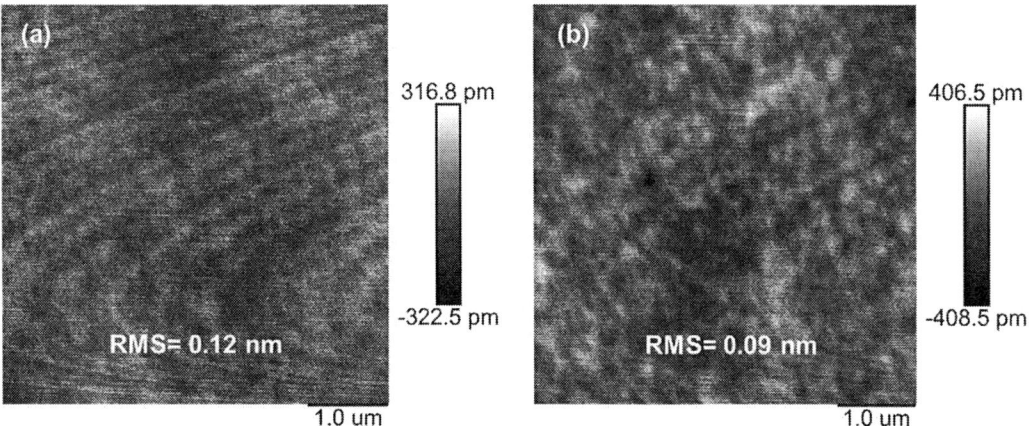

Fig. 1 The surface morphology of (a) sample A, (b) sample B (Scan area is 5 × 5 μm² for all the images).

III. RESULTS AND DISCUSSION

According to previous work, oxidation of SiC under ultrahigh temperature and low oxygen partial pressure may result in the active oxidation [13]. This can lead to a rough surface of the sample. The Atomic force microscopy was employed to analyze the surface morphology of SiO_2 in sample A and B. Fig. 1 shows the results and the root-mean-square (RMS) roughness values of SiO_2 surface for sample A and B are 0.12 nm and 0.09 nm, respectively. The results reveal that ultrahigh temperature NO oxidation will not lead to a rough surface. Then, SE was employed to determine the refractive index of the SiO_2. The refractive index values (the wavelength ranges between 300 and 900 nm) of the samples were shown in Fig. 2(a) which were extracted from the SE data. The n values for sample A and B are from 1.44 to 1.47, which are within the range of values given by reference [14]. However, sample A has a larger n value compared with sample B. The result indicates that SiO_2 in sample A has a denser structure.

TABLE II. T_{OX} AND $E_{CRITICAL}$ FOR SAMPLE A AND B

Sample	Tox	$E_{critical}$
A	41.29 nm	8.60 MV/cm
B	41.66 nm	7.56 MV/cm

Fig. 2(b) shows the current–electric field (*I–E*) characteristics of sample A and B. The thickness of the SiO_2

(T_{OX}) was determined by SE. The oxide electric field (E_{OX}) can be calculated by:

$$Eox = V_G / Tox \qquad (1)$$

where V_G are the voltage on SiO_2.

The T_{OX} and oxide critical electric field ($E_{critical}$) of sample A and B were summarized in Table II. The results show that sample B has a larger leakage current and a lower $E_{critical}$ which indicates a high density of traps. The ultrahigh temperature NO oxidation method can effectively reduce the creation of traps in the oxide which suppress the leakage current and increase the oxide critical electric field. The $E_{critical}$ of sample A and B are smaller than 9 MV/cm. Note that, the T_{OX} of samples are around 40 nm which should be responsible for this. And the sample A exhibits typical *I–E* characteristics of Fowler–Nordheim tunneling current at a sufficiently high-oxide field (about 7 MV/cm).

Fig. 3 shows the high(100 kHz)-low(quasi-static) C-V curve of samples measured at room temperature. The flatband voltage shift (ΔV_{fb}) is determined by:

$$\Delta V_{FB} = V_{FB,theory} - V_{FB,exp} \qquad (2)$$

where $V_{FB,theory}$ and $V_{FB,exp}$ are the theoretical and experimental value of flatband voltage, respectively. And the $V_{FB,exp}$ can be calculated via the high-low(quasi-static) *C-V* data. Then the density of effective fixed dielectric charge (Q_{eff}) was determined as:

$$Q_{eff} = (C_{OX} \cdot \Delta V_{FB}) / q \qquad (3)$$

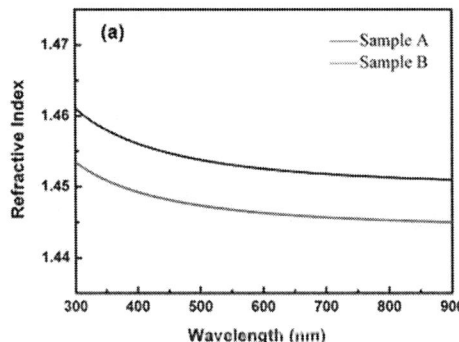
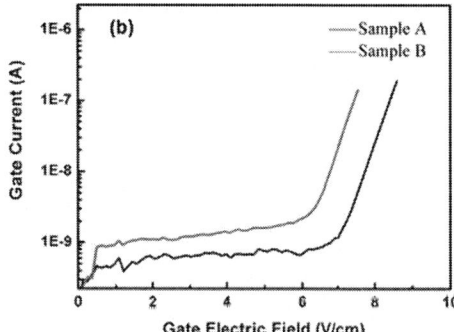

Fig. 2 (a) Extracted refractive index of the SiO_2 for sample A and B as a function of wavelength. (b) I–E characteristics of sample A and B.

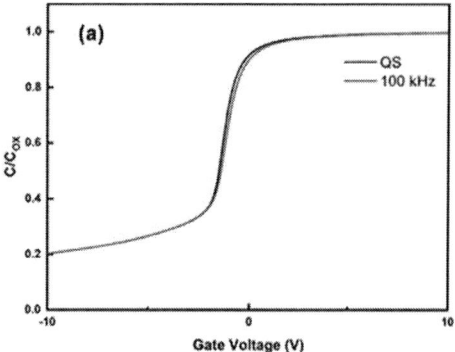 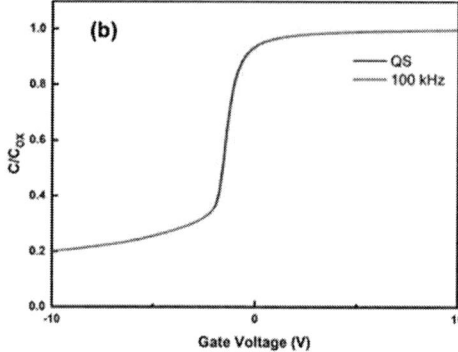

Fig. 3 High frequency-quasi-static C-V curves of SiC MOSCAP: (a) sample A (ultrahigh temperature nitric-oxide oxidation); (b) sample B.

Where C_{OX} is the oxide capacitance per unit area [15]. The value of Q_{eff} for sample A and sample B is 7×10^{11} cm^{-2} and 8.3×10^{11} cm^{-2}, respectively. This indicates that ultrahigh temperature NO oxidation can effective lower the Q_{eff}.

Fig 4(a) shows the D_{it} extracted from C–V characteristics for SiC MOS structures of sample A and B. The result indicates that the method slightly reduces the improvement effect on the SiO$_2$/SiC interface compared with standard NO passivation method. This is also proofed by the C-V data. Compared with Fig 3(b), the shift of 100 kHz curve is more obvious in Fig 3(a). And the explanation is as follows:

The oxidation rate is large under the 1450 °C which leads to the duration of the oxidation process less than half hour. Besides, the SiO$_2$ in sample A formed without a post-oxidation annealing process. Therefore, the SiO$_2$/SiC interface of sample A is not well passivated.

The output characteristics of sample A 4H-SiC MOSFETs were shown in Fig. 4(b) (Vg = 6 V, 10 V, 15 V). The results prove that the SiC MOSFETs with SiO$_2$ gate formed by the method can work normally.

IV. CONCLUSION

A novel thermal oxidation process combined nitric-oxide and high temperature is reported to improve the SiO$_2$ gate quality for 4H-SiC MOSFETs. The SiO$_2$ was grown on 4H-SiC wafer using the method with a smooth surface. No active oxidation was found. The SE data shows that the SiO$_2$ film has a larger refractive index value which means a denser structure. Besides, the sample with the SiO$_2$ has a lower leakage current,

Q_{eff} and a higher oxide critical electric field. These proofs that the method can effectively reduce the creation of traps in the oxide and lower the Q_{eff}. Finally, the 4H-SiC MOSFETs with SiO$_2$ gate formed through the method can work properly which proves that the method is compatible with existing SiC manufacturing processes. In the next step, The method will be improved to better reduce interface state density.

ACKNOWLEDGMENT

This work was supported by Scientific and Technology Project of State Grid Corporation of China "Research on the Control Electrode Technology of Silicon Carbide Chips Based on Polycrystalline Silicon Material", the project number is "5500-202399656A-3-2-ZN".

REFERENCES

[1] B. J. Baliga, Power semiconductor device figure of merit for highfrequency applications, IEEE Electron Device Lett., 1989, 10 (10), 455–457.

[2] A. Castellazzi, A. Fayyaz, G. Romano, L. Yang, M. Riccio, and A. Irace, SiC power MOSFETs performance, robustness and technology maturity, Microelectron. Rel., 2016, 58, 164–176.

[3] V. V. Afanasev, M. Bassler, G. Pensl, and M. Schulz, Intrinsic SiC/SiO$_2$ interface states, Phys. Status Solidi A, 1997, 16 (1), 321–337.

[4] A. Siddiqui, H. Elgabra, and S. Singh, The current status and the future prospects of surface passivation in 4H-SiC transistors, IEEE Trans. Device Mater. Rel., 2016, 16 (3), 419–428.

[5] J. Rozen, S. Dhar, S. K. Dixit, V. V. Afanas'ev, F. O. Roberts, H. L. Dang, S. Wang, S. T. Pantelides, J. R. Williams, and L. C. Feldman, Increase in oxide hole trap density associated with nitrogen incorporation at the SiO$_2$/SiC interface, J. Appl. Phys., 2008, 103, 124513.

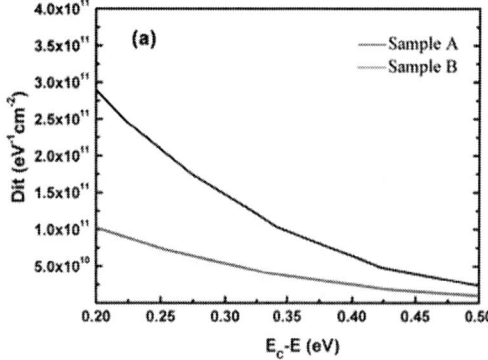 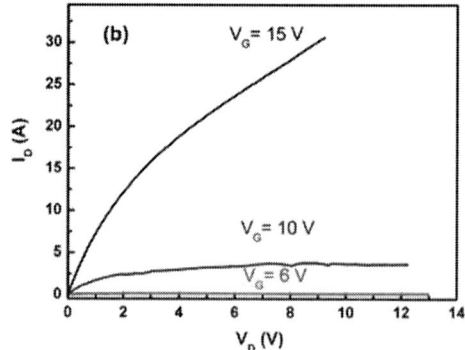

Fig. 4 (a) Distribution of interface state density near the conduction band edge for SiC MOS structures of sample A and B. (b) Output characteristic curves of sample A 4H-SiC MOSFETs.

[6] Y. Katsu, T. Hosoi, Y. Nanen, T. Kimoto, T. Shimura, and H. Watanabe, Impact of NO annealing on flatband voltage instability due to charge trapping in SiC MOS devices, Mater. Sci. Forum, 2016, 858, 599.

[7] T. Kimoto, Material science and device physics in SiC technology for high-voltage power devices, Jpn. J. Appl. Phys., 2015, 54, 040103.

[8] D. Okamoto, H. Yano, K. Hirata, T. Hatayama, and T. Fuyuki, Improved inversion channel mobility in 4H-SiC MOSFETs on Si face utilizing phosphorus-doped gate oxide, IEEE Electron Device Lett., 2010, 31(7), 710–712.

[9] D. Okamoto, M. Sometani, S. Harada, R. Kosugi, Y. Yonezawa, and H. Yano, Improved channel mobility in 4H-SiC MOSFETs by boron passivation, IEEE Electron Device Lett., 2014, 35(12), 1176–1178.

[10] D. J. Lichtenwalner, L. Cheng, S. Dhar, A. Agarwal, and J. W. Palmour, High mobility 4H-SiC (0001) transistors using alkali and alkaline Earth interface layers, Appl. Phys. Lett., 2014, 105(18), 182107.

[11] Z. Luo, C. Wan, Z. Jin and H. Xu, Effects of sequential annealing in low oxygen partial-pressure and NO on 4H-SiC MOS devices, Semicond. Sci. tech., 2021, 36, 045021.

[12] Y. Zhang, H. Yuan, J. Guo, H. Yang, Y. Zhou, F. Du, Y. Liu, K. Liu, C. Han, Q. Song and X. Tang, Long-Term lifetime evolution mechanism of 4H-SiC MOSFETs under nitric oxide annealing, IEEE Trans. Electron Devices, 2024, 71(12), 7682-7688.

[13] T Hosoi, Y Katsu1, K Moges, D Nagai, M Sometani, H Tsuji, T Shimura and H Watanabe, Passive–active oxidation boundary for thermal oxidation of 4H-SiC(0001) surface in O_2/Ar gas mixture and its impact on SiO_2/SiC interface quality, Appl. Phys. Express, 2018, 11, 091301.

[14] O. J. Guy, T. E. Jenkins, M. Lodzinski, A. Castaing, S. P. Wilks, P. Bailey and T. C. Q. Noakes, Ellipsometric and MEIS studies of 4H-SiC/Si/SiO2 and 4H-SiC/SiO_2 interfaces for MOS devices," Mat. Sci. Forum, 2007, 509, 556-557.

[15] L M Lin and P T.Lai, Improved high-field reliability for a SiC metal–oxide–semiconductor device by the incorporation of nitrogen into its HfTiO gate dielectric, J Appl Phys, 2007, 102(5), 054515.

[16] Y. Jia, H Lv, X Tang, C Han, Q Song, Y Zhang, Y Zhang, S Dimitrijev, J Han, D Haasmann, Influence of various NO annealing conditions on N-type and P-type 4H-SiC MOS capacitors, J. Mater. Sci., Mater. Electron., 2019, 30(11), 10302-10310.

The Method for Chip Sorting Based on SiC MOSFET Parameters for Parallel Current Sharing under a Symmetrical Layout

Yi Li
State Key Laboratory of Alternate Electrical Power System
with Renewable Energy Sources
North China Electric Power University
Beijing, China

BoWen Tian
State Key Laboratory of Alternate Electrical Power System
with Renewable Energy Sources
North China Electric Power University
Beijing, China

Bin Zhao
State Key Laboratory of Alternate Electrical Power System
with Renewable Energy Sources
North China Electric Power University
Beijing, China

Peng Sun*
State Key Laboratory of Alternate Electrical Power System
with Renewable Energy Sources
North China Electric Power University
Beijing, China

ZhiBin Zhao
State Key Laboratory of Alternate Electrical Power System
with Renewable Energy Sources
North China Electric Power University
Beijing, China

Xu Cheng
Zhuhai Power Supply Bureau
Guangdong Power Grid
Guangdong, China

Yong Chen
Zhuhai Power Supply Bureau
Guangdong Power Grid
Guangdong, China

XingYu Pei
Zhuhai Power Supply Bureau
Guangdong Power Grid
Guangdong, China

JianBiao Li
Zhuhai Power Supply Bureau
Guangdong Power Grid
Guangdong, China

HongYuan Wu
Zhuhai Power Supply Bureau
Guangdong Power Grid
Guangdong, China

Abstract—**Parallel connection of multiple SiC MOSFET chips is a crucial approach to meeting the demands of higher-power applications. However, due to variations in chip parameters, imbalanced current distribution occurs among the parallel devices. This paper systematically investigates current balancing methods for parallel SiC MOSFETs to address this issue. First, the static parameters of 30 single-chip SiC MOSFET devices were tested, and a statistical analysis was conducted to evaluate the parameter dispersion affecting parallel current distribution. Then, the Pearson correlation coefficients between the gate-source voltage and transconductance under different current levels were established. The gate-source voltage at which the Pearson correlation coefficient reaches -0.7 was defined as the current-balancing voltage, based on which a chip sorting criterion was proposed. Subsequently, chip selection was** optimized using the current-balancing voltage as the sorting indicator. Finally, an experimental platform for multiple discrete devices in parallel was constructed to verify the effectiveness of the proposed method.

Keywords—*SiC MOSFET; multi-chip parallel connection; current balancing; current-balancing voltage; chip selection*

I. INTRODUCTION

In recent years, third-generation wide-bandgap semiconductor devices, represented by silicon carbide (SiC), have rapidly developed due to their advantages in high frequency, high voltage, high efficiency, and high-temperature resistance[1-3]. These devices have been widely applied in

Supported by science and technology project of China Southern Power Grid under Grant 030400KC23090015(GDKJXM20231031).

various aspects of clean energy integration, transmission, and utilization. To control chip costs and yield rates, the current rating of commercially mass-produced SiC chips is generally limited to below 150 A. Therefore, utilizing multiple discrete devices in parallel or adopting multi-chip parallel packaging in power modules has become a crucial approach to meeting higher power application demands[4-6]. However, in parallel applications, current imbalance can lead to uneven electrical and thermal stress, which, in severe cases, may cause excessive current stress on certain devices, resulting in their failure. This, in turn, can compromise the reliability of other parallel-connected devices and even the entire converter system[7-8].

Parallel current imbalance can be categorized into transient dynamic current imbalance during switching events and steady-state current imbalance after device conduction. Among these, steady-state current imbalance has a relatively minor impact on device operational reliability, as the positive temperature coefficient of the on-state resistance enables self-regulated current balancing. In contrast, dynamic current imbalance is influenced by the threshold voltage variation among chips, which has a negative temperature coefficient. When a device carries a higher current, its junction temperature rises, further reducing the threshold voltage. This, in turn, leads to an increase in device current and junction temperature, eventually resulting in overcurrent failure or thermal runaway. Therefore, particular attention must be paid to the issue of dynamic current imbalance in parallel-connected devices.

Various methods have been studied to mitigate dynamic current imbalance. Research on chip parameter sorting mainly focuses on selecting chips based on key parameters affecting parallel current distribution, such as threshold voltage and transconductance. Fuji Electric proposed a method to suppress transient current imbalance by selecting chips with similar threshold voltages for parallel operation[9]. However, threshold voltage only reflects the turn-on point of the device at low current levels and fails to characterize the current rise rate at high currents, resulting in limited accuracy. References proposed a hierarchical clustering sorting method based on the Lance and Williams Distance of transfer curves, which considers both threshold voltage and transconductance in transient current distribution[10-11]. This approach achieves more balanced current sharing after chip selection. However, obtaining the transfer curve requires testing chips across the entire current range from low to high, leading to a time-consuming process. Additionally, storing the large amounts of test data poses a significant challenge, especially for wafer-level sorting.

To address the aforementioned issues, this study builds upon previous research and conducts a comprehensive analysis of traditional chip sorting parameters, revealing their application limitations and proposing directions for optimization. The gate-source voltage corresponding to a Pearson correlation coefficient of -0.7 with transconductance is defined as the current-sharing voltage. Using this voltage as the key sorting parameter enables simultaneous consideration of both threshold voltage and transconductance. Subsequently, under a symmetrical layout, the current-sharing voltage is employed as a control parameter for chip sorting. Finally,

experimental results validate the effectiveness of the proposed method.

II. CHIP PARAMETER STATISTICS

This paper first selects 30 single-chip SiC MOSFETs from the same batch produced by a well-known international manufacturer as the research subject. The transfer characteristics of these 30 chips are tested using the static characteristic parameter tester B1505.

A. Chip Parameter Statistical Indicators

To more accurately describe the dispersion of chip parameters, the range (D_{ra}), relative range (D_{rra}), and coefficient of variation (D_{cv}) are used to characterize the dispersion of chip parameters, as shown in Equations (1) to (3).

$$D_{ra} = x_{max} - x_{min} \tag{1}$$

$$D_{rra} = \frac{x_{max} - x_{min}}{\frac{1}{n}\sum_{i=1}^{n} x_i} \tag{2}$$

$$D_{cv} = \frac{\sqrt{\frac{1}{n-1}\left(\sum_{j=1}^{n}\left(x_j - \frac{1}{n}\sum_{i=1}^{n} x_i\right)^2\right)}}{\frac{1}{n}\sum_{i=1}^{n} x_i} \tag{3}$$

where, x_{max} and x_{min} represent the maximum and minimum measured values within the sample, respectively; x_i and x_j denote the measured values of the parameter for the i-th and j-th chips, respectively; and n is the total number of samples.

B. Threshold Voltage and Transconductance

Under test conditions of 25 °C ambient temperature and a drain-source voltage of 20 V, the statistical distribution of threshold voltage (V_{th}) and maximum transconductance (g_{fsmax}) for 30 SiC MOSFETs is summarized in Table I. The results indicate that while the variation in threshold voltage among the devices is relatively small, the difference in maximum transconductance is significantly larger.

TABLE I. CHIP PARAMETER STATISTICS

Definition	V_{th}	g_{fsmax}
Maximum value	3.86V	13.41S
Minimum value	3.65V	12.00S
Mean value	3.76V	12.57S
Range	0.21V	1.41S
Relative range	5.64%	11.22%
Coefficient of variation	1.57%	2.52%

The following section will propose the key chip parameters for chip sorting based on the statistical analysis data of the chip parameters shown in Table I.

III. OPTIMIZATION AND SELECTION OF CHIP PARAMETERS

A. Optimization Analysis of Chip Sorting Parameters

The traditional chip sorting method is based on threshold voltage selection. The following section first analyzes the limitations of the chip sorting parameters selected by the traditional sorting method and proposes directions for improving chip sorting parameters.

The threshold voltage-based sorting method uses threshold voltage as the chip sorting parameter. However, threshold voltage only characterizes the turn-on voltage of the chip and cannot ensure that the chip has a consistent rise rate during the current increase phase, i.e., the same transconductance. The drain current during the turn-on process can be expressed as:

$$i_{dk} = g_{fsk}\left(V_{gsk} - V_{th}\right) \tag{4}$$

where, g_{fsk} represents the transconductance corresponding to a specific drain current; V_{gsk} represents the gate-source voltage corresponding to the specific drain current; and V_{th} is the threshold voltage of the chip.

From Equation (4), it can be seen that the transconductance at the threshold voltage is close to 0, meaning the gate-source voltage cannot effectively characterize the transconductance at this point. As the drain current increases, the transconductance gradually increases. When the drain currents of parallel chips are equal, the product of transconductance and gate-source voltage is also equal, meaning that when the gate-source voltage of the parallel chips is consistent, the transconductance of the parallel chips is also consistent. To verify the effectiveness of the theoretical analysis, the Pearson Correlation Coefficient is introduced to measure the linear correlation between gate-source voltage and transconductance. The formula for calculating the Pearson Correlation Coefficient is:

$$r = \frac{\sum (V^i_{gsj} - v_{avj})(g^i_{fsj} - g_{avj})}{\sqrt{\sum (V^i_{gsj} - V_{avj})^2 \sum (g^i_{fsj} - g_{avj})^2}} \tag{5}$$

where, V^i_{gsj} is the gate-source voltage corresponding to the drain current j for the i-th chip; g^i_{fsk} is the transconductance corresponding to the drain current j for the i-th chip; V_{avj} is the average gate-source voltage corresponding to the drain current j for all parallel chips; and g_{avj} is the average transconductance corresponding to the drain current j for all parallel chips.

The Pearson correlation coefficient ranges from -1 to 1, indicating both the strength and direction of the linear relationship between two variables. A coefficient of 1 represents a perfect positive correlation, meaning the two variables increase or decrease together in a completely linear manner. A coefficient of -1 signifies a perfect negative correlation, where one variable increases as the other decreases in a completely linear fashion. A coefficient of 0 indicates no linear correlation between the variables.

By fitting the measured data, the fitting equation at a drain current of 5 mA is given as:

$$g_{fs} = -0.0069V_{gs} + 0.0409 \tag{6}$$

The Pearson correlation coefficient between the two variables is -0.399, indicating that the gate-source voltage corresponding to the threshold voltage cannot form an effective correlation with the transconductance. At a drain current of 5 A, the obtained fitting equation is:

$$g_{fs} = -0.557V_{gs} + 8.578 \tag{7}$$

The Pearson correlation coefficient between the two variables reaches -0.966, indicating a strong negative correlation between the gate-source voltage and transconductance at this current level. Therefore, when the drain current rises to a certain level, the gate-source voltage corresponding to a single data point can be used as a key chip parameter. Sorting based on this key parameter enables simultaneous sorting with respect to transconductance.

B. Requirements for the Selection of Chip Sorting Parameters

As indicated by the preceding analysis, sorting based on the gate-source voltage at a single operating point under a specific drain current level enables simultaneous classification of chips in terms of both gate-source voltage and transconductance. This approach significantly reduces the complexity of parameter testing and data storage. While selecting a lower drain current can shorten the measurement time during sorting, excessively low current levels lead to a weaker correlation between gate-source voltage and transconductance. Therefore, a further analysis is conducted to investigate the variation in gate-source voltage among paralleled chips under different drain current levels, in order to determine the optimal current level for chip sorting.

The gate-source voltages of the paralleled chips under the same drain current are sorted in ascending order, with the chip exhibiting the lowest gate-source voltage assigned as Rank 1 and the highest as Rank 30. An analysis was conducted on the ranking variations under different drain current levels. The results show that as the drain current increases from 1 A to 5 A, the overall ranking of gate-source voltages among the chips remains largely stable. Specifically, when the drain current increases from 1 A to 2 A, 9 chips exhibit a change in their ranking; from 2 A to 3 A, 4 chips change; from 3 A to 4 A, 7 chips change; and from 4 A to 5 A, only 2 chips (Chip No. 3 and Chip No. 15) show a change, with their gate-source voltage rankings switching positions.

At a drain current of 4 A, the measured data is fitted, and the resulting fitting equation is:

$$g_{fs} = -0.494V_{gs} + 7.241 \tag{8}$$

Meanwhile, the Pearson correlation coefficient between the two variables is -0.974, indicating a strong negative correlation between the gate-source voltage and transconductance at this current level. This suggests that sorting based on gate-source voltage at this point can simultaneously achieve effective sorting with respect to transconductance.

In statistics, a Pearson correlation coefficient below -0.7 indicates a significant negative correlation. Therefore, when the drain current of a chip reaches a certain level and the Pearson correlation coefficient between the gate-source voltage and transconductance falls below -0.7, the corresponding voltage can be defined as the "current-sharing voltage" V_{ct} and used for chip sorting. In the above analysis, at a drain current of 4 A, the Pearson correlation coefficient between gate-source voltage and transconductance is -0.974, which meets the criterion for selecting V_{ct}. Thus, this voltage can be effectively used as a sorting parameter.

C. Chip Sorting Method

Based on the analyses presented in the previous two sections, this section proposes a method for achieving current balancing among parallel chips using the current-sharing voltage parameter. Under a symmetrical layout, the circuit parameters of the paralleled chips are identical, and the parasitic inductance of the power source terminals is equal. In this case, the voltage difference at the source terminals of the paralleled chips is zero, meaning that balanced current distribution can be achieved simply by ensuring consistency in the current-sharing voltage among the selected chips.

IV. EXPERIMENTAL VERIFICATION AND COMPARATIVE ANALYSIS

To facilitate the quantitative evaluation of current distribution characteristics among parallel chips in the following sections, the definition of the maximum dynamic current imbalance factor, α, is introduced.

$$\alpha = \frac{2\left|i_{\text{peak}x} - i_{\text{peak}y}\right|}{i_{\text{peak}x} + i_{\text{peak}y}} \quad (9)$$

In the equation, $i_{\text{peak}x}$ and $i_{\text{peak}y}$ represent the peak drain current values at the second turn-on for chips Q_x and Q_y, respectively.

The proposed sorting method is experimentally validated using a test platform consisting of multiple paralleled SiC MOSFETs, as shown in Fig. 1. The selected devices are the same SiC MOSFET chips analyzed in the previous theoretical study. The current distribution characteristics among the parallel chips are examined during the turn-on phase of the second pulse in a double-pulse test.

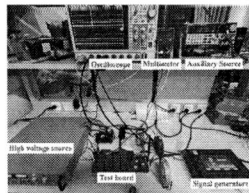

(a) Experimental equipment   (b)Test board
Fig. 1 Double pulse experiment platform

To verify whether the Pearson correlation coefficient between gate-source voltage and transconductance reaching -0.7 meets the requirements for chip sorting, the gate-source voltage and transconductance of 30 selected 1.2 kV/100 A chips were analyzed at a drain current of 3 A. The resulting

Pearson correlation coefficient was -0.753, which satisfies the criterion for selecting the current-sharing voltage.

Under the symmetrical layout, the parasitic inductance of the power source terminals on each chip's test board is identical. The current-sharing voltage for Chip No. 4 and Chip No. 20 is consistent at 7.45V, which serves as the experimental group based on current-sharing voltage sorting. Chip No. 3 and Chip No. 4, having the same threshold voltage of 3.83V, form the control group based on threshold voltage sorting. The current distribution among the paralleled chips under both sorting methods is evaluated to validate the effectiveness of the proposed method.

The experimental results are shown in Fig. 2. The sorting method based on current-sharing voltage significantly improved the current distribution of paralleled SiC MOSFETs, especially during the switching transient process. Compared to the traditional threshold voltage-based sorting method, the maximum dynamic current imbalance was reduced from 2.17 A to 0.26 A, and the maximum dynamic current imbalance factor decreased from 6.98% to 1.15%. These results validate the superiority of the sorting method based on current-sharing voltage in achieving current balancing.

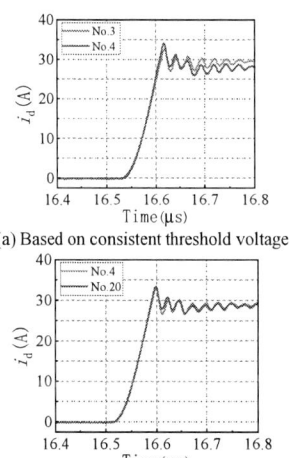

(a) Based on consistent threshold voltage

(b) Based on consistent current balancing voltage
Fig. 2 Parallel current distribution of two sorting methods under symmetrical layout

V. CONCLUSION

To address the issue of uneven current distribution in the parallel application of SiC MOSFETs, this paper proposes a sorting method based on current-sharing voltage under a symmetrical layout. The method uses current-sharing voltage as the sorting criterion, and under the premise of a symmetrical circuit layout, it ensures consistent current-sharing voltage among the chips, thereby achieving balanced current distribution in parallel circuits. Experimental results show that this method reduces the maximum current imbalance from 2.17 A to 0.26 A, and the imbalance factor from 6.98% to 1.15%, significantly improving current balancing. Additionally, both theoretical analysis and experimental verification indicate that sorting based solely on threshold voltage cannot ensure consistent transconductance among the chips, with a Pearson correlation coefficient of -0.399 between threshold voltage and transconductance. In contrast, the current-sharing voltage exhibits a stronger negative correlation with transconductance, with a Pearson coefficient reaching -0.7, making it a suitable sorting criterion. Compared with traditional sorting methods, the sorting method based on current-sharing voltage not only enhances current balancing but also simplifies the chip sorting process, thereby improving its feasibility for industrial applications.

ACKNOWLEDGMENT

This work was supported by science and technology project of China Southern Power Grid under Grant 030400KC23090017(GDKJXM20231033).

REFERENCES

[1] Z. Zeng et al., "Changes and challenges of photovoltaic inverter with silicon carbide device," Renew. Sustain. Energy Rev., vol. 78, pp. 624–639, Oct. 2017.

[2] H. Lee, V. Smet, and R. Tummala, "A review of SiC power module pack aging technologies: Challenges, advances, and emerging issues," IEEE J. Emerg. Sel. Topics Power Electron., vol. 8, no. 1, pp. 239–255, Mar. 2020.

[3] S. Kuang, R. Na, X. Hongyi. "A Recent Review on Silicon Carbide Power Devices Technologies", in Proceedings of the CSEE, vol. 40, pp. 1741–1753. March 2020.

[4] Z. Zeng, X. Zhang, and X. Li, "Layout-dominated dynamic current imbalance in multichip power module: Mechanism modeling and com parative evaluation," IEEE Trans. Power Electron., vol. 34, no. 11, pp. 11199–11214, Nov. 2019.

[5] Zhao Z, Sun P, Cai Y, et al. "Analytical Model for Predicting the Junction Temperature of Chips Considering the Internal Electrothermal Coupling inside Silicon Carbide MOSFET modules", in IET Power Electronics, vol. 3, no. 3, pp. 436-444, February 2019.

[6] K. Puschkarsky, H. Reisinger, T. Aichinger, W. Gustin, and T. Grasser, "Understanding BTI in SiC MOSFETs and its impact on circuit oper ation," IEEE Trans. Device Mater. Rel., vol. 18, no. 2, pp. 144–153, Jun. 2018.

[7] G. Wang, J. Mookken, J. Rice, and M. Schupbach, "Dynamic and static behavior of packaged silicon carbide mosfets in paralleled applications," in Proc. IEEE Appl. Power Electron. Conf. Exposit., Mar. 2014, pp. 1478 1483.

[8] J.Lv,C.Chen,B.Liu,Y.Yan,andY.Kang,"Adynamiccurrentbalancing method for paralleled SiC MOSFETs using monolithic Si- RC snubber based on adynamiccurrent sharing model," IEEE Trans. Power Electron., vol. 37, no. 11, pp. 13368–13384, Nov. 2022.

[9] Fuji Power MOSFET[J]. Application Note FA5504N, Fuji Electric, 2014, http://pdf.datasheet. Directory /datasheets0 /fuji_electric/FA5504.pdf.

[10] Ke Junji, Zhao Zhibin, Sun Peng, et al. Chips Classification for Suppressing Transient Current Imbalance of Parallel-Connected Silicon Carbide MOSFETs[J]. IEEE Trans Power Electron, 2020, 35(4): 3963-72.

[11] Zhao Bin, Yu Qiuping, Sun Peng, et al. Device Screening Strategy for Suppressing Current Imbalance in Parallel-Connected SiC MOSFETs. IEEE Transactions on Device and Materials Reliability, 2021, 21(4): 556-568.

Charaterization of A Nonlinear Conductivity Encapsulant for Electric Field Reduction in High-Voltage Power Module Packaging

Meiyu Wang [a, b*], Peng Gao [a], Yiting Han [a], Haidong Yan [c, *]

a College of Electronic Information and Optical Engineering, Nankai University, Tianjin 300350, China

b Shenzhen Research Institute of Nankai University, Shenzhen 518083, China

c College of Electrical Engineering, Zhejiang University, Hangzhou 311200, China

* Corresponding author: haidong_yan@zju.edu.cn

Abstract—Electrical insulation, particularly the partial discharge at triple points, remains a challenge in high-voltage SiC power module packaging. To address this issue, a field-dependent conductivity (FDC) insulation material was employed as a coating at the triple points of the direct-bond-copper (DBC) substrate prior to the encapsulation with silicon gel. The FDC material was developed by incorporating 0.1 wt.% -1.0 wt.% metal nanoparticle fillers into a silicon gel matrix. Experimental results have validated that the FDC material exhibits nonlinear conductivity characteristics, specifically an exponential increase with electric field. Notably, the FDC material with filler content as low as 0.5 wt.% has a low viscosity of 500 mPa·s. This ensures uniform potting during the module packaging process and contributes to excellent electric field homogenization effects. The results of 2D and 3D model simulation reveal that, when subjected to a high voltage of 15 kV, the electric field is reduced by 47.6%, i.e., dropping from 50.2 kV/mm to 26.3 kV/mm in comparison to the sample encapsulated solely with silicon gel. These findings suggest that the proposed FDC material presents a viable solution for enhancing the insulation performance in the packaging of high-voltage power modules.

Key words: high voltage module packaging, electric field, field-dependent conductivity, nonlinear conductivity

I. INTRODUCTION

Wide bandgap (WBG) semiconductor devices, such as silicon carbide (SiC) and gallium nitride (GaN), exhibit broad prospects in high-power, high-frequency, and high-temperature applications due to their superior blocking voltage, faster switching speed, and better high-temperature performance when compared to traditional silicon devices [1, 2]. The higher blocking voltages are necessary in order to pursue high power density, which results in the concentration of electric field (E-field) at the triple point, such as the junction of ceramics, metals, and encapsulant materials [3]. A partial discharge occurrence

can happen when the E-field strength exceeds the insulation material's breakdown threshold, which will cause great issue to the insulation material of the high-voltage power module. Despite the current state-of-the-art strategies, such as modifying the geometry, thickening the insulating metal substrate, or implementing multi-layer stacking technology, have mitigated the problem to a certain degree, these approaches not only increase manufacturing costs but also may introduce additional thermal resistance and reliability concerns [4, 5].

In this study, an innovative approach using a field-dependent conductivity (FDC) insulation material is proposed to reduce the E-field concentration for high-voltage power modules. More precisely, the FDC material was synthesized by incorporating 0.1 wt.%-1.0 wt.% metal nanoparticle fillers into a silicon gel matrix [6]. The nonlinear electrical conductivity arises from the tunneling effect and the jump conduction mechanism between nanoparticles prior to the filler content reaching the permeability threshold. The nonlinear electrical conductivities of the FDC materials were validated through measurements under diverse E-field excitations. Subsequently, the FDC material was applied as a coating at triple points of a direct-bond-copper (DBC) substrate. The 2D and 3D E-field simulation results vividly demonstrate that the FDC materials exhibit excellent E-field homogenization effects [7]. In comparison to the samples filled with silicon gel, the E-field strength of the sample coated with 0.5 wt.% FDC material was reduced by 47.6%, i.e., dropping from 50.2 kV/mm to 26.3 kV/mm. The findings indicate that the proposed FDC material is a promising solution for increasing partial discharge initial voltage (PDIV) for high-voltage module packaging [8].

II. FDC MATERIAL AND SAMPLE FABRICATION

The FDC materials were synthesized by incorporating metal nanoparticle fillers into the silicon gel (Wacker SilGel-612) matrix. To ensure compatibility with industrial potting procedures, a low nanofiller content ranging from 0.1 wt.% to 1.0 wt.% was employed to maintain a low viscosity of about 500 mPa·s. The material fabrication process involves precisely weighing the silicon gel and nanofillers, then mixing in a high-speed mixer (Thinky ARE-310) at a rotational speed of 2000 r/min for 30 min to guarantee homogeneous dispersion. For the samples used in electrical conductivity measurement, the mixture was poured into a Teflon mold and cured at 230°C for 30 min. Fig. 1 presents scanning electron microscope (SEM) images of the FDC material with nanofiller content ranging from 0.1 wt.% to 1.0 wt.%. The microstructural analysis reveals that the nanofillers were dispersed uniformly in the silicon gel matrix.

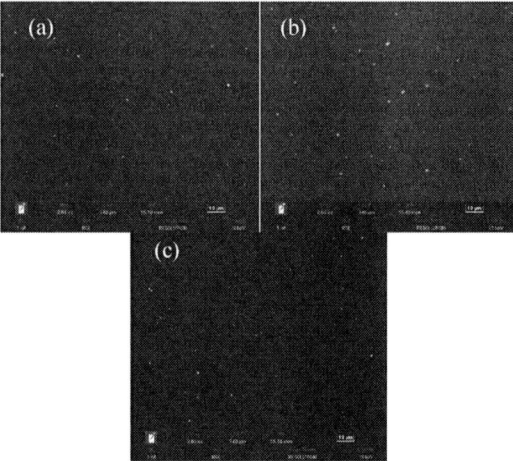

Fig. 1. SEM images for filler concentrations of (a) 0.1wt%, (b) 0.5wt% and (c) 1.0 wt%. Black area is silicon gel, white dot is nanofiller.

III. Nonlinear Conductivity Characterization of FDC material

The electrical conductivities of the FDC materials with nanofiller contents ranging from 0.1 wt.% to 1.0 wt.% were measured at voltage (U) spanning from 1 kV to 10 kV. During the testing process, an ultralow current (I) of less than 1 mA was restricted to prevent self-heating effects on temperature. The electrical conductivity (σ) can be calculated as follows:

$$\sigma = \frac{I}{U} \times \frac{t}{A} \tag{1}$$

where A is the area perpendicular to the current path, and t is the distance between two electrodes parallel to the current path. Here, t was fixed at 0.5 mm; therefore, the E-field varied from 2 kV/mm to 20 kV/mm when the applied voltage ranged from 1 kV to 10 kV.

Fig. 2 depicts the measured relationship between the electrical conductivity and the applied E-field for FDC materials with nanofiller contents ranging from 0.1 wt.% to 1.0 wt.%. The results indicate that the conductivity of silicon gel (Wacker SilGel-612) remains nearly constant, unaffected by the E-field. In contrast, FDC materials exhibit distinct nonlinear conductivity characteristics in relation to the E-field. After a switching E-field (E_b), the conductivity of FDC materials increases exponentially with the rise in the E-field. This behavior is crucial for alleviating the E-field concentration and, consequently, increasing the PDIV. Moreover, as the filler content increases, the nonlinearity of the conductivity first increases and then decreases [7]. Specifically, when the fillers content is 0.5 wt.%, the nonlinearity reaches its maximum value.

When the doping concentration of nano-metallic particles approaches approximately 0.5 wt%, the system lies near the percolation threshold of the electrical connectivity phase transition. At this critical concentration, the quantum tunneling effect through the silicone gel insulating matrix is maximized, resulting in the most pronounced nonlinear characteristics in the I–V curves [9]. Below 0.5 wt%, the increased interparticle spacing exponentially reduces tunneling probability, leading to low overall conductivity dominated by linear responses. Near 0.5 wt%, however, the number of tunneling junctions between particles surges, and localized electric field concentration forms conductive pathways, generating a strong nonlinear conductivity peak. Beyond this threshold, continuous conductive networks gradually develop within the composite, favoring Ohmic transport mechanisms and diminishing field-induced nonlinearity [10].According to the classical percolation-tunneling composite model, the power-law exponent of nonlinear conductivity exhibits a maximum near the percolation threshold, with non-universal behavior commonly observed in metal/insulator composites [11].Similar nonlinear conductivity peaks have been reported in carbon nanotube/polymer and ZnO/polymer systems within the 0.1–2.0 wt% range, further validating the universality of this critical concentration [12]. Thus, the synergistic interplay between percolating networks and quantum tunneling effects at ~0.5 wt% maximizes nonlinear responses, providing theoretical insights into the pivotal role of nanoparticle concentration in such systems.

It is worth highlighting that in contrast to the high viscosity of 1350 mPa·s associated with 60 wt.% SiC fillers in the studies by Li and Liang *et al.* [13, 14], the FDC material proposed in this study has a significant lower viscosity of 500 mPa·s. This lower viscosity ensures uniform potting during the module packaging process.

The measurement results demonstrate that a very low filler content, such as 0.5 wt.%, can yield the maximum nonlinear conductivity. In order to more accurately describe the law of the FDC, the measured results were fitted using an exponential function as follows:

$$\sigma(E) = \sigma_0 \left(1 + \left(\frac{E}{E_b}\right)^\alpha\right) \tag{2}$$

979-8-3315-1110-4/25 $31.00 © 2025 IEEE

where σ_0 is the electrical conductivity at low E-field, E_b is the switching E-field, and α is the nonlinear coefficient. The fitting curves for FDC materials with different nanofiller contents are presented in Fig. 2.

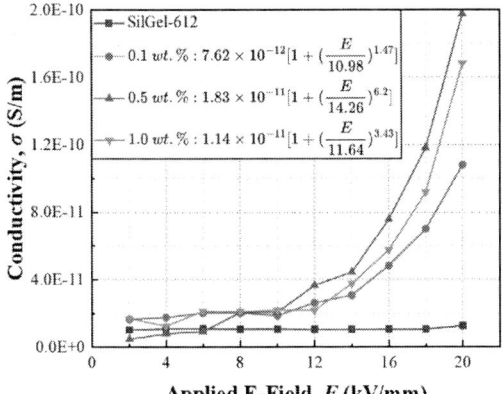

Fig. 2. The measured electrical conductivity versus applied E-field of FDC materials with different nanofillers contents.

IV. ELECTRIC FIELD SIMULATION

To validate the effect of the nonlinear conductivity performance on the E-field distribution, simulations were carried out using COMSOL Multiphysics software [15, 16]. Unlike the sole used of a 2D model in Zhang's study, this study compares 2D and 3D models to account for the effect of space on the E-field distribution. Fig. 3 presents the 2D and 3D schematic diagrams of the DBC structures employed in the simulation. In Fig. 3 (a), (c), and (d), the blue portion represents the internal FDC coating at the triple points, while in Fig. 3 (b) and (e), the green portion depicts the external SilGel-612 encapsulant. Regarding the excitations, one of the top-side Cu plates is connected to a high voltage of 15 kV, and the other two Cu plates are grounded. The electrical properties of the materials used in the COMSOL simulation are listed in Table I.

Table I. Material Parameters Used in E-field Simulations

Material	Conductivity (S/m)	Relative Permittivity
Copper	5.81×10^7	1
Alumina	1.0×10^{-12}	9.9
Silicone Gel (SilGel-612)	1.0×10^{-13}	2.7
0.1 wt.% FDC	$7.62 \times 10^{-12}[1+(E/10.98)^{1.47}]$	2.7
0.5 wt.% FDC	$1.83 \times 10^{-11}[1+(E/14.26)^{6.2}]$	2.7
1.0 wt.% FDC	$1.1410^{-11}[1+(E/11.64)^{3.43}]$	2.7

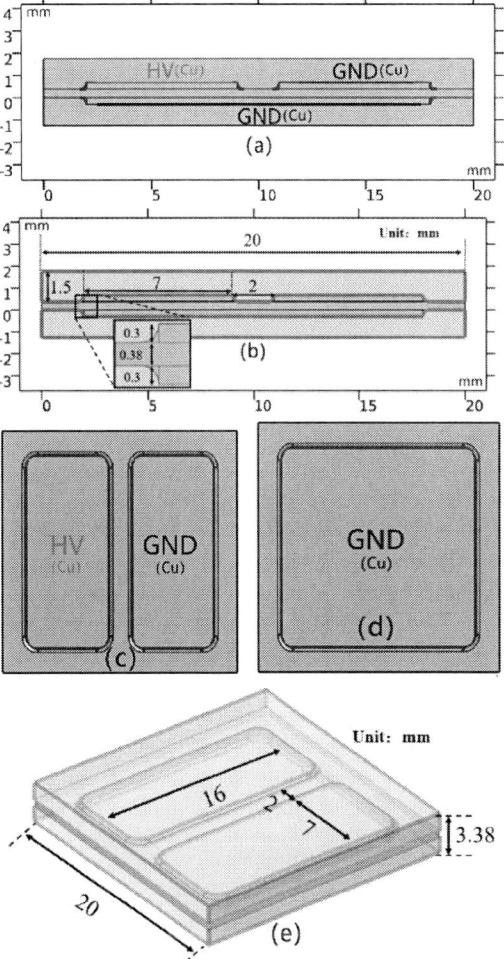

Fig. 3. Schematic of the DBC structure used in the E-field simulation: cross-sectional 2D model showing (a) FDC coating and (b) SilGel-612 encapsulant; 3D model showing (c) FDC coating on top-side, (d) FDC coating on bottom-side, and (e) SilGel-612 encapsulant.

Fig. 4 and Table II illustrate the simulation results of the 2D and 3D models incorporating diverse insulation assemblies. For the sample filled with SilGel-612 both internally and externally, the peak E-field strength reaches 69.8 kV/mm in the 2D model and 50.2 kV/mm in the 3D model. Evidently, the 3D model aligns more closely with the actual scenario. Specifically, the E-field strength is reduced by 28.6% due to the homogenization effect within the 3D space. For the 3D structures that are coated with FDC materials internally, the maximum E-field first decreases and then increases as the nanofiller content varies from 0.1 wt.% to 1.0 wt.%.

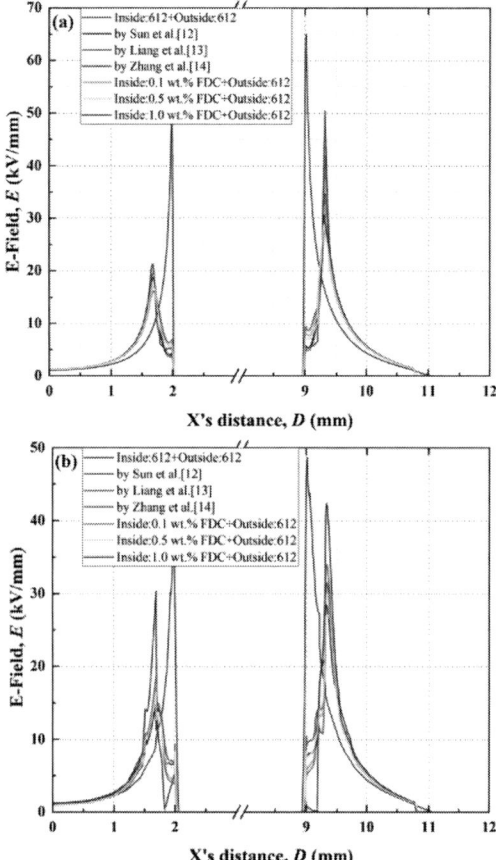

Fig. 4. The E-field simulation resuts of (a) 2D and (b) 3D DBC srtucture with different insulation assemby.

In the case of the structure coated with FDC material at nanofiller content of 0.5 wt.%, the maximum E-field reaches the lowest value of 26.3 KV/mm among all samples under investigation. This value is notably 23.9 KV/mm lower than that of the sample with SilGel-612, and 11.4 KV/mm, 8.6 kV/mm, and 20.4 KV/mm lower than the results reported by Liang et al. [14], Zhang et al. [17], and Sun et al. [18], respectively. The simulation results clearly demonstrate that the FDC coating effectively mitigates the E-field concentration through the mechanism of E-field grading by nonlinear conductivity. The findings indicate great potential for enhancing the PDIV for high-voltage module.

Table II. The peak E-field strength of the simulation results with different inside insulation materials

Inside material	2D model	3D model
SilGel-612	69.8 kV/mm	50.2 kV/mm
Sun et al. [12]	50.3 kV/mm	46.7 kV/mm
Liang et al. [13]	42.0 kV/mm	37.7 kV/mm
Zhang et al. [14]	34.7 kV/mm	34.9 kV/mm
0.1 wt.% FDC	32.5 kV/mm	31.0 kV/mm
0.5 wt.% FDC	26.5 kV/mm	26.3 kV/mm
1.0 wt.% FDC	29.2 kV/mm	28.9 kV/mm

V. CONCLUSIONS

To address the issue of E-field concentration in high-voltage power modules, a field-dependent conductivity (FDC) insulation material was developed. This was achieved by incorporating metal nanoparticle fillers into a silicon gel matrix. The nonlinear conductivity properties of the FDC materials were experimentally verified through measurements within an E-field spanning from 2 kV/mm to 20 kV/mm. The results clearly show that after a certain switching E-filed, the conductivities increase exponentially with the increasing E-field. This characteristic is of great significance as it effectively alleviates E-field concentration. Moreover, as the filler content increases from 0.1 wt.% to 1.0 wt.%, the nonlinearity of the conductivity first increases and then decreases. Notably, when the fillers content is 0.5 wt.%, the nonlinearity reaches its maximum value. This low filler content results in a low viscosity of 500 mPa·s, ensuring uniform potting during the module packaging process.

Furthermore, the FDC material was applied as a coating at triple points of a DBC substrate at a bias voltage of 15 kV. The 2D and 3D E-field simulation results vividly demonstrate that the FDC materials exhibit excellent E-field homogenization capabilities. For instance, when coated with the 0.5 wt.% FDC material, the E-field was reduced by 47.6%, i.e., dropping from 50.2 kV/mm to 26.3 kV/mm in comparison to the sample filled solely with silicon gel. The findings suggest that the proposed FDC material holds great promise as a viable solution for increasing PDIV in high-voltage module packaging applications.

REFERENCES

[1] C. M. DiMarino, B. Mouawad, C. M. Johnson, D. Boroyevich, and R. Burgos, "10-kV SiC MOSFET Power Module With Reduced Common-Mode Noise and Electric Field," *Ieee Transactions on Power Electronics,* vol. 35, no. 6, pp. 6050-6060

[2] B. Zhang *et al.*, "Electrical Properties of Silicone Gel for WBG-Based Power Module Packaging at High Temperatures," *Ieee Transactions on Dielectrics and Electrical Insulation,* vol. 30, no. 2, pp. 852-861

[3] Y. Zhou, L. Sang, X. Tang, H. Shi, and Ieee, "Design and Research on Package Insulation of Highvoltage Silicon Carbide Module," in *IEEE Workshop on Wide Bandgap Power Devices and Applications in Asia (WiPDA Asia)*, Huazhong Univ Sci & Technol, Wuhan, PEOPLES R CHINA, 2021

[4] X. Su, S. Zhang, F. Zhao, P. Liu, Z. Peng, and Ieee, "Experimental Study on the Effect of Thermal Aging on Epoxy Composites Used for UHV Dry Bushing," in *12th International Conference on the Properties and Applications of Dielectric Materials (ICPADM)*, Xian Jiaotong Univ, Xian, PEOPLES R CHINA, 2018

[5] Q. Chi, M. Yang, T. Zhang, and C. Zhang, "Investigation of electrical and mechanical properties of silver-hexagonal boron nitride/EPDM

composites," *Journal of Materials Science-Materials in Electronics,* vol. 30, no. 14, pp. 13321-13329, Jul 2019

[6] Z. Zhang *et al.*, "Packaging of a 10-kV Double-Side Cooled Silicon Carbide Diode Module With Thin Substrates Coated by a Nonlinear Resistive Polymer-Nanoparticle Composite," *Ieee Transactions on Power Electronics,* vol. 37, no. 12, pp. 14462-14470, Dec 2022

[7] C. Leopold, T. Augustin, T. Schwebler, J. Lehmann, W. V. Liebig, and B. Fiedler, "Influence of carbon nanoparticle modification on the mechanical and electrical properties of epoxy in small volumes," *Journal of Colloid and Interface Science,* vol. 506, pp. 620-632, Nov 15 2017.

[8] W. Pan *et al.*, "Nonlinear Materials Applied in HVDC Gas Insulated Equipment: From Fundamentals to Applications," *Ieee Transactions on Dielectrics and Electrical Insulation,* vol. 28, no. 5, pp. 1588-1603, Oct 2021

[9] D. Zhang *et al.*, "Field-dependent nonlinear electrical response characteristics in polymer dielectrics with sodium alginate scaffold," *Advanced Composites and Hybrid Materials,* vol. 7, no. 5, p. 162, 2024/10/02 2024

[10] D. Toker, D. Azulay, N. Shimoni, I. Balberg, and O. Millo, "Tunneling and percolation in metal-insulator composite materials," *Physical Review B,* vol. 68, no. 4, p. 041403, 07/25/ 2003

[11] I. Balberg, D. Azulay, D. Toker, and O. Millo, "PERCOLATION AND TUNNELING IN COMPOSITE MATERIALS," *International Journal of Modern Physics B,* vol. 18, no. 15, pp. 2091-2121

[12] A. Can-Ortiz, L. Laudebat, Z. Valdez-Nava, and S. Diaham, "Nonlinear Electrical Conduction in Polymer Composites for Field Grading in High-Voltage Applications: A Review," *Polymers,* vol. 13, no. 9

[13] J. Li, Y. Liang, Y. Mei, X. Tang, and G.-Q. Lu, "Packaging Design of 15 kV SiC Power Devices With High-Voltage Encapsulation," *Ieee Transactions on Dielectrics and Electrical Insulation,* vol. 29, no. 1, pp. 47-53, Feb 2022

[14] Y. Liang, G. Zhu, G.-Q. Lu, and Y.-H. Mei, "Reliable epoxy/SiC composite insulation coating for high-voltage power packaging," *Journal of Materials Science-Materials in Electronics,* vol. 33, no. 26, pp. 20508-20517, Sep 2022

[15] C. F. Bayer, E. Baer, U. Waltrich, D. Malipaard, and A. Schletz, "Simulation of the Electric Field Strength in the Vicinity of Metallization Edges on Dielectric Substrates," *Ieee Transactions on Dielectrics and Electrical Insulation,* vol. 22, no. 1, pp. 257-265, Feb 2015

[16] Q. Wang, X. Chen, C. Dai, A. Paramane, M. Awais, and N. Ren, "Experimental and Finite Element Analysis of Epoxy-Based Composites for Packaging Materials to Reduce Electric Field and Power Loss Under AC and DC Conditions," *Ieee Transactions on*

Components Packaging and Manufacturing Technology, vol. 12, no. 1, pp. 11-26, Jan 2022

[17] Z. Zhang, K. D. T. Ngo, and G.-Q. Lu, "Characterization of a Nonlinear Resistive Polymer-Nanoparticle Composite Coating for Electric Field Reduction in a Medium-Voltage Power Module," *Ieee Transactions on Power Electronics,* vol. 37, no. 3, pp. 2475-2479, Mar 2022

[18] K.-B. Sun, Y.-H. Mei, Z.-B. Shuai, and L. Li, "Insulation and Reliability Enhancement by a Nonlinear Conductive Polymer-Nanoparticle Coating for Packaging of High-Voltage Power Devices," *Ieee Transactions on Dielectrics and Electrical Insulation,* vol. 30, no. 6, pp. 2514-2521, Dec 2023

[19] L. Mu, B. Wang, J. Hao, Z. Fang, and Y. Wang, "Study on material and mechanical characteristics of silicone rubber shed of field-aged 110 kV composite insulators," *Scientific Reports,* vol. 13, no. 1, p. 16889, 2023/10/06 2023

[20] X. Liu, Z. Ma, D. Auhl, and F. Xia, "Effects of Nanoparticles and Surface Modification on Thermal, Mechanical, and Electrical Properties of Composites from Liquid Silicone Rubber with Expanded Graphite," *ACS Omega,* vol. 10, no. 16, pp. 16370-16383, 2025/04/29 2025

[21] C. Tang, R. Chen, J. Zhang, X. Peng, B. Chen, and L. Zhang, "A review on the research progress and future development of nano-modified cellulose insulation paper," *IET Nanodielectrics,* vol. 5, no. 2, pp. 63-84, 2022/06/01 2022

[22] M. Zielecka and A. Rabajczyk, "Silicone Nanocomposites with Enhanced Thermal Resistance: A Short Review," *Materials,* vol. 17, no. 9

Configuration Selection for Degradation Trajectory Prediction of Power Modules Based LSTM Model

1st Yichi Zhang
AAU Energy
Aalborg University
Aalborg, Denmark
yzhang@energy.aau.dk

2nd Yi Zhang
Department of Electrical and
Electronic Engineering
Hong Kong Polytechnic
University
Hong Kong, China
yiz@ieee.org

3rd Jie Kong
AAU Energy
Aalborg University
Aalborg, Denmark
jiek@energy.aau.dk

4th Jiahong Liu
AAU Energy
Aalborg University
Aalborg, Denmark
jliu@energy.aau.dk

5th Bo Yao
AAU Energy
Aalborg University
Aalborg, Denmark
ybo@energy.aau.dk

6th Huai Wang
AAU Energy
Aalborg University
Aalborg, Denmark
hwa@energy.aau.dk

Abstract—This paper investigates a data-driven approach for degradation trajectory prediction aimed at reducing reliability testing time, specifically employing iterative sequence-to-sequence prediction based on the long short-term memory (LSTM) model. It provides a comprehensive understanding of the application of the data-driven method to the scenarios analyzed and the related details, which involves data processing and the hyperparameter selection process. The degradation data support is from 18 samples under three test conditions in the power cycling test. The study considers the impact of different configurations (i.e., hyperparameters) of deep learning models on the prediction results, namely input/output features, the data down-sampling coefficient, the number of network layers, the number of hidden layer units, and the lengths of input and output sequences. Moreover, two indicators, the prediction accuracy and the degree of testing time reduction, are defined to quantify the prediction analysis performance. Finally, the sensitivity analysis quantifies the contribution of each of the six factors to both predicted performance metrics.

Index Terms—degradation trajectory prediction, long short-term memory model, configuration, power cycling test, power module

I. INTRODUCTION

As power electronic systems become ubiquitous across modern applications, from electric vehicles and renewable-energy converters to industrial automation and smart grids, their long-term reliability has emerged as a paramount concern, demanding rigorous design, testing, and qualification protocols [1], [2], especially for one of the key components, power modules. Currently, its reliability quantification primarily relies on testing, with power cycling tests serving as the standard method for assessing package reliability in power modules [3]. However, the conventional approach, continuing tests until a predefined failure criterion is met or complete failure occurs, is highly time-consuming. This issue becomes even more pronounced under lower test stress conditions and as the number of test samples increases, highlighting the urgent need to reduce reliability testing time as a key research focus.

Lifetime prediction provides an effective solution by forecasting future degradation trajectories based on early aging data [4]. This requires the development of models capable of accurately capturing aging trends. Existing modeling approaches can be broadly categorized into analytical and data-driven methods. For the first one, some empirical models [5], such as the exponential or a power law function. However, the ambiguous interaction between variations in device characteristics and the effects of test conditions limits these models' generalizability across applications. Moreover, given the complexity of failure mechanisms, data-driven approaches are gaining increasing attention.

In power cycling tests, health indicators are represented as time-series data, making deep learning techniques, particularly recurrent neural networks (RNNs) and their variants, widely adopted due to their effectiveness in sequential data analysis [6]–[8]. The work in [6] is based on an RNN and adds both boundary constraints that the remaining lifetime should be monotonically decreasing and that no negative values are possible. Studies [7] and [8], on the other hand, are based on the long short-term memory (LSTM) model, the latter of which is based on bidirectional LSTM. These promising predictive results demonstrate the potential of the data-driven methods.

However, the existing studies typically focus on presenting the configuration that yields good prediction results, yet they often lack an explanation of the selection process and the rationale behind it. This omission not only limits the generalizability of the method but also undermines confidence in its applicability. Moreover, most research relies on limited databases, typically consisting of only 4 to 8 samples. To address these limitations, this paper conducts a comprehensive analysis of the degradation trajectory of the power modules based on the sequence-to-sequence iteration prediction approach. This study offers deeper insights into data-driven reliability modeling, which involves all aspects of the prediction process, including data preprocessing, choice of configuration (i.e., hyperparameters), and quantification of prediction performance. Additionally, the analysis is based on richer data, 18 samples under three test conditions.

II. DATABASE PREPARATION

A. Power cycling test

Fig. 1 illustrates the platform used for conducting power cycling tests. The test conditions and corresponding sample numbers are detailed in Table I. Throughout the test, two common health indicators, on-state voltage (V_{ce}) and thermal resistance (R_{th}), are monitored. The

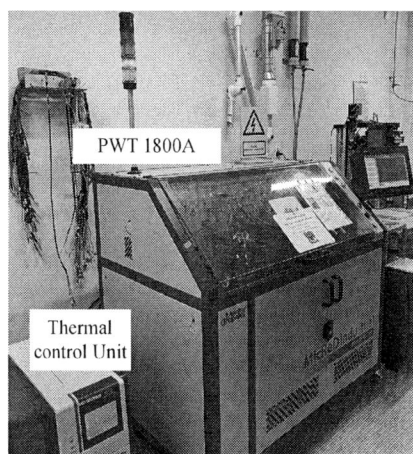

Fig. 1. Test platform of power cycling test.

experiment is a second-level power cycling test, where aging primarily occurs at the topside interconnection, leading to a noticeable increase in on-state voltage [9]. In contrast, thermal resistance remains nearly constant, reinforcing on-state voltage as the primary indicator for subsequent prediction analysis. Moreover, all tests are conducted until the device reaches the destruct limit, which indicates the complete failure. As the number of cycles increases, power cycling current, on-state voltage, maximum junction temperature, and minimum junction temperature are recorded, respectively.

TABLE I
TEST CONDITIONS AND SAMPLE NUMBERS

No.1	Test condition	Sample number	Sample index
1	ΔT_j=100 K, T_{jmax}=150 °C	6	1-6
2	ΔT_j=90 K, T_{jmax}=150 °C	6	7-12
3	ΔT_j=100 K, T_{jmax}=125 °C	6	13-18

B. Data processing

The data needs to be preprocessed before subsequent predictive analysis can be carried out, mainly for the on-state voltage. This process consists of two steps, the first is temperature decoupling, it is because the on-state voltage is strongly temperature-dependent, which can lead to a part of the increase not caused by aging but by temperature rise. Following decoupling, the data undergoes additional filtering. The results of one sample

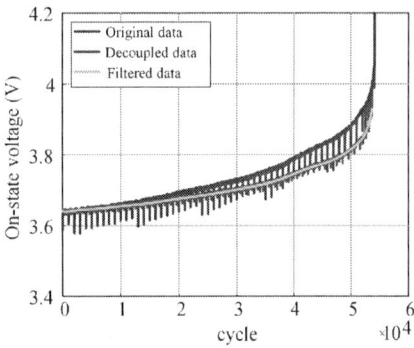

Fig. 2. On-state voltage with the number of cycles with data pre-processing of one sample.

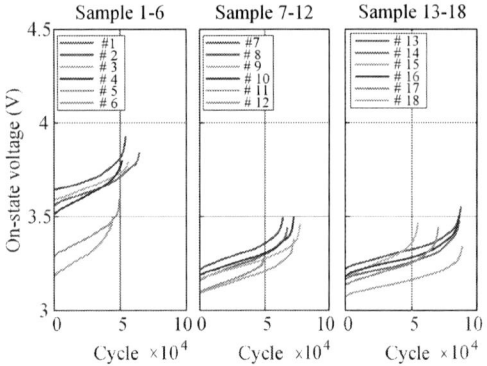

Fig. 3. Decoupling data of all tested samples.

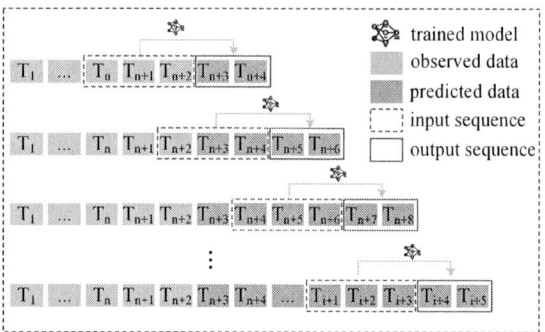

Fig. 4. Degradation trajectory prediction based on sequence-to-sequence iteration.

Firstly, because the on-state voltage is decoupled, it has been decoupled to the reference maximum temperature (T_{jmaxr}), which is 150°C, 150°C, and 125°C for the three test conditions of Table I. For the junction temperature fluctuation, it increases with the cycle. However, its update will increase the prediction variables, and the determination of the degree of aging and lifetime assessments is only derived based on the on-state voltage, so here we only introduce it and use its initial value (ΔT_{j0}), but do not do the update. And the power cycling current (I_{heat}) in the test is supposed to be kept constant [3].

III. TRAJECTORY PREDICTION BASED ON LSTM MODEL

A. Prediction strategy

For the sequence-to-sequence-based prediction strategy, as shown in Fig. 4, the entire prediction process follows an iterative approach. where T represents the data point, the blue one means that the data is known, i.e., the actual test value, and the yellow one means the predicted value. It is also important to note that the actual lengths of the selected input and output sequences are not 3 and 2, as shown in the figure, which is only a schematic representation. Through continuous iteration, the future trajectory can be generated. Furthermore, when the output window comprises a single data point, then it is the common sequence-to-point prediction strategy, which is adopted in [8].

B. Prediction expectations and configuration consideration

For degradation trajectory prediction in power cycling tests, two primary expectations are prediction

are shown in Fig. 2. The spikes in the raw data are caused by the suspension of the power cycling test for thermal resistance measurement. Since aging is an irreversible process, the data is filtered to prevent subsequent neural networks from learning irrelevant, noisy information, thereby enhancing the accuracy of aging trend analysis. Fig. 3 presents the results for all samples. It is evident that, under identical test conditions, both the on-state voltage and the device lifetime exhibit substantial variation, and the test conditions are overall clearly affecting the lifetime of the device.

Next, all data variables and the corresponding variation with the cycle index of each sample are detailed in Table II. These data are then partially or fully selected, even transformed, for the next predictive analysis. The last three columns of data remain constant.

TABLE II
DATA VARIABLES AND THE CORRESPONDING VARIATION WITH CYCLE INDEX

Cycle index	Filtered on-state voltage	Maximum junction temperature (reference)	Junction temperature fluctuation	Power cycling current
1	$V_{\text{ce-filtered1}}$	T_{jmaxr}	ΔT_{j0}	I_{heat}
2	$V_{\text{ce-filtered2}}$	T_{jmaxr}	ΔT_{j0}	I_{heat}
3	$V_{\text{ce-filtered3}}$	T_{jmaxr}	ΔT_{j0}	I_{heat}
$\vdots$	$\vdots$	$\vdots$	$\vdots$	$\vdots$

accuracy and the degree of test time reduction. In other words, achieving accurate predictions with less early data is expected. For model configuration, there are six main factors to consider.

For the first one, we consider three scenarios, as detailed in the table below: Case 1: focuses exclusively on the measured on-state voltage data, which is common in existing studies [7], [8]. Case 2: accounts not only for aging but also for the influences of current and temperature on the on-state voltage and test conditions. Case 3: aging is driven primarily by an increase in package connection resistance, with the degradation rate closely tied to the test conditions, where R is defined as the ratio of $V_{\text{ce-filtered}}$ to I_{heat}.

TABLE III
THREE CONSIDERATIONS FOR INPUT/OUTPUT FEATURES

No.1	Input features	Output feature
Case 1	$V_{\text{ce-filtered}}$	$V_{\text{ce-filtered}}$
Case 2	$V_{\text{ce-filtered}}$, I_{heat}, T_{jmaxr}, ΔT_{j0}	$V_{\text{ce-filtered}}$
Case 3	ΔR, T_{jmaxr}, ΔT_{j0}	ΔR

Next, the data should be downsampled considering that the full dataset is large, on the order of 5×10^4 to 10^5 points, as shown in Fig.3. For the LSTM architecture, there are two main hyperparameters: the number of layers and the number of hidden units. In a multi-layer configuration, the hidden-unit count is halved at each successive layer (for example, a three-layer network might use $256 \rightarrow 128 \rightarrow 64$ units). The input and output sequence lengths correspond to the number of points in each segment, with the total window size (input + output) constrained by the length of the down-sampled data. It is not denied that hyperparameters such as learning rate, regularization coefficient, and number of training iterations also influence prediction quality; however, tuning them all at once imposes prohibitive computational costs. Therefore, other related configurations are not considered.

We evaluate prediction performance using two quantification metrics. For prediction accuracy, uniformly quantified as the root-mean-square error (RMSE) of the on-state voltage of each sample from the respective dataset, which means that the predicted resistance value needs to be inverted to a voltage. Further considering the presence of multiple samples, this value is averaged again as shown in Equation (2). For the degree of test time reduction, we assume that only the initial input sequence is available at prediction. Equation (3) formalizes this definition, the resulting metric is also averaged over all samples in the respective dataset. The target is the lower $\overline{\text{RMSE}}$ and $\overline{\text{RT}}$.

$$\text{RMSE}_i = \sqrt{\frac{1}{n_i} \sum_{j=1}^{n_i} \left(\hat{V}_{\text{ce}-\text{filter}}^{(i,j)} - V_{\text{ce}-\text{filter}}^{(i,j)} \right)^2} \quad (1)$$

$$\overline{\text{RMSE}} = \frac{1}{M} \sum_{i=1}^{M} \text{RMSE}_i \quad (2)$$

where M is the number of samples of the respective dataset, n_i is the total number of data point of the ith sample, $V_{\text{ce,filter}}^{(i,j)}$ is the true value of the jth data point of the ith sample, $\hat{V}_{\text{ce,filter}}^{(i,j)}$ indicates the responding predicted value.

$$\text{RT}_i = \frac{L_{\text{input}}^{(i)}}{L_{\text{total}}^{(i)}} \quad (3)$$

$$\overline{\text{RT}} = \frac{1}{M} \sum_{i=1}^{M} \text{RT}_i \quad (4)$$

where $L_{\text{input}}^{(i)}$ is the input length of the ith sample, $L_{\text{total}}^{(i)}$ is the total length of the ith sample.

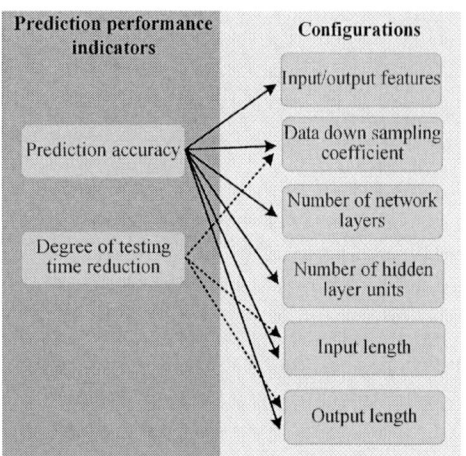

Fig. 5. Two prediction expectations and six model configurations.

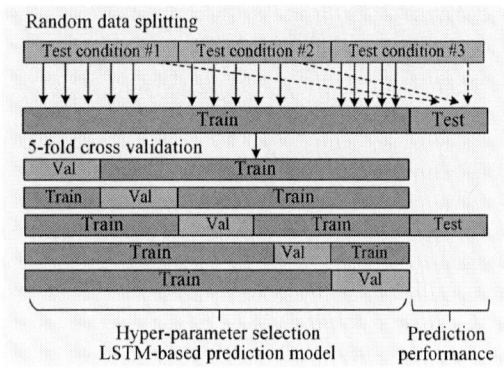

Fig. 6. Data split and dataset division.

IV. PREDICTION ANALYSIS

A. Optimal parameter acquisition

Before evaluating how different configuration choices affect prediction performance, the optimal hyperparameters are first determined using a grid-based search process. This approach is intended chiefly to generate results for each configuration, thereby enabling subsequent sensitivity analysis; alternative methods, such as random search or Bayesian optimization, may also be employed. Table IV summarizes the parameter search space and step, where the first factor corresponds to the three cases in Table III. Output length is defined as a proportion of the input length, preventing impractical

pairings, such as an input length of 10 with an output length of 400. It should be noted that the search space presented here is merely illustrative; researchers should tailor it to their specific requirements. In the absence of prior knowledge, a broader range may be selected, but doing so will inevitably increase the computational effort required to identify the optimal parameters.

TABLE IV
HYPERPARAMETER SEARCH SPACE

Factor	Search range	Search step
Input/output feature	[1, 3]	1
Data down sampling coefficient	[50, 200]	50
Number of network layers	[1, 3]	1
Number of hidden units	[64, 512]	64
Input length	[20, 100]	20
Output length ratio	[0.2, 1]	0.2

Next is the data splitting, one sample from each test condition is randomly selected to form the test dataset; the remaining samples constitute the train dataset. During hyperparameter tuning, five-fold cross-validation is performed on the train dataset. Moreover, in light of the two objective functions, the optimality is selected when the sum of their squared values is minimum. End-users may tailor the relative weighting of the two objective functions to suit their priorities, for example, by acquiring additional test data to enhance predictive accuracy. In this study, equal weights were applied. The corresponding configurations are listed in Table V. Note that optimal parameters are chosen based on performance on a five-fold validation set. Once selected, the model is retrained on the full train dataset, and the predictions are then generated for the test samples. The resulting forecasts are shown in Fig. 7. It can be seen that accurate predictions are realized and that the observations used for all three samples do not exceed 10% overall, and sample 16 is even less than 5%.

TABLE V
OPTIMAL CONFIGURATION SELECTION

Factor	Value
Input/output feature	Case 3 of Table III
Data down sampling coefficient	50
Number of network layers	3
Number of hidden units	512
Input length	80
Output length ratio	1

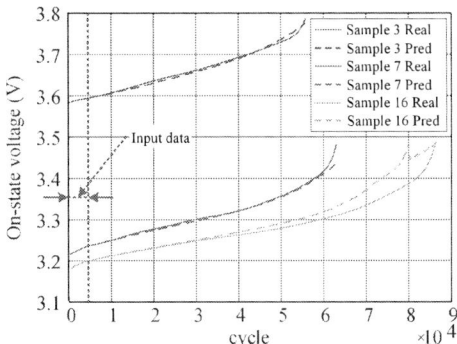

Fig. 7. Degradation trajectory prediction of three samples from the test dataset.

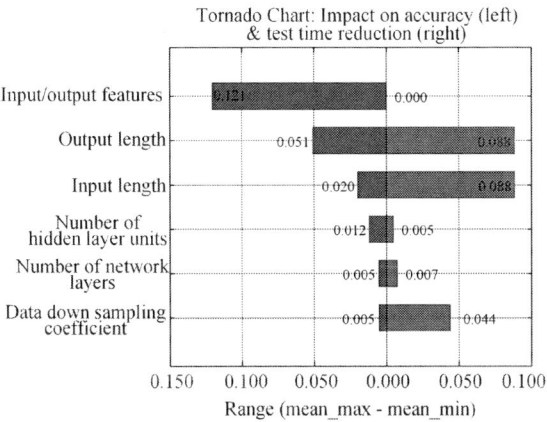

Fig. 8. Impact of configurations on the two performance metrics on the test dataset.

B. Impact of different configuration selection

Fig. 8 presents the impact of different configurations on the two performance metrics across all test dataset configurations based on the Tornado chart. It shows that feature selection exerts the greatest impact on prediction error, followed by the input and output length. A widely held explanation is that sequence length dictates the number of iterations; each iteration introduces incremental error, so more iterations yield greater cumulative error. Moreover, test-time reduction is governed primarily by sequence length, with the downsampling coefficient exerting a secondary influence.

V. CONCLUSION

This paper comprehensively presents the process of degradation trajectory prediction of power modules when the LSTM model is adopted and the prediction strategy is based on the sequence-to-sequence iteration. It involves the preprocessing of data, the construction of data variables, six common configurations, and the selection of optimal parameters. Moreover, the impact of configuration on the prediction performance is analyzed. These results show that input–output feature selection has the greatest influence on prediction error, whereas the network architecture exerts a relatively weak effect. Although sequence length markedly impacts test-time reduction, its benefits must be weighed against any degradation in predictive accuracy. Together, these findings highlight the critical importance of careful feature selection in data-driven analyses.

ACKNOWLEDGEMENT

This research has been supported by the Project of Artificial Intelligence for Next-Generation Power Electronics (AI-Power).

REFERENCES

[1] S. Yang, D. Xiang, A. Bryant, P. Mawby, L. Ran, and P. Tavner, "Condition monitoring for device reliability in power electronic converters: A review," *IEEE Power Electron. Mag.*, vol. 25, no. 11, p. 2734–2752, Nov. 2010.

[2] H. Wang and F. Blaabjerg, "Power electronics reliability: State of the art and outlook," *IEEE J. Emerg. Sel. Topics Power Electron.*, vol. 9, no. 6, pp. 6476–6493, Dec. 2021.

[3] ECPE, "324—qualification of power modules for use in power electronics converter units in motor vehicles," *ECPE*, 2025.

[4] S. Zhao, S. Chen, F. Yang, E. Ugur, B. Akin, and H. Wang, "A composite failure precursor for condition monitoring and remaining useful life prediction of discrete power devices," *IEEE Trans. Ind. Informat.*, vol. 17, no. 1, pp. 688–698, Jan. 2021.

[5] J. W. McPherson, *Reliability physics and engineering: time-to-failure modeling.* Springer, 2018.

[6] Z. Lu, C. Guo, M. Liu, and R. Shi, "Remaining useful lifetime estimation for discrete power electronic devices using physics-informed neural network," *Sci. Rep.*, vol. 13, no. 1, p. 10167, Jun. 2023.

[7] W. Li, B. Wang, J. Liu, G. Zhang, and J. Wang, "IGBT aging monitoring and remaining lifetime prediction based on long short-term memory (LSTM) networks," *Microelectronics Reli.*, vol. 114, pp. 1–8, Nov. 2020.

[8] A. Vaccaro, D. Biadene, and P. Magnone, "Remaining useful lifetime prediction of discrete power devices by means of artificial neural networks," *IEEE Open J. Power Electron.*, vol. 4, pp. 978–986, Nov. 2023.

[9] Y. Zhang, Y. Zhang, S. Zhao, B. Yao, and H. Wang, "Physics-based modeling of packaging-related degradation of IGBT modules," in *Proc. IEEE Appl. Power Electron. Conf. Expo.*, 2023, pp. 2463–2468.

A Review and Analysis of Grid-Forming Technologies in Renewable Energy Power Systems

Zhicheng Liu, Dezheng Zhang, Yehan Fu, Yuying He, and Li Zhang

College of Electrical and Power Engineering, Hohai University, Nanjing, China

Abstract— **High renewable penetration poses challenges to grid stability, especially in low-inertia environments. In this scenario, grid-forming control gained increasing attention for improving voltage/frequency support capability. This paper focuses on the grid-forming control techniques, and systematically reviews the evolution of droop control, virtual synchronous generator (VSG) control, matching control, virtual oscillator control (VOC), and hybrid synchronous control (HSC). A comparison highlights each strategy's unique benefits and drawbacks in enhancing stability and dynamic response. Finally, this review offers key insights for theoretical optimization and practical design of inverter control technologies in modern power systems.**

Keywords—Grid-forming, virtual synchronous generator, droop control, matching control, virtual oscillator control.

I. INTRODUCTION

With the rapid advancement of renewable energy technologies and power electronics, modern power systems demand enhanced grid stability [1]. The grid-connected behavior of power converters is primarily dictated by their control strategies, which fall into two categories: grid-following and grid-forming approaches [2].

Grid-following control relies on a phase-locked loop (PLL) to track the grid phase, ensuring precise power injection but lacking inertial support [3]. In contrast, grid-forming control autonomously establishes voltage and frequency, enabling robust transient responses in low-inertia grids [2]. Fig. 1 illustrates the distinctions between these two types in terms of synchronization methods, inertia support, transient voltage response, and adaptability to both strong and weak grids.

Over the years, researchers have developed various grid-forming control strategies. Through mathematical modeling and dynamic simulation of core characteristics, such as inertia, damping, and grid frequency regulation of synchronous generators, early researchers proposed two basic grid-forming technologies: droop control and virtual synchronous generator (VSG). As the demand for active support in modern power systems increased, the technological development has now surpassed the simple paradigm of mimicking synchronous machines, forming three new technologies: matching control, virtual oscillator control (VOC), and hybrid synchronous control (HSC).

This paper reviews grid-forming control technologies, and explores their distinct characteristics. A comparative analysis of various grid-forming strategies follows, highlighting their advantages and limitations. Finally, key

challenges and future directions are discussed, offering insights for advancing grid-forming converter research.

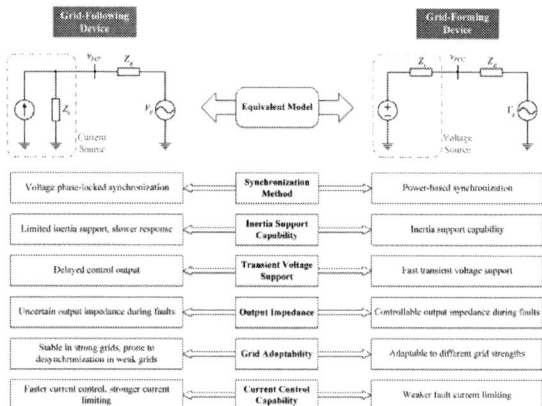

Fig. 1 Comparison of Functions and Features of Grid-Following and Grid-Forming Devices

II. GRID-FORMING CONTROL TECHNOLOGIES

This section reviews the five common control strategies, discussing their basic principles and practical applications.

A. Droop Control

The droop control strategy is widely applied to regulate voltage-source converters (VSCs) to ensure grid stability and reliability [4]. Droop Control simulates the natural droop characteristics of synchronous generators, enabling autonomous regulation of active power-frequency (*P-f*) and reactive power-voltage (*Q-V*) [5]. In traditional droop control, frequency and voltage change linearly with active and reactive power, respectively. The frequency ω and voltage V control are

$$\begin{cases} \omega = \omega_0 + K_{pf}(P_0 - P) \\ V = V_0 + K_{qv}(Q_0 - Q) \end{cases} \quad (1)$$

Fig. 2 shows the system structure of a grid-forming inverter with droop control [6]. The inverter is connected to the grid and PCC through an LC filter, where L and C represent the filter inductance and capacitance. According to Thévenin's theorem, the AC grid at the PCC can be viewed as an ideal voltage source V_g in series with the grid impedance L_g [7]. ω_0 and V_0 are the reference values for frequency and voltage, respectively, and the power deviation is obtained by the difference between the output power reference values $P_0(Q_0)$ and the measured values. The power deviation multiplied by the droop coefficients $K_{pf}(K_{qv})$ constitutes the frequency (voltage) adjustment term [8].

This work is supported in part by the National Key Research and Development Program of China under Grant 2024YFB2408600, in part by National Natural Foundation of China under Grant 524036711, in part by the Natural Science Foundation of Jiangsu Province under Grant SBK2023045379, and in part by the Special Fund of Jiangsu Province for the Transformation of Scientific and Technological Achievements under Grant BA2023108.

979-8-3315-1110-4/25 $31.00 © 2025 IEEE

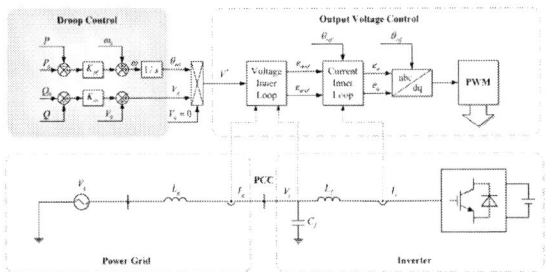

Fig. 2. The configuration of the GFM inverter with droop control.

Droop control can achieve distributed power sharing and load distribution without relying on complex communication, making it modular and plug-and-play [9]. Additionally, its fast response speed ensures that the system can quickly adjust when faced with sudden load changes.

However, droop control has limitations in dynamic and static stability. To address the issue of low-frequency negative impedance instability caused by the voltage-current inner-loop structure under strong grids [10], Ref. [11] adopts a virtual impedance control method to statically reshape the converter port impedance characteristics. Ref. [12],[13] propose replacing the voltage loop in the voltage-current dual inner-loop control with virtual admittance control to dynamically adjust virtual impedance. It is worth noting that traditional droop control still has inherent defects, such as weak inertia and limited reactive power distribution accuracy. To overcome these issues, Ref [9] proposes a combined strategy of cooperative virtual output impedance and adaptive clipping control, which significantly improves frequency fluctuation control and system dynamic recovery in off-grid mode.

B. Virtual Synchronous Generator Control

The VSG strategy is a leading grid-forming control method that mimics the operating characteristics of traditional synchronous generators, offering voltage-source features similar to or exceeding those of synchronous machines [14]. In Fig. 3, comparative analysis of VSG and conventional generators across seven operational parameters reveals that VSG's adjustable impedance/inertia parameters enable enhanced adaptability and dynamic responsiveness, including effective frequency/voltage regulation and DC component control. In contrast, conventional generators demonstrate inherent limitations in parameter flexibility and regulation speed, highlighting VSG's potential for improving power system stability and operational flexibility.

Figure 4 shows the overall block diagram of grid-connected virtual synchronous control, while Equation (2) describes its frequency and voltage regulation dynamics.

$$\begin{cases} \dfrac{d\omega_0}{dt} = \dfrac{1}{J}\left[P_{ref} - P_e - D\left(\omega_{ref} - \omega_0\right)\right] \\ V_0 = V_{ref} + K_q\left(Q_{ref} - Q_e\right) \end{cases} \quad (2)$$

The frequency dynamic equation converts the difference between the reference power P_{ref} and the actual electrical power P_e into an angular frequency adjustment using the virtual inertia constant J and damping coefficient D. This dynamic process simulates the rotor motion of a synchronous generator: the power deviation $P_{ref} - P_e$ frequency changes, while the damping term $D\left(\omega_{ref} - \omega_0\right)$ suppresses frequency oscillations. Meanwhile, the voltage equation uses a reactive power control gain K_q to linearly map the reactive power deviation $Q_{ref} - Q_e$ to the output voltage V_0, achieving rapid voltage tracking.

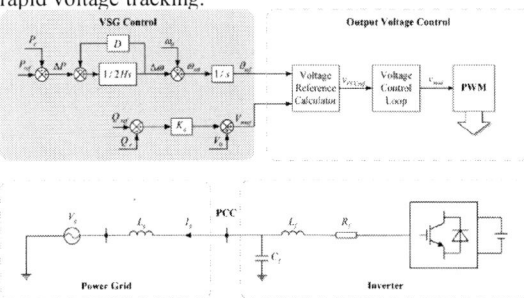

Fig. 4. The configuration of the GFM inverter with VSG.

The technical schemes for virtual synchronous generator control can be divided into grid-following based and grid-forming based types. In 2007, Ref. [15] introduced a virtual synchronous machine scheme (VISMA) that indirectly imparted the characteristics of a synchronous generator to renewable energy systems by directly controlling the current through filter inductance. On this basis, by adjusting the current loop control command, it was possible to emulate the primary frequency and voltage regulation characteristics of synchronous machines. However, since this control method essentially adds virtual synchronous control to grid-following converters, it cannot overcome issues such as poor damping characteristics and the inability to operate in islanded mode that stem from the inherent current-source nature of renewable energy units.

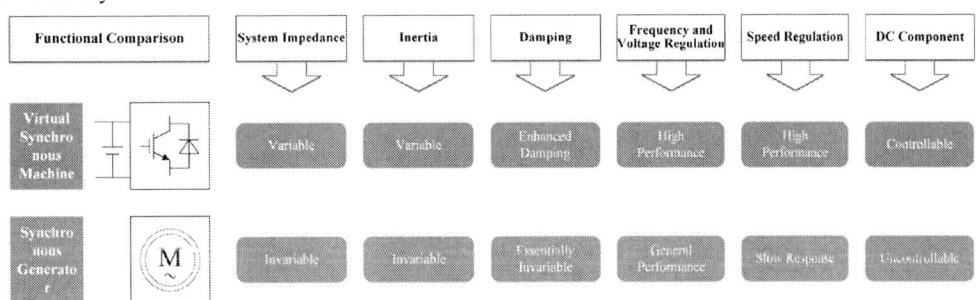

Fig. 3. Functional Comparison of VSG and Synchronous Generators.

TABLE I. KEY COMPARISON BETWEEN SYNCHRONVERTER AND CONVENTIONAL VSG TECHNOLOGIES

Characteristic	Synchronverter	VSG
Control Objective	Track power commands (PQ/VF control)	Emulate the electromechanical dynamic characteristics of synchronous generators
Inertia Support	None	Provide inertia via a virtual rotor equation (requires energy storage or capacitors)
Frequency/Voltage Regulation	Relies on external controller (e.g., droop control)	Integrated primary frequency and voltage regulation (similar to the autonomous response of synchronous generators)
Dynamic Response	Fast (millisecond-level)	Relatively slower (second-level)
Weak Grid Stability	Medium	Strong

For grid-forming virtual synchronous generator control, two main developmental approaches have emerged. One approach is the "synchronverter" control scheme proposed in 2010 [16]. Its core idea is to replicate the rotor motion and electromagnetic equations of synchronous generators through mathematical modeling, enabling the inverter to mimic the dynamic behavior of synchronous generators (e.g., inertia, damping, frequency, and voltage regulation) [17]. Today, grid-forming VSG control technologies encompass all inverter control algorithms that emulate synchronous generator characteristics, including synchronverter and VSG controls along with subsequent improvements (such as improved VSG [18]-[21], and adaptive VSG [22]-[25]. Table 1 presents a key comparison between traditional synchronverter control and current conventional VSG technology. Overall, synchronverters are more suitable for scenarios requiring rapid response, whereas VSGs are better suited for weak grids that demand both inertia and stability.

Compared with VSG technology, synchronverter control, although enhancing grid stability by emulating synchronous machine characteristics, may have its inherent fast response capability compromised due to its inertia properties [26]. This challenge necessitates balancing grid stability with rapid responsiveness during design.

Beyond precisely mimicking the electromagnetic characteristics of synchronous machines, some researchers have proposed an alternative approach: directly replicating the active support function of synchronous machines. One proposed control method for distributed energy resources emulates the primary frequency regulation and inertia functions of synchronous machines, enabling distributed generation units to operate in both grid-connected and islanded modes while achieving a smooth transition between the two, thereby embodying all the features of a virtual synchronous machine [27], [28]. In this approach, the grid-side converter of renewable energy units adopts a grid-forming control strategy, with a more flexible selection of outer-loop control methods. Consequently, this has become the mainstream direction in recent research and applications.

C. Matching Control

Matching control is a promising grid-forming control strategy that effectively balances energy exchange between the source and the grid [29]. This approach was first introduced by researchers who established a correspondence between the DC-link capacitor voltage of power converters and the rotor speed of synchronous machines (SMs) [30], [31]. By emulating the electromechanical energy transfer of synchronous

machines, matching control enables synchronization with the power grid.

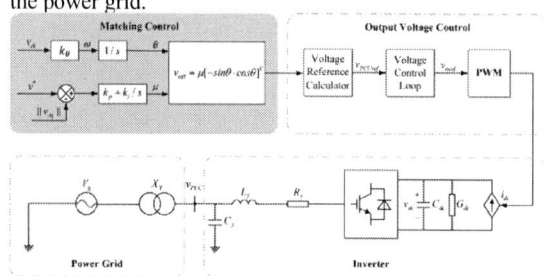

Fig. 5. The configuration of the GFM inverter with matching control.

The control structure of matching control is illustrated in Fig. 5 [32], showcasing its mechanism of energy balancing and synchronization through an analog to conventional synchronous machine dynamics.

As illustrated in Fig. 5, the 'Inverter' section represents the DC-side dynamics of the system, where i_{dc} denotes the source current, G_{dc} and C_{dc} correspond to the DC conductance and capacitance, respectively, v_{dc} represents the voltage across the DC-link capacitor, and i_x is the DC current at the input switching stage. The transient behavior of the DC-link voltage can be mathematically expressed as

$$C_{dc}\dot{v}_{dc} = i_{dc} - G_{dc}v_{dc} - i_x \qquad (3)$$

In conventional matching control, the control coefficient k_θ is employed to establish a relationship between the AC angular frequency ω and the DC-link voltage v_{dc}, i. e.,

$$k_\theta = \frac{\omega^*}{v^*_{dc}} \qquad (4)$$

where ω^* and v^*_{dc} represent the reference values for the system's angular frequency and DC-link voltage, respectively. The system's angular frequency is expressed as

$$\omega = k_\theta v_{dc} \qquad (5)$$

In this equation, ω and v_{dc} denote the system angular frequency and DC-link voltage, respectively.

Subsequently, the reference value for the output voltage, $v_{\alpha\beta}$, is defined as

$$v_{\alpha\beta} = \mu\begin{bmatrix}-\sin\theta & \cos\theta\end{bmatrix}^T \qquad (6)$$

Here, θ represents the AC output voltage phase angle, which is determined by the DC-link voltage. The AC voltage amplitude μ is obtained through a proportional-integral (PI) controller, which regulates the deviation between the AC-side voltage reference v^* and the measured output voltage amplitude $\|v_{dq}\|$.

By combining (3) and (5), the system dynamics can be formulated to emulate the synchronous machine's electromechanical equation, expressed as

$$J_v \dot{\omega} = T_{vm} - T_{ve} - D_v \omega \tag{7}$$

In this formulation, the virtual inertia coefficient and virtual damping coefficient are given by

$$J_v = C_{dc}/k_\theta{}^2, T_{ve} = i_x/k_\theta \tag{8}$$

Similarly, the virtual mechanical torque and virtual electromagnetic torque are defined as

$$T_{vm} = i_{dc}/k_\theta, T_{ve} = i_x/k_\theta \tag{9}$$

This approach achieves establishes a direct coupling between the DC voltage source and the AC voltage frequency, making it particularly suitable for Inverter-Based Resources (IBRs). Moreover, recognizing the limited flexibility and constrained frequency regulation performance of traditional Matching control, previous studies have proposed an improved frequency control strategy to enhance its adaptability and effectiveness in dynamic grid conditions.

Matching control offers advantages such as fast response speed and strong adaptability. However, its high dependency on DC capacitor capacity limits its applicability to systems where multiple wind turbines share a common DC bus or DC grid networks, making it unsuitable for single wind turbine control [33]. To address this issue, scholars have proposed a modification in the context of doubly-fed wind turbine systems, where the DC voltage signal is utilized as the input for the inertia transfer control loop, thereby increasing the active power output capability of the DC-side prime mover [34],[35]. This improved approach is commonly referred to as "Inertial Synchronous Control."

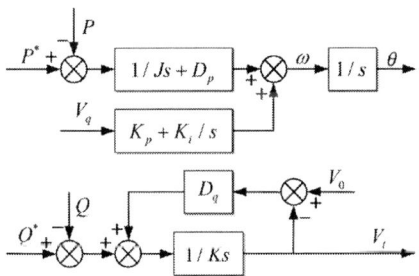

Fig. 6. Hybrid-Synchronous Control

D. Other Control Strategies

In recent years, VOC and HSC have emerged as key research focuses in grid-connected inverters due to their superior synchronization performance in communication-free environments.

a). Virtual Oscillator Control

VOC is an advanced nonlinear control strategy [36] that utilizes the synchronized oscillatory behavior of coupled nonlinear oscillators to achieve autonomous voltage synchronization in parallel inverter systems [37].

Common VOC implementations include the Van der Pol Oscillator (VdP), Dead-Zone Oscillator (DZO), and Andronov-Hopf Oscillator (AHO) [38]. Tab. 2 compares key performances of VdP, DZO, and AHO [39]-[42]. AHO excels in dispatchability, enabling dynamic grid reconfiguration, whereas VdP and DZO lack real-time scheduling capabilities. In harmonic suppression, AHO achieves a high-frequency noise attenuation rate, outperforming VdP and DZO.

For fault ride-through, AHO sustains continuous operation during grid disturbances, enhancing transient stability. DZO improves fault tolerance indirectly through system stability enhancements, whereas VdP lacks this capability entirely. Grid stability tests further show that AHO achieves superior dynamic regulation accuracy and environmental adaptability, while VdP and DZO are better suited for steady-state analysis and fundamental oscillation studies. Overall, AHO has strong engineering potential in complex grid applications.

Recent research has focused on harmonic suppression and grid adaptability optimization [43]. Future work will advance novel oscillator designs, including dVOC [44], AH-VOC [45], and uVOC [46], along with multi-machine system stability analysis and control strategy innovation. These developments will further integrate VOC into modern power systems .

b). Hybrid-Synchronous Control

The Hybrid-synchronous control combines PLL and PSL, as shown in Fig. 6. By tuning the proportional gain K_p and integral gain K_i, HSC improves transient stability and mitigates power loop failures during grid faults.

TABLE II. COMPARISON OF KEY CHARACTERISTICS OF THREE VIRTUAL OSCILLATOR CONTROL STRATEGIES

Characteristics	VdP	DZO	AHO
Basic Principle	Based on nonlinear domain theory; primarily used for analyzing system oscillations.	Based on voltage-dependent nonlinearity for grid stability analysis.	Uses adaptive feedback mechanisms for system stability and dynamic adjustments.
Application Scenarios	Theoretical analysis and simulation verification of oscillator behavior.	Suppression of grid harmonics and support for grid stability in fault conditions.	Intelligent power control and fault ride-through; suitable for dynamic grid conditions.
Adjustment Capability	No real-time adjustment capability; primarily for theoretical validation.	Does not support real-time adjustment but indirectly enhances grid stability.	Real-time adjustable; dynamic power distribution and system optimization capability.
Harmonic Suppression	Limited harmonic suppression capability.	Strong harmonic suppression, reducing high-order harmonics effectively.	Dynamic harmonic suppression; ideal for high-quality power requirements.
Fault Ride-Through	Not supported; suitable mainly for stable system analysis.	Not supported; contributes partially to stability enhancement.	Supports fault ride-through, enhancing grid stability under fault conditions.
Grid Stability	Limited stability capability; theoretical validation only.	Enhances grid stability by reducing harmonics and optimizing voltage.	Adaptively enhances transient stability and overall system performance under fault conditions.

TABLE III. RENEWABLE ENERGY GRID-FORMING CONTROL STRATEGIES

Characteristic	Droop Control	VSG	Matching Control	VOC	Hybrid-Synchronization
Synchronization	Power Synchronization $(P-\omega)$	Power Synchronization $(P-\omega)$	Inertia Synchronization $(U_{dc}-\omega)$	Extreme Ring Oscillation	Multi-Mode Synchronization
Inertia/Damping Properties	None/Weak	Strong	Configurable	Present	Strong and Adjustable
Response Speed	Fast	Slow	Moderate	Fast	Adjustable
Control Complexity	Low	Medium	Medium-High	High	High
PLL Dependence	Dependent	Partially Dependent	No Dependence	No Dependence	No Dependence
Power Quality	Average	Good	Good	Average	Good

The foundation of HSC was laid in 2020, when researchers first proposed simulating synchronous generator behavior by integrating PLL and PSL in a synchronization controller [47]. Concurrently, studies [48] explored HSC's transient stability enhancements through PLL and VSG control integration. Ref. [49] examined its performance under various operating scenarios, offering theoretical guidance for parameter selection. However, optimal parameter design remains a challenge, especially considering grid impedance variations, load fluctuations, and fault conditions.

III. COMPARISON

To systematically compare different grid integration control techniques for renewable energy systems, this section examines the core advantages and limitations of several major control strategies. Tab. 3 summarizes the distinct characteristics of these technologies across seven key dimensions.

Quantitative analysis demonstrates that VSG technology exhibits significant advantages in inertial support, though its power response speed is relatively slower. Matching control achieves faster response but provides no inertial support. Hybrid synchronization control achieves a balanced performance with moderate inertia and response speed, albeit with higher algorithm complexity.

VSG achieves self-synchronization by emulating synchronous machine characteristics without requiring phase-locked loops. Under weak grid conditions, it demonstrates superior power angle stability and power quality compared to conventional voltage source converters.

In high penetration scenarios, VSG excels in frequency regulation compliance rate and harmonic performance due to its high inertia and moderate response speed, making it particularly suitable for weak grids or microgrids. Matching control is preferable for applications demanding faster response, while hybrid synchronization control shows potential in complex power systems despite its technical challenges.

IV. CONCLUSION

Grid-forming technology, crucial for renewable integration, confronts three key challenges under high renewable penetration: system inertia deficiency, multi-timescale oscillation risks, and device-control-communication conflicts. Moreover, the issues such as the high-overload capacity for WBG devices in GFM inverters, the digital twins for system testing, and system transient stability improvements, are to be further studied. Future priorities include standard alignment and adaptive impedance technologies, et al., which are promising for advancing intelligent power systems.

REFERENCES

[1] Y. Bao, J. Pan, K. Wang et al., "An improved grid-connected control method combining GFM and GFL merits based on virtual synchronous generator," *J. Energy Storage*, vol. 112, Art. No. 115483, 2025.

[2] R. H. Lasseter, Z. Chen, and D. Pattabiraman, "Grid-forming inverters: A critical asset for the power grid," *IEEE J. Emerg. Sel. Top. Power Electron.*, vol. 8, no. 2, pp. 925–935, Jun. 2020.

[3] J. Rocabert, A. Luna, F. Blaabjerg et al., "Control of power converters in AC microgrids," *IEEE Trans. Power Electron.*, vol. 27, no. 11, pp. 4734–4749, Nov. 2012.

[4] Y. Zhang, B. Yang, Z. Li et al., "Frequency stability control strategy of power systems based on VSC-HVDC," *Electr. Eng. Technol.*, vol. 50, no. 13, pp. 35–40, 2023.

[5] P. Unruh, M. Nuschke, P. Strauß et al., "Overview on grid-forming inverter control methods," *Energies*, vol. 13, no. 10, Art. No. 2589, May 2020.

[6] D. Pan, X. Wang, F. Liu et al., "Transient stability impact of reactive power control on grid-connected converters," in *Proc. IEEE Energy Convers. Congr. Expo. (ECCE)*, 2019, pp. 4311–4316.

[7] H. Zhang, "Transient instability mechanism and stability enhancement strategy of droop-controlled grid-forming inverters," Ph.D. dissertation, Huazhong Univ. Sci. Technol., Wuhan, China, 2023.

[8] J. X. Xu, W. Liu, S. Liu ., "Current status and development trends of grid-forming control technology for power system converters," *Power Grid Technol.*, vol. 46, no. 9, pp. 3586–3595, 2022.

[9] H. He, H. Xiao, and L. Zhang, "Transient stability analysis and adaptive limiting method for grid-forming converters considering current limitation," *Autom. Electr. Power Syst.*, vol. 49, no. 2, pp. 30–40, 2025.

[10] W. Du, Z. Chen, K. P. Schneider et al., "A comparative study of two widely used grid-forming droop controls on microgrid small-signal stability," *IEEE J. Emerg. Sel. Top. Power Electron.*, vol. 8, no. 2, pp. 963–975, Jun. 2020.

[11] J. M. Guerrero, J. C. Vasquez, J. Matas et al., "Hierarchical control of droop-controlled AC and DC microgrids—A general approach toward standardization," *IEEE Trans. Ind. Electron.*, vol. 58, no. 1, pp. 158–172, Jan. 2011.

[12] C. Liu, X. Cai, R. Li et al., "Optimal short-circuit current control of the grid-forming converter during grid fault condition," *IET Renew. Power Gener.*, vol. 15, no. 10, pp. 2185–2194, Oct. 2021.

[13] Y. Zhang, C. Zhang, R. Yang et al., "Current-constrained power-angle characterization method for transient stability analysis of grid-forming voltage source converters," *IEEE Trans. Energy Convers.*, vol. 38, no. 2, pp. 1338–1349, Jun. 2023.

[14] J. Driesen and K. Visscher, "Virtual synchronous generators," in *Proc. IEEE Power Energy Soc. Gen. Meeting*, 2008, pp. 1–3.

[15] H. P. Beck and R. Hesse, "Virtual synchronous machine," in *Proc. 9th Int. Conf. Electr. Power Qual. Utilisation (EPQU)*, 2007, pp. 1–6.

[16] Q. C. Zhong and G. Weiss, "Synchronverters: inverters that mimic synchronous generators," *IEEE Trans. Ind. Electron.*, vol. 58, no. 4, pp. 1259–1267, Apr. 2011.

[17] Q. C. Zhong, P. L. Nguyen, Z. Ma et al., "Self-synchronized synchronverters: inverters without a dedicated synchronization unit," *IEEE Trans. Power Electron.*, vol. 29, no. 2, pp. 617–630, Feb. 2014.

[18] S. Chen, Y. Sun, H. Han et al., "A modified VSG control scheme with virtual resistance to enhance both small-signal stability and transient synchronization stability," *IEEE Trans. Power Electron.*, vol. 38, no. 5, pp. 6005–6014, May 2023.

[19] K. Koiwa, K. Inoo, T. Zanma et al., "Virtual voltage control of VSG for overcurrent suppression under symmetrical and asymmetrical voltage dips," *IEEE Trans. Ind. Electron.*, vol. 69, no. 11, pp. 11177–11186, Nov. 2022.

[20] P. Ge, C. Tu, F. Xiao et al., "Design-oriented analysis and transient stability enhancement control for a virtual synchronous generator," *IEEE Trans. Ind. Electron.*, vol. 70, no. 3, pp. 2675–2684, Mar. 2023.

[21] M. Chen, D. Zhou, and F. Blaabjerg, "Enhanced transient angle stability control of grid-forming converter based on virtual synchronous generator," *IEEE Trans. Ind. Electron.*, vol. 69, no. 9, pp. 9133–9144, Sep. 2022.

[22] H. Wu and X. Wang, "Control of grid-forming VSCs: a perspective of adaptive fast/slow internal voltage source," *IEEE Trans. Power Electron.*, vol. 38, no. 8, pp. 10151–10169, Aug. 2023.

[23] X. Hou, Y. Sun, X. Zhang et al., "Improvement of frequency regulation in VSG-based AC microgrid via adaptive virtual inertia," *IEEE Trans. Power Electron.*, vol. 35, no. 2, pp. 1589–1602, Feb. 2020.

[24] S. P. Me, S. Zabihi, F. Blaabjerg et al., "Adaptive virtual resistance for postfault oscillation damping in grid-forming inverters," *IEEE Trans. Power Electron.*, vol. 37, no. 4, pp. 3813–3824, Apr. 2022.

[25] L. Huang, C. Wu, D. Zhou et al., "A power-angle-based adaptive overcurrent protection scheme for grid-forming inverter under large grid disturbances," *IEEE Trans. Ind. Electron.*, vol. 70, no. 6, pp. 5927–5936, Jun. 2023.

[26] H. Liu, S. Yu, D. Sun et al., "Review on grid-forming converter control techniques and principles," *Proc. CSEE*, vol. 45, no. 1, pp. 277–297, 2025.

[27] F. Gao and M. R. Iravani, "A control strategy for a distributed generation unit in grid-connected and autonomous modes of operation," *IEEE Trans. Power Del.*, vol. 23, no. 2, pp. 850–859, Apr. 2008.

[28] K. Sakimoto, Y. Miura, and T. Ise, "Stabilization of a power system with a distributed generator by a virtual synchronous generator function," in *Proc. 8th Int. Conf. Power Electron. – ECCE Asia*, 2011, pp. 1498–1505.

[29] H. Luo, Y. Zhu, Y. Yang, Y. Peng, and L. Zhang, "Modified Matching Control with Enhanced Frequency Support Capability," in *Proc. 2024 CPSS & IEEE International Symposium on Energy Storage and Conversion (ISESC)*, Xi'an, China, 2024, pp. 246-251, doi: 10.1109/ISESC63657.2024.10785458.

[30] T. Jouini, C. Arghir, and F. Dörfler, "Grid-friendly matching of synchronous machines by tapping into the DC storage," *IFAC-PapersOnLine*, vol. 49, no. 22, pp. 192–197, 2016.

[31] C. Arghir, T. Jouini, and F. Dörfler, "Grid-forming control for power converters based on matching of synchronous machines," *Automatica*, vol. 95, pp. 273–282, Sep. 2018.

[32] X. Xiong, C. Wu, F. Blaabjerg, "An improved synchronization stability method of virtual synchronous generators based on

frequency feedforward on reactive power control loop," *IEEE Trans. Power Electron.*, vol. 36, no. 8, pp. 9136–9148, 2021.

[33] J. L. Li, Z. Y. Ding, H. T. Liu et al., "Research on grid-forming energy storage converters and control strategies, " *Power Generation Technol.*, vol. 43, no. 5, pp. 679–686, 2022.

[34] S. Sang, C. Zhang, X. Cai et al., "Voltage-source control of full-power conversion wind turbines (I): Control architecture and weak grid stability analysis," *Proc. CSEE*, vol. 41, no. 16, pp. 5604–5616, 2021.

[35] Y. Qin, H. Wang, Z. Yang et al., "Voltage-source control of full-power conversion wind turbines (II): Grid fault ride-through control and protection," *Proc. CSEE*, vol. 43, no. 2, pp. 530–543, 2023.

[36] M. A. Awal and I. Husain, "Unified virtual oscillator control for grid-forming and grid-following converters," *arXiv preprint*, 2020.

[37] M. Lu, "Virtual oscillator grid-forming inverters: state of the art, modeling, and stability," *IEEE Trans. Power Electron.*, vol. 37, no. 10, pp. 11579–11591, Oct. 2022.

[38] S. A. Aghdam and M. Agamy, "Virtual oscillator-based methods for grid-forming inverter control: A review," *IET Renew. Power Gener.*, vol. 16, no. 5, pp. 835–855, May 2022.

[39] M. A. Awal, H. Yu, I. Husain et al., "Selective harmonic current rejection for virtual oscillator controlled grid-forming voltage source converters," *IEEE Trans. Power Electron.*, vol. 35, no. 8, pp. 8805–8818, Aug. 2020.

[40] V. Gurugubelli, A. Ghosh, A. K. Panda et al., "Implementation and comparison of droop control, virtual synchronous machine, and virtual oscillator control for parallel inverters in standalone microgrid," *Int. Trans. Electr. Energy Syst.*, vol. 31, no. 5, Art. No. e12859, May 2021.

[41] V. Gurugubelli, A. Ghosh, and A. K. Panda, "Comparison of deadzone and VanderPol oscillator controlled voltage source inverters in islanded microgrid," in *Proc. IEEE 2nd Int. Conf. Smart Technol. Power, Energy Control (STPEC)*, 2021, pp. 1–6.

[42] M. Lu, S. Dhople, and B. Johnson, "Benchmarking nonlinear oscillators for grid-forming inverter control," *IEEE Trans. Power Electron.*, vol. 37, no. 9, pp. 10250–10266, Sep. 2022.

[43] Y. Tu, J. H. Su, X. Z. Yang et al., "A novel wireless parallel scheme for inverters based on virtual oscillator control," *Proc. CSEE*, vol. 36, no. 15, pp. 4184–4192, 2016.

[44] G. S. Seo, M. Colombino, I. Subotic et al., "Dispatchable virtual oscillator control for decentralized inverter-dominated power systems: Analysis and experiments," in *Proc. IEEE Appl. Power Electron. Conf. Expo. (APEC)*, 2019, pp. 561–566.

[45] M. Lu, S. Dutta, and B. Johnson, "Self-synchronizing cascaded inverters with virtual oscillator control," *IEEE Trans. Power Electron.*, vol. 37, no. 6, pp. 6424–6436, Jun. 2022.

[46] M. A. Awal and I. Husain, "Unified virtual oscillator control for grid-forming and grid-following converters," *IEEE J. Emerg. Sel. Top. Power Electron.*, vol. 9, no. 4, pp. 4573–4586, Dec. 2021.

[47] L. Harnefors, J. Kukkola, M. Routimo et al., "A universal controller for grid-connected voltage-source converters," *IEEE J. Emerg. Sel. Top. Power Electron.*, vol. 9, no. 5, pp. 5761–5770, Oct. 2021.

[48] P. Hu, W. Jiang, Y. Yu et al., "Transient stability improvement of grid-forming voltage source converters considering current limitation," *Sustain. Energy Technol. Assess.*, vol. 54, Art. No. 102839, 2022.

[49] H. Gong and X. Wang, "Design-oriented analysis of grid-forming control with hybrid synchronization," in *Proc. Int. Power Electron. Conf. (IPEC-Himeji - ECCE Asia)*, 2022, pp. 440–446.

Switching Oscillation Suppression Based on Embedded SiC Power Module With Low Parasitic Inductance for CLLC Resonant Converter

Jiarui Zhang
Zhejiang University
College of Electrical
Engineering
Hangzhou, China
3200102822@zju.edu.cn

Bodong Li
Zhejiang University
College of Electrical
Engineering
Hangzhou, China
bodong_li@zju.edu.cn

Xinnan Sun
Zhejiang University
College of Electrical
Engineering
Hangzhou, China
sxnan@zju.edu.cn

Jiahui Wang
Zhejiang University
College of Electrical
Engineering
Hangzhou, China
3170105809@zju.edu.cn

Liwen Jia
Zhejiang University
College of Electrical
Engineering
Hangzhou, China
3190105626@zju.edu.cn

Kelin Chen
Zhejiang University
College of Electrical
Engineering
Hangzhou, China
3210103075@zju.edu.cn

Feng Jiang
Zhejiang University
College of Electrical
Engineering
Hangzhou, China
jiangfeng@zju.edu.cn

Min Chen
Zhejiang University
College of Electrical
Engineering
Hangzhou, China
calim@zju.edu.cn

Abstract—The application of SiC MOSFET in CLLC converter has received more and more attention due to its high performance. However, the parasitic inductance of the conventional TO-247 package will result in driving voltage interference and switching oscillation. This paper analyzed the influence mechanism of parasitic inductance on voltage oscillation. An embedded SiC power module with low parasitic inductance was used to suppress voltage oscillation. Experimental results show that compared with TO-247 package, the module can reduce driving voltage interference by 28.7% and drain-to-source voltage oscillation by 34.4%.

Keywords—Embedded package, TO-247 package, parasitic inductance, SiC MOSFET, CLLC resonant converter.

I. INTRODUCTION

CLLC resonant converter has gained significant attention in vehicle-to-grid (V2G) systems and DC microgrid owing to its superior performance characteristics, including high efficiency, high power density, and full-range zero-voltage-switching (ZVS) [1] [2] [3] [4]. The application of SiC MOSFET in CLLC converter has become a research focus for advantages such as high voltage, high temperature, high frequency, and reduced loss. But in high-frequency applications of SiC power devices, parasitic parameters of MOSFET and PCB wiring can cause more serious voltage and current oscillations than Si power devices[5]. Switching oscillation will increase voltage stress, lead to EMI problems, cause extra losses and severe heating. And the interference caused by parasitic parameters on driving signals may cause false turn-on, which can directly damage the MOSFET in the bridge circuit and make it difficult to identify the reason of the fault.

Significant research efforts have focused on developing advanced packaging technologies with minimized parasitic inductance. With the rapid development of advanced substrate technology, embedded packaging technology has emerged[6] [7]. This innovative approach integrates power dies within substrates, and the dies interconnected with outside circuits by redistribution layers (RDL), vias, leads, etc to achieve substantial reductions in electrical path lengths.

In this paper, an embedded SiC half-bridge power module with low parasitic inductance is applied in CLLC converter to suppress swiching oscillation. In Section II, the influence of parasitic inductance on the drive voltage and drain-to-source voltage is analyzed. In Section III, the low parasitic inductance characteristic of the embedded module is introduced, and the advantages over TO-247 in reducing PCB driving circuit wiring inductance is demonstrated. In Section IV, experimental results show the effect of the embedded module on switching oscillation suppression compared to TO-247. In section V, conclusions are drawn.

II. ANALYSIS OF SWITCHING OSCILLATION IN CLLC

Fig.1 shows the equivalent circuit of a typical MOEFET with driver circuit considering parasitic parameters. L_g, L_d and L_s represent the gate, drain and source parasitic inductance of the package. L_{wir} represents the parasitic inductance of the PCB wiring in the driving circuit. When MOSFET is turned off, the displacement current of C_{ds} flowing through L_g and L_{wir} will disturb V_{gs}, which may lead to false turn-on. The rapid current change in the power circuit can also cause significant voltage oscillations on L_{cs}, but it can be avoided by Kelvin source connection. In power loop circuit, L_d and L_s will cause drain-to-source oscillations during the hard switching process.

979-8-3315-1110-4/25 $31.00 © 2025 IEEE

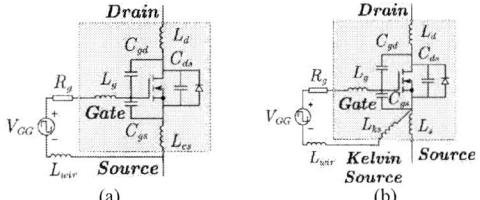

(a)   (b)

Fig. 1 MOSFET equivalent circuit considering parasitic parameters. (a) Without Kelvin source. (b) With Kelvin source.

Fig.2 shows the equivalent circuit of CLLC considering parasitic parameters. L_{rp} represents the parasitic inductance of packaging and PCB wiring. In CLLC resonant converters, the secondary MOSFET can achieve both soft turn-on and soft turn-off under appropriate synchronous rectification strategy[8]. However, the primary MOSFET usually work in ZVS turn-on but hard turn-off. Therefore, the drain-to-source voltage oscillation is most severe when the primary MOSFETs are turned off. Fig.3 shows the oscillation waveforms of drain-to-source voltage of the primary and secondary MOSFET of CLLC in three different modes: below-resonant-frequency mode (BRFM), resonant-frequency mode (RFM) and above-resonant-frequency mode (ARFM). v_{ds1} and v_{ds7} represent the drain-to-source voltage of the primary side S_1 and the secondary side S_7. All three modes indicate that when the primary MOSFET is turned off, there will be significant high-frequency oscillation needed to be suppressed.

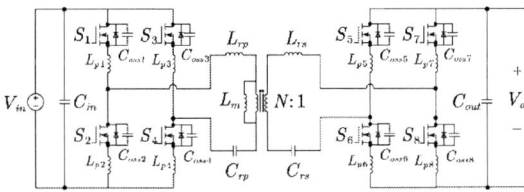

Fig. 2 Topology of CLLC resonant converter considering parasitic inductance and parasitic capacitance.

In order to suppress v_{ds} switching oscillation, the influence of parasitic parameters on the amplitude and frequency of oscillation is analyzed below. Fig.4 shows that the oscillation of the primary hard turn-off starts within the dead time. At the beginning of the dead time, all MOSFETs are turned off, the polarity of i_p remains consistent, charging C_{oss2} and C_{oss3},

meanwhile discharging C_{oss1} and C_{oss4}. After v_{ds1} drops to 0 then creating the conditions for ZVS, the equivalent circuit is shown in Fig.5(a). i_p flows through the body diodes of S_1 and S_4, clamping v_{ds1} and v_{ds4} to 0 then short circuiting C_{oss1} and C_{oss4}. S_2 and S_3 are turned off, equivalent to an open circuit. Since there is no high-frequency oscillation in the primary resonant current, the resonant current can be assumed to be constant during the dead time. Assuming that each MOSFET has the same parasitic parameters, the equivalent circuit is simplified as Fig.5(b).

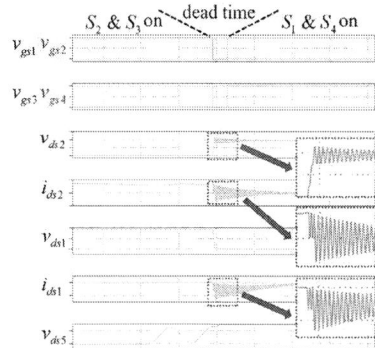

Fig. 4 Hard turn-off oscillation during dead time.

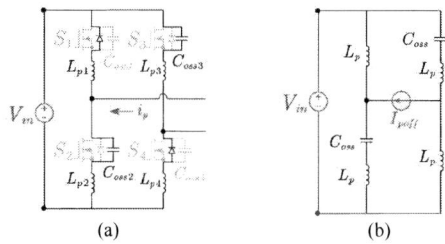

(a)   (b)

Fig. 5 Equivalent circuit within the dead time after S_2 and S_3 are turned off. (a) Without approximation. (b) Assuming the resonant current remains constant.

According to the superposition principle, the v_{ds2} oscillation response caused by the excitation of the input voltage source and the turn-off current can be obtained as follows.

$$\begin{cases} v_{ds2v} = V_{in} - V_{in}\cos(\dfrac{t}{\sqrt{2L_pC_{oss}}}) \\ v_{ds2c} = I_{poff}\sqrt{\dfrac{L_p}{2C_{oss}}}\sin(\dfrac{t}{\sqrt{2L_pC_{oss}}}) \end{cases} \quad (1)$$

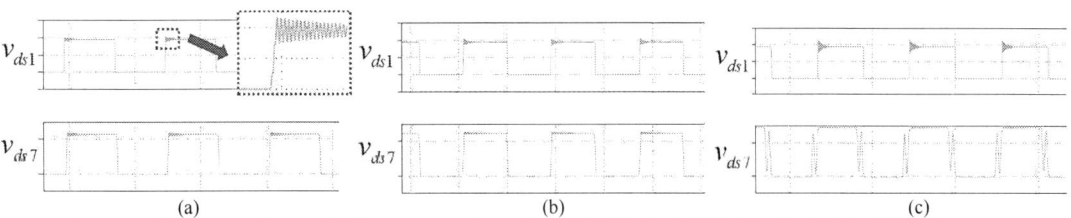

(a)   (b)   (c)

Fig. 3 The switching oscillation waveforms of drain-to-source voltage. (a) ARFM ($f_s > f_r$). (b) RFM ($f_s = f_r$). (c) BRFM ($f_s < f_r$).

V_{in} represents the DC input voltage of CLLC converter. I_{poff} represents the resonant current which is assumed to be equal to turn-off current. v_{ds2v} and v_{ds2c} represent the response of v_{ds2} under independent excitation of input voltage and turn-off current. Fig.6 shows the effect of L_p on the relationship between oscillation amplitude and turn-off current at $C_{oss1}=300$pF. The calculation result indicates that the oscillation amplitude caused by the turn-off current is proportional to $\sqrt{L_p}$, and a smaller L_p is effective in suppressing v_{ds} switching oscillation.

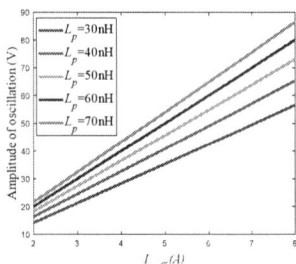

Fig. 6 The influence of L_p on the amplitude of switch oscillation.

III. COMPARISON OF PARASITIC INDUCTANCE

A. Parasitic Characteristic of Embedded Module

A face-down embedded half-bridge module with low parasitic inductance has been proposed for high-frequency applications[6]. The structure of the module is given in Fig.7. The overall design of Kelvin source connection, die face-down, and Cu connecting blocks reduce the parasitic inductance.

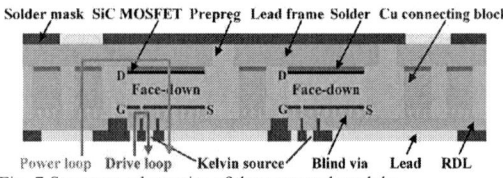

Fig. 7 Structure schematics of the proposed module.

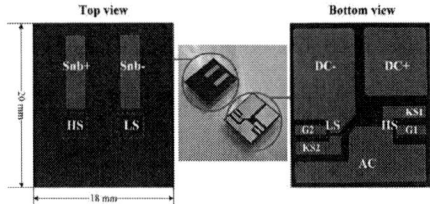

Fig. 8 Layout of the proposed module.

The parasitic inductances of the proposed embedded module were extracted. Benefiting from the face-down placement, the inductances of drive circuits are only 22.41pH and 23.36pH, and the inductances of commutation path and the discharge loop provided by snubber capacitor are 151.80pH and 153.19pH. The total parasitic inductance is 257.10pH, which is significantly lower than many new packaging

methods aimed to reduce parasitic inductance as shown in TABLE I.

TABLE I
COMPARISON OF PARASITIC INDUCTANCE

Package	Total parasitic inductance
Proposed module	0.257nH
Power module based on SKiN structure[9]	1.4nH
Face-up embedded module with vias[10]	2.496nH
Press-pack package based on LTCC interposer[11]	4.3nH
Power chip on bus power module[12]	8nH
Wire-bonded package[13]	>10nH

B. Parasitic Inductance of PCB Wiring

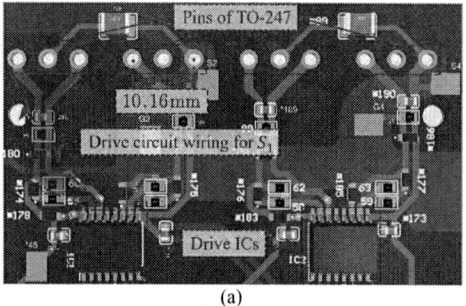

(a)

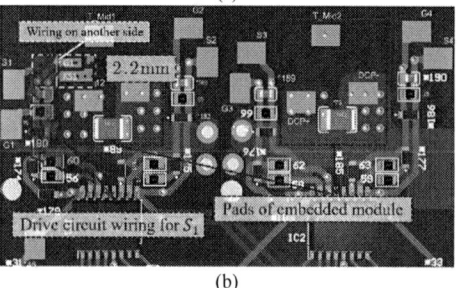

(b)

Fig. 9 PCB wiring of driver circuit for (a) TO-247 and (b) embedded module.

Due to the closer distance between the gate pin and source pin of the embedded module compared to TO247, the parasitic inductance caused by PCB wiring in drive circuit is smaller. TABLE II lists the PCB wiring inductance for each MOSFET driver circuit in two different packages, calculated as follows:

$$L_{wiring} = k_1 l \left(\ln \frac{2l}{w+t} + k_2 \frac{w+t}{l} + k_3 \right) \qquad (2)$$

l, w, and t represent the length, width, and thickness of the wiring. k_1, k_2, and k_3 are coefficients based on experience.

TABLE II
PCB WIRING INDUCTANCE OF DRIVING CIRCUIT

Packaging	PCB wiring inductance			
	S_1	S_2	S_3	S_4
TO-247	42.8nH	46.7nH	45.1nH	42.5nH
embedded module	40.4nH	42.9nH	39.8nH	42.4nH

The average PCB wiring inductance of driving circuit for each TO-247 packaged MOSFET is calculated to be 44.3nH, and the embedded module's is calculated to be 41.4nH, which is beneficial for reducing drive signal interference.

IV. EXPERIMENTAL VERIFICATION

Two 3kW CLLC prototypes were produced with the MOSFET packaged in TO-247 and embedded modules as a comparison. C2M0080120D in TO-247 package is selected, which uses the same MOSFET die (CPM2-1200-0080B) with the embedded module. Two prototypes share the same resonant tank. Parameters are shown in TABLE III.

As shown in Fig.11. the prototype with embedded modules can operate in RFM, ARFM, and BRFM at rated power. Primary MOSFET can achieve ZVS soft turn-on but hard turn-off. As shown in Fig.12, compared to TO-247, the embedded module reduces the interference voltage of the drive turn off circuit from 2.02V to 1.44V, and reduces the maximum oscillation amplitude of the turn off drain source voltage from 26.52V to 17.40V.

The experimental results indicate the advantages of low parasitic inductance and shorter drive circuit wiring length due to module packaging structure can suppress driving voltage interference and drain-to-source voltage oscillation.

(a)

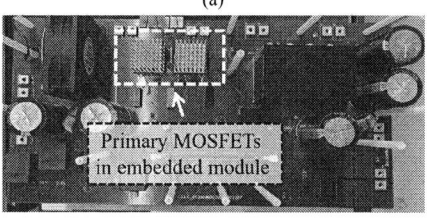

(b)

Fig. 10 The main circuit of CLLC prototype with primary MOSFETs in (a) TO-247 and (b) embedded module.

TABLE III
THE PARAMETERS OF CONVERTER

Parameters	Values
Input voltage, V_{in}	390V
Nominal output voltage, V_{out}	260V
Nominal input power, P_{nom}	3kW
Transformer turn ratio, n	3:2
Magnetizing inductance, L_m	79.53μH
Resonant inductance, L_{rp}, L_{rs}	20.93μH, 9.3μH
Resonant capacitance, C_{rp}, C_{rs}	100nF, 225nF

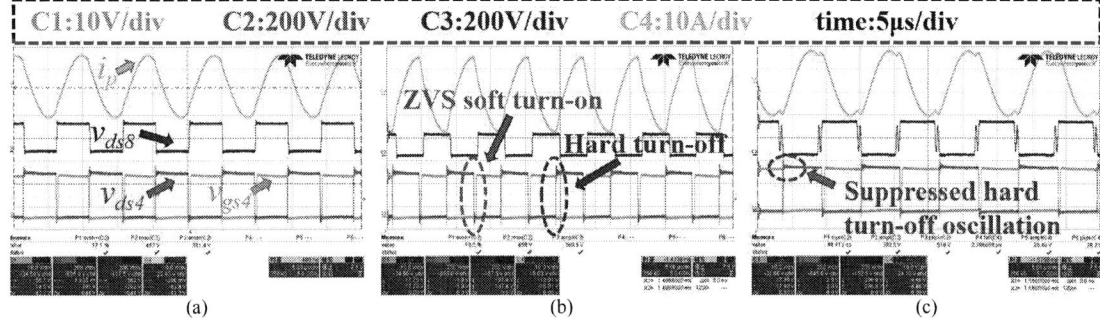

(a)   (b)   (c)

Fig. 11 The working waveforms of 3kW CLLC prototype based on embedded module. (a) RFM ($f_s = f_r$). (b) ARFM ($f_s > f_r$). (c) BRFM ($f_s < f_r$).

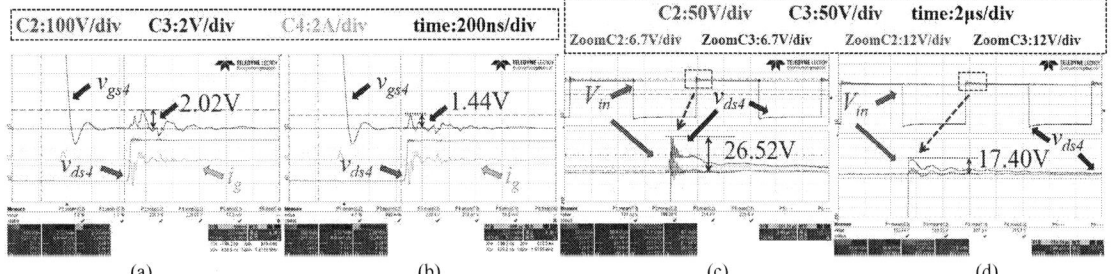

(a)   (b)   (c)   (d)

Fig. 12 The waveforms of switching oscillation. (a) TO-247 drive voltage interference. (b) Embedded module drive voltage interference.

(c) TO-247 drain-to-source voltage oscillation. (d) Embedded module drain-to-source voltage oscillation.

V. CONCLUSION

This paper analyzed the effect of parasitic inductance on driving voltage interference and switching oscillation in CLLC converter. An embedded SiC power module with low parasitic inductance was used to build CLLC prototype. The experiment results proved that the module can reduce driving voltage interference by 28.7% and drain-to-source voltage oscillation by 34.4% compared to TO-247 package.

ACKNOWLEDGMENT

The authors would like to thank PLEXIM Inc. for the support of PLECS.

REFERENCES

[1] N. Chen et al., "Synchronous rectification based on resonant inductor voltage for CLLC bidirectional converter," *IEEE Trans. Power Electron*, vol. 37, no. 1, pp. 547–561, Jan. 2022.

[2] B. Li, M. Chen, X. Sun, J. Wang, and F. Jiang, "A hybrid control for smooth power direction transition of bidirectional resonant CLLC converter with wide voltage gain," *IEEE Trans. Power Electron.*, vol. 39, no. 12, pp. 15898–15914, Dec. 2024.

[3] M. Chen et al., "A coupled inductor scheme for CLLC bidirectional converter and optimized current detection method, " *IEEE Trans. Power Electron.*, vol. 37, no. 10, pp. 11546-11551, Oct. 2022.

[4] M. Chen, L. Jia, B. Li, D. Zhang and F. Jiang, "A novel CTTC structure and optimization design method for CLLC bidirectional resonant converter, " *IEEE Trans. Power Electron.*, early access, Jan. 31, 2025.

[5] S. Yin et al., "An accurate subcircuit model of SiC half-bridge module for switching-loss optimization," *IEEE Trans. Ind. Appl.*, vol. 53, no. 4, pp. 3840–3848, Jul./Aug. 2017.

[6] X. Sun et al., "Design and evaluation of a face-down embedded SiC power module with low parasitic inductance and low thermal resistance, " *IEEE Trans. Power Electron.*, vol. 38, no. 3, pp. 2799-2804, March 2023.

[7] X. Sun et al., "A novel substrate-embedded SiC power module with integrated liquid cooling, " *IEEE Trans. Power Electron.*, early access, April. 21, 2025.

[8] B. Li, M. Chen, X. Wang, N. Chen, X. Sun, and D. Zhang, "An optimized digital synchronous rectification scheme based on time-domain model of resonant CLLC circuit," *IEEE Trans. Power Electron.*, vol. 36, no. 9, pp. 10933–10948, Sep. 2021.

[9] C. Chen, F. Luo, and Y. Kang, "A review of SiC power module packaging: Layout, material system and integration," *CPSS Trans. Power Electron. Appl.*, vol. 2, no. 3, pp. 170–186, Sep. 2017.

[10] T.-C. Chang, C.-C. Lee, C.-P. Hsieh, S.-C. Hung, and R.-S. Cheng, "Electrical characteristics and reliability performance of IGBT power device packaging by chip embedding technology," *Microelectron. Rel.*, vol. 55, no. 12, pp. 2582–2588, Oct. 2015.

[11] N. Zhu, H. A. Mantooth, D. Xu, M. Chen, and M. D. Glover, "A solution to press-pack packaging of SiC MOSFETS," *IEEE Trans. Ind. Electron.*, vol. 64, no. 10, pp. 8224–8234, Oct. 2017.

[12] P. Beckedahl, S. Buetow, A. Maul, M. Roeblitz, and M. Spang, "400 A, 1200 V SiC power module with 1nH commutation inductance," in *Proc. 9th Int. Conf. Integr. Power Electron. Syst.*, 2016, pp. 1–6.

[13] Y. Xu, I. Husain, H. West, W. Yu, and D. Hopkins, "Development of an ultra-high density power chip on bus (PCoB) module," in Proc. *IEEE Energy Convers. Congr. Expo.*, 2016, pp. 1–7.

Modeling and Design of the Planar Magnetic Integration for Dual-stage EMI Filters

Haiyan Liang
College of Mechatronics and Control
Engineering
Shenzhen University
Shenzhen, China
2310295004@email.szu.edu.cn

Yitao Liu
College of Mechatronics and Control
Engineering
Shenzhen University
Shenzhen, China
liuyt@szu.edu.cn

Zijian Lu
College of Mechatronics and Control
Engineering
Shenzhen University
Shenzhen, China
2016110091@email.szu.edu.cn

Abstract—**A planar magnetic integration method of dual-stage electromagnetic interference (EMI) filters is proposed. Common-mode (CM) inductors, CM and differential-mode (DM) capacitors are integrated into a single planar EIE-type magnetic core, DM inductors are provided by the leakage inductance of CM inductors. A planar magnetic integrated dual-stage EMI filter (IDSEF) is built and tested on the PLECS simulation platform to verify the feasibility of the proposed method. A GaN single-phase inverter and a test platform are developed. Compared with a discrete single-stage EMI filter (DSSEF), the discrete dual-stage EMI filter (DDSEF) achieves weight and volume reductions of 22.57% and 24.42%, respectively. Furthermore, the proposed IDSEF achieves additional reductions of 33.64% in weight and 29.16% in volume relative to the DDSEF. The proposed IDSEF delivers comparable EMI attenuation performance to the DDSEF while significantly reducing the weight and volume of the filter and enhancing the power density of the inverter system.**

Keywords—*Conducted EMI, EMI filter, Planar magnetic integration, Dual-stage.*

I. INTRODUCTION

Wide-bandgap (WBG) materials have gained widespread attention and application in the power electronics industry due to their exceptional performance. These materials are better suited to the high-frequency and high-efficiency requirements of inverter systems but also intensify electromagnetic interference (EMI) issues [1, 2]. To maintain grid stability, the International Special Committee on Radio Interference (CISPR) has established relevant standards for the electromagnetic compatibility (EMC) of inverter systems. Many researchers have focused on EMI filter design to optimize the EMC of inverter systems [3, 4]. In [5], common-mode (CM) and differential-mode (DM) inductors are integrated into a single core using a nested magnetic ring structure, effectively reducing both volume and copper wire usage. In [6], the inductors of single-stage and dual-stage EMI filters were integrated into a single core structure, optimizing volume in both structures. The integrated structure of the dual-stage EMI filter demonstrated superior noise suppression performance but did not incorporate capacitors. Flexible copper foil has been used instead of copper wire to achieve the integration of magnetic components, and dielectric materials inserted between copper foil layers further integrate the filter capacitors. However, this method imposes high manufacturing requirements on the flexible tape material [7]. In [8, 9], printed circuit board (PCB) approach was employed to achieve a magnetic integrated structure using planar coils. Additionally, filter capacitors were further integrated between planar coil layers, enhancing system power density. However, further

This work was supported partially by the Shenzhen Science and Technology Program (JCYJ20230808104910021), partially by the Guangdong Provincial Natural Science Foundation (2025A1515012166).

modeling and optimization of leakage inductance and high-frequency parasitic parameters are still required.

This paper proposes a planar magnetic integration design method to integrate dual-stage passive EMI filter with planar EIE-type core. A comparative analysis is conducted among a discrete single-stage EMI filter (DSSEF), a discrete dual-stage EMI filter (DDSEF), and the proposed integrated dual-stage EMI filter (IDSEF) with identical parameters. The proposed planar magnetic integration structure significantly optimizes the filter design, leading to substantial reductions in the weight and volume of the IDSEF. The feasibility and effectiveness of the proposed method and structure were validated through both simulations and experiments.

II. PLANAR MAGNETIC INTEGRATION STRUCTURE OF THE DUAL-STAGE EMI FILTER

A planar EIE-type core is used to achieve the planar magnetic integration design of the IDSEF. Planar magnetic integrated inductors are wound using flat copper foil on a planar core. Depending on the operating current, the planar coils are designed with different trace spacing l, trace width w, and copper thickness to meet voltage withstand requirements. By inserting planar dielectric materials between the layers of planar coils, planar capacitors can be formed, enabling the planar magnetic integration of both inductors and capacitors.

$$C = \frac{\varepsilon_0 \varepsilon_r S}{d}, \varepsilon_0 = 8.85 \times 10^{-12} \text{ F/m} \quad (1)$$

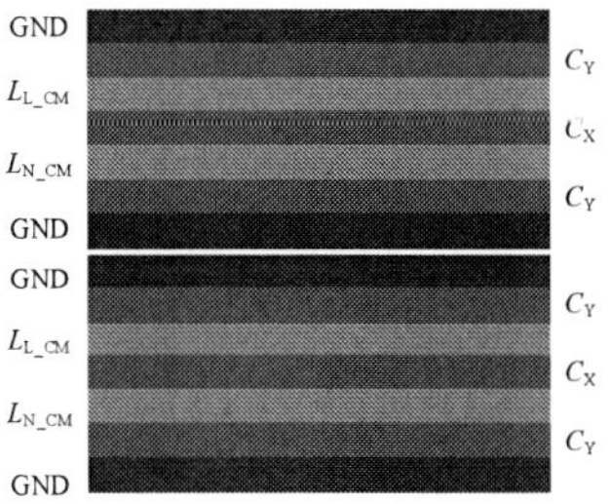

Fig. 1. Layered structure of the planar coil in the IDSEF.

The stacking order of the coils are carefully designed to minimize the equivalent parasitic capacitance between adjacent turns of the same conductor. According to (1), the parasitic capacitance between the windings of the same conductor can be reduced by increasing d, decreasing ε_r or S, where ε_0 denotes the vacuum permittivity, ε_r is the relative permittivity of the dielectric material, S represents the relative area between coil layers, while d denotes the relative distance between them. Interleaving the planar coils of the L and N lines to form a sandwich structure to increase the relative distance between the planar coils of the same conductor, thereby reducing parasitic capacitance. Additionally, both sides of the planar coils are covered with FR-4 insulation material, which further lowering parasitic capacitance. By inserting dielectric materials between the interleaved L-line and N-line planar coils, as shown in Fig. 1, the required DM capacitance can be formed. Additionally, copper foils are placed on the topmost and bottommost layers of the planar coil stack and are connected to ground via conductors. The introduced grounding layers facilitate the formation of the CM capacitance required for the EMI filter between the L-line and N-line planar coils.

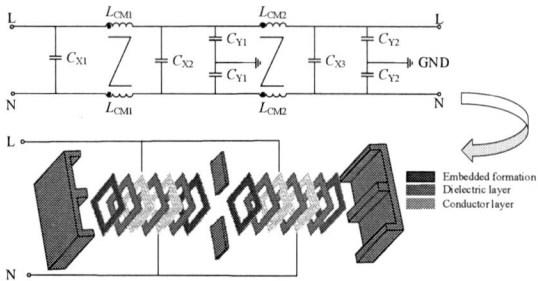

Fig. 2. Planar magnetic integrated structure of the dual-stage EMI Filter.

Based on the stacking design of the planar coils, the planar structure of the dual-stage EMI filter is obtained, as shown in Fig. 2. CM inductance is formed by the coupling of planar coils on the magnetic core, while the leakage inductance is utilized as DM inductance. Additionally, dielectric materials are inserted between the L-line planar coils, the N-line planar coils, and the grounding layers to form the required CM and DM capacitances, thereby achieving a fully integrated planar inductor-capacitor structure.

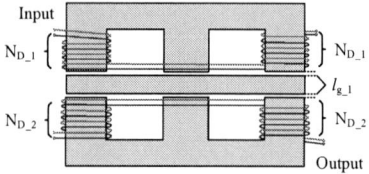

Fig. 3. Winding structure of the dual-stage EMI Filter.

Based on Ampere's law in electromagnetics, the winding structure of the dual-stage EMI filter is designed, as shown in Fig. 3. Planar windings required for two CM inductors are wound on the side columns of the core, where the red traces represent the L-line and the blue traces represent the N-line. The leakage inductance of the CM inductors provides the necessary DM inductance. The upper and lower E-type core side columns are wound with L_{CM1} and L_{CM2}, respectively. In the upper section, the left and right side columns each have N_{D_1} turns, while N_{D_2} turns in the lower section. Since $L_{CM1}=L_{CM2}$, $N_{D\_1}=N_{D\_2}$. The winding directions of the upper and lower side-column windings are kept consistent to ensure that the magnetic flux cancels out within the I-type core. Additionally, appropriate air gaps are introduced between E-type cores and I-type core to prevent magnetic saturation during circuit operation.

Under CM excitation, the magnetic flux distribution of the proposed EMI filter is shown in Fig. 4. The CM noise-induced magnetic flux ϕ_{L1}、ϕ_{N1} reinforce each other within the E-type core. This results in high impedance to CM noise, effectively functioning as a CM inductor. Under DM

excitation, the magnetic flux generated by the L-line and N-line cancels out within the core, as illustrated in Fig. 5.

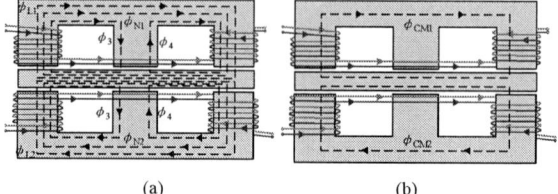

(a)   (b)

Fig. 4. Magnetic flux distribution under CM excitation. (a) Magnetic flux distribution. (b) Equivalent magnetic flux path.

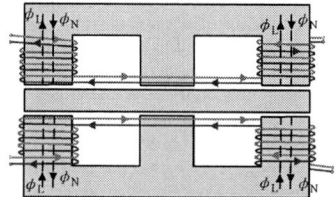

Fig. 5. Magnetic flux distribution under DM excitation.

III. PARAMETER DESIGN OF A PLANAR MAGNETIC INTEGRATED DUAL-STAGE EMI FILTER

The equivalent magnetic circuit of the proposed EMI filter is shown in Fig. 6, where R_s、R_c、R_y represent the reluctances of the side legs, center leg, and yoke of the E-core, respectively. R_{gs}、R_{gc}、R_i denote the reluctances of the air gaps in the side legs, center leg of the E-core, and the I-core, respectively. The reluctances of each part can be obtained from (2), where l_s、l_c、l_y represent the average lengths of the side legs, center leg, and yoke of the E-core, respectively. l_{g_1}、l_i denote the average lengths of the air gap and the I-core, respectively. A_s、A_c、A_y、A_i represent the cross-sectional areas of the side legs, center leg, yoke of the E-core, and the I-core, respectively. μ_r、μ_0 denote the relative permeability of the core material and the vacuum permeability, respectively. To facilitate the analysis of the magnetic coupling between the left and right side legs, a further simplification leads to the model shown in Fig. 7.

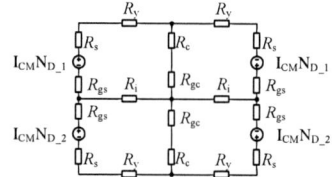

Fig. 6. Magnetic circuit of the IDSEF.

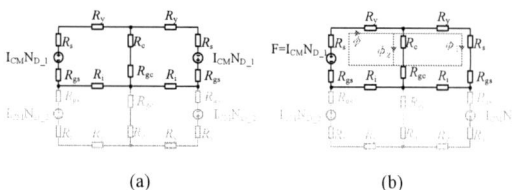

(a)   (b)

Fig. 7. Simplified magnetic circuit model. (a) Simplified magnetic circuit. (b) Magnetic coupling analysis.

According to (3), the equivalent self-inductance reluctance R_{d_s} and the equivalent mutual inductance reluctance R_{d_m} can

979-8-3315-1110-4/25 $31.00 © 2025 IEEE   845

be obtained, where R_1 and R_2 are defined in (4). ϕ、ϕ_1、ϕ_2 represent the total magnetic flux, the flux flowing through the side legs and center leg, respectively.

$$\begin{cases} R_s = \dfrac{l_s}{\mu_r \mu_0 A_s}, R_c = \dfrac{l_c}{\mu_r \mu_0 A_c}, R_y = \dfrac{l_y}{2\mu_r \mu_0 A_y} \\ R_{gs} = \dfrac{l_{g\_1}}{\mu_0 A_s}, R_{gc} = \dfrac{l_{g\_1}}{\mu_0 A_c}, R_i = \dfrac{l_i}{2\mu_r \mu_0 A_i} \end{cases} \quad (2)$$

$$\begin{cases} R_{d\_s} = R_1 + (R_1 \,\square\, R_2) \\ R_{d\_m} = \dfrac{F}{\phi_1} = \dfrac{F}{\phi}\square\dfrac{R_2}{R_1 + R_2} = \dfrac{R_1(R_1 + 2R_2)}{R_2} \end{cases} \quad (3)$$

$$\begin{cases} R_1 = R_y + R_s + R_{gs} + R_i \\ R_2 = R_c + R_{gc} \end{cases} \quad (4)$$

$$K_{d\_m} = \frac{R_{d\_s}}{R_{d\_m}} = \frac{R_2}{R_1 + R_2} \quad (5)$$

$$L_{CM} = \frac{2N_{D\_1}^2}{R_{d\_s}} + 2M = \frac{2N_{D\_1}^2 \left(K_{d\_m} + 1 \right)}{R_{d\_s}} \quad (6)$$

The magnetic coupling coefficient K_{d_m} of the both side legs of the E-core can be calculated according to (5). Furthermore, the relationship between the number of turns of the planar coil and the inductance can be determined according to (6).

Based on the analysis method in [10], the cores selected for the proposed EMI filter are ELP 42/10/28 (N95) and I 42/4/28 (N95). The dimensions are shown in Fig. 8. The initial permeability is 3000, with a saturation magnetic flux density of 525 mT at 25°C and 410 mT at 100°C.

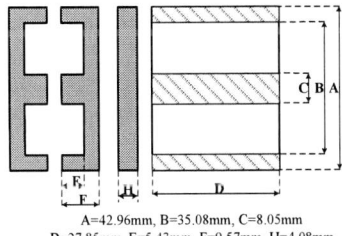

A=42.96mm, B=35.08mm, C=8.05mm
D=27.85mm, E=5.43mm, F=9.57mm, H=4.08mm

Fig. 8. Core dimensions.

The reluctance of the air gaps should satisfy (7). Taking ten times the reluctance of the I-core as the threshold, the final air gap length is selected as l_{gs}=0.1mm.

$$\begin{cases} R_{gs} + R_s + R_y \gg R_i \\ R_{gc} + R_c \gg R_i \end{cases} \quad (7)$$

Based on the original EMI noise of the inverter circuit, the required insertion loss of the EMI filter is determined. Using the cutoff frequency method, the filter parameters are obtained. Combined with the magnetic circuit analysis, the number of turns for the planar coils is determined to be $N_{D\_1} = N_{D\_2} = 20$. The equivalent area of the planar coil is calculated by (8). Zirconia ceramic is selected as the dielectric material to integrate the required capacitance, while FR-4 is used as the

insulating material. The parameters of the proposed planar magnetic integrated structure are shown in TABLE I.

$$S = 2w\mathrm{N}(C + D) + 4w^2 \mathrm{N}^2 + 4wl\mathrm{N}(\mathrm{N} - 1) \quad (8)$$

TABLE I. Structural Parameters of the IDSEF

Parameters		Value
$N_{D\text{-}1}$ /$N_{D\text{-}2}$	Number of layers	4
	Turns per layer	5
Inductive coil	Copper thickness	0.35 mm
	Wiring width	1.2 mm
	Clearance	0.5 mm
C_Y /C_X	Dielectric thickness	0.2 mm/0.1 mm
	ε_r	26
Insulation layer	Dielectric thickness	1 mm
GND	Copper thickness	0.035 mm

IV. SIMULATION ANALYSIS AND EXPERIMENTAL VERIFICATION

In this section, a 300W single-phase inverter circuit and an experimental test platform are constructed to verify the feasibility and effectiveness of the design method for the planar magnetic integrated dual-stage EMI filter (IDSEF). The output voltage is AC 122 V/50 Hz, and the switching frequency is 200 kHz. The EMI standard is set to EN55022A.

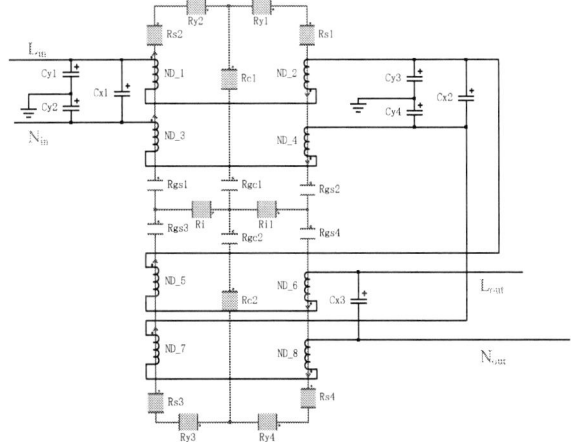

Fig. 9. Simulation circuit of the planar magnetic integrated dual-stage EMI filter (IDSEF).

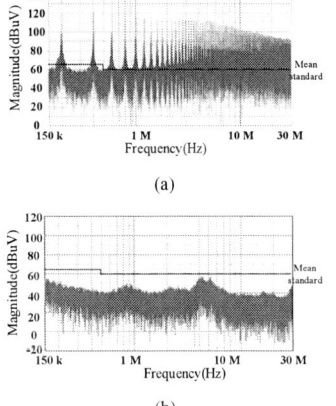

(a)

(b)

Fig. 10. Noise measurements of the simulation circuit. (a) Without filter.

(b)With IDSEF.

A. Simulation Verification

The IDSEF is constructed on the PLECS simulation platform, as shown in Fig. 9. EMI measurements are conducted for the simulated circuit with and without the filter, with the results shown in Fig. 10. After the planar magnetic integrated filter is connected, there is a significant suppression of EMI within the test frequency range, validating the effectiveness of the planar magnetic integrated structure.

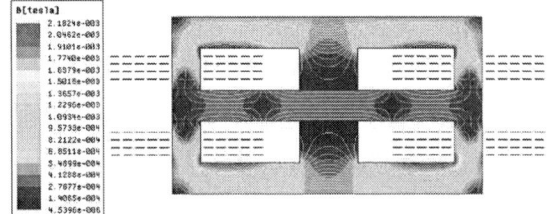

Fig. 11. Magnetic flux distribution of IDSEF.

The magnetic obtained from finite element simulation software is shown in Fig. 11. The maximum magnetic flux density in the core is only 2.2 mT, which is much lower than the saturation magnetic flux density of 525 mT at 25°C, as well as 410 mT at 100°C.

B. Experiment Results

The feasibility of the proposed structure of IDSEF is verified through experiments, with the system parameters consistent with the simulation. A comparison is made between the DSSEF, the DDSEF, and the proposed IDSEF, as shown in Fig. 12. The weight and volume of the three types of filters are compared in TABLE II. Due to the small inductance value, the weight and volume of the DDSEF are reduced by 22.57% and 24.42%, respectively, compared to the DSSEF. The IDSEF reduces the weight and volume by 33.64% and 29.16%, respectively, compared to the DDSEF.

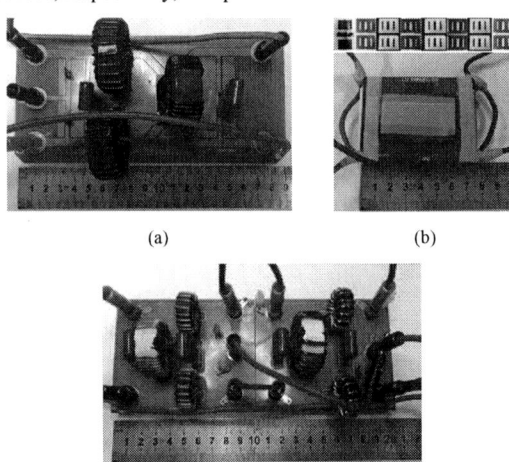

Fig. 12. Size comparison between the DSSEF, IDSEF and DDSEF. (a)DSSEF. (b) IDSEF. (c) DDSEF.

TABLE II. Size and Weight Comparison

Parameters	DSSEF	IDSEF	DDSEF
Volume (cm^3)	167.9	89.9	126.9
Weight (g)	362.4	186.2	280.6

CM and DM insertion losses of the three types of EMI filters are measured separately based on the differences in CM and DM EMI conduction paths, as shown in Fig. 13, Fig. 14, and Fig. 15. The DSSEF achieves a maximum CM EMI attenuation of up to 70 dB, and exhibits relatively high DM EMI suppression at low frequencies. The insertion loss curves of IDSEF and DDSEF exhibit similar trends. However, due to parasitic effects present in the planar magnetic integration structure at high frequencies, oscillations are observed in CM insertion loss. Additionally, the omission of the DM inductor design in the planar magnetic integration structure leads to a faster decay in DM insertion loss. Nevertheless, the insertion losses of all three EMI filters meet the design targets.

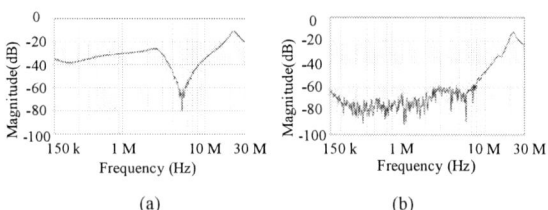

(a)   (b)

Fig. 13. The insertion loss of DSSEF. (a) CM. (b) DM.

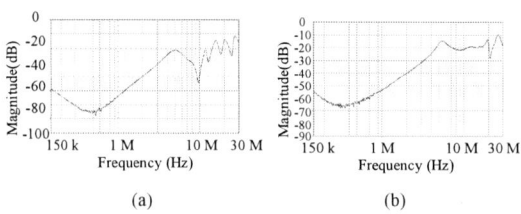

(a)   (b)

Fig. 14. The insertion loss of IDSEF. (a) CM. (b) DM.

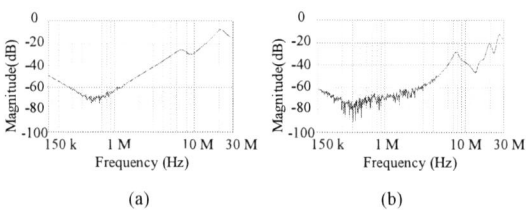

(a)   (b)

Fig. 15. The insertion loss of DDSEF. (a) CM. (b) DM.

The EMI measurements with DSSEF, IDSEF and DDSEF are shown in Fig. 16, Fig. 17, and Fig. 18. The DSSEF can reduce both CM and DM noise to below the standard limits. The IDSEF exhibits a spike while suppressing CM noise, slightly exceeding the standard value, but overall, it significantly suppresses CM noise and meets the requirements. Due to the omission of DM inductance design in the magnetic integration, its ability to suppress DM noise is somewhat limited, and further improvements in the design of DM inductance are needed. Due to the inclusion of additional DM inductance, the DDSEF with discrete components exhibits better DM noise suppression performance than the IDSEF. All three filters show significant noise suppression effects, with stable filtering performance for both CM and DM noise in the low-frequency range. The planar magnetic integrated structure greatly optimizes the filter design, significantly reducing the weight and volume of the filter. However, the design of the DM inductance needs further consideration. The DSSEF and DDSEF are less affected by parasitic parameters, but their weight and volume cannot be further optimized, making it difficult to improve the power density of the filters.

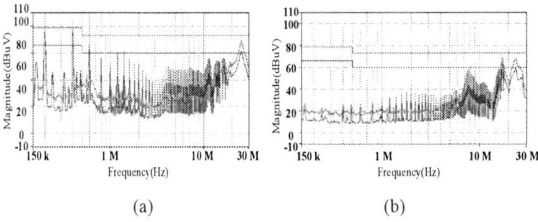

Fig. 16. EMI with DSSEF. (a) CM EMI. (b) DM EMI.

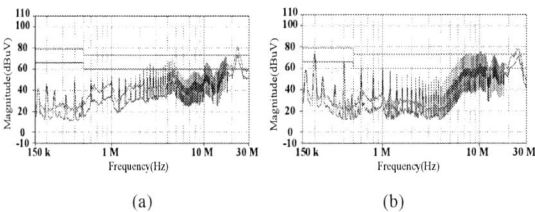

Fig. 17. EMI with IDSEF. (a) CM EMI. (b) DM EMI.

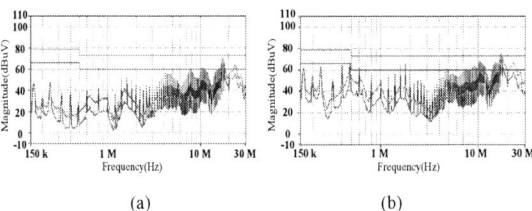

Fig. 18. EMI with DDSEF. (a) CM EMI. (b) DM EMI.

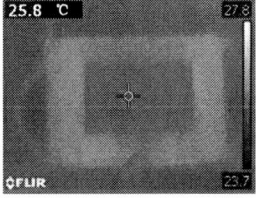

Fig. 19. Magnetic core temperature in experiments at rated power.

Fig. 19 shows the thermal imaging of the IDSEF after 10 minutes of operation at rated power. Due to the excellent surface area-to-volume ratio of the planar core, the core dissipates heat quickly, which helps reduce power losses. The filter shows no significant temperature rise, with the highest temperature reaching 27.8°C. The saturation magnetic flux density of the core does not significantly decrease, guaranteeing the reliability of the planar magnetic integrated structure during operation at rated power.

V. CONCLUSION

In this paper, the planar magnetic integrated dual-stage EMI filters are designed, and the planar magnetic integration structure is analyzed and modeled in detail. The effectiveness and feasibility of the proposed structure are validated through simulation and experimental test. The DSSEF, IDSEF and DDSEF are fabricated for comparative evaluation. Simulation and experimental results demonstrate that all three filters achieve significant EMI suppression. The DDSEF provides superior DM EMI attenuation compared to IDSEF, while the planar magnetic integration structure significantly reduces the weight and volume of filters. Specifically, compared to DSSEF, the weight and volume of DDSEF are reduced by 22.57% and 24.42%, respectively, while IDSEF achieves

additional reductions of 33.64% in weight and 29.16% in volume relative to the discrete-component design.

REFERENCES

[1] Z. Ma, Y. Pei, L. Wang, Q. Yang, Z. Qi, and G. Zeng, "An Accurate Analytical Model of SiC MOSFETs for Switching Speed and Switching Loss Calculation in High-Voltage Pulsed Power Supplies," *IEEE Trans. Power Electron.*, vol. 38, no. 3, pp. 3281-3297, 2023.

[2] F. Zhou, W. Xu, F. Ren, Y. Xia, L. Wu, T. Zhu, *et al.*, "1.2 kV/25 A Normally off P-N Junction/AlGaN/GaN HEMTs With Nanosecond Switching Characteristics and Robust Overvoltage Capability," *IEEE Trans. Power Electron.*, vol. 37, no. 1, pp. 26-30, 2022.

[3] A. L. Julian and G. Oriti, "Novel Common Mode Voltage Elimination Methods in Three-Phase Four-Wire Grid-Connected Inverters," *IEEE Transactions on Industry Applications*, vol. 59, no. 1, pp. 1044-1053, 2023.

[4] P. Zhou, X. Pei, K. Zhang, and Y. Shan, "Improved EMI Behavioral Modeling Method of Three-Phase Inverter Based on the Noise-Source Phase Alignment," *IEEE Trans. Power Electron.*, vol. 37, no. 8, pp. 9333-9344, 2022.

[5] R. Lai, Y. Maillet, F. Wang, S. Wang, R. Burgos, and D. Boroyevich, "An Integrated EMI Choke for Differential-Mode and Common-Mode Noise Suppression," *IEEE Trans. Power Electron.*, vol. 25, no. 3, pp. 539-544, 2010.

[6] Y. Liu, S. Jiang, W. Liang, H. Wang, and J. Peng, "Modeling and Design of the Magnetic Integration of Single- and Multi-Stage EMI Filters," *IEEE Trans. Power Electron.*, vol. 35, no. 1, pp. 276-288, 2020.

[7] S. Jiang, W. Wang, P. Wang, and D. Xu, "A Fully Integrated Common-Mode Choke Design Embedded With Differential-Mode Capacitances," *IEEE Trans. Power Electron.*, vol. 37, no. 5, pp. 5501-5513, 2022.

[8] M. Li, Z. Ouyang, and M. A. E. Andersen, "High-Frequency LLC Resonant Converter With Magnetic Shunt Integrated Planar Transformer," *IEEE Trans. Power Electron.*, vol. 34, no. 3, pp. 2405-2415, 2019.

[9] S. Wang and C. Xu, "Design Theory and Implementation of a Planar EMI Filter Based on Annular Integrated Inductor–Capacitor Unit," *IEEE Trans. Power Electron.*, vol. 28, no. 3, pp. 1167-1176, 2013.

[10] W. G. Hurley, T. Merkin, and M. Duffy, "The Performance Factor for Magnetic Materials Revisited: The Effect of Core Losses on the Selection of Core Size in Transformers," *IEEE Power Electron. Mag.*, vol. 5, no. 3, pp. 26-34, 2018.

A Method to Decrease the Submodule Capacitor Voltage Fluctuations in Voltage-Source Modular Multilevel Converter with Wide-Bandgap Power Devices

1st Qian Kang
School of Electrical Beijing Jiaotong University
Beijing, China
22110470@bjtu.edu.cn

2nd Tiancong Shao*
School of Electrical Beijing Jiaotong University
Beijing, China
shaotc@bjtu.edu.cn

3rd Trillion Zheng
School of Electrical Beijing Jiaotong University
Beijing, China
tqzheng@bjtu.edu.cn

4th Yuqing Geng
School of Electrical Beijing Jiaotong University
Beijing, China
23121408@bjtu.edu.cn

5th Yaqi Li
School of Electrical Beijing Jiaotong University
Beijing, China
22126203@bjtu.edu.cn

6th Zhitong Bai
School of Electrical Beijing Jiaotong University
Beijing, China
20221218@bjtu.edu.cn

7th Xiaofeng Yang
School of Electrical Beijing Jiaotong University
Beijing, China
xfyang@bjtu.edu.cn

Abstract—**The Voltage-Source Modular Multilevel Converter (VMMC) improves conventional Modular Multilevel Converter (MMC) by reducing submodule count, offering a compact solution for grid-forming converters (GFCs) in high-voltage applications. This paper proposes a high-frequency energy balancing method to suppress submodule capacitor voltage fluctuations in VMMC, improving WBG performance and enabling a more efficient, lightweight design. It describes the operational principle of the topology, and analyses the submodule capacitor voltage fluctuation model in detail to reveal the impact mechanism between energy balancing frequency and the fluctuation. Analytical modeling and simulations validate its effectiveness, providing an opportunity to optimize WBG-based grid-forming converters in high-voltage applications.**

Keywords—Voltage-Source Modular Multilevel Converter (VMMC); Wide-Bandgap (WBG) Power Devices; Energy Balancing; Voltage Fluctuation

I. INTRODUCTION

Grid-forming converters (GFCs) play a crucial role in future power systems by enabling stable and resilient renewable energy integration. Among various GFC topologies, the Modular Multilevel Converter (MMC) has gained significant attention due to its high scalability, superior power quality, and fault-tolerant capabilities. However, conventional MMCs face challenges in applications with strict space and weight constraints, such as offshore wind power DC transmission, photovoltaic DC collection, and medium-voltage DC power supplies for data centers. These emerging applications require more compact and efficient multilevel converter designs to enhance power density and overall system performance [1].

This work was supported in part by the National Natural Science Foundation of China under Grant 52377165.

To address these challenges, researchers have introduced wide-bandgap (WBG) power devices, such as silicon carbide (SiC) MOSFETs, into MMC-based GFCs, replacing conventional silicon-based insulated-gate bipolar transistors (Si IGBTs). The adoption of SiC MOSFETs enables higher switching frequencies, improving dynamic performance and reducing submodule capacitor voltage fluctuations [2-3]. Furthermore, hybrid MMC topologies integrating both SiC MOSFETs and Si IGBTs have been explored to optimize the trade-off between switching loss reduction and high-frequency operation. Advanced modulation strategies have been developed to refine switching dynamics, mitigate dv/dt effects, and minimize electromagnetic interference (EMI) and losses [4-8].

Despite these advancements, conventional MMCs still suffer from 2nd harmonic-frequency power pulsations in submodule capacitors, which blocks the full utilization of WBG power devices in high-voltage grid-forming converters. To address this issue, Reference [9] proposed a Voltage-Source Modular Multilevel Converter (VMMC) topology, which reduces the number of submodules by 25% while offering greater control flexibility over energy balancing. However, VMMC still encounters submodule capacitor voltage fluctuation issues, which are influenced by the energy balancing current frequency.

To address this challenge, this paper proposes a high-frequency energy balancing method to suppress submodule capacitor voltage fluctuations in VMMC, enhancing WBG performance and enabling a more efficient, lightweight design. Furthermore, by optimizing energy balancing control, the proposed method provides greater flexibility in managing submodule energy flow, which is essential for GFC applications. The VMMC topology and its operational principles are analyzed, and the impact of energy-balancing frequency on submodule capacitor voltage fluctuations is

979-8-3315-1110-4/25 $31.00 © 2025 IEEE

systematically studied. Finally, simulation results validate the effectiveness of the proposed method.

II. TOPOLOGY AND OPERATING PRINCIPLE

A. Converter Topology

Fig. 1 shows the topology of the VMMC, which consists of three phase units: Phase A, Phase B, and Phase C. Each phase unit includes an upper bridge arm, a lower bridge arm, and a High-Frequency Energy Balancing Current Source (HF-EBCS). Each phase unit's upper and lower arms consist of submodule links (SML) and semiconductor switch links (SSL). For easy presentation, in this paper, the submodules are acted by the half-bridge submodule (HBSM). SML_{jP} and SML_{jN} ($j=A, B, C$) consist of N cascaded submodules, while SSL_{jP} and SSL_{jN} consist of N semiconductor switches connected in series. SML and SSL are connected in series to form the upper or lower arm. Each phase's upper and lower arms have two terminals: the SML terminal and the SSL terminal. The SML terminals of the upper and lower arms are connected to the positive and negative DC sides, respectively, while the SSL terminals form the AC-side output terminal u_j. Each phase's HF-EBCS branch consists of an SML_{jB} and an AC inductor L_j in series. The SML_{jB} is formed by cascading N submodules. Each phase's HF-EBCS branch connects to the junction points of the upper and lower arms' SML and SSL.

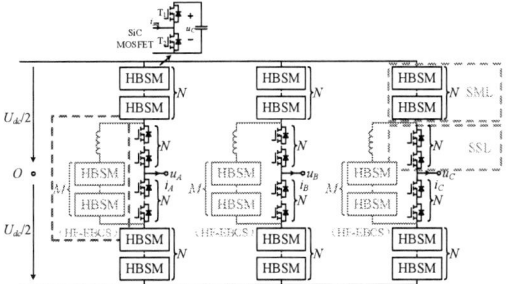

Fig. 1. Topology of VMMC with High Frequency Energy Balance Current Source (HF-EBCS)

B. Operating Principle

The VMMC controls the switching of the SSL to connect the upper and lower arm submodule strings during the positive and negative half-cycles of the fundamental wave. This shapes the AC-side voltage. The conduction modes of the VMMC are analyzed using Phase A as an example. Based on the VMMC topology, the AC-side voltage u_A and current i_A of Phase A are expressed as:

$$\begin{cases} u_A = U_m \sin(\omega t) \\ i_A = I_m \sin(\omega t - \varphi) \end{cases} \quad (1)$$

Where U_m and I_m are the amplitudes of the AC voltage and current, respectively. ω is the power frequency angular frequency, and φ is the power factor angle. The voltage modulation ratio m, which defines the relationship between the AC and DC voltage, is given by:

$$m = \frac{U_m}{u_{dc}/2} \quad (2)$$

As shown in Fig. 2(a), when $u_A>0$ (positive half-cycle of the fundamental wave $[0, \pi]$), the upper arm SSL_{AP} conducts, and the lower arm SSL_{AN} is off. The upper arm SML_{AP} connects to

the AC side, shaping the AC voltage (red line). The HF-EBCS connects to the lower arm SML_{AN}, forming the HF-EBCS loop with the upper arm SML_{AP} (blue line). Voltage balance is maintained by controlling the HF-EBCS to generate balancing currents. In Fig. 2(b), when $u_A < 0$ (negative half-cycle of the fundamental wave $[\pi, 2\pi]$), the upper arm SSL_{AP} is off, and the lower arm SSL_{AN} conducts. The lower arm SML_{AN} connects to the AC side, shaping the AC voltage (red line). The HF-EBCS connects to the upper arm SML_{AP}, forming the HF-EBCS loop with the lower arm SML_{AN} (blue line). Voltage balance is maintained by controlling the HF-EBCS to generate balancing currents.

From the conduction modes of the VMMC, it is clear that within one fundamental wave cycle, the submodule capacitors of Phase A form a DC voltage path (yellow line in the figure). Therefore, the submodule capacitors serve as DC-side support capacitors, eliminating the need for additional ones.

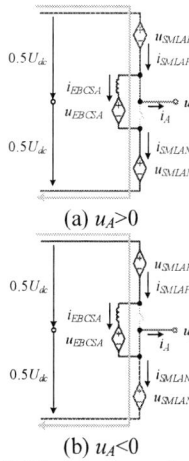

(a) $u_A>0$

(b) $u_A<0$

Fig. 2. VMMC phase A conduction mode

III. ANALYSIS OF SUBMODULE CAPACITOR VOLTEGE FLUCTUATION

Stable VMMC operation requires submodule capacitor voltage balance. HF-EBCS regulates the balancing current i_{EBCS} via SML, ensuring energy balance per fundamental cycle. Since VMMC operates in two modes per cycle, f_{EBCS} must be at least 100Hz to maintain energy balance in each half-cycle. If f_{EBCS} increases, it should be a multiple of 100Hz.

$$\begin{cases} f_{EBCS,\min} = 100\text{Hz} \\ f_{EBCS} = n f_{EBCS,\min} \end{cases} \quad (3)$$

Where n is a positive integer, i.e., $n=1,2,3,\cdots$

Taking the HF-EBCS current as a square wave, the positive amplitude I_{EBCSP_Square} and the negative amplitude I_{EBCSN_Square} can be derived as:

$$\begin{cases} I_{EBCSP\_Square} = \dfrac{4-m\pi}{2(\pi-2m)} I_m \cos\varphi \\ I_{EBCSN\_Square} = \dfrac{4-m\pi}{\pi-2m} \dfrac{m}{\pi} I_m \cos\varphi \end{cases} \quad (4)$$

By considering the VMMC topology and mode changes, the energy fluctuations in the upper arm, lower arm, and EBCS branch can be analyzed. Since the upper and lower arms alternate in AC voltage shaping and energy balancing during each half-cycle, their energy fluctuation patterns are similar. Therefore, the analysis focuses on the upper arm. The energy variation in the upper arm and EBCS branch is expressed as the integral of voltage and current.

$$\begin{cases} E_{SMLAP}(t) = \int_0^t u_{SMLAP}(t) i_{SMLA}(t) dt \\ E_{EBCSA}(t) = \int_0^t u_{EBCSA}(t) i_{EBCSA}(t) dt \end{cases} \quad (5)$$

Where E_{SMAP} and E_{EBCSA} represent the energy of the upper arm and the HF-EBCS, respectively, u_{SMLAP} and u_{EBCSA} are their voltages, and i_{SMLA} and i_{EBCSA} are their currents.

Numerical calculations show the maximum and minimum energy deviation values at different HF-EBCS current frequencies for a phase angle $\varphi=0$, as shown in Fig. 3. In Fig. 3(a), as the HF-EBCS current frequency increases, the energy deviation in the HF-EBCS branch submodule capacitors decreases. The decrease is faster at lower frequencies and slower at higher frequencies. In Fig. 3(b), as the HF-EBCS current frequency increases, the energy deviation in the upper arm submodule capacitors also decreases.

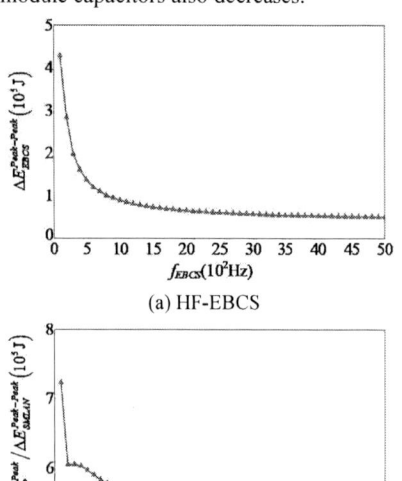

(a) HF-EBCS

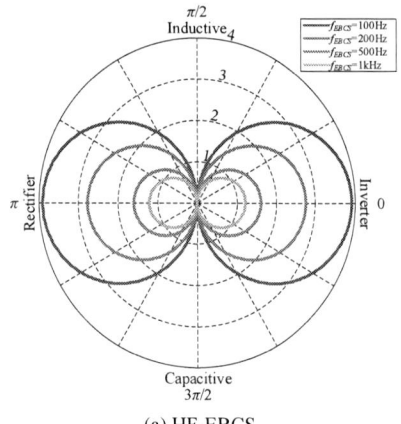

(b) Upper arm or lower arm

Fig. 3. Relationship between HF-EBCS current frequency and energy deviation

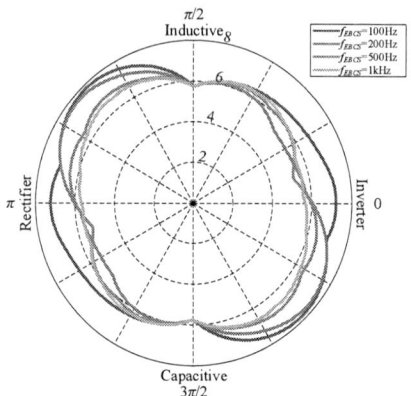

(b) Upper arm or lower arm

Fig. 4. Relationship between energy deviation and phase angle

Fig. 4 illustrates the energy deviation versus phase angle φ for different HF-EBCS current frequencies (100 Hz, 200 Hz, 500 Hz, and 1 kHz). In Fig. 4(a), the HF-EBCS branch submodule string's energy deviation varies with φ, peaking at $\varphi=0$ or π (purely resistive load in inverter or rectifier mode). As φ changes, the deviation shifts, but under purely reactive power conditions, it is zero, achieving energy balance without control. Fig. 4(b) shows similar trends for the upper and lower arm submodule strings, where energy deviation remains constant across all frequencies in purely reactive conditions, ensuring balance without the HF-EBCS branch.

The energy fluctuations of the upper and lower arms, as well as the HF-EBCS, are obtained from the above analysis, which allows determining the submodule capacitor voltage fluctuations:

$$E = \frac{1}{2} C u^2 \quad (6)$$

In summary, increasing the i_{EBCS} frequency helps reduce submodule capacitor voltage fluctuations in the VMMC.

IV. CASE STUDY

To verify the impact of HF-EBCS current frequency on VMMC submodule capacitor voltage fluctuations, a single-phase VMMC simulation model was built in MATLAB/Simulink using HBSM submodules. The system operates with a DC bus voltage of 800V, 4 submodules per SML, submodule capacitance of 4.28mF, an AC inductance of 5mH, a voltage modulation ratio of 0.85, and a power factor angle of 0.

Fig. 5 shows the AC-side output voltage in VMMC, where submodule capacitor voltages combine to form a multilevel sinusoidal waveform. To achieve the same voltage levels, the traditional MMC requires 16 submodules per phase, while VMMC needs only 12, reducing submodules by 25%.

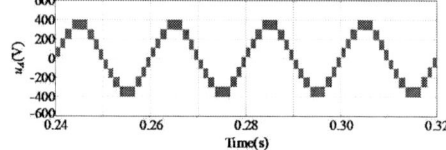

Fig.5 The AC output voltage waveforms in VMMC

Fig. 6 and Fig. 7 show the simulation results of the EBCS current and submodule capacitor voltage for EBCS current frequencies of 100Hz and 500Hz. The energy balance control allows the EBCS to produce the desired square wave current at these frequencies. Meanwhile, the VMMC keeps the submodule capacitor voltage dynamically balanced, fluctuating around 100V. Comparing the voltage waveforms at 100 Hz and 500Hz, the submodule capacitor voltage fluctuates by 7V at 100Hz and 3V at 500Hz. As the EBCS current frequency increases, the voltage fluctuation of the submodule capacitors decreases significantly.

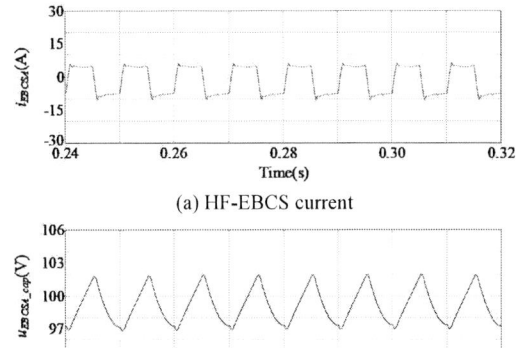

(a) HF-EBCS current

(b) HF-EBCS submodule capacitor voltage

Fig. 6: HF-EBCS current with a 100Hz square waveform

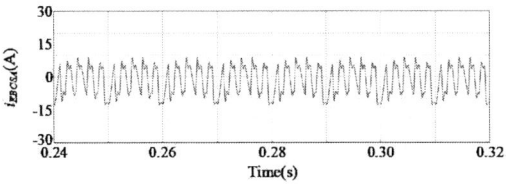

(a) HF-EBCS current waveform

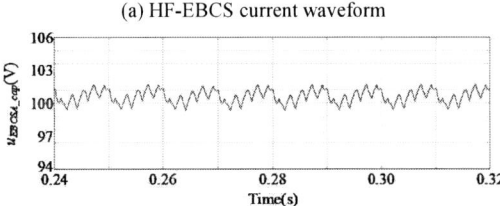

(b) HF-EBCS submodule capacitor voltage

Fig. 7: HF-EBCS current with a 500Hz square waveform

V. CONCLUSION

This paper proposes a high-frequency energy balancing method to mitigate submodule capacitor voltage fluctuations in the VMMC. Simulation results show that increasing the energy balancing current frequency significantly reduces voltage fluctuations, from 7V at 100Hz to 3V at 500Hz. This improvement enables a more compact and efficient converter design. The proposed approach enhances the advantages of WBG devices and provides an opportunity to optimize grid-forming converters' operation with lightweight, high-performance power systems.

REFERENCES

[1] R. Pan, D. Liu, S. Liu, J. Yang, L. Kou and G. Tang, "Stability Comparison Between Grid-forming and Grid-following Based Wind Farms Integrated MMC-HVDC," in Journal of Modern Power Systems and Clean Energy, vol. 11, no. 4, pp. 1341-1355, July 2023.

[2] K. Jing, L. Lin and T. Yin, "Analysis of the Flying-Capacitor Modular Multilevel Converter Based on SiC MOSFET," 2020 IEEE 1st China International Youth Conference on Electrical Engineering (CIYCEE), Wuhan, China, 2020, pp. 1-5.

[3] P. Guicharrousse, M. Rishad Ahmed, P. Wheeler and P. Zanchetta, "New approach for comparing Modular Multilevel Converter submodule losses considering IGBT and SiC MOSFET devices," IECON 2022 – 48th Annual Conference of the IEEE Industrial Electronics Society, Brussels, Belgium, 2022, pp. 1-6.

[4] T. Yin, L. Lin, C. Xu, D. Zhu and K. Jing, "A Hybrid Modular Multilevel Converter Comprising SiC MOSFET and Si IGBT With Its Specialized Modulation and Voltage Balancing Scheme," in IEEE Transactions on Industrial Electronics, vol. 69, no. 11, pp. 11272-11282, Nov. 2022.

[5] T. Yin, C. Xu, L. Lin and K. Jing, "A SiC MOSFET and Si IGBT Hybrid Modular Multilevel Converter With Specialized Modulation Scheme," in IEEE Transactions on Power Electronics, vol. 35, no. 12, pp. 12623-12628, Dec. 2020.

[6] T. Yin, L. Lin, X. Shi and K. Jing, "A Si/SiC Hybrid Full-Bridge Submodule for Modular Multilevel Converter With its Control Scheme," in IEEE Journal of Emerging and Selected Topics in Power Electronics, vol. 11, no. 1, pp. 712-721, Feb. 2023.

[7] C. Xu, J. He and L. Lin, "Research on Capacitor-Switching Semi-Full-Bridge Submodule of Modular Multilevel Converter Using Si-IGBT and SiC-MOSFET," in IEEE Journal of Emerging and Selected Topics in Power Electronics, vol. 9, no. 4, pp. 4814-4825, Aug. 2021.

[8] X. Li et al., "Simple Switching Strategies for dv/dt Reduction in SiC-Device-Based Modular Multilevel Converters," in IEEE Transactions on Power Electronics, vol. 38, no. 2, pp. 1485-1493, Feb. 2023.

[9] Q. Kang, T. Shao, T. Q. Zheng. "Research on a Voltage-Type Modular Multilevel Converter Topology with Reduced Number of Submodules," 2023 China Power Electronics and Energy Conversion Congress - CPEEC , Guang Zhou, China: 10-13 Nov. 2023.

Junction Temperature Control and Thermal Stability Enhancement Method for Power Devices Based on Vapor Phase-Change Principle

1st Ruya Song
School of Electrical Engineering and Automation
Hefei University of Technology
Hefei, China
songruya@mail.hfut.edu.cn

2nd Shuang Zhao
School of Electrical Engineering and Automation
Hefei University of Technology
Hefei, China
shuang.zhao@hfut.edu.cn

3rd Jinxiao Wei
School of Electrical Engineering and Automation
Hefei University of Technology
Hefei, China
jxwei@hfut.edu.cn

4th Waleed Alhosaini
Jouf University
Saudi Arabia
wsalhosaini@ju.edu.sa

5th Lijian Ding
School of Electrical Engineering and Automation
Hefei University of Technology
Hefei, China
ljding@hfut.edu.cn

Abstract—The total equivalent thermal capacitance of the junction-to-heatsink path in power devices significantly influences the junction temperature swing (ΔT_j). As the equivalent thermal capacitance increases, ΔT_j exhibits a decreasing trend. Based on this observation, this study proposes a junction temperature regulation method for power devices by modulating the thermal capacitance parameter using a liquid-to-gas phase change mechanism. A theoretical model of phase-change cooling is established to elucidate its regulatory mechanism on junction temperature, and experiments are conducted to quantify the operational boundaries of phase-change cooling. Experimental results demonstrate that this method optimizes the heat transfer process, improves the thermal network parameters of the device, and effectively controls both junction temperature fluctuations and average junction temperature, thereby enhancing the calculated thermal cycling lifetime of the device.

Keywords—thermal capacitance, junction temperature, phase change, thermal network

I. INTRODUCTION

With the rapid development of power semiconductor technology, power electronic devices and systems play an increasingly critical role across various fields. Enhancing the reliability of power semiconductor devices is of paramount engineering significance for ensuring the overall stability and continuous operational reliability of power electronic systems. One effective approach to improving the reliability of power devices is the active regulation of their junction temperature [1].

At different operational levels, the highest-level loss control primarily involves system-level regulation, where the junction temperature of power semiconductor devices is managed by limiting their load [2]. At the intermediate level, junction temperature regulation is achieved by adjusting the switching frequency of power electronic converters to modulate switching losses [3]. At the lowest level, junction temperature control is implemented by modifying the gate drive circuit and the switching trajectory of the device, directly affecting the loss distribution and thereby regulating the junction temperature [4].

Overall, current methods primarily mitigate device loss fluctuations by adjusting the switching frequency and system load. Although these approaches can reduce junction temperature fluctuations to some extent, they largely overlook the power processing capacity and output power quality of the converter system.

This study proposes a junction temperature regulation method for power devices by modulating the thermal capacitance parameter based on a liquid-to-gas phase change mechanism. The theoretical analysis reveals the underlying mechanism by which phase-change cooling optimizes the junction temperature of power devices. Without compromising the output characteristics of the device, the proposed phase-change heat sink enhances the thermal spreading angle, thereby optimizing the thermal network of the power device and enabling precise control over both junction temperature fluctuations and the absolute average junction temperature.

II. PRINCIPLE OF PHASE-CHANGE COOLING

As shown in Fig. 1., the vapor phase-change cooling device consists of a shell, columns, a vapor chamber, a capillary structure, and working fluid. The shell and columns, typically made of aluminum or copper, provide thermal conductivity and mechanical support. The vapor chamber reduces atmospheric pressure to facilitate phase change at lower temperatures, while the capillary structure, made using micro-grooves or similar techniques, drives the liquid's return flow. High-purity deionized water is used to minimize corrosion.

When the shell is heated, thermal energy transfers to the capillary structure, where the liquid rapidly evaporates under near-vapor conditions, absorbing latent heat and cooling the heated area. The vapor spreads across the chamber, enabling uniform heat distribution, and condenses at cooler regions, releasing heat. Capillary pressure drives the liquid back to the heated zone, maintaining a continuous evaporation-condensation cycle.

979-8-3315-1110-4/25 $31.00 © 2025 IEEE

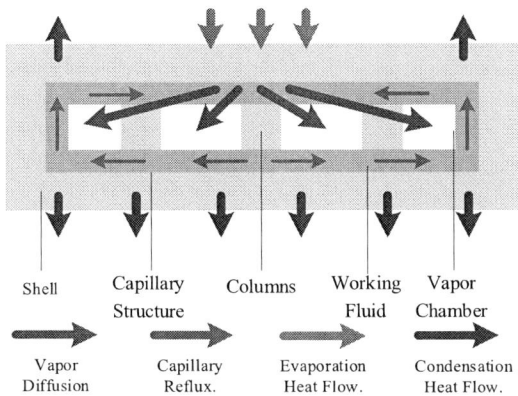

Shell | Capillary Structure | Columns | Working Fluid | Vapor Chamber

Vapor Diffusion | Capillary Reflux. | Evaporation Heat Flow. | Condensation Heat Flow.

Fig. 1. Vapor chamber heat dissipation.

This phase-change mechanism achieves much higher heat transfer efficiency than conventional conduction or convection and is unaffected by gravity, allowing operation in various orientations. This study applies vapor phase-change cooling to optimize the thermal management of power devices.

III. STRUCTURE AND MODELING OF THE VCPCH SYSTEM

This study applies the phase-change heat transfer method to power device cooling and proposes a vapor chamber phase-change heat sink (VCPCH) for power devices. As shown in Fig. 2., the vapor chamber containing the liquid working fluid is installed at the drain of the discrete device. A thermally conductive insulating pad is used to ensure both thermal conduction and electrical insulation between the discrete device and the primary heat sink.

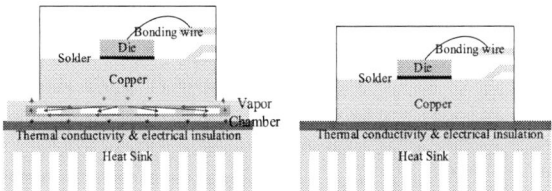

Fig. 2. Vapor chamber phase-change heat sink vs. conventional heat sink.

Structurally, the thermal network of a power device using VCPCH has one additional thermal layer compared to a conventional heat sink, corresponding to the vapor chamber itself. The junction-to-heatsink thermal impedance of the device with VCPCH is expressed by (1), while that of the conventional heat sink is given by (2).

$$Z_{th(j-c)} = \{\{[(Z_{RthDtci} \| Z_{CthDtci} + Z_{RthVC}) \| Z_{CthVC} + Z_{RthDcu}]$$

$$\| Z_{CthDcu} + Z_{RthDs}\} \| Z_{CthDs} + Z_{RthDd}\} \| Z_{CthDd}$$

$$Z_{Rth} = R_{th} \tag{1}$$

$$Z_{Cth} = \frac{1}{j\omega C_{th}}$$

$$Z_{th(j-c)} = \{\{[(Z_{RthDtci} \| Z_{CthDtci} \| Z_{RthDcu}) \| Z_{CthDcu} + Z_{RthDs}]$$

$$\| Z_{CthDs} + Z_{RthDd}\} \| Z_{CthDd} \tag{2}$$

When using a conventional heat sink, the thermal spreading effect occurs between the power device and the thermally conductive insulating layer. The spreading angle between the copper lead frame and the insulating layer is described by (3). The equivalent heat transfer area of the thermal insulation layer is expressed by (4).

$$\alpha_{Dcu} = \tan^{-1}(\lambda_{Dcu}/\lambda_{Dtci}) \tag{3}$$

$$A_{Dtci} = (a_{Dcu} + 2h_{Dtci}\tan\alpha_{Dcu})(b_{Dcu} + 2h_{Dtci}\tan\alpha_{Dcu}) \tag{4}$$

With the VCPCH, the enhanced thermal spreading effect increases the equivalent heat transfer area compared to a conventional heat sink. This also enlarges the effective heat transfer area of the insulating pad, thereby reducing its equivalent thermal resistance while increasing its equivalent thermal capacitance, as described by (5). Consequently, the VCPCH effectively regulates both the junction temperature fluctuation and the average junction temperature of the power device.

$$\begin{aligned} A'_{Dtci} &> A_{Dtci} \\ R'_{thDtci} &< R_{thDtci} \\ C'_{thDtci} &> C_{thDtci} \end{aligned} \tag{5}$$

IV. VCPCH ANALYSIS BASED ON FINITE ELEMENT SIMULATION

A finite element simulation model is developed using COMSOL Multiphysics to analyze the thermal transfer process under different cooling methods. With an ambient and initial temperature of 300 K and a power loss of 50 W, the steady-state junction temperature distribution for conventional and VCPCH cooling is shown in Fig. 3.(a) and Fig. 3.(b), respectively.

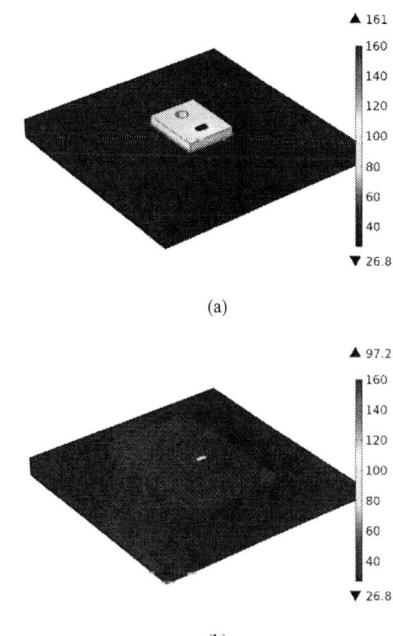

(a)

(b)

Fig. 3. Steady-state junction temperature distribution.

The results indicate that the overall temperature of the power device with a conventional heat sink is higher than that with the VCPCH. The temperature distribution of the chip, solder layer, and copper lead frame, along with their corresponding in-plane temperature profiles using the conventional heat sink, are illustrated in Fig. 4. The maximum temperatures of the chip, solder, and copper lead frame are 161°C, 145°C, and 116°C, respectively. Significant in-plane temperature non-uniformities are observed, with the center temperature being higher than the edges. The temperature differences between the center and the edge for the chip, solder, and copper lead frame are 5°C, 10°C, and 30.1°C, respectively.

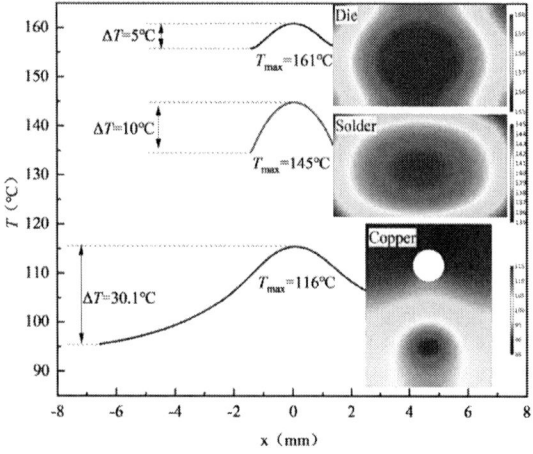

Fig. 4. The internal temperature distribution of devices with traditional cooling methods.

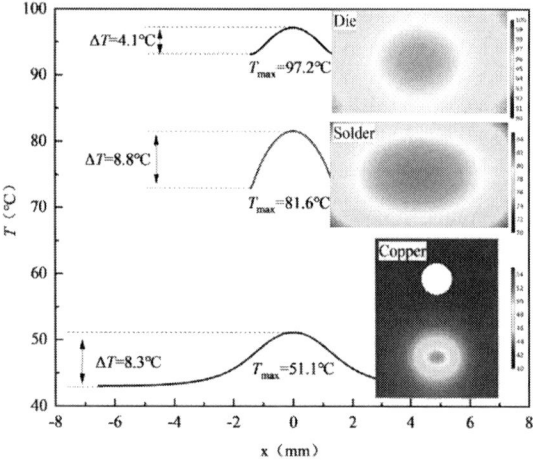

Fig. 5. Internal temperature distribution of devices during phase change cooling.

For the VCPCH, the temperature distribution and corresponding in-plane temperature profiles of these three layers are shown in Fig. 5. Compared to the conventional heat sink, the VCPCH not only reduces the maximum temperatures of the chip, solder layer, and copper lead frame but also improves the in-plane temperature uniformity across these layers.

Additionally, the vertical temperature gradient at the geometric center of the chip under different cooling methods is compared in Fig. 6. The results show that with the VCPCH, the temperature gradient across the thermally conductive insulating layer decreases significantly from 84.5°C/mm to 16.3°C/mm—an 80.7% reduction—accounting for only 19.3% of that observed with the conventional cooling method. This reduction in temperature gradient further enhances the thermal uniformity within the power device.

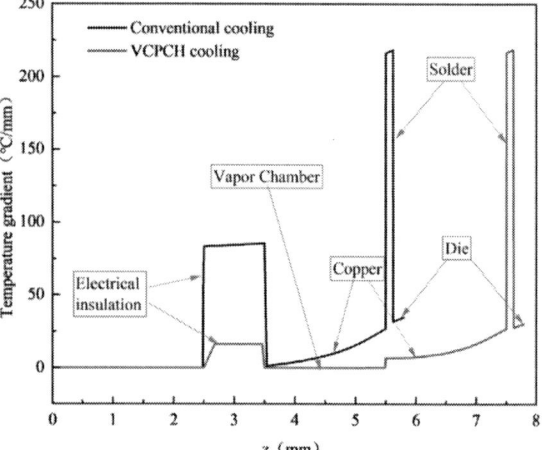

Fig. 6. Temperature gradient distribution.

The finite element analysis demonstrates that the VCPCH optimizes the heat transfer process by enhancing thermal spreading, effectively increasing the equivalent heat transfer area. This improvement plays a crucial role in regulating the junction temperature of power devices.

V. T_j CONTROL EXPERIMENT BASED ON VCPCH

This section experimentally verifies the effectiveness of the vapor chamber phase-change heat sink (VCPCH) in controlling the junction temperature fluctuation and average junction temperature of power devices. Using a C3M0160120D SiC MOSFET, a temperature-sensitive electrical parameter (TSEP)-based junction temperature monitoring circuit is constructed (Fig. 7.). The experiment simulates load-induced temperature fluctuations with a 10 s current cycle, 8.5 A current amplitude, and 15 V gate voltage.

Four VCPCHs with identical widths (26 mm) and thicknesses (2 mm) but different lengths (40 mm, 45 mm, 50 mm, and 55 mm) are tested. The relationship between the peak-to-valley values of the on-state resistance and the chamber size is shown in Fig. 8.

A power semiconductor static tester is used to calibrate the relationship between the junction temperature , drain current , on-state resistance , and gate-source voltage. The variation of R_{dson} with T_j at a drain current of 8.5 A is shown in Fig. 9.

By using the R_{dson}-T_j calibration curve, the junction temperature distribution under load-induced temperature fluctuation is derived as a function of the vapor chamber size, as shown in Fig. 10. When the chamber size is 26 mm × 45 mm, the best temperature control is achieved, reducing junction temperature fluctuation by 8.5°C and the average junction temperature by 4.2°C. Further increases in chamber size yield minimal improvement.

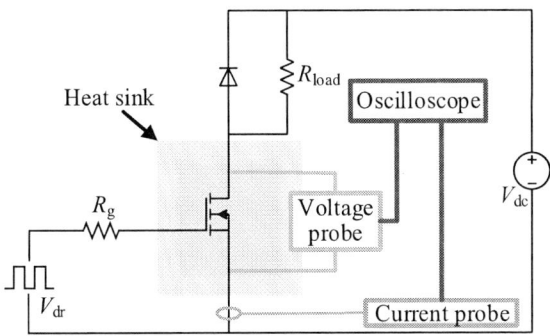

（a）

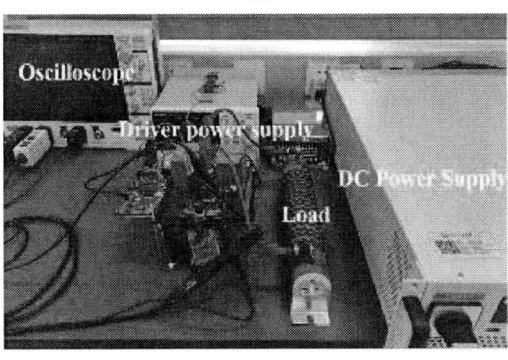

（b）

Fig. 7. Experimental principles.

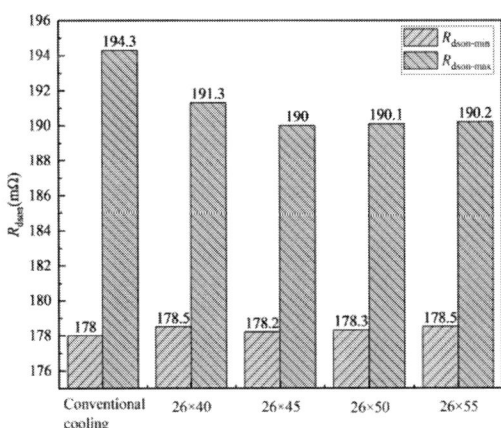

Fig. 8. The relationship between R_{dson} and vapor chamber area.

According to the Coffin–Manson model, the thermal cycle lifetime of the power device with the VCPCH is 4.45 times longer than that with a conventional heat sink. The results demonstrate that the VCPCH enhances thermal capacity and reduces thermal resistance, effectively mitigating junction temperature fluctuations and improving the device's estimate cycle lifetime.

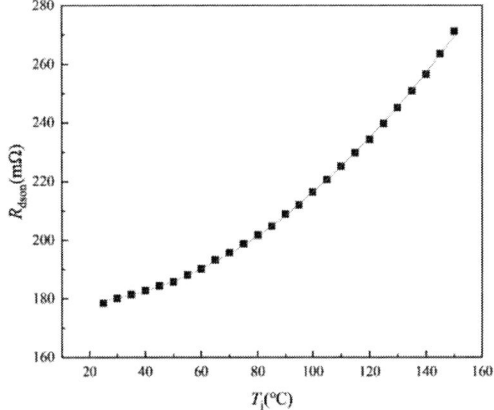

Fig. 9. The relationship between R_{dson} and T_j under 8.5A current.

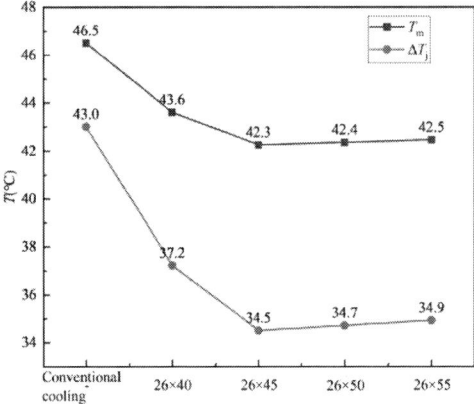

Fig. 10. The relationship between T_j and vapor chamber area.

VI. CONCLUSION

This study proposes a vapor chamber phase-change cooling method to control junction temperature in power devices. The method enhances thermal capacity and reduces thermal resistance, effectively suppressing junction temperature fluctuations and lowering the average junction temperature. Experimental results show that the VCPCH optimizes heat transfer by increasing the effective thermal spreading area, improving thermal management, and extending the estimate cycle lifetime of power devices.

REFERENCES

[1]. L. Ding, R. Song, S. Zhao, J. Wang, and H. A. Mantooth, "Active peltier effect heat sink for power semiconductor device thermal stability enhancement," *IEEE Trans. Power Electron.*, vol. 38, no. 9, pp. 11507-11520, Sept. 2023.

[2]. J. Lemmens, J. Driesen and P. Vanassche, "Thermal management in traction applications as a constraint optimal control problem," in *Proc. VPPC*, 2012, pp. 36-41.

[3]. L. Wei, J. McGuire and R. A. Lukaszewski, "Analysis of PWM frequency control to improve the lifetime of PWM inverter," *IEEE Trans. Ind. Appl.*, vol. 47, no. 2, pp. 922-929, Mar.2011.

[4]. Z. Li *et al.*, "Active gate delay time control of Si/SiC hybrid switch for junction temperature balance over a wide power range," *IEEE Trans Power Electron.*, vol. 35, no. 5, pp. 5354-5365, May 2020.

A self-clamped L-shaped trench gate SiC MOSFET with Improved Breakdown and Short-Circuit Reliability

Xiaobo Cao[a,b], Jing Liu[b], Shaowei Zhang[b], Qian Zhang[b] and Zhonggang Yin[a]

[a] Department of Electrical Engineering, Xi'an University of Technology, Xi'an 710048, China

[b] Department of Electronic Engineering, Xi'an University of Technology, Xi'an 710048, China

A novel self-depleting L-shaped channel SiC trench MOSFET (SCL-TMOS) has been proposed and analyzed using a TCAD simulator. By extending the p-base region to the bottom of the trench gate through deep P-well injection, the SCL-TMOS better shields the gate oxide from high electric fields in the off state. This enhancement allows for an increased breakdown voltage without significantly degrading output characteristics. Additionally, the introduction of the self-depletion mechanism reduces the peak current of the device during short-circuit conditions, improving the short-circuit tolerance. Compared to conventional SiC trench MOSFETs, the peak electric field value in the gate oxide is reduced by 64.9%. Furthermore, the breakdown voltage has increased by 13.7%. The capacitance of HW-TMOS has been optimized by converting the gate-drain capacitance (C_{gd}) at the trench gate sidewall into gate-source capacitance. The short-circuit robustness time of SCL-TMOS is improved by 66.7% compared to C-TMOS. Overall, SCL - TMOS meets the high demands for power devices in modern electronic equipment.

Keywords —SiC trench MOSFET, breakdown voltage, peak electric field, short circuit (SC)

I. Introduction

The rapid development of renewable energy systems, electric vehicles (EVs), and industrial electrification demands power electronic converters with higher efficiency, power density, and reliability [1]. Silicon (Si)-based insulated gate bipolar transistors (IGBTs), while dominant in high-voltage applications (>1.2 kV), face inherent limitations in switching frequency and high-temperature operation due to material properties. [2-3]Silicon carbide (SiC) MOSFETs, distinguished by their superior electrical properties including high critical breakdown field, fast switching capability, and low on-resistance, are positioning themselves as leading candidates for next-generation power switching applications.[4-5]

SiC trench MOSFETs are increasingly adopted for power applications due to low conduction losses and high channel density, making structural design a prominent research focus [6-8]. However, electric field crowding at trench corners during blocking state causes gate oxide degradation. This critical issue drives investigations into diverse electric field shielding techniques. One approach involves integrating a P+ shielding layer connected to the source electrode at the trench bottom to suppress high electric fields in that region. [9-10] However, the introduction of the P+ region creates an undesired JFET area, leading to increased on-state losses, while the shielding effectiveness of the P+ layer remains limited. The cross-sectional view of the structure is shown in Fig. 1(a).To further alleviate electric field concentration at trench corners, researchers have proposed an L-shaped gate trench MOSFET (LSG-MOS) design. [11-12] This structure enhances protection capability by deepening the P-Well implantation depth, enabling partial encapsulation of the gate trench to redistribute peak electric fields. However, the threshold voltage of this structure is higher than that of traditional structures.

Another notable point is that, compared to Si IGBTs, SiC MOSFETs exhibit higher power density and unsaturated drain current due to short channel effects. Additionally, the input capacitance (C_{iss}) of SiC MOSFETs is lower, resulting in faster switching speeds (in the nanosecond range) and extremely high short-circuit current rise rates (di/dt) exceeding 10^9 A/s. This leads to the concentration of short-circuit energy being released in a shorter time, exacerbating thermal stress and resulting in poorer short-circuit capability. A critical challenge in the design optimization of SiC MOSFETs arises from the inherent trade-off

979-8-3315-1110-4/25 $31.00 © 2025 IEEE

between achieving low on-state losses and ensuring robust short-circuit ruggedness.[13-14]

This paper presents an optimized design based on the L-shaped trench gate MOSFET, introducing a 1200 V SiC MOSFET with a self-clamping channel structure (SCL-TMOS) to address the aforementioned two issues, requiring only an additional ion implantation process. For a clearer analysis, we conducted Sentaurus TCAD numerical simulations on both LSG-MOS and SCL-MOS to evaluate the advantages of the proposed structure. The models involved in this work include incomplete ionization, SRH recombination, Auger recombination, as well as traps and fixed charges along the SiC/SiO2 interface. These experimentally validated models have been rigorously verified and consistently implemented in our previous research works.[15]

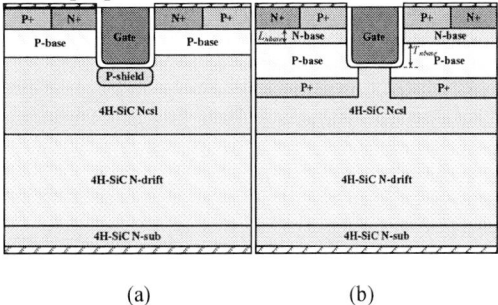

(a)   (b)

Fig. 1. Cross-sectional view of (a) C-TMOS (b) SCL-TMOS

TABLE I
MAIN PARAMETERS OF HW-TMOS AND C-TMOS

Device parameter	C-TMOS	SCL-TMOS
Oxide thickness T_{ox}	50nm	50nm
Trench depth T_{gate}	1.2μm	1.2μm
Trench width W_{gate}	1.0μm	1.0μm
N-base doping N_{N_base}	5×10^{17}cm^{-3}	-
CSL doping N_{CSL}	1×10^{16}cm^{-3}	1×10^{16}cm^{-3}
P-base doping N_{P_base}	2×10^{17}cm^{-3}	2×10^{17}cm^{-3}
P-shield doping N_{P_shield}	1×10^{18}cm^{-3}	1×10^{18}cm^{-3}
Wcell pitch W_{cellf}	4,0μm	4.0μm

II. DEVICE STRUCTURES AND CHARACTERISTICS

The proposed structure in Fig. 1(b) achieves dual functional enhancements: (i) An L-shaped channel formed by deep P-Well implantation provides corner protection for gate oxide, and (ii) the N-base region introduced N-type implantation enables threshold voltage tuning and intrinsic current-limiting through self-pinch-off during short-circuit conditions.[16]

The conventional structures utilize a P-shield layer implanted at the trench bottom. However, this approach requires relatively high doping concentration in the P-type region to effectively protect the gate oxide bottom. Furthermore, the P-shield layer must be electrically connected to the source terminal for optimal shielding performance, which significantly complicates the device layout design.

First, a deep P-Well implantation is employed to form an L-shaped channel, effectively protecting the gate oxide corners and thereby enhancing the device's gate oxide reliability. Second, through an additional N-type implantation step, an N-base region is introduced to modulate the threshold voltage. This design prevents excessively high threshold voltage (reducing switching losses) and incorporates a self-pinch-off mechanism. Under extreme conditions such as short-circuit events, the Nbase region becomes fully depleted, eliminating the current path and suppressing abnormal current surges, ultimately improving the device's short-circuit robustness.

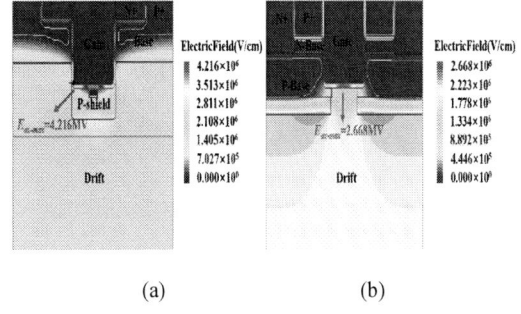

(a)   (b)

Fig. 2. Comparison of the off-state electric field distributions of (a)C-TMOS (b) SCL-TMOS at the Breakdown voltage

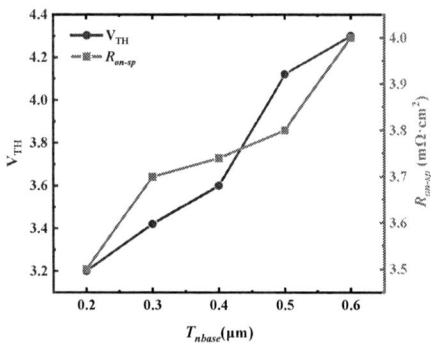

Fig. 3. Effect of distance between the N-base and the bottom of the trench distance on threshold voltage and specific on-resistance

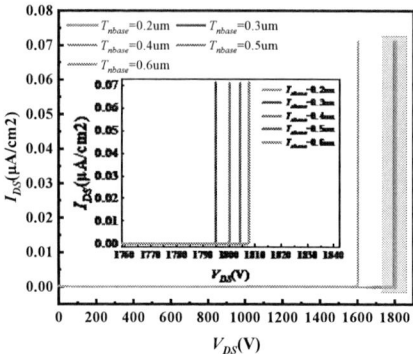

Fig. 4. Effect of distance between the N-base and the bottom of the trench distance on breakdown voltage

The SCL-TMOS structure utilizes deep P-Well implantation to form an L-shaped channel that encapsulates the gate bottom within the P-Well, thereby effectively reducing the electric field concentration at the gate oxide corners and enhancing dielectric reliability. As evidenced by the breakdown electric field distributions in Fig. 2, the conventional structure exhibits a peak electric field intensity of 4.381 MV/cm, whereas the SCL-TMOS achieves a significantly lower peak field of 2.668 MV/cm. This represents a remarkable reduction of 1.713 MV/cm compared to the conventional design.

The distance between the N-base region and trench bottom (T_{nbase}) serves as a critical parameter governing the device's threshold voltage (V_{th}), specific on-resistance ($R_{on,sp}$), and breakdown voltage. Fig. 3 demonstrates the dependence of SCL-MOS V_{th} and s $R_{on,sp}$ on T_{nbase} variations. With increasing T_{nbase}, the effective channel length extends, leading to elevated threshold voltage and increased overall resistance due to reduced carrier mobility along the elongated conduction path. However, decreasing Tnbase reduces the breakdown voltage, which shifts the source potential distribution toward the drain side. Interestingly, this potential redistribution alleviates electric field crowding at the gate oxide bottom, as more field lines are attracted by the base region through charge compensation effects.

Fig. 4 demonstrates the forward blocking characteristics under varying T_{nbase} conditions, revealing a 23% reduction in breakdown voltage when T_{nbase} decreases from 1.2 μm to 0.6 μm. Correspondingly, Fig. 5 shows a 38% decrease in the peak electric field intensity at the gate oxide interface with reduced T_{nbase}, confirming the field redistribution mechanism.

Another critical design parameter is the width of the Nbase region (L_{nbase}). Reducing L_{nbase} decreases the channel conduction area, thereby increasing the channel resistance (R_{ch}). However, the narrowed L_{nbase} facilitates intrinsic current limiting during short-circuit events through enhanced carrier scattering and velocity saturation effects, which significantly improves the short-circuit withstand capability.

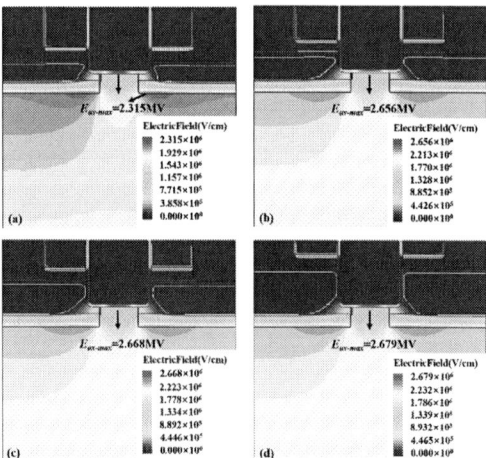

Fig. 5. The electric field distribution of SCLMOS at the breakdown voltage: (a) T_{nbase} = 0.2 μm, (b) T_{nbase} = 0.3 μm, (c) T_{nbase} = 0.4 μm, (d) T_{nbase} = 0.6 μm.

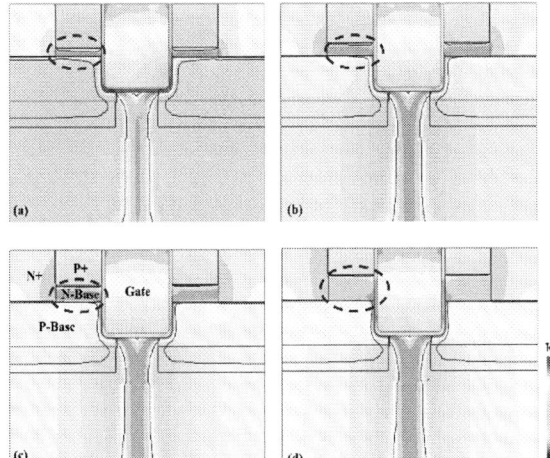

Fig. 6. The current density distribution of the device in a short-circuit condition at V_{DS} = 800 V and t_{SCW} = 3 μs (a) L_{nbase}=0.1μm, (b) L_{nbase}=0.2μm, (c) L_{nbase}=0.3μm, (d) L_{nbase} =0.35μm

Fig. 6 illustrates the current density distributions under short-circuit conditions with V_{DS}=800 , t_{SCW}=3 μs comparing devices with varying L_{nbase}. The narrower L_{nbase} configuration demonstrates a 42% reduction in peak current density compared to the baseline design, effectively suppressing current surge through conduction area limitation.[17] This suppression mechanism originates from the partial depletion of the Nbase region by the adjacent P+ source and P-well under high drain-source bias, which dynamically modulates the effective conduction path cross-section during fault events.

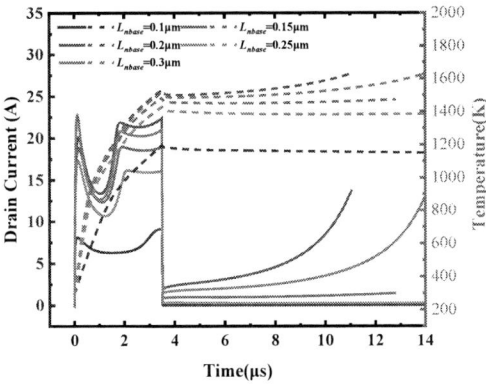

Fig.7. Short-circuit characteristics comparison with varying Lnbase at V_{DS}=800 and T_j=125℃.

As demonstrated in Fig. 7, structures with L_{nbase} = 0.1 μm achieve 66% lower peak short-circuit current compared to those with L_{nbase} = 0.3 μm, while maintaining $R_{on,sp}$ within 15% of the design target.

III. RESULTS AND DISCUSSION

Fig. 8 compares the output characteristics between C-TMOS and SCLMOS, showing a slight increase of 8% in $R_{on,sp}$ for SCL-MOS. This slight improvement is attributed to the trade-off between channel resistance increase (due to L-shaped topology) and drift region optimization.

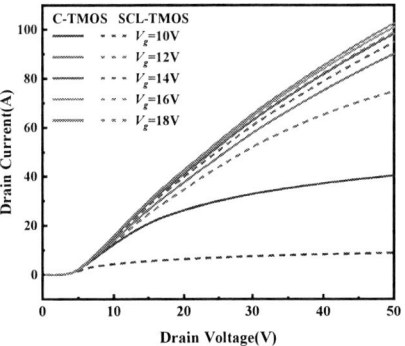

Fig. 8 Comparison of the on-state characteristics of C-TMOS and SCL-TMOS.

Fig. 9 demonstrates the gate charge characteristics, where SCL-MOS achieves 52% lower gate-drain charge (Q_{gd}) compared to C-TMOS, owing to the reduced gate-drain overlap area enabled by the L-shaped gate architecture.

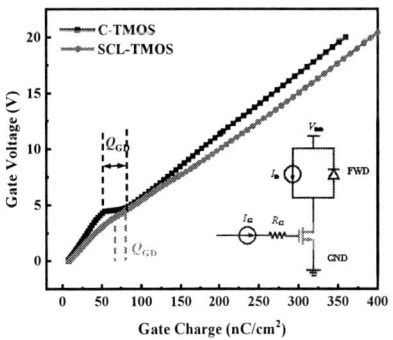

Fig. 9 Comparison of the gate charge characteristics of C-TMOS and SCL-TMOS.

Fig. 10 further validates the capacitance reduction, showing a 70.2% decrease in gate-drain capacitance (C_{gd}) for SCL-MOS. This enhanced frequency response positions SCL-

MOS as a superior candidate for high-frequency applications. Figure 11 shows the short-circuit characteristics of C-TMOS and SCL-TMOS at V_{DS} = 800 V, with the SCLMOS demonstrating a short-circuit withstand time of up to 35 µs.

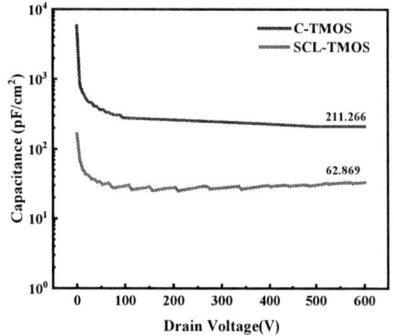

Fig. 10 Comparison of the gate-diain capacitor characteristics of C-TMOS and SCL-TMOS.

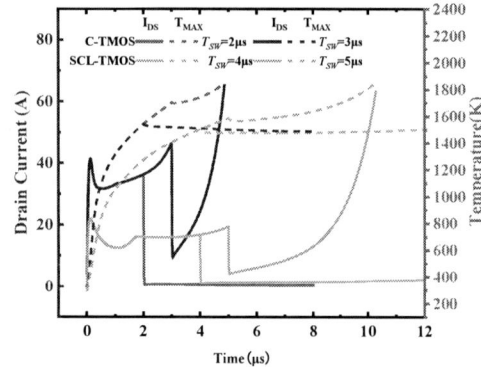

Fig. 10 Comparison of the short-circuit characteristics of C-TMOS and SCL-TMOS.

参考文献：

[1]. X. Ding et al., "Analytical and experimental evaluation of SiC-inverter nonlinearities for traction drives used in electric vehicles," IEEE Trans. Veh. Technol., vol. 67, no. 1, pp. 146–159, Jan. 2018, doi:10.1109/TVT.2017.2765670.

[2]. J. W. Palmour, "Silicon carbide power device development for industrial markets," in Proc. IEEE IEDM., 2014, pp. 1.1.1–1.1.8.

[3]. X. She, A. Q. Huang, Ó. Lucía and B. Ozpineci, "Review of Silicon Carbide Power Devices and Their Applications," in *IEEE Transactions on*

Industrial Electronics, vol. 64, no. 10, pp. 8193-8205, Oct. 2017, doi: 10.1109/TIE.2017.2652401

[4]. M. Imaizumi and N. Miura, "Characteristics of 600, 1200, and 3300 V planar SiC-MOSFETs for energy conversion applications," IEEE Trans. Electron Devices, vol. 62, no. 2, pp. 390–395, Feb. 2015, doi: 10.1109/TED.2014.2358581.

[5]. S. Parashar, A. Kumar, and S. Bhattacharya, "High power medium voltage converters enabled by high voltage SiC power devices," in Proc. Int. Power Electron. Conf., Niigata, Japan, May 2018, pp. 3993–4000 doi: 10.23919/IPEC.2018.8506674.

[6]. W. Chen, Y. Zhou, Y. Xiao, H. Zhang, Y. Huang and Z. Han, "Asymmetric Trench SiC MOSFET With Integrated Three Channels for Improved Performance and Reliability," in *IEEE Transactions on Device and Materials Reliability*, vol. 25, no. 1, pp. 95-100, March 2025, doi: 10.1109/TDMR.2024.3510782.

[7]. D. Kim, S. Y. Jang, A. J. Morgan and W. Sung, "Influence of P+ Body on Performance and Ruggedness of 1.2 kV 4H-SiC MOSFETs," in *IEEE Transactions on Electron Devices*, vol. 71, no. 12, pp. 7659-7665, Dec. 2024, doi: 10.1109/TED.2024.3474615.

[8]. B. -Y. Tsui *et al*., "A Recessed Source Contact Technology to Reduce the Specific On-Resistance of Power MOSFET on 4H-SiC," in *IEEE Electron Device Letters*, vol. 45, no. 10, pp. 1930-1932, Oct. 2024, doi: 10.1109/LED.2024.3437372.

[9]. J. Tan, J. A. Cooper, and M. R. Melloch, "High-voltage accumulation layer UMOSFET's in 4H-SiC," IEEE Electron Device Lett., vol. 19, no. 12, pp. 487–489, Dec. 1998, doi: 10.1109/55.735755.

[10]. J. Wei, M. Zhang, H. Jiang, H. Wang and K. J. Chen, "Dynamic Degradation in SiC Trench MOSFET With a Floating p-Shield Revealed With Numerical Simulations," in IEEE Transactions on Electron Devices, vol. 64, no. 6, pp. 2592-2598, June 2017, doi: 10.1109/TED.2017.2697763.

[11]. Q. Song *et al*., "4H-SiC Trench MOSFET With L-Shaped Gate," in *IEEE Electron Device Letters*, vol. 37, no. 4, pp. 463-466, April 2016, doi: 10.1109/LED.2016.2533432.

[12]. X. Zhou, R. Yue, J. Zhang, G. Dai, J. Li and Y. Wang, "4H-SiC Trench MOSFET With Floating/Grounded Junction Barrier-controlled Gate Structure," in *IEEE Transactions on Electron Devices*, vol. 64, no. 11, pp. 4568-4574, Nov. 2017, doi: 10.1109/TED.2017.2755721.

[13]. D. Kim, S. DeBoer, S. Y. Jang, A. J. Morgan, and W. Sung, "A comparison of short-circuit failure mechanisms of 1.2 kV 4H-SiC MOSFETs and JBSFETs," in Proc. IEEE 9th Workshop Wide Bandgap Power Devices Appl. (WiPDA), Redondo Beach, CA, USA, Nov. 2022, pp. 54–57, doi: 10.1109/WIPDA56483.2022.9955302.

[14]. X. Deng et al., "Short-circuit capability prediction and failure modeof asymmetric and double trench SiC MOSFETs," IEEE Trans.Power Electron., vol. 36, no. 7, pp. 8300–8307, Jul. 2021, doi:10.1109/TPEL.2020.3047896.

[15]. Cao, X.; Liu, J.; An, Y.; Ren, X.; Yin, Z. A Novel 4H-SiC SGT MOSFET with Improved P+ Shielding Region and Integrated Schottky Barrier Diode. *Micromachines* **2024**, *15*, 933. https://doi.org/10.3390/mi15070933

[16]. X. Li *et al.*, "A Novel SiC MOSFET With Embedded Auto-Adjust JFET With Improved Short Circuit Performance," in *IEEE Electron Device Letters*, vol. 42, no. 12, pp. 1751-1754, Dec. 2021, doi: 10.1109/LED.2021.3124526.

[17]. Y. Chen, C. Li, Y. Wu and Z. Zheng, "Improved HF-FOM and SC Ruggedness of Split-Gate 4H-SiC MOSFET With P+ Buffer," in *IEEE Electron Device Letters*, vol. 45, no. 7, pp. 1269-1272, July 2024, doi: 10.1109/LED.2024.3401043.

Degradation of Planar-Gate SiC MOSFETs Under Repetitive Short-Circuit Stress in Different Gate Bias

Yifan Wu
Department of Electrical Engineering
Tsinghua University
Beijing, China
wu-yf21@mails.tsinghua.edu.cn

Chi Li
Department of Electrical Engineering
Tsinghua University
Beijing, China
chi.li.2014@ieee.org

Jianwei Liu
Department of Electrical Engineering
Tsinghua University
Beijing, China
winer32@tsinghua.edu.cn

Zedong Zheng
Department of Electrical Engineering
Tsinghua University
Beijing, China
zzd@tsinghua.edu.cn

Abstract—The lifetime and degradation of silicon carbide (SiC) metal-oxide-semiconductor field-effect transistors (MOSFETs) have garnered increasing attention from device users and power electronics system designers. This work establishes an experimental platform for repetitive short-circuit tests of SiC MOSFETs. Under varying gate voltage (U_{GS}) conditions, comprehensive experimental investigations are conducted on the degradation characteristics of static parameters under repetitive short-circuit stresses, including zero gate voltage drain current (I_{DSS}), threshold voltage (V_{TH}), on-state resistance ($R_{DS(on)}$), and breakdown voltage (BV), and gate-source leakage current (I_{GSS}). The experimental results reveal the degradation trends and patterns of these parameters as the number of short-circuit pulses increases.

Keywords—silicon carbide (SiC) MOSFET, repetitive short-circuit (RSC), static parameters, degradation characteristics

I. INTRODUCTION

Silicon carbide (SiC) MOSFETs demonstrate superior performance in high-frequency and high-temperature working situations with low switching losses, minimal switching oscillations, and negligible current tail [1]. Nevertheless, their reliability is less than mature silicon (Si) IGBTs due to higher wafer defect densities, immature fabrication processes, and reduced chip area [2]. This reliability limitation is particularly critical under short-circuit faults, impeding their widespread adoption. Current research on SiC MOSFET's short-circuit reliability focuses on two domains: the one is faster protection circuits preventing irreversible failure during single fault event [3], the other is accelerated degradation induced by short-circuit stresses.

Recently, lifetime and aging issues of SiC MOSFETs have gained significant attention from device users and power electronics system designers [4]. To explore aging patterns under short-circuit faults, understand the failure mechanisms, and find ways to reduce degradation, researchers increasingly use repetitive short-circuit (RSC) tests to evaluate degradation of characteristic parameters. RSC test is a kind of accelerated aging testing method. Many studies have conducted such tests under various short-circuit pulse durations (t_{SC}) [5], drain-source voltages (U_{DS}) [6], case temperature (T_{CASE}) [7], gate-source voltages (U_{GS}) [8], and different device structures [9]. However, limited RSC test results for SiC MOSFETs are available in literature. This makes it difficult to identify general patterns or draw conclusions with practical value for real-world applications.

This work establishes an RSC test platform for SiC MOSFETs. Experimental investigations were conducted on the degradation characteristics of key static parameters under a wide range of gate-source voltages (U_{GS}), including zero gate voltage drain current (I_{DSS}), threshold voltage (V_{TH}), on-state resistance ($R_{DS(on)}$), breakdown voltage (BV), gate-source leakage current (I_{GSS}). The tests reveal degradation trends of these parameters versus increasing RSC pulse counts across varyin U_{GS} levels.

II. EXPERIMENT SETTINGS

Because the parameter degradation of SiC MOSFETs caused by single short-circuit events is too small for precise measurement, researchers typically employ RSC tests to evaluate accumulated degradation under multiple short-circuit pulses. This study establishes an RSC test platform, as shown in Fig. 1, to evaluate static parameter degradation accumulation in SiC MOSFETs under RSC stresses. The circuit schematic and gate-drive control logic are illustrated in Fig. 2, with equipment specifications detailed in TABLE I. In this study, C2M0040120D (1200V/40mΩ planar-gate SiC MOSFET produced by Wolfspeed) is selected as the device under test (DUT). After each RSC test phase, static parameters I_{DSS}, V_{TH}, $R_{DS(on)}$, BV, and I_{GSS} were measured using a static parameter tester (HUSTEC-1600A-MT).

This study configures U_{GS} across a broad range (20V, 18V, 16V, 14V, 10V, and 8V) in the RSC tests to reveal universal degradation patterns of SiC MOSFET static parameters under short-circuit stress. To closely emulate practical operating scenarios, other test conditions were carefully selected: DC bus voltage U_{DS} = 800V, representing the most common application scenario for 1200V-rated SiC MOSFETs; short-circuit duration t_{SC} = 2μs, matching the typical response time of commercial desaturation (DESAT) protection circuits.

To prevent irreversible thermal runaway in the DUT, the interval between the repeated short-circuit pulses (t_{INT}) was carefully controlled. Based on the Wolfspeed's thermal network model and single-pulse short-circuit test results at U_{GS} = 20V, MATLAB/Simulink simulations were conducted to estimate the junction temperature dynamics during fault events. The results in Fig. 3 show that the peak junction temperature reaches 563.4K at t = 2μs, and the temperature decayed to 300.7K within 50ms after the fault. This value is very close to the initial temperature (300K). Therefore, based on the thermal simulations, t_{INT} was set to 1s in the RSC tests,

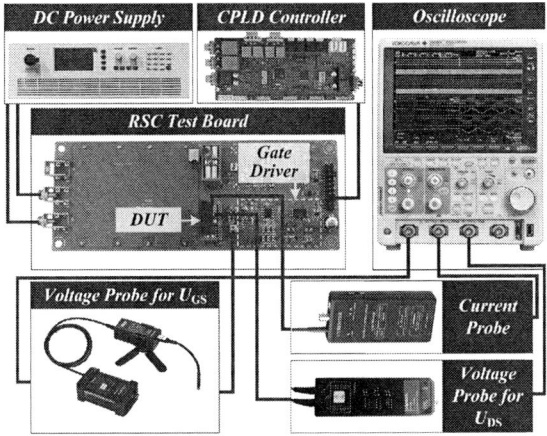

Fig.1 Repetitive short-circuit test platform for SiC MOSFETs

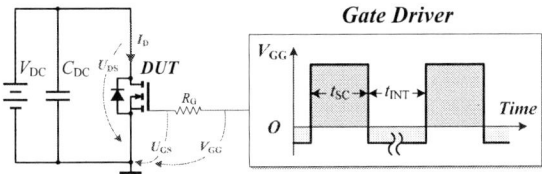

Fig.2 Schematic circuit diagram of the RSC test platform and the control logic of the gate driver.

TABLE I. EQUIPMENTS USED IN RSC TESTS

Equipment	Type	Parameters
Oscilloscope	YOKOGAWA DLM3024	2.5GS/s,200MHz
Voltage Probe for U_{DS}	CYBERTEK DP6150	1500V_{pk},70MHz
Voltage Probe for U_{GS}	CYBERTEK OP6031A	30V_{pk},150MHz
Current Probe	CYBERTEK CP9121SA	I_{pk}=12kA,30MHz
DC Power Supply	Ainuo AN532250-20	2250V,20A,15kW

ensuring that the DUTs can cool completely to the baseline conditions between short-circuit pulses.

III. RESULTS AND DISCUSSIONS

A. Zero Gate Voltage Drain Current (I_{DSS})

Fig. 4 shows the changes in zero gate voltage drain current (I_{DSS}) for SiC MOSFETs during RSC tests. These tests were done using varying U_{GS} from 8V to 20V. In Fig. 4, it can be clearly seen that I_{DSS} steadily grows larger as the device undergoes more short-circuit pulses, from the first pulse all the way to 2000 pulses. This upward pattern happens consistently across every tested U_{GS} level. Furthermore, the way I_{DSS} increases follows a two-stage process. During the beginning stages of RSC tests, the current rises quickly. Then, as the tests continue to higher pulse counts, the current still increases but at a much slower speed. Most notably, the U_{GS} does not produce any meaningful effect on how I_{DSS} changes. Whether we use high or low gate voltages, the current follows almost identical growth patterns. This independence from U_{GS} levels shows that I_{DSS} degradation is not dominated by electrical stresses at the gate of SiC MOSFET. Instead, the primary

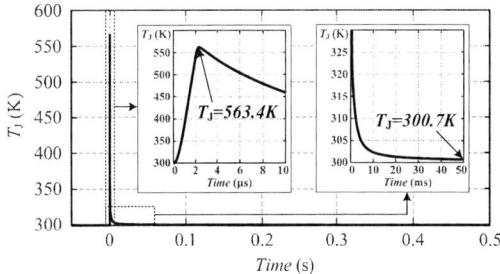

Fig.3 Schematic circuit diagram of the RSC test platform and the control logic of the gate driver.

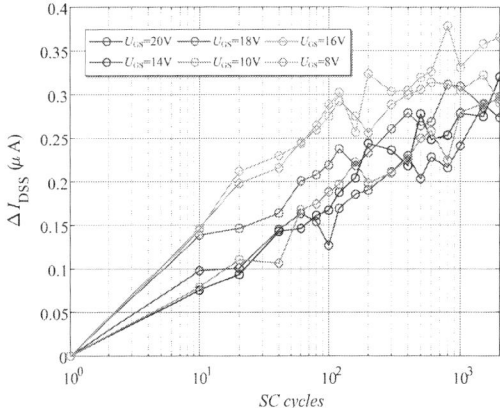

Fig.4 Degradation characteristics curves of I_{DSS} under the conditions of different U_{GS}.

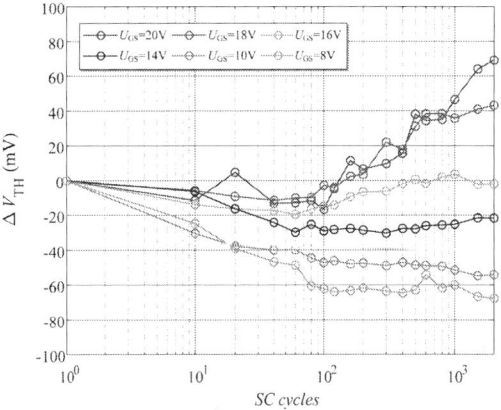

Fig.5 Degradation characteristics curves of V_{TH} under the conditions of different U_{GS}.

causes appear to be physical stresses and high heat dissipation during short-circuit events.

B. Threshold Voltage (V_{TH})

Fig. 5 presents the threshold voltage (V_{TH}) degradation characteristics of the tested SiC MOSFET under RSC stresses across varying U_{GS} from 8V to 20V. Quantitative degradation values within three pulse-count ranges (0-80, 80-2000, and 0-2000 pulses) are summarized in TABLE II.

TABLE II. EQUIPMENTS USED IN RSC TESTS

	U_{GS}	0~80	80~2000	0~2000
DUT1	20V	-11.7mV	80.9mV	69.2mV
DUT2	18V	-9.9 mV	53.1mV	43.2mV
DUT3	16V	-16.6mV	14.7mV	-1.9mV
DUT4	14V	-25.3mV	3.7mV	-21.6mV
DUT5	10V	-44.5mV	-9.5mV	-54.0mV
DUT6	8V	-60.6mV	-7.0mV	-67.6mV

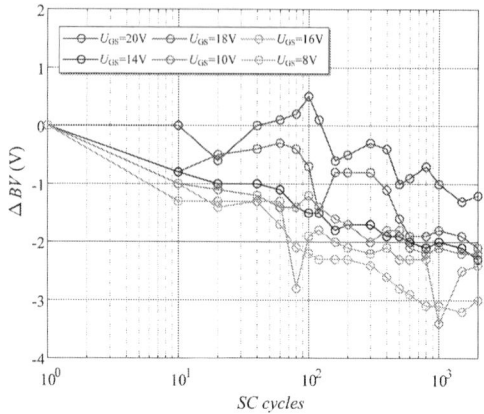

Fig.7 Degradation characteristics curves of *BV* under the conditions of different U_{GS}.

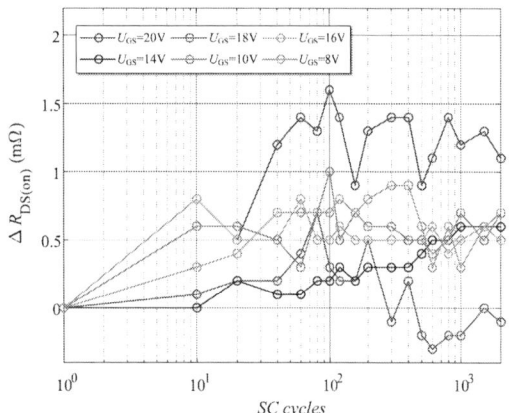

Fig.6 Degradation characteristics curves of $R_{DS(on)}$ under the conditions of different U_{GS}.

Within the range of initial 0-80 pulse, V_{TH} exhibits a consistent negative drift across all voltage conditions, with degradation rates amplifying proportionally to elevated U_{GS} levels. Notably, minimal degradation is observed at U_{GS} = 20V and 18V, indicating near complete suppression of early-stage parameter shifts. Transitioning to the 80-2000 pulse region, the degradation pattern undergoes a fundamental inversion: V_{TH} demonstrates positive drift at higher U_{GS} = 20V, 18V, 16V, stagnates at U_{GS} = 14V, and exhibits minor negative drift at lower U_{GS} = 10V and 8V. Particularly, the slope of degradation characteristics maintains a positive correlation with U_{GS} value throughout both region.

In summary, a unified U_{GS}-dependent principle governs the full 0-2000 pulse region: increasing U_{GS} systematically shifts V_{TH} degradation toward positive drift. However, optimal degradation suppression requires opposing gate bias strategies in each region: higher U_{GS} = 20V and 18V minimizes early-stage V_{TH} shifts, whereas medium U_{GS} = 14V can effectively suppress the late-phase V_{TH} shifts. The U_{GS}-dependent pattern indicates distinct underlying degradation mechanisms. Notably, V_{TH} degradation can be affected by gate bias, but I_{DSS} degradation cannot be controlled. It proves that different failure processes operate in different parts of the devices. These discoveries highlight the need for better reliability prediction methods that can realize the separate monitoring of degradation from heat and electrical stresses at gate.

C. On-State Resistance ($R_{DS(on)}$)

Fig. 6 characterizes the on-state resistance ($R_{DS(on)}$) degradation in SiC MOSFETs under RSC stress across U_{GS} ranging from 8V to 20V. In the experimental results, $R_{DS(on)}$ demonstrates an increasing trend throughout the 0-2000 pulse range under all tested U_{GS} conditions. This upward trend reflects cumulative damage to the devices from repetitive thermal-electrical stresses. It is worth noting that the $R_{DS(on)}$ degradation also shows different trends between the first 100 pulses and beyond that range. In the first 100 pulses, $R_{DS(on)}$ demonstrates a nearly monotonic increase. However, beyond 100 pulses, the resistance enters a relatively stabilized plateau, maintaining near-constant values despite continued stress application. This plateau behavior may suggest the defect saturation mechanisms. Besides, the results confirm no meaningful correlation between U_{GS} level and $R_{DS(on)}$ degradation characteristics. Both the initial rise rate and the plateau values remain U_{GS}-independent. This independence further indicates that SiC MOSFETs' $R_{DS(on)}$ degradation during short-circuit faults is primarily governed by the bulk semiconductor or the packaging degradation mechanisms rather than gate-field-related effects.

D. Breakdown Voltage (BV)

Fig. 7 shows the breakdown voltage (*BV*) degradation characteristics of SiC MOSFETs under RSC stresses across U_{GS} from 8V to 20V. In the test results, *BV* exhibits a nearly consistent monotonic decline throughout the 0-2000 pulse range. This degradation also follows a predictable pattern: the initial stage is dominated by an accelerated rate of reduction, and it transitions to progressively slower degradation at higher pulse counts. Additionally, the results confirm negligible correlation between U_{GS} levels and *BV* degradation characteristics, indicating that the BV stability is governed primarily by bulk semiconductor properties rather than the gate-field-related effects. Particularly, the quantitative evaluation reveals very modest degradation magnitudes. The maximum observed *BV* shift after 2000 short-circuit pulses shows less than 0.5% deviation from the initial *BV*. Such minimal decreases have negligible operational impact under practical application scenarios. In summary, these results show that while *BV* changes in a predictable way during RSC stresses, its small amount of degradation will cause no real

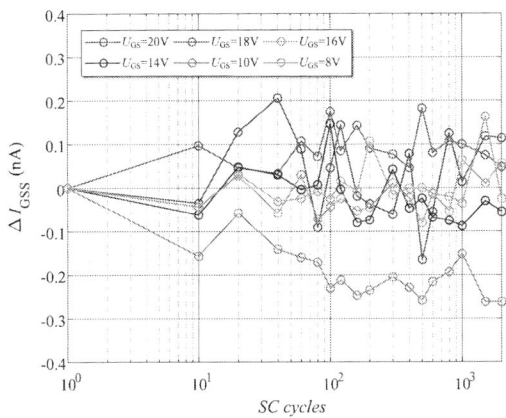

Fig.8 Degradation characteristics curves of I_{GSS} under the conditions of different U_{GS}.

reliability problems for SiC MOSFETs in power electronics systems.

E. Gate-Source Leakage Current (I_{GSS})

Fig. 8 shows the degradation characteristics of gate source leakage current (I_{GSS}) under RSC stresses across varying U_{GS} from 8V to 20V. For U_{GS} = 20V, 16V, 14V, and 8V, I_{GSS} oscillates near the baseline values throughout 0-2000 pulses, exhibiting no measurable degradation. Conversely, at U_{GS} = 10V and 18V, I_{GSS} demonstrates progressive drift. Critically, no clear degradation pattern appears as the short-circuit pulses increase, and no U_{GS}-dependent pattern appear either. This randomness probably comes from manufacturing differences, rather than direct damage from electrical and thermal stresses.

F. Summary

Comprehensive repeated short-circuit (RSC) tests of SiC MOSFETs confirm stable degradation progression in zero gate voltage drain current (I_{DSS}), threshold voltage (V_{TH}), on-state resistance ($R_{DS(on)}$), and breakdown voltage (BV). Among these parameters, BV exhibits minimal reduction, with measured below 0.5% of its initial value. Gate-source leakage current (I_{GSS}) shows negligible degradation across all test conditions. Crucially, V_{TH} alone shows significant sensitivity to gate-source voltage (U_{GS}) variations, exhibiting distinct degradation patterns across pulse-count regions. In contrast, the degradation characteristics of I_{DSS}, $R_{DS(on)}$, BV, and I_{GSS} remain uncorrelated with U_{GS} changes. This selective voltage dependence indicates that gate-oxide-related parameters are uniquely sensitive to gate-field effects during electrical stress, whereas other characteristics follow bulk semiconductor degradation mechanisms.

IV. CONCLUSION

This study established a repeated short-circuit (RSC) degradation test platform for SiC MOSFETs and conducted comprehensive tests on Wolfspeed C2M0040120D devices under varying U_{GS}. The experimental investigation evaluated degradation characteristics of critical static parameters: zero gate voltage drain current (I_{DSS}), threshold voltage (V_{TH}), on-state resistance ($R_{DS(on)}$), breakdown voltage (BV), and gate-source leakage current (I_{GSS}). Key findings reveal marked and stable degradation progression in V_{TH}, I_{DSS}, $R_{DS(on)}$, and BV with increasing pulse counts, while I_{GSS} exhibits negligible parameter drift. Crucially, V_{TH} degradation demonstrates significant U_{GS} dependence following a consistent pattern: higher U_{GS} will shift degradation toward positive drift while amplifying degradation rates. However, this U_{GS}-sensitive behavior shows distinct characteristics across two pulse-count regions (0-80 pulses and 80-2000 pulses), exhibiting fundamentally different degradation dynamics between these ranges. Conversely, all other parameters maintain U_{GS}-insensitive degradation characteristics, showing no correlation with U_{GS} variations.

REFERENCES

[1] T. Kimoto, "Material science and device physics in SiC technology for high-voltage power devices," Jpn. J. Appl. Phys., vol. 54, no. 4, p. 040103, Mar. 2015, doi: 10.7567/JJAP.54.040103.

[2] L. Ceccarelli, P. D. Reigosa, F. Iannuzzo, and F. Blaabjerg, "A survey of SiC power MOSFETs short-circuit robustness and failure mode analysis," Microelectron. Reliab., vol. 76–77, pp. 272–276, Sep. 2017, doi: 10.1016/j.microrel.2017.06.093.

[3] A. Anurag, S. Acharya, N. Kolli, and S. Bhattacharya, "Gate Drivers for Medium-Voltage SiC Devices," IEEE J. Emerg. Sel. Top. Ind. Electron., vol. 2, no. 1, pp. 1–12, Jan. 2021, doi: 10.1109/JESTIE.2020.3039108.

[4] S. Pu, F. Yang, B. T. Vankayalapati, and B. Akin, "Aging Mechanisms and Accelerated Lifetime Tests for SiC MOSFETs: An Overview," IEEE J. Emerg. Sel. Top. Power Electron., vol. 10, no. 1, pp. 1232–1254, Feb. 2022, doi: 10.1109/JESTPE.2021.3110476.

[5] J. Sun, J. Wei, Z. Zheng, Y. Wang, and K. J. Chen, "Short Circuit Capability and Short Circuit Induced V_\mathrmTH Instability of a 1.2-kV SiC Power MOSFET," IEEE J. Emerg. Sel. Top. Power Electron., vol. 7, no. 3, pp. 1539–1546, Sep. 2019, doi: 10.1109/JESTPE.2019.2912623.

[6] D. Pappis and P. Zacharias, "Failure modes of planar and trench SiC MOSFETs under single and multiple short circuits conditions," in 2017 19th European Conference on Power Electronics and Applications (EPE'17 ECCE Europe), Sep. 2017, p. P.1-P.11. doi: 10.23919/EPE17ECCEEurope.2017.8099206.

[7] H. Du, P. Diaz Reigosa, L. Ceccarelli, and F. Iannuzzo, "Impact of Repetitive Short-Circuit Tests on the Normal Operation of SiC MOSFETs Considering Case Temperature Influence," IEEE J. Emerg. Sel. Top. Power Electron., vol. 8, no. 1, pp. 195–205, Mar. 2020, doi: 10.1109/JESTPE.2019.2942364.

[8] J. Wei et al., "Comprehensive Analysis of Electrical Parameters Degradations for SiC Power MOSFETs Under Repetitive Short-Circuit Stress," IEEE Trans. Electron Devices, vol. 65, no. 12, pp. 5440–5447, Dec. 2018, doi: 10.1109/TED.2018.2873672.

[9] Y. Li et al., "Gate Bias Dependence of VTH Degradation in Planar and Trench SiC MOSFETs Under Repetitive Short Circuit Tests," IEEE Trans. Electron Devices, vol. 69, no. 5, pp. 2521–2527, May 2022, doi: 10.1109/TED.2022.3142237.

Low Roughness and Shape Reforming 4H-SiC Trench Process optimized by CCP Etching with High Temperature Annealing and Sacrificial Oxidation

Qiongyang Zhuang
Shenzhen Pengjin High-Tech Company, Ltd
Shenzhen, China
qiongyang.zhuang@pjht.com.cn

Xixi Luo
Shenzhen Pinghu Laboratory
Shenzhen, China
luoxixi@phlab.com.cn

Yu Chen
Shenzhen Pengjin High-Tech Company, Ltd
Shenzhen, China
yu.chen@pjht.com.cn

Caixin Gu
Shenzhen Pengjin High-Tech Company, Ltd
Shenzhen, China
caixin.gu@pjt.com.cn

Lei Song
Shenzhen Pengjin High-Tech Company, Ltd
Shenzhen, China
lei.song@pjt.com.cn

Kaiju Liao
Shenzhen Pengjin High-Tech Company, Ltd
Shenzhen, China
kaiju.liao@pjt.com.cn

Qin Hu
Shenzhen Pinghu Laboratory
Shenzhen, China
huqin@phlab.com.cn

Jiamin Tian
Shenzhen Pinghu Laboratory
Shenzhen, China
tianjiamin@phlab.com.cn

Yidan Chen
Shenzhen Pinghu Laboratory
Shenzhen, China
chenyidan@phlab.com.cn

Gang Chen
Shenzhen Pinghu Laboratory
Shenzhen, China
chengang@phlab.com.cn

Jinliang He
Shenzhen Pengjin High-Tech Company, Ltd
Shenzhen, China
jinliang.he@pjht.com.cn

Abstract—**This paper focuses on the evaluation of two subsequent process steps—post-trench processing (PTPs) and sacrificial oxidation (SAC)—following the etching of 4H silicon carbide (4H-SiC) trenches. Through hydrogen (H2) annealing followed by sacrificial oxidation, exceptionally low trench sidewall roughness of 0.24 nm and well-rounded, micro-trench-free SiC trenches were achieved. This study also presents a comparison study of SiC annealing processes using argon (Ar) and hydrogen (H2), as well as an examination of the roughness associated with the SiC sacrificial oxidation process. These findings contribute to enhancing the electrical performance of trench MOSFETs.**

Keywords—SiC trench, low roughness, sacrificial oxidation, argon (Ar) annealing, hydrogen (H2) annealing

I. INTRODUCTION

Recently, The SiC trench MOSFET (UMOSFET) has gained attention due to its smaller cell pitch and higher electron mobility in the trench sidewall, resulting in lower on-state resistance compared to traditional planar MOSFETs. However, SiC trench sidewalls fabricated using capacitive coupled plasma(CCP) etch techniques which induced the poor surface smoothness of sidewall and micro-trenching issues. The poor surface smoothness of sidewall and micro-trenching issues are exist which has significantly degrade device performance and reliability. Moreover, the high electric field is easily to concentrate at the trench corners near the gate oxide. This may cause reduction in breakdown voltage under high-voltage conditions. However, it's very hard to improve the trench quality by only optimizing the capacitive coupled plasma(CCP) etch condition. Therefore, it is very necessary to explore into reducing sidewall roughness through high-temperature annealing and sacrificial oxidation methods [1].

II. EXPERIMENTAL DETAILS

This paper focuses on the evaluation of two subsequent process steps—post-trench processing (PTPs) and sacrificial oxidation (SAC)—following the etching of 4H silicon carbide (4H-SiC) trenches, with an emphasis on maintaining low sidewall roughness as well as reshaping the profile of the trenches for SiC Trench-MOSFETs [2].

For better comparison, all trenches were etched on 8 inches n-type 4H-SiC substrates with 4°-off orientation. PETEOS was grown as hardmask and SiC trench was etched by capacitively coupled plasma (CCP) dry etch. All trenches have a shape of Figure 1.

Fig. 1. tilted cross-section of the as-etched SiC trench.

Scanning Electron Microscope(SEM) and Atomic Force Microscope(AFM) were utilized to estimate surface roughness of the as-etched trenches as in Figure 2a and 2b.

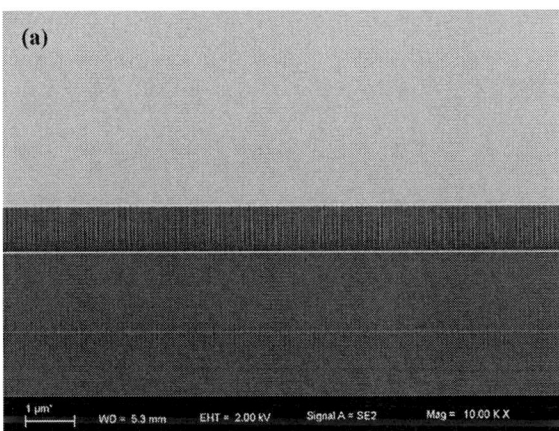

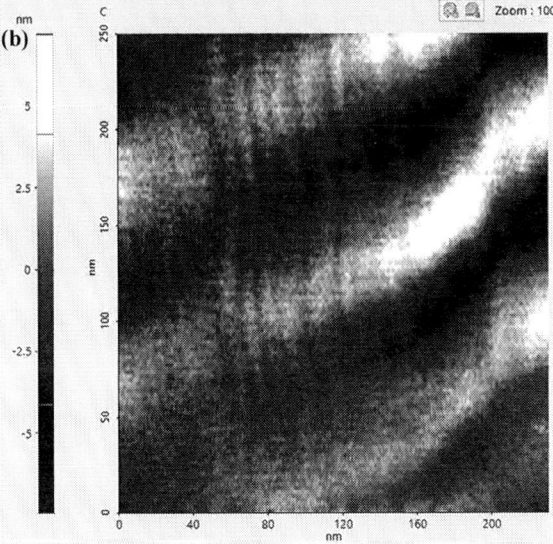

Fig. 2. (a)SEM view and (b) AFM view of the sidewall of as-etched SiC trench.

Obvious surface roughness can be identified from the SEM image and a Ra of about 1.647nm was obtained by AFM. It is then shown that the CCP dry etch damaged the SiC surface making it no longer suitable for a SiC Trench MOSFETs as the channel will be located on the sidewall.

In order to recover the SiC surface, post-trench treatment involved subjecting the wafer to a hydrogen (H2) or argon (Ar) ambient from 1300°C to 1500°C. All process were conducted under a chamber pressure of 20 mBar and the process time varied from 1.5min to 5min.

After the high temperature annealing process, followed by a sacrificial oxidation step to a thickness of approximately 70nm or 140nm at 1100°C.

III. RESULTS AND DISCUSSION

According to the SEM profile, the high anneal treatments both in H2 ambient, as in Figure 3a and 3b, or Ar ambient, as in Figure 4a and 4b, effectively reduced the roughness and bunching along the trench sidewalls, which were caused by the CCP trench etching process.

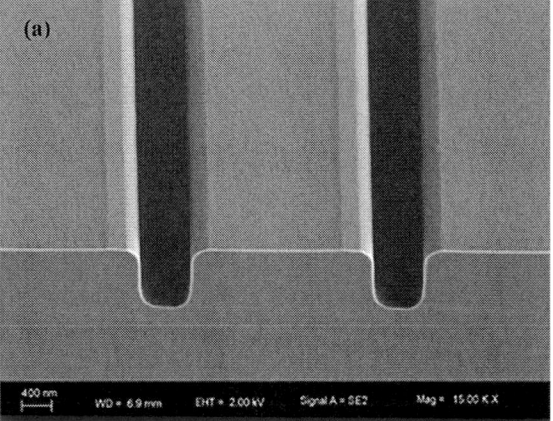

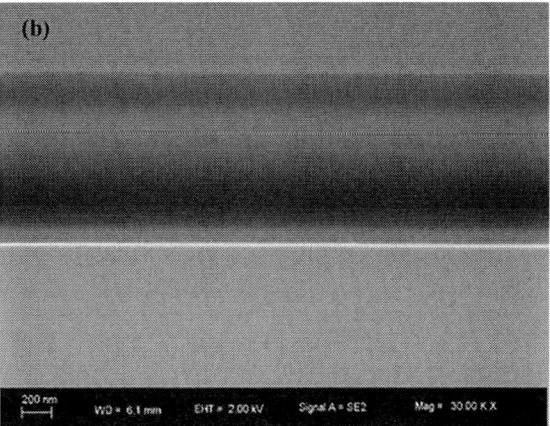

Fig. 3. SEM view (a) corss-section and (b) sidewall of post-anneal SiC trench after H2 annealing at 1400°C for 1.5min and 70nm sacrificial oxidation.

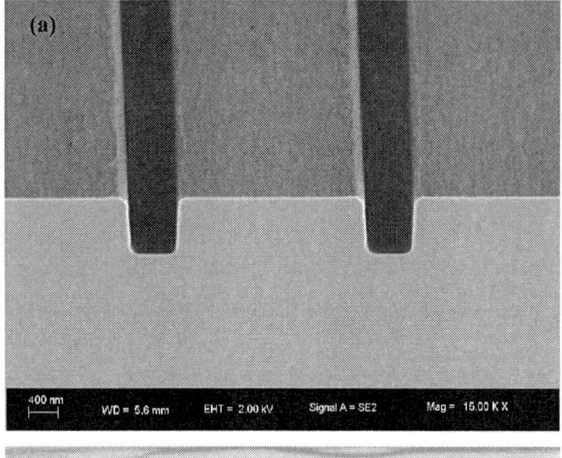

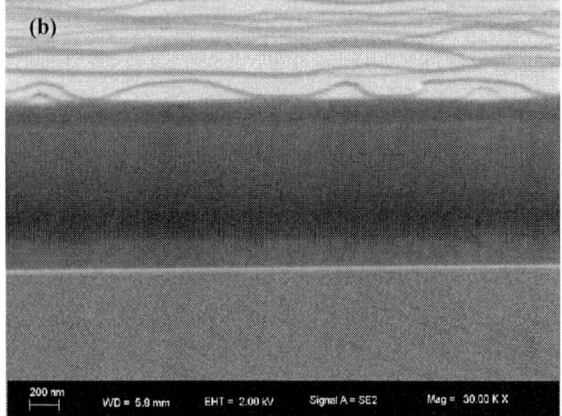

Fig. 4. SEM view (a) corss-section and (b) sidewall of post-anneal SiC trench after Ar annealing at 1400°C for 1.5min and 70nm sacrificial oxidation.

However, although shown a sidewall smoothing effect, Ar annealed SiC trench indicated a strong top surface damage which is not desirable for device reliability concern. These top surface damage seem getting worse when temperature and process time increases, as in Figure 5.

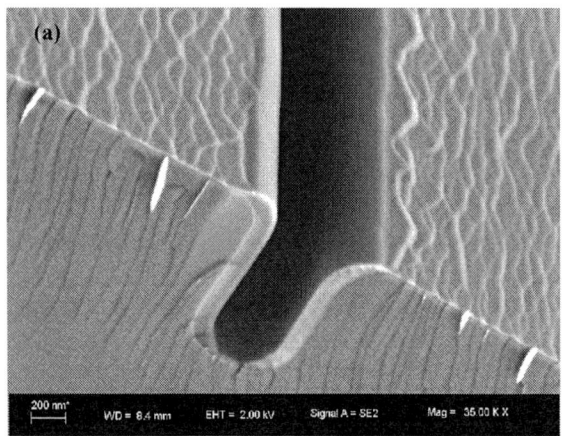

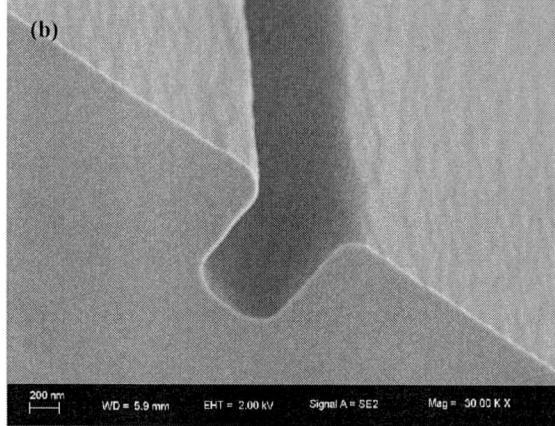

Fig. 5. SEM view tilted corss-section of post-anneal SiC trench after Ar annealing at (a)1500°C for 5min (b)1450°C for 5min (c)1400°C for 5min and (d)1400°C for 1.5min.

On the other hand, H2 annealing indicated a much better smoothing effect as well as an excellent edge rounding capability [3]. As shown in Figure 6, higher process temperature and longer process time gives better edge rounding effect.

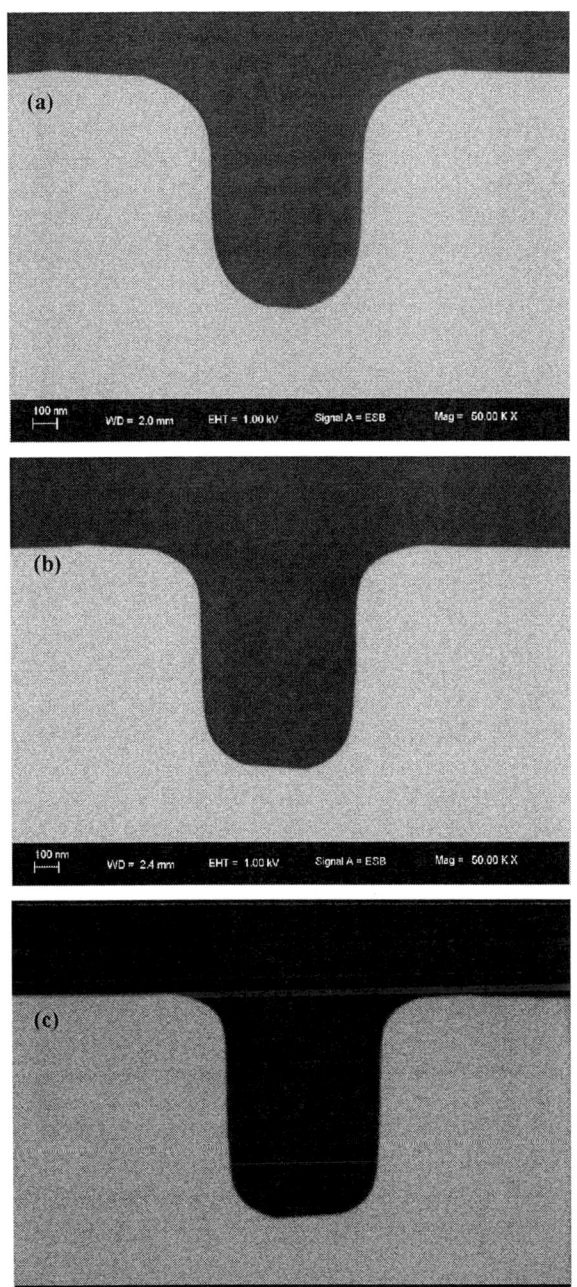

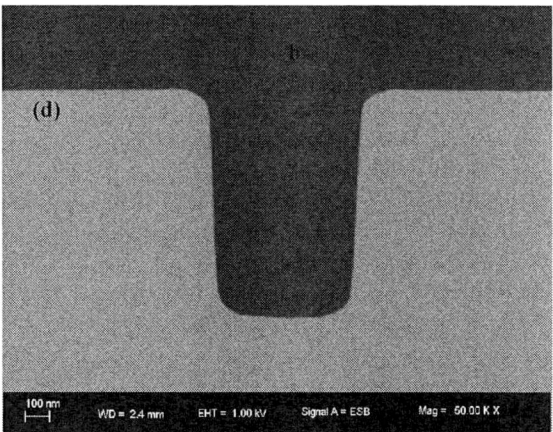

Fig. 6. SEM view corss-section of post-anneal SiC trench after H2 annealing at (a)1450°C for 1.5min (b)1400°C for 1.5min (c)1350°C for 5min and (d)1300°C for 5min.

Although SEM images shown no obvious sidewall and top surface roughness in all H2 annealed samples, as in Figure 7. The AFM roughness measurement results did indicated a trench top and sidewall roughness getting worse when H2 anneal temperature increases, as in Table I.

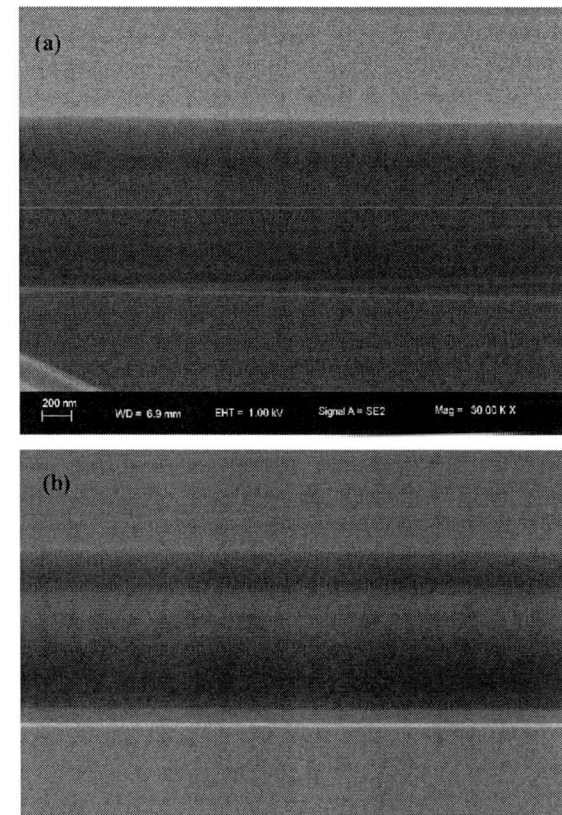

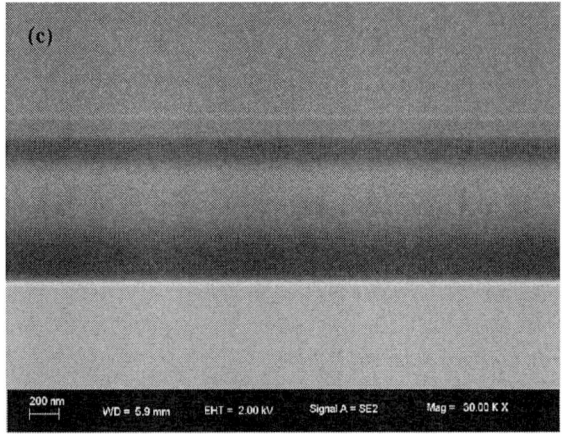

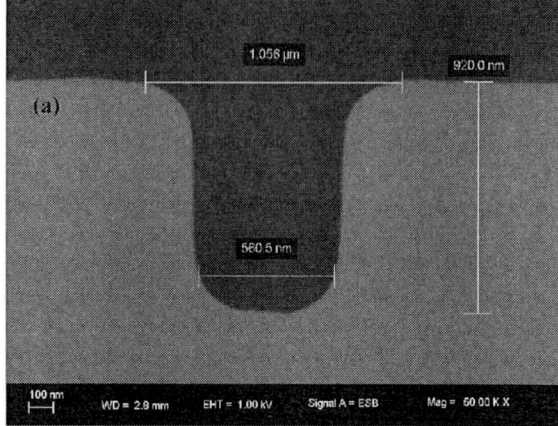

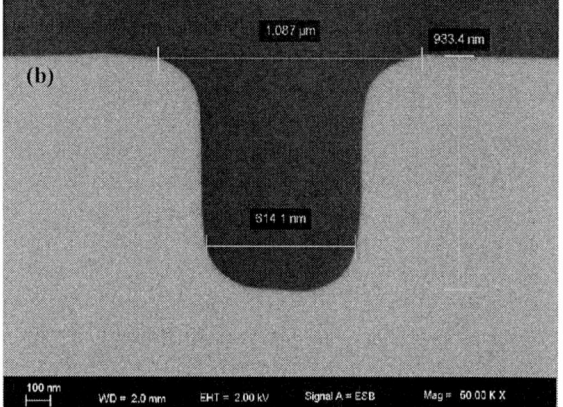

Fig. 7. SEM view sidewall of post-anneal SiC trench after H2 annealing at (a)1450°C for 1.5min (b)1400°C for 1.5min (c)1350°C for 5min and (d)1300°C for 5min.

TABLE I.   ROUGHNESS OF SiC TRENCH WITH DIFFERENT H2 ANNEALING

H2 anneal condition	1450°C 1.5min	1400°C 1.5min	1350°C 5min	1300°C 5min
Sidewall Ra (nm)	1.216	0.622	0.313	0.244
Top surface Ra (nm)	0.227	0.153	0.142	0.039

Subsequently, dry oxidation was grown at 1100°C to estimate the surface smoothing effect of the sacrificial oxidation. For 120min oxidation time, a sidewall oxide thickness of about 70nm was achieved. For 300min oxidation time, a sidewall oxide thickness of about 140nm was achieved.

After removing the sacrificial oxidation, a trench widen effect was observed since sidewall oxidation removed parts of the SiC. However, the trench depth kept the same since the trench bottom oxidation rate much slower than the sidewall, as in Figure 8.

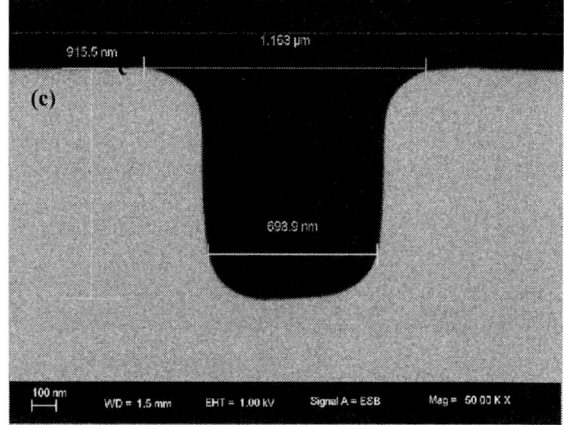

Fig. 8. SEM view cross-section of post-anneal SiC trench after H2 annealing at 1400°C 1.5min and (a) no (b)70nm (c)140nm sacrificial oxidation.

However, AFM measurement indicates limited sidewall and top surface roughness improvement of the sacrificial oxidation process since the surface has been very well smoothed with H2 annealing, as in Table II.

TABLE II.   ROUGHNESS OF SiC TRENCH WITH DIFFERENT SAC-OXIDATION

H2 anneal &Sac-oxide condition	1400°C 1.5min 0nm oxide	1400°C 1.5min 70nm oxide	1400°C 1.5min 140nm oxide
Sidewall Ra (nm)	0.622	0.445	0.341
Top surface Ra (nm)	0.153	0.051	0.143

IV. CONCLUSION

In conclusion through the H2 anneal then sac-oxidation two-steps process, exceptionally low trench sidewall roughness of 0.24nm and well rounded micro-trench-free SiC trenches were achieved. This study also presents a comparison study of SiC Ar and H2 annealing process and the roughness study of SiC sac-oxidation process, which are all conducive to enhancing the electrical performance of Trench-MOSFETs.

REFERENCES

[1] Kawada Y , Tawara T , Nakamura S I ,et al.Shape Control and Roughness Reduction of SiC Trenches by High-Temperature Annealing[J].Japanese Journal of Applied Physics, 2009, 48(11):116508.1-116508.6.DOI:10.1143/JJAP.48.116508.

[2] Zheng C , Wang Z , Jiao S ,et al.Low Roughness SiC Trench Formed by ICP Etching with Sacrificial Oxidation and Ar Annealing Treatment[J].2021 IEEE Workshop on Wide Bandgap Power Devices and Applications in Asia (WiPDA Asia), 2021:354-357. DOI:10.1109/ wipdaasia 51810. 2021.9656026.

[3] Takatsuka A , Tanaka Y , Yano K ,et al.Shape Transformation of 4H-SiC Microtrenches by Hydrogen Annealing[J].Japanese Journal of Applied Physics, 2009, 48(4issue1):041105-041105-3.DOI:10.1143/JJAP.48.041105.

Investigation of L-FER ESD Protection Capability on E-mode *p*-GaN HEMT

Junye Wu[a], Yitian Gu[a], Chao Feng[a], Danfeng Mao[a], Yanlin Wu[a], Haolin Hu[a], Wei Zeng[a], David Zhou[a] and Yuxi Wan[a]

[a] Shenzhen Pinghu Laboratory, Shenzhen 518100, China,
davidzhou@phlab.com.cn; wanyuxi@phlab.com.cn

Abstract—In this work, several lateral field effect rectifiers (L-FERs) were investigated as electrostatic discharge (ESD) protection circuit for p-GaN AlGaN/GaN HEMTs. Four different ESD protection circuits were fabricated simultaneously for comparison, based on a Complementary Metal Oxide Semiconductor (CMOS)-compatible 6-inch GaN-on-Si platform. Human Body Model (HBM) ESD single-pulse stress testing was performed, following the JS-001-2023 testing standard. Notably, the device pass a 30th stress cycle test, demonstrating the robustness of HBM ESD capability within specification.

Keywords—Electrostatic discharge, E-mode, p-GaN HEMT, Human body model

I. INTRODUCTION

AlGaN/GaN based high electron mobility transistors (HEMTs) have been regarded as a promising candidate in high-performance power applications due to high electron mobility and sheet density of the 2-dimensional electron gas (2DEG)[1-2]. The *p*-GaN gate HEMT structure is used to achieve enhancement-mode (E-Mode) operation, which has garnered considerable research attention. However, p-GaN HEMT shows weak gate ESD capability due to lack of efficient discharge channel in its gate structure. Therefore, considerable effort has been devoted to find effectively ways to improve the degradation or failure ESD event [3-6], which is often occurs in the manufacturing and practical usage of semiconductor devices and could cause catastrophic failure or non-catastrophic damage. On-chip integration of ESD protection circuits into p-GaN HEMTs is both convenient and effective, while also reducing fabrication costs[7].

In this work, based on a CMOS-compatible 6-inch GaN-on-Si platform, four ESD protection circuits were simultaneously fabricated to investigate their protective effects on p-GaN gate AlGaN/GaN HEMT devices. A ESD protection circuit comprising six L-FERs was subjected to Human Body Model (HBM) ESD and survived until 30th stress cycle, demonstrating achievement of HBM ESD protection capability within specification.

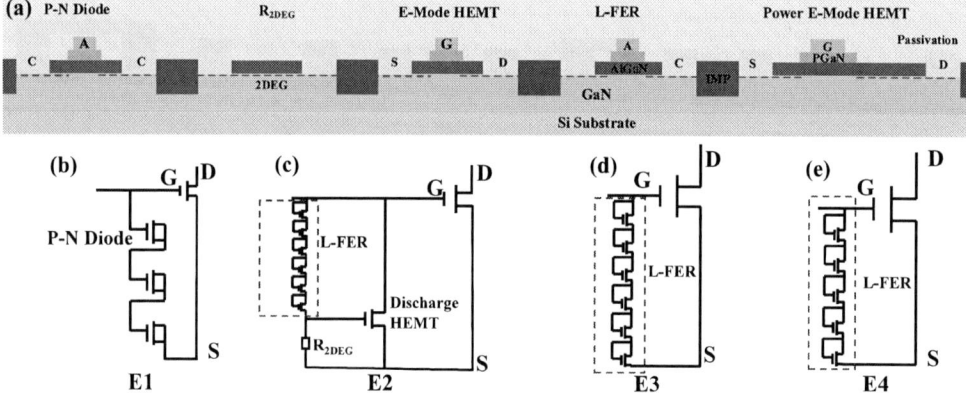

Fig.1 (a) The demonstration of components within the monolithic integrated 6inch GaN-Si platform, (b)-(e) Simplified schematic diagram of the four ESD circuit.

Table 1: The physical parameters of the Power E-mode HEMT

Parameter	Definition	Value
L_G	Gate length	1.4 μm
L_{GS}	Gate-to-source distance	0.75 μm
L_{GD}	Gate-to-drain distance	18 μm
W_G	Gate width	116 mm

II. STRUCTURE AND MECHANISM

As shown in Fig. 1(a), based on a 6-inch CMOS-compatible platform, we proposed a

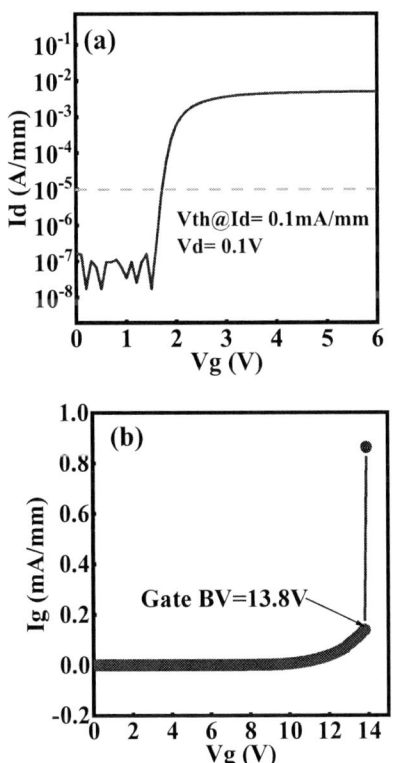

Fig.2 (a) Transfer (I_D-V_G) and (b) forward gate leakage current (I_G) characteristics of the Power E-mode HEMT without ESD protection circuit.

650V power E-Mode p-GaN HEMT platform with multiple functional components isolated by ion-implantation including P-N diodes, 2DEG resistors, discharge HEMTs and L-FERs. Notably, all the components acquire no additional process. Table 1 shows the key device parameters of the protected E-Mode p-GaN HEMT. Simplified schematics of four gate(G)- source(S) ESD protection circuit configurations are presented in Fig. 1(b)-(e). In addition, the second design (E2) was previously proposed in the previous work[3] which serves as a benchmark for comparison with other configurations. All DC electrical characterization measurements were performed under ambient temperature conditions (25°C) using a Keysight B1505A semiconductor parameter analyzer.

III. RESULT AND ANALYSIS

Fig. 2 presents static I-V characteristics of the power E-Mode p-GaN HEMT without ESD protection circuit. Transfer characteristics curve reveals a threshold voltage (V_{TH}) of 1.7 V, extracted at $I_D = 0.1$ mA/mm and $V_D = 0.1$ V. Forward gate leakage current (I_G) characteristics exhibit a gate breakdown voltage (BV) of 13.8V at $I_G = 0.14$ mA/mm.

Forward and reverse gate leakage current of the power E-mode HEMT with different ESD protection circuit are shown in Fig. 3. Only the E2 exhibited a sharp increase in reverse gate leakage current, attributable to the 2DEG Resistance (R_{2DEG}) and discharge HEMT in the circuit between the G and S terminals during reverse conduction. In contrast, the E1 structure demonstrated identical forward/reverse characteristics to the power E-Mode p-GaN HEMT w/o ESD protection circuits, due to the series-connected diodes containing both reverse-biased Schottky and p-n junctions at the anode, which remained non-conductive under all bias conditions and thus failed to provide ESD protection. Both E2 and E3 configurations showed a marked increase in forward I_G beyond 7.5 V, indicating ESD circuit activation at this threshold. The E4 structure, featuring one fewer series-connected L-FER compared to

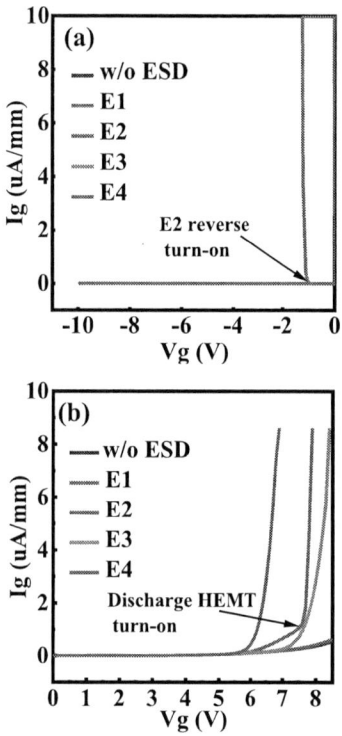

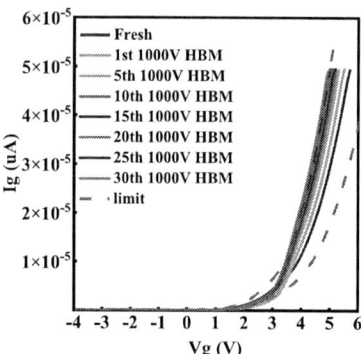

Fig.4 IG-VG characteristics of the Power E-mode HEMT with E3 protection circuit after HBM test.

pulse HBM ESD stress testing at 1000 V. Post-stress I_G-V_G characteristics were analyzed using a specification of leakage current exceeding 10% of the fresh device baseline. The tested devices exhibited failure upon the 30th stress cycle, featuring the robustness of HBM ESD capability within specification.

Fig.3 (a) Reverse and (b) forward gate leakage current (IG) characteristics of the Power E-mode HEMT with different ESD protection circuit.

E3, initiated ESD protection functionality at a reduced gate voltage of V_G = 6 V. Compared with the E2 configuration, the E3 and E4 structures exhibit simplified designs while maintaining ESD protection functionality. Critically, these configurations remain inactive under reverse voltage conditions, thus preventing unintended activation of the HEMT as a switching controller due to excessive reverse current. The E3 structure initiates forward conduction at V_G = 7.5 V, ensuring no interference with device operation at its nominal working voltage of 6 V.

HBM ESD robustness evaluation was performed on devices incorporating the E3 structure in accordance with the JS-001-2023 testing standard [8]. As demonstrated in Fig. 4, the devices underwent repeated single-

IV. CONCLUSION

This study implements four monolithically integrated ESD protection circuits for a E-mode *p*-GaN HEMT based on a CMOS-compatible 6-inch GaN-on-Si platform to systematically evaluate their ESD protection capability. The ESD protection circuit with six L-FER in series was characterized under Human Body Model (HBM) ESD stress conditions following industry-standard qualification protocols. The tested devices exhibited failure upon the 30th stress cycle, enabling the robustness of HBM ESD capability within specification.

REFERENCE

[1] K. J. Chen et al., "GaN-on-Si power technology: Devices and applications," *IEEE Trans. Electron Devices*, vol. 64, no. 3, pp. 779–795, Mar. 2017, doi: 10.1109/TED.2017.2657579.

[2] W. Chen, K.-Y. Wong, and K. J. Chen, "Single-chip boost converter using monolithically integrated AlGAN/GAN

lateral field-effect rectifier and normally off HEMT," *IEEE Electron Device Lett.*, vol. 30, no. 5, pp. 430–432, May 2009, doi: 10.1109/LED.2009.2015897.

[3] C. Zhou et al., "On-chip gate ESD protection for AlGaN/GaN E-mode power HEMT delivering >2kV HBM ESD capability," *IEEE 7th Workshop Wide Bandgap Power Devices Appl. (WiPDA)*, 2019, pp. 175–176, doi: 10.1109/WiPDA46397.2019.8998945.

[4] J. Sun et al., "Characteristics and evaluation approaches of human-body-model electrostatic discharge across Schottky p-GaN gate HEMTs," *IEEE Trans. Ind. Electron.*, vol. 71, no. 3, pp. 3113–3121, 2024, doi: 10.1109/TIE.2023.3265036.

[5] P.-Y. Huang et al., "Comprehensive Study of Human-Body-Model Electrostatic Discharge on p-GaN Gate Power HEMT with AlGaN Barrier Spacers," 2024 36th International Symposium on Power Semiconductor Devices and ICs (ISPSD), pp.538-541, 2024, doi: 10.1109/ISPSD59661.2024.10579584.

[6] J. Sun et al., "Self-Protection Mechanism of Schottky-Type p-GaN Gate HEMTs Under Forward Gate ESD Stress," 2024 36th International Symposium on Power Semiconductor Devices and ICs (ISPSD), pp.271-274, 2024, doi: 10.1109/ISPSD59661.2024.10579656.

[7] Y. Shi et al., "Experimental Investigation and Model Analysis on a GaN Electrostatic Discharge Clamp," IEEE J. ELECTRON DEVI., vol.10, pp. 976-982, 2022, doi: 10.1109/JEDS.2022.3218020.

[8] ESDA/JEDEC Joint Standard for Electrostatic Discharge Sensitivity Testing - Human Body Model (HBM) Device Level, ANSI/ESDA/JEDEC JS-001-2023, 2023.

2025 IEEE Workshop on Wide Bandgap Power Devices and Applications in Asia (WiPDA Asia)

Study on 4H-SiC Trench Gate Dual-Mode Composite Transistors (T-DCT) Structure to reduce R_{on} and Enhance SCWT

Wenyu Xi
Department of Electronic Engineering
Xi'an University of Technology
Xian, China
343872706@qq.com

Cailin Wang*
Department of Electronic Engineering
Xi'an University of Technology
Xian, China
wangcailin@xaut.edu.cn

Lei Guan
Department of Electronic Engineering
Xi'an University of Technology
Xian, China
842347439@qq.com

Abstract—In order to solve of high specific on-resistance and limited short-circuit withstand capability of conventional 4H-SiC Vertical U-groove MOSFET (VUMOS), a Trench-gate Dual-mode Composited Transistor (T-DCT) structure with VUMOS and BJT in parallel is proposed. The static characteristics, switching behavior, and short-circuit robustness of the T-DCT are systematically evaluated by TCAD and compared with that of VUMOS. Simulation results show that the T-DCT achieves a 50% reduction in R_{on} while exhibiting enhanced short-circuit withstand time (SCWT) of 7.0μs, a 29.6% improvement compared to the 5.4μs SCWT of the VUMOS.

Keywords—4H-SiC;VUMOS;bipolar-junction transistor (BJT)

I. INTRODUCTION

4H-SiC exhibits significant advantages in power device applications due to its superior critical breakdown electric field, high electron saturation mobility, and exceptional thermal conductivity coefficient[1,2], which exceptional properties have positioned 4H-SiC MOSFET as a prominent focus in wide-bandgap semiconductor device research and commercialization[3], and extensive application prospects at transportation, aerospace, new energy systems, and defense technologies[4,5]. Nevertheless, the inherent trade-off between specific on-resistance(R_{on}) and breakdown voltage in unipolar devices, coupled with intrinsically limited short-circuit withstand time (SCWT)，has imposed fundamental constraints on their widely application.

In order to improve the performance of 4H-SiC MOSFET by reducing on-resistance and improving short-circuit withstand capability, this paper proposes a Trench-gate Dual-mode Composited Transistor (T-DCT) in which integrates a Vertical U-groove MOSFET (VUMOS) and a bipolar junction transistor (BJT) in parallel, hence, the conductivity modulation effect is introduced in T-DCT during conduction to effectively reduce the R_{on}. Concurrently, the reduced cell channel density leads to a lower saturation current, thereby exhibiting enhanced SCWT. The static characteristics, switching behavior, and short-circuit performance of the T-DCT are systematically investigated by TCAD simulator and compared with that of VUMOS.

II. DEVICE STRUCTURE AND MECHANISM

A. Device Structure

4H-SiC VUMOS structure incorporating a current spreading layer (CSL) and p-floating region is illustrated in Fig. 1(a). Compared to conventional silicon VUMOS devices, a higher doping concentration and a deeper n-type CSL

positioned beneath the p-body region is incorporated to reduce the resistance of the conduction region, which accelerate rapid current spreading into the n-drift region during conduction. Additionally, a p-floating region is formed at the trench bottom to mitigate high electric fields, thereby reducing the electric field intensity within the gate oxide layer at the trench bottom.

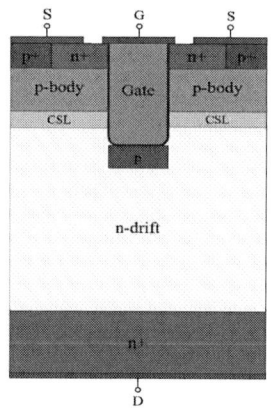

(a)4H-SiC VUMOS structure with current spreading layer and p-floating region

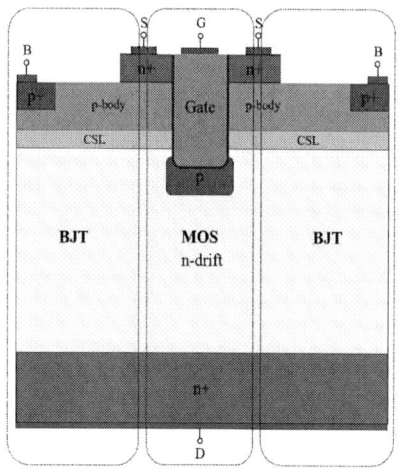

(b)4H-SiC T-DCT structure

Fig. 1. 4H-SiC T-DCT and 4H-SiC VUMOS structures with current spreading layer and p-floating region

This work is supported by Shaanxi Province "Two Chain" Integration Key Project of China (Grant No. 2021LLRH-02), and Qin Chuang yuan Scientist and Engineer Team Building Project (Grant No.2023KXJ-189).

979-8-3315-1110-4/25 $31.00 © 2025 IEEE

The 4H-SiC DCT structure is shown in Fig. 1(b), in the existing 4H-SiC VUMOS cell structure, the n+ source region and p+ ohmic contact region are first separated, and then a control electrode B is set on the etched surface of p- body region, thus a BJT is introduced in parallel with the VUMOS to form a double-control-electrode (i.e. G and B) which separately controls the turn-on and turn-off of the VUMOS and BJT. This approach reduces the R_{on} and on-state loss of the T-DCT, decreases channel density to increase SCWT, and eliminates the influence of the body diode on device characteristics. Additionally, a p-floating region with a side arc shape at the trench bottom are formed by multiple ion implantation steps, thereby reducing the electric field strength in the gate oxide layer at the trench corners and solving gate oxide breakdown issues.

B. Mechanism

The control signal timing of T-DCT is shown in Fig. 2. When $V_{DS} > 0$, $V_{GS}=V_{BS}=0$, no inversion layer channel at the surface of the p-body region along the vertical sidewalls of the trench. Under this bias condition, the p-body/CSL junction becomes reverse-biased and supports the applied forward voltage.

When $V_{DS}>0$ $V_{GS}>V_{th}$, an inversion layer channel forms at surface of the p-body region along the vertical sidewalls of the trench, enabling electron injection from the source through the channel into the n⁻ drift region. Simultaneously, BJT is triggered conduction by applying an enough base current ($I_B=I_{BM}$) to B of T-DCT, the emitter junction(i.e. the p-body/n+ source junction) is forward-biased, allowing n+ emitter region (i.e. the n+ source region of T-DCT) electrons inject into the p-base region (i.e. the p-body region of T-DCT). While the minority carriers will be recombined to produce hole current, majority electrons are swept into the n⁻ drift region and then into n+ collector regions (i.e. the n+ drain region of T-DCT) as collector current. A large base current induces a strong conductivity modulation effect across the p-body and n- drift regions, effectively lowering the conduction resistance.

The VUMOS and BJT in parallel interaction forms parallel conduction paths, thereby the T-DCT operates in dual mode and has enhanced conductivity modulation during conduction, which enables R_{on} reduction.

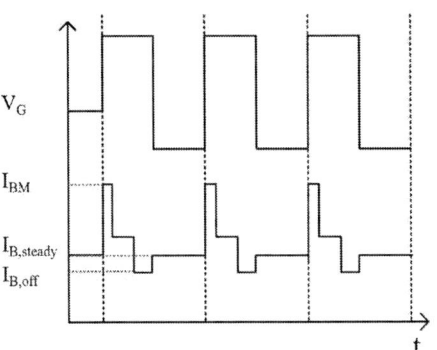

Fig. 2. The control signal timing of T-DCT

When $V_{GS}<V_{th}$, inversion layer channel at the surface of the p-body region along the vertical sidewalls of the trench gradually disappears, resulting in T-DCT turn-off.

As shown in Fig. 2, a negative base current is applied to B of T-DCT prior to MOS channel turn-off to rapidly extract the stored charge within the device, enabling accelerated turn-off transition. And higher the base current, the faster of T-DCT turns off.

III. CHARACTERISTIC ANALYSIS

The structural models of the T-DCT and VUMOS with CSL and p-floating region are established by TCAD simulator, and the static, switching and short circuit characteristics are simulated and compared. Key structural parameters are listed in Table 1.

TABLE I.   MAIN STRUCTURAL PARAMETERS OF TWO DEVICES S

Regions and parameters		VUMOS	T-DCT
p-floating region	concentration	2.5×10^{17} cm⁻³	2.5×10^{17} cm⁻³
	thickness	0.6μm	0.6μm
trench	depth	1.4μm	1.4μm
between S and B	space	-	5μm
n+ source region	width	2μm	6μm

A. Breakdown Characteristic

Fig. 3 shows the breakdown characteristic curves of the proposed T-DCT and VUMOS with same vertical structural parameters. At 10 nA of leakage current, the V_{BR} of VUMOS and T-DCT is 2181 V and 2228 V respectively.

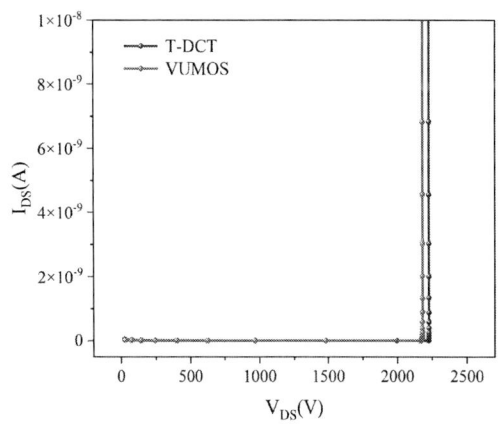

Fig. 3. Breakdown characteristic curves of two devices

B. Conduction Characteristic

Fig. 4 shows the conduction characteristic curves of the proposed T-DCT and VUMOS. At the rated current of 60 A, the R_{on} for T-DCT and VUMOS are 11.7 mΩ and 23.5 mΩ respectively, demonstrating a 50% reduction in R_{on} achieved by the T-DCT. And two curves intersect. Below the crossover point of curves, the T-DCT exhibits lower on-resistance; and above the crossover point of curves, the T-DCT exhibits lower saturation current compared to VUMOS. This shows that T-DCT has lower conduction loss and longer SCWT.

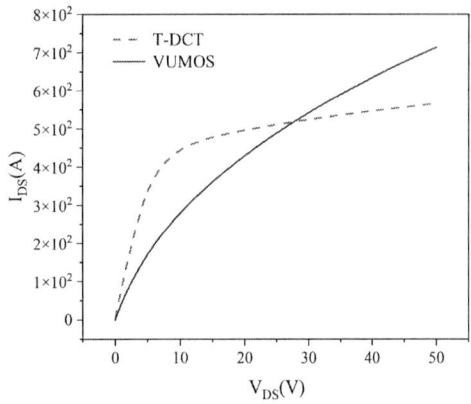

Fig. 4. Conduction characteristic curves of two devices

Fig. 5 shows the current density distributions of the proposed T-DCT and VUMOS during conduction. Seen from Fig. 5(a), electrons flow through the MOS channel into the CSL, establishing a spreading current distribution into the drain to form the drain current. Seen from Fig. 5(b), when the base current activates the BJT, both holes in the p-base region and electrons in the n-collector region coexist as charge carriers, inducing the conductivity modulation effect. The device from unipolar mode into bipolar mode, which significantly reduces the on-state resistance through carrier density enhancement.

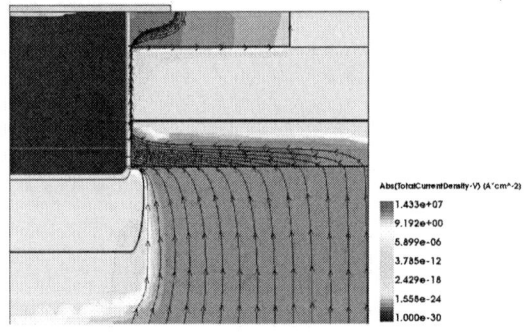

(a)4H-SiC VUMOS with p-floating structure

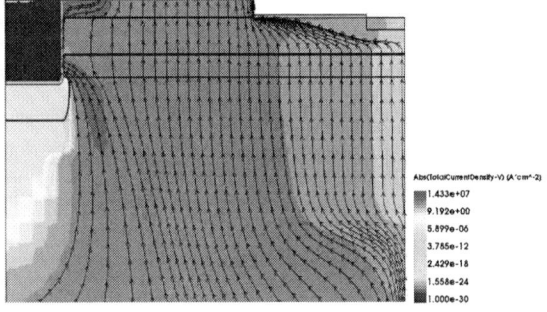

(b)4H-SiC T-DCT structure

Fig. 5. Current density distributions of the two devices

C. Switching Characteristic

Fig. 6 (a)shows the turn-on characteristics of the proposed T-DCT and VUMOS. Under the test conditions of V_{DD}=1000

V, R_G=2Ω, R_L=16.6Ω, V_{GS}=+18V, VUMOS exhibits a turn-on time of 75.6 ns with corresponding switching energy loss of 0.51 mJ. In contrast, the proposed T-DCT demonstrates superior performance under identical voltage and load conditions with additional control parameters (I_{BM}=7A and $I_{B,steady}$=2A), achieving reduced turn-on time of 59.1ns and lower energy consumption of 0.37mJ, indicating excellent turn-on characteristics.

Fig. 6 (b) shows the turn-off characteristics of proposed T-DCT and VUMOS. Under conditions of V_{DD}=1000V, I_{DS}=60A, R_G=2Ω, R_L=16.6Ω, V_{GS}=-2V and I_D=60A, VUMOS shows a turn-off time of 85.5 ns with energy dissipation of 0.24 mJ. The proposed T-DCT, operating under identical voltage and load conditions with additional a reverse bias current $I_{B,off}$=-5A, achieves improved turn-off performance characterized by shortened turn-off time of 74.0ns while maintaining comparable energy loss of 0.26mJ.

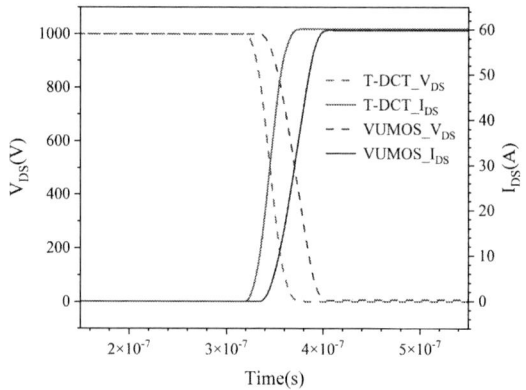

(a) Turn-on curve

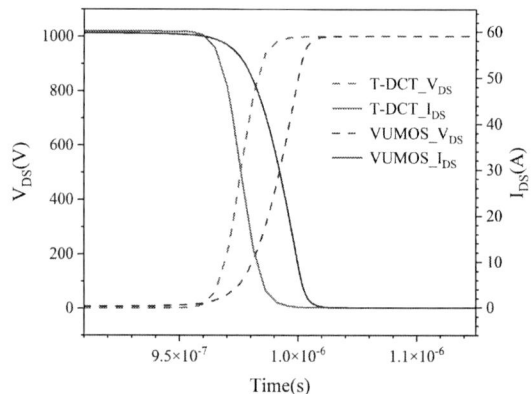

(b) Turn-off curve

Fig. 6. Switching characteristics of the two devices

D. Short-circuit characteristic

Fig.7 shows the short-circuit current curves of the proposed T-DCT and VUMOS under the condition of 1000V bus voltage and -5A base current . The VUMOS exhibits a SCWT of 5.4μs, while the T-DCT demonstrates a SCWT of 7μs, representing a 29.6% improvement in robustness. If the reverse base current is further increased, the SCWT of the T-DCT will increase.

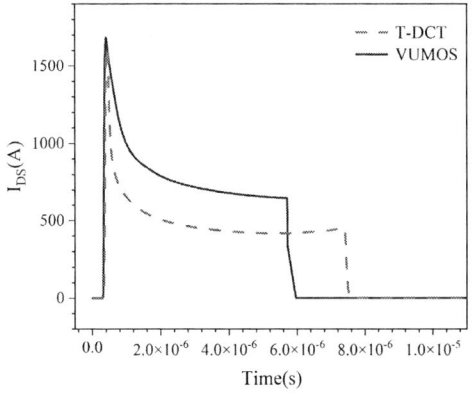

Fig. 7. Comparison of short-circuit current curves of two devices

This significantly enhanced performance of T-DCT can be attributed to the lower saturation current combined with its negative base current extraction capability under short-circuit fault conditions.

IV. KEY PARMETERS OPTIMIZATION

A. Space between S and B

Fig.8 shows the influences of the space, S_B, between n+ source and p+ regions on the conduction characteristics of T-DCT. The on-state resistance, R_{on}, increases slightly and the saturation current, I_{Dsat}, reduces observably with increase of S_B. When S_B increases from 5μm to 7μm, the R_{on} increases from 11.7 mΩ to 13.7 mΩ under I_{DS}=60A condition, while I_{Dsat} accordingly decreases. If S_B decreases to 3μm, R_{on} reduces to 10.6 mΩ·cm² at I_{DS}=60A accompanied by I_{Dsat} observably increases. By trading off the lower R_{on} and lower I_{Dsat}, Optimal performance is achieved when S_B is 5μm.

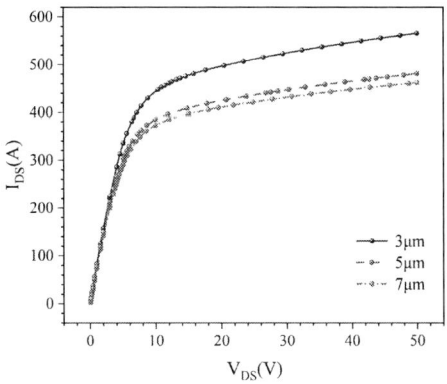

Fig. 8. Distance between S and B electrodes

B. Thickness of n+ source region

Fig.9 shows the influences of the n+ source region widths ,W_{n+}, on the conduction characteristics of T-DCT. The results show that increase of W_{n+} lead to increased R_{on} and observably reduced I_{Dsat}. When W_{n+} increases from 6μm to 7μm, the R_{on} increases from 11.7 mΩ to 13.5 mΩ under I_{DS}=60A condition. If W_{n+} decreases to 5μm, R_{on} reduces to 10.8 mΩ·cm² at I_{DS}=60A accompanied by I_{Dsat} observably increases. Optimal performance is achieved at W_{n+}=6μm, with the lower R_{on} and favorable I_{Dsat}.

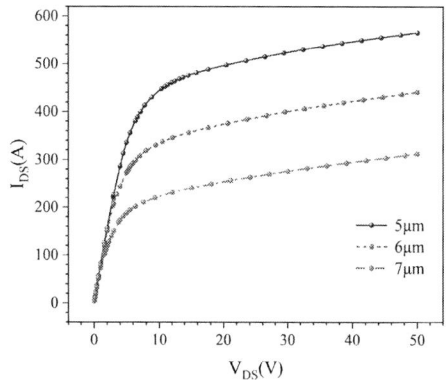

Fig. 9. Thickness of n+ source region

V. CONCLUSIONS

By independently controlling the source and p-body region of the 4H-SiC VUMOS, the induced BJT activates a conductivity modulation effect during conduction. The simulation results show that, compared to the VUMOS with a SCWT of 5.4μs, the T-DCT demonstrates a 29.6% improvement in SCWT (7.0μs) and a 50% reduction in R_{on}.

REFERENCES

[1] Yu Q, Chen W, Huang J, et al. A Novel 4H-SiC Power MOSFET with Source-side Poly-Si/SiC Heterojunctions for Single-Event Effects Hardening[J]. Micro and Nanostructures, 2024: 208064.

[2] Lin W-C, Yu W-C, Chen B-R, et al. Investigation of the time dependent gate dielectric stability in SiC MOSFETs with planar and trench gate structures[J]. Microelectronics Reliability, 2023, 150: 115141.

[3] Guan L, Wang C. Study on the Characteristics and Mechanism of High-Voltage 4H-SiC Dual-Mode Composite Transistor (DCT)[C]. 2024 3rd International Symposium on Semiconductor and Electronic Technology (ISSET), 2024: 134-137.

[4] Bai Z, Tang X, Xie S, et al. Investigation on single pulse avalanche failure of 1200-V SiC MOSFETs via optimized thermoelectric simulation[J]. 2021, 68(3): 1168-1175.

[5] Zhang L, Yang D, Ren L, et al. The Method of the SiC MOSFET Replacing the Si IGBT in the Traditional Power Electronics Converter without Redesigning the Main Circuit and the Driver Circuit[J]. Energy Engineering, 2021, 118(4): 1155-1170.

Super-Junction IGBT with Adaptive Hole Channel around Stepped Trench Gate for Low On-State Voltage and Low Turn-Off Loss

Xuelei Zhou
College of Electrical Engineering
Zhejiang University
Hangzhou, China
zhouxuelei@zju.edu.cn

Hengyu Wang
College of Electrical Engineering
Zhejiang University
Hangzhou, China
wanghengyu@zju.edu.cn

Yifan Wang
Electric Power Research Institute
State Grid Zhejiang Electric Power Co.,ltd
Hangzhou, China
39627979@qq.com

Kuang Sheng
College of Electrical Engineering
Zhejiang University
Hangzhou, China
shengk@zju.edu.cn

Abstract—This paper presents a novel super-junction insulated gate bipolar transistor (SJ-IGBT) with an adaptive hole channel around the stepped trench gate. During on-state operation, the hole channel is off due to the potential barrier caused by the positive gate bias. As a result, hole extraction from the drift region is effectively suppressed. Conductance modulation in both the n-pillar and p-pillar regions are enhanced. And hence, the on-state voltage (V_{ON}) can be largely reduced. During the turn-off transient, the hole channel is on as the gate bias is at a low level. As a result, the hole channel region facilitates rapid extraction of holes from the P-pillar, which leads to shorter turn-off time and reduced turn-off loss (E_{OFF}). The performance of the proposed SJ-IGBT and conventional super-junction IGBT structures are studied and compared through TCAD simulation. The results show that the trade-off between V_{ON} and E_{OFF} have been significantly improved. In addition, this paper analyzes the internal depletion region variation during turn-off and components of turn-off loss of SJ-IGBT.

Keywords—insulated gate bipolar transistor (IGBT), super-junction (SJ), on-state voltage, turn-off loss

I. INTRODUCTION

The insulated gate bipolar transistor (IGBT) is one of the most important power devices for efficient and compact power conversion in various applications. An essential research direction of IGBT lies in improving the trade-off between turn-off loss (E_{OFF}) and on-state voltage (V_{ON}). Super-junction (SJ) IGBTs leverage N-pillar and P-pillar regions to achieve lower V_{ON} and E_{OFF} compared to conventional IGBTs [1]. Specifically, the floating P-pillar SJ-IGBT (FP-SJ-IGBT) exhibits lower V_{ON} than the conventional SJ-IGBT (Con-SJ-IGBT) by isolating the P-pillar region from the P-base region, enhancing conductance modulation in the drift region during forward conduction. However, FP-SJ-IGBTs suffers from increased turn-off time and turn-off loss. Recent advancements have achieved improved V_{ON}-E_{OFF} trade-offs through structural designs featuring variable P-pillar region connections — separated during on-state and connected during turn-off [2]- [6]. Additionally, enhancements in emitter region design have accelerated electron extraction during turn-off, reducing E_{OFF} [7], [8]. However, those devices incorporate additional

This paper is supported by State Grid Zhejiang Electric Power Co.,Ltd. Technology Project (5211DS21N00A).

components, such as dual emitters, P-MOS, or dual gates with separate gate signals. The inclusion of more components will substantially increase the difficulty of device fabrication as well as the complexity of device control. In addition, the mechanism of SJ-IGBT during turn-off and factors affecting the E_{OFF} have not been fully investigated.

This paper proposes a new 1.2kV SJ-IGBT with an adaptive hole channel (AHC-SJ) around the stepped trench gate to improve the V_{ON}-E_{OFF} trade-off. TCAD simulations have been conducted to study the device mechanism and performance. Furthermore, comparative investigation of static and dynamic charateristics with other conventional SJ-IGBTs are performed. Further, the internal mechanism and loss of SJ-IGBT during turn-off have been deeply investigated.

II. STRUCTURE AND MECHANISM

Fig. 1 illustrates the schematic cross-sectional structures of Con-SJ-IGBT, FP-SJ-IGBT, and the proposed SJ-IGBT. The proposed SJ-IGBT introduces stepped gate corners and a nearby P-type hole channel, facilitating P-pillar region to P-base region connection during turn-off. Additionally, the N-type carrier storage layer is positioned above the N-pillar region alongside the hole channel region. Key device parameters used in TCAD simulations are listed in Table I [9]. Physical models including Effective Intrinsic Density (OldSlotboom), the Shockley – Read – Hall recombination, Auger recombination, impact ionization models, high-field saturation mobility, and Philips unified mobility are used.

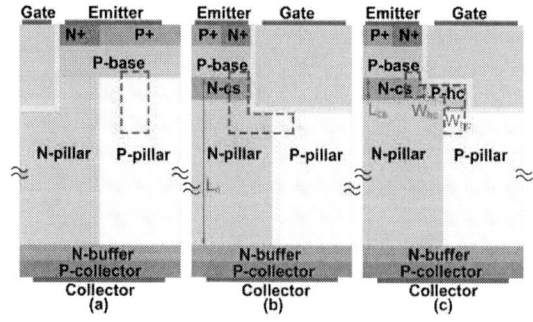

Fig. 1. Schematic cross-sectional structures of (a) the Con-SJ-IGBT, (b) the FP-SJ-IGBT and (c) the AHC-SJ-IGBT.

TABLE I. KEY DEVICE PARAMETERS

Parameters	Definitions	Value
N_{base} (cm^{-3})	P-base doping	2×10^{17}
L_n (μm)	N-pillar length	95
N_p (cm^{-3})	P-pillar&N-pillar doping	4×10^{15}
W_p (μm)	N-pillar&N-pillar width	2
N_{buffer} (cm^{-3})	N-buffer doping	5×10^{16}
W_{hc} (μm)	P-hc width	0.5
N_{hc} (cm^{-3})	P-hc doping	2×10^{17}
L_{cs} (μm)	N-cs length	0.5
N_{cs} (cm^{-3})	N-cs doping	5×10^{16}

During forward blocking, FP-SJ-IGBT concentrates electric fields at trench gate corners and causing premature device breakdown. In AHC-SJ-IGBT, stepped gate design and hole channel region at trench gate corners mitigate electric field aggregation, achieving breakdown voltages comparable to Con-SJ-IGBT.

In on-state, since the gate voltage of the AHC-SJ is high, the hole channel next to the gate is off to form a depletion region, blocking the connection between the P-pillar region and the P-base region. The hole channel region and the N-type carrier storage layer form a hole barrier that prevents holes from being extracted, thus enhancing the level of conductance modulation in the drift region. Fig. 2(a) shows schematic diagrams of the energy band from the P-base region to the P-pillar region during forward conduction. The AHC-SJ forms a potential barrier to prevent the hole from moving. Therefore, AHC-SJ is able to obtain the same high level of conductance modulation as FP-SJ, resulting in a lower V_{ON}.

During the turn-off transient, the hole channel of AHC-SJ is on due to low gate voltage, and the holes are able to move from the P-pillar region to the P-base region quickly through the hole channel, thus its turn-off speed is faster. Fig. 2(b) shows schematic diagrams of the energy band from the P-base region to the P-pillar region during turn-off transient. AHC-SJ is the same as Con-SJ in that the hole current can enter the P-base region from the P-pillar region more easily, which reduces the turn-off time and turn-off loss. FP-SJ has a large potential barrier in the energy band, so it needs to accumulate holes increasing potential at the bottom of the gate in order to conduct, which increases the turn-off time and turn-off loss. In addition, FP-SJ-IGBT forms a structure similar to a carrier storage layer because the top region of the N-pillar is not depleted from the P-pillar region. Therefore, the depletion region expands for a longer time during turn-off, which can lead to a larger Miller plateau with larger E_{OFF}.

III. SIMULATION RESULTS AND DISCUSSION

Fig. 3 presents breakdown characteristics of Con-SJ, FP-SJ, and AHC-SJ. AHC-SJ achieves a breakdown voltage of approximately 1475 V, slightly lower than Con-SJ's 1493 V due to higher doping concentration in the carrier storage layer. FP-SJ's BV is 1248 V, with AHC-SJ improving by around 230 V, attributed to hole channel region design and stepped trench gate alleviating electric field aggregation. In addition, the three devices have similar leakage currents in blocking state, indicating that the introduction of the new structure does not increase the leakage current.

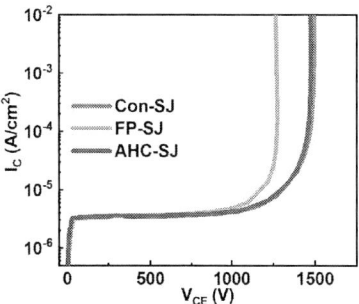

Fig. 3. Breakdown characteristics of the SJ-IGBTs.

Forward conduction characteristics in Fig. 4 show AHC-SJ achieving low V_{ON} similar to FP-SJ, significantly lower than Con-SJ due to enhanced conductance modulation. At I_C=100 A/cm^2, the V_{ON} of AHC-SJ is 1.312 V, and FP-SJ's V_{ON} is 1.225 V, Con-SJ's V_{ON} is 2.889 V. Comparing with Con-SJ, AHC-SJ's V_{ON} is 54.6% lower.

Fig. 5 illustrates the carrier distributions of the three devices at I_C=100 A/cm^2 in positive conduction. The left panel shows the electron density distribution and the right panel shows the hole density distribution. It can be seen that Con-SJ has lower hole density and electron density in part of the drift region, which indicates a lower level of conductance modulation, resulting in larger V_{ON}. Con-SJ has partial unipolar conduction due to the connection of the P-pillar region to the P-base region, where the P-pillar region holes are easily extracted. In contrast, it can be seen that FP-SJ and AHC-SJ have high hole density and electron density in the whole drift region, which indicates a higher level of conductance modulation, resulting in lower V_{ON}. The hole channel of the AHC-SJ is closed at V_G=15 V in forward conduction and prevents hole extraction together with the N-type carrier storage layer.

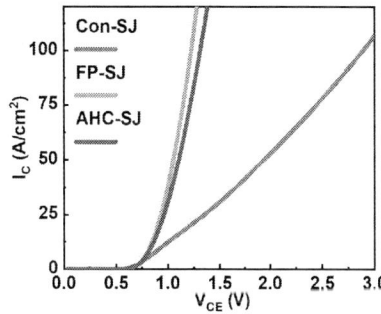

Fig. 4. Forward conduction characteristics of the SJ-IGBTs at V_G=15V.

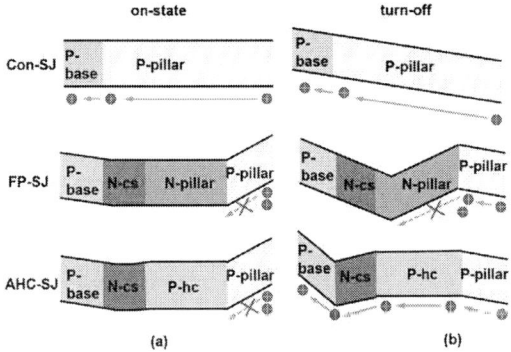

Fig. 2. During (a) on-state operation conduction and (b) turn-off transient, schematic diagrams of energy band for Con-SJ, FP-SJ and AHC-SJ in the red square in Fig. 1.

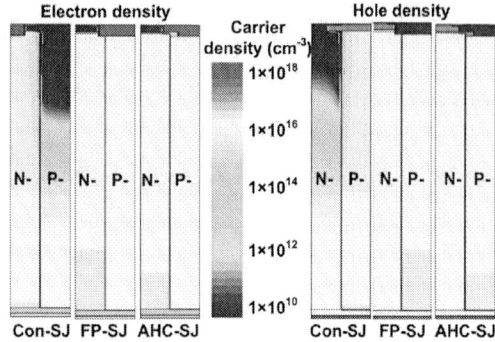

Fig. 5. Electron density and hole density distribution in the on-state with $I_C=100A/cm^2$.

Fig. 6(b) shows the inductive turn-off waveform for the circuit shown in Fig. 6(a). Fig. 6(b) divides the turn-off transient into three phases: a slow rise in V_{CE} in the first phase, a rapid rise in V_{CE} in the second phase and a fall in I_C in the third phase. Fig. 7 shows the gate voltage and the electron current of the emitter current during turn-off. The turn-off loss mainly starts from the rise of voltage and ends with the fall of current..

For SJ-IGBT turn-off, it is divided into three main parts:

When entering the Miller platform period, the voltage starts to rise slowly and then sharply. Voltage slowly rising stage is also to the unipolar conductive transition process, the voltage rises rapidly for the depletion zone expansion process. Fig. 8 shows the hole density distribution of the devices during these two phases. Considering the symmetry of the super-junction structure, the higher electron density is on the other side of the higher hole density.

The first part: Voltage rises slowly. Particularly for SJ-IGBTs, a transverse electric field is generated between the N-pillar region and P-pillar region during turn-off, which leads to a slow rise in V_{CE}. The minority carrier on both sides of the depletion region are extracted. The SJ-IGBT begins to switch from bipolar conduction to unipolar conduction. Since the top part of the Con-SJ is unipolar conductive when it is forward conducting, a transverse electric field is already formed at the beginning of this stage, and thus the voltage can rise faster, as shown in Fig. 8(a). In FP-SJ, there is no corresponding P-region next to the N-cs region, and thus the depletion region expands between the N-cs region and the P-base region at the initial stage. It takes longer time for the FP-SJ to make the depletion region expands to P-pillar region to generate a transverse electric field between N-pillar and P-pillar, as in Fig. 8(b). For the AHC-SJ, in the early stage of this phase, a transverse electric field is formed between the P-hp region and the N-cs region, and thus the AHC-SJ could start unipolar conductivity earlier than the FP-SJ as shown in Fig. 8(c). Therefore, compared to FP-SJ, AHC-SJ is able to effectively reduce the time of this phase, which efficiently reducing E_{OFF1}.

The second part: Voltage rise rapidly. The narrow depletion region between the N-pillar region and the P-pillar region, formed during the first part, gradually widens from top to bottom, as shown on the right side of Fig. 8. At this stage the device has entered the unipolar type of conduction, with a low level of electrons and a a high level of holes in the P-pillar region. Therefore AHC-SJ could effectively removes

the holes from the P-pillar region through the P-hp region as shown in Fig. 2(b), thus realizing a rapid voltage rise. In Con-SJ, the depletion layer extension is slow due to the electron current injection at the emitter, as illustrated by the red dashed circle in Fig. 7. Therefore, AHC-SJ is able to achieve the fastest depletion region extension and the fastest voltage rise, which results in the reduction of E_{OFF2}.

The third part: Current drops. It is known that the IGBT turn-off current is mainly due to the electron current of the MOS, the hole current (emitter side) and the electron current (collector side) caused by the excess carrier sweeping out due to the depletion layer expansion [10]. As illustrated by the red dashed circle in Fig. 7, when the current starts to decrease, MOS electron current of Con-SJ and FP-SJ still exists. Thus the current decreases slowly due to influence of the gate voltage. The emitter current of AHC-SJ is only hole current in Fig. 7. Therefore, the AHC-SJ has the fastest current drop and has the smallest E_{OFF3}.

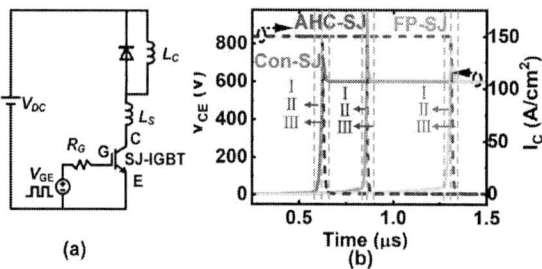

Fig. 6. (a) Simulation circuit with $R_G=10\Omega$, stray inductance $L_S=10nH$, and inductive load $L_C=1mH$. (b) Inductive turn-off waveforms at 300K for different SJ-IGBTs.

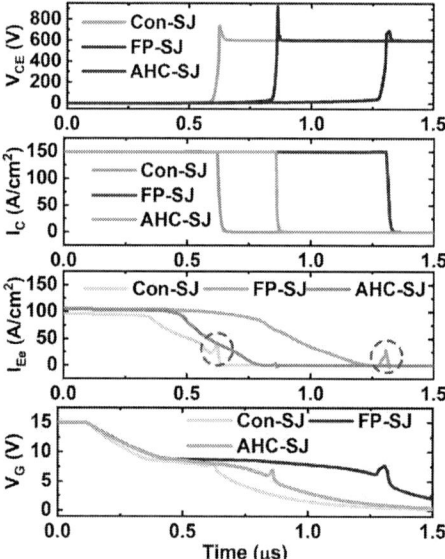

Fig. 7. Detailed turn-off waveforms at 300K for SJ-IGBTs.

Fig. 9 shows the power waveforms at turn-off for the three devices. Overall, in terms of turn-off loss, the AHC-SJ has the lowest turn-off loss in both the second and third part. The first part of the loss is unavoidable due to the bipolar conductivity during on-state, but there is a huge improvement compared to the FP-SJ. The turn-off loss, E_{OFF} is calculated uniformly from 2% V_{DC} to 1% I_L at a load

979-8-3315-1110-4/25 $31.00 © 2025 IEEE

current I_L of 150 A/cm². The E_{OFF} of the Con-SJ is 1.69 mJ/cm² and that of the FP-SJ is 3.47 mJ/cm². The E_{OFF} of the AHC-SJ is 1.59 mJ/cm², which is 54% lower than that of the FP-SJ.

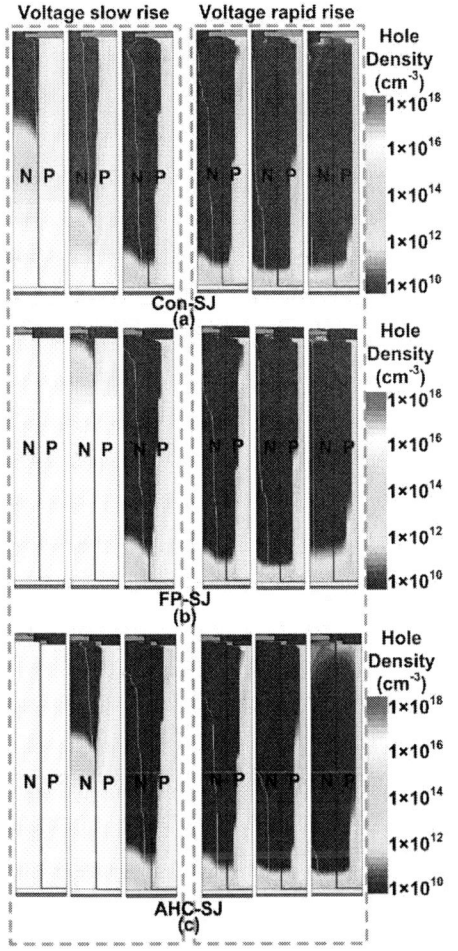

Fig. 8. Hole density distribution of Con-SJ, FP-SJ and AHC-SJ in turn-off transient when V_{CE} rise.

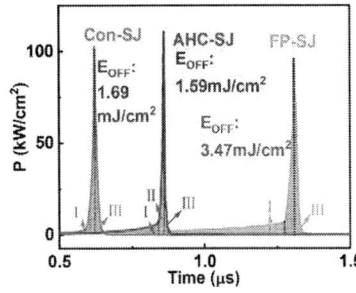

Fig. 9. Turn-off loss power waveforms for Con-SJ, FP-SJ and AHC-SJ.

Fig. 10 demonstrates V_{ON}-E_{OFF} trade-off by varying doping concentration in the P-collector region. The AHC-SJ achieves a superior trade-off compared to Con-SJ and FP SJ. At V_{ON} of 1.41 V, AHC-SJ's E_{OFF} is 1.2 mJ/cm², 58% lower than FP-SJ's 2.9 mJ/cm². At E_{OFF} of 1.92 mJ/cm², AHC-SJ's V_{ON} is 1.27 V, 49% lower than Con-SJ's 2.54 V.

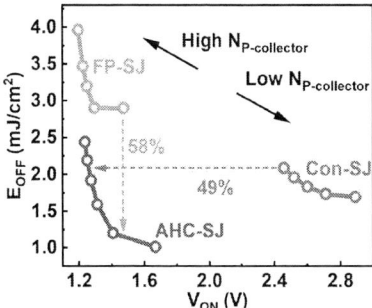

Fig. 10. E_{OFF}-V_{ON} trade-off relationship for different SJ-IGBTs.

IV. CONCLUSION

This paper proposes and simulates a novel SJ-IGBT with an adaptive hole channel. The proposed SJ-IGBT features an adaptive hole channel around the stepped trench gate. Such a channel is off during on-state and on during turn-off process. Therefore, the hole extraction can be largely suppressed or facilitated, respectively. The simulation results show that the V_{ON} of AHC-SJ-IGBT can be reduced by 49% compared to that of Con-SJ-IGBT, while the E_{OFF} of AHC-SJ-IGBT is 55% lower than that of FP-SJ-IGBT. In addition, this paper investigates the turn-off mechanism of the SJ-IGBT and explains the internal reason for the voltage rise of the SJ-IGBT during turn-off.

REFERENCES

[1] Vinod Kumar Khanna, "Novel IGBT Design Concepts, Structural Innovations, and Emerging Technologies," in Insulated Gate Bipolar Transistor IGBT Theory and Design , IEEE, 2003, pp.499-544.

[2] J. Huang, H. Huang and X. B. Chen, "Simulation Study of a Low ON-State Voltage Superjunction IGBT With Self-Biased PMOS," in IEEE Transactions on Electron Devices, vol. 66, no. 7, pp. 3242-3246, July 2019.

[3] M. Huang, B. Gao, Z. Yang, L. Lai and M. Gong, "A Carrier-Storage-Enhanced Superjunction IGBT With Ultralow Loss and On-State Voltage," in IEEE Electron Device Letters, vol. 39, no. 2, pp. 264-267, Feb. 2018.

[4] J. Wei, S. Zhang, X. Luo, D. Fan and B. Zhang, "Low Switching Loss and EMI Noise IGBT With Self-Adaptive Hole-Extracting Path," in IEEE Transactions on Electron Devices, vol. 68, no. 5, pp. 2572-2576, May 2021.

[5] J. Wei, M. Zhang and K. J. Chen, "Design of Dual-Gate Superjunction IGBT towards Fully Conductivity-Modulated Bipolar Conduction and Near-Unipolar Turn-Off," 2020 32nd International Symposium on Power Semiconductor Devices and ICs (ISPSD), Vienna, Austria, 2020, pp. 498-501.

[6] W. Saito and S. -I. Nishizawa, "Switching Noise-Loss Trade-Off Improvement of SJ-IGBTs," 2022 IEEE 34th International Symposium on Power Semiconductor Devices and ICs (ISPSD), Vancouver, BC, Canada, 2022, pp. 53-56.

[7] L. Li et al., "650V Planar Anode Gate Super-junction IGBT with Superior Von-Eoff Trade-off," 2022 IEEE 16th International Conference on Solid-State & Integrated Circuit Technology (ICSICT), Nangjing, China, 2022, pp. 1-3

[8] Z. Wu, Y. He, X. Ge and D. Liu, "A Fast Switching Superjunction IGBT with Segmented Anode NPN," 2021 IEEE 16th Conference on Industrial Electronics and Applications (ICIEA), Chengdu, China, 2021, pp. 633-636

[9] TCAD Sentaurus Device Manual, Synopsys, Inc., Mountain View, CA, USA, 2013.

[10] Y. Onozawa, M. Otsuki and Y. Seki, "Investigation of carrier streaming effect for the low spike fast IGBT turn-off," 2006 IEEE International Symposium on Power Semiconductor Devices and IC's, Naples, Italy, 2006, pp. 1-4.

Effect of Voltage Probes on the Characterisation of Switching Processes in Wide-bandgap Semiconductor Devices

Yishun Yan
School of Electrical Engineering & Automation
Harbin Institute of Technology
Harbin, China
24S006068@stu.hit.edu.cn

Lurenhang Wang
School of Electrical Engineering & Automation
Harbin Institute of Technology
Harbin, China
23S006077@stu.hit.edu.cn

Xuchong Cai
School of Electrical Engineering & Automation
Harbin Institute of Technology
Harbin, China
24S006020@stu.hit.edu.cn

Mingcheng Ma
School of Electrical Engineering & Automation
Harbin Institute of Technology
Harbin, China
24B906005@stu.hit.edu.cn

Yanchen Pan
School of Electrical Engineering & Automation
Harbin Institute of Technology
Harbin, China
24b906019@stu.hit.edu.cn

Dianguo Xu
School of Electrical Engineering & Automation
Harbin Institute of Technology
Harbin, China
xudiang@hit.edu.cn

Abstract—The high switching speed of wide-bandgap (WBG) semiconductor devices puts higher requirements on the measurement accuracy and intrusion effects of voltage probes. This paper analyses the influence of input capacitance and parasitic inductance of voltage probes on the characterization of switching processes from the point of view of measurement accuracy and the intrusion effect of probes. It analyses the reason why the latest opto-isolated probes measure the gate-source voltage affecting the switching process. The GaN HEMT-based double-pulse experiment verifies the correctness of the theoretical analysis. The performance of different voltage probes and different methods for measuring gate-source voltage and drain-source voltage of WBG devices is systematically compared. The results indicate that conventional high-voltage differential probes fail to meet the testing requirements due to insufficient bandwidth and excessive parasitic inductance. Opto-isolated probes demonstrate superior performance with high bandwidth and low input capacitance. However, forward-direction measurements are critical to avoid the reverse connection's grounding capacitance impact on switching. This study provides guidance on probe selection and measurement methodologies for accurate characterization of dynamic characteristics in WBG semiconductor devices.

Keywords—voltage probes, wide-bandgap semiconductor devices, measurement

I. INTRODUCTION

Third-generation wide-bandgap (WBG) semiconductor devices can completely meet the diversified demands for switching devices in power electronics with their excellent high switching speed and low conduction loss characteristics [1]-[3]. However, the extremely high switching speed challenges the measurement limits of oscilloscopes and probes, which seriously affects the dynamic characterization of third-generation semiconductor power devices, hampers the research of gate driving strategies, and restricts the exhaustive application of power devices [4]. On the one hand, the high switching speed puts higher requirements on the probe bandwidth, while the parasitic inductance in the measurement probe triggers oscillation problems, resulting in switching edges that cannot be accurately captured, which seriously affects the measurement accuracy of voltage overshoot and switching loss [5], [6]; on the other hand, due to the very low junction capacitance of WBG power devices, the intrusion effect of the probe reshapes the impedance of the

circuit under test [7], affecting the switching transients and even leading to instability. Regarding the measurement-induced parasitic effects, references [8] and [9] proposed some generalized methods for high-speed and high-precision measurement of SiC devices. In addition, for the same voltage probe, the measurement accuracy and the size of the intrusion effect are not the same for different measurement methods. Therefore, it is necessary to investigate the factors affecting the measurement accuracy, reduce the intrusion effect of the probe, and achieve the in-situ accurate characterization of the switching characteristics of WBG semiconductor devices, to give full play to the excellent characteristics of the third-generation WBG semiconductor devices [10].

In this paper, the influence of the parasitic inductance of the probe on the measurement accuracy and the load effect of the probe are analyzed from the perspectives of self-inductance and mutual inductance. The performance of different voltage probes and different measurement methods for measuring gate-source and drain-source voltages of WBG semiconductor devices is systematically compared.

II. THEORETICAL ANALYSIS

The principles of the double-pulse test, which includes parasitic parameters, are shown in Fig. 1. Q_H and Q_L are GaN devices. To achieve a full range of dynamic characterization of the switching process of the power device needs to simultaneously measure the gate-source voltage of Q_L, the drain-source voltage of Q_L, and the drain-source voltage of Q_H.

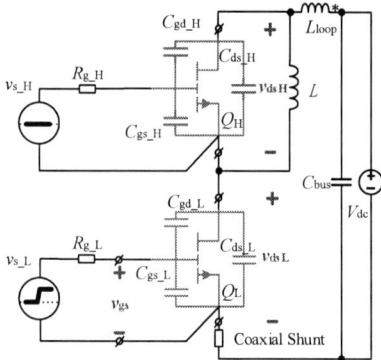

Fig. 1. Principles of the double-pulse test.

979-8-3315-1110-4/25 $31.00 © 2025 IEEE

Typical circuit models for passive and differential probes are shown in Fig. 2 and Fig. 3. If the buffer input impedance is considered to be infinite, both passive and differential probes can be approximated as equivalent to the circuit shown in Fig. 4.

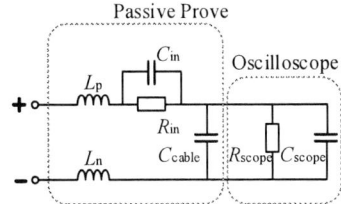

Fig. 2. Typical circuit models for passive probes.

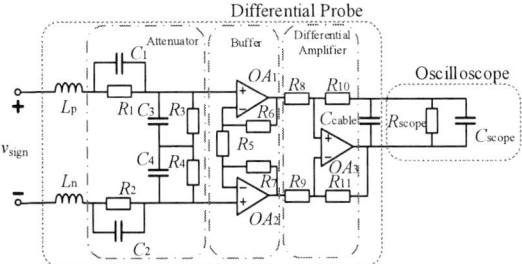

Fig. 3. Typical circuit models for differential probes.

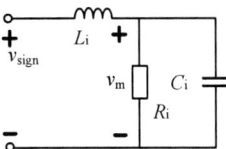

Fig. 4. Voltage probe simplified equivalent circuit.

A. Voltage Probe Bandwidth Requirements

The ringing frequency due to LC oscillation in the circuit is higher than the bandwidth required to measure the switching time. The ringing frequency is given by the following equation:

$$f_r = \frac{1}{2\pi\sqrt{L_{\text{loop}} C_{\text{oss}}}} \tag{1}$$

The output junction capacitance C_{oss} of GaN SYSTEM GS66508B is about 60 pF. The parasitic inductance of the power loop L_{loop} is 10 nH, so f_r is 205.5 MHz. To guarantee a measurement accuracy of not less than 98%, the bandwidth of the probe should be more than five times the bandwidth of the measured signal. Therefore, a voltage probe with a bandwidth of not less than 1 GHz and able to withstand higher voltages is required.

B. Voltage probe intrusion effect

The voltage probe test circuit is shown in Fig. 5. v_c is the signal source. R is the signal source resistor. C is the load capacitor. v_{sign} is the target signal. R_i and C_i are the input resistor and input capacitor of the voltage probe, respectively. Measuring voltage with a voltage probe is equivalent to introducing extra resistance, capacitance, and inductance into the circuit under test, which will reshape the original circuit impedance, i.e., there is an intrusion effect of the voltage

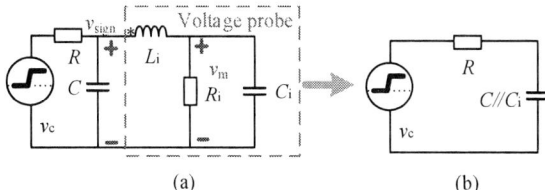

Fig. 5. Voltage probe test circuit.

probe. The target signal is altered by the addition of the probe. The voltage probe input resistance is of the order of MΩ level, so the effect of the probe input resistance can be approximately ignored. As is shown in Fig. 5 (b). The degree of change a in the target signal due to the probe intrusion effect can be expressed as:

$$a = \frac{C_i}{C_s} \tag{2}$$

It can be seen that the smaller C_i is, the smaller the intrusion effect of the probe is. The equivalent load capacitance corresponding to v_{gs} and v_{ds} can be approximated by the input capacitance C_{iss} and output capacitance C_{oss} of the device, respectively, and is usually much larger than C_i. Therefore, when measuring v_{ds} with the voltage probe, the effect on the switching process is extremely small. However, when using a voltage probe to measure v_{gs}, although the effect on v_{gs} is very small, it is amplified by the transconductance of the device, which significantly affects the drain current and thus the switching process.

C. Impact of Probe Inductance

The front end of the voltage probe will inevitably have parasitic inductance L_i due to the measurement leads, etc.

On the one hand, considering the role of self-inductance, when v_{sign} changes very fast, the parasitic inductance L_i resonates with the input capacitance C_i, and the resonance frequency is:

$$f_s = \frac{1}{2\pi\sqrt{L_i C_i}} \tag{3}$$

Typically, C_i is extremely small and although the parasitic inductance L_i varies depending on lead length, this resonant frequency will still be far higher than the device switching ringing. Therefore, the effect of parasitic inductive self-inductance is often not observed in low bandwidth probes.

On the other hand, considering the role of mutual inductance, the drain current i_d slope during GaN switching is as high as 10 A/ns, and the EMI problem is extremely serious. The measurement loop and the power loop inductance are coupled to each other, so the measurement result v_m will have oscillations with the same frequency as the power loop's ringing frequency. v_m can be expressed as:

$$v_m = v_{\text{sign}} - L_i \frac{di_i}{dt} - M \frac{di_d}{dt} \tag{4}$$

$$M = k\sqrt{L_i L_{\text{loop}}} \tag{5}$$

Where i_i is the internal current of the probe, and k is the coupling coefficient. The placement of the probe during measurement affects the coupling coefficient, which affects the magnitude of the mutual inductance voltage, significantly affecting the measurement results.

D. Proper use of Opto-isolated probes

Opto-isolated probes are widely used in WBG semiconductor testing due to their high bandwidth, low input capacitance, and high common-mode rejection ratio. Because of the coaxial cable at the front end of their probes, there is minimal interference with measurements due to external magnetic fields. Fig. 6 shows the coaxial cable schematic. However, the coaxial cable makes the positive and negative terminals of the opto-isolated probes lose their symmetry like a differential probe. Therefore, when using an opto-isolated probe to measure v_{gs}, the center conductor should be connected to the gate, at which time the capacitance of the probe to the ground is C_c, otherwise the capacitance of the probe to the ground is C_s, which is much larger than C_c, and seriously affects the switching process. Define the center conductor connected to the gate for forward measurement, and vice versa for reverse measurement.

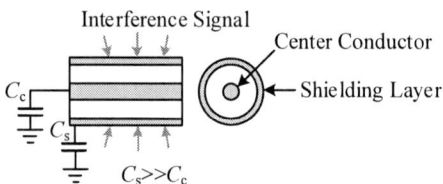

Fig. 6. The coaxial cable schematic.

III. EXPERIMENTAL VERIFICATION

A double-pulse test platform, as shown in Fig. 7, is established to compare the measurement accuracy and intrusion effect of gate-source and drain-source voltages of WBG semiconductor devices using different voltage probes and different measurement methods. The oscilloscope used is Tektronix MSO68B with a bandwidth of 2.5 GHz. i_d is measured using a coaxial shunt SSDN-414-10 with a bandwidth of 2 GHz. The parameters of the voltage probes are shown in TABLE I. The power device is a GaN SYSTEM GS66508B. The turn-on gate resistance is 10 Ω, and the turn-off gate resistance is 4.7 Ω. Tests are performed at a bus voltage of 400 V and a load current of 20 A. The intrusion effect of the probes is reflected by i_d. To avoid additional errors introduced by simultaneous measurements with multiple probes, only one voltage probe was used for each experiment. All probe timings have been calibrated.

A. Measurement of the gate-source voltage

v_{gs} measured using different voltage probes are shown in Fig. 8. From the figure, it can be seen that using a high-voltage differential probe to measure v_{gs} slows down the turn-on process. Low voltage differential probe, high voltage

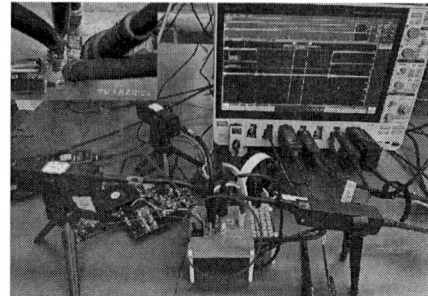

Fig. 7. Double-pulse test platform.

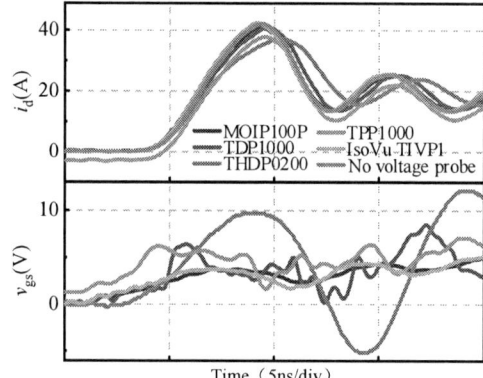

Fig. 8. v_{gs} measured using different voltage probes.

differential probe, and passive probe have oscillations, and since i_d does not oscillate abnormally, this indicates that oscillations do not actually occur on the gate-source side, but rather that there are oscillations within the probe, resulting in poor measurement accuracy. In addition, v_{gs} waveforms measured by the low-voltage differential probe and the passive probe have both low-frequency and high-frequency oscillations. The low-frequency oscillation frequency is the same as the switch ringing frequency, which confirms the previous analysis. The high-frequency oscillations cannot be observed in the high-voltage differential probe because the bandwidth is only 200 MHz. At the same time, since both the coaxial shunt and passive probe share a common ground through the oscilloscope, a large ground loop is formed. This causes ground potential drift in the oscilloscope and introduces measurement errors. Therefore, the simultaneous use of multiple passive probes should be avoided during measurements.

Fig. 9 shows the waveforms of v_{gs} measured in different ways using the opto-isolated probes. As can be seen from i_d waveforms, the different ways of measuring v_{gs} do not affect the switching process, but they do affect the

TABLE I. TYPICAL VOLTAGE PROBE PARAMETERS

Probe Type	Probe Model	Bandwidth (Hz)	Voltage class (V)	Input impedance	Isolated or not
Passive probe	TPP1000	1 G	300V	3.9 p//10 M	N
High Voltage Differential Probe	THDP0200	200 M	1500	<2 p//10 M	Y
Low Voltage Differential Probe	TDP1000	1 G	42	<1 p//1 M	Y
Opto-isolated probe	IsoVu TIVP1	1 G	2500	2.4 p//40 M	Y
Opto-isolated probe	SigOFIT MOIP100P	1 G	2500	1 p//20.52 M	Y

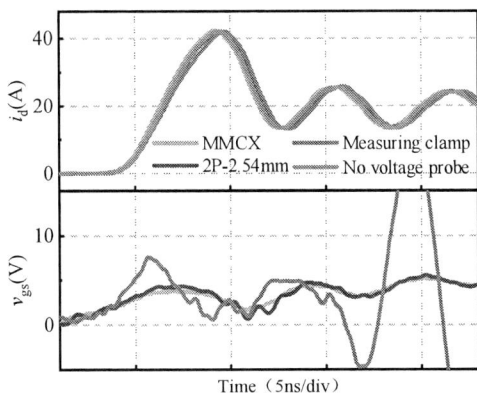

Fig. 9. v_{gs} measured in different ways by Opto-isolated probe.

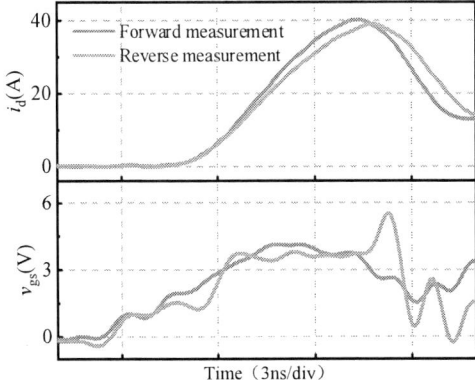

Fig. 10. v_{gs} measured in forward and reverse directions.

measurement accuracy, with the use of the MMCX giving the best results. The introduction of additional measurement parasitic inductance causes strong oscillations inside the probe when measured with a measuring clamp.

Fig. 10 shows the waveforms of v_{gs} measured in forward and reverse directions using the opto-isolated probe. It can be found that v_{gs} rises in the reverse direction is slower than that in the forward direction, which is due to the increased capacitance of the reverse measurement probe to the ground, and due to the transconductance relationship of the device turn-on process, the drain current also rises slower, seriously affecting the switching process.

B. Measurement of the drain-source voltage

The waveforms of v_{dsL} during turn-on and turn-off measured by different voltage probes are shown in Figs. 11 and Fig. 12, respectively. During turn-on process, the high-voltage differential probe exhibited an anomalous voltage undershoot of -69.9 V in the measured v_{dsL} waveform, which should not exist theoretically. During turn-off process, the opto-isolated probe measured a v_{dsL} overshoot of 18.6 V, while the high-voltage differential probe produced erroneous oscillations with a measured overshoot of 130.0 V.

The waveforms of v_{dsH} during turn-on and turn-off measured by different voltage probes are shown in Figs. 13 and Fig. 14, respectively. Similar measurement artifacts were observed, with the high-voltage differential probe showing incorrect undershoot (-154.1 V) and overshoot (149.9 V). Compared to v_{dsL} measurements, the voltage measurement

errors were more pronounced in v_{dsH} measurements due to the poor common-mode rejection ratio (CMRR) of the high-voltage differential probe. The conventional high-voltage differential probe, with its limited bandwidth and significant measurement parasitic inductance, fails to meet the testing requirements for WBG semiconductor devices. In contrast, the opto-isolated probe demonstrates superior measurement accuracy and high CMRR, enabling precise characterization of both voltage transition timing and amplitude. The measurement results from both models of opto-isolated probes demonstrate good overall consistency.

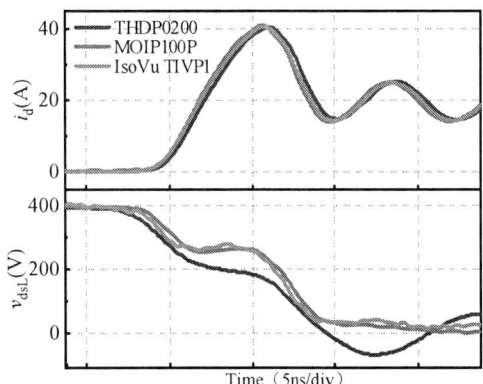

Fig. 11. v_{dsL} measured using different voltage probes at turn on process.

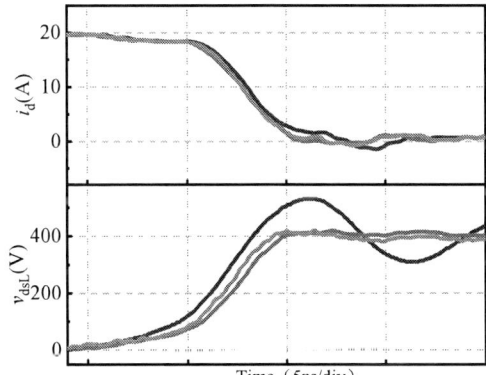

Fig. 12. v_{dsL} measured using different voltage probes at turn off process.

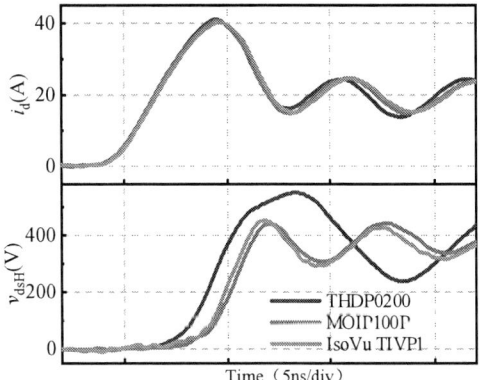

Fig. 13. v_{dsH} measured using different voltage probes at turn on process.

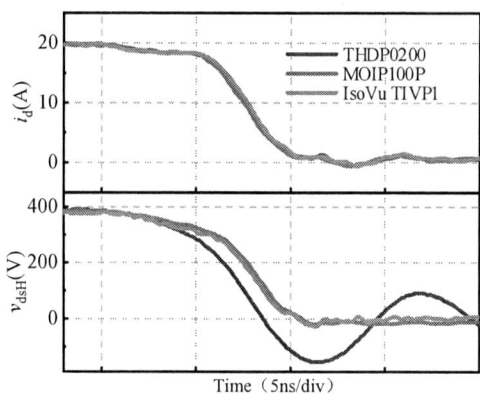

Fig. 14. v_{dsH} measured using different voltage probes at turn off process.

IV. CONCLUSION

The correct selection and use of voltage probes is crucial for the accurate measurement of high-frequency transient voltages in WBG semiconductor devices. WBG semiconductor testing requires voltage probes with high accuracy, high voltage, high bandwidth, and low parasitic effects. Traditional high-voltage differential probes are no longer able to meet the testing requirements due to their low bandwidth and large parasitic inductance. The passive probe has no isolation function, and multi-point grounding will lead to oscilloscope potential drift. When measuring, the parasitic inductance at the test point should be reduced as much as possible to improve the measurement accuracy, and avoid using lead measurement. Opto-isolated probes have good performance in WBG semiconductor testing, but it should be noted that the test needs to be in the forward direction of the measurement, otherwise, the intrusion effect of the probe will seriously affect the switching process. These findings provide practical guidelines for selecting and deploying voltage probes to enable accurate in-situ dynamic characterization of WBG semiconductor devices.

REFERENCES

[1] J. P. Kozak *et al.*, "Stability, Reliability, and Robustness of GaN Power Devices: A Review," in *IEEE Transactions on Power Electronics*, vol. 38, no. 7, pp. 8442-8471, July 2023.

[2] K. J. Chen *et al.*, "GaN-on-Si Power Technology: Devices and Applications," in *IEEE Transactions on Electron Devices*, vol. 64, no. 3, pp. 779-795, March 2017.

[3] R. Ramachandran and M. Nymand, "Experimental Demonstration of a 98.8% Efficient Isolated DC–DC GaN Converter," in *IEEE Transactions on Industrial Electronics*, vol. 64, no. 11, pp. 9104-9113, Nov. 2017.

[4] C. F. Tong et al.,"Challenges in switching waveforms measurement for a high-speed switching module,"in Proc. IEEE ECCE, Sep. 2015, pp. 6175–6179.

[5] M. Grubmüller, B. Schweighofer and H. Wegleiter, "Development of a Differential Voltage Probe for Measurements in Automotive Electric Drives," in *IEEE Transactions on Industrial Electronics*, vol. 64, no. 3, pp. 2335-2343, March 2017

[6] Z. Zeng, X. Zhang, and L. Miao,"Inaccuracy and instability: Challenges of SiC MOSFET transient measurement intruded by probes,"in IEEE ECCE Asia, May 2019, pp. 1–6.

[7] Z. Zeng, X. Zhang, F. Blaabjerg and L. Miao, "Impedance-Oriented Transient Instability Modeling of SiC mosfet Intruded by Measurement Probes," in IEEE Transactions on Power Electronics, vol. 35, no. 2, pp. 1866-1881, Feb. 2020.

[8] H. Sakairi, T. Yanagi, H. Otake, N. Kuroda and H. Tanigawa, "Measurement Methodology for Accurate Modeling of SiC MOSFET Switching Behavior Over Wide Voltage and Current Ranges," in *IEEE Transactions on Power Electronics*, vol. 33, no. 9, pp. 7314-7325, Sept. 2018

[9] Z. Zhang, B. Guo, F. F. Wang, E. A. Jones, L. M. Tolbert and B. J. Blalock, "Methodology for Wide Band-Gap Device Dynamic Characterization," in *IEEE Transactions on Power Electronics*, vol. 32, no. 12, pp. 9307-9318, Dec. 2017

[10] C. F. Tong *et al.*, "Challenges in switching waveforms measurement for a high-speed switching module," *2015 IEEE Energy Conversion Congress and Exposition (ECCE)*, Montreal, QC, Canada, 2015, pp. 6175-6179

Performance improvement of SiC n-LTT by semi-through via structure

Yulei Zhang
Department of Electronic Engineering
Xi'an University of Technology
Xi'an, China
zhangyulei_zl@163.com

Xi Wang*
Department of Electronic Engineering
Xi'an University of Technology
Xi'an, China
wangxii@xaut.edu.cn

Xuhui Pu
Department of Electronic Engineering
Xi'an University of Technology
Xi'an, China
puxuhuii@163.com

Jichao Hu
Department of Electronic Engineering
Xi'an University of Technology
Xi'an, China
jchu@xaut.edu.cn

Hongbin Pu
Department of Electronic Engineering
Xi'an University of Technology
Xi'an, China
puhongbin@xaut.edu.cn

Yuan Yang
Department of Electronic Engineering
Xi'an University of Technology
Xi'an, China
yangyuan@xaut.edu.cn

Abstract—Compared to Si thyristors, SiC thyristors exhibit superior performance. As an essential category of SiC thyristors, SiC Light-Triggered Thyristors (LTTs) with a blocking voltage of 15 kV or below typically employ a p-type long-base structure. In this work, n-type long-base SiC LTTs were simulated and analyzed, revealing that p-type substrate SiC n-LTT exhibits high conduction resistance, while n-type substrate SiC n-LTT encounters difficulties in turning on. To address these issues, this work proposes a semi-through via structure for n-type long-base SiC LTTs to reduce conduction resistance, with numerical simulations conducted using TCAD software. The proposed structure effectively mitigates the high conduction resistance in p-type substrate SiC n-LTT and resolves the turn-on difficulty in n-type substrate SiC n-LTT, with a detailed analysis of these challenges. Simulation results indicate that the n-type substrate SiC n-LTT with a semi-through via structure can be successfully turned on. At an anode current density of 100 A/cm², it exhibits a forward conduction voltage drop of 4.0 V, which is lower than the 4.2 V observed in the p-type substrate SiC n-LTT. In addition, its on-state resistance is lower than that of the p-type long-base SiC LTT, thereby significantly improving the device's overall conduction performance.

Keywords—SiC, thyristor, via structure, on-resistance

I. INTRODUCTION

Compared to silicon (Si), silicon carbide (SiC), as a third-generation semiconductor material, offers several advantages, including a wide bandgap, high critical breakdown electric field strength, high carrier saturation drift velocity, high thermal conductivity, and excellent chemical stability. These properties make SiC highly suitable for manufacturing high-power power electronic devices. Thanks to its superior material characteristics, SiC thyristors exhibit higher voltage resistance, greater current density, and better high-temperature tolerance than their silicon counterparts[1]. In recent years, with the continuous maturation of SiC material manufacturing technologies and the gradual improvement of device fabrication processes, theoretical and experimental research on SiC thyristors has made continuous breakthroughs[2-5]. As one of the important categories of SiC thyristors, SiC LTTs offer advantages such as optical isolation, simple drive, and strong electromagnetic interference resistance. Despite the numerous excellent characteristics of SiC materials, due to process limitations, SiC LTTs with a blocking voltage of 15 kV or below generally adopt p-type long-base regions with relatively low carrier mobility, and the use of n-type long-base regions has not significantly enhanced device performance[6,7]. In SiC, for blocking voltages below

15 kV, the substrate must be sufficiently thick to support the wafer. In 2022, the first experimental demonstration of a waffle-substrate n-channel IGBT in 4H-SiC is reported[8], The feasibility of this approach is demonstrated by fabricating a 10-kV class n-channel IGBT. This also presents an improved approach for the realization of n-type long-base Light Triggered Thyristors with blocking voltages below 15 kV.

In this work, a SiC n-type long-base LTT with a semi-through via structure on an n-type substrate is proposed. By creating semi-through vias on the bottom surface of the substrate, the adverse effects of the substrate on the device's conduction characteristics are effectively mitigated, leading to a significant improvement in the device's performance.

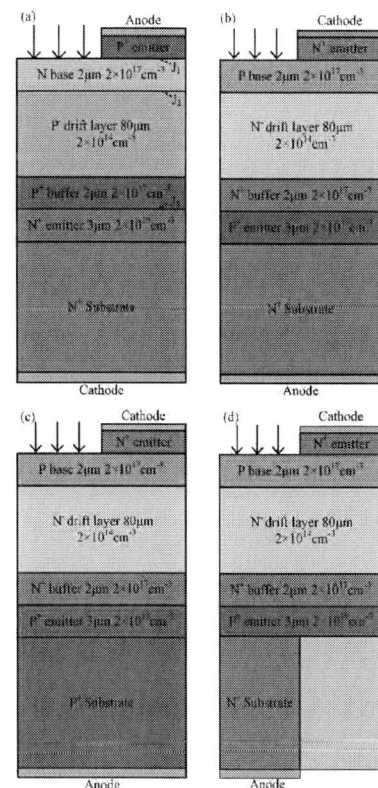

Fig. 1. Structure of the 4H-SiC thyristor. (a) n-substrate SiC p-LTT, (b) n-substrate SiC n-LTT, (c) p-substrate SiC n-LTT, (d) Semi-through via SiC n-LTT.

II. DEVICE STRUCTURE

The 10 kV SiC LTT structures are depicted in Fig. 1. The three thyristors share similar structural parameters, including layer thickness and doping concentration, differing only in doping type. The cross-sectional schematic of the n-substrate SiC p-LTT is shown in Fig. 1(a), while the SiC n-LTT structures are illustrated in Figs. 1(b) and 1(c). Additionally, the SiC n-LTT with a semi-through via structure on an n-type substrate is presented in Fig. 1(d). The top emitter region has a thickness of 3.0 μm and a doping concentration of 2.0×10^{19} cm^{-3}. Directly beneath it, the short base region has a thickness of 2.0 μm and a doping concentration of 2.0×10^{17} cm^{-3}. The blocking base region, responsible for voltage blocking capability, has a thickness of 80 μm and a doping concentration of 2.0×10^{14} cm^{-3}. Below the blocking base, a buffer layer is introduced with a thickness of 2.0 μm and a doping concentration of 2.0×10^{17} cm^{-3}. Finally, the bottom emitter region has a thickness of 3.0 μm and a doping concentration of 2.0×10^{19} cm^{-3}.

III. RESULTS AND DISCUSSION

To compare the SiC p-LTT and the SiC n-LTT with an n-type long base region, three different SiC LTT structures were simulated in this work: the n-substrate SiC p-LTT, as shown in Fig. 1(a), the SiC n-LTT with an n-type substrate, as shown in Fig. 1(b), and the SiC n-LTT with a p-type substrate, as shown in Fig. 1(c). The numerical simulations of these three structures were performed using simulation software, incorporating various physical models, including the bandgap model, carrier mobility model, recombination model, impact ionization model, incomplete ionization model, and complex refractive index model[9-11]. The electrical characteristics of the proposed novel SiC LTT with a semi-through via structure were investigated, and the simulation results were analyzed and discussed.

The forward I-V characteristics of the SiC p-LTT, SiC n-LTT with an n-type substrate, and SiC n-LTT with a p-type substrate under ultraviolet illumination (365 nm, 1.0 W/cm²) at an anode bias of 10 V are presented in Fig. 2. As seen in Fig. 2, under a 10 V anode bias, the n-type substrate SiC n-LTT remains non-conducting, while at an anode current density of 100 A/cm², the forward conduction voltage drop of the SiC p-LTT is 3.4 V, whereas that of the p-type substrate SiC n-LTT is 4.2 V. The SiC p-LTT exhibits lower conduction resistance compared to the p-type substrate SiC n-LTT, while the n-type substrate SiC n-LTT only reaches an anode current density of 0.003 A/cm² under a 10 V anode bias, indicating that it remains in an off-state. Compared to the n-type substrate SiC n-LTT and p-type substrate SiC n-LTT, the SiC p-LTT demonstrates a lower conduction resistance and reduced forward voltage drop at an anode current density of 100 A/cm².

Due to the significant disparity in electron and hole mobility in SiC, with electron mobility at 1040 cm²/V·s and hole mobility at 118 cm²/V·s, electrons exhibit a significantly higher mobility than holes. Therefore, theoretically, the SiC p-LTT is expected to have a higher conduction resistance than the SiC n-LTT. However, as shown in Fig. 2, which presents the forward I-V characteristics of the three different structures, this expected trend is not observed—the SiC p-LTT does not exhibit higher conduction resistance than the SiC n-LTT. To elucidate this discrepancy, this study analyzes the reasons behind the failure of the n-type substrate SiC n-LTT to turn on properly and the higher conduction resistance of the p-type

substrate SiC n-LTT compared to the SiC p-LTT. The electric potential distributions of the n-type substrate SiC n-LTT and p-type substrate SiC n-LTT are illustrated in Fig. 3. Under an anode bias, the p-type substrate SiC n-LTT is already in conduction, with the potential gradually decreasing from the anode to the cathode. In contrast, the n-type substrate SiC n-LTT maintains a high potential at the n-type substrate, while the n-type long base region remains at a low potential, preventing the SiC LTT from turning on.

Structural analysis of the n-type substrate SiC n-LTT and p-type substrate SiC n-LTT reveals that the n-type substrate SiC n-LTT remains non-conducting under a 10 V bias because the n-type substrate forms a pn junction with the p-emitter region, effectively introducing a reverse-biased pn junction diode in series with the SiC LTT. Furthermore, Al is commonly employed as the p-type dopant in SiC devices, and due to its high ionization energy of 0.24–0.26 eV in SiC[12,13], the injection efficiency of the pn junction between the p-emitter and the n-buffer layer is significantly reduced. Consequently, the n-type substrate SiC n-LTT fails to turn on under a 10 V anode bias, as represented by the equivalent structure in Fig. 4(a). Conversely, the higher conduction resistance of the p-type substrate SiC n-LTT under a 10 V bias is attributed to the p-type substrate, which effectively introduces a large series resistance into the SiC LTT. This results in an elevated conduction resistance for the p-type substrate SiC n-LTT under a 10 V anode bias, preventing the device from exhibiting the low conduction resistance characteristic of the SiC n-LTT, as depicted by the equivalent structure in Fig. 4(b).

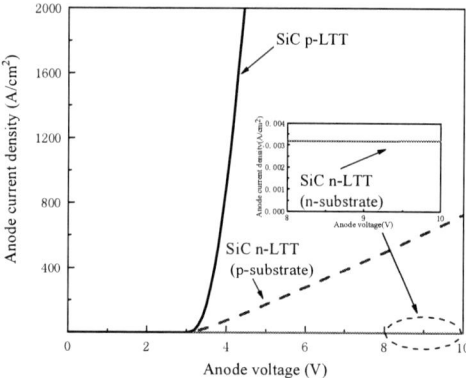

Fig. 2. Forward turn-on curves of SiC LTTs.

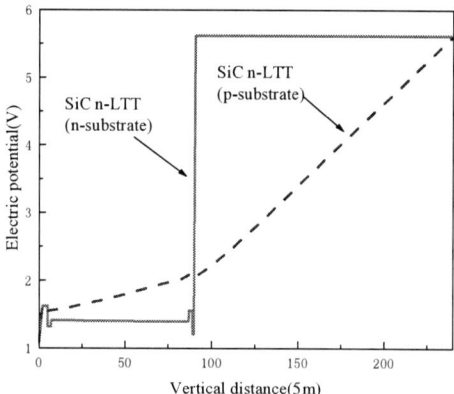

Fig. 3. Electric potential of SiC n-LTT with an n-substrate and a p-substrate.

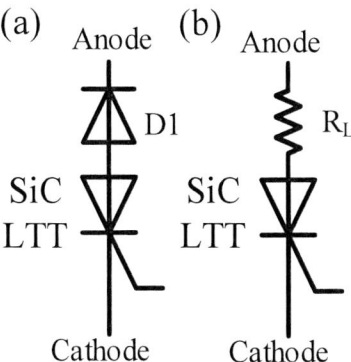

Fig. 4. Equivalent structures (a) SiC n-LTT with an n-substrate, (b) SiC n-LTT with a p-substrate.

Through the simulation analysis of the SiC n-LTT with an n-type substrate and the SiC n-LTT with a p-type substrate, it was found that the non-conducting behavior of the n-type substrate SiC n-LTT and the high on-state resistance of the p-type substrate SiC n-LTT result from structural limitations in both devices. To address these issues, this paper proposes an SiC n-LTT with a semi-through via structure on an n-type substrate, as illustrated in Fig. 1(d). A via is introduced into the n-type substrate, and metal is used to partially replace the n-substrate, thereby mitigating the adverse impact of the substrate on the device's conduction characteristics.

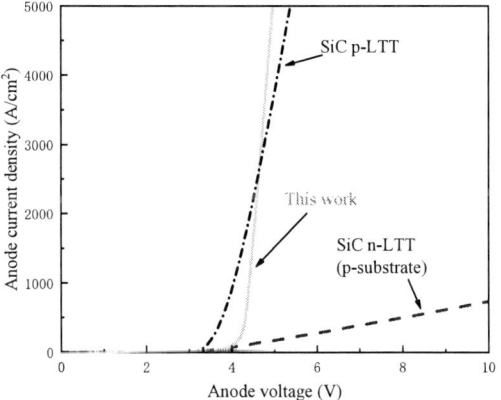

Fig. 5. Forward turn-on curve.

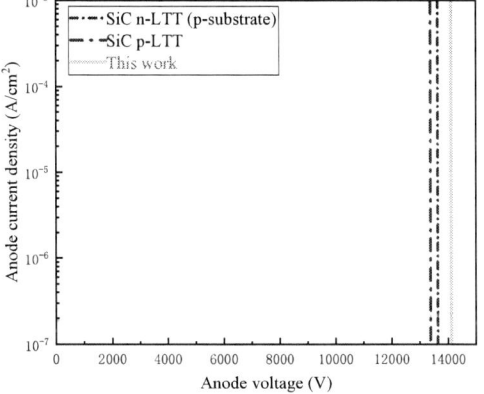

Fig. 6. Device forward blocking voltage.

The forward I-V characteristics of the SiC n-LTT with a semi-through via structure, measured under a 10 V anode bias and illuminated with 365 nm ultraviolet light at an intensity of 1.0 W/cm², are shown in Fig. 5. As observed, at an anode current density of 100 A/cm², the forward conduction voltage drop of the SiC n-LTT with a semi-through via structure is 4.0 V, while that of the p-type substrate SiC n-LTT is 4.2 V, and that of the SiC p-LTT is 3.4 V. Although the semi-through via structure reduces the forward conduction voltage drop compared to the p-type substrate SiC n-LTT, it remains 0.6 V higher than that of the SiC p-LTT. This is attributed to the high ionization energy of Al in SiC, which leads to a significantly low injection coefficient at the junction between the anode p-emitter and the n-buffer layer. Furthermore, the pn junction between the n-type substrate and the p-emitter region remains present, as the semi-through via structure does not eliminate this junction. Consequently, the SiC n-LTT with a semi-through via structure requires a higher turn-on voltage than the SiC p-LTT.

The on-state resistance of the SiC n-LTT with an n-type substrate and a semi-through via structure, as shown in Fig. 5, is lower than that of both the SiC p-LTT and the SiC n-LTT with a p-type substrate after the device turns on. This is because the electron mobility in SiC is significantly higher than the hole mobility, leading to a lower on-state resistance in the n-type long base region compared to the p-type long base region. The semi-through via structure of the SiC n-LTT effectively mitigates the issue of increased on-state resistance caused by the p-type substrate and prevents the SiC n-LTT from failing to turn on due to the n-type substrate. As a result, the on-state resistance of the SiC n-LTT with a semi-through via structure is significantly reduced.

Furthermore, the forward blocking characteristics of the three different SiC LTT structures are shown in Fig. 6. The forward blocking voltage of the SiC n-LTT with a semi-through via structure and an n-type substrate is 14127 V, whereas the SiC p-LTT and the SiC n-LTT with a p-type substrate exhibit forward blocking voltages of 13392 V and 13632 V, respectively. The proposed semi-through via structure in the SiC n-LTT effectively reduces the on-state resistance while maintaining excellent forward blocking capability.

IV. CONCLUSION

A 10 kV-class 4H-SiC LTT featuring an n-type substrate and an innovative semi-through via structure is proposed in this work. The introduced structure effectively addresses several critical limitations of conventional SiC LTTs, including the high on-state resistance of devices with p-type long-base regions, the increased conduction resistance caused by p-type substrates in n-LTTs, and the failure of n-type substrate n-LTTs to turn on under low anode bias. For LTTs with blocking voltages up to 15 kV, this study systematically investigates the conduction behavior associated with different substrate and base region configurations. It is found that the p-type substrate introduces considerable series resistance, severely degrading the conduction performance, while the pn junction between the n-type substrate and the p-emitter remains reverse-biased, which inhibits carrier injection and prevents device turn-on. To overcome these issues, a novel n-type substrate n-LTT incorporating a semi-through via structure is developed, and its static characteristics are evaluated through numerical simulations. The results show that the proposed device exhibits a forward conduction voltage drop of 4.0 V at an anode current density of 100 A/cm², which is 0.6 V higher than that of the p-type LTT and 0.2 V lower than that of the n-LTT with a p-type substrate. During forward conduction, the pn junction between the n-type substrate and the p-emitter remains reverse-biased, resulting in a higher turn-on voltage compared to the p-type LTT. However, once the device is turned on, the semi-through via structure effectively reduces the on-state resistance due to enhanced electron transport in the n-type base, thereby improving overall conduction performance.

ACKNOWLEDGMENT

This work was supported in part by the National Natural Science Foundation of China under Grants 62004161, 62174135, and 52377197, in part by the Scientific Research Project of Shaanxi Provincial Education Department under Grant 23JP101, and in part by the Shaanxi Innovation Capability Support Project under Grant 2021TD-25.

RFERENCES

[1] Q. Zhanga,A. Agarwal, C. Capell, L. Cheng,M. O'Loughlin, A. Burk, et al. "SiC super GTO thyristor technology development: Present status and future perspective," 2011 IEEE Pulsed Power Conference, Chicago, IL, USA, 2011, pp. 1530-1535.

[2] Z. Li, K. Zhou, L. Zhang, X. Xu, L. Li, J. Li, and G. Dai, "A simple multistep etched termination technique for 4H-SiC GTO thyristors," Solid-State Electronics, vol. 151, pp. 1-5, 2019.

[3] H. Long, H. Xu, H. Wang, N. Ren, Q. Guo and K. Sheng, "Single-Mask Implantation-Free Technique Based on Aperture Density Modulation for Termination in High-Voltage SiC Thyristors," IEEE Transactions on Electron Devices, vol. 68, no. 3, pp. 1181-1184, March 2021.

[4] T. Yang, X. Li, Y. Wang, P. Yao and R. Yue, "12.5 kV SiC Gate Turn Off Thyristor With Trench-Modulated JTE Structure," IEEE Transactions on Electron Devices, vol. 69, no. 3, pp. 1258-1264, March 2022.

[5] X. Wang, H. Pu, Q. Liu, L. An, X. Tang and Z. Chen, "Demonstration of 4H-SiC Thyristor Triggered by 100-mW/cm2 UV Light," IEEE Electron Device Letters, vol. 41, no. 6, pp. 824-827, June 2020.

[6] I. G. Ivanov, A. Henry, and E. Janzén, "Ionization energies of phosphorus and nitrogen donors and aluminum acceptors in 4H silicon carbide from the donor-acceptor pair emission," Phys. Rev. B, vol. 71, no. 24, pp. 241201, Jun. 2005.

[7] C. Darmody and N. Goldsman, "Incomplete ionization in aluminum-doped 4H-silicon carbide," J. Appl. Phys., vol. 126, no. 14, pp. 145701, 2019.

[8] M. Alam, N. Opondo, D. T. Morisette and J. A. Cooper, "Demonstration of a 10-kV Class Waffle-Substrate n-Channel IGBT in 4H-SiC," IEEE Transactions on Electron Devices, vol. 69, no. 10, pp. 5683-5688, Oct. 2022.

[9] O. S. Saadeh, H. A. Mantooth and J. C. Balda, "The modeling and characterization of Silicon carbide gate turn off thyristors," 2012 IEEE Energy Conversion Congress and Exposition (ECCE), Raleigh, NC, USA, 2012, pp. 3589-3594.

[10] Wang, Jun. Design, characterization, modeling and analysis of high voltage silicon carbide power devices. North Carolina State University, 2010.

[11] Y. Sui, J. A. Cooper, X. Wang, and G. G. Walden, "Design, simulation, and characterization of high-voltage SiC p-IGBTs, " Solid-State Electron., vols. 600–603, pp. 1191–1194, Jan. 2008.

[12] M. E. Levinshtein, T. T. Mnatsakanov, S. N. Yurkov, A. G. Tandoev, S.-H. Ryu, and J. W. Palmour, "High-voltage silicon-carbide thyristor with an n-type blocking base," Phys. Semicond. Devices, vol. 50, pp. 404–410, Mar. 2016.

[13] Y. Huang, R. Wang, Y. Zhang, D. Yang, and X. Pi, "Compensation of p-type doping in Al-doped 4H-SiC," J. Appl. Phys., vol. 131, no. 18, p. 185703, 2022.

Comparative Investigation of Gate Oxide Degradation in 1.2 kV Planar, Double-Trench, and Asymmetric-Trench SiC MOSFETs

Dingkun Zhao
College of Electrical and Information Engineering
Hunan University
Changsha, China
zhaodk@hnu.edu.cn

Xin Yang*
College of Electrical and Information Engineering
Hunan University
Changsha, China
xyang@hnu.edu.cn

Abstract—SiO₂ is commonly employed as the gate-oxide material in SiC MOSFETs, but its tendency to degrade poses a significant challenge to device reliability. Therefore, monitoring the gate oxide's condition is crucial. Many degradation precursors are influenced by junction temperature, suggesting that a composite indicator combining threshold voltage and transconductance may offer a more dependable metric. However, most existing research concentrates on conventional planar and trench gate designs, leaving the comparative analysis of planar, double-trench, and asymmetric-trench SiC MOSFETs largely unexplored. This study systematically evaluates the three structures and analyzes the shift characteristics of their transconductance and threshold voltage. The findings serve as a reference for understanding the degradation characteristics of various gate structures during gate oxide aging, thereby allowing for more accurate assessments of gate oxide reliability.

Keywords—Aging precursor, gate-oxide, degradation, SiC MOSFETs, gate structure

I. INTRODUCTION

Silicon dioxide (SiO₂) is widely used as the primary gate-oxide material in silicon (Si) and silicon carbide (SiC) MOSFETs. However, the gate oxide in SiC MOSFETs undergoes more significant degradation than in Si-based devices. This heightened susceptibility arises from the thinner gate-oxide layer typically utilized in SiC MOSFETs to achieve the required threshold voltage and transconductance levels [1]. As a result, under equivalent gate-bias conditions, the gate oxide in SiC MOSFETs experiences a stronger electric field, leading to greater stress. Moreover, the SiC/SiO₂ interface often exhibits lower quality compared to the Si/SiO₂ interface, with the interface and near-oxide trap densities being approximately two to three orders of magnitude higher [2]. These defect sites promote charge trapping, gradually altering the electrical characteristics of SiC MOSFETs over time. Therefore, ensuring the reliability of the gate oxide is paramount, highlighting the importance of monitoring its degradation in SiC MOSFETs.

To effectively monitor the health of SiC MOSFET gate oxides, a suitable precursor must reliably indicate oxide degradation while remaining stable against temperature fluctuations and packaging deterioration. Additionally, this precursor should be straightforward to measure consistently.

Existing literature has examined various precursors, with threshold voltage often highlighted as a leading indicator of gate-oxide degradation [3]. However, threshold voltage is strongly influenced by junction temperature, making it difficult to distinguish genuine degradation from temperature-induced variations. Parameters derived from switching characteristics, such as the gate plateau voltage, plateau duration, and turn-on delay time, provide valuable insights into gate-oxide integrity. However, the high sensitivity of these parameters to noise, stemming from the rapid switching behavior of SiC MOSFETs, presents a significant measurement challenge. Although the use of an external gate resistance can enhance measurement resolution, this technique compromises the high-speed switching performance which is a key advantage of SiC MOSFET technology. The gate leakage current (I_{GSS}) is a known marker for degradation [4], but its usefulness for ongoing condition monitoring is limited, since it generally remains stable until the device approaches the end of its lifespan. To address these challenges, some studies have proposed using a composite precursor that combines transconductance with threshold voltage [5], allowing aging effects to be distinguished more reliably from temperature-induced variations. However, much of the existing research has been conducted on traditional planar and trench gate structures. Comparative analyses of threshold voltage and transconductance effects across planar, double trench, and asymmetric trench SiC MOSFETs remain limited. This work seeks to address that gap through a systematic evaluation of these gate structures, ultimately validating the broader applicability of these degradation indicators.

This paper consists of the following sections: Section II introduces the devices under test (DUTs) that feature distinct structural configurations. Section III, outlines the foundational details regarding the effects of gate-oxide degradation in SiC MOSFETs. Section IV, explores how threshold voltage and transconductance are affected by gate oxide degradation across various gate structures. Section V, offers a comprehensive conclusion to summarize the findings of the study.

II. DEVICE UNDER TEST

Commercially available discrete SiC MOSFETs are classified into planar and trench gate designs, each offering distinct structural and performance characteristics. Trench gate

---

This work was supported in part by the National Natural Science Foundation of China under Grant U24A20155, and in part by the Science and Technology Innovation Program of Hunan Province, China under grant 2023PT1009 and 2024RC1036. *(Corresponding author: Xin Yang.)*

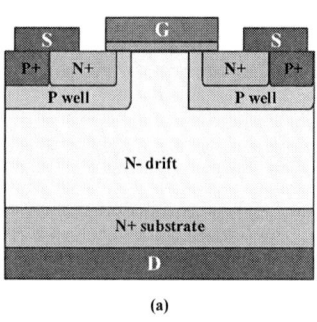 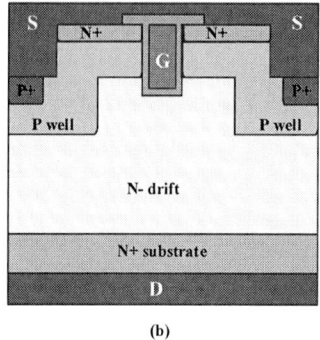 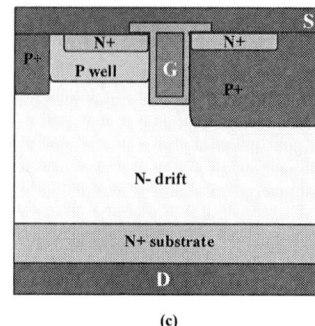

(a) (b) (c)

Fig. 1. Schematic cross-sectional views of SC MOSFETs. (a) Planar gate SC MOSFET (J-MOSFET), (b) Double trench SC MOSFET (DT-MOSFET), (c) Asymmetric trench SC MOSFET (AT-MOSFET).

MOSFETs typically achieve higher channel mobility and lower ON-resistance than planar gate structures (D-MOSFET). Within trench-based architectures, the double trench (DT-MOSFET) and asymmetric trench (AT-MOSFET) configurations stand out. DT-MOSFETs incorporate p-implanted source trenches to mitigate the electric field at the trench bottom [as shown in Fig. 1(b)] [6], while AT-MOSFETs employ adjacent p-well regions to generate a JFET effect that reduces oxide stress at trench corners [as shown in Fig. 1(c)] [7]. However, trench etching may introduce a high density of interface states, and the gate oxide remains susceptible to degradation under high OFF-state electric fields.

TABLE I. PARAMETERS OF THE DUTS

Vendor	Device Type	Voltage Rating	Current Rating	$R_{DS(on)}$
A	D-MOSFET	1200 V	30 A	75 mΩ
B	DT-MOSFET	1200 V	31 A	80 mΩ
C	AT-MOSFET	1200 V	36 A	60 mΩ

Previous research has mainly addressed gate-oxide degradation in planar and conventional trench MOSFETs. This study expands the scope by analyzing SiC MOSFETs from three commercial vendors, encompassing various planar and trench-based designs. All devices are housed in standard TO-247 packages, ensuring consistent parasitic inductances. These devices share comparable voltage ratings, current handling capabilities, and ON-resistance levels, providing a robust foundation for evaluating threshold voltage and transconductance as indicators of gate-oxide health. To mitigate sample variability, three units of each device type were tested. Table I summarizes the detailed specifications.

III. GATE-OXIDE DEGRADATION IN SIC MOSFETS

Despite the rapid advancements in SiC technology and its increasing adoption in the electric vehicle market, ensuring reliable gate-oxide performance remains a considerable challenge. Several interrelated factors contribute to this issue.

A primary concern is the elevated defect density at the SiC/SiO$_2$ interface, which far exceeds that of conventional Si/SiO$_2$ systems. These defects largely result from carbon-containing molecules generated during SiC oxidation. Although hydrogen passivation can mitigate these defects, it cannot completely eliminate them. Additionally, the thin gate-oxide layer commonly used in SiC MOSFETs is more susceptible to Fowler–Nordheim tunneling under high electric fields. This tunneling introduces extra electrons into the oxide, which, in conjunction with a higher abundance of interface trapped charges at the SiC/SiO$_2$ interface compared to silicon-based systems, exacerbates Coulombic scattering, reduces carrier mobility, and ultimately degrades device performance.

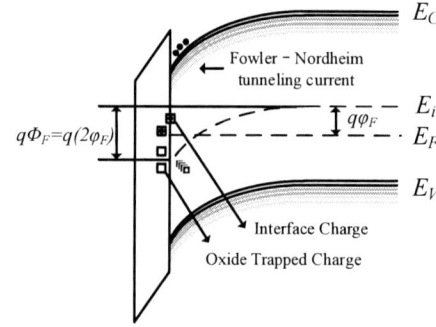

Fig. 2. Electron tunneling into oxide traps and interface traps under positive electric field.

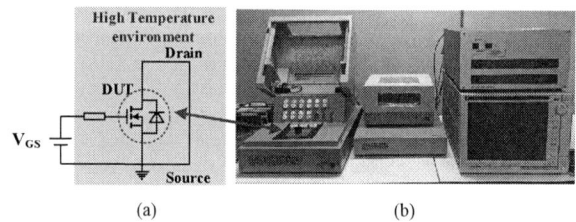

(a) (b)

Fig. 3. Illustration of the measure-stress-measure procedure for precursors monitoring. (a) Circuit schematic of HTGB test. (b) Keysight B1506A curve tracer.

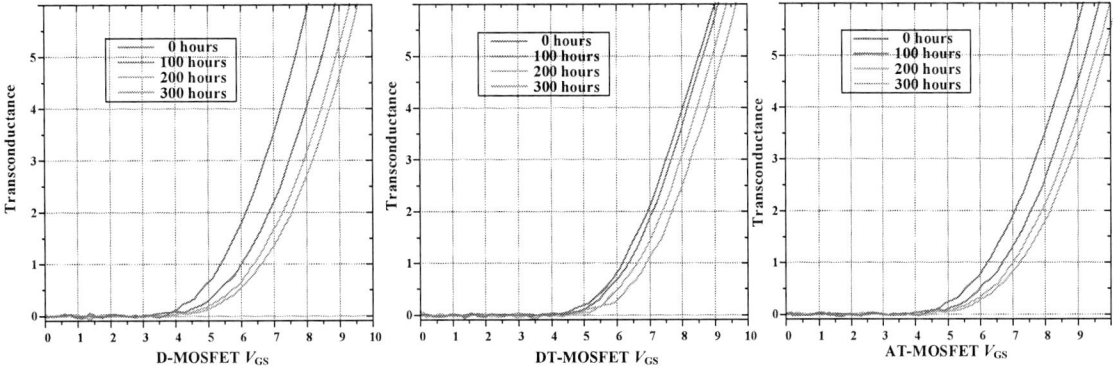

Fig. 4. Transconductance curves of different SiC MOSFET devices: (a) D-MOSFET, (b) DT-MOSFET, (c) AT-MOSFET

Bias conditions introduce additional complexity to these challenges. As shown in Fig. 2, when a gate-source bias is applied, the electric field causes the Fermi level at the SiO_2/SiC interface to shift. An increase in the gate-source bias enhances this electric field, further raising the Fermi level and promoting electron accumulation near the conduction band edge of the SiC material. Some electrons tunnel into pre-existing defects, causing threshold voltage instability and bias temperature instability (BTI). Over time, these processes generate additional defects in the oxide layer, eventually leading to time-dependent dielectric breakdown (TDDB), a critical failure mechanism.

Defects that undermine gate-oxide reliability generally fall into four categories: oxide-trapped charges (Q_{ot}), interface-trapped charges (Q_{it}), fixed charges (Q_f), and mobile ion charges (Q_m). Of these, oxide and interface-trapped charges are the primary contributors to reduced channel mobility, threshold voltage shifts, and BTI. Interface and near-interface traps can also lead to unpredictable threshold voltage variations, heightening the risk of crosstalk-induced conduction errors.

HTGB testing is a widely recognized accelerated aging method used to assess the impact of gate-oxide degradation on the performance and reliability of MOSFETs. This technique involves subjecting the gate oxide to a strong electric field and elevated temperatures. By applying a high gate source voltage greater than the standard operating voltage but safely below the breakdown threshold along with prolonged thermal stress, HTGB creates a controlled environment for observing aging phenomena. Elevated temperatures, typically set around 150°C, accelerate degradation, enabling the identification of long-term reliability issues.

Standard procedures, such as those specified in JEDEC JESD22-A108D, recommend a delay before parameter measurements after removing the bias. This precaution ensures that only permanent degradation effects are measured, eliminating transient shifts that could obscure the actual aging trends. As illustrated in Fig. 3, the circuit design employed for the HTGB test is depicted schematically. In this study, a gate–source voltage of 30 V was chosen, as prior research [8] indicates that this level effectively induces observable aging

patterns within a reasonable test duration, while avoiding dielectric breakdown.

To ensure stable test conditions, a large external gate resistance is used. This resistance suppresses voltage spikes at the test's outset and prevents gate current overshoots, maintaining a stable applied voltage throughout the experiment. Since gate leakage currents remain low over the device's lifespan, the resistance minimally affects the effective gate source voltage. Periodically, such as every 100 hours, the test is paused for static characterization using precise measurement tools like a Keysight 1506A curve tracer. These regular intervals of static characterization allow for a detailed analysis of gate-oxide aging progression and its eventual impact on device behavior, thereby ensuring a comprehensive understanding of long-term reliability and performance.

IV. EFFECT OF GATE-OXIDE DEGRADATION ON PRECURSORS

A composite precursor that integrates both transconductance and threshold voltage offers a more robust means of distinguishing aging-related effects from temperature-induced variations. This method enables temperature-independent monitoring of gate-oxide degradation. To confirm its general applicability, this study evaluated three distinct SiC MOSFET designs, each featuring a unique gate structure, under HTGB testing conditions. These tests simulate accelerated gate-oxide degradation, providing a comparative analysis of the composite precursor's performance across different device architectures.

Transconductance, which represents the relationship between drain current and gate–source voltage, serves as a practical indicator of gate-oxide healthy. Under aging stress conditions, all three types of SiC MOSFETs, D-MOSFET, DT-MOSFET, and AT-MOSFET are exhibit a consistent rightward shift in their transconductance curves, as shown in Fig. 4(a–c). This uniform trend indicates that aging leads to a reduction in channel mobility or an increase in interface traps, both of which require a higher gate–source voltage to achieve the same drain current. The rightward shift thus provides a

979-8-3315-1110-4/25 $31.00 © 2025 IEEE

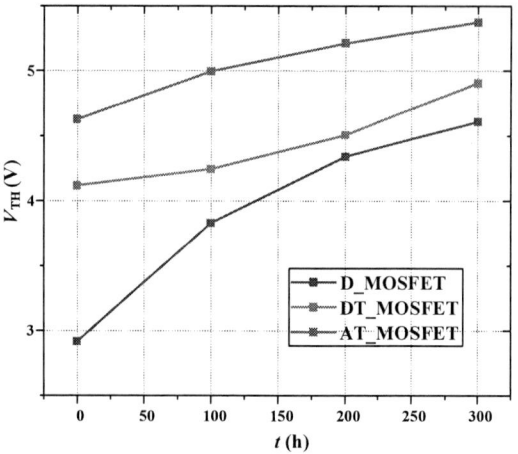

Fig. 5. Threshold voltage variation over time for different SiC MOSFET devices

clear and quantifiable signal of gate-oxide degradation across different device architectures.

Threshold voltage is widely regarded as a critical indicator of gate-oxide degradation. However, its strong temperature dependence necessitates careful compensation methods. Research indicates that multiple factors influence threshold voltage shifts, all of which must be accounted for when assessing degradation [9],

$$V_{th0} = \frac{\sqrt{4\varepsilon_s kTN_A ln(\frac{N_A}{n_i})}}{C_{ox}} + \frac{2kT}{q}ln(\frac{N_A}{n_i}) \qquad (1)$$

where the gate-oxide capacitance per unit area C_{ox}, the dielectric constant ε_s, the p-type well doping concentration (N_A), the intrinsic carrier concentration of SiC n_i, and the junction temperature T. This relationship also depends on fundamental constants such as the Boltzmann constant k and the elementary charge q. The threshold voltage shift can also be calculated as,

$$\Delta V_{th} = V_{th0} - \frac{Q_{ot}}{C_{ox}} + \frac{qN_{it}}{C_{ox}} \qquad (2)$$

The threshold voltage shift in SiC MOSFETs primarily results from two charge categories: interface trapped-charges and oxide-trapped charges. Among these, interface-trapped charges have a more pronounced impact, as indicated by the interface trap density N_{it} in equation (2) .

As shown in Fig. 5, gate oxide degradation significantly affects the threshold voltage across different SiC MOSFET structures. After 300 hours of electrical and thermal stress, the threshold voltage shift reaches approximately 1.7 V in the D-MOSFET, while both the DT-MOSFET and AT-MOSFET exhibit a shift of about 0.8 V. The results indicate that threshold voltage and transconductance exhibit clear changes across all device types, and can effectively reflect the degradation of the gate oxide. Moreover, both parameters are

capable of distinguishing degradation caused by device aging from variations induced by temperature, making them suitable for degradation diagnosis and health monitoring across different SiC MOSFET structures.

V. CONCLUSION

During the aging process, the transconductance curves of the D-MOSFET, DT-MOSFET, and AT-MOSFET all exhibit a gradual rightward shift with increasing gate voltage, accompanied by a corresponding rise in threshold voltage. These consistent trends across different gate structures indicate that degradation is occurring in the gate oxide layer. By jointly analyzing the variations in transconductance and threshold voltage, it is possible to effectively distinguish degradation induced by device aging from changes caused by temperature fluctuations. This finding demonstrates that both parameters are reliable indicators of gate oxide reliability in various SiC MOSFET architectures, providing a solid foundation for accurate degradation diagnosis and long-term health monitoring.

REFERENCES

[1] T.-T. Nguyen, A. Ahmed, T. V. Thang, and J.-H. Park, "Gate oxide reliability issues of SiC MOSFETs under short-circuit operation," *IEEE Trans. Power Electron.*, vol. 30, no. 5, pp. 2445–2455, May 2015.

[2] M. Nawaz, "On the evaluation of gate dielectrics for 4H-SiC based power MOSFETs," *Act. Passive Electron. Compon.*, vol. 2015, pp. 1–12, Jan. 2015.

[3] H. Luo, F. Iannuzzo, and M. Turnaturi, "Role of threshold voltage shift in highly accelerated power cycling tests for SiC MOSFET modules," *IEEE J. Emerg. Sel. Topics Power Electron.*, vol. 8, no. 2, pp. 1657–1667, Jun. 2020.

[4] P. Wang, J. Zatarski, A. Banerjee, and J. S. Donnal, "Condition monitoring of SiC MOSFETs based on gate-leakage current estimation," *IEEE Trans. Instrum. Meas.*, vol. 71, pp. 1–10, Jan. 2022.

[5] M. Farhadi, B. T. Vankayalapati, R. Sajadi, and B. Akin, "Gate-Oxide Degradation Monitoring of SiC MOSFETs Based on Transfer Characteristic With Temperature Compensation," *IEEE Trans. Transp. Electrific.*, vol. 10, no. 1, pp. 1837–1849, March 2024.

[6] T. Nakamura *et al.*, "High performance SiC trench devices with ultralow ron," *IEDM Tech. Dig.*, Washington, DC, USA, Dec. 2011, pp. 26.5.1–26.5..

[7] D. Peters *et al.*, "The new CoolSiC trench MOSFET technology for low gate oxide stress and high performance," *Proc. Int. Exhib. Conf. Power Electron., Intell. Motion, Renew. Energy Manage.*, Nuremberg, Germany, May 2017, pp. 1–7.

[8] D. J. Lichtenwalner *et al.*, "Gate oxide reliability of SiC MOSFETs and capacitors fabricated on 150 mm wafers," *Mater. Sci. Forum*, vol. 963, pp. 745–748, Jul. 2019.

[9] N. Stojadinovic *et al.*, "Effects of electrical stressing in power VDMOSFETs," *Proc. IEEE Conf. Electron Devices Solid-State Circuits*, Kowloon, Hong Kong, Mar. 2003, pp. 291–296.

2025 IEEE Workshop on Wide Bandgap Power Devices and Applications in Asia (WiPDA Asia)

Understanding the Role of Buffer Traps in GaN HEMTs: A Simulation Study on Dynamic R_{on}

Haiyang Li[1,†], Mengqi Fan[1,†], Xinyue Dai[1], Xiaoping Wang[1], David Zhou[1], Danfeng Mao[1], Yan Wang[2], Yuxi Wan[1]

[1]Shenzhen Pinghu Laboratory
Shenzhen, China
[2]Tsinghua University
Beijing, China
wangxiaoping@phlab.com.cn
†These authors equally contributed to this work.

Abstract—This work employs calibrated Technology Computer-Aided Design (TCAD) simulations to analyze buffer trap behavior under varying stress amplitude (V_{stress}) and duration (t_{stress}) conditions and its impact on dynamic on-resistance (R_{on}) degradation. Two distinct degradation phases are identified: (I) a V_{th}-shift dominant phase; (II) a current collapse dominant phase without additional V_{th} shift. The two phases arise from the time-dependent, region-specific electron trapping within buffer layer during off-state drain stress. These findings provide insights into buffer trap dynamics and their impact on dynamic R_{on} degradation.

Keywords—p-GaN HEMT, TCAD simulation, buffer traps, dynamic R_{on}

I. INTRODUCTION

The rapid advancement of GaN HEMT power devices has been driven by their superior material properties, including high breakdown field and electron mobility [1]. Carbon doping in GaN buffer layers, which introduces acceptor-like traps, is crucial for suppressing leakage current in power GaN HEMTs. While extensive research has explored the impacts of these traps on vertical breakdown characteristics [2-4], capacitance-voltage (C-V) behavior [5-6] and dynamic on-resistance (R_{on}) [7-8], few studies have employed Technology Computer-Aided Design (TCAD), which is a powerful tool to uncover the microscopic mechanism of trap behaviors. Moreover, the temporal-spatial evolution of these buffer traps across various drain bias conditions remains insufficiently studied. In this work, we conduct a detailed simulation study of buffer trap behavior under a wide range of off-state drain stress voltage and time, offering deeper insights into their role in dynamic R_{on} in GaN HEMTs.

The paper is organized as follows: First we briefly describe the device simulation method and the calibration process. Next, we analyze trap dynamics and their correlation with electrical characteristics—specifically threshold voltage shift and dynamic on-resistance degradation—under two type of conditions: (1) short stress pulses (down to 1 μs) with varying amplitudes, and (2) fixed stress amplitude with varying durations. Finally, we summarize the key findings and conclusions.

II. DEVICE SIMULATION AND CALIBRATION

Fig. 1 shows the device structure of the 650 V p-GaN HEMT with Schottky gate contact, fabricated on a 6-inch wafer. In our TCAD simulations, the two-dimensional electron gas (2DEG) in the AlGaN/GaN heterostructure is modeled by accounting for both spontaneous and piezoelectric polarization, as well as surface donors at the AlGaN/passivation interface. For the p-GaN layer, incomplete Mg ionization is considered. Carrier transport is described using the drift-diffusion model, incorporating radiative and

Shockley–Read–Hall (SRH) recombination. The mobility model includes high-field saturation effects. Carbon-related acceptor-like traps, are assumed in the GaN buffer layer with an energy level of $E_{t,acc} = E_V + 0.9$ eV [9] and a uniform density of $N_{t,acc} = 1 \times 10^{18}$ cm^{-3}. The calibration methodology follows a similar approach to that described in [10-11]. As shown in Fig. 2, the calibration achieves excellent agreement between simulations (transfer characteristics and C-V curves) and experimental measurements.

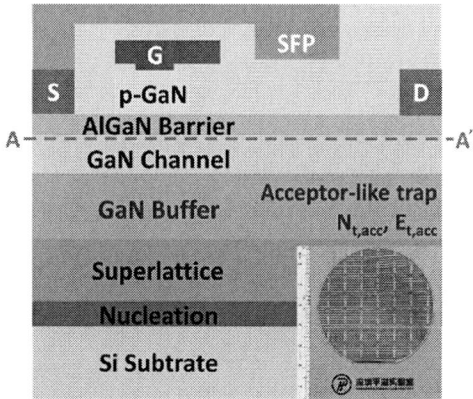

Fig. 1. Schematic of the 650 V p-GaN HMET. Inset: Fabricated 6-inch wafer.

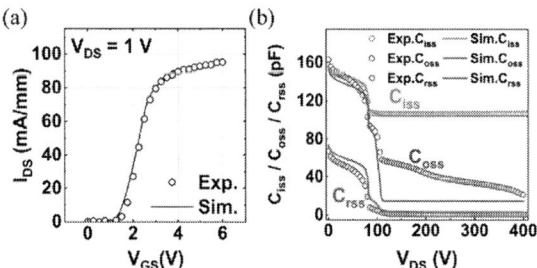

Fig. 2. Excellent agreement has been achieved between the experimental and simulated results of p-GaN HEMT: (a) transfer (b) off-state C-V characteristics. $E_{t,acc} = E_V + 0.9$ eV and $N_{t,acc} = 1 \times 18$ cm^{-3} are adopted to reproduce the measured results.

The $C_{rss} - V_{DS}$ characteristics exhibits high sensitivity to the $E_{t,acc}$ and $N_{t,acc}$ of buffer traps. Fig. 3 shows the $C_{rss} - V_{DS}$ curves for varying $E_{t,acc}$ and $N_{t,acc}$ values of the buffer traps. The curve shifts leftward as $E_{t,acc}$ decreases and $N_{t,acc}$ increases. The shift occurs because with lower trap energy level and higher trap density enhance electron trapping. The accumulation of trapped electrons in the buffer layer elevates the conduction band energy of the GaN channel layer, accelerating electron depletion in the channel layer. Therefore,

979-8-3315-1110-4/25 $31.00 © 2025 IEEE   898

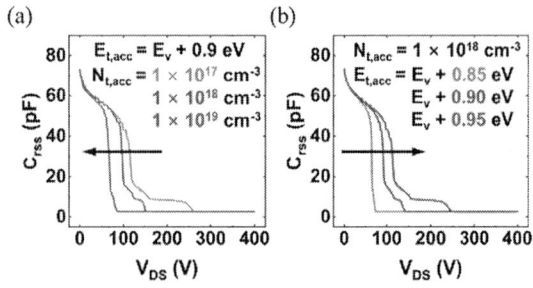

Fig. 3. (a) $N_{t,acc}$ and (b) $E_{t,acc}$ dependence of $C_{rss} - V_{DS}$ characteristics.

in the following discussion, $N_{t,acc}$ and $E_{t,acc}$ is reasonably fixed at 1×10^{18} cm^{-3} and $E_V + 0.9$ eV, respectively, to match the measured C_{rss} curve.

Fig. 4 shows the V_{GS} and V_{DS} bias sequence used to simulate the dynamic R_{on}. During the stress phase, the device is subjected to a V_{DS} stress with amplitude V_{stress} and pulse width of t_{stress}. Afterward, the on-resistance is measured at $V_{GS} = 6$ V and $V_{DS} = 1$ V. The dynamic R_{on} is normalized to that of the fresh device. V_{stress} ranges from 50 V to 400 V while t_{stress} is varied from 1 μs to 100 s.

III. RESULTS AND DISCUSSION

A. Trap behavior under varying V_{stress}

Fig. 5 compares the electron trapping distribution in the fresh device with that in post-stress devices subjected to different 1 μs V_{stress} pulses. Relative to the fresh device, the V_{DS}-induced electron trapping predominantly occurs at the buffer layer surface, especially near the gate region. At higher V_{stress}, electron trapping extends deeper into the buffer layer beneath the drain-side gate region, driven by enhanced buffer leakage under high drain bias.

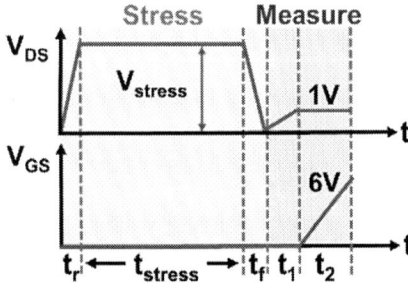

Fig. 4. Sketch of waveforms applied to the gate and drain terminal to monitor dynamic R_{on}. $t_r = t_f = 100$ ns, $t_1 = 1$ s, $t_2 = 6$ s.

Fig. 6 (a) and (b) presents the transfer curves of post-stress devices with the fresh device as a reference. With increased stress amplitude, the threshold voltage (V_{th}) shifts positively, accompanied by current collapse. The threshold voltage shift (ΔV_{th}) in Fig. 6 (c) agrees with the measured results [12].

The positive V_{th} shift results from trapped electrons accumulating beneath the gate region, lowering the electrostatic potential (Fig. 7) — a mechanism analogous to the body bias effect in MOSFETs. Meanwhile, the current collapse arises from a reduction in two-dimensional electron

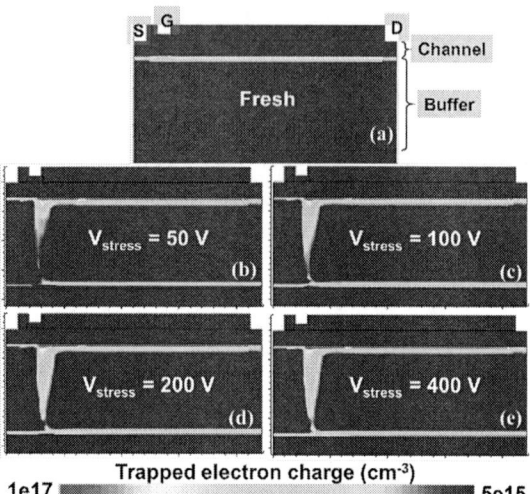

Fig. 5. Simulated electron trapping distribution in GaN buffer layer of (a) fresh device and post-stress devices with fixed $t_{stress} = 1$ μs and varying V_{stress}: (b) 50 V (c) 100 V (d) 200 V (e) 400 V.

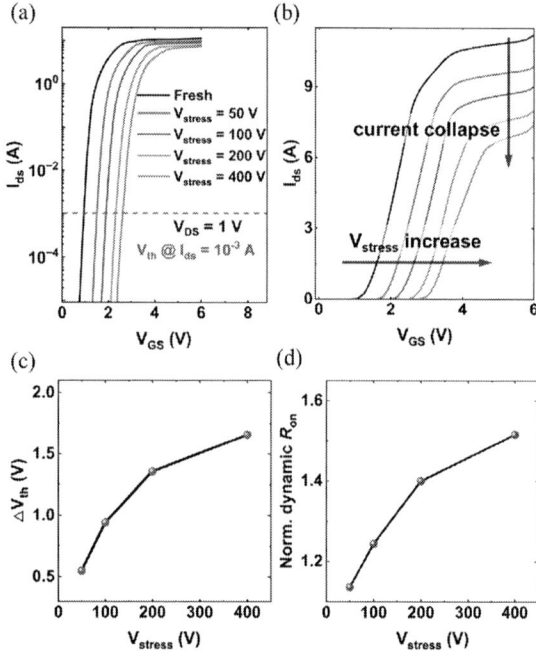

Fig. 6. Simulated post-stress Id-Vg curves with fixed $t_{stress} = 1$ μs and varying V_{stress}: (a) log scale (b) linear scale. Corresponding (c) V_{th} shift and (d) dynamic R_{on}.

gas (2DEG) density due to electron trapping under the access region.

B. Trap behavior under varying t_{stress}

Fig. 8 presents the post-stress electron trapping distribution immediately following a $V_{stress} = 200$ V pulse for different t_{stress} duration. Between $t_{stress} = 1$ μs and $t_{stress} = 100$ μs, no additional electron trapping is observed. However, from $t_{stress} = 10$ ms to 1 s, additional electron trapping emerges at the upper surface of the buffer layer and near the drain-side gate corner (as indicated by the arrow).

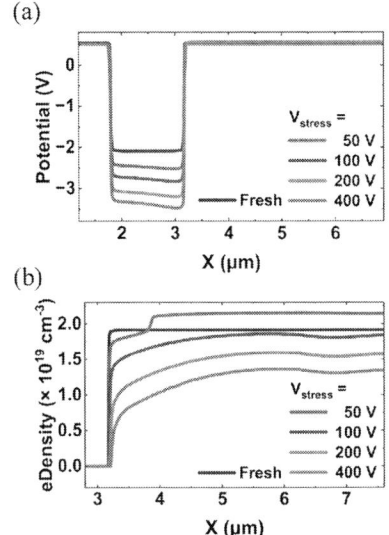

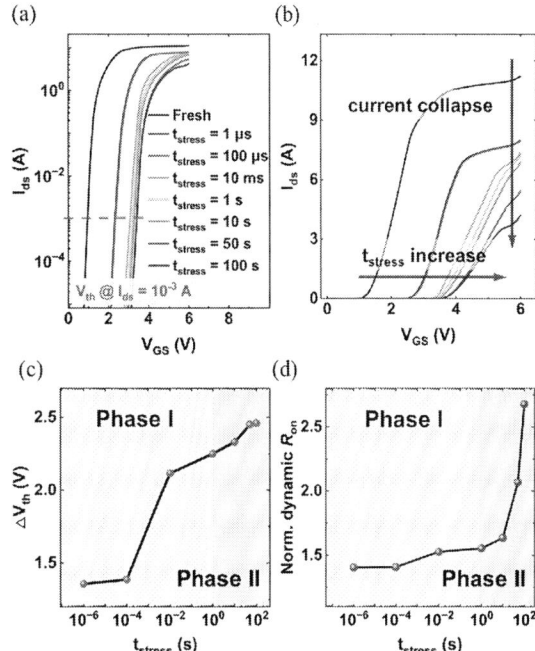

Fig. 7. Simulated zero-bias electrostatic potential distribution channel (cutline at AA' in Fig. 1) after different V_{DS} stress. t_{stress} = 1 μs.

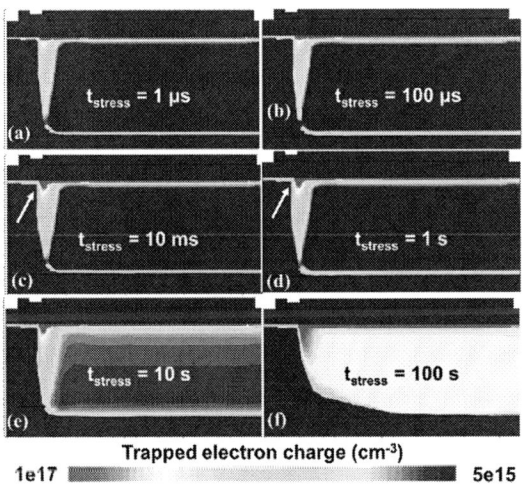

Fig. 8. Simulated post-stress electron trapping distribution in GaN buffer layer with fixed V_{stress} = 200 V and varying t_{stress}: (a) 1 μs (b) 100 μs (c) 10 ms (d) 1 s (e) 10 s (f) 100 s.

When t_{stress} exceeds 10 s, severe electron trapping occurs in the buffer layer below the access region. For t_{stress} = 100 s, becomes pronounced and extends throughout the entire buffer layer.

The corresponding post-stress transfer curves are shown in Fig. 9 (a) and (b). The V_{th} degradation and dynamic R_{on} degradation over t_{stress} are shown in Fig. 9 (c) and (d), where two distinct phases can be clearly identified: (I) from a V_{th}-shift dominant phase; (II) a current collapse-dominant phase without additional V_{th} shift.

To elucidate the relationship between trapped charges evolution and electrical characteristics degradation in these two phases, we continuously monitor both the electrostatic potential distribution along the channel and the electron density in the access region, as shown in Fig. 10. During phase

Fig. 9. Simulated post-stress Id-Vg curves with fixed V_{stress} = 200 V and varing t_{stress}: (a) log scale (b) linear scale. Corresponding (c) V_{th} shift and (d) dynamic R_{on}.

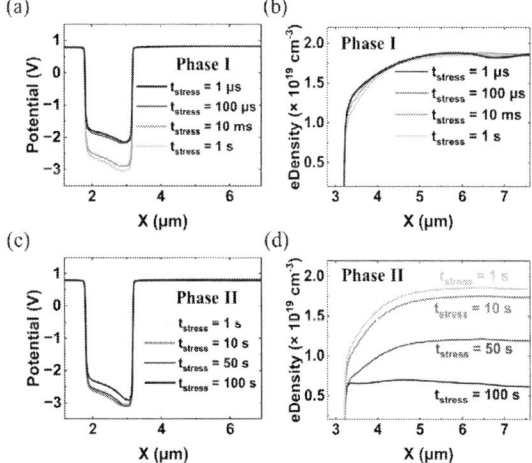

Fig. 10. Simulated zero-bias (a) electrostatic potential distribution and (b) access region electron density along the channel (cutline at AA' in Fig. 1) after different t_{stress} duration. V_{stress} = 200 V. (a)(b): Phase I; (c)(d): Phase II.

I, as shown in Fig. 10 (a-b), the electrostatic potential under gate region is pulled down by the accumulated trapped electron, while the access region remains unaffected. During phase II, the variation in electrostatic potential under the gate region is minimal (consistent with the negligible V_{th} shift) while a significant reduction in 2DEG density in the access region is observed. This reduction is attribute to the accumulation of trapped electrons in the buffer layer beneath the access region.

C. Dynamic R_{on} under varying t_{stress} & V_{stress}

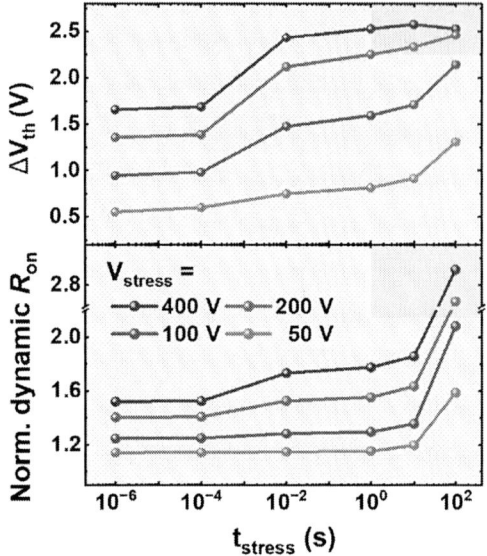

Fig. 11. Simulated (a) V_{th} shift and (b) dynamic R_{on} with both varying t_{stress} and V_{stress}. Blue shade: Phase I; Pink shade: Phase II.

Fig. 11 summarizes the V_{th} shift and dynamic R_{on} degradation under various off-state V_{DS} stress conditions, with the Phase I and Phase II highlighted in blue and pink shading, respectively. At relatively low V_{stress}, the device exhibits only Phase I degradation. The transition to Phase II degradation occurs exclusively under higher V_{stress} levels combined with prolonged stress durations.

D. Conclusion

In this work, we investigate the buffer trap behavior under different V_{stress} and t_{stress} conditions using calibrated TCAD simulation. Electron trapping within the buffer layer exhibits two distinct time-dependent, region-specific characteristics. Under short off-state V_{DS} stress, trapped electron primarily accumulates at the surface of the buffer layer, and further electron trapping occurs beneath the gate region, leading to a positive V_{th} shift. For longer off-state V_{DS} stress, trapped electrons extend from gate region toward drain and penetrate deeper into the buffer layer, resulting in 2DEG reduction. These findings provide a deeper understanding of buffer trap behavior in GaN HEMT, and its impact on dynamic R_{on} degradation.

REFERENCES

[1] M. Meneghini, C. d. Santi, I. Abid, M. Buffolo, M. Cioni, R. A. Khadar, et al., "GaN-based power devices: Physics, reliability, and perspectives," Journal of Applied Physics, vol. 130, no. 18, Art. no. 18, Nov. 2021, doi: 10.1063/5.0061354.

[2] M. J. Uren, S. Karboyan, I. Chatterjee, A. Pooth, P. Moens, A. Banerjee, et al., "'Leaky Dielectric' Model for the Suppression of Dynamic R_{ON} in Carbon-Doped AlGaN/GaN HEMTs," IEEE Trans. Electron Devices, vol. 64, no. 7, pp. 2826–2834, Jul. 2017, doi: 10.1109/TED.2017.2706090.

[3] M. Borga, C. D. Santi, S. Stoffels, B. Bakeroot, X. Li , M. Zhao, et al., "Modeling of the Vertical Leakage Current in AlN/Si Heterojunctions for GaN Power Applications," IEEE Trans. Electron Devices, vol. 67, no. 2, pp. 595–599, Feb. 2020, doi: 10.1109/TED.2020.2964060.

[4] D. Cornigli, S. Reggiani, E. Gnani, A. Gnudi, G. Baccarani, P. Moens, et al., "Numerical investigation of the lateral and vertical leakage currents and breakdown regimes in GaN-on-Silicon vertical structures," in 2015 IEEE International Electron Devices Meeting (IEDM), Washington, DC, USA: IEEE, Dec. 2015, p. 5.3.1-5.3.4, doi: 10.1109/IEDM.2015.7409633.

[5] M. Cioni , A. Chini , N. Zagni, G. Verzellesi, G. Giorgino, G. Cappellini, et al., "On the Dynamic R_{ON} , Vertical Leakage and Capacitance Behavior in pGaN HEMTs With Heavily Carbon-Doped Buffers," IEEE Electron Device Lett., vol. 45, no. 8, pp. 1437–1440, Aug. 2024, doi: 10.1109/LED.2024.3417313.

[6] G. Verzellesi, L. Morassi, G. Meneghesso, M. Meneghini, E. Zanoni, G. Pozzovivo, et al., "Influence of Buffer Carbon Doping on Pulse and AC Behavior of Insulated-Gate Field-Plated Power AlGaN/GaN HEMTs," IEEE Electron Device Lett., vol. 35, no. 4, pp. 443–445, Apr. 2014, doi: 10.1109/LED.2014.2304680.

[7] M. Cioni, N. Zagni, F. Iucolano, M. Moschetti, G. Verzellesi, and A. Chini, "Partial Recovery of Dynamic R_{ON} Versus OFF-State Stress Voltage in p-GaN Gate AlGaN/GaN Power HEMTs," IEEE Trans. Electron Devices, vol. 68, no. 10, pp. 4862–4868, Oct. 2021, doi: 10.1109/TED.2021.3105075.

[8] S. Yang, J. Du, and S. Li, "Characterization of Dynamic R_{on} in GaN HEMTs," in TENCON 2024 - 2024 IEEE Region 10 Conference (TENCON), Singapore, Singapore: IEEE, Dec. 2024, pp. 456–460, doi: 10.1109/TENCON61640.2024.10902940.

[9] J. L. Lyons, A. Janotti, and C. G. Van De Walle, "Carbon impurities and the yellow luminescence in GaN," Applied Physics Letters, vol. 97, no. 15, p. 152108, Oct. 2010, doi: 10.1063/1.3492841.

[10] P. V. Raja, J.-C. Nallatamby, N. DasGupta, and A. DasGupta, "Trapping effects on AlGaN/GaN HEMT characteristics," Solid-State Electronics, vol. 176, p. 107929, Feb. 2021, doi: 10.1016/j.sse.2020.107929.

[11] W.-C. Cheng, P.-H. Lin, C.-H. Lin, Y.-W. Lien, C.-Y. Chen, C.-H. Lee, et al., "Developing Physics-Based TCAD Model for AlGaN/GaN Power HEMTs," in 2023 IEEE Workshop on Wide Bandgap Power Devices and Applications in Asia (WiPDA Asia), Hsinchu, Taiwan: IEEE, Aug. 2023, pp. 1–5, doi: 10.1109/WiPDAAsia58218.2023.10261931.

[12] H. Wang, J. Wei, R. Xie, C. Liu, G. Tang, and K. J. Chen, "Maximizing the Performance of 650-V p-GaN Gate HEMTs: Dynamic RON Characterization and Circuit Design Considerations," IEEE Trans. Power Electron., vol. 32, no. 7, pp. 5539–5549, Jul. 2017, doi: 10.1109/TPEL.2016.2610460.

A 3×1 Silicon Carbide Bidirectional Switch Power Module with Balanced Inductance During Current Commutation

Zhiwei Jiao
State Key Laboratory of Intelligent
Power Distribution Equipment and
System
Hebei University of Technology
Tianjin, China
2572475217@qq.com

Lei Ming
State Key Laboratory of Intelligent
Power Distribution Equipment and
System
Hebei University of Technology
Tianjin, China
minglei@hebut.edu.cn

Yufeng Cao
Beijing Keytone Electronic Relay Corp.
Beijing, China
yfcao9@163.com

Tao Xu
State Key Laboratory of Intelligent
Power Distribution Equipment and
System
Hebei University of Technology
Tianjin, China
2040687037@qq.com

Zihang Gu
State Key Laboratory of Intelligent
Power Distribution Equipment and
System
Hebei University of Technology
Tianjin, China
lwj_qhb@sina.com

Zhen Xin
State Key Laboratory of Intelligent
Power Distribution Equipment and
System
Hebei University of Technology
Tianjin, China
xzh@hebut.edu.cn

Abstract—**Bidirectional switch is an inevitable device in many topologies, such as current-source converter and matrix converter. The performance of these converters are however deeply affected by the current commutation process, which requires balanced inductance, especially for the fast-switched silicon carbide converters. To address this issue, this paper proposes an optimized 3×1 silicon carbide bidirectional switch power module. This paper firstly analyzes the commutation loop and parasitic inductance rules of the power module. Based on this, the circuit topology and geometric packaging structure of the bidirectional switch module are presented, which is designed with the standard substrate size of EasyPACK 1B. By adopting a symmetric layout for the module, a 3×1-phase silicon carbide-based bidirectional switch power module with balanced inductance during current commutation is developed. The effectiveness of the proposed method has been verified by the double-pulse test.**

Keywords—*bi-directional power switches, SiC Power MOSFETs, Balanced Inductance, Layout Design*

I. INTRODUCTION

A bidirectional switch is an active electronic component composed of semiconductor switching transistors that facilitates bidirectional current flow in its on-state while blocking voltage in both directions during its off-state. It is widely used in fields，such as electric vehicles, new-energy power generation, rail transit, and smart grids. It is a basic component of circuit topologies such as current-source converters, matrix converters, high-frequency link inverters, and three-phase Vienna PFC rectifiers[1].

Traditional bidirectional switches typically utilize silicon-based IGBTs with anti-parallel diodes, which suffer from significant conduction losses, low switching speeds, substantial switching losses, and low temperature resistance. As a representative wide-bandgap semiconductor material, SiC MOSFETs are demonstrated superior characteristics, including high voltage tolerance, fast switching capability, low switching losses, and excellent high-temperature resistance, making them suitable for high-temperature, high-

This study was supported in part by the National Nature Science Foundation of China under Grant 52207196 and in part by the Overseas Returnees Funding Program of Hebei Province under Grant C20230316 and funded by Hebei Natural Science Foundation under Grant E2024202265

frequency applications and higher power scenarios[2]. Furthermore, SiC MOSFET-based bidirectional switch(SiC-BDS) topologies can utilize the body diode of the MOSFET without requiring additional anti-parallel diodes, thereby reducing module size and enhancing power density. Given the current challenges in monolithic bidirectional switch technology, such as complex manufacturing processes and low technological maturity, the bidirectional switch topology employing two SiC MOSFET in common-source configuration emerges as the optimal solution that balances superior performance with existing technological readiness[3].

The parasitic inductance inherent in power module packaging structures, particularly within high-frequency switching loops, induces detrimental voltage overshoot and electromagnetic interference that critically compromises system reliability and operational efficiency. During the device turn-off process, the parasitic inductance's combined action and the current change rate will generate a voltage overshoot. This voltage overshoot is applied to the power chip, increasing the voltage stress on the power chip, raising the requirements for the chip's voltage-withstand level, and limiting the further increase of the switching frequency[4]. The imbalance of parasitic inductances between commutation loops will lead to uneven voltage stress distribution among different phases, thereby degrading reliability and shortening the operational lifespan of the module [5].

Regarding the optimization of the parasitic inductance of power modules, there have been some studies, mainly focusing on optimizing the substrate layout and improving the packaging structure. The concept of P-cell and N-cell has been proposed to shorten the length of the commutation loop by placing switch pairs together in [6]; The research in [7] discusses that the parasitic inductance of the module can be reduced by integrating decoupling capacitors inside the module; Eric Vagnon proposed a three-dimensional press-fit chip stacking scheme in 2008. In this structure, two power chips are vertically connected through copper blocks, which greatly shortens the length of the commutation loop and effectively reduces the parasitic inductance of the commutation loop[8]. There is currently a lack of low-cost parasitic inductance optimization solutions specifically targeting current-source lateral commutation schemes.

Currently, most of the existing bidirectional switch power modules are single-bidirectional switch packages and nine-in-one bidirectional switch power modules used in three-by-three matrix converters. In response to the requirements of high power density, low parasitic inductance, and multi-topology applications of SiC-BDS modules, this paper proposes a packaging scheme for a 3×1-phase SiC-BDS power module to achieve balanced parasitic inductance between each commutation loop and meets the needs of current-source lateral commutation topology applications.

II. PROPOSED MODULE STRUCTURE AND LAYOUT

A. Analysis Of the Parasitic Inductance

The parasitic inductance induced by the packaging structure within the power module's switching loop poses significant challenges to device integrity and system reliability during high-frequency switching operations. For SiC MOSFET power modules, internal parasitic inductance limits their high-frequency applications[9].

Therefore, it is extremely important to accurately analyze the internal parasitic inductance in power modules. Parasitic inductance in the commutation loop of a power module is mainly introduced by conductive components such as power chips, copper layers on ceramic substrates, terminals, and bonding wires. Based on existing calculation methods, conductors can be categorized into two types: one is cylindrical conductors, mainly consisting of bonding wires and terminals, which are usually arranged in parallel; the other is plate-like conductive bodies, mainly consisting of power chips and substrate copper layers.

For multiple parallel conductors, their inductance can be determined using the following formula:

$$L_W = \frac{1}{N}\frac{\mu_0 l}{2\pi}\left[\ln\left(\frac{4l}{d}\right)-1\right] \qquad (1)$$

where N is the number of interconnected conductors; μ_0 is the vacuum permeability; l and d are the length and diameter of the conductor, respectively.

According to (1), the parasitic inductance in the power module can be reduced by increasing the number of parallel bonding wires and terminals with the same current direction.

For a rectangular conductor, the inductance calculation formula is:

$$L_C = \frac{\mu_0 l}{2\pi}\left[\ln(\frac{2l}{b+h})+\frac{1}{2}\right] \qquad (2)$$

where μ_0 represents the magnetic constant; l, b, and h represent the length, width, and height of the conductor, respectively.

According to (2), the parasitic inductance of the rectangular conductor can be minimized by increasing its current path area and shortening the current path length.

B. Introduction of Module Topology

The traditional module packaging scheme is optimized for voltage-source vertical commutation methods with low-inductance packaging considerations. Since vertical commutation involves current transfer between upper and lower arms, the conventional approach integrates both arms as a unified package to minimize the commutation loop path

length, thereby reducing parasitic inductance in the commutation loop. However, such packaging schemes are not applicable to current-source topologies employing horizontal commutation methods[10-11]. Different commutation approaches are illustrated in Figure 1.

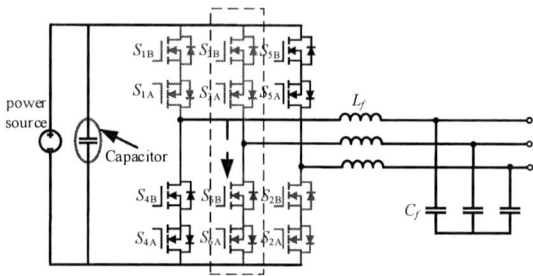

(a) Voltage-source topology with vertical commutation

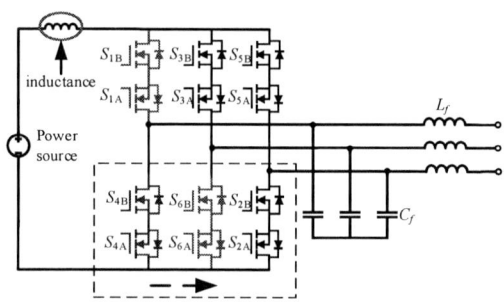

(b) Current-source topology with horizontal commutation
Fig. 1. Comparison of Different Commutation Approaches

In current-source circuit topologies, the internal commutation method within the circuit uses lateral commutation, meaning commutation occurs between different bridge arms. To meet the requirements of balanced parasitic parameters between different commutation loops, reduced parasitic parameters in the commutation loops, and multi-topology applications, a three-in-one packaging solution for bidirectional switches is proposed to address these requirements.

As shown in Figure 2, this circuit represents the circuit topology diagram of the 3×1 SiC-BDS power module proposed for packaging in this paper. This module serves as the fundamental building block for topologies such as current-source converters (CSCs), matrix converters, and PFC rectifiers.

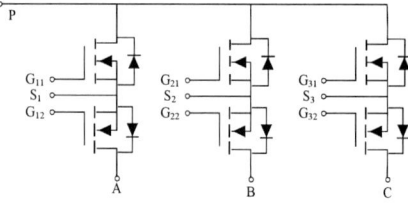

Fig. 2. 3×1-phase bidirectional switch power module circuit topology diagram

C. Module Structure and Layout Design

The majority of currently deployed bidirectional power switch modules utilize planar interconnection packaging techniques. Research on bidirectional power switch modules frequently employs 3D packaging. The 3D packaging form of power modules involves complex manufacturing processes and has low technological maturity. Among the planar interconnection packaging forms, the EasyPACK packaging proposed by Infineon Technologies offers advantages such as simplified integration, reduced inductance, and plug-and-play capability. Consequently, this paper will adopt the EasyPACK packaging form.

First, we introduce the structure and manufacturing process of this module. This module adopts the EasyPACK packaging form. The cross-sectional view is shown in Figure 3. Six SiC MOSFETs form a 3×1 bidirectional switch power module. Each phase adopts a common-source connection of two power chips. Direct Bonding Copper (DBC) is used, and the alumina ceramic in it is used for insulation and heat conduction. The DBC board uses the EasyPACK 1B standard size of 39mm × 32.5mm, and the thicknesses of the upper and lower copper layers and the ceramic layer are 0.3mm and 0.38mm respectively. Inter-chip electrical interconnections and chip-to-substrate coupling are achieved through 0.254 mm (10 mil) diameter aluminum bond wires ultrasonically welded to copper layers. The power terminals are directly bonded to the corresponding positions on the DBC board using a terminal bonder. Finally, the outer shell and the substrate are combined and sealed with silica gel.

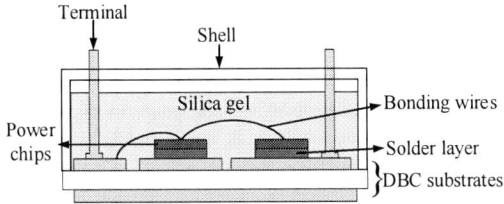

Fig. 3. Cross-sectional diagram of the 3×1-phase silicon carbide-based bidirectional switch power module

Secondly, based on the packaging form of planar interconnection, to balance and reduce the parasitic inductance of different commutation loops, a compact, symmetrical layout is proposed to arrange the three bidirectional switches symmetrically around the center. For each bidirectional switch in the three-phase configuration, the drain terminal of the upper switch device is interconnected with the central copper layer embedded in the upper substrate metallization. Simultaneously, the source terminals of both upper and lower switch devices establish direct electrical continuity through multiple aluminum bonding wires, forming a low-impedance current path between the power devices. The three-phase symmetrical layout ensures consistent parasitic parameters in the commutation loops of the 3×1 bidirectional switch power module, which improves the reliability of the long-term stable operation of the Power Module. Through strategic implementation of parallel pin connections at the power terminals, the power loop parasitic inductance is effectively mitigated by employing the inductance cancellation principle, where opposing magnetic fluxes generated through the parallel conductors achieve mutual cancellation based on Maxwell's equations. The final layout and the position of the terminal settings are shown in Figure 4.

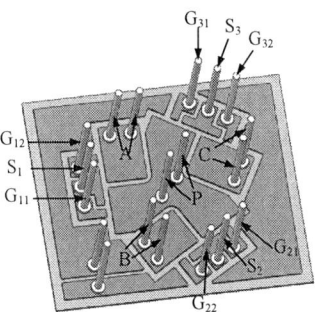

Fig. 4. 3×1-phase SiC-BDS module substrate layout and terminal position diagram.

The optimization of parasitic inductance in this power module is mainly carried out from the following three aspects, which include the parasitic inductance of bonding wires, terminals, and that generated by the DBC layout. This work reduces gate-loop parasitic inductance in bidirectional switches through optimized DBC layout and magnetic flux cancellation. By shortening current paths and arranging anti-parallel conductors to generate opposing fluxes per Faraday's law, the design achieves 62% lower loop inductance than conventional structures, as validated by electromagnetic simulations. Therefore, as shown in Figures 4 and 5, the design of the gate copper layer adopts this method.

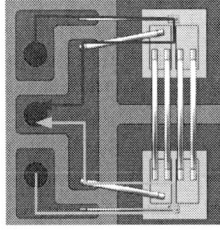

Fig.5. Power module gate circuit schematic diagram.

III. SIMULATION AND EXPERIMENT

A. Analysis of the Simulation

a. Electrical Simulation

The parasitic inductances of the three-phase SiC-BDS power module were extracted using ANSYS Q3D software based on the aforementioned packaging layout configuration. As illustrated in Figure 6, the parasitic inductance of the module consists of two components, namely, parasitic inductance of the power loop and the gate drive loop.

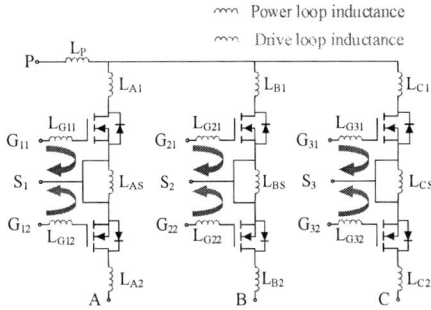

Fig. 6. Schematic diagram of module parasitic inductance.

In ANSYS Q3D, the parasitic inductance of the commutation loop is extracted using frequency scanning. The simulation results for the parasitic inductance are shown in Figure 7. When the sampling frequency reaches 100 MHz, the parasitic inductance between different phase approaches 20.2 nH. The imbalance level of power loop commutation inductance is calculated to be 0.2%. Similarly, the parasitic inductance in the gate driver loop converges to 14.15nH. According to simulation results of other gate driver loop parasitic inductances in this module, the inductance imbalance level in the driver circuit is determined as 0.15%.

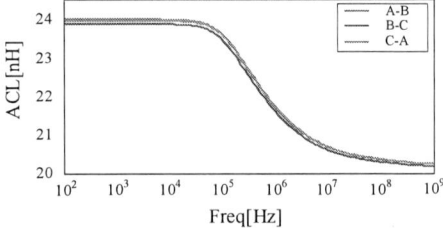

(a) Simulation results of the parasitic inductance in the power loop.

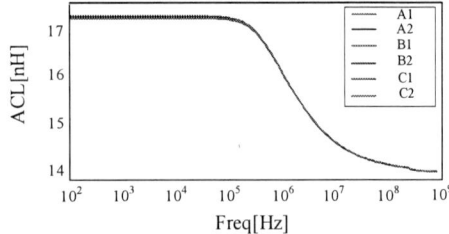

(b) Simulation results of the parasitic inductance in the drive loop.

Fig. 7. Simulation Results of Parasitic Inductance.

According to the simulation results, the commutation inductance of each commutation loop is consistent in this layout of the power module, ensuring uniform voltage stress on each phase SiC power chip during operation. Additionally, the parasitic inductance of each SiC power chip's drive circuit remains consistent, ensuring consistent switching times and losses across all power chips.

b. Thermal Simulation

The substrate was imported into COMSOL Multiphysics for finite element simulation, where material parameters and thermal boundary conditions were configured. Thermal simulation validation was conducted on the substrate, as shown in Figure 8. The heat distribution on the substrate is uniform, with the highest temperature of the chip reaching 110°C, which meets the cooling requirements.

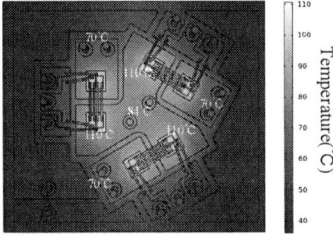

Fig. 8. 3×1-phase silicon carbide-based bidirectional switch power module thermal distribution

B. Module Fabrication and Processing

The WM2A075120B SiC power device, rated at 1200 V and 33 A, was employed as the power semiconductor component in the module, and the module packaging process is shown in Figure 9.

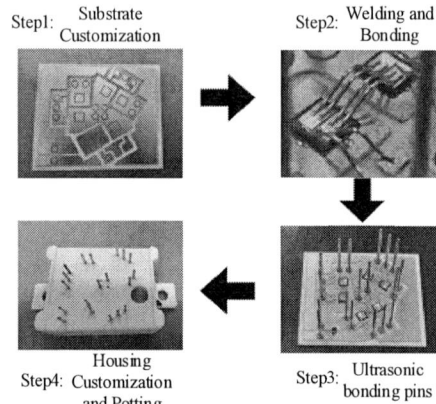

Fig. 9. Fabrication process of silicon carbide-based bidirectional switch power module in EasyPACK packaging format

C. Test of the Static and Dynamic Parameter

The static parameter characterization of the power module was performed using Keysight B1506A Power Device Analyzer, with comprehensive measurements including threshold voltage, on-state resistance, leakage current, and body diode forward voltage drop of the embedded SiC MOSFET dies. Subsequently, these measured values will be compared with those of the bare power chip. It is concluded that after the Packaging Process, the static parameters of the power chip remain normal.

For the bidirectional switch module, a double-pulse test circuit is constructed to perform dynamic parameter testing. The schematic diagram of the double-pulse test for this module is shown in Figure 10. Considering its unique lateral commutation method, two of the three-phase bridge legs are configured as a half-bridge structure of a conventional double-pulse test circuit, with the P terminal serving as the midpoint of the test circuit. The gate and source terminals of the upper switch in the upper bridge arm are shorted, while the lower switch is connected to a 15V voltage signal. In the lower bridge arm, the upper switch is connected to the double-pulse control signal, and the lower switch is connected to a 15V voltage signal.

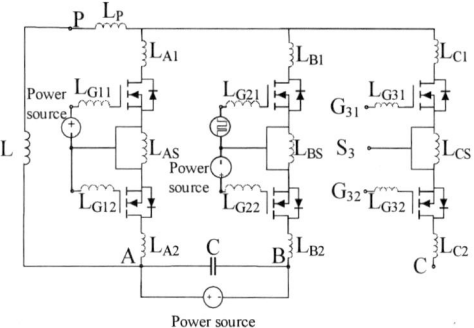

Fig. 10. Schematic diagram of the double-pulse test for this module

979-8-3315-1110-4/25 $31.00 © 2025 IEEE

Based on the double-pulse test schematic, a double-pulse test platform is constructed for this module, including a DC voltage source, signal generator, load inductor, voltage probe, oscilloscope, current clamp, and the device under test (DUT) module. The double-pulse test platform is shown in Figure 11.

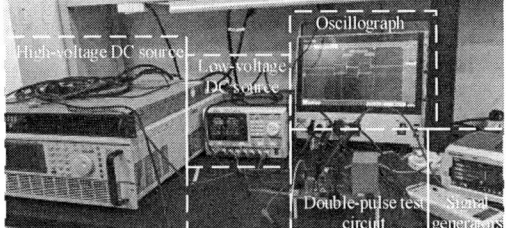

Fig. 11. Bidirectional switch module double-pulse test platform

Under the conditions of a DC bus voltage of 400 V and a load current of 15 A, experimental investigations were conducted on the A-phase to B-phase, B-phase to C-phase, and C-phase to A-phase, respectively. During testing, the pulse width was set to 10 μs, and the turn-on and turn-off waveforms of the bidirectional switch power module were measured. The experimental waveforms are shown in Figure 12. It can be observed that during turn-off, the drain-source voltage of the module rises to 400V, with voltage overshoots of 55V, 58V, and 63V in the three commutation circuits, respectively. The voltage stress across the three commutation circuits is evenly distributed, meeting the design requirements.

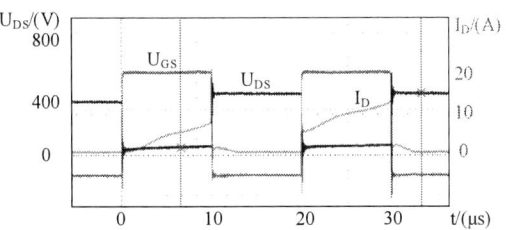

(a) A-B phase switching waveform

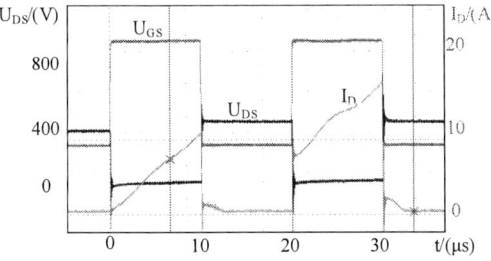

(b) B-C phase switching waveform

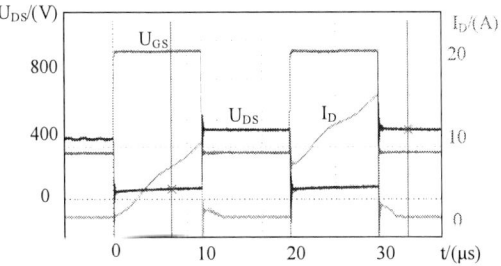

(c) C-A phase switching waveform

Fig. 12. Dual-pulse test waveform diagram

IV. CONCLUSION

To address the challenges of achieving high power density, minimizing parasitic inductance, and ensuring multi-topology compatibility in SiC-BDS modules, this paper presents a novel 3×1 SiC-BDS power module incorporating commutation inductance balancing through symmetrical layout design. The proposed architecture effectively balances parasitic inductance distribution across commutation loops via geometric symmetry optimization. Parasitic parameter characterization was performed using ANSYS Q3D, revealing exceptional inductance balance with a 0.2% imbalance ratio in power commutation loops and 0.15% in driver loops. A complete manufacturing workflow for the SiC-BDS module is established, culminating in prototype development and experimental validation through double-pulse testing.

REFERENCES

[1] G. Wang, H. Wen, W. Liu and F. Li, "Physical Structure, Characteristics, and Applications of Monolithic Bidirectional Switches: A Comprehensive Review," in IEEE Transactions on Power Electronics, vol. 40, no. 7, pp. 9187-9199, July 2025.

[2] D. Bisi, "GaN Bidirectional Switches: The Revolution is Here," in IEEE Power Electronics Magazine, vol. 12, no. 1, pp. 29-36, March 2025.

[3] A. Kanale et al., "Comparison of the Capacitances and Switching Losses of 1.2 kV Common-Source and Common-Drain Bidirectional Switch Topologies," 2021 IEEE 8th Workshop on Wide Bandgap Power Devices and Applications (WiPDA), Redondo Beach, CA, USA, 2021, pp. 112-117.

[4] D. Jia et al., "A Multi-terminal Silicon Carbide Power Module with Low Parasitic Inductance," 2024 IEEE 10th International Power Electronics and Motion Control Conference (IPEMC2024-ECCE Asia), Chengdu, China, 2024, pp. 2039-2044.

[5] L. Qiao et al., "Performance of a 1.2kV, 288A full-SiC MOSFET module based on low inductance packaging layout," 2017 IEEE Applied Power Electronics Conference and Exposition (APEC), Tampa, FL, USA, 2017, pp. 3038-3042.

[6] S. Li, L. M. Tolbert, F. Wang, and F. Z. Peng, "Stray Inductance Reduction of Commutation Loop in the P-cell and N-cell-Based IGBT Phase Leg Module," IEEE Transactions on Power Electronics, vol. 29, no. 7, pp. 3616–3624, Jul. 2014.

[7] Y. Ren, X. Yang, F. Zhang, L. Tan and X. Zeng, "Analysis of a low-inductance packaging layout for Full-SiC power module embedding split damping," 2016 IEEE Applied Power Electronics Conference and Exposition (APEC), Long Beach, CA, USA, 2016, pp. 2102-2107.

[8] E. Vagnon, J. C. Crebier, Y. Avenas and P. O. Jeannin, "Study and realization of a low force 3D press-pack power module," 2008 IEEE Power Electronics Specialists Conference, Rhodes, Greece, 2008, pp. 1048-1054.

[9] S. Liang, Y. Liang, P. Sun, J. Gong, M. Zou and Z. Zeng, "Design and Demonstration of 3D-Stacked Packaging SiC Power Module with Low Parasitic Inductance and Low Thermal Resistance," 2023 IEEE 2nd International Power Electronics and Application Symposium (PEAS), Guangzhou, China, 2023, pp. 1715-1719.

[10] R. Amorim Torres, H. Dai, W. Lee, B. Sarlioglu and T. Jahns, "Current-Source Inverter Integrated Motor Drives Using Dual-Gate Four-Quadrant Wide-Bandgap Power Switches," in IEEE Transactions on Industry Applications, vol. 57, no. 5, pp. 5183-5198, Sept.-Oct. 2021.

[11] L. Ming et al., "A SiC-Si Hybrid Module for Direct Matrix Converter With Mitigated Current Spikes," in IEEE Journal of Emerging and Selected Topics in Power Electronics, vol. 10, no. 4, pp. 3805-3817, Aug. 2022

2025 IEEE Workshop on Wide Bandgap Power Devices and Applications in Asia (WiPDA Asia)

An Experimental Study on Single Pulse Avalanche Characteristics of Si/SiC Hybrid Switch

Hangzhi Liu*
Anhui Provincial Key Laboratory of Power Electronics and Motion Control
Anhui University of Technology
Ma'anshan, China
hangzhi_liu@ahut.edu.cn

Yuming Zhou
Anhui Provincial Key Laboratory of Power Electronics and Motion Control
Anhui University of Technology
Ma'anshan, China
ymzhou@ahut.edu.cn

Abstract—**This paper presents a comprehensive evaluation of the single-pulse avalanche behavior of the Si/SiC hybrid switch through experimental testing. Based on the fundamental operating principles of the hybrid switch, both nondestructive and destructive single-pulse avalanche tests are performed under the optimal gate turn-on and turn-off delay time between the two device branches. The branch that primarily determines the unclamped inductive switching (UIS) characteristics and consequently limits the single-pulse avalanche withstand capability of the Si/SiC hybrid switch is identified, and the dominating mechanisms are revealed. Additionally, the effect of gate turn-off delay time on the UIS performance of the hybrid switch is extensively investigated and analyzed through experimental tests. The findings of this study provide valuable evidences for the long-term reliable application of the Si/SiC hybrid switch in power conversion systems.**

Keywords—hybrid switch, Si/SiC, single-pulse avalanche

I. INTRODUCTION

Thanks to the outstanding electrical and thermal properties of Silicon Carbide (SiC), SiC-based MOSFETs have garnered significant interest in both academic and industrial circles in recent years. These MOSFETs outperform traditional Silicon (Si) IGBTs in various performance [1]. However, the high cost of both SiC materials and the fabrication process have hindered the widespread commercial adoption of SiC MOSFETs, especially when the current ratings are scaled to match those of Si IGBTs. To overcome these challenges, the Si/SiC hybrid switch has been introduced, typically combining a high-current Si IGBT with a lower-current SiC MOSFET. This hybrid configuration has been the subject of extensive research by numerous groups [1-3].

Since the introduction of the Si/SiC hybrid switch, plenty of studies have shown that its performance is comparable with that of SiC MOSFETs rated for the same voltage and current, all while significantly reducing overall power device costs [1-3]. These studies have covered diverse topics such as optimal switching patterns, current sharing mechanisms, gate control strategy optimization, simulation model development, and the selection of appropriate chip sizes [4-7]. In addition to improving the basic performances in practical applications, the robustness of the individual devices within the Si/SiC hybrid switch, particularly the SiC MOSFET under short-circuit (SC) and unclamped inductive switching (UIS) conditions, has been a major focus of research over the past decades [8-11]. SC and UIS testing have become standard methods for assessing the robustness of the hybrid switch under extreme stress conditions. For example, in [12],

experimental study revealed that the SC capability of the hybrid switch is primarily constrained by the auxiliary SiC MOSFET branch. This study also founds that factors like gate driving voltage and dc-link voltage have a more significant effect on the SC performance than case temperature. Additionally, failure modes and mechanisms were thoroughly identified. However, similar attention has not yet been given to the UIS characteristics of the hybrid switch. The specific device branch responsible for limiting UIS robustness has yet to be determined, and how physical factors relevant to the device's basic performance improvement influence UIS avalanche characteristics of hybrid switch remains largely unexplored.

The purpose of the study conducted in this paper is to address this problem. The structure of this paper is outlined as follows: Section II introduces the experimental setup for the UIS testing of the Si/SiC hybrid switch. Section III presents the single-pulse avalanche tests performed on the hybrid switch with the optimized gate delay time. It also identifies which branch governs UIS capability of the hybrid switch and explores the impact of the gate control pattern on the avalanche behavior of Si/SiC hybrid switch. Finally, the conclusions of this paper are drawn in Section IV.

II. EXPERIMENTAL TESTING SETUP

The UIS test circuit diagram for the hybrid switch is presented in Fig. 1, along with the corresponding waveforms showing avalanche failure. In the built experimental testing platform, the bank capacitance, C, is composed of four parallel capacitors, each rated at 420 µF, while the main inductor, L, is set at 0.19 mH. The hybrid switch is formed by connecting a 20 A Si IGBT and an 11 A SiC MOSFET in parallel, with both components being commercially available products rated at 1200 V.

As there is no anti-parallel diode across the inductor, when the current through the hybrid switch increases linearly and reaches the maximum avalanche current, it no longer rises due to the abrupt switching off of the hybrid switch, but instead decreases linearly. This decreasing of the current through the hybrid switch coupled with the main inductor induces a voltage (V_{br}) across the hybrid switch, triggers avalanche breakdown of the hybrid switch and allows current to continue flowing. As the hybrid switch enters avalanche breakdown mode, the energy stored in the load inductor during the current ramp-up phase is dissipated through the hybrid switch. As soon as the energy consumption exceeds thermal limits, device failure occurs, and thus the current flowing through the hybrid switch doesn't decreasing with a rate of $(V_{br}-V_{DC})/L$ any more but rises suddenly.

This work was supported in part by the Opening Project of Key Laboratory of Power Electronics and Motion Control of Anhui Higher Education Institutions under Grant PEMC24004, in part by Anhui University of Technology Young Teachers Research Fund under Grant QZ202412, and in part by Scientific Research Startup Fund for Introduced Talents of Anhui University of Technology under Grant QD202340.

979-8-3315-1110-4/25 $31.00 © 2025 IEEE

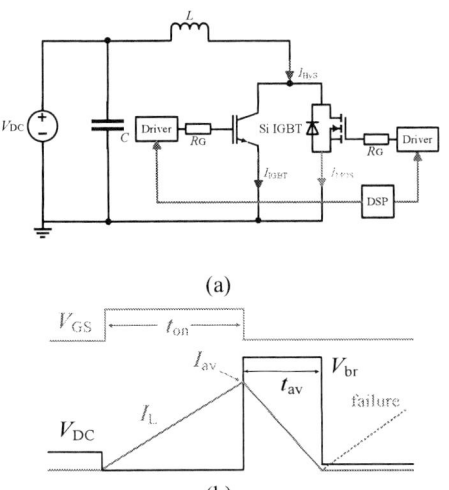

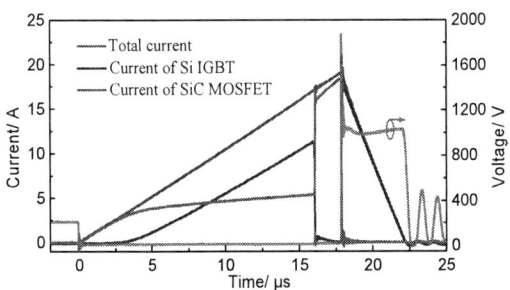

Fig. 1 (a) Schematic diagram of UIS testing circuit for the hybrid switch and (b) the ideal testing waveforms

Fig. 2 The tested waveforms of the hybrid switch under single pulse avalanche condition

An experiment test is then conducted on the hybrid switch at room temperature using the experimental platform, with the current and voltage waveforms shown in Fig. 2. The dc-link voltage is set to be 200 V for the single-pulse UIS test to ensure a complete current conduction path during the avalanche mode. The gate drive duration for the SiC MOSFET is set to be 19.2 μs, resulting in a maximum avalanche current of approximately 15 A. As shown in Fig. 2, when the Si IGBT is switched off, the total current is diverted to the SiC MOSFET branch and continues to increase linearly over time until the SiC MOSFET is turned off. As the current begins to decrease linearly, the hybrid switch enters avalanche mode, with the voltage across the switch reaching around 1000 V. Upon closer inspection of Fig. 2, it becomes apparent that once the hybrid device enters avalanche mode, the avalanche current abruptly shifts from the SiC MOSFET branch to the Si IGBT, whose gate signal is turned off 1.8 μs earlier than the internal SiC MOSFET. This observation indicates that for the Si/SiC hybrid switch configuration, it is not the auxiliary SiC MOSFET but the main Si IGBT that enters avalanche mode.

III. RESULTS AND DISCUSSIONS

In this section, UIS measurements are then carried out. During these tests, we capture the current and voltage waveforms of both device branches and concentrate our analysis on the avalanche current within the hybrid assembly. Inspired by the distinctive conduction and commutation behaviors observed under avalanche conditions, we proceed

to perform single-pulse avalanche trials on discrete IGBT and SiC MOSFET. These comparative tests yield clear evidence for identifying which branch limits the UIS performance and governs the avalanche characteristics of the combined switch. Finally, how crucial factors, such as the T_{off_delay}, impact the hybrid switch's single-pulse avalanche ruggedness is also experimentally analyzed.

A. Dominate Branch for Single Pulse Avalanche Capability of Hybrid Switch

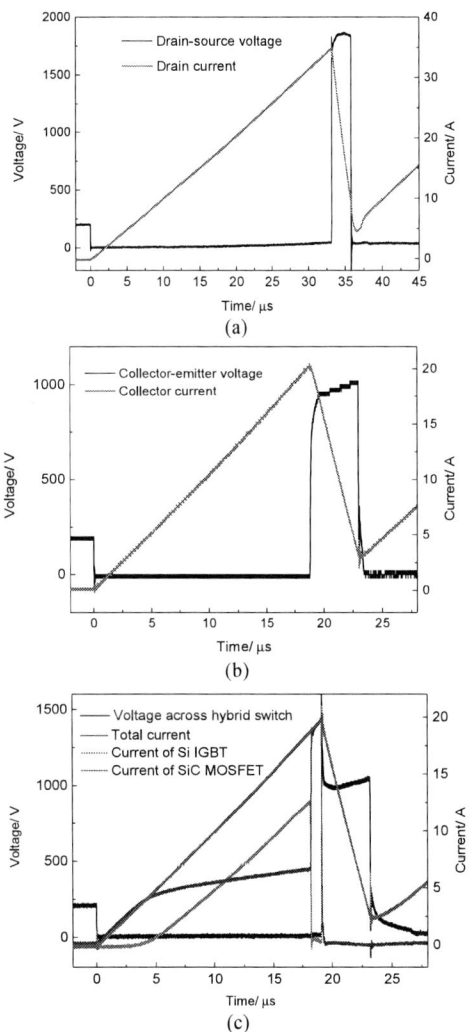

Fig.3 Waveforms of the discrete (a) SiC MOSFET, (b) Si IGBT and (c) the hybrid switch under single pulse avalanche condition with thermal failures

UIS tests are performed on the individual SiC MOSFET and Si IGBT, which together form the hybrid switch, using the specially designed experimental setup. The results of these tests are presented in Fig. 3. The experimental tests are conducted by progressively increasing the gate turn-on duration for each device, allowing the energy stored in the inductor to rise steadily to a level that induces thermal breakdown failure in the discrete SiC MOSFET, IGBT, and Si/SiC hybrid switch during the avalanche breakdown phase. The results indicate that a gate turn-on duration of approximately 20 μs is required for both the Si/SiC hybrid switch and the discrete IGBT, whereas the discrete SiC MOSFET requires a much longer gate duration of 33.1 μs to reach destructive failure. This longer duration results in a peak avalanche current of 34.7 A, which is around 15 A higher than that observed in both the hybrid switch and discrete IGBT.

The critical avalanche energy (E_{cr}) represents the energy dissipated by the SiC MOSFET, IGBT, and Si/SiC hybrid switch during avalanche breakdown just before failure, and is calculated by integrating the product of the total current and the corresponding voltage. Testing data indicates that E_{cr} of the discrete SiC MOSFET is 98.6 mJ, approximately 1.2 times greater than that of the other two devices. In contrast, E_{cr} of the IGBT and the Si/SiC hybrid switch are nearly identical to be 45 mJ. What's more significant is that the avalanche breakdown voltage of the hybrid switch during avalanche breakdown mode is almost equal to that of the individual IGBT forming the branch inside the hybrid switch. These further accounts for the conclusion that it is not the auxiliary SiC MOSFET, but the Si IGBT branch, that determines the UIS characteristics and limits the single-pulse avalanche capability of the Si/SiC hybrid switch.

B. Impact of Gate Turn-off Delay on Single Pulse Avalanche Capability of Hybrid Switch

In practical operation, the chosen gate turn-off delay for the internal SiC MOSFET not only shifts the power loss distribution within the hybrid switch but also upsets the thermal balance between its two branches. Furthermore, because load current and temperature typically fluctuate, using a fixed gate turn-off delay on the auxiliary SiC MOSFET cannot reliably maintain the optimal power conversion efficiency.This study extensively investigates the impact of varying the gate signal turn-off delay time of the smaller SiC MOSFET branch on the UIS avalanche performance of the hybrid switch by means of experiment testing.

With the peak avalanche current held constant and the gate-pulse width of the SiC MOSFET fixed, Fig. 4 plots how varying the IGBT's turn-off delay (i.e. the interval between the MOSFET's gate-off command and the actual drop of collector current) influences the total energy dissipated in avalanche. As the turn-off delay is increased from 0.9 μs to 1.8 μs, the measured avalanche energy falls in a nearly linear fashion. This behavior can be attributed to the fact that, during the charging of the series inductor, a well-defined amount of magnetic energy is stored and must be extracted when the current commutates off the internal MOSFET. In our hybrid arrangement, that stored inductive energy is partially shed as IGBT turn-off losses. Specifically, through the tail-current region, and thus does not contribute to the voltage–current product used to calculate avalanche energy in the MOSFET. As the turn-off delay lengthens, two effects coincide to reduce the avalanche energy: first, a greater fraction of the inductor's stored energy is bled off during the IGBT's extended turn-off interval; second, the diminishing tail current—caused by the recombination and sweep-out of excess carriers in the IGBT's drift region—becomes a progressively smaller contributor to total energy loss. Beyond approximately 1.8 μs, the curve plateaus, signifying that virtually all injected carriers have been removed and the residual tail current is negligible. In this regime, further extension of the gate-off interval cannot appreciably lower the avalanche energy, as there is essentially no remaining

charge to dissipate in the IGBT's tail region. Importantly, while the turn-off delay exerts a pronounced influence on the extracted avalanche energy, it has a negligible impact on the peak avalanche voltage or the device's single-pulse avalanche withstand capability. This decoupling indicates that, once carriers are fully cleared, the MOSFET's avalanche ruggedness is dictated primarily by its intrinsic blocking voltage and channel-junction characteristics, rather than by dynamic gate-timing effects. From a design standpoint, selecting a turn-off delay in the vicinity of 1.8 μs therefore optimizes energy dissipation without compromising voltage-withstand performance, ensuring both robust avalanche operation and minimized thermal stress in the hybrid switch.

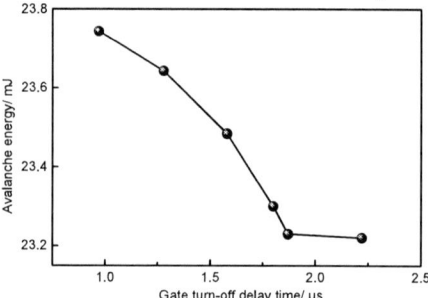

Fig. 4 The avalanche energy E_{av} of the hybrid switch at varied gate turn-off delay time

IV. CONCLUSION

This work systematically investigates the unclamped inductive switching (UIS) behavior of Si/SiC hybrid configuration, combining an auxiliary SiC MOSFET with a main silicon IGBT, through both non-destructive and destructive single-pulse avalanche tests. In the non-destructive UIS trials, we observe that the hybrid device's avalanche current is carried almost exclusively by the Si IGBT: once the inductor-stored energy forces the device into avalanche, the MOSFET remains below its breakdown voltage and essentially "hands off" the current to the IGBT. Conversely, in the high-stress (destructive) UIS experiments, the hybrid switch's critical-energy threshold—i.e. the maximum single-pulse avalanche energy prior to device failure—matches almost exactly that of the standalone Si IGBT. Likewise, the peak avalanche voltage at which breakdown occurs in the hybrid string aligns with the IGBT's own avalanche voltage, confirming that the Si leg dictates both energy-withstand and voltage-withstand limits.

Together, these findings make clear that the single-pulse avalanche ruggedness and UIS robustness of the Si/SiC hybrid topology are fundamentally constrained by the silicon IGBT branch, not the wide-bandgap MOSFET. Moreover, by systematically varying the gate turn-off delay of the IGBT, we demonstrate that while longer delays noticeably reduce the energy dissipated in avalanche—by allowing more of the inductor's stored energy to be bled off during the extended turn-off tail region—they do not alter either the avalanche breakdown voltage or the ultimate single-pulse energy capability of the hybrid switch. This decoupling implies that, beyond a certain delay threshold, all excess carriers in the IGBT have been cleared and further timing adjustments yield diminishing returns in energy reduction without impacting voltage ruggedness. From a practical standpoint, these insights provide clear guidance for optimizing gate-drive timing in power converters: by selecting a turn-off delay bordering the carrier-clearing threshold, designers can minimize thermal stress and switching losses in avalanche events without sacrificing voltage-withstand reliability. In doing so, the Si/SiC hybrid architecture can be confidently deployed for long-term, high-reliability applications in renewable-energy inverters, motor drives, and other demanding power-conversion systems.

REFERENCES

[1] A. Q. Huang, X. Song and L. Zhang, "6.5 kV Si/SiC hybrid power module: An ideal next step?," 2015 IEEE International Workshop on Integrated Power Packaging (IWIPP), Chicago, IL, USA, 2015, pp. 64-67.

[2] D. Woldegiorgis, M. M. Hossain, Z. Saadatizadeh, Y. Wei and H. A. Mantooth, "Hybrid Si/SiC Switches: A Review of Control Objectives, Gate Driving Approaches and Packaging Solutions," in IEEE Journal of Emerging and Selected Topics in Power Electronics, vol. 11, no. 2, pp. 1737-1753, April 2023.

[3] J. He, R. Katebi and N. Weise, "A Current-Dependent Switching Strategy for Si/SiC Hybrid Switch-Based Power Converters," in IEEE Transactions on Industrial Electronics, vol. 64, no. 10, pp. 8344-8352, Oct. 2017.

[4] J. Wang, Z. Li, X. Jiang, C. Zeng and Z. J. Shen, "Gate Control Optimization of Si/SiC Hybrid Switch for Junction Temperature Balance and Power Loss Reduction," in IEEE Transactions on Power Electronics, vol. 34, no. 2, pp. 1744-1754, Feb. 2019.

[5] Z. Li, J. Wang, L. Deng, Z. He, X. Yang, B. Ji. Z. J. Shen, "Active Gate Delay Time Control of Si/SiC Hybrid Switch for Junction Temperature Balance Over a Wide Power Range," in IEEE Transactions on Power Electronics, vol. 35, no. 5, pp. 5354-5365, May 2020.

[6] Y. Fu, Z. Ma and H. Ren, "A Low Cost Compact SiC/Si Hybrid Switch Gate Driver Circuit for Commonly Used Triggering Patterns," in IEEE Transactions on Power Electronics, vol. 37, no. 5, pp. 5212-5223, May 2022.

[7] X. Song, L. Zhang and A. Q. Huang, "Three-Terminal Si/SiC Hybrid Switch," in IEEE Transactions on Power Electronics, vol. 35, no. 9, pp. 8867-8871, Sept. 2020.

[8] J. Wei et al., "Review on the Reliability Mechanisms of SiC Power MOSFETs: A Comparison Between Planar-Gate and Trench-Gate Structures," in IEEE Transactions on Power Electronics, vol. 38, no. 7, pp. 8990-9005, July 2023.

[9] P. Steinmann, S. Ganguly, B. Hull, at al., "Improvements to the Analytical Model to Describe UIS Events," in IEEE Transactions on Electron Devices, vol. 69, no. 7, pp. 3848-3853, July 2022.

[10] J. Wang and X. Jiang, "Review and analysis of SiC MOSFETs' ruggedness and reliability", IET Power Electronics, vol. 13, no. 3, pp. 445-455, 2020.

[11] H. Liu, D. He, Q. Guo and Y. Zhou, "UIS Characterization of Hybrid Si/SiC Device under Single Pulse Avalanche Mode," 2024 IEEE 10th International Power Electronics and Motion Control Conference (IPEMC2024-ECCE Asia), Chengdu, China, 2024, pp. 1856-1859.

[12] J. Wang, X. Jiang, Z. Li and Z. J. Shen, "Short-Circuit Ruggedness and Failure Mechanisms of Si/SiC Hybrid Switch," in IEEE Transactions on Power Electronics, vol. 34, no. 3, pp. 2771-2780, March 2019.

Substrate Coupling Considerations for Monolithic Integration of High-Voltage Power Transistors with Low-Voltage Devices and Circuits in GaN-on-Si Technology

Rui (Ray) Yao[*]
Department of Electrical Engineering and Electronics
The University of Liverpool
Liverpool, UK

Miao Cui
Department of Electrical and Electronic Engineering, SAT
Xi'an Jiaotong-Liverpool University
Suzhou, China

Zhao Wang
Department of Communications and Networking, SAT
Xi'an Jiaotong-Liverpool University
Suzhou, China

Sang Lam
Department of Electrical and Electronic Engineering, SAT
Xi'an Jiaotong-Liverpool University
Suzhou, China
s.lam.cn@ieee.org

Stephen Taylor
Department of Electrical Engineering and Electronics
The University of Liverpool
Liverpool, UK

([*]also with
School of Advanced Technology (SAT)
Xi'an Jiaotong-Liverpool University
Suzhou, China)

Abstract— The device area dependence of substrate coupling in GaN-on-Si technology is investigated in situations of integrating a high-voltage large-sized transistor on the same chip. It is found by computational electromagnetic (EM) investigation that $|S_{21}|$ as a measure of the EM coupling increases minimally by ≈7.7 dB/decade or less, when the device area increases by size expansion along the separation direction. This implies that $|S_{21}|$ at 30 MHz would increase by at least 31 dB from -77 dB (for 2×2 μm^2 area) to -46 dB (or -25 dB in the worst case), when a typical power transistor of 40-mm channel width is placed at 300 μm away from a device that is 10 thousand times smaller in area.

Keywords—substrate coupling, power transistor integration, device area, GaN-on-Si technology, computational EM

I. INTRODUCTION

Gallium nitride (GaN) is an important wide-bandgap semiconductor [1]-[2] with its great promise for making high-voltage power electronic devices operating at higher frequencies beyond sub-megahertz (MHz) as in traditional power semiconductor devices in silicon (Si) technology. Such GaN power electronic devices can offer superior circuit performance in high frequency electronic power conversion systems [1],[3] for common applications in such as electric vehicles and renewable energy technology. In particular, GaN-on-Si technology has been an attractive option for monolithic integration [3]-[6] of high-voltage power transistors and low-voltage electronic circuits (such as voltage references [7] and temperature sensing [8]). However, signal integrity is of paramount importance in such power integrated circuits (ICs) in which the switching transients of high voltage levels can easily interfere with the small signals of low-voltage circuits built on the same substrate.

To reduce the manufacturing cost and complexity as well as the parasitic circuit elements in the device and circuit packaging, it is certainly desirable to monolithically integrate the high-voltage power semiconductor devices with the low-voltage circuits on the same chip. By sharing the same Si semiconductor substrate however, the high-voltage switching transients can easily couple through the substrate and then interfere with the electronic circuits in low voltage operation.

To make things even worse, high-voltage power transistors usually occupy a much larger chip area [9] and hence likely larger capacitive coupling to the substrate. As a result, non-negligible crosstalk voltage may be generated in other circuits on the same chip despite large enough separation distance. So, there is engineering importance to investigate substrate coupling for considerations of the monolithic integration of high-voltage power transistors in GaN-on-Si technology.

II. SUBSTRATE COUPLING IN GaN-ON-Si TECHNOLOGY

With signal integrity considerations in power ICs that have high-voltage power transistors integrated on the same chip, there have been various research works on the substrate coupling in GaN-on-Si technology. In one early experimental study [10], the current-voltage characteristics of a low-voltage GaN-based high electron mobility transistor (HEMT) were measured when a high voltage bias was applied to an adjacent electrode that was on the same chip. There was no investigation into the dependence on the electrode chip area though. There was also theoretical work on suppressing the crosstalk in monolithic GaN devices by using electric-field decoupling structure [11]. The electric fields and the device voltages (for calculating the crosstalk current) were computed by TCAD simulations which give no frequency dependence results. There was no investigation into the device area impact on the crosstalk. We also recently reported experimental measurements [12] and computational electromagnetic (EM) results [13]-[14] on substrate coupling in GaN-on-Si technology. However, the device structures (for modelling the aggressor and victim transistors) occupy more or less the same chip area. There is a gap of knowledge about how much a large device area might worsen the substrate coupling in GaN-on-Si power ICs. Such a gap is to be filled up by this work.

III. 3D COMPUTATIONAL ELECTROMAGNETIC (EM) INVESTIGATION INTO SUBSTRATE COUPLING

To investigate the crosstalk through substrate coupling from a high-voltage power transistor in GaN-on-Si technology, we performed three-dimensional (3D) computational EM studies by the finite element method (FEM). Maxwell's equations were solved computationally to

979-8-3315-1110-4/25 $31.00 © 2025 IEEE

find out the electric field in the GaN-on-Si device structure and then to compute the frequency-dependent voltage at the victim end. Compared with TCAD simulations which computationally solve Poisson's equations, S-parameters are obtained in our EM investigation to find out the substrate coupling. The S-parameter data can reveal more clearly the EM coupling through the substrate at high frequencies of multi-MHz and beyond. Such frequency dependence results are not available from TCAD simulations [11]. The electric field distribution details in the GaN-on-Si structure cannot be determined from experimental investigation.

Fig. 1 shows the GaN-on-Si device structure with key dimensions of the geometrical design included. Table I lists the essential material parameters with the corresponding values adopted in our 3D computational EM investigation up to 100 MHz. Both the device structure and the material properties are close as much as possible to the actual fabricated devices [15]. A small discrepancy is the AlGaN barrier layer of which the thickness should be 25 nm and the electrical conductivity (σ) is not exactly known in the experimental work. The EM simulation considerations of this small discrepancy were described in [13] where more details can also be found about the EM simulation settings (including the 50-Ω lumped ports) adopted in this work. It is worth pointing out that the aluminium (Al) metal strips, which cover the AlGaN barrier layer and GaN buffer layer, respectively represent terminals (e.g. drain or gate) of two separate GaN transistors sharing the same substrate. So, the Al metal strip area corresponds to the device area of the GaN transistors.

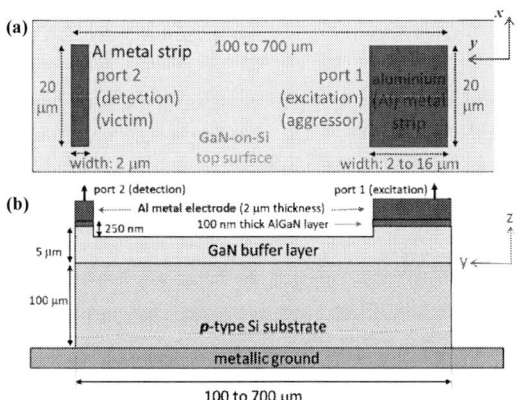

Fig. 1. A top view in (a) and a schematic cross-sectional diagram in (b) showing the GaN-on-Si device structure for investigating the substrate coupling when the agressor (excitation port) has much larger device area.

TABLE I. MATERIAL PARAMETERS AND CORRESPONDING VALUES OF THE GAN-ON-SI DEVICE STRUCTURE ADOPTED FOR 3D COMPUTATIONAL EM INVESTIGATION UP 100 MHZ

material	conductivity σ (S/m)	relative permittivity ε_r
Al metal electrode	3.8×10^7	1
AlGaN barrier layer	5000	9.2
GaN buffer layer	200	8.9
p-type Si substrate	2000	11.9

Fig. 2 shows the computed electric field distribution in cases with increasing electrode area (port 1) by increasing the Al metal strip width. It can be seen that an electrode (as the

aggressor) of a doubled device area extends the region of strong electric field laterally in GaN buffer layer. On one hand, the electric field is not significantly stronger (as the applied voltage to the aggressor electrode remains the same). On the other hand, the electric field in the Si substrate remains very weak compared with that in the GaN layer.

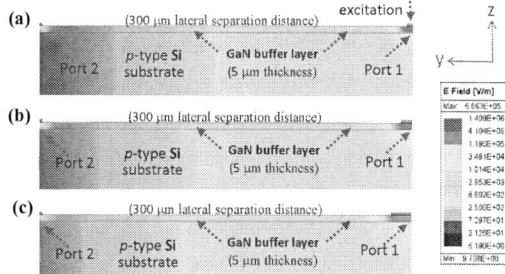

Fig. 2. Electric field distribution of the GaN-on-Si device structure, with doubling widths (hence doubling device area) of the agressor electrode (excitation port) (a) 4 μm, (b) 8 μm, and (c) 16 μm, for 300 μm separation.

In Fig. 3, the vector plots of the electric field distribution near the excitation port (as the aggressor) show more clearly the considerably stronger electric field that penetrates laterally (along the y-axis) in the GaN buffer layer, eventually giving the dominant crosstalk voltage at the detection port (as the victim). The electric field direction (indicated by the arrows) and the strength (indicated by the colour) have no significant difference regardless of the doubled widths hence the doubled device areas of the electrode at port 1.

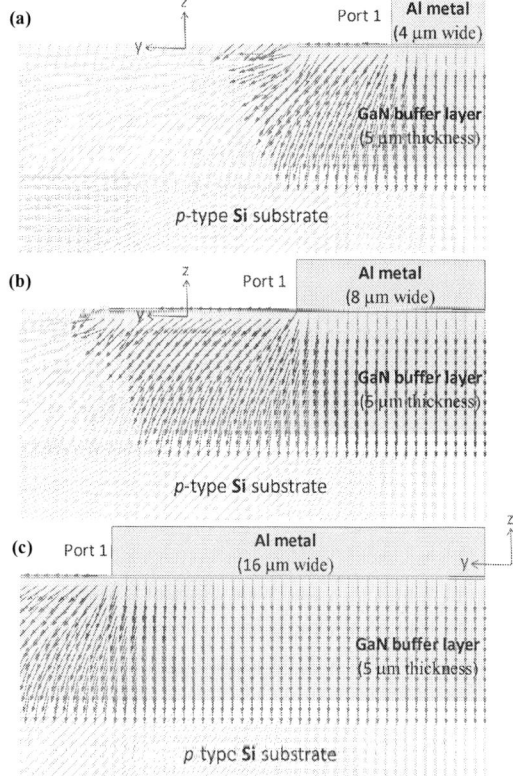

Fig. 3. Vector plot of the electric field distribution near the excitation port (as the aggressor), with doubled widths (hence the doubled device area) of the agressor electrode: (a) 4 μm, (b) 8 μm, and (c) 16 μm.

Looking more closely at the region right underneath the Al metal electrode at port 1 (Fig. 3), the electric field almost entirely points downwards (along the negative z-axis), with the resistive p-type Si substrate connected to a metallic ground (which corresponds to the grounded bottom surface of the GaN-on-Si wafer or chip). With the electrical conductivity (σ) being much larger in the AlGaN barrier layer and the resistive Si substrate than that in the GaN buffer layer, the electric field is much weaker (indicated by the pale green arrows in Fig. 3) in these two layers. In contrast, the electric field is strong in the GaN buffer layer, as indicated by the red arrows in Fig. 3. These differences in the electric field strength can be easily understood by the fact that most of the applied voltage at the excitation port drop across the GaN buffer layer (with smaller σ) which acts more like a capacitor while the AlGaN barrier layer and the Si substrate act more like resistors. When it is away from the excitation metal electrode (port 1), there are more lateral components (along the y-axis) of the electric field in the GaN buffer layer, and the electric field still remains stronger than that in the resistive Si substrate. Such electric field distribution results have implications to the high frequency coupling path and strength.

IV. ELECTRIC FIELD PROFILES & S-PARAMETER RESULTS

Fig. 4 shows the electric field magnitude profiles in the GaN buffer layer for a lateral separation distance of 300 μm. The electric field overall increases slightly when the device area of the aggressor electrode is doubled by doubling the Al metal strip width of the excitation port (port 1). Such behaviour is similar for an even larger separation distance (e.g. 700 μm). This implies that there is a likely increase in EM coupling through the GaN buffer layer when the metal strip width is increased and hence the increased device area.

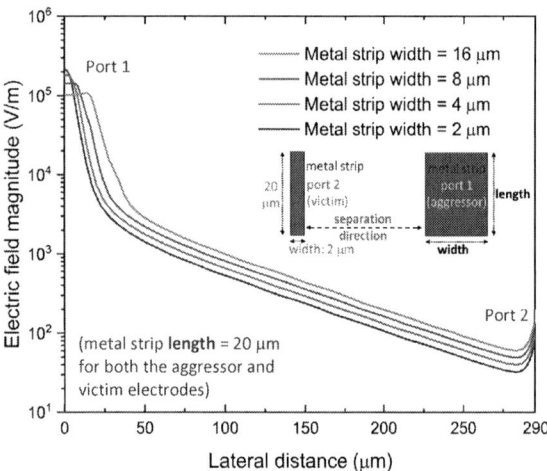

Fig. 4. Profiles of electric field (magnitude) in the middle-depth of the GaN buffer layer as a function of the lateral distance, for differing device area of the agressor electrode (port 1) as shown in the inset.

Fig. 5 and Fig. 6 show the S-parameter results from 3D EM simulations, specifically $|S_{21}|$ as a measure of the frequency-dependent EM coupling in the GaN-on-Si device structure for two very different lateral separation distances (300 μm and 700 μm). The frequency spans from 1 MHz to 100 MHz, covering adequately the high frequency operation

of even advanced GaN-based power ICs [3]. The frequency-dependent $|S_{21}|$ results agree with the inference from the electric field profiles (Fig. 4). In particular, $|S_{21}|$ increases by $\approx$ 2.3 dB or less when the Al metal strip width (and hence the aggressor's device area) is doubled, with the width extending along the negative y-axis (Fig. 1). It is equivalently 7.7 dB/decade in the dependence of $|S_{21}|$ on the device area, as $\log_2 10 = 3.332$ (= 7.7/2.3). Besides, when the Al metal strip width is doubled further and further (i.e. with the size expansion along the separation direction), the incremental change in $|S_{21}|$ diminishes. Such a trend overall is more or less the same for a separation distance of 300 μm (in Fig. 5) and for a much larger distance of 700 μm (in Fig. 6). However, there is still a significant difference at low frequencies, especially below 20 MHz. For example, with a 300-μm separation distance, $|S_{21}|$ drops steadily to about -59 dB at 1 MHz for a metal strip width of 16 μm (Fig. 5); but with a 700-μm separation distance, $|S_{21}|$ drops rapidly to about -68 dB also at 1 MHz for the same metal strip width (Fig. 6). Such a big difference of 9 dB at 1 MHz narrows to < 3 dB at 100 MHz.

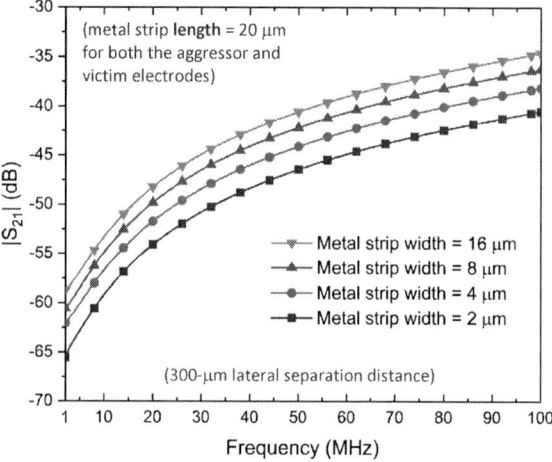

Fig. 5. $|S_{21}|$ as a measure of EM coupling in the GaN-on-Si structure, showing small steps of deteriotion ($\approx$ 2.3 dB or less in $|S_{21}|$) when the device area is doubled for the agressor electrode (port 1), with 300 μm separation.

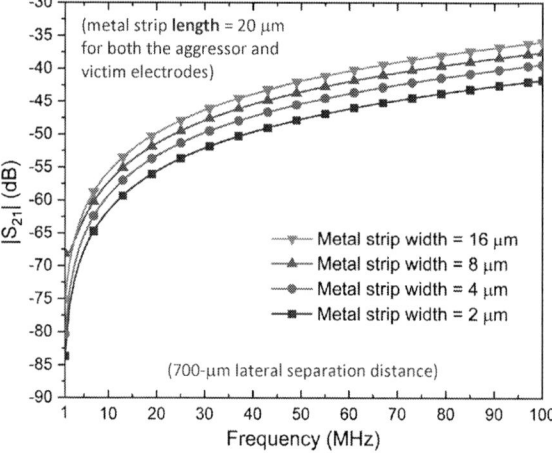

Fig. 6. $|S_{21}|$ as a measure of EM coupling in the GaN-on-Si structure for a much larger physical separation distance of 700 μm, showing more or less the same small steps of deteriotion ($\approx$ 2.3 dB or less in $|S_{21}|$) when the device area is doubled for the agressor electrode (port 1).

Fig. 7 and Fig. 8 show the $|S_{21}|$ results as a function of the lateral separation distance and of the Al metal strip width (and hence the device area). It confirms that the EM coupling in the GaN-on-Si structure increases by about 2.3 dB when the Al metal strip width is doubled at various lateral separation distances between the aggressor and victim electrodes. As shown especially in Fig. 7, a 300-μm lateral separation distance seems to be adequate for effectively lowering the substrate coupling. It gives a steep decrease in $|S_{21}|$ with the lateral distance but without using considerably more chip area due to larger physical separation.

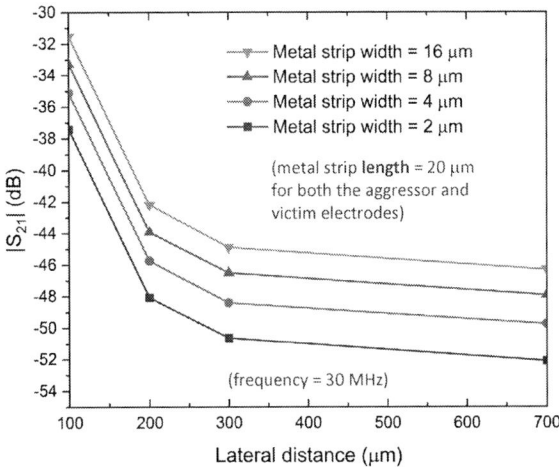

Fig. 7. $|S_{21}|$ at 30 MHz of the GaN-on-Si device structure as a function of the lateral separation distance, revealing almost the same the EM coupling dependence on the device area, even increasing the lateral separation distance indefinitely beyond about 300 μm.

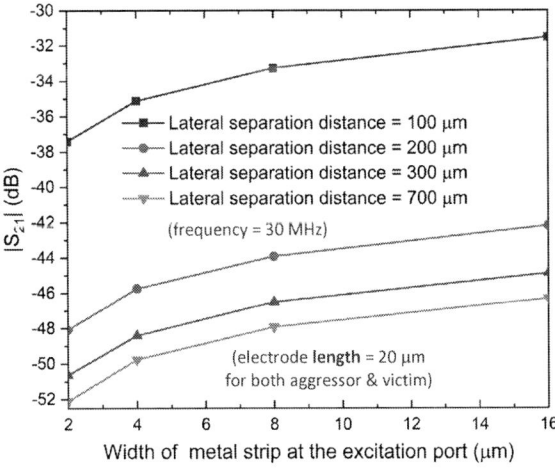

Fig. 8. $|S_{21}|$ at 30 MHz of the GaN-on-Si device structure as a function of the width of the agressor electrode (port 1), revealing that the EM coupling does not increase sharply when further doubling the device area of the agressor electrode.

Looking at the substrate coupling at 30 MHz for a lateral separation distance of 300 μm (shown in Fig. 7 or Fig. 8), $|S_{21}|$ ≈ -50.8 dB when both the aggressor (port 1) and the victim (port 2) have the device area of 20 μm (length) by 2 μm (width). By shrinking the aggressor and victim device area by 10 times (i.e. getting to 2 μm × 2 μm), $|S_{21}|$ is estimated to be -66.2 dB (= -50.8 dB – 2*7.7 dB) based on the trend shown in

Fig. 8. Considering a typical GaN power transistor as in [7] with its gate width being 40 mm and gate length 1 μm (as the minimum feature size), the drain terminal device area as the equivalent aggressor area is estimated to be about 4×10^{-8} m². This is four orders of magnitude of a victim area of 2 μm × 2 μm. This translates to $|S_{21}|$ = -35.4 dB (= -66.2 dB – 4*7.7 dB) for the substrate coupling between the high-voltage power transistor and a transistor of the minimum size in the 1 μm GaN-on-Si technology (which would give a drain area of 2 μm × 2 μm estimated by scalable CMOS layout design rule).

Apart from changing the device area by changing the width (i.e. along the *y*-axis direction shown in Fig. 1), it was also investigated about the changes in the length of the metal strip. The results are shown in Fig. 9. $|S_{21}|$ increases by ≈3.9 dB when the metal strip length (and hence the victim device area) is doubled, extending along the *x*-axis (Fig. 1). It is equivalently 13 dB/decade in the dependence of $|S_{21}|$ on the device area changes by size expansion perpendicular to the aggressor-victim separation direction. Such different dependence on the device area change can be attributed to the EM coupling path through the GaN buffer layer as explained in [16]. Such results imply some needed strategy in the IC layout of a high-voltage power transistor to be oriented adequately with respect to the low-voltage sensitive devices on the same GaN-on-Si chip. Note that the aforementioned estimation of $|S_{21}|$ should be revised to -76.8 dB (= -50.8 dB – 2*13 dB) when both the aggressor and victim device area being 2 μm × 2 μm).

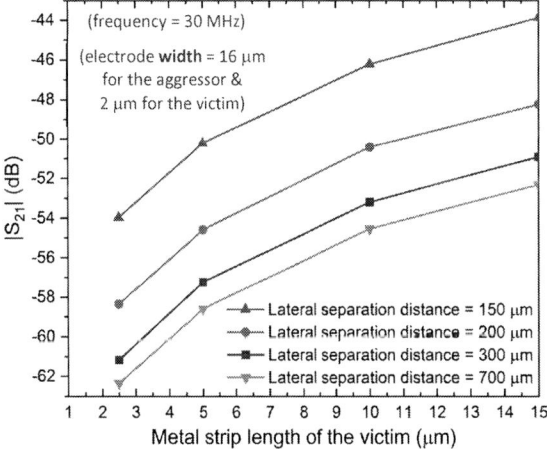

Fig. 9. $|S_{21}|$ at 30 MHz of the GaN-on-Si device structure as a function of the length of the victim electrode (port 2), revealing that the EM coupling increases more when further doubling the device area of the victim electrode by changing the length perpendicular to separation direction.

V. CONCLUSION

We have reported computational EM investigation into the device area dependence of substrate coupling in GaN-on-Si technology. The 3D EM simulation results have revealed that the substrate coupling does not increase sharply when further increasing the device area occupied by a power transistor as the aggressor, if the size expansion is along the separation direction. In such a case, $|S_{21}|$ increases by about 7.7 dB/decade or less when the aggressor's device area is increased in the width (namely the size along the same direction of the lateral separation). This implies that by keeping the victim transistor small enough (i.e. occupying much less chip area), the crosstalk voltage can be minimised with an aggressor power device placed 300 μm or more away on the same chip. In contrast, when the device area is increased but by the size expansion perpendicular to the separation direction, the substrate coupling increases considerably at 13 dB/decade. These substrate coupling trends are valid for changes in device area of either the aggressor or victim transistors in GaN-on-Si technology.

The device area dependence results are tied to the fact that the coupling electric field still penetrates dominantly through the GaN buffer layer, regardless of the indefinite increase in the aggressor device area. By adopting some isolation trench designs [14],[16], $|S_{21}|$ is expected to be lowered further by 10 dB to effectively suppress the substrate coupling. In the technology down-scaling, substrate coupling may also be improved by the transistors' smaller device area and by the suitable layout orientation of especially the power transistors. The results obtained from computational EM investigation inform the design and fabrication of GaN-on-Si power ICs when integrating a high-voltage large-size power transistor on the same chip.

ACKNOWLEDGMENT

The authors acknowledge the support from the Departments of Electrical and Electronic Engineering (EEE) and of Communications and Networking (CAN), School of Advanced Technology (SAT) of XJTLU as well as Department of Electrical Engineering and Electronics, The University of Liverpool. This work is supported in part by PGRS funding (FOSA2406036) of XJTLU. Both R. Yao and S. Lam sincerely acknowledge the technical and administrative support by Mr. Yubin Gu and other colleagues of Academic Enhancement Team in MITS for the arrangements of floating licenses of Ansys HFSS within the campus network of XJTLU.

REFERENCES

[1] T. J. Flack, B. N. Pushpakaran, and S. B. Bayne, "GaN technology for power electronic applications: a review," *Journal of Electronic Materials*, vol. 45, no. 6, pp. 2673–2682, June 2016.

[2] J. Millán, P. Godignon, X. Perpiñà, A. Pérez-Tomás and J. Rebollo, "A survey of wide bandgap power semiconductor devices," *IEEE Transactions on Power Electronics*, vol. 29, no. 5, pp. 2155-2163, May 2014.

[3] T.-W. Wang, Y.-Y. Kao, S.-H. Hung, Y.-H. Wen, T.-H. Yang, S.-Y. Li, K.-H. Chen, K.-L. Zheng, Y.-H. Lin, S.-R. Lin, and T.-Y. Tsai, " Monolithic GaN-based driver and GaN switch with diode-emulated GaN technique for 50-MHz operation and sub-0.2-ns deadtime control," *IEEE Journal of Solid-State Circuits*, vol. 57, no. 12, pp. 3877-3888, December 2022.

[4] Kevin J. Chen, Oliver Häberlen, Alex Lidow, Chun lin Tsai, Tetsuzo Ueda, Yasuhiro Uemoto, and Yifeng Wu, "GaN-on-Si power technology: devices and applications," *IEEE Transactions on Electron Devices*, vol. 64, no. 3, pp. 779-795, March 2017.

[5] H. W. Then, M. Radosavljevic, N. Desai, R. Ehlert, V. Hadagali, K. Jun, P. Koirala, N. Minutillo, R. Kotlyar, A. Oni, M. Qayyum, J. Rode, J. Sandford, T. Talukdar, N. Thomas, H. Vora, P. Wallace, M. Weiss, X. Weng, and P. Fischer, "Advances in research on 300mm Gallium Nitride-on-Si(111) NMOS transistor and silicon CMOS integration," *Digest of 2020 IEEE International Electron Devices Meeting (IEDM)*, pp. 581-584, 2020.

[6] J. Wei, Z. Zheng, G. Tang, H. Xu, G. Lyu, L. Zhang, J. Chen, M. Hua, S. Feng, T. Chen, and K. J. Chen, "GaN power integration technology and its future prospects," *IEEE Transactions on Electron Devices*, vol. 71, no. 3, pp. 1365-1382, March 2024.

[7] Rongming Chu, Andrea Corrion, Mary Chen, Ray Li, Danny Wong, and Daniel Zehnder, "1200-V normally off GaN-on-Si field-effect transistors with low dynamic on-resistance," *IEEE Electron Device Letters*, vol. 32, no. 5, pp. 632-634, May 2011.

[8] A. Li, Y. Shen, Z. Li, I. Z. Mitrovic, H. Wen, S. Lam and W. Liu, "A monolithically integrated 2-transistor voltage reference with a wide temperature range based on AlGaN/GaN technology," *IEEE Electron Device Letters*, vol. 43, no.3, pp. 362-365, March 2022.

[9] A. Li, Y. Shen, Z. Li, F. Li, R. Sun, I. Z. Mitrovic, H. Wen, S. Lam and W. Liu, "A 4-transistor monolithic solution to highly linear on-chip temperature sensing in GaN power integrated circuits," *IEEE Electron Device Letters*, vol. 44, no. 2, pp. 333-336, February 2023.

[10] Qimeng Jiang, Zhikai Tang, Chunhua Zhou, Shu Yang, and Kevin J. Chen, "Substrate-coupled cross-talk effects on an AlGaN/GaN-on-Si smart power IC platform," *IEEE Transactions on Electron Devices*, vol. 61, no. 11, pp. 3808-3813, November 2014.

[11] R. Sun, J. Lai, W. Chen, C. Liu, F. Wang, J. Zhou, Z. Li, and B. Zhang, "Crosstalk suppression in monolithic GaN devices based on inverted E-field decoupling," *IEEE Transactions on Electron Devices*, vol. 68, no. 4, pp. 1542 - 1549, April 2021.

[12] Miao Cui and Sang Lam, "Use of DC probes for multi-MHz measurements of crosstalk and substrate coupling in gallium nitride power integrated circuits," *2024 IEEE 36th International Conference on Microelectronic Test Structures (ICMTS)*, Edinburgh, United Kingdom, April 2024, pp. 1-5.

[13] Rui Ray Yao, Miao Cui, Zhao Wang, Sang Lam, Stephen Taylor, "Electromagnetic investigation of substrate coupling in power integrated circuits in GaN-on-Si technology", *2024 IEEE 11th Workshop on Wide Bandgap Power Devices & Applications (WiPDA)*, pp.1-5, Dayton, Ohio, USA, November 2024.

[14] Zijin Jiang, Rui Ray Yao, Miao Cui, Zhao Wang, Sang Lam, Stephen Taylor, "On the design and effectiveness of isolation trenches to suppress substrate coupling in power integrated circuits in GaN-on-Si technology", *2024 IEEE Workshop on Wide Bandgap Power Devices and Applications in Europe (WiPDA Europe)*, pp.1-5, September 2024.

[15] M. Cui, Y. Cai, S. Lam, W. Liu, C. Z. Zhao, I. Z. Mitrovic, S. Taylor, and P. R. Chalker, "Characterization of transient threshold voltage shifts in enhancement- and depletion-mode AlGaN/GaN metal-insulator-semiconductor (MIS)-HEMTs," *2018 IEEE International Conference on Electron Devices and Solid State Circuits (EDSSC 2018)*, Shenzhen, China, June 2018, pp. 1-2.

[16] R. R. Yao, Z. Jiang, M. Cui, Z. Wang, S. Lam, and S. Taylor, "Grounded isolation trenches in GaN-on-Si power integrated circuits: an electromagnetic study for trench filling considerations," *The 16th IEEE International Conference on Electron Devices and Solid-State Circuit Conference (EDSSC 2025)*, Yinchuan, Ningxia, China, 13th-15th June 2025.

Analysis and compensation method of transient unbalanced current in parallel connection of SiC MOSFET

1st Yong Chen
Zhuhai Power Supply Bureau of Guangdong Power Grid Co., Ltd. DC Power Distribution and Consumption Technology Research Center of Guangdong Power Grid Co., Ltd.
Zhuhai, China
35665035@qq.com

2nd Xu Cheng
Zhuhai Power Supply Bureau of Guangdong Power Grid Co., Ltd. DC Power Distribution and Consumption Technology Research Center of Guangdong Power Grid Co., Ltd.
Zhuhai, China
664157102@qq.com

3rd Xingyu Pei
Zhuhai Power Supply Bureau of Guangdong Power Grid Co., Ltd. DC Power Distribution and Consumption Technology Research Center of Guangdong Power Grid Co., Ltd.
Zhuhai, China
31977550@qq.com

4th Jianbiao Li
Zhuhai Power Supply Bureau of Guangdong Power Grid Co., Ltd. DC Power Distribution and Consumption Technology Research Center of Guangdong Power Grid Co., Ltd.
Zhuhai, China
zhlijianbiao@126.com

5th Hongyuan Wu
Zhuhai Power Supply Bureau of Guangdong Power Grid Co., Ltd. DC Power Distribution and Consumption Technology Research Center of Guangdong Power Grid Co., Ltd.
Zhuhai, China
whyqsxddc@163.com

6th Bin Zhao
Advanced Power Transmission Institute North China Electric Power University
Beijing, China
binzhaocj@163.com

Abstract—This paper proposes a comprehensive solution to the problem of transient current imbalance caused by chip parameter mismatch and package parasitic parameter coupling in parallel operation of silicon carbide (SiC) MOSFET. First, by analyzing the nonlinear characteristics of threshold voltage (V_{th}) and transconductance (g_{fs}), a chip screening method based on cubic polynomial fitting is established, which significantly reduces the influence of parameter dispersion on dynamic current sharing. Secondly, a mathematical model of multi-chip parallel connection of inductance and circuit layout is constructed, revealing the mechanism of mutual inductance effect aggravating unbalanced current under the discrete layout of drain bus and power source bus. Finally, a branch impedance compensation method is proposed. By adding compensation on the source side, the common branch impedance coupling and mutual inductance effect are suppressed, and the reliability of the parallel system is significantly improved.

Keywords— SiC MOSFET; transient current imbalance; parameter mismatch; package parameters; branch impedance compensation

I. INTRODUCTION

With the rapid development of new energy generation and power electronic systems, silicon carbide (SiC) MOSFET has gradually become the core device of high-power power electronic equipment. In practical applications, although the chip parameters and circuit design are as consistent as possible, transient current imbalance will inevitably occur due to the device itself, packaging parasitic parameters, etc., which will not only cause local overheating and increased power loss, but also may cause device failure.

The current silicon carbide MOSFET production process is not yet mature, and the parameters of the same batch of chips are significantly dispersed [1]. In 2019, Alessandro Borghese's team at the University of Naples showed through Monte Carlo simulation analysis that gate resistance mismatch affects the current distribution of parallel chips: chips with larger gate resistance have slower turn-on speeds and larger turn-off currents [2]. In the same year, Ke Junji's team at North China Electric Power University measured the parameters of 30 silicon carbide MOSFETs from the same batch and found that the threshold voltage dispersion was the highest. Chips with higher transconductance had higher dynamic currents due to the high current rise rate. Studies have shown that threshold voltage and transconductance mismatch are the main causes of dynamic imbalance in parallel chips [3]. The packaging parameters mainly include the parasitic parameters (stray inductance) and packaging structure of the multi-device parallel power module. In 2014, Li H of Aalborg University in Denmark analyzed two discrete devices in parallel and confirmed that source stray inductance mismatch is the main cause of dynamic imbalance [4]. In 2020, Cheng Zhao of Xi'an Jiaotong University further quantified the analysis through theoretical modeling and found that the dynamic current is most sensitive to source inductance mismatch, but not to drain/gate inductance mismatch [5].

In 2020, Ke Junji proposed a chip screening method based on the transfer curve distance coefficient [6]. In 2017, Miao Wang's team in the United States proposed a dual power terminal layout (as shown in Figure 1-1 (a)) [7]. In 2019, Shao Weihua's team at Chongqing University designed a circularly symmetrical layout power module (as shown in Figure 1-1 (b)) [8].

Existing studies have shown that the transfer curve method has mathematical significance but lacks physical parameter correlation, and ignores the complex circuit topology coupling effect when multiple chips are connected in parallel. It is urgent to build a mathematical model that includes parasitic parameter coupling and mutual inductance, and propose a dynamic unbalanced current control method that takes into account packaging optimization.

This paper systematically studies the influence mechanism of chip parameters on the dynamic unbalanced current of parallel SiC MOSFETs, and proposes a dynamic current screening method based on chip parameter statistical analysis.

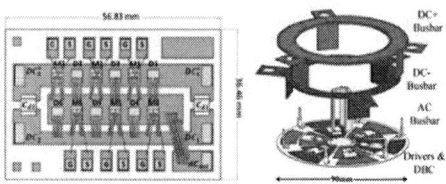

Figure .1-1. Different layouts of parallel power modules (a) Schematic diagram of dual power terminal layout (b) Schematic diagram of circular layout

II. INFLUENCE AND REGULATION OF CHIP PARAMETERS ON PARALLEL CURRENT SHARING OF SiC MOSFETs

Chip parameter mismatch can lead to unbalanced dynamic/static current between parallel chips, which can cause overcurrent failure in severe cases and threaten the reliability of high-current applications. To solve this problem, it is necessary to obtain chips with similar parameters through pre-screening. In view of the high testing cost of unpackaged SiC MOSFET, this study selected TO247-4 packaged discrete devices for experiments and proposed a chip screening method based on physical significance to effectively suppress the impact of parameter mismatch on parallel current distribution.

A. Impact of Chip Parameters on Dynamic Current Sharing Maintaining the Integrity of the Specifications

The immaturity of SiC MOSFET production technology leads to significant dispersion of chip parameters, including threshold voltage, transconductance, etc. The dispersion of parasitic capacitance has little effect on parallel current distribution [3]. Subsequent analysis focuses on the discrete statistics of threshold voltage and transconductance parameters.

$$i_{\mathrm{d}} = g_{\mathrm{fs}}(v_{\mathrm{gs}} - V_{\mathrm{th}}), v_{\mathrm{gs}} > V_{\mathrm{th}} \quad (1\text{-}1)$$

From formula (1-1), it can be seen that the parallel dynamic current distribution of SiC MOSFET is mainly determined by the threshold voltage (V_{th}) and transconductance (g_{fs}). Let g be the transconductance coefficient. According to the structure of SiC MOSFET, when MOSFET works in the saturation region, due to the channel cutoff, the switching transient dynamic current can be expressed as:

$$i_{\mathrm{d}} = g(v_{\mathrm{gs}} - V_{\mathrm{th}*})^3, v_{\mathrm{gs}} > V_{\mathrm{th}*} \quad (1\text{-}2)$$

B. Chip screening method based on dynamic current sharing

The difference in the cubic polynomial fitting transconductance coefficients between the first chip m1 and the remaining m-1 chips in the parallel chip can be expressed as:

$$d_{11} = \sum_{i=2}^{m} \left| g_{m_1} - g_{m_i} \right| \quad (1\text{-}3)$$

$$\Delta d_1 = \frac{m \cdot (d_{11} + d_{12} + \cdots + d_{1(m-1)})}{g_{m_1} + g_{m_2} + \cdots g_{m_m}} \quad (1\text{-}4)$$

$$\Delta d_2 = \frac{m \cdot (d_{21} + d_{22} + \cdots + d_{2(m-1)})}{V_{\mathrm{th}m_1*} + V_{\mathrm{th}m_2*} + \cdots + V_{\mathrm{th}m_{m*}}} \quad (1\text{-}5)$$

The Δd_{d} of the group with the closest chip parameters should take the minimum value.

C. Experimental validation of the chip screening method

The maximum current imbalance between parallel chips is defined as:

$$\alpha_{xy} = \frac{2 \cdot \max(|i_{\mathrm{d}x} - i_{\mathrm{d}y}|)}{i_{\mathrm{d}x} + i_{\mathrm{d}y}} \quad (1\text{-}6)$$

This chapter proposes a dynamic current-sharing screening method for parallel chips based on fitting threshold voltage and transconductance coefficient with cubic polynomial. The effectiveness is verified by two sets of experiments. The new method reduces the maximum dynamic unbalanced current from 12.72% to 2.81%, and the peak current is only 0.88A, which is significantly better than the traditional threshold voltage screening method.

III. EFFECT OF PACKAGING PARAMETERS ON PARALLEL CURRENT SHARING OF SiC MOSFETs

A. Impact of packaging parameters on dynamic current sharing

There are two common circuit layouts for power modules with multiple SiC MOSFET chips connected in parallel, namely, the drain bus and the power source bus lead out from one side or from both sides [9, 10], as shown in Figure 2-1. The only difference in the layout is the location of the drain and power source bus, and the rest of the structure is exactly the same. Based on this, the equivalent circuit models of the two layouts are shown in Figure 2-1.

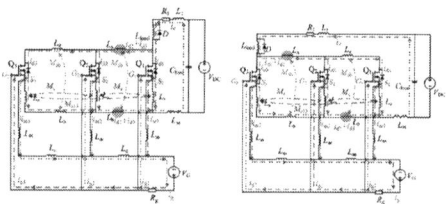

Figure .2-1 . Dynamic circuit model of two layouts of SiC MOSFET multi-chip parallel power module (a) The drain and power source meet at one side. (b) The drain and power source meet at two sides.

In Figure 2-1, the drain current id2/id3 forms a common branch impedance coupling through the drain parasitic inductance La and the power source parasitic inductance Lb between the parallel chips Q1-Q2; there is a common branch impedance coupling phenomenon [11]. This switching transient, the drain current of the parallel chip can be expressed as:

$$i_{\mathrm{d}j} = g_{\mathrm{fs}}(v_{\mathrm{gs}j} - V_{\mathrm{th}}), v_{\mathrm{gs}j} > V_{\mathrm{th}}, j = 1,2,3 \quad (2\text{-}1)$$

The gate-source voltage and power source voltage of the parallel chips can be expressed as:

$$\begin{bmatrix} v_{\mathrm{gs}1} \\ v_{\mathrm{gs}2} \\ v_{\mathrm{gs}3} \end{bmatrix} = \begin{bmatrix} 1 & -R_g & -(L_a+L_{ba}) & -(M_a+L_{ba}) & -(M_{a13}+L_{ba}) \\ 1 & -R_g & -(M_a+L_{ba}) & -(L_a+L_b-M_{ab}+L_{ba}) & -(M_a+L_b-M_{ab}+L_{ba}) \\ 1 & -R_g & -(M_{a13}+L_{ba}) & -(M_a+L_b-M_{ab}+L_{ba}) & -(L_a+2L_b-2M_{ab}+L_{ba}) \end{bmatrix} \begin{bmatrix} V_G \\ i_g \\ i_{d1} \\ i_{d2} \\ i_{d3} \end{bmatrix} \quad (2\text{-}2)$$

$$\begin{bmatrix} v_{s12} \\ v_{s13} \\ v_{s23} \end{bmatrix} = \begin{bmatrix} -L_s + M_s & L_s - M_s + L_b - M_{ab} & (1-k_s)M_s + L_b - M_{ab} \\ -L_s + k_s M_s & L_b - M_{ab} & L_s + 2L_b - 2M_{ab} - k_s M_s \\ (k_s - 1)M_s & -L_s + M_s & L_s - M_s + L_b - M_{ab} \end{bmatrix} \begin{bmatrix} \dot{i}_{d1} \\ \dot{i}_{d2} \\ \dot{i}_{d3} \end{bmatrix} \quad (2\text{-}3)$$

According to formula (2-1), the dynamic unbalanced current between parallel chips can be expressed as:

$$\begin{bmatrix} \Delta i_{d1} \\ \Delta i_{d13} \\ \Delta i_{d23} \end{bmatrix} = g_{fs} \begin{bmatrix} -L_s & L_s & 0 \\ -L_s & 0 & L_s \\ 0 & -L_s & L_s \end{bmatrix} + \begin{bmatrix} M_s & -M_s & (1-k_s)M_s \\ k_s M_s & 0 & -k_s M_s \\ (k_s-1)M_s & M_s & -M_s \end{bmatrix} +$$
$$\begin{bmatrix} 0 & L_b & L_b \\ 0 & L_b & 2L_b \\ 0 & 0 & L_b \end{bmatrix} - \begin{bmatrix} 0 & M_{ab} & M_{ab} \\ 0 & M_{ab} & 2M_{ab} \\ 0 & 0 & M_{ab} \end{bmatrix} \begin{bmatrix} \dot{i}_{d1} \\ \dot{i}_{d2} \\ \dot{i}_{d3} \end{bmatrix} \quad (2\text{-}5)$$

From formula (2-5), we can know that the dynamic unbalanced current of SiC MOSFET parallel connection is composed of four parts. The effects of each part are shown in Table 2-1.

Table 2-1 Drain bus point and power source bus point are located on one side

Factors affecting unbalanced current	Effects
Part I (Power Source Parasitic Inductance)	inhibition
Part II (Mutual inductance of power sources between parallel chips)	Aggravation
Part III (Common branch impedance coupling between parallel chips)	Aggravation
Part IV (Mutual inductance on the drain side and power source side between parallel chips)	inhibition

When the drain side confluence point and the power source confluence point of the parallel chips are located on both sides, the gate-source voltage of the parallel chips and the dynamic unbalanced current between the parallel chips can be expressed as:

$$\begin{bmatrix} \Delta i_{d1} \\ \Delta i_{d13} \\ \Delta i_{d23} \end{bmatrix} = g_{fs} \begin{bmatrix} -L_s & L_s & 0 \\ -L_s & 0 & L_s \\ 0 & -L_s & L_s \end{bmatrix} + \begin{bmatrix} M_s & -M_s & (1-k_s)M_s \\ k_s M_s & 0 & -k_s M_s \\ (k_s-1)M_s & M_s & -M_s \end{bmatrix} +$$
$$\begin{bmatrix} 0 & L_b & L_b \\ 0 & L_b & 2L_b \\ 0 & 0 & L_b \end{bmatrix} + \begin{bmatrix} M_{ab} & 0 & 0 \\ 2M_{ab} & M_{ab} & 0 \\ M_{ab} & M_{ab} & 0 \end{bmatrix} \begin{bmatrix} \dot{i}_{d1} \\ \dot{i}_{d2} \\ \dot{i}_{d3} \end{bmatrix} \quad (2\text{-}6)$$

It can be seen from formula (2-6) that when the drain side bus point and the power source bus point of the parallel chip are on the same side, the effects of each part are shown in Table 2-2.

Table 2-2 Drain bus point and power source bus point are located on both sides

Factors affecting unbalanced current	Effects
Part I (Power Source Parasitic Inductance)	inhibition
Part II (Mutual inductance of power sources between parallel chips)	Aggravation
Part III (Common branch impedance coupling between parallel chips)	Aggravation
Part IV (Mutual inductance on the drain side and power source side between parallel chips)	Aggravation

B. Experimental analysis

Figure 2-2 shows the circuit layout when the drain side and the power source junction are located on one side and two sides. The experimental results show that the maximum dynamic unbalanced current under the two layouts are 16.88A and 30.17A respectively (as shown in Figure 2-3).

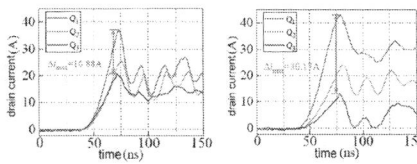

Figure .2-2 . Parallel current distribution under two layouts (a) The drain and power source meet at one side. (b) The drain and power source meet at two sides.

IV. SILICON CARBIDE MOSFET POWER MODULE PARALLEL CURRENT SHARING CONTROL

A. Theoretical Analysis of Dynamic Current Sharing Impedance Compensation Method

The following is an example of branch impedance compensation design using the layout where the drain and power source confluence points are located on one side. At the instant of SiC MOSFET switching, the dynamic unbalanced current between the mth device and the nth device can be expressed as:

$$\Delta i_{dmn} = g_{fs} \frac{n^2 - (2m-1)n + m^2 - m}{2} L_{s12} \frac{di}{dt} \quad (3\text{-}5)$$

From formula (3-5), we can see that in order to eliminate the dynamic unbalanced current between parallel devices, the compensation inductance of the power source side of the mth device is:

$$\Delta L_{sm} = \frac{n^2 - (2m-1)n + m^2 - m}{2} L_{s12} \quad (3\text{-}6)$$

B. Simulation Study on Branch Impedance Compensation Method

In order to further verify the practical effectiveness of the branch impedance compensation method, a dedicated simulation circuit was built in software. The simulation process specifically used a silicon carbide MOSFET in a standard TO-247-4 package configuration. Before the branch impedance compensation, the branch impedances Z1, Z2, and Z3 were initially equal by design. Compensation was performed precisely according to formula (3-6) through iterative adjustments. The comparative simulation results before and after the branch impedance compensation are shown in Figures 3-1 (a) and 3-1 (b) respectively. The simulation results are closely consistent with the theoretical analysis framework. It can be clearly found that the parasitic parameter mismatch significantly causes a large unbalanced current between parallel silicon carbide MOSFET devices, particularly during the critical device turn-on process and the subsequent current rise process after the device is turned on. The proposed branch impedance compensation method can effectively and consistently suppress the dynamic unbalanced current caused by the common branch impedance coupling phenomenon.

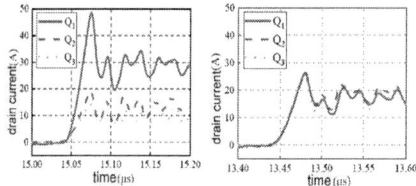

Figure .3-1.   SiC MOSFET dynamic current distribution before branch impedance compensation (a) The drain and power source meet at one side. (b) The drain and power source meet at two sides.

C. Experimental verification

In order to verify the effectiveness of the branch impedance compensation method, a double pulse test platform was designed and built. In the dynamic unbalanced current test, the current peak of Q_1 at the moment of opening reached 47.11A (exceeding 57.03% of the device rated current of 30A), which was significantly higher than 13.90A of Q_2 and 8.45A of Q_3, resulting in the dynamic unbalance of Q_1-Q_2, Q_1-Q_3, and Q_2-Q_3 of 54.43%, 69.61%, and 24.44%, respectively. After compensating the branch inductance Ls1/Ls2 on the power source side of Q_1 and Q_2, the dynamic current unbalance between the three groups of chips was significantly reduced to 0.31%/0.45%/0.14%. The results show that the branch impedance compensation method can effectively suppress the dynamic unbalanced current caused by the common branch impedance coupling, verifying its engineering feasibility.

V. SUMMARIZE

In view of the problem of unbalanced dynamic current distribution of parallel SiC MOSFET, the main research results of this paper are summarized as follows:

The key chip parameters that affect the dynamic current distribution of parallel SiC MOSFET are extracted, the cubic polynomial fitting model of the transfer characteristic curve is established, and a screening method to reduce the dispersion of parallel chip parameters is proposed. The influence mechanism of packaging parameters on the dynamic current distribution of parallel SiC MOSFET is revealed, which provides theoretical guidance for the research of unbalanced current control methods of parallel SiC MOSFET. A branch impedance compensation method for paralleling multiple discrete devices of SiC MOSFET to suppress dynamic unbalanced current is proposed, and the effectiveness of the branch impedance compensation method is verified by simulation and experiment. The parallel current sharing characteristics of SiC MOSFET are significantly improved.

ACKNOWLEDGMENTS

This work was supported by science and technology project of China Southern Power Grid under Grant 030400KC23090015 (GDKJXM20231031).

REFERENCES

[1] Sun P, Zhao Z, Cai Y, et al. Parallel SiC MOSFET chip screening based on switch energy balancing[J]. Proceedings of the CSEE, 2019, 39(19): 5613-23+889.

[2] Borghese A, Riccio M, Fayyaz A, et al. Statistical Analysis of the Electrothermal Imbalances of Mismatched Parallel SiC Power MOSFETs [J]. IEEE J Emerg Sel Topics Power Electron, 2019, 7(3): 1527-38

[3] Ke J, Zhao Z, Sun P, et al. Influence of Device Parameters Spread on Current Distribution of Paralleled Silicon Carbide MOSFETs[J]. Journal of Power Electronics, 2019, 19(4): 1054-67.

[4] Li H, Munknielsen S, Pham C, et al. Circuit mismatch influence on performance of paralleling silicon carbide MOSFETs[C]. European Conference on Power Electronics & Applications. Lappeenranta, Finland:

IEEE, 2014: 1-8.

[5] Hu J, Alatise O, Gonzalez J A O, et al. The Effect of Electrothermal Nonuniformities on Parallel Connected SiC Power Devices Under Unclamped and Clamped Inductive Switching[J]. IEEE Trans Power Electron, 2016, 31(6): 4526-35.

[6] Ke J, Zhao Z, Sun P, et al. Chips Classification for Suppressing Transient

Current Imbalance of Parallel-Connected Silicon Carbide MOSFETs[J].

IEEE Trans Power Electron, 2020, 35(4): 3963-72.

[7] Wang M, Luo F, Xu L. A Double-End Sourced Wire-Bonded Multichip SiC MOSFET Power Module With Improved Dynamic Current Sharing

[J]. IEEE J Emerg Sel Topics Power Electron, 2017, 5(4): 1828-36.

[8] Shao W, Li R, Zeng Z, et al. Design and evaluation of SiC multichip power module with low and symmetrical inductance[C]. Institution of Engineering and Technology, Milwaukee, WI, USA: IEEE, 2019:3573-

3577.

[9] Chen Z, Yao Y, Boroyevich D, et al. A 1200-V, 60-A SiC MOSFET Multichip Phase-Leg Module for High-Temperature, High-Frequency Applications[J]. IEEE Trans Power Electron, 2014, 29(5): 2307-20.

[10] Xu F, Han T J, Jiang D, et al. Development of a SiC JFET-Based SixPack Power Module for a Fully Integrated Inverter[J]. IEEE Trans Power Electron, 2013, 28(3): 1464-78.

[11] Cittanti D, Iannuzzo F, Hoene E, et al. Role of parasitic capacitances in power MOSFET turn-on switching speed limits: A SiC case study[C]. 2017 IEEE Energy Conversion Congress and Exposition (ECCE). Cincinnati, OH, USA: IEEE, 2017: 1387-94.

979-8-3315-1110-4/25 $31.00 © 2025 IEEE

Design and fabrication of a SiC trench MOSFET with multi-step p-type shielding and multiple CSL layers

Wei Chen
Dept of Integrated Power Systems and Device Technology
JFS Laboratory
Wuhan, China
chenwei@jfslab.com.cn

Fei Guo
Dept of Integrated Power Systems and Device Technology
JFS Laboratory
Wuhan, China
carry_lee@jfslab.com.cn

Yangyang Wu
Dept of Integrated Power Systems and Device Technology
JFS Laboratory
Wuhan, China
wuyangyang@jfslab.com.cn

Kuan Wang
Dept of Integrated Power Systems and Device Technology
JFS Laboratory
Wuhan, China
wangkuan@jfslab.com.cn

Zhijie Cheng
Dept of Integrated Power Systems and Device Technology
JFS Laboratory
Wuhan, China
chengzhijie@jfslab.com.cn

Jun Yuan
Dept of Integrated Power Systems and Device Technology
JFS Laboratory
Wuhan, China
yuanjun@jfslab.com.cn

Rong Zhang
School of Electrical and Electronic Engineering
Huazhong University of Science and Technology
Wuhan, China
rongzhang@hust.edu.cn

Guoqing Xin
School of Electrical and Electronic Engineering
Huazhong University of Science and Technology
Wuhan, China
guoqingxin@hust.edu.cn

Zhiqiang Wang
School of Electrical and Electronic Engineering
Huazhong University of Science and Technology
Wuhan, China
zhiqiangwang@hust.edu.cn

Abstract—This study presents the design and fabrication of a SiC trench MOSFET featuring a multi-step p-type shielding region (P+SLD) and multiple current spreading layers (CSL) (MPC-TMOS). Compared with the conventional SiC trench MOSFET with single P+SLD (P-TMOS), the proposed structure employs a self-aligned process to form the multi-step P+SLD without additional masks. Therefore, there are only minor changes in the fabrication process and cost does not increase significantly. Numerical 2D-simulation has been conducted to compare the electrical characteristics of the MPC-TMOS and P-TMOS. The results indicate comparable specific on-resistance ($R_{on,sp}$) and breakdown voltage (BV) for both structures. However, the MPC-TMOS demonstrates superior gate oxide reliability, with a peak electric field ($E_{ox,peak}$) of 2.89 MV/cm at 1200 V drain voltage—a 12.69% reduction compared to the 3.31 MV/cm observed in the P-TMOS. Experimental tests of the MPC-TMOS samples show a BV of 1700 V and an on-state resistance (R_{on}) of 50.7 mΩ. Additionally, the influence of key structural parameters is analyzed, offering insights for further device optimization.

Keywords—SiC trench MOSFET, multi-step P+SLD, multiple CSL, gate oxide protection, device fabrication

I. Introduction

Silicon carbide (SiC)-based metal-oxide-semiconductor field-effect transistors (MOSFETs) are widely recognized for their exceptional performance in high-voltage, high-power, and high-efficiency applications. This is attributed to the material's outstanding characteristics, such as an elevated critical breakdown field, superior thermal conductivity, and high electron saturation velocity [1-2]. Compared with the silicon (Si)-based counterparts, SiC MOSFETs enhance overall system efficiency and minimize thermal management requirements by reducing heat sink size and weight.

Consequently, they facilitate the development of more compact, energy-efficient power electronic systems.

SiC trench MOSFET has gained significant research interest as it allows higher channel density and enables a more space-efficient layout, thus further reducing the specific on-resistance ($R_{on,sp}$) [3]. Nevertheless, a critical drawback is the exposure of the trench-bottom gate oxide to intense electric fields, which accelerates oxide degradation and may cause catastrophic failure, significantly undermining long-term reliability.

To address this issue, multiple gate oxide shielding techniques have been explored. Rohm's double-trench MOSFET (DTMOS) design employs dual source trenches to alleviate electric field stress [7-8]. Infineon, on the other hand, introduced an asymmetric trench MOSFET incorporating a P+ shielding layer (P+SLD), which demonstrates enhanced gate oxide reliability [9-10]. Alternative approaches include integrating a p-type epitaxial layer below the gate trench [11–13] or implementing a heavily doped p-type shielding region under the gate (P-TMOS) to suppress the peak electric field within the gate oxide layer ($E_{ox,peak}$) [14].

In this paper, a SiC trench MOSFET featuring multi-step P+SLD and multiple current spreading layers (CSL) (MPC-TMOS) is proposed and fabricated. Compared with the P-TMOS, the multi-step P+SLD of the MPC-TMOS is able to effectively optimizes the electric field distribution at the trench bottom, which enhances gate oxide reliability while maintains output performance. The test results of the manufactured MPC-TMOS samples demonstrate the promising potential of this structure. Additionally, key parameters of the PGP-TMOS are also studied to facilitate further optimization.

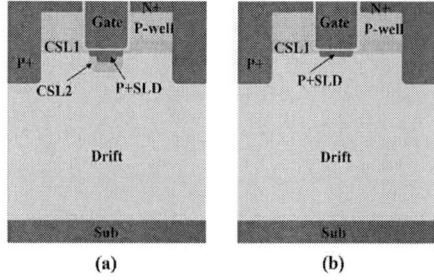

(a) **(b)**

Fig. 1 Schematic cross-section view of the (a) MPC-TMOS and (b) P-TMOS

TABLE I.   DEVICE STRUCTURE PARAMETERS

Parameters	PGP-TMOS	FP-TMOS
Cell width (μm)	7.5	7.5
Trench width (μm)	1.5	1.5
Trench depth (μm)	0.9	0.9
Drift thickness (μm)	11	11
CSL1 thickness (μm)	1	1
P+ region thickness (μm)	1.5	1.5
1st-step P+SLD thickness (μm)	0.2	0.2
1st-step P+SLD width (μm)	1.1	1.1
2nd-step P+SLD thickness (μm)	0.3	\
2nd-step P+SLD width (μm)	0.7	\
Drift concentration (cm⁻³)	8×10^{15}	8×10^{15}
P+ concentration (cm⁻³)	2×10^{18}	2×10^{18}
CSL1 concentration (cm⁻³)	4×10^{16}	4×10^{16}
CSL2 concentration (cm⁻³)	4×10^{16}	\
P+SLD concentration (cm⁻³)	2×10^{18}	2×10^{18}

II. DEVICE STRUCTURE AND SIMUTATION

Fig. 1 shows the schematic cross-section view of the proposed MPC-TMOS and the conventional P-TMOS. Both structures utilize wafers with two epitaxial layers (drift and CSL1) and employ P+SLD beneath the gate trench along with deep P+ regions on either side for gate oxide protection. However, the MPC-TMOS incorporates a multi-step P+SLD design, which redistributes electric field concentration at the trench corners, thereby reducing peak oxide field ($E_{ox,peak}$). Furthermore, an additional CSL2 layer is integrated to maintain conduction performance without compromising fabrication efficiency—the self-aligned process enables this enhanced structure with no additional masks and only a minor cost increase. Key simulation parameters for both devices are summarized in Table I.

Numerical 2D-simulations have been used to compare the performance of the P-TMOS and the MPC-TMOS. As shown in Fig. 2(a), both structures exhibit nearly identical blocking capabilities, with BV of 1727 V for the MPC-TMOS and 1733 V for the P-TMOS. Their output characteristics also demonstrate minimal variation. However, the MPC-TMOS achieves a 12.69% reduction in the $E_{ox,peak}$, decreasing from 3.31 MV/cm to 2.89 MV/cm. Notably, the multi-step P+SLD in the MPC-TMOS redistributes the electric field, with peak values of 2.88 MV and 2.18 MV at its two corners, compared to 2.61 MV at the single P+SLD corner of the P-TMOS. This

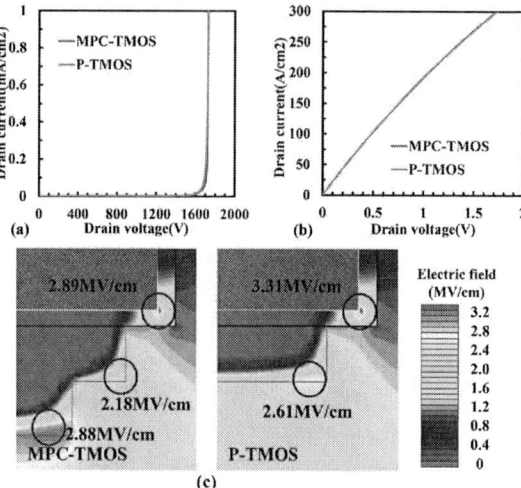

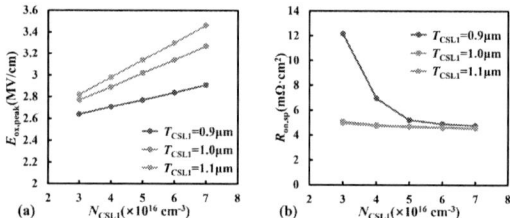

Fig.2 (a) Blocking and (b) conduction I-V characteristics. (c)Electric field contours of the two structures at the drain bias of 1200V.

Fig. 3 Influence of the T_{CSL1} and N_{CSL1} on performance of the MPC-TMOS. (a) $E_{ox,peak}$, (b)$R_{on,sp}$.

modified electric field distribution effectively lowers the $E_{ox,peak}$. Therefore, the gate oxide reliability is enhanced.

The doping concentration (N_{CSL1}) and thickness (T_{CSL1}) of the CSL1 layer primarily influence the $E_{ox,peak}$ and specific on-resistance ($R_{on,sp}$), as illustrated in Figs. 3(a) and 3(b). Within the simulated parameter range, the $E_{ox,peak}$ exhibits a gradual increase with both parameters. Conversely, the $R_{on,sp}$ decreases with elevated N_{CSL1} and T_{CSL1}. Notably, when the T_{CSL1} is below 1 μm, the N_{CSL1} significantly affects the $R_{on,sp}$. However, this dependence becomes negligible when the T_{CSL1} reaches or exceeds 1 μm. This phenomenon indicates that under these conditions, the channel resistance and drift region resistance dominate the total device resistance, while the JFET region contribution becomes relatively insignificant. Consequently, variations in CSL1 parameters no longer substantially impact the overall resistance.

Fig. 4(a) illustrates the impact of P+SLD region depth (T_{PS}) on the BV, $E_{ox,peak}$, and $R_{on,sp}$ of the MPC-TMOS, while maintaining a constant first-step P+SLD thickness of 0.2μm. In this configuration, the CSL2 thickness is consistently maintained at 0.1μm greater than that of the P+SLD. As T_{PS} increases, the specific on-resistance ($R_{on,sp}$) of the device remains essentially constant. This stability arises because the width of the JFET region is sufficiently large that minor variations in its length have negligible impact on JFET resistance. However, T_{PS} significantly affects blocking characteristics. On the one hand, the peak electric field in

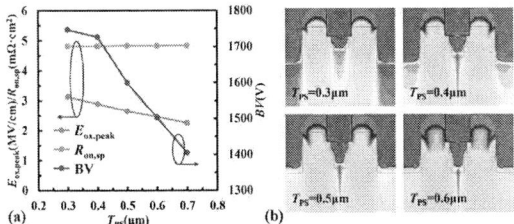 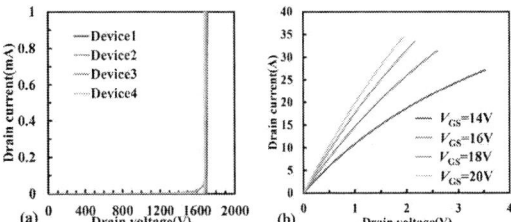

Fig. 4 (a) Influence of the T_{PS} on performance of the MPC-TMOS. (b) Impact generation rate at breakdown when T_{PS}=0.3 μm, 0.4 μm, 0.5 μm and 0.6 μm, respectively.

Fig. 7 Tested (a) blocking I-V curves and (b) conduction I-V curves at different V_{GS} of the samples.

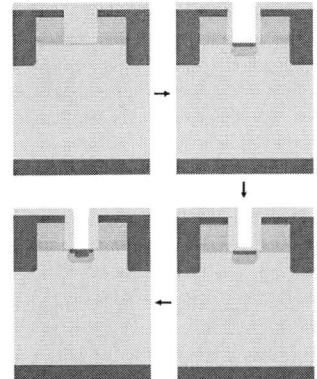

Fig. 5 Process flow to form the CSL2 and multi-step P+SLD.

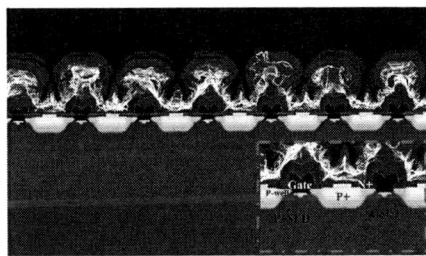

Fig. 6 SEM micrograph of the fabricated MPC-TMOS.

the gate oxide gradually decreases due to the enhanced field shielding effect from the deeper P+SLD. On the other hand, the BV decreases markedly, particularly when T_{PS} exceeds 0.4 μm. As revealed by the impact generation rate in Fig. 4(b), the breakdown point shifts from the P+ region bottom to the P+SLD bottom with increasing T_{PS}. This phenomenon occurs because: (1) the electric field shielding effect provided by the P+ regions on the P+SLD weakens with larger T_{PS} values; and (2) the P+SLD, being surrounded by higher-concentration CSL2, reaches critical breakdown electric field at lower drain voltage than the P+ regions.

III. DEVICE FABRICATION

The key process steps for forming the CSL2 and multi-step P+SLD are depicted in Fig. 5. The sequence begins with etching an ion implantation window, followed by the first implantation to create the CSL2 and the initial P+SLD layer. Next, plasma-enhanced chemical vapor deposition (PECVD) is applied to deposit thicker silicon dioxide spacers on both sidewalls. Finally, a self-aligned process creates a narrower

implantation window, enabling the formation of the second P+SLD step through higher-energy ion implantation.

The MPC-TMOS has been fabricated on wafers with CSL layer. The thickness of the CSL layer is 1 μm and the concentration is 4×10^{16} cm^{-3}. Beneath the CSL layer is the drift layer with a thickness of 11 μm and a concentration of 8×10^{15} cm^{-3}. The micrograph of the scanning electron microscope (SEM) of the MPC-TMOS is shown in Fig. 6. The tested blocking curve and output curve are shown in Fig. 7(a) and Fig. 7(b), respectively. BV of the samples is about 1700 V, which is consistent with the simulation results. The on-state resistance (R_{on}) is 50.7 mΩ at the condition of V_{GS}=20 V and I_{DS}=10 A.

IV. CONCLUSION

In this work, a novel SiC trench MOSFET incorporating multi-step P+SLD and multiple current spreading layers is proposed and fabricated. Compared with conventional P-TMOS designs, the proposed MPC-TMOS demonstrates enhanced gate oxide reliability while maintaining comparable output characteristics. The proposed structure is achieved through a self-aligned fabrication process requiring no additional masks. The MPC-TMOS were fabricated on wafers featuring a CSL layer. Experimental results indicate that BV of the samples is approximately 1700 V and R_{on} is 50.7 mΩ. In addition, the impact of the key parameters on the performance of the structure is also verified by numerical analysis to provide guidance for further optimization.

REFERENCES

[1] T. Kimoto, "Material science and device physics in SiC technology for high-voltage power devices," *Jpn. J. Appl. Phys.* vol. 54, no.4, p.040103, Mar. 2015.

[2] M. Östling, "Silicon Carbide Power Devices: Evolution, Applications, and Future Opportunities," *IEEE Electron Devices Magazine*, vol. 2, no. 4, pp. 30-35, Dec. 2024.

[3] D. Peters, R. Siemieniec, T. Aichinger, T. Basler, R. Esteve, W. Bergner, D. Kueck, "Performance and ruggedness of 1200V SiC — Trench — MOSFET," in *ISPSD 2017*, Sapporo, Japan, 2017, pp. 239-242.

[4] X. Deng, W. Huang, X. Li, X. Li, C. Chen, Y. Wen, J. Ding, W. Chen, Y. Sun, B. Zhang, "Investigation of Failure Mechanisms of 1200V Rated Trench SiC MOSFETs under Repetitive Avalanche Stress," *IEEE Trans. Power Electron.*, vol. 37, no. 9, pp. 10562-10571, Mar. 2022.

[5] J. Li, A. Shekhar, W. D. van Driel and G. Zhang, "A Review on Gate Oxide Failure Mechanisms of Silicon Carbide Semiconductor Devices," *IEEE Trans. Electron Devices*, vol. 71, no. 12, pp. 7230-7243, Dec. 2024.

[6] M. Chaturvedi, S. Dimitrijev, D. Haasmann, H. A. Moghadam, P. Pande and U. Jadli, "Comparison of Commercial Planar and Trench SiC MOSFETs by Electrical Characterization of Performance-Degrading Near-Interface Traps," *IEEE Trans. Electron Devices*, vol. 69, no. 11, pp. 6225-6230, Nov. 2022.

[7] Y. Nakano, R. Nakamura, H. Sakairi, S. Mitani, T. Nakamuraet, "690V, 1.00 mΩcm^2 4H-SiC Double-Trench MOSFETs," *Mater. Sci. Forum*, vol. 717-720, pp. 1069-1072, May 2012.

[8] T. Nakamura, Y. Nakano, M. Aketa, R. Nakamura, S. Mitani, H. Sakairi, Y. Yokotsujiet, "High Performance SiC Trench Devices with Ultra-low Ron," in *IEDM 2011*, Washington, DC, 2011, pp. 26.5.1-26.5.3.

[9] D. Peters, T, Basler, B. Zippelius, T, Aichinger, W, Bergner, R. Esteve, D. Kueck, R. Siemieniec, "The New CoolSiC™ Trench MOSFET Technology for Low Gate Oxide Stress and High Performance," in *Proc. PCIM Eur. 2017 Int. Exhibit. Conf. for Power Electron. Intell. Motion Renew. Energy and Energy Manage.*, Nuremberg, Germany, 2017, pp. 1-7.

[10] R. Siemieniec, D. Peters, R. Esteve, W. Bergner, D. Kück, T. Aichinger, T. Basler, B. Zippeliuset, "A SiC Trench MOSFET Concept Offering Improved Channel Mobility and High Reliability," in *EPE'17 ECCE Europe*, Warsaw, 2017, pp. 1-13.

[11] W. Chen, J. Yuan, F. Guo, Y. Wu, Z. Chen, K. Wang, G. Xin, Z. Wang, "4H-SiC Trench MOSFET with Half-Surrounded Gate for Improved Gate Oxide Reliability," in *SSLCHINA: IFWS 2023*, Xiamen, China, 2023, pp. 31-34.

[12] Y. Wu, J. Yuan, K. Wang, F. Guo, W. Chen, Z. Cheng, S. Xu, G. Xin, Z. Wang, "A Novel SiC Trench MOSFET with Unilateral P Buried Layer for Improved Oxide Electric Field and Switching Loss," in *SSLCHINA: IFWS 2023*, Xiamen, China, 2023, pp. 35-38.

[13] J. Yuan, Z. Cheng, F. Guo, K. Wang, W. Chen, Y. Wu, S. Xu, G. Xin, Z. Wang, "A Trench and Field Limiting Rings Co-assisted JTE Termination With N-P-N Sandwich Epitaxial Wafers for 4H-SiC Devices," *IEEE Electron Device Lett.*, vol. 45, no. 8, pp. 1425-1428, June 2024.

[14] Y. Kagawa, N. Fujiwara, K. Sugawara, R. Tanaka, Y. Fukui, Y. Yamamoto, N. Miura, M. Imaizumi, S. Nakata, S. Yamakawa, "4H-SiC Trench MOSFET with Bottom Oxide Protection," *Mat. Sci. For.*, Vol. 778-780, pp. 919-922, Feb. 2014.

A Novel Integrated Power Module Package Method with SiC MOSFETs and Energy Absorber in Solid-state Circuit Breaker Application

Dongxin Jin
College of Intelligent Robotics and Advanced Manufacturing
Fudan University
Shanghai, China
24110860043@m.fudan.edu.cn
Eaton (China) Investment Co., Ltd
dongxinjin@eaton.com

Jie Gong
Eaton Research Labs
Eaton (China) Investment Co., Ltd
Shanghai, China
jiegong3@eaton.com

Yuchen Wang
Eaton Research Labs
Eaton (China) Investment Co., Ltd
Shanghai, China
yuchenwang2@eaton.com

Cheng Luo
Eaton Research Labs
Eaton (China) Investment Co., Ltd
Shanghai, China
chengluo@eaton.com

Xiaojun Dong
Eaton Research Labs
Eaton (China) Investment Co., Ltd
Shanghai, China
xiaojundong@eaton.com

Guangyin Lei
College of Intelligent Robotics and Advanced Manufacturing
Fudan University
Shanghai, China
guangyinlei@fudan.edu.cn

Abstract—**Solid-state circuit breaker (SSCB) has emerged as a cutting-edge circuit protection solution in direct current (DC) power distribution system, showing significant supercities in tripping speed, reliability and intelligence. Compared to conventional mechanical circuit breaker, the SSCB is built up with power semiconductor and energy absorber devices, which basically determine static and dynamic characteristics of SSCB. In this paper, an integrated power module consisted of Silicon Carbide Metal Oxide Semiconductor Field Effect Transistor (SiC MOSFET) and Transient Voltage Suppressor (TVS) is proposed. This new device can play a pivotal role to optimize footprint, tripping capability and cost in SSCB application. The topology and electrical analysis of the power module are discussed. Finally, experimental results verify the feasibility and advancements of the new device, further, these findings offer valuable insights for practical engineering design.**

Keywords—**SSCB, integrated package, SiC MOSFET, power module, stray inductance**

I. INTRODUCTION

With the increasing penetration of renewable energies, the DC power distribution systems have witness remarkable growth, and this trend paves the way to adoption of new DC circuit protection devices [1]. The solid-state circuit breaker (SSCB) is one of the most promising solutions for DC circuit protection, which offers a typical response time in a range of tens of microseconds [2]. There are many topologies of SSCB, particularly the pure SSCB utilizes the power semiconductor devices to conductor the nominal current as well as interrupt the fault current [3]. As shown in Fig. 1, the core components for pure SSCB are power semiconductor and energy absorber, additionally there are also assisted parts including driver, sensing, controller circuit and cooling system [4],[5].

Wide-bandgap (WBG) semiconductor such as silicon carbide (SiC) and gallium nitride (GaN) is gradually emerged in SSCB application due to their notable enhancements in operating voltage, temperature and speed [6]. SiC MOSFET is extensively investigated for SSCB implementation, however, it is difficult to trade-off between the turn off speed and peak voltage of energy absorber because the existing stray inductance between these two devices [7],[8]. A faster turn - off speed, facilitated by the driver, can effectively reduce the peak fault current and potentially alleviate the thermal stress on the SiC MOSFET, nevertheless, this comes at the cost of increased voltage stress across the device, which can severely undermine its reliability [9]. To tackle this issue, low inductance design by optimizing circuitry layout and component mapping between these two kinds of devices can be effective, however, when devices are paralleled to meet higher current requirements, it becomes difficult to satisfy these design constraints and even results in unexpected gate-source voltage oscillation by inductance mismatch [10]-[12]. Transient voltage suppressor (TVS) outperforms metal oxide varistor (MOV) in terms of reducing clamping voltage and extending electric service life, but the parasitic inductance and capacitance inherent in off-the-shelf power TVS packages can significantly degrade the dynamic performance of the SSCB [13]-[15].

This paper presents a novel integrated power module solution with SiC MOSFET and power TVSs, which is a unique package design that can be used in bidirectional SSCB implementation at exceeding 800V DC power distribution application. Section II elaborates on the topology and operational principles of the power module. Subsequently, Section III delves into the design considerations regarding gate drivers and layout, aiming to enhance the tripping capability, and also illustrates the corresponding simulation verification. Section IV demonstrates the experimental results and advancements of the power module tripping test. Finally, Section V summarizes the research findings and draws conclusions.

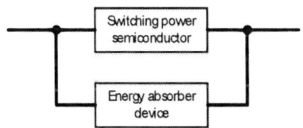

Fig. 1. General structure of pure SSCB

II. INTEGRATED POWER MODULE DESIGN

The proposed integrated power module is consisted with SiC MOSFETs and power TVS. The innovation of this integrated package method lies in design a user-friendly module for constructing SSCB, thus it overcomes all the limitations associated with traditional SSCB with power semiconductors and energy absorbers separately.

A. Overview of the Proposed Power Module

As illustrated in Fig. 2, the proposed power module features a bidirectional configuration. In this setup, the SiC MOSFET and TVS are first connected in parallel and subsequently arranged in a reverse - series connection.

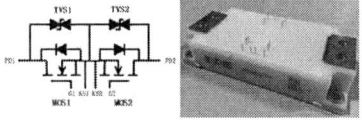

Fig. 2. Proposed integrated power module with SiC MOSFET and TVS

This design exhibits electrical symmetry within the circuit, and also maintains geometric symmetry in the power module layout. With power terminals positioned on two opposing edges and driver signals located in the central region, it facilitates enhanced paralleling capabilities to form a perfect symmetric circuit and reduce connection materials. Moreover, the overall footprint of SSCB is significantly reduced.

B. Operating Principal of the Power Module

The current flow of the power module at on-state and turn-off conditions are shown in Fig. 3 and Fig. 4 respectively. There are multiple SiC MOSFETs in parallel and TVS in series due to high current and voltage requirement in SSCB operation. When the SSCB is in on-state, all the SiC MOSFETs are turning on, and the current flows in the channels of MOSFETs. Especially, the body diodes are not activated if the current is relatively small.

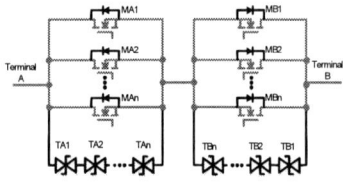

Fig. 3. Current flow path at SSCB on-state

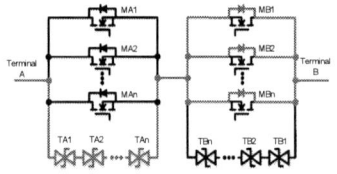

Fig. 4. Current flow path during SSCB turn-off

When the SSCB is turning off by simultaneously switching off all SiC MOSFETs, the current is commuted from channels of MOSFETs to the TVSs in one side, and the body diodes take over the current from MOSFET channel in the other side. The circuit tripping or turn-off energy is absorbed by TVSs at the same time. The theoretical time

sequence of voltage and current of the power module in SSCB application are illustrated in Fig. 5.

- At t_0: short circuit or overcurrent happens.

- During $t_0 \sim t_1$: fault current is rising rapidly, and the SSCB trips at t_1, the SiC MOSFET begins to turn off.

- At t_2: the voltage across power module starts to rise after a short time for gate-source capacitor to discharge.

- During $t_2 \sim t_3$: the SiC MOSFET enters the Miller plateau, the drain-source voltage continues to rise.

- At t_3: the voltage of power module or SSCB is equal to that of the external power source, and the fault current begins to go down.

- At t_4: the voltage of power module reaches the breakdown voltage of TVS, so the current starts to commute from SiC MOSFET to TVS.

- During $t_4 \sim t_5$: Miller plateau ends, gate-source voltage and current of SiC MOSFET goes down, concurrently the drain-source voltage experience overshoot, which can be attributed to the stay inductance between SiC MOSFET and TVS and high current ramp. The drain-source voltage and current of SiC are illustrated as (1) and (2), respectively, where g and V_{TH} are the trans-conductivity and gate threshold voltage of SiC MOSFET, and L_M is the stray inductance between the SiC MOSFET and TVS [11].

$$i_{DS} = gf(v_{GS} - V_{TH}) + C_{DS}\frac{d(i_{DS})}{dt} - C_{GD}\frac{d(v_{GD}-v_{DS})}{dt}$$
$$\approx gf(v_{GS} - V_{TH}) \qquad (1)$$

$$v_{DS} = V_S - L_M\frac{d(i_{DS})}{t} \qquad (2)$$

- During $t_5 \sim t_6$: the TVS takes over all fault current and absorbs the fault energy, simultaneously, the voltage of power module is higher than that of external power source to drive the fault current to zero.

- At t_6: the voltage across power module eventually goes to the same to the external power source, so the SSCB trips the circuit fault successfully.

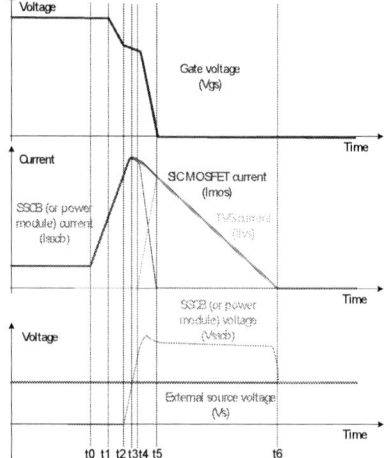

Fig. 5. Diagram of time sequence of voltage and current in the power module when SSCB turns off

III. INFLUENCES OF PARASITIC PARAMETERS ON TRIPPING CAPABILITY

The SSCB is designed to trip circuit fault by turning off power semiconductor and clamping voltage with energy absorber, so tripping capability of SSCB is determined by turn-off capability of the power module. Consequently, parasitic parameters in the power module layout and driver circuit loop can have considerable impact on the static and dynamic performance of SSCB.

A. Analysis of Stray Inductance among Components inside the Power Module Package

In order to achieve large current and voltage capability, it is common to implement SiC MOSFET dies in parallel and TVS chips in series to build up an integrated power module. The theoretical circuit model with stray inductance of the power module with 2 SiC MOSFET in parallel and 2 TVSs in series is shown in Fig. 6. Since the topology is the power module is symmetric in dual directions, analysis work can be carried out in only unidirectional as an example.

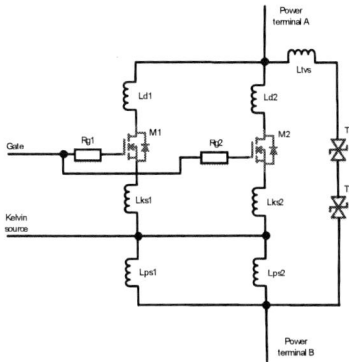

Fig. 6. Circuit of the power module with stray inductance in one direction

The power loop stray inductance can be classified as TVS inductance (L_{tvs}), Kevin source loop inductance (L_{ks1} and L_{ks2}) and drain-source inductance (L_{d1}, L_{d2}, L_{ps1}, and L_{ps2}). Since there can be many chips inside a module, the inductance mismatch is a very common phenomenon, which can result in degradation of tripping capability in terms of unexpected gate-source oscillation, excessive drain-source voltage overshoot and current/voltage significant unbalance. It is challenging to mathematically model an accurate circuit, thus, simulation in LTSPICE can be an effective approach to investigate and understand the behavior of the power module and influences of these parameters.

B. Simulation Results and Comparison Analysis

Simulation conditions and number are shown in Table 1. There are two 1700V/20mohm SiC MOSFET dies in parallel and TVS stack with 1200V clamping voltage in the simulation.

TABLE 1. SIMULATION CONDITIONS

Sim. No.	N0	N1	N2	N3	N4
Condition	*Normal*	*ΔRg*	*ΔLtvs*	*ΔLks*	*ΔLp*
Rg1 (Ω)	5	10	5	5	5
Rg2 (Ω)	5	10	5	5	5
Lks1 (nH)	1	1	1	2	1
Lks2 (nH)	1	1	1	1	1
Lp1 (nH)	5	5	5	5	10
Lp2 (nH)	5	5	5	5	5
Ltvs (nH)	15	15	30	15	15

The simulation results are shown in Fig. 7~Fig. 10. All the simulations are conducted with DC 800V as external power source and 580A turn-off current.

To compare the results between N0 and N1, the higher gate resistor can make turn off speed of SiC MOSFET slower, so a lower current ramp result in smaller drain-source voltage overshoot. However, large turn-off resistor may result in high thermal stress of the SiC MOSFET, and can be harmful to switch performance when the capacitor load is pre-charged by SSCB chopping.

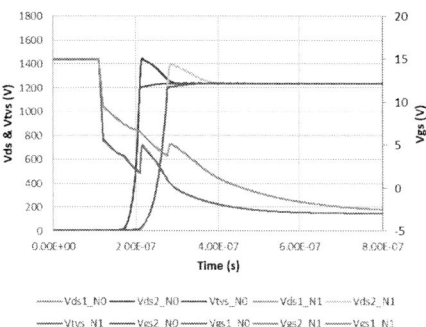

Fig. 7. V_{ds}, V_{tvs} & V_{gs} comparison @ different gate resistors

The drain-source voltage differences between N0 and N2 demonstrates the critical impact of stray inductance between the power module and energy absorber on SSCB voltage stress. The integrated power module with SiC MOSFET and TVS can significantly reduce the stray inductance in the current commutation loop compared to the traditional SSCB with external connections among individual components, in a way, the voltage boundary of SiC MOSFET is extended.

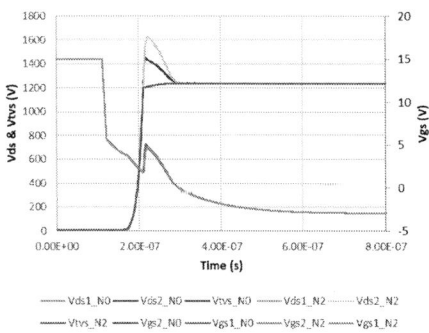

Fig. 8. V_{ds}, V_{tvs} & V_{gs} comparison @ different TVS stray inductances

The next two groups of comparisons are focusing on the stray inductance difference between the paralleled SiC MOSFETs. As shown in the results of N0 and N3, very small Kevin source inductance difference can lead to both gate-source and drain-source voltage difference. Moreover, the gate-source voltage difference can even result in gate voltage oscillation and turn-off failure. This inductance is both in the driver and power loop of the SiC MOSFET, and the interference between driver and power is unwanted and harmful. Therefore, the package design should make sure these inductances symmetric and as small as possible.

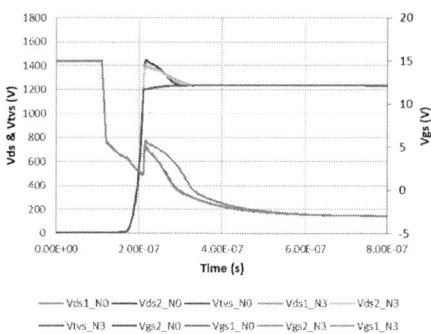

Fig. 9. V_{ds}, V_{tvs} & V_{gs} comparison @ different Kevin loop inductances

The final comparison group takes power loop inductance difference of paralleled SiC MOSFETs into consideration. By analyzing the drain-source voltage waveforms, it is evident that the die with larger power inductance suffers higher voltage stress. As power loop inductance is more likely encountered with unbalance in practical design even in the integrated power module package, the layout and chip mapping optimization are of paramount importance in engineering development.

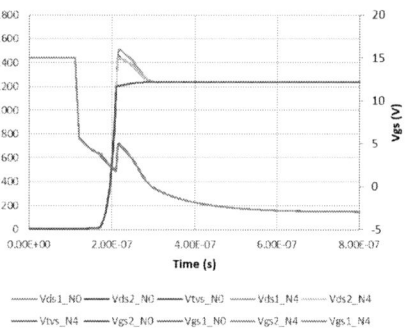

Fig. 10. V_{ds}, V_{tvs} & V_{gs} comparison @ different power loop inductances

IV. EXPERIMENTAL RESULTS

The experimental test focus on tripping capability validation with of the proposed power module. The test platform is shown in Fig. 11. Firstly, the capacitor bank is charged by DC source and close the DC breaker. Then, the SSCB with the proposed power module is closed to make the fault short circuit current. Finally, the SSCB trips at certain current setting by controller.

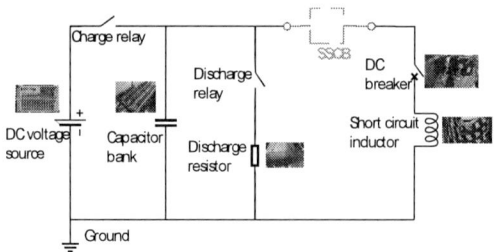

Fig. 11. Tripping Test platform

The testing results are illustrated in Fig. 12. (a)~(c), where the SSCB current, SiC MOSFET current, TVS current, drain-

source voltage, gate-source voltage and external DC source (capacitor bank) voltage are in ch6, ch5, M1, ch4, ch1 and ch3, respectively.

As shown in Fig. 12. (a), SSCB with the new power module is verified at DC 800V external voltage and 700A peak current condition. The peak drain-source voltage is 1298V, which is much lower than breakdown voltage of 1700V SiC MOSFET devices.

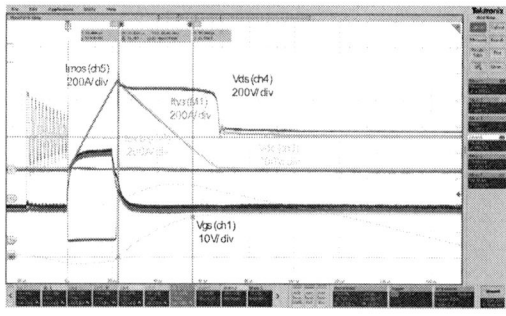

(a) New power module @ DC 800V external source and 700A peak

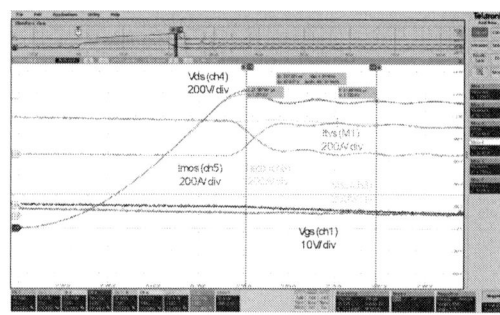

(b) New power module @ DC 100V external source and 70A peak

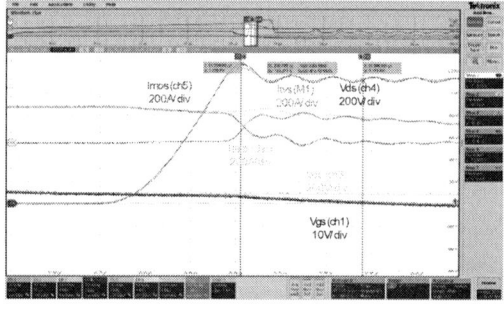

(c) traditional SSCB @ DC 100V external source and 70A peak

Fig. 12. Tripping test results with new power module and traditional SSCB

The zoom in waveform to investigate the current commutation inside the power module is shown in Fig. 12. (b). At DC 100V and 70A peak condition, the current commutation from SiC MOSFET to TVS takes tens of nanoseconds, which generates very small heating energy on SiC MOSFET. Besides, the drain-source peak voltage is only 1226V.

The final waveform in Fig. 12. (c) illustrated the tripping result of the traditional SSCB, which utilizes the same SiC MOSFETs package layout as the new power module and externally mounted individual TVSs in parallel. By comparing to the new power module, the drain-source peak voltage of the traditional one is 1257V, which is 31V higher than the new solution. It is evident that the integrated power module can considerably reduce stray inductance and alleviate the voltage stress of SSCB.

V. CONCLUSION

This paper has successfully proposed and validated a novel integrated power module with SiC MOSFET and TVSs. The unique power module has proven its feasibility for bidirectional SSCB application in exceeding DC 800V power distribution scenarios. The critical design rules lie in parasitic parameters optimization in terms of minimizing stray inductances and symmetric layout. This characteristic presents a promising opportunity to decrease the rated breakdown voltage requirements of power semiconductors. Consequently, it has the potential to reduce both the cost of the SSCB and its on-state resistance, thereby enhancing the overall economic efficiency and performance of the device. Nevertheless, there are still some challenges to overcome, for instance, thermal dissipation, electrical life cycles and vibration in engineering design and industrialization.

REFERENCES

[1] S. Beheshtaein, R. M. Cuzner, M. Forouzesh, M. Savaghebi and J. M. Guerrero, "DC Microgrid Protection: A Comprehensive Review," in IEEE Journal of Emerging and Selected Topics in Power Electronics, 2019.

[2] Z. J. Shen, Z. Miao and A. M. Roshandeh, "Solid state circuit breakers for DC micrgrids: Current status and future trends," 2015 IEEE First International Conference on DC Microgrids (ICDCM), Atlanta, GA, USA, 2015, pp. 228-233.

[3] X. Song, P. Cairoli, Y. Du and A. Antoniazzi, "A Review of Thyristor Based DC Solid-State Circuit Breakers," in IEEE Open Journal of Power Electronics, vol. 2, pp. 659-672, 2021.

[4] R. Rodrigues, Y. Du, A. Antoniazzi and P. Cairoli, "A Review of Solid-State Circuit Breakers," in IEEE Transactions on Power Electronics, vol. 36, no. 1, pp. 364-377, Jan. 2021.

[5] Z. J. Shen, G. Sabui, Z. Miao and Z. Shuai, "Wide-Bandgap Solid-State Circuit Breakers for DC Power Systems: Device and Circuit Considerations," in IEEE Transactions on Electron Devices, vol. 62, no. 2, pp. 294-300, Feb. 2015.

[6] A. Y. Liu, B. Xingwen Li, C. W. Lei, D. Zhaozi Zhang and E. C. Gao, "A Review of Research Progress on Low-Voltage DC Solid-State Circuit Breakers," 2024 IEEE 10th International Power Electronics and Motion Control Conference (IPEMC2024-ECCE Asia), Chengdu, China, 2024, pp. 3551-3556.

[7] Z. Wang, X. Shi, Y. Xue, L. M. Tolbert, F. Wang and B. J. Blalock, "Design and Performance Evaluation of Overcurrent Protection Schemes for Silicon Carbide (SiC) Power MOSFETs," in IEEE Transactions on Industrial Electronics, vol. 61, no. 10, pp. 5570-5581, Oct. 2014.

[8] Y. Ren, X. Yang, F. Zhang, F. Wang, L. M. Tolbert and Y. Pei, "A Single Gate Driver Based Solid-State Circuit Breaker Using Series Connected SiC MOSFETs," in IEEE Transactions on Power Electronics, vol. 34, no. 3, pp. 2002-2006, March 2019.

[9] X. Song, Y. Nawafleh and M. D. Rahman, "A SiC MOSFET Based Multi-Port Solid State Circuit Breaker for DC Protection," 2024 IEEE Energy Conversion Congress and Exposition (ECCE), Phoenix, AZ, USA, 2024, pp. 714-719.

[10] X. Wu et al., "Mitigating Gate Oscillations of Parallel SiC MOSFETs for Enhanced Performance in DC Solid-State Circuit Breaker," in IEEE Journal of Emerging and Selected Topics in Power Electronics, vol. 12, no. 2, pp. 1822-1833, April 2024.

[11] D. Jin, N. Qi, J. Ouyang and C. Luo, "Analysis and Reduction of Turn-on Gate-source Voltage Oscillation on Paralleled SiC MOSFETs Application," 2022 IEEE Energy Conversion Congress and Exposition (ECCE), Detroit, MI, USA, 2022, pp. 1-7.

[12] N. Qi, D. Jin, X. Ge and C. Luo, "Influence of Gate Driver Loop Inductance on SiC MOSFET Module Turn-on Gate Voltage Oscillation in High Power Application," 2022 IEEE Transportation Electrification Conference and Expo, Asia-Pacific (ITEC Asia-Pacific), Haining, China, 2022, pp. 1-5.

[13] J. Hayes, K. George, P. Killeen, B. McPherson, K. J. Olejniczak and T. R. McNutt, "Bidirectional, SiC module-based solid-state circuit breakers for 270 Vdc MEA/AEA systems," 2016 IEEE 4th Workshop on Wide Bandgap Power Devices and Applications (WiPDA), Fayetteville, AR, USA, 2016, pp. 70-77.

[14] X. Song, Y. Du and P. Cairoli, "Survey and Experimental Evaluation of Voltage Clamping Components for Solid State Circuit Breakers," 2021 IEEE Applied Power Electronics Conference and Exposition (APEC), Phoenix, AZ, USA, 2021, pp. 401-406.

[15] K. Askan, M. Bartonek and F. Stueckler, "Bidirectional Switch Based on Silicon High Voltage Superjunction MOSFETs and TVS Diode Used in Low Voltage DC SSCB," PCIM Europe 2019; International Exhibition and Conference for Power Electronics, Intelligent Motion, Renewable Energy and Energy Management, Nuremberg, Germany, 2019, pp. 1-8.

2025 IEEE Workshop on Wide Bandgap Power Devices and Applications in Asia (WiPDA Asia)

Dry Oxidation and SiO₂ Deposition Strategies for High-Performance Gate Dielectrics in 3.3 kV SiC Power MOSFETs

Zijian Hu
Zhejiang University
Electrical Engineering Department
HangZhou, China
12310046@zju.edu.cn

Hongyi Xu*
Hangzhou Silicon Magic
Semiconductor Technology Co.,Ltd
HangZhou, China
charles.xu@silicon-magic.com

Na Ren
Zhejiang University
Electrical Engineering Department
Hangzhou Innovation Center
HangZhou, China
ren_na@zju.edu.cn

Kuang Sheng
Zhejiang University
Electrical Engineering Department
HangZhou, China
shengk@zju.edu.cn

Abstract—**This study presents the static electrical characteristics of a 3300 V silicon carbide planar MOSFET employing an optimized JFET region design. Devices with dry oxide and SiO₂ deposited by low pressure chemical vapor deposition followed by NO annealing were utilized in this study. The MOSFETs with SiO₂ deposition as the gate dielectric achieved a mobility of 20 cm²/Vs, representing an enhancement over the 16 cm²/Vs achieved for the dry oxide, albeit with a thicker thickness. With the gate voltage fixed at 20 V, a specific on-resistance of 21 mΩcm² was measured, and the fabricated MOSFETs demonstrated a drain voltage of 4375 V and a leakage current of 1 mA without exhibiting any avalanche breakdown.**

Keywords— Silicon Carbide, MOSFETs, 3300V, specific resistance, gate oxide, mobility

I. INTRODUCTION

Silicon carbide (SiC) has emerged as a revolutionary wide-bandgap semiconductor for next-generation power electronics, distinguished by its superior intrinsic characteristics including a wide bandgap (3.3 eV), exceptionally high critical breakdown field (3 MV/cm), and outstanding thermal conductivity (4.9 W/cm·K) [1].These inherent advantages enable SiC metal oxide semiconductor field effect transistors (MOSFETs) to achieve significantly enhanced performance compared to conventional silicon-based devices in high-voltage applications exceeding 3 kV [1]. Particularly in industrial converter systems and traction drive applications, 3300 V power devices have become increasingly critical for energy conversion efficiency [2], [3], [4], [5].

Nevertheless, the development of SiC MOSFETs in low-to-medium voltage ranges faces fundamental limitations due to inherent material constraints, particularly poor channel mobility. This challenge becomes more pronounced in high-voltage devices where thicker gate oxide layers are required to withstand elevated electric fields. Conventional thermal oxidation processes with subsequent nitric oxide (NO) annealing, while being the industry standard, demonstrate an inverse relationship between oxide thickness and channel mobility [1]. Recent advancements in deposition techniques have shown that low-pressure chemical vapor deposition (LPCVD) of SiO₂ as gate dielectric can effectively enhance channel mobility to 80 cm²/V·s, representing a 40% improvement over thermal oxide counterparts [6], [7], [8], [9]. In this study, we present a 3300 V SiC MOSFET fabricated using LPCVD-derived gate oxide, demonstrating enhanced channel mobility while maintaining robust gate oxide integrity. The device architecture incorporates optimized doping profiles and junction termination techniques to address high-field reliability challenges.

II. FABRICATION OF MOSFET

The fabrication of the devices was conducted on 6-inch 4H-SiC wafers with N+ substrate and a 33 μm 2.5×10¹⁵ cm⁻³ doped N-type epitaxial layer. The optimized JFET parameters were achieved through TCAD simulation with 3.5 ×10¹⁶ cm⁻³ and 1 μm depth by implantation with nitrogen at high energy. Following the annealing activation with carbon cap, the gate oxide was formed by thermally grown in dry oxygen (named Device A) and SiO₂ deposition by LPCVD with tetraethoxysilane (TEOS) as precursor (named Device B). Then the devices were annealed in NO ambient at 1250 °C for the improvement of the channel mobility. Polysilicon gates were applied in the devices, and the front ohmic contacts were formed by self-alignment. The ohmic contacts located on the posterior surface undergo laser annealing, while the wafer thickness is maintained at 350 μm. Fig.1 depicts the SEM cross-sections of the fabricated MOSFETs.

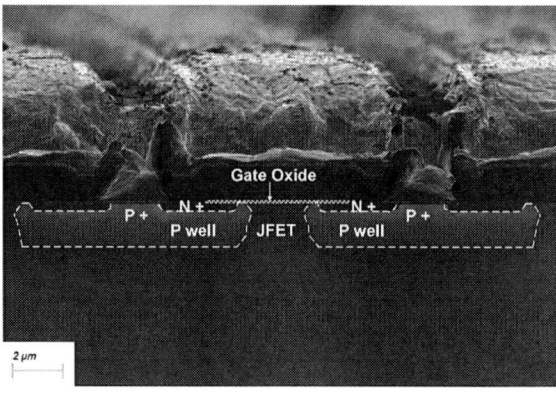

Fig. 1. The SEM cross-section of the fabricated MOSFETs.

III. RESULTS AND DISCUSSION

A. Characterication of the fabricated Gate Oxide

As illustrated in Fig.2, the *I-V* characteristics of the MOSFETs' gate oxide are depicted while the drain voltage was maintained at 0 V. The thickness of the gate oxide is

979-8-3315-1110-4/25 $31.00 © 2025 IEEE

derived from the *C-V* data that has been tested for both device A and device B, which have a thickness of 44.3 nm and 48.6 nm, respectively. It is evident from the figure that the gate leakage current remains on the order of 10^{-11} when the gate voltage is maintained below 25 V. This outcome demonstrates that a voltage withstand capability approaching that of the thermally grown oxide can be achieved by utilizing LPCVD deposited SiO_2 as the gate dielectric layer.

Fig.3 depicts the gate characteristics and the field-effect mobility of the lateral MOSFETs with various gate oxide process. The channel length and width of the lateral MOSFETs were 150 μm and 300 μm, respectively. The drain voltage was 0.1 V during the measurement. The peak field-effect mobility of Device B is observed to be 20 cm²/Vs, which is higher than that of Device A, despite the thickness of Device B being greater than that of Device A. The mobility of the MOSFETs is affected by different scatter mechanisms, including bulk mobility, Coulomb scattering, phonon scattering and surface roughness scattering. The low field mobility is predominantly influenced by Coulomb scattering, which is significantly impacted by traps situated at the SiC/SiO₂ interface, and the roughness scattering dominates the high field mobility [1]. The sub-threshold swings (SS) of Device A and Device B are 910 mV/decade and 630 mV/decade, which were extracted from the gate characteristics of the fabricated lateral MOSFETs, while the drain current from 10^{-10} to 10^{-9} A. The lower interface state density (D_{it}) at the SiC/SiO₂ interface of device B can be attributed to the fact that the deposited SiO₂ process can effectively avoid the C-related defects introduced at the SiC/SiO₂ interface during the oxidation of SiC. The following formula defines the **SS.** Where e is the elementary charge, T is the temperature in Kelvin, k is Boltzmann's constant, C_{ox} is the oxide capacitance, and C_D is the semiconductor capacitance.

$$SS = \frac{dV_{gs}}{d\log_{10} I_d} = \frac{kT}{e} ln10 \frac{C_{ox} + C_D + e^2 D_{it}}{C_{OX}} \qquad [1]$$

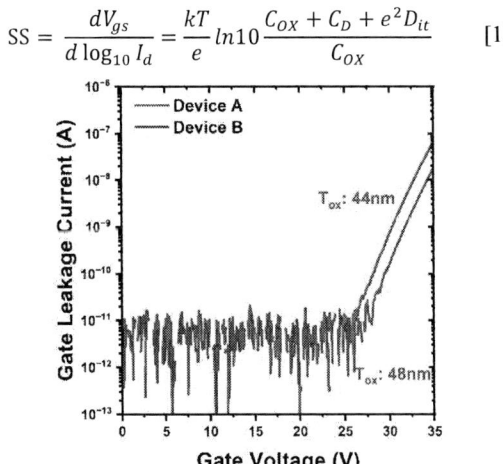

Fig. 2. *I-V* characteristics of the MOSFETs' gate oxide.

As demonstrated in Fig.3, the field-effect mobility of Device A exhibits a decline when the gate voltage is elevated from 15 V to 25 V. Conversely, the mobility of Device B demonstrates an upward trend in response to an increase in gate voltage. This phenomenon can be attributed to the observation that the SiC/SiO₂ interfacial roughness of device B is less pronounced than that of device A. This suggests that the process of depositing SiO₂ is effective in mitigating the issue of roughness enhancement in the high temperature

oxidation process of SiC. This is a significant finding, as it pertains to enhancing mobility at elevated fields, particularly in scenarios where the gate oxide thickness necessitates augmentation at higher voltage levels.

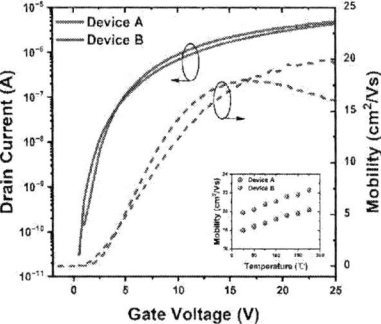

Fig. 3. The gate characteristics and the field-effect mobility of the fabricated MOSFETs. The peak value of the field-effect mobility of the Device B is 20 cm²/Vs which is higher than that of Device A. The mobility increases with the temperature increases which contribute a lower channel resistance at elevated temperature.

B. Characterication of the fabricated MOSFETs

Fig.4 depicts the transfer curves of the MOSFETs were measured under constant gate and drain voltages. The threshold voltage (V_{th}) of the Device A is 2.55 V and that of the Device B is 1.95 V. The gate voltage is defined as the V_{th} when the drain current reaches to 1 mA. The higher fixed charge at the SiC/SiO₂ interface of the Device B results in a smaller V_{th}, which is a consequence of the NO annealing.

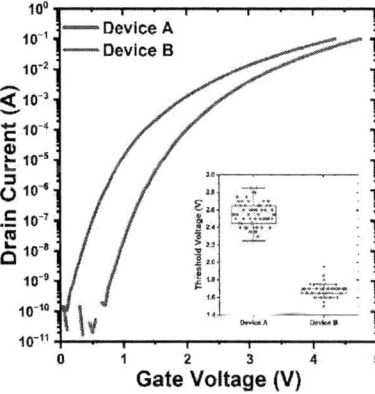

Fig. 4. The gate characteristics of the fabricated MOSFETs.

Fig.5 depicts the output I-V curves of the MOSFETs at V_{gs} = 16 V and V_{gs} = 20 V. The specific resistance ($R_{on,sp}$) of Device A is 21 mΩcm², while that of Device B is 25 mΩcm². It is evident that the design parameters of both devices are identical, however, Device B exhibits a higher channel mobility. Meanwhile, with the temperature increases from 25 °C to 175 °C with step is 50 °C. As the temperature increases, the discrepancy in the $R_{on,sp}$ of devices B and A diminishes and the $R_{on,sp}$ of device B exhibits a further reduction above 125 °C. In order to explore the difference between the two types of devices, the resistance of the MOSFETs is decomposed into substrate resistance, drift resistance, JFET resistance, channel resistance and contact resistance. The parameters used in the resistance calculation are combined with theoretical and experimental values, where the channel mobility is depicted in Fig.3 and the channel length is extracted from Fig.1, while the ohmic contact resistivity is extracted from the CTLM test structure. The ohmic contact resistivity for Device A and Device B are determined to be 3.3×10^{-4} Ωcm^2 and 2.0×10^{-2} Ωcm^2, respectively.

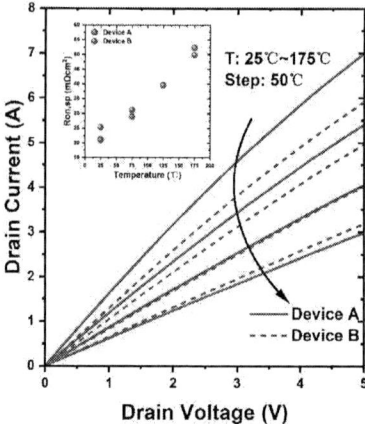

Fig. 5. The I-V characteristics of the fabricated MOSFETs' with $V_{gs}= 20$ V and the temperature increases from 25 °C to 175 °C with step is 50 °C. As the temperature increases, the discrepancy in the $R_{on,sp}$ of devices B and A diminishes and the $R_{on,sp}$ of device B exhibits a further reduction above 125 °C.

As illustrated in Figure 6, the resistance analysis of Devices A and Device B reveals that, given the identical drift region parameter, the drift, substrate and JFET resistances are presumed to be equivalent. The channel resistance of Device B, measured at 1.2 $m\Omega cm^2$, is lower than that of Device A, which is 1.37 $m\Omega cm^2$, attributable to the higher channel mobility exhibited by Device B. Device B exhibits a higher source ohmic contact resistance (4.5 $m\Omega cm^2$) than Device A. This is attributable to the higher ohmic contact resistivity exhibited by Device B. The poor source contact is contributed to the ohmic process, which can be excluded by optimizing process flow. As demonstrated in Fig.6, the drift resistance of the devices is 17.1 $m\Omega cm^2$, which is the predominant component of the 3300 V SiC planar MOSFETs resistance. Optimization of the MOSFET structure, such as through the implementation of trench or super junction MOSFETs, has been identified as a strategy to reduce drift resistance. Consequently, in the context of SiC MOSFETs with a voltage below 1700 V, the drift resistance will be superseded by the channel resistance as the predominant component.

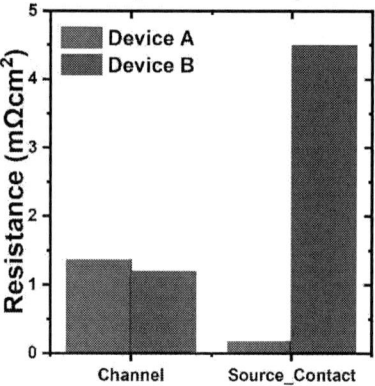

Fig. 6. The Channel and Source contact resistance comparation for Device A and B at 25 °C. The elevated ohmic contact resistance of Device B gives rise to an augmented output resistance of Device B.

Fig.7 depicts the $R_{on,sp}$ of the Device A and Device B at different JFET width. The $R_{on,sp}$ of the MOSFETs decreases with increasing width due to a smoother path for the current to flow when the device is on. Fig.7 also demonstrates that increasing the width of the JFET has a diminishing effect on reducing the device resistance, with the $R_{on,sp}$ gradually saturating. However, as the JFET width increases, the gate oxide electric field (E_{ox}) of the MOSFET device also increases, thereby decreasing the reliability of the device. In this paper, the most optimal JFET width parameter selected is 2.8 µm. Through TCAD simulation, it was determined that the E_{ox} is below 4 MV/cm, thereby ensuring the long-term reliability of the devices.

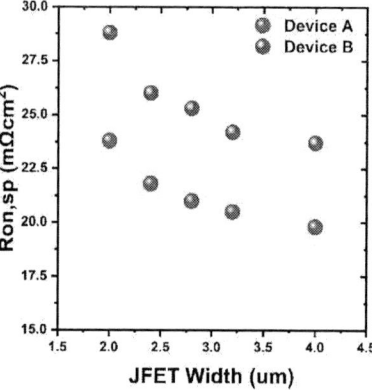

Fig. 7. The $R_{on,sp}$ of the fabricated MOSFETs at different JFET width. With the increasement of the JFET width the $R_{on,sp}$ decreases.

As demonstrated in Fig.8, the block curves of devices A and B are presented. During the measurement, the gate voltage was maintained at 0 V. The drain voltage of device A attains a maximum of 4375 V, which is designated as the block voltage when the drain current is 1 mA. The block voltage of device B is 4236 V. The field-limiting guard rings (FLRs) terminal was implemented in the MOSFETs. The PN diodes were fabricated with the designed FLRs as MOSFETs, and as can be seen from Fig. 8, the block curves of the diodes. Device B exhibits a lower block voltage than Device A, which is attributed to the high fixed charge density at the interface between SiC and SiO₂, which contributes to the leakage current increase though the applied voltage is low.

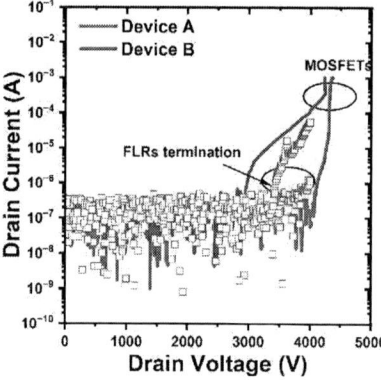

Fig. 8. The block characteristics of the fabricated MOSFETs with the gate voltage kept at 0 V and the block characteristics of the FLRs termination structure at PN diodes were included.

As demonstrated in Fig. 9, a plot of specific on-resistance versus blocking voltage is presented for SiC MOSFET documented in the extant literature[10], [11], [12], [13], [14], [15], [16], [17], [18], [19], [20], [21], [22], [23], [24] at 3300 V level. The parameters of Device A and Device B that were tested are indicated by pentagrams in the plot. It can thus be concluded that the 3300 V SiC planar MOSFET reported in this paper exhibit superior performance.

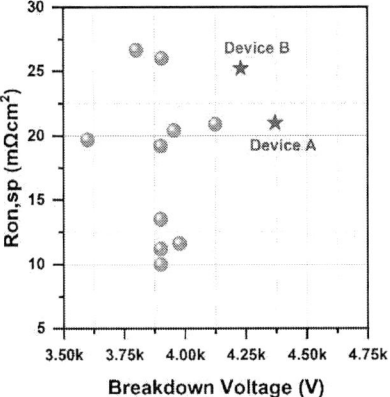

Fig. 9. The benchmark of the fabricated 3300 V SiC planar MOSFETs and the data reported in other literature is included as comparison.

IV. CONCLUSION

In this study, 3300V SiC planar MOSFETs with different gate oxide processes have been fabricated. The MOSFETs' block voltage has been optimized to 4375V, and the specific resistance is $21m\Omega cm^2$ with the gate oxide formed by thermal oxide. The investigation focused on the gate oxide process, which involves the deposition of SiO_2 followed by NO annealing. The primary objective was to enhance mobility, irrespective of the thickness of the oxide layer. This approach was found to be effective in enhancing the performance of high voltage SiC MOSFETs with trench or super junction structures.

ACKNOWLEDGMENT

This work was supported in part by the National Natural Science Foundation of China under Grant 52377200 and in part by the "Leading Goose" R&D Program of Zhejiang under Grant 2024C01113.

REFERENCES

[1] T. Kimoto and J. A. Cooper, Fundamentals of Silicon Carbide Technology: Growth Characterization Devices and Applications, Piscataway, NJ, USA: Wiley, 2014.

[2] S. Narasimhan, A. Kanale, S. Bhattacharya, and J. Baliga, "Performance evaluation of 3.3 kV SiC MOSFET and schottky diode for medium voltage current source inverter application," in *2021 IEEE 8th Workshop on Wide Bandgap Power Devices and Applications (WiPDA)*, Redondo Beach, CA, USA: IEEE, Nov. 2021, pp. 366–371. doi: 10.1109/WiPDA49284.2021.9645089.

[3] Z. Guo, H. Li, and X. Dong, "A self-voltage balanced hybrid three-level MV inverter using 3.3-kV SiC MOSFET module with false-trigger-proof design," *IEEE J. Emerg. Sel. Top. Power Electron.*, vol. 10, no. 6, pp. 6854–6864, Dec. 2022, doi: 10.1109/JESTPE.2021.3114623.

[4] M. Sakai, S. Toyoshima, K. Wada, M. Furumai, T. Tsuno, and Y. Mikamura, "3.3 kV SiC power module with low switching loss".

[5] G. Fortes, P. Ladoux, J. Fabre, and D. Flumian, "Characterization of a 300 kW IIsolated DC-DC converter using 3.3 kV SiC-MOSFETs," 2021.

[6] J. H. Moon, I. H. Kang, H. W. Kim, O. Seok, W. Bahng, and M.-W. Ha, "TEOS-based low-pressure chemical vapor deposition for gate

oxides in 4H–SiC MOSFETs using nitric oxide post-deposition annealing," *Curr. Appl. Phys.*, vol. 20, no. 12, pp. 1386–1390, Dec. 2020, doi: 10.1016/j.cap.2020.09.003.

[7] M. W. Lim *et al.*, "Pre-deposition interfacial oxidation and post-deposition interface nitridation of LPCVD TEOS used as gate dielectric on 4H–SiC," *Mater. Sci. Forum*, vol. 1004, pp. 535–540, Jul. 2020, doi: 10.4028/www.scientific.net/MSF.1004.535.

[8] Lumei Wei, M.-L. Locatelli, S. Diaham, C. D. Pham, G. Grosset, and L. Dupuy, "High temperature dielectric properties of SiO_2 films deposited by TEOS PECVD," in *2016 IEEE International Conference on Dielectrics (ICD)*, Montpellier, France: IEEE, Jul. 2016, pp. 1061–1064. doi: 10.1109/ICD.2016.7547801.

[9] K. Tachiki, M. Kaneko, and T. Kimoto, "Mobility improvement of 4H-SiC (0001) MOSFETs by a three-step process of H_2 etching, SiO_2 deposition, and interface nitridation," *Appl. Phys. Express*, vol. 14, no. 3, p. 031001, Mar. 2021, doi: 10.35848/1882-0786/abdcd9.

[10] S. Li, Y. Chen, H. Liu, R. Huang, Q. Liu, and S. Bai, "Simulation, fabrication and characterization of 3300V/10A 4H-SiC power DMOSFETs," in *2018 15th China International Forum on Solid State Lighting: International Forum on Wide Bandgap Semiconductors China (SSLChina: IFWS)*, Shenzhen: IEEE, Oct. 2018, pp. 1–4. doi: 10.1109/IFWS.2018.8587331.

[11] K. Hamada, S. Hino, and T. Kitani, "Low on-resistance SiC-MOSFET with a 3.3-kV blocking voltage".

[12] G. Song, Y. Wang, X. Chen, and C. Li, "Investigation on electrical characteristic of 3.3kV SiC MOSFET with integrated SBD," 2021.

[13] K. Hamada *et al.*, "Investigation of cell structure and doping for low-on-resistance SiC metal–oxide–semiconductor field-effect transistors with blocking voltage of 3300 V," *Jpn. J. Appl. Phys.*, vol. 52, no. 4S, p. 04CP03, Apr. 2013, doi: 10.7567/JJAP.52.04CP03.

[14] K. Kawahara *et al.*, "Impact of embedding schottky barrier diodes into 3.3 kV and 6.5 kV SiC MOSFETs," *Mater. Sci. Forum*, vol. 924, pp. 663–666, Jun. 2018, doi: 10.4028/www.scientific.net/MSF.924.663.

[15] X. Chen *et al.*, "Different JFET designs on conduction and short-circuit capability for 3.3 kV planar-gate silicon carbide MOSFETs," *IEEE J. Electron Devices Soc.*, vol. 8, pp. 841–845, 2020, doi: 10.1109/JEDS.2020.3010951.

[16] L. Fursin, X. Q. Li, X. Huang, K. Zhu, W. Simon, and A. Bhalla, "Development of a high-performance 3,300V silicon carbide MOSFET," *Mater. Sci. Forum*, vol. 924, pp. 770–773, Jun. 2018, doi: 10.4028/www.scientific.net/MSF.924.770.

[17] R. Huang *et al.*, "Design and fabrication of a 3.3 kV 4H-SiC MOSFET," *J. Semicond.*, vol. 36, no. 9, p. 094002, Sep. 2015, doi: 10.1088/1674-4926/36/9/094002.

[18] W. Ni *et al.*, "Design and fabrication of 3300V 100mΩ 4H-SiC MOSFET with stepped p-body structure," in *2019 16th China International Forum on Solid State Lighting & 2019 International Forum on Wide Bandgap Semiconductors China (SSLChina: IFWS)*, Shenzhen, China: IEEE, Nov. 2019, pp. 50–53. doi: 10.1109/SSLChinaIFWS49075.2019.9019791.

[19] X. Huang, L. Fursin, A. Bhalla, W. Simon, and J. C. Dries, "Design and fabrication of 3.3kV SiC MOSFETs for industrial applications," in *2017 29th International Symposium on Power Semiconductor Devices and IC's (ISPSD)*, Sapporo: IEEE, May 2017, pp. 255–258. doi: 10.23919/ISPSD.2017.7988908.

[20] K. Wada *et al.*, "Blocking characteristics of 2.2 kV and 3.3 kV-class 4H-SiC MOSFETs with improved doping control for edge termination," *Mater. Sci. Forum*, vol. 778–780, pp. 915–918, Feb. 2014, doi: 10.4028/www.scientific.net/MSF.778-780.915.

[21] T. Tsuji, H. Shiomi, N. Ohse, Y. Onishi, and K. Fukuda, "3300V-class 4H SiC implantation-epitaxial mosfets with low specific on-resistance of 11.6mΩcm² and high avalanche withstanding capability," *Mater. Sci. Forum*, vol. 858, pp. 962–965, May 2016, doi: 10.4028/www.scientific.net/MSF.858.962.

[22] B. Powell, K. Matocha, S. Chowdhury, and C. Hundley, "3300V SiC DMOSFETs fabricated in high-volume 150 mm CMOS fab," *Mater. Sci. Forum*, vol. 924, pp. 731–734, Jun. 2018, doi: 10.4028/www.scientific.net/MSF.924.731.

[23] H. Kono, M. Furukawa, K. Ariyoshi, T. Suzuki, Y. Tanaka, and T. Shinohe, "14.6 mΩcm² 3.3 kV DIMOSFET on 4H-SiC (000-1)," *Mater. Sci. Forum*, vol. 778–780, pp. 935–938, Feb. 2014, doi: 10.4028/www.scientific.net/MSF.778-780.935.

[24] A. Bolotnikov *et al.*, "3.3kV SiC MOSFETs designed for low on-resistance and fast switching," in *2012 24th International Symposium on Power Semiconductor Devices and ICs*, Bruges, Belgium: IEEE, Jun. 2012, pp. 389–392. doi: 10.1109/ISPSD.2012.6229103.

2025 IEEE Workshop on Wide Bandgap Power Devices and Applications in Asia (WiPDA Asia)

Low Parasitic Repackaging and Integration of Multiple GaN HEMT Devices

Yue Chen, Mingrui Zou, Dongjun Jiang, Senhao Liang, Jiakun Gong, Zheng Zeng

State Key Laboratory of Power Transmission Equipment Technology, School of Electrical Engineering, Chongqing University

Chongqing, China

Email: yuechen@stu.cqu.edu.cn, zoumingrui@cqu.edu.cn, dongjun@stu.cqu.edu.cn, liangsenhao@cqu.edu.cn, gongjiakun@cqu.edu.cn, zengerzheng@cqu.edu.cn

Abstract—The wide bandgap semiconductor GaN HEMT device performs faster switching speed, which brings the advantage of high switching frequency and the challenge of switching stability for high power density of power converter. Besides, the current rating of the GaN HEMT discrete device is insufficient for industrial applications, so as that the high capacity multi-chip parallel GaH HEMT power module with innovative packaging structure is pursued. However, the traditional wire-bonding packaging with high parasitic is not suitable for the GaH HEMT power module with voltage overshooting and ringing issues. Due to the low threshold voltage of GaN HEMT, the unbalanced current sharing also challenge the multi-chip power module. In this paper, a low parasitic repackaging concept is proposed for a 650 V / 120 A GaN HEMT power module integrated with multiple GaN HEMT discrete devices and gate driver. A novel self-powered isolated gate driver is employed to simplify the parallel paths and shorten the power loop. Compared to the packaging with conventional layout, the measured parasitic inductance of the proposed multi-chip GaN HEMT power module is 4.75 nH, which is reduced by 46.7% benefit from the optimized symmetric layout. Furthermore, the experiments also present the overshooting voltage is only 7.3% under 400 V and 80 A during turn-off transient. The experimental results also ensure the outstanding current sharing performance of the proposed power module layout.

Keywords—GaN HEMT, repackaging power module, low parasitic inductance, current sharing, integrated gate driver

I. INTRODUCTION

With the emergence of new semiconductor materials and breakthroughs in chip fabrication technology, wide bandgap semiconductor power devices, such as GaN HEMT, have made great progress. Compared with traditional silicon-based devices, the GaN HEMT device have a high mobility of electrons in the two-dimensional electron gas region, which can reduce the on-resistance of the device and thus lower switching losses [1]–[3]. High electron mobility also allows the device to have lower input and output capacitance under the same rated current conditions, these lower junction capacitance and higher electron saturation rate allows the device to have faster switching speeds, thus enabling higher switching frequencies and power densities to be achieved [4]–[5]. These advantages of GaN HEMT devices are extremely attractive for high power supply devices such as motor drives. However, the current ratings of GaN HEMT discrete devices do not meet high power requirements, and multiple chips in parallel are required to increase the current rating.

Higher switching speed of GaN HEMT devices results in higher di/dt and dv/dt, which makes the power module more sensitive to parasitic parameters. To avoid false triggering or even breakdown of the device caused by voltage overshoot, it is necessary to design favorable module packaging and layout [6]–[8]. At the same time, due to the low threshold voltage of

GaN HEMT, if the gate driving signals between parallel chips are not synchronized enough, it is difficult to maintain a consistent dynamic current distribution among the chips in parallel, which will lead to local overheating and even device damage [9]–[11]. Therefore, it is also crucial to design the gate drive circuit with excellent performance.

At present, some researches have been carried out on the multichip parallel connection of GaN HEMT power modules. In terms of packaging and layout of power module, APEIL proposed a GaN HEMT multichip power module applying a conventional wire bond package, which has a rated voltage of 650 V and a rated current of 150 A. The inductance of the power loop is between 5.1 nH and 5.6 nH [12]. A study by a team from the University of Stuttgart, Germany, demonstrated an integrated gate-driven GaN HEMT power module with a power loop inductance of 2.35 nH by embedding the GaN HEMT device into a PCB [13]. Literature [14] presents a 650 V / 60 A bondless GaN HEMT power module with a low power loop inductance of 0.68 nH by realizing a vertical commutation loop through a hybrid package of PCB + DBC. In the design of gate driving circuit, Reference [15] proposes a driving circuit with a vertical loop structure, which reduces the thickness of the board layer by using ultra-thin flexible PCBs to make the loop have a smaller area, thereby reducing the parasitic inductance of the driving circuit. Literature [16] uses a single gate driver to construct a complete driving circuit to simultaneously drive 12 parallel devices, so that each device's gate circuit has a different parasitic inductance. A method of driving resistance compensation design is proposed to solve the problem of transient current equalization. Literature [17] adopts a packaging structure combining IMS substrate and PCB, which constructs four parallel driving circuits with a single gate driver on the PCB. Brass pins are used to connect and drive the four chips located on the IMS substrate. By isolating the gate circuit from the power circuit, mutual coupling is avoided from affecting the driving circuit, thus making the driving circuit more stable in performance. Above studies have demonstrated different options for the development of GaN HEMT power modules. In general, the development of new packaging and device layout suitable for GaN HEMT to optimize the parasitic inductance of the power module, as well as designing more advanced gate driving circuits to achieve current sharing of parallel devices, are key to leveraging the unique advantages of GaN HEMT [18]–[22].

In this paper, we propose a secondary encapsulated 650 V / 120 A GaN HEMT half-bridge power module benchmarked against the DCM standard package, using GaN HEMT discrete devices for four-chip paralleling, and verifying the excellent performance of the power module through testing. The paper is structured as follows. The layout optimization of the proposed GaN HEMT power module is presented in Section II. Section III demonstrates the vertical structure

979-8-3315-1110-4/25 $31.00 © 2025 IEEE   933

design and implementation of the power module. Comprehensive experimental results are displayed in Section IV. Section V concludes the paper.

II. Low-Parasitic-Oriented Optimal Layout of Proposed GaN HEMT Power Module

In order to improve the reliability of the GaN HEMT power module and ensure the current balance between parallel chips，the layout of power module is optimized in this paper. Fig. 1 depicts the circuit diagram of the power module containing key parasitic inductors.

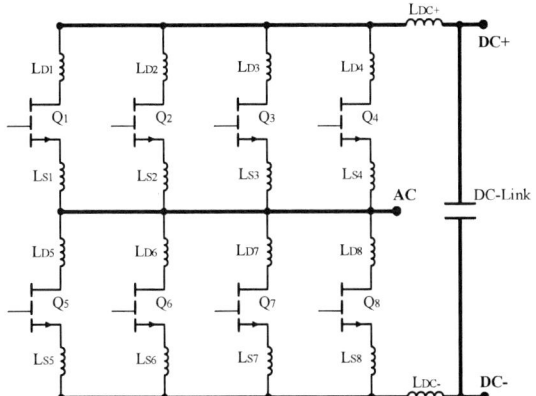

Fig. 1. Circuit diagram of half bridge with four GaN HEMT chips in parallel with parasitic inductances.

In this paper, a four-chip parallel structure is designed, each chip contains parasitic inductors L_D and L_S, and the complete power loop also contains parasitic inductors L_{DC+} and L_{DC-}. Ignoring unnecessary related factors, and according to the classical equation of PCB inductance [23] − [24], the inductance of power circuit can be calculated as

$$L_{loop} \cong \mu \frac{h}{w} l \ , \tag{1}$$

where μ is the permeability of the FR4 material, h is the height, l is the length and w is the width of the power path in PCB, it is an accurate approximation if $h \gg w$. So, the optimization of the parasitic inductors is based on the scheme of shortening the length of the loop as much as possible, and the symmetric design of both sides of the chip and the overlapping power paths of the top and bottom layers are used to realize the magnetic field phase cancellation, so as to reduce the parasitic inductance of the circuit.

The power circuit is circulated in the power motherboard PCB, which is aligned with the DCM standard package, with two DC+ terminals and one DC− terminal on one side, and AC terminals on the other side of the power motherboard PCB. The GaN HEMT chips are located in the bottom layer of the power motherboard PCB, and the through-hole is used to connect the top layer of the PCB with the bottom layer, constituting a complete power circuit. The component layout of the power motherboard PCB and the current path are shown in Fig. 2. In Fig. 2, the blue arrow indicates the current path in the bottom copper layer of the PCB, and the red arrow indicates the current path in the top copper layer of the PCB.

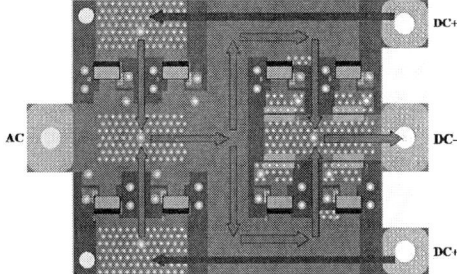

Fig. 2. Diagram of proposed power PCB current path.

The majority of GaN HEMT multichip power modules are in the long column layout [12]−[16]. This paper extracts this feature, designs a traditional module with this layout, and sets it to have the same power terminal as the proposed module to eliminate the influence of other factors on parasitic inductance. The component layout of the traditional module and the current path are shown in Fig. 3. Finally, the traditional module is compared with the proposed module.

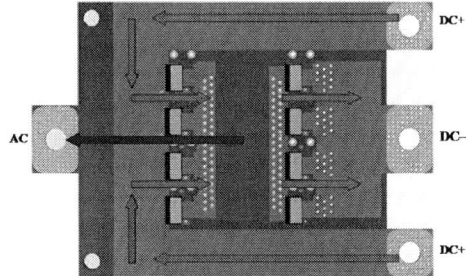

Fig. 3. Diagram of traditional power PCB current path.

In order to verify the effectiveness of the power motherboard PCB in reducing parasitic inductance, this paper extracts the parasitic parameters through ANSYS Q3D finite element simulation software and compares it with the traditional layout. The simulation results show that the parasitic inductance of the power circuit of the power motherboard PCB proposed in this paper is 4.72 nH, and the parasitic inductance of the power circuit of the PCB under the traditional layout is 8.85 nH, which is reduced by 46.7%.

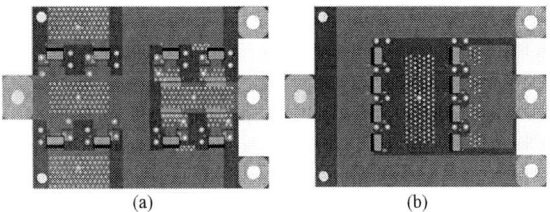

(a)   (b)

Fig. 4. Schematic diagram of the GaN power module with (a) proposed layout and (b) traditional layout.

Fig. 5. Diagram of the power module magnetic field intensity distribution.

III. CONFIGURATION OF MULTI-CHIP GaN HEMT POWER MODULE WITH REPACKAGING CONCEPT

A. Vertical Implementation of the Power Modulel Structure

The proposed multi-chip power module is created by the discrete GaN HEMT device GS66508T from Infineon, with all the electrodes of the chip located on the bottom surface and a heat sink metal substrate on the top surface. Its specific electrical parameters are shown in TABLE I.

TABLE I. SPECIFICATIONS OF EMPLOYED GAN HEMT DEVICE GS66508T

Specification	Value	Unit
Drain-source voltage V_{DS}	650	V
Gate-source voltage V_{GS}	−10 to +7	V
Continous conduction drain current I_{ds}	30	A
On-resistance $R_{ds(on)}$	50	mΩ

This paper proposes a secondary encapsulated power module adapted to the lateral structure of GaN HEMT devices, and the components of the power module include the bottom DBC for heat dissipation, the middle power motherboard PCB for DCM standard package, and the top signal daughterboard PCB for optimizing the electrical performance of the module. the bottom structure is the DBC, which is connected to the top heat dissipation substrate of the inverted GaN HEMT chips. The middle layer structure of the power module is the motherboard PCB, and the electrodes at the bottom of the GaN HEMT chips are soldered to the bottom layer of the motherboard PCB, i.e., the GaN HEMT chips are located between the DBC and the power motherboard PCB, and on the top layer of the power motherboard PCB are the driver circuit, the decoupling ceramic capacitors, and the contact terminals that are connected to the signal daughterboard PCB. Located on the signal daughter PCB are the PWM signal ports and the necessary power supply circuits. This hybrid package achieves low parasitic inductance and high insulation with good heat dissipation. The side view of the power module is shown in Fig. 6. The 3D model design is shown in Fig. 7.

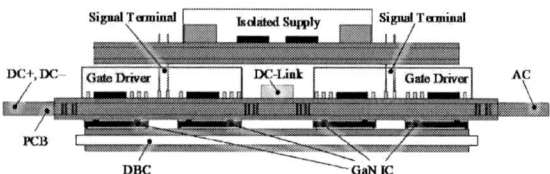

Fig. 6. Configuration of repackaged GaN HEMT power module.

Fig. 7. Exploded view of repackaged GaN HEMT power module.

B. Gate Driver Circuit Integration of the Power Module

The module uses a new-type self-powered isolated gate drive circuitry, a circuit chipset that simultaneously transmits the PWM signal and gate bias power through an external transformer, eliminating the need for an external gate drive auxiliary bias power supply or high side bootstrap. This greatly simplifies system design and reduces electromagnetic interference (EMI) by reducing common mode (CM) capacitance. It also allows floating switches to be driven anywhere in the switching power supply topology. The power module proposed in this paper is a four-chip parallel structure, i.e., the same PWM signal needs to be transmitted synchronously to the gate of the four chips, therefore, this paper adopts a single isolated emission IC plus a transformer as the front half of the driver circuit structure is located in the signal sub-panel PCB, and the circuit from the secondary side pins of the transformer to the input pins of the four driver reception ICs transmit only the signal, and there are no other limitations except to ensure that the four transmission paths have as equal a length as possible for synchronous driving. The specific structure of the drive circuit and the layout of the signal daughter board PCB are shown in Fig. 8 and Fig. 9.

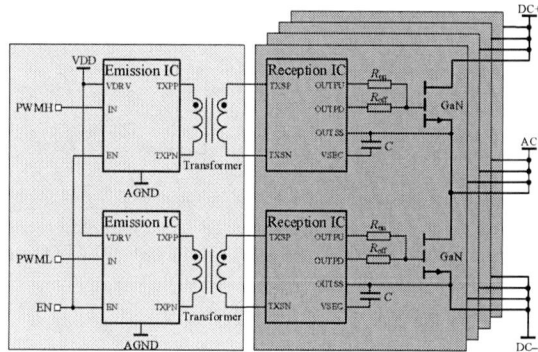

Fig. 8. Schematic diagram of integrated gate driver in repackaged GaN HEMT power module.

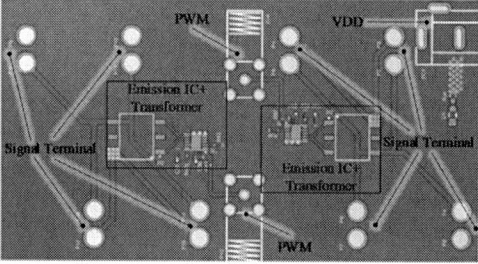

Fig. 9. Structure diagram of signal daughter-board in PCB.

Four driver reception ICs together with the corresponding driving resistors and other components are located on the top layer of the power motherboard PCB as the second half of the driver circuit structure to drive the four GaN HEMT chips synchronously. The second half of the driver circuit and the GaN HEMT chips form a very small loop area, and the good design of the driver circuit achieves low parasitic inductance. In addition, the decoupling capacitor is also located on the top layer of the power motherboard PCB, between the DC+ and DC− terminals.

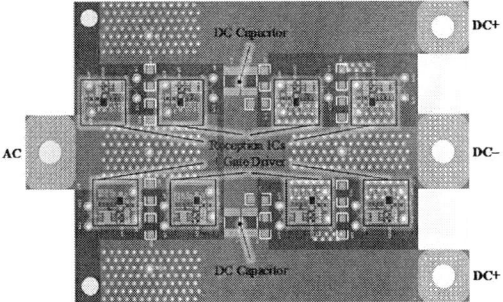

Fig. 10. Structure diagram of power motherboard PCB

IV. EXPERIMENTAL VALIDATIONS

A. Measurement and Verification of Parasitic Inductance

To verify the correctness of the simulation result, we use KEYSIGHT's 4294A Impedance Analyzer to test the parasitic inductance of the proposed module. The results show that the parasitic inductance is 4.75 nH, the error between the simulation results and the actual results is 0.6%.

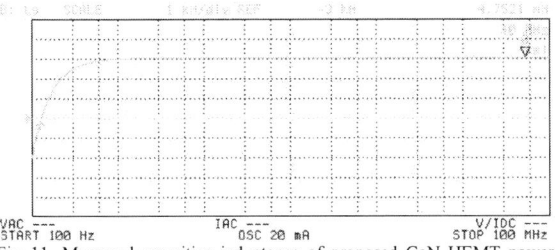

Fig. 11. Measured parasitice inductance of proposed GaN HEMT power module by using Impedance Analyzer.

B. Current Sharing Test of the Power Module

To confirm the advances of the proposed GaN HEMT power module, the double-pulse test rig is set up, as shown in Fig. 15 and Fig. 15. The matched busbar with low parasitic inductance is also carefully designed and prepared for the GaN HEMT power module. Both the power module and the capacitor busbar are fabricated by SMT process, and are assembled by screws. The signal generator is employed to generate the trigger pulses, while the oscilloscope is used to capture the waveforms. The optical-fiber isolated voltage probe and Rogowski coil are utilized to measure the transient switching trajectories. The load inductor is 560 µH. Meanwhile, the turn-on and -off resistances for the gate driver are 10 Ω and 0 Ω, respectively.

Fig. 12. Prototype of proposed GaN HEMT power module.

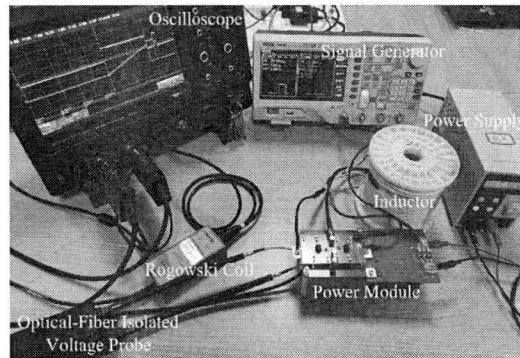

Fig. 13. Double test rig for proposed GaN HEMT power module.

In order to verify the chip equalization performance of the module proposed in this paper, the load current waveforms of the four chips are tested separately and compared experimentally with the traditional layout module. Under the setting parameter of load current 20 A, the experimental results show that the waveforms of the four chips of the proposed module have high consistency during the turn-on process, the difference between the maximum value and the minimum value is 0.6 A, and the difference of the traditional module is 1.7 A. Compared with the traditional layout, the difference of measured load current is reduced by 64.7%, has a smaller difference from the set parameter of 20 A. The experimental results show that the module has a good performance in terms of equalizing current performance.

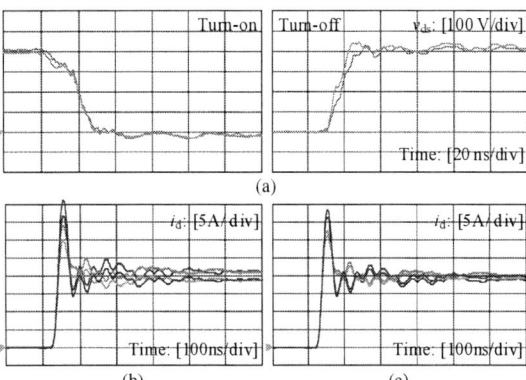

Fig. 14. Experimental current sharing of paralleled four chips with (a) voltage waveform of proposed chip layout(red) and traditional chip layout(green), (b) traditional chip layout and (c) proposed chip layout.

C. Voltage Overshoot and Load Test of the Power Module

Under different dc-link voltages, the measured turn-on and -off waveforms are displayed in Fig. 15. The load current is 80 A for the power module and 20 A for each chip. It can be found that, benefit from the low parasitic design, the overshooting voltage during turn-off is only 30 V, which is 7.3% of the nominal dc-link voltage. Besides, no significant overshooting is observed in the gate loop. As a result, the fabricated GaN HEMT power module has clear and excellent switching performances.

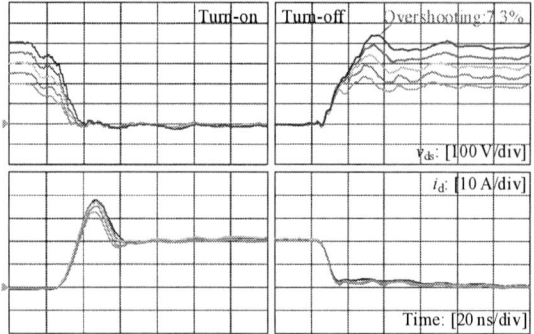

Fig. 15. Experimental waveforms of double-pulse test under different voltage conditions.

Under 400 V dc-link voltage but different currents, the measured turn-on and turn-off waveforms of total load current and single chip current are displayed in Fig. 15.

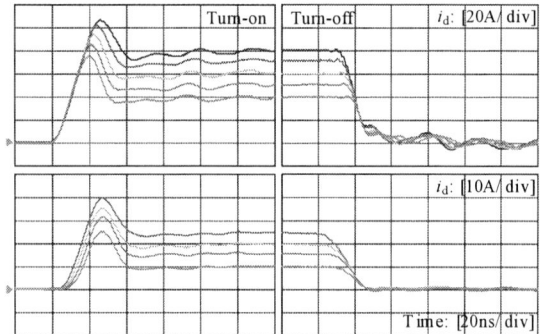

Fig. 16. Experimental waveforms of double-pulse test under different current conditions.

It can be seen from the figure that under different current conditions, the rising slope of current has high consistency, and the overshoot is basically the same, which indicates that the current characteristics also have good performance.

V. ONLUSIONS

In this paper, a four-chip parallel GaN power half-bridge module with integrated gate driver is proposed, fabricated and tested. The novel self-powered isolated gate driver circuit is designed to mitigate the complexity associated with the parallel connection of chips. The power motherboard PCB is globally optimized and compared with the traditional layout package, and the simulation results show that the parasitic inductance of the module is reduced by 46.7%. The bus voltage overshoot is only 7.3% under 400 V, 80 A double-pulse experimental condition, which verifies that the module has more stable and reliable switching performance. The comparison of the current-averaging characteristics through the double-pulse experiment shows that the chip layout of the proposed module has better chip current-averaging performance than the traditional chip layout. The thermal management of the proposed module will be further design and optimize in the future research.

ACKNOWLEDGMENT

The authors would like to thank the financial supports from the National Natural Science Foundation of China under Grant 52177169 and the Chongqing Research Program of

Basic Research, the Natural Science Foundation of Chongqing under Grants CSTB2024NSCQ-JQX0016, and the Graduate Research and Innovation Foundation of Chongqing, China under Grant CYB240023.

REFERENCES

[1] J. P. Kozak, R. Zhang, M. Porter, Q. Song, J. Liu and B. Wang, "Stability, reliability, and robustness of GaN power devices: A review, " *IEEE Trans. Power Electron.*, vol. 38, no. 7, pp. 8442-8471, July 2023.

[2] E. A. Jones, F. F. Wang and D. Costinett, "Review of commercial GaN power devices and GaN-based converter design challenges, " *IEEE J. Emerg. Sel. Top. Power Electron.*, vol. 4, no. 3, pp. 707-719, Sept. 2016.

[3] M. Meneghini, C. D. Santi, I. Abid, et al, "GaN-based power devices: Physics, reliability, and perspectives," *J. Appl. Phys.* , vol. 130, no. 18, pp. 181101, Nov. 2021.

[4] K. J. Chen, O. Häberlen, A. Lidow, C. L. Tsai, T. Ueda and Y. Uemoto, "GaN-on-Si power technology: Devices and applications," *IEEE Trans. Electron. Devices*, vol. 64, no. 3, pp. 779-795, March 2017.

[5] A. Letellier, M. R. Dubois, J. P. Trovao and H. Maher, "Gallium nitride semiconductors in power electronics for electric vehicles: Advantages and challenges," in *IEEE VPPC*, 2015, pp. 1-6.

[6] B. Sun, Z. Zhang and M. A. E. Andersen, "Switching transient analysis and characterization of GaN HEMT," in *IGBSG*, 2018, pp. 1-4.

[7] Q. Xin, H. Peng, Q. Yue, J. Chen and L. Tong, "Investigate of gate driver loop impedance on switching transients of both power device side and gate driver side," in *IEEE PEAC*, 2022, pp. 1467-1472.

[8] E. Gurpinar, B. Ozpineci and S. Chowdhury, "Design, analysis, comparison, and experimental validation of insulated metal substrates for high-power wide-bandgap power modules," *J. Electron. Packag.*, vol. 142, no. 4, pp. 1-10, Dec. 2020.

[9] S. Lu, T. Zhao, R. Burgos and G. Q. Lu, "Packaging of (650 V, 150 A) GaN HEMT with low parasitics and high thermal performance," in *ICEP*, 2021, pp. 39-40.

[10] L. Kou and J. Lu, "Applying GaN HEMTs in conventional housing-type power modules," in *IEEE ECCE*, 2020, pp. 4006-4011.

[11] F. Luo, Z. Chen, L. Xue, P. Mattavelli, D. Boroyevich and B. Hughes, "Design considerations for GaN HEMT multichip halfbridge module for high-frequency power converters," in *IEEE APEC*, 2014, pp. 537-544.

[12] B. Passmore, S. Storkov, B. McGee, J. Stabach, G. Falling and A. Curbow, "A 650 V/150 A enhancement mode GaN-based half-bridge power module for high frequency power conversion systems," in *IEEE ECCE*, 2015, pp. 4520-4524.

[13] S. Moench, R. Reiner, P. Waltereit, F. Benkhelifa, J. Hückelheim and D. Meder, "PCB-embedded GaN-on-Si half-bridge and driver ICs with on-package gate and DC-link capacitors," *IEEE Trans. Power Electron.*, vol. 36, no. 1, pp. 83-86, Jan. 2021.

[14] A. I. Emon, H. Carlton, J. Harris, A. Krone, M. Hassan and A. Mirza, "A 650V/60A gate driver integrated wire-bondless multichip GaN module," in *IEEE PEDG*, 2021, pp. 1-6.

[15] H. Kong, L. Jia, L. Wang, Y. Yao, F. Yang and H. Cui, "A flexible-PCB on DPC GaN power module with ultralow parasitic inductance," *IEEE Trans. Power Electron.*, vol. 40, no. 4, pp. 5241-5251, April 2025.

[16] P. Han, P. Liu, Q. Huang, Z. Chen and A. Q. Huang, "A 650 V, 2.1 mohm GaN Half-bridge Power Module for 400V EV Traction Inverter Application," in *IEEE ECCE*, 2022, pp. 1-6.

[17] M. T. Tran, D. D. Tran, K. Deepak, G. E. Martin, O. Bay and M. E. Baghdadi, "A high performance GaN power module with parallel packaging for high current and low voltage traction inverter applications," *IEEE J. Emerg. Sel. Top. Power Electron.*, vol. 13, no. 1, pp. 1188-1209, Feb. 2025.

[18] A. B. Jørgensen, S. Bęczkowski, C. Uhrenfeldt, N. H. Petersen, S. Jørgensen and S. M. Nielsen, "A fast-switching integrated full-bridge power module based on GaN eHEMT devices," *IEEE Trans. Power Electron.*, vol. 34, no. 3, pp. 2494-2504, March 2019.

[19] A. I. Emon, H. Carlton, J. Harris, A. Krone, A. Mirza and M. Hassan, "Design and optimization of 650 V/60 A double-sided cooled multichip GaN module," in *IEEE APEC*, 2021, pp. 2313-2317.

[20] Y. Yan, L. Zhu, J. Walden, Z. Liang, H. Bai and M. H. Kao, "Packaging a top-cooled 650 V/150 A GaN power modules with insulated Thermal Pads and Gate-Drive Circuit," in *IEEE APEC*, 2021, pp. 2345-2350.

[21] B. Li, X. Yang, K. Wang, H. Zhu, L. Wang and W. Chen, "A compact double-sided cooling 650 V/30 A GaN power module with low parasitic parameters," *IEEE Trans. Power Electron.*, vol. 37, no. 1, pp. 426-439, Jan. 2022.

[22] C. Laurant, J. Delaine, P. Périchon, B. Thollin, C. Lanneluc and A. Izoulet, "Very low parasitic inductance double side cooling power modules based on ceramic substrates and GaN devices," in *IEEE ECTC*, 2020, pp. 1402-1407.

[23] E. Gurpinar, F. Iannuzzo, Y. Yang, A. Castellazzi and F. Blaabjerg, "Design of Low-Inductance Switching Power Cell for GaN HEMT Based Inverter," *IEEE Trans. Ind. Appl.*, vol. 54, no. 2, pp. 1592-1601, March-April 2018.

[24] M. T. Thompson, "Inductance Calculation Techniques-Part I: Classical Methods," *Power Control Intell. Motion.*, vol. 25, no. 12, pp. 40-45, 1999, Accessed: Jan. 29, 2024.

Impact of Electroluminescence Spectrum Sampling on SiC MOSFET Junction Temperature and Current Sensing

Yuting Jin
College of Electrical Engineering
Zhejiang University
Hangzhou, China
yutingjin@zju.edu.cn

Hao Huang
College of Electrical Engineering
Zhejiang University
Hangzhou, China
12310083@zju.edu.cn

Jingyang Hu
College of Electrical Engineering
Zhejiang University
Hangzhou, China
jingyanghu@zju.edu.cn

Shengjie Luo
College of Electrical Engineering
Zhejiang University
Hangzhou, China
22310083@zju.edu.cn

Haoze Luo
College of Electrical Engineering
Zhejiang University
Hangzhou, China
haozeluo@zju.edu.cn

Wuhua Li
College of Electrical Engineering
Zhejiang University
Hangzhou, China
woohualee@zju.edu.cn

Abstract—The electroluminescence (EL) effect in SiC MOSFETs offers a non-invasive method for sensing key operating states, such as junction temperature and current. However, the performance of such sensing systems is influenced by the channel attributes of the spectral sampling unit, which affect accuracy, bandwidth, and interference immunity. Additionally, the optical coupling method directly impacts the spectral preprocessing requirements and the difficulty of module integration. This work investigates the impact of electroluminescence spectral sampling on device state sensing in detail using a 1200V SiC module operating in a Buck topology and a partial least squares (PLS) sensing model as a case.

Keywords—Electroluminescence effect, SiC MOSFET, state sensing, spectrum sampling

I. INTRODUCTION

With the widespread adoption of SiC MOSFETs in fields such as renewable energy, electric vehicles, rail traction, and power transmission systems, the operating conditions of SiC devices have become increasingly stringent in terms of temperature, voltage, and current, leading to significantly higher reliability requirements [1]. On the one hand, this trend underscores the growing importance of monitoring the operating state of SiC MOSFETs; on the other hand, the challenges of thermal isolation under high temperatures and electrical isolation under high voltages present significant difficulties for state monitoring solutions.

Traditional thermal measurement methods based on physical heat conduction, such as thermistors with negative temperature coefficients (NTC), struggle to simultaneously fulfill the fast response and insulation requirements for junction temperature measurement. Similarly, conventional current measurement techniques [2] based on electromagnetic principles, such as coaxial resistors, current transformers, Hall sensors, and Rogowski coils [3], often face difficulties in resisting temperature drift and electromagnetic interference. Electroluminescence, an intrinsic characteristic of the body diode in SiC MOSFET, provides critical information about the device's operating states via the visible light spectrum emitted during reverse conduction. Leveraging light as the medium to transmit electrical and thermal information effectively addresses the challenges of thermal isolation and electrical insulation, thereby opening new pathways for state sensing in SiC devices. Early research has explored the significant correlation of spectral characteristics, such as light intensity and wavelength redshift, to device states [4-5]. Later studies introduced advanced multispectral [6] and hyperspectral [7] acquisition methods, along with improved analytical algorithms, to enhance signal-to-noise ratios and the accuracy of spectral measurement and analysis under practical operating conditions.

However, the influence of spectral sampling and processing methods on state sensing has yet to be systematically investigated. Different optical coupling methods can impact the integration complexity and baseline of the spectrum, while the number and characteristics of sampling channels affect sensing accuracy, anti-interference capability, and cost. This study investigates the impact of spectral sampling on partial least squares (PLS) state extraction results by taking a 1200V SiC module operating under synchronous rectification in a Buck topology as an example.

The rest of this paper is organized as follows. Section II briefly introduces the electroluminescence phenomenon and characteristics of SiC MOSFET, and explains the analysis method, experimental basis and algorithm flow of the spectral sampling impact on state sensing. Sections III and IV analyze the influence of optical coupling mode and sampling channel characteristics respectively. Section V provides a summary of this work.

II. ELECTROLUMINESCENCE PRINCIPLE AND ANALYSIS PROCESS OF SAMPLING IMPACT

Taking the Buck topology as an example, when the lower SiC device operates in reverse conduction, the body diode emits a blue-green light, which is the electroluminescence effect of the SiC MOSFET, as shown in Fig. 1. The electroluminescence spectrum exhibits two characteristic peaks, generated by direct band-to-band recombination and trap-assisted recombination, respectively. The intensity and wavelength of these peaks show varying sensitivities to the device's junction temperature and current. Since the device only emits light during the dead time under actual synchronous rectification conditions, more than two channels

---

This work was supported by the National Key Research and Development Program of China under Grant 2022YFE0138400, in part by Zhejiang Provincial Natural Science Foundation of China under Grant LR24E070001, and in part by the Joint Foundation for Basic Research on Railways under Grant U2368206.

are required for different wavelengths to ensure an adequate signal-to-noise ratio and sensing accuracy.

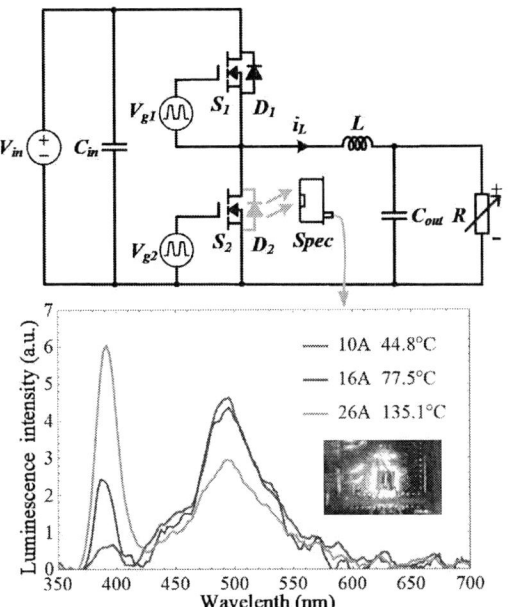

Fig. 1. Electroluminescence phenomena of SiC MOSFETs in Buck topology.

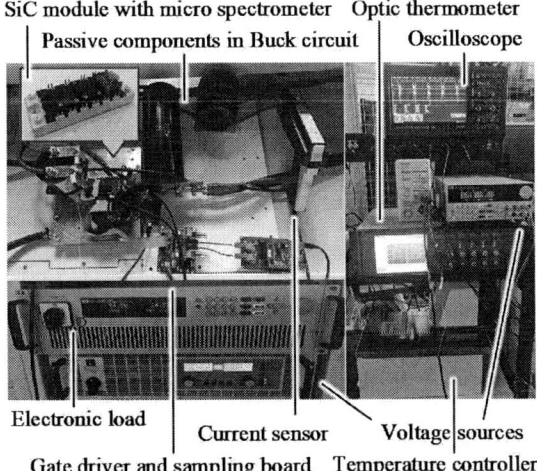

Fig. 2. Experimental platform setup.

In this work, the optical measurement equipment, spectral analysis method and experimental data from previous research [7] are adopted as the foundation to analyze the impact of electroluminescence spectrum sampling on state sensing. The experimental platform is shown in Fig. 2. A miniaturized spectrometer with 288 visible-light channels was used to measure the spectra of a 1200V SiC power module under different operating conditions, including turn-off voltages ranging from 50V to 300V, currents from 10A to 26A, and junction temperatures from 44.8°C to 153.2°C. A fundamental dataset consisting of 1120 samples was constructed. The analysis process for investigating the impact of spectral sampling on state sensing is illustrated in Fig. 3. To account

for potential measurement variations between different spectrometers or experiments, adjustments are made to the same fundamental dataset to simulate different optical coupling methods and channel characteristics.

Under each sampling condition, the dataset will be preprocessed using convolution smoothing and normalization techniques, followed by the training of a partial least squares (PLS) regression model. Savitzky-Golay convolution smoothing is a filtering method based on local polynomial least squares fitting in the time domain, which can effectively retain the spectral characteristics while removing spectral measurement noise. Unlike principal component analysis (PCA), which solely reduces data dimensionality without considering target variables, PLS explicitly maximizes the covariance between spectral features (X: I_{EL}) and device states (Y: T_j, I). This characteristic proves critical in scenarios where subtle spectral shifts (e.g., 0.1–0.3 nm wavelength drifts) carry significant state information. The detailed procedures of data preprocessing and model construction have been thoroughly described in [7] and will not be reiterated here. The junction temperature and current sensing mean absolute errors (MAE) corresponding to the basic sampling and regression processes are 3.11°C and 0.81A respectively. By applying the test set data to the optimal model under each sampling condition, the quantitative impact of different spectral sampling configurations on state sensing can be obtained, providing a foundation for further analysis.

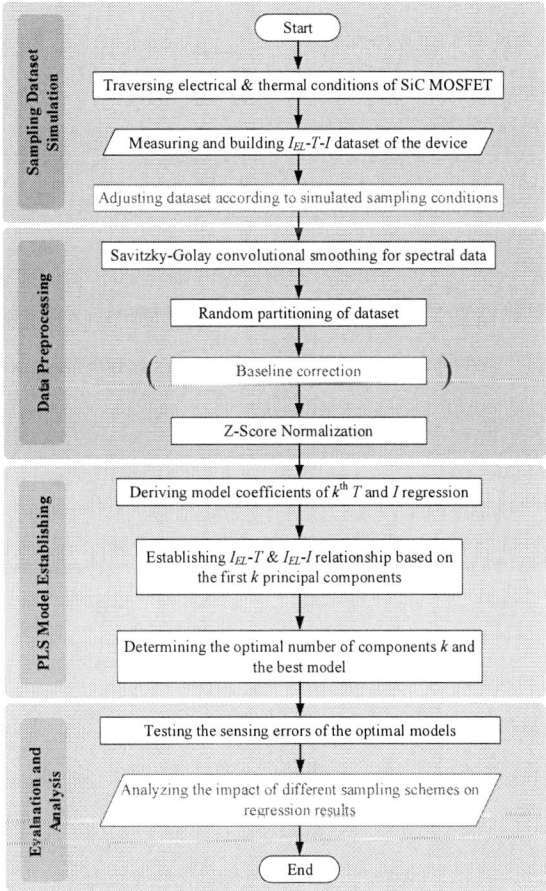

Fig. 3. Analysis flowchart of the Spectral Sampling Impact.

III. Impact Analysis of Optical Coupling Method

Optical measurement setups can typically be classified into two coupling modes: free-space coupling and fiber coupling. In free-space coupling, the electroluminescence emitted by the device passes through a spatial medium within the module, such as silicone gel, and is directly fed into the measurement system. This approach offers higher integration but imposes stricter requirements on the spectrometer's size and is more susceptible to environmental influences [8]. In contrast, fiber coupling directs the luminescence through an optical fiber to be measured and processed at a remote location. This method provides more stable measurements [9] but results in a more complex system.

In this work, free-space coupling was employed for experimental measurements. The baseline of the measured spectra (i.e., the horizontal line at the initial state) was found to fluctuate with operating conditions, as shown in Fig. 4. And the baseline correction of the fundamental dataset can effectively simulate the stable spectral data that would have been obtained using fiber coupling.

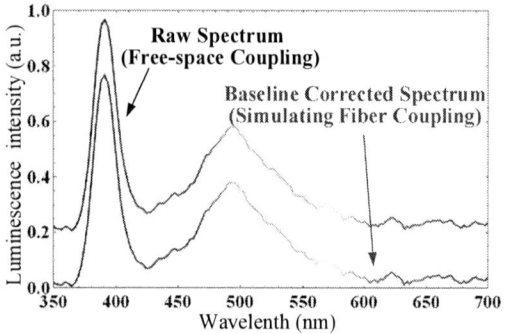

Fig. 4. Electroluminescence spectra corresponding to different coupling modes.

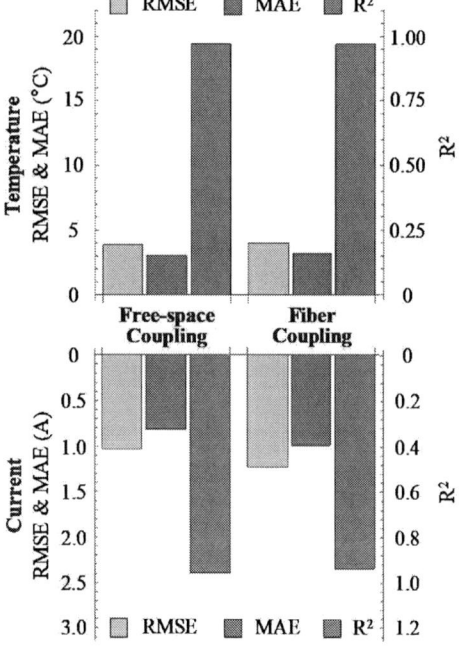

Fig. 5. Sensing results of different coupling modes.

The sensing performance of these two optical coupling methods is shown in Fig. 5. Counterintuitively, free-space coupling exhibits slightly better sensing performance than fiber coupling for both junction temperature and current. The root mean square errors (RMSE) and mean absolute errors (MAE) are smaller, while the coefficients of determination (R^2) are larger.

This phenomenon can be attributed to the fact that baseline drift in the measured spectra during module operation is primarily caused by the dark current of the photodetector. The dark current is positively correlated with the spectrometer's temperature, as shown in Fig. 6 [10]. The temperature-dependent behavior of dark current in CMOS-based spectrometers follows fundamental semiconductor physics principles. As derived from Shockley-Read-Hall recombination theory [11], the dark current density J_{dark} can be expressed as:

$$J_{dark}(T) = q \cdot n_i^2(T) \cdot \frac{W}{\tau_{eff}(T)} \left(e^{\frac{qV}{k_B T}} - 1 \right) \quad (1)$$

where q denotes the elementary charge, $n_i(T)$ is the intrinsic carrier concentration, W is the depletion region width, $\tau_{eff}(T)$ is the effective carrier lifetime, and V the applied bias voltage. For practical photodetectors operating in the weak inversion regime ($V \approx 0$), (1) simplifies to a temperature-dominated form [12]:

$$I_{dark}(T) = A \cdot T^{\frac{3}{2}} \cdot e^{-\frac{E_g}{2k_B T}} + I_{leakage} \quad (2)$$

where A represents a device-specific proportionality constant, E_g is the effective bandgap energy of the detector material, and $I_{leakage}$ denotes temperature-independent leakage components from surface states.

This implies that the baseline is not detrimental to spectral analysis but rather serves as a variable related to the sensor temperature, which is determined by the device junction temperature, ambient temperature and the thermal network of the integrated module, as illustrated in Fig. 7. Thus, free-space coupling effectively integrates an additional "temperature sensor" within the module. When baseline information is incorporated into the PLS model, it enhances the sensing accuracy over that of fiber coupling.

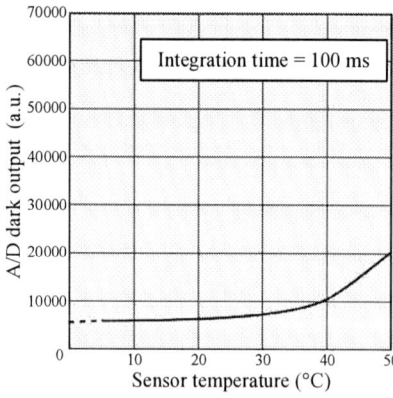

Fig. 6. Relationship between the dark current output of the micro-spectrometer and the sensor temperature [10].

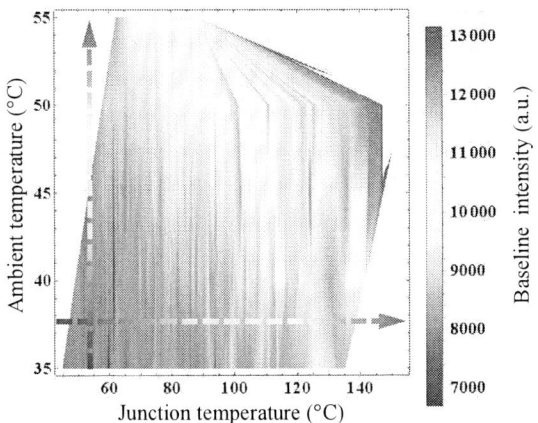

Fig. 7. Dependence of baseline intensity on temperature.

IV. IMPACT ANALYSIS OF SAMPLING CHANNELS

The number of sampling channels and their wavelengths determine the accuracy and anti-interference capability of the perception system, while also influencing the cost and integration complexity of the optical path. Therefore, achieving an appropriate state sensing accuracy with as few channels as possible is desirable. The sampling conditions for n channels can be obtained by retaining only the mth channel for every p channels in the sample spectra from the fundamental dataset, as shown in Fig. 8. The relationship between n and p can be expressed as:

$$n = \lceil 288 / p \rceil \qquad (3)$$

It is worth noting that the choice of m also impacts the perception performance since the electroluminescence signal-to-noise ratio varies across adjacent wavelength channels.

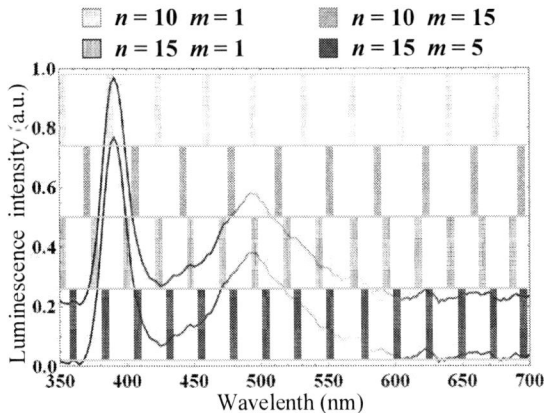

Fig. 8. Schematic diagram of different channel numbers and sequences.

The influence of different channel amounts and sequences on sensing results is illustrated in Fig. 9. The correlation between junction temperature and current with spectral wavelength is fundamentally consistent. When the number of channels drops below 50, the impact of channel sequence on the model becomes markedly significant, indicating that the spectral information is no longer complete. When the number

of channels is reduced to 10, the RMSE of junction temperature and current perception under the optimal channel sequence is 2.85 times and 1.95 times higher, respectively, compared to the worst-case channel sequence. The results indicate that junction temperature sensing exhibits a higher sensitivity to channel sequence compared to current sensing.

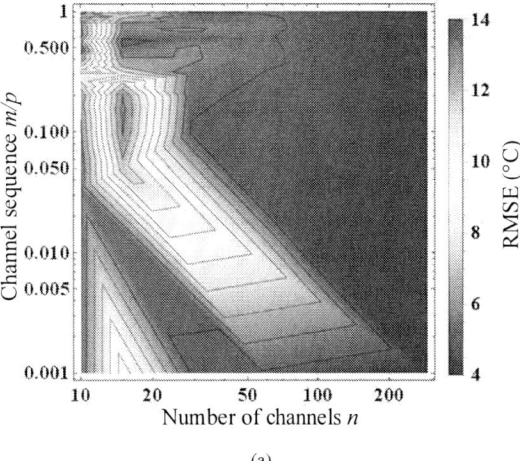

(a)

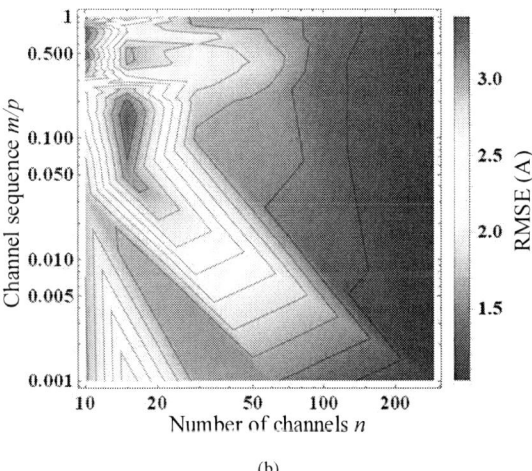

(b)

Fig. 9. Sensing results under different channel numbers and sequences. (a) Temperature model. (b) Current model.

V. CONCLUSION

This work investigates the impact of electroluminescence sampling and processing methods on junction temperature and current sensing in SiC MOSFETs. In terms of optical coupling, unlike conventional application scenarios, the free-space coupling mode demonstrates higher sensing accuracy than fiber coupling by additionally capturing baseline drift characteristics, which include contributions from dark current and ambient temperature variations. Regarding sampling channels, spectral information becomes incomplete when the number of channels falls below 50. Furthermore, the junction temperature exhibits higher sensitivity to channel sequencing than current. The proposed analysis provides valuable insights for designing integrated sensing solutions in power modules.

REFERENCES

[1] M. H. Nguyen and S. Kwak, "Enhance Reliability of Semiconductor Devices in Power Converters," *Electronics*, vol. 9, no. 12, pp. 1-37, Dec. 2020.

[2] J. P. Kozak, R. Zhang, J. Liu, K. D. T. Ngo and Y. Zhang, "Degradation of SiC MOSFETs Under High-Bias Switching Events," *IEEE J. Emerg. Sel. Topics Power Electron.*, vol. 10, no. 5, pp. 5027-5038, Oct. 2022.

[3] Y. Shi, Z. Xin, P. C. Loh and F. Blaabjerg, "A Review of Traditional Helical to Recent Miniaturized Printed Circuit Board Rogowski Coils for Power-Electronic Applications," *IEEE Trans. Power Electron.*, vol. 35, no. 11, pp. 12207-12222, Nov. 2020.

[4] J. Winkler, J. Homoth and I. Kallfass, "Utilization of Parasitic Luminescence from Power Semiconductor Devices for Current Sensing," in *Proc. PCIM Europe 2018; International Exhibition and Conference for Power Electronics, Intelligent Motion, Renewable Energy and Energy Management*, 2018, pp. 1-8.

[5] H. Luo, J. Mao, C. Li, F. Iannuzzo, W. Li and X. He, "Online Junction Temperature and Current Simultaneous Extraction for SiC MOSFETs With Electroluminescence Effect," *IEEE Trans. Power Electron.*, vol. 37, no. 1, pp. 21-25, Jan. 2022.

[6] L. A. Ruppert and R. W. De Doncker, "Multispectral Electroluminescence Sensing of SiC MOSFETs for Junction Temperature and Current Extraction," in *Proc. PCIM Europe 2024; International Exhibition and Conference for Power Electronics, Intelligent Motion, Renewable Energy and Energy Management*, Nürnberg, Germany, 2024, pp. 1698-1706.

[7] Y. Jin, S. Ye, Q. Wu, H. Luo, W. Li and X. He, "Junction Temperature and Current Synchronous Sensing for SiC MOSFETs Based on Electroluminescence Hyperspectral," in *Proc. 2023 11th International Conference on Power Electronics and ECCE Asia (ICPE 2023 - ECCE Asia)*, Jeju Island, Republic of Korea, 2023, pp. 896-901.

[8] H. A. Zain *et al.*, "Review of Microbottle Resonators for Sensing Applications," *Micromachines*, vol. 14, no. 4, pp. 734-746, 2023.

[9] J. Oh *et al.*, "Metasurfaces for Free-Space Coupling to Multicore Fibers," in *Journal of Lightwave Technology*, vol. 42, no. 7, pp. 2385-2396, Apr. 2024.

[10] Hamamatsu Photonics K.K., " Fingertip-sized, ultra-compact spectrometer head supporting high sensitivity, " C12880MA Microspectrometer Datasheet, Japan, Mar. 2024.

[11] S. M. Sze and K. K. Ng, "Physics of Semiconductor Devices," 3rd ed. Hoboken, NJ: Wiley, 2006, pp. 78-82.

[12] E. R. Fossum, "Modeling the temperature dependence of dark current in CMOS image sensors," *IEEE Trans. Electron Devices*, vol. 59, no. 8, pp. 1989-1995, Aug. 2012.

Research on Threshold Voltage Instability of SiC MOSFETs at High Temperature

1st Xu Cheng
Zhuhai Power Supply Bureau of
Guangdong Power Grid Co., Ltd.
DC Power Distribution and
Consumption Technology Research
Center of Guangdong Power Grid Co.,
Ltd.
Guangdong, China
664157102@qq.com

2nd Yong Chen
Zhuhai Power Supply Bureau of
Guangdong Power Grid Co., Ltd.
DC Power Distribution and
Consumption Technology Research
Center of Guangdong Power Grid Co.,
Ltd.
Guangdong, China
35665035@qq.com

3rd Xingyu Pei
Zhuhai Power Supply Bureau of
Guangdong Power Grid Co., Ltd.
DC Power Distribution and
Consumption Technology Research
Center of Guangdong Power Grid Co.,
Ltd.
Guangdong, China
31977550@qq.com

4th Jianbiao Li
Zhuhai Power Supply Bureau of
Guangdong Power Grid Co., Ltd.
DC Power Distribution and
Consumption Technology Research
Center of Guangdong Power Grid Co.,
Ltd.
Guangdong, China
zhlijianbiao@126.com

5th Hongyuan Wu
Zhuhai Power Supply Bureau of
Guangdong Power Grid Co., Ltd.
DC Power Distribution and
Consumption Technology Research
Center of Guangdong Power Grid Co.,
Ltd.
Guangdong, China
whyqsxddc@163.com

6th Cong Chen
School of Electrical and Electronic
Engineering,
North China Electric Power University.
Beijing, China
chencong@ncepu.edu.cn

Abstract—**Silicon carbide MOSFETs have a promising future in electric vehicles, aerospace and other fields due to their high-temperature resistance and low-loss characteristics, but their high-temperature reliability is severely constrained by threshold voltage instability. In this paper, in response to the shortcomings of existing studies at ultra-high temperatures (150~275°C) and wide gate voltage ranges, a bias temperature instability (BTI) experimental platform is constructed to systematically investigate the effects of gate bias and high temperature on the threshold voltage instability of silicon carbide MOSFETs. It is found that: ΔV_{TH} is negatively correlated with temperature in the range of 150~175°C, and positively correlated with temperature in the range of 175~275°C This is due to the competition between the bidirectional tunneling process of charges in the traps and the thermal activation process of the traps in the different temperature intervals; under negative gate bias (NBTI), the trap-assisted tunnelling (TAT) is triggered by the high field strength and the high temperature can activate the deep mobile ions in the oxide layer, leading to an anomalous temperature dependence of the threshold voltage drift. The dominant roles of the host trap (forbidden band mid/valence band) and the donor trap (conduction band) on the drift direction are further verified by Sentaurus TCAD simulations. In this study, the synergistic degradation path between gate voltage and temperature at ultra-high temperatures is clarified for the first time, and the application design guideline based on the temperature turning point is proposed, which provides theoretical support and experimental basis for the optimization of the reliability of silicon carbide devices under extreme operating conditions.**

Keywords—Silicon Carbide MOSFET, Bias Temperature Instability, High Temperature, Gate Bias, Threshold Voltage Instability

I INTRODUCTION

Compared with traditional silicon (Si) devices, silicon carbide (SiC) materials show significant advantages in high-temperature and high-voltage scenarios due to their wide-bandwidth, high breakdown field strength, and high thermal conductivity properties, which enable SiC MOSFETs to be used in high-temperature and high-voltage scenarios. However, the reliability of the gate oxide layer of SiC MOSFETs at high temperatures is still a key bottleneck that restricts the performance of SiC MOSFETs in practical applications. Studies have shown that when the temperature exceeds 200°C, the gate-oxygen interface state traps increase significantly, leading to threshold voltage (V_{TH}) drift, which in turn triggers the risk of rising on-resistance or mis-conductivity, and seriously affects the device stability and system safety [1]. Therefore, in-depth investigation of the V_{TH} instability mechanism of SiC MOSFETs at high temperatures is of great significance to promote their application in extreme environments.

Since the 20th century, several research organizations around the world have conducted systematic studies around the high-temperature reliability of SiC MOSFETs: Cree experimentally found that the V_{TH} of 1200V SiC MOSFETs in the range of -187~300°C decreases rapidly (-187~100°C) and then decreases slowly (100~300°C) with the increase in temperature, which is attributed to the trap de-trapping and the competing mechanism of surface potential change [2]; the University of Warwick compared the on-resistance changes of three commercial SiC MOSFETs at 350°C, and found that high temperature significantly exacerbates V_{TH} drift, and the PBTI effect is particularly prominent after 200°C [3]; the U.S. Army Research Laboratory revealed the role of gate bias in regulating the interface trap charge under high-temperature stress, and pointed out that the positive gate pressure accelerates the electron trap accumulation [4]; Sandia National Laboratories found through HTGB experiments that the V_{TH} drift rate of plastic packaged devices reached 0.3%/h at 225°C and -20V negative gate voltage, which was much higher than that of metal packaged devices [5]; Xi'an University of Electronic Science and Technology found through PBTI experiments that the V_{TH} drift at 25V gate voltage increased exponentially with the temperature, and at 275°C, it was as high as 12mV at 275°C [6];

979-8-3315-1110-4/25 $31.00 © 2025 IEEE

Existing studies mostly focus on the gate oxygen degradation mechanism below 200°C, while the multi-physical field coupling effect at ultra-high temperatures (>200°C) lacks systematic analysis. To address the above issues, this paper designs a high-temperature gate bias (HTGB) experimental platform based on TO-247 packaged SiC MOSFETs to realize multi-group ($\pm$ 20V gate voltage, different heating rates) long-term aging experiments in the range of 150~275℃. Through infrared thermal imaging and K-type thermocouple calibration, the junction temperature measurement error is ensured to be <3℃; Sentaurus TCAD is used to establish a two-dimensional planar gate SiC MOSFET model, coupled with quantum corrected mobility, interfacial trap density spectra, and nonequilibrium carrier transport equations to simulate the V_{TH} drift and leakage current degradation behavior under PBTI/NBTI stresses; the combination of experimental and simulation data reveals the gate pressure, temperature, and leakage current degradation behavior under PBTI/NBTI. In addition, the coupling mechanism of gate pressure, temperature and stress duration is revealed to quantify the degradation contribution of F-N tunneling effect under high field strength.

II. Threshold Voltage Instability Measurement Method for Silicon Carbide MOSFETs at High Temperature

In this paper, we focus on the permanent threshold voltage drift, and based on the reliability test standards JEP-184 [7] and AEC-Q101 [8] for silicon carbide MOSFETs, we developed a high-temperature gate bias experimental platform for TO-247 package devices, and measured the changes in the static parameters, such as threshold voltage and leakage current, before and after aging, in order to evaluate the threshold voltage of silicon carbide MOSFETs under high temperature instability.

A. Development of high-temperature gate bias experimental platform

In this paper, the parameters of the device are measured using the current (HCSMU) and power (HPSMU) modules of Agilent B1505A power device curve tracer and its supporting Agilent N1259A test fixture, which are used to test the threshold voltage (V_{TH}), the transfer curve, the output curve, the on-resistance (R_{ON}) and the gate leakage current (I_{GSS}).

A physical diagram of the BTI experimental setup is shown in Figure 1. The DC source in Figure 1 provides the voltage to the device under test and the heating table is used to control the temperature during the stressing period.The BTI test board is connected to the Agilent N1259A fixture through a measurement cable and the Agilent B1505A curve tracer is used to measure the static parameters of the device.The boat switches on the BTI test board are double-knife, double-throw switches for switching between different states.

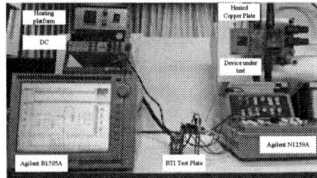

Fig. 1. Experimental platform physical picture

In this paper, TO packaged devices are unpacked and mounted on a heated plate, and the temperature of the chip surface is monitored with an infrared thermal imager. The temperature of the chip surface is approximated as the chip junction temperature. In which the chip surface is sprayed with a black body calibration paint to obtain the infrared imager readings. And a temperature sensor (type K thermocouple) was used to assist in monitoring the temperature of the heated table surface.

The device was heated and stabilized at a range of different temperatures and the readings of the IR thermographer and temperature sensor were recorded, this calibration was carried out from 150°C to 275°C with an average step size of 25°C. The relationship between the two readings is highly linear, indicating that all that is required to bring the device junction temperature up to the required temperature in this setup is to determine the temperature of the hot plate surface, which in turn determines the temperature at which the heating table needs to be set.

B. Experimental setup for high-temperature grid bias

The whole experimental procedure is as follows: first characterize the static parameters of the device at room temperature before starting to apply the stress; then apply high-temperature stress to the device under test, wait for 5 minutes for the device to reach thermal equilibrium and then start to pressurize the device; after 1 hour of high-temperature gate bias stress, stop applying the high-temperature stress, and output 0 V from the DC source to carry out the 30-minute recovery process; finally, measure the static constants of the device again at room temperature to evaluate the Finally, the static constants of the device were measured again at room temperature to evaluate the parameter degradation caused by the high temperature gate bias stress. The above process was repeated five times for each group of experiments, i.e., the cumulative stress time for each group was 5 hours.

A flat-gate silicon carbide MOSFET (D-MOSFET) with a voltage and current rating of 1200V/32A from a company was used for the experiments. In this study, operating conditions higher than the device operating gate voltage were chosen to accelerate the degradation process. Aging tests were conducted for four different positive gate bias stresses and five different negative gate bias stresses and six different temperatures.

Three silicon carbide MOSFETs were tested in each set of experiments in this study to determine the consistency of the changes in electrical parameters. Comparison of the results for all samples shows that these changes in aging parameters are reproducible, and only representative experimental results are provided below.

III. Threshold Voltage Instability Study of Silicon Carbide MOSFET under High Temperature Gate Bias

A. Effect of different gate voltages on threshold voltage instability

Threshold voltage instability of silicon carbide MOSFETs under positive gate bias (PBTI) stress is a central challenge for their high-temperature applications. In this section, the degradation law of the device under the coupling of a wide gate voltage range (+20~+35V) and ultra-high temperature (150~275°C) is systematically investigated experimentally. The experimental results show that the threshold voltage drift (ΔV_{TH}) exhibits a complex nonlinear relationship with gate voltage and temperature. As shown in Fig. 2(a), ΔV_{TH} shows a monotonically increasing trend with the increase of gate voltage in the range of +20~+30V gate voltage, and the drift curves at each temperature are highly overlapped, indicating that the gate oxygen degradation is mainly dominated by the trapping effect of electron traps in this interval. However, the threshold voltage drift increases sharply when the gate voltage rises to +35 V, and the device undergoes gate oxygen breakdown rapidly at temperatures above 200 °C, indicating that the high field strength triggers Fowler-Nordheim (F-N) tunneling and trap-assisted tunneling (TAT) mechanisms. By analyzing the leakage current degradation, it is found that the gate leakage current under +35 V stress decreases with stress time, which is attributed to the relaxation of the interfacial electric field by the accumulation of electron traps, and the enhancement of the local TAT current ultimately leads to the breakdown of the oxide layer.

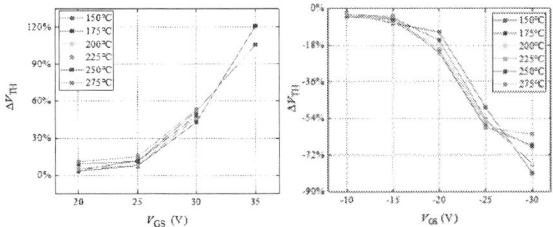

Fig. 2. Relationship between ΔVTH and gate bias at different temperatures：（a）PBTI,(b)NBTI

Different from PBTI, the threshold voltage of silicon carbide MOSFETs under negative gate bias (NBTI) stress exhibits a negative drift characteristic, and the degradation mechanism is closely related to hole traps and removable ions. As shown in Fig. 2(b), ΔV_{TH} increases with the absolute value of gate voltage in the range of -10~-30V gate voltage, but exhibits a differentiated pattern in different temperature intervals. Under -10~-25V stress, the threshold voltage decreases rapidly during the initial hour and then tends to saturate, indicating that the filling of hole traps mainly occurs at the beginning of the stress. The leakage current analysis shows that the F-N tunneling starts to activate at -25 V but does not trigger breakdown, while the leakage current increases significantly under -30 V stress and the device undergoes early failure at 150 °C, indicating that the high field strength together with TAT accelerates the oxide wear.

B. Effect of Temperature on Threshold Voltage Instability

The effect of temperature on the threshold voltage drift under PBTI shows a significant interval characteristic, as shown in Fig. 3(a). In the range of 150~175°C, ΔV_{TH} decreases with increasing temperature, which is due to the fact that high temperature accelerates the electron de-trapping process and counteracts the trap trapping effect; whereas, when the temperature exceeds 175°C, ΔV_{TH} turns to a positive temperature dependence, indicating that the high temperature activates more deep electron traps, and the rate of trap generation exceeds the rate of de-trapping. This phenomenon is especially obvious at +30V gate voltage, and ΔV_{TH} at 150°C is even higher than the value at 250°C. Further analysis of the energy band model reveals that electron bi-directional tunneling dominates in the low-temperature interval (<175°C), while thermal activation of traps becomes a key factor in the high-temperature interval (>175°C). At extreme gate voltages (+35 V), the temperature increase exacerbates the reduction of

the TAT equivalent barrier, leading to a sharp deterioration in the reliability of the oxide.

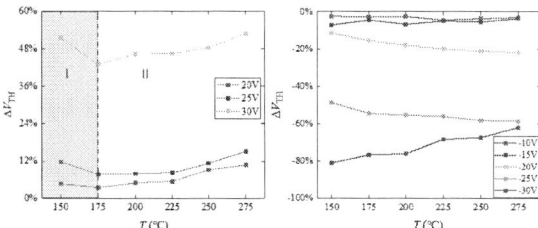

Fig. 3. Relationship between ΔVTH and temperature in (a)PBTI,(b)NBTI

The effect of temperature on NBTI presents contradictory characteristics, as shown in Fig. 3(b). Under -20~-25 V gate pressure, ΔV_{TH} increases with temperature, which is consistent with the expectation that high temperature promotes the generation of hole traps; however, under -30 V stress, ΔV_{TH} decreases with temperature instead. The energy band analysis reveals that the high temperature promotes the migration of mobile ions (e.g., Na^+) from the deeper part of the oxide layer to the interface, which partially neutralizes the charge effect of the hole traps and thus mitigates the threshold voltage drift. This phenomenon is verified in the leakage current variation: the compensating effect of movable ions at high temperatures suppresses the local enhancement of the TAT current, which makes the lifetime of the oxide layer at 275 °C better than that at 150 °C instead.

C. Degradation mechanism explanation

Comparison of the degradation mechanisms of PBTI and NBTI reveals that the electron traps under positive gate bias are concentrated near the interface, while the hole traps are more deeply distributed with negative gate bias and are regulated by movable ions. Experiments further reveal that the threshold voltage drift of NBTI is generally larger than that of PBTI, which is consistent with the characteristic that hole traps are more difficult to be detrapped.

Based on the above results, this paper proposes design guidelines for high-temperature applications: under positive gate bias, the temperature should be strictly controlled below 175°C to avoid thermal activation of the traps; while under negative gate bias, gate voltages above -20 V can be used safely, but localized failures induced by field strengths of -30 V or more should be prevented.

IV. SIMULATION ANALYSIS OF THRESHOLD VOLTAGE INSTABILITY OF SILICON CARBIDE MOSFETs

In order to verify the threshold voltage instability of the device, Synopays' Sentaurus TCAD software is used to firstly establish the metacellular structure model of the device, and then add a variety of physical models that affect the electrical performance of the device to establish a device model that can respond to the actual device characteristics. Finally, a degradation model is added to simulate the degradation of the threshold voltage of the device, and the effects of different trap types, different energy levels and different stress conditions on the threshold voltage instability of the device are simulated and analyzed.

A. Device simulation modeling

In this study, the structure of the device is built using the SDE module of the device structure editing tool in TCAD software, and firstly, the N-channel silicon carbide

MOSFEET half-cell model with planar gate structure is established, and the substrate of the device is an N-type doped silicon carbide material, the oxide layer is a silicon dioxide material, and the thickness of the gate oxygen layer is 50 nm, the source and drain are made of aluminum material, and the gate is made of a heavily doped polysilicon material.Forbidden band width modeling:The ΔE_g is different in different forbidden band narrowing models, in this study the OldSlotboom model is used with the ΔE_g equation:

$$\Delta E_g = E_{ref}\left[\ln(\frac{N_{tot}}{N_{ref}}) + \sqrt{\left(\ln(\frac{N_{tot}}{N_{ref}})\right)^2 + 0.5} \right] \quad (1)$$

Where N_{tot} is the total doping concentration, $N_{ref} = 1 \times 10^{17} \text{cm}^{-3}$ and $E_{ref} = 9 \times 10^{-3}\text{eV}$

*(b) Incomplete ionization model:*In silicon carbide, the dopant is incompletely ionized, so the incomplete ionization model needs to be considered. At a certain temperature, the actual ionized impurity concentration is:

$$N_D = \frac{N_D}{1 + g_D \exp(\frac{E_{F,n} - E_D}{kT})} \quad (2)$$

$$N_A = \frac{N_A}{1 + g_A \exp(\frac{E_{F,p} - E_A}{kT})} \quad (3)$$

where N_A and N_D are the doping concentrations of the P-type and N-type materials, respectively, g_A and g_A are their simplicity factors, E_A and E_D are their ionization activation energies, respectively, $E_{F,P}$ and $E_{F,N}$ are the Fermi energy levels of the P-type and N-type semiconductors, respectively, and k is the Boltzmann constant.

(c) Mobility modeling: In considering the doping-related mobility model this study uses the Arora model

For the channel region of the MOSFET, the IALMob model, which includes the contributions from Coulomb impurity scattering, phonon scattering, and surface roughness scattering, is used in this study, with Eq:

$$\frac{1}{\mu_{IALMob}} = \frac{1}{\mu_C} + \frac{1}{\mu_{ph}} + \frac{D}{\mu_{sr}} \quad (4)$$

where μ_C, μ_{ph}, and μ_{sr} represent the mobility for Coulomb impurity scattering, phonon scattering, and surface roughness scattering, respectively, and D is a quantity related to the interfacial distance.

In high electric fields, the carrier drift velocity is no longer proportional to the electric field but saturates to a finite velocity. The Cannali model from the software was chosen in this study with the following equation:

$$\mu_H = \frac{(\alpha+1)\mu_L}{\alpha + \left[1 + (\frac{(\alpha+1)\mu_L F_{hfs}}{v_{sat}})^\beta\right]^{1/\beta}} \quad (5)$$

$$v_{sat} = v_{sat}(\frac{300K}{T})^{v_{sat\,exp}} \quad (6)$$

$$\beta = \beta_0(\frac{T}{300K})^{\beta_{exp}} \quad (7)$$

where F_{hfs} is the driving electric field, α is the fitting parameter defaulted to 0 in the software. v_{sat} is the saturation drift velocity, and β is the correction factor, both of which are temperature dependent. The other relevant fitting coefficients use the default values in the software.

*(d) Interface Charge Modeling:*In order to make the threshold voltage and transfer curve of the device as similar as possible to that of the actual device, a certain energy level of uniformly distributed charge at the interface of silicon carbide/silicon dioxide is set as the main trap. After setting the structural parameters of the device and adding the above physical model, the simulation results of the device are shown in Figure 4.

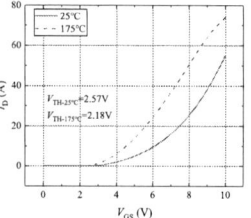

Fig. 4. Transfer curve simulation results

The negative temperature characteristics of the device transfer curve can be obtained from Figure 4, and the threshold voltages of the device at 25°C and 175°C are 2.57V and 2.18V, respectively, which are basically the same as 2.5V and 2.2V in the datasheet. It shows that the simulation model established in this study can reflect the nature of the device.

B. Degenerate model setup

The reaction-diffusion (R-D) model was used in the simulation to simulate the degradation behavior of the threshold voltage. The model assumes that when a gate voltage is applied, it initiates a field-dependent reaction at the silicon carbide/silicon dioxide interface, which creates a trap by breaking the passivated silicon-hydrogen (Si-H) bonds. The newly released hydrogen diffuses away from the interface, leaving a positively charged interfacial state (Si+), which is responsible for the threshold voltage drift.The R-D model can be divided into two phases, the reaction phase and the diffusion phase. The initial trap density depends on the breaking of the Si-H bond i.e. the reaction phase and the subsequent trap density depends on the movement of H i.e. the diffusion phase [9].

CONCLUSION

This study systematically investigates the gate oxygen layer reliability of silicon carbide MOSFETs under high temperature and DC bias stress, focusing on the threshold voltage degradation mechanism and its influencing factors. The study concludes that the gate oxygen reliability of SiC MOSFETs is synergistically affected by multiple factors: in the positive gate voltage scenario, high temperature does not necessarily exacerbate the drift, but ultra-high temperatures (>275°C) significantly reduce the stability of the gate oxygen, which needs to be combined with the optimization of the cooling system; in the negative gate voltage scenario, the gate overstress is the key limiting factor, and the crosstalk needs to be tightly controlled to prevent the misconductivity. This study provides theoretical basis and engineering guidance for the reliability design of silicon carbide devices in extreme environments.

ACKNOWLEDGMENTS

This work was supported by science and technology project of China Southern Power Grid under Grant 030400KC23090017(GDKJXM20231033).

REFERENCES

[1] Sheng Zhen, Ren Na, Xu Hongyi. Review and prospect of silicon carbide power device technology[J]. Chinese Journal of Electrical Engineering,2020,40(06):1741-1753.

[2] Cheng L, Agarwal A K, Dhar S, et al. Static performance of 20 A, 1200 V 4H-SiC power MOSFETs at temperatures of− 187 C to 300 C[J]. Journal of electronic materials, 2012, 41(5): 910-914.

[3] Hamilton D P, Jennings M R, Pérez-Tomás A, et al. High-temperature electrical and thermal aging performance and application considerations for SiC power DMOSFETs[J]. IEEE Transactions on Power Electronics, 2016, 32(10): 7967-7979.

[4] Lelis A J, Green R, Habersat D B. High-temperature reliability of SiC power MOSFETs[J]. Materials Science Forum. Trans Tech Publications Ltd, 2011, 679: 599-602.

[5] Kaplar R, DasGupta S, Marinella M J, et al. Degradation mechanisms and characterization techniques in silicon carbide MOSFETs at high temperature operation[J]. Proc. Electrical Energy Storage Applications and Technologies, 2012: 121-124.

[6] Bai Z Q, Tang X Y, Han C, et al. The Influence of Temperature Storage on Threshold Voltage Stability for SiC VDMOSFET[J]. Materials Science Forum. Trans Tech Publications Ltd, 2019, 954: 144-150.

[7] JEP 184, Guideline for evaluating Bias Temperature Instability of Silicon Carbide Metal-Oxide-Semiconductor Devices for Power Electronic Conversion[S]. JEDEC Solid State Technology Association, Mar 2021.

[8] AEC-Q101-Rev-E, Failure mechanism based stress test qualification for discrete semiconductors in automotive applications[S]. Automotive Electronics Council Component Technical Committee, Mar 2021.

[9] Franchi J, Domeij M, Lee K. 1200 V SiC MOSFETs with Stable V_{TH} under High Temperature Gate Bias Stress[J]. Materials Science Forum. Trans Tech Publications Ltd, 2019, 963: 753-756.

A High-Isolation X-ray Power Supply with Multi-Transformer Series Configuration*

1st Ziyang An
ZJU-UIUC Institute
Zhejiang University
Haining, China
ziyang1.23@intl.zju.edu.cn

2nd Ye Tian
Electric Power Research Institute
State Grid Fujian Electric Power Co.
Fujian, China
tianbobing@outlook.com

3rd Jie Ming
School of Automation
Wuhan University of Technology
Wuhan, China
jming@whut.edu.cn

4th Yu Dou
College of Electrical Engineering
Zhejiang University
Hangzhou, China
yudou@intl.zju.edu.cn

5th Chushan Li
ZJU-UIUC Institute
Zhejiang University
Haining, China
chushan@intl.zju.edu.cn

Abstract—**Micro-focus computed tomography (CT) systems demand high-voltage power supplies capable of delivering both high efficiency and robust voltage isolation to achieve high-resolution imaging. Conventional solutions, such as flyback converters, often exhibit insufficient isolation and efficiency under stringent operational requirements. This study proposes a novel design integrating multi-transformer series isolation with LLC resonant topology to overcome these limitations. The proposed architecture enhances power conversion efficiency and voltage isolation, critical for stable operation in high-voltage environments. Additionally, the multi-transformer isolation strategy effectively distributes high-voltage stress across cascaded stages, achieving an isolation voltage. This design offers a reliable and efficient solution for micro-focus CT systems, significantly advancing the performance of high-voltage power supplies in precision imaging applications.**

Index Terms—**micro-focus CT, LLC resonant converter, multi-transformer isolation.**

I. INTRODUCTION

The rapid development of high-voltage power supply technology has significantly impacted a wide range of industrial and scientific applications that require precise control and stable operation. High-voltage power supplies are crucial in systems such as capacitor charging, X-ray generators, electrostatic precipitation, and other pulsed power supply applications [1-4], where they ensure the stability, efficiency, and regulation of power to various components. These applications demand high-performance power supplies that can maintain voltage and current stability under varying loads while ensuring proper isolation and safety.

In particular, high-voltage power supplies play a critical role in micro-focus Computed Tomography (CT) [5, 6], a technology that requires precise and stable power delivery to achieve high-resolution imaging. The power supply system

in micro-focus CT is responsible for not only providing stable power but also meeting the high-voltage requirements essential for accurate imaging. Proper design and optimization of the power supply are key to ensuring that the CT system operates with the necessary precision and efficiency, resulting in superior imaging quality and stability.

Figs. 1 and 2 illustrate the overall power supply system architecture for a microfocus CT system. This power system is composed of several key modules, including the front-end power supply, filament supply, cathode supply, and grid supply. Each of these modules fulfills a specific function in supporting the high-precision operation of the CT system. Among these, the front-end power supply serves as the core module, providing power to all other units while ensuring voltage regulation and isolation, especially for the high-voltage components. This function is critical in microfocus CT applications, where stable and precise high-voltage output directly impacts imaging accuracy and system reliability.

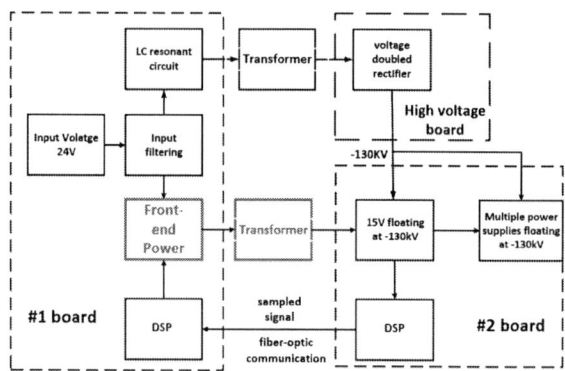

Fig. 1. Power system block diagram

This work was supported in part by National Key Research and Development Program of China 2024YFF1401201, and in part by Grants from the Power Electronics Science and Education Development Program of Delta Group. DREG2024001

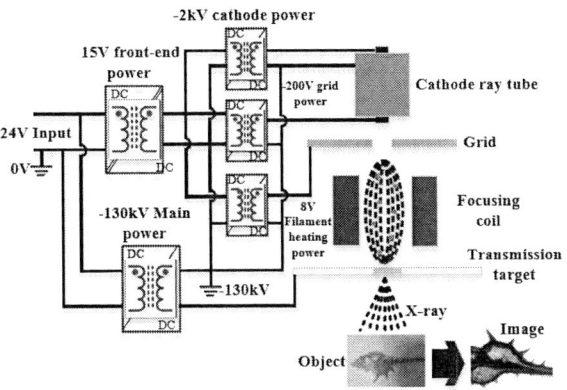

Fig. 2. Electrical connection diagram

II. SYSTEM DESIGN: POWER SUPPLY TOPOLOGY DESIGN AND HIGH-VOLTAGE ISOLATION SCHEME

The high voltage power supply design includes two key aspects: first, the LLC resonant converter topology, and second, the series stack isolation transformer scheme.

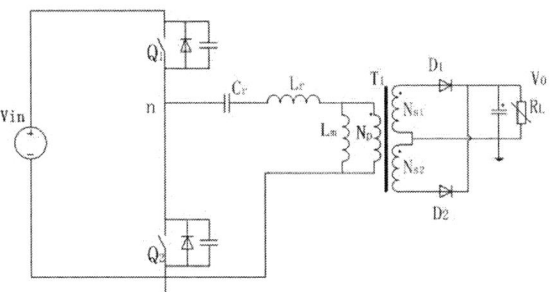

Fig. 3. LLC topology

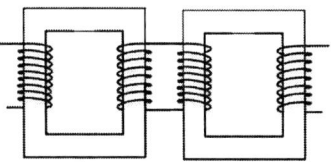

Fig. 4. Series connection of transformer

Traditional power supply designs, such as those using fly-back topologies, face challenges in high-voltage applications like micro-focus CT. While these designs work for lower voltage systems, they struggle to handle the high-voltage demands of micro-focus CT, leading to issues such as high switching losses and low efficiency. Therefore, there is a need for new high-voltage power supply solutions that ensure high efficiency, enhanced voltage isolation, and improved stability [7], which are essential for the evolving requirements of modern micro-focus CT systems. Recent advances have introduced LLC resonant converters [8-10], known for their high efficiency and reduced switching losses. Despite these improvements, achieving the required high-voltage isolation remains a challenge. To address this, multi-transformer series isolation designs have been proposed, which enhance voltage isolation while maintaining efficiency. This approach aims to provide a more reliable solution for the high-voltage power supplies used in micro-focus CT. This research aims to address these challenges by introducing a novel high-voltage power supply design based on a multi-transformer series isolation structure combined with LLC resonant topology. These innovations address key challenges in terms of efficiency, voltage isolation, and stability, which are essential for micro-focus CT applications. Table 1 presents the key design specifications for the front-end power supply. These specifications, including output voltage, current capacity, and efficiency, highlight the critical performance requirements that the power supply must meet to support the operation of the micro-focus CT system.

Fig. 3 illustrates the LLC resonant topology employed in our system and Fig. 4 illustrates how the multi-transformer series isolation is configured. The transformer shown in Fig.3 is a simplified representation, the actual implementation uses the series stacked transformers depicted in Fig. 4. In this configuration, we first determine the specifications of the resonant inductor, resonant capacitor, and magnetizing inductance to ensure that the converter achieves efficient zero-voltage soft-switching at the target frequency. Simultaneously, to satisfy the −130 kV isolation requirement, we develop a multi-transformer series stacking scheme including selecting appropriate core materials, defining each transformer's turns ratio, and arranging inter-winding insulation layers to guarantee safe, reliable operation while keeping the overall footprint compact.

A. Parameter Design Process

The converter's magnetizing inductance is denoted L_m, its resonant inductance L_r, resonant capacitance C_r, and turns-ratio n. The two-element LC resonant frequency is $f_r = 1/2\pi\sqrt{L_r C_r}$ while the three-element LLC resonant frequency is $f_m = 1/2\pi\sqrt{(L_r + L_m)C_r}$. The converter's switching frequency is f_s. To achieve high-frequency operation and soft-switching, the circuit's operating frequency must satisfy $f_m < f_s < f_r$. Based on the half-bridge structure, the turns ratio n of the transformer is first determined according to (1).

TABLE I
PARAMETER OF THE FRONT-END POWER

Parameter	Value
Output Voltage	15V
Efficiency	80%
Isolation Voltage	130kV
Input Voltage	24V DC
Output Power	30W

$$n = \frac{0.5 \times V_{in}}{V_{out} + V_D} \quad (1)$$

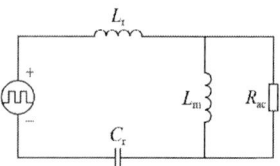

Fig. 5. equivalent circuit of the LLC resonant converter

Fig. 5 shows the AC equivalent circuit of the LLC resonant converter. In the Fig. 5, R_{ac} represents the AC equivalent load of a voltage-mode full-wave rectifier circuit, which is the actual load reflected from the secondary side to the primary side: $R_{ac} = n^2 \cdot \frac{8}{\pi^2} \cdot R_L$.

Based on the equivalent circuit, the input impedance Zin and voltage gain G of the LLC resonant converter can be derived.

$$Z_{in} = Q \cdot R_{ac} \left[\frac{x^2 k^2 Q}{x^2 k^2 Q^2 + 1} + j \left(x - \frac{1}{x} - \frac{xk}{x^2 k^2 Q^2 + 1} \right) \right] \quad (2)$$

$$|G| = \frac{1}{\sqrt{\left[1 + \frac{1}{k} \left(1 - \frac{1}{x^2} \right) \right]^2 + Q^2 \cdot \left(x - \frac{1}{x} \right)^2}} \quad (3)$$

In this case, k is the coefficient, defined as $k = L_m / L_r$; x is defined as $x = f_s / f_r$; and the quality factor of the series resonant circuit Q is $Q = \sqrt{L_r / C_r} / R_{ac}$.

To determine the maximum and minimum frequencies of the LLC resonant converter, as well as the corresponding Q value, it is important to note that the maximum gain of the LLC occurs at the boundary between the capacitive and inductive regions, where the impedance becomes purely resistive. At this point, the imaginary component of the input impedance is zero. According to (4), the maximum quality factor Q_{max} can be calculated.

$$Q_{max} = \sqrt{\frac{1}{k(1 - x^2)} - \frac{1}{(k - x)^2}} \quad (4)$$

Q_{max} is the quality factor at the critical point between the capacitive and inductive regions. To prevent the resonant tank from entering the capacitive region, the range of Q is set between 0.9 and 0.95. At this point, the LLC achieves its maximum gain, corresponding to full load operation with low input voltage and high output voltage. Therefore, the critical points are G_{max} , Q_{max} and f_{min}.

$$f_{min} = \frac{f_r}{\sqrt{k \left(1 - \frac{1}{G_{max}^2} \right) + 1}} \quad (5)$$

By substituting the obtained minimum frequency f_{min} into (4), the result can be obtained.

$$Q_{max} = \frac{1}{k \cdot G_{max}} \cdot \sqrt{k + \frac{G_{max}^2}{G_{max}^2 - 1}} \quad (6)$$

The minimum gain Q_{min} occurs at high voltage input and low voltage output, while the maximum switching frequency f_{max} is achieved under no-load conditions (when $Q = 0$).

$$f_{max} = \frac{f_r}{\sqrt{k \cdot \left(1 - \frac{1}{G_{min}} \right) + 1}} \quad (7)$$

According to (8) and (9), the maximum and minimum gains can be obtained.

$$G_{max} = 2 \cdot n \cdot \frac{V_{out} + V_D}{V_{inmin}} \quad (8)$$

$$G_{min} = 2 \cdot n \cdot \frac{V_{out} + V_D}{V_{inmax}} \quad (9)$$

As mentioned earlier, the value of k determines that the larger the value, the lower the conduction and switching losses of the MOSFET near the resonant frequency. Considering all factors, the typical range for k is taken to be between 2.5 and 6. Based on the previous equations, the required design values for C_r, L_r and L_m can be obtained.

$$C_r = \frac{1}{2\pi f_r R_\alpha Q} \quad (10)$$

$$L_r = \frac{1}{4\pi^2 C_r f_r^2} \quad (11)$$

$$L_m = k \cdot L_r \quad (12)$$

Next, the minimum number of turns on the primary side of the transformer can be calculated using (13), where ΔB represents the change in magnetic flux density, and A_e represents the effective cross-sectional area of the core.

$$N_p = \frac{n \cdot (V_{out} + V_D)}{2 \cdot f_{min} \cdot \Delta B \cdot A_e} \quad (13)$$

B. Resonant Parameter Calculation

Based on the above design process, the resonant parameters are designed with the specific requirements of this paper as an example. The design specifications are referenced in Table 1. Considering the high-isolation transformer scheme, the transformer core uses the UY15 core. The specifications not listed in the table are as follows:

- Resonant frequency f_r: 25 kHz
- Input voltage range V_{in}: 23~25 V
- K value: 6
- Secondary diode forward voltage drop V_D: 1 V
- Effective cross-sectional area A_e: 192 mm^2
- Magnetic flux density variation ΔB: 0.2 T

According to the previous design process, the calculation results are shown in Table II.

TABLE II
PARAMETER CALCULATION RESULTS

Parameter	Value
C_r	$3.3\,\mu F$
L_r	$12\,\mu H$
L_m	$72\,\mu H$
N_p	15
N_s	20

For the series transformer isolation scheme, the first transformer uses a 1:1 turns ratio with 15 turns on the primary side and 15 turns on the secondary side. The second transformer uses a 15:20 turns ratio, with 15 turns on the primary side and 20 turns on the secondary side.

III. SIMULATION RESULTS

Based on the previous theoretical analysis, simulations were conducted for the LLC resonant converter topology and the series transformer isolation scheme. The design was simulated using MATLAB Simulink and ANSYS MAXWELL, respectively.

A. Topology Simulation

This paper verifies the feasibility of the proposed scheme by separately building the traditionally used flyback topology and the previously calculated LLC resonant converter.

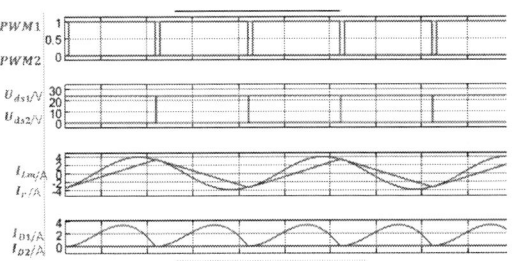

Fig. 6. simulation results of LLC

Fig. 6 shows the waveform of the LLC resonant converter under calculated parameters, highlighting the soft-switching behavior. The waveform clearly demonstrates the voltage and current characteristics at the primary and secondary sides, where the resonant topology helps minimize switching losses by allowing current to flow naturally during zero-voltage switching (ZVS) conditions. Then, by referencing the traditional flyback topology for sampling in the front-end power supply, simulations were conducted based on the corresponding parameters. The results were then validated by comparing the efficiency with that of the LLC resonant converter. Fig. 7 illustrates the efficiency comparison between the LLC resonant converter and the traditional flyback converter under various operating conditions. The graph clearly shows that the flyback converter suffers from significantly lower efficiency, particularly under higher load conditions, where its efficiency drops below 65%. This is due to high switching losses and

inadequate soft-switching capabilities. In contrast, the LLC resonant converter outperforms the flyback converter across all load conditions, maintaining efficiency above 66% even at the highest load (106 Ω). LLC's superior performance is attributed to its ability to reduce switching losses and its optimal power conversion characteristics, making it more suitable for high-voltage applications such as micro-focus CT.

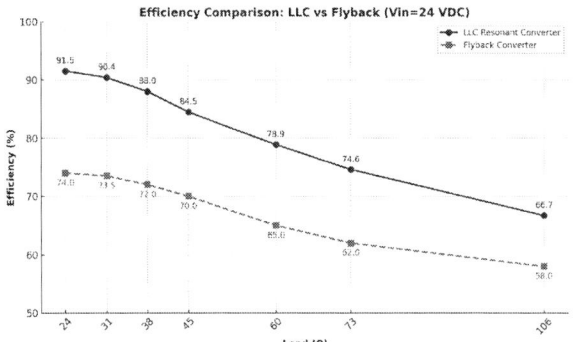

Fig. 7. Efficiency comparsion of flyback and LLC

B. Transformer Simulation

In this study, ANSYS Maxwell software is used to simulate the transformer design. The objective of the simulation is to verify whether the designed series-connected transformers can meet the high isolation and voltage withstand requirements necessary for the product. Specifically, the transformers must support a voltage difference of up to 130 kV between the primary and secondary sides. Given that two transformers are connected in series, each transformer only needs to withstand 65 kV. To ensure that the voltage is evenly distributed between the two transformers, high-voltage isolation resistors are used to divide the voltage. This approach is illustrated in Fig. 8.

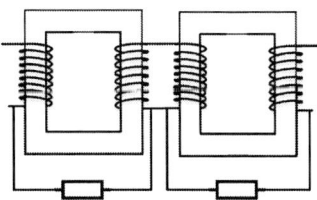

Fig. 8. Series connection of transformer with high-voltage resistors

In Maxwell, it is difficult to model the series connection of two transformers directly, so only a single transformer is modeled. The simulation focuses on observing the electric field distribution around the transformer to verify whether it can withstand the required 65 kV. For this modeling, it is essential to determine the following parameters: the transformer core material properties, the winding insulation material properties, and the insulating oil material properties surrounding the transformer.

TABLE III
RELATIVE PERMITTIVITY

Parameter	Value
magnetic core	12
insulating oil	1.63
casing	2.1

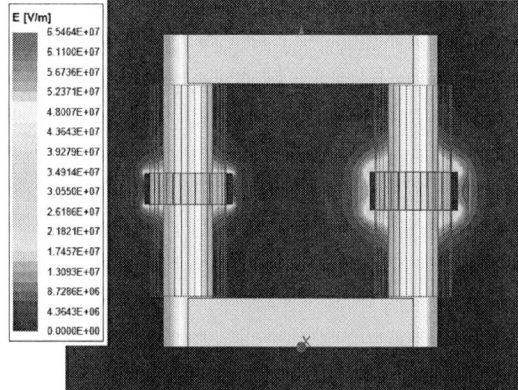

Fig. 9. Electric field distribution of transformer

Fig. 9 presents how the electric field is distributed around the transformer, providing insight into the potential stress points and areas requiring enhanced insulation. The field distribution indicates that the multi-transformer series isolation effectively mitigates high-voltage concentrations, reducing the electrical stress on individual components and improving the safety and efficiency of the overall system. The results demonstrate that the proposed isolation design achieves the necessary high-voltage performance while maintaining a manageable electric field, which is crucial for the reliability and durability of the system under operating conditions.

IV. SUMMARY AND PROSPECTS

This study introduces an innovative power supply architecture combining an LLC resonant topology with a multi-transformer series isolation configuration, specifically tailored for high-voltage applications such as micro-focus computed tomography (CT) systems. The LLC resonant converter, renowned for its soft-switching capability, effectively minimizes switching losses and enhances overall efficiency, addressing the inherent limitations of conventional flyback converters in high-voltage environments. Complementing this topology, the multi-transformer isolation design strategically distributes voltage stress across cascaded stages, significantly improving voltage isolation robustness while mitigating the risk of component failure.

The synergy between these two designs ensures stable and reliable operation under stringent high-voltage requirements, offering a scalable solution for applications demanding both precision and safety. By eliminating the dependency on single-transformer isolation and leveraging resonant characteristics,

this architecture sets a new benchmark for high-performance power supply systems in medical imaging and industrial high-voltage scenarios. Future efforts will focus on further optimizing component integration and expanding the design's applicability to broader voltage and power ranges.

ACKNOWLEDGMENT

This work was supported in part by National Key Research and Development Program of China 2024YFF1401201, and in part by Grants from the Power Electronics Science and Education Development Program of Delta Group. DREG2024001, and we would like to express our gratitude for this support.

REFERENCES

[1] Z. Zhao, L. Zhuang, and R. Zhang, "A high voltage pulse power supply based on piezoelectric transformer," in *Proc. 2022 IEEE 5th Int. Electrical and Energy Conf. (CIEEC)*, Nanjing, China, 2022, pp. 3158–3163, doi: 10.1109/CIEEC54735.2022.9846444.

[2] G. L. Piazza, R. L. Alves, C. H. I. Font, and I. Barbi, "Resonant circuit model and design for a high frequency high voltage switched-mode power supply," in *Proc. 2009 Brazilian Power Electronics Conf. (COBEP)*, Bonito-MS, Brazil, 2009, pp. 236–231, doi: 10.1109/COBEP.2009.5347683.

[3] G. Yinghui et al., "Development on high-power and high power density capacitor charging power supply," in *Proc. 2014 17th Int. Symp. on Electromagnetic Launch Technology*, La Jolla, CA, USA, 2014, pp. 1–6, doi: 10.1109/EML.2014.6926049.

[4] D.-W. Liu, J.-L. Yao, H.-Y. Ni, C. Pan, and W.-X. Yan, "A portable X-ray flaw detection equipment with high voltage for transmission line," in *Proc. 2022 IEEE Int. Conf. on High Voltage Engineering and Applications (ICHVE)*, Chongqing, China, 2022, pp. 1–4, doi: 10.1109/ICHVE53725.2022.9961550.

[5] G. Cao, L. Lee, Y. Z. Liu, and Z. et al., "Respiratory-gated micro-CT using a carbon nanotube based microfocus field emission x-ray source," *Medical Imaging: Physics of Medical Imaging, Int. Society for Optics and Photonics*, 2008.

[6] Intibayn A. et al., "High-power microfocus X-ray installation," in *Proc. 7th Mediterranean Conf. on Embedded Computing (MECO)*, Budva: IEEE, 2018, pp. 1–4, doi: 10.1109/MECO.2018.8405985.

[7] Son S. H. et al., "Development of 80-kW high-voltage power supply for X-ray generator," *IEEE Trans. on Industrial Electronics*, vol. 70, no. 4, pp. 3652–3662, 2023, doi: 10.1109/TIE.2022.3181663.

[8] Wang H., Chen Y., Liu Y. F., et al., "A passive current sharing method with common inductor multiphase LLC resonant converter," *IEEE Trans. on Power Electronics*, vol. 32, no. 9, pp. 6994–7010, 2017, doi: 10.1109/TPEL.2016.2626312.

[9] Qu J., Gao F., Zhao H., et al., "Design and analysis of LLC resonant converter for X-ray high-voltage power," in *Proc. 2019 IEEE 4th Adv. Inf. Technology, Electronic and Automation Control Conf. (IAEAC)*, Chengdu: IEEE, 2019, pp. 505–510, doi: 10.1109/IAEAC47372.2019.8997738.

[10] Yoon C. O., Kim J. W., Park M. H., et al., "Improving the light-load efficiency capability of LLC series resonant converter using impedance analysis," *IEEE Trans. on Power Electronics*, vol. 32, no. 9, pp. 7056–7067, 2017, doi: 10.1109/TPEL.2016.2629517.

A High-speed Dynamic Gate Driver with Low Oscillation for GaN HEMTs

Xuetong Zhou
State Key Laboratory of Materials for Integrated Circuits
Shanghai Institute of Microsystem and Information Technology,
Chinese Academy of Sciences
Shanghai, China
zhouxuetong@mail.sim.ac.cn

Li Zheng*
State Key Laboratory of Materials for Integrated Circuits
Shanghai Institute of Microsystem and Information Technology,
Chinese Academy of Sciences
Shanghai, China
zhengli@mail.sim.ac.cn

Xinhong Cheng*
State Key Laboratory of Materials for Integrated Circuits
Shanghai Institute of Microsystem and Information Technology,
Chinese Academy of Sciences
Shanghai, China
xh_cheng@mail.sim.ac.cn

Lingyan Shen
State Key Laboratory of Materials for Integrated Circuits
Shanghai Institute of Microsystem and Information Technology,
Chinese Academy of Sciences
Shanghai, China
shenly@mail.sim.ac.cn

Abstract—GaN HEMTs are extremely sensitive to the parasitic inductance of driving loop due to the short switching time and the low maximum allowed gate voltage. However, for conventional gate driver, it's impossible to achieve high switching speed and low voltage overshoot simultaneously. In this paper, a dynamic gate driver with suppressed gate voltage overshoot is proposed. The dynamic three-stage switching process realized by the high-speed and low-speed branches enables fast charging/discharging during the first stage and low oscillation during the last stage. Therefore, the proposed dynamic gate driver achieves short switching time of 9.76 ns and low voltage overshoot of 0.60 V at the same time, which are reduced by 8.36% and 34.78% compared with the conventional gate driver. The switching loss of the buck converter with the proposed dynamic gate driver is reduced by 18.5% and the drain voltage oscillation during dead time is much reduced.

Keywords—*GaN HEMTs, dynamic gate driver, parasitic inductance.*

I. INTRODUCTION

Benefitting from the high switching speed and low conduction resistance, GaN HEMTs are suitable for high-frequency power converters. Since the switching time of GaN HEMTs is as short as nanoseconds, the unavoidable parasitic inductance of driving loop leads to unexpected gate-source voltage (v_{gs}) oscillation[1]. However, for commercial enhancement-mode GaN HEMTs, the threshold voltage V_{TH} is around 1.5 V and the maximum allowed v_{gs} is limited to only 7 V. Thus, the allowed v_{gs} oscillation amplitude is only 1~2 V and the v_{gs} overshoot must be limited to prevent false turn-on, false turn-off or even breakdown.

The conventional gate driver for GaN HEMTs consists of a commercial gate driver IC and 1 or 2 gate resistances (R_g). These components, along with the parasitic inductance ($L_{driving}$) and the input capacitance (C_{iss}) of GaN HEMTs, form a second-order system. In order to minimize the v_{gs} overshoot,

the value of R_g must be increased, which inevitably leads to higher switching time and higher switching loss.

Switching output branches via MOS circuits, dynamic gate driver can overcome the tradeoff issue between switching speed and voltage oscillation by adjusting circuit parameters during switching periods. R_g and driving voltage ($v_{driving}$) are the most common choices. For dynamic-R_g gate drivers, R_g is designed sufficiently low at the beginning of turn-on periods to guarantee the fast rise of v_{gs}. As v_{gs} increases, R_g will gradually increase to a higher value to suppress the voltage overshoot occurred at the end of rising stage[2]. For dynamic-$v_{driving}$ gate drivers, $v_{driving}$ is initially maintained at a relatively low level at the beginning and subsequently increased at the end of turn-on period[3][4].

In terms of fundamental principles, dynamic-R_g gate drivers increase the second-order system's damping factor, while dynamic-$v_{driving}$ gate drivers decreases the bias of v_{gs} oscillation. Nevertheless, since the switching time of GaN HEMTs is extremely low, the adjustment of R_g and $v_{driving}$ must be completed in nanoseconds. Such stringent requirement is why most dynamic gate drivers rely on MOS circuits — MOS devices offer rapid switching capabilities, enabling precise and timely control of R_g and $v_{driving}$.

In this paper, a dynamic gate driver implemented by discrete components on PCB is proposed. The proposed gate driver achieves low switching time and low v_{gs} overshoot at the same time, making it suitable for GaN-based applications. The paper is organized as follows. In Section II, the principle of the proposed driver is analyzed in detail. The performances of the proposed driver are discussed according to the SPICE simulation. The experiment results of double pulse test (DPT) and buck converter test are shown in Section III.

II. DYNAMIC GATE DRIVER FOR GAN HEMTS

A. Operation Principle of the Proposed Gate Driver

Fig. 1 gives the diagram of the proposed dynamic gate driver. The proposed driver consists of two branches: a high-speed branch containing R_{fast}, C_{iso}, D_1, D_2 and a low-speed branch containing R_{slow}. The typical component values are listed in Table I.

Fig. 2 gives the waveforms of the proposed driver during the turn-on period, which can be divided into 3 stages.

This work was supported by the National Key Research and Development Program of China (Grant No. 2022YFB3604300, 2022YFB3604301, 2022YFB3604303), National Natural Science Foundation of China (Grant No. 11705263), the Science and Technology Commission of Shanghai Municipality (Grant No. 23511102602), Youth Innovation Promotion Association CAS, Autonomous deployment project of State Key Laboratory of Materials for Integrated Circuits (No. SKLJC-Z2024-C02) and Shanghai Post-doctoral Excellence Program (Grant No. 2024697). (*Co-corresponding authors: Li Zheng and Xinhong Cheng*)

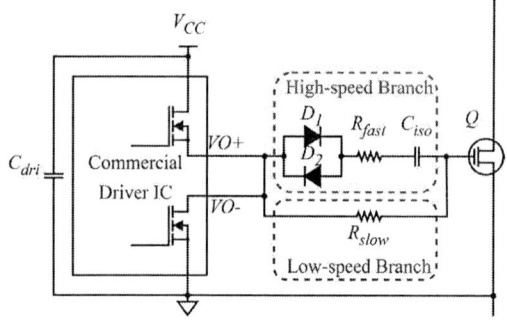

Fig. 1. Diagram of the proposed dynamic gate driver.

TABLE I. TYPICAL PARAMETERS OF THE PROPOSED DYNAMIC GATE DRIVER

Parameters	Values
High-speed resistance R_{fast}	1 Ω
Low-speed resistance R_{slow}	20 Ω
Isolated capacitance C_{iso}	100 nF
Diode forward voltage V_F	0.5 V

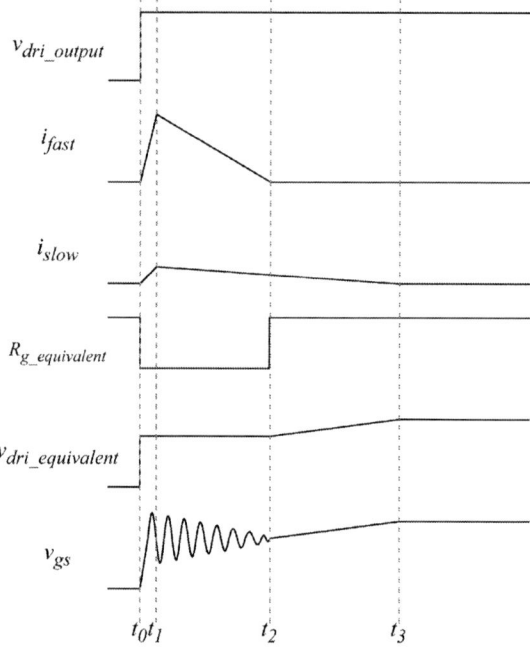

Fig. 2. Timing diagram of the proposed dynamic gate driver during turn-on period.

- Stage 1 ($t_0 \sim t_1$): The C_{iss} of GaN HEMT is charged by the high-speed branch to the voltage value of $V_{CC} - V_F$, where V_{CC} is the supply voltage of driver IC, V_F represents to the forward voltage of D_1 and D_2.

- Stage 2 ($t_1 \sim t_2$): The current through high-speed branch decreases, while the low-speed branch gradually plays a dominant role in the charging process.

- Stage 3 ($t_2 \sim t_3$): The C_{iss} is finally charged by the low-speed branch to the voltage value of V_{CC}.

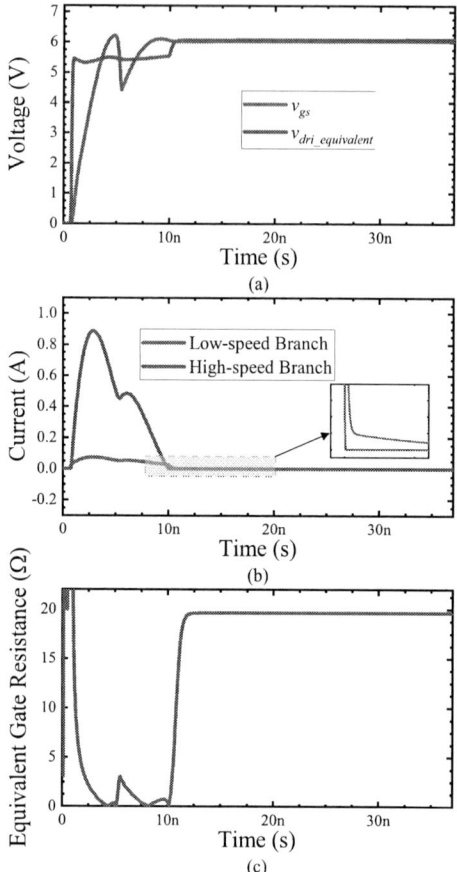

Fig. 3. Simulated waveforms of the proposed dynamic gate driver during turn-on period.

Similarly, during turn-off periods, the C_{iss} firstly discharges through the high-speed branch from V_{CC} to V_F in stage 1, and finally discharges to 0 through the low-speed branch in stage 3.

Since the value of R_{fast} is set to 1~2 Ω to achieve high switching speed, the v_{gs} oscillation occurs at stage 1 and the v_{gs} oscillates on the $V_{CC} - V_F$ bias. Thus the v_{gs} overshoot can be reduced.

The proposed driver is firstly simulated in LTspice. The GS-065-018-2 devices from GaN Systems are adopted and the parasitic inductance of driving loop $L_{driving}$ is set to 5 nH. As shown in Fig. 3, the simulated waveforms are consistent with the analysis. During stage 1, the equivalent gate resistance ($R_{g\_equivalent}$) decreases to 1 Ω, and the equivalent driving voltage ($v_{dri\_equivalent}$) is around 5.5 V, which is the value of $V_{CC} - V_F$. Once the current of the high-speed branch drops to 0, the $R_{g\_equivalent}$ increases to 20 Ω and the $v_{dri\_equivalent}$ rises to 6 V.

The simulated waveforms of the proposed dynamic gate driver and the conventional gate driver are compared in Fig. 4. Compared with the conventional driver with a 1-Ω R_g, the turn-on time and the turn-off time are slightly increased by 0.32 ns and 0.38 ns, while a 0.53 V reduction in v_{gs} overshoot is achieved, which closely approximates the value of V_F. Meanwhile, compared with the conventional driver with a 2-Ω R_g, the v_{gs} overshoot is reduced by 0.09 V and the turn-on time and turn-off time are reduced by 0.22 ns and 2.43 ns.

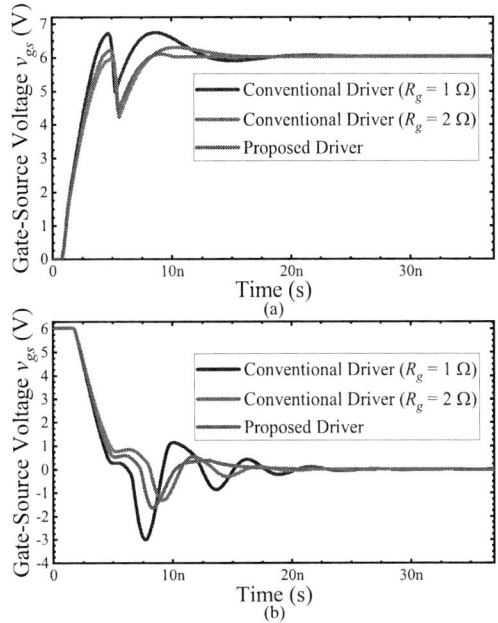

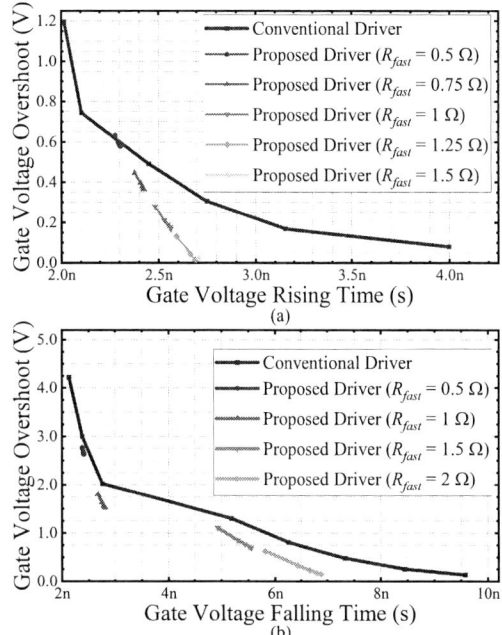

Fig. 4. Simulated waveforms of different gate drivers.
(a) Turn-on period. (b) Turn-off period.

Fig. 5. Simulation data distribution of switching time and gate voltage overshoot of different gate drivers.
(a) Turn-on period. (b) Turn-off period.

B. Influence of the Components in the Proposed Dynamic Gate Driver

Fig. 6 shows the influence of R_{fast}, R_{slow}, C_{iso} and V_F. The switching speed is mainly determined by the value of R_{fast}. A smaller R_{fast} leads to smaller switching time during stage 1, whereas inevitably increases the v_{gs} overshoot. R_{slow} determines the charging or discharging speed of C_{iss} during stage 3. The value of C_{iso} must be over 1000 times higher than C_{iss}, otherwise the voltage of C_{iso} will be neglected, which will slow down the switching speed. V_F determines the bias of v_{gs} oscillation during stage 1. A larger V_F can reduce the v_{gs} overshoot, while it's more likely to cause false turn-on during turn-off stage.

C. Tradeoff between Switching Speed and Overshoot

The simulation data of the proposed dynamic gate driver with different component parameters is plotted in Fig. 5, which is a scatter plot with switching time as x axis and v_{gs} overshoot as y axis. The proposed gate driver demonstrates enhanced comprehensive capabilities, outperforming traditional designs in both switching speed and stability. For a 5-nH parasitic inductance, the proposed driver can raise the to the driving voltage without v_{gs} overshoot in 2.7 ns.

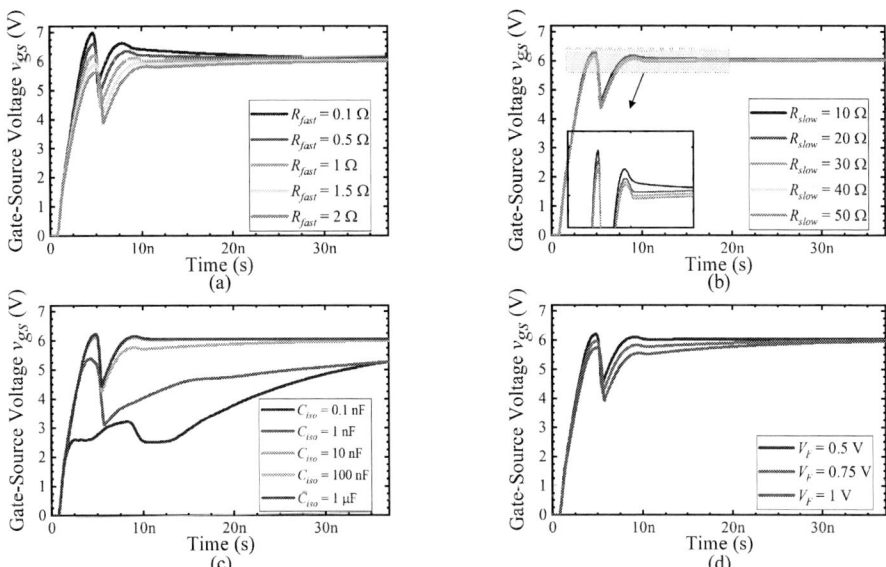

Fig. 6. Influence of the components in the proposed dynamic gate driver during turn-on periods.

III. EXPERIMENT VERIFICATION

To verify the superior performances of the proposed dynamic gate driver, the DPT and buck converter test are adopted.

A. Double Pulse Test Results

The detail DPT data is given by Table II. Compared with the conventional gate driver with an 2-Ω R_g, the turn-on time of the proposed driver is increased by 8.8%, whereas the v_{gs} overshoot is reduced by 34.8%. Meanwhile, compared with the conventional gate driver with an 4-Ω R_g, the v_{gs} overshoots are similar, while the switching time is reduced by 8.3%. Fig. 7 gives the switching time and v_{gs} overshoot of different drivers in DPT. Obviously, the comprehensive performances of the proposed dynamic gate driver is better than the conventional driver.

TABLE II.   DPT RESULTS OF DIFFERENT GATE DRIVERS

Parameters		Conventional Driver		Proposed Driver
		Rg = 2 Ω	Rg = 4 Ω	
Turn-on period	v_{gs} rising time	8.79 ns	10.65 ns	9.76 ns
	v_{gs} overshoot	0.92 V	0.51 V	0.60 V
Turn-off period	v_{gs} falling time	8.25 ns	9.38 ns	9.70 ns
	v_{gs} overshoot	1.41 V	0.89 V	0.77 V

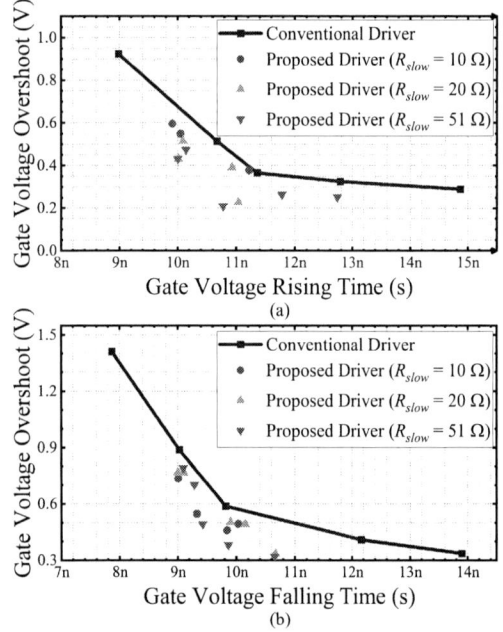

Fig. 7.   Experiment data distribution of switching time and gate voltage overshoot of different gate drivers.
(a) Turn-on period. (b) Turn-off period.

TABLE III.   EXPERIMENT RESULTS OF BUCK CONVERTERS WITH DIFFERENT GATE DRIVERS

Parameters	With Conventional Gate Driver	With Proposed Dynamic Gate Driver
Efficiency of Buck converter	97.3%	97.8%
Peak-peak value of v_{ds} oscillation during dead time	4.36 V	4.04 V

B. Buck Converter Test Resultss

To further validate the performance of the proposed dynamic gate driver in practical applications, the gare drivers are evaluated in a buck converter and the experiment results are given by Table III. With the proposed driver, the efficiency of the buck converter reaches 97.8%, which is higher than the converter with conventional drivers. Additionally, as shown in Fig. 8, the drain-source voltage v_{ds} oscillation during dead time is significantly reduced by the proposed driver, which indicates that the parasitic effect of driving loop is suppressed[5][6][7].

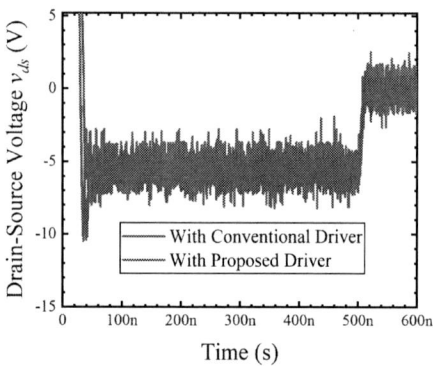

Fig. 8.   Drain-Source voltage Waveforms of the buck converters with different gate drivers during dead time.

IV. CONCLUSSION

This study presents a dynamic gate driver architecture featuring adaptive switching control, which demonstrates significant improvements in both switching speed and gate-source voltage stability. Through comprehensive validation encompassing analytical modeling, SPICE simulations, double-pulse testing (DPT), and buck converter prototyping, the proposed solution exhibits 8.36% faster switching transients while maintaining v_{gs} overshoots of 0.6 V during turn-on periods. This cost-effective solution not only improves the reliability of GaN-based power converters but also enables operation at higher switching frequency beyond MHz. These findings highlight the potential for advancing GaN-based power converters through innovative driver solutions.

REFERENCES

[1] E. A. Jones, F. F. Wang, D. Costinett, Review of Commercial GaN Power Devices and GaN-Based Converter Design Challenges. IEEE Journal of Emerging and Selected Topics in Power Electronics, 2016, 4(3): 707-719. doi: 10.1109/JESTPE.2016.2582685.

[2] W. J. Zhang, J. Yu, W. Cui et al. A Smart Gate Driver IC for GaN Power HEMTs With Dynamic Ringing Suppression. IEEE Transactions on Power Electronics, 2021, 36(12): 14119-14132. doi: 10.1109/TPEL.2021.3089679.

[3] Y. Yang, Y. Wen, Y. Gao. A Novel Active Gate Driver for Improving Switching Performance of High-Power SiC MOSFET Modules. IEEE Transactions on Power Electronics, 2019, 34(8): 7775-7787. doi: 10.1109/TPEL.2018.2878779.

[4] X. Ming, X. Li, Z. Zhang et al. A GaN HEMT Gate Driver IC with Programmable Turn-on dV/dt Control. 2020 32nd International Symposium on Power Semiconductor Devices and ICs (ISPSD), 2020: 98-101. doi: 10.1109/ISPSD46842.2020.9170152.

[5] K. Wang, X. Yang, L. Wang et al. Instability Analysis and Oscillation Suppression of Enhancement-Mode GaN Devices in Half-Bridge

Circuits. IEEE Transactions on Power Electronics, 2018, 33(2): 1585-1596. doi: 10.1109/TPEL.2017.2684094.

[6] J. Chen, X. Du, Q. Luo et al. A Review of Switching Oscillations of Wide Bandgap Semiconductor Devices. IEEE Transactions on Power Electronics, 2020, 35(12): 13182-13199. doi: 10.1109/TPEL.2020.2995778.

[7] Y. Tian, R. Pan, L. Shen et al. Monolithic Integration of p-GaN HEMT with Anti-parallel Lateral Rectifier to Reduce the Negative Resistance Effect. IEEE Transactions on Power Electronics, doi: 10.1109/TPEL.2025.3528180.

Research on An Anti-Offset Wireless Power Transfer System With Auxiliary Resonant Circuit

Youzheng Wang[1], Shengxiu Xu[1], Shuyu Wang[2], Hongchen Liu[2], Longnv Li[1], Gaojia Zhu[1] and Yunhui Mei[1]

[1]School of Electrical Engineering, Tiangong University, Tianjin, China
[2]School of Electrical Engineering and Automation, Harbin Institute of Technology, Harbin, China
E-mail:youzhengwang@tiangong.edu.cn

Abstract— Aiming at the issue of unstable output voltage and rapid efficiency degradation caused by coil offset during the wireless power transfer (WPT) process, an anti-offset WPT system with auxiliary resonant circuit is presented. The WPT system employs the high-order compensation network and an auxiliary resonant circuit composed of an auxiliary coupling coil and a resonant capacitor. The coupler of the system adopts the traditional four-coil BP coupler. By adjusting the resonant inductance value of the LCC network, the magnetic coupler can accommodate different scenarios with varying equivalent mutual inductance (MI) fluctuation requirements. Finally, a 1-kW prototype was developed to validate the strong anti-offset characteristics of the proposed WPT system.

Index Terms—Wireless power transfer, auxiliary resonant circuit, anti-offset characteristics, BP magnetic coupler.

I. INTRODUCTION

Wireless power transfer (WPT) technology has attracted much attention due to its automation, security, convenience and advantages in various situations such as underwater [1] and mine [2].

WPT systems must be optimized to ensure efficient and reliable energy transfer. This optimization includes achieving accurate power control across a wide load range, minimizing reactive power in the resonant circuit to reduce device stress, implementing soft-switching to enhance efficiency, and other critical factors. The aforementioned optimizations are based on the assumption that the coupler is well-aligned. However, in practical WPT applications, such as WPT system for autonomous underwate vehicles, precise alignment of parking positions is often challenging, leading to variations in mutual inductance (MI). These variations can significantly impact output power and system efficiency [3]. Thus, WPT systems should possess anti-offset within a specified range of coil misalignment to accommodate different application requirements.

Currently, research on the strong anti-offset characteristic for WPT systems primarily focuses on two aspects: compensation topology design [4], [5] and magnetic coupler optimization [6], [7], [8]. Regarding compensation topology, scholars have employed two main approaches to enhance WPT system misalignment tolerance: topology parameter optimization and hybrid compensated topology. These methods can achieve relatively stable output without complex control mechanisms. However, in cases of large-range offset, the topology parameter optimization method tends to introduce reactive power, thereby reducing system efficiency. While hybrid compensated topology exhibits strong anti-offset characteristic, it can impose stringent requirements on the parameters of circuit and couplers. For instance, the hybrid compensated topology proposed in [6] incorporates numerous passive components. Presently, scholars both domestically and internationally have designed a family of typical couplers with certain anti-offset feature [9], [10], [11], [12], such as double-D (DD), double-D quadrature (DDQ), tripolar (TP), and bipolar (BP). The self-inductance and mutual inductance (MI) coefficients of these couplers are relatively small in specific misalignment directions, but they struggle to accommodate misalignment in other directions. [13] introduced the concept of the hybrid coupler, which not only demonstrates high misalignment tolerance in both X-axis and Z-axis directions but also offers greater universality. However, for the four-coil BP coupler within the hybrid coupler, the coupling MI parameters of the two overlapping coils on the same side of the transmitter and receiver cannot be individually designed, leading to multiple iterations in the coupler design process and complicating the design methodology.

With the purpose of solving the above problems, the paper proposes an anti-offset

WPT system with an auxiliary resonant circuit. The WPT system employs the high-order compensation network and an auxiliary resonant circuit composed of an auxiliary coupling coil and a resonant capacitor. The coupler of the system adopts the traditional four-coil BP coupler. The demands for the parameters design of the BP coupler are reduced, and the difficulty of the design of the BP coupler is simplified. And it can realize the equivalent MI of the coupler fluctuation is small under the large offset distance. Moreover, by adjusting the value of resonant inductance value of LCC network, the coupler can be adapted to the occasions requiring different equivalent MI fluctuation rates.

II. THEORETICAL ANALYSIS OF THE PROPOSED WPT SYSTEM

The anti-offset WPT system with auxiliary resonant circuit proposed in this paper is depicted in Fig.1, which is mainly composed of the high frequency single-phase full-bridge inverter, high-order compensation topology, four-coil BP coupler and the uncontrollable rectifier. Four power switches MOSFETs (Q_1~Q_4) and DC voltage V_{in} constitute the high frequency single-phase full-bridge inverter. The high-order compensation topology consists of the primary compensation capacitors C_P, C_{P1} and C_{PA}, the primary compensation inductor L_{P1}, and the receiving compensation capacitor C_S. The magnetic coupler employs the traditional four-coil BP coupler. The auxiliary resonant circuit is composed of one coil in the BP coupler and a high-frequency resonant capacitor C_{SA}. D_1 to D_4 and the output electrolytic capacitor C_O constitute the rectifier. L_P, L_{PA}, L_S and L_{SA} are the self inductance of the BP coupler respectively. M_{PS}, M_{PASA}, M_{PPA}, M_{SSA}, M_{PSA} and M_{SPA} are MI between four coils respectively. Owing to the structure of BP coil, cross coupling MI M_{PSA} and M_{SPA} are much smaller than M_{PS}, M_{PASA}, M_{PPA} and M_{SSA}, so M_{PSA} and M_{SPA} can be approximately equal to 0.

Based on the Kirchhoff's voltage law, the two-port matrix equation for ac input and ac output can be derived as follows:

$$\begin{bmatrix} \dot{V}_{AB} \\ \dot{V}_{ab} \\ 0 \\ 0 \end{bmatrix} = j\omega \begin{bmatrix} L_P + L_P' & M_{PS} & 1/\omega^2 C_{P1} & 0 \\ M_{PS} & L_S & 0 & |M_{SSA}| \\ 1/\omega^2 C_{P1} & 0 & L_{PA}' & M_{PASA} \\ 0 & |M_{SSA}| & M_{PASA} & L_{SA}' \end{bmatrix} \begin{bmatrix} \dot{i}_P \\ \dot{i}_S \\ \dot{i}_{PA} \\ \dot{i}_{SA} \end{bmatrix} (1)$$

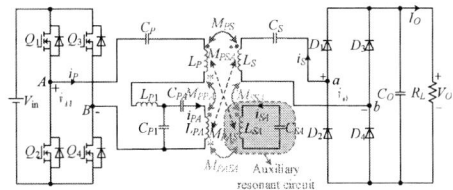

Fig.1 The structure of the proposed WPT topology.

Where $L_P' = L_{P1} + L_{PA} - \dfrac{1}{\omega^2 C_{P1}}$, $L_{PA}' = L_{P1} + \dfrac{1}{\omega^2 C_{P1}} - \dfrac{1}{\omega^2 C_{PA}}$,

$$L_{SA}' = L_{SA} - \frac{1}{\omega^2 C_{SA}} .$$

During normal operation, the system operates in a resonant state, and the resonant frequency ω satisfies Eq.(2).

$$\omega = \frac{1}{\sqrt{L_{P1}C_{P1}}} = \frac{1}{\sqrt{(L_{PA} - L_{P1})C_{PA}}} = \frac{1}{\sqrt{L_{SA}C_{SA}}} (2)$$

By substituting Eq.(2) into Eq.(1), the following relationships can be derived.

$$\begin{cases} \dot{i}_{PA} = -\dfrac{|M_{SSA}|}{M_{PASA}} \dot{i}_S \\ \dot{i}_{SA} = \dfrac{L_{P1}}{M_{PASA}} \dot{i}_P \end{cases} (3)$$

By substituting Eq.(3) into Eq.(1), Eq.(4) can be derived.

$$\begin{bmatrix} \dot{V}_{AB} \\ \dot{V}_{ab} \end{bmatrix} = j\omega \begin{bmatrix} L_P & M_{PS} + \dfrac{L_{P1}|M_{SSA}|}{M_{PASA}} \\ M_{PS} + \dfrac{L_{P1}|M_{SSA}|}{M_{PASA}} & L_S \end{bmatrix} \begin{bmatrix} \dot{i}_P \\ \dot{i}_S \end{bmatrix}$$

$$(4)$$

Thus, the relationships governing the equivalent MI of the magnetic coupler can be derived.

$$M_{EQ} = M_{PS} + \frac{L_{P1}|M_{SSA}|}{M_{PASA}} (5)$$

The magnetic coupler depicted in Fig. 1 employs a typical four-coil BP coupler. The two independent coils L_P and L_{PA} (or L_S and L_{SA}) on the same side are typically mechanically fixed, ensuring that the MIs M_{PPA} and M_{SSA} remain nearly constant and are unaffected by coil offset. Additionally, these two independent coils on the same side are usually designed with identical dimensions and turns, resulting in nearly equal values of M_{PS} and M_{PASA}, and any coil offset affects both similarly. Thus, the expression for the equivalent MI can be formulated as follows:

$$M_{EQ} = M_{PS} + \frac{L_{P1}\left|M_{SSA}\right|}{M_{PS}} \qquad (6)$$

Eq.(6) indicates that, when the operating frequency ω of the system is fixed, the product $L_{A1} \times M_{SSA}$ remains constant, and M_{EQ} becomes a function of M_{PS}, as illustrated in Fig. 2. Assuming that M_{EQref} represents the equivalent mutual MI in the well-aligned state of the magnetic coupler and allows an acceptable fluctuation ratio λ, the permissible misalignment range for M_{PS} is $[M_{PSmin}, M_{PSmax}]$.

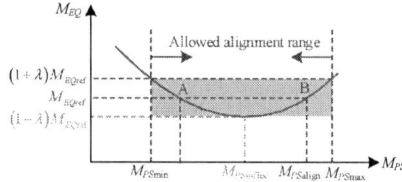

Fig.2 Curve of M_{EQ} versus M_{PS}.

To maximize the tolerance to coil offset, the minimum value of M_{EQ} at the curve inflection point $M_{PSinflec}$ should be set to $(1-\lambda)M_{EQref}$ and substituted into Eq.(6) to obtain Eq.(7).

$$M_{P\,\mathrm{Sin\,flec}} = \sqrt{L_{P1}\left|M_{SSA}\right|} = \left(\frac{1-\lambda}{2}\right)M_{EQref} \quad (7)$$

Furthermore, when the coils experiences offset, MI typically decreases. Thus, $M_{PSalign}$ should be designed to correspond to point B in case of well-aligned conditions. Similarly, this can be achieved by substituting Eq.(6).

$$M_{P\,\mathrm{Salign}} = \frac{1+\sqrt{\lambda(2-\lambda)}}{2}M_{EQref} \qquad (8)$$

Combining (7) with (8), the expression of L_{P1} can be acquire.

$$L_{P1} = \frac{(1-\lambda)^2 M_{P\,\mathrm{Salign}}^2}{\left(1+\sqrt{\lambda(2-\lambda)}\right)^2 \left|M_{SSA}\right|} \qquad (9)$$

From Eq.(9), it can be observed that the value of L_{P1} is determined by the values of $M_{PSalign}$ and λ.

For the magnetic coupler of a given size and distance, i.e., while $M_{PSalign}$ is specified, if λ is known, Eq.(9) allows for the flexible design of L_{P1}. This flexibility in selecting L_{P1} enables the BP coupler to optimize M_{SSA} by minimizing cross-coupling MIs M_{PSA} and M_{SSA}. To further simplify the BP coupler design, the MI between two coils overlapping the transmitter can be set to zero, which is achieved by ensuring that the overlapping area of coils L_P and L_{PA} is half of their respective areas. Given that the two independent coils on the same side are typically designed with

identical dimensions and turns, Ampère's theorem dictates that when the overlap area is half of each coil's area, the total magnetic flux linkage between two coils is zero, thereby eliminating any coupling effect. Finally, the value of L_{P1} is determined using Eq.(9), and C_{P1}, C_{PA}, and C_{SA} are calculated according to Eq.(2).

The expression of load current I_O can be expressed as follows:

$$I_O = \frac{8V_{\mathrm{in}}}{\pi^2 \omega\left[M_{PS} + \left(L_{A1}M_{SSA}/M_{PS}\right)\right]} \quad (11)$$

The M_{EQ} of the magnetic coupler fluctuates with the coil offset $(1\pm\lambda)M_{EQref}$ range.

III. EXPERIMENT VALIDATION

A. Experimental Setup

To validate the aforementioned analysis, an experimental 1kW prototype is constructed based on Fig. 1, as illustrated in Fig.3. The parameter values of the passive components for the system are listed in Table I.

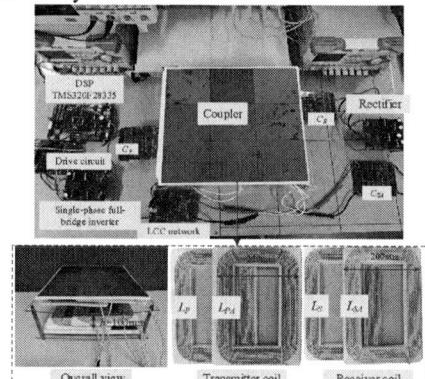

Fig.3 Experimental prototype.

TABLE I
THE PARAMETER VALUES OF THE PASSIVE COMPONENTS

Parameter	Value	Parameter	Value
L_P	52.23µH	L_{PA}	53.13µH
C_P	67.33nF	C_S	74.20nF
C_{P1}	502.28nF	L_S	47.25µH
L_{P1}	6.98µH	L_{SA}	48.37µH
C_{PA}	75.97nF	C_{SA}	72.48nF

B. Performance of the Strong Anti-Offset Features

Fig.4(a) and 4(b) present the experimental waveforms of the coupler under well-aligned conditions and with 150mm offset along the X-axis, respectively, when the output power is 1kW (R_L=8.6Ω). Despite 150mm offset along the X-axis, the system maintains an output current I_O close to 10.8 A, indicating strong offset in the X-axis direction. Additionally, the phase difference between i_p and v_{AB} is minimal, suggesting that the system's reactive power is nearly zero and that switches Q_1 to

Q_4 can achieve soft-switching. To verify the load-independent constant-current output characteristics of the system, Fig.5(a) and Fig.5(b) show the experimental waveforms at an output power of 500 W ($R_L = 4.3\Omega$).

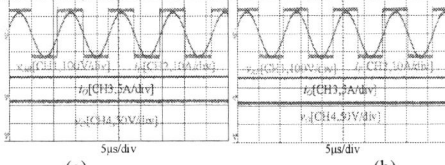

(a)   (b)

Fig.4 Experimental waveforms when the output power is 1kW. (a) Well-Aligned. (b) 150 mm offset along X-axis.

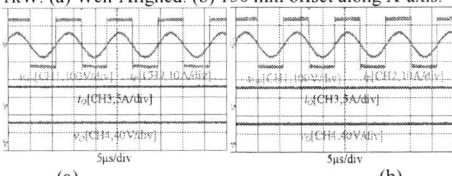

(a)   (b)

Fig.5 Experimental waveforms when the output power is 500W. (a) Well-Aligned. (b) 150mm offset along X-axis.

Fig.6(a) and 6(b) present the experimental waveforms of the coupler under -10 mm and 50 mm misalignments along the Z-axis, respectively, when the output power is 1kW. Fig.9 illustrates the curves of I_O variation with misalignment distance and load resistance R_L when the coupler is misaligned along the X-axis and Z-axis directions. The experimental results are in good agreement with the theoretical predictions.

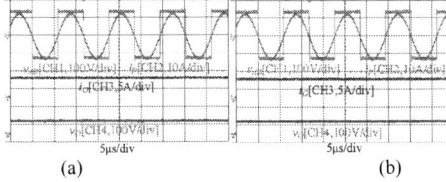

(a)   (b)

Fig.8 Experimental waveforms under -10mm and 50mm offset along the Z-axis when the output power is 1kW. (a) -10 mm offset along Z-axis. (b) 50mm offset along Z-axis.

IV. CONCLUSION

This paper introduces an anti-offset WPT system with auxiliary resonant circuit. The proposed system can achieve strong offset characteristic by integrating an auxiliary resonant circuit, high-order compensation network, and BP coupler. The scheme proposed in this paper can use various types of WPT systems and has a certain degree of universality. Experimental results confirm that the system maintains high misalignment tolerance along both the X-axis and Z-axis directions.

REFERENCES

[1] C. Cai, S. Wu, Z. Zhang, L. Jiang and S. Yang, "Development of a Fit-to-Surface and Lightweight Magnetic Coupler for Autonomous Underwater Vehicle Wireless Charging Systems," *IEEE Trans. Power Electron.*, vol. 36, no. 9, pp. 9927-9940, Sept. 2021.

[2] Y. Wang, H. Liu, P. Wheeler and F. Wu, "Implementation and Analysis of an Efficient Soft-Switching Battery Wireless Charger with Re-Configurable Rectifier," *IEEE Trans. Ind. Electron.*, vol. 71, no. 5, pp. 4640-4651, May 2024.

[3] A. Hossain, P. Darvish, S. Mekhilef, K. S. Tey and C. W. Tong, "A New Coil Structure of Dual Transmitters and Dual Receivers With Integrated Decoupling Coils for Increasing Power Transfer and Misalignment Tolerance of Wireless EV Charging System," *IEEE Trans. Ind. Electron.*, vol. 69, no. 8, pp. 7869-7878, Aug. 2022.

[4] H. Liu, Y. Wang, H. Yu, F. Wu and P. Wheeler, "A Novel Three-Phase Omnidirectional Wireless Power Transfer System With Zero-Switching-Loss Inverter and Cylindrical Transmitter Coil," *IEEE Trans. Power Electron.*, vol. 38, no. 8, pp. 10426-10441, Aug. 2023.

[5] Y. Chen, B. Yang , Q. Li, H. Feng, X. Zhou, Z. He and R. Mai., "Reconfigurable Topology for IPT System Maintaining Stable Transmission Power Over Large Coupling Variation," *IEEE Trans. Power Electron.*, vol. 35, no. 5, pp. 4915-4924, May 2020.

[6] X. Qu, Y. Yao, D. Wang, S. -C. Wong and C. K. Tse, "A Family of Hybrid IPT Topologies With Near Load-Independent Output and High Tolerance to Pad Misalignment," *IEEE Trans. Power Electron.*, vol. 35, no. 7, pp. 6867-6877, July 2020.

[7] L. Zhao, D. J. Thrimawithana and U. K. Madawala, "Hybrid Bidirectional Wireless EV Charging System Tolerant to Pad Misalignment," *IEEE Trans. Ind. Electron.*, vol. 64, no. 9, pp. 7079-7086, Sept. 2017.

[8] Y. Chen, B. Yang, Z. Kou, Z. He, G. Cao and R. Mai, "Hybrid and Reconfigurable IPT Systems With High-Misalignment Tolerance for Constant-Current and Constant-Voltage Battery Charging," *IEEE Trans. Power Electron.*, vol. 33, no. 10, pp. 8259-8269, Oct. 2018.

[9] M. Budhia, J. T. Boys, G. A. Covic and C. Huang, "Development of a Single-Sided Flux Magnetic Coupler for Electric Vehicle IPT Charging Systems," *IEEE Trans. Ind. Electron.*, vol. 60, no. 1, pp. 318-328, Jan. 2013.

[10] A. Zaheer, H. Hao, G. A. Covic and D. Kacprzak, "Investigation of Multiple Decoupled Coil Primary Pad Topologies in Lumped IPT Systems for Interoperable Electric Vehicle Charging," *IEEE Trans. Power Electron.*, vol. 30, no. 4, pp. 1937-1955, April 2015.

[11] S. Kim, G. A. Covic and J. T. Boys, "Tripolar Pad for Inductive Power Transfer Systems for EV Charging," *IEEE Trans. Power Electron.*, vol. 32, no. 7, pp. 5045-5057, July 2017.

[12] Y. Yao, Y. Wang, X. Liu, Y. Pei and D. Xu, "A Novel Unsymmetrical Coupling Structure Based on Concentrated Magnetic Flux for High-Misalignment IPT Applications," *IEEE Trans. Power Electron.*, vol. 34, no. 4, pp. 3110-3123, April 2019.

[13] W. Zhao, X. Qu, J. Lian and C. K. Tse, "A Family of Hybrid IPT Couplers With High Tolerance to Pad Misalignment," *IEEE Trans. Power Electron.*, vol. 37, no. 3, pp. 3617-3625, March 2022.

Research on Low Speed Power Boosting Technology for High Speed Maglev Trains

Zheyi Zheng
School of Electrical Engineering
Southwest Jiaotong University
Cheng du, China
zhengzheyi@my.swjtu.edu.cn

Fuao Chen
School of Electrical Engineering
Southwest Jiaotong University
Cheng du, China
chenfuao@my.swjtu.edu.cn

Xiaojun Zhang
School of Electrical Engineering
Southwest Jiaotong University
Cheng du, China
yiquebkx@my.swjtu.edu.cn

Haoyun Wang
School of Electrical Engineering
Southwest Jiaotong University
Cheng du, China
wanghaoyun_swjtu@163.com

Yang Chen
School of Electrical Engineering
Southwest Jiaotong University
Cheng du, China
yangchen@swjtu.edu.cn

Ruikun Mai*
School of Electrical Engineering
Southwest Jiaotong University
Cheng du, China
mairk@swjtu.edu.cn

Abstract—This paper introduces a method for using switched capacitors to enhance the system power of high-speed maglev trains operating under low-speed conditions. High-speed maglev trains generate electricity through the cogging effect to meet the power demand of the trains. In the low-speed operation condition, the voltage induced by the train is too low, so the traditional power supply method requires additional equipment for auxiliary power supply. Different from the traditional power supply method, the approach proposed in this paper can solve the problem of difficult power supply for high-speed maglev trains during low-speed operation. Firstly, the principle of power generation by the cogging effect of high-speed maglev trains is discussed, and then the factors affecting power supply are discussed. Then, based on the fundamental wave approximation model, switched capacitors are used to tune the circuit to increase the output power. Moreover, even under the condition of variable frequency, it can still track the circuit frequency and dynamically tune the circuit to enhance the circuit power. Finally, a simulation model is established on MATLAB/Simulink and experiments are carried out to verify the effectiveness of the proposed method.

Keywords—High-speed maglev train, switched control capacitor, dynamic tuning, power enhancement

I. INTRODUCTION

Maglev trains theoretically overcome the two major drawbacks of high-speed railways, namely wheel-rail friction drive and pantograph-catenary sliding current collection. However, a key challenge that impedes the rapid development of high-speed maglev trains is the issue of "safe, reliable, and efficient" power collection for the onboard auxiliary power supply system. This system serves as the sole energy source for critical equipment, including control, communication, and life support systems within the carriage. The safe, reliable, and efficient operation of the auxiliary power supply system is crucial to the advancement of high-speed maglev trains.

There are two primary types of suspension systems for maglev trains: the EMS (Electromagnetic Suspension) system and the EDS (Electrodynamic Suspension) system [1]-[3]. The auxiliary power supply system of the TR08 maglev train, which utilizes the EMS system, employs a "high-speed dynamic/low-speed contact" power supply mode. Specifically, when operating at high speeds (>100 km/h), the harmonic magnetic field generated by the interaction between the linear motor and the collector coil is used to generate "dynamic" power (i.e., tooth-slot power generation), As shown in Fig. 1.

At low speeds (≤100 km/h), electricity is supplied through "contact" between the collector shoe and the power rail via friction. However, the contact-based power supply method during low-speed operation is subject to several limitations, including significant wear of the collector shoes, poor current collection from exposed power rails in harsh environments, and susceptibility to electric shock [4]-[6].

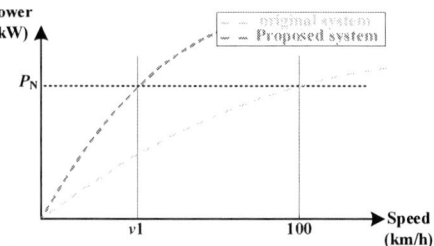

Fig. 1. The proposed system meets the speed requirement for power compared to the original system

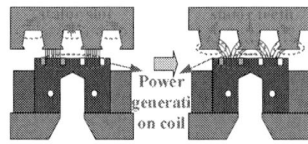

Fig. 2. Principle of linear generator reluctance generation

The strong inductance of the vehicle-mounted collector coil induces a phase difference between the current and voltage, resulting in waveform distortion and limiting the efficient transfer of electrical power. Maintaining the circuit in a resonant state can mitigate this issue [7]. The approach of combining fixed-frequency control with dynamic reactive power parameter tuning can facilitate reactive power adjustment; however, it is constrained by the fixed frequency [8]-[9]. Capacitive arrays can provide discrete dynamic tuning, but they are bulky and costly [10]-[11]. Moreover, the use of non-adjustable inductance circuits [12] has become largely irrelevant. Currently, dynamic tuning technologies for maglev trains predominantly achieve rectification through PWM converters [13]-[14], or impedance matching between the generator coil and the load circuit via series compensation capacitors [15]. Nevertheless, these methods have not effectively addressed the challenge of converting limited power generation into usable electrical energy, particularly for trains operating at low speeds.

In contrast to traditional power supply methods, the approach proposed in this study employs switched capacitors for continuous dynamic tuning. This method effectively addresses the power supply challenges faced by high-speed maglev trains when operating at low speeds.

II. THEORETICAL ANALYSIS

A. The principle of power generation by the cogging effect

As shown in Fig. 2, The principle of cogging effect power generation for high-speed maglev trains was explored. The main magnetic pole of a long stator linear synchronous motor incorporates a generator slot, where the generator coil is positioned. The excitation magnetic field induces magnetic reluctance due to the tooth-slot effect. When the generator coil aligns with the stator teeth, the system experiences maximum magnetic reluctance and minimum magnetic flux. As the train moves, the magnetic flux within the linear generator coil alternates. According to Faraday's law of electromagnetic induction, the changing magnetic flux generates an induced electromotive force. Consequently, as the train progresses, an induced voltage is produced. The faster the train travels, the higher the induced voltage, while the slower the speed, the lower the induced voltage. In accordance with the characteristics of synchronous motors, the frequency of the stator current also varies with the operating speed. The angular frequency of the stator current can be determined using the (1) and (2).

$$f = \frac{v}{2\tau_s} \tag{1}$$

$$\omega = 2\pi f = \frac{\pi v}{\tau_s} \tag{2}$$

In the formula, f is the stator frequency, v is the running speed of the train, τ_s is the stator pole distance.

The variation curves of induced voltage and system frequency with speed are shown in Fig. 3.

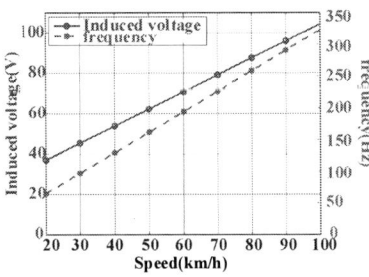

Fig. 3. Curve graph of induced voltage and frequency variation with speed

B. Switch-controlled capacitor

Based on the aforementioned analysis, it can be concluded that under low-speed operating conditions, both the frequency and induced voltage of the system are relatively low. The pronounced inductive characteristics of the collector coil result in excessive reactive power, which in turn leads to reduced output power and insufficient power generation during low-speed operation. This issue is particularly prominent under low-speed conditions and significantly impacts the overall energy conversion efficiency of the system. To mitigate this problem, the present study proposes compensating for the strong inductive effects of the collector coil by incorporating variable capacitors. Specifically, by dynamically tuning the circuit, the reactive power within the system can be effectively reduced, thereby optimizing the power factor and enhancing the efficient utilization of available energy. This approach not only improves power output efficiency under low-speed conditions but also ensures the stability and efficiency of the system across various operating conditions.

At low frequencies, a large compensation capacitance is required, and the capacitance value must be continuously adjusted in response to the time-varying system speed and frequency. Consequently, switching capacitors present an ideal solution. As illustrated in the Fig. 4, the capacitance ratio C_{sc}/C_a in relation to the phase angle offset can be derived using the appropriate formula. It is evident that by altering the phase angle offset α, the equivalent capacitance C_{sc} can be effectively controlled and modulated.

The typical circuit diagram of SCC is shown in Fig. 5 (a), where Sa and Sb are two back-to-back MOSFETs, and Ca is a parallel capacitor. SCC uses PWM (Pulse Width Modulation) method to change the external equivalent capacitance value to alter the output current. Fig. 5 (b) shows the voltage U_{ab} which across Terminals A and B. This voltage is always zero at the instants when the stitches are turning ON and turning OFF, thus achieving the ZVS condition. The equivalent calculation of switch capacitor is shown in (3).

$$C_{SC} = \frac{\pi C_a}{2\pi - 2\alpha + \sin 2\alpha} \tag{3}$$

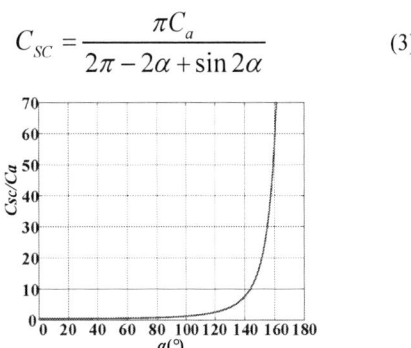

Fig 4　Capacitance ratio C_{sc}/C_a versus phase angle shift α.

III. THE PROPOSED METHOD AND ITS CONTROL STRATEGY

The schematic diagram of the original system is shown in Fig. 6. The obvious inductive characteristics of the collector coil L_S in the circuit lead to the generation of excessive reactive power, resulting in a decrease in output power and insufficient power generation during low-speed operation.

A. The proposed system with SCC

[15] proposed using a compensation capacitor to compensate for the strong inductiveness of the collector coil. However, it was only implemented with a fixed capacitor. The problem with this approach is that reactive power compensation can only be achieved at a certain frequency, and it is difficult to implement due to the large capacitance value of the compensation capacitor. Ultimately, the achieved effect is not satisfactory. By using switched capacitors, the above problems can be avoided. Reactive power compensation can be achieved at different frequencies, and it can be realized with a relatively small capacitance value.

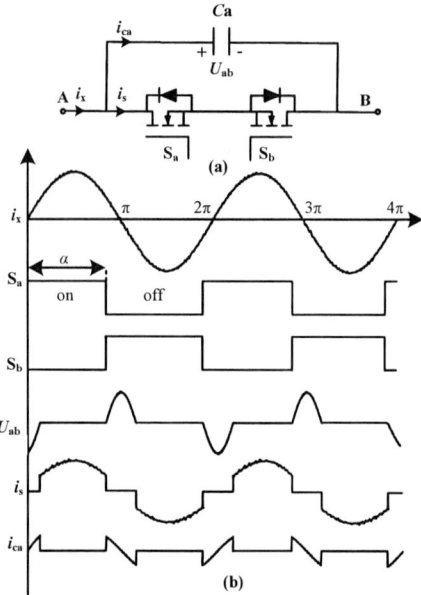

Fig. 5. (a)SCC Structure. (b)Typical waveforms.

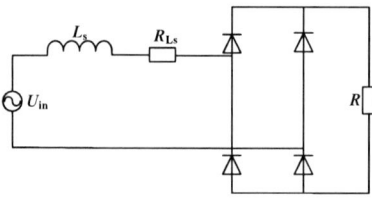

Fig. 6. Original system schematic diagram

As shown in Fig. 4, theoretically, the use of switched capacitors can increase the capacitance value to infinity. However, when the control angle α is between 140° and 180°, the curve is very steep, and the capacitance value changes abruptly. Generally, a capacitor is connected in series in the switched capacitor structure to make the curve gentler. This also reduces the magnification of the capacitance value and adds a new component. Since the value of the collector coil in the original system structure is already fixed and cannot be changed, and the system is in a low-frequency state, the required compensation capacitance value is as high as the millifarad level. Therefore, switched capacitors are composed by connecting three microfarad-level capacitors in parallel. In this way, not only can the capacitance value be increased, but also the adjustment in the smooth section of the curve can be achieved.

B. Control strategy

As shown in Fig. 7, the generation mechanism of the PWM driving signal for the Switched Capacitor (SCC) module relies on the collaborative effect of a dual closed-loop control strategy. Specifically, the red control path conducts real-time sampling of the input current through a high-precision current sensor. Then, the signal conditioning circuit extracts the zero-crossing characteristic points of the current, providing a phase reference for the synchronous triggering of the switching

tubes. This ensures that the switching state is changed when the current naturally crosses zero to reduce losses. The blue control path, on the other hand, employs the Phase-Locked Loop (PLL) technology to track the frequency of the input current signal. Through a dynamic frequency-locking algorithm, it obtains the current operating frequency of the system in real time. By combining the circuit parameters with the mathematical model established in (3), it deduces the control angle α that matches the current working condition. After being processed by the Pulse Width Modulation (PWM) logic unit, this angular parameter generates a PWM driving waveform with a specific duty cycle and phase difference, precisely controlling the turn-on and turn-off timing of the switching tubes (such as MOSFETs). Through the phase synchronization and parameter coupling of the two control signals, the dynamic adjustment of the equivalent capacitance value of the switched capacitor is achieved. This, in turn, drives the resonant frequency of the system to shift towards the target operating point, ultimately achieving an adaptive tuning process based on the real-time working conditions. This control strategy, by combining the hardware triggering mechanism of current zero-crossing detection with the software algorithm of PLL frequency tracking, effectively improves the response speed and control accuracy of the tuning process, providing a reliable technical guarantee for the stable power output of the maglev system in the low-frequency band.

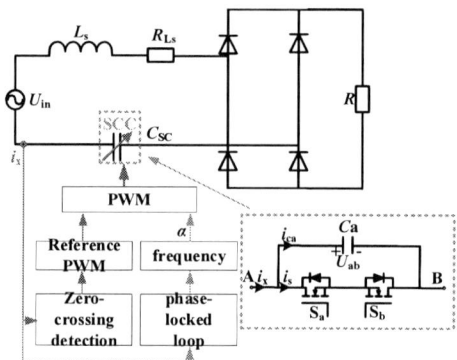

Fig. 7. The block diagram of the proposed control strategy

IV. SIMULATION AND EXPERIMENTAL VERIFICATION

In this section, the power improvement method proposed in Section III is verified through simulations and experiments.

A. Simulation

Finally, a simulation model was developed using MATLAB/Simulink to verify the effectiveness of the proposed strategy. The simulation was carried out according to the actual situation of the cogging effect. As shown in the Fig. 9, the output powers of the original system and the proposed system under the actual induced voltage and frequency are compared. It can be seen that after the speed reaches 20 km/h, this method can effectively improve the output power of the system.

Fig. 8 provides us with a detailed comparison of the current waveforms before and after tuning. Before tuning, the system's current waveform exhibited obvious distortion and discontinuity. The distortion and discontinuity of the current waveform not only lead to a decrease in the system's efficiency but may also trigger a series of power quality issues, such as harmonic pollution, voltage fluctuations, etc., posing a serious threat to the stable operation of the system and the normal operation of the equipment. However, after tuning using the proposed strategy, the system's current waveform has significantly improved. As can be clearly seen from Fig. 8, both the distortion and discontinuity of the current waveform have been greatly reduced, and the waveform has become smoother and more regular. This indicates that the proposed strategy effectively adjusts the circuit parameters of the system, bringing the operation of the system closer to the ideal state. Meanwhile, there has also been a significant increase in the current amplitude. This change means that the system can output more power, thereby improving the overall performance and efficiency of the system.

In conclusion, the verification through the MATLAB/Simulink simulation model fully demonstrates that the proposed strategy has remarkable effectiveness and superiority in enhancing the system's output power and improving the current waveform. This provides a solid theoretical basis and reliable technical support for the application of this strategy in practical engineering projects.

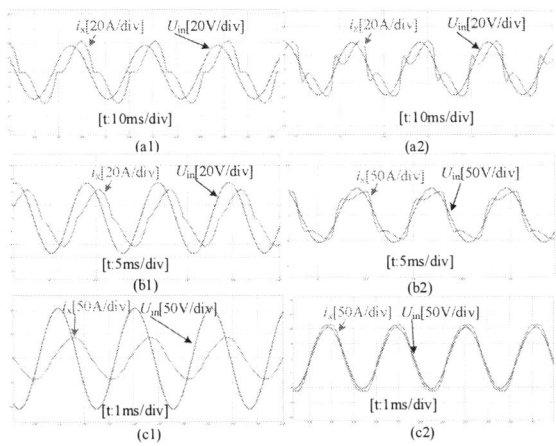

Fig. 8. (a1) and (a2), (b1) and (b2), (c1) and (c2), are voltage and current waveforms before and after tuning at speeds of 20km/h, 40km/h, and 100km/h, respectively.

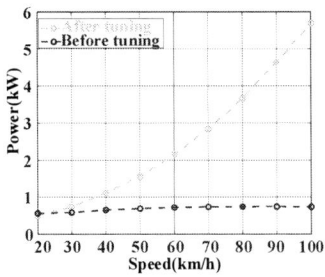

Fig. 9. Comparison of output power before and after simulation tuning

B. Experiment

In this study, experimental methods were employed to verify the effectiveness of the proposed strategy. The experimental setup is shown in the Fig. 10, and the experimental parameters are listed in TABLE I. .

During the experimental design process, taking into account the actual situation of the overcurrent limitation of the current sampling board, in order to ensure the safety and feasibility of the experiment, a comparative experimental method of applying the same input voltage at different frequencies was adopted to conduct a low-power experiment. This experimental design can effectively verify the correctness of the theory while avoiding the risk of overcurrent in the current sampling board.

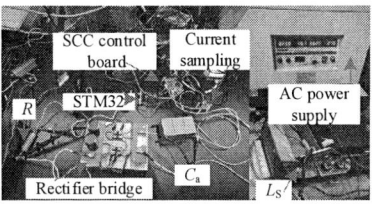

Fig. 10. Experimental setup

TABLE I.   THE EXPERIMENTAL PARAMETERS

Symbol	Value
f	64.5-323Hz
U_{in}	8V
L_S	1.1mF
C_a	150μF
C_{SC}	0.221-5.518mF

The experiment aims to verify the consistency between the theoretical analysis and the simulation results. Through the collection and analysis of experimental data, it can be clearly observed from Fig. 11 of the experimental results that after tuning, the phase relationship between the input voltage and current waveforms has been significantly improved. Specifically, the phase difference between the voltage and current waveforms has decreased, approaching the ideal in-phase state, which indicates that the power factor of the system has been improved. At the same time, the current amplitude has also increased significantly, which means that the system can absorb more electrical energy, thereby increasing the output power of the system. As shown in Fig. 12, the output power of the system has been significantly enhanced after tuning. This result is consistent with the theoretical analysis and simulation results, fully verifying the effectiveness and feasibility of the proposed strategy.

In conclusion, through the design of the experimental scheme and data analysis, this experiment has successfully verified the theoretical and simulation results, providing a solid experimental basis for the application of the proposed strategy in practical engineering projects.

V. CONCLUSION

This paper proposes a novel method for enhancing the system power when the train is operating at a low speed, which is based on switched capacitors. By detecting the frequency and phase of the current flowing through the capacitor, the phase difference that needs to be compensated under the current working condition is calculated. Then, the signal of the switching tube is controlled to compensate for the strong inductance in the circuit, so as to improve the power transmission capability of the system. This method not only avoids the limitation of relying on external equipment during the low-speed stage of the train but also fills the research gap in power regulation under low-frequency working conditions. Compared with the traditional auxiliary power supply scheme, this study provides a new power supply path that is lightweight, efficient, and has engineering adaptability for the maglev system under low-speed working conditions.

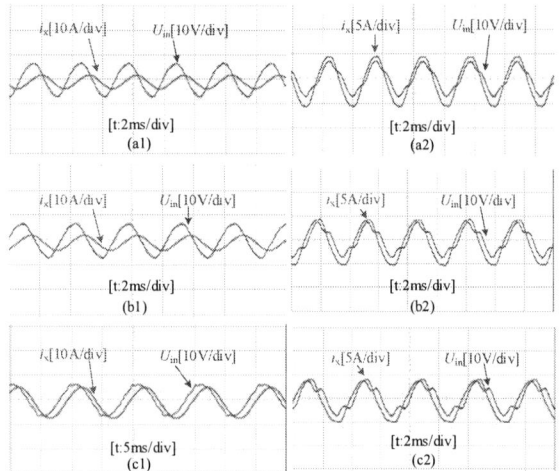

Fig. 11. (a1) and (a2), (b1) and (b2), (c1) and (c2), are voltage and current waveforms before and after tuning at speeds of 20km/h, 40km/h, and 100km/h, respectively.

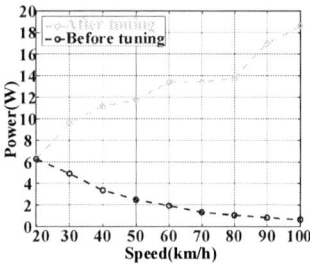

Fig. 12. Comparison of the output power before and after experimental tuning

REFERENCES

[1] Diekmann A, Hahn W, Kunze K, "The support magnet cladding with integrated IPS pick-up coil of Transrapid vehicles," Maglev: International Conference on Magnetically Levitated Systems & Linear Drives. 2006.

[2] Lyu Gang, Guo Xilin, "Calculation of Power Generation Characteristics of Linear Harmonic Generator for Electrodynamic Suspension Maglev Train (电动磁浮列车用直线谐波发电机发电特性计算)," Xinan Jiaotong Daxue Xuebao/Journal of Southwest Jiaotong University, vol 58, no 4, pp 783-791, August 2023.
吕刚,郭曦临.电动磁浮列车用直线谐波发电机发电特性计算[J].西南交通大学学报,2023,58(04):783-791.

[3] Fischperer Rolf, "Power Supply System for a Long-stator Drive for a Magnetic Levitation Train," U.S. Patent 5,569,987, issued October 29, 1996.

[4] Wang Lu, Chen Min, Xu De-Hong, "Engineering design of contactless emergency power supply in maglev," Proceedings of the Chinese Society of Electrical Engineering, vol. 27, no 18, pp 67-70, June 25, 2007.
王璐,陈敏,徐德鸿.磁浮列车非接触紧急供电系统的工程化设计[J].中国电机工程学报,2007,(18):67-70.

[5] Maki N, Tatsumi T, Iwahana T, "Methods and characteristics of train power source system utilizing the flux produced by track coils," The transactions of the Institute of Electrical Engineers of Japan. B, vol. 101, no. 1, pp. 33-40, 1981.

[6] Yu Jiaqi, Zhou Lingyun, Liu Shunpan, Wang Zhoulong, Mai Ruikun, "Research on Inductive Power Transfer Method for Electrodynamic Suspension Maglev Train Based on Collector Coil Reuse (基于集电线圈复用的电动磁浮列车感应式电能传输技术研究)," Diangong Jishu Xuebao/Transactions of China Electrotechnical Society, vol 9, no 4, pp 976-986, 2024.
余嘉淇,周凌云,刘顺攀,王州龙,麦瑞坤.基于集电线圈复用的电动磁浮列车感应式电能传输技术研究[J].电工技术学报,2024,39(04):976-986.

[7] Liu Hao, Li Zhenjie, Tian Yuhong, Song Wenlong, "Dynamic Tuning Method for Wireless Charging SystemBased on the Magnetic Flux Controllable Inductor," Journal of Harbin University of Science & Technology, vol. 28, no. 5, 2023.
刘浩,李振杰,田育弘,宋文龙.采用磁通可控可变电感的无线充电动态调谐[J].哈尔滨理工大学学报,2023,28(05):11-18.

[8] Su Yugang, Tang Chunsen, Sun Yue, Wang Zhihui, "Load adaptive technology of contactless power transfer system," Diangong Jishu Xuebao/Transactions of China Electrotechnical Society, vol 24, no 1, pp 153-157, January 2009.
苏玉刚,唐春森,孙跃,王智慧.非接触供电系统多负载自适应技术[J].电工技术学报,2009,24(01):153-157.

[9] Mai Ruikun, Lu Liwen, Li Yong, He Zhengyou, "Dynamic resonant compensation approach based on minimum voltage and maximum current tracking for IPT system," Diangong Jishu Xuebao/Transactions of China Electrotechnical Society, vol 30, no 19, pp 32-38, October 5, 2015.
麦瑞坤,陆立文,李勇,何正友.一种采用最小电压与最大电流跟踪的IPT系统动态调谐方法[J].电工技术学报,2015,30(19):32-38.

[10] Sun Yue, Wu Jing, Wang ZhiHui, Tang ChunSen, "Frequency stabilization control method for ICPT system based on capacitor array," Dianzi Keji Daxue Xuebao/Journal of the University of Electronic Science and Technology of China, vol 43, no 1, pp 54-59, January 2014.
孙跃,吴静,王智慧,唐春森.ICPT系统基于电容阵列的稳频控制策略[J].电子科技大学学报,2014,43(01):54-59.

[11] He Zhengyou, Li Yong, Mai Ruikun, Li Yanling, "Dynamic compensation strategy of inductive power transfer system with inductive-resistive load," Xinan Jiaotong Daxue Xuebao/Journal of Southwest Jiaotong University, vol 49, no 4, pp 569-575 and 589, August 1, 2014.
何正友,李勇,麦瑞坤,李砚玲.考虑阻感性负载IPT系统的动态补偿技术[J].西南交通大学学报,2014,49(04):569-575+589.

[12] Yang MinSheng, Wang YaoNan, "Transferred power regulating method with a dynamically detuning inductor for ICPT pickups," Dianji yu Kongzhi Xuebao/Electric Machines and Control, vol 16, no 1, pp 72-78, January 2012.
杨民生,王耀南.感应耦合电能传输系统动态解谐传输功率控制[J].电机与控制学报,2012,16(01):72-78.

[13] Yamamoto T, Murai T, Hasegawa H, Youshioka H, Fujiwara S, Hatsukade S, "Development of distributed-type linear generator with damping control," Quarterly Report of RTRI, vol. 41, no.2, pp. 83-88, 2000.

[14] T. Murai, Y. Sakamoto and H. Hasegawa, "High Power Factor Converter Control by Instantaneous Single-Phase Current for a Maglev System Linear Generator,"2007 Power Conversion Conference Nagoya, Nagoya, Japan, 2007, pp. 1158-1163.

[15] W. Ying, L. Weiguo, H. Hongyun, L. Zongjian, X. Yang and L. Da, "Research on Contactless Power Supply of High Speed Maglev Train Based on MCR-WPT," 2019 14th IEEE Conference on Industrial Electronics and Applications (ICIEA), Xi'an, China, 2019, pp. 2297-2302.

Current Overshoot and Oscillation Suppression in SiC MOSFETs Through Variable Gate Capacitance During Turn-on Transient

Xuchong Cai
School of Electrical Engineering & Automation
Harbin Institute of Technology
Harbin, China
24S006020@stu.hit.edu.cn

Yishun Yan
School of Electrical Engineering & Automation
Harbin Institute of Technology
Harbin, China
24S006068@stu.hit.edu.cn

Yanchen Pan
School of Electrical Engineering & Automation
Harbin Institute of Technology
Harbin, China
24B906019@stu.hit.edu.cn

Mingcheng Ma
School of Electrical Engineering & Automation
Harbin Institute of Technology
Harbin, China
24B906005@stu.hit.edu.cn

Binbo Xu
School of Electrical Engineering & Automation
Harbin Institute of Technology
Harbin, China
2022112922@stu.hit.edu.cn

Dianguo Xu
School of Electrical Engineering & Automation
Harbin Institute of Technology
Harbin, China
xudiang@hit.edu.cn

Abstract—Wide-bandgap devices, such as silicon carbide MOSFETs (SiC MOSFETs), are increasingly being adopted in high-frequency and high-power-density conversion systems due to their ability to achieve higher switching speeds and lower switching losses compared to conventional silicon-based devices. However, the high switching speeds may induce pronounced current overshoots and high-frequency oscillations, potentially compromising system reliability and inducing electromagnetic interference (EMI) issues. To address these challenges, this paper proposes an innovative gate driver circuit with variable gate capacitance for SiC MOSFETs, featuring dynamic adjustment of gate capacitance during turn-on transients. The proposed solution incorporates an additional circuit module that is autonomously activated during the Miller plateau, effectively suppressing both current overshoot and switching oscillations. Experimental validation under 400 V, 20 A operating conditions reveals that the proposed method achieves a 23% current overshoot reduction with simultaneous oscillation suppression during the turn-on transients when compared with conventional gate driver implementations.

Keywords—silicon carbide (SiC) MOSFET, Variable gate Capacitance, Current Overshoot, Switching Losses

I. INTRODUCTION

The prevailing trend in modern power electronic systems is toward achieving higher power density and efficiency. SiC MOSFETs exhibit significant advantages over traditional Si devices[1], including faster switching speeds, lower switching losses, and reduced on-state resistance. These merits have enabled their widespread adoption in wind power systems[2], motor drive systems[3-5], electric-vehicle powertrains[6], and other energy-critical applications. However, excessively high switching speeds tend to induce current overshoot and high-frequency oscillations, which exacerbate electromagnetic interference (EMI) issues, degrade device reliability, and may even cause permanent damage[7].To address the problem of current overshoot and oscillation during turn-on transients, some active gate driver (AGD) strategies have been extensively investigated according to their control variables, including variable gate resistance[8-10], variable gate current[11-13], and variable gate voltage methods[14-17]. The following sections will provide detailed analyses of the implementation

approaches and corresponding advantages and limitations for each category of AGD strategies.

A. Variable gate resistance

Among these, the variable gate resistance method is the most widely implemented. As reported in [8], this method identifies switching stages by monitoring the gate-source voltage and dynamically adjusts the gate resistance. However, the feedback signals require amplification and processing to enable resistance adjustment, thereby increasing system complexity. Reference [9] has further segmented the switching process into multiple phases and employs high-speed Field-Programmable Gate Arrays (FPGAs) to dynamically regulate both phase durations and gate resistances, effectively suppressing voltage and current overshoots and oscillations. Reference [10] has proposed a relatively simple variable gate resistance driver circuit for SiC MOSFETs, demonstrating a 30%-40% reduction in turn-on losses, but it introduces additional current overshoot during the turn-on transients.

B. Variable gate current

The variable gate current method utilizes an adjustable current source to adjust the gate drive current[11]. In [12], complementary switches are designed to adjust the charging rate during the switching process, reducing the gate charge current during the Miller plateau while enhancing the gate charge current in other stages to optimize the switching trajectory of the SiC MOSFET. Since current sources are typically implemented by MOSFETs operating in saturation mode, the core challenges of this approach lie in the design of stage identification algorithms and high-bandwidth, high-slew-rate current sources. Reference [13] proposes an advanced digitally controlled gate current source-based active gate driver for SiC MOSFETs, which enables precise adjustment of switching trajectories and achieves a reduction in switching losses, but introduces implementation difficulty due to complex real-time digital signal processing algorithms.

C. Variable gate voltage

The variable gate voltage method generally lowers the driving voltage during the Miller plateau[14]. However, the brief duration of the Miller plateau presents significant

979-8-3315-1110-4/25 $31.00 © 2025 IEEE

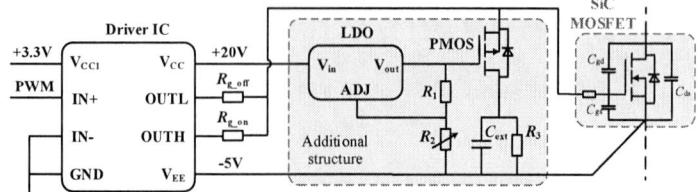

Fig.2 Schematic diagram of the gate driver circuit topology structure with variable gate capacitance

challenges for implementing variable gate voltage methods in SiC MOSFET applications. Reference [15] proposes a microcontroller unit (MCU) based solution that dynamically adjusts the output voltage of a digital-to-analog converter (DAC) to alter the intermediate voltage, thereby achieving adaptive voltage switching during switching transitions. Additionally, complex topologies often fail to deliver expected performance due to discrepancies between simulation assumptions and practical constraints. For instance, while an advanced topology in [16] demonstrates effectiveness in simulations through multi-stage control strategies, its experimental validation remains unfeasible due to implementation complexity. Reference [17] presents a novel four-level active gate driver topology achieving a switching loss reduction. However, the four-level implementation requires additional devices and complex multi-stage sequencing, significantly increasing system complexity while introducing reliability problems.

In this paper, a simplified active gate driver circuit for SiC MOSFETs is proposed based on variable gate capacitance. By introducing an additional circuit structure equipped with autonomous detection capabilities, the proposed method selectively reduces the gate charging rate during the Miller plateau, thereby effectively mitigating current overshoot and oscillations in the turn-on transient. Experimental results demonstrated that the proposed driver circuit achieves a 23% reduction in current overshoot with significant oscillation suppression while maintaining nearly identical turn-on losses characteristics compared with conventional gate driver implementations.

II. ANALYSIS OF WORKING PRINCIPLE

A. Double-pulse test circuit

The schematic diagram of the double-pulse test circuit is shown in Fig. 1, where R_g represents the gate resistor, the upper SiC MOSFET is connected in the form of a gate-source short-circuit configuration, so it can be equivalent to a diode D for freewheeling current path, the load inductance L is connected in parallel with the equivalent diode. C_{gs}, C_{gd}, and C_{ds} are the parasitic capacitances of the SiC MOSFET, and L_{ss}, L_s, L_d, and L_g are the parasitic inductances of the SiC MOSFET's Kelvin source, source, drain, and gate, respectively. V_{drive} is generated by the driver IC. L_{loop} is the parasitic inductance of the power loop, and V_{dc} is the bus voltage, with C_{bus} stabilizing the bus voltage.

B. Circuit Topology Structure Analysis

The gate driver circuit with variable gate capacitance comprises three main components: a driver IC providing basic driving capability, a gate resistor for turn-on speed limitation, and an additional dynamic compensation structure. The additional structure includes a P-channel MOSFET, an adjustable low dropout regulator (LDO) to control the P-channel MOSFET, two resistors R_1 and R_2 for LDO output

voltage adjustment, an additional compensation capacitor C_{ext} for gate charging rate modulation, and a discharge resistor R_3 for ensuring capacitor reset safely. This configuration enables dynamic adjustment of the effective gate-source capacitance during the Miller plateau. Fig.2 shows the schematic diagram of the gate driver circuit topology structure with variable gate capacitance.

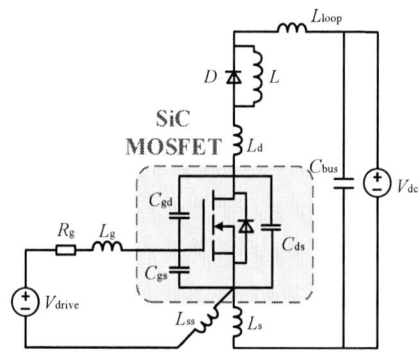

Fig.1 Double-pulse test circuit schematic diagram

C. Working principles

Fig. 3 shows the theoretical waveforms of the gate source voltage v_{gs}, drain source voltage v_{ds}, and drain current i_d in the SiC MOSFET turn-on transition. Based on the variations in gate-source voltage v_{gs}, drain-source voltage v_{ds}, and drain current i_d during the turn-on transient, the turn-on transition can be divided into four stages, including the turn-on delay stage, the current rise stage, the drain-source voltage drop stage, and the gate-source voltage rise stage.

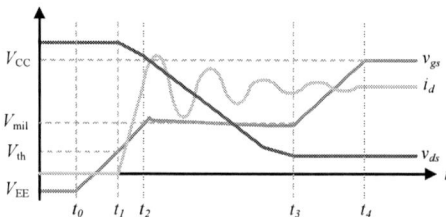

Fig. 3. Theoretical turn-on waveforms of the SiC MOSFET

1) Turn-on delay stage [t_0, t_1]: Upon receiving a PWM signal, the driver IC transitions its output voltage from V_{EE} to V_{CC}, initiating the gate charging stage. The driver IC charges the gate-source parasitic capacitor C_{gs} of the SiC MOSFET through the gate resistor R_g, and the gate-source voltage v_{gs} gradually rises to the threshold voltage V_{th}. During this stage, the SiC MOSFET remains in the off-state. Consequently, the p-channel MOSFET in the additional circuit remains off. This renders the additional structure inactive during this stage. The drain-source voltage v_{ds} remains at the bus voltage V_{dc}, and the drain current i_d remains zero.

979-8-3315-1110-4/25 $31.00 © 2025 IEEE

2) Current rising stage [t_1, t_2]: When the gate-source voltage v_{gs} reaches the threshold voltage V_{th}, initiating channel formation, the SiC MOSFET enters conduction. During this stage, the drain current i_d increases rapidly from zero, while the drain-source voltage v_{ds} experiences inductive voltage drop due to the power loop parasitic inductance L_{loop}. Concurrently, the gate-source voltage v_{gs} continues to rise. Since the gate-source voltage of the P-channel MOSFET remains higher than its conduction threshold voltage V_{thp}, the additional circuit remains inactive

3) Drain-source voltage dropping stage [t_2, t_3]: At t_2, marking the commencement of the Miller plateau, the drain current i_d approaches steady-state levels, meaning the current overshoot is about to occur. During this stage, the gate drive current is diverted to discharge the gate-drain parasitic capacitor C_{gd}, causing the drain-source voltage v_{ds} to drop rapidly, while the gate-source voltage v_{gs} remains at the Miller plateau voltage V_{mil}.

From this stage, the additional structure exerts its effect through control of gate charging rate. The LDO provides a bias voltage V_p to the gate of the P-channel MOSFET that is approximately equal to the Miller plateau voltage V_{mil}. This synchronization enables precise timing for implementing additional structural interventions. The effects of inherent device characteristics such as the P-channel MOSFET threshold voltage V_{thp} and the turn-on delay time t_{on} will be thoroughly examined in subsequent sections to quantify their influence on transient performance. As the gate-source voltage of the SiC MOSFET increases, the gate-source voltage of the P-channel MOSFET gradually falls below its threshold voltage V_{thp}, initiating conduction in the P-channel MOSFET, establishing a parallel charging path. Consequently, a significant portion of the current originally dedicated to charging the gate-source parasitic capacitor C_{gs} is diverted into the capacitor C_{ext} within the additional structure. This current redistribution effectively slows down the gate charging rate by introducing an additional capacitive load, thereby dampening the current overshoot and oscillations, as detailed in Fig. 4.

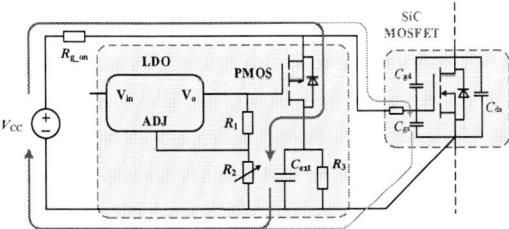

Fig. 4. Working principle of the additional structure during the drain-source voltage dropping stage

4) Gate-source voltage rising stage [t_3, t_4]: At t_3, marking the completion of the drain-source voltage transition, the drain-source voltage v_{ds} approaches zero with full channel conduction. During the drain-source voltage dropping stage, sustained charging elevates the capacitor C_{ext} voltage, progressively reducing its gate current shunting capacity. This transition redirects the majority gate current to C_{gs} charging, completing channel enhancement shown in Fig. 5.

During the turn-off transients, the P-channel MOSFET maintains conduction initially, as illustrated in Fig. 6. The driver IC simultaneously discharges capacitor C_{ext} through the

additional structure and the gate-source parasitic capacitance C_{gs}.

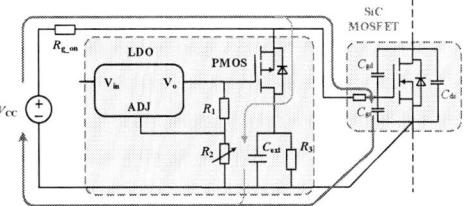

Fig. 5. Working principle of the additional structure during the gate-source voltage rising stage

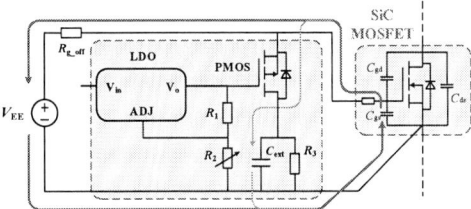

Fig. 6. Working principle of the additional structure during the initial stage of turn-off transients

When the gate-source voltage of the SiC MOSFET decreases, causing the gate-source voltage of the P-channel MOSFET to increase beyond the threshold voltage V_{thp}, the P-channel MOSFET transitions into the off-state. The additional capacitor C_{ext} discharges through a reset path formed by resistor R_3, as depicted in Fig. 7, thereby reinitializing the circuit for the next turn-on transition.

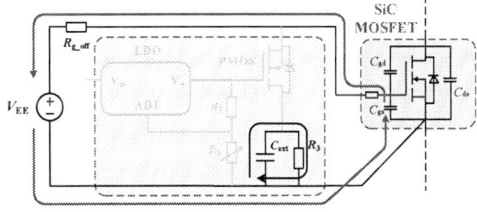

Fig. 7. Working principle of the additional structure during the reset process

Compared to closed-loop gate drivers needing active sensing and feedback control, the proposed structure connects the gate of the SiC MOSFET to the source of the P-channel MOSFET in the additional structure, enabling self-activation as the gate-source voltage v_{gs} of the SiC MOSFET increases naturally without external triggering signals. This gate driver circuit eliminates the need for complex digital control components, minimizes the components in the driver loop, and prevents instability from feedback signal propagation delays, collectively reducing complexity and cost.

D. Parametric design

The performance optimization of the proposed gate driver requires coordinated control of two critical parameters: the bias voltage V_p and the capacitance C_{ext}.

The P-channel MOSFET BSS84LT1G by Onsemi is specifically selected for its optimal switching speed. According to the datasheet of BSS84LT1G, the turn-on time t_{on} is 16 ns, and the threshold voltage V_{thp} is -1.7 V. The SiC MOSFET SCTL35N65G2V by STMicroelectronics is selected as the power switching device. According to the datasheet of SCTL35N65G2V, its Miller plateau voltage V_{mil} measures approximately 9 V. The rise rate of the gate-source voltage v_{gs} for SCTL35N65G2V is about 0.3V/ns during the turn-on transient. Therefore, the output voltage of the LDO can be calculated by the following formula:

$$V_{out} = V_{mil} + V_{thp} - k * t_{on} = 2.5V \qquad (1)$$

The LDO LM317 by Texas Instruments is selected for voltage regulation of the P-channel MOSFET. According to the datasheet of LM317, R_1 is chosen to be 240Ω, and R_2 is a variable resistor.

Larger capacitance C_{ext} values effectively suppress current overshoot and oscillations during the turn-on transient, but extend the turn-on duration, leading to a significant increase in turn-on losses. Conversely, smaller capacitance C_{ext} values minimally affect turn-on speed but provide limited current overshoot and oscillations suppression effectiveness. To strike an optimal balance between switching loss and transient overshoot suppression, an additional capacitance value of 200 pF was selected. R_3 is chosen to be 10kΩ to discharge the capacitor C_{ext} in the additional structure.

III. EXPERIMENTAL VALIDATION

A. Experimental Test Platform

To evaluate the performance of the proposed SiC MOSFET gate driver circuit with variable gate capacitance, a double-pulse test platform was constructed as shown in Fig. 8. The gate driver IC UCC5350SBDR by Texas Instruments was selected. Other components used in the driver circuit were described in Section II.

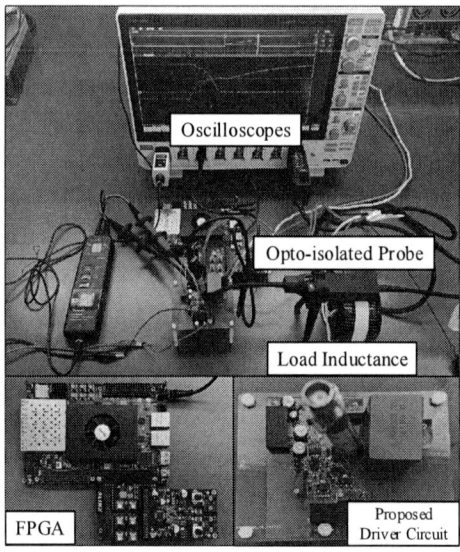

Fig. 8. Double-pulse experimental test platform

The measurement system utilizes a Tektronix MSO68B oscilloscope with a sampling rate of 12.5 GS/s. The gate-source voltage v_{gs} is monitored using an IsoVu™ TIVP1 isolated probe with a bandwidth of 1 GHz, while the drain-source voltage v_{ds} is acquired via the Tektronix THDP0200 high-voltage differential probe. The drain current i_d quantification is achieved through the Cybertek CSD050A coaxial shunt.

B. Comparative analysis of experimental results

Fig. 9 compares the turn-on transients between the proposed active gate driver (AGD) and conventional gate driver (CGD) under 400 V DC bus voltage, 20 A load current, and 20 Ω gate resistance conditions. The CGD exhibits a peak current overshoot of 14.4 A with significant oscillations, while the AGD reduces this value to 9.8 A via variable gate capacitance, achieving a 4.6 A suppression.

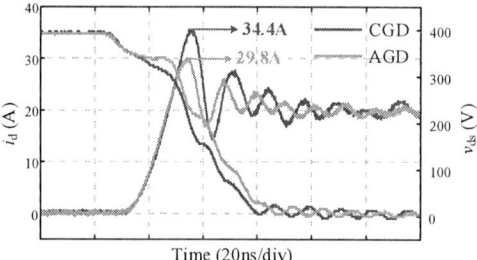

Fig. 9. Experimental results of CGD and AGD for i_d and v_{ds}

The optimization effects of bias voltage V_p are quantified in Fig. 10 through parametric sweeps conducted at three distinct levels: V_p = 2.5 V, 3.5 V, and 6 V. When V_p is set to 2.5 V, the current overshoot reaches its minimum value of 9.8 A, which is attributed to the full utilization of the addition capacitance path. As V_p is progressively increased to 3.5 V and 6 V, the capacitance modulation effect becomes less pronounced, leading to current overshoot escalations of 11.2 A and 14.2 A respectively. This performance degradation primarily stems from the growing temporal misalignment between the p-channel MOSFET activation timing and the critical Miller plateau phase.

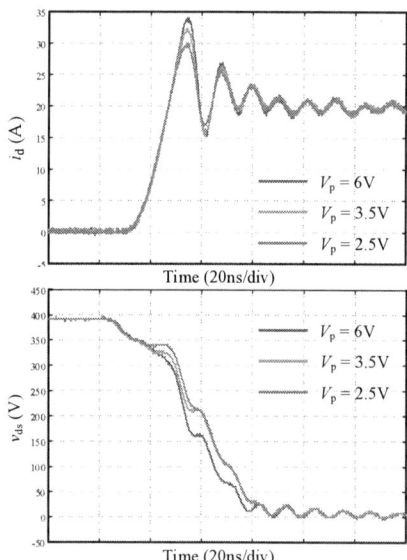

Fig. 10. Experimental results of AGD for i_d and v_{ds} with different V_p values

The selection of capacitance C_{ext} inherently involves balancing current overshoot suppression against turn-on losses. This design trade-off is experimentally validated in Fig. 11 through comparative evaluations of three capacitance configurations: C_{ext} = 200 pF, 520 pF, and 1 nF. Under 200 pF operation, the current overshoot is reduced to 9.8 A with only a marginal loss increase. The 520 pF configuration achieves further overshoot reduction to 8.8 A, but incurs a 7 ns extension in turn-on time. The maximum 1 nF capacitance provides the strongest overshoot suppression of 6.3 A at the cost of 12 ns additional turn-on time.

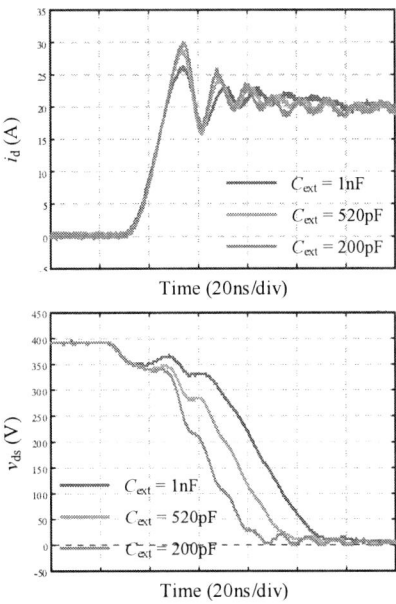

Fig. 11. Experimental results of AGD for i_d and v_{ds} with different C_{ext} values

Fig. 12 provides experimental validation of gate-source voltage v_{gs} sensitivity to parameter variations within the additional structure. Across various V_p and C_{ext} combinations, the gate-source voltage v_{gs} waveforms exhibit strong consistency, demonstrating negligible influence of additional structure parameter deviations on gate voltage characteristics. A slight voltage drop in the gate-source voltage v_{gs} during the Miller plateau occurs exclusively when $C_{ext} = 1$ nF.

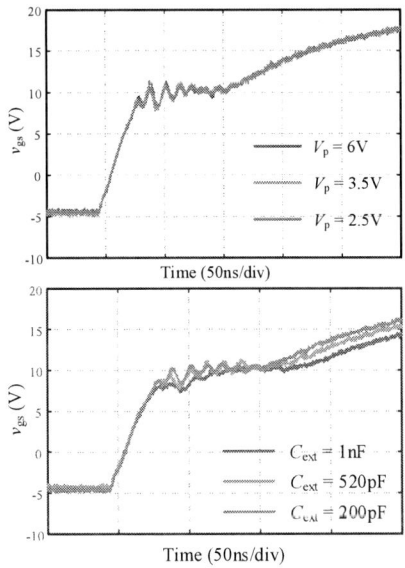

Fig. 12. Experimental results of AGD for v_{gs} with different C_{ext} and V_p values

C. Loss analysis

Balancing current overshoot suppression with turn-on loss reduction represents a core optimization challenge in gate driver design. Comprehensive validation requires implementing a dual-metric assessment framework that simultaneously evaluates current overshoot and turn-on losses.

Fig. 13 demonstrates the inverse correlation between turn-on losses and current overshoot characteristics. Experimental results confirm that optimized selection of the capacitance C_{ext} and bias voltage V_p achieves precise control over the loss-overshoot compromise. Under optimal parameters of $V_p = 2.5$ V, $C_{ext} = 200$ pF, comparative testing exposes critical CGD deficiencies. Under equivalent overshoot conditions, the CGD approach exhibits 66% higher turn-on losses, while with identical gate resistance configurations, it demonstrates 23% greater current overshoot magnitudes compared to the proposed method. These quantified results conclusively validate the proposed method's superior capability in achieving balanced performance optimization.

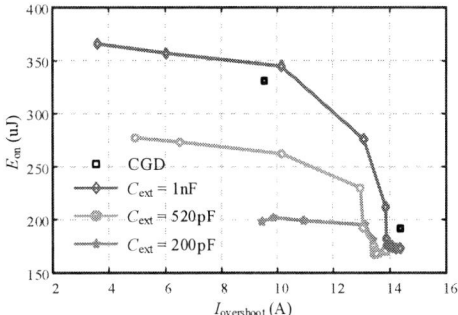

Fig. 13. Experimental results for turn-on losses and current overshoot

IV. CONCLUSION

The proposed SiC MOSFET driver circuit with variable gate capacitance achieves effective current overshoot and oscillation suppression through additional structure. Under 400 V, 20 A testing conditions, experimental comparisons reveal that the proposed driver reduces current overshoot by 23% and mitigates turn-on transient oscillations compared to conventional implementations. This topology achieves effective current overshoot mitigation during turn-on transients while suppressing high-frequency oscillations, and eliminates reliance on complex digital control algorithms or high-frequency pulse-width modulation signals requiring nanosecond-scale temporal resolution.

REFERENCES

[1] J. Millán, P. Godignon, X. Perpiñà, A. Pérez-Tomás and J. Rebollo, "A Survey of Wide Bandgap Power Semiconductor Devices," in IEEE Transactions on Power Electronics, vol. 29, no. 5, pp. 2155-2163, May 2014.

[2] Y. Zhang, X. Peng, X. Yuan, Y. Li and K. Wang, "A Unidirectional Cascaded High-Power Wind Converter With Reduced Number of Active Devices," in IEEE Access, vol. 11, pp. 10902-10911, 2023.

[3] B. Zhang, S. Wang, Y. Lai and Y. Yang, "Modeling and Prediction of Low-frequency Radiated EMI for a SiC Motor Drive System," in IEEE Transactions on Industrial Electronics, vol. 71, no. 9, pp. 10210-10220, Sept. 2024.

[4] A. Marzoughi, R. Burgos and D. Boroyevich, "Investigating Impact of Emerging Medium-Voltage SiC MOSFETs on Medium-Voltage High-Power Industrial Motor Drives," in IEEE Journal of Emerging and

Selected Topics in Power Electronics, vol. 7, no. 2, pp. 1371-1387, June 2019.

[5] A. B. Mirza, K. Choksi, S. S. Vala, A. Anwar and F. Luo, "Investigation of Reflected Wave Phenomenon in SiC-Based Two-Level Split-Phase Inverter-Fed Motor Drives," in IEEE Transactions on Power Electronics, vol. 40, no. 4, pp. 5768-5786, April 2025.

[6] H. Xu et al., "SiC MOSFETs: The Inevitable Trend for 800V Electric Vehicle Air Conditioning Compressors," in IEEE Transactions on Vehicular Technology, vol. 74, no. 2, pp. 2620-2634, Feb. 2025.

[7] T. Moaz, N. Rajagopal, C. DiMarino and M. Fish, "EMI Mitigation for SiC MOSFET Power Modules Using Integrated Common-Mode Screen," in IEEE Open Journal of Power Electronics, vol. 4, pp. 873-886, 2023.

[8] A. P. Camacho, V. Sala, H. Ghorbani and J. L. R. Martinez, "A Novel Active Gate Driver for Improving SiC MOSFET Switching Trajectory," in IEEE Transactions on Industrial Electronics, vol. 64, no. 11, pp. 9032-9042, Nov. 2017.

[9] G. Engelmann, T. Senoner and R. W. De Doncker, "Experimental investigation on the transient switching behavior of SiC MOSFETs using a stage-wise gate driver," in CPSS Transactions on Power Electronics and Applications, vol. 3, no. 1, pp. 77-87, March 2018.

[10] Y. Teng, Q. Gao, Q. Zhang, J. Kou and D. Xu, "A Variable Gate Resistance SiC MOSFET Drive Circuit," IECON 2020 The 46th Annual Conference of the IEEE Industrial Electronics Society, Singapore, 2020, pp. 2683-2688.

[11] Z. Zhang, W. Eberle, P. Lin, Y. -F. Liu and P. C. Sen, "A 1-MHz High-Efficiency 12-V Buck Voltage Regulator With a New Current-Source

[12] S. Zhao, X. Zhao, Y. Wei, Y. Zhao and H. A. Mantooth, "A Review of Switching Slew Rate Control for Silicon Carbide Devices Using Active Gate Drivers," in IEEE Journal of Emerging and Selected Topics in Power Electronics, vol. 9, no. 4, pp. 4096-4114, Aug. 2021.

[13] Y. Sukhatme, V. K. Miryala, P. Ganesan and K. Hatua, "Digitally Controlled Gate Current Source-Based Active Gate Driver for Silicon Carbide MOSFETs," in IEEE Transactions on Industrial Electronics, vol. 67, no. 12, pp. 10121-10133, Dec. 2020.

[14] N. Idir, R. Bausiere and J. J. Franchaud, "Active gate voltage control of turn-on di/dt and turn-off dv/dt in insulated gate transistors," in IEEE Transactions on Power Electronics, vol. 21, no. 4, pp. 849-855, July 2006.

[15] S. Zhao, X. Zhao, A. Dearien, Y. Wu, Y. Zhao and H. A. Mantooth, "An Intelligent Versatile Model-Based Trajectory-Optimized Active Gate Driver for Silicon Carbide Devices," in IEEE Journal of Emerging and Selected Topics in Power Electronics, vol. 8, no. 1, pp. 429-441, March 2020.

[16] X. Li, X. Ding and Y. Yang, "A SiC MOSFETs Switching Trajectory Optimization Strategy Based on an Active Gate Drive Circuit," 2023 26th International Conference on Electrical Machines and Systems (ICEMS), Zhuhai, China, 2023, pp. 5119-5125.

[17] H. B. Ekren, D. A. Philipps, G. Lyng Rødal and D. Peftitsis, "Four Level Voltage Active Gate Driver for Loss and Slope Control in SiC MOSFETs," 2022 IEEE 13th International Symposium on Power Electronics for Distributed Generation Systems (PEDG), Kiel, Germany, 2022, pp. 1-6.

Gate Driver," in IEEE Transactions on Power Electronics, vol. 23, no. 6, pp. 2817-2827, Nov. 2008.

979-8-3315-1110-4/25 $31.00 © 2025 IEEE

An Active Clamped Resonant Ultra-High Frequency Quasi-Square Wave Gate Driver for SiC MOSFET

Zhiqing Liang
SKL-EQCEP
Hunan University
Changsha,China
lzq@hnu.edu.cn

Zhixing He*
SKL-EQCEP
Hunan University
Changsha,China
hezhixing@hnu.edu.cn

Haoyi Sheng
SKL-EQCEP
Hunan University
Changsha,China
ShengHaoyi@hnu.edu.cn

Renfeng Guan
SKL-EQCEP
Hunan University
Changsha,China
grf@hnu.edu.cn

Zhenyuan Ou
SKL-EQCEP
Hunan University
Changsha,China
ouzhenyuan@hnu.edu.cn

Yang Liu
SKL-EQCEP
Hunan University
Changsha,China
liuyang24@hnu.edu.cn

Zongjian Li
SKL-EQCEP
Hunan University
Changsha,China
lzjq1@hnu.edu.cn

Jun Wang
SKL-EQCEP
Hunan University
Changsha,China
junwang@hnu.edu.cn

Abstract—The SiC MOSFETs are hard to be driven at MHz due to their large gate capacitances, this paper presents an active-clamped resonant high-frequency quasi-square-wave driver circuit which can drive SiC MOSFETs in MHz switching power amplifiers. The proposed driver can operate over a wide frequency range and duty cycle range as its parameters do not need to be modified according to resonant switching frequency. And it has ns-level rise/decline time with the use of high frequency resonant network, which means no deviation in the output duty cycle. Meanwhile, the clamp structure of the proposed circuit can reduce the sensitivity of the high-frequency driver to parasitic parameters, and has a strong anti-interference capability. Finally, a prototype of the proposed active clamped resonant gate drive (RGD) and a Class-E RGD were built and applied to drive a 13.56MHz/1.5kW Class-E power amplifier for comparison. The experimental results strongly demonstrate the advantages of the proposed driver. The proposed driver can drive SiC MOSFET in quasi-square-wave over 6.78~20.12 MHz, with a duty cycle of 35%~65% and t_r/t_f about 3.9~6.2 ns. And it improves the efficiency of the Class-E power amplifiers from 84% to 87% comparing with the Class-E RGD.

Keywords—Ultra-high frequency gate driver, SiC MOSFET, MHz, Power amplifier

I. INTRODUCTION

With the rapid development of power electronics technology and wide-bandgap semiconductor devices, the demand for high power density, high efficiency and miniaturization is growing. The ultra-high frequency (UHF) power amplifier from several MHz to tens of MHz is gradually becoming a research hotspot and showing great potential for application in many fields, such as plasma etching which require a specific frequency (13.56 MHz), induction heating, radio frequency energy transfer and wireless energy transfer, etc. In the design of high-power UHF circuits, voltage stress and device heat dissipation level are often the main factor limiting the power capacity. Thanks to improvements in manufacturing process, SiC MOSFETs now offer usable UHF performance. They are becoming a promising high-power, high-frequency device thanks to their superior voltage rating and heat dissipation capability compared to GaN switching devices. Using SiC MOSFET in MHz switching power amplifiers is a highly feasible way to increase the power level of individual machines. However, SiC MOSFET have higher requirements for driver circuits for its larger gate capacitance and higher drive voltage, which brings higher gate losses.

In order to minimize the gate loss, resonant gate drive (RGD) technology has been widely used. RGD can be categorized into single-switch drive[4-10] and bridge drive[14-17] according to the circuit structure, and sinusoidal wave drive (SWD), trapezoidal wave drive (TWD) and quasi-square wave drive (QSWD) according to the output waveform. The more widely used single-switch RGD is the Class-E RGD[5-9], which can output sinusoidal waveform or rectified trapezoidal waveform with a fixed duty cycle. The disadvantages are slow switching speed, circuit parameters need to be tuned according to the operating frequency, high parameter-frequency coupling, impedance matching, and a narrow operating frequency range. In order to avoid gate impedance matching and rectification, [1] proposes a Class-Φ2 RGD to take advantage of low switching stress and use the device terminal voltage as the output, which can generate a trapezoidal wave drive with a fixed duty cycle of 20 MHz, but the amplitude of the drive voltage is sensitive to parasitic parameters. In order to realize the adjustable duty cycle, a RGD based on an adjustable active clamp source that can output a trapezoidal wave drive is proposed in [10]. By varying the voltage difference, an adjustable duty cycle at 12MHz-15MHz is achieved, but it requires dual power supplies and output rectification.

The earlier application of bridge RGD is the Inductive half-bridge RGD[14], which can output sinusoidal drive. The advantage is the simplicity of the circuit, and the disadvantages are the narrow range of operating frequency, fixed duty cycle, slow driving speed, and high correlation between the driving voltage amplitude and L and C need to be precisely designed. By adding a resonant cavity, Multi-resonant RGD with totem-pole and Frequency doubler RGD with quasi-square and sinusoidal waveforms are introduced in [15-16], respectively, which realize faster driving speed and lower loss of internal buffer stage. The disadvantage is that it can only operate at a fixed switching frequency and duty cycle and the parameters are complicated to design. By adding a switch, [17] introduced a bridge quasi-square wave RGD with a bidirectional resonant path. It can work in a wide frequency range and has a wide duty cycle adjustment capability, fast

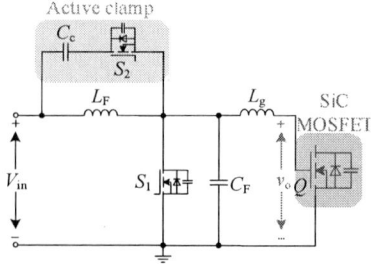

Fig. 1. The proposed gate-drive circuit.

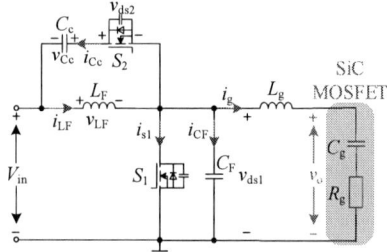

Fig. 2. Equivalent circuit of the proposed gate driver.

driving speed, but the driving signal logic requirements are more stringent, and the control is hard to realize at MHz high frequency.

In summary, the existing single switch RGDs and bridge RGDs are difficult to combine the ability of wide operating frequency range, wide modulation of duty cycle, fast driving speed and simple control realization at the same time under tens of MHz operating frequency. There is still room for improvement of high-performance ultra-high frequency SiC MOSFET driving circuit. Therefore, this paper proposes an active clamped resonant driver circuit with a wide range of quasi-sinusoidal waveforms with variable frequency and duty cycle output for tens of MHz.

The rest of the paper is organized as follows. Section II discusses the working principle of the proposed active clamp driver. Section III analyzes the steady state relational characteristics and the dynamic switching process. Section IV gives experimental validation and comparison experiments between the proposed driver and an existing driver used to drive a Class E power amplifier. Finally, brief conclusions are drawn in Section V.

II. TOPOLOGY AND OPERATION

A. Circuit Configuration and Description

An active clamped SiC MOSFET high-frequency resonant driver circuit based on GaN HEMTs is proposed, as shown in Fig.1. The proposed active clamp driving circuit consists of a main switch S_1, an input inductor L_F, a shunt resonant capacitor C_F, and an active clamp circuit consisting of an auxiliary switch S_2 and a clamp capacitor C_c. Q is the target switching of the proposed driver. S_1 and S_2 conduct alternately to produce a quasi-square-wave driving output at the two ends of S_1.

The target SiC MOSFET is equivalent to the gate capacitance C_g and gate resistance R_g in series. The gate inductance L_g is the non-negligible gate line inductance at UHF. Therefore, the circuit structure can be equivalent to Fig.2, where the marking is the reference positive direction.

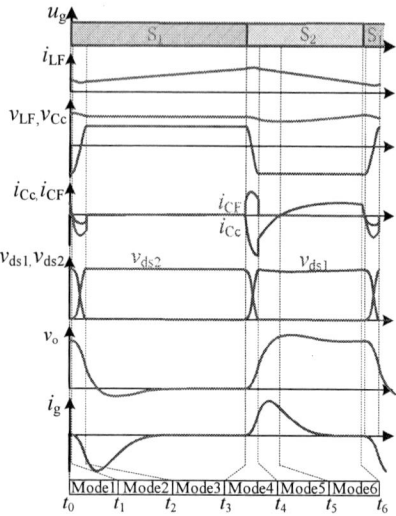

Fig. 3. The main working waveforms.

B. Operation Principles

The operating waveform of the proposed gate driver is shown in Fig.3. The operation of the proposed circuit can be divided into six sub-modes in a single operating cycle, according to the changing state of the switch and the changing direction of current flow. The current loop in each mode is given in Fig.4, which identifies the charging and discharging of C_c, C_F and the switch output capacitors C_{S1} and C_{S2}, with green indicating charging and blue indicating discharging.

Mode1[$t_0 \sim t_1$, Fig.4(a)]: At t_0, S_1 is turned on and S_2 is turned off. However, since i_{Cc} is still positive at this time, C_{s2} does not start charging immediately but keeps the body diode of S_2 on first. v_{ds2} keeps unchanged, C_c continues to charge, and the output voltage v_o is still clamped at $V_{in}+v_{Cc}$. At this stage, C_F and C_{s1} are also added to the resonant network to participate in the resonance until i_{Cc} is resonated to 0, and the body diode of S_2 cuts off to enter the next mode.

Mode2[$t_1 \sim t_2$, Fig.4(b)]: At t_1, i_{Cc} resonates through zero, and S_2 cuts off. C_{s2} starts charging, and C_{s1} and C_F discharge. At this mode, V_{in}, L_F, C_c and C_F provide energy for the resonant network composed of C_{s2} and load gate RLC. a small capacitance C_{s2} is equivalent to a small capacitance in series with C_c, so that at this time, the resonant network has a very high resonant frequency, and v_{ds2} rises rapidly, while v_{ds1} and v_o fall rapidly. Until v_{ds1} falls to zero at t_2, into the next mode.

Mode3[$t_2 \sim t_3$, Fig.4(c)]: At t_2, v_{ds1} drops to zero and is clamped at $v_{ds1}=0$ by the body diode of S_1, and v_{ds2} is clamped at $V_{in}+v_{Cc}$ to hold. C_c is isolated outside the current loop neither charging nor discharging, and v_{LF} is clamped at V_{in}. V_{in} is charged to the L_F through S_1, and the current i_{LF} rises linearly, and the increase of i_{LF} at this stage is:

$$\Delta i_{LF(+)} = \frac{1}{L_F}\int_{t_2}^{t_3} V_{in} dt \quad (1)$$

Mode4[$t_3 \sim t_4$, Fig.4(d)]: At t_3, S_1 is turned off and S_2 is turned on. C_{s1} and C_F are charged and C_{s2} is discharged. The resonant network composition in this mode is the same as that of Mode2, with v_{ds1} and v_o rising rapidly and v_{ds2} falling rapidly. Until v_{ds2} drops to 0 and is clamped at zero by the body

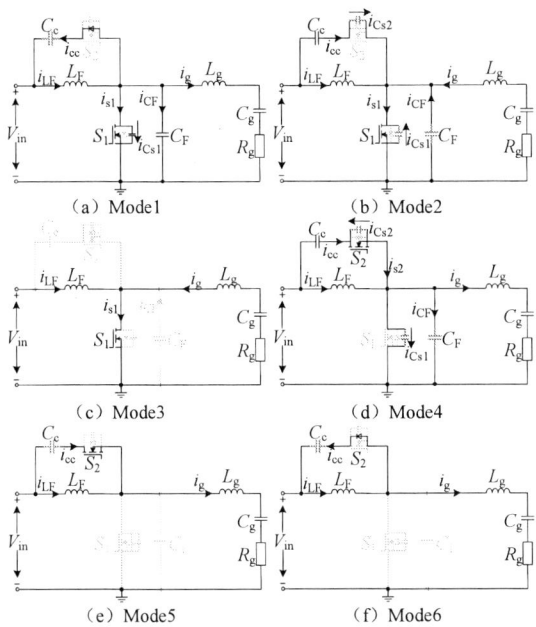

Fig. 4. Equivalent circuit of each mode.

diode of S_2 at the moment t_4, entering the next mode.

Mode5[$t_4 \sim t_5$, Fig.4(e)]: At t_4, v_{ds2} decreases to 0 and the body diode of S_2 conducts. C_c and L_F discharges together with V_{in} energize the gate *RLC* network. Due to the clamping effect of C_c, v_{Cc} decreases only slightly and v_{ds1} is always clamped at $V_{in}+v_{Cc}$.

Mode6[$t_5 \sim t_6$, Fig.4(f)]: At t_5, the i_{Cc} resonates through zero, and S_2 changes from body diode conduction to MOS switch conduction. The resonant network composition in this mode is the same as in Mode5, except that C_c becomes charged, v_{Cc} is slightly elevated, and v_{ds1} is still clamped at $V_{in}+v_{Cc}$. Both in this mode and Mode5, v_{LF} is -v_{Cc}, and i_{LF} decreases by the amount of i_{LF} reduction:

$$\Delta i_{LF(-)} = \frac{1}{L_F} \int_{t_4}^{t_6} v_{Cc}(t) dt \qquad (2)$$

III. CHARACTERIZATION AND ANALYSIS OF THE DRIVER

The proposed circuit is formed by adding an active clamp circuit to a single switch Class E RGD. The high voltage stress of the Class E circuit is clamp-limited and used to drive the target switching device.

The clamping effect eliminates the need to adjust the circuit parameters according to frequency. This results in a large operating bandwidth and reduced sensitivity to parasitic parameters, making the device highly versatile. The inclusion of S_2 forms a half-bridge structure. The large clamp capacitor and the power supply work together to provide voltage to both ends of the bridge arm, making it easy to adjust the duty cycle and the drive voltage amplitude. The double switch uses a complementary mode of operation. The presence of input inductance at the midpoint of the bridge arm without the risk of straight-through and acceptable with deadband operation, making the control simple to implement. The large clamp capacitor in series with S_2 and L_F in parallel increases the resonance frequency of the SiC MOSFET gate charging and

discharging process, resulting in faster driving speeds.

A. Steady-state analysis

Assumptions: (1)The clamp capacitance C_c is so large that the clamp capacitance voltage can be regarded as a constant value V_{Cc}. (2)The input inductance L_F is large enough to make i_{LF} constant positive.The duration of the charging and discharging processes in Mode1, Mode2 and Mode4 is much less than T_s. (3)As the duration of Mode1, Mode2 and Mode4 is extremely short, these three modes can be ignored in the analysis of the steady-state relationship. In this case, the circuit is considered to have only two operating states, Mode3 (S_1ON, S_2OFF) and Modes5 and 6 (S_1OFF, S_2ON).

Once the circuit has reached a steady state, the change in inductor current i_{LF} over one cycle is balanced. According to the volt-second balance principle of the input inductor L_F, the steady-state voltage value of the clamp capacitor C_c is approximately known to be:

$$V_{Cc} = \frac{V_{in}D_1}{(1-D_1)} \qquad (3)$$

Then, the amplitude V_g and the duty cycle D_g of the output drive voltage v_o exist in relation to the input voltage V_{in} and the D_1 of the S_1 duty cycle:

$$V_g = V_{Cc} + V_{in} = \frac{V_{in}}{(1-D_1)} \qquad (4)$$

$$D_g = 1 - D_1 \qquad (5)$$

From the power conservation it can be concluded that the average value of the current flowing through the input inductor L_F is:

$$I_L = \frac{f_s Q_g V_g}{V_{in}} = \frac{f_s Q_g}{1-D_1} \qquad (6)$$

The maximum and minimum value of the input inductor current i_{LF} is:

$$I_{L\max} = \frac{f_s Q_g}{1-D_1} + \frac{V_{in}}{2L_F}D_1 T_s \qquad (7)$$

$$I_{L\min} = \frac{f_s Q_g}{1-D_1} - \frac{V_{in}}{2L_F}D_1 T_s \qquad (8)$$

Then, the input inductance L_F should be taken to satisfy:

$$L_F > \frac{(1-D_1)^2}{\omega^2 C_g^{\ 2} R_g} D_1 T_s \qquad (9)$$

To ensure that the input inductor current is always positive, the actual value of L_F is recommended to be 4~8 times of the calculated value.

B. Dynamic switching time analysis

The rising process of the driving voltage v_o occurs in Mode4, and the corresponding equivalent circuit of the turn-on process is shown in Fig.5. S_1 turn-off is regarded as a disconnection, and S_2 is equivalent to L_{s2} in series with r_{s2}. C_{F1}

in Fig.5 is the parallel value of C_F and C_{s1}, i.e..

$$C_{F1}=C_F+C_{s1} \qquad (10)$$

Similarly the equivalent circuit of the turn-off process corresponding to the falling process Mode2 of the drive voltage v_o is shown in Fig.5.

To analyze the rise time t_r and the fall time t_f uniformly, Fig.5(a) and Fig.5(b) can be represented uniformly as Fig.5(c). Note that the current directions in the figures are all in the positive reference direction. The drive turn-on process corresponds to $i_{s1}\equiv0$, and the drive turn-off process corresponds to $i_{s2}\equiv0$. According to the theorems of KVL and KCL, the equations of Fig.5(c) can be written and organized to obtain (11).

$$\begin{cases} \dfrac{d}{dt}v_{ds1}(t)=\dfrac{i_{CF1}(t)}{C_{F1}} \\[2mm] \dfrac{d}{dt}v_{ds2}(t)=\dfrac{i_{Cs2}(t)}{C_{s2}} \\[2mm] \dfrac{d}{dt}v_{LF}(t)=-\dfrac{i_{CF1}(t)}{C_{F1}} \\[2mm] \dfrac{d}{dt}v_{Lg}(t)=\dfrac{i_{CF1}(t)}{C_{F1}}-\dfrac{R_g}{L_g}v_{Lg}(t)-\dfrac{1}{C_g}i_g(t) \\[2mm] \dfrac{d}{dt}i_g(t)=\dfrac{v_{Lg}(t)}{L_g} \\[2mm] \dfrac{d}{dt}i_{LF}(t)=\dfrac{v_{LF}(t)}{L_F} \\[2mm] \dfrac{d}{dt}i_{s1}(t)=\dfrac{v_{ds1}(t)}{L_{s1}}-\dfrac{r_{s1}}{L_{s1}}i_{s1}(t) \\[2mm] \dfrac{d}{dt}i_{s2}(t)=\dfrac{v_{ds2}(t)}{L_{s2}}-\dfrac{r_{s2}}{L_{s2}}i_{s2}(t) \\[2mm] \dfrac{d}{dt}i_{CF1}(t)=(\dfrac{v_{LF}(t)}{L_F}+\dfrac{d}{dt}i_{s2}(t)-\dfrac{d}{dt}i_{s1}(t)-\dfrac{v_{Lg}(t)}{L_g})/\ (1+\dfrac{C_{s2}}{C_{F1}}) \\[2mm] \dfrac{d}{dt}i_{Cc}(t)=(\dfrac{d}{dt}i_{s1}(t)+\dfrac{v_{Lg}(t)}{L_g}-\dfrac{v_{LF}(t)}{L_F}-\dfrac{d}{dt}i_{s2}(t))/(1+\dfrac{C_{F1}}{C_{s2}}) \end{cases} \quad (11)$$

The initial values of the corresponding switching processes for (11) set of differential equations are respectively:

$$t_r: \begin{cases} v_{ds1}(0)=0 \\ v_{ds2}(0)=V_{in}+V_{Cc} \\ v_{LF}(0)=V_{in} \\ v_{Lg}(0)=0 \\ i_g(0)=0 \\ i_{LF}(0)=I_{Lmax} \\ i_{s2}(0)=0 \\ i_{CF1}(0)=-C_{s2}I_{Lmax}/(C_{F1}+C_{s2}) \\ i_{Cc}(0)=C_{F1}I_{Lmax}/(C_{F1}+C_{s2}) \end{cases} \quad t_f: \begin{cases} v_{ds1}(0)=V_{in}+V_{Cc} \\ v_{ds2}(0)=0 \\ v_{LF}(0)=-V_{Cc} \\ v_{Lg}(0)=0 \\ i_g(0)=0 \\ i_{LF}(0)=I_{Lmin} \\ i_{s1}(0)=I_{Lmin} \\ i_{Cc}(0)=0 \\ i_{CF1}(0)=0 \end{cases} \quad (12)$$

According to (11) and (12), the fall time t_f and rise time t_r of the drive voltage satisfying (13) can be solved.

$$\begin{cases} v_{ds1}(t_r)-v_{Lg}(t_r)=V_{in}+V_{Cc} \\ v_{ds1}(t_f)-v_{Lg}(t_f)=0 \end{cases} \qquad (13)$$

Once the switching devices S and Q are selected, the values of L_s, r_s, C_g, and R_g are determined. By varying the

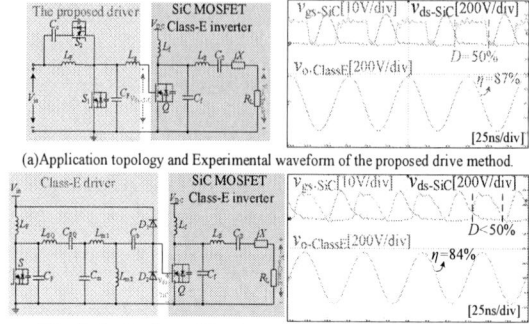

(a)Application topology and Experimental waveform of the proposed drive method.

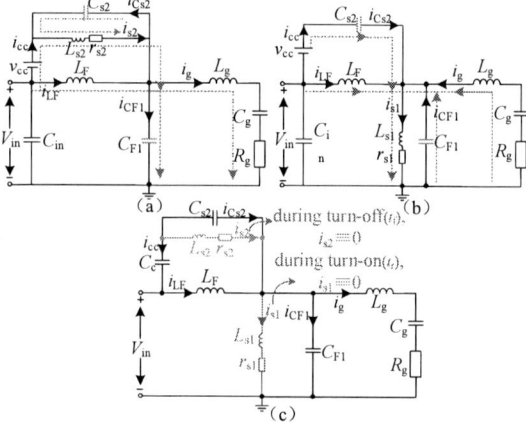

(b)Application topology and Experimental waveform of Class-E drive method.

Fig. 10. Topology and experimental waveforms of the proposed driver and the driver in [5] driving the 13.56MHz Class E power amplifier.

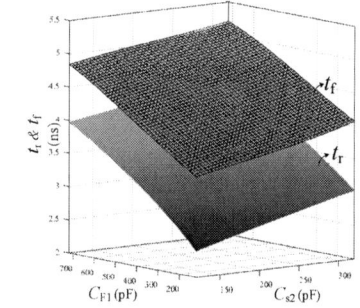

Fig. 5. Dynamic equivalent circuit. (a)Turn-on process(t_r), (b)Turn-on process(t_f), (c)Unified analysis circuits (t_r&t_f).

Fig. 6. Switching time variation curve with C_{F1}, C_{s2}.

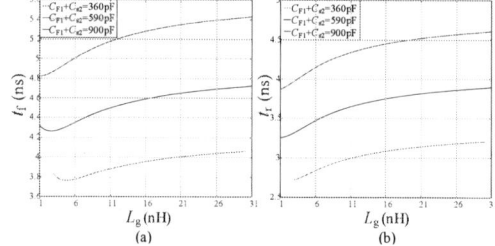

Fig. 7. Switching time versus gate inductance L_g for different capacitance values. (a)Fall time. (b)Rise time.

values of the shunt capacitor C_{F1}, the switching output capacitance C_{s2}, and the gate inductance L_g, the curves of the switching time versus C_{F1}, C_{s2}, and L_g are obtained as

Fig. 8. Experimental platform of the proposed driver.

(D_1=35%, 50% and 65%) at operating frequencies of 6.78MHz, 13.56MHz and 20.12MHz. The test results are shown in Fig.9.

As shown in Fig.9, the proposed driver circuit can output quasi-square-wave drive signal at frequencies from 6.78MHz to 20.12MHz. The output drive duty cycle D_g is consistent with 1-D_1. Experiments were conducted to change the duty

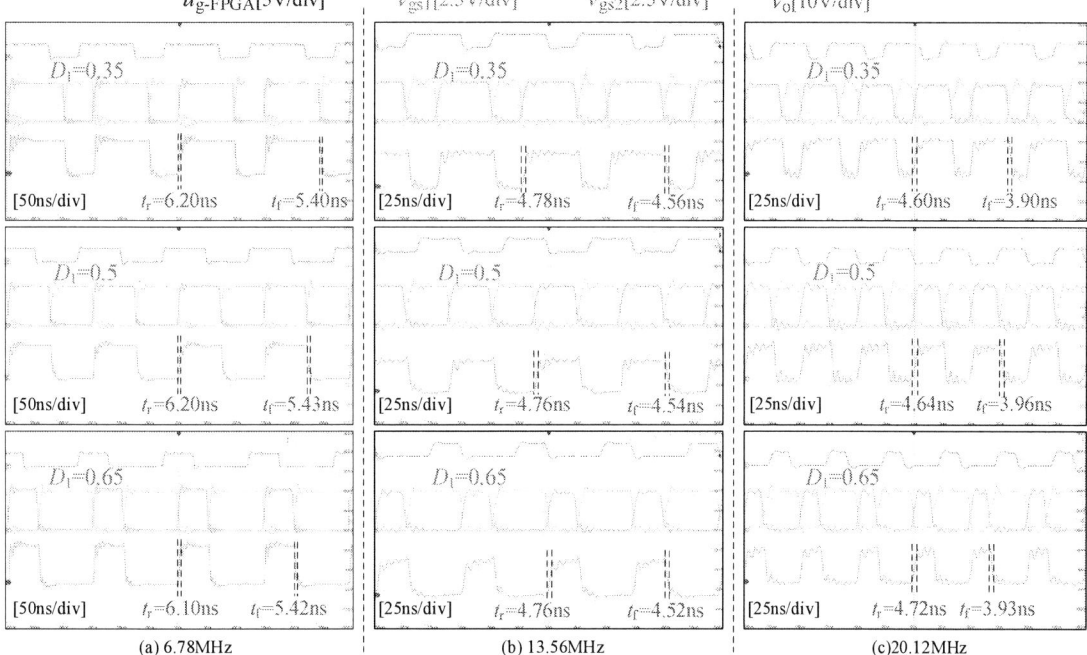

Fig. 9. Experimental waveforms of the proposed gate driver when the duty cycle of S_1 is 35%, 50% and 65% under different switching frequencies of 6.78MHz,13.56MHz and 20.12MHz.

shown in Fig.6 and Fig.7.As shown in Fig.6 and Fig.7, there is a clear positive relationship between the t_r and t_f of the drive voltage v_o and both C_{F1} and C_{s2}, as well as the gate inductance L_g. This means that by reducing either the gate inductance L_g or the capacitance sum of C_{F1} and C_{s2}, we can both shorten the resonance period and increase the switching speed. However, since the PCB line inductance and the MOSFET package inductance are comprised in L_g. They are the main cause of overshoot voltage and gate oscillation, which can easily damage the switching components. Therefore, the gate loop should be optimized for the PCB layout to be as short as possible, reducing the impact of the gate inductance L_g. The rise and fall time of the drive voltage v_o are generally only adjusted by changing the total capacitance value ($C_{F1}+C_{s2}$). It is generally believed that the rise and fall time of the drive voltage v_o should be less than $0.2T_s$. However, longer or shorter switching time at MHz switching frequency will lead to a significant increase in the drive power consumption. Therefore, the switching speed and the drive power consumption need to be balanced according to the actual requirements of the design.

IV. EPERIMENTAL

To demonstrate the practicality of the proposed active clamp driver circuit, an experimental prototype was established, and the experimental platform is shown in Fig.8. The main parameters of the prototype are shown in Table I. The prototype was tested under three operating conditions

cycle of v_{gs1} and v_{gs2} by adjusting the duty cycle of the FPGA control signal. This enabled the drive duty cycle D_g to be adjusted within the range of 35% to 65%. The proposed driver has a fast switching speed, t_r and t_f are less than $0.1T_s$ ($\approx$ 3.9~5.6ns). In the case of high dv/dt brought about by high switching speed, the clamping effect also restrained the oscillation of the output driving voltage v_o to a certain extent. It is also optional to increase the capacitance of C_{F1} and C_{s2} and to further enhance the clamping effect of the circuit by sacrificing a certain switching speed in order to better suppress oscillations.

TABLE I. COMPONENTS OF THE DRIVER PROTOTYPE

Symbol	Implementation	
	Description	*Value*
f_s	Switching frequency	6.78MHz~20.12MHz
D_1	Duty cycle of S_1	35%, 50%, 65%
$S_1\&S_2$	Main & auxiliary switch	GS61004B
driver IC	Driver chip for $S_1\&S_2$	LMG1020
C_{in}	Input capacitance	24.5μF
C_c	Clamp capacitance	30μF
L_F	Input inductance	500nH
C_F'	Parallel resonant capacitor	188pF
L_g	Gate inductance	15nH
Q	Target switch	AIMZH120R020M1T (C_g=2667pF,R_g=2.2Ω)

In order to further show the advantages of the proposed driver circuit, Table II lists the comparison between the proposed driver circuit and some other UHF driver circuits. In addition, a comparative experiment was conducted by applying the Class E RGD in [5-9] and the proposed driver to drive the same 13.56MHz/1.5kW Class E power amplifier. The comparative topology and experimental waveforms are shown in Fig.10. As we know, Class E power amplifiers are generally recommended to work at the optimal operating point of 50% duty cycle. However, since the circuit parameters of the commonly used Class E RGD (Fig. 10(c)) are strongly coupled to the frequency. It requires precise impedance matching, otherwise it is easy to deviate from the rated operating point during operation, and its driving voltage is a sinusoidal waveform obtained by diode rectification, which can result in output duty cycle deviation and slow switching speed. From Fig. 10(d), the driving voltage of Class E RGD output rises and falls slowly, and the duty cycle deviates from 50%.From Fig. 10(b) and (d), it can be seen that the proposed driver circuit has a faster switching speed and a duty cycle that is more in line with the 50% theoretical value of the optimal duty cycle for Class E. This allows the efficiency of the class-E power amplifier to be increased from 84% to 87% with the proposed driver. The advantages of the proposed driver are further verified through comparative experiments.

TABLE II.   COMPARISON OF THE PROPOSED RGD WITH OTHER RGDS

Features	Circuit				
	[5]	*[4]*	*[10]*	*[15]*	*This*
Structure	Class-E RGD	Class-Φ2 RGD	Tunable Class-E RGD	Frequency doubler RGD	Active clamped RGD
Num.of switches/ diodes	1 2	1 0	1 2	2 0	2 0
Num.of passive devices	8	3	3	3	3
Frequency （MHz）	12.56-14.56	20	12-15	13.56	**6.78-20.12**
Parameter-to-f_s decoupling	Weak	Weak	Weak	Weak	**Strong**
Duty ratio adjustment	×	×	√	×	√
Drive Waveform	TWD fast	QSWD faster	TWD fast	SWD Slow	**QSWD fastest**

V. CONCLUSION

This paper presents a high-speed active clamped quasi-square wave driver circuit for driving SiC MOSFET operating at several MHz to tens of MHz. It is features with high operating frequency with wide range, flexible and controllable duty cycle, ns-level rise/fall time, etc. The switching speed can also be fine-tuned by adjusting the capacitance value. In addition, thanks to the clamping effect that reduces the coupling between the driving waveform and the parasitic parameters, the circuit has good versatility and anti-interference performance. This paper analyzed the operating modes, steady state relationship and dynamic switching process of the proposed circuit, and verified the feasibility and correctness of the analysis of the proposed active clamp driver circuit based on the experimental platform.

REFERENCES

[1] Guo D. Topology and regulation method for optimizing the efficacy of Class E wireless energy transmission [D]. Xiamen University,2022.

[2] Ziyue Dang. High power density silicon carbide drive technology based on gallium nitride [D]. Huazhong University of Science and Technology,2021.

[3] Liu P. Isolated resonant driver circuit design for SiC MOSFETs [D]. Southeast University,2017.

[4] H. Jedi, T. Salvatierra, A. Ayachit and M. K. Kazimierczuk, "High-Frequency Single-Switch ZVS Gate Driver Based on a Class Φ2 Resonant Inverter," in *IEEE Trans. Ind. Electron.*, vol. 67, no. 6, pp. 4527-4535.

[5] J. Xu, Z. Tong and J. Rivas-Davila, "1 kW MHz Wideband Class E Power Amplifier," *2021 IEEE 22nd Workshop on Control and Modelling of Power Electronics (COMPEL)*, Cartagena, Colombia, 2021, pp. 1-6.

[6] C. Chen, H. Wang, G. Ning, X. Chen and M. Su, "Cascaded Resonant RF Drive Design and Soft-Switching Optimization of High-Power Class-E Power Amplifiers," *2022 4th International Conference on Smart Power & Internet Energy Systems (SPIES)*, Beijing, China, 2022, pp. 1539-1543.

[7] H. Wang *et al.*, "Design Method of Impedance Matching Network for High Power Cascaded Class-E Inverter," in *IEEE J. Emerging Sel. Top. Power Electron*, vol. 6, no. 1, pp. 327-337, Jan. 2025.

[8] H. Yogi, X. Wei, H. Sekiya and T. Hikihara, "Design of 6.78 MHz SiC MOSFET Class-E Inverter with a Class-Φ High-Speed Driver," *2019 IEEE Energy Conversion Congress and Exposition (ECCE)*, Baltimore, MD, USA, 2019: 375-379.

[9] Y. Sun, R. Sugano, X. Wei, T. Hikihara and H. Sekiya, "High-speed driver for SiC MOSFET based on class-E inverter," *2017 IEEE International Symposium on Circuits and Systems (ISCAS)*, Baltimore, MD, USA, 2017, pp. 1-4 .

[10] W. Liu, Y. Zhu and M. Liu, "A Tunable Active Clamping Source Based High Frequency Single-Switch Resonant Gate Driving Circuit with Duty Cycle and Frequency Modulation Capability," *2024 IEEE Wireless Power Technology Conference and Expo (WPTCE)*, Kyoto, Japan, 2024, pp. 23-27.

[11] Z. Ye, Z. Tong, L. Gu and J. Rivas-Davila, "A High Frequency Resonant Gate Driver for SiC MOSFETs," *2021 IEEE 22nd Workshop on Control and Modelling of Power Electronics (COMPEL)*, Cartagena, Colombia, 2021, pp. 1-5.

[12] ZHAO Qinglin, CHEN Lei, YUAN Jing, et al. A resonant drive circuit for GaN devices[J]. Power Automation Devices, 2019, 39(04): 114-118.

[13] Hao Peng. Research on high-performance resonant drive of silicon carbide MOSFET [D]. Huazhong University of Science and Technology,2021.

[14] I. D. de Vries, "A resonant power MOSFET/ IGBT gate driver," *APEC. Seventeenth Annual IEEE Applied Power Electronics Conference and Exposition*, Dallas, TX, USA, 2002, pp. 179-185.

[15] F. Hattori *et al.*, "Frequency Doubler Gate Drive Circuit Suitable for High-Frequency Applications," in *IEEE J. Emerging Sel. Top. Power Electron*, vol. 10, no. 1, pp. 617-631, Feb. 2022.

[16] L. Gu, Z. Tong, W. Liang and J. Rivas-Davila, "A Multiresonant Gate Driver for High-Frequency Resonant Converters," in *IEEE Trans. Ind. Electron.*, vol. 67, no. 2, pp. 1405-1414, Feb. 2020.

[17] N. Teerakawanich and C. M. Johnson, "A new resonant gate driver with bipolar gate voltage and gate energy recovery," *2013 Twenty-Eighth Annual IEEE Applied Power Electronics Conference and Exposition (APEC)*, Long Beach, CA, USA, 2013, pp. 2424-2428.

[18] J. V. P. S. Chennu, R. Maheshwari and H. Li, "New Resonant Gate Driver Circuit for High-Frequency Application of Silicon Carbide MOSFETs," in *IEEE Trans. Ind. Electron.*, vol. 64, no. 10, pp. 8277-8287, Oct. 2017.

[19] H. Fujita, "A resonant gate-drive circuit capable of high-frequency and high-efficiency operation," *2009 IEEE 6th International Power Electronics and Motion Control Conference*, Wuhan, China, 2009, pp. 351-357.

[20] F. Hattori, H. Umegami, and M. Yamamoto, "Multi-resonant gate drive circuit of isolating-gate GaN HEMTs for tens of MHz," *IET Circuits, Devices Syst.*, vol. 11, no. 3, pp. 261–266, Jun. 2017.

48V-to-0.9V Voltage Regulator with GaN Devices and Integrated Magnetic Design

Zikang Li
*School of Electrical
Engineering and Automation
Harbin Institute of
Technology,*
Harbin, China
24s006041@stu.hit.edu.cn

Jingyang Tan
*School of Electrical
Engineering and Automation
Harbin Institute of
Technology,*
Harbin, China
23s006036@stu.hit.edu.cn

Yijie Wang
*School of Electrical
Engineering and Automation
Harbin Institute of
Technology,*
Harbin, China
wangyijie@hit.edu.cn

Dianguo Xu
*School of Electrical
Engineering and Automation
Harbin Institute of
Technology,*
Harbin, China
xudiang@hit.edu.cn

Abstract—In recent years, the development of artificial intelligence (AI) has been increasingly rapid, posing a huge challenge to the power supply of data centers. Enhancing key performance metrics such as energy conversion efficiency, power density and control bandwidth constitutes pivotal research directions for next-generation voltage regulator modules (VRMs). GaN devices have smaller sizes and higher power densities, making it possible to create more compact power electronics systems. A GaN-based, low-voltage and high-current converter of a multi-phase interleaved current-doubler rectifier is proposed in this article, with an input voltage of 48 V, an output voltage of 0.9 V, and an output current of 120 A. By integrating magnetic design, the power density of the module is enhanced. The structure of the traditional dual-EI magnetic core is optimized to integrate transformers and output inductors. A comprehensive time-domain analysis method combining magnetic and electrical circuits is employed to conduct small-signal modeling of the converter. The final prototype has a current density of 0.2A/mm^2 and a peak efficiency of 89%.

Keywords—wide bandgap device, voltage regulator module (VRM), magnetically integrated design, Small-signal modeling

I. INTRODUCTION

The rapid advancement of artificial intelligence (AI) in recent years has driven exponential growth in parameter scales and computational demands of large AI models, escalating from billions to trillillions of parameters. As the core supporting component of XPU, the innovation of VRMs is not only related to the improvement of chip performance, but also directly affects the efficiency and stability of the entire AI system. Voltage regulation modules (VRMs) with high efffciency, high power density, and high control bandwidth are needed to support future microprocessors[1].

Due to their superior electrical properties, GaN-based components can be designed with smaller physical dimensions while still handling significant amounts of power. This enables the creation of more compact power electronics systems[2][3]. In the XPU power supply environment with a high step-down ratio and high-current output, the two-stage power supply solution with a 12V intermediate bus has been widely accepted and implemented. However, problems still exist, such as the intermediate bus decoupling capacitors limiting the overall efficiency and power density, and the complexity of the overall loop control. In contrast, the single-stage VRM has the advantages of high-power desimplicity and flexibility in application. VRMs based on this architecture are developing rapidly.

Low-profile direct power converter was proposed in [4][5]. Planar magnetic design adapts to the power transmission structure of vertical power supply, and the soft - switching of the clamping capacitor further improves the overall efficiency of the system. However, the output current of this structure is discontinuous, and the transformation ratio is difficult to adjust. [6] presents a Direct-Step-Down STC (DSDSTC) based on the design of resonant switched-tank converter and buck converter. By combining the two - stage structure, the number of devices is reduced, and higher power density and efficiency are achieved. However, this structure also makes its control method more complex. A resonant converter based on an autotransformer was proposed in [7][8]. While impro-ving the transformation ratio, it utilizes magnetic integration design to achieve a higher power density. However, this structure has a narrow input voltage variation range and a limited transformation ratio. The magnetic integration design based on multi-phase inductors[9] can achieve negative coupling of the output inductors, greatly reducing the volume of magnetic components. Combined with the implementation of the Trans-Inductor Voltage Regulator (TLVR), the transient response capability of the converter is further enhanced.

In this paper, a low-voltage and high-current converter based on GaN devices is proposed. The magnetic integration structure of the traditional current-doubler rectifier is improved to achieve higher efficiency and power density. Four transformers and the output inductors of each phase are integrated into a planar magnetic structure. The magnetic core contains five legs, including four cylinders around which the primary and secondary windings are wound (hereinafter referred to as winding legs), and a square column for storing magnetic field energy (hereinafter referred to as the leakage leg). And the small-signal modeling of the converter is carried out by using the comprehensive time-domain analysis method of magnetic circuit and circuit. The paper is structured as follows. Section II introduces the design of the converter topology and magnetic structure Section III obtains the small-signal model through the comprehensive time-domain analysis of the magnetic circuit and electrical circuit for the afore-mentioned converter. Section IV provides the prototype parameters and related experimental waveforms. Section V summarizes the main viewpoints of this article.

II. OPERATION MECHANISMS OF THE TOPOLOGY

The topology of the multiphase current doubler rectifier is shown in Fig. 1. This topology is composed of two identical parallel-connected parts. The input and output of the converter are isolated by transformers, with the primary side employing a half-bridge inverter circuit and the secondary side adopting a center-tapped rectifier circuit.

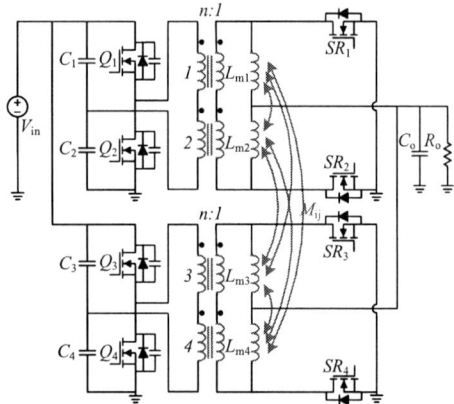

Fig. 1. Topology of the multiphase current doubler rectifier.

Fig. 2 shows the key waveforms of the multiphase current doubler rectifier. Q_x and SR_x are controlled by a set of complementary signals, and the duty cycle of Q_x should be less than 50% to avoid short-circuit faults. The outputs of each channel have a successive 90° phase delay. When Q_1 is turned on and SR_1 is turned off, current will go through the transformers and SR_2, where the current of SR_2 is the sum of the I_{Lm1} and I_{Lm2}. The voltage across L_{m2} is clamped to the output voltage V_o, and the I_{Lm1} increases. Based on the voltage variation across L_{mx}, the relationship between output and input voltage can be achieved as

$$\frac{V_o}{V_{in}} = \frac{D}{2n} \quad (1)$$

Here, n represents the turns ratio of the transformer, and D represents the duty cycle of Q_x. In order to prevent the duty cycle from being too low and the lamination of PCB windings from being overly complicated, the turns ratio of the transformer is finally chosen as 4.

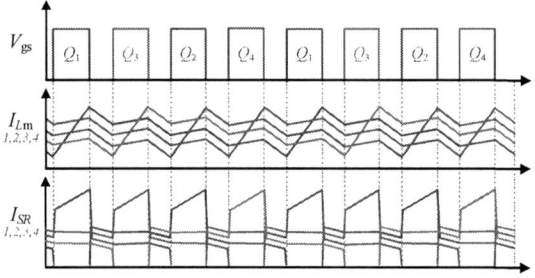

Fig. 2. Key waveforms of the multiphase current doubler rectifier.

III. SMALL-SIGNAL MODEL ASSOCIATED WITH THE MAGNETIC INTEGRATED STRUCTURE

A. Magnetic Integrated Structure of the Converter

The magnetic integrated structure of the converter is shown in Fig. 3. All the transformers and the output inductors of the converter share one magnetic core, with the excitation inductance of the transformer serving as output inductor. The magnetic core consists of four winding legs and a leakage leg.

The primary windings and the secondary windings are wound around the winding legs to fulfill the functions of the transformer. The leakage leg optimizes the magnetic flux path. Two primary windings with opposite winding directions are connected in series to achieve a 180° phase interleaving. All the secondary windings are connected in parallel to enhance the current-carrying capacity at the output terminal and possess expandability.

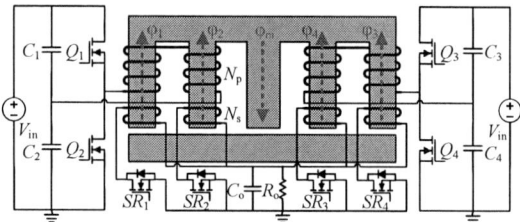

Fig. 3. Magnetic integrated structure of the converter.

All the secondary windings have the same winding method, and the four-phase output inductors achieve negative coupling. This structure can cause the magnetic fluxes going through the winding legs to cancel each other out, reducing the DC magnetic flux and lowering the saturation limit of the magnetic core. The magnetic fluxes going through the leakage leg are superimposed, and the variation of the AC magnetic flux is decreased, which is beneficial for reducing magnetic losses. Fig. 4 shows the magnetic flux density simulation results of the magnetic integrated structure of the converter. When the output current is 150A, the average magnetic flux density of the winding legs is 0.16T, and that of the leakage leg is 0.14T. The implementation of negative coupling can also reduce the transient output inductance of the converter, enabling a faster response speed when the load changes.

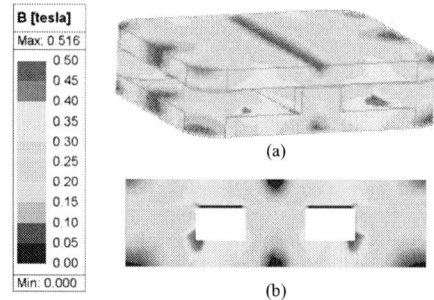

Fig. 4. Magnetic field simulation results of the integrated core. (a) 3D perspective. (b) longitudinal section perspective

The magnetic core in this design is made of ferrite material DMR51W which is suitable for high frequency. The reluctance model of the magnetic integrated structure is shown in Figure 5. Due to the high permeability of this material, most of the reluctance in the magnetic circuit is composed of air gap reluctance. Therefore, the reluctance of the magnetic material is ignored for approximate calculation during magnetic circuit analysis. When the magnetic integrated structure is implemented, the equivalent value of the single-phase magnetic reluctance is

$$R_x = R_{lg} + \left(\frac{R_{lg}}{3} // \frac{R_{lgk}}{4} // R_m // R_{mk} \right) \quad (x = 1, 2, 3, 4) \quad (2)$$

R_{lg} represents the reluctance of the air gap in the winding leg, R_m represents the reluctance of the air gap in the leakage leg, R_{lgk} and R_{mk} are the equivalent reluctance characterizing the diffused magnetic flux of the air gap. The existence of this part may cause the actual inductance value to be smaller compared to the theoretical value.

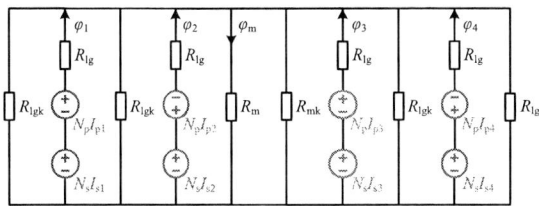

Fig. 5. Reluctance model of the magnetic integrated structure.

B. Converter Modeling with Comprehensive Time-domain Analysis of Magnetic Circuit and Electrical Circuit

This design adopts a magnetic integration structure. The traditional circuit modeling method cannot intuitively analyze the equivalent magnetic circuit model. Therefore, a comprehensive time-domain analysis method combining magnetic circuit and electric circuit is used to construct the small-signal model of the converter. During a working cycle, this topology has five main working modes. The four phases are sequentially excited by the primary voltage (Mode 1~4), and there is also a secondary freewheeling state between the two aforementioned modes (Mode 5). The magnetic flux of each phase and the output voltage under different modes are shown in Table I. Due to the split magnetic flux structure of the primary winding, the actual turns ratio of the magnetically integrated transformer during the excitation stage is N_p:0.5 N_s. V_{px} and V'_{px} are the voltages across the primary windings of each phase. The state - space equations of the converter under the average condition of a single switching cycle can be achieved as

$$
\begin{cases}
\dfrac{d\varphi_1}{dt} = \dfrac{D(V_{p1} - V'_{p2})}{N_p} - \dfrac{V_o}{N_s} \\[2mm]
\dfrac{d\varphi_2}{dt} = \dfrac{D(V'_{p1} - V_{p2})}{N_p} - \dfrac{V_o}{N_s} \\[2mm]
\dfrac{d\varphi_3}{dt} = \dfrac{D(V_{p3} - V'_{p4})}{N_p} - \dfrac{V_o}{N_s} \\[2mm]
\dfrac{d\varphi_4}{dt} = \dfrac{D(V'_{p3} - V_{p4})}{N_p} - \dfrac{V_o}{N_s} \\[2mm]
C_o\dfrac{dV_o}{dt} = i_{L\_avg} - \dfrac{V_o}{R}
\end{cases}
\tag{3}
$$

The primary windings of the converter are connected in series, thus the conditions that should be met are

$$
\begin{cases}
V_{p1} + V'_{p1} = V_{p3} + V'_{p3} = \dfrac{V_{in}}{2} \\[2mm]
V_{p2} + V'_{p2} = V_{p4} + V'_{p4} = -\dfrac{V_{in}}{2}
\end{cases}
\tag{4}
$$

Ideally, the output currents of the four phases are equal. When the magnetic circuits of the four phases are completely symmetrical (including the diffused magnetic circuits of the air gaps), except that the phases are successively staggered by 90°, the changes in the magnetic fluxes of each phase are

exactly the same, and the excitation voltages satisfy the following relationship.

$$
\begin{cases}
V_{p1} = -V_{p2} = V_{p3} = -V_{p4} \\[2mm]
V'_{p1} = -V'_{p2} = V'_{p3} = -V'_{p4}
\end{cases}
\tag{5}
$$

The output current can be expressed as

$$
i_{L\_avg} = 4i_{s\_avg} = \dfrac{4\varphi_x R_x}{N_s} \quad (x = 1, 2, 3, 4)
\tag{6}
$$

By combining (4)(5) and (6), (3) can be simplified to

$$
\begin{cases}
\dfrac{d\varphi_1}{dt} = \dfrac{DV_{in}}{2N_p} - \dfrac{V_o}{N_s} \\[2mm]
C_o\dfrac{dV_o}{dt} = \dfrac{4\varphi_1 R_1}{N_s} - \dfrac{V_o}{R}
\end{cases}
\tag{7}
$$

The steady-state operating point of the system can be achieved as

$$
\begin{cases}
V_o = \dfrac{N_s D V_{in}}{2N_p} \\[2mm]
\varphi_1 = \dfrac{N_s V_o}{4R_1 R}
\end{cases}
\tag{8}
$$

Add a small perturbation to the relevant parameters near the steady-state operating point, and further simplify it to obtain the linearized state-space equation.

$$
\begin{cases}
u_{in} = V_{in} + \hat{u}_{in} \\
d = D + \hat{d} \\
u_o = V_o + \hat{u}_o \\
\phi_1 = \varphi_1 + \hat{\phi}_1
\end{cases}
\tag{9}
$$

$$
\begin{cases}
\dfrac{d\hat{\phi}_1}{dt} = \dfrac{V_{in}}{2N_p}\hat{d} + \dfrac{D}{2N_p}\hat{u}_{in} - \dfrac{1}{2N_s}\hat{u}_o \\[2mm]
C_o\dfrac{d\hat{u}_o}{dt} = \dfrac{4R_1}{N_s}\hat{\phi}_1 - \dfrac{1}{R}\hat{u}_o
\end{cases}
\tag{10}
$$

Perform the Laplace transform on the above equation. The transfer function from the duty cycle D of the control signal to the output voltage V_o can be achieved as

$$
\begin{bmatrix} s & \dfrac{1}{N_s} \\[2mm] -\dfrac{4R_1}{N_s C_o} & s + \dfrac{1}{RC_o} \end{bmatrix}
\begin{bmatrix} \phi_1(s) \\[2mm] u_o(s) \end{bmatrix}
=
\begin{bmatrix} \dfrac{D}{2N_p} \\[2mm] 0 \end{bmatrix} u_{in}(s)
+
\begin{bmatrix} \dfrac{V_{in}}{2N_p} \\[2mm] 0 \end{bmatrix} d(s)
\tag{11}
$$

$$
L_{equ} = \dfrac{N_s^2}{4R_1}
\tag{12}
$$

$$
\dfrac{u_o(s)}{d(s)} = \dfrac{1}{C_o L_{equ} s^2 + \dfrac{L_{equ}}{R}s + 1} \dfrac{N_s}{N_p}\dfrac{V_{in}}{2}
\tag{13}
$$

The transfer function (13) of this converter is similar to that of the traditional Buck converter. The difference is that the transformer turns ratio in this topology will affect the gain.

It indicates that the essence of this topology is an isolated Buck converter. L_{equ} is the equivalent inductance, and its magnitude is affected by the specific magnetic integration structure.

TABLE I.   VARIABLES AND EXPRESSIONS IN DIFFERENT MODES

Modes	Variables and expressions				
	$d\varphi_1/dt$	$d\varphi_2/dt$	$d\varphi_3/dt$	$d\varphi_4/dt$	$C_o dV_o/dt$
Mode 1 [0~D]	$\dfrac{V_{p1}}{N_p}$	$\dfrac{V'_{p1}}{N_p}-\dfrac{2V_o}{N_s}$	$-\dfrac{V_o}{N_s}$	$-\dfrac{V_o}{N_s}$	
`Mode 2 [0.25~ 0.25+D]	$-\dfrac{V_o}{N_s}$	$-\dfrac{V_o}{N_s}$	$\dfrac{V_{p3}}{N_p}$	$\dfrac{V'_{p3}}{N_p}-\dfrac{2V_o}{N_s}$	
Mode 3 [0.5~ 0.5+D]	$\dfrac{V'_{p2}}{N_p}-\dfrac{2V_o}{N_s}$	$\dfrac{V_{p2}}{N_p}$	$-\dfrac{V_o}{N_s}$	$-\dfrac{V_o}{N_s}$	$i_L-\dfrac{V_o}{R}$
Mode 4 [0.75~ 0.75+D]	$\dfrac{V_o}{N_s}$	$\dfrac{V_o}{N_s}$	$-\dfrac{V'_{p4}}{N_p}-\dfrac{2V_o}{N_s}$	$\dfrac{V_{p4}}{N_p}$	
Mode 5 [others]	$-\dfrac{V_o}{N_s}$				

IV. EXPERIMENTAL RESULTS

The prototype with magnetic integration structure is shown in Fig. 6. The key component list of the prototype is shown in Table II. In this design, the high-voltage switches adopt the third-generation semiconductor GaN device, which has lower driving losses and switching losses, and is beneficial to the improvement of the converter's efficiency. The selected chip integrates the power half-bridge and its driver, effectively reducing the volume of the peripheral circuits and meeting the requirements of power density.

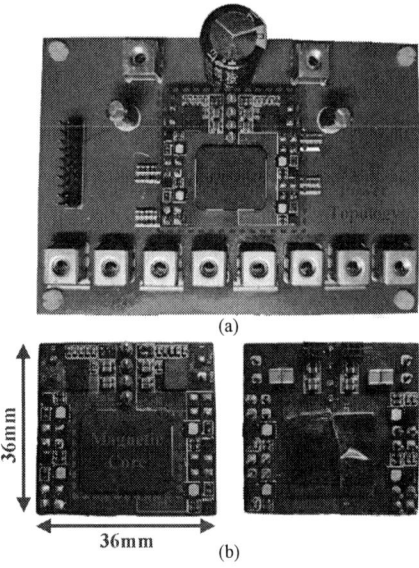

Fig. 6. experimental platform of proposed magnetic integration structure. (a) The power topology is installed on the test base plate. (b) Top and bottom view of the power prototype with a footprint of 36 mm × 36 mm and the height is 4.5 mm.

The prototype can work for a wide range of input and output voltages and offer high output current. The operation of it is defined for the nominal condition, i.e., 48-V input, 0.9-V output and 500kHz frequency. Because of the integration of transformers and coupled inductors, the prototype has a very small footprint. The overall height of power topology is only 4.5 mm. Finally, the current density is 0.2A/mm² at 120 A output current.

TABLE II.   KEY COMPONENT LIST OF THE PROTOTYPE

Component	Parameters
Pri. switch & gate driver	Innoscience ISG3201(2x 100V, 3.2mΩ GaN with Half Bridge Driver)
Sec. switch	Infineon IQE004NE1LM7CGSC (15V, 0.45mΩ)
Sec. gate driver	TI UCC27614DSGR (1-channel low-side, 10/10A)
Magnetic Core	DMEGC DMR51W High-frequency Ferrite Core

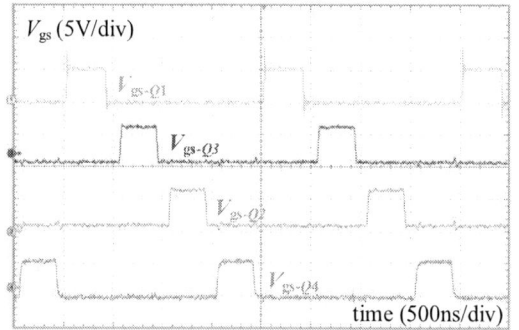

Fig. 7. The driving signals of the primary side switches Q_x with 90° phase differences

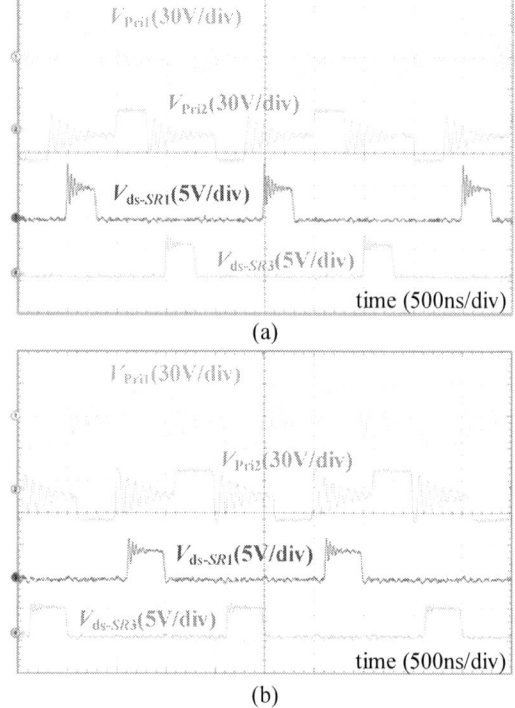

Fig. 8. Measured waveform at 48-V input and 0.9-V output. (a) Light-load condition (I_o = 10A). (b)Full-load condition (I_o = 120A).

Fig. 7 and Fig. 8 show the waveforms measured under different working conditions. In Fig. 7, V_{gs} is the gate voltage of Q_x. In Fig. 8, V_{prix} is the terminal voltage of the primary winding of the transformer, and V_{ds-SRx} is the drain-source voltage of SR_x. Since the prototype is hard switching, ringing can be observed during the switching process of Q_x, and the leakage energy of the transformer is dissipated. As the output current increases, the ringing becomes more severe. However, the peak voltage of the ringing is 60V at the full-load condition of 120 A, which leaves more than 40% margin for ISG3201. The experimental results show that the ringing iswithin an acceptable range and will not affect the operation of the converter.

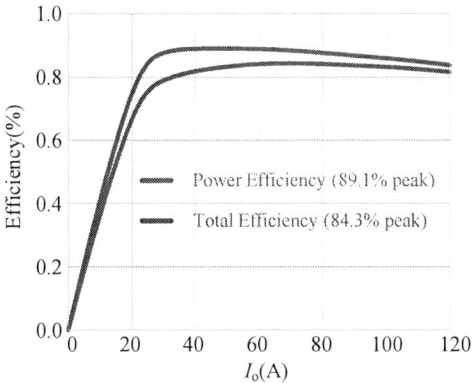

Fig. 9. Measured efficiency with 48-V input and 0.9-V output (natural cooling).

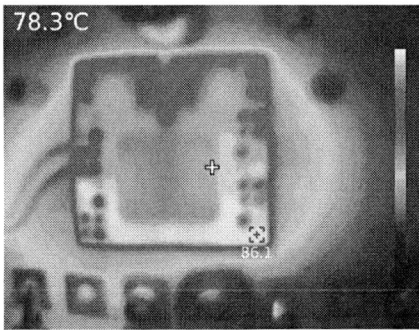

Fig. 10. Thermal image of the prototype at 48-V input and 0.9-V/120-A output. The thermal image was measured at 20°C ambient temperature, under a thermal steady state without heatsinks.

Measured efficiency with 48-V input and 0.9-V output is shown in Fig. 9. When the gate drive losses are not included, the peak efficiency of the topology can achieve 89%, and the full-load efficiency can achieve 84% The peak efficiency appears at 40A output current, which is approximately 30% of the load. Due to the high output current, winding loss and SR loss are the main sources of overall system loss. In this prototype, the secondary winding has only 4 layers and the copper thickness is 1oz. By increasing the number of layers in parallel and the copper thickness of the winding, the prototype may further improve the efficiency.Fig. 10 shows the thermal image of the prototype at 48-V input and 0.9-V/120-A output. The highest temperature of the system is 86 °C.

V. CONCLUSION

In this work, a GaN-based, low-voltage and high-current converter of a multi-phase interleaved current-doubler rectifier is proposed. The magnetic integration was optimized to improve power density and reduce loss. A comprehensive time-domain analysis method combining magnetic circuits and electrical circuits is adopted to conduct small-signal modeling for the converter, and it is analyzed that its essence is an isolated Buck converter, which provides a theoretical basis for the control of the topology. The current density of the final prototype is 0.2 A/mm², and the peak efficiency is 90%.

ACKNOWLEDGMENT

The authors would like to thank DMEGC for providing the magnetic core samples.

REFERENCES

[1] M. Chen, S. Jiang, J. A. Cobos and B. Lehman, "Design considerations for 48-V VRM: architecture, magnetics, and performance Tradeoffs," International Symposium on 3D Power Eectronics Integration and Manufacturing (3D-PEIM), Miami, FL, USA, 2023, pp. 1-9.

[2] P. Wang, Y. Chen, G. Szczeszynski, S. Allen, D. M. Giuliano and M. Chen, "MSC-PoL: hybrid GaN–Si multistacked switched-capacitor 48-V pwrSiP VRM for chiplets," IEEE Transactions on Power Electronics, vol. 38, no. 10, pp. 12815-12833, 2023.

[3] X. Xu, Y. Guan, Y. Wang and D. Xu, "A single-stage LLC resonant GaN-based DC-DC converter with switched capacitor," International Conference on Electrical Machines and Systems (ICEMS), Chiang Mai, Thailand, 2022, pp. 1-5.

[4] Figueroa, P. Mazariegos, J. Goicoechea, A. Castro and J. A. Cobos, "Low-profile direct power converter: 350A/48V-1V with planar matrix transformer using standard PCB and commercial cores," IEEE Applied Power Electronics Conference and Exposition (APEC), Long Beach, CA, USA, 2024, pp. 2172-2177.

[5] J. A. Cobos, P. Mazariegos, A. Figueroa, A. Castro and Á. Cobos, "500A stacked direct power converter with standard PCB transformer," IEEE Applied Power Electronics Conference and Exposition (APEC), Long Beach, CA, USA, 2024, pp. 925-930.

[6] S. Y. Sim, X. Zhang, J. Jiang, K. Wei and C. Huang, "A 94.7% efficiency direct-step-down switched-tank-based 48V to 1V-3.3V hybrid converter with constant-resonant-time closed-loop control," IEEE Applied Power Electronics Conference and Exposition (APEC), Long Beach, CA, USA, 2024, pp. 1344-1350.

[7] Z. Li, H. Wu, Y. Zhang, Z. Wang, Y. Song and Y. Xing, "Hybrid resonant converter-based 8:1 bus converter for 48V-to-1V power system," IEEE 2nd International Power Electronics and Application Symposium (PEAS), Guangzhou, China, 2023, pp. 53-57.

[8] H. Wu, Y. Zhang and Z. Li, "Hybrid resonant converter-based 8:1 bus converter with 3.5 kW/in3 and 98.6%-efficient for 48 V data-center power systems," IEEE Transactions on Power Electronics, vol. 39, no. 1, pp. 36-41, 2024.

[9] F. Zhu, X. Lou and Q. Li, "An improved hybrid-coupled inductor structure with flux reduction and integrated controllable coupling function," IEEE Transactions on Power Electronics, vol. 39, no. 1, pp. 1103-1114, 2024.

[10] X. Lou and Q. Li, "Multiphase Half-Bridge Current-Doubler Rectifier: A 93.1%-efficiency single-stage 48V voltage regulator with 1.04 kW/in3 power density," IEEE Applied Power Electronics Conference and Exposition (APEC), Orlando, FL, USA, 2023, pp. 1975-1981.

[11] X. Lou, F. Zhu, Z. Li, Q. Li and F. C. Lee, "Single-stage 48 V/1.8 V coupled-transformer voltage regulator (CTVR)," IEEE Energy Conversion Congress and Exposition (ECCE), Nashville, TN, USA, 2023, pp. 2333-2339.

[12] X. Lou and Q. Li, "Single-stage 48 V/1.8 V converter with a novel integrated magnetics and 1000 W/in3 power density," IEEE Transactions on Industrial Electronics, vol. 71, no. 7, pp. 6601-6611, 2024.

A Bidirectional-Signal Transmission Method for Gate Drive Application using Single Isolation Transformer

Junru Lin
the College of Electrical Engineering
Zhejiang University
Hangzhou, China
linjr@zju.edu.cn

Junming Zhang
the College of Electrical Engineering
Zhejiang University
Hangzhou, China
zhangjm@zju.edu.cn

Abstract—**Gate drivers with galvanically isolation capability are essential for switching devices in high-voltage applications. The conventional magnetic isolated gate drive uses separate isolation channels to transmit power, gate drive signal and fault feedback signal, thus increasing both the overall volume and the cost. In order to simplify the gate drive circuit, a novel self-powered gate driver with a single isolation transformer that supports bidirectional signal transmission is proposed in this paper. Experimental results verify that the proposed gate driver exhibits very short transmission delay of approximately 65ns, and supports gate driver signal frequency of 100 kHz with duty cycle ranging from 3% to 97%. Additionally, the feedback signal is capable of distinguishing between different types of faults.**

Keywords—*bidirectional signal transmission, gate drive, single isolation channel, amplitude modulation*

I. INTRODUCTION

In recent years, Insulated Gate Bipolar Transistors (IGBTs) and high-voltage wide bandgap semiconductors have been extensively used in diverse applications, including renewable energy, electric vehicles, and industrial power electronics. These power devices typically operate under high-voltage conditions, thus the galvanically isolation of the gate drives (GD) is necessary to ensure the safety and reliability.

Among the isolated gate drives, optical coupler isolation usually has limitations of high signal delay and reduced reliability [1][2], while capacitive isolation features short transmission delay and better longevity, but is susceptible to EMI interference, especially in high dv/dt environments [3]. Magnetic isolation also has a short propagation delay, and has become the mainstream choice of isolated gate drive [4]. However, to reduce the magnetic core size, the isolation transformer is difficult to transmit low-frequency signals, especially for chip scale transformers, thus the gate drive signal should be properly modulated to match the transmission characteristics of the transformer.

Besides, fault detection and protection are usually required for gate drive circuit to protect the power devices under abnormal operation condition, such as short circuit and over current [5] – [7]. These protection circuits on the secondary side of the GD, as well as the push-pull output stage for the GD PWM signal, all require an isolated low voltage power supply. As a result, conventional isolated GD often requires three separate isolation channels for signal and power transmission, thereby leading to a large overall footprint and higher system cost.

To address these limitations, integrating signal and power transmission within a single isolation channel has emerged as a promising solution to improve system integration, reliability,

This work was supported by National Key Research and Development Program of China under Grant 2021YFB2401600.

and reduce the cost. In previous studies on self-powered isolated GD, [8] and [9] extracted the edges of the GD signal to transmit both signal and power through single pulse transformer, thereby minimizing the transformer's volt-second product and physical size. However, the transmitted power may not sufficient to support the operation of protection circuits. In [10], the GD signal is modulated by high-frequency carrier for on-off key control, yet the transmitted power will decrease sharply as the duty cycle approached 0%, potentially leading to the failure of protection circuits. Furthermore, none of these designs supported the reverse transmission of the feedback fault signals, which is crucial for protection and monitoring.

Recent advancements have demonstrated the feasibility of bidirectional signal communication and forward power delivery over a single isolated channel. In [11] and [12], an isolated DC-DC converter is adopted to transmit power, while GD signals are transmitted via frequency modulation (FM) of the converter's switching control signals. Meanwhile, fault signals are reverse transmitted by modulating the voltage amplitude on the secondary side of the isolation transformer. However, FM demodulation typically requires integration circuits, which introduce inherent propagation delays and restrict the duty cycle range. For instance, in [12], the propagation delay of the GD signal is up to 400 ns, limiting the available duty cycle range to 7.5%–92.5% at 100 kHz PWM input. Even with the switching frequency increased to 50 MHz, the propagation delay is still over 100 ns [13]. Additionally, the propagation delays of rising and falling edges are asymmetric, which may have an adverse effect on the control of the power converter. In wide-bandgap semiconductor applications, GD with short and symmetrical propagation delay is essential.

To overcome these challenges, a novel isolated GD is proposed in this paper, which supports bidirectional signal transmission and forward power transfer via a single isolation transformer. The proposed GD achieves a propagation delay of less than 100 ns and supports almost full duty cycle range of 3%–97% at a 100 kHz PWM input. The protection circuit on the secondary side remains functional at any duty cycle. Section II introduces the operating principle and schematic of proposed GD, along with a detailed timing analysis. Section III presents experimental validation, and Section IV concludes this work.

II. OPERATION PRINCIPLE AND CIRCUIT DESCRIPTION

A. Proposed Bidirectional Transmission Method

The operation principle of the proposed GD is illustrated in Fig. 1. The energy required by the secondary side circuit is transferred via an isolated DC-DC converter sharing the same isolation transformer. Meanwhile, the GD signal and the feedback fault signal are transmitted together through the same

979-8-3315-1110-4/25 $31.00 © 2025 IEEE

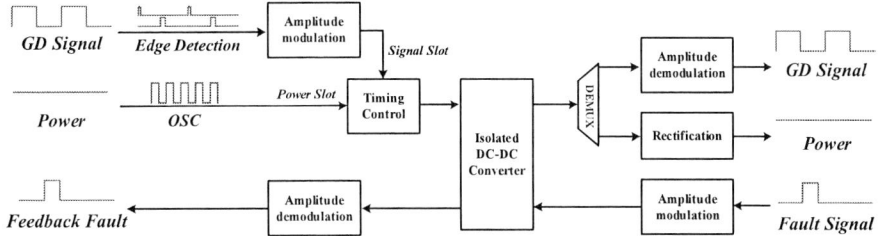

Fig. 1. The block diagram of the proposed GD.

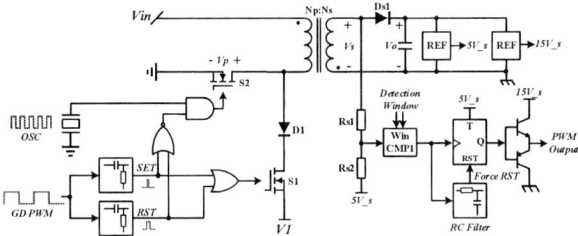

Fig. 2. GD forward transmission circuit.

magnetic isolation channel of the converter. In order to ensure that the power transmission is not affected by the duty cycle of the GD PWM signal, while also guarantee a short transmission delay, the amplitude modulation (AM) technique is adopted. The GD PWM signal is edge-extracted by a differentiator circuit, and the timing control circuit assigns the isolation channel to transmit the GD edge signal. During this period, the input voltage of DC-DC converter is also synchronously modulated to represent the edges of the PWM signal. By sampling and amplitude demodulating the voltage of the isolation transformer's secondary winding, the edges of the GD PWM signal can be identified, thus the PWM signal can be reconstructed by a flip-flop. The reconstructed GD signal will be further amplified by the push-pull output stage.

The energy transferred by DC-DC converter powers the secondary side amplitude demodulation circuit, push-pull circuit, and the fault protection circuit. When a fault is detected (such as overheating, overcurrent, or short circuit), the amplitude modulation circuit on the secondary side will briefly changes the voltage of the transformer's secondary winding. On the primary side, the feedback fault signal can be restored using a sampling and amplitude demodulation circuit

B. Circuit Schematics

Forward transmission of power and GD signal

The proposed GD forward transmission circuit is shown in Fig. 2. A flyback converter is selected to transfer power and provide galvanic isolation between the primary side and secondary side. The control signal of the switching device S2 is generated by an oscillator with a fixed frequency and duty cycle.

When the flyback converter operates in continuous conduction mode (CCM), the output voltage V_o on the secondary side is given as:

$$V_o = \frac{N_S D}{N_p(1-D)}V_i - V_d \qquad (1)$$

where N_s/N_p is the turns ratio of the transformer, D is the duty cycle, V_i is the input voltage, and V_d is the forward voltage drop of diode D_{s1}.

The GD signal is processed by two differentiating circuit

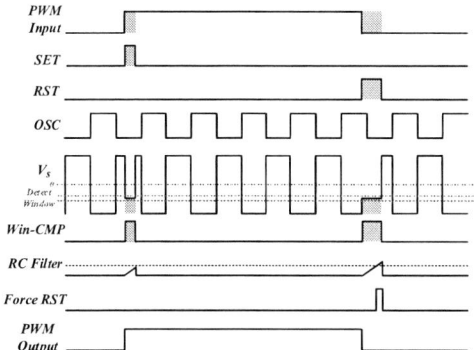

Fig. 3. Key waveform of forward transmission.

to separately extract the rising and falling edges, generating corresponding pulse signals *SET* and *RST*. The pulse width of *RST* is wider than the pulse width of *SET*, and both two signals is used to control the switching device *S1* with a OR gate. Meanwhile, a timing control circuit composed of a NOR gate and an AND gate ensures that *S2* remains off when *S1* is on. Fig. 3 shows the key waveforms of the GD forward transmission. During the power transmission period, when *S2* is on, the secondary side diode D_{s1} is off, and the voltage of the transformer's secondary winding V_s is given by (2):

$$V_s = -\frac{N_s}{N_p}V_i \qquad (2)$$

When *S2* turns off, diode D_{s1} is on, and in this case, we have:

$$V_s = V_o + V_d = \frac{N_s D}{N_p(1-D)}V_i \qquad (3)$$

During the signal transmission period, *S1* turns on, *S2* is off, and the secondary side diode D_{s1} is off. Therefore, the equivalent input voltage of the flyback converter is V_i-V_1-V_d, thus voltage of the transformer's secondary winding V_s is given by (4):

$$V_s = -\frac{N_s}{N_p}(V_i - V_1 - V_d) \qquad (4)$$

On the secondary side, two non-isolated voltage regulators *REF1* and *REF2* are adopted to provide the required supply voltages: 15V for the push-pull stage, and 5V for the demodulation and protection circuits. The voltage of the transformer's secondary winding V_s is sampled and DC biased through two sampling resistors R_{s1} and R_{s2}. Based on the analysis of V_s in (2) - (4), a window comparator (Win-CMP) can set appropriate detection window to restore the edge pulses of the GD signal. The original drive signal is then reconstructed using a T flip-flop. In addition, since the pulse width of *RST* is wider than that of *SET*, an RC filter circuit is used to extract the *RST* signal and asynchronously clear the T flip-flop, which

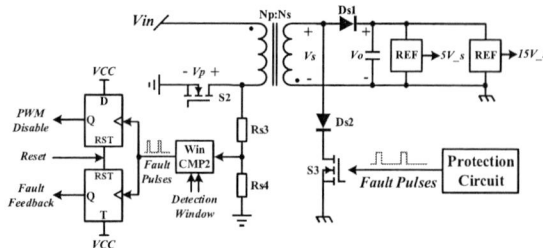

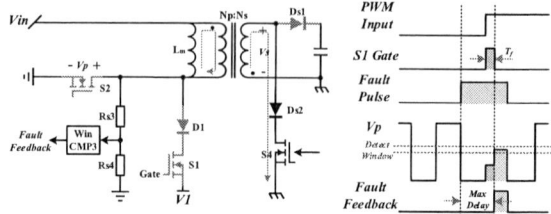

Fig. 4. GD reverse transmission circuit.

Fig. 6. Schematic of circuit operating state and key waveforms during reverse transmission.

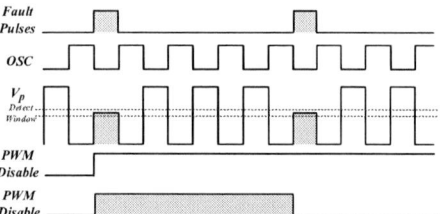

Fig. 5. Key waveform of reverse transmission.

ensures the gate drive signal is correctly reset in each cycle.

Reverse transmission of fault signal

The proposed GD reverse transmission circuit is shown in Fig. 4. When a fault occurs, the corresponding protection circuit will immediately generate the first narrow pulse to turn on $S3$. During the conduction period of $S3$, when the switching device $S2$ on the primary side is turned off, D_{s2} will conduct, and the voltage of the transformer's secondary winding V_s is clamped to V_d. At this time, the voltage across the switching device $S2$ V_p is given by (5):

$$V_p = \frac{N_p}{N_s}(V_i + V_d) \qquad (5)$$

V_p is sampled by two resistors R_{s3} and R_{s4}. By performing amplitude demodulation using another window comparator with appropriate detection window according to (5), the feedback fault pulse can be reconstructed. The first narrow pulse will instantly disable the transmission of the GD signal, and after a period of time, a second narrow pulse is sent to the primary side. Different types of faults correspond to different time intervals between two pulses. The demodulation circuit on the primary side outputs a high level during the interval between the two narrow pulses, and the duration of the high level can indicate different fault types. Corresponding waveforms are shown in Fig. 5.

C. Duty Cycle Range and Propagation Delay Analysis

For the forward-transmitted GD signal, amplitude modulation is employed, and since amplitude variations can be detected instantaneously, this modulation method theoretically introduces no propagation delay. However, the logic gates and comparators used in the modulation circuit and demodulation circuit will introduce additional transmission delays. Therefore, total delay time of forward transmission T_{fd} is given by (6):

$$T_{fd} = \sum T_{dn} \qquad (6)$$

where T_{dn} is the delay introduced by each logic gates and comparators.

When a fault occurs, $S3$ turns on. However, $Ds2$ can only conduct when $S1$ and $S2$ both turn off. Only under this

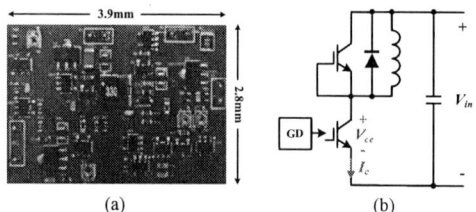

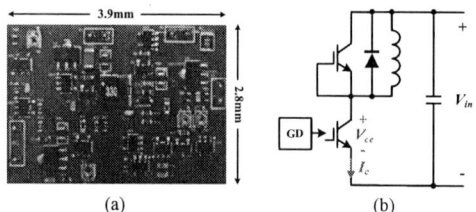

Fig. 7. (a) The GD prototype hardware. (b) Circuit of double-pulses test.

TABLE I.   PROTOTYPE KEY PARAMETERS

Parameters	Values
Transformer inductance	33uH (1:1)
Switching frequency	2MHz
Input voltage	20V
GD signal output	0V (low)/15V (high)

condition can the feedback fault signal be transmitted through the isolation transformer to the primary side, as shown in Fig. 6. The range of the fault signal reverse transmission delay T_{rd} is given by (7):

$$T_{rd} \in \left(0, \frac{T_s}{2} + T_f\right] \qquad (7)$$

where T_s is the switching period of the flyback converter, and T_f is the max width of the forward transferred *SET* and *RST* pulses.

The duty cycle of the GD PWM signal is constrained by the width of width of the forward transferred *SET* and *RST* pulses. When *SET* and *RST* overlap, the GD signal cannot be correctly reconstructed. As a result, the actual available duty cycle range is given by (8):

$$D \in [(T_{SET} + T_{RST})f_s, 1 - (T_{SET} + T_{RST})f_s] \qquad (8)$$

In this paper, T_{SET} is set to about 100ns, while T_{RST} is set to about 200ns. If the frequency of the GD PWM signal f_s is 100 kHz, the available duty cycle range is from 3% to 97%, which is sufficient to meet the requirements of most application scenarios.

III. EXPERIMENTAL VERIFICATIONS

A GD prototype was built to verify the effectiveness of the proposed method. The key parameters of the GD prototype are listed in Table I, and Fig. 7(a) shows the prototype hardware.

A. Forward Transmission Function Verification

Fig. 8 shows the measured forward transmission waveforms after modulation. When an edge of the forward-transmitted PWM signal arrives, the isolation channel is allocated

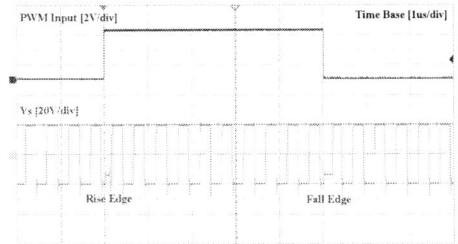

Fig. 8. Measured forward transmission waveform.

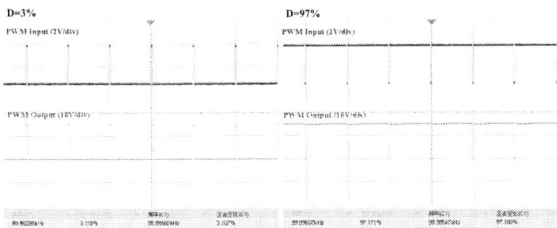

Fig. 9. Available GD duty cycle test.

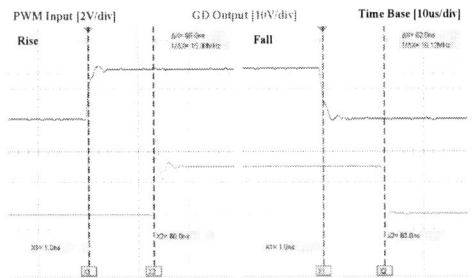

Fig. 10. Propagation delay.

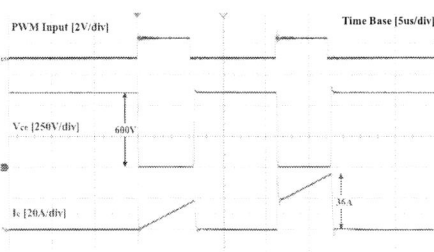

Fig. 11. Double-pulses test waveform.

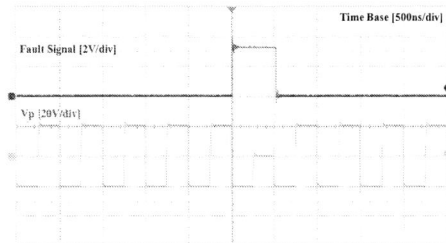

Fig. 12. Measured reverse transmission waveform.

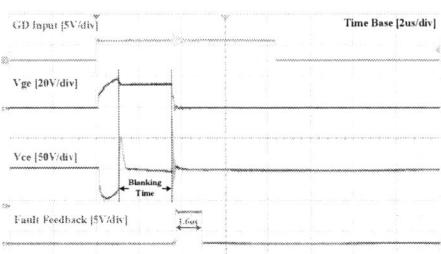

Fig. 13. Short circuit protection test waveform.

for transmitting the gate drive signal. The AM narrow pulses are transmitted to the secondary side through the isolation transformer and restored to the original GD signal after amplitude demodulation.

To verify the performance of the proposed isolated gate driver, experimental testing was conducted on the prototype using a 100 kHz PWM signal. Fig. 9 shows the input and output waveforms of the GD when D=3% and 97% respectively. Clearly the gate drive can operate properly under a wide duty cycle range. Fig. 10 shows the propagation delay of the GD, where the rising edge delay is measured at 65ns, and the falling edge delay is 62ns—both short and nearly symmetrical.

The double-pulses test was conducted on the proposed GD prototype, and the corresponding circuit schematic is shown in Fig. 7(b). The test bus voltage is 600V, and the IGBT models used are IKW25N120T2. The key waveform of the multipulse test is presented in Fig. 11, verifying the GD's capability to operate under high voltage and high current condition.

B. Reverse Transmission Function Verification

Fig. 12 shows the waveforms of the voltage across the switching device S3 during reverse signal transmission period. When the fault occurs, a significant voltage drop can be observed at the detection point on the primary side. The demodulation circuit can detect this amplitude change and reconstruct the fault pulses.

A short-circuit protection experiment was conducted to validate the reverse transmission performance of the isolated

GD. The IGBT is turned on under short circuit condition. The waveform is shown in Fig. 13. After a blanking time of 2.5μs (desaturation), the short-circuit protection circuit performs turn-off action and transmits the feedback fault signal to the primary side. The transmission of fault signal will not terminate the forward transmission GD signal. The demodulation circuit on the primary side output a high-level pulse with a duration of approximately 1.6μs, and the width of this pulse is variable to represent different types of faults.

IV. CONCLUSION

This paper proposes an isolation method for GD application, which realizes the bidirectional transmission of the GD signal and the fault feedback signal, as well as the power transmission, significantly simplifying the circuit and reducing the costs. The proposed GD features low and symmetrical propagation delays. Experimental results verify that GD supports a duty cycle range of 3%–97% under a 100 kHz PWM input. Short-circuit protection tests further verify the driver's capability to transmit fault signals in reverse, and different types of faults can be distinguished by the pulse width of the feedback signal.

REFERENCES

[1] A. Smith and K. Lenz, "New communication and isolation technology for integrated gate driver IC solutions suitable for IGBT and Si/SiC MOSFETs: Gate drive units, intelligent integrated drivers," in *2018 Conf. Proc. IEEE Appl. Power Electron. Conf. Expo.*, San Antonio, TX, USA, 2018, pp. 2986-2991

[2] H. Fujita, "A resonant gate-drive circuit with optically isolated control signal and power supply for fast-switching and high-voltage power semiconductor devices", *IEEE Trans. Power Electron.*, vol. 28, no. 11, pp. 5423-5430, Nov. 2013.

[3] R. Zhang, X. Li, J. Ding, S. Chen, H. Yang and H. Guo, "Review of IGBT Intelligent Gate Drive and Protection Strategies," *IEEE Trans. Power Electron.*, vol. 39, no. 6, pp. 7392-7403, June 2024

[4] J. M. V. Bikorimana and A. V. Bossche, "Less-conventional low-consumption galvanic separated MOSFET-IGBT gate drive supply", *Act. Passive Electron. Compon.*, vol. 2017, 2017.

[5] J. Henn et al., "Intelligent gate drivers for future power converters", *IEEE Trans. Power Electron.*, vol. 37, no. 3, pp. 3484-3503, Mar. 2022.

[6] Y. Chen, W. Li, F. Iannuzzo, H. Luo, X. He and F. Blaabjerg, "Investigation and classification of short-circuit failure modes based on three-dimensional safe operating area for high-power IGBT modules", *IEEE Trans. Power Electron.*, vol. 33, no. 2, pp. 1075-1086, Feb. 2018.

[7] C. H. Van Der Broeck, S. Kalker and R. W. De Doncker, "Intelligent monitoring and maintenance technology for next-generation power electronic systems", *IEEE Trans. Emerg. Sel. Topics Power Electron.*, vol. 11, no. 4, pp. 4403-4418, Aug. 2023.

[8] J. Yu et al., "A high frequency isolated resonant gate driver for SiC power MOSFET with asymmetrical ON/OFF voltage", *Proc. IEEE Appl. Power Electron. Conf. Expo.*, pp. 3247-3251, 2017.

[9] J. Garcia, E. Gurpinar and A. Castellazzi, "Impulse transformer based secondary-side self-powered gate-driver for wide-range PWM operation of SiC power MOSFETs", Proc. IEEE 4th Workshop Wide Bandgap Power Devices Appl., pp. 59-63, 2016.

[10] J. Garcia, S. Saeed, E. Gurpinar, A. Castellazzi and P. Garcia, "Self-powering high frequency modulated SiC power MOSFET isolated gate driver", IEEE Trans. Ind. Appl., vol. 55, no. 4, pp. 3967-3977, Jul./Aug. 2019.

[11] A. Seidel, M. Costa, J. Joos and B. Wicht, "Isolated 100% PWM gate driver with auxiliary energy and bidirectional FM/AM signal transmission via single transformer," in *2015 Conf. Proc. IEEE Appl. Power Electron. Conf. Expo.*, Charlotte, NC, USA, 2015, pp. 2581-2584.

[12] S. Song, H. Peng, C. Jiang, Q. Tong and Y. Kang, "A Novel Single Isolation Channel Gate Driver with Bidirectional-Signal and Forward-Power Transmission," *IEEE Trans. Power Electron.*, vol. 39, no. 6, pp. 6580-6585, June 2024.

[13] Z. Guo and H. Li, "A Fiber-Optic-Less 50-MHz Single Transformer Isolated Gate Driver with Fault Feedback for 10-kV SiC MOSFETs", *IEEE Trans. Ind. Electron.*, vol. 71, no. 9, pp. 10854-10863, Sept. 2024.

2025 IEEE Workshop on Wide Bandgap Power Devices and Applications in Asia (WiPDA Asia)

A Dual-Voltage 650 V and 100 V GaN Integrated Platform Featuring High-Performance Monolithic Components

Yanlin Wu[1], Junye Wu[1], Zuoheng Jiang[1], Danfeng Mao[1], Keping Wu[1], Chao Feng[1], Jiawei Chen[1],
David Zhou[1, z], and Yuxi Wan[1, z]

[1] Shenzhen Pinghu Laboratory, Shenzhen 518100, China

[z]E-mail: davidzhou@phlab.com.cn; wanyuxi@phlab.com.cn;

Abstract—This paper presents two integrated platforms developed by PH laboratory: a 650 V platform and a 100 V platform. Both platforms feature high-performance e-mode GaN HEMTs and a variety of passive components, including 2DEG resistors, metal resistors, p-GaN stack capacitors, M/2DEG capacitors, and MIM capacitors. The 650 V e-mode GaN HEMT demonstrates a V_{TH} of 1.3 V, an R_{ON} of 10.8 $\Omega\cdot$mm, and a BV of 843 V, while the 100 V e-mode GaN HEMT exhibits a V_{TH} of 1 V, an R_{ON} of 3.1 $\Omega\cdot$mm, and a BV of 174 V. Passive components are characterized: 2DEG resistors show a contact resistance of 0.38 $\Omega\cdot$mm (650 V) and 0.3 $\Omega\cdot$mm (100 V), with sheet resistances of 347.5 Ω/sq (650 V) and 403 Ω/sq (100 V). Metal resistors achieve a low sheet resistance of 17 Ω/sq, ideal for precision applications. The p-GaN stack capacitors reach maximum densities of 162 nF/cm² (650 V) and 115 nF/cm² (100 V), while MIM capacitors achieve 20 nF/cm² (650 V) and 589.7 nF/cm² (100 V), enabling significant area savings. These platforms offer superior performance and design flexibility, making it highly suitable for compact, high-power-density applications.

Keywords—Monolithic integration, e-mode, 650 V integrated platform, 100 V integrated platform, passive component.

I. INTRODUCTION

Over the past two decades, GaN transistors have been extensively developed, emerging as strong candidates for next-generation power switching applications [1]-[2]. A critical challenge in their evolution has been the realization of enhancement-mode (e-mode) operation. GaN HEMTs, which leverage two-dimensional electron gas (2DEG) conduction at the AlGaN/GaN interface, are inherently depletion-mode (d-mode) devices. The gate control of d-mode GaN HEMTs is notably more complex, with an inherent risk of unintended turn-on. To address these limitations and achieve e-mode operation, researchers have explored various strategies, including fluorine ion treatment, recessed gate technology, p-GaN gate technology, and cascode configurations [3]-[7]. Among these, p-GaN gate HEMTs and cascode switches have been commercialized.

Compared to cascode switches, the p-GaN gate HEMT, as a single-chip solution, offers distinct advantages in monolithic integration. As opposed to multi-chip systems assembled on PCBs, monolithic integration significantly reduces interconnect parasitic effects, enabling higher operating frequencies and fully leveraging GaN's high-frequency characteristics [8]-[13]. Additionally, the increased switching frequency reduces the size requirements for passive components. As a result, the layout area occupied by passive components is minimized, leading to a more compact chip size [14]-[16]. Such compact, high-power-density chips are particularly advantageous for applications in drones, electric bicycles, and similar scenarios.

Currently, leading companies like EPC and Navitas have launched 100 V and 650 V GaN monolithic integrated chips [17]-[18]. These chips integrate GaN power devices with driver and other modules, significantly enhancing both performance and functionality. Research institutions like IMEC and foundries such as TSMC have developed 200 V and 650 V integration platforms, providing IC designers with tools to create high-performance monolithic integrated devices [19]-[21]. However, detailed parameters of passive components in these integrated platforms remain underreported.

This paper presents the PH laboratory's advanced 650 V and 100 V integrated pilot platform, which demonstrates industry-leading performance in both power devices and passive

979-8-3315-1110-4/25 $31.00 © 2025 IEEE

components. The platform offers circuit designers a comprehensive selection of passive components, including metal-insulator-metal (MIM) capacitors, metal/2DEG (M/2DEG) capacitors, p-GaN capacitors, 2DEG resistors, and metal resistors. Notably, the MIM and p-GaN capacitors feature high capacitance density, enabling significant area reduction and space savings. The platform's diverse resistor options, with their wide resistance range, provide enhanced design flexibility while minimizing area wastage from repeated winding. These diverse passive components significantly expand the possibilities for circuit design.

II. DEVICE CHARACTERIZATION

Fig. 1 illustrates the integrated platform developed by PH Laboratory. This platform, based on p-GaN gate technology, includes 650 V e-mode GaN HEMTs, 100 V e-mode HEMTs, 2DEG resistors, metal resistors, p-GaN stack capacitors, MIM capacitors, and M/2DEG capacitors. Table 1 compares the PH laboratory's integrated platform with those from IMEC and TSMC. While the IMEC platform covers 200 V and 650 V voltage levels, and TSMC focuses on a 650 V platform, the PH Laboratory platform supports both 650V and 100V power devices, catering to a broader design range of 100–650 V. Passive components are fundamental to integrated circuit design, and the PH Laboratory platform offers five types of passive components, significantly expanding design options for IC designers. This diversity surpasses the reported passive components of the other two platforms.

Fig. 2 presents the I-V characteristics of the 650 V e-mode GaN HEMT. The transfer characteristics curve reveals a positive threshold voltage (V_{TH}) of 1.3 V, extracted at $I_D = 1$ mA/mm and $V_D = 1$ V. From the output characteristics curve, the on-resistance (R_{ON}) is determined to be 10.8 $\Omega \cdot$mm. The 650 V e-HEMT exhibits a breakdown voltage (BV) of 843 V, defined at $I_D = 1$ µA/mm.

Fig. 3 shows the I-V characteristics of the 100 V e-mode GaN HEMT. The transfer characteristics curve indicates a positive V_{TH} of 1 V, extracted at $I_D = 0.1$ mA/mm and $V_D = 0.1$ V. The output characteristics curve yields an R_{ON} of 3.1 $\Omega \cdot$mm. The 100 V e-HEMT

demonstrates a BV of 174 V, defined at $I_D = 1$ µA/mm.

Fig. 4(a) includes an inset depicting the test schematic for the transfer length method (TLM) of the 2DEG resistor. By measuring resistances across different gap spacings, the contact resistance and sheet resistance are extracted. For the 650 V platform, the contact resistance (R_C) is 0.38 $\Omega \cdot$mm, and the sheet resistance ($R_{SH-2DEG}$) is 347.5 Ω/sq. For the 100 V platform, the R_C is 0.3 $\Omega \cdot$mm, and the $R_{SH-2DEG}$ is 403 Ω/sq. The total resistance of the 2DEG resistor in the integrated platform is calculated as $R_{2DEG} = 2 R_C + R_{sh-2DEG} \times$ L, where L is length of a resistor.

As shown in Fig. 5, the sheet resistance of the metal resistor is extracted using the TLM method. The metal sheet resistance (R_{SH-M}) is 17 Ω/sq, significantly lower than that of the 2DEG resistor, making it suitable for precision resistor applications in circuit design. Since the metal is conductive, the contact resistance is negligible. The total resistance of the metal resistor in the integrated platform is calculated as $R_M = R_{sh-M} \times$ L, where L is length of a resistor.

Fig. 6 illustrates the structure of a p-GaN stack capacitor, which consists of a Schottky gate/p-GaN junction capacitance and a p-GaN/-AlGaN/GaN heterojunction capacitance. When $V_G < V_{TH}$, the 2DEG channel is not formed, and the p-GaN/AlGaN/GaN heterojunction capacitance is negligible, with the gate capacitance primarily determined by the Schottky junction capacitance and fringing capacitance [22]-[23]. When $V_G > V_{TH}$, the 2DEG channel gradually forms, and the gate capacitance increases. The maximum gate capacitance corresponds to the point where the channel is fully formed. As the gate voltage further increases, the Schottky junction becomes reverse-biased, the depletion region widens, and the Schottky capacitance decreases, leading to a reduction in the gate capacitance. For the 650 V platform, the p-GaN stack capacitor achieves 111 nF/cm² at a 6 V bias, reaching a maximum density of 162 nF/cm². For the 100 V platform, the p-GaN stack capacitor achieves 87 nF/cm² at a 6 V bias, reaching a maximum density of 115 nF/cm².

Additionally, Table 2 summarizes the capacitance values of the capacitors on both platforms. The M/2DEG capacitors achieve 13.1 nF/cm² (650 V) and 37.7 nF/cm² (100 V), while the MIM capacitors attain 20 nF/cm² (650 V platform) and 589.7 nF/cm² (100 V platform). The p-GaN capacitor and the MIM capacitor on the 100 V platform exhibit high capacitance density, enabling a reduction in capacitor size and saving layout area.

III. CONCLUSION

This work demonstrates two advanced integrated platforms: a 650 V platform and a 100 V platform, both featuring e-mode GaN HEMTs based on p-GaN gate technology. Each platform incorporates a comprehensive suite of passive components, including 2DEG resistors, metal resistors, p-GaN stack capacitors M/2DEG capacitors, and MIM capacitors. The resistors offer a wide resistance range, enabling versatile circuit design while minimizing area inefficiencies caused by repeated winding in traditional designs. Additionally, the p-GaN and the MIM capacitors exhibit high capacitance density, significantly reducing component size and optimizing layout area. With superior performance, design flexibility, and a broad voltage range, these platforms are well-suited for compact, high-power-density applications in next-generation power electronics.

REFERENCES

[1] H. Amano, et al., "The 2018 GaN power electronics roadmap," J. Phys. D, Appl. Phys., vol. 51, no. 16, Apr. 2018, Art. no. 163001.

[2] K. J. Chen, et al., "GaN-on-Si power technology: Devices and applications," IEEE Trans. Electron Devices, vol. 64, 64, no. 3, pp. 779–795, Mar. 2017.

[3] Y. Cai, Y. Zhou, K. J. Chen, and K. M. Lau, "High-performance enhancement-mode AlGaN/GaN HEMTs using fluoride-based plasma treatment," IEEE Electron Device Lett., vol. 26, no. 7, pp. 435–437, Jul. 2005.

[4] T. Kikkawa, et al., "600 V JEDEC-qualified highly reliable GaN HEMTs on Si substrates," in IEDM Tech. Dig., Dec. 2014, pp. 2.6.1–2.6..

[5] T. Oka and T. Nozawa, "AlGaN/GaN recessed MIS-gate HFET with high-threshold-voltage normally-off operation for power electronics applications," IEEE Electron Device Lett., vol. 29, no. 7, pp. 668–670, Jul. 2008.

[6] J. Wei, et al., "Low on-resistance normally-off GaN double-channel metal–oxide–semiconductor high-electron-mobility transistor," IEEE Electron Device Lett., vol. 36, no. 12, pp. 1287–1290, Dec. 2015.

[7] X. Hu, et al., "Enhancement mode AlGaN/GaN HFET with selectively grown pn junction gate," Electron. Lett., vol. 36, no. 8, p. 753, Apr. 2000.

[8] D. Reusch and J. Strydom, "Understanding the effect of PCB layout on circuit performance in a high-frequency gallium-nitride-based point of load converter," IEEE Trans. Power Electron., vol. 29, no. 4, pp. 2008–2015, Apr. 2014.

[9] K. Wang, L. Wang, X. Yang, X. Zeng, W. Chen, and H. Li, "A multiloop method for minimization of parasitic inductance in GaN-based high-frequency DC–DC converter," IEEE Trans. Power Electron., vol. 32, no. 6, pp. 4728–4740, Jun. 2017

[10] O. Trescases, S. K. Murray, W. L. Jiang, and M. S. Zaman, "GaN power ICs: Reviewing strengths, gaps, and future directions," in IEDM Tech. Dig., San Francisco, CA, USA, Dec. 2020, pp. 27.4.1–27.4.4.

[11] D. Kinzer, "Monolithic GaN power IC technology drives wide bandgap adoption," in IEDM Tech. Dig., San Francisco, CA, USA, Dec. 2020, pp. 27.5.1–27.5.4.

[12] K. J. Chen et al., "Planar GaN power integration—The world is flat," in IEDM Tech. Dig., San Francisco, CA, USA, Dec. 2020, pp. 27.1.1–27.1.4.

[13] C.-L. Tsai et al., "Smart GaN platform: Performance & challenges," in IEDM Tech. Dig., San Francisco, CA, USA, Dec. 2017, pp. 737–740.

[14] Y. Guan, Y.Wang, D. Xu, andW.Wang, "A 1 MHz half-bridge resonant DC/DC converter based on GaN FETs and planar magnetics," IEEE Trans. Power Electron., vol. 32, no. 4, pp. 2876–2891, Apr. 2017.

[15] R. C. N. Pilawa-Podgurski, "Emerging circuit techniques to utilize wide-bandgap semiconductors in compact, lightweight, and efficient power converters," in IEDM Tech. Dig., San Francisco, CA, USA, Dec. 2021, pp. 5.6.1–5.6.4.

[16] M.-J. Liu and S. S. H. Hsu, "A miniature 300-MHz resonant DC–DC converter with GaN and CMOS integrated in IPD technology," IEEE Trans. Power Electron., vol. 33, no. 11, pp. 9656–9668, Nov. 2018.

[17] EPC23104 Datasheet, EPC, El Segundo, CA, USA, 2024. [Online]. Available: https://www.epc-co.com.

[18] NV6115 Datasheet, Navitas Semi, Torrance, CA, USA, 2022. [Online]. Available: www. navitassemi.com.

[19] X. Li, et al., "GaN-on-SOI: Monolithically integrated all-GaN ICs for power conversion," in IEDM Tech. Dig., San Francisco, CA, USA, Dec. 2019, pp. 4.4.1–4.4.4.

[20] X. Li, et al., "Integration of 650 V GaN power ICs on 200 mm engineered substrates," IEEE Trans. Semicond. Manuf., vol. 33, no. 4, pp. 534-538, Nov. 2020.

[21] M. H. Kwan, et al., "CMOS-compatible GaN-on-Si field-effect transistors for high voltage power applications," in IEDM Tech. Dig., San Francisco, CA, USA, Dec. 2014, pp. 17.6.1–17.6.4.

[22] G. Tang, et al., "High-capacitance-density p-GaN gate capacitors for high-frequency power integration," IEEE Electron Device Lett., vol. 39, no. 9, pp. 1362–1365, Sep. 2018.

[23] J. Wei, et al., "GaN power integration technology and its future prospects," IEEE Trans. Electron Devices, vol. 71, no. 3, pp. 1365–1382, Mar. 2025.

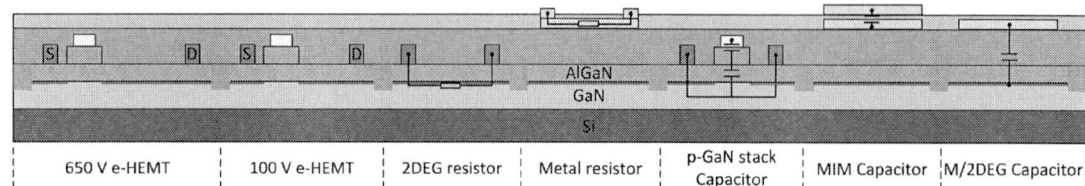

Fig. 1: Schematic cross-section of PH laboratory's GaN power integration platform. The platform features monolithic co-integration of 650-V e-HEMTs, 100-V e-HEMTs, 2 types of resistors, three types of capacitors.

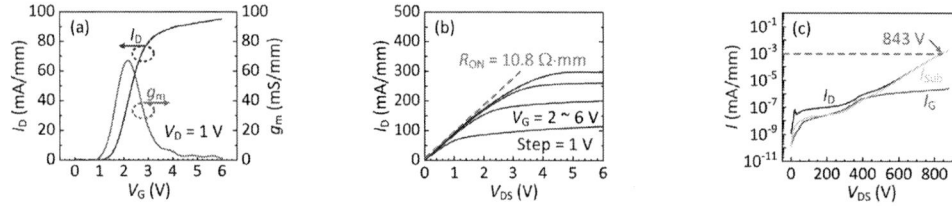

Fig. 2: (a) Transfer, (b) output, and (c) breakdown voltage characteristics of 650 V GaN HEMT. V_{TH} is 1.3 V extracted at I_D = 1 mA/mm and V_D = 1 V. R_{ON} is 10.8 Ω·mm. The breakdown voltage is 843 V at I_D = 1 μA/mm.

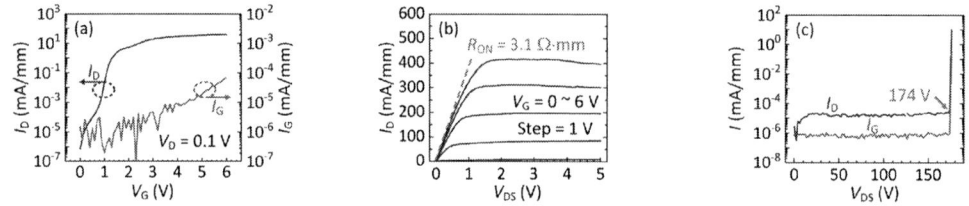

Fig. 3: (a) Transfer, (b) output, and (c) breakdown voltage characteristics of 100 V GaN HEMT. V_{TH} is 1 V extracted at I_D = 0.1 mA/mm, V_D = 0.1 V. R_{ON} is 3.1 Ω·mm. The breakdown voltage is 174 V at I_D = 1 μA/mm.

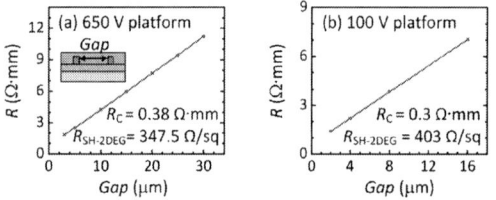

Fig. 4: Contact resistance and sheet resistance extracted by TLMs for (a) 650 V platform and (b) 100 V platform.

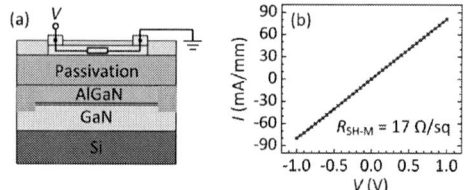

Fig. 5: (a) TLM pattern utilized for extracting the sheet resistance of the metal. (b) The sheet resistance is 17 Ω/sq.

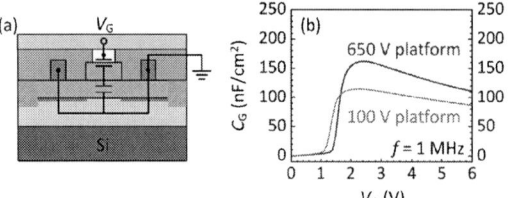

Fig. 6: (a) Cross-sectional schematic of the p-GaN stack capacitor structure. (b) C-V characteristics showing capacitance densities of 111 nF/cm² (650 V platform) and 87 nF/cm² (100 V platform) at V_G = 6 V.

Table 1: Comparison of GaN platform among IMEC, TSMC and PH laborotary.

	IMEC	TSMC	PH lab.
Voltage level	650 V, 200 V	650 V	100 V, 650 V
E/D HEMT	E, D	E, D	E
Capacitor	MIM	/	MIM, M/2DEG, p-GaN
Resistor	2DEG	/	2DEG, metal

Table 2: Capacitance of 650 V and 100 V GaN platform.

Capacitor	650 V platform	100 V platform
M/2DEG (nF/cm²)	13.1	37.7
MIM (nF/cm²)	20	589.7
p-GaN (nF/cm²)	> 111	> 87

Exploring the Soft-Switching Benefits of TZCM Mode in Three-Level DC-DC Converters Using Wide Bandgap Power Devices

Zhigang Yao
Energy Research Institute
Nanyang Technological University
Singapore
zhigang.yao@ntu.edu.sg

Jingrui Liu
School of Electrical and Electronic Engineering
Nanyang Technological University
Singapore
jingrui001@e.ntu.edu.sg

Sankun Yao
School of Big Data and Automation
Chongqing Chemical Industry Vocational College
China
dsj20242045@cqcivc.edu.cn

Bac-Bien Ngo
School of Electrical and Electronic Engineering
Nanyang Technological University
Singapore
ngob0002@e.ntu.edu.sg

Ziheng Xiao
Energy Research Institute
Nanyang Technological University
Singapore
ziheng.xiao@ntu.edu.sg

Yi Tang
School of Electrical and Electronic Engineering
Nanyang Technological University
Singapore
yitang@ntu.edu.sg

Abstract—**Wide bandgap power devices such as SiC MOSFETs and GaN HEMTs offer low conduction resistance and reduced switching losses. However, under high-frequency hard-switching operation, their switching losses remain significant. To improve efficiency, soft-switching techniques have been developed, including the well-known triangular current mode (TCM) and the emerging trapezoidal current mode (TZCM). The two modes enable zero-voltage switching (ZVS) without requiring additional auxiliary components, and effectively eliminate turn-on losses and body diode reverse recovery losses. But, a detailed comparison of their soft-switching benefits has not yet been conducted. This paper aims to clarify the respective advantages and application boundaries of continuous conduction mode (CCM), TCM, and TZCM. Through theoretical analysis and comparison, the advantageous operating ranges of the three modes are identified. Experimental results confirm that each operating mode has its own advantageous operating range.**

Keywords—Three-level DC-DC converter, trapezoidal current mode (TZCM), soft-switching, SiC MOSFET

I. INTRODUCTION

Soft-switching techniques are being increasingly adopted to reduce switching losses and electromagnetic interference (EMI) in high-frequency power electronic converters [1], [2]. Zero-current switching (ZCS) is primarily used to reduce turn-off losses of power semiconductor devices, while zero-voltage switching (ZVS) is effective in minimizing turn-on losses and body diode reverse recovery losses of power semiconductor devices. Since the turn-on and reverse recovery losses are typically greater than turn-off losses in most power devices, such as SiC MOSFETs and GaN HEMTs, ZVS soft-switching is more commonly employed [3], [4], [5], [6]. These popular power converter topologies, including LLC resonant converters, dual active bridge (DAB) converters, and buck/boost DC-DC converters operated in TCM, all take advantage of ZVS soft-switching [7], [8], [9], [10].

Benefiting from reduced voltage stress on power switches and smaller passive component requirements, three-level DC-DC converters are widely used in battery storage systems, photovoltaic (PV) generation, fuel cells, and various DC loads

[11], [12]. Traditionally, these converters operate in continuous conduction mode (CCM), which leads to hard-switching behavior, resulting in significant switching losses and EMI issues [13]. To address these drawbacks, several studies[14], [15], [16] have investigated soft-switching techniques, particularly zero-voltage switching. Conventional methods for achieving ZVS typically require additional auxiliary components and circuits, which increase hardware complexity and system size. For example, in [14] and [17], auxiliary inductors, capacitors, and power switches are introduced to create resonant transitions that enable ZVS.

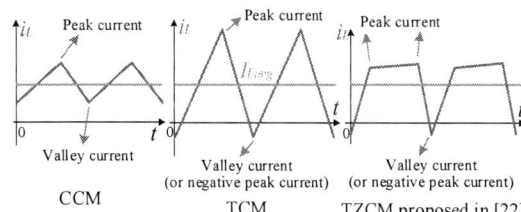

Fig. 1. Inductor current waveforms in CCM, TCM and TZCM.

To realize ZVS without the need for auxiliary circuits, triangular current mode (TCM) has been explored in both two-level and three-level converters [15], [18], and some control methods are studied for the four-switch buck-boost converter [19], [20], [21]. However, TCM results in excessive peak currents—often exceeding twice the average inductor current—which imposes significant current stress and increases turn-off losses in the switching devices. To mitigate this issue, recent studies [22], [23] have proposed trapezoidal current mode (TZCM) for three-level converters, achieving ZVS and efficiencies exceeding 99%. Moreover, the principles of TCM and TZCM can be extended to multilevel DC-DC converter topologies [24], [25]. The inductor current waveforms in CCM, TCM and TZCM are shown in Fig. 1.

However, a detailed comparison among the three operating modes—CCM, TCM, and TZCM—has not yet been conducted, and the soft-switching benefit region of TZCM remains underexplored. This paper presents a comparative analysis of their basic operating principles and respective advantages in three-level DC-DC converters. The ZVS beneficiary regions for each mode are identified, and experimental results verify that each has its own advantageous operating range.

II. COMPARISON OF OPERATING PRINCIPLE

The series-capacitor and flying-capacitor three-level DC-DC converters are shown in Fig. 2. Both they work in CCM, TCM, and TZCM. Compared with the traditional modulation of CCM and TCM, the new TZCM not only reduces the turn-off and conduction losses of the power switches by lowering the peak and RMS currents, but also realizes ZVS under near-CRM at D=0.5, where the ZVS cannot be realized in the traditional modulation because the inductor current ripple is equal to zero.

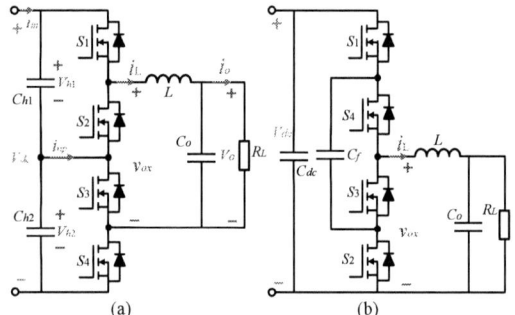

Fig. 2. Typical three-level DC-DC converters. (a) Series-capacitor three-level DC-DC converter. (b) Flying-capacitor three-level DC-DC converter.

The modulation methods of the three working modes for three-level DC-DC converters are shown in Fig. 3. The modulation methods of CCM and TCM modes are the same, except that the ripple current in TCM is larger. A fixed negative peak current is realized through an adjusted switching frequency control to enable ZVS soft-switching. For the modulation of TZCM, the duty cycles D_1 and D_4 of the two main switches are different, resulting in a 2-1-0 switching output voltage level sequence. In TZCM, every switching cycle has two positive peak currents and one negative peak current.

Estimating the power losses of power devices requires calculating values of the peak/valley and RMS currents. In CCM or TCM, the peak-to-peak value of the triangular inductor current can be expressed as

$$\Delta i_L = \begin{cases} \dfrac{D(1-2D)V_{dc}}{2Lf_{sw}}, 0 \le D \le 0.5 \\ \dfrac{(1-D)(2D-1)V_{dc}}{2Lf_{sw}}, 0.5 < D \le 1 \end{cases} \quad (1)$$

By fixing the average output current I_{Lavg}, the peak and valley currents in CCM can be calculated as

$$\begin{cases} I_p = I_{Lavg} + 0.5\Delta i_L \\ I_v = I_{Lavg} - 0.5\Delta i_L \end{cases} \quad (2)$$

To meet the ZVS condition in TCM, the valley current is fixed at a negative value of I_{p-}, so the positive peak current can be calculated as

$$I_{p+} = 2I_{Lavg} - I_{p-} \quad (3)$$

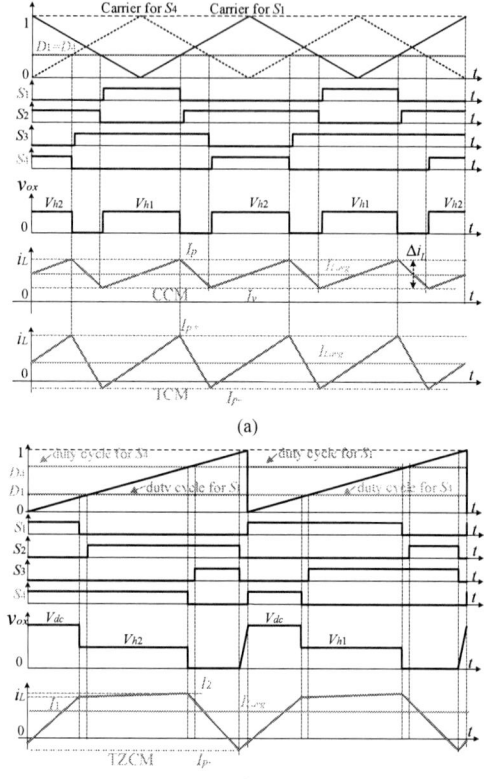

(a)

(b)

Fig. 3. Modulation methods for three-level DC-DC converters. (a) Conventional modulation method for CCM and TCM. (b) Trapezoidal current modulation method for TZCM.

In TZCM, the valley current is also fixed at a negative value of I_{p-} to meet the ZVS condition, referring to [22], the two peak currents I_1 and I_2 can be expressed as

$$\begin{cases} I_1 = \Delta i_{L1} - |I_{p-}| = \dfrac{4\left(I_{Lavg} + |I_{p-}|\right)(V_{dc} - V_o)D_1}{\left(D_1 + D_4 - D_1^2 - D_4^2\right)V_{dc}} - |I_{p-}| \\ I_2 = \Delta i_{L2} - |I_{p-}| = \dfrac{4\left(I_{Lavg} + |I_{p-}|\right)V_o(1-D_4)}{\left(D_1 + D_4 - D_1^2 - D_4^2\right)V_{dc}} - |I_{p-}| \end{cases} \quad (4)$$

Using the current RMS calculation method for segmented line segments, the inductor current RMS values are (5), (6), and (7) for CCM, TCM, and TZCM modes, respectively.

$$I_{Lrms\_CCM} = \sqrt{\frac{I_p^2 + I_p I_v + I_v^2}{3}} = \sqrt{I_{Lavg}^2 + \frac{\Delta i_L^2}{12}} \quad (5)$$

$$I_{Lrms\_TCM} = \sqrt{\frac{I_{p+}^2 + I_{p+}I_{p-} + I_{p-}^2}{3}} \quad (6)$$

$$I_{Lrms\_TZCM} = \sqrt{\begin{array}{l} \dfrac{I_1^2 + I_1 I_{p-} + I_{p-}^2}{3}D_1 + \dfrac{I_1^2 + I_1 I_2 + I_2^2}{3}(D_4 - D_1) \\ + \dfrac{I_2^2 + I_2 I_{p-} + I_{p-}^2}{3}(1 - D_4) \end{array}} \quad (7)$$

Based on expressions (2), (3) and (4), the peak currents can be calculated and compared, as shown in Fig. 4. The relationship between the peak inductor current and the duty cycle shows that TZCM's peak current is less than TCM's during $0.3 < D < 0.7$. This is the suitable working range for TZCM because of the lower peak current, turn-off losses, and switching current stress.

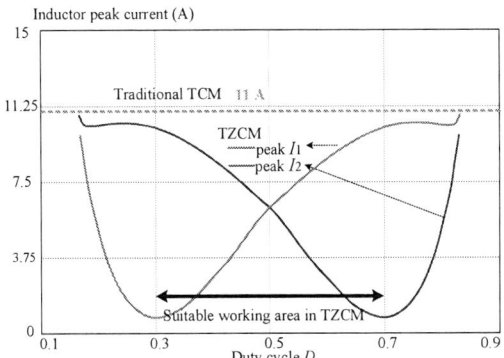

Fig. 4. Relationship between the inductor peak current and the duty cycle D in TZCM and TCM with V_{dc}=600 V, I_{Lavg}=5 A and I_{p-}=-1 A, where duty cycle D=0.5(D_1+D_4).

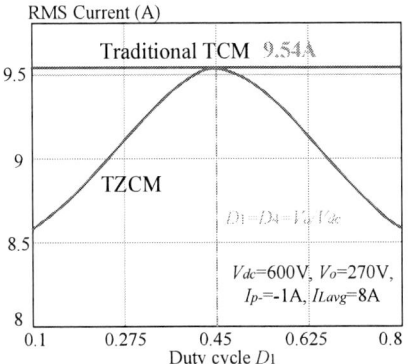

Fig. 5. Comparison of the inductor RMS current between TZCM and TCM, with V_{dc}=600 V, V_o=270V I_{Lavg}=8 A and I_{p-}=-1 A.

Based on expressions (5), (6) and (7), the inductor RMS current can be calculated, as shown in Fig. 5. Compared to traditional modulation, the TZCM mode realizes ZVS soft-switching in near-CRM and has smaller inductor RMS currents. This means it will have lower conduction losses in power devices. It is evident that the smaller the duty cycle D_1, the lower the inductor's RMS current. Benefiting from the reduced peak and RMS current, using the TZCM mode is advantageous for mitigating conduction losses and turn-off losses in power semiconductor switches.

III. COMPARISON OF POWER LOSSES IN DIFFERENT MODES

In CCM, TCM and TZCM, the power losses of power semiconductor switches mainly consist of conduction loss and switching loss. Neglecting the body diode conduction loss generated during the deadband process, the conduction losses for per complementary switch pair can be simply approximated as (8), (9), and (10) for CCM, TCM, and TZCM modes, respectively.

$$P_{cond\_CCM} = R_{DSon}I_{Lrms\_CCM}^2 \tag{8}$$

$$P_{cond\_TCM} = R_{DSon}I_{Lrms\_TCM}^2 \tag{9}$$

$$P_{cond\_CCM} = R_{DSon}I_{Lrms\_TZCM}^2 \tag{10}$$

In CCM mode, the two main switches S_1 and S_4 have turn-on and turn-off losses. In contrast, the two complementary switches S2 and S3 have no turn-on or turn-off losses, but have body diode reverse recovery losses. So, the switching losses for per complementary switch pair can be calculated as

$$P_{sw\_CCM} = \left(E_{on\_I_v} + E_{off\_I_p} + E_{rr\_I_v}\right)f_{sw\_CCM} \tag{11}$$

In TCM mode, both the two main switches and the complementary switches can achieve ZVS soft-switching. This results in the TCM mode having no turn-on losses and no body diode reverse recovery losses, leaving only turn-off losses. Thus, the switching losses for per complementary switch pair in TCM can be calculated as

$$P_{sw\_TCM} = \left(E_{off\_i_{p+}} + E_{off\_i_{p-}}\right)f_{sw\_TCM} \tag{12}$$

In TZCM mode, although the two main switches and the complementary switches can achieve ZVS soft-switching, there are two positive peak currents and one negative peak current. as same as TCM, there are no turn-on losses and no body diode reverse recovery losses in TZCM. However, there are three peak currents that produce turn-off losses. Therefore, the switching losses for per complementary switch pair in TCM can be calculated as

$$P_{sw\_TZCM} = \left(E_{off\_I_1} + E_{off\_I_2} + E_{off\_ip-}\right)f_{sw\_TZCM} \tag{13}$$

Based on expressions (8)-(13), the conduction and switching losses of power semiconductor switches can be calculated using data from the datasheet or experimental results using double pulse test. The respective optimized working intervals can also be compared and selected.

Usually, E_{on} of the power semiconductor devices is much greater than E_{off} for most SiC MOSFETs and GaN HEMTs [4], [26], [27], [28]. For the operating modes in TZCM and TCM, eliminating turn-on losses is more effective, so it is preferable to choose the power semiconductor devices with E_{on} much greater than E_{off}, such as that of Fig. 6. Since the ripple of the inductor current increases relative to the CCM mode, only this power device characteristic with $E_{on} > E_{off}$ can be beneficial in reducing the total switching losses by using TCM or TZCM.

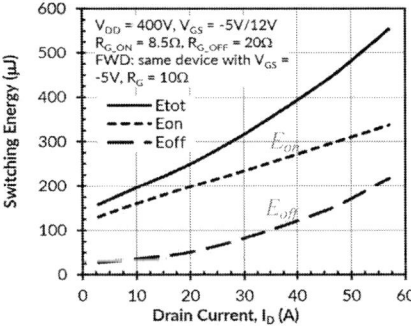

Fig. 6. A typical power semiconductor device with E_{on} much greater than E_{off}, which is a screenshot from SiC MOSFET UF3C065030K4S [29].

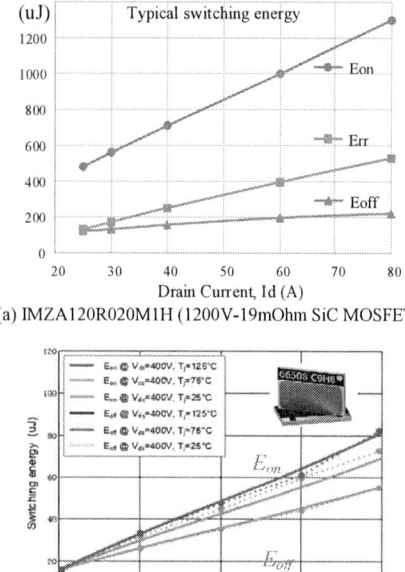

(a) IMZA120R020M1H (1200V-19mOhm SiC MOSFET)

(b) GS66508T (650V-50mOhm GaN HEMTs) in [4].

Fig. 7. Switching energy data from the datasheet and test results.

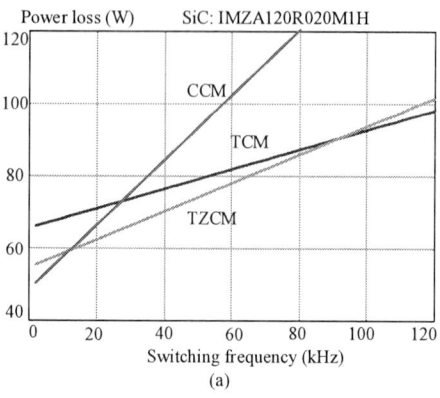

(a)

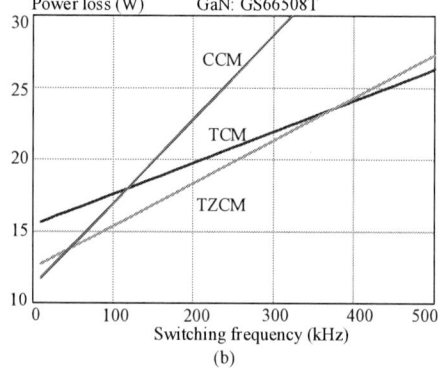

(b)

Fig. 8. Power loss comparison in CCM, TCM and TZCM. (a) SiC MOSFET IMZA120R020M1H with I_{Lavg}=40 A, (b) GaN HEMT GS66508T with I_{Lavg}=10 A.

Using wide bandgap power devices as a study case to demonstrate the characteristics of the three operating modes. The switching energy data of SiC MOSFET and GaN HEMT is shown in Fig. 7, obtained from the datasheet and test results in [4], [30]. By fitting these switching energy data and calculating using Eqs. (11)-(13), the switching losses can be obtained. Combined with the calculation of conduction loss using Eqs. (8)-(10), the total loss of the power semiconductor device can be obtained at different switching frequencies, as shown in Fig. 8.

The working condition in Fig. 8(a) are based on the SiC MOSFET IMZA120R020M1H (R_{on}=0.03Ω @125°C) [31], which has a rated voltage of 800 V and a rated current of 40 A for each switch, i.e., V_{dc}=1600 V and V_o=800 V for three-level DC-DC converter. The peak-to-peak value of the inductor current ripple is 40% of the average current (I_{Lavg}=40 A) in CCM mode, and the negative peak current is -2 A in TCM and TZCM modes. For Fig. 8(b), the power device is GaN GS66508T(R_{on}=0.11Ω @125°C) [30] , the peak-to-peak value of the inductor current ripple is 40% of the average current (I_{Lavg}=10 A) in CCM mode, and the negative peak current is -1 A in TCM and TZCM modes.

Fig. 8 shows the total conduction and switching losses for each complementary switch pair. The losses in all three modes increase with the switching frequency due to the presence of switching losses. However, they have different dominance intervals: 1) CCM has an advantage only at a very low switching frequency due to its low conduction losses and high switching losses. 2) TCM has a slight advantage at a very high switching frequency, provided that the turn-off energy satisfies $E_{off\_ip+} < E_{off\_l1} + E_{off\_l2}$. 3) TZCM has an advantage at the suitable middle switching frequency due to its lower conduction and switching losses compared to TCM and CCM, respectively.

IV. EXPERIMENTAL RESULTS

To compare the losses of the three operating modes, a three-level DC-DC converter prototype is built as shown in Fig. 7. The type of power switch is SiC MOSFET UF3C065030K4S.

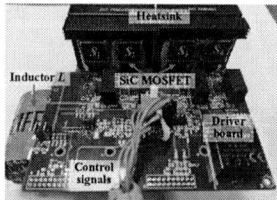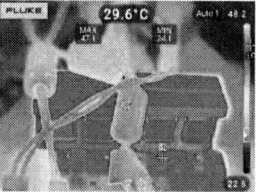

Fig. 9. Experimental prototype of the series-capacitor three-level DC-DC converter and its thermal test.

Fig. 10 shows the waveform comparison in CCM, TCM, and TZCM for a three-level DC-DC converter. The peak current value in TCM is more than twice the average inductor current. The two peak currents in the TZCM are greater than in the CCM but less than in the TCM. Both TCM and TZCM have negative peak currents. Additionally, the CCM waveforms exhibit more pronounced interference than the TCM and TZCM waveforms.

The basic waveforms in TZCM for the three-level DC-DC converter with a 1.2kW load are shown in Fig. 11, which is used to test the efficiency of the converter by changing the load and the switching frequency.

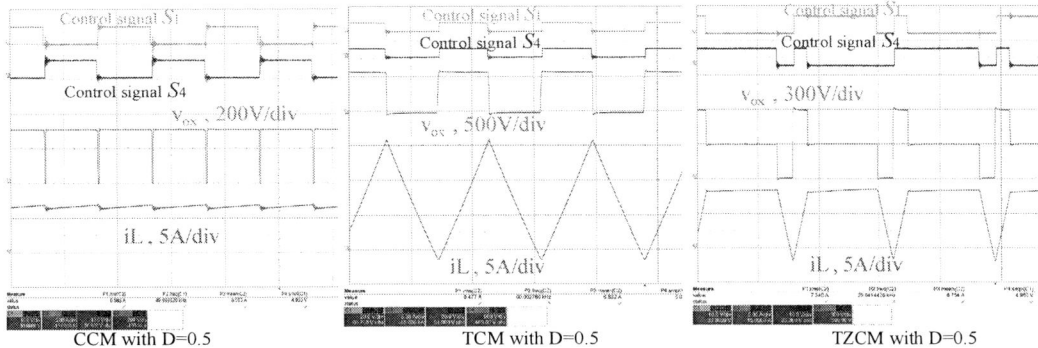

Fig. 10. Comparison of key waveforms in CCM, TCM, and TZCM for three-level DC-DC converter with V_{dc}=600 V.

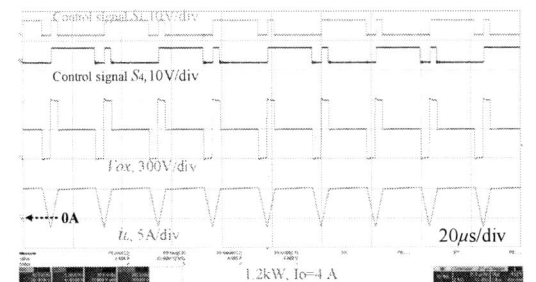

Fig. 11. Basic waveforms in TZCM for three-level DC-DC converter with a 1.2kW load.

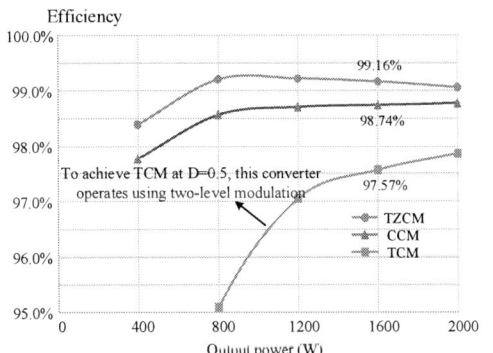

Fig. 12. Efficiency comparison in TZCM, CCM and TCM with V_{dc}=600 V and D=0.5.

The efficiency results for TZCM, CCM and TCM with V_{dc}=600 V and D=0.5 are shown in Fig. 12. The order of their efficiency is TZCM > CCM > TCM. The efficiency of the converter in TCM is lower than that of TZCM and CCM due to the large inductor current ripple, which is caused by the two-level modulation because the inductor current ripple in the conventional three-level modulation is almost zero. In practice, the TCM is not suitable for this three-level converter around D=0.5, whereas the TZCM mode just solves this difficulty in achieving high efficiency soft-switching at a duty cycle of 0.5.

V. CONCLUSION

This paper compares the advantages of CCM, TCM and TZCM in a three-level DC-DC converter. For wide bandgap power devices with E_{on}>E_{off}, TZCM and TCM are preferable to CCM because they can achieve zero-voltage switching (ZVS) soft-switching to eliminate switching losses caused by E_{on} and E_{rr}. Although both TCM and TZCM can achieve ZVS soft-switching, they have own advantageous operating region. CCM only has an advantage at a very low switching frequency. TCM has a slight advantage at a very high switching frequency provided that the turn-off energy satisfies E_{off_ip+}< E_{off_I1} + E_{off_I2}. TZCM has an advantage at the suitable middle switching frequency due to its lower conduction and switching losses compared to TCM and CCM, respectively. Experimental results show differences in waveforms and efficiencies, where the efficiency in TZCM is higher when the duty cycle is close to 0.5.

REFERENCES

[1] Y. Chen and D. Xu, "Review of Soft-Switching Topologies for Single-Phase Photovoltaic Inverters," *IEEE Trans. Power Electron.*, vol. 37, no. 2, pp. 1926-1944, Feb. 2022.

[2] Z. Yao and S. Lu, "A Simple Approach to Enhance the Effectiveness of Passive Currents Balancing in an Interleaved Multiphase Bidirectional DC–DC Converter," *IEEE Trans. Power Electron.*, vol. 34, no. 8, pp. 7242-7255, Aug. 2019.

[3] J. A. Anderson, C. Gammeter, L. Schrittwieser and J. W. Kolar, "Accurate Calorimetric Switching Loss Measurement for 900 V 10 mΩ SiC mosfets," *IEEE Trans. Power Electron.*, vol. 32, no. 12, pp. 8963-8968, Dec. 2017.

[4] R. Hou, Y. Shen, H. Zhao, H. Hu, J. Lu and T. Long, "Power Loss Characterization and Modeling for GaN-Based Hard-Switching Half-Bridges Considering Dynamic on-State Resistance," *IEEE Trans. Transp. Electrif.*, vol. 6, no. 2, pp. 540-553, June 2020.

[5] M. Guacci *et al.*, "On the Origin of the Coss -Losses in Soft-Switching GaN-on-Si Power HEMTs," *IEEE Journal of Emerging and Selected Topics in Power Electronics*, vol. 7, no. 2, pp. 679-694, June 2019.

[6] Q. Song, R. Zhang, Q. Li and Y. Zhang, "Origin of Soft-Switching Output Capacitance Loss in Cascode GaN HEMTs at High Frequencies," *IEEE Transactions on Power Electronics*, vol. 38, no. 11, pp. 13561-13566, Nov. 2023.

[7] F. Meng, S. Yong, J. Wu, Y. Shu and Y. Chen, "ZVS Implementation Analysis and Optimization Design of LLC Converter Based on TDM-DC," *IEEE Trans. Power Electron.*, doi: 10.1109/TPEL.2025.3579730.

[8] Z. Yao, T. Jiang, T. Sun, Z. Xiao, W. Chen and Y. Tang, "Smooth Optimal Trajectory and Voltage Balancing Control for T-Type Three-Level LLC Resonant Converter in Alternating Mode," *IEEE Trans. Power Electron.*, vol. 40, no. 7, pp. 9278-9288, July 2025.

[9] G. Park, H. Kim, B. -G. Cho and S. Cui, "ZVS-Enhanced and RMS-Current-Minimized Optimal Modulation Scheme of Dual-Active Bridge Converter With Comprehensive ZVS Analysis," *IEEE Trans. Power Electron.*, vol. 40, no. 7, pp. 9004-9018, July 2025.

[10] G. Yu, S. Yadav, J. Dong and P. Bauer, "Revisiting the Reverse Switched Current of Buck, Boost, and Buck–Boost Converters in Voltage-Mode TCM–ZVS Control Considering Parasitic Resistances," *IEEE Trans. Power Electron.*, vol. 39, no. 7, pp. 8254-8268, July 2024.

[11] Z. Yao et al., "Current Ripple Prediction and ZVS-Based Variable Switching Frequency Control for Interleaved Multiphase Three-Level DC–DC Converter," *IEEE Trans. Power Electron.*, vol. 40, no. 4, pp. 5109-5119, April 2025.

[12] M. Harasimczuk et al., "Common-Leg Coupled Inductor Configuration in a Three-Level Interleaved DC–DC Medium Voltage SiC-Based Converter," *IEEE Trans. Power Electron.*, vol. 39, no. 8, pp. 9694-9704, Aug. 2024.

[13] S. Dusmez, A. Hasanzadeh and A. Khaligh, "Comparative Analysis of Bidirectional Three-Level DC–DC Converter for Automotive Applications," *IEEE Trans. Ind. Electron.*, vol. 62, no. 5, pp. 3305-3315, May 2015.

[14] J. Wan, F. Liu, K. -Z. Liu and Y. Li, "An Efficient Soft-Switching Buck Converter With Parasitic Resonance Suppression in Auxiliary Circuit," *IEEE Trans. Ind. Electron.*, vol. 70, no. 2, pp. 1367-1377, Feb. 2023.

[15] Z. Yao and S. Lu, "Voltage Self-Balance Mechanism Based on Zero-Voltage Switching for Three-Level DC–DC Converter," *IEEE Trans. Power Electron.*, vol. 35, no. 10, pp. 10078-10087, Oct. 2020.

[16] YAO Zhigang, ZHANG Yuxin, LIU Tao, LU Shuai, "ZVS Control of an Interleaved Three-phase Three-level DC-DC Converter," *Proceedings of the CSEE*, vol. 40, no. 13, pp. 4256–4266, Jul. 2020. DOI: 10.13334/j.0258-8013.pcsee.190969. (in chinese).

[17] Z. Yan, J. Zeng, Z. Guo, R. Hu and J. Liu, "A Soft-Switching Bidirectional DC–DC Converter With High Voltage Gain and Low Voltage Stress for Energy Storage Systems," *IEEE Trans. Ind. Electron.*, vol. 68, no. 8, pp. 6871-6880, Aug. 2021.

[18] C. Marxgut, F. Krismer, D. Bortis and J. W. Kolar, "Ultraflat Interleaved Triangular Current Mode (TCM) Single-Phase PFC Rectifier," *IEEE Trans. Power Electron.*, vol. 29, no. 2, pp. 873-882, Feb. 2014.

[19] J. Fang, X. Ruan, X. Huang, R. Dong, X. Wu and J. Lan, "A PWM Plus Phase-Shift Control for Four-Switch Buck-Boost Converter to Achieve

ZVS in Full Input Voltage and Load Range," *IEEE Trans. Ind. Electron.*, vol. 69, no. 12, pp. 12698-12709, Dec. 2022.

[20] G. Yu et al., "Three-Mode Variable-Frequency ZVS Modulation for Four-Switch Buck+Boost Converters With Ultra-High Efficiency," *IEEE Trans. Power Electron.*, vol. 38, no. 4, pp. 4805-4819, April 2023.

[21] J. Liao, G. Qiu, Y. Huang and V. Khadkikar, "Lagrange-Multiplier-Based Control Method to Optimize Efficiency for Four-Switch Buck–Boost Converter Over Whole Operating Range," *IEEE Trans. Ind. Electron.*, vol. 71, no. 1, pp. 822-833, Jan. 2023.

[22] Z. Yao et al., "A Trapezoidal Current Mode to Reduce Peak Current in Near-CRM for Three-Level DC–DC Converter," *IEEE Trans. Power Electron.*, vol. 39, no. 8, Aug. 2024.

[23] Z. Yao, X. He, M. Liu, J. Liu, Z. Xiao and Y. Tang, "Fixed Switching Frequency Control Using Trapezoidal Current Mode to Achieve ZVS in Three-Level DC–DC Converters," *IEEE Trans. Ind. Electron.*, vol. 72, no. 4, April 2025.

[24] D. Chou, Y. Lei and R. C. N. Pilawa-Podgurski, "A Zero-Voltage-Switching, Physically Flexible Multilevel GaN DC–DC Converter," *IEEE Trans. Power Electron.*, vol. 35, no. 1, pp. 1064-1073, Jan. 2020.

[25] Z. Yao et al., "Generalized Trapezoidal Current Mode-Based Zero-Voltage Switching for Multilevel DC–DC Converters," *IEEE Trans. Power Electron.*, vol. 40, no. 7, pp. 8956-8961, July 2025.

[26] A. Hu and J. Biela, "Fast and Accurate Data Sheet Based Analytical Switching Loss Model for a SiC MOSFET and Schottky Diode Half-Bridge," *IEEE Open Journal of Power Electronics*, vol. 5, pp. 1684-1696, 2024.

[27] H. Gui, J. Sun and L. M. Tolbert, "Charge Pump Gate Drive to Reduce Turn-ON Switching Loss of SiC MOSFETs," in *IEEE Transactions on Power Electronics*, vol. 35, no. 12, pp. 13136-13147, Dec. 2020.

[28] Z. Gu et al., "Comparative Study on High-Temperature Electrical Properties of 1.2 kV SiC MOSFET and JBS-Integrated MOSFET," in *IEEE Transactions on Power Electronics*, vol. 39, no. 4, pp. 4187-4201, April 2024.

[29] Onsemi's datasheet of UF3C065030K4S, www.onsemi.cn.

[30] GaN Systems' datasheef of GS66508T, www.gansystems.com.

[31] Infineon's datasheet of IMZA120R020M1H, www.infineon.com.

Design and Current-Sharing Study of the TL-Boost Power Unit Based on Discrete Devices in Parallel

Jianing Wang
School of Electrical Engineering and Automation
Hefei University of Technology
Hefei, China
jianingwang@hfut.edu.cn

Honghong Li
School of Electrical Engineering and Automation
Hefei University of Technology
Hefei, China
2023170551@mail.hfut.edu.cn

Shaolin Yu*
The Institute of Energy, Hefei Comprehensive National Science Center(Anhui Energy Laboratory)
Hefei, China
yusl@ie.ah.cn

Donglei Zhang
School of Electrical Engineering and Automation
Hefei University of Technology
Hefei, China
2022170536@mail.hfut.edu.cn

Baolong Yan
School of Electrical and Information Engineering
Anhui University of Science and Technology
Huainan, China
2023201832@aust.edu.cn

Zhenchun Xia
School of Electrical Engineering and Automation
Hefei University of Technology
Hefei, China
2023170526@mail.hfut.edu.cn

Weina Mao
School of Electrical Engineering and Automation
Hefei University of Technology
Hefei, China
2023170557@mail.hfut.edu.cn

Abstract—The TL-Boost scheme utilizing parallel discrete devices has gained significant attention due to its cost-effectiveness and flexible power scalability. However, asymmetric design of interconnection structures in main power units could compromise the consistency of parasitic parameters across branches, resulting in unbalanced current distribution among parallel devices. This paper establishes a parallel parasitic parameter model based on practical commutation paths of TL-Boost converters, proposing an engineer-friendly multi-device parallel power unit architecture with practical applicability. Field-circuit co-simulation reveals the mapping relationship between structural parameters and current-sharing characteristics, enabling quantitative evaluation of current distribution patterns. Furthermore, a prototype was developed and experimentally validated through double-pulse testing, demonstrating excellent current equalization performance. The proposed simulation methodology provides technical support for implementing discrete devices in medium/high-power converter applications.

Keywords — Parallel discrete devices, TL-Boost, Parasitic parameters, Current balancing, Double-pulse test.

I. INTRODUCTION

Non-isolated photovoltaic inverters have been widely adopted in high-voltage, high-power applications due to their high efficiency and compact form factor. In such systems, TL-Boost circuits - typically serving as front-stage boost converters - effectively elevate voltage levels while reducing switching losses[1]. For medium/high-power TL-Boost converters employing parallel discrete devices, this approach not only enhances current-carrying capacity but also offers superior power scalability. Moreover, it demonstrates notable cost-effectiveness in practical implementation.

Current imbalance among parallel-connected devices remains a critical challenge, causing unequal voltage stress and power dissipation across devices. This imbalance not only compromises operational reliability but may also induce derated system operation through the weakest-link effect[2]. Primary factors affecting current distribution include device-to-device variations (e.g., threshold voltage, on-state resistance) and parasitic parameter discrepancies in power loops[3]. Particularly for SiC devices with their faster switching speeds compared to Si-based counterparts, meticulous consideration of both intrinsic parameter variations and parasitic parameter distributions becomes imperative during power unit design. Current industrial practice involves device selection from identical production batches with parameter consistency screening to mitigate the impact of intrinsic device variations. Consequently, achieving symmetrical parasitic parameter design in parallel branches emerges as the decisive factor for current balancing among parallel devices.

Reference [4] analyzes the parallel technology of Si IGBT discrete devices in new energy vehicles and proposes that the symmetrical design of the main power loop is the key to achieving current sharing, though no actual prototype structure is provided. References [5] and [6] address the issue of current imbalance by adjusting bond wire lengths or introducing additional parasitic parameters via carefully designed copper clips to balance branch parasitic parameters. However, this approach sacrifices device switching speed and causes additional voltage overshoot. Some researchers have developed a circumferentially symmetric layout for power modules[7], where the physical space structure is entirely symmetric, meaning that the parasitic parameters between the external power unit branches are perfectly symmetric. However, this structure is complex and has low space utilization, making it difficult to widely apply in engineering practice..

The symmetric layout of the external power loop is an important method for achieving current sharing, and evaluating the current-sharing performance of the layout

---

This work was supported by Anhui Province Key Research and Development Program Project, Project Number: JZ2024AKKG0057.

through reasonable and accurate simulation methods is a crucial basis for iterative design of power unit layouts. Reference [8] established a parasitic parameter model considering the self-inductance and mutual inductance effects of parallel branches for multi-SiC MOSFET parallel power modules, focusing on the impact of mutual inductance coupling effects on switching transient electrical characteristics. However, the influence of parasitic resistance on current distribution was ignored. Reference [9] only considers self-inductance and self-resistance along the current paths, without considering the coupling effects between parasitic parameters. Additionally, parasitic parameters are extracted at a single frequency point, without considering the complex conditions across a broad frequency range.

This paper establishes a parallel-type parasitic parameter model for the actual commutation path and proposes a TL-Boost power unit structure based on multi-discrete device parallelization, which is applicable in practical engineering. Through co-simulation of the field and circuit, the mapping relationship between the current sharing characteristics is established, and the current distribution characteristics under this structure are evaluated. Finally, a prototype of the discrete device parallel-type TL-Boost is built and subjected to double pulse testing. The test results demonstrate that the proposed power unit structure exhibits good current sharing characteristics. Additionally, the accuracy of the proposed simulation evaluation method is also verified.

II. TL-BOOST TOPOLOGY COMMUTATION LOOP AND PARASITIC PARAMETER MODELING ANALYSIS

A. Power Conversion Circuit Analysis under Synchronous Modulation

When the two switches are controlled by the same driving signal, turning on and off simultaneously, it is defined as synchronous modulation. The TL-Boost topology often uses synchronous modulation to suppress common-mode currents in the circuit[10]. As shown in Fig. 1, under continuous conduction mode (CCM), the circuit operates in two modes:

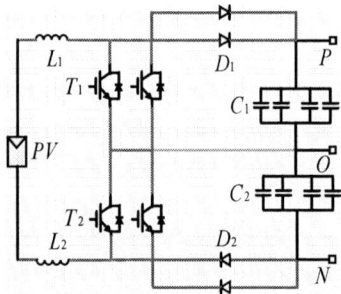

Fig. 1. Current path under synchronous modulation

Mode 1 (Current path in red): The switches T1 and T2 are both turned on simultaneously, while diodes D1 and D2 are off. In this mode, the DC source provides energy to inductors L1 and L2, and the current flowing through the switches is equal to the input current, exhibiting a linear rise.

Mode 2 (Current path in blue): The switches T1 and T2 are both turned off simultaneously, each experiencing half of the input voltage, while diodes D1 and D2 are on. In this mode, the DC source and inductors work together to provide energy to the capacitor and load. The current flowing through the switches is zero, and the inductor current decreases linearly.

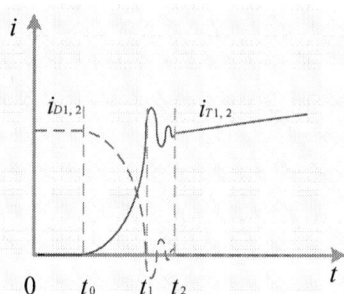

Fig. 2. Current analysis of switch states.

Assuming the circuit transitions from Mode 2 to Mode 1, due to the presence of large inductance, the current in the loop cannot change instantaneously. As shown in Fig. 2, the current switches from the path through diodes D1 and D2 to the path through switches T1 and T2. In this transition, the current through diodes D1 and D2 decreases, while the current through switches T1 and T2 increases. The magnitude of the switching transient is equal but in the opposite direction, maintaining the same rate of change. This creates a power conversion loop between diodes D1, D2, switches T1, T2, and the conductive paths. Due to the mutual resonance between the parasitic inductance of the power loop and the junction capacitance of the switches, the current waveform experiences oscillations at the beginning of the switching process.

B. Parasitic Parameter Network Model

Based on the previous section's analysis of the power conversion loop under synchronous modulation, each current path in the loop corresponds to a conductor segment, which can be represented by self-inductance and self-resistance. Due to the proximity effect, there exists mutual inductance and mutual resistance between any two paths. Therefore, the parasitic parameter model must include both inductive and resistive elements. Additionally, for the subsequent experimental verification, full-voltage double-pulse testing will be used, taking into account the external load and DC source connection method. Ultimately, the established power loop considers a parasitic parameter model that includes certain self-inductance, mutual inductance, self-resistance, and mutual resistance, as shown in Fig. 3.

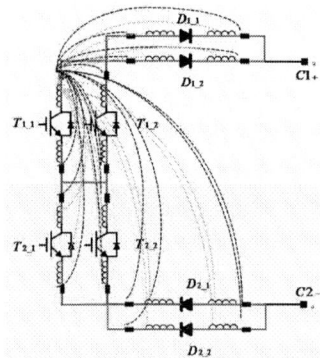

Fig. 3. Parasitic parameter network with mutual inductance and mutual resistance.

The parasitic parameter network indicates that in the main power loop, each switch IGBT's collector and emitter, as well as the current paths of the anode and cathode of the diodes, all have self-inductance and self-resistance. Additionally, there will be mutual resistance and mutual inductance coupling between different self-resistances and self-inductances, collectively forming a complex and large parasitic parameter network. If the impact of the parasitic parameters from the switch IGBT's driving circuit is also considered, the parasitic parameter network becomes even larger.

Due to the complexity of this parasitic parameter network, it is difficult to directly analyze its influence on device parallel current-sharing through analytical formulas. Therefore, this paper proposes using a Q3D-Simplorer-based field-circuit co-simulation method for research.

III. POWER UNIT PCB STRUCTURE LAYOUT

A. Parameters and Structure of the Actual Device

The electrical parameters of the developed TL-Boost prototype device are shown in Table I.

TBALE II. ELECTRICAL PARAMETERS OF PROTOTYPE DEVICE

Parameter	Material
Input Voltage Range/V	500~1500
Full-load Input Voltage/V	800~1350
Rated Output Voltage/V	1300
Maximum Input Current/A	75
Switching Frequency/kHz	40
Number of Parallel Devices	2
Cooling Method	Forced-air cooling

During the design of the device, considering constraints such as electrical clearance and creepage distances, and the minimal area required to accommodate all components, a PCB size limitation of 300 mm × 220 mm was set based on the board's mechanical strength and the practical assembly space within the overall system. The main goal is to achieve high power density, meeting the demands of compact system design.

However, pursuing higher power density leads to increased thermal concentration. To improve cooling efficiency and facilitate practical assembly, the pins of switching devices (IGBT and diodes) are bent and placed into positioning slots, allowing their back surfaces to directly contact heat sinks.

With the PCB dimensions, device types, and quantities already fixed, the following subsection discusses how to optimally place the devices within the limited space to achieve uniform parasitic parameters and maximize current-sharing performance.

B. Optimization of Stray Inductances

In the power module, both stray inductances and stray resistances influence device switching performance. However, since stray inductances and resistances exhibit similar distribution characteristics, this analysis primarily focuses on stray inductances.

Device placement and orientation directly affect the current entry and exit points, influencing current paths. Given fixed PCB dimensions due to power density and cost constraints, achieving symmetric layouts for the components within the confined area intuitively ensures consistency in the parasitic inductances of each branch, thereby promoting

uniform current sharing among parallel devices. Furthermore, inspired by laminated busbar designs, the proposed structure exploits mutual inductance coupling principles by arranging current paths on different PCB layers in opposite directions, partially canceling stray inductances. The following discussion details the optimization of loop inductances.

Loop inductances in commutation paths can adversely affect circuit performance in several ways. When switching devices operate at high speeds, the energy stored in the inductance rapidly transfers into the output capacitance () of the device being turned off, causing voltage overshoot and electromagnetic interference due to high-frequency oscillations[11] Film capacitors and act as critical points in the commutation loop and significantly influence loop inductances. Taking capacitor placement as an example, this paper analyzes the impact of positive-negative terminal spacing and orientation on PCB busbar loop inductance. Fig. 4 compares three capacitor placement schemes and examines how adding additional capacitors influences loop inductances in each configuration.

Subsequently, a 3D model is constructed in Ansys Q3D, setting constant values for the width, thickness, and inter-layer spacing of rectangular busbars. Fig. 5 shows simulated parasitic inductances corresponding to different capacitor placement schemes. From Fig. 5 Layout 2 yields the highest inductance due to minimal mutual inductance cancellation area, while Layout 3 achieves the lowest inductance owing to maximal mutual inductance cancellation. Thus, Layout 3 is selected for capacitor placement. Furthermore, comparing Layout 3 and 4 demonstrates that increasing the number of film capacitors reduces loop inductances.

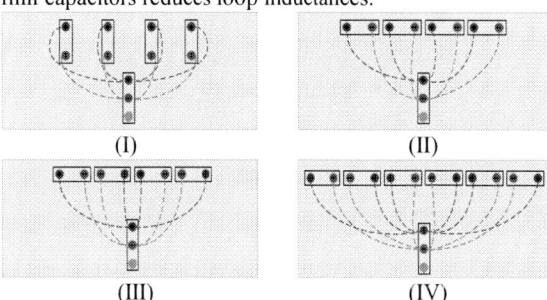

Fig. 4. Depicts capacitor layouts.

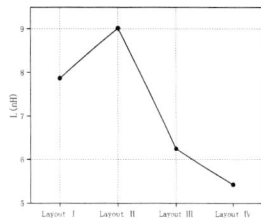

Fig. 5. Parasitic inductances for various capacitor placement schemes.

Based on these findings, a practical PCB layout structure for the triple parallel TL-Boost topology power module is designed as shown in Fig. 6. Fig. 6(a) illustrates the corresponding topology diagram for the actual main power loop PCB structure, and Fig. 6(b) depicts the three-dimensional structural layout of the main power loop and driver circuits.

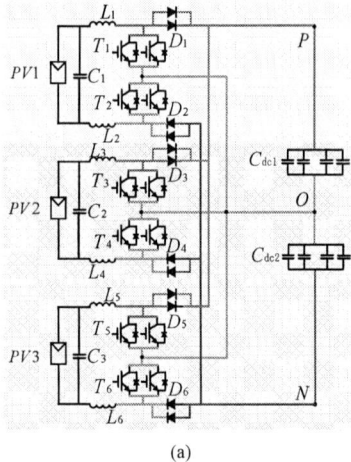

(a)

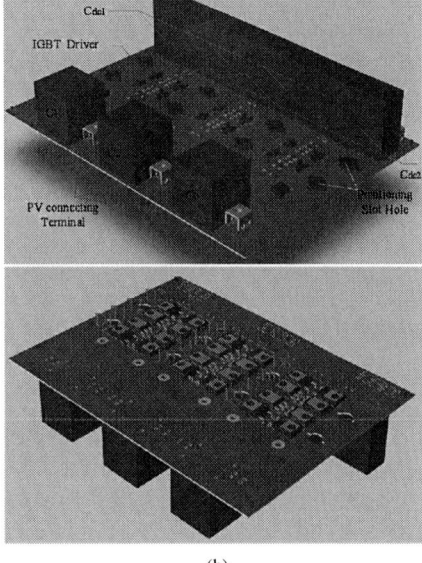

(b)

Fig. 6. Actual power unit topology and PCB structure:(a) Three-phase parallel TL-Boost topology diagram,(b) 3D structure of PCB layout.

The three-phase TL-Boost power unit adopts a left-middle-right placement, with the positive terminal P, the zero voltage terminal O, and the negative terminal N connected to the corresponding terminals on one side of the PCB. On the opposite side, the input terminals for the three-phase TL-Boost are located. Taking the layout of the first TL-Boost structure as an example, the switch devices T1 and T2 use the TO247-3PLUS package, with the model DGQ75N120CTH1B, $= 1200V$, $= 75A$. Two single devices are connected in parallel and their pins are bent to align with the PCB plane. The back lead frame is facing upwards to facilitate the attachment of the heat sink. The freewheeling diodes D1 and D2 use the TO-247-2 package, with the model 75F120WT fast recovery diodes, $= 75A$, $= 1200V$, with bent leads for installation. The thin-film capacitors C1 and C2 are both made up of four capacitors connected in parallel, with the model C3D2K146KB30382, each capacitor having a nominal value of 14μF. C1 and C2 are placed alternately. The energy storage inductors are not installed as they were not involved in the simulation or actual testing. The DC buffer

capacitors are placed at each input terminal. The IGBT driver circuit uses the 2FSC0110T12C1 dual-channel driver, which can provide a single-channel drive power of 1W (= 85°C).

IV. FIELD-CIRCUIT CO-SIMULATION WIDE-FREQUENCY DOMAIN CURRENT-SHARING SIMULATION ANALYSIS BASED ON Q3D-SIMPLORER

After completing the layout and routing work in Cadence, the 3D entity model of the PCB current path is extracted in Q3D. In Q3D, material properties and source/sink locations are set. Due to the characteristics of the Q3D software, only one sink can be set for a single conductor net, while multiple sources can be set. To accurately extract the branch parasitic parameters, some conductors need to have reverse excitation settings. In this case, the magnitude of the extracted mutual inductance and mutual resistance remains unchanged, but the signs are reversed. However, when building the peripheral circuit in Simplorer later, controlling the current flow direction at the same terminal name for self-inductance and self-resistance will correct the sign issue of mutual inductance and mutual resistance.

Inductance and resistance are physical quantities that vary with frequency. Extracting parasitic parameters at a single frequency point cannot accurately reflect the actual operating conditions with various switching frequencies. Therefore, the Q3D simulation frequency is set from 0 to 10 MHz, using an exponential form for point extraction, covering the range from DC to the device's edge equivalent frequency. This approach can effectively reflect the parasitic parameter values of parasitic inductance and parasitic resistance in different frequency domains under the entire operating condition.

Through the co-simulation between Ansys Simplorer and Q3D, a full-voltage double-pulse circuit is built in Simplorer, as shown in Fig. 7. The simulation results show the current waveform of parallel devices during dynamic switching under the influence of the parasitic parameters of the current path. In the Fig. 7, the film capacitors account for the actual capacitive ESL and ESR, with these values obtained by measuring several capacitors of the model C3D2K146KB30382 using a vector network analyzer. The switch IGBT deviates from the traditional method of parameterizing devices using datasheets. Instead, the device's spice model from Infineon's official website is imported into Simplorer for modeling, better considering the actual parasitic parameters inside the device.

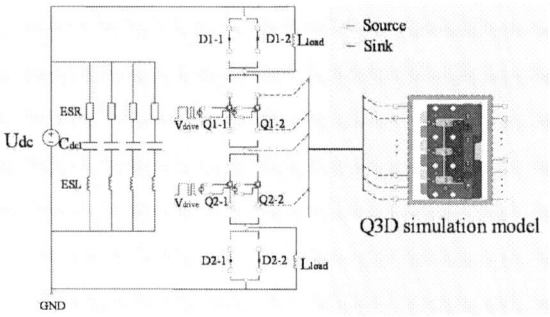

Fig. 7. Simplorer peripheral full-voltage double-pulse circuit.

The double-pulse simulation waveform diagram shows the current characteristics of the upper switch IGBT-T1, the lower switch T2, and diodes D1 and D2 during the dynamic turn-on process and the static process after full conduction. The diagram also includes a zoomed-in section to compare the current waveform trends of the parallel devices during the turn-on process. This comparison allows for the evaluation of the current PCB layout and routing performance, facilitating iterative optimization and design to achieve the most balanced current distribution among the parallel devices.

For ease of analysis, the current imbalance is quantified. If, during a certain process, the currents through two parallel devices are I1 and I2, the current imbalance at this point is given by equation (1):

$$I_N = \frac{I_1 - I_2}{(I_1 + I_2)/2} \qquad (1)$$

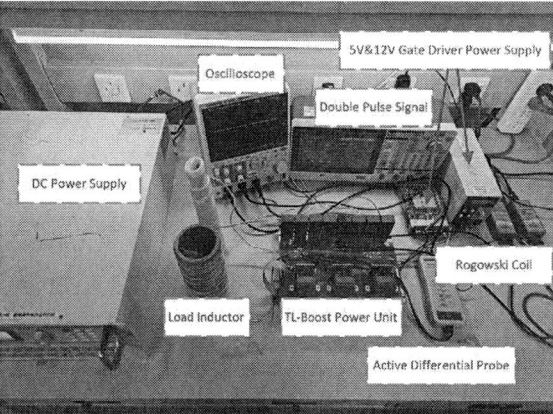

Fig. 8. Full-voltage double-pulse simulation result waveform:(a) D1 current waveform,(b) D2 current, (c) Collector Current of T1, (d) Collector Current of T2.

Fig. 8 shows the current waveforms of the switches and diodes under the combined coupling effect of parasitic parameters from both the power loop and the driver circuit. From the final simulation results, it can be seen that, except for the dynamic turn-on process of the upper switch D1 and T1, where the conduction times of the individual devices are inconsistent, leading to a large current overshoot and a significant current imbalance, the current imbalance of the remaining devices during the dynamic turn-on and static conduction processes is all below 5%, demonstrating good current-sharing performance.

V. PERFORMANCE TESTING

A. Experimental Platform Setup

To verify the current-sharing characteristics of the TL-Boost power unit, a double-pulse test platform was established as illustrated in Fig. 9. The DC-side voltage of 750 V was supplied by a 30 kW DC power source. The durations of the double pulses were set to 20 μs and 10 μs, respectively. Initially, the signal generator produced a double-pulse signal ranging from 0 V to 5 V, feeding the gate driver board. Subsequently, the gate driver converted this voltage to a gate drive signal varying from -8 V to +15 V to control the IGBTs. Two load inductors, each with an inductance of 90 μH, were employed. Moreover, device voltages and currents were measured using active differential voltage probes and Rogowski coils, respectively, with the waveforms acquired by a Yokogawa oscilloscope (500 MHz, 2.5 GS/s).

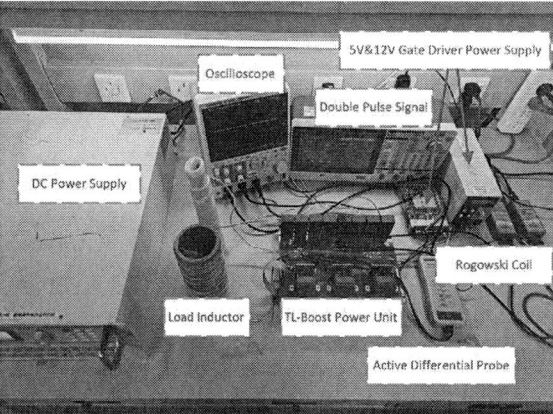

Fig. 9. Double-pulse test platform.

B. Analysis of Experimental Results

The diodes and IGBTs utilized in the experiments were sourced from the same manufacturer and were carefully screened by static test equipment to minimize variations caused by intrinsic device dispersion. The measured current waveforms of each branch are illustrated in Fig. 10.

During the experiment, additional parasitic parameters were introduced due to practical factors such as wiring and the positioning angle of the Rogowski coils, resulting in a dynamic current imbalance increase of approximately 5–10% and a static current imbalance increase of around 1–3% for the IGBTs. Furthermore, due to diode placement, measurements were susceptible to parasitic inductances from adjacent branches, causing the current imbalance to increase by nearly 10%. Despite these increases, the observed current imbalances remained well below acceptable engineering application standards. These results validate the effective current-sharing capability of the proposed discrete-device parallel TL-Boost power module..

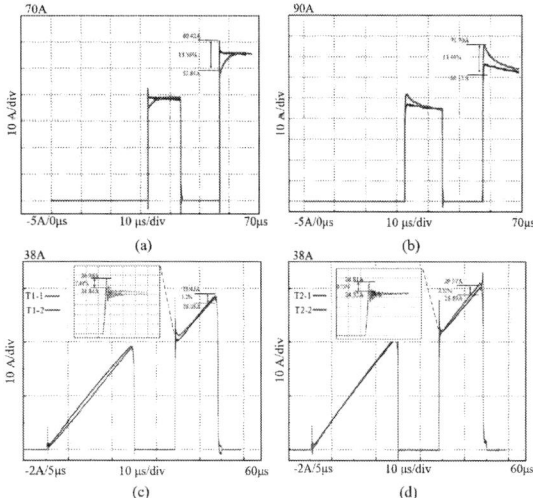

Fig. 10. Experimental waveforms obtained from full-voltage double-pulse tests. (a) D1. (b) D2. (c) T1. (d) T2.

VI. CONCLUSION

The use of discrete semiconductor devices in parallel configurations has notable advantages in terms of system cost and power scalability. Based on this concept, a discrete-device parallel TL-Boost power module was developed in this paper. A parallel-connected parasitic parameter model was constructed considering the actual commutation paths of the topology. Furthermore, a co-simulation combining field and circuit analyses was utilized to establish the mapping relationship between parasitic parameters and current-sharing characteristics, thus enabling the evaluation of current distribution performance. Finally, a practical prototype of the discrete-device parallel TL-Boost power unit was assembled and evaluated using double-pulse testing. Experimental results demonstrated the superior current-sharing characteristics of the developed power unit and confirmed the accuracy of the proposed simulation and evaluation method..

REFERENCES

[1] Zeng Jiang, Huang Zhonglong, Qiu Guobi. Coordination Control of Three-level Boost-inverter Considering Neutral-point Potential Balance [J]. Electrical Drive,2020,50(04):38-44.

[2] Zeng Z, Zhang X, Zhang Z. Imbalance current analysis and its suppression methodology for parallel SiC MOSFETs with aid of a differential mode choke[J]. IEEE Transactions on Industrial Electronics, 2020, 67(2): 1508-1519.

[3] H. Li, S. Zhao, X. Wang, et al. Parallel Connection of Silicon Carbide MOSFETs — Challenges, Mechanism, and Solutions[J]. IEEE Transactions on Power Electronics, 2023, 38(8):9731-9749.

[4] Liu Dan, Xiao Zhe, Liu Weiwen. Parallel Connection Analysis of Power Discrete IGBTs in Electric Vehicle Application [J]. Power Electronics, 2011,45(12):66-68.

[5] S. Beczkowski, A. Bjorn Jorgensen, H. Li , et al, Switching current imbalance mitigation in power modules with parallel connected SiC MOSFETs[J]. 2017 19th European Conference on Power Electronics and Applications (EPE'17 ECCE Europe),2017, pp. P.1–P.8.

[6] L. Wang et al., Cu clip-bonding method with optimized source inductance for current balancing in multichip SiC MOSFET power module, IEEE Trans. Power Electron., vol. 37, no. 7, pp. 7952–7964, Jul. 2022.

[7] Shao W, Li R, Zeng Z, et al. Design and evaluation of SiC multichip power module with low and symmetrical inductance [C]. Institution of Engineering and Technology, Milwaukee, WI, USA: IEEE, 2019:3573-3577.

[8] Zhang B, Wang S. Parasitic Inductance Modeling and Reduction for Wire-Bonded Half-Bridge SiC Multichip Power Modules[J]. IEEE Transactions on Power Electronics, 2021, 36(5):5892- 5903.

[9] G. Chang et al., "Optimization and Validation of Current Sharing in IGBT Modules With Multichips in Parallel," IEEE Transactions on Power Electronics, vol. 39, no. 12, pp. 15672-15681, Dec. 2024.

[10] Xie Yong, Cheng Gan, Fang Yu,et al. Research on Two Key Technical Problems in TL-Boost Photovoltaic System [J]. Power Electronics,2016,50(01

[11] A. Bhargava, D. Pommerenke, K. W. Kam, F. Centola, and C. W. Lam, "Dc-dc buck converter emi reduction using pcb layout modification," IEEE Transactions on Electromagnetic Compatibility, vol. 53, no. 3, pp. 806–813, 2011.

2025 IEEE Workshop on Wide Bandgap Power Devices and Applications in Asia (WiPDA Asia)

Study on New Structures of EST to inhibit snapback effect and enhance MCC

Wuhua Yang
Department of Electronic Engineering
Xi'an University of Technology
Xian, China
yangwuhua@xaut.edu.cn

Jia Liping
Department of Electronic Engineering
Xi'an University of Technology
Xian, China
Jia_LiPing56@163.com

Guo Jiarui
Department of Electronic Engineering
Xi'an University of Technology
Xian, China
1540738346@qq.com

Zhang Chao
Department of Electronic Engineering
Xi'an University of Technology
Xian, China
zhangchao4403@xaut.edu.cn

Shen Sihao
Department of Electronic Engineering
Xi'an University of Technology
Xian, China
1549347775@qq.com

Wang Cailin
Department of Electronic Engineering
Xi'an University of Technology
Xian, China
wangcailin@xaut.edu.cn

Abstract—This paper proposes two novel structures, namely the Improved planar Emitter Switched Thyristor with the introduction of the CS layer and the Trench-planar Emitter Switched Thyristor (TP-EST) combining planar gate and trench gate, which address the snapback effect in the traditional planar-gate EST structure. Based on the above-mentioned two devices, the p$^+$ well region in the structure is replaced with a shallow p^{++} layer to increase the maximum controllable current.

Keywords—Emitter Switched Thyristor (EST), snapback effect, Maximum Controllable Current(MCC)

I. INTRODUCTION

In the high-power and high-voltage field, the emitter switched thyristor (EST), which is formed by combining the MOS and thyristor structures, has a better trade-off between the on-state voltage drop and the withstanding voltage[1-2]. However, there are two problems for the EST in high-power application: (1) the snapback effect during the turn-on of the EST[3]; (2) the lower maximum controllable current (MCC) during the turn-off of the EST[4].

II. BACKGROUND

Fig. 1 shows the turn-on characteristic curve of the conventional 6.5kV P-EST in the inductive load. The "hump" in the V$_{AK}$ curve with the increase of J$_A$ reflects the snapback effect of EST,which is caused by the transition of working state from the IGBT mode into thyristor mode.

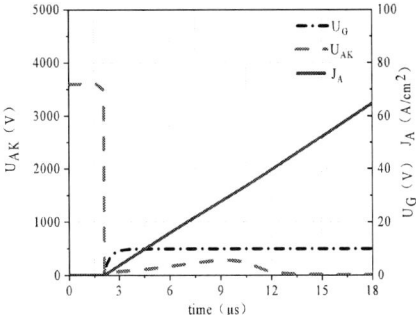

Fig. 1 Turn-on characteristic curve of the Con. P-EST

When the gate voltage of the two gates of the Con. P-EST drops to zero, the electron channel beneath the gate disappears, resulting in the cutoff of electron current and turn-off of the device. Fig. 2 shows the turn-off characteristic

curve of the device, with a turn-off current density of approximately 85A/cm². During the turn-off, as the gate voltage pulse width increases, the maximum controllable current density of the Con. P-EST is limited to about 198.1 A/cm² by the parasitic thyristor latch-up.

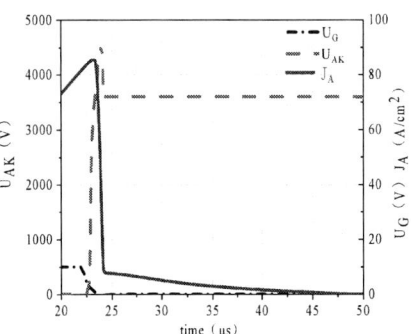

(a) Turn-off characteristic curve

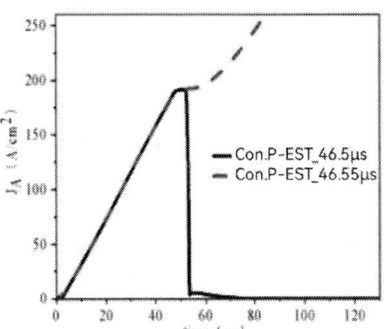

(b) Maximum controllable current density
Fig. 2 Turn off chatacteristics curve of the Con. P-EST

For the Con. P-EST, the snapback phenomenon can cause a larger turn-on loss and impact on turn-on robustness, and the parasitic thyristor latch-up seriously affects the MCC of the device.

III. IMPROVED PLANAR-GATE EST

Fig. 3 shows the structure of the improved P-EST. The improved P-EST structure introduces a CS layer to accelerate the carrier accumulation at the cathode-side. The n-CS layer is formed by the retrograde doping process, and the overall device process is compatible with that of IGBTs.

979-8-3315-1110-4/25 $31.00 © 2025 IEEE

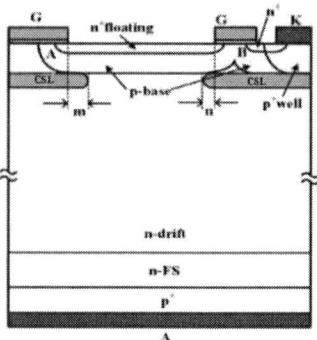

Fig.3 The improved P-EST structure

Fig. 4 shows on-state characteristics curve of the improved P-EST.Fig. 5 shows turn-on characteristic curve of the improved P-EST. During the conduction process, the voltage corresponding to the conversion of the device from the IGBT mode to the EST mode is approximately 14.9V, and the snapback effect has been improved by approximately 92.2%.There is no obvious " hump " in the V_{AK} curve, indicating that the snapback effect of the device has been significantly improved, which greatly reduces the turn-on loss of the device.

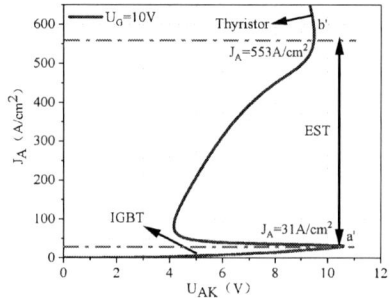

Fig. 4 On-state characteristics curve of the improved P-EST

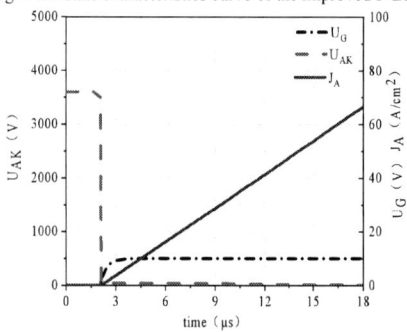

Fig. 5 Turn-on characteristic curve of the improved P-EST

As shown in Fig. 6, at t=2.5μs, the electron channels beneath the two gates are turned on. At t=7μs, the electron is injected into the drift region through the left channel, and the device operates in the IGBT mode. Until t=11μs, the thyristor structure is triggered, and with the current density increasing, the main thyristor structure continues to expand. Compared to Con. P-EST, the CS layer of improved P-EST accelerate the conversion of the device from IGBT mode to EST mode, improving the snapback effect.

The paper is supported by the Teachers' Doctoral Research Funding of Xi＇an University of Technology of China (Grant No.103-451121007), and the Scientific Research Program of Shaanxi Provincial Education Department of China (Grant No.24JK0566).

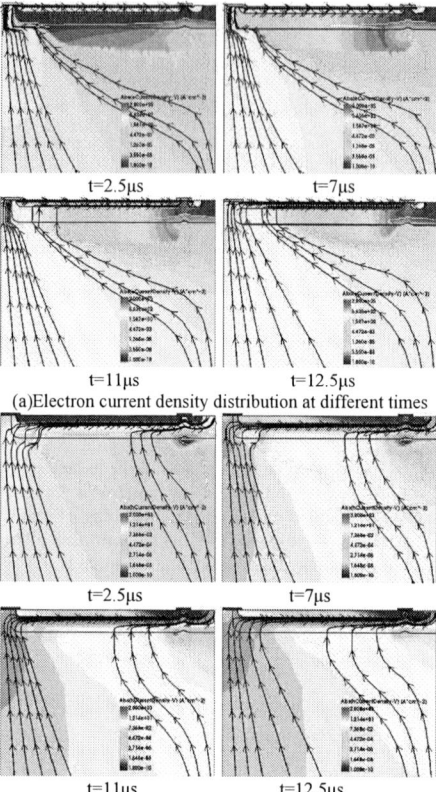

(a)Electron current density distribution at different times

(b)Hole current density distribution at different times
Fig. 6 The electron and hole current density distribution of the improved P-EST during the turn-on

Fig. 7 and Fig. 8 shows the influences of CS layer doping concentration N_{CS} and CS layer thickness h_{CS} on the device performance of improved P-EST, respectively. It can be seen that the increase of N_{CS} or h_{CS} can better improve the snapback effect due to the higher the hole barrier at the p-base/CS layer junction. However, the higher N_{CS} or h_{CS} can increase the electric field strength at the p-base/CS layer junction, leading to decrease of the blocking voltage U_{BR} of the device[5].

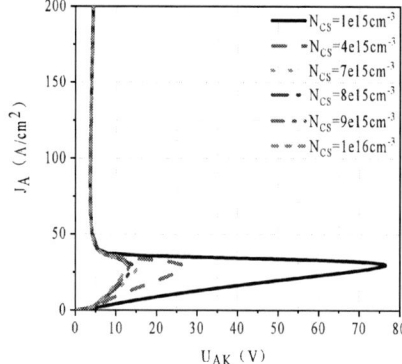

(a) Influence of CS layer doping concentration N_{CS} on on-state characteristic

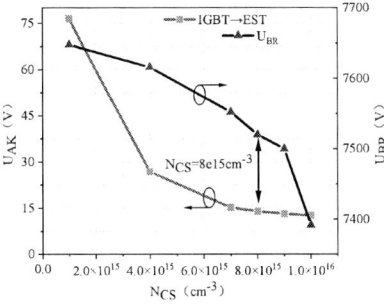

(b) Influence of N_{CS} on the turning point voltage of on-state curves and the blocking voltage U_{BR}

Fig.7 Influence of CS layer doping concentration N_{CS} on the device performance of the improved

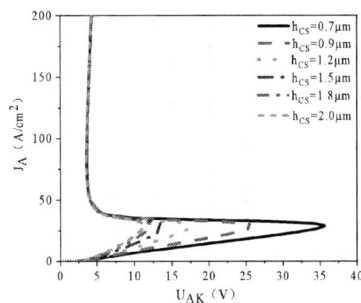

(a)Influence of CS layer doping thickness hCS on on-state characteristics.

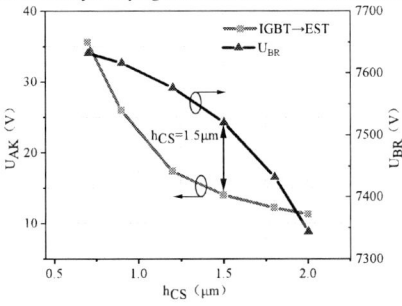

(b)Influence of hCS on the turning point voltage of the on-state curves and the blocking voltage

Fig.8 Influence of CS layer doping thickness hcs on the device performance of the improved P-EST

For the improved P-EST, the snapback effect is significantly improved. However, the maximum controllable current is not increased. Without changing other parameters, to increase $J_{A,MAX}$, the original p^+ well region in the structure is replaced with a shallow p^{++} layer. Fig. 9 shows the structure of the improved P - EST with a shallow p^{++} layer.

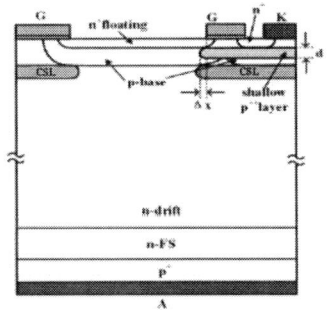

Fig. 9 The improved P-EST structure with shallow p^{++} layer

After introducing the shallow p^{++} layer, the maximum controllable current of the improved P-EST structure has been significantly increased. $J_{A,MAX}$ can reach at least 232.7A/cm^2. Compared with the improved P-EST without the shallow p^{++} layer, the maximum controllable current density has increased by 30%. Compared with the Con. P-EST structure, the maximum controllable current density has increased by 17.5%.

IV. TRENCH-PLANAR EST

Fig. 10 shows the proposed TP-EST structure. The TP - EST changes the left-side planar gate of the Con. P-EST into a trench gate, positioning the channel on the sidewall of the gate. Electrons enter the drift region along the channel. Meanwhile, the resistance of the JFET region is also reduced.

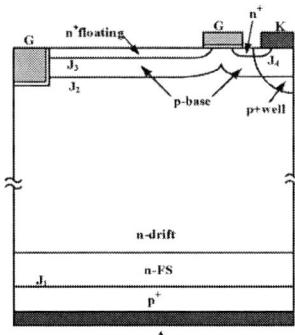

Fig.10 TP-EST structure

Fig. 11 shows on-state characteristics curve of the TP-EST. Fig. 12 shows turn-on characteristic curve of the TP-EST. During the conduction process of the TP-EST structure, the voltage corresponding to the transition from the IGBT mode to the EST mode is approximately 4.8V. The snapback phenomenon has been improved by approximately 97.5%, so there is no obvious "hump" in the V_{AK} curve.

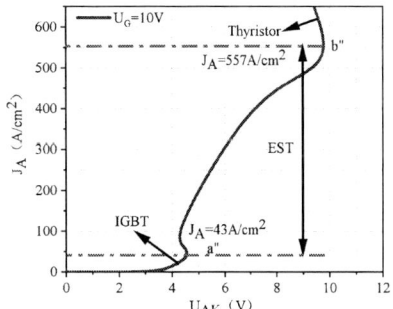

Fig. 11 On-state characteristics curve of TP-EST

Fig.12 Turn-on characteristic curve of TP-EST

Fig. 13 shows the distribution of electron and hole current densities during the turn-on process of the TP-EST. Compared with the Con. P-EST, the hole concentration of the TP-EST increases. At t=12.5μs, the main thyristor structure continues to expand, and the device operates in the EST mode.

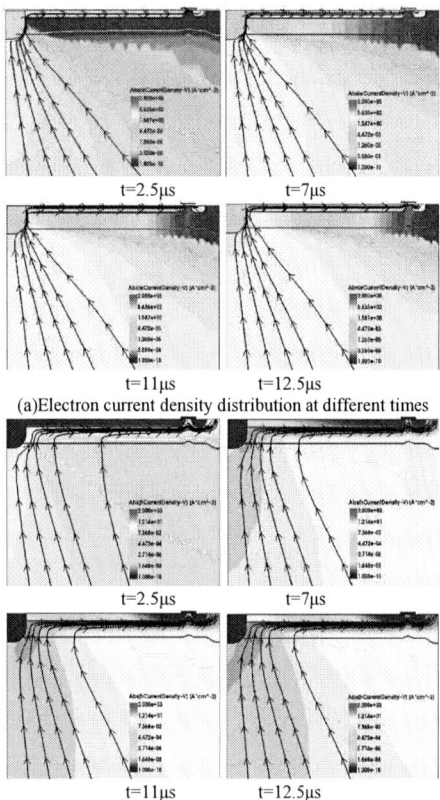

(a)Electron current density distribution at different times

(b)Hole current density distribution at different times

Fig.13 The electron and hole current density

Fig. 14 shows the influences of trench depth on the device performance of TP-EST. It can be seen that with the increase of trench depth, the forward blocking voltage of the device decreases,due to the increase of the electric field at the trench bottom. Meanwhile, the J3 junction between the n$^+$ floating region and the p-base region conducts rapidly, accelerating the conversion of the device from IGBT mode to EST mode (thyristor mode), improving the snapback effect.

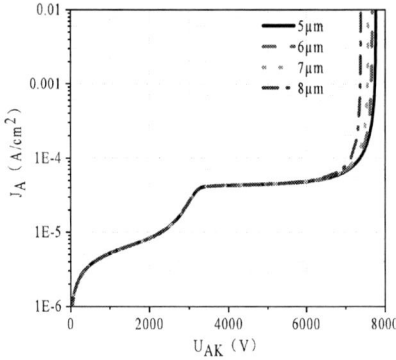

(a)Influence of trench depth on blocking voltage U_{BR}

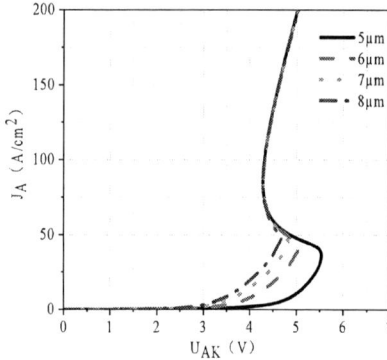

(b)Influence of trench depth on on-state characteristic

Fig.14 Effects of different trench depths on static characteristics

Fig. 15 shows the structure of the TP-EST with a shallow p^{++} layer. Compared with the TP-EST structure, the maximum controllable current of the TP-EST structure with a shallow p^{++} layer reaches 280A/cm^2.

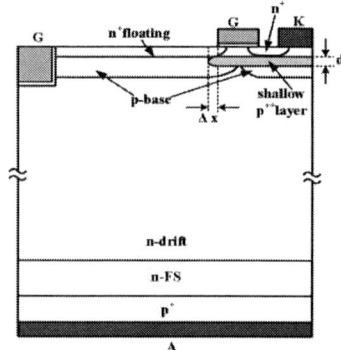

Fig. 15 TP-EST structure with shallow p++ layer

V. CONCLUSIONS

Fig. 16 shows the comparison of the conduction characteristics of the three structures. At room temperature, Fig. 17 presents the turn-on characteristic curves of the three structures. During the conduction process of the Con. P-EST, the voltage corresponding to the transition from the IGBT mode to the EST mode is approximately 190V, while that of the improved P-EST is about 14.9V, and for the TP-EST structure, it is 4.8V. Compared with the Con. P-EST, the snapback effect during the conduction process of the improved P-EST is improved by approximately 92.2%, and that of the TP-EST is improved by approximately 97.5%. When comparing the TP-EST structure with the improved P-EST, the improvement is about 67.8%. Therefore, during the turn-on process, compared with the Con. P-EST, after the voltage across the anode and cathode of the improved P-EST structure and the TP-EST structure drops, no "hump" appears on the curve, reducing the power consumption generated during the device's turn-on process.

979-8-3315-1110-4/25 $31.00 © 2025 IEEE

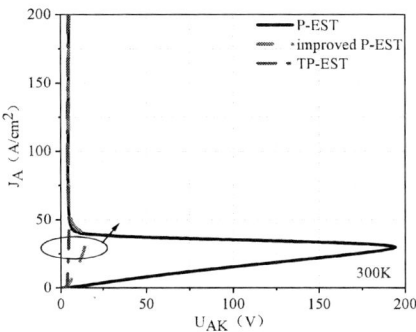

Fig.16 Comparison of static characteristics of three EST structures

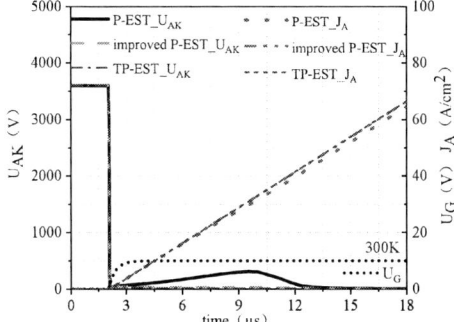

Fig. 17 Comparison of the turn-on characteristic of three EST structures

Fig. 18 presents a comparative analysis of the maximum controllable current density for three types of devices at varying temperatures. To enhance the maximum controllable current capability, the original p^+ well region in both improved P-EST and TP-EST device structures was replaced with a shallow p^{++} layer. This structural modification resulted in significant performance improvements across all tested temperatures. At room temperature, the improved P-EST structure achieved a maximum controllable current density of 232.7 A/cm^2 , representing a 17.5% enhancement compared

to the conventional design. More notably, the optimized TP-EST configuration demonstrated superior performance with a maximum controllable current density reaching 280 A/cm^2 , corresponding to a substantial improvement of approximately 41.3%. These enhancements in current handling capability are attributed to the improved carrier modulation and reduced parasitic resistance achieved through the implementation of the shallow p^{++} layer substitution.

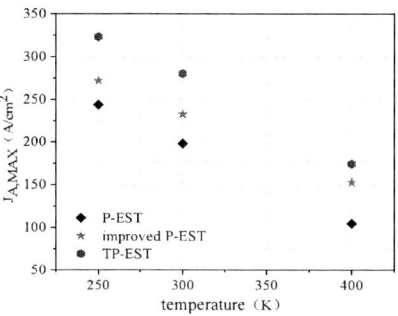

Fig. 14 Comparison of MCC of three EST structures

REFERENCES

[1] Liu, X., Zhang, Y., & Li, Z. (2020). Improved Performance of Power Devices with Novel Structures. IEEE Transactions on Electron Devices, 67(8), 3345 - 3352.

[2] Li, M., & Wang, Y. (2024). Novel Design and Performance Evaluation of Power Electronic Devices. International Journal of Electrical Power & Energy Systems, 164, 108653.

[3] Yang, J., & Sun, L. (2022). Analysis and Optimization of EST Devices for High - Power Applications. IEEE Journal of Emerging and Selected Topics in Power Electronics, 10(3), 2567 - 2574.

[4] Wang, H., Chen, Q., & Zhao, X. (2021). A New Method to Enhance the Current Capability of Thyristor - Based Devices. Journal of Power Electronics, 21(4), 987 - 994.

[5] Zhang, L., & Liu, S. (2023). Research on the Influence of Doping and Structure on the Characteristics of Power Semiconductor Devices. Journal of Semiconductor Technology and Science, 13(2), 189 - 196.

A Transient Interaction Mechanism Analysis Method for the Grid-forming Voltage Support Device Integrated into LCC-HVDC System

Yanlin Song
College of Electrical Engineering
Beijing Jiaotong University
Beijing, China
24121331@bjtu.edu.cn.cn

Zhichang Yang
China Electric Power Research
Institute
Beijing, China
yangzhichang@epri.sgcc.com

Hong Li*
College of Electrical Engineering
Zhejiang University
Zhejiang, China
hong_li@zju.edu.cn

Abstract—The sending end of the line-commutated converter based high-voltage direct current (LCC-HVDC) system will undergo a "low-then-high" transient voltage fluctuation when faults occur at the receiving end. Consequently, reactive power compensation devices are often employed to enhance the transient voltage stability of LCC-HVDC sending-end systems. However, there is insufficient analysis regarding the transient voltage support mechanisms for these reactive power compensation devices integrated into LCC-HVDC systems. To address this problem, this paper develops an equivalent model for a grid-forming voltage support device (GVSD) integrated into LCC-HVDC system, illustrating its reactive power compensation process during the "low-then-high" transient process. On this basis, an analysis method based on the short-circuit ratio (SCR) is proposed to clarify the transient voltage support mechanism for GVSD. Finally, the accuracy of the proposed method is confirmed through simulations.

Keywords—*LCC-HVDC system, voltage support, transient capability, SCR*

I. INTRODUCTION

With the rapid development of new energy sources, the conventional power grid, which has primarily relied on synchronous generators, is gradually evolving into a new type of grid primarily powered by renewable energy sources, leading to significant changes in the grid operational characteristics [1]. In this case, line-commutated converter based high-voltage direct current (LCC-HVDC) systems have become popular for long-distance, high-capacity power transmission due to their high efficiency and cost-effectiveness. However, it is believed that LCC-HVDC systems consume reactive power due to line commutation, which can lead to transient voltage stability problems, such as insufficient voltage support capability [2]. Research has shown that when severe grounding faults occur at the receiving end of LCC-HVDC system, the sending-end grid experiences a distinctive "low-then-high" transient voltage fluctuation [3]. To address this problem, reactive power compensation devices with active voltage support capabilities, such as the static synchronous compensator (STATCOM), have been introduced to improve the transient voltage support stability of LCC-HVDC systems [4]. Therefore, to enable efficient and fast reactive power compensation for the post-fault LCC-HVDC system, it is critically necessary to characterize the transient overvoltage mechanism of LCC-HVDC system, and evaluate the transient voltage support

capabilities of reactive power compensation devices when integrated into such a system.

Recent studies on the voltage support capabilities of reactive power compensation devices mainly focus on enhancing control strategies, while lacking in-depth theoretical analysis of their voltage support mechanisms. Typically, mathematical equivalent models of STATCOM are created based on voltage phasor diagrams, which clarify the transient voltage support process from a voltage perspective [4], [5]. However, they failed to establish the mathematical connection between HVDC systems and STATCOM, hindering the analysis of STATCOM transient voltage support mechanism when integrated into HVDC systems. In this case, models have been developed for various reactive power compensation devices integrated into HVDC systems in [6], which elucidates their voltage support mechanisms through their respective reactive power-voltage (Q-V) relationships.

Besides, the short-circuit ratio (SCR) has been extensively used to quantify the voltage support strength of post-fault LCC-HVDC systems [7], [8]. Some researchers have proposed different improved versions of SCR to assess the voltage support strength of HVDC systems, developing mathematical models and formulas for transient overvoltage in HVDC systems, and conducting quantitative analysis of the key influencing factors [9]-[11]. Additionally, recent studies are exploring the application of SCR to assess the transient voltage support capabilities of reactive power compensation devices when integrated into grid systems [12], [13].

However, most current research primarily concentrates on either the transient overvoltage mechanisms of HVDC systems, or the voltage support capabilities of reactive power compensation devices through separate models, failing to develop a comprehensive analysis method to clarify the transient interaction mechanisms for reactive power compensation devices integrated into HVDC systems. Therefore, this paper presents a transient voltage equivalent model for a grid-forming voltage support device (GVSD) integrated into LCC-HVDC system, illustrating its reactive power compensation process. Furthermore, a SCR-based analysis method is proposed to clarify the transient voltage support mechanism of GVSD. Finally, the accuracy of the proposed method is confirmed through simulations.

II. BASIC PRINCIPLE OF GVSD

A. Main Circuit of GVSD

The main circuit topology of GVSD is shown in Fig. 1, which mainly consists of a cascaded H-bridge multilevel inverter, a transformer, and energy storage modules.

---

* This work was supported in part by the National Science Fund of China under Grant 52237008 and the National Science Fund for Distinguished Young Scholars of China under Grant 52325704.
* The corresponding author is Hong Li (hong_li@zju.edu.cn).

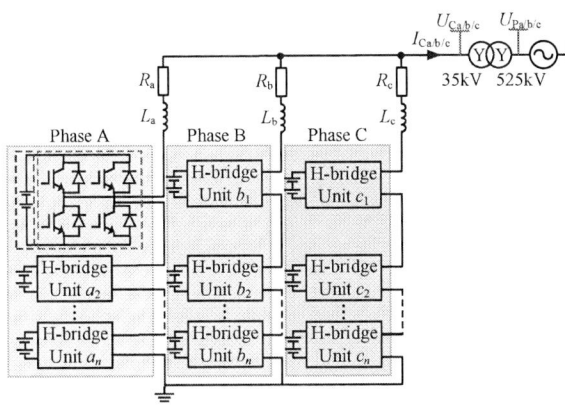

Fig. 1. The main circuit of GVSD.

The cascaded H-bridge multilevel inverter employs star connections, where each phase contains 16 power modules using IGBTs as switching components. The DC side of the power modules is connected to energy storage modules composed of several series-connected DC voltage sources. U_{Ca}, U_{Cb}, U_{Cc} are the three-phase output voltages of GVSD; U_{Pa}, U_{Pb}, U_{Pc} are the three-phase output voltages of the grid side; I_{Ca}, I_{Cb}, I_{Cc} are the three-phase output currents of GVSD; L_a, L_b, L_c are the three-phase filter inductors; R_a, R_b, R_c are the three-phase equivalent resistances.

B. Control Circuit of GVSD

The control circuit for GVSD in this paper includes a target voltage formation link and voltage/current control loops, as shown in the structural diagram in Fig. 2.

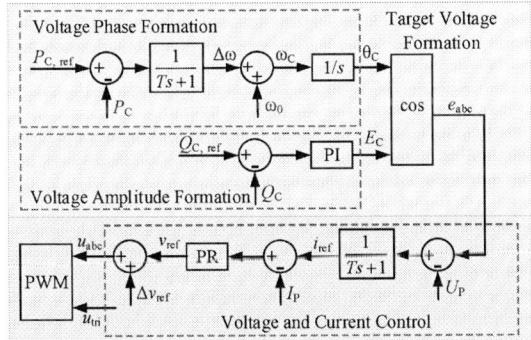

Fig. 2. The control circuit of GVSD.

Since GVSD operates without a phase-locked loop (PLL), it relies on a target voltage formation link to generate a grid-synchronized output voltage. This link comprises a voltage amplitude formation part and a voltage phase formation part.

The power-frequency (P-f) control is employed in the voltage amplitude formation part to emulate synchronous generator characteristics, which adjusts the phase deviation between GVSD and the point of common coupling (PCC) to control its active power output; while the reactive power-voltage (Q-V) control is employed in the voltage amplitude formation part, which adjusts the voltage amplitude of GVSD to control its reactive power output. The mathematical formulations of these control strategies are given below [11]:

$$P_C = \frac{E_C V_{PCC}}{X_C} \sin(\theta_C - \theta_{PCC}) \tag{1}$$

$$Q_C = \frac{E_C^2 - E_C V_{PCC} \cos(\theta_C - \theta_{PCC})}{X_C} \tag{2}$$

Where P_C and Q_C are respectively the active and reactive power output of GVSD; E_C is the developed voltage amplitude; V_{PCC} is the voltage of PCC; θ_C is the phase of developed voltage; θ_{PCC} is the phase of PCC; X_C is the reactance between the transformer and the gird.

III. TRANSIENT VOLTAGE SUPPORT MECHANISM OF GVSD

A. Transient Voltage Equivalent Models for LCC-HVDC system

The equivalent model for LCC-HVDC sending-end system is established as shown in Fig. 3. Where U_{er} and U_{dr} are respectively the equivalent AC and DC voltages of the sending end; R_r and X_r are the equivalent line resistance and reactance; U_{pr} is the bus voltages of the sending end, with the rated value U_{prN}; I_{pr} and I_{dr} are the line currents of the AC and DC sides; Q_{cr} is the three-phase reactive power compensation capacity at rated voltages. $S_{acr} = P_r + jQ_r$ is the transmission power of the sending end, in which P_r and Q_r are respectively its active and reactive transmission power.

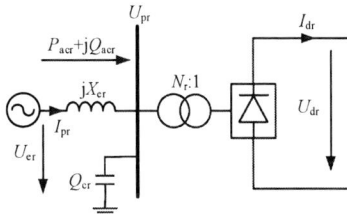

Fig. 3. The sending-end equivalent model of LCC-HVDC system.

In a steady-state condition, the direction of power transfer from the AC side to the bus side is defined as the positive direction. Consequently, the sending-end bus voltage U_{pr} and the equivalent AC voltage U_{er} are connected as follows:

$$\dot{U}_{er} - \dot{U}_{pr} = (R_r + jX_r)\dot{I}_{pr} \tag{3}$$

Considering the sending-end bus voltage as a reference axis, denoted as $U_{pr} = U_{pr}\angle 0°$, a transient voltage fluctuation characterized by an initial low voltage followed by a high voltage is observed at the sending-end bus when disturbances occur at the LCC-HVDC receiving end. The voltage phasor diagrams of "low-voltage" and "high-voltage" stages during this transient process are illustrated in Fig. 4.

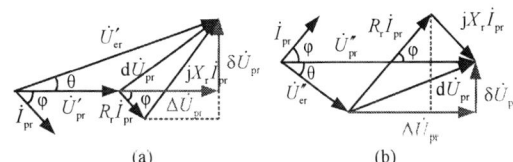

Fig. 4. The transient voltage phasor diagrams of LCC-HVDC sending-end system. (a) The "low-voltage" stage. (b) The "high-voltage" stage.

Where $\Delta U_{pr} = R_r I_{pr} \cos\varphi + X_r I_{pr} \sin\varphi \approx Q_r X_r / U_{pr}$ is the quadrature component of the sending-end voltage drop due to reactive power transmission, and $\delta U_{pr} = X_r I_{pr} \cos\varphi - R_r I_{pr} \sin\varphi \approx P_r R_r / U_{pr}$ is the longitudinal component of the sending-end voltage drop due to active power transmission; θ is the phase deviation between the equivalent AC voltage and the bus voltage; φ is the phase deviation between the equivalent AC voltage and the sending-end line current.

During this "low-then-high" process, the relationship between the actual reactive power transmitted at the sending end Q'_{cbp} and the reactive power at the rated condition Q_{cbp} can be expressed as below:

$$Q'_{cbp} = \left(\frac{U'_{pr}}{U_{prN}}\right)^2 Q_{cbp} \tag{4}$$

The relational expression for the "low-voltage" stage is shown as follows:

$$U_{er} = \sqrt{(U'_{pr} + \Delta U_{pr})^2 + \delta U_{pr}^2} \tag{5}$$

For simplification, the quadrature voltage component δU_p, which exerts a negligible influence on the voltage amplitude, can be ignored. This leads to the simplified relationship between the bus voltage and the AC voltage as below:

$$U'_{pr} = U_{er} - \Delta U_{pr} \tag{6}$$

Then the transient voltage expression of sending-end bus during the "low-voltage" stage is obtained from (4) and (6):

$$
\begin{aligned}
U'_{pr} &= \frac{(1 - \sqrt{1 + 4\dfrac{Q_{cbp} X_r}{U_{prN}^2}})}{2\dfrac{Q_{cbp} X_r}{U_{prN}^2 U_{er}}} \\
&= \frac{U_{prN}^2 U_{er} - U_{pN} U_{er}\sqrt{U_{prN}^2 + 4Q_{cbp} X_r}}{2Q_{cbp} X_r}
\end{aligned} \tag{7}
$$

Similarly, the relational expression for the "high-voltage" stage is also obtained as follows:

$$U''_{pr} = \sqrt{(U_{er} + \Delta U_{pr})^2 + \delta U_{pr}^2} \tag{8}$$

Then the simplified relationship between the bus voltage and the AC voltage are as below after ignoring the quadrature voltage component δU_p:

$$U''_{pr} = U_{er} + \Delta U_{pr} \tag{9}$$

Then the transient voltage expression of sending-end bus during the "high-voltage" stage is obtained from (4) and (9):

$$
\begin{aligned}
U''_{pr} &= \frac{(1 - \sqrt{1 - 4\dfrac{Q_{cbp} X_r}{U_{prN}^2}})}{2\dfrac{Q_{cbp} X_r}{U_{prN}^2 U_{er}}} \\
&= \frac{U_{prN}^2 U_{er} - U_{prN} U_{er}\sqrt{U_{prN}^2 - 4Q_{cbp} X_r}}{2Q_{cbp} X_r}
\end{aligned} \tag{10}
$$

On this basis, SCR is introduced to quantitatively characterize the severity of transient voltage fluctuation in LCC-HVDC system, thereby evaluating the transient voltage support capability of GVSD. The defining equation of SCR is as below:

$$\text{SCR} = \frac{S_{sc}}{P_{dN}} \tag{11}$$

Where S_{sc} is the short-circuit capacity at sending-end bus; P_{dN} is the rated DC power.

When the line resistance R is ignored, the sending-end AC system can be represented equivalently as a voltage source containing only reactance. This leads to the expression for the equivalent reactance:

$$X_r = 3\frac{U_{prN}^2}{S_{sc}} \tag{12}$$

The relationship between SCR and the equivalent reactance is then obtained based on (11) and (12):

$$\text{SCR} = \frac{3U_{prN}^2}{X_r P_{dN}} \tag{13}$$

Finally, by substituting (13) into (8) and (10), the relationships between the transient bus voltage and SCR for "low-voltage" and "high-voltage" stages can be derived respectively in (14) and (15):

$$U'_{pr} = \frac{U_{er} - U_{er}\sqrt{1 + \dfrac{12Q_{cbp}}{S_{dN} \cdot \text{SCR}}}}{\dfrac{6Q_{cbp}}{S_{dN} \cdot \text{SCR}}} \tag{14}$$

$$U''_{pr} = \frac{U_{er} - U_{er}\sqrt{1 - \dfrac{12Q_{cbp}}{S_{dN} \cdot \text{SCR}}}}{\dfrac{6Q_{cbp}}{S_{dN} \cdot \text{SCR}}} \tag{15}$$

Based on the equations derived above, the relationships between the transient bus voltage at the sending end and SCR for the two transient stages are established, as shown in Fig. 5.

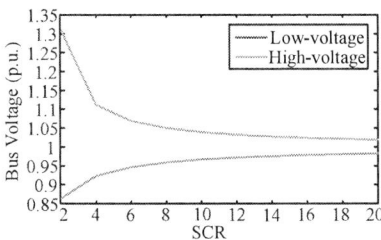

Fig. 5. Relationships between the transient bus voltage and SCR.

This indicates that for both transient stages, an increase in SCR corresponds to a reduction in the transient bus voltage deviation, measured relative to 1.0 per-unit. It can be therefore concluded that the transient voltage deviation of LCC-HVDC system exhibits an inverse relationship with SCR.

B. The Equivalent Model for GVSD Integrated into LCC-HVDC system

Fig. 6 shows the equivalent model for GVSD integrated into LCC-HVDC system. GVSD is linked to the sending-end bus of LCC-HVDC system, thereby providing voltage support to this bus.

Then the equivalent model of GVSD integrated into LCC-HVDC system is derived, as shown in Fig. 7. GVSD is equivalently represented as a series configuration that includes a transformer, line reactance, and a controllable voltage source. Where U_{PCC} is the voltage of PCC; U_C is the equivalent voltage of GVSD; X_L and R are respectively the equivalent reactance and resistance of the transmission line between PCC and GVSD; I_C is the line current flowing through X_L and R.

Furthermore, since phase A, B and C of GVSD are symmetrical, the equivalent model shown in Fig. 7 can be further simplified by considering only one phase, resulting in a single-phase equivalent model for GVSD shown in Fig. 8.

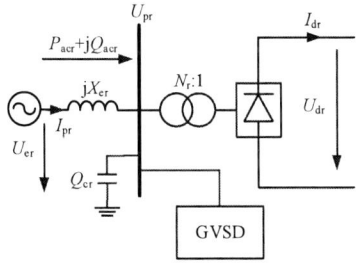

Fig. 6. The diagram for GVSD integrated into LCC-HVDC system.

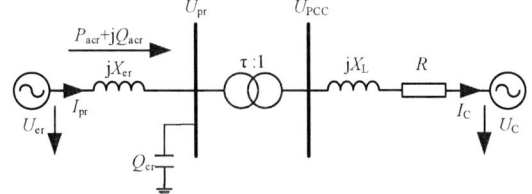

Fig. 7. The equivalent model for GVSD integrated into LCC-HVDC system.

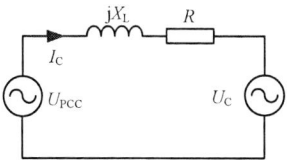

Fig. 8. A single-phase equivalent model for GVSD.

C. Transient Voltage Support Mechanism of GVSD

Based on Fig. 8, the corresponding voltage phasor diagrams of GVSD integrated into LCC-HVDC system for "low-voltage" and "high-voltage" stages is derived in Fig. 9. Similarly, the direction from PCC to GVSD is specified to be the positive direction, and the equivalent voltage of GVSD is considered as a reference axis, that is, $U_C = U_C \angle 0°$.

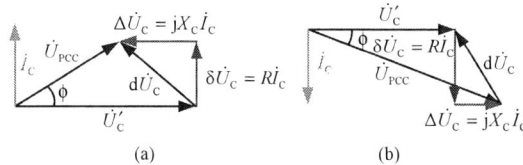

Fig. 9. The voltage phasor diagrams of GVSD. (a) The "low-voltage" stage. (b) The "high-voltage" stage.

Similarly, the voltage phasor diagrams in Fig. 9 can be simplified as follows by ignoring the quadrature voltage component δU_e:

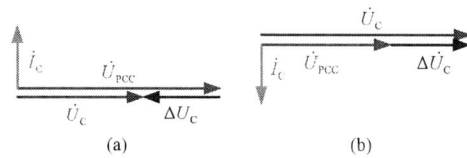

Fig. 10. The simplified voltage phasor diagrams of GVSD. (a) The "low-voltage" stage. (b) The "high-voltage" stage.

Therefore, the reactive power consumed by GVSD can be expressed as follows:

$$Q_m = \text{Im}\left(U_{PCC} I_C\right) = U_{PCC} \frac{U_{PCC} - U_C}{X_C} \quad (16)$$

In the "low-voltage" stage, $U_{PCC} > U_C$, $Q_m > 0$, GVSD provides reactive power to LCC-HVDC system; and in the "high-voltage" stage, $U_{PCC} < U_C$, $Q_m < 0$, GVSD absorbs the surplus reactive power from LCC-HVDC system.

Therefore, by combining Fig. 4 and Fig. 10, it can be seen from Fig. 11 that GVSD effectively minimizes the line current I_{pr} of LCC-HVDC system by injecting a current I_C with the same amplitude but opposite phase to counteract the q-axis current component I_{Qpr}, which reduces the reactive power transmission on LCC-HVDC system and fulfills its transient voltage support requirement.

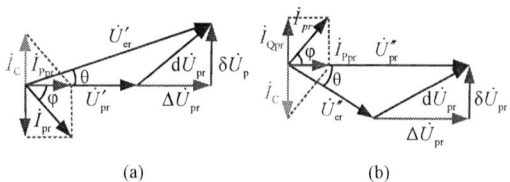

(a)　　　　　　(b)

Fig. 11. The voltage phasor diagrams for GVSD integrated into LCC-HVDC system. (a) The "low-voltage" stage. (b) The "high-voltage" stage.

By integrating (14), (15) and (16), it is evident from (17) and (18) that GVSD provides reactive power to reduce the reactive power transmitted Q_{cbp} in the sending end, resulting in the transient voltage support for LCC-HVDC system.

$$U'_{pr} = \frac{U_{er} - U_{er}\sqrt{1 + \dfrac{12(Q_{cbp} - |Q_{m}|)}{S_{dN} \cdot SCR}}}{\dfrac{6(Q_{cbp} - |Q_{m}|)}{S_{dN} \cdot SCR}} \quad (17)$$

$$U''_{pr} = \frac{U_{er} - U_{er}\sqrt{1 - \dfrac{12(Q_{cbp} - |Q_{m}|)}{S_{dN} \cdot SCR}}}{\dfrac{6(Q_{cbp} - |Q_{m}|)}{S_{dN} \cdot SCR}} \quad (18)$$

Based on the relationships above, the relationships between the transient bus voltage at the sending end and SCR for the two transient stages, respectively, with and without GVSD, are derived in Fig. 12.

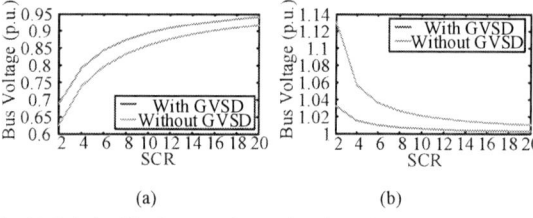

(a)　　　　　　(b)

Fig. 12. Relationships between the transient bus voltage and SCR. (a) The "low-voltage" stage. (b) The "high-voltage" stage.

This indicates that, during both transient stages, the integration of GVSD can reduce the transient bus voltage deviation at different values of SCR. It is therefore concluded that GVSD adjusts the reactive power capability of LCC-HVDC system by either consuming or compensating reactive power, which suppresses the transient voltage fluctuations at different values of SCR and enhances the transient voltage stability of LCC-HVDC system.

IV. VERIFICATION OF GVSD TRANSIENT VOLTAGE SUPPORT CAPABILITY

The GVSD simulation model has been developed on RTDS and is integrated into a standard LCC-HVDC system test case. In this model, LCC-HVDC system is configured in a bipolar arrangement, featuring a rated reactive power capacity of 12,000 MVA on the rectifier side and 7,500 MVA on the inverter side, and GVSD is designed with a rated reactive power capacity of 120 MVA.

A. Effect of SCR on Transient Voltage Support Capability of LCC-HVDC system

With GVSD not integrated into LCC-HVDC system, a three-phase grounding fault is applied at the receiving end system at 0.8s. The transient voltage fluctuation curves of the sending-end bus corresponding to different SCR are shown in Fig. 13.

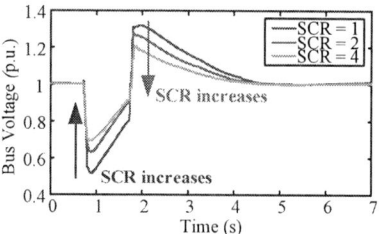

Fig. 13. Transient voltage curves at different SCR without GVSD.

It can be seen that, as SCR increases, there is a gradual reduction in the severity of transient voltage fluctuations at the sending-end bus. This indicates an inverse relationships between the transient voltage deviation and SCR, aligning with the conclusions drawn from theoretical analysis.

B. The Transient Voltage Support Capability of GVSD Integrated into LCC-HVDC system

Similarly, with GVSD integrated into LCC-HVDC system, a three-phase grounding fault is applied at its receiving end at 0.8s. The transient voltage fluctuation curves of the sending-end bus corresponding to different SCR are shown in Fig. 14.

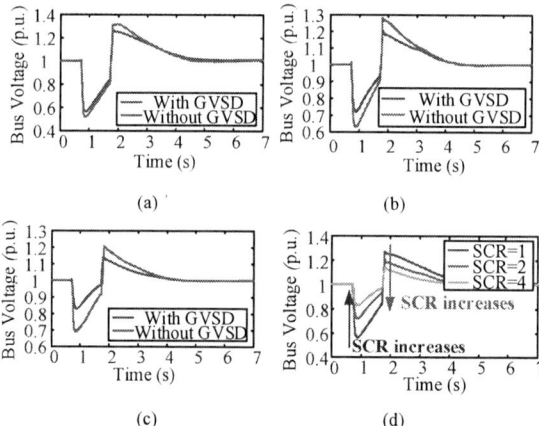

(a)　　　　　　(b)

(c)　　　　　　(d)

Fig. 14. Transient voltage curves at different SCR with or without GVSD. (a) SCR = 1. (b) SCR = 2. (c) SCR = 4. (d) Different SCR with GVSD.

It can be seen that, compared with conditions without GVSD, the integration of GVSD can suppresses the transient voltage fluctuation at different SCR, aligning with the conclusions drawn from the theoretical analysis.

V. CONCULISIONS

This paper investigates the transient voltage support mechanism of GVSD integrated into LCC-HVDC system from its operational principles, reactive power transmission and voltage support strength. Additionally, a SCR-based analysis method is proposed to clarify the transient voltage

979-8-3315-1110-4/25 $31.00 © 2025 IEEE

support capabilities of GVSD. The specific conclusions obtained in this paper are as follows:

1) The transient voltage deviation of LCC-HVDC system exhibits an inverse relationship with SCR; specifically, as SCR increases, the transient bus voltage deviation decreases.

2) The same amplitude but opposite phase current is provided by GVSD in response to the voltage deviation between GVSD and LCC-HVDC sending-end bus. This mechanism effectively reduces the reactive power transmitted within LCC-HVDC system, thereby resulting in the transient voltage support to LCC-HVDC system.

3) The integration of GVSD reduces the transient bus voltage deviation at different values of SCR, enhancing the voltage support strength of LCC-HVDC system.

REFERENCES

[1] Y. Peng, L. Feng, J. Qirong, and M. Hangyin. "Large-disturbance stability of power systems with high penetration of renewables and inverters: Phenomena, challenges, and perspectives," in *Journal of Tsinghua University* (Science and Technology), vol. 61, no.5, pp. 403-414, Apr. 2021.

[2] W. Qiunan, J. Hongyang, L. Dong, and S. Huaping, "Review on Reactive Voltage Control of LCC-HVDC Converter Station," in *Power Capacitor and Reactive Power Compensation*, vol. 44, no. 2, pp. 10-17, Apr. 2023.

[3] L. Dongyan, "Transient Voltage Stability Assessment and Realization in High Renewable Penetration Power Grids," in *Electrical Equipment and Economy*, early access, Feb. 2025, doi: 10.3969/j.issn.1673-8845.2025.02.100.

[4] H. Julin, Z. Bo, and H. Yibin, "Research on the Mechanism of Transient Overvoltage in Photovoltaic Grid-Connected HVDC Systems," in *Heibei Eletric Power*, vol. 43, no. 6, pp. 44-49, Dec. 2024.

[5] W. Shuqiang, "Mathematical Model and Simulation of Distribution Systems STATCOM," in *2013 Electrical Engineering Society of China*, Chengdu, China, 2013, pp. 236-249.

[6] X. Fan, and Z. Youbin, "Analysis of the Impact of Different Reactive Power Compensation Devices on HVDC System Rectifier Station Power System Automation," in *2019 4th International Conference on Power and Renewable Energy* (ICPRE), Chengdu, China, 2019, pp. 254-258.

[7] T. Lan et al. "LCC-HVDC's Systematical Impact on Voltage Stability: Theoretical Analysis and a Practical Case Study," in *IEEE Transactions on Power Systems*, vol. 38, no. 2, pp. 1663-1675, Mar. 2023.

[8] Y. Lin, S. Huadong, X. Shiyun, Z. Bing, Z. Jian, ang L. Zonghan, "Overview of Strength Quantification Indexes of Power System With Power Electronic Equipment," in *Proceedings of the CSEE*, vol. 42, no. 2, pp. 499-515, Jun. 2022.

[9] W. Feng, L.Tianqi, D. Yuanyuan, Z. Qi, and L. Xingyuan, "Calculation Method and Influencing Factors of Transient Overvoltage Caused by HVDC Block," in *Power System Technology*, vol. 40, no.10, pp. 3059-3065, Oct. 2016.

[10] F. Li et al. "Short Circuit Ratio Index Considering HVDC Transient Characteristics," in *2023 6th Asia Conference on Energy and Electrical Engineering* (ACEEE), Chengdu, China, 2023, pp. 211-216.

[11] Z. Cao, X. Liu, Z. Suo, S. Teng and B. Gao, "Transient Voltage Evaluation of HVDC Transmission System Based on Improved Short Circuit Ratio," in *2024 IEEE 14th International Conference on CYBER Technology in Automation, Control, and Intelligent Systems* (CYBER), Copenhagen, Denmark, 2024, pp. 611-616.

[12] L. Beihua et al. "Voltage-supported new energy field station construction scheme based on distributed regulator--(I) Mechanism analysis," in *Automation of Electric Power Systems*, early access, Mar. 2025, doi: 10. 7500/AEPS20240807001.

[13] H. Yuan et al. "Small Signal Stability Analysis of Grid-Following Inverter-Based Resources in Weak Grids With SVGs Based on Grid Strength Assessment," in *2021 IEEE 1st International Power Electronics and Application Symposium* (PEAS), Shanghai, China, 2021, pp. 1-6.

2025 IEEE Workshop on Wide Bandgap Power Devices and Applications in Asia (WiPDA Asia)

A Numerical Model of SiC MOSFET for Electro-Thermal Characteristics Based on TCAD

1st Yujie Zhang
National Key Laboratory of
Electromagnetic Energy
Naval University of Engineering
Wuhan,China
34312019001@nue.edu.cn

2nd Yongle Huang
National Key Laboratory of
Electromagnetic Energy
Naval University of Engineering
Wuhan,China
nudt_mse_501@163.com

3rd Yifei Luo
National Key Laboratory of
Electromagnetic Energy
Naval University of Engineering
Wuhan,China
yfluo23@nue.edu.cn

Abstract—**SiC semiconductor devices are replacing Si-based semiconductor devices due to their bandwidth, high breakdown field strength, and high thermal conductivity. Due to the differences in the physical and geometrical structures of Si and SiC power devices, it is more difficult to accurately characterize the dynamic and static properties of SiC devices than those of conventional Si devices. In this paper, a metameric simulation model is constructed through the TCAD simulation platform to accurately characterize the spatial and temporal evolution of multi-physical fields and semiconductor physical processes inside SiC MOSFETs, which can be used to guide the practical application of SiC MOSFETs. The simulation results of the characteristic curves of the device at different temperatures are in good agreement with the manual. The switching experiments, avalanche breakdown experiments and gate breakdown experiments at different temperatures are designed, and the errors between the simulation results and the experimental results are within the acceptable range, which verifies the accuracy of the model.**

Keywords—*SiC MOSFET, TCAD, breakdown boundary, wide-bandwidth*

I. INTRODUCTION

The current mainstream silicon (Si) power semiconductor devices are limited by the physical limit of the material, it is difficult to meet the power electronic system of large-capacity power conversion device high-speed, high-frequency and high-temperature and other extreme operating conditions. Silicon carbide (SiC) devices as a representative of the wide bandwidth semiconductor devices due to the bandwidth, breakdown field, electron saturation speed, thermal conductivity and other aspects of the Si has obvious advantages, so that the device operating temperature, switching frequency, loss and other performance has been greatly improved, is gradually replacing the Si-based semiconductor devices.

The main failure mechanisms of Si MOSFETs are thermal failure due to current accumulation caused by $\triangle T$, electrical failure due to parasitic bipolar junction transistor (BJT) conduction, electrical failure due to the diode body effect, electrical failure due to the Kirk effect, secondary decomposition, the Egawa effect, and failure due to dynamic avalanches occurring during reverse recovery. There are other failure mechanisms such as hot carrier injection, negative bias temperature instability, thermo-mechanical stress and gate reliability. Among them the presence of parasitic BJTs significantly affects the robustness of Si MOSFETs.

The main failure mechanisms of SiC MOSFETs are thermal runaway due to BJT locking, early breakdown of the gate oxide due to the oxidized electric field at the gate trench exceeding the critical electric field value, thermal failure due

to self-heating effect, early failure due to cracks in the gate oxide, gate oxide failure due to melting of the top aluminum caused by the self-heating effect, and gate oxide failure due to thermal stresses and hot electron injection. Failure of the gate oxide due to thermal stress and hot electron injection. Among them, the gate oxide significantly affects the robustness of SiC MOSFETs [2].

The high intrinsic temperature of SiC MOSFETs, the almost complete absence of BJT opening at room temperature, and the high SiC electro-thermal instability triggering conditions relative to material physical boundary failures. Therefore, the limiting capability of SiC MOSFETs is mainly affected by the physical boundary relative to Si MOSFETs. Due to the differences in the physical and geometrical structures of Si and SiC power devices, it is difficult for Si-based models to accurately describe the dynamic and static characteristics of SiC devices, and it is more difficult to accurately characterize SiC devices than conventional Si devices. Therefore, in order to guide the application of SIC MOSFETs in power electronics, a simulation model that can accurately characterize the spatial and temporal evolution of its internal multiphysics fields and semiconductor physical processes is needed.

In terms of SiC MOSFET device modeling research, the current domestic and international studies are mainly divided into behavioral, physical, semi-physical, and numerical models [3].

Behavioral model is a kind of model that fits the actual operating state of the device based on the existing test characteristic curve of SiC MOSFET, and its advantages are fast calculation speed, easy implementation, and the ability to characterize the periodic steady state solution of physical quantities at the device port. The disadvantage is that the model is not very accurate for its switching process, and it is difficult to characterize the internal physical field of the device.

The physical model is an analytical model based on the physical basis of semiconductor devices, which is an appropriate simplification of the physical equations related to semiconductor physics, and has the advantage of being able to better reproduce the device port characteristics. Its disadvantage is that the accurate characterization of the model requires more complex calibration, and the model construction is more difficult.

Numerical modeling can realize accurate calculation of multi-physical fields inside the device through material parameters, physical model definition and finite element meshing. The numerical model can portray the time-varying and spatial distribution of the physical fields inside the power semiconductor device during the operation process, and its solution can realize the spatial accuracy at the micron level

Innovative Group Project of Hubei Natural Science Foundation.

979-8-3315-1110-4/25 $31.00 © 2025 IEEE

and the temporal accuracy at the nanosecond level. The advantages are high model accuracy, accurate solution, and the ability to consider multiple physical field effects. The disadvantages are poor model convergence, slow solution speed, and complicated model construction.

In summary, in order to characterize the spatial and temporal evolution of multi-physical fields and semiconductor physical processes inside SiC MOSFETs, a numerical model needs to be constructed to provide guidance for its practical application.

II. MODELING

TCAD (Technology Computer-Aided Design) is a software application widely used in the semiconductor field. The TCAD numerical model simulation relies on the precise resolution of the device's physical fields, with the relevant simulation parameters being of paramount importance. In this study, several physical models are employed to accurately simulate the behavior of the SiC MOSFET under high current density conditions. The Shockley-Read-Hall (SRH) model is used to account for carrier recombination, while the Auger mobility model addresses high current density effects. Additionally, the OldSlotboom model is employed to consider bandgap narrowing, and the incomplete ionization model is used to simulate the incomplete activation of dopant impurities. Furthermore, the Okuto collisional ionization model and the HighFieldSaturation mobility degradation model are incorporated to capture high-field effects.

The TCAD numerical model incorporates both structural and process parameters. Specifically, the structural parameters of the SiC MOSFET can be observed using SEM or FIB imaging. Meanwhile, the process parameters can be extracted through semiconductor metrology and characterization curves.

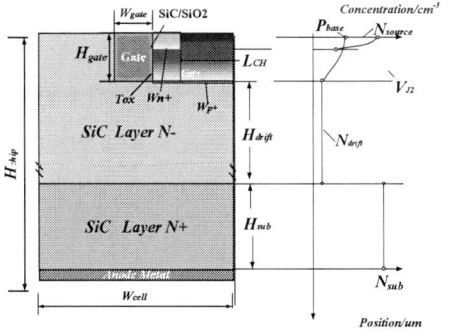

Fig. 1. SiC MOSFET numerical model structure and process parameters

The physical dissection method is used to remove the SiC MOSFET device package shell, and the SiC MOSFET chip is peeled off from the module by heating the backing plate to remove the bonding filaments on the surface, so that the die structure of the device can be obtained under an optical microscope. After PN junction coloring, the cross-section distribution of the device can be obtained by cutting along the active array using the FIB method, and the dimension information in the figure can be directly obtained based on the measurement tool.

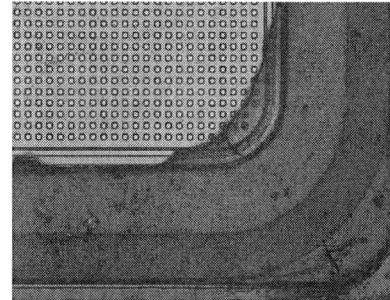

Fig. 2. Surface morphology of SiC MOSFET module

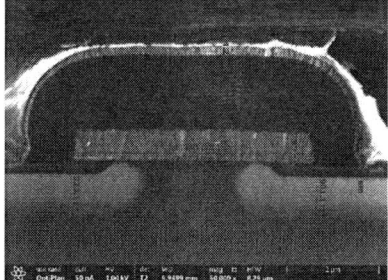

Fig. 3. Longitudinal sectional view of SiC MOSFET cell

Based on PN junction staining information, the junction depth of the P-base region of the device can be obtained, and the doping concentration of the P-base region of the device can be calibrated by threshold voltage. Drift region concentration, slowly varying doping peak valley values, etc. can be directly obtained based on SPR. Build a TCAD numerical model based on the obtained parameters.

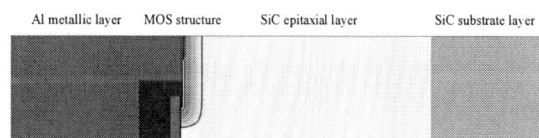

Fig. 4. SiC MOSFET TCAD Model

III. VERIFICATION MODEL

The model's accuracy was verified through four distinct aspects: characteristic curves, switching waveforms, gate breakdown boundary, and avalanche breakdown boundary.

A. Characteristic Curves and Switching Waveforms

For the transmission characteristic curves, the Vds is fixed at 20 V, with temperatures set to 25°C, 125°C, or 150°C. Similarly, for the output characteristic curves, the Vds of 20 V is applied, with temperatures set to 25°C, 125°C, or 150°C. Under the same conditions, the TCAD model is used to scan the Vg and Vd, and the Id change curves are recorded. Double-pulses experiments were conducted, and the waveforms were recorded. The experimental conditions included a DC voltage of 900 V, an AC current of 250 A, a resistance of 1 Ω, and temperatures of 125 °C and 150 °C. The comparison shows that the characteristic curves at different temperatures closely match the experimental data. The maximum deviation of the transmission characteristic curves is approximately 8%, and the errors of the output characteristic curves are less than 5%. Furthermore, the simulated and experimental switching waveforms at varying temperatures demonstrate a high degree of agreement. Key parameters, such as switching time, rate of change, and spikes, exhibit errors below 10%.

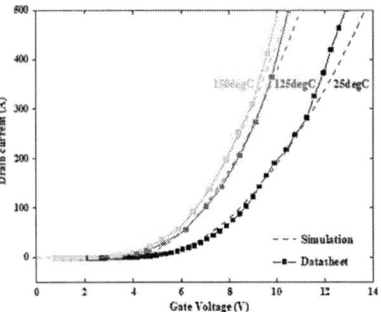

(a)Transmission characteristic curves at different temperatures (V_{ds}=20V)

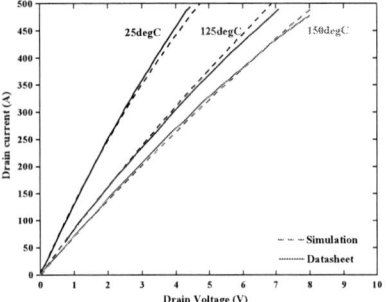

(b) Output characteristic curves at different temperatures (V_{gs}=18V)

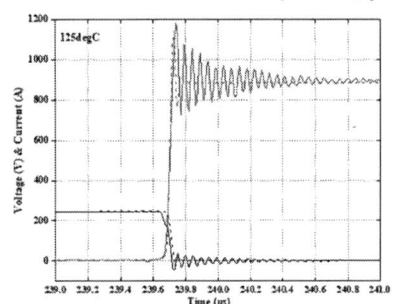

(c) Shutdown waveforms at 125 degC, 900V, 250A, 1Ω

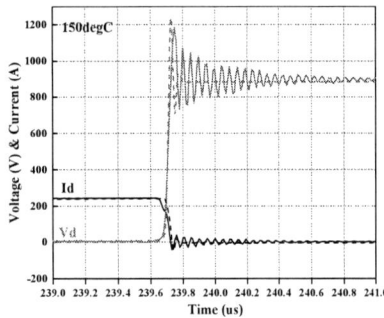

(d) Shutdown waveforms at 150 degC, 900V, 250A, 1Ω

Fig. 5.   Electrical characterization of the SiC MOSFET TCAD model

B. Avalanche Breakdown Boundary

To verify the accuracy of the model, avalanche breakdown experiments were conducted. Using a power semiconductor device dynamic tester, dynamic characteristic tests were conducted on SiC MOSFETs under conditions of 125degC

and 1000V, and the variation of voltage spikes with switch current was recorded. When the switch current is below 1100A, the peak of the turn off voltage of SiC MOSFET increases linearly with the switch current; When the switch current is higher than 1100A, the turn off voltage peak of SiC MOSFET reaches about 1800V, and then the voltage peak remains basically unchanged with the increase of current, indicating that the device has undergone dynamic avalanche breakdown and entered the turn off self clamping mode, with an avalanche breakdown voltage of about 1800V.

The TCAD model simulation circuit structure is a half bridge inductive diode clamp circuit, with a stray inductance of 32nH, a driving resistance of 1 Ω, a switching pulse width of 50us, gate voltages of -2V and 18V, a switching voltage range of 900V~1500V, a switching current range of 250A~1000A, and a switching temperature range of 150 °C~200 °C. The simulation results show that the turn off voltage peak of the SiC MOSFET will eventually enter the clamp, at which point the PN junction electric field strength exceeds 2.0MV/cm, and local avalanche breakdown occurs. The avalanche breakdown voltage is about 1780V, which is less than 3% different from the experimental results. The simulation and experimental results of the avalanche breakdown voltage of the device at different temperatures are shown in Figure 8, indicating that the TCAD model can accurately obtain the avalanche breakdown voltage of the device at different temperatures.

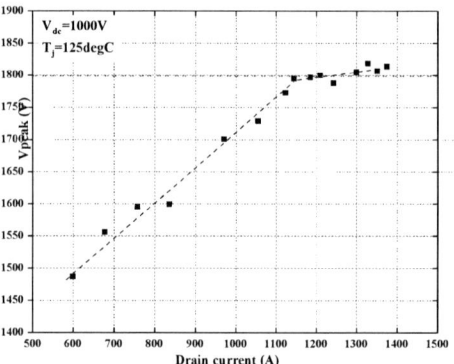

Fig. 6.   SiC MOSFET dynamic characterization test turn-off voltage spike Vpeak with switching current trend: Vdc=1000V, T=125degC, Rg=1Ω.

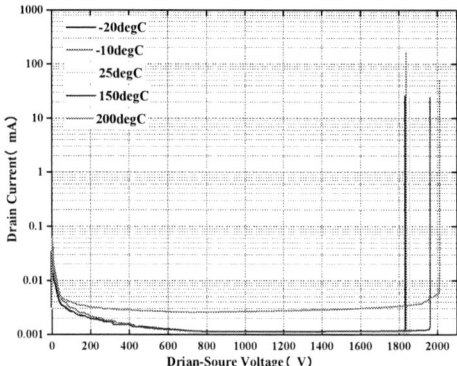

Fig. 7.   SiC MOSFET breakdown voltage simulation at different temperatures: -20degC~200degC.

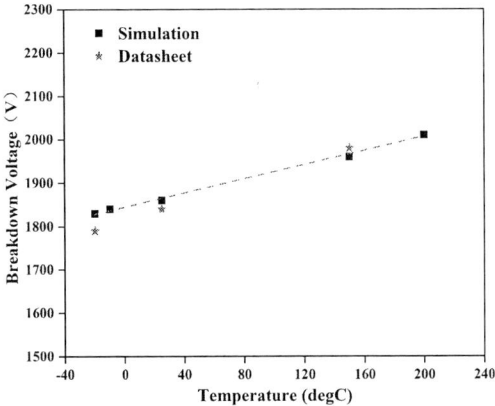

Fig. 8. Avalanche breakdown voltage diagram of SIC MOSFET

C. Gate Breakdown Boundary

To verify the accuracy of the model, gate breakdown experiments were conducted. Using the voltage linear scanning method, an electrical breakdown test was conducted on the gate source electrode of SiC MOSFET at room temperature (with extremely short drain source), and the variation curve of gate leakage current Igss with gate source voltage Vgs was collected. The voltage scanning rate is 0.5V/s. When the forward scan of the gate source voltage exceeds about 33V, the gate leakage current of SiC MOSFET increases exponentially, with a maximum value exceeding 2mA (instrument protection threshold); When the negative scan of the gate source voltage exceeds about 18V, the gate leakage current of SiC MOSFET also shows a rapid increase phenomenon. The experimental results show that the gate withstand voltage range of SiC MOSFET is around -18V~33V.

A simulation study was conducted on the variation of gate leakage current with bias voltage based on the TCAD model. The results show that within the range of -20V to 30V, the gate leakage current remains constant at approximately 20~50nA, and the gate operates normally; When the gate bias voltage is scanned forward to about 30V, the gate oxide layer breaks down, and the drain current increases exponentially with the voltage; When the gate bias voltage is scanned in the negative direction, the same phenomenon exists, and the deflection threshold is about -20V. The maximum deviation between simulation and experimental results is about 3V, and the TCAD model can accurately obtain the gate breakdown voltage of the device.

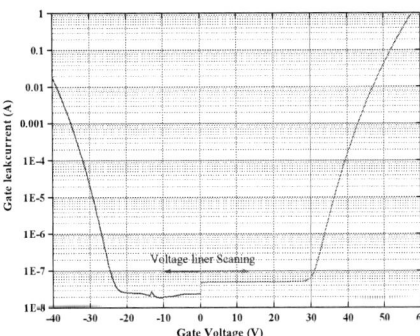

Fig. 9. SiC MOSFET gate voltage linear sweep: V_g=-40V/+60V, V_{ds}=0V, T=150degC, R_g=1Ω.

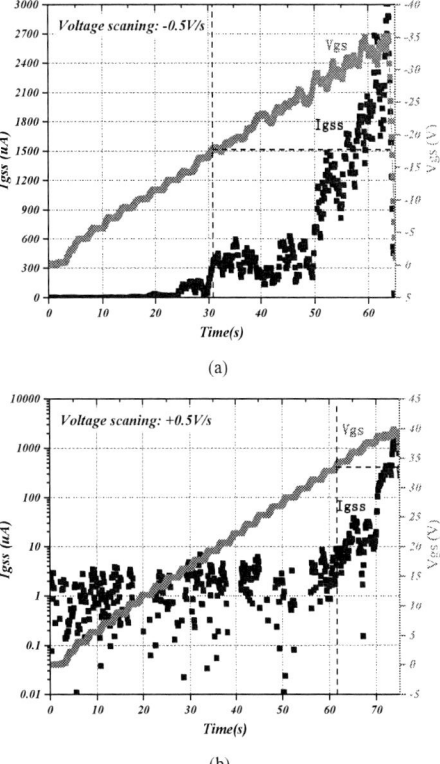

(a)

(b)

Fig. 10. Leakage current variation curves during voltage linear scan of the SiC MOSFET: 0.5V/s, room temperature, (a)-negative scan, (b)-positive scan.

Through the TCAD model, 900V-1300V, 500A, The simulation results of the 150degC transient switch are shown in Figures 11 and 12. The simulation model can accurately characterize the evolution process of multiple physical fields inside the device and perform simulation analysis under different operating conditions.

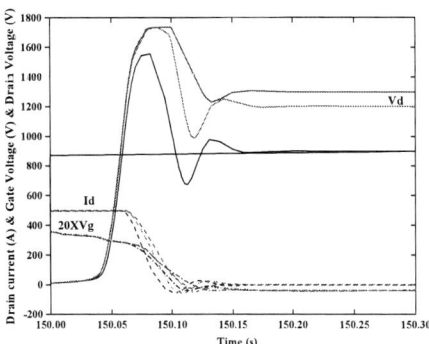

Fig. 11. SiC MOSFET transient switching characteristic curves: 900V-1300V, 500A , 150 degC, R_g=1Ω

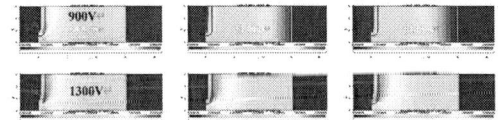

Fig. 12. The evolution process of multiple physical fields in SiC MOSFET turn off transient cells:: 900V-1300V, 500A , 150 degC, R_g=1Ω

IV. CONCLUSION

In summary, the numerical model constructed using the TCAD simulation platform demonstrates a high degree of agreement between the characteristic curves obtained from simulation and manual measurements at different temperatures, as well as between the switching waveforms from simulation and experiments. Under the conditions of 125 °C and 1000 V, the discrepancy between the dynamic breakdown voltage of the SiC MOSFET and the experimental results is less than 3%. Furthermore, the maximum deviation observed between the simulated and experimental gate withstand voltages of the SiC MOSFET is approximately 3 V. These findings verify the accuracy of the TCAD model and affirm its capability to study the characteristics of the SiC MOSFET under various operating conditions. Furthermore, the model can serve as a framework for guiding the practical applications of the SiC MOSFET.

REFERENCES

[1] R. Tambone, A. Ferrara, R. Siemieniec, A. Wood, F. Magrini and R. J. E. Hueting, "Ruggedness of Silicon Power MOSFETs—Part I: Cell Structure Design Related Failure: A Review," in IEEE Transactions on Electron Devices, vol. 71, no. 6, pp. 3445-3457, June 2024.

[2] J. Wei et al., "Review on the Reliability Mechanisms of SiC Power MOSFETs: A Comparison Between Planar-Gate and Trench-Gate Structures," in IEEE Transactions on Power Electronics, vol. 38, no. 7, pp. 8990-9005, July 2023.

[3] X. Li, Y.F. Luo, Z.N. Shi, R.T. Wang, F. Xiao. A physically based improved circuit model for SiC MOSFET [J]. Journal of Electrotechnology, 2022, 37 (20): 5214-5226.(in Chinese)

2025 IEEE Workshop on Wide Bandgap Power Devices and Applications in Asia (WiPDA Asia)

The Method to Evaluate the SOA of Drain-Source Voltage for SiC MOSFET

Fengming Yang
School of Electrical and Electronic
Engineering
Huazhong University of Science and
Technology
Wuhan, 430074, China
yangfengming@hust.edu.cn

Xin Li*
National Key Laboratory of Electromagnetic
Energy
Naval University of Engineering
Wuhan 430033, China
xinlee@nue.edu.cn

Yifei Luo
National Key Laboratory of Electromagnetic
Energy
Naval University of Engineering
Wuhan 430033, China
yfluo23@nue.edu.cn

Lin Liang
School of Electrical and Electronic
Engineering
Huazhong University of Science and
Technology
Wuhan, 430074, China
lianglin@hust.edu.cn

Yongle Huang
National Key Laboratory of Electromagnetic
Energy
Naval University of Engineering
Wuhan 430033, China
xinlee@nue.edu.cn

Abstract—**Silicon Carbide (SiC) MOSFETs have superior material characteristics, including high switching frequency and high-temperature. They are widely used in high-density power electronic system. In order to access its full potential, especially in high-frequency and high-temperature applications, in this paper, a method to evaluate the safe operating area (SOA)defined by the drain-source voltage peak during switching transient is proposed. Firstly, theoretical treatment on the switching transient and the failure mechanism during turn-off transient for SiC MOSFET is conducted. Then, temperature dependency of the drain-source overshoot is derived. Finally, double pulse test platform is designed for different temperature conditions, and the SOA defined by the drain-source voltage overshoot has been investigated. The conclusion of this paper implies that the device can be operated on a higher switching speed under high temperature conditions, which will provide some guidance on accessing the full advantage of the high frequency and high temperature characteristics of SiC MOSFET.**

Keywords—*SOA, SiC MOSFET, drain-source voltage peak, high frequency, reliability.*

I. INTRODUCTION

Silicon carbide (SiC) metal oxide semiconductor field effect transistors (MOSFETs) can be employed in the high-density power system due to its superior material characteristics. It has been considered the perfect substitute of Si insulated gate bipolar transistor (IGBT) [1]. However, higher switching frequency increases the voltage and current change rate in the circuit. Higher dv/dt and di/dt slew rate applied to the lower junction capacitance could result in a small damping ratio, which is more susceptible to oscillation peak. These side effects become the major issue that limit its full potential in high frequency.

In order to access the full potential of SiC MOSFET, several existing research works investigate this problem from different perspective, including oscillation mechanism and modeling [2-5], influence factors [6-9], and suppression methods [10-13]. In terms of the mechanism of the oscillation, plenty of studies generally agree that the rapid change of voltage and current in the commutation loop during switching transient should be responsible for the oscillation in drain source loop. Simultaneously, two feedback networks are

This work was supported by National Natural Science Foundation of China under Grant 52407231, National Key Laboratory of Electromagnetic Energy under Project 61422172320102, and the Innovative Group Project of

Hubei Natural Science of Foundation under 2025AFA045. (Corresponding author: Xin Li)

sensing the drain-source voltage and changing the gate-source voltage [2], cause the gate-source voltage oscillation. The second topic that previous research focus on is the influence of the parasitic parameters in the voltage and current oscillation. The common source inductance and the gate inductance play more important role in the gate-source oscillation, they should be eliminated as much as possible to achieve a stable switching operation [3]. In order to mitigate the side effects during the switching transient, several effective methods have been approached to mitigate the switching ringing and other side effects, including RC snubber [7],[9], optimizing the parasitic parameters in both gate and power loop [10-11], and active gate drive (AGD) techniques [12].

In order to evaluate the potential of SiC MOSFET more precisely, this paper proposed a method to describe the SOA of drain-source voltage during switching transient. The contribution of this paper lies in two aspects. First, theoretical analyzing of the distortion mechanism and its temperature dependency in the drain-source loop of SiC MOSFET during switching transient are carried out. Second, this paper explores the SOA of drain-source voltage for the SiC MOSFET based on the double pulse test. The rest of this article is organized as follows. In Section II, the failure mechanism, and the temperature dependency of drain-source voltage overshoot are illustrated. Then, the experiment detail is illustrated in Section III. Based on that, the SOA evaluation method is depicted. Finally, the conclusions are drawn for summarizing this article in Section IV.

II. THE THEORETICAL ANALYZING FOR THE BOUNDARY OF DRAIN SOURCE VOLTAGE

A. SiC MOSFET Switching Transient Analyzing

The half-bridge SiC MOSFET circuit is used to accurately evaluate the SiC MOSFET switching transient characteristics through double pulse test (DPT), as shown in Fig.1. Q_H and Q_L are high side and low side SiC MOSFET in the half-bridge circuit respectively. V_{dc} is the dc bus voltage. C_{dc} is paralleled at both ends of V_{dc} to improve the power supply quality. L_{load} is the inductive load. The parasitic inductance of SiC MOSFET are L_s, L_d and L_g, which represent the parasitic inductance on source, drain and gate terminal, respectively. The internal parasitic capacitance includes gate-source capacitance C_{gs}, gate-drain capacitance C_{gd} and drain-source capacitance C_{ds}. The gate and source terminal are connected to a gate driver through gate resistance R_g and gate inductance

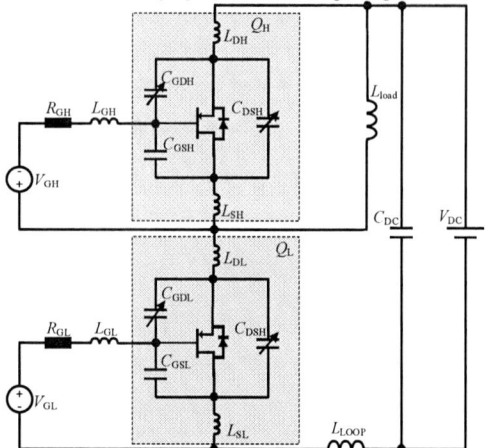

Fig. 1. Half-bridge circuit schematic of SiC MOSFET

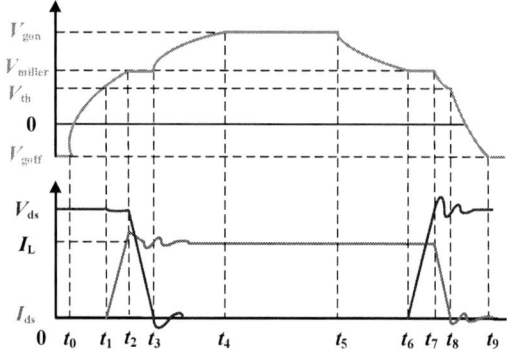

Fig. 2. Typical waveform of SiC MOSFET switching trasnient

L_g. In the half-bridge circuit, Q_H and Q_L are complementary, and the turn-on transient processes of Q_H and Q_L are similar. Therefore, the switching transient behavior is analyzed by taking the Q_L as an example.

The typical waveform of SiC MOSFET switching transient can be depicted in Fig.2. The turn-on transient can be divided into four sequential intervals, including turn-on delay interval (t_0-t_1), current rising interval (t_1-t_2), voltage falling interval (t_2-t_3) and complete turn-on interval (t_3-t_4). Similarly, the turn off transient also can be divided into four sequential intervals, including turn-off delay interval (t_5-t_6), voltage rising interval(t_6-t_7), current falling interval (t_7-t_8) and complete turn-off interval (t_8-t_9). SiC MOSFET high-speed switching increases the voltage and current change rate in the circuit, which generated higher dv/dt and di/dt. The high dv/dt and di/dt generated during the switch transient, combined with the parasitic parameters such as common source inductance, output capacitance, etc. could result in the voltage oscillation.

B. The Failure Mechanism under the Voltage Oscillation in Dain-Source Loop

During the turn-off transient, the gate electrode of SiC MOSFET is connected to the source via the gate resistance to discharge its input capacitance. The gate voltage decreases exponentially with time. This stage is called turn-off delay (t_5-t_6) before the drain voltage begins to increase.

When the gate voltage drops to V_{miller} at t_6, the voltage rising interval begins. The average change rate of the drain source voltage during this stage is determined by the discharge rate of the capacitance of gate-drain (C_{gd}). In the current falling stage (t_7-t_8), the gate source voltage drops from V_{miller} to V_{th}. The average change rate of the drain current during this stage can be approximately expressed as

$$\frac{dI_D}{dt} = \frac{V_{SS} - \left(\frac{I_L}{2g_m} + V_{TH} \right)}{\frac{R_G C_{iss}}{g_m} + L_S} \tag{1}$$

where, V_{ss} is the negative voltage supply of the gate, R_G is the gate resistance, I_L is the load current, V_{TH} is the threshold of SiC MOSFET, C_{iss} is the input capacitance of SiC MOSFET, L_s is the parasitic inductance of source terminal, g_m is the transconductance of SiC MOSFET, and it is given by

$$g_m = \frac{W \mu_0 C_{OX}}{L} (V_{GS} - V_{TH}) \tag{2}$$

where, W is the channel width of SiC MOSFET, L is the channel length, μ_0 is mobility of the carriers, C_{OX} is the gate-oxide capacitor, V_{GS} is the gate bias voltage source.

Moreover, due to the existence of switching loop stray inductances (L_{loop}), there is a voltage drop on L_{loop}, the voltage across drain and source terminal will be reshaped, which cause the overshoot of V_{ds}. The value of the drain source voltage overshoot can be expressed as

$$V_{overshoot} = -L_{loop} \frac{dI_D}{dt} \tag{3}$$

This is the reason introducing the positive spike during turn off transient. It will lead to breakdown failure if the drain-source voltage oscillation peak exceeds the breakdown voltage of SiC MOSFET.

C. The Temperature Dependency Analyzing for Voltage Overshoot in Dain-Source Loop

According the above analyzing, the overshoot voltage between drain source terminal is proportional to the current change rate di/dt. Hence, it is necessary to investigate the temperature dependency of di/dt.

The temperature dependency of di/dt can be evaluated by taking derivative of Eq.(1) with respect to temperature. The result is shown as

$$\frac{d^2 I_D}{dt dT} = K_1 \frac{d\mu_0}{dT} - K_2 \frac{dV_{TH}}{dT} \tag{4}$$

Where, K_1 and K_2 are given in Appendix. From Eq.(4), it can be seen that there are two temperature sensitive parameters: μ_0 and V_{TH}. According reference [14], [15], these two parameters show a negative coefficient of the temperature. Combined with the μ_0 is larger than the coefficient of $d\mu_0/dT$, hence, the temperature dependency of change rate of drain current is dominated by the temperature dependency of threshold voltage. Therefore, during turn-off transient, due to the change of di/dt polarity, the temperature dependency of di/dt exhibits a decline trend with the rising temperature. It can be concluded that the overshoot peak decreases with the increment of temperature.

III. The SOA Evaluation Method of Dain Source voltage for SiC MOSFET

In this section, the detail for the double-pulse platform under low-temperature and high-temperature conditions is exhibited. Based on the experimental platform, the variation ranges of drain oscillation voltage of SiC MOSFET under different switching voltages and temperature conditions are conducted. Then, the safe operating area of SiC MOSFET

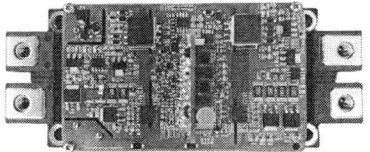

Fig. 3.   The 1700V/300A SiC MOSFET module and its driver board

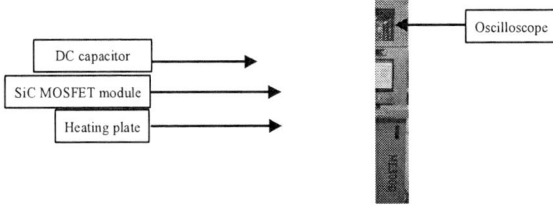

Fig. 4.   The DPT exprimental platform for room and high temperature

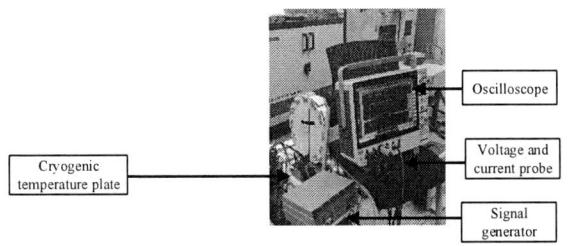

Fig. 5.   The DPT exprimental platform for cryogenic temperature

TABLE I.   TYPE NUMBERS OF MAIN EQUIPMENT IN DPT TEST

Equipment	Type Number
Oscilloscope	MDO3054
DC power supply	JP12006D
Differential voltage probe	THDP0100
Current probe	CWTMini-HF3B
SiC MOSFET module	YS17CB300EM
Heating plate	JF966C
Microcontroller	STM32F429IGT6

defined by the drain source voltage peak is analyzed.

A. Expremental platform Setup

A SiC MOSFET half-bridge module with a rated voltage 1700V and a current of 300 A is used as the device to be tested, as shown in Fig.3. The double pulse test platforms for different temperature are shown as Fig.4 and Fig.5. The auxiliary supply provides a 24V for driving SiC MOSFET module, and a microcontroller board provides the double pulse signal for the driver board. A MDO3054 oscilloscope of Tektronix is utilized to capture the waveforms. The bus voltage is measured by a differential voltage probe THDP0100 of Keysight. Drain-source current during switching transient is measured by Rogowski coil probe CWTMini-HF3B. Besides, the high voltage dc supply is JP12006D, and a JF966C model heating plate to increase the

temperature of the SiC MOSFET module. The main equipment is shown in Table I.

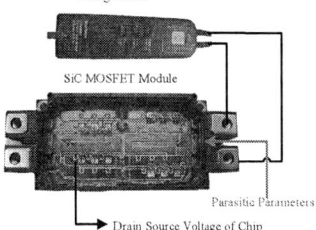

Fig. 6.   The connection between the voltage probe and SiC MOSFET moddule in the exprimental platform

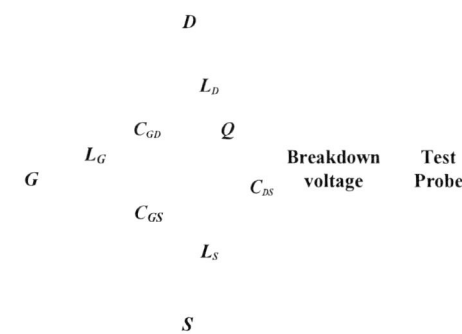

Fig. 7.   The relationship between the probe tesing and the chip voltage for SiC MOSFET.

B. The Drain Source Voltage Testing Detail

The experimental temperature was set at -12°C, 25°C and 175°C, 200°C respectively. Before each test, setting the heating plate to specified working temperature, and until its temperature was stable for 10 minutes to make sure the whole module has entered a thermal steady state. Then, a double pulse test was set and the waveforms can be captured by the oscilloscope, and the experimental waveforms including V_{ds}, I_{ds}, and V_{gs} were recorded.

Due to the package of SiC MOSFET module, it is not very easy to evaluate the drain source voltage of the chip. Fig.6 and Fig.7 show the relationship between the probe testing voltage and the inner breakdown voltage of SiC MOSFET. According the overshoot mechanism during the turn off transient in Section II, the overshoot voltage between drain source terminal has proportional relationship to the current change rate di/dt in the specific system. Hence, the value of parasitic inductance of the testing system can be obtained through double pulse test. Then the voltage across the chip can be obtained by eliminating the voltage drop across the parasitic inductor due to the change current from the bus voltage.

C. Expremental Results

The experimental results and the SOA are carried out under different switching voltage and temperature conditions, as shown in Fig.8-Fig.11. The boundary of drain-source voltage spike is the breakdown voltage between two terminals, which is 1700V.

The experimental results show that the turn-off voltage spike of SiC MOSFET increases approximately linearly with switching voltage and current. Same tendency can be seen in different temperature conditions. Under room temperature, as shown in Fig.9, when the drain-source voltage reaches its boundary, the maximum switching current of SiC MOSFET

lines: bus voltage, drain source voltage boundary, and the desaturation current.

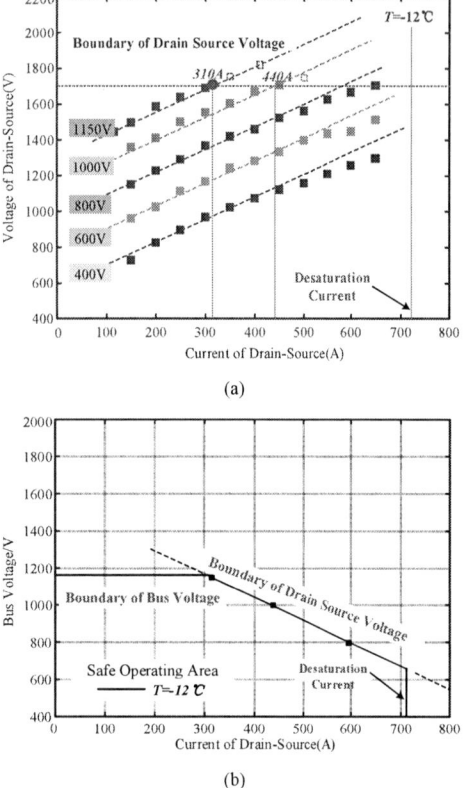

(a)

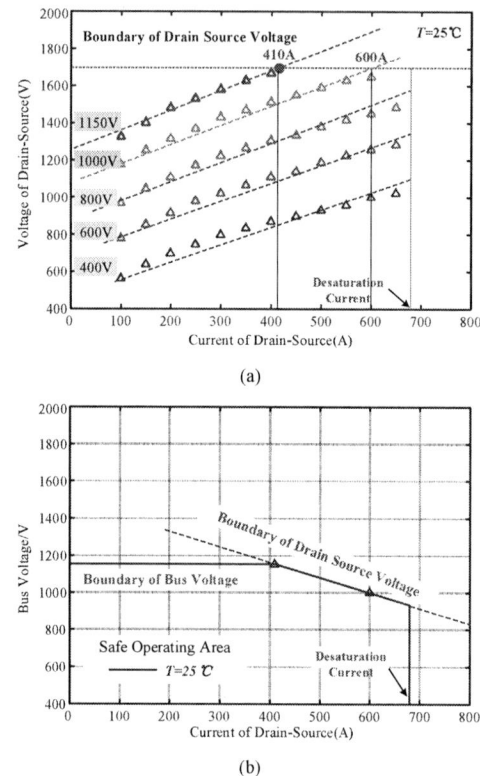

(a)

(b)

Fig. 8. The exprimental result (a) and the SOA (b) under -12°C temperature

(b)

Fig. 9. The exprimental result (a) and the SOA (b) under 25°C temperature

is 410A with 1150V bus voltage. In comparison, the maximum switching current of SiC MOSFET is 600A with 1000V bus voltage.

From temperature perspective, the safe operating area is gradually expanded due to the temperature increasing. Under the condition of -12°C, as shown in Fig.8, the maximum current of SiC MOSFET at the voltage rated 1150V and 1000V rated are approximately 310A and 440A respectively. At 175°C, as shown in Fig.10, the maximum switching current of SiC MOSFET under 1150V rated and 1000V rated voltage is about 495A and 655A respectively. When the temperature increases to 200°C, as shown in Fig.11, under same condition, these value are about 510A and 668A respectively.

However, the desaturation current of SiC MOSFET has been observed to be 450A and 385A at 175°C and 200°C, respectively. Hence, the desaturation protection is triggered at 175°C and 200°C prior to the occurrence of the maximum switching current capacity of the SiC MOSFET.

The preceding analysis indicates that the safe operating area defined by the drain-source voltage peak gradually increases with the increasing of temperature. The main reason is because as the temperature increases, the threshold voltage and the Miller plateau of the SiC MOSFET decrease, resulting in a slower switching rate of the SiC MOSFET, which leads to a lower turn-off spike voltage.

D. The SOA Defined by Drain-Source Voltage Oscillation

The SOA of SiC MOSFET defined by the drain-source voltage peak under different temperature conditions are shown in Fig.12. The SOA of SiC MOSFET consists of three

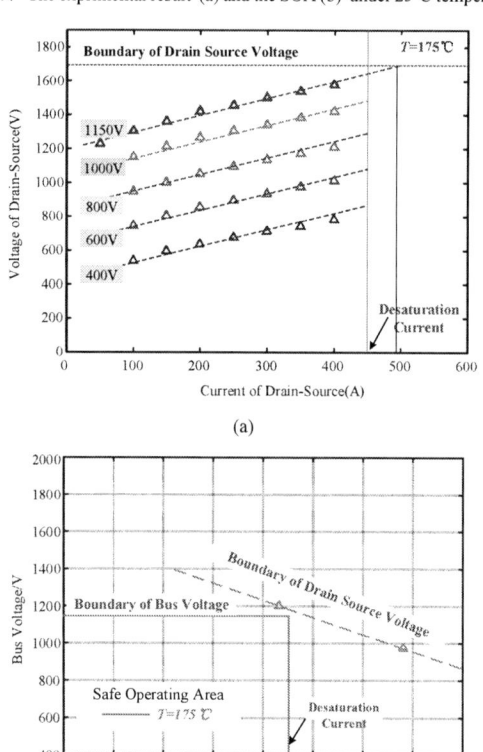

(a)

(b)

Fig. 10. The exprimental result (a) and the SOA (b) under 175°C temperature

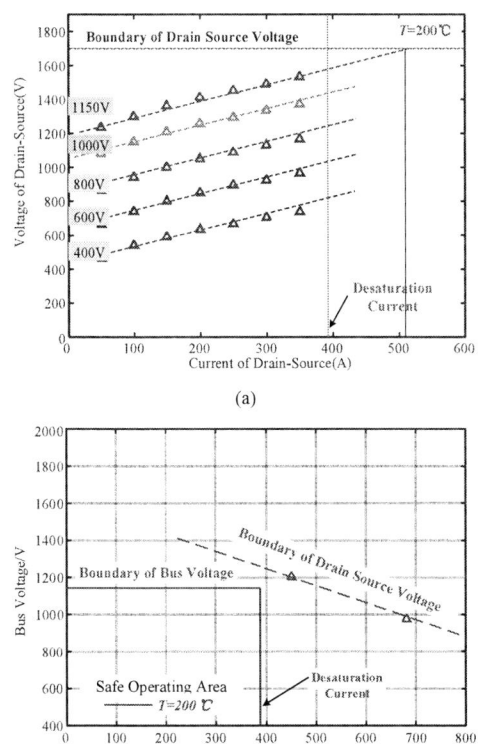

(a)

(b)

Fig. 11. The exprimental result (a) and the SOA (b) under 200°C temperature

From the Fig.12, it can be seen that the boundary of drain source voltage undergoes an expansion as the temperature increases. Simultaneously, there has been a gradual decline trend on the desaturation current. Due to the boundary of drain source voltage is far from the desaturation current boundary, the SOA under high temperature is determined by the desaturation current.

There are some conclusions can be drawn through the analysis above. Firstly, the SiC MOSFET has a narrow gate source boundary, a slower switching speed and a higher current density are recommended. Then, a widen gate source voltage allows the device is operated on a higher switching speed with a lower current density to ensure that the device does not cause failure due to overheat. In practice, this conclusion can be used as a guide to find the optimal power point to access the full advantage of the high frequency and high temperature characteristics of SiC MOSFET.

IV. CONCLUSION

In order to evaluate the potential of SiC MOSFET more precisely, this paper proposes a method to describe the SOA of drain-source voltage during switching transient. Firstly, a theoretical treatment on the switching transient and the failure mechanism during turn-off transient for SiC MOSFET have been presented. Then, the temperature dependency of the drain-source overshoot has been derived. Finally, double pulse test platform is designed for different temperature conditions, and the SOA defined by the drain-source voltage has been investigated. The key findings are summarized as follows:

The temperature dependency analyzing of *di/dt* during turn-off transient exhibits a decline trend with the rising of

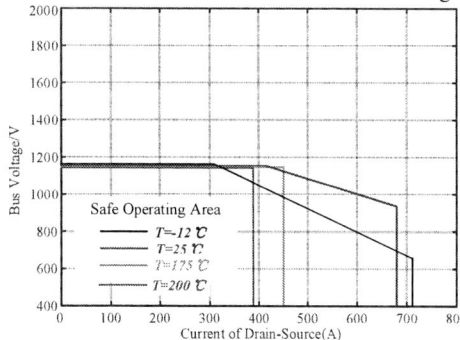

Fig. 12. The SOA defined by drain-source voltage oscillation peak

temperature, which caused the overshoot peak decreases with the increment of temperature. Based on the analytical results, the SOA defined by the drain-source voltage peak has a wide trend due to the increasing temperature. This implies that the device can be operated on a higher switching speed under high temperature conditions. In practice, this conclusion can be used as a guide to access the full advantage of the high frequency and high temperature characteristics of SiC MOSFET.

APPENDIX

The detail about the temperature dependency of di/dt during turn-off transient is shown as follow:

According Eq.(1), assuming

$$
\begin{cases}
M = V_{SS} - \left(\dfrac{I_L}{2g_m} + V_{TH} \right) \\
N = \dfrac{R_G C_{iss}}{g_m} + L_S
\end{cases}
\tag{5}
$$

taking derivative of Eq. (1) with the temperature, which can be expressed as

$$
\frac{d^2 I_D}{dt dT} = \frac{N \dfrac{dM}{dT} - M \dfrac{dN}{dT}}{N^2}
\tag{6}
$$

where,

$$
\begin{cases}
\dfrac{dM}{dT} = -\left(-\dfrac{I_L}{2g_m^2} \dfrac{dg_m}{dT} + \dfrac{dV_{TH}}{dT} \right) \\
\dfrac{dN}{dT} = -\dfrac{R_G C_{iss}}{g_m^2} \dfrac{dg_m}{dT}
\end{cases}
\tag{7}
$$

by substituting Eq. (2) to Eq. (7), it can be obtained:

$$
\begin{cases}
\dfrac{dM}{dT} = \dfrac{I_L}{2g_m^2} \left[\dfrac{d\mu_0}{dT} (V_{GS} - V_{TH}) - \dfrac{W \mu_0 C_{OX}}{L} \dfrac{dV_{TH}}{dT} \right] - \dfrac{dV_{TH}}{dT} \\
\dfrac{dN}{dT} = -\dfrac{R_G C_{iss}}{g_m^2} \left[\dfrac{d\mu_0}{dT} (V_{GS} - V_{TH}) - \dfrac{W \mu_0 C_{OX}}{L} \dfrac{dV_{TH}}{dT} \right]
\end{cases}
\tag{8}
$$

the value of K_1 and K_2 can be obtained from Eq.(6)- Eq.(8).

$$
\begin{cases}
K_1 = \dfrac{V_{GS} - V_{TH}}{N^2} \left[N \dfrac{I_L}{2g_m^2} - M \dfrac{R_G C_{iss}}{g_m^2} \right] \\
K_2 = \dfrac{1}{N^2} \left[N \left(\dfrac{I_L}{2g_m^2} \dfrac{W \mu_0 C_{OX}}{L} - 1 \right) + M \dfrac{R_G C_{iss}}{g_m^2 L} \dfrac{W \mu_0 C_{OX}}{L} \right]
\end{cases}
\tag{9}
$$

REFERENCES

[1] J. Chen, X. Du, Q. Luo, X. Zhang, P. Sun, and L. Zhou, "A Review of Switching Oscillations of Wide Bandgap Semiconductor Devices," IEEE Trans. Power Electron., vol. 35, no. 12, pp. 13182–13199, Dec. 2020, doi: 10.1109/TPEL.2020.2995778.

[2] W. Zhang, X. Wang, M. S. A. Dahidah, G. N. Thompson, V. Pickert, and M. A. Elgendy, "An Investigation of Gate Voltage Oscillation and its Suppression for SiC MOSFET," IEEE Access, vol. 8, pp. 127781–127788, 2020, doi: 10.1109/ACCESS.2020.3008940.

[3] X. Yang, M. Xu, Q. Li, Z. Wang, and M. He, "Analytical Method for RC Snubber Optimization Design to Eliminate Switching Oscillations of SiC MOSFET," IEEE Trans. Power Electron., vol. 37, no. 4, pp. 4672–4684, Apr. 2022, doi: 10.1109/TPEL.2021.3127516.

[4] Y. Xiao, H. Shah, T. P. Chow, and R. J. Gutmann, "Analytical modeling and experimental evaluation of interconnect parasitic inductance on MOSFET switching characteristics," in Nineteenth Annual IEEE Applied Power Electronics Conference and Exposition, 2004. APEC '04., Anaheim, CA, USA: IEEE, 2004, pp. 516–521. doi: 10.1109/APEC.2004.1295856.

[5] Y. Wu, N. He, L. Yu, D. Xu, S. Igarashi, and T. Fujihira, "Effectiveness Analysis of SiC MOSFET Switching Oscillation Damping," in 2020 IEEE 9th International Power Electronics and Motion Control Conference (IPEMC2020-ECCE Asia), Nanjing, China: IEEE, Nov. 2020, pp. 20–27. doi: 10.1109/IPEMC-ECCEAsia48364.2020.9367969.

[6] Y. Wu, S. Yin, H. Li, and W. Ma, "Impact of RC Snubber on Switching Oscillation Damping of SiC MOSFET With Analytical Model," IEEE J. Emerg. Sel. Topics Power Electron., vol. 8, no. 1, pp. 163–178, Mar. 2020, doi: 10.1109/JESTPE.2019.2953272.

[7] Zezheng Dong, Xinke Wu, Kuang Sheng, and Junming Zhang, "Impact of common source inductance on switching loss of SiC MOSFET," in 2015 IEEE 2nd International Future Energy Electronics Conference (IFEEC), Taipei, Taiwan: IEEE, Nov. 2015, pp. 1–5. doi: 10.1109/IFEEC.2015.7361607.

[8] N. Qi, D. Jin, X. Ge, and C. Luo, "Influence of Gate Driver Loop Inductance on SiC MOSFET Module Turn-on Gate Voltage Oscillation in High Power Application," in 2022 IEEE Transportation Electrification Conference and Expo, Asia-Pacific (ITEC Asia-Pacific), Haining, China: IEEE, Oct. 2022, pp. 1–5. doi: 10.1109/ITECAsia-Pacific56316.2022.9941889.

[9] T. Liu, R. Ning, T. Wong, and Z. J. Shen, "Modeling and Analysis of SiC MOSFET Switching Oscillations," IEEE J. Emerg. Sel. Topics Power Electron., pp. 1–1, 2016, doi: 10.1109/JESTPE.2016.2587358.

[10] Y. Han, H. Lu, Y. Li, and J. Chai, "Open-Loop Gate Control for Optimizing the Turn-ON Transition of SiC MOSFETs," IEEE J. Emerg. Sel. Topics Power Electron., vol. 7, no. 2, pp. 1126–1136, Jun. 2019, doi: 10.1109/JESTPE.2018.2848900.

[11] Y. Sugihara, K. Nanamori, M. Yamamoto, and Y. Kanazawa, "Parasitic Inductance Design Considerations to Suppress Gate Voltage Oscillation of Fast Switching Power Semiconductor Devices," in 2018 International Power Electronics Conference (IPEC-Niigata 2018 - ECCE Asia), Niigata: IEEE, May 2018, pp. 2789–2795. doi: 10.23919/IPEC.2018.8507433.

[12] M. R. Ahmed, R. Todd, and A. J. Forsyth, "Predicting SiC MOSFET Behavior Under Hard-Switching, Soft-Switching, and False Turn-On Conditions," IEEE Trans. Ind. Electron., vol. 64, no. 11, pp. 9001–9011, Nov. 2017, doi: 10.1109/TIE.2017.2721882.

[13] L. Yu, Y. Wu, A. U. Ibrahim, D. Xu, S. Igarashi, and T. Fujihira, "Suppression Switching Ringing of SiC-MOSFET Inverters with Combined Design of DC Bus Snubber and Gate Drive," in 2021 IEEE 12th International Symposium on Power Electronics for Distributed Generation Systems (PEDG), Chicago, IL, USA: IEEE, Jun. 2021, pp. 1–7. doi: 10.1109/PEDG51384.2021.9494224.

[14] X. Lu, L. Wang, Q. Yang, F. Yang, Y. Gan, and H. Zhang, "Investigation and Comparison of Temperature-Sensitive Electrical Parameters of SiC mosfet at Extremely High Temperatures," IEEE Trans. Power Electron., vol. 38, no. 8, pp. 9660–9672, Aug. 2023, doi: 10.1109/TPEL.2023.3267472.

[15] Z. Ni, S. Zheng, M. S. Chinthavali, and D. Cao, "Investigation of Dynamic Temperature-Sensitive Electrical Parameters for Medium-Voltage SiC and Si Devices," IEEE J. Emerg. Sel. Topics Power Electron., vol. 9, no. 5, pp. 6408–6423, Oct. 2021, doi: 10.1109/JESTPE.2021.3054018.

GaN-based 1.5 MHz Synchronous Buck Converter with Partial Soft-switching Control Scheme

Zeqi Yang
School of Electrical Engineering
Hebei University of Technology
Tianjin, China
15031348977@163.com

Yuan Liu
School of Electrical Engineering
Hebei University of Technology
Tianjin, China
ly620304@outlook.com

Zhe Zhang
School of Electrical Engineering
Hebei University of Technology
Tianjin, China
zhezhang@outlook.com

Abstract—This paper presents a systematic design methodology for a GaN-based synchronous Buck converter. The study introduces an analog hysteresis control scheme that enables precise and continuous tracking of perturbation current. Furthermore, based on the proposed control scheme, this work establishes analytical models for the quantitative assessment of power losses during both hard-switching and soft-switching operations, and proposes a partial soft-switching control scheme. Finally, an experimental prototype with 1.5 MHz switching frequency was developed, the efficiency is improved by 3% with the proposed partial soft-switching control scheme.

Keywords—MHz-level switching frequency, loss analysis, variable-frequency control, partial soft-switching

I. INTRODUCTION

High efficiency and high power density have consistently been the pivotal performance criteria for power electronics devices and systems [1]. In recent years, the wide-bandgap (WBG) semiconductor technology represented by GaN has experienced significant progress and these emerging power devices exhibit superior characteristics, including faster switching speeds, reduced on-resistance and lower output capacitance [2]. The GaN-based synchronous Buck converter stands out for its simplicity and efficiency, positioning it as an attractive option power conversion [3]. Its near-Critical Conduction Mode (near-CRM) facilitates soft-switching without auxiliary circuits, enabling zero-voltage turn-on characteristics and improved transient response [4]. This paper proposes analog hysteresis control scheme and implements a discretized approximation of the variable-frequency inductor current waveform. Consequently, analytical expressions for losses in hard-switching and soft-switching modes under variable switching frequency operation are formulated. Subsequently, a partial soft-switching control scheme aimed at achieving higher efficiency is introduced.

"This work is supported by National Key Research and Development Program of China under Grant 2024YFB2504900."

II. GaN-BASED SYNCHRONOUS BUCK CONVERTER

A. Synchronous Buck Converter Based on Analog Hysteresis Control

The topology shown in Fig. 1 depicts a synchronous Buck converter, consisting of three primary components: the control loop, the gate driver circuit, and the power stage.

B. Analog Hysteresis Control Scheme

Reference [5] proposed an analog control scheme utilizing quasi-square-wave zero-voltage switching (QSW-ZVS), which employs a straightforward voltage loop compensator to regulate boundary voltage, facilitating soft-switching in synchronous DC-DC converters under wide input-voltage and switching-frequency variations. Reference [6] quantitatively analyzed the zero-voltage switching (ZVS) boundary conditions for near-CRM synchronous Buck converters, deriving reverse current and dead-time parameters from device datasheets. Building upon these findings, this

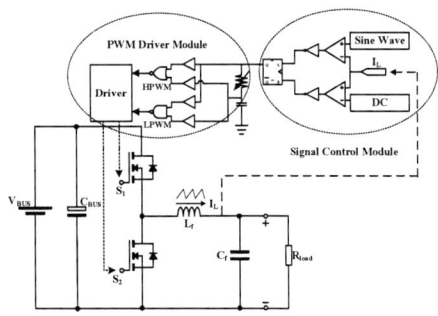

Fig. 1 Converter Topology

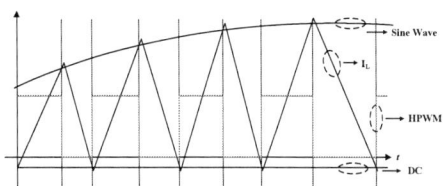

Fig. 2 Principle of Analog Hysteresis Control Scheme

paper proposes an analog hysteresis control scheme that achieves soft-switching via precise hysteresis boundary voltage design. The operational principle of this control scheme is illustrated in Fig. 2.

III. CONVERTER SWITCHING LOSS ANALYSIS

With the increasing switching frequencies of power electronic systems, switching losses in semiconductor devices escalate significantly. Developing an accurate switching loss model is essential for estimating semiconductor device losses, serving as the foundation for converter-level loss analysis and efficiency enhancement. Existing switching loss models can be classified into three categories: physics-based models, behavioral models [7], and analytical modele [8],[9]. This work implements a segmented linear approximation model, which accounts for the effects of loop common-source inductance (L_S). A refined linearization analysis is conducted, and the improved switching loss model is illustrated in Fig. 3.

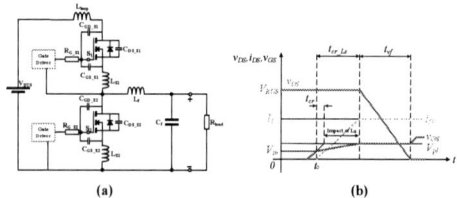

(a)   (b)

Fig. 3 (a). Considering the equivalent circuit model of the loop common-source inductance L_S. (b). The impact of the loop common-source inductance L_S on the device turn-on process.

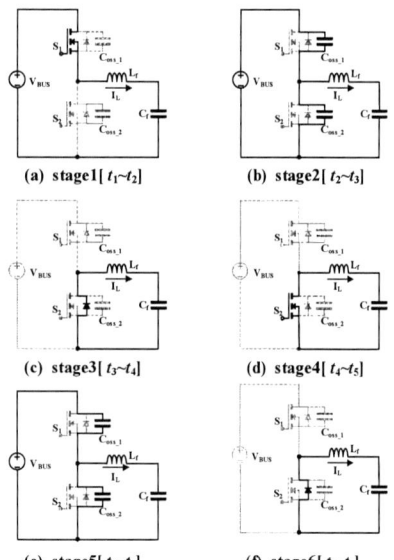

(a) stage1[t_1~t_2]   **(b) stage2[t_2~t_3]**

(c) stage3[t_3~t_4]   **(d) stage4[t_4~t_5]**

(e) stage5[t_5~t_6]   **(f) stage6[t_6~t_7]**

Fig. 4 The transient process of hard-switching operation

Through parameterizing L_S as an equivalent resistance (R_{Ls}) during switching transient and reevaluating the average gate drive current (I_G), the accuracy of loss estimation can be enhanced significantly. The equivalent resistance R_{Ls} is expressed as (1):

$$R_{Ls} = \frac{L_{CS} \cdot g_m}{C_{GS}} \qquad (1)$$

where g_m is the transconductance parameter, which is determined by calculating the slope of the V_{GS}-I_D transfer characteristic across the threshold voltage $V_{GS(th)}$ to plateau voltage $V_{GS(pl)}$ operating region.

A. Analysis of Hard-Switching Losses

Switching losses in power converters primarily comprise three components: output capacitance energy loss (E_{OSS}), commutation overlap energy loss ($E_{ON,overlap}$ and $E_{OFF,overlap}$), and conduction energy loss (E_{cond}). The output capacitance energy loss is influenced by the nonlinear output capacitance (C_{OSS}) characteristics of the power transistors in the half-bridge circuit. The commutation overlap energy loss depends on both gate current dynamics and switching transition duration during turn-on/turn-off events. The conduction energy loss is determined by the RMS current magnitude through the semiconductor devices and their effective conduction period. The hard-switching process is illustrated in Fig. 4.

All power semiconductor devices exhibit a highly nonlinear voltage-dependent capacitance characteristics between drain-source terminals, as illustrated in Fig. 4. This capacitance, denoted as the effective C_{OSS}, can be modeled by fitting the device characterization curves from the device datasheet. The energy dissipation resulting from the charging and discharging of C_{OSS} is defined as E_{OSS}, which is expressed as (2):

$$E_{oss} = \int_0^{V_{BUS}} C_{oss}(v_{DS}) \cdot v_{DS} dv_{DS}$$
$$= \frac{1}{2} \cdot C_{OSS,er} \cdot V_{BUS}^2 \qquad (2)$$

where $C_{OSS,er}$ is the equivalent capacitance derived from the nonlinear C_{OSS} characteristics at a specific bus voltage (V_{BUS}).

During the turn-on switching transition of GaN HEMTs, non-negligible commutation energy loss emerges in addition to C_{OSS}-related energy loss. This energy dissipation, defined as $E_{ON\_overlap}$, is expressed as (3):

$$E_{ON,overlap} = \frac{1}{2} \cdot V_{BUS} \cdot I_L \cdot \left(t_{cr,Ls} + t_{vf}\right) \qquad (3)$$

During the turn-off switching transition of GaN HEMTs, the commutation overlap energy loss is expressed as (4):

979-8-3315-1110-4/25 $31.00 © 2025 IEEE   1029

$$E_{OFF,overlap} \approx \frac{\frac{1}{12} \cdot (I_L \cdot t_{cf})^2}{C_{OSS,S1}(0V) + C_{OSS,S2}(V_{BUS})} \quad (4)$$

As illustrated in Fig. 4 (c) and (f), under hard-switching operation, the synchronous rectifier transistor S_2 manifests reverse conduction characteristics during dead-time intervals. In GaN-based low-voltage converters, the higher reverse conduction voltage drop V_r of S_2 contributes significantly to the total losses and cannot be neglected. The reverse conduction energy loss is expressed as (5):

$$E_{SD} = [I_{Lmax} \cdot V_{SD1} \cdot t_{SD1}] \\ + [I_{Lmin} \cdot V_{SD2} \cdot t_{SD2}] \quad (5)$$

where V_{SD1} and V_{SD2} represent the reverse conduction voltage drop of the GaN HEMTs, which can be determined according to the I_{SD}-V_{SD} transfer characteristic. The reverse conduction time t_{SD1} and t_{SD2} are determined by the effective dead time and the voltage-current commutation time, which are expressed as (6) and (7):

$$t_{SD1} = t_{dead} - t_{vf} - \frac{1}{2} \cdot t_{cr\_Ls} - \frac{1}{2} \cdot t_{off} \quad (6)$$

$$t_{SD2} = t_{dead} - t_{cf} - t_{vr} - \frac{1}{2} \cdot t_{on} \quad (7)$$

The gate charge-related energy dissipation can be expressed as (8):

$$E_G = Q_G \cdot (V_{drv,on} - V_{drv,off}) \quad (8)$$

where $V_{drv,on}$ and $V_{drv,off}$ respectively represent the positive voltage and negative voltage of the gate-drive circuit.

The conduction energy loss is expressed as (9) and (10):

$$E_{cond,S1} = (\frac{I_{Lmin}^2}{2} + \frac{I_{Lripple}^2}{6}) \cdot R_{DS,on} \cdot \frac{1}{2 \cdot f_{sw}} \\ + \frac{I_{Lripple} \cdot I_{Lmin}}{2} \cdot R_{DS,on} \cdot \frac{1}{2 \cdot f_{sw}} \quad (9)$$

$$E_{cond,S2} = (\frac{I_{Lmax}^2}{2} + \frac{I_{Lripple}^2}{6}) \cdot R_{DS,on} \cdot \frac{1}{2 \cdot f_{sw}} \\ - \frac{I_{Lripple} \cdot I_{Lmax}}{2} \cdot R_{DS,on} \cdot \frac{1}{2 \cdot f_{sw}} \quad (10)$$

B. Analysis of Soft-Switching Losses

In soft-switching operation, the inductor current exhibits negative polarity in Stages 5-6 of Fig. 4. During this interval, the negative inductor current is reverse-conducted through the upper transistor S_1 following the turn-off event of S_2. Consequently, the reverse

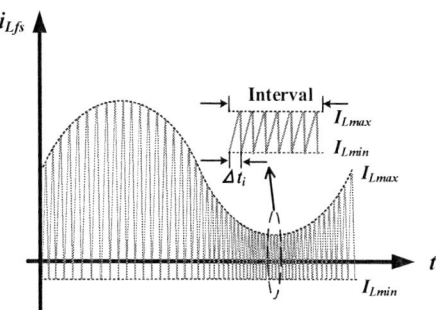

Fig. 5 Schematic diagram of segmented approximation of variable-frequency I_L

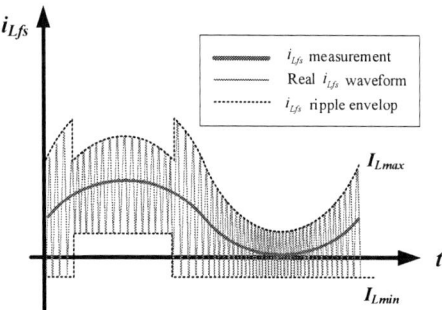

Fig. 6 Partial soft-switching control scheme

conduction energy loss of S_1 and S_2 must be separately expressed as (11) and (12):

$$E_{SD,S2} = |I_{Lmin}| \cdot V_{SD2} \cdot t_{SD2} \quad (11)$$

$$E_{SD,S1} = I_{Lmax} \cdot V_{SD1} \cdot t_{SD1} \quad (12)$$

C. Segmented Approximation of Variable-Frequency Inductor Current

As illustrated in Fig. 2, the inductor current exhibits variable switching frequency characteristics due to the dynamic hysteresis voltage boundary. To accurately estimate switching loss, it is essential to determine the switching event density within each modulation signal cycle. For simplicity, the sinusoidal modulation signal is discretized into infinitesimal intervals via linear approximation techniques, as illustrated in Fig. 5. Assuming quasi-static frequency conditions within each interval, the effective switching period is expressed as (13):

$$\Delta t_i = \frac{(k_1 + k_2) \cdot (I_{Lmax} - I_{Lmin})}{k_1 \cdot k_2} \quad (13)$$

TABLE I
PARAMETERS OF THE EXPERIMENTAL PROTOTYPE

Fundamental parameter	Typical value	Fundamental parameter	Typical value
V_{BUS}	10 V	L_{CS}	0.1 nH
I_{OUT}	5±2 A	$R_{DS,on}$	16 mΩ
f_{OUT}	10 kHz	$R_{G,on}$	0.9 Ω
f_{sw}	1.5 MHz	$V_{GS(th)}$	1.7 V
t_{dead}	10 ns	$V_{GS(pl)}$	3.5 V
C_{OSS}	260 pF	Q_{GS}	1.5 nC
C_{GS}	230 pF	Q_{OSS}	9 nC

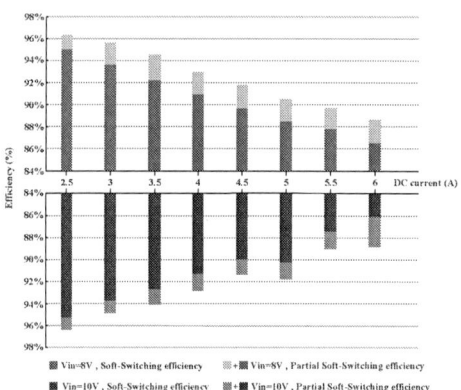

Fig. 8 Efficiency comparison of the converter under soft-switching control scheme and partial soft-switching control scheme.

$$P_{loss} = E_{total} \cdot f_{module} \qquad (15)$$

D. Partial Soft-Switching control scheme

The soft-switching operation introduces elevated inductor current ripple ($I_{Lripple}$), leading to higher conduction loss compared to hard-switching operation. As illustrated in Fig. 5, the switching frequency reaches its minimum and the inductor current ripple reaches its maximum at the peak of I_{Lmax}. Consequently, the conduction loss in this region can significantly exceed those in hard-switching operation. To mitigate this compromise, a partial soft-switching control scheme is proposed, enabling soft-switching operation at high-frequency regions and hard-switching operation at low-frequency regions to enhance converter efficiency. The proposed scheme is illustrated in Fig. 6.

IV. EXPERIMENTAL VERIFICATION

An experimental prototype with 1.5 MHz switching frequency was developed, and the critical parameters are summarized in Table I. Analytical loss model predictions indicate that under hard-switching operation, the total switching loss reaches 3.58 W with conversion efficiency η=84.7%. In contrast, soft-switching operation demonstrates Ploss=2.01 W, achieving an enhanced efficiency of 90.8%. Furthermore, the experimental confirms the operational advantages of the proposed partial soft-switching control scheme. The experimental waveforms are shown in Fig. 7.

To comprehensively validate the proposed partial soft-switching control scheme, Fig. 8 compares the conversion efficiency characteristics between soft-switching and partial soft-switching operations under varying load conditions. Quantitative analysis reveals that the partial soft-switching control scheme can effectively improve the efficiency of the converter by about 3% compared with the soft-switching control

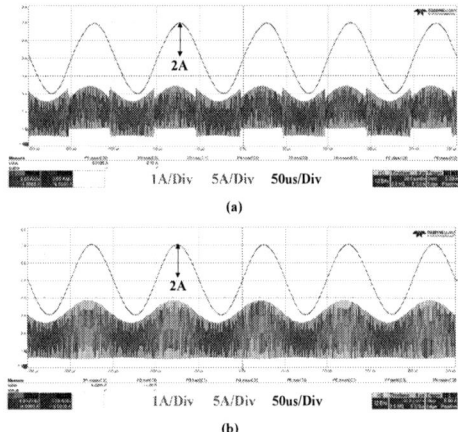

(a)

(b)

Fig. 7 Experimental waveform：Inductor current (blue waveform), filter current (red waveform). (a) Patial soft-switching control scheme.(b) Soft-switching control scheme.

where k_1 and k_2 represent the positive current slew rate and negative current slew rate of the inductor current respectively.

Within each discretized time interval, it is assumed that the energy loss generated by each switching operation is the same and is expressed as E_i. The cumulative switching energy loss over the complete sinusoidal modulation signal period is obtained through integration of the instantaneous energy loss density, which is expressed as (14):

$$E_{total} = \int_0^{\frac{1}{f_{module}}} E_i \cdot \frac{1}{\Delta t_i} dt \qquad (14)$$

where f_{module} represents the frequency of sinusoidal modulation signal.

The total power dissipation (P_{loss}) within a single cycle of the sinusoidal modulation signal is expressed as (15):

scheme, which verifies the feasibility of the partial soft-switching control scheme.

V. CONCLUSION

This paper analyzes the analog hysteresis control scheme of the synchronous Buck converter. The mathematical analytical models of power loss for hard-switching and soft-switching were established respectively. An estimation method for energy loss under variable-frequency control was proposed, and on this basis, a partial soft-switching control scheme was proposed. Finally, an experimental prototype was developed based on the proposed control scheme. Experimental results clearly demonstrate that the proposed analog hysteresis control scheme enables the achievement of complete soft-switching. In comparison to the hard-switching condition, the converter's efficiency of the converter is increased by 6.1%. Furthermore, the efficacy of the proposed partial soft-switching control scheme has been substantiated. Under identical operating conditions, the efficiency of the converter employing the partial soft-switching control scheme surpasses that of the soft-switching control scheme by 3%.

REFERENCES

[1] Z. Yu, J. He and Z. Li, "Design of a Novel GaN-Based 4-MHz ZVS Active-Clamping Synchronous Buck Converter," *2020 IEEE 9th International Power Electronics and Motion Control Conference (IPEMC2020-ECCE Asia)*, Nanjing, China, 2020, pp. 1077-1082, doi: 10.1109/IPEMC-ECCEAsia48364.2020.9368034.

[2] T. Fogec, J. Bačmaga, I. Krois and A. Barić, "Characterization of GaN-Based Synchronous Buck Converter Operating in MHz-Range," *2021 44th International Convention on Information, Communication and Electronic Technology (MIPRO)*, Opatija, Croatia, 2021, pp. 90-95, doi: 10.23919/MIPRO52101.2021.9596956.

[3] Martinez A, Torres F, Marin J, Rojas CA, Gak J, Rommel M, May A, Wilson-Veas AH, Miguez M, Rossi C, et al. Analysis and Design of an SiC CMOS Three-Channel DC-DC Synchronous Buck Converter for High-Temperature Applications. *Applied Sciences*. 2024; 14(21):9789. https://doi.org/10.3390/app14219789

[4] Z. Yao, X. He, Z. Xiao, Z. He, Y. Jiang and Y. Tang, "Full-Range ZVS Control Method Based on Near-CRM for T-Type Three-Level Inverter," *2023 IEEE Energy Conversion Congress and Exposition (ECCE)*, Nashville, TN, USA, 2023, pp. 2987-2992, doi: 10.1109/ECCE53617.2023.10362405.

[5] Vazquez, K. Martin, M. Arias and J. Sebastian, "A very simple analog control for QSW-ZVS source/sink dc-dc converters with seamless mode transition," *2019 IEEE Applied Power Electronics Conference and Exposition (APEC)*, Anaheim, CA, USA, 2019, pp. 220-226, doi: 10.1109/APEC.2019.8721813.

[6] T. Konjedic, L. Korošec, M. Truntič, C. Restrepo, M. Rodič and M. Milanovič, "DCM-Based Zero-Voltage Switching Control of a Bidirectional DC–DC Converter With Variable Switching Frequency," in IEEE Transactions on Power Electronics, vol. 31, no. 4, pp. 3273-3288, April 2016, doi: 10.1109/TPEL.2015.2449322.

[7] H. Xu, J. Wei, R. Xie, Z. Zheng, J. He and K. J. Chen, "Incorporating the Dynamic Threshold Voltage Into the SPICE Model of Schottky-Type p-GaN Gate Power HEMTs," in IEEE Transactions on Power Electronics, vol. 36, no. 5, pp. 5904-5914, May 2021, doi: 10.1109/TPEL.2020.3030708.

[8] Y. Zhang, C. Chen, Y. Xie, T. Liu, Y. Kang and H. Peng, "A High-Efficiency Dynamic Inverter Dead-Time Adjustment Method Based on an Improved GaN HEMTs Switching Model," in IEEE Transactions on Power Electronics, vol. 37, no. 3, pp. 2667-2683, March 2022, doi: 10.1109/TPEL.2021.3112694.

[9] Z. Qi et al., "An Accurate Datasheet-Based Full-Characteristics Analytical Model of GaN HEMTs for Deadtime Optimization," in IEEE Transactions on Power Electronics, vol. 36, no. 7, pp. 7942-7955, July 2021, doi: 10.1109/TPEL.2020.3044083.

Design and Demonstration of a Novel SiC Trench MOSFET with Periodically Grounded P Shield Island Based on Secondary Epitaxy Process

Yangyang Wu
Dept of Integrated Power Systems and Device Technology
Hubei Jiufengshan Laboratory
Wuhan, China
wuyangyang@jfslab.com.cn

Fei Guo
Dept of Integrated Power Systems and Device Technology
Hubei Jiufengshan Laboratory
Wuhan, China
carry_lee@jfslab.com.cn

Kuan Wang
Dept of Integrated Power Systems and Device Technology
Hubei Jiufengshan Laboratory
Wuhan, China
wangkuan@jfslab.com.cn

Wei Chen
Dept of Integrated Power Systems and Device Technology
Hubei Jiufengshan Laboratory
Wuhan, China
chenwei@jfslab.com.cn

Zhijie Cheng
Dept of Integrated Power Systems and Device Technology
Hubei Jiufengshan Laboratory
Wuhan, China
chengzhijie@jfslab.com.cn

Yuan Jun
Dept of Integrated Power Systems and Device Technology
Hubei Jiufengshan Laboratory
Wuhan, China
yuanjun@jfslab.com.cn

Rong Zhang
School of Electrical and Electronic Engineering
Huazhong University of Science and Technology,
Wuhan, China
rongzhang@hust.edu.cn

Guoqing Xin
School of Electrical and Electronic Engineering
Huazhong University of Science and Technology,
Wuhan, China
guoqingxin@hust.edu.cn

Zhiqiang Wang
School of Electrical and Electronic Engineering
Huazhong University of Science and Technology,
Wuhan, China
zhiqiangwang@hust.edu.cn

Abstract—This paper proposes a novel silicon carbide (SiC) trench MOSFET with periodically grounded P shield islands (SiC PSI-MOS) based on a secondary epitaxy process. The proposed SiC PSI-MOS structure utilizes a patterned P shield island formed through etching, ion implantation, and secondary epitaxy, enabling deeper shielding with lower energy implantation while maintaining gate oxide reliability. In the blocking state, the P shield region and P shield island effectively shield the drain potential, reducing the gate oxide electric field to a safe level (1.07 MV/cm at 1200 V). In the on-state, the drift between the P shield and the island creates additional current paths, lowering specific on-resistance (4.9 mΩ·cm² at a 7.5 μm cell pitch). In addition, a feasible fabrication process is demonstrated, with experimental results confirming the device's output and transfer characteristics. The SiC PSI-MOS offers an optimized trade-off between gate oxide reliability and conduction performance.

Keywords—Silicon carbide, MOSFET, Gate oxide reliability

I. INTRODUCTION

Silicon carbide (SiC) trench MOSFET have attracted researchers and industry with its superior on-resistance and lower switching capacitance than planar gate MOSFETs [1]. Unfortunately, SiC trench MOSFETs have severe gate trench bottom oxide strong electric field events [2]. A common solution is to design a P shield region at the gate trench bottom to reduce the electric field strength [3-6]. Although a deeper P shield region provides better protection, it requires higher performance of the ion implantation equipment and high energy ion implantation leads to lateral straggle and lattice damage, which ultimately adversely affects the design and reliability of SiC MOSFETs [7-9]. In addition, the P shield region needs to be shorted to the source to avoid degradation of dynamic R_{ON} during the switching operation [10].

In this paper, a novel SiC trench MOSFET with periodically grounded P shield island based on secondary epitaxy process (SiC PSI-MOS) is proposed and discussed. The advantages of SiC PSI-MOS are demonstrated by Silvaco TCAD, involving in the concentration-dependent mobility, high-field saturation mobility, concentration-dependent SRH lifetime, auger recombination, incomplete ionization of impurities and anisotropic impact ionization models.

II. DETICE STRUCTURE AND MECHANISM

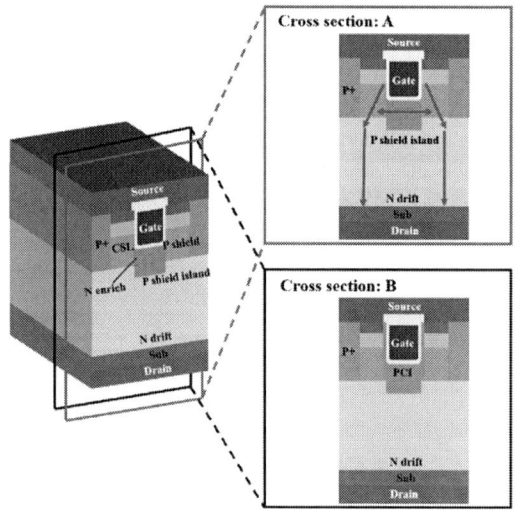

Fig. 1. SiC trench MOSFET with periodically grounded P shield island.

Fig. 1 shows the cross-section view of SiC PSI-MOS. The proposed structure forms a patterned P shield island by etching alignment mark (ALG), patterned mask, ion implantation and secondary epitaxy. As shown in section B of Fig. 1, a P-type connecting implant (PCI) is formed on both sides and the bottom of the gate trench through a self-aligned implantation process. This region connects the P shield region, P shield island, and the P well (P+ source) to electrically ground the P shield structures. Furthermore, an N-type doped region (N-enrich region) is introduced between the P shield and P shield island to prevent pinch-off by the depletion region. The process based on secondary epitaxy achieves a deeper P shield island region under low-energy conditions. The proposed SiC PSI-MOS not only ensures the reliability of the gate oxide but also provides an additional current path through the gap between the P shield region and the P shield island to enhance conduction capability. The key structure parameters of the SiC PSI-MOS are shown in Table I.

TABLE I.   KEY STRUCTURE PARAMETERS OF SiC PSI-MOS

Parameters	Value	Unit
cell pitch	7.5	μm
drift region thickness	10	μm
drift region doping	1×10^{16}	cm^{-3}
gate trench width	1.5	μm
gate trench deep	0.9	μm
oxide thickness	80	nm
P well doping	1×10^{17}	cm^{-3}
P+ source doping	5×10^{18}	cm^{-3}
P+ source thickness	1.5	μm
channel length	0.5	μm
P shield doping	5×10^{18}	μm
P shield width	1.1	μm
P shield thickness	0.2	μm
P shield island doping	1×10^{18}	cm^{-3}
P shield island width	1.5	μm
P shield island thickness	1	μm
N enrich doping	1×10^{18}	cm^{-3}

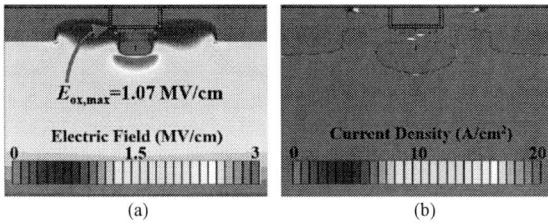

Fig. 2. (a) Electric field strength distribution in blocked state at V_{ds}=1200 V and (b) current density distribution of in conduction state at I_{ds}=300 A/cm^2 of SiC PSI-MOS.

In the blocked state, the P shield region at the trench bottom and the P shield island in the drift region can shield most of the potential from the drain electrode, ensuring that the gate oxide electric field is at a safe value, as illustrated in Fig. 2 (a).

In the on-state, the absence of a P-doped region between the P shield region and the P shield island enables dual conduction mechanisms: (1) a wider vertical current path in the vertical direction, reducing the specific on-resistance ($R_{on,sp}$), and (2) an additional horizontal current path between the P shield region and P shield island, further minimizing $R_{on,sp}$. This optimized carrier transport is illustrated in Fig. 2(b).

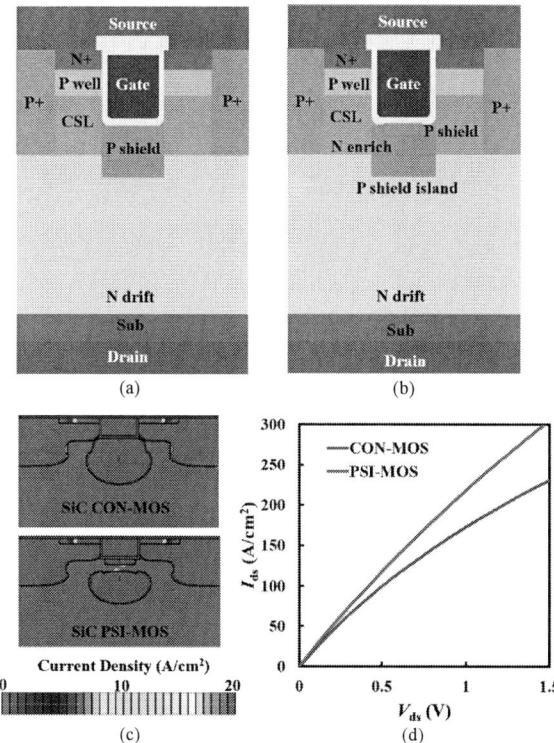

Fig. 3. Schematic structure of (a) SiC CON-MOS and (b) SiC PSI-MOS, (c) current density distribution in the on-state, and (d) comparison of output characteristics.

To demonstrate the advantages of the proposed structure, we conducted TCAD-based comparisons of the conduction capabilities between the SiC PSI-MOS and conventional structures. Fig. 3 (a) and (b) illustrate the schematic diagrams of the conventional P-shield SiC MOSFET (SiC CON-MOS) for comparison and the proposed SiC PSI-MOS structure in this work, respectively. The black lines in Fig. 3 (c) depict the depletion region boundaries of the SiC CON-MOS and SiC PSI-MOS under the on-state. In the SiC PSI-MOS, owing to the P shield island region providing a wider current path, the $R_{on,sp}$ of the SiC PSI-MOS is 4.9 mΩ·cm^2 (V_{gs}=20 V & I_{ds}=300 A/cm$_2$), which is 34.7% lower than that of the conventional structure (7.5 mΩ·cm^2), as shown in Fig. 3 (d).

III. FABRICATION

Additionally, a feasible fabrication process for the SiC PSI-MOS is proposed, as illustrated in Fig. 4, which comprises the following steps: (a) preparing the epitaxial wafer, (b) performing ion implantation to form the P shield island region, (c) conducting secondary epitaxy to form the CSL layer, (d) performing ion implantation to form the P+ source, N+ source, P well, and etching the gate trench, (e) utilizing a self-aligned process at the bottom of the trench to implant and form the N enrich region and P shield region, or (f) performing ion implantation to form the PCI region, (g) depositing and etching the polysilicon, and (h) fabricating the surface electrodes.

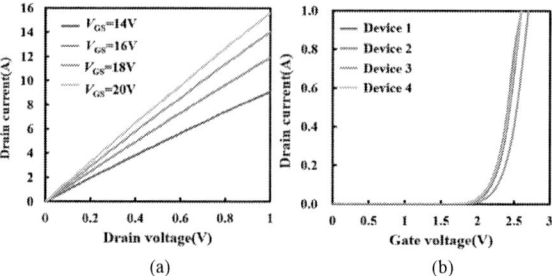

(a)   (b)

Fig. 6. Output characteristics and transfer characteristics of the fabricated SiC PSI-MOS.

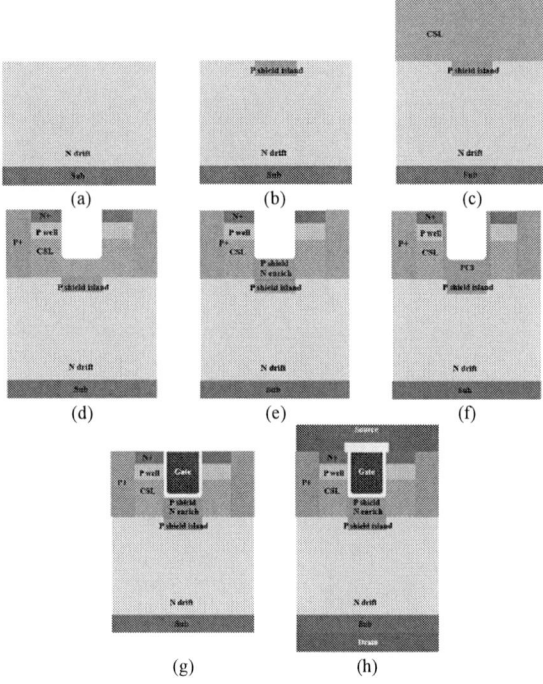

Fig. 4. SEM micrograph of (a) cross section A and (b) cross section B of the fabricated SiC PSI-MOS.

Fig. 5 shows the SEM images of the SiC PSI-MOS fabricated using the aforementioned method. The prepared device exhibits a conduction cross-section and a connecting cross-section that ensures the short connection between the P shield/P shield island regions and the source electrode, as illustrated in Fig. 5 (a) and (b), respectively. Notably, the P shield island region exhibits a slight tilt after the secondary epitaxy.

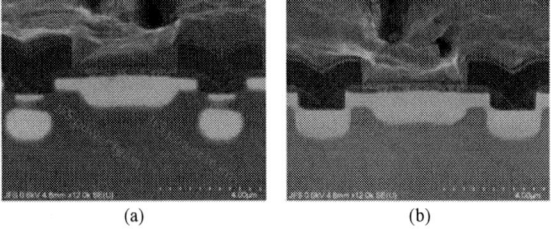

Fig. 5. SEM micrograph of (a) cross section A and (b) cross section B of the fabricated SiC PSI-MOS.

After wafer fabrication, wafer-level electrical characterization was performed on the SiC PSI-MOS devices. Fig. 6 presents the typical output and transfer characteristics of the fabricated SiC PSI-MOS. The device demonstrated a R_{on} of 60 m$\Omega\cdot$with a cell pitch of 7.5 μm (V_{gs} = 20 V, V_{ds} = 1 V), with a threshold voltage of 2.5 V, exhibiting excellent uniformity.

IV. CONCLUTION

A novel SiC trench MOSFET with periodically grounded P shield island based on secondary epitaxy process is proposed and characterized by simulation. The proposed new structure shows an excellent $R_{on,sp}$ and $E_{ox,max}$ trade-off. The proposed novel structure will be a strong contender for future high power density power electronics applications due to its excellent conduction capability and high reliability.

REFERENCES

[1] Xuan Li, Yifan Wu, Zhao Qi, Zhen Fu, Yanning Chen, Wenmin hang, Quan Zhang, Hanqing Zhao, Xiaochuan Deng, Bo Zhang, "An In-Depth Investigation into Short-Circuit Failure Mechanisms of State-of-the-Art 1200 V Double Trench SiC MOSFETs," in IEEE Transactions on Power Electronics, doi: 10.1109/TPEL.2024.3431296.

[2] A. K. Agarwal, R. R. Siergiej, S. Seshadri, M. H. White, P. G. McMullin, A. A. Burk, L. B. Rowland, C. D. Brandt, and R. H. Hopkins, "A critical look at the performance advantages and limitations of 4H-SiC power UMOSFET structures," Proc. 8th Int. Symp. Power Semiconductor Devices ICs, Waikoloa, HI, USA, May 1996, pp. 119-122.

[3] Yasuhiro Kagawa, Nobuo Fujiwara, Katsutoshi Sugawara, Rina Tanaka, Yutaka Fukui, Yasuki Yamamoto, Naruhisa Miura, Masayuki Imaizumi, Shuhei Nakata, Satoshi Yamakawa, "4H-SiC trench MOSFET with bottom oxide protection," Mater. Sci. Forum, vols. 778–780 (2014), pp. 919-922.

[4] Dethard Peters, Ralf Siemieniec, Thomas Aichinger, Thomas asler, Romain Esteve, Wolfgang Bergner, Daniel Kueck, "Performance and ruggedness of 1200 V SiC-Trench-MOSFET," Proc. 29th Int. Symp. Power Semiconductor Devices IC's (ISPSD), Sapporo, Japan, May 2017, pp. 239-242.

[5] T. Hiyoshi, K. Uchida, M. Sakai, M. Furumai, T. Tsuno and Y. Mikamura, "Gate oxide reliability of 4H-SiC V-groove trench MOSFET under various stress conditions," 2016 28th International Symposium on Power Semiconductor Devices and ICs (ISPSD), Prague, Czech Republic, 2016, pp. 39-42, doi: 10.1109/ISPSD.2016.7520772.

[6] Jun Yuan, Zhijie Cheng, Fei Guo, Kuan Wang, Wei Chen, Yangyang Wu, Shaodong Xu, Guoqing Xin, Zhiqiang Wang, "A Trench and Field Limiting Rings Co-Assisted JTE Termination With N-P-N Sandwich Epitaxial Wafers for 4H-SiC Devices," in IEEE Electron Device Letters, vol. 45, no. 8, pp. 1425-1428, Aug. 2024, doi: 10.1109/LED.2024.3416959.

[7] Julietta Weiße, Constantin Csato, Martin Hauck, Jürgen Erlekampf, Shavkat Akhmadaliev, Mathias Rommel, Heinz Mitlehner, Michael Rüb, Michael Krieger, Anton Bauer et al., "Impact of Al-Ion Implantation on the Formation of Deep Defects in n-Type 4H-SiC," 2018 22nd International Conference on Ion Implantation Technology (IIT), Würzburg, Germany, 2018, pp. 66-69, doi: 10.1109/IIT.2018.8807980.

[8] Jiashu Qian, Tianshi Liu, Jake Soto, Mowafak M. Al-Jassim, Robert Stahlbush, Nadeemullah Mahadik, Limeng Shi, Michael Jin, Anant K. Agarwal, "A Comparison of Ion Implantation at Room Temperature and Heated Ion Implantation on the Body Diode Degradation of Commercial 3.3 kV 4H-SiC Power MOSFETs," 2022 IEEE 9th Workshop on Wide Bandgap Power Devices & Applications (WiPDA), Redondo Beach, CA, USA, 2022, pp. 49-53, doi: 10.1109/WiPDA56483.2022.9955255.

[9] Huan Ge, Wanli Zhao, Tao Zhu, Jingrong Yan, Rui Wang, Zheyang Li, Rui Jin, "Low to Medium Dose Room and Elevated Temperature Implants in 4H-SiC Devices," 2023 20th China International Forum on Solid State Lighting & 2023 9th International Forum on Wide Bandgap Semiconductors (SSLCHINA: IFWS), Xiamen, China, 2023, pp. 117-121, doi: 10.1109/SSLChinaIFWS60785.2023.10399700.

[10] Jin Wei, Meng Zhang, Huaping Jiang, Hanxing Wang, Kevin J. Chen, "Dynamic Degradation in SiC Trench MOSFET With a Floating p-Shield Revealed With Numerical Simulations," in IEEE Transactions on Electron Devices, vol. 64, no. 6, pp. 2592-2598, June 2017, doi: 10.1109/TED.2017.2697763.

AUTHOR INDEX

Afanasenko, Valentyna 147
Al-Haddad, Kamal 246, 393, 644
Alhosaini, Waleed 853
An, Wenbo 609
An, Ziyang 949
Bai, Jinyang 398
Bai, Song 462
Bai, Zhitong 849
Bao, Dingyuan 26
Bao, Guojun 430
Barón, Kevin Muñoz 147
Ben, Hongqi 133
Cai, Xuchong 885, 968
Cailin, Wang 1006
Cao, Hanlin 11
Cao, Jiaying 21, 164
Cao, Junhou 462, 659
Cao, Liqiang 52
Cao, Pei 692
Cao, Pingyu 11, 232
Cao, Ruibo 52
Cao, Wenping 71, 355
Cao, Xiaobo 857
Cao, Ying 410
Cao, Yufeng 902
Chan, Ian Yj 582
Chang, Given Shucheng 813
Chang, Liuchen 328, 468
Chang, Yifei 21, 164
Chang, Yongqi 717
Chao, Zhang 1006
Chen, Bingxin 450
Chen, Cai 118, 177, 280, 520
Chen, Cen 577, 630
Chen, Cong 944
Chen, Dawei 430
Chen, Fuao 963
Chen, Gang 867
Chen, Haobin 221
Chen, Jian 32, 524
Chen, Jiawei 990
Chen, Jia-Xiang 562
Chen, Jiaxiang 704
Chen, Jiwen 190
Chen, Junyang 762
Chen, Kelin 480, 839
Chen, Min 47, 62, 211, 480, 839
Chen, Pengyu 544

Chen, Tiwei 713
Chen, Wei 920, 1033
Chen, Xiao 109, 201
Chen, Xingye 122
Chen, Yang 963
Chen, Yenan 104
Chen, Yidan 867
Chen, Yifeng 151
Chen, Yong 314, 324, 416, 474, 753, 817, 916, 944
Chen, Yongang 491
Chen, Yongxin 355
Chen, Yu 867
Chen, Yue 933
Chen, Zihao 404
Chen, Zilong 343, 485
Chen, ... 688
Cheng, Haifeng 138
Cheng, Haoyuan 67
Cheng, Huaihao 753
Cheng, Ji 221
Cheng, Jinpeng 339
Cheng, Xinhong 550, 954
Cheng, Xu 275, 324, 416, 474, 753, 817, 916, 944
Cheng, Yujie 138
Cheng, Zhijie 920, 1033
Cheng, Zizhen 747
Chi, Jialiang 304
Chi, Qingguo 696
Chiu, Yi 762
Chu, Hao 118
Cui, Miao 11, 232, 911
Cui, Qian 343, 485
Cui, Wentao 206
Cui, Xinchun 572
Cui, Yingxin 731
Dai, Haohao 295
Dai, Jianxun 169
Dai, Siqi 177
Dai, Xinyue 898
Dai, Yuxing 246, 328, 393, 468, 644
Deng, Chaofan 726
Deng, Gaoqiang 644
Deng, Xiaochuan 295, 737
Diao, Lijun 491
Ding, Desheng 462
Ding, Lijian 445, 667, 788, 853
Ding, Xiang-Jin 562
Ding, Xiangjin 704

Dong, Jintong....................128
Dong, Ruixiao....................328, 468
Dong, Xiaojun....................924
Dong, Xiaonan....................186
Dou, Chong....................343, 485
Dou, Wenzheng....................16
Dou, Yu....................949
Du, Yifei....................47, 211
Duan, Yuhan....................21, 164
Fahlbusch, Sebastian....................638
Fan, Meng-Qi....................562
Fan, Mengqi....................898
Fan, Minfan....................582
Fan, Shanzhen....................700
Fan, Wendi....................104
Fang, Liu....................445
Fang, Mingzhu....................788
Fang, Yin....................260
Feng, Chao....................873, 990
Feng, Hao....................339
Feng, Siyuan....................280
Feng, Yu....................445
Fu, Hao....................462, 659
Fu, Minfan....................544
Fu, Xichen....................566
Fu, Yehan....................833
Fu, Yu....................308, 334
Gao, Mingyang....................83
Gao, Peng....................767, 822
Gao, Shanshan....................133, 376
Gao, Tianyu....................717
Ge, Xinglai....................741
Geng, Guifeng....................678, 798
Geng, Yuqing....................849
Gong, Jiakun....................933
Gong, Jie....................924
Gong, Kaixiang....................314, 416
Gong, Shoulai....................731
Gong, Xiaowu....................366, 516, 709, 726
Gu, Caixin....................867
Gu, Yitian....................873
Gu, Yu....................264
Gu, Zihang....................382, 902
Guan, Guolian....................495
Guan, Hao....................21, 164
Guan, Jiajia....................118, 280, 520
Guan, Lei....................877
Guan, Renfeng....................974
Guan, Yueshi....................528
Guang, Yang....................160
Guangzhou, Jiahang Wang....................366
Guangzhou, Xi Jiang....................366

Gui, Qingzhong....................424
Guo, Fei....................920, 1033
Guo, Gaofu....................713
Guo, Suxia....................71
Guo, Yanao....................731
Hai, Dong....................26
Han, Jisheng....................731
Han, Shouhui....................684
Han, Yiting....................767, 822
Han, Yu....................304
Han, Zhiyun....................398, 450, 684
Hao, Xiamin....................757
Hao, Yue....................655
He, Daozhen....................673
He, Feng....................757
He, Hailong....................290
He, Jinliang....................867
He, Quanbo....................371
He, Shuiyuan....................491
He, Wenzhi....................450
He, Xiaomin....................91, 440
He, Yanjie....................485
He, Yanjing....................516, 709
He, Yingfeng....................813
He, Yuanheng....................673
He, Yuying....................606, 609, 833
He, Zhixing....................974
Ho, Carl Ngai Man....................260
Hou, Chunyao....................181
Hou, Fengze....................52
Hu, Cungang....................71, 355
Hu, Haipeng....................788
Hu, Haolin....................572, 704, 873
Hu, Jichao....................91, 890
Hu, Jingyang....................939
Hu, Pei....................87
Hu, Qiang....................32
Hu, Qin....................867
Hu, Qingmao....................424
Hu, Qingrong....................366
Hu, Tingwen....................524
Hu, Xiaofei....................500
Hu, Yaoyu....................614
Hu, Yifan....................290
Hu, Yirui....................269
Hu, Zhuangzhuang....................138
Hu, Zijian....................929
Hua, Mengyuan....................704
Huang, Baoying....................349, 588
Huang, Hao....................26, 939
Huang, Huolin....................169
Huang, Jun....................211

Huang, Kai	83
Huang, Lei	659
Huang, Qian	295
Huang, Yongle	1017, 1022
Huang, Zhi-Ying	762
Hui, Xiaoshuang	142, 156
Huo, Yiting	588
Iannuzzo, Francesco	26
Jayamaha, Shan	260
Ji, Kai	128
Ji, Mingming	109
Ji, Runyang	792
Ji, Yanfei	190
Jia, Liwen	480, 839
Jia, Pengyu	190, 195, 777, 783
Jia, Ziqi	290
Jiang, Dong	420
Jiang, Dongjun	933
Jiang, Feng	47, 211, 480, 839
Jiang, Haonan	655
Jiang, Junsong	71
Jiang, Runquan	678, 798
Jiang, Shaoyan	566
Jiang, Xi	516, 655, 709, 726
Jiang, Yu	290
Jiang, Yuteng	450
Jiang, Zepeng	741
Jiang, Zuoheng	990
Jiao, Teng	562, 572, 704
Jiao, Zhiwei	382, 902
Jiarui, Guo	1006
Jie, Huamin	717
Jin, Dongxin	924
Jin, Rui	77, 757, 813
Jin, Yuting	26, 939
Jingfei, Wang	160
Jun, Yuan	1033
Kallfass, Ingmar	147
Kang, Qian	849
Kang, Yong	118, 177, 280, 520
Kang, Yuhui	156
Kefan, Yu	160
Kong, Hang	100
Kong, Jie	827
Kong, Liudan	21, 164
Kong, Weixuan	577
Lai, Jia-Jun	62
Lai, Jiankun	516
Lai, Wei	236
Lam, Sang	11, 911
Lan, Jianyu	480
Lan, Xin	16, 151

Lei, Guangyin	924
Lei, Yun	169
Lei, Zhengzi	609
Li, Binbin	456
Li, Bingru	57
Li, Bodong	480, 839
Li, Chao	77
Li, Chi	863
Li, Chuangye	667
Li, Chushan	949
Li, Dawei	299
Li, Fang	600
Li, Haiyang	898
Li, Hao	410
Li, Helong	122, 445, 667, 788
Li, Hong	500, 594, 624, 673, 772, 1011
Li, Honghong	215, 226, 808, 1000
Li, Jianbiao	753, 817, 916, 944
Li, Jie	47, 211
Li, Longnv	57, 319, 959
Li, Mingfu	62
Li, Ningbo	382
Li, Qingmin	688, 700
Li, Shouxiang	308, 334
Li, Shuanglong	349
Li, Tianxi	280
Li, Wuhua	26, 939
Li, Xianfeng	688
Li, Xiangdong	655, 726
Li, Xiao	269, 619
Li, Xin	151, 436, 506, 512, 808, 1022
Li, Xing	38
Li, Xu	295, 737
Li, Xuan	295, 737
Li, Xuebao	77, 410
Li, Xuefei	491
Li, Yanjun	624
Li, Yanzuo	709, 726
Li, Yaqi	849
Li, Yi	817
Li, Yitong	614
Li, Zeyu	692
Li, Zhen	398
Li, Zheyang	813
Li, Zhihui	700
Li, Zhucheng	713
Li, Zicong	506
Li, Zikang	980
Li, Zixiao	491
Li, Zongjian	974
Liang, Haiyan	844
Liang, Lin	1022

Liang, Mei	190
Liang, Senhao	933
Liang, Shiwei	393, 644
Liang, Yidi	594, 772
Liang, Zhiqing	974
Liao, Aojie	624
Liao, Hui	186
Liao, Kaiju	867
Liao, Yiyang	614
Lin, Junru	985
Lin, Xinpeng	704
Lin, Xiyuan	544
Lin, Xuanyu	630
Lin, Yen-Liang	762
Liping, Jia	1006
Liu, Baihan	177
Liu, Bin	349, 588
Liu, Boyang	304
Liu, Chang	71, 731
Liu, Chao	388
Liu, Chaohui	221
Liu, Chuang	388
Liu, Chuyuan	319
Liu, Decai	813
Liu, Guoyou	524
Liu, Hangzhi	907
Liu, Hongchen	959
Liu, Hui	264, 420
Liu, Jiahong	827
Liu, Jianwei	863
Liu, Jing	857
Liu, Jingrui	994
Liu, Ming	556
Liu, Mingjun	195, 777, 783
Liu, Pan	21, 164
Liu, Qifan	709
Liu, Qingchang	582
Liu, Qingsong	398, 684
Liu, Shuyu	339
Liu, Sinuo	264
Liu, Siyang	462, 659
Liu, Wei	95, 556
Liu, Yan	562, 572, 704
Liu, Yang	974
Liu, Yi	802
Liu, Yipeng	177
Liu, Yitao	844
Liu, Yu	236
Liu, Yuan	1028
Liu, Yushan	269, 619
Liu, Zhaocheng	410
Liu, Zheng	684

Liu, Zhicheng	802, 833
Liu, Zhiqiang	667
Liu, Zhongyue	692
Liu, Zixuan	361, 649
Lou, An	181
Lu, Bohang	630
Lu, Jie	71
Lu, Zijian	844
Luan, Aozu	181
Luo, Anjing	713
Luo, Cheng	924
Luo, Haoze	26, 939
Luo, Jian	83
Luo, Shengjie	939
Luo, Xixi	867
Luo, Yifei	506, 512, 1017, 1022
Lv, Jiahui	1
Lv, Jianwei	177
Ma, Hao	349, 588
Ma, Hongbo	114
Ma, Jun	704
Ma, Mingcheng	534, 885, 968
Mai, Ruikun	963
Man, Lichang	133
Mao, Danfeng	873, 898, 990
Mao, Weina	215, 226, 1000
Mao, Yanfang	792
Mei, Yun-Hui	57, 319
Mei, Yunhui	802, 959
Mi, Tianhe	572
Ming, Jie	949
Ming, Lei	382, 902
Mo, Guorui	709
Mo, Zhili	550
Ngo, Bac-Bien	994
Ning, Jianping	241, 250
Ning, Puqi	142, 156
Ning, Wenjie	614
Niu, Chunping	290
Niu, Jiaxuan	474
Niu, Yukun	275, 314, 416
Nuerdebieke, Yeerzhati	299
Ou, Zhenyuan	974
Ou, Zhujian	792
Ouyang, Runze	366
Pan, Siyu	349, 588
Pan, Xiang	215
Pan, Yanchen	885, 968
Pan, Yunbin	62
Pang, Zhuo	762, 767
Pei, Xingyu	753, 817, 916, 944
Pei, Xuejun	678, 798

Peng, Bo	91
Peng, Wensong	737
Pu, Hongbin	890
Pu, Xuhui	890
Qi, Bin	87
Qi, Jingjing	284
Qi, Xianbin	122, 445, 788
Qi, Yuwen	491
Qi, Zhiyuan	404
Qin, Peng	445, 788
Qiu, Kai	195
Qiu, Maohang	114
Qu, Vickie	582
Ran, Li	339
Ren, Hanwen	688, 692, 700
Ren, Na	929
Ruan, Jiabin	566
Ruan, Xinbo	255
Sang, Xikun	376
Sang, Zihan	410
Shangguan, Miaomiao	236
Shangguan, Xu	594, 772
Shao, Shuai	181, 206
Shao, Tiancong	87, 849
Shao, Zhe	456
Shen, Lingyan	954
Sheng, Haoyi	974
Sheng, Kuang	6, 67, 173, 881, 929
Sheng, Tsung-Huan	762
Shi, Mingxin	624, 673
Shi, Qianru	582
Shi, Zenan	506, 512
Shi, Ziliang	62
Shu, Peng	410
Shu, Zhou	717
Sihao, Shen	1006
Solomakha, Oleksandr	147
Song, Jian	52
Song, Lei	867
Song, Qingwen	404, 516
Song, Ruya	853
Song, Wensheng	32, 524
Song, Xuanting	246, 393, 644
Song, Xuhui	275, 314, 416
Song, Yanlin	1011
Su, Yan	122
Sun, Hao	62
Sun, Jiahua	539
Sun, Peng	594, 772, 817
Sun, Weifeng	462, 659
Sun, Xinnan	47, 211, 839
Sun, Xizhi	534

Sun, Xuchen	118, 280
Sun, Yuhan	87
Sun, Zhen	241, 250
Tan, Jingyang	980
Tan, Kun	71, 355
Tang, Jiuyang	21, 164
Tang, Tao	524
Tang, Xi	71, 355
Tang, Xiaoyan	516
Tang, Yi	994
Tang, Yihui	122
Tao, Ze	512
Taylor, Stephen	911
Thomas, Rony	638
Tian, Bowen	817
Tian, Jiamin	867
Tian, Ye	430, 949
Tian, Yufei	550
Udrea, Florin	371
Wan, Fayu	663
Wan, Yu-Xi	562
Wan, Yuxi	572, 704, 873, 898, 990
Wang, Cailin	877
Wang, Ce	6
Wang, Changdong	717
Wang, Chaojun	206
Wang, Chen	361, 649
Wang, Chenghai	696
Wang, Chenlu	659
Wang, Chenyi	577
Wang, Denggui	138
Wang, Dewen	495, 747
Wang, Feng	700
Wang, Hao	236
Wang, Haodong	577
Wang, Haoyun	963
Wang, Hengyu	6, 67, 173, 371, 881
Wang, Huai	827
Wang, Jiahui	839
Wang, Jian	688, 700
Wang, Jianing	38, 215, 226, 436, 808, 1000
Wang, Jun	246, 328, 393, 468, 634, 644, 974
Wang, Junbo	655
Wang, Kai	87
Wang, Kuan	920, 1033
Wang, Laili	100, 495, 747
Wang, Lei	77
Wang, Lu	319
Wang, Lurenhang	534, 885
Wang, Meiyu	762, 767, 822
Wang, Nianzheng	100
Wang, Peng	572

Wang, Pengfei 582
Wang, Qian 83, 634
Wang, Qidong 52
Wang, Runze 38, 226, 436, 808
Wang, Ruoyin 721
Wang, Shuyu 959
Wang, Tiefu 67
Wang, Wei 688, 700
Wang, Xi 890
Wang, Xiahao 38, 215, 226, 436
Wang, Xiao-Ping 562
Wang, Xiaoping 898
Wang, Xulong 361, 649
Wang, Yan 898
Wang, Yao 241, 250
Wang, Yifan 881
Wang, Yijie 128, 133, 376, 528, 980
Wang, Ying 366, 709
Wang, Youzheng 57, 319, 959
Wang, Yuanfeng 539
Wang, Yuchen 924
Wang, Yudong 343
Wang, Yulin 328, 468
Wang, Yuwei 634
Wang, Zhao 911
Wang, Zhi 450
Wang, Zhiqiang 304, 920, 1033
Wang, Zhiyuan 456
Wang, Zhongjie 109, 186, 201
Wang, Zicheng 630
Wang, Ziyang 32
Wang, Zuoxing 624
Wangliang, ...43
Wei, Jiaxing 462, 659
Wei, Jinxiao 445, 853
Wei, Mingbo 594
Wei, Suhang 177
Wei, Yuqi 343, 485
Wei, Zhaoxiang 462, 659
Wei, Zheng 614
Wen, Yi 737
Wu, Enyou 280
Wu, Hongfei 138
Wu, Hongyuan 753, 817, 916, 944
Wu, Junye 873, 990
Wu, Keping 990
Wu, Min 314
Wu, Shunqing 388
Wu, Tuanzhuang 462, 659
Wu, Yangyang 920, 1033
Wu, Yanlin 873, 990
Wu, Yi 290

Wu, Yifan 863
Wu, Yucheng 47, 211
Wu, Yue 181
Wu, Yunjie 236
Wu, Yuzhen 255
Xi, Wenyu 877
Xia, Runze 6
Xia, Zhenchun 38, 215, 226, 436, 1000
Xiao, Chuanwei 566
Xiao, Xiangan 118
Xiao, Ziheng 994
Xiaoguang, Wei 43, 160
Xie, Dong 741
Xie, Lihong 255
Xie, Minglei 450
Xin, Guoqing 920, 1033
Xin, Sixiao 450
Xin, Zhen 382, 902
Xing, Wenbin 95
Xing, Yimei 195, 777, 783
Xinling, Tang 43, 160
Xiong, Zhuofan 619
Xu, An 709
Xu, Aoxue 83
Xu, Binbo 968
Xu, Dianguo 133, 376, 528, 534, 885, 968, 980
Xu, Hongyi 929
Xu, Jun 600
Xu, Ke 57
Xu, Lei 355
Xu, Shengxiu 959
Xu, Shikang 737
Xu, Tao 382, 902
Xu, Wenjie 747
Xu, Xiaohui 128
Xu, Xiaoyi 792
Xu, Xingque 566
Xu, Xunjin 62
Xu, Yihao 232
Xu, Zhiliang 741
Xue, Fei 232
Xue, Lingxiao 109, 186, 201
Xue, Tangman 696
Xue, Yao 600
Xue, Zedong 404
Yan, Baolong 38, 436, 808, 1000
Yan, Haidong 221, 767, 822
Yan, Na 52
Yan, Yishun 534, 885, 968
Yan, Yuxin 334
Yan, Zhangzhe 138
Yan, Zhaoheng 709, 726

Yang, Dongsheng	606
Yang, Fengming	1022
Yang, Fengtao	747
Yang, Jiajun	142, 156
Yang, Jingli	717
Yang, Lei	539
Yang, Mao-Jin	562
Yang, Maojin	704
Yang, Wuhua	1006
Yang, Xiaodong	91
Yang, Xiaofeng	849
Yang, Xiaolei	462
Yang, Xin	424, 894
Yang, Xiong	62
Yang, Xu	474
Yang, Yahong	480
Yang, Ying	361, 649
Yang, Yingkun	767
Yang, Yuan	1, 95, 890
Yang, Yun	250
Yang, Zeqi	388, 1028
Yang, Zhichang	1011
Yang, Zhihao	91
Yang, Zhiqing	122, 445, 788
Yany, Yateng	700
Yao, Bo	827
Yao, Jianguang	792
Yao, Rui Ray	911
Yao, Sankun	994
Yao, Siyi	544
Yao, Zhigang	994
Yin, Shuangyan	717
Yin, Xunran	717
Yin, Zhonggang	857
You, Shuzhen	655
You, Xiangan	52
Yu, Jinfeng	264
Yu, Jinhui	91
Yu, Kanghua	634
Yu, Shaolin	38, 215, 226, 436, 808, 1000
Yu, Zhe	638
Yu, Zheyuan	314
Yuan, Hao	404, 516
Yuan, Jun	920
Yuan, Song	366, 516, 709, 726
Yue, Hao	32, 524
Yujie, Du	43
Zalinge, Harm Van	232
Zeng, Chunhong	713
Zeng, Wei	873
Zeng, Zheng	933
Zeng, Zhijie	430

Zeng, Zhongming	713
Zha, Xian-Hu	562
Zha, Xianhu	704
Zhai, Fan	201
Zhan, Xinbin	516, 709
Zhang, Baoshun	713
Zhang, Bo	737
Zhang, Changhai	696
Zhang, Chao	468
Zhang, Chi	67
Zhang, Chuanqi	393
Zhang, Congcong	544
Zhang, Dao-Hua	562, 704
Zhang, Daohua	572
Zhang, Dezheng	833
Zhang, Donglei	1000
Zhang, Fan	275, 314, 416, 474
Zhang, Haining	284
Zhang, Haitao	440
Zhang, Haochen	528
Zhang, Haoyu	304
Zhang, Huanyu	713
Zhang, Jiarui	480, 839
Zhang, Jiayu	169, 741
Zhang, Jincheng	655
Zhang, Junming	206, 985
Zhang, Junzhao	420
Zhang, Kai	284
Zhang, Kuang	500
Zhang, Li	606, 609, 713, 833
Zhang, Maosheng	221
Zhang, Minmin	1
Zhang, Peng	77
Zhang, Ping	232
Zhang, Qian	857
Zhang, Qingchun	21, 164
Zhang, Rong	920, 1033
Zhang, Ruihao	663
Zhang, Shaowei	857
Zhang, Taohui	26
Zhang, Tiandong	696
Zhang, Tongyu	495
Zhang, Xiangqian	186
Zhang, Xiaodong	713
Zhang, Xiaojun	963
Zhang, Xiaolu	275, 314, 416
Zhang, Xiaotian	324
Zhang, Xingye	704
Zhang, Xinle	696
Zhang, Xinyu	308
Zhang, Xuelun	62
Zhang, Yajing	349, 588, 600

Zhang, Yakun.................................221
Zhang, Yaodong...............................324
Zhang, Yi....................................827
Zhang, Yichi.................................827
Zhang, Yifan.................................177
Zhang, Yihang................................190
Zhang, Yiheng................................777
Zhang, Yujie................................1017
Zhang, Yukun............................343, 485
Zhang, Yulei.................................890
Zhang, Yuming...........................404, 516
Zhang, Zewei.................................762
Zhang, Zhaofu.................................71
Zhang, Zhe.........................87, 388, 1028
Zhang, Zhihao...........................109, 201
Zhang, Ziyang................................151
Zhao, Bin...............................817, 916
Zhao, Dengrui................................713
Zhao, Dingkun................................894
Zhao, Fangwei............................87, 600
Zhao, Haiqiang...............................762
Zhao, Hang...................................236
Zhao, Hanqing................................737
Zhao, Jianyu.................................169
Zhao, Kepeng.................................232
Zhao, Kexin..................................474
Zhao, Mingzhi................................495
Zhao, Ning....................................16
Zhao, Shuang............................241, 853
Zhao, Yao....................................304
Zhao, Yixiang................................500
Zhao, Yuanzhi................................366
Zhao, Yucheng...........................308, 334
Zhao, Yunxuan...........................709, 726
Zhao, Zhenyu.................................717
Zhao, Zhibin.................................817
Zhao, Zhiqiang..........................495, 747
Zheng, Hong..................................721
Zheng, Hongjun...............................544
Zheng, Li...............................550, 954
Zheng, Trillion..............................849
Zheng, Yuze.........................275, 314, 416
Zheng, Zedong................................863
Zheng, Zexiang...............................177
Zheng, Zheyi.................................963
Zheng, Zhongshu..............................609
Zheng, Zijie............................246, 644
Zhi, Shuaiqing...............................534
Zhong, Linhai................................709
Zhongkang, Lin...............................160
Zhou, Chengyuan..............................173
Zhou, David.........................873, 898, 990

Zhou, Han.................................47, 211
Zhou, Jianjun................................138
Zhou, Liang...................................83
Zhou, Peng..............................678, 798
Zhou, Qiang..................................619
Zhou, Shaoze.................................614
Zhou, Xiang..................................299
Zhou, Xinyu..................................659
Zhou, Xuelei.................................881
Zhou, Xuetong...........................550, 954
Zhou, Yuming.................................907
Zhou, Zekun..................................737
Zhou, Zhenning...............................577
Zhu, Gaojia.........................57, 319, 959
Zhu, Hexin...................................295
Zhu, Lixuan..................................430
Zhu, Shuangxi................................520
Zhu, Tao.....................................813
Zhu, Xiaoyong................................721
Zhu, Ye.......................................62
Zhu, Zhanshan................................667
Zhu, Zhengyun.................................83
Zhuang, Huizhu..........................594, 772
Zhuang, Qiongyang............................867
Zong, Yujian.................................206
Zou, Liang..........................398, 450, 684
Zou, Liwen...................................186
Zou, Mingrui.................................933
Zou, Wen.....................................606
Zou, Yang....................................173
Zou, Yongzhou...........................246, 644
Zou, Zhili...................................713
Zou, Zhixiang................................430
Zuo, Qingyuan................................169

IEEE
445 Hoes Lane
Piscataway, NJ 08854-4141

ISBN 979-8-3315-1110-4